Genetics

Genetics

G. D. ELSETH
BRADLEY UNIVERSITY

KANDY D. BAUMGARDNER
EASTERN ILLINOIS UNIVERSITY

ADDISON-WESLEY PUBLISHING COMPANY

Reading, Massachusetts / Menlo Park, California / London
Amsterdam / Don Mills, Ontario / Sydney

Sponsoring Editor: Nancy J. Kralowetz
Developmental Editors: Katharine Gregg and Nancy Shapiro
Production Manager: Martha K. Morong
Production Editor: Laura R. Skinger
Designer: Rita Naughton
Cover Designer: Patricia O'Hare Williams
Art Coordinator: Kristin Belanger
Illustrator: Oxford Illustrators

Library of Congress Cataloging in Publication Data

Elseth, G. D. (Gerald D.), 1936-
Genetics

Includes index.
1. Genetics. I. Baumgardner, Kandy D., 1946-
II. Title.
QH430.E38 1984 575.1 83-9293
ISBN 0-201-03953-2

ABCDEFGHIJ-DO-8987654

Genetics, the science of heredity, deals with the factors responsible for the similarities and differences between generations. These factors affect form and function at many different levels, from the molecules of cells to populations of organisms. The concepts of genetics are thus fundamental to all the biological sciences and play a central role in the study of modern biology.

The purpose of this book is to provide an introduction to genetics that is readable and challenging to students and broad enough in scope to serve as the textbook in a one-quarter, one-semester, or two-quarter general genetics course. In this text, we are giving what we feel is a balanced coverage of the major areas of genetics: cytogenetics, molecular genetics, population genetics, quantitative genetics, and transmission genetics, but in a fashion such that the component parts can be dissected out for individual course needs. The concepts and techniques of these areas are presented along with sufficient historical background to show how many of the important principles have evolved. Introductory topics, such as basic Mendelism, probability, and mitosis and meiosis, are explained in full, and many practical applications to fields such as agriculture and medicine are given. The book is thus suited for use in courses where students have varied backgrounds and interests.

The book consists of seventeen chapters, which are arranged into seven discrete sections to provide flexibility in the order of their use. The chapters start with a description of Mendel's experiments and the application of Mendel's rules of inheritance. This historical introduction is followed by a consideration of the rules of counting and probability and their application in the analysis of genetic crosses. Probability is introduced early in the book to maximize its use in problem solving. Five chapters (Chapters 5 through 8 and Chapter 12) deal with the concepts of cytogenetics and transmission genetics. The emphasis throughout these chapters is on the chromosomal basis of inheritance. Six chapters (Chapters 3 and 4, Chapters 9 through 11, and Chapter 17) cover topics related to the molecular basis of inheritance, and are more chemically oriented than the others. Although DNA, RNA, and proteins are considered at length, chemical notation and jargon are kept to a minimum since the text does not presuppose a course in organic chemistry. The remaining four chapters (Chapters 13 through 16) discuss population and quantitative genetics. Quantitative inheritance, selection, and breeding principles are covered in greater depth in these chapters than in most introductory texts, and thus serve as a useful source of information for students specializing in animal and plant breeding. Other useful features of the book include an extensive glossary of terms and frequent internal summaries within each chapter, which help the student in reviewing important terms and concepts.

Although genetics is a comparatively young science, having developed primarily in the twentieth century, the growth of knowledge in this field has been phenomenal. Progress has been particularly impressive at the molecular level. As more discoveries are made, more techniques are developed that enable geneticists to probe even deeper into the molecular basis of inheritance. Research in modern genetics has therefore become a mushrooming industry that continues to open up exciting new areas of inquiry. The area of genetic engineering is a well-publicized example, in which recent technological advances have served to broaden our understanding of genetics at the molecular level and have provided new approaches to the synthesis of drugs and other chemicals and to pollution control. These new technologies also promise to have a major impact on the agricultural and medical sciences by providing tools for crop and livestock improvement and cures for genetic diseases. Since these new technologies have potential applications in all areas of genetics, we have included a detailed account of the current accomplishments and future prospects of genetic engineering in the last chapter, after the more traditional topics have been discussed.

Genetics is an analytical branch of science in which principles are expressed in quantitative terms. A certain amount of mathematics is therefore unavoidable. Despite the quantitative nature of the subject, the mathematical operations used in elementary genetics are not difficult, but require only the elements of algebra and a basic understanding of probability. A problem approach is used in this book to help the student develop the needed analytic skills and to provide the student with an opportunity to apply these skills to the analysis of genetic experiments. Graded sets of supplementary problems and review questions are included at the end of each chapter with answers given at the end of the text. The problems vary in character and degree of difficulty, and range from exercises that provide necessary repetition of basic skills to questions that are designed to challenge the student and test his or her comprehension of the subject material. Great care has been taken in the development of the problems to provide a useful teaching aid and to illustrate the wide range of applications of this area of study. Since students entering a beginning genetics course usually have little experience with problem solving, the book includes several numerical examples in the body of the text and a number of solved example problems that are set off from the text proper. The solved example problems extend and amplify basic principles and help to familiarize the student with the logical sequence of steps that can be used in finding solutions to problem situations.

ACKNOWLEDGMENTS

This project could not have been completed without the assistance of many people. We wish to thank our reviewers, Glenn Bewley, North Carolina State University; L. Herbert Bruneau, Oklahoma State University; Ronald P. Cantrell, Purdue University; David P. Campbell, California State Polytechnic University, Pomona; Thomas C. Gray,

University of Kentucky; Robert E. Goodwill, University of Kentucky; Nancy Z. Hartung, College of St. Thomas; Barbara Hollar, Mercy College; Becky B. Johnson, Oklahoma State University; Clifford Johnson, University of Florida; J. Spencer Johnston, Texas A & M University; David Knauft, University of Florida; Larry Leamy, California State University at Long Beach; Margaret Y. Menzel, Florida State University; and Henry Schaffer, North Carolina State University for their critical reading of the manuscript. Their valuable comments and suggestions helped to shape the text into its final form. Any errors that remain are the authors' responsibility. We also gratefully acknowledge the assistance of Ms. Connie Huber, who took responsibility for obtaining permissions and artwork and helped in many other ways. Credits for illustrations and tables from individuals or publications are given as these materials appear in the text. We especially wish to thank those people who provided photographs and other illustrative material. Finally, we are grateful to the talented staff of Addison-Wesley for their enthusiasm, competence, and support, particularly to Ms. Nancy Kralowetz, editor, Mr. Dick Morton and Ms. Kristin Belanger and staff, art, and Ms. Laura Skinger and staff, production department.

Abridged Contents

Contents

I Basic Mendelian Inheritance

II Nucleic Acids and Chromosomes

Chapter 6 Allelic Genes on Autosomes and Sex Chromosomes 195

Chapter 7 Linked Genes and Chromosome Mapping in Eukaryotes 235

VII Genetic Engineering

Part I

Basic Mendelian Inheritance

Chapter 1

Mendelian Principles

Many of the basic principles of genetics have their roots in the experiments performed by a nineteenth-century Austrian monk, Gregor Mendel (Fig. 1.1). Mendel published the results of these experiments in 1866 in a paper entitled *Experiments in Plant Hybridization.* The publication of this paper marked the birth of genetics as a science. The principles stated in the paper still stand today as the cornerstone of modern genetics.

Mendel's work is considered a classic in experimental design and interpretation. His **particulate theory** of inheritance proposed that the hereditary factors, which we call **genes,** are transmitted intact from one generation to the next. In his experiments, Mendel observed that certain traits can be "hidden" from view in the offspring. However, the factors that express these traits are not destroyed, so that these same characteristics may reappear unchanged in subsequent generations.

Mendel's concept of heredity is in contrast with the then prevailing but now defunct *blending theory* of inheritance, which regarded the characteristics of the progeny as a diluted mixture of the traits of the parents. According to the blending theory, the hereditary factors received by the progeny from their parents did not remain distinct, but rather blended together and lost their separate identities in the process.

Figure 1.1. (a) Gregor Mendel, Austrian monk, in 1865. (b) Monastery garden at Brno in Moravia (now part of Czechoslovakia) where Mendel's experiments were conducted. *Source*: From the Mendelianum of the Moravian Museum, Brno, Czechoslovakia.

The Mendelian Approach to Genetic Analysis

Mendel's success in establishing the basic rules of inheritance occurred for several reasons. One of the most important was his clear understanding of the kind of experimental design needed to answer the questions he was asking about the hereditary process. This clarity of understanding was reflected in his choice of an experimental organism.

Mendel chose to work with the garden pea *(Pisum sativum)* because (1) it possesses traits that are easily recognizable and appear in widely contrasting forms, and (2) it lends itself well to controlled crosses. Mendel realized that he had to be able to identify unambiguously the hereditary characteristics of each individual. He therefore limited his analysis to seven different traits, each expressed in a pair of contrasting forms (Fig. 1.2). These traits were seed coat texture (round or wrinkled), cotyledon, or seed leaf, color (yellow or green), flower color (purple

Figure 1.2. The seven pairs of contrasting traits of the garden pea studied by Mendel.

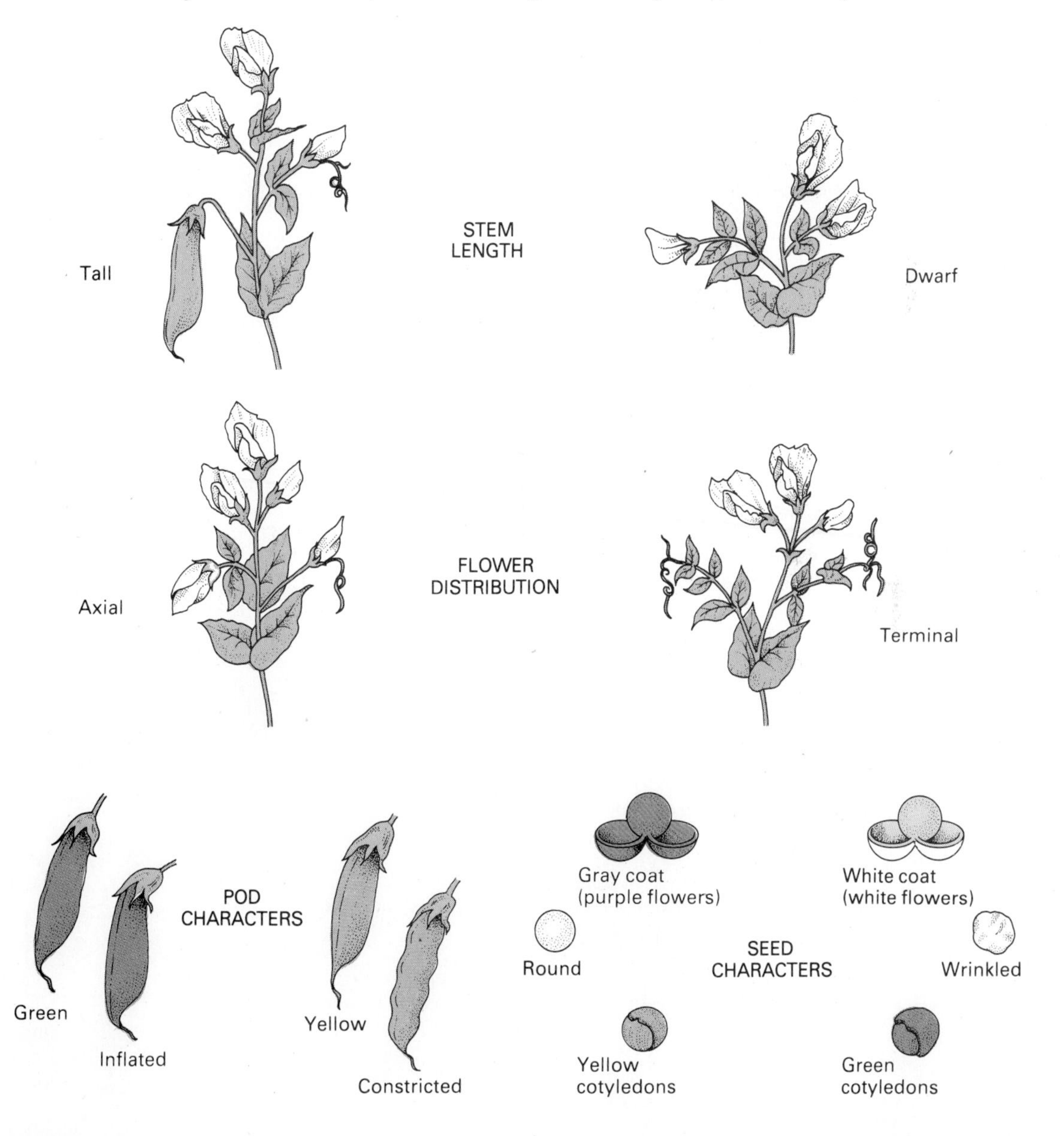

or white), pod shape (inflated or constricted), pod color (green or yellow), flower location (axial or terminal), and stem length (long or short). The alternative characters of each trait were clearly distinguishable, leaving little room for error in identification, and retained their contrasting appearances from one generation to the next.

Mendel knew that he had to have complete control over the kinds of plants being mated. The pea plant is naturally self-fertilizing, but it can be forced to cross-pollinate if the anthers (or male parts) of the flower are removed and the pistil (or female part) is then dusted with the pollen from another source (Fig. 1.3). By using this technique, Mendel was able to dictate the precise sequence of matings in his experiments.

Individual plants that continue to self-pollinate over many generations eventually lose their genetic variability, so that the offspring are all the same types. This knowledge enabled plant breeders even before Mendel's time to develop so-called **true-** or **pure-breeding** lines. When a plant of a pure-breeding line is self-fertilized (**selfed**) or crossed with another plant of the same line, the offspring are identical in appearance to the parent or parents. By using established pure-breeding lines, Mendel was able to begin each sequence of matings by crossing plants that were pure for the contrasting alternatives of a trait (for example, a pure-breeding long-stemmed plant could be crossed with a pure-breeding short-stemmed variety). Mendel then recorded the appearance of the offspring. He allowed the offspring to self-pollinate or mated them to one of the parental lines in a **backcross** to determine the pattern of inheritance in the following generation.

Figure 1.3. (a) The reproductive parts of the pea flower. The garden pea has a *perfect flower*, because male and female parts are both contained within the same flower. The anthers are the pollen-bearing (or male) part, and the pistil is the female part. (b) Controlled mating between purple-flowered and white-flowered plants. The anthers of the purple-flowered parent had previously been removed to prevent self-fertilization. Then, the pollen from the white-flowered parent is dusted onto the pistil of the purple-flowered parent.

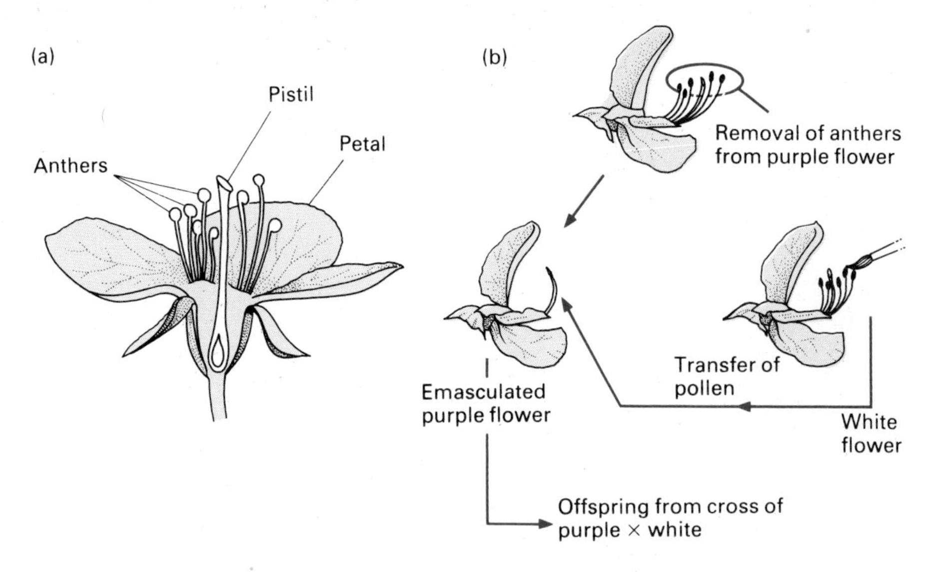

Mendel kept the different generations of plants distinct so that he could record the traits of each generation separately. To ensure statistical accuracy, he analyzed large numbers of plants from each cross. Mendel pursued this careful experimental plan for eight years.

A second reason for Mendel's success in elucidating the principles of heredity was his ability to work with numbers and to recognize basic quantitative relationships. Because of this ability, he was able to interpret experimental data using a mathematical model with direct applications in biology, even though he knew nothing about the physical basis of inheritance. In fact very little was known at that time about the process of cell division and the subcellular structures, such as chromosomes. Mendel's principles therefore described the transmission of genes in abstract mathematical terms. The importance of Mendel's work went completely unrecognized for 34 years, because few scientists understood the biological significance of these principles at the time they were proposed.

In 1900 three botanists, Hugo de Vries (Dutch), Carl Correns (German), and Erich von Tschermak-Seysenegg (Austrian), simultaneously discovered Mendel's paper on heredity, recognized its importance, and cited it in their own publications. Scientists by 1900 knew much more about the structure of cells and the behavior of cells during reproduction. They were better able to grasp the biological meaning of Mendel's principles, and these principles were quickly accepted as the foundation of genetics. In the following sections we will discuss the two main principles proposed by Mendel: the Principle of Genetic Segregation and the Principle of Independent Assortment.

Principle of Genetic Segregation

Mendel began his experiments by crossing true-breeding plants that differed in only one pair of contrasting characters such as coat texture or color. By limiting his analysis to single character differences, Mendel avoided the overwhelming complexity that plagued the results of earlier studies, in which several characters were permitted to vary. He then analyzed the appearance of the *hybrid* offspring of this first cross between true-breeding lines, termed the F_1 (for first filial generation). Since these F_1 progeny were hybrid for a single pair of alternative characters, they are also called **monohybrids** (*mono* = one). The F_1 monohybrids were then permitted to self-pollinate, as they normally would, to produce a second generation of offspring, the F_2 (or second filial generation). The appearance of these progeny too was analyzed. Mendel repeated this same sequence of crosses with the remaining six pairs of alternative traits.

Segregation Ratios with Dominance

The results of Mendel's experiments on the seven pairs of contrasting characters are summarized in Table 1.1. In every case, all of the F_1 exhibited the characteristic of just one of the parental plants. For example, when true-breeding plants with round seeds were crossed with true-breeding plants with wrinkled seeds, all of the F_1 seeds were round. Similarly, pure-breeding long-stemmed plants crossed with short-stemmed plants produced all long-stemmed progeny, and so on. The

Table 1.1. Summary of the results of Mendel's experiments with seven traits of the garden pea.

PARENTAL TRAITS	F_1	F_2 (NUMBER) DOMINANT	F_2 (NUMBER) RECESSIVE	F_2 (RATIO)
(1) seed coat texture (round vs. wrinkled)	all round	5474 round	1850 wrinkled	2.96:1
(2) cotyledon color (yellow vs. green)	all yellow*	6022 yellow	2001 green	3.01:1
(3) flower color (purple vs. white)	all purple	705 purple	224 white	3.15:1
(4) pod shape (inflated vs. constricted)	all inflated	882 inflated	299 constricted	2.95:1
(5) pod color (green vs. yellow)	all green*	428 green	152 yellow	2.82:1
(6) flower location (axial vs. terminal)	all axial	651 axial	207 terminal	3.14:1
(7) stem length (long vs. short)	all long	787 long	277 short	2.84:1

* Even though two entirely different traits (cotyledon color and pod color) may exhibit the same contrasting characters (yellow vs. green), the character that is dominant in one case is not necessarily the dominant one in the other. A dominance relation holds constant for any one trait, but is not necessarily the same for a different trait.

characters in the progeny definitely did not look like a blend of the parental characteristics. Mendel termed the character that appeared in the F_1 the **dominant** trait. The alternative that failed to appear in the F_1 was called the **recessive** trait.

Unlike the F_1, the F_2 was composed of both dominant and recessive types (see Table 1.1). But in each of the seven experiments, plants that expressed the dominant trait were approximately three times more frequent than plants that expressed the recessive trait. That is, both characters reappeared in the F_2, but at a ratio of about three to one. Therefore, three-fourths of the F_2 expressed the dominant trait and one-fourth expressed the recessive trait.

The Basic Model: Interpretation of Results. Mendel interpreted the results of his crossing sequence in the following manner: Each pair of alternative characters is determined by a pair of hereditary factors, or **genes.** (The term *gene* was originally coined by W. L. Johannsen in 1903 to mean a Mendelian factor, so we will use this term from now on.) These genes are passed intact from one generation to the next in the sex cells (or **gametes**) of the parents. They do not blend together in the F_1 hybrids. Indeed, they are not altered in any fashion by being present together in a hybrid form, as witnessed by the reappearance of the recessive type in the F_2. Dominance is thus considered to have a masking effect on the expression of the recessive gene; it in no way alters the integrity of the recessive gene.

The genes that control a pair of alternative characters are known as **allelic genes,** or simply **alleles.** In modern terminology, we say that the basic unit of inheritance is the gene, which exists in alternative allelic forms. The allelic forms of each gene are usually symbolized by the uppercase and lowercase of the same letter (such as *A* and *a*), with the uppercase symbol designating the dominant allele. In certain situations, which we will discuss later, symbols such as A_1 and A_2 or *A* and *A′* might be employed to indicate different alleles of the same gene.

Mendel also proposed that every individual, whether true-breeding or hybrid, has two copies of each gene. In a true-breeding line, the parents have a gene constitution (or **genotype**) consisting of either two *A* genes or two *a* genes. The expression of these genes constitutes their **phenotype.** Thus a genotype of *AA* is used to designate the genes of an individual that breeds true for the dominant trait. A genotype of *aa* is used to designate the true-breeding recessive form. Individuals in which both genes of a pair are alike (either *AA* or *aa*) are called **homozygous** for that trait.

Another important feature of Mendel's model is that during reproduction, only one member of each pair of genes is included in a single gamete. The F_1 monohybrids are formed by the union of two gametes from two separate true-breeding lines. In the cross, they receive an *A* allele from the parent that breeds true for the dominant trait and an *a* allele from the parent that breeds true for the recessive trait. The F_1 monohybrids therefore all have the genotype *Aa*. Since the hybrids carry contrasting allelic genes, they are called **heterozygous** for that trait. The *Aa* heterozygote expresses the phenotype of the dominant parent, thus indicating that the presence of only one dominant *A* allele is sufficient to completely mask the appearance of the recessive character. Both the parental *AA* homozygote and the F_1 heterozygote will produce the dominant phenotype. The dominant phenotype is therefore only an outward manifestation of the presence of at least one dominant allele. It is not a very reliable indication of an individual's genotype.

The three important features of the model, namely that (1) each form of a trait is determined by a specific form of a gene (called an allele), (2) every individual carries two copies of the gene for a given trait, and (3) only one member of each pair of genes in an individual is transferred in a gamete to the next generation, are illustrated by the following sequence.

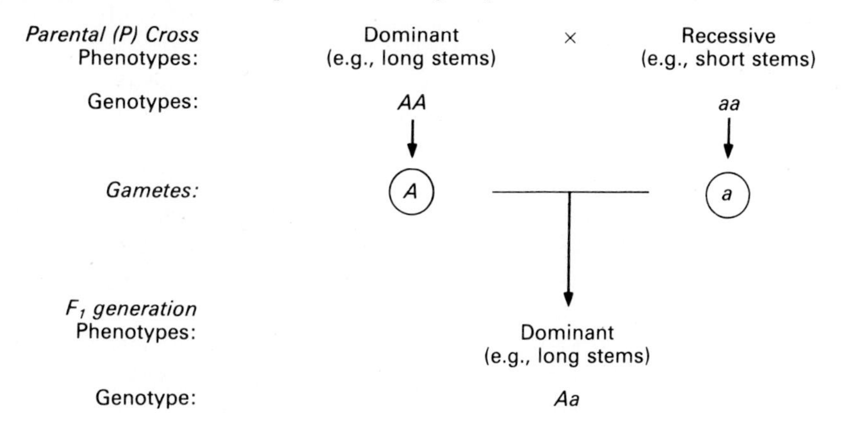

If the F_1 reproduce, each gamete will contain just one of the allelic genes—either *A* or *a*—chosen at random from the pair of alleles present in the heterozygote. This is Mendel's *Principle of Genetic Segregation,* which is formally stated as follows:

RULE 1.1

Principle of genetic segregation

In the formation of gametes, the members of a pair of alleles separate (or segregate) cleanly from each other, so that only one member is included in each gamete.

The two types of gametes of an F_1 heterozygote, designated Ⓐ and ⓐ, are thus equally likely to occur. The proportion of Ⓐ and ⓐ gametes is therefore one-half of each type:

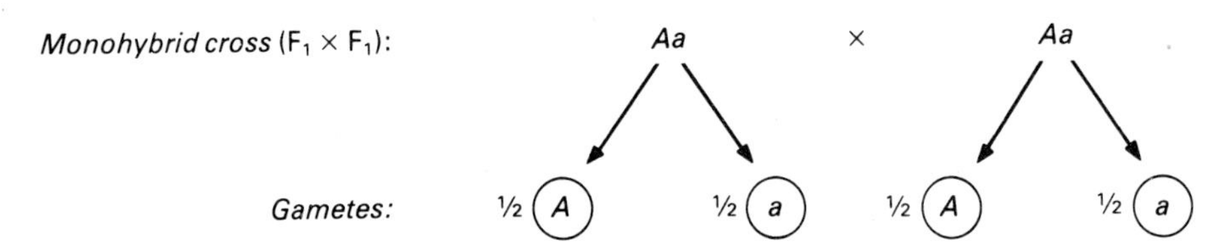

The process of fertilization is random. All possible unions of gametes from the male and female portions of the plant must therefore be taken into account to determine the genotypic and phenotypic composition of the F_2 zygotes. One way to make sure that all possible combinations are included is to set them up in the form of a checkerboard, called a *Punnett square:*

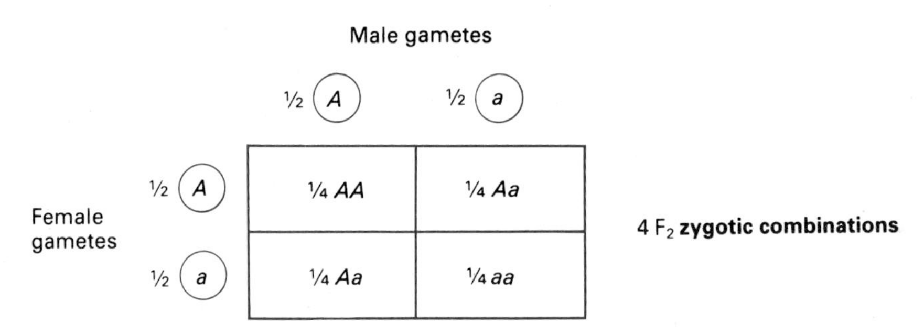

The Punnett square is merely a device to show all of the possible ways in which gametes can combine during fertilization. The four entries within the square represent the four possible F_2 zygotic combinations. The actual numbers of each type of zygote produced by the cross are not included, since they may vary greatly from one experiment to another. But more instructive than the actual numbers is the **relative frequency** or **proportion** of each offspring class. The proportion of each class is given in the square. We can see that of the total number of F_2 zygotes produced, we expect ¼ *AA*, ¼ + ¼ = ½ *Aa*, and ¼ *aa*. The F_2 genotypic ratio of 1 *AA* : 2 *Aa* : 1 *aa* expresses itself as a phenotypic ratio of three-fourths dominant (¼ *AA* + ½ *Aa*) to one-fourth recessive (¼ *aa*). The phenotypic ratio can be further abbreviated as 3 $A-$: 1 aa, where $A- = AA + Aa$ and is a shorthand way of designating the dominant phenotype.

It is not necessary to write the proportion ½ in front of each gamete in a Punnett square. The fact that the gametes occur with equal likelihood is understood. As you become more proficient at predicting the results of genetic crosses, you can eventually dispense with the checkerboard altogether and visualize the fertilization events as follows.

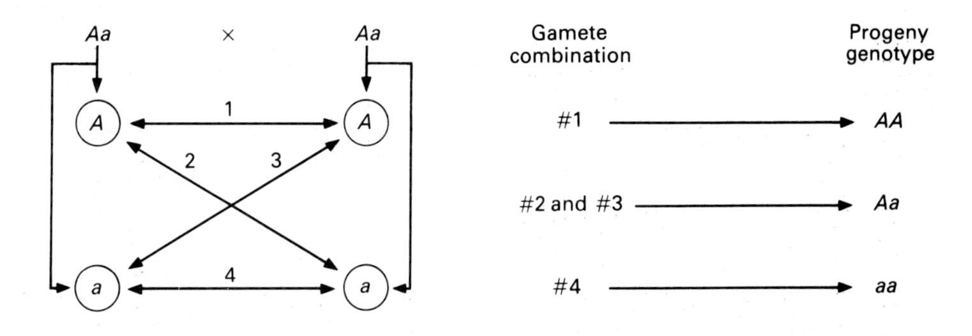

EXAMPLE 1.1. The standard (wild-type) coat color in mink is dark black-brown. Several other coat colors are known. One is a blue-gray (or platinum) color. When a wild-type mink was crossed with a mink of platinum color, all the offspring were wild type. Intercrossing the F_1 mink resulted in an F_2 consisting of 33 wild-type and 10 platinum animals. **(a)** Explain these results in terms of a single pair of genes. **(b)** How many wild-type mink in the F_2 are expected to be heterozygous?

Solution. **(a)** Wild-type color is evidently dominant to platinum, since only wild type showed up in the F_1 and approximately three-fourths of the F_2 expressed this color. Let P be the gene for wild-type coloration, and let p be the gene for platinum. The results of the cross in terms of genotypes can then be summarized as follows:

P: wild type × platinum
(PP) *(pp)*

F_1: wild type
(Pp)

F_2: 33 wild type + 10 platinum
(P−) *(pp)*

(b) The mating $Pp \times Pp$ is expected to yield $\frac{1}{4}\,PP + \frac{1}{2}\,Pp + \frac{1}{4}\,pp = \frac{3}{4}\,PP$ or $Pp + \frac{1}{4}\,pp$. Of the ¾ PP or Pp, ⅓ are PP and ⅔ are Pp. Thus, two out of every three dominant offspring in the F_2 should, on the average, be heterozygous. The number of heterozygotes expected among the offspring is therefore ⅔(33) = 22.

Progeny Testing: Detecting Heterozygotes. At this point we can test the genetic interpretations given for Mendel's results by predicting what would happen if the F_2 plants are subjected to further mating experiments. For example, if the F_2 plants with the recessive phenotype are allowed to self-pollinate, their F_3 progeny should all show the recessive phenotype, since the cross *aa* × *aa* yields only *aa* offspring. On the other hand, selfing of the F_2 plants with the dominant phenotype should yield an F_3 of all dominant offspring or part dominant and part recessive offspring, depending on which of the dominant F_2 plants are being selfed. This outcome is shown as follows:

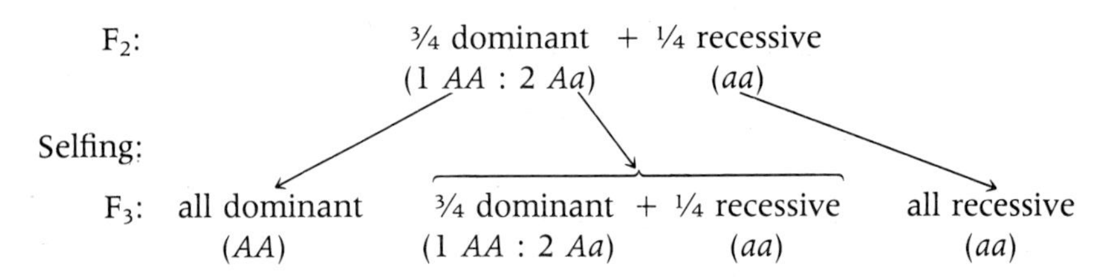

One-third of the dominant F_2 plants (one-fourth of the total F_2) are of the pure-breeding genotype *AA*. When selfed, these plants should produce only dominant progeny in the F_3. The other two-thirds of the dominant F_2 individuals, with the *Aa* genotype, should yield a progeny ratio of three-fourths dominant to one-fourth recessive when selfed. Mendel's data on F_3 generations confirmed these predictions.

In the preceding example, a procedure known as **progeny testing** was used to verify Mendel's predictions. This procedure uses genetic crosses to evaluate the genotype of an individual based on the phenotypes of its offspring. Progeny tests are used extensively in breeding programs to determine whether a dominant individual is homozygous for the dominant allele or is a heterozygous carrier of the recessive allele. One application is in cattle breeding, where progeny tests provide a means for detecting the presence of a recessive allele for dwarfism in certain breeds of beef cattle. Because of their size, cattle with this trait are of little economic value. Since a single heterozygous bull can spread the allele to several herds, it presents a practical problem to beef breeders. One approach is to test promising young bulls by breeding them to known carriers of the allele (this can be easily and quickly performed using artificial insemination). If all the progeny in the test are normal, then the bull breeds true for the normal allele, and the cross is *AA* × *Aa*. None of that bull's future offspring is expected to exhibit dwarfism, since none can receive two recessive alleles. If the bull himself is a carrier, however, the cross is *Aa* × *Aa*, and some of the progeny in the test will show the recessive trait. One-fourth of the future offspring from such a cross are then expected to be dwarfs. Bulls that produce dwarf calves in such crosses can then be eliminated as breeding stock.

Progeny testing is also useful in plant genetics. Some examples in maize (corn) are shown in Fig. 1.4. An ear of corn is in many respects an ideal subject for analysis of genetic ratios. Each kernel represents a separate individual and is the result of an independent fertilization event. Therefore, as many as one thousand progeny from a single cross are conveniently located on one ear. The kernels must be planted to express the characters of the mature plant. But many character

Figure 1.4. *Zea mays* (maize, or corn) showing (a) 3 : 1 ratio of purple to white and (b) 1 : 1 ratio of purple to white. Each kernel on an ear represents an individual act of fertilization. Therefore, as many as 1000 progeny from a cross are conveniently located on one ear. The 3 : 1 ratio arose from a selfing of an F_1 monohybrid. The 1 : 1 ratio was the result of a testcross of the F_1 monohybrid.

differences, such as those affecting seed shape and seed color, can be identified by a direct examination of the kernel phenotypes. Progeny ratios can then be obtained by merely counting the number of each type of kernel on individual ears of corn. For instance, Fig. 1.4 shows two kernel phenotypes, purple and white, with purple being dominant over white. If a dominant plant is heterozygous, self-pollination will yield a phenotypic ratio among the kernels of 3 purple : 1 white. The dominant plant can be further tested by crossing it to one that normally produces white kernels. (This cross is equivalent to $Aa \times aa$.) A progeny ratio of 1 purple : 1 white is then expected, since half of the kernels will receive the recessive allele from both parents. If the dominant plant had been of a true-breeding line (AA), neither mating would have produced white kernels.

As we have seen, two types of matings can be used for progeny testing. In one type, a dominant individual suspected of being heterozygous is mated to a known carrier of the recessive allele ($A- \times Aa$), or to itself ($A- \times A-$) if it is normally a self-fertilizing plant. This approach is generally taken when individuals homozygous for the recessive allele are inviable, or infertile, or otherwise uneconomic to produce. In the second type of mating, the dominant individual is crossed with a recessive homozygote ($A- \times aa$). Any dominant × recessive mating such as this is called a **testcross,** since it is used so often to test whether dominant

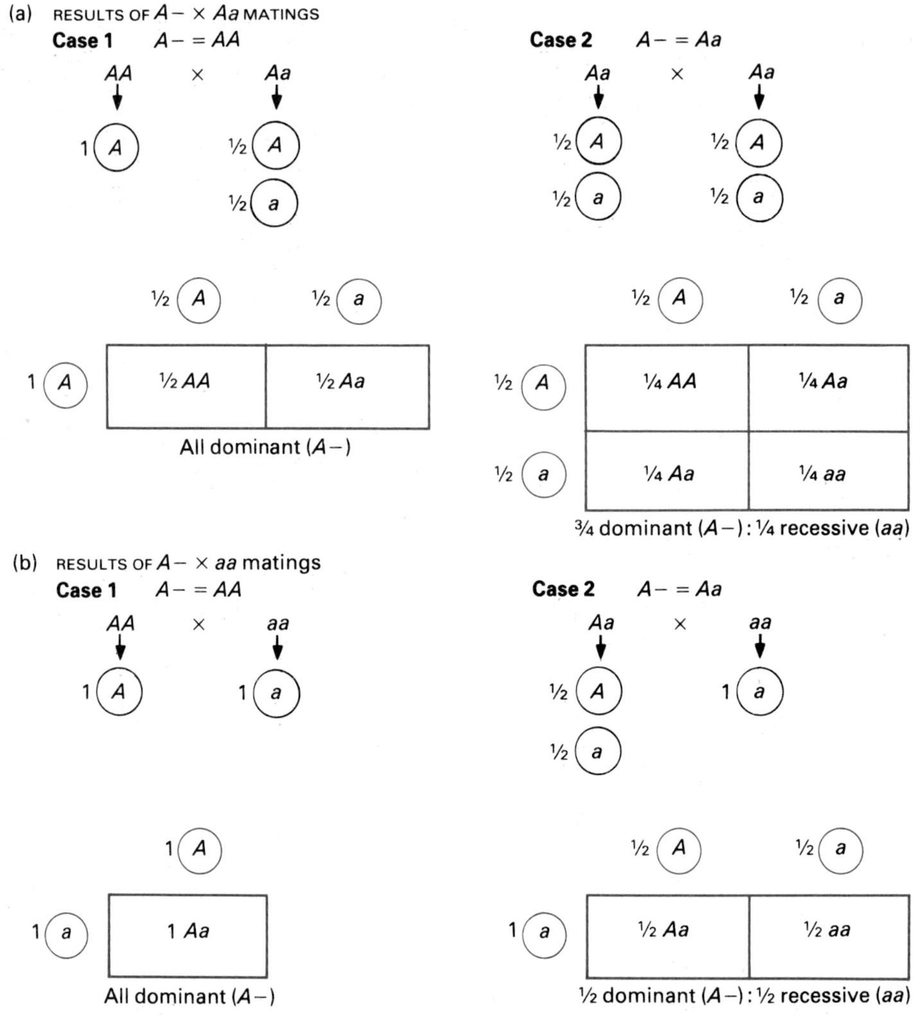

Figure 1.5. Progeny tests. These crosses provide the criteria used to distinguish heterozygotes from dominant homozygotes. In (a) a dominant individual of unknown genotype is mated to a known heterozygote. The unknown genotype is judged to be heterozygous if recessive offspring are produced. In (b) the dominant individual is mated to one that is recessive. Again, the unknown genotype is judged to be heterozygous if recessive offspring are produced.

individuals in the cross are heterozygous or homozygous. The criteria for distinguishing heterozygotes from homozygotes in the two types of matings are shown in Fig. 1.5.

EXAMPLE 1.2. In cattle, the polled (hornless) condition ($H-$) is dominant over the presence of horns (hh). A breeder who wishes to establish a pure-breeding herd of hornless cattle purchases a polled bull. To test whether the bull is a carrier of the recessive gene, he is mated to horned cows. The first mating produces an offspring that develops horns. **(a)** What is the genotype of the bull? **(b)** What percentage of offspring from such matings is expected to be polled? **(c)** Suppose that the bull is mated only to cows that are known carriers of the recessive gene. What percentage of such matings is expected to give rise to horned calves?

Solution. **(a)** The bull must be heterozygous (*Hh*), since the recessive offspring (*hh*) must have received an *h* allele from both parents. None of the offspring could be horned if the bull had the *HH* genotype. **(b)** The mating is *Hh* × *hh*. One-half (or 50 percent) of the offspring of such matings will be heterozygous (*Hh*), or polled. **(c)** The mating would then be *Hh* × *Hh*. One-fourth (or 25 percent) of the offspring of such matings, on the average, will be horned (*hh*).

It should be apparent to you by now that geneticists must be able to predict the outcome of several different crosses in order to draw the proper conclusions from the results of a cross. As a student of genetics, you too must become familiar with the genotypic and phenotypic ratios expected from these various matings. The results of the six basic crosses with allelic genes discussed in this section are summarized in Table 1.2.

Dominant/Recessive Inheritance in Humans. Dominant/recessive inheritance is known to occur in all organisms subjected to genetic analysis. In humans, many inherited disorders are recessive to the normal condition. A case in point is the fatal and untreatable condition known as *Tay-Sachs disease.* This is a disorder in lipid (fat) metabolism that affects infants who are homozygous for the Tay-Sachs allele (that is, *aa* in genotype). Although normal at birth, these infants undergo rapid and progressive neurological deterioration leading to mental retardation, blindness, paralysis, and finally death by the age of three or four years. Individuals with Tay-Sachs disease experience these symptoms because of their inability to produce an enzyme known as hexosaminidase A. The absence of this enzyme results in the accumulation of a specific fatty substance (ganglioside GM_2) within

Table 1.2. Summary of crosses involving a single gene pair. The *A* allele is assumed to be dominant over *a* in every case.

CROSS	GAMETES		PROGENY	
	PARENT 1	PARENT 2	GENOTYPIC RATIO	PHENOTYPIC RATIO
AA × *AA*	Ⓐ	Ⓐ	1 *AA*	1 dominant
AA × *Aa*	Ⓐ	Ⓐ and ⓐ	½ *AA* : ½ *Aa*	1 dominant
AA × *aa*	Ⓐ	ⓐ	1 *Aa*	1 dominant
Aa × *Aa*	Ⓐ and ⓐ	Ⓐ and ⓐ	¼ *AA* : ½ *Aa* : ¼ *aa*	¾ dominant : ¼ recessive
Aa × *aa*	Ⓐ and ⓐ	ⓐ	½ *Aa* : ½ *aa*	½ dominant : ½ recessive
aa × *aa*	ⓐ	ⓐ	1 *aa*	1 recessive

nerve cells. The accumulation, in turn, leads to the deterioration of the central nervous system.

Tay-Sachs disease is most common among Jewish people of eastern European origin (Ashkenazi Jews). It is estimated that as many as 1 out of 40 persons of such descent is a heterozygous carrier of the defective gene. Since the condition is fatal before the reproductive age, affected infants can result only from marriages between heterozygotes ($Aa \times Aa$). When offspring from all such marriages are totaled, about one-fourth of the siblings (brothers and sisters) of affected individuals are also born with the disease. This ratio is in agreement with the basic model for simple recessive inheritance.

EXAMPLE 1.3. Pigmentation in humans is determined by a dominant allele, *A*, and albinism (lack of pigment) is determined by the recessive allele, *a*, for the same gene. A couple who are both normally pigmented have their first child, and it is albino. **(a)** What are the possible genotypes of the parents? **(b)** If the couple goes on to have four other children, and two of them are albino, would this result invalidate the genotypes you proposed in (a)?

Solution. **(a)** Both parents are normally pigmented and are therefore either *AA* or *Aa* in genoype. However, their first child is albino, so that child is genotypically a recessive homozygote (*aa*). Since the child must have inherited a recessive gene from each parent, both parents must be heterozygous (*Aa*). **(b)** If both parents are *Aa*, we should expect three-fourths of the children from the monohybrid cross ($Aa \times Aa$) to have normal pigmentation. If a ratio of one normal to one albino occurs among the four children, this finding does not necessarily invalidate the proposed genotypes. The expected ratio is 3:1, but a family of four children is too small a sample size to ensure that the expected ratio will be observed. In tossing a coin, you would expect the outcome to be heads one-half of the time and tails the other half. You would not conclude that the coin is biased if you obtained seven heads and three tails in ten tosses. But if in 1000 tosses you got 700 heads and 300 tails, you might very well question whether heads and tails are really equally likely outcomes for that particular coin. Mendel analyzed hundreds, even thousands, of offspring before drawing any conclusions as to the precise ratios involved.

A frequent misconception among beginning students of genetics is that dominant traits are better or more desirable than their recessive counterparts, and that they are more abundant in a population as a consequence. While many recessive traits in humans are detrimental and rare, several dominant traits fall into this category as well. These traits include achondroplasia (a form of dwarfism), Huntington's chorea (a progressive neurological disease), and dyslexia (a type of reading disability), to name a few. It should be apparent from even this small list that dominance implies nothing about the desirability or adaptivity of a trait.

Modifications of Monohybrid Ratios

After the discovery of Mendel's work, interest in genetics grew as many researchers entered the field. It quickly became apparent that Mendel's segregation principle explained a wide variety of inheritance patterns, even when the typical 3:1 or 1:1

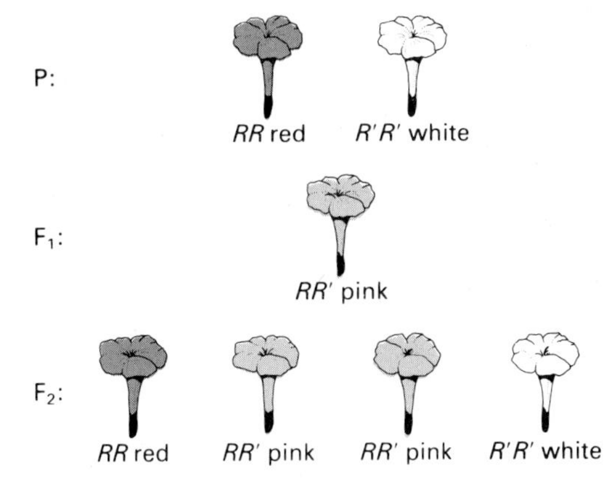

Figure 1.6. Incomplete dominance of alleles for flower color in four-o'clocks. A cross between red and white plants gives pink F_1, because the red allele does not completely mask the white allele. When intercrossed, these F_1 yield a 1 : 2 : 1 phenotypic ratio in the F_2, instead of the classical 3 : 1 ratio found if dominance is complete.

phenotypic ratios were not observed. In such cases, segregation of alleles still occurs, but the results are modified by the way in which genotypes are expressed and organized into phenotypic classes. In the next section, we will discuss some of the more common departures from the expected results for the basic dominant/recessive form of inheritance.

Dominance Relationships. In our previous discussion of dominance, we considered only complete dominance, in which the dominant allele completely masks the expression of the recessive allele. Not all alleles show complete dominance. For example, flower color in carnations, snapdragons, and four-o'clocks can be red (*RR*), pink (*RR'*) or white (*R'R'*) (Fig. 1.6). The flowers of the heterozygotes in these species are intermediate in color; in contrast, heterozygotes in the garden pea have the same flower color as the dominant homozygotes. Since the pink color gives a diluted appearance to the flowers of the heterozygotes, neither allele appears to mask the other completely. The *R* and *R'* alleles are thus said to exhibit **incomplete dominance.** Note that the intercross between heterozygotes (*RR'* × *RR'*) yields a ratio of 1 red (*RR*) : 2 pink (*RR'*) : 1 white (*R'R'*), instead of the familiar 3:1 ratio, so that the phenotypic ratio is identical to the genotypic ratio.

We can readily explain how incomplete dominance operates in flower color if we assume that the alternative state, recessiveness, is due to a complete or partial loss of gene activity. Suppose, for example, that a red flower pigment is produced from colorless precursors through a complex series of chemical reactions, each catalyzed by a different enzyme. Furthermore, assume that the amount of enzyme catalyzing the last reaction is under the control of gene *R*, so that a one-to-one relationship exists between the activity of the enzyme and the number of *R* alleles per genotype. (Possible mechanisms for this gene–enzyme relationship are discussed in later chapters.) We can represent the stepwise conversion of precursors to pigment in the following manner:

$$\text{Chemical precursors} \rightarrow \rightarrow \rightarrow \rightarrow \rightarrow \text{Colorless substrate} \xrightarrow[\quad]{\substack{R\text{ gene} \\ \downarrow \\ \text{enzyme}}} \text{Red pigment}$$

Observe that when the *R'* allele lacks function, so that a functional enzyme can only be formed in the presence of the *R* allele, *R'R'* homozygotes cannot produce red pigment. The flowers of $R'R'$ homozygotes will then appear white.

Flower pigments are complex substances. Ordinarily, several different chemical steps are required in order to synthesize them from simple precursor materials. Despite the complexity of the process, the overall rate of pigment synthesis tends to be limited by the rate of the slowest reaction (the rate-limiting step). If the reaction in question is the rate-limiting step, then any increase in the activity of the enzyme under the control of gene R should result in a corresponding increase in the amount of red pigment and, hence, in the intensity of flower color. This allelic affect is shown in Fig. 1.7 by a plot of the observed phenotypic effect against the genotype. At low levels of gene (and enzyme) activity, the amount of red pigment produced is directly proportional to the number of active R alleles in the

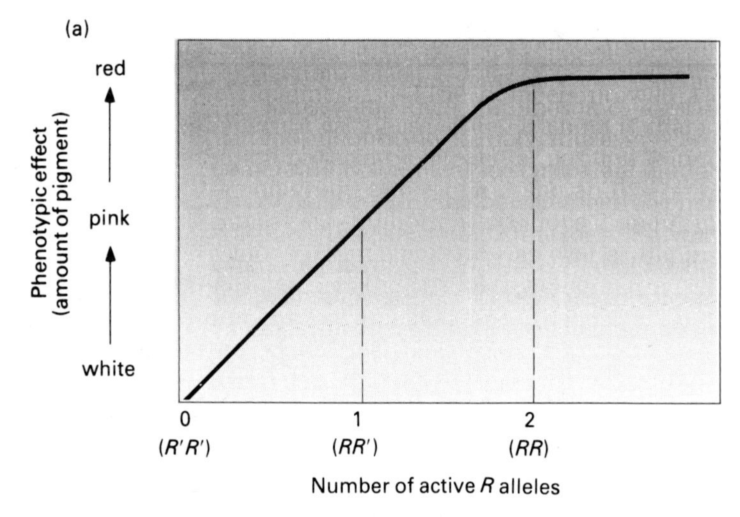

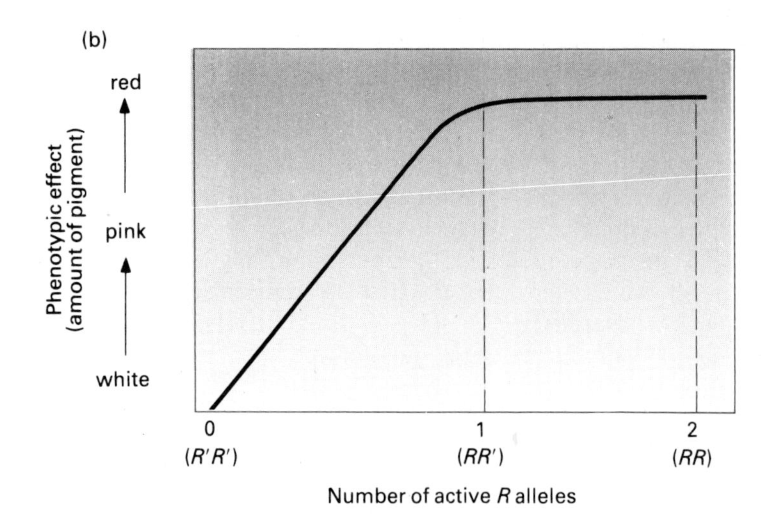

Figure 1.7. The distinction between complete and incomplete dominance. The graphs show a postulated increase in the degree of flower pigmentation corresponding to an increase in the number of active R alleles. It is assumed that the R allele controls a rate-limiting chemical step involved in the synthesis of red pigment. The amount of red pigment will rise up to a maximum level at which point another chemical reaction becomes the limiting step in pigment production. (a) Incomplete dominance, in which the effect of RR' falls below the maximum and is intermediate between that of $R'R'$ and RR. (b) Complete dominance, in which both RR and RR' have effects at the plateau level and are indistinguishable from one another in phenotype.

genotype. In this region of gene activity, heterozygotes show phenotypic effects that are intermediate to those of the *RR* and *R'R'* homozygotes (Fig. 1.7a). The *R* allele would then exhibit incomplete dominance.

Increases in pigment intensity cannot continue indefinitely. Ultimately, a plateau is reached, the point at which some other chemical reaction among those involved in pigment formation cannot keep pace and becomes the rate-limiting step. If the presence of a single *R* allele is sufficient to raise the rate of the reaction under its control to this level, additional *R* alleles would be ineffectual, and heterozygotes will express a flower color that is indistinguishable from that of the *RR* homozygotes (Fig. 1.7b). The *R* allele would then show complete dominance over the *R'* allele.

In this model of flower color, dominance is not a masking effect at all; rather, it is the result of a difference in allele activity. We see then that the degree of dominance can vary along a continuous scale, ranging from no dominance when heterozygotes are exactly intermediate in character, through various forms of partial dominance, to complete dominance at the other extreme when heterozygotes and dominant homozygotes are indistinguishable in appearance.

In addition to quantitative differences in the *amount* or degree of phenotypic effect, alleles may also show qualitative differences in the *kind* or nature of the effect produced. Such qualitative differences in allele expression form the basis of another dominance relation, called **codominance.** Two alleles are codominant if both are fully functional and express themselves individually when in the heterozygous condition. In codominance, as in incomplete dominance, heterozygotes are distinguishable on the basis of phenotype. But unlike incomplete dominance, a dilution effect does not occur. Heterozygotes show instead an unblended mixture of the separate effects of both alleles.

One example of codominance involves the inheritance of coat color in cattle. In the Shorthorn breed, homozygotes are either red (*RR*) or white (*R'R'*). Heterozygotes express a reddish-gray or roan color. Close examination of a roan coat shows it to be a mixture of red and white hairs. Each hair is either red or white; no single hair is an intermediate color, as would be expected with incomplete dominance. Since the *R* and *R'* alleles are both fully expressed in the heterozygote, they are codominant. Furthermore, intercrosses between heterozygotes give a 1 red : 2 roan : 1 white phenotypic ratio. Because the phenotypic ratio is again identical with the genotypic ratio, codominance cannot be distinguished from incomplete dominance when the only information available is the ratio of phenotypes.

A well-known example of codominance in humans occurs in the ABO blood group system. Alleles I^A and I^B show codominance, so that people with the I^AI^A genotype have type A blood, people with the I^BI^B genotype have type B blood, and the heterozygotes (I^AI^B) have type AB blood, expressing both alleles in full.

Lethal Genes. Another modification of typical Mendelian ratios occurs when a gene has such an extremely deleterious effect during development that it is lethal in the homozygous condition. Such a **lethal gene** occurs in mice. Two alleles that affect coat color in mice are *A* and A^Y, where *AA* mice have the normal wild-type color typical of mice in natural populations, AA^Y mice have yellow coats, and the A^YA^Y condition is lethal. The double dose of the A^Y allele results in death during early development. When yellow mice are bred together, a modified phenotypic ratio of two yellow to one wild type is found among the liveborn offspring. This modified ratio is in contrast to the 3:1 ratio expected from classical Mendelism. The modified ratio can be explained in the following way:

Parents:		$AA^Y \times AA^Y$		
Progeny:	Ratio at fertilization	1 AA normal	: 2 AA^Y yellow	: 1 A^YA^Y dies
	Ratio at birth	1 AA normal	: 2 AA^Y yellow	

Litter sizes from crosses of yellow mice are only three-fourths as large as typical mouse litters, supporting the theory that one-fourth of the fertilized eggs die during embryonic development. When a lethal gene is involved, it is customary to state the phenotypic ratio among the live offspring. In this instance, the ratio would be 2:1 (or $\frac{2}{3} : \frac{1}{3}$), rather than 3:1.

Many kinds of lethal genes are known. Some cause death early in embryonic development, as with the gene for coat color in mice just described. Another example of this type of lethal gene is an allele *c* in chickens that results in the so-called creeper condition. Creepers have short, crooked legs, giving them a squatty appearance, and are genetically heterozygous (*Cc*). Homozygotes for the *c* allele die on or about the fourth day of development. In contrast to the creeper condition, other lethal genes have been discovered that exert their effects later in development, but before or immediately after birth. A gene in humans that is lethal in homozygotes causes achondroplastic dwarfism in heterozygotes. A gene in cattle that produces a short-legged form of cattle in heterozygotes (known as the Dexter breed) causes the lethal "bulldog" calf condition in homozygotes. The calves have a highly abnormal appearance, somewhat resembling a bulldog, and are usually aborted at six to eight months of development.

Some of the better-known genetic disorders in humans are caused by genes that exert a lethal effect at some point after birth. We have already considered one example of this type, Tay-Sachs disease. Another example is *sickle-cell anemia*, a severe hemolytic disease characterized by anemia, jaundice, and episodes of severe pain (sickle-cell crises), usually localized in the joints, chest, and abdomen. Two alleles, Hb^A and Hb^S, are involved. Individuals who have the Hb^SHb^S genotype suffer from severe anemia as a result of problems associated with the abnormal sickle shape of their red blood cells (Fig. 1.8). The sickled red blood cells are less able to pass through small blood vessels, where they get trapped and obstruct the flow of blood to the surrounding tissues. Circulatory blockage that is widespread and prolonged can lead to sickle-cell crises, accompanied by organ and tissue damage. Some of the symptoms that can occur in this disease are summarized in Fig. 1.9. The many seemingly unrelated effects that can develop from a single genetic cause (sickled red blood cells) illustrate an important concept in genetics: *A single gene can influence more than one character.* The multiple phenotyic effect of

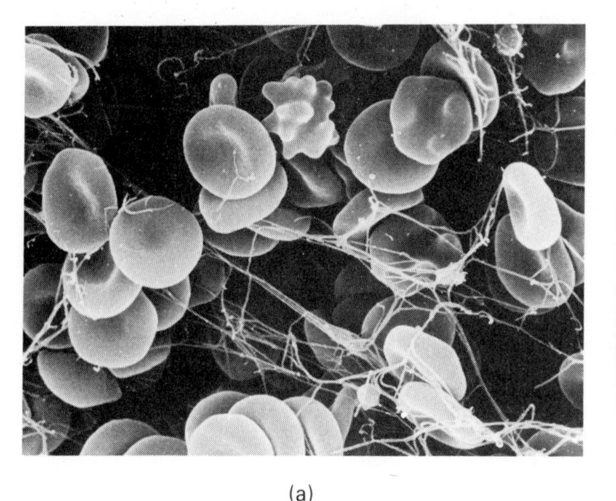

(a)

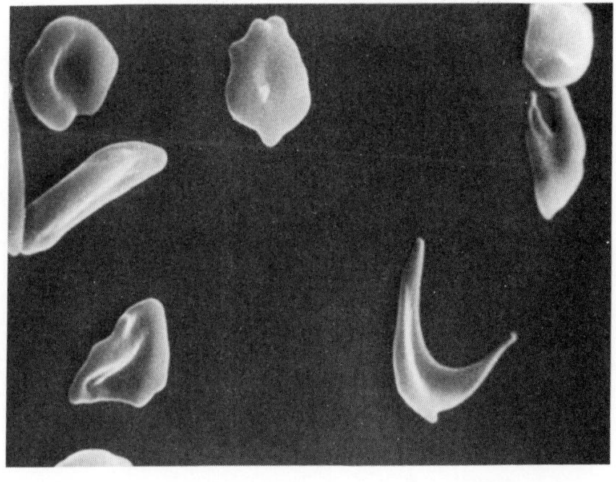

(b)

Figure 1.8. Normal (a) and sickled (b) red blood cells. The normal cells are disk-shaped, whereas cells from individuals who possess the Hb^SHb^S genotype have collapsed and assume the shape of a sickle. The sickled cells tend to clump together and clog blood vessels, impairing the transport of oxygen. (Both photographs were taken with a scanning electron microscope.) *Source*: (a) Courtesy of Philips Scientific and Analytical Equipment, Einhaven, The Netherlands; (b) Courtesy of Dr. Marion Barnhart, Professor of Physiology, Wayne State University School of Medicine, Detroit, Michigan.

a single gene, called **pleiotropy,** is a manifestation of the involvement of a single gene product in several different processes during development.

While a few homozygotes for the sickle-cell gene are only mildly affected and may live for a number of years, many die before reaching reproductive age. Heterozygotes (Hb^AHb^S), on the other hand, have enough functional red blood cells to remain in generally good health. Only under conditions of low oxygen tension (for example, at high altitudes) do such people suffer ill effects. They are carriers of the sickle-cell allele, however, and are said to have sickle-cell trait. Marriages between heterozygotes result in a ratio of 1 Hb^SHb^S (sickle-cell anemia): 2 Hb^SHb^A (sickle-cell trait) : 1 Hb^AHb^A (completely normal) at birth. This ratio becomes modified later in life, approaching 2:1 after the death of many of the affected individuals.

The lethal genes that we have described so far exert their effect early enough in life to eliminate, or at least drastically reduce, the reproductive potential of affected homozygotes. But this is not always the case. For example, Huntington disease (also known as Huntington's chorea) is a nervous disorder in humans, marked by spasmodic uncontrolled movements and mental deterioration. The age of onset is highly variable. Symptoms most often begin to appear between 30 and 40 years, but they sometimes do not appear for the first time until after 60 years of age. This means that most individuals who have the causative gene do not manifest any symptoms until middle age or older. By then, many of these individuals have already passed on the lethal gene to half of their children, on the average.

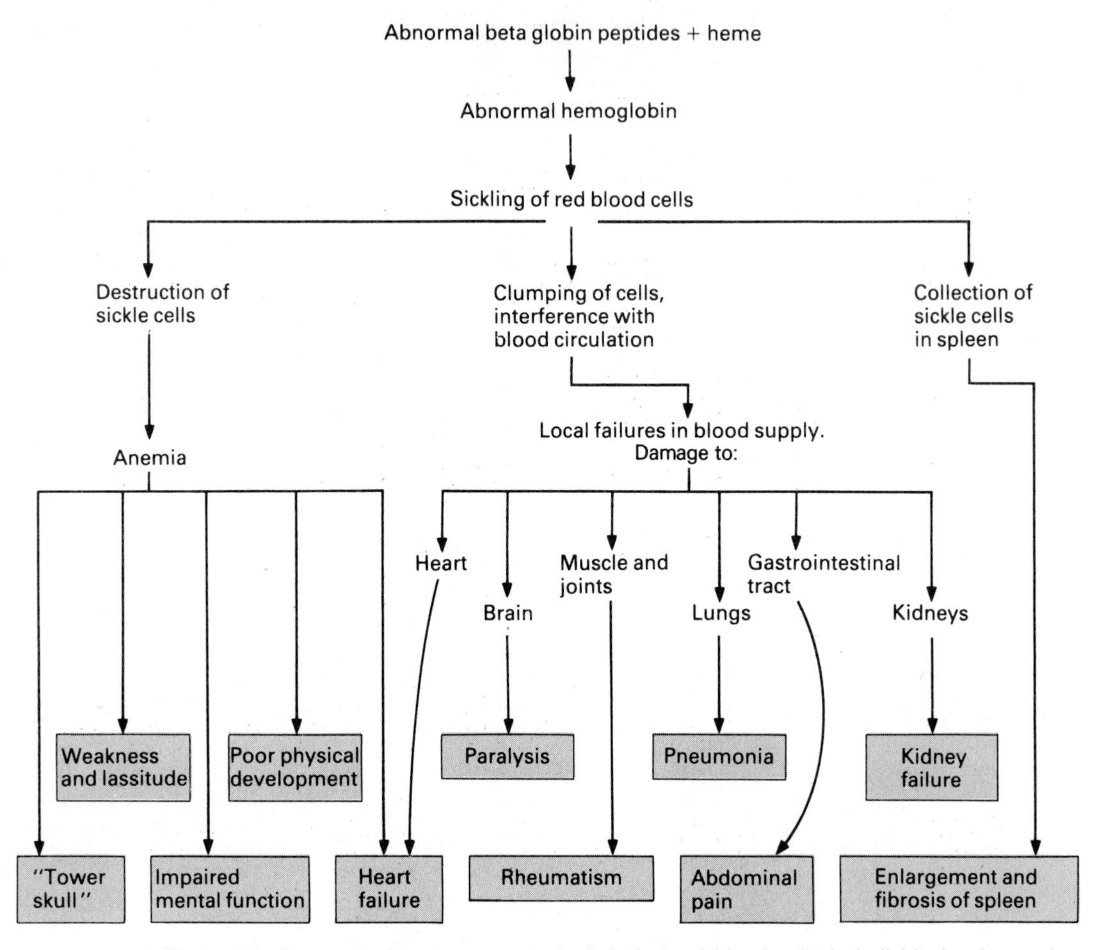

Figure 1.9. Some possible consequences of sickled red blood cells in individuals who are homozygous for the sickle cell gene. Multiple phenotypic effects that arise from a single genetic cause are known as pleiotropic effects. *Source*: Adapted from Neel and Schull, *Human Heredity*, University of Chicago Press, Copyright 1954 by The University of Chicago.

TO SUM UP

1. According to the basic Mendelian model of inheritance, genes are assumed to occur in pairs within the cells of higher organisms. The different members of a pair of genes are known as alleles.
2. The Principle of Genetic Segregation states that only one member of a pair of genes can enter a particular gamete. Fertilization then restores the genes to pairs. This principle holds for all sexually reproducing organisms.
3. Six different kinds of crosses are possible for a single pair of genes. These crosses and the offspring they produce are: $AA \times AA$ = all AA; $AA \times Aa = \frac{1}{2}\, AA : \frac{1}{2}\, Aa$;

$AA \times aa$ = all Aa; $Aa \times Aa$ = ¼ AA : ½ Aa : ¼ aa; $Aa \times aa$ = ½ Aa : ½ aa; and $aa \times aa$ = all aa. Of the crosses, only $Aa \times Aa$ and $Aa \times aa$ yield both dominant and recessive progeny.

4. Different degrees of dominance can be explained by differences in the activity of alleles. Quantitative differences (or differences in the amount of phenotypic effect) can lead to incomplete (or partial) dominance. Qualitative differences (or differences in the kind of phenotypic effect) by two equally functional alleles can lead to codominance. When the product of an allele affects several developmental processes, the allele can produce many different phenotypic effects. Such an allele is said to exhibit pleiotropy.
5. When dominance is complete, heterozygotes are indistinguishable in appearance from dominant homozygotes. Heterozygotes can be detected by progeny testing, in which a dominant individual of unknown genotype is usually mated to either a known heterozygote or a recessive homozygote. In either case, recessive offspring can be produced only when the dominant individual is heterozygous. A cross to determine whether a dominant individual is a heterozygote that involves a dominant × recessive mating is referred to as a testcross.
6. A number of factors transform basic Mendelian genotypic ratios into non-Mendelian phenotypic ratios. For example, the classical 3:1 monohybrid ratio expected for complete dominance can change to a 1:2:1 ratio with incomplete dominance or codominance. When a gene alters the phenotype of heterozygotes and is lethal in its homozygous state, the expected ratio changes from 3:1 to 2:1.

Principle of Independent Assortment

Mendel did not restrict his analysis to the inheritance of single alternative traits. He also determined the ratios produced when two separate character differences are considered simultaneously in the same cross. For example, in one of his crosses he studied the inheritance of both seed coat texture (round vs. wrinkled) and seed leaf color (yellow vs. green). The purpose of these experiments was to determine the influence that the inheritance of one trait might have on the transmission of another. Different traits are specified by different genes. If the alleles governing one trait are *A* and *a*, we might designate the alleles for a completely different trait as *B* and *b*. Different genes, such as *A* and *B* or *a* and *b*, are referred to as **nonallelic genes,** or simply **nonalleles.** Alleles, by contrast, are alternative forms of the same gene. In modern terminology, we could say that Mendel's experiments with the inheritance of two different traits concern the combined behavior of nonallelic genes during gamete formation and the effect, if any, of one gene pair on the segregation of the other.

Two Gene Pairs: Segregation Ratios with Dominance

Consider the cross of a true-breeding round, yellow-seeded pea plant with one that, when selfed, characteristically produces wrinkled and green seeds. Let *a* be the recessive allele for wrinkled seeds, and let *A* be the dominant allele for round seeds. Similarly, let *b* represent the recessive allele for green, and let *B* represent the dominant allele for yellow. Members of the allelic pair *A* and *a* are nonalleles of *B* and *b*. The cross of round, yellow-seeded plants with the wrinkled, green-seeded ones can be symbolized as *AABB* × *aabb*. According to Mendel's Principle of Genetic Segregation, *each gamete will contain only one member of each pair of alleles.* A gamete must always contain one representative of each pair of genes that the parent possesses, so that the union of gametes through fertilization restores the proper number of genes in the next generation. The following sequence illustrates this point.

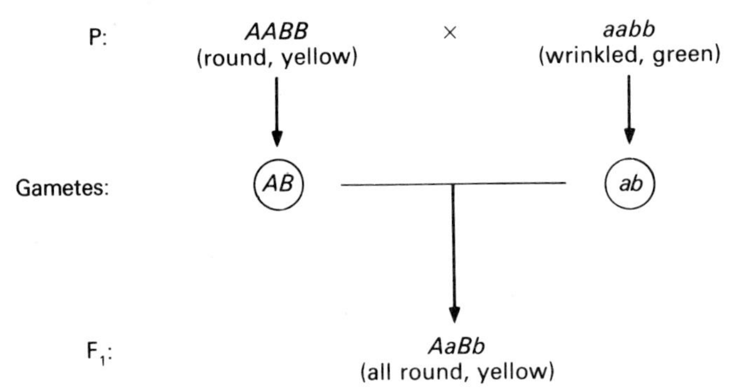

The F_1 produced from this cross are called **dihybrids,** because they are hybrid for two different pairs of contrasting characters.

Mendel continued the crossing sequence by allowing the dihybrid F_1 plants to self-pollinate, resulting in matings of the form *AaBb* × *AaBb*. Since a gamete receives only one of each pair of alleles, four kinds of gametes will be formed by the F_1: (*AB*), (*Ab*), (*aB*), and (*ab*). What we do not know at this point is the expected proportion of each type. Are all gamete types equally likely to occur (1/4 or 25 percent of each) or are certain gamete types formed more frequently than others? For example, is there perhaps a tendency for gametes to contain all dominant or all recessive genes, so that (*AB*) and (*ab*) gametes are more likely than (*Ab*) and (*aB*) types? After all, the original parents were totally dominant and totally recessive.

The simplest hypothesis, which turns out to be the correct one, is that nonallelic genes segregate independently of each other. The four gene combinations are then equally likely to occur in gametes, and the predicted results of a dihybrid cross can be obtained by using a Punnett square. Figure 1.10 illustrates the 16 different ways in which gametes in a dihybrid cross can combine during fertilization. Of these 16 F_2 zygotic combinations, there are only nine different genotypes, since some genotypes occur more than once. The F_2 genotypes and their proportions are 1/16 *AABB*, 2/16 *AABb*, 1/16 *AAbb*, 2/16 *AaBB*, 4/16 *AaBb*, 2/16 *Aabb*, 1/16 *aaBB*, 2/16 *aaBb*, and 1/16 *aabb*, yielding a ratio of 1:2:1:2:4:2:1:2:1. Because of dominance, the phenotypic ratio is 9 round, yellow (1 *AABB* + 2 *AABb* + 2 *AaBB* + 4 *AaBb*) : 3 round, green (1 *AAbb* + 2 *Aabb*) : 3 wrinkled, yellow (1 *aaBB* + 2 *aaBb*) : 1 wrinkled, green (1 *aabb*). If we use a dash (-) to indicate that the second gene can be either dominant or recessive and still allow the first gene to produce a dominant phenotype for that gene pair, the phenotypic ratio

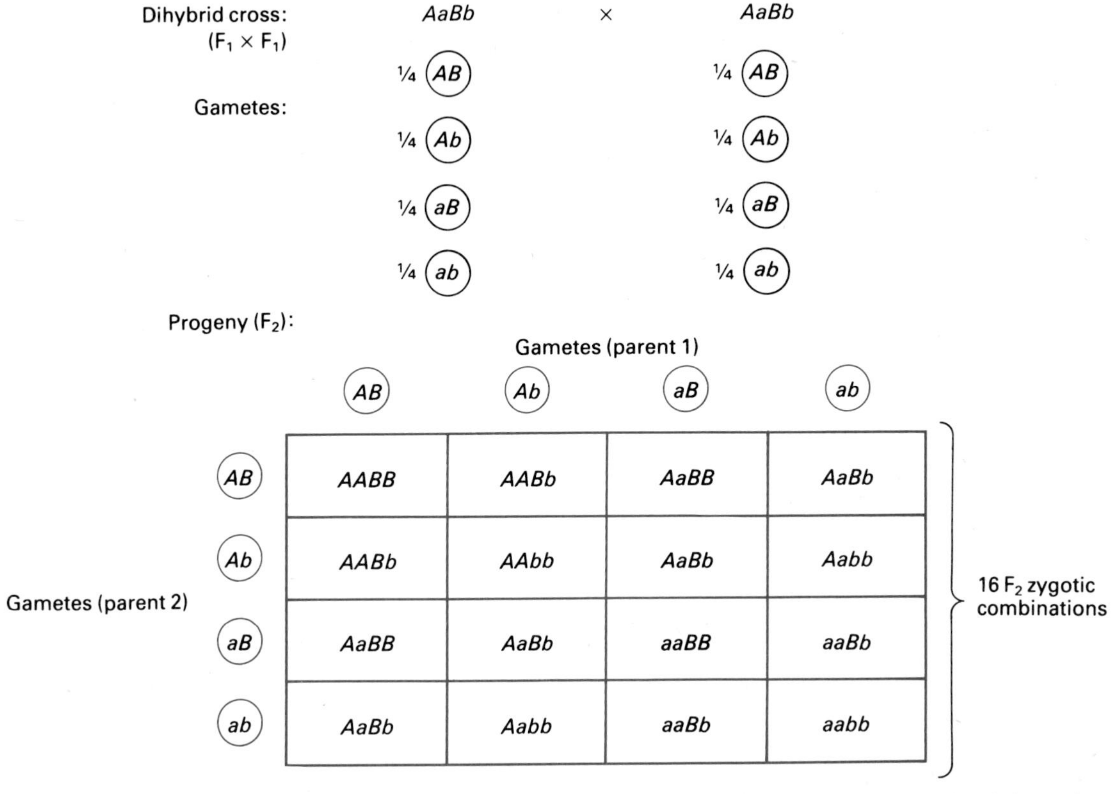

Figure 1.10. The independent assortment of nonallelic genes in a dihybrid cross. With independent assortment, all gamete types from both dihybrid parents are equally frequent. Each type makes up 1/4 of the total gametes of that parent. The 16 ways in which gametes can then combine during fertilization are illustrated in the Punnett square. Each possible zygotic combination constitutes 1/16 of the total F_2 generation.

can be expressed as 9 *A-B-* : 3 *A-bb* : 3 *aaB-* : 1 *aabb*. The F_2 ratios of genotypes and phenotypes that result from the cross *AaBb* × *AaBb* are summarized in Table 1.3.

The assumption that nonallelic genes associate independently during gamete formation was verified by Mendel's analysis of F_2 results. His actual dihybrid cross (*AaBb* × *AaBb*) for these traits produced 556 offspring:

315 round, yellow	101 wrinkled, yellow
108 round, green	32 wrinkled, green

The ratio of 315 to 101 or of 315 to 108 is about 3 to 1, and the ratio of 315 to 32 is about 9 to 1. The overall phenotypic ratio in Mendel's experiment is therefore 9 round, yellow : 3 round, green : 3 wrinkled, yellow : 1 wrinkled, green. If we inspect the data more closely, however, we find that each trait, when considered separately of the other, produces a 3:1 phenotypic ratio. The total number of

Table 1.3 Genotypic and phenotypic ratios resulting from the cross *AaBb* × *AaBb*.

GENOTYPE	GENOTYPIC PROPORTIONS	PHENOTYPE	PHENOTYPIC PROPORTIONS
AABB	1/16	Round, yellow	9/16
AABb	2/16		
AaBB	2/16		
AaBb	4/16		
AAbb	1/16	Round, green	3/16
Aabb	2/16		
aaBB	1/16	Wrinkled, yellow	3/16
aaBb	2/16		
aabb	1/16	Wrinkled, green	1/16

round-seeded plants produced is 315 + 108 = 423, and the total number of wrinkled-seeded plants produced is 101 + 32 = 133, for a ratio of 423 : 133, or about 3:1. Similarly, the total number of yellow-seeded plants produced is 315 + 101 = 416, and that of green-seeded plants is 108 + 32 = 140, again yielding a ratio of about 3:1. The 9:3:3:1 phenotypic ratio is thus seen to be the product of the two separate 3:1 ratios:

(3 round + 1 wrinkled)(3 yellow + 1 green) =
9 found, yellow + 3 round, green + 3 wrinkled, yellow + 1 wrinkled, green.

When the proportions from a cross involving two or more traits can be obtained by multiplying together the proportions expected for each trait alone, then the genes for the different traits are behaving independently during gamete formation. (The multiplication law of mathematical probability, which applies in this situation, is discussed in the next chapter.) Each gamete is formed in the F_1 as though we had randomly chosen an *A* or *a* allele and then randomly picked a *B* or *b* allele to go into the gamete with it. Whether the *A* or *a* allele is chosen has no bearing on whether the *B* or *b* allele is picked. Each possible combination of nonallelic genes is equally likely. This is Mendel's *Principle of independent assortment* stated formally as follows:

RULE 1.2 **Principle of independent assortment**

Nonallelic genes segregate into gametes independently of each other, producing equal proportions of all possible gamete types.

EXAMPLE 1.4. Coat color and spotting pattern in the Cocker Spaniel breed of dogs depend on two gene pairs that segregate independently. A black coat color (*B*-) is dominant to red (*bb*) and solid color (*S*-) is dominant to white spotting (*ss*). A red and white spotted female (*bbss*) produces a litter of five pups: two red and white spotted, one black and white spotted, one solid red, and one solid black. Deduce the genotype and phenotype of the male parent in this cross.

Solution. The five pups have the following genotypes: red and white spotted = *bbss*, black and white spotted = *B-ss*, solid red = *bbS-*, and solid black = *B-S-*. Observe that the solid black pup has at least one *B* and at least one *S* allele. Since the mother is *bbss*, the *B* and *S* alleles had to come from the male parent. Based on this information, we can deduce that the genotype of the father is *B-S-*. The red and white spotted pups must have received both a *b* and an *s* allele from each parent. The father must therefore also carry a copy of both the *b* and *s* alleles. We now know that the male parent must be *BbSs* in genotype and solid black in color.

Testcross Results. Experimental geneticists in the early 1900s showed that the Principle of Independent Assortment does not apply to all gene pairs. That is, not all nonallelic genes segregate independently during gamete formation. We will discuss how gamete proportions are determined for genes that do not assort independently in Chapter 7. For now, we need an experimental method to determine whether or not genes are segregating independently. A 9:3:3:1 ratio is certainly one indication of independent assortment. But there is a more direct way to confirm independent assortment. That method is a testcross involving a dihybrid (such as *AaBb*) and a recessive homozygote. The recessive parent (called the testcross parent) produces only the doubly recessive type of gamete. The phenotypic ratio among the offspring will therefore depend solely on the ratio of gametes produced by the dihybrid parent. If nonallelic gene pairs segregate into gametes independently, then the four gamete types will be equally frequent, as will the four offspring classes produced upon fertilization. The predicted results are shown in Fig. 1.11.

When Mendel performed a testcross of a dihybrid with a doubly recessive homozygote, he obtained results of 55 round, yellow : 51 round, green : 49 wrinkled, yellow : 52 wrinkled, green. These results are sufficiently close to the theoretical 1:1:1:1 ratio to confirm independent assortment. If the nonallelic gene pairs had not assorted independently, then some ratio other than 1:1:1:1 would have resulted.

Ratios without Complete Dominance

Both incomplete dominance and codominance can modify dihybrid phenotypic ratios, just as they modified the results of monohybrid crosses. These modifying factors can act on either one or both gene pairs, giving rise to deviations from the classical 9:3:3:1 ratio.

If either one or both gene pairs in a dihybrid cross show incomplete dominance or codominance, the phenotypic ratio will more closely resemble the genotypic ratio. For example, the shape of radishes is determined by an incompletely dominant pair of alleles, where *LL* = long, *LL'* = oval, and *L'L'* = round. Radish color is also determined by an incompletely dominant pair of alleles, where *RR* = red, *RR'* = purple, and *R'R'* = white. When oval, purple dihybrids are intercrossed, the genotypic ratio in the progeny is expected to be 1 *LLRR* : 2 *LLRR'* : 1 *LLR'R'* : 2 *LL'RR* : 4 *LL'RR'* : 2 *LL'R'R'* : 1 *L'L'RR* : 2 *L'L'RR'* : 1 *L'L'R'R'*. Because of the lack of complete dominance in each allelic pair, the expected phenotypic ratio is 1 long, red : 2 long, purple : 1 long, white : 2 oval, red : 4

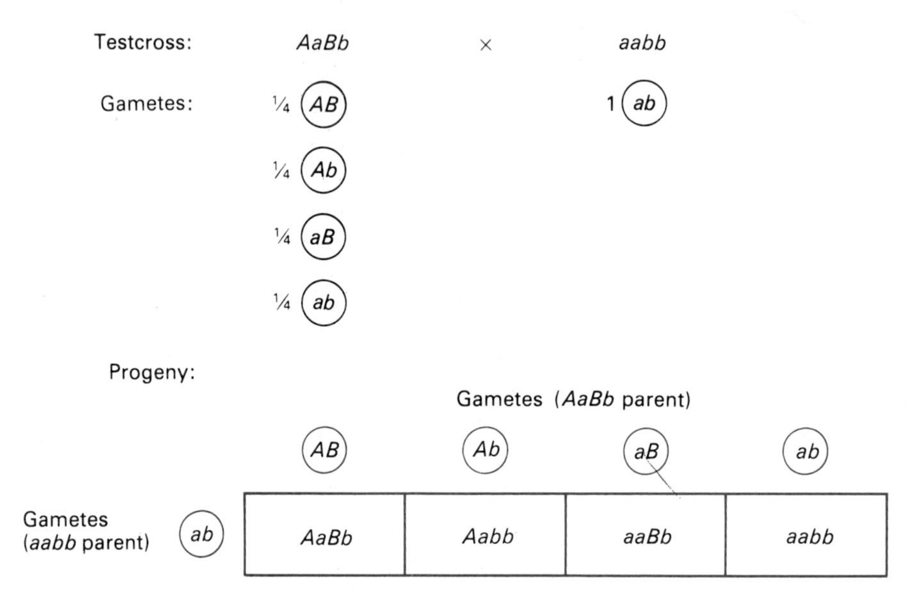

Progeny genotypic ratio = progeny phenotypic ratio = 1:1:1:1

Figure 1.11. The independent assortment of nonallelic genes in a testcross in which the doubly dominant parent is dihybrid. The ways that gametes combine during fertilization are shown in the Punnett square. Note that the proportions of zygotic combinations are the same as the frequencies of the gamete types produced by the heterozygous parent. Since each zygotic combination expresses both a different genotype and a different phenotype, the genotype and phenotype ratios are both 1:1:1:1.

oval, purple : 2 oval, white : 1 round, red : 2 round, purple: 1 round, white, or 1:2:1:2:4:2:1:2:1. Note that an important aspect of incomplete dominance (or codominance) is an increase in the number of phenotypic classes, so that they equal the number of genotypic classes.

There are many cases in which dominance is complete in one gene pair but not in the other. In cattle, the allele for the hornless (or polled) condition (*H*) is completely dominant over that for horns (*h*). But recall that the alleles for color are codominant, so that RR = red, RR' = roan, and $R'R'$ = white. A cross between heterozygous hornless, roan cattle ($HhRR' \times HhRR'$) is expected to yield 3 hornless, red (1 $HHRR$ + 2 $HhRR$) : 6 hornless, roan (2 $HHRR'$ + 4 $HhRR'$) : 3 hornless, white (1 $HHR'R'$ + 2 $HhR'R'$) : 1 horned, red (1 $hhRR$) : 2 horned, roan (2 $hhRR'$) : 1 horned, white (1 $hhR'R'$), for a ratio of 3:6:3:1:2:1.

EXAMPLE 1.5. In tomatoes, the degree of pubescence (hairy covering) for fruit and for stems and leaves varies independently. The texture of the fruit covering may be smooth (*P*-) or peach (*pp*). Stems and leaves can be either hairy (*HH*), with scattered hairs (*HH'*), or hairless (*H'H'*). Suppose that seeds from a particular cross are planted and yield 50 tomato plants with the following characteristics: 9 smooth, hairy; 19

smooth, scattered hairs; 8 smooth, hairless; 3 peach, hairy; 7 peach, scattered hairs; and 3 peach, hairless. What are the most likely genotypes of the parents?

Solution. The ratio of smooth (9 + 19 + 8 = 36) to peach (3 + 7 + 3 = 13) is about 3:1. Such a ratio is expected for a monohybrid cross with complete dominance. Thus, both parents probably have the *Pp* genotype. The ratio of hairy (9 + 3 = 12) to scattered hairs (19 + 7 = 26) to hairless (8 + 3 = 11) is about 1:2:1. This is the result expected for a monohybrid cross with incomplete dominance. We can assume that both parents are probably *HH′* in genotype. The overall genotype of the parents in the cross is most likely *PpHH′*. (Another way to arrive at this same conclusion is by recognizing that the numbers approximate a 3:6:3:1:2:1 ratio overall.)

Two or More Genes Affecting the Same Character

Up to this point, we have dealt with simple genetic patterns in which each trait is determined by a single gene pair. We will now look at some situations in which a trait is determined by two or more gene pairs. The products of the nonallelic genes act together in this case to produce the phenotype. Effects produced by multiple genes are quite common among organisms, because even seemingly simple traits tend to be formed through a complex sequence of developmental steps, with different genes controlling different parts of the overall process.

One consequence of the effects produced by multiple genes is that more phenotypic classes may occur than can be accounted for by the action of only one pair of alleles. A classic example is the inheritance of comb shape in domestic breeds of chickens. The Leghorn breed has a "single" comb (Fig. 1.12). Wyandottes have a structurally different type of comb, a "rose" comb, whereas Brahmas have "pea" combs. When Wyandottes and Brahmas are crossed, the F_1 hybrids all have a "walnut" type of comb, whose shape is different from that of either parent.

Figure 1.12. Comb shapes in domestic breeds of chickens: (a) single, (b) pea, (c) rose, and (d) walnut. Comb shape is determined by two independently assorting gene pairs.

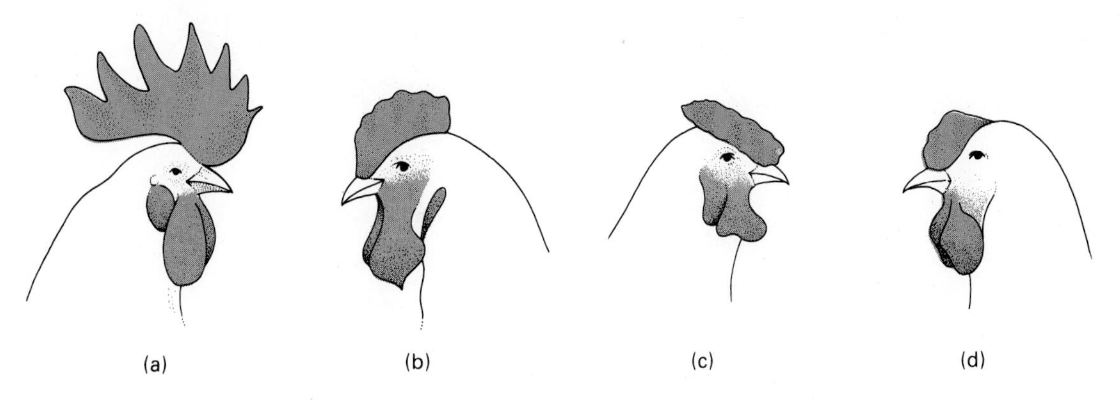

Intercrosses among the F_1 give *four* kinds of F_2 progeny—too many to be accounted for by a single pair of genes. Moreover, the progeny occur in a ratio of 9 walnut : 3 rose : 3 pea : 1 single, which is a classical Mendelian *dihybrid* ratio with complete dominance in each allelic pair. Using the symbols *R,r* and *P,p* to denote the allelic pairs, the genotypes and their corresponding phenotypes are *R-P-* = walnut, *R-pp* = rose, *rrP-* = pea, and *rrpp* = single. The Wyandotte breed is *RRpp*, the Brahmas are *rrPP*, Leghorns are *rrpp*, and the F_1 walnuts are *RrPp*. In this manner, one character is able to exhibit several novel phenotypes because two gene pairs, rather than one, specify the trait.

Inheritance of Coat Color. One trait for which several genes are responsible is coat color in mammals. Various genes are known to affect the amount, kind, and localization of melanin pigment in animal coats. Two of these gene pairs are the *A,a* alleles, which determine the distribution pattern of color in hairs, and the *B,b* alleles, which determine the color of pigment produced. The dominant *A* allele is responsible for a pigment pattern called *agouti*, which yields a grayish appearance. This protective coloration is characteristic of many mammals in wild populations. It is produced by a band of yellow located toward the tip of each colored hair shaft (Fig. 1.13). The recessive *a* allele produces a solid-colored shaft, without the yellow band. (The lethal A^Y allele for yellow color in mice, which was mentioned in an earlier section, is another variant form of the *A* gene.)

The black-colored hair of many wild animals is formed in the presence of the dominant *B* allele. The recessive *b* allele yields a brown color. When the genotypes

Figure 1.13. Individual hairs showing color patterns common to many mammals. The agouti (wild-type) pattern consists of a dark-colored hair shaft (either black or brown), with a yellow band located toward the tip. The yellow bands give the wild-type coat a grayish appearance. The nonagouti (or plain) pattern lacks the yellow band.

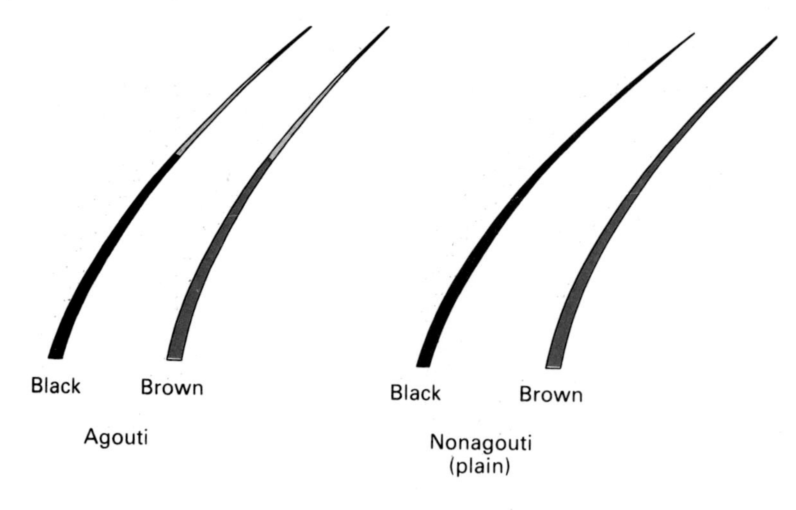

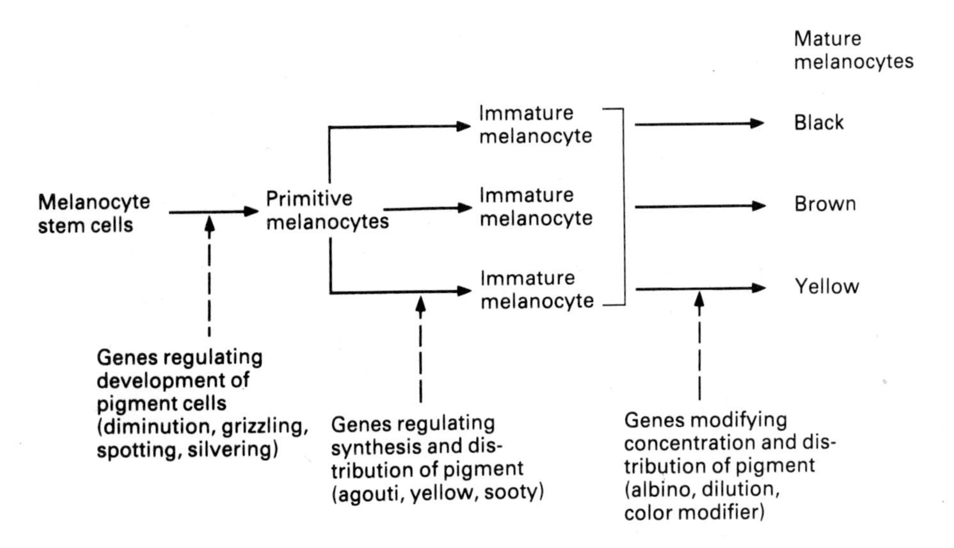

Figure 1.14. Relationship of various known genes to the development of coat color in certain mammals. *Source*: From S. Wright, "Genic Interaction," in *Methodology in Mammalian Genetics*, W. J. Burdette, ed., Holden Day, San Francisco, 1963.

for both gene pairs are varied, four distinguishable coat color patterns result:

A-B-, agouti black
A-bb, agouti brown (or cinnamon)
aaB-, plain black
aabb, plain brown

The involvement of two gene pairs rather than one in the inheritance of these color patterns is verified by the finding that all four traits appear among the offspring of a cross between two dihybrid agouti black animals. The ratio in the offspring is 9 *A-B-* (agouti black) : 3 *A-bb* (cinnamon) : 3 *aaB-* (black) : 1 *aabb* (brown).

Many interesting variations to this basic scheme exist, particularly among domestic animals. For example, the agouti pattern is absent from horses except in a wild ancestral stock known as the Prejvalski horse. The agouti gene is replaced by another type of dominant *A* gene in domestic breeds. This gene is responsible for limiting the distribution of hair color produced by the *B,b* alleles to certain parts of the body. Instead of agouti black, the bay color occurs in *A-B-* animals. Bay horses are reddish brown with black manes, tails, and lower legs. In contrast, *aaB-* animals are plain black, *A-bb* animals are a full reddish brown color (referred to as chestnut), and *aabb* animals are liver-colored (or liver chestnut).

Several other genes that affect coat color are known. Some produce spotting, and others alter the color or intensity of shading. The relationship of these various genes to the development of coat color is shown in Fig. 1.14. A few will be described in later sections.

Interaction between Nonallelic Genes

Dominance, or interaction among allelic genes, is not the only way in which genes can interact to produce different phenotypic ratios. When two or more gene pairs affect the development of the same character, the interaction between nonallelic genes (called *epistatic* interaction) can also modify the phenotypic ratios. The term **epistasis** was originally used to describe the masking of the expression of one or both members of a pair of alleles by a nonallelic gene. This definition is similar to dominance, except that dominance involves the relationship between alleles of the same gene, whereas epistasis involves an interaction between entirely different genes. As with dominance, the conventional definition of epistasis, which implies an inhibiting effect, should not be taken literally. Only in certain limited cases does a member of one gene pair actually act to suppress the action of another gene.

There are three basic types of epistatic interactions. They can be recognized, at least in part, by the nature of their departure from the classical 9:3:3:1 dihybrid ratio. Listed below are the three types of epistatic interactions, together with some examples of the types of modified groupings they produce in the 9:3:3:1 ratio.

Complementary gene interaction	$9:7 = 9:(3 + 3 + 1)$
	$9:3:4 = 9:3:(3 + 1)$
Modifier gene interaction	$13:3 = (9 + 3 + 1):3$
	$12:3:1 = (9 + 3):3:1$
Duplicate gene interaction	$15:1 = (9 + 3 + 3):1$

Note that a typical result of epistatic interaction is to reduce the number of phenotypic classes. The modified dihybrid ratios are summarized in Table 1.4. The three forms of interactions are described in the sections that follow.

Complementary Gene Interaction. Complementary genes act together in the same biochemical or developmental pathway. The function of one gene is needed for the expression of the other.

An example of complementary gene interaction is provided by common white clover. Some strains of white clover are high in cyanide (HCN) content, and others test negatively for the substance. Cyanide is associated with richer vegetative growth in white clover and, despite its usual toxicity, does not seem to harm cattle that eat these plants. High cyanide production has been shown to be the result of one dominant allele at each of two pairs of genes (*A-B-*). If there is not at least one dominant allele at each pair of genes, the plants fail to produce HCN in detectable amounts. A dihybrid cross (*AaBb* × *AaBb*) gives rise to a 9:7 phenotypic ratio:

$$\underbrace{9\ A\text{-}B\text{-}}_{\text{Dominant allele at each gene pair yields high HCN}} : \underbrace{(3\ A\text{-}bb + 3\ aaB\text{-} + 1\ aabb)}_{aa \text{ and/or } bb \text{ results in no HCN produced}} = 9 \text{ HCN positive} : 7 \text{ HCN negative}$$

Table 1.4 Interactions between nonallelic genes that produce modifications in the classical Mendelian phenotypic ratio of 9:3:3:1. The two gene pairs assort independently and exhibit complete dominance.

Type of interaction	Phenotypic ratio			
	A–B–	*A–bb*	*aaB–*	*aabb*
None (classical ratio)	9	3	3	1
Complementary gene interaction	9	7		
	9	3	4	
Modifier gene interaction	13		3	
	12		3	1
Duplicate gene interaction	15			1

In this case, a testcross (*AaBb* × *aabb*) would give a ratio of 1 HCN positive : 3 HCN negative.

In white clover, cyanide is produced from the substance cyanogenic glucoside through a reaction that is catalyzed by a specific enzyme (enzyme β). Another enzyme (enzyme α) is responsible for the formation of cyanogenic glucoside from precursors. The fact that two different enzymes as well as two different genes are involved in cyanide production suggests that the *A* and *B* genes may function by controlling the formation of enzymes α and β. One possible sequence of biochemical reactions might be the following:

$$\text{Precursor} \xrightarrow[\text{enzyme } \alpha]{\text{Gene } A \downarrow} \text{cyanogenic glucoside} \xrightarrow[\text{enzyme } \beta]{\text{Gene } B \downarrow} \text{cyanide}$$

We can now see why genes *A* and *B* must both be present in their functional form (that is, as *A-B-*) in order for both reactions to occur and for the precursor compound to be transformed sequentially into HCN. Plants lacking a functional *A* gene, as in the case of *aaB-* and *aabb*, also lack enzyme α and are unable to produce cyanogenic glucoside. Without the substrate needed for the second reaction, such plants cannot form HCN. Similarly, plants lacking a functional *B* gene are HCN negative. In this case, the plants are missing enzyme β and are unable to convert cyanogenic glucoside into cyanide.

Experimental tests of the above model have been performed by measuring cyanide production in extracts from the leaves of plants following the administra-

Table 1.5. Tests of leaf extracts for cyanide production.

PHENOTYPE	LEAF EXTRACT ALONE	LEAF EXTRACT AND GLUCOSIDE	LEAF EXTRACT AND ENZYME β
A-B-	+	+	+
aaB-	−	+	−
A-bb	−	−	+
aabb	−	−	−

+ = cyanide production
− = no cyanide production

tion of either enzyme β or cyanogenic glucoside. Results of such tests are summarized in Table 1.5. Observe that *A-bb* and *aabb* plants lack enzyme β. Moreover, neither *aaB-* nor *aabb* genotypes can produce cyanogenic glucoside. Both observations are in agreement with the model.

The same mechanism can account for a 9:3:4 ratio in a dihybrid cross merely by assuming a third distinguishable phenotype. For example, if we were to define the phenotype in white clover on the basis of whether plants can produce both cyanide and cyanogenic glucoside (+ +), cyanogenic glucoside but not cyanide (+ −), or neither compound (− −), a dihybrid cross (*AaBb* × *AaBb*) would then produce a ratio of 9 + + (*A-B-*) : 3 + − (*A-bb*) : 4 − − (*aaB-* and *aabb*).

A more familiar example of the 9:3:4 ratio occurs in the inheritance of coat color in certain mammals. Recall that coat color depends on alleles at the *A* and *B* color genes. But in order to be fully colored, an individual must also possess an active *C* gene. An animal that is *cc* in genotype is unable to produce the melanin pigment, regardless of which alleles are at its other color genes. Such an animal appears albino, with pure white hair and pink eyes. Crosses between *AaBBCc* agouti animals, for example, thus produce an offspring ratio of 9 agouti (*A-BBC-*) : 3 plain black (*aaBBC-*) : 4 albino (3 *A-BBcc* and 1 *aaBBcc*).

EXAMPLE 1.6. While performing some of their early experimental work in genetics, Bateson and Punnett made the surprising discovery that crossing two true-breeding white-flowered varieties of the sweet pea gave rise to an F_1 with all purple flowers. Intercrossing the F_1 yielded an F_2 ratio of 9 purple : 7 white. Explain these results and propose how two white-flowered plants could have all purple-flowered offspring.

Solution. Flower color in sweet peas is determined by two gene pairs with complementary effects. Only plants with the *A-B-* genotype have purple flowers. The flowers of plants with other genotypes (*A-bb*, *aaB-*, and *aabb*) are white. Because a 9:7 ratio is produced in the F_2, the purple-flowered F_1s must be *AaBb* in genotype. Two possible matings between homozygous parental varieties can produce *AaBb* offspring: *AABB* × *aabb* and *AAbb* × *aaBB*. Obviously, the first of these matings could not have occurred in the present situation, since one of the parents (*AABB*) would then have to possess purple flowers. But the second possibility can account for the results observed in the cross. Both *AAbb* and *aaBB* plants would have white flowers. The crossing sequence is then as follows:

P: *AAbb* × *aaBB*
(white) (white)

F_1: *AaBb*
(purple)

F_2: 9 *A-B-* : (3 *A-bb* + 3 *aaB-* + 1 *aabb*) = 9 purple : 7 white

Modifier Gene Interaction. Modifier genes suppress or otherwise alter the activity of other genes. An example of a modifier gene that suppresses the activity of other genes is involved in the inheritance of feather color in chickens. Two of the gene pairs that affect feather color are *I,i* (where *I-* inhibits the expression of color and *ii* gives no inhibition) and *C,c* (where *C-* expresses color in the absence of inhibition and *cc* does not). In this case, *I* is the dominant modifier gene, with an inhibitory effect, and *c* is a recessive gene that fails to carry out a necessary step in pigment production (similar to the albino gene in mammals). One or the other of these genes is responsible for the lack of color in different breeds of chickens. The White Leghorn breed is white because it is homozygous for the dominant *I* gene, even though its other color genes are functional. The White Silkie breed lacks color because it is homozygous for the *c* gene, even though no inhibition is present. A mating between these breeds (*IICC* × *iicc*) produces an F_1 of genotype *IiCc*. The F_1 is also white because of the inhibiting action of the *I* gene. But intercrosses of the F_1s (*IiCc* × *IiCc*) yield a ratio of 13:3.

(9 *I-C-* + 3 *I-cc*) : 3 *iiC-* : 1 *iicc* = 13 white : 3 colored

I- gives inhibition and no color | *ii* and *C-* give color | *cc* fails to give color

A testcross of *IiCc* × *iicc* produces a ratio of 3 white : 1 colored.

This example illustrates a case in which a modifier gene prevents the expression of other genes. There are also modifier genes that have only a partial effect, some enhancing and others detracting from the final phenotype. An example of a modifier gene that detracts from the final phenotype is a dilution gene, *D′*, which controls the intensity of coat color in mammals. In horses, the dilution gene is incompletely dominant, yielding progressively lighter colors with increasing dose (Fig. 1.15). One dose of the *D′* allele (*DD′*) in horses that otherwise have the potential for bay yields dun (or buckskin), with a black mane and tail, while two doses of the *D′* allele (*D′D′*) yields pearl white (called perlino). On the other hand, one dose of the *D′* allele in horses that otherwise have the potential for chestnut yields palomino (a golden color with white mane and tail) while two doses of the *D′* allele results in ivory white (called cremello).

EXAMPLE 1.7. The Angus breed of cattle is black, whereas the Jersey breed is a color called black-and-red. In crosses between the two breeds, the F_1 are always black.

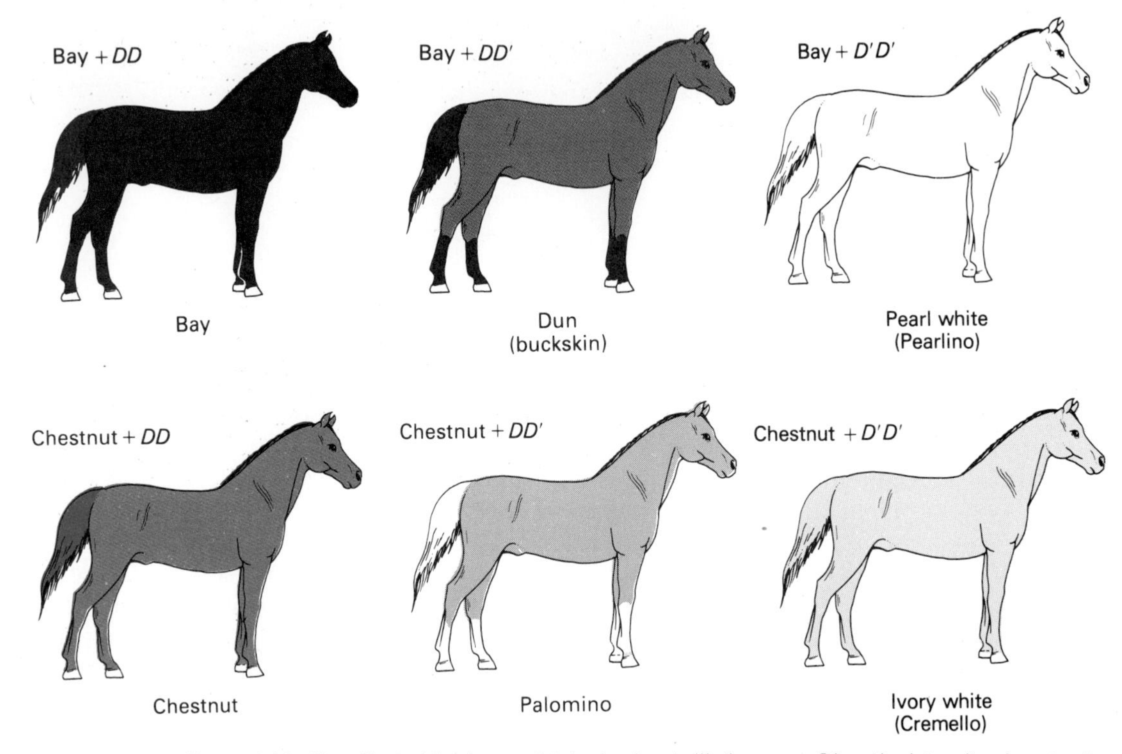

Figure 1.15. The effect of the incompletely dominant dilution gene D' on the intensity of coat color in horses.

Intercrossing these F_1 gives three colors among the F_2: black, black-and-red, and red. One such group of F_2 consisted of 122 black : 29 black-and-red : 9 red. Based on this phenotypic ratio, explain the genetic basis of this trait.

Solution. Since three colors are involved, we might postulate that a single gene pair with incomplete dominance or codominance is responsible for the trait. Thus, red could be *aa*, black *AA*, and black-and-red *Aa*. The $F_1 \times F_1$ crosses would then give an F_2 ratio of 1 red : 2 black-and-red : 1 black. But examination of the observed ratio in the F_2 reveals that it is considerably different from 1:2:1, suggesting that the one gene pair hypothesis is incorrect. (Always try the simplest explanation first; if it does not explain the situation, as in this case, then try an explanation based on a more complex mechanism.) The numbers 122:29:9 approximately form a 12:3:1 ratio, which indicates that two gene pairs and a modifier gene effect are operating. If black is *A-B-* and *A-bb*, then black-and-red is *aaB-*, and red is *aabb*. Thus, the Angus breed is genotypically *AAbb*, the Jersey breed is *aaBB*, the F_1 dihybrids are *AaBb*, and the red F_2 are *aabb*.

Duplicate Gene Interaction. Duplicate genes are separate gene pairs that appear to be identical in function. Traits for which duplicate genes exist are then inherited in the pattern expected for four genes of a kind rather than two.

The shape of the seed capsule in the flower of shepherd's purse is an example. If a dominant allele of either the *A* or *B* gene is present, the shape of the seed capsule is triangular. If only recessive genes are present (*aabb*), the shape is ovoid. A dihybrid cross (*AaBb* × *AaBb*) will therefore yield a 15:1 phenotypic ratio:

$$\underbrace{(9\ A\text{-}B\text{-} + 3\ A\text{-}bb + 3\ aaB\text{-})}_{\substack{A\text{- and/or } B\text{-} \\ \text{gives triangular}}} : \underbrace{1\ aabb}_{\substack{\text{All recessive} \\ \text{gives ovoid}}} = 15 \text{ triangular} : 1 \text{ ovoid}$$

A testcross of *AaBb* × *aabb* results in 3 triangular : 1 ovoid.

As in several of the other examples, in this case a testcross gives a 3:1 ratio. This is the same phenotypic ratio as would be obtained from a monohybrid cross with complete dominance. It often happens that various types of crosses yield the same overall phenotypic ratio even though the genetic effects differ. The ratio alone then does not necessarily imply that a particular genetic mechanism is operating. Before making a final decision on the kind of gene action involved, we must be careful to conduct several kinds of crosses and to note the ratios obtained from each. The ratios obtained must be correlated with the types of crosses performed.

EXAMPLE 1.8. In the Duroc Jersey breed of swine, color can be red, sandy, or white. A given cross of red × white yields all red in the F_1, as does a certain cross of sandy × sandy. Mating the F_1 offspring from the two parental crosses gives 9 red : 6 sandy : 1 white in the F_2. Explain these results in terms of an acceptable genetic model.

Solution. This is an example of a trait determined by duplicate dominant genes with a cumulative effect. Let us assume that two pairs of genes are responsible for body color and that the two pairs contribute cumulatively to the intensity of coloration. Let *aabb* = white, *A-bb* and *aaB-* = sandy, and *A-B-* = red. The cross results can then be explained as follows:

P_1: *AABB* × *aabb* (red) (white) P_2: *AAbb* × *aaBB* (sandy) (sandy)

F_1: *AaBb* (red) × *AaBb* (red)

F_2: 9 *A-B-* : (3 *A-bb* + 3 *aaB-*) : 1 *aabb* = 9 red : 6 sandy : 1 white

TO SUM UP

1. The Principle of Genetic Segregation describes the behavior of alleles, and the Principle of Independent Assortment describes the behavior of nonallelic genes. The latter principle states that members of different gene pairs behave independently of each other during gamete formation. In other words, nonallelic genes do not influence each other during their segregation into gametes. As a result, all of the different gamete genotypes are equally frequent in number.
2. A dihybrid cross in which both gene pairs show complete dominance typically gives a 9:3:3:1 phenotypic ratio in the progeny. A dihybrid cross with incomplete dominance (or codominance) yields a phenotypic ratio of 1:2:1:2:4:2:1:2:1, which is the same as the genotypic ratio.

3. A testcross of 1:1:1:1 is the most direct indication of independent assortment. Thus, in addition to testing for the presence of a recessive gene in the heterozygous condition, a testcross can be used to test for independent assortment of nonalleles.
4. The expression of a genotype containing two or more pairs of genes depends on a number of factors. These factors include dominance and nonallelic gene (or epistatic) interactions. Lack of complete dominance tends to increase the number of phenotypic classes above that number seen with complete dominance. In contrast, epistasis tends to reduce the number of phenotypic classes.
5. Certain modifications of the basic dihybrid ratio of 9:3:3:1 should be immediately recognized as indicating epistasis. These modifications include 9:7 and 9:3:4 for complementary gene interaction, 13:3 and 12:3:1 for modifier genes with a dominant inhibitory effect, and 15:1 for the interaction of duplicate genes. The associated testcross ratios, such as 2:1:1 and 3:1, can easily be confused with monohybrid ratios. You should therefore exercise caution in deducing genetic mechanisms from phenotypic ratios alone.
6. Most traits are formed through a complex series of developmental steps, with different genes controlling different steps in the overall process. Therefore, not only can many characters be affected by a single gene, but many genes can affect a single character. Because of this multiple gene effect, genetic crosses that involve character differences for a single trait can sometimes give rise to phenotypic ratios that would be expected for the independent segregation of two or more gene pairs.

Questions and Problems

1. Define the following terms, and distinguish between the members of paired terms:
   pure-breeding
   dominant–recessive
   alleles
   genotype–phenotype
   homozygous–heterozygous
   testcross
   incomplete dominance–codominance
   epistasis
   pleiotropy
   independent assortment
2. Determine the genotypic ratio expected for each of the following crosses: **(a)** *AA* × *aa*, **(b)** *AA* × *Aa*, **(c)** *Aa* × *Aa*, **(d)** *AAbb* × *aaBB*, **(e)** *AaBB* × *AaBb*.
3. Sheep can be either black or white in color. These colors are determined by a single pair of alleles. A white ram (male) and white ewe (female) produce a black lamb. Which color is dominant? Explain.
4. Wire-haired texture versus smooth hair are alternative charcteristics in dogs. A mating between two wire-haired dogs results in a litter of 3 wire-haired and 2 smooth-haired pups. **(a)** Which trait is dominant? **(b)** If the same parents were to produce several other litters, what ratio of wire-haired to smooth-haired pups is expected among the collective progeny?
5. In guinea pigs, coat color can be black or white. Matings between black guinea pigs sometimes yield 3 black : 1 white and sometimes yield all black offspring. Crosses of black × white sometimes produce all black progeny and sometimes produce a ratio of 1 black : 1 white. White × white crosses produce only white offspring. Explain these results, and indicate which trait is dominant. Write the genotypes of the parents and the offspring in each cross mentioned.
6. In tomatoes, purple stems are dominant to green stems. Several crosses are made involving parental plants with known phenotypes but unknown genotypes. The results of these crosses are given below:

PARENTAL STRAINS	PROGENY	
	PURPLE	GREEN
Purple × Green	121	0
Purple × Purple	97	31
Green × Green	0	101
Purple × Green	63	67
Purple × Purple	76	0

Letting *A* represent the dominant allele and *a* the recessive allele, designate the genotypes of the parents in each cross.

7. Members of the Holstein breed of dairy cattle normally have a black and white spotted coat. On occasion, calves are born that have a recessive red and white spotted coat. A dairy farmer purchases a prized black and white spotted bull. To the farmer's dismay, the bull produces a calf with the recessive coloration when bred to one of his black and white cows. **(a)** What is the genotype of the farmer's bull? (Use *R* and *r* for the color alleles.) **(b)** What phenotypic ratio is expected in the offspring if the bull is mated to red and white spotted cows?
8. Matings between palomino horses yield an overall phenotypic ratio among their offspring of 1 chestnut : 2 palomino : 1 cremello. Predict the phenotypic ratios produced by the following crosses: **(a)** palomino × chestnut, **(b)** palomino × cremello, **(c)** cremello × cremello.
9. Many traits that appear to be completely dominant when viewed at a superficial level actually show incomplete dominance on closer examination. One such case is that of round vs. wrinkled seeds in pea plants. The wrinkled seeds convert less of their sugar content into starch than the round seeds do. As a consequence, less water is retained when the seeds mature, and they take on a wrinkled or shrunken appearance. Microscopic examination of the seed contents reveals that despite their round external shape, heterozygous seeds contain starch granules that are intermediate in size and amount. The granules in heterozygotes are distinctly smaller and less numerous than those in dominant homozygotes. Suppose that a heterozygous plant self-pollinates. Give the phenotypic ratios expected among the seeds of this plant in terms of **(a)** external appearance and **(b)** size and quantity of starch granules.
10. A normally pigmented couple marry and have several offspring. Both the husband and wife in this case come from families in which one of the parents is albino. What fraction of the offspring of this couple is expected to have normal pigmentation?
11. Matings between members of the Mexican Hairless breed of dogs produce smaller than usual litters consisting of both hairless and haired offspring at a ratio of 2:1. Occasional puppies are born dead with certain structural abnormalities. Provide a genetic explanation for these results.
12. Thalassemia is a type of human anemia that appears most frequently among Mediterranean populations. This disorder occurs in two forms. The severe form known as thalassemia major occurs among homozygotes for an aberrant gene. The milder form called thalassemia minor occurs among heterozygotes for the gene. **(a)** What type of dominance appears to be involved in this trait? **(b)** If a mildly affected woman marries a normal man, what fraction of their offspring is expected to be normal? Will any of the children of this couple be severely affected?
13. Platinum-colored foxes produce a progeny phenotypic ratio of 1 silver : 2 platinum : 1 white. The white pups that are produced by such matings die shortly after birth. **(a)** Explain these results by defining the genotypes for the silver, platinum, and white foxes. Use gene symbols starting with the first letter of the alphabet. **(b)** How might fox breeders produce the maximum number of platinum offspring from their foxes, without the losses resulting from the birth of white pups?
14. Lethal traits are not restricted to animals but occur in plants as well. For example, a dominant gene in corn, designated in this problem as *G*, is needed for chlorophyll production. The *G-* plants are green, while the *gg* plants lack chlorophyll and are yellowish white in color, since chlorophyll no longer masks the yellow carotenoid pigments. They are also incapable of photosynthesis and cannot survive independently. Early seedling development does occur in these albino plants, provided by food reserves stored within the seed. But the seedlings die after a period of about two weeks, once these reserves have been exhausted. Suppose that a heterozygous *Gg* plant is selfed. **(a)** What will be the phenotypic ratio among the seedlings? Among the mature plants? **(b)** What fraction of the progeny is expected to carry the *g* allele at maturity?
15. Suppose that a geneticist crosses two true-breeding pea plants: one with purple flowers and long stems and the other with white flowers and short stems. **(a)** Describe the appearance of the F_1 (see Table 1.1 for information

concerning dominance and recessiveness in these traits). **(b)** If the F_1 individuals are selfed, what phenotypes will appear in the F_2, and in what proportions?

16. The fruit of the watermelon can be either short (*A*-) or elongate (*aa*), and its skin can be either green (*B*-) or striped (*bb*). A cross is made between a short, striped variety and a long, green variety. Four different phenotypes appear among the offspring, with approximately 25 percent of the F_1 plants producing fruits that are long and striped. What are the genotypes of the parental varieties in this cross?

17. What phenotypic ratio is expected from the testcross *AaBb* × *aabb* if the corresponding ratio from the dihybrid cross *AaBb* × *AaBb* is (a) 1:2:1:2:4:2:1:2:1, (b) 3:6:3:1:2:1, (c) 9:3:4, (d) 12:3:1, (e) 9:7, (f) 13:3, (g) 15:1, (h) 9:6:1?

18. Coat color (*C*-) in rabbits is dominant to albino (*cc*), and short hair (*L*-) is dominant to long (*ll*). Suppose that in a series of crosses, parents of known phenotypes but unknown genotypes produce the results shown in the table at the bottom of the page. What are the most likely genotypes of the parents in each cross?

19. Susceptibility of corn to a particular plant disease depends on two independently assorting gene pairs. Plants with an *A-B-* genotype are highly resistant to infection, *A-bb* plants are moderately resistant, *aaB-* plants are slightly resistant, and *aabb* plants are unresistant (highly susceptible to infection). Suppose that you have access to moderately resistant and slightly resistant plants. Describe how you would proceed to develop a true-breeding strain of corn that is highly resistant to the disease.

20. Recall that mice with at least one *A* gene and one *B* gene (designated *A-B-*) are agouti black, *aaB-* mice are plain black, *A-bb* mice are agouti brown (or cinnamon), and *aabb* mice are plain brown. **(a)** An agouti black female and a cinnamon male produce offspring, approximately 3/8 of which are agouti black, 3/8 are cinnamon, 1/8 are black, and 1/8 are brown. What are the genotypes of the parents? **(b)** In repeated matings, a black male and a cinnamon female produce a total of 34 agouti black and 36 cinnamon offspring. What are the genotypes of the parents?

21. The Blue Andalusian variety of chickens has slate-blue feathers edged with black. This variety is produced when birds with splashed-white feathers (white with scattered slate-blue flecks) are mated to birds with solid black feathers. Matings between Blue Andalusians produce a ratio in the progeny of approximately 1 black : 2 Blue Andalusian : 1 splashed white. **(a)** What appears to be the genetic basis for feather color in the Blue Andalusian fowl? **(b)** Would it be possible to develop a true-breeding variety of Blue Andalusians? Why or why not?

22. Suppose that Blue Andalusian fowl (see Problem 21) that also show the creeper condition are mated. (Recall that creepers are heterozygous for the recessive lethal *c* gene.) What phenotypes are expected among the live offspring, and in what proportions will they appear?

23. Another character difference in chickens is normal versus frizzled feathers. Homozygous *FF* birds have normal straight feathers, whereas *F'F'* homozygotes show an extreme frizzled condition, with abnormally brittle feathers that curve upward and forward. Heterozygotes (*FF'*) are mildly frizzled, with only slightly abnormal feathers. Suppose that mildly frizzled Blue Andalusian fowl are mated (see Problem

	NUMBER OF OFFSPRING			
PHENOTYPES OF PARENTS	COLORED, SHORT	COLORED, LONG	ALBINO, SHORT	ALBINO, LONG
Colored, short × colored, short	74	26	24	8
Colored, long × albino, short	34	31	33	32
Colored, short × colored, long	49	47	15	17
Colored, long × albino, short	37	32	0	0
Colored, short × albino, short	76	26	0	0

21). (a) How many different phenotypes will be produced among the offspring with respect to both feather shape and color? (b) In what ratio will the phenotypic classes appear?

24. Fruit color in summer squash can be white, yellow, or green. A parental cross of true-breeding white- and yellow-fruited varieties results in an F_1 of all white-fruited plants. Selfing the F_1 gives an F_2 of 12 white : 3 yellow : 1 green. **(a)** How many gene pairs appear to be responsible for fruit color in these crosses? Give the reasons for your answer. **(b)** Designate genotypes for the white- and yellow-fruited parental varieties and for the F_1, and give the general genetic formulas for the different phenotypic classes in the F_2. Use gene symbols starting with the first letter of the alphabet.

25. Two independently assorting genes, *D* and *E*, interact in a complementary fashion to produce normal hearing in humans. Individuals who are homozygous for either or both of the recessive *d* or *e* alleles (*D-ee*, *ddE-*, or *ddee*) are born deaf. **(a)** Suppose that a couple who are both *DdEe* in genotype marry and have a number of children. What proportion of their offspring is expected to have normal hearing? **(b)** Show how a deaf couple in this case could have children with normal hearing.

26. We mentioned in this chapter that matings between the White Leghorn and White Silkie breeds of chickens yield an F_1 that is all white and an F_2 that has a ratio of 13 white : 3 colored. When White Leghorns and White Silkies are mated to members of another white breed of chickens, known as White Wyandottes, they yield the following results:

P: White Wyandotte × White Leghorn
F_1: all white
F_2: 13 white : 3 colored

P: White Wyandotte × White Silkie
F_1: all colored
F_2: 9 colored : 7 white

(a) Is the lack of color in the White Wyandotte breed due to a dominant or recessive gene effect? **(b)** What type of epistatic gene interaction is involved in the White Wyandotte × White Leghorn cross? What type is involved in the White Wyandotte × White Silkie cross? **(c)** Propose the genotype for each breed in these crosses, including in the genotype all genes that affect feather color. Work out the results of each cross using your designated symbols.

27. Seed coat color in a particular type of plant can be either red or white. When true-breeding red-seeded and white-seeded varieties are crossed, their offspring (the F_1) all have red seeds. Intercrossing the F_1 yields an F_2 in which 6 percent of the seeds produced are white. **(a)** How many gene pairs are responsible for seed coat color in these crosses? **(b)** Explain the results of the crosses by designating genotypes for the true-breeding parents and for the F_1 and F_2 plants. Use gene symbols starting with the first letter of the alphabet.

28. Several genes in addition to those described in the text affect coat color in horses. The recessive *e* allele, for example, reduces the intensity of the darker pigments without affecting the lighter red and yellow pigments. Thus while *A-bbDDE-* horses are chestnut in color, *A-bbDDee* horses are sorrel, with a light mane and tail. Another gene, designated *G*, acts in a dominant manner to replace the original hair color with gray. This graying effect often occurs early in life, so that the phenotype of mature *G-* horses is gray, regardless of the presence of other color genes. Horses with a *gg* genotype retain their original color. **(a)** Suppose that crosses are made between gray horses that are *AAbbDDEeGg* in genotype. What colors are expected among the offspring of these matings, and in what proportions will they occur? **(b)** A gray mare is bred to a sorrel stallion. The first mating results in a chestnut foal that later becomes gray. The second mating produces a sorrel colt that retains its color to maturity. What are the genotypes of the stallion and mare in this cross?

29. Three independently assorting genes (*A*, *C*, and *R*) work together in a complementary fashion to produce kernel color in corn. Kernels with at least one dominant allele for each of these three genes (*A-C-R-*) show color; all others are white. Give the ratio of colored to colorless kernels for each of the following crosses:

(a) *AACCRr* × *AACCRr*
(b) *AaCcrr* × *AaCcRR*
(c) *AaCcRr* × *aaccrr*
(d) *AaCcRr* × *AaCcRr*

Chapter 2

Chance and Mendelian Ratios

The results of a genetic cross can be likened to the results of a game of chance, such as coin tossing. One or the other allele in a monohybrid (*Aa*) individual is included in each gamete *at random,* producing two equally likely outcomes: Ⓐ and ⓐ. The toss of a coin also produces two equally likely outcomes: heads (H) and tails (T). When two hybrids are involved in a cross (*Aa* × *Aa*), the alleles in the gametes combine at random to produce three genotypes, in the ratio of 1 *AA* : 2 *Aa* : 1 *aa*. Similarly, when two coins labeled 1 and 2 are tossed together, heads and tails combine at random so that three results—heads on both; head on one, tail on the other; and tails on both—occur in a ratio of 1 (H_1H_2) : 2 ($H_1T_2 + H_2T_1$) : 1 (T_1T_2). The basic laws of probability can be used to predict the outcome of any chance event, such as the toss of a coin. It follows that these laws must also apply to the random processes of gamete formation and fertilization.

Even though the laws of probability give predictions about the likely results of genetic crosses, there is no guarantee that any particular outcome will necessarily occur. Because they are based on random processes, genetic ratios often fail to conform precisely to the predicted results. A precise fit may be lacking because of chance deviations from the predicted values (called random sampling error) or because the initial expectations were incorrect. Statistical tests must be used to determine whether random deviations or incorrect expectations explain the lack of conformity between observation and prediction.

In the sections that follow, we will first consider the mathematics of counting. We will then define the rules of probability and discuss their applications to the analysis of results in genetic crosses. We will also explain a statistical procedure known as the chi-square test, which evaluates the significance of the differences that exist between observed and expected ratios. The chi-square test allows us to decide objectively whether the departures of the actual results from those predicted represent real genetic differences or are merely statistical accidents.

Basis of Genetic Algebra

The alternative results of a single trial are known as outcomes or simple events. For example, the toss of a single coin is a trial that is associated with two possible simple events: H and T. Similarly, gamete production in an *Aa* heterozygote is associated with the simple events Ⓐ and ⓐ. Simple events can also be defined for the throw of a single die. In this case, however, there are six possible outcomes, corresponding to the six faces of a cube.

But most geneticists deal with more complex events, such as crosses that involve two or more pairs of genes rather than just one pair. Let us consider crosses with the dihybrid *AaBb,* for example. Distribution of genes to gametes in the *AaBb* individual results from the simultaneous occurrence of two simple events: one involving the *A,a* allele pair and the other associated with the *B,b* pair. The toss of two different coins, such as a penny and a nickel, also combines two simple events into a complex event. In our discussion, we will be concerned mainly with these more complex events, which are the result of two or more simple events.

In how many ways can these more complex events occur? Gamete formation in the *AaBb* individual yields four alternative outcomes: (*AB*), (*Ab*), (*aB*), and (*ab*). The combined toss of a penny (P) and a nickel (N) also yields four possibilities: H_PH_N, H_PT_N, T_PH_N, and T_PT_N. In these examples, we can quite easily determine the number of ways in which the events can occur by simply listing the alternatives. But many situations involve a larger number of possible outcomes, so that listing them becomes unfeasible. In the sections that follow, we will learn some rules for calculating the number of possible outcomes when it is not feasible to count them.

Computing the Ways That Events Can Occur

Three basic rules are useful in genetics for calculating the number of possible outcomes. These rules are the addition rule, the multiplication rule, and the rule of combinations. We will begin by discussing the addition rule.

RULE 2.1 **Addition rule**

If event 1 can happen in n_1 ways and event 2 can happen in n_2 ways and these two events are mutually exclusive (only one of them can occur), then either event 1 or event 2 can happen in $n_1 + n_2$ ways.

The words *either* and *or* constitute a key phrase that signals the use of the addition rule. For example, suppose that we are interested in the event of getting one head and one tail upon the toss of a penny and a nickel. We can represent this event in general as HT. In order to obtain the event HT, *either* H_PT_N *or* T_PH_N is a satisfactory outcome. So there are two ways in which this event can occur.

We can readily extend Rule 2.1 to include any number of simple events. Let us consider the cross *AaBb* × *AaBb*. If the desired event is a double dominant phenotype (*A-B-*), in how many ways can this event occur? There is one way of producing *AABB*, two ways each of producing *AaBB* and *AABb*, and four ways of producing *AaBb* (see the Punnett square in Fig. 1.10). The answer is therefore 1 *AABB* or 2 *AaBB* or 2 *AABb* or 4 *AaBb* = 1 + 2 + 2 + 4 = 9 total ways. As you can see, we were already using the addition rule when we added together all the genotypes that give a certain phenotype. In many instances, the addition rule is used intuitively. We do not always need to state that it is being used.

RULE 2.2

Multiplication rule

If event 1 can happen in n_1 ways and event 2 can happen in n_2 ways, and these events are independent (that is, the outcome of one does not affect the outcome of the other), then they can happen together in $n_1 \times n_2$ ways, which is the product of the number of ways that each event can occur alone.

The word *and* signals the use of the multiplication rule. For example, suppose that we are interested in the total number of outcomes possible from the toss of a penny and a nickel. We have already calculated that there are four possible results: H_PH_N, H_PT_N, T_PH_N, and T_PT_N. But we can also arrive at this number directly by multiplying the number of alternative outcomes for the penny by the number of alternative outcomes for the nickel. The number of outcomes for the penny (2, H or T), *and* for the nickel (2, H or T) *together* then equals 2 × 2, or 4.

As a genetic example, let us consider gamete formation in the dihybrid *AaBb*. We already know that this genotype can form four kinds of gametes. We can calculate this result directly by using the multiplication rule. If event 1 is the presence of a member of the *A,a* gene pair in a gamete, then event 1 can occur in two ways (*A* or *a*), so $n_1 = 2$. If event 2 is the presence of a member of the *B,b* gene pair in a gamete, then event 2 can also occur in two ways (*B* or *b*), so $n_2 = 2$. Using the multiplication rule, we can calculate that the number of ways in which the two nonallelic genes can occur together in a gamete is 2 × 2 = 4 different ways.

In the example just given, involving two gene pairs, it may seem unnecessary to use the multiplication rule to calculate the possibilities. But when the genotype comprises several gene pairs, the number of possible gamete types can be very large. In such cases, listing all the gamete types can be extremely tedious and often unnecessary, since we are frequently interested only in how many kinds of gametes are possible and not in their specific gene constitutions. As an example, let us consider gamete formation in the tetrahybrid *AaBbCcDd*. The number of different kinds of gametes produced by this genotype can be calculated by extending the multiplication rule to four events: $n_1 \times n_2 \times n_3 \times n_4$ = 2 (*A* or *a*) × 2 (*B* or *b*) × 2 (*C* or *c*) × 2 (*D* or *d*) = 16 gamete types.

Permutations and Combinations. Suppose that a study is being made of a large sample of families with five children. Of interest is the sex composition of each family of children. A few of the families are composed of all boys, and a few are composed of all girls. But most are a mixture of boys and girls. Each child can be a boy or a girl (that is, there are two possible outcomes per child), and the sex determination of each child can be considered a separate event. Using the multiplication rule, we can therefore calculate that $2 \times 2 \times 2 \times 2 \times 2 = 2^5 = 32$ different types of families are possible, when we consider both the sex and the birth order of each child. Some of these possible outcomes (listing the children in age order, beginning with the oldest) are BBBBB, GGGGG, BBBBG, GBBBB, BGBBB, BBGBB, BBBGB, BBBGG, GGBBB, BBGGB, . . ., BBGGG, . . ., BGGGG, and so on.

Although 32 possibilities exist if the birth order is considered, we might find it more informative to group these outcomes into classes based only on overall sex composition, ignoring the birth order. In this case, six events are possible: 5 boys; 4 boys and 1 girl; 3 boys and 2 girls; 2 boys and 3 girls; 1 boy and 4 girls; and 5 girls. Consider those families with 3 boys and 2 girls. How many of the 32 possible outcomes are included in this group? They can be listed as BBBGG, BBGGB, BGGBB, GGBBB, BGBGB, BBGBG, BGBBG, GBGBB, GBBGB, and GBBBG. There are a total of 10 different orders or sequences into which we can arrange the 3 boys and 2 girls. Thus, the event 3 boys, 2 girls can occur in 10 ways.

We have applied the addition rule to find that there are 10 ways in which the desired event can happen. Since these outcomes are mutually exclusive, their numbers were added to get the total of 10. There is nothing theoretically wrong with this approach. But if the number of alternative possibilities is very large, listing them would be a very tedious procedure. We therefore need a formula that can be used to calculate directly the number of orders into which objects can be arranged, without having to list all the different possibilities. The theory of mathematical probability includes formulas by which such calculations can be made. They are called *permutation formulas,* since another word for an arrangement is a permutation. There are 10 sequences, or permutations, of 3 boys and 2 girls in families with five children. Permutation formulas make use of factorials (!), which you have probably encountered previously. We will review the calculation of factorials here and learn how permutation formulas can be used to determine the number of possible outcomes, when each sequence is considered a separate outcome.

Imagine five balls: one red, one green, one blue, one yellow, and one orange. If we arrange these balls in a row, in how many different ways could they be ordered? Some possibilities are Ⓡ Ⓖ Ⓑ Ⓨ Ⓞ, Ⓞ Ⓨ Ⓑ Ⓖ Ⓡ, Ⓑ Ⓞ Ⓖ Ⓨ Ⓡ, Ⓨ Ⓡ Ⓑ Ⓖ Ⓞ, Ⓨ Ⓖ Ⓑ Ⓞ Ⓡ, Ⓖ Ⓞ Ⓨ Ⓑ Ⓡ, and so on. The list includes a very large number of permutations—120 in all. We can calculate this number in the following manner. The first position in the row can be filled by any of five balls, so there are five choices for the first spot. Once that position is filled, there are only four balls left to work with. The number of choices for the second position is therefore four. The third spot can then be filled by any of the three balls remaining. By the time the fourth position is reached, there are only two balls left to choose from, and the last position receives the one remaining ball. Since the filling of each position can be considered a separate event, and we want to fill all the positions together, we multiply $5 \times 4 \times 3 \times 2 \times 1$ to get 120 as the total number of alternative ways the balls can be arranged.

The product $5 \times 4 \times 3 \times 2 \times 1$ is 5 factorial, symbolized 5!. You may recall that any number factorialized is that number multiplied by one less than that number multiplied by two less than that number, and so on. Symbolically,

$$N! = N(N-1)(N-2)(N-3)\ldots(3)(2)(1) \tag{2.1}$$

Thus, the number of permutations or sequences of N distinct objects is N factorial.

EXAMPLE 2.1. A geneticist wishes to make a series of genetic crosses using the fruit fly, *Drosophila melanogaster,* as the experimental organism. The geneticist has seven female flies and seven male flies, all of which are genetically distinct. The geneticist places the seven females into separate bottles and then selects a different male for each. How many different crosses are possible, assuming that each fly is mated only once?

Solution. The first male selected can be mated with any one of seven different females, the second male with any one of the remaining six females, the third male with any one of the remaining five females, and so on, until all males and females are paired off. This gives us $7 \times 6 \times 5 \times 4 \times 3 \times 2 \times 1 = 7! = 5040$ different possible crossing sequences.

One problem with the discussion so far is that the five different colored balls are not really analogous to the five children. While the five balls are distinguished by their five different colors, there are only two distinct kinds of children: boys and girls. Each of the five children cannot have a different sex. A better analogy would be for the five balls to have only two colors: red and green. If three of them are red and two are green, in how many ways can we arrange the five balls in a row? The arrangements are Ⓡ Ⓡ Ⓡ Ⓖ Ⓖ, Ⓡ Ⓡ Ⓖ Ⓖ Ⓡ, Ⓡ Ⓖ Ⓖ Ⓡ Ⓡ, Ⓖ Ⓖ Ⓡ Ⓡ Ⓡ, Ⓡ Ⓖ Ⓡ Ⓖ Ⓡ, Ⓡ Ⓡ Ⓖ Ⓡ Ⓖ, Ⓡ Ⓖ Ⓡ Ⓡ Ⓖ, Ⓖ Ⓡ Ⓖ Ⓡ Ⓡ, Ⓖ Ⓡ Ⓡ Ⓖ Ⓡ, and Ⓖ Ⓡ Ⓡ Ⓡ Ⓖ, for a total of 10. This total is far less than the 120 sequences possible when the five balls are each of a different color.

The permutation formula can be modified to account for the fact that not all objects are different from each other. We begin by noting that if the three red balls could be distinguished from one another, there would be $3! = 6$ ways in which they could be ordered among themselves in any one of the 10 arrangements. The orders are $R_1R_2R_3$, $R_1R_3R_2$, $R_2R_1R_3$, $R_2R_3R_1$, $R_3R_1R_2$, and $R_3R_2R_1$. Similarly, there are $2! = 2$ ways that the two green balls could be ordered in each of the 10 arrangements: G_1G_2 and G_2G_1. Multiplying 10 by both of these factorials gives us the total number of ways in which five distinguishable objects can be ordered: $(10)(3!)(2!) = 5!$, or 120 ways. Thus, in the case in which the balls are one of

Table 2.1. Birth orders in families with five children (B = boy, G = girl).

Event:						
Boys	$x_1 = 5$	$x_1 = 4$	$x_1 = 3$	$x_1 = 2$	$x_1 = 1$	$x_1 = 0$
Girls	$x_2 = 0$	$x_2 = 1$	$x_2 = 2$	$x_2 = 3$	$x_2 = 4$	$x_2 = 5$
List of orders	BBBBB	BBBBG	BBBGG	BBGGG	BGGGG	GGGGG
		BBBGB	BBGGB	BGBGG	GBGGG	
		BBGBB	BGGBB	BGGBG	GGBGG	
		BGBBB	GGBBB	BGGGB	GGGBG	
		GBBBB	BGBGB	GGGBB	GGGGB	
			BBGBG	GGBBG		
			BGBBG	GBBGG		
			GBGBB	GBGBG		
			GBBGB	GBGGB		
			GBBBG	GGBGB		
Calculated no. of orders	$\frac{5!}{5!0!^*} = 1$	$\frac{5!}{4!1!} = 5$	$\frac{5!}{3!2!} = 10$	$\frac{5!}{2!3!} = 10$	$\frac{5!}{1!4!} = 5$	$\frac{5!}{0!5!} = 1$

* Recall that 0! = 1.0.

two colors rather than of five, we can alter the original permutation formula by dividing 5! by the product (3!)(2!):

$$\frac{5!}{3!2!} = \frac{5 \times 4 \times 3 \times 2 \times 1}{3 \times 2 \times 1 \times 2 \times 1} = 10$$

The general rule for this procedure is as follows:

RULE 2.3

Rule of combinations

The number of possible permutations of N total objects, in which x_1 are of one type and x_2 are of another type, is equal to $N!/x_1!x_2!$, where $x_1 + x_2 = N$.

The equation $N!/x_1!x_2!$ is often referred to as the *combination formula;* it is only one of several permutation formulas that exist. It is the only one we will use in calculations. It is a mathematical method that can be used to calculate the number of ways in which an event can occur in any situation in which the total number of objects (N) is divided into two classes (red and green, or boy and girl, for example). In Table 2.1 the formula is used to compute the different birth orders in each type of family with five children.

EXAMPLE 2.2. What fraction of all the families with five children will consist of two boys and three girls?

Solution. Since each child can be a boy or girl (two possibilities per child), there are $2 \times 2 \times 2 \times 2 \times 2 = 32$ different types of families with five children, considering both the sex and the birth order of the children. Of the 32 possibilities, $5!/2!3! = 10$ possibilities will consist of 2 boys and 3 girls. The fraction with 2 boys and 3 girls is then equal to $10/32$, or $5/16$ of the total.

Most uses of permutations in genetics involve cases such as we have just considered, in which the total objects are of just two different kinds. Usually, the members of one class are referred to as successes, and the members of the other class are called failures. For example, we could call red balls successes and green balls failures. The number of successes desired is x_1 and the number of failures (the remainder) is x_2. It makes no difference which class of objects is labeled successes and which failures, since both terms appear in the denominator of the formula. Situations such as these, in which each object can be categorized as either a success or a failure, are called **binomial events,** because there are only two classes of objects (*bi* = two).

Applications to the Analysis of Genetic Crosses

Many features describing the expected outcome of a genetic cross can be predicted mathematically, eliminating the need to work through the cross mechanically. The multiplication rule is particularly useful in this regard. For example, the number of different types of gametes produced by a given genotype is equal to the product of the possibilities for each gene pair. Hence, an *AaBbCcDd* individual can produce $2 \times 2 \times 2 \times 2 = 2^4 = 16$ gamete types. But an *AaBBCcdd* individual can produce only 2 (*A* or *a*) × 1 (*B*) × 2 (*C* or *c*) × 1 (*d*) = 4 gamete types. The number of different gene combinations in the gametes depends not so much on the total number of gene pairs considered as on the number of gene pairs that are heterozygous.

Fertilization gives rise to zygotic combinations, which become the entries in a Punnett square. In general, the number of zygotic combinations produced by a given cross is equal to the product of the number of gamete types from the female parent times the number of gamete types from the male parent. Thus, the dihybrid cross *AaBb* × *AaBb* yields $4 \times 4 = 16$ zygotic combinations in the offspring. The cross *AaBbCc* × *AaBbCc* gives rise to $2^3 = 8$ gamete types from each parent, for a product of $8 \times 8 = 64$ zygotic combinations. The cross *AaBbCcDd* × *AaBbCcDd* yields $16 \times 16 = 256$ zygotic combinations. But the cross *AaBbCcDd* × *AaBBCcdd* gives rise to only $16 \times 4 = 64$ zygotic possibilities.

The number of different genotype classes among the offspring of a cross is usually less than the number of zygotic combinations, since some genotypes may appear more than once in the Punnett square. In calculating the number of genotypic classes, it is helpful to rewrite the cross, separating the different gene pairs into individual crosses written in parentheses. For example, the cross *AaBbCc* × *AaBbCc* can be written as

$$(Aa \times Aa)(Bb \times Bb)(Cc \times Cc).$$

This notation is useful for calculating the results of a cross when each gene pair behaves in an independent manner. It has no special biological meaning. We can see that the number of genotypic classes produced from the *Aa* × *Aa* cross is three (*AA*, *Aa*, and *aa*). Similarly, there are three genotypic classes produced from the *Bb* × *Bb* cross (*BB*, *Bb*, and *bb*), and three from the *Cc* × *Cc* cross (*CC*, *Cc*, and *cc*). Applying the multiplication rule, we calculate the total number of genotypic classes from the overall cross as $3 \times 3 \times 3 = 3^3 = 27$.

Table 2.2. The number of classes of gametes, zygotic combinations, genotypes, and phenotypes for the tetrahybrid cross *AaBbCcDd* × *AaBbCcDd*.

Cross				*AaBbCcDd* × *AaBbCcDd*					
No. of gamete classes				$2^4 = 16$		$2^4 = 16$			
No. of zygotic combinations				$16 \times 16 = 256$					
	(*Aa* × *Aa*)		(*Bb* × *Bb*)		(*Cc* × *Cc*)		(*Dd* × *Dd*)		
No. of genotypic classes	3	×	3	×	3	×	3		= 81
No. of phenotypic classes if dominance is									
complete	2	×	2	×	2	×	2		= 16
not complete	3	×	3	×	3	×	3		= 81
complete for *A,a,* and *B,b* only	2	×	2	×	3	×	3		= 36

The number of phenotypic classes can also be obtained by this method. If dominance is complete, each separate monohybrid cross produces two phenotypic classes in the progeny. The number of phenotypic classes from the overall cross will then be $2 \times 2 \times 2 = 2^3 = 8$. Note, however, that if dominance is not complete, there will be $3^3 = 27$ possible phenotypes among the offspring, so that the phenotypic classes are numerically equal to the genotypic classes.

As a further example, a complete analysis of a tetrahybrid cross is given in Table 2.2. Notice that at this point we are not interested in the proportions of the various classes, only in the number of different classes. In the next section of this chapter, which is concerned with probability, the proportion of each gamete type, genotype, and phenotype will be discussed.

EXAMPLE 2.3. Among the offspring of the cross *AaBbCCdd* × *AabbccDd,* determine the number of genotypic and phenotypic classes expected if dominance is (a) complete or (b) incomplete.

Solution. Rewrite the cross as a series of single-gene pair crosses, and multiply the expected possibilities from each:

	(*Aa* × *Aa*)		(*Bb* × *bb*)		(*CC* × *cc*)		(*dd* × *Dd*)	
No. of genotypes	3	×	2	×	1	×	2	= 12
No. of phenotypes if dominance is								
complete	2	×	2	×	1	×	2	= 8
incomplete	3	×	2	×	1	×	2	= 12

The third rule, the rule of combinations, can be used to analyze situations in which the offspring from a cross are divided into two classes. The disease galactosemia provides an example. This disease is caused by a recessive gene, *g*. Individuals of genotype *gg* are unable to metabolize the milk sugar galactose and must be placed on a milk-free diet. Let us consider the case of carrier parents (*Gg* × *Gg*). If these parents have six children, in how many ways can the event four normal, two affected be obtained? We will arbitrarily designate the normal offspring as successes and the affected offspring as failures. It is now clear that we have a binomial situation, in which the total objects considered is $N = 6$, the number of successes is $x_1 = 4$, and the number of failures is $x_2 = 2$. The number of arrangements of six objects when four are successes and two are failures is $6!/4!2! = 15$. There are therefore 15 sequences or permutations in which the desired event can occur. (You might want to verify this finding by listing the 15 arrangements.)

TO SUM UP

1. We have discussed three rules that govern the calculation of the ways in which events can happen. These rules are the addition rule, the multiplication rule, and the rule of combinations, which governs the arrangements of a group of objects composed of two separate classes.
2. The addition rule is used in "either/or" situations. If there are two or more alternative outcomes that can satisfy the desired event, then the total number of ways the event can occur is equal to the sum of the number of acceptable alternative outcomes.
3. The multiplication rule is used when the desired outcome includes the simultaneous occurrence of two or more events. The number of ways that both events can occur together is the product of the number of ways that each could occur separately, $n_1 \times n_2$.
4. Some situations involve a total number of objects (N) partitioned into two groups, called successes and failures. The desired event in such a case is the occurrence of a specified number of successes (x_1). The number of ways that this event can occur is given by the combination formula $N!/x_1!x_2!$, where $x_1 + x_2 = N$.
5. Problems concerning binomial combinations can also be worked out by applying the addition rule and summing all the possible arrangements that yield a satisfactory outcome. The combination formula is the method generally preferred, however, because it provides a more accurate and more rapid calculation when large numbers of possible arrangements are involved.

Probability and Genetics

We are all familiar with probability statements. These statements are typically given as the chance or odds that some event will occur. Perhaps the most common probability statement is that of the weather forecaster predicting the chance of precipitation. For example, the weather forecaster may say that the chance of rain on a given day is 30 percent. This statement means that out of a large number of days similar to the one in question with respect to weather conditions, it will rain on 30 percent of those days. Thus, out of 100 such days, rain will occur on $0.3 \times 100 = 30$ of them, on the average. Despite the numerical prediction, however, there is no way of knowing whether today is one of those days. Probability statements are useful in describing the average characteristics of a group or population. They are not as meaningful in characterizing any one single item or individual. For this reason, probability statments are often associated with some degree of uncertainty.

Probability statements are frequently used in genetics when groups or populations of gametes and offspring are being studied. The concept of probability is particularly helpful in making precise predictions when the organism being studied produces large numbers of offspring. It is also used on occasion to provide information about single individuals. Genetic counselors, for example, make extensive use of probability in calculating the risk of inheriting a genetic disorder. For instance, a couple who are both carriers of the recessive gene for cystic fibrosis would be advised that each time they conceive a child, there is a 1 in 4 chance of the child inheriting the disease. When applied to a single child, this information is only of limited value as a relative measure of risk, since any one child either will or will not be affected. The information is more meaningful when the offspring of many carrier parents are considered together as a group. Twenty-five percent of the children in this group are expected to have cystic fibrosis.

Usually, though, the prospective parents are not interested in the risk for the whole group of offspring, only in their own. Information based on probability figures, if misunderstood, can then result in some mistaken notions. Some couples might incorrectly interpret the 1 in 4 risk figure as meaning that the disease will not strike until the fourth child. We know from our study of the random events of gamete formation and fertilization that this is simply not true. There are many possible orders in which events can occur. The oldest child could be affected, or the youngest, or any one in between. Moreover, more than one child may be affected, or none at all. Another false notion is that a 1 in 4 risk means that if the first child is affected, then the next several children are less likely to have the disease. In fact, each child has the same probability of having the disorder, regardless of the phenotypes of any previous children. Advice from a genetic counselor is often greeted with mixed emotions and misunderstanding, because of the difficulties most people have in grasping the meaning of a probability statement.

Fundamental Rules of Probability

In our discussion of combinations, we were concerned with the number of ways in which an event can occur. Now we can calculate the ratio of this number to the total number of possible outcomes. This ratio is the **relative frequency** of the event in question. For example, in tossing two coins, 1 and 2, the desired event might be getting a head and a tail, HT. This event can occur in two ways: H_1T_2 or H_2T_1. Since there are four possible outcomes in all (H_1H_2, H_1T_2, H_2T_1, T_1T_2), the relative frequency of the event HT is $2/4 = 1/2 = 50$ percent. This relative frequency is the same as saying that a proportion of $1/2$ of all two-coin tosses are expected to result in the event HT.

Now that we can express the number of ways an event can occur as a ratio or proportion of the total, we can define the **probability** of an event, symbolized $P(E)$, as the theoretical relative frequency:

$$P(E) = \frac{\text{Number of ways that event } E \text{ can occur}}{\text{Total number of possible outcomes}}. \tag{2.2}$$

Thus, in the case of the coin tossing experiment, the probability of the event HT is simply $1/2$.

One must be careful to distinguish between the theoretical relative frequency (or probability) and the empirically determined proportion of successes. When the coin tossing experiment is performed, the relative frequency actually observed may or may not be 0.5. Say the coins are tossed 10 times, with three tosses giving H_1T_2 and three giving H_2T_1. The observed frequency of HT is then $3/10 + 3/10 = 6/10$, or 0.6. This value is close to the theoretical value, but not equal to it. If this experiment of tossing the coins 10 times is repeated several times, and the observed relative frequency of HT is calculated each time, we will probably obtain an array of different values. Some of these values may be close to 0.5; some may not be. The observed variability may be particularly great in this case since the number of tosses is not very large.

Now suppose the experiment is redesigned so that the pair of coins is tossed 100 times. If this 100-toss procedure is repeated several times, more of the observed HT frequencies will probably be closer to 0.5, the theoretical value. In mathematical terms, this result is referred to as the *Law of Large Numbers.* This law states that as the sample size becomes very large, the observed relative frequency of an event approaches its theoretical value (probability). The expected or theoretical value should therefore be a good predictor of what is actually observed if the sample size is large. The converse is also true. If the sample size is small, the observed relative frequency will often not be the same as the expected frequency. This tendency results solely from the small number of trials; it is not an indication that anything is wrong with the theory itself.

We used the addition, multiplication, and combination rules to determine the number of ways that events can occur. These same rules are used in calculating probabilities. As before, "either/or" situations call for addition, the occurrence of two or more events together implies multiplication, and multiple binomial trials can be analyzed with the aid of the combination formula.

RULE 2.4

Addition rule of probability

If E_1 and E_2 are mutually exclusive (alternative) events, the probability that either of them will happen is calculated as the sum of their individual probabilities. Symbolically, $P(E_1 \text{ or } E_2) = P(E_1) + P(E_2)$.

We have already used this rule to calculate the probability of the event HT in coin tossing. There are four possible alternative outcomes when two coins are tossed. Since these four outcomes are equally likely, the probability of each possible event is $1/4$. The probability of HT can then be calculated as $P(\text{HT}) = P(\text{H}_1\text{T}_2 \text{ or } \text{H}_2\text{T}_1) = P(\text{H}_1\text{T}_2) + P(\text{H}_2\text{T}_1) = 1/4 + 1/4 = 1/2$. In throwing two dice, there are $6 \times 6 = 36$ possibilities, each with a probability of $1/36$. Therefore, the chance of obtaining a total of seven showing on the two dice is $P(\text{seven}) = P(1{,}6 \text{ or } 2{,}5 \text{ or } 3{,}4 \text{ or } 4{,}3 \text{ or } 5{,}2 \text{ or } 6{,}1) = P(1{,}6) + P(2{,}5) + P(3{,}4) + P(4{,}3) + P(5{,}2) + P(6{,}1) = 1/36 + 1/36 + 1/36 + 1/36 + 1/36 + 1/36 = 6/36 = 1/6$. An alternative way of viewing the situation is to reason that since there are six satisfactory outcomes out of a total of 36 possible outcomes, the probability of the event is simply $6/36$, or $1/6$.

RULE 2.5

Multiplication rule of probability

If E_1 and E_2 are independent events, the probability that they will both happen is the product of their individual probabilities. Symbolically, $P(E_1 \text{ and } E_2) = P(E_1)P(E_2)$.

In tossing a pair of coins, say a penny and a nickel, the chance of getting a head on the penny is $\frac{1}{2}$, and the chance of getting a tail on the nickel is also $\frac{1}{2}$. Thus, the probability that both these events will occur together is $P(H_P \text{ and } T_N) = P(H_P)P(T_N) = (\frac{1}{2})(\frac{1}{2}) = \frac{1}{4}$. Similarly, the probability that a gamete will receive the *A* allele from an *Aa* individual is $\frac{1}{2}$, as is the probability of receiving the *B* allele from a *Bb* individual. The chance that an *AB* gamete will be produced from an *AaBb* individual is then $P(A \text{ and } B) = P(A)P(B) = (\frac{1}{2})(\frac{1}{2}) = \frac{1}{4}$.

As a final example of the use of the multiplication rule, suppose that both the phenotype and sex of an offspring are specified in the event. Let us consider a couple, both of whom are carriers of the recessive gene *g* for galactosemia ($Gg \times Gg$). What is the probability that their first child will be a galactosemic girl? Event 1 is that the child is galactosemic. The probability of this event is $\frac{1}{4}$. Event 2 is that the child is a girl; its probability is $\frac{1}{2}$. The outcome of interest is for the resulting zygote to be both galactosemic and female. The probability of the combined event is $P(gg \text{ and girl}) = P(gg)P(\text{girl}) = (\frac{1}{4})(\frac{1}{2}) = \frac{1}{8}$. We multiplied in this case because *both* events must occur together for the combined event to occur. If we had added instead, we would be calculating the probability that the child would be either galactosemic *or* girl, which was not the event we were asked to consider.

EXAMPLE 2.4. A husband and wife are both carriers of the recessive gene *g* for galactosemia. They would like to have two children, with the oldest being a girl and the youngest a boy. What is the chance that they will have a girl first and then a boy, where neither child is affected with galactosemia?

Solution. Since the parents are both heterozygous (*Gg*), each child has a $\frac{1}{4}$ chance of being affected with the disorder and a $\frac{3}{4}$ chance of being normal. Each child also has a $\frac{1}{2}$ chance of being a girl and a $\frac{1}{2}$ chance of being a boy. The probability that the first child is an unaffected girl is therefore $(\frac{3}{4})(\frac{1}{2}) = \frac{3}{8}$. Similarly, the chance that the second child is an unaffected boy is $(\frac{3}{4})(\frac{1}{2}) = \frac{3}{8}$. The combined probability that the couple will have a girl as the oldest child and a boy second, neither of whom is affected with galactosemia, is then $(\frac{3}{8})(\frac{3}{8}) = \frac{9}{64}$.

EXAMPLE 2.5. A man, whose sister is affected with cystic fibrosis (a disease caused by a recessive gene), marries a woman whose brother is affected with the same disease. What is the chance that their first child will have cystic fibrosis?

Solution. The solution to this problem illustrates the general procedure that is employed to determine the probability that a mating between two individuals of dominant phenotype produces an offspring with the homozygous recessive phenotype. The chance of a recessive offspring is equal to the product of three probabilities: (probability that the female parent is heterozygous) (probability that the male parent is heterozygous) (probability that the offspring is homozygous recessive if both parents are carriers), where the last probability is $1/4$. We therefore need to determine the probability that each parent is a carrier. In this particular situation, we reason as follows.

All individuals not otherwise designated are assumed to be phenotypically normal; this category includes the man and the woman and both sets of parents. The fact that the man's sister is affected indicates that both his parents are heterozygous (*Aa*). The man is therefore either *AA* or *Aa*. Since he is known to be unaffected by the disease, he cannot be *aa*. Both possible kinds of *A*- offspring can be produced from the cross *Aa* × *Aa*, in a ratio of 2 *Aa* : 1 *AA*. There is thus a $2/3$ chance that the man is a carrier, *conditional* upon the premise that he is normal (dominant) in phenotype. The same reasoning applies to the woman. The conditional probability that she is a carrier, given her normal (dominant) phenotype, is $2/3$.

If both the man and the woman are carriers, then the chance of having an affected child is $1/4$. Since the probability that both are carriers is $(2/3)(2/3) = 4/9$, the overall probability that this couple will have an affected child is then $(4/9)(1/4) = 1/9$.

Conditional probabilities are used extensively in human genetics. Genetic counselors often meet with prospective parents who come from families having a history of a particular recessive disease. These individuals want to know if they carry the causative allele in the heterozygous condition. Breeding experiments, such as testcrosses, are obviously not applicable to humans. With some disorders there are biochemical tests that can directly determine whether an individual with a normal phenotype is heterozygous or homozygous. But with most inherited diseases, there is no way to make this determination with certainty. The best that can be done is to calculate the probability that an individual is a carrier. This is a **conditional probability,** because it is given (or known) that the individual in question is unaffected. That individual must therefore be either homozygous dominant or a heterozygous carrier of the recessive allele. If the parents of an individual can themselves be established as carriers (through an affected sibling of the individual in question, for example), then the chance that the individual is a heterozygous carrier is taken as $2/3$.

Calculating Mendelian Ratios

The rules of probability allow us to predict the proportions of gamete, genotypic, and phenotypic classes. We often apply the addition rule without even realizing that we are using it. For example, when we stated that three-fourths of the progeny from a monohybrid cross have the dominant phenotype, we have added the $2/4$ *Aa* and $1/4$ *AA* classes to obtain $3/4$. The event *A*- can be satisfied by either *Aa* or *AA*. Since they are alternative possibilities for any single individual, their probabilities have been summed. A somewhat less obvious case involves the dihybrid cross *AaBb* × *AaBb*. This cross yields 16 zygotic combinations, of which $9/16$ are *A-B-*, $3/16$ are *A-bb*, $3/16$ are *aaB-*, and $1/16$ are *aabb*. Suppose that normal hearing requires the presence of at least one dominant member of each pair of genes. What fraction of the offspring are expected to be deaf? The outcomes that satisfy this event are *A-bb*, *aaB-*, and *aabb*. Therefore, the chance that a single offspring from such a cross will be deaf is $P(A\text{-}bb \text{ or } aaB\text{- or } aabb) = P(A\text{-}bb) + P(aaB\text{-}) + P(aabb) = 3/16 + 3/16 + 1/16 = 7/16$.

The multiplication rule is used quite extensively to calculate the expected proportion of each gamete type produced by each member of a cross. Let us consider the genotype *AaBbCcDd*. What proportion of all gametes from individuals with this genotype are expected to be (*ABCd*)? As long as the gene pairs are known to segregate independently, we can work with each pair individually and then combine the results through the multiplication rule. When each gene pair is considered separately, the chance of a gamete's receiving the designated gene is ½. The probability of getting an (*ABCd*) gamete is then $P(A \text{ and } B \text{ and } C \text{ and } d) = P(A)P(B)P(C)P(d) = (1/2)(1/2)(1/2)(1/2) = (1/2)^4 = 1/16$.

The genotypic and phenotypic ratios expected from a specific cross can also be determined by applying the multiplication rule. When nonallelic genes assort independently, each mating can be treated as a series of crosses, with one cross for each gene pair. These individual expected values are then multiplied to obtain the final result. To illustrate this procedure, consider the cross *AaBb* × *AaBb*. Recall that by using the Punnett square method, we determined that this cross yields a genotypic ratio of 1:2:1:2:4:2:1:2:1. This ratio can be obtained directly by considering each nonallelic gene pair separately, and then combining the results through multiplication:

$$Aa \times Aa \rightarrow 1/4\ AA : 1/2\ Aa : 1/4\ aa$$
$$Bb \times Bb \rightarrow 1/4\ BB : 1/2\ Bb : 1/4\ bb$$

When the results are combined, we get the following: $(1/4\,AA + 1/2\,Aa + 1/4\,aa)(1/4\,BB + 1/2\,Bb + 1/4\,bb) = (1/4)(1/4)AABB + (1/4)(1/2)AABb + (1/4)(1/4)AAbb + (1/2)(1/4)AaBB + (1/2)(1/2)AaBb + (1/2)(1/4)Aabb + (1/4)(1/4)aaBB + (1/4)(1/2)aaBb + (1/4)(1/4)aabb = (1/16)AABB + (2/16)AABb + (1/16)AAbb + (2/16)AaBB + (4/16)AaBb + (2/16)Aabb + (1/16)\,aaBB + (2/16)aaBb + (1/16)aabb$.

The phenotypic ratio of 9:3:3:1 for this cross can be obtained in the same manner: $(3/4\ A\text{-} + 1/4\ aa)(3/4\ B\text{-} + 1/4\ bb) = (3/4)(3/4)A\text{-}B\text{-} + (3/4)(1/4)A\text{-}bb + (1/4)(3/4)aaB\text{-} + (1/4)(1/4)aabb = (9/16)A\text{-}B\text{-} + (3/16)A\text{-}bb + (3/16)aaB\text{-} + (1/16)aabb$.

The multiplication rule is especially useful when we are interested in the proportion of only one class, not the entire ratio. For example, suppose the cross is *AaBbCcDd* × *AabbCcDD*. What proportion of the progeny of this cross is expected to be *A-B-C-D-* in phenotype? To answer this question, we multiply ¾ (*A-*) × ½ (*B-*) × ¾ (*C-*) × 1 (*D-*), to get 9/32. This approach obviously saves time. If the cross had been written out in a Punnett square, we would have had to write down and sort through a total of 4 × 2 × 4 × 2 = 64 zygotic combinations to obtain the desired result!

Binomial Probability

Earlier in this chapter, we learned that the combination formula could be used to calculate the number of ways that a binomial event could occur. We found that N total objects (or events) could be partitioned into x_1 successes and x_2 failures in a total of $N!/x_1!x_2!$ different ways. In this section, we will consider the probability of a binomial event.

If five children comprise three boys and two girls, there are $5!/3!2! = 10$ different ways that this event can occur. One possibility is the birth order BBBGG. Since the sex of each child is independent of the sex of each of the preceding children, the probability of this event is $(1/2)(1/2)(1/2)(1/2)(1/2) = (1/2)^5 = 1/32$. It should be apparent that this is also the probability of any one of the other possible arrangements, since each arrangement contains three boys, with a probability of $(1/2)^3$, and two girls, with a probability of $(1/2)^2$, for a combined probability of $(1/2)^3(1/2)^2 = (1/2)^5$. The overall probability of three boys and two girls occurring in any order can then be calculated by multiplying the total number of ways that three boys and two girls can be ordered by the probability of any one individual arrangement. This calculation gives us $(5!/3!2!)(1/2)^3(1/2)^2 = 10(1/32) = 5/16$.

These calculations would apply equally well to the probability of obtaining, say, three heads and two tails in five tosses of a coin. The same approach could be used for calculating the probability of having three *A*- and two *aa* offspring in a family of five from the cross *Aa* × *Aa*. In this case, however, the chance of getting each particular sequence of three *A*- and two *aa* is $(3/4)^3(1/4)^2$ rather than $(1/2)^3(1/2)^2$.

Since individual probabilities can vary from case to case, we will use the symbol p to represent the *probability of each success.* The value of p in genetics problems is often 1/2 or 3/4, but in general it could be any value from 0 to 1. We will use the symbol q to represent the *probability of each failure.* The value of q in genetics problems is often 1/2 or 1/4, but it also can take on any value from 0 to 1. The p and q values must add up to one, however: $p + q = 1$. For example, if the chance of each success is 3/4, then the chance of each failure has to be 1/4, since $3/4 + 1/4 = 1$. In general then, if the desired event is three successes and two failures, we can write the probability of any one particular order as p^3q^2. The total probability of this event occurring in any order is then $(5!/3!2!)p^3q^2$. The specific values to be inserted for p and q depend on the problem being considered.

We can now generalize our formula to include any number of successes and failures:

RULE 2.6

Binomial rule of probability

If p is the probability of success in a single trial, and $q = 1 - p$ is the probability of failure in that trial, then the probability of exactly x_1 successes and x_2 failures in N total trials is calculated as

$$P(x_1,x_2) = \frac{N!}{x_1!x_2!}p^{x_1}q^{x_2} \qquad (2.3)$$

This equation is know as the *binomial formula.* In the equation, $N!/x_1!x_2!$ is the number of ways the desired event can occur, and $p^{x_1}q^{x_2}$ is the probability of any one particular arrangement. For example, what is the probability of obtaining 4 heads and 2 tails in 6 tosses of a coin? In this case, the total is $N = 6$, the number of successes desired is $x_1 = 4$, the number of failures is $x_2 = 2$, and the chance of each success (head) and each failure (tail) is 1/2. There are $6!/4!2! = 15$ ways the event can occur, each with a probability of $(1/2)^4 (1/2)^2 = 1/64$. Therefore, the probability of obtaining 4 heads and 2 tails in 6 tosses is

$$P(4,2) = \frac{6!}{4!2!}(1/2)^4(1/2)^2 = 15(1/16)(1/4) = 15/64 = 0.23$$

EXAMPLE 2.6. A husband and wife are both carriers of the recessive gene for galactosemia. What is the chance that among 5 of their children, 4 are normal and 1 is affected with galactosemia?

Solution. The situation clearly conforms to the binomial formula. The values of the symbols in the formula are $N = 5$, $x_1 = 4$ normal children, $x_2 = 1$ galactosemic child, p (the chance of a child's being normal) $= 3/4$, and q (the chance of a child's having galactosemia) $= 1/4$. The overall probability of the event 4 normal children and 1 galactosemic child is then

$$P(4,1) = \frac{5!}{4!1!}(3/4)^4(1/4)^1 = 5(81/256)(1/4) = 405/1024 = 0.40.$$

When the probability $P(x_1,x_2)$ is calculated for all possible values of x_1 and x_2, we obtain what is known as the *binomial probability distribution*. The binomial probability distribution for sex phenotype in families with four children ($N = 4$, $p = 1/2$) is given below for illustrative purposes. The distribution is plotted in Fig. 2.1(a) in the form of a histogram, or bar graph.

Event:	$x_1 = 4$	$x_1 = 3$	$x_1 = 2$	$x_1 = 1$	$x_1 = 0$	boys
	$x_2 = 0$	$x_2 = 1$	$x_2 = 2$	$x_2 = 3$	$x_2 = 4$	girls
Probability:	1/16	4/16	6/16	4/16	1/16	

Figure 2.1. Histograms showing binomial distributions for (a) $N = 4$, $p = 1/2$ and (b) $N = 5$, $p = 3/4$. The abscissa gives the number of successes (x_1) and the ordinate shows the proportion of the population expected to have that number of successes. Note that binomial distributions are symmetrical if $p = q = 1/2$ (graph a), but are asymmetrical if $p \neq q$ (graph b).

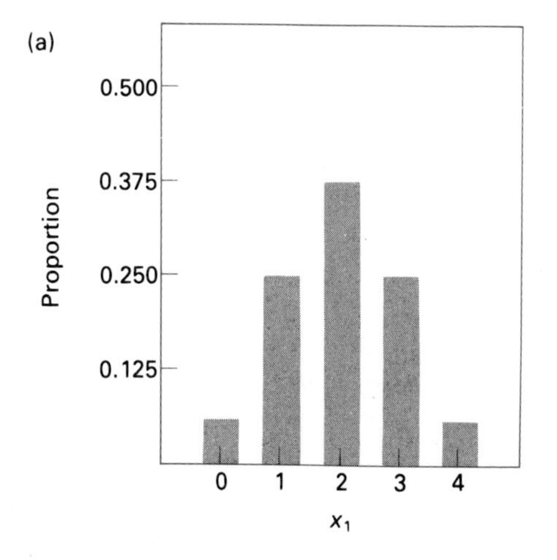

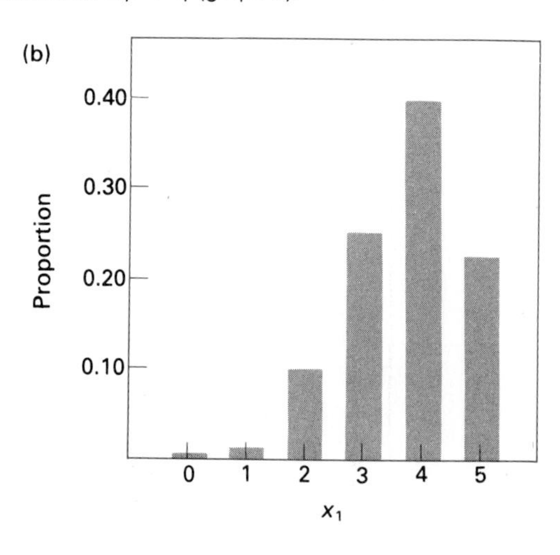

Five outcomes are possible in families with four children. The probabilities of these outcomes add to one: 1/16 + 4/16 + 6/16 + 4/16 + 1/16 = 16/16 = 1. (Since one of these outcomes *must* occur in each family with four children, the sum of the probabilities of all possible outcomes must add up to one.) We can therefore learn a great deal by examining a complete probability distribution such as this one, because it shows the relative proportions of families that are expected to fall into each category. A probability distribution therefore tells us which events are fairly common and which are rare. In this case, only 1/16 (or 6.25 percent) of the families are expected to be made up of all boys, and the same percentage is expected for families to be made up of all girls. At the other extreme, 6/16 (or 37.5 percent) of the families are expected to be made up of 2 boys and 2 girls. An equal number of boys and girls is statistically the overall expectation (or average) result, since each sex has an equal probability of occurring. The chance of this event is indeed higher than that of any other, but it may surprise you to notice that although 6/16 is the highest proportion, it is not even 50 percent.

TO SUM UP

1. We can calculate genotypic and phenotypic proportions directly by using the addition and multiplication rules of probability, thus avoiding the time-consuming Punnett square method. The addition rule of probability is used when the final event can occur in two or more alternative ways. In contrast, if the overall event involves the occurrence of two or more simple events together, then the multiplication rule of probability is the one to use.
2. Binomial probabilities occur when the total number of objects (or events) is divided into successes and failures. The probability of x_1 successes in N total trials is calculated as

$$P(x_1,x_2) = \frac{N!}{x_1 x_2!} p^{x_1} q^{x_2}.$$

The symbols p and q stand for the chance of a success and a failure, respectively, in each individual trial. The term $N!/x_1!x_2!$ gives the number of ways that the event can occur, while the product $p^{x_1}q^{x_2}$ gives the probability of each possible way. By multiplying the two terms together, we get the probability of the binomial event occurring in any possible sequence of successes and failures.
3. The binomial probability distribution lists all possible events and their probabilities. This distribution is useful in that it gives the relative proportions expected in each category.

Tests of Significance: Determining "Goodness-of-Fit"

In genetics, as in other branches of science, the observed numerical results of an experiment are often compared with those expected on the basis of some hypothesis. Suppose an experiment is conducted with the fruit fly, *Drosophila*, to determine whether or not two gene pairs assort independently. A testcross, *RrSs* × *rrss*, yields 500 progeny in a ratio of 142 *RrSs* : 138 *Rrss* : 115 *rrSs*: 105 *rrss*. If the nonallelic gene pairs assort independently, the expected testcross ratio is 1:1:1:1. We must therefore decide if 142:138:115:105 is close enough to a 1:1:1:1 ratio for us to conclude that independent assortment is operating for these gene pairs. In statistical terms, we are being asked to judge the "goodness-of-fit" of the observed ratio to the expected ratio.

Based on the hypothesis of independent assortment, the expected numbers in each class are ¼(500) = 125 *RrSs*, (¼)500 = 125 *Rrss*, (¼)500 = 125 *rrSs*, and (¼)500 = 125 *rrss*. Comparing the observed values with these expected values, we note that there is some deviation in each of the four classes:

	CLASS			
	RrSs	*Rrss*	*rrSs*	*rrss*
Observed (*O*)	142	138	115	105
Expected (*E*)	125	125	125	125
Deviations (*O* − *E*)	+17	+13	−10	−20

We must decide whether these deviations are *insignificant* or *significant* by comparison to chance occurrence. If the deviations are insignificant, we are saying that we believe they are merely a result of random sampling error. In that case, we do not consider that the hypothesis is wrong. Rather, we are concluding that the differences between the observed and expected values reflect the chance deviations that tend to occur to varying degrees when experiments are performed. The hypothesized 1:1:1:1 ratio is then accepted, and we conclude that the gene pairs do assort independently. On the other hand, if the deviations are significant, we are saying that they are too large to be accounted for merely on the basis of chance alone. We must then reject the hypothesis as being correct. Further experimentation or a new hypothesis is then needed.

In determining whether the deviations are insignificant or significant, we base our judgment on the likelihood that random sampling error alone would result in departures from the expected values that are as large as or larger than those observed. The calculations of a statistical test are used to give the probability value associated with a set of deviations. The larger the probability value, the more likely it is that the deviations are merely due to random error. For example, suppose that the calculations of a statistical test give a probability of 0.40. This probability value indicates that every time the testcross *RrSs* × *rrss* is performed under the conditions employed in this experiment, there is a 40 percent chance that random sampling error alone will result in deviations at least as great as those actually found. By contrast, a probability value of 0.01 would indicate that there is only a 1 percent chance that random sampling error alone could cause departures as large as or larger than those observed.

We need to decide how large a probability value we require in order to be confident that random sampling error is the sole cause of the deviations. A cutoff probability value, termed the *level of significance,* must be decided upon prior to running a statistical test. Typically, 0.05 or 5 percent is the value chosen; this helps to minimize the chance of accepting a wrong hypothesis without overly increasing the chance of rejecting a correct one. Thus, if the statistical test gives a probability value above 5 percent, we conclude that random sampling error is indeed the cause of the deviations. We can then call the deviations insignificant and accept the hypothesis (in a statistical sense, we fail to reject the hypothesis). But if the statistical test gives a probability value below 5 percent, then we cannot attribute the observed deviations to chance alone. In that case, we call the deviations significant and reject the hypothesis.

The Chi-Square Test

The chi-square (χ^2) test is a statistical test designed to evaluate the significance of deviations between observed and expected values in two or more categories. The calculations involved in the test transform the deviations in the various categories into a single χ^2 value. This value is a standardized measure of the magnitude of the deviations. Each χ^2 value has an associated probability that gives the chance that repeated experimentation would result in deviations as large as or larger than those observed.

The chi-square value is calculated in the following manner: First, find the difference between the observed (O) and expected (E) values in each class. Next, square each of these deviations in the format $(O - E)^2$. Since a squared term is always positive, this step eliminates the negative sign in front of some of the deviations. After squaring the deviations, divide each squared deviation by the expected value of that class, in the format $(O - E)^2/E$. The division puts each squared deviation in proportion to the expected value for its class. The reason for this procedure can best be demonstrated by an example. Suppose there is one class with a deviation, $(O - E)$, of 20, yielding a squared deviation, $(O - E)^2$, of 400. This appears to be a large deviation. If the expected value (E) is small (e.g., 50), the deviation is indeed large relative to the expected size for the class. But if the expected value for the class is itself large (e.g., 1000), then the relative squared deviation, $(O - E)^2/E$, may actually be minor. The chi-square test thus compensates for different expected values by dividing each squared deviation by the expected value for that class.

The steps of the chi-square test so far for an hypothesis (H_O) of a 1:1:1:1 ratio are summarized below:

	CLASSES			
	RrSs	*Rrss*	*rrSs*	*rrss*
Observed (O)	142	138	115	105
Expected (E)	125	125	125	125
Deviations ($O - E$)	+17	+13	−10	−20
$(O - E)^2$	289	169	100	400
$(O - E)^2/E$	$\frac{289}{125} = 2.31$	$\frac{169}{125} = 1.35$	$\frac{100}{125} = 0.80$	$\frac{400}{125} = 3.20$

We are now ready to determine the value of χ^2. The formula for the χ^2 value is

$$\chi^2 = \sum (O - E)^2/E \tag{2.4}$$

Table 2.3. Probabilities of different chi-square values for degrees of freedom from 1 to 30.

	PROBABILITIES									
df	0.99	0.90	0.80	0.70	0.50	0.30	0.20	0.10	0.05	0.01
1	0.000	0.016	0.064	0.15	0.46	1.07	1.64	2.71	3.84	6.64
2	0.02	0.21	0.45	0.71	1.39	2.41	3.22	4.61	5.99	9.21
3	0.12	0.58	1.00	1.42	2.37	3.67	4.64	6.25	7.82	11.35
4	0.30	1.06	1.65	2.20	3.36	4.88	5.99	7.78	9.49	13.28
5	0.55	1.61	2.34	3.00	4.35	6.06	7.29	9.24	11.07	15.09
6	0.87	2.20	3.07	3.83	5.35	7.23	8.56	10.65	12.59	16.81
7	1.24	2.83	3.82	4.67	6.35	8.38	9.80	12.02	14.07	18.48
8	1.65	3.49	4.59	5.53	7.34	9.52	11.03	13.36	15.51	20.09
9	2.09	4.17	5.38	6.39	8.34	10.66	12.24	14.68	16.92	21.67
10	2.56	4.87	6.18	7.27	9.34	11.78	13.44	15.99	18.31	23.21
15	5.23	8.55	10.31	11.72	14.34	17.32	19.31	22.31	25.00	30.58
20	8.26	12.44	14.58	16.27	19.34	22.78	25.04	28.41	31.41	37.57
25	11.52	16.47	18.94	20.87	23.34	28.17	30.68	34.38	37.65	44.31
30	14.95	20.60	23.36	25.51	29.34	33.53	36.25	40.26	43.77	50.89

where the symbol Σ indicates summation. The sum of the $(O - E)^2/E$ values is $2.31 + 1.35 + 0.80 + 3.20 = 7.66$. Thus, $\chi^2 = 7.66$.

The deviations have now been transformed into a χ^2 value. The probability of getting a χ^2 value that is equal to or greater than the one obtained is found by consulting Table 2.3. Within the body of the table are representative chi-square values, separated into columns. Each column gives the χ^2 values associated with a certain probability value (P value). This is the probability of obtaining a χ^2 value at least as large as the one observed simply as a result of random sampling error. The P values shown are 0.99, 0.90, 0.80, 0.70, 0.50, 0.30, 0.20, 0.10, 0.05, and 0.01. Of course, χ^2 values and P values between any of those given in the table are possible, but we do not have room to list them all here.

To the left of the columns of χ^2 values is a column marked *degrees of freedom* (df). The df in a chi-square test is one less than the number of classes into which the data have been divided. In our case, df $= 4 - 1 = 3$. In a statistical analysis, the df represents the number of classes that can be freely or independently filled. If we place 125 of the 500 offspring into the *RrSs* class, another 125 into the *Rrss* class, and 125 into the *rrSs* class, we have no choice but to place the remaining 125 into the *rrss* class. There is freedom in filling all classes except for the last one filled, which must be composed of the remaining offspring. A helpful analogy is to consider a race with four runners. Once the first, second, and third places have been won, the last runner to cross the finish line has no choice but to be awarded fourth place.

Our χ^2 value is 7.66. We next locate the row in the table in which df $= 3$ and follow this row across to the χ^2 values that bracket 7.66. These values are 6.25 and 7.82. Our calculated χ^2 value is between 6.25 and 7.82, which corresponds to a P value between 0.10 and 0.05. The probability that random sampling error alone would produce a χ^2 value as large as or larger than 7.66, conditioned on the truth of the hypothesis being tested, is between 5 and 10 percent. Since the probability value is above the 5 percent significance level, we conclude that random sampling error is the cause of the deviations between the observed and expected values. The deviations are not regarded as significant and we accept the hypothesis of a 1:1:1:1 ratio. By accepting the hypothesis, we are saying that the observed data statistically conform to a 1:1:1:1 ratio, indicating independent assortment of the nonallelic gene pairs.

EXAMPLE 2.7. A dihybrid cross that involves differences in kernel color yields the following results:

KERNEL COLOR	NUMBER OF KERNELS
Purple	354
Red	132
White	154
Total	640

If we assume a 9 purple : 3 red : 4 white phenotypic ratio, check the goodness-of-fit by means of a chi-square test. Use a 0.05 level of significance for acceptance or rejection of the hypothesis.

Solution. The expected numbers of purple, red, and white kernels under the stated hypothesis are $(9/16)(640) = 360$ purple, $(3/16)(640) = 120$ red, and $(4/16)(640) = 160$ white. The chi-square value is then calculated as follows:

$$\begin{aligned}\chi^2 &= \frac{(354 - 360)^2}{360} + \frac{(132 - 120)^2}{120} + \frac{(154 - 160)^2}{160} \\ &= \frac{(-6)^2}{360} + \frac{(12)^2}{120} + \frac{(-6)^2}{160} \\ &= 0.1 + 1.2 + 0.2 \\ &= 1.5\end{aligned}$$

In this case, df = 3 − 1 = 2. The calculated χ^2 value is between 1.39 and 2.41 in the chi-square table, which corresponds to a P value between 0.5 and 0.3. Since this probability is greater than 0.05, we do not regard the deviations of observed values from expected values as significant, and we accept the hypothesis of a 9:3:4 ratio.

A few words of caution are necessary on using the chi-square test. The test is valid only when whole numbers are used; it will not work with values that are expressed in fractional or percent form. The test is also very sensitive to small sample sizes. As a rule of thumb, the expected value for each class should be greater than or equal to 5 for the test to be accurate. Finally, the decision to accept or reject the hypothesis must not be taken as proof that the proposed genetic ratio is the true one or not. A statistical test can only support or fail to support a proposed hypothesis; it can never actually prove that the hypothesis is true or false. Thus, by failing to reject a proposed hypothesis, we do not necessarily mean that there is nothing wrong with it, only that we have failed to detect anything wrong with it.

Questions and Problems

1. Define the following terms:
random sampling error
independent events
mutually exclusive events
conditional probability
permutations
degrees of freedom

2. Determine the number of ways of obtaining each of the following events:
(a) An ace *or* a king upon drawing a single card from an ordinary deck of 52 cards.
(b) An ace *and* a king upon drawing two cards, each from a different deck.
(c) Four aces from a single deck, when four cards are drawn without replacement.
(d) An even number with the roll of an honest die.
(e) Numbers adding to eight when two dice are rolled simultaneously.

3. Calculate the probability of obtaining each of the outcomes listed in Problem 2.

4. Seven coins are tossed. **(a)** In how many ways can you obtain five heads and two tails? **(b)** What is the probability of getting four heads and three tails?

5. A husband and wife are both carriers of the recessive gene for galactosemia. They have two children. Calculate the probability of each of the following events:
(a) The first child is galactosemic.
(b) Both children are galactosemic.
(c) The first child is galactosemic, and the second is not.
(d) Only one child is galactosemic.

6. In cattle, the polled condition (*H*-) is completely dominant to the horned condition (*hh*). In contrast, coat color shows codominance, with *RR* being red color, *RR'* roan, and *R'R'* white. Calculate the probability of each of the following events:
(a) A polled offspring from the cross *HhRR'* × *HhRR'*.
(b) A polled, roan offspring from the cross *HhRR'* × *HhRR'*.
(c) A horned, red offspring from the cross *HhRR'* × *hhRR*.

7. How many different genotypic classes in gametes can be produced by an individual that is heterozygous for **(a)** three pairs of genes, **(b)** four pairs of genes, **(c)** five pairs of genes, **(d)** *n* pairs of genes?

8. How many different genotypic classes in the progeny can be produced by selfing a plant that is heterozygous for **(a)** three pairs of genes, **(b)** four pairs of genes, **(c)** five pairs of genes, **(d)** *n* pairs of genes?

9. How many different progeny genotypic classes can be produced by a testcross in which one parent is heterozygous for **(a)** three pairs of genes, **(b)** four pairs of genes, **(c)** five pairs of genes, **(d)** *n* pairs of genes?

10. Assume five gene pairs that assort independently. The genes designated by capital letters show complete dominance over their respective recessive alleles. Calculate the probability of each of the following events:
(a) An *AbcDE* gamete from an individual of genotype *AaBbccDdEe*.
(b) An *AaBbCCDdee* offspring from the cross *AaBbCCddEe* × *AaBbCcDDee*.
(c) An *A-bbC-D-ee* offspring from the cross *AaBbCCddEe* × *AabbCcDdee*.
(d) An *A-B-C-D-E-* offspring from the cross *AabbCcddEe* × *aaBbccDdee*.

11. Determine the probability that in a family of six children **(a)** four are girls and two are boys, **(b)** the four eldest are girls and the two youngest are boys, **(c)** at least four are girls, **(d)** all are of the same sex.

12. Albinism is a recessive trait. A normal couple have an albino girl. Calculate the probability that **(a)** the next child is albino, **(b)** of the next four children, the first is albino and the next three are normal, **(c)** of the next four children, one is albino, **(d)** of the next four children, all are normal girls, **(e)** of the next four children, two are albino boys.

13. The creeper condition in chickens is the heterozygous expression of a recessive lethal gene, *c*. Creepers have deformed wings and legs that are shorter than those of normal (*CC*) birds. **(a)** What is the probability that a chick produced from a creeper × creeper mating will be normal? **(b)** What is the chance that out of four chicks produced from creeper × creeper matings, three are normal?

14. Fruit color in summer squash is determined by two interacting gene pairs, *A,a* and *B,b*. Color can be white (*A-B-* or *A-bb*), yellow (*aaB-*), or green (*aabb*). Fruit shape in this plant is controlled by two other gene pairs, *C,c* and *D,d*. Shape may be disk (*C-D-*), sphere (*C-dd* or *ccD-*), or elongate (*ccdd*). **(a)** How many different genotypes are responsible for the white, sphere-shaped phenotype? **(b)** If a tetrahybrid white, disk-shaped plant is selfed, what fraction of the offspring will be yellow and sphere-shaped?

15. In dogs, black color (*B-*) is dominant to red (*bb*), and short hair (*L-*) is dominant to long (*ll*). A mating between dihybrid black, short-haired dogs yields a litter of five pups. What is the probability that two of the offspring will have red, short hair?

16. A small population consists of one female of genotype *AaBb* and two males, one of genotype *AaBB* and the other of genotype *aabb*. The female produces a single egg, which is then fertilized by either one of the males with equal likelihood. What is the chance that the single offspring is of genotype *Aabb*?

17. Determine the number of degrees of freedom (df) involved in testing deviations from the following ratios: **(a)** 3:1, **(b)** 1:2:1, **(c)** 9:3:3:1, **(d)** 3:6:3:1:2:1, **(e)** 1:2:1:2:4:2:1:2:1.

18. When a purple-flowered plant is selfed, it produces an F_1 consisting of 60 purple-flowered and 40 white-flowered offspring. **(a)** Is this result consistent with the 3:1 ratio expected from a monohybrid cross? Check this hypothesis by means of a chi-square test. **(b)** What other explanation might account for these results?

19. In tomatoes, red fruit (*A-*) is dominant to yellow (*aa*), and a tall stem (*B-*) is dominant to dwarf (*bb*). A series of matings between dihybrid tomato plants yields the following offspring: 557 tall, red-fruited plants, 187 tall, yellow-fruited plants, 192 dwarf, red-fruited plants, and 64 dwarf, yellow-fruited plants. Use the chi-square test to check these data for the hypothesis that independent assortment has occurred (i.e., a 9:3:3:1 ratio). What conclusion do you reach?

20. Two of the cross results obtained by Mendel are shown below, along with the corresponding theoretical ratios:

CROSS	RESULTS	HYPOTHESIS
Green pod × yellow pod	428 : 152	3:1
Round, yellow seeds × wrinkled, green seeds	31:26:27:26	1:1:1:1

Check the results of each cross for goodness-of-fit by means of a chi-square test. Use a 0.05 level of significance for acceptance or rejection of the hypothesis.

Suggested Readings / Part I

Chapter 1

Atwood, S. S. and J. T. Sullivan. Inheritance of a cyanogenetic glucoside and its hydrolyzing enzyme in *Trifolium repens*. *J. Hered.* 34:311–320, 1943.

Carlson, E. A. *The Gene: A Critical History*, Saunders, Philadelphia, 1973.

Dunn, L. C. *A Short History of Genetics*, McGraw-Hill, New York, 1965.

Gardner, E. J. *History of Biology*, Burgess, Minneapolis, 1972.

Mendel, G. Experiments in plant hybridization. Reprinted in J. A. Peters (ed.), *Classical Papers in Genetics*, Prentice-Hall, Englewood Cliffs, N.J., 1959.

Searle, A. G. *Comparative Genetics of Coat Colour in Mammals*, Logos Press, London, 1968.

Sturtevant, A. H. *A History of Genetics*, Harper & Row, New York, 1965.

Chapter 2

Batschelet, E. *Introduction to Mathematics for Life Scientists*, Springer-Verlag Berlin, Heidelberg, 1975.

Elandt-Johnson, R. C. *Probability Models and Statistical Methods in Genetics*, Wiley, New York, 1971.

Grossman, S. I. and J. E. Turner. *Mathematics for the Biological Sciences*, Macmillan, New York, 1974.

Mosimann, J. E. *Elementary Probability for the Biological Sciences*, Appleton-Century-Crofts, New York, 1968.

Nucleic Acids and Chromosomes

Chapter 3

Nature of the Genetic Material

So far in our study of genetics, we have regarded genes as the fundamental units of heredity. We defined genes as units of function, each acting in the expression of a particular trait, and treated them as units of transmission that are passed from parents to their offspring according to demonstrable rules. We will now consider the chemistry of the genetic material and describe genes in molecular terms.

The concept of a gene as a chemical substance has largely taken form since the early 1940s with the development of an area of genetic research known as molecular genetics. The great success of investigators in this field of genetics can be attributed in part to their choice of experimental organisms. At first, molecular geneticists concentrated their studies on two main experimental subjects: bacteria and the viruses of bacteria (known as bacteriophages, or phages). Bacteria and phages share several features that make them attractive for genetic analysis. These features include a short reproductive cycle, large numbers of progeny, and ease of manipulation. The small size of these microorganisms also permits the investigator to work with large numbers of individuals. In more recent years, researchers in molecular genetics have extended their investigations to the genes of higher organisms.

Because of fundamental differences that exist in the structure and complexity of these groups of organisms, we begin this chapter with a brief comparison of the two major cell types: bacteria and the cells of higher organisms.

Prokaryotes and Eukaryotes

Much of our knowledge of the form and structure of cells has been obtained from observations using the electron microscope. With this instrument it is possible to investigate structures as small as 0.4 nanometer (nm) in size, constituting a 500-fold increase in resolving power over that obtained with an ordinary light microscope. These cytologic studies show that cells can be classified on the basis of their structural complexity into two major types: **prokaryotes** and **eukaryotes.**

Figure 3.1. Structure of a prokaryotic cell. (a) Generalized bacterial cell showing major structural components. (b) Electron micrograph of bacterial cells. *Source:* (b) From D. P. Snustad, *et al., J. Virology,* 10:126, 1972. Original photograph courtesy of D. P. Snustad and the American Society for Microbiology.

(a)

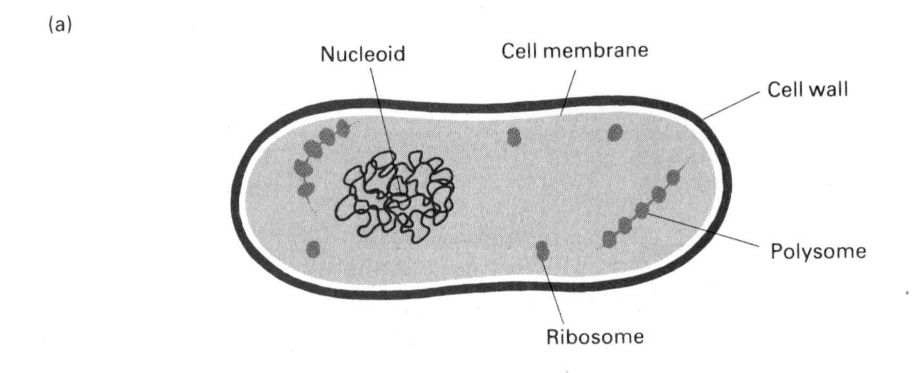

(b)

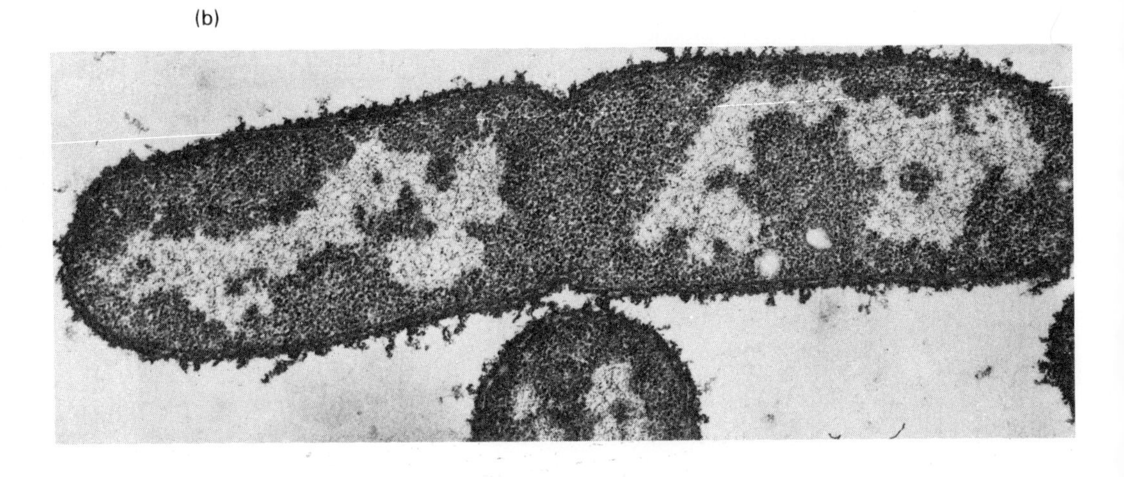

The Prokaryotic Cell

Prokaryotes consist of bacteria and blue-green algae (*cyanobacteria*), which lack the distinct nucleus and certain other membrane-bounded compartments that are characteristic of cells in higher organisms. Prokaryotic cells are relatively small, on the order of 1 to 2 micrometers (μm) in length, and are typically enclosed in a rigid cell wall (Fig. 3.1). Underlying the cell wall is the plasma membrane, consisting of lipid and protein only a few molecules thick. Dispersed throughout the cytoplasm are numerous tiny particles known as **ribosomes.** These particles, each about 20 to 30 nm in diameter, contain RNA (ribonucleic acid) and proteins and function in protein synthesis. Instead of a nucleus bounded by a membrane, the interior of the cell contains an irregularly folded molecule of DNA (deoxyribonucleic acid) in a complex with proteins and RNA. This structure, called a **nucleoid,** serves as a chromosome in the prokaryotic cell.

Bacterial Growth. Bacterial cells increase in number by binary fission. This is a doubling process, in which a single cell divides to form two daughter cells. These two cells then give rise to four, the four give rise to eight, and so on. Each division cycle therefore doubles the number of cells in a population in the geometric progression $1 \rightarrow 2 \rightarrow 4 \rightarrow 8 \rightarrow 16 \rightarrow 32 \rightarrow 64$, and so on. Each term in this progression can be represented as 2^x, where x is the number of division cycles (cell doublings) that have occurred. Progressing at this rate, it would take only 20 division cycles for a single cell to produce a population of 2^{20}, or about one million cells. If each cell doubled every 30 minutes, the population would attain this size in only 600 minutes, or 10 hours. Generation times of as short as 20 to 30 minutes and concentrations of as high as 10^9 bacteria per milliliter are not uncommon in liquid growth medium.

Bacteria are also able to grow on solid medium. When highly diluted samples of liquid cultures are placed on a nutrient agar medium, individual cells are trapped on the gel surface and divide to form separate macroscopic colonies (Fig. 3.2). Each colony constitutes a **clone** of some 10^7 or more cells that originated from the same genetic ancestor.

Eukaryotic Cells

Eukaryotic cells are the cells of organisms other than bacteria and blue-green algae. Eukaryotic cells, particularly those of higher plants and animals, are much larger than prokaryotic cells and more complex in structure. A distinctive feature of eukaryotic cells is the presence of a well-developed, complex nucleus that is separated from the cytoplasm by a nuclear envelope (Fig. 3.3). The nuclear envelope is composed of two membranes that are joined around openings called *nuclear pores.* Dispersed throughout the nucleus of a nondividing cell is a network of fibers known as **chromatin.** Chromatin consists of threadlike strands of DNA in a complex with a number of specialized proteins. Just before cell division, the chromatin fibers become organized into discrete bodies to form the chromosomes of the eukaryotic cell. Also suspended in the nucleus are one or more *nucleoli* (singular, **nucleolus**), which function in the synthesis of ribosomes.

The cytoplasm of eukaryotic cells is characterized by the presence of a network of membranous channels known as the *endoplasmic reticulum* (or ER). Certain of these channels known as the rough ER are attached to numerous ribosomes and function in the synthesis and transport of proteins. Other specialized structures in the cytoplasm include *lysosomes* (for digestion), *Golgi bodies* (for secretion), *mitochondria* (for cellular respiration), and, in the case of plants, *chloroplasts* (for photosynthesis). Mitochondria and chloroplasts both function by producing the cell's main energy currency, adenosine triphosphate (ATP). They are semiautonomous organelles, in that they arise by growth and division and possess their own unique DNA, RNA, and ribosomes.

Figure 3.2. Growing bacterial colonies on solid medium. A highly diluted sample of a bacterial culture is placed on nutrient agar medium. Each individual viable cell grows into a separate colony. The number of viable cells in the original culture is calculated as the number of colonies multiplied by the dilution factor. In this case, for example, if 100 colonies appear on the plate, then the number of viable cells in the original liquid culture is estimated to be $100(10^7) = 10^9$/ml.

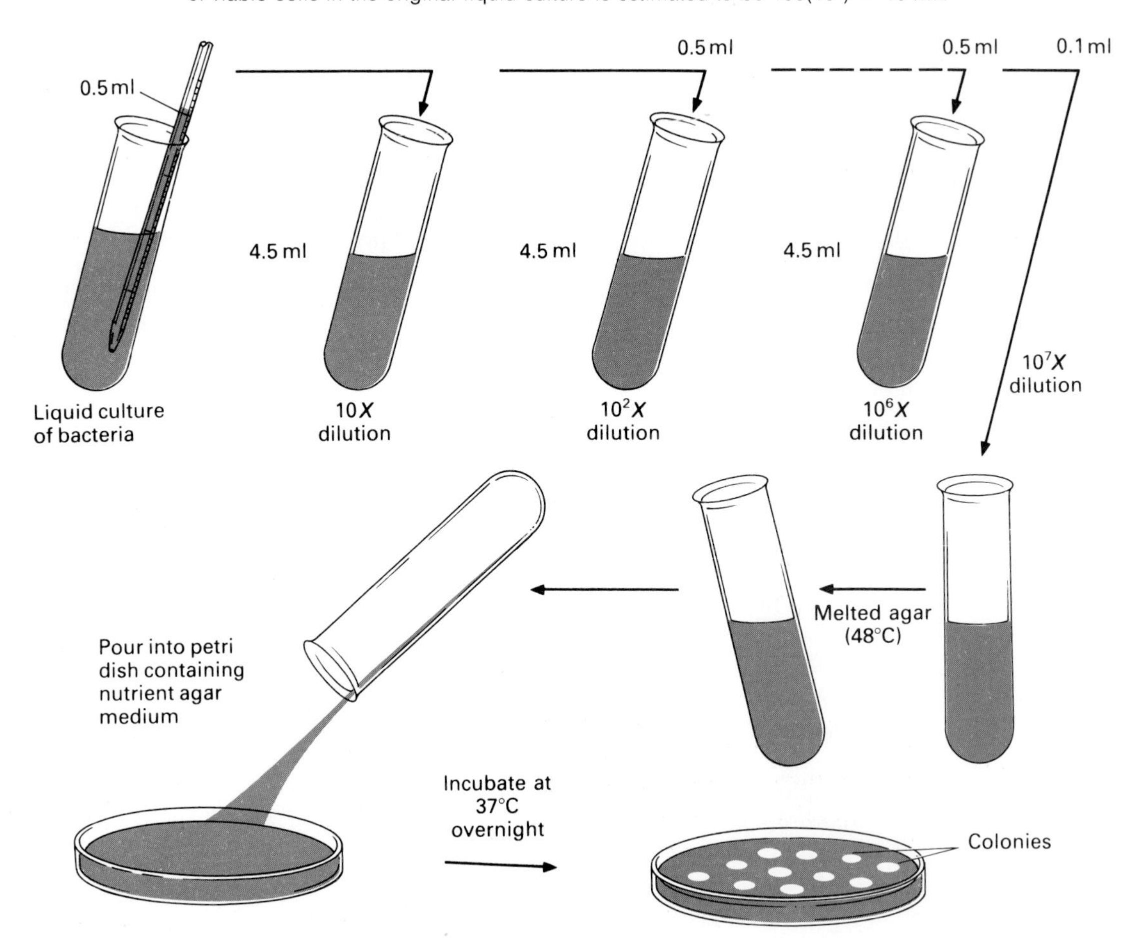

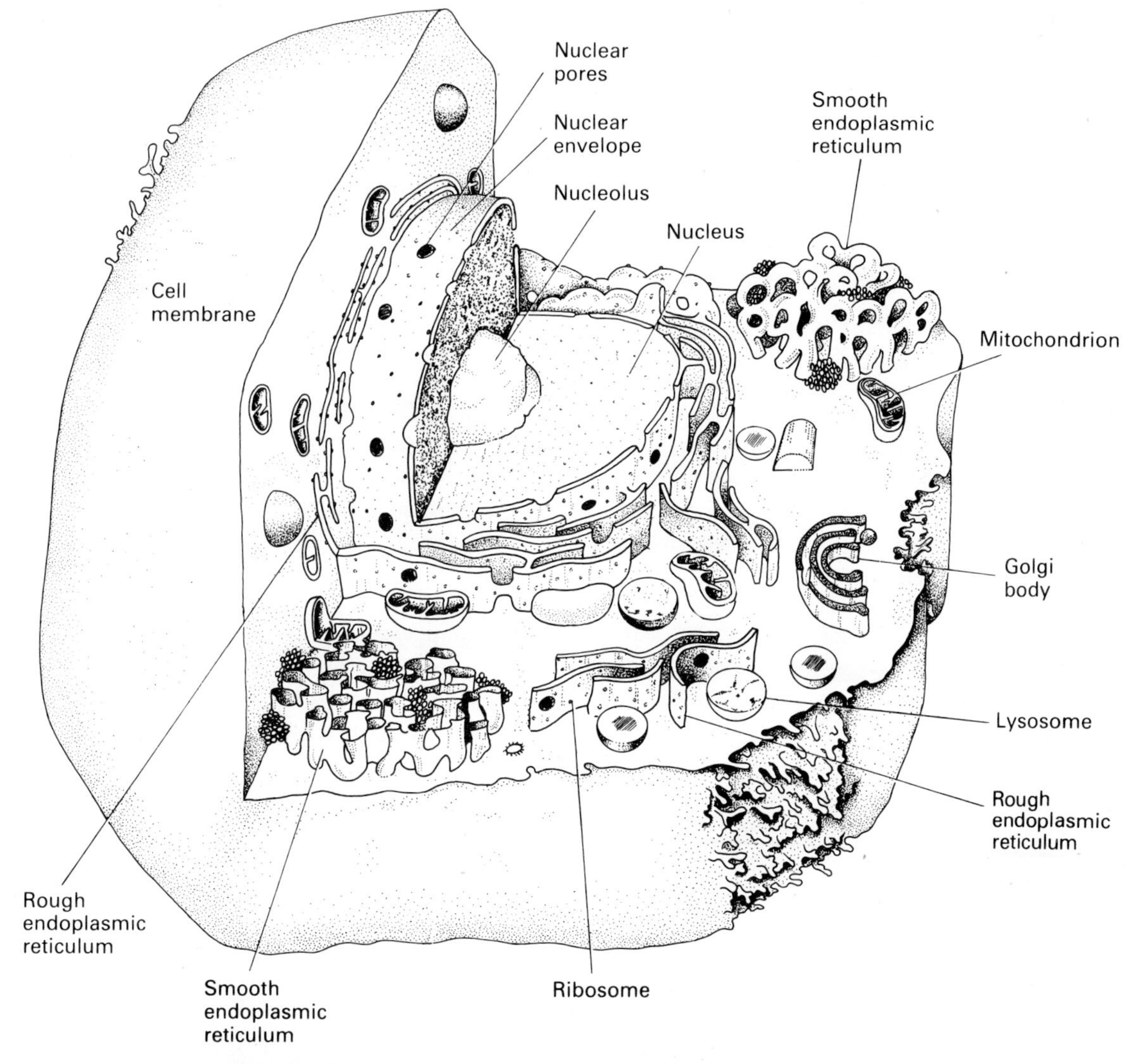

Figure 3.3. Structure of a eukaryotic cell. Generalized mammalian cell showing major structural components.

DNA as the Genetic Material

Any given class of molecules must meet certain basic requirements if it is to serve as the genetic material. The requirements that are essential to function as a gene include: (1) the ability to store genetic information in a stable form and transfer this information to other parts of the cell, (2) the ability to duplicate the information accurately during cell division, and (3) the capability of (infrequent) variation in the form of mutation. Of all the molecules within a cell, only DNA, RNA, and proteins are sufficiently complex to possibly meet these requirements and qualify as candidates for genes. This led to disagreement among scientists over which class of molecules serves as the genetic material. Several experiments conducted during the 1940s and early 1950s finally proved that genes are composed of DNA. Two experiments dealing with genetic transformation of bacteria and the replication of bacterial viruses were particularly important in establishing the genetic role of DNA and are considered in some detail in the following sections. We will see later that RNA is also capable of serving as the genetic material, but it does so only in certain viruses that lack DNA.

Genetic Transformation

Unlike eukaryotes, gene exchange in bacteria typically involves a *one-way transfer* of genetic material from one strain (the *donor* strain) to another strain (the *recipient* strain). **Genetic transformation** was the first such mechanism of gene exchange to be discovered in bacteria. In genetic transformation, the genetic material from a donor cell is taken up directly by a recipient cell in a process that does not involve direct contact of the cells or mediation by any vector (transmitting agent), such as a virus. Small fragments of genetic material from a donor, each containing just one or two genes, are adsorbed directly to receptor sites on the outer surface of the recipient cell. The fragments move through the cell wall into the cell interior, where some are incorporated into the genetic material of the recipient.

The usual source of transforming genetic material in nature is dead cells that lyse, releasing their contents. In the laboratory, large amounts of genetic material from a specified donor strain can be isolated and purified, thereby making transformation more efficient than it probably is in nature. Even so, only certain kinds of bacteria have a cell wall structure that is permeable to the transforming agent. Since the types of bacteria that are able to undergo transformation are limited, most studies have involved just three species: *Streptococcus* (formerly *Diplococcus*) *pneumoniae, Bacillus subtilis,* and *Hemophilus influenzae.*

Discovery of the Transforming Agent.

Transformation was first discovered in 1928 by F. Griffith while working with *S. pneumoniae,* commonly known as pneumococcus. The virulent (infective) form of this organism causes pneumonia in humans and is lethal when injected into laboratory mice. Virulence in pneumococcus is a genetically determined characteristic; those pneumococcal cells that are enclosed in a polysaccharide capsule are virulent. We can detect the presence of the capsule by observing the appearance of colonies produced by the multiplication of individual pneumococcal cells. Cells that have the capsule produce colonies with a smooth texture and are said to have the S (for smooth) phenotype. Mutant cells that have lost their virulence lack a capsule and produce colonies with a rough (or type R) appearance.

Griffith's experiments and results are summarized in Fig. 3.4. Control experiments showed that as expected, livc cells of the S type are lethal to mice (Fig. 3.4a), whereas live cells of the R type or dead cells of the S type are not lethal if only one type is injected (Fig. 3.4b and c). If a mixture of live R and heat-killed S cells is injected into the same mouse, however, the mouse dies (Fig. 3.4d). Furthermore, live pneumococcal cells extracted from the dead mouse produce smooth colonies and are lethal when used subsequently to infect other mice, as

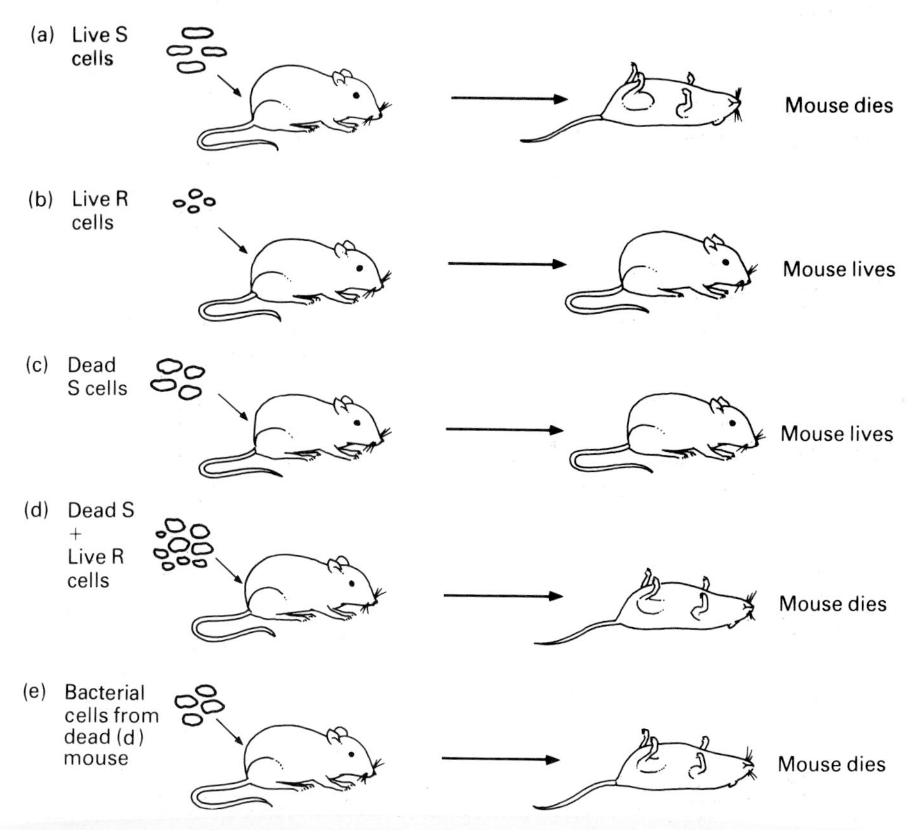

Figure 3.4. Griffith's original demonstration of transformation. (a) Mice die after injection of virulent S bacteria. Mice survive infection by (b) nonvirulent R bacteria and (c) heat-killed S bacteria. Mice die following injection with a mixture of dead S and live R bacteria (d), indicating that some of the R cells have been transformed to the virulent S condition. Pneumococcal cells isolated from the dead mouse in (d) are lethal when injected into another mouse (e), indicating that the transformation event is a permanent genetic change.

shown in Fig. 3.4(e). Griffith concluded that something in the debris of the dead S cells had somehow transformed the live R cells into virulent S types, and that this transformation results in a permanently inherited change in genotype. Just what this substance was remained a mystery, since the experimental results of Griffith gave no clue as to the chemical nature of the transforming agent.

In 1944, a team of three investigators—O. Avery, C. M. MacLeod, and M. McCarty—reported on results of experiments designed to identify the transforming agent. They isolated the different classes of molecules found in the debris of killed S cells and tested each class separately for transforming activity. Tests of purified polysaccharide, DNA, RNA, and protein fractions isolated from dead S cells revealed that only DNA could bring about transformation of R cells to type S (Fig. 3.5a). These results were made even more definitive by experiments using enzymes that degrade specific kinds of molecules. These enzymes include DNase (deoxyribonuclease), which degrades DNA but not RNA or protein; RNase (ribonuclease),

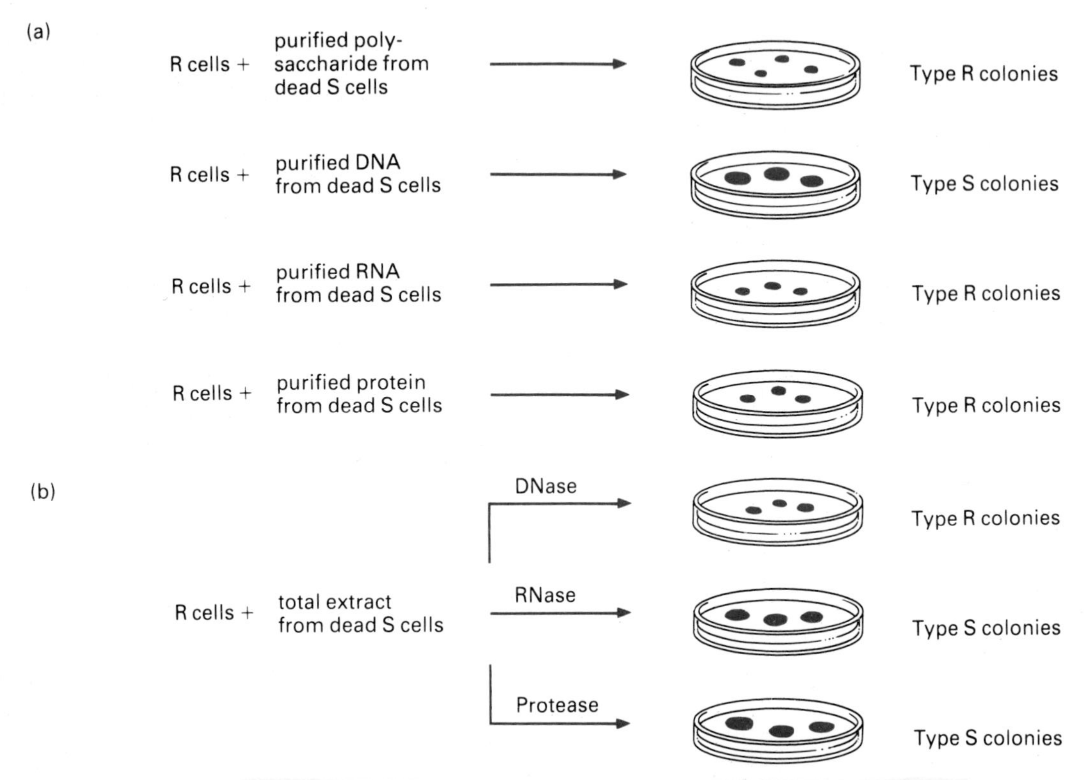

Figure 3.5. (a) Avery, MacLeod, and McCarty tested purified classes of molecules isolated from the debris of heat-killed S cells for transforming activity. Only the DNA from dead S cells transformed the R cells to type S. (b) When the total extract from dead S cells was treated with a specific degrading enzyme, the transforming activity was destroyed only by DNase. This outcome confirmed that the transforming agent is composed of DNA.

which degrades only RNA; and proteases, which degrade proteins. In separate experiments, the investigators treated extracts from dead S cells with either DNase, RNase, or proteases; they then mixed each extract with live R cells and injected the mixture into mice. RNase and proteases had no effect on the transformation of R cells to type S; in contrast, DNase totally eliminated all transforming activity (Fig. 3.5b).

The efficiency of transformation is very low. For a given gene, the proportion of potentially recipient bacteria that are actually transformed is never greater than 10 percent and is usually considerably lower. One reason for the low efficiency of transformation is that usually only a minority of cells are in a physiological state that permits them to undergo the transformation process. Cells that are able to be transformed are said to be competent. **Competence** represents the capacity of bacteria to take up DNA from the medium, the extent of which varies with growth conditions and the phase of growth of recipient cells.

Other factors that limit the efficiency of transformation for a given gene include (1) the number of DNA binding sites on the bacterial surface and (2) the size of the DNA molecules involved in transformation. When investigators extract the

DNA from donor cells and subsequently purify it, the single DNA molecule of each bacterial nucleoid breaks down into fragments ranging in size from 1⁄200 to 1⁄500 of its total length. Thus, any competent bacterium that makes effective contact with a single fragment of DNA has less than a 1 in 200 chance of being transformed for a particular gene.

Replication of Bacterial Viruses

Of the many different types of viruses that have been discovered, those that infect bacteria **(bacteriophages)** have been used the most extensively in genetic research. In contrast to bacteria, phages have no cellular structure and, like animal and plant viruses, are not classified as either prokaryotes or eukaryotes. In fact, some scientists do not even consider them to be organisms, since they are unable to metabolize or reproduce entirely on their own. They are obligate parasites of bacteria in that they are completely dependent on the bacterial cell's metabolic machinery for growth.

Bacteriophages have a comparatively simple structure, which consists of a single molecule of DNA or RNA surrounded by a protein coat (Fig. 3.6). Phages can exist in this form indefinitely while they are stored in the laboratory. In order to reproduce, however, phage particles must be mixed with a population of bacterial cells. They can then begin their life cycle, which is summarized in Fig. 3.7.

Figure 3.6. Representative bacteriophage strains of *Escherichia coli* that contain DNA. Phages T2, T4, and T6 (the T-even phages) are structurally similar and are among the best understood of all genetic subjects (a). They are composed of a polyhedral-shaped protein head, which contains the chromosome, and a complex protein tail apparatus, ending in six tail fibers. (b) Phage λ is smaller than the T-even phages and has a simpler tail apparatus that ends in only one tail fiber. (c) Phage φX174 consists of a protein capsule containing the chromosome, but no tail structure.

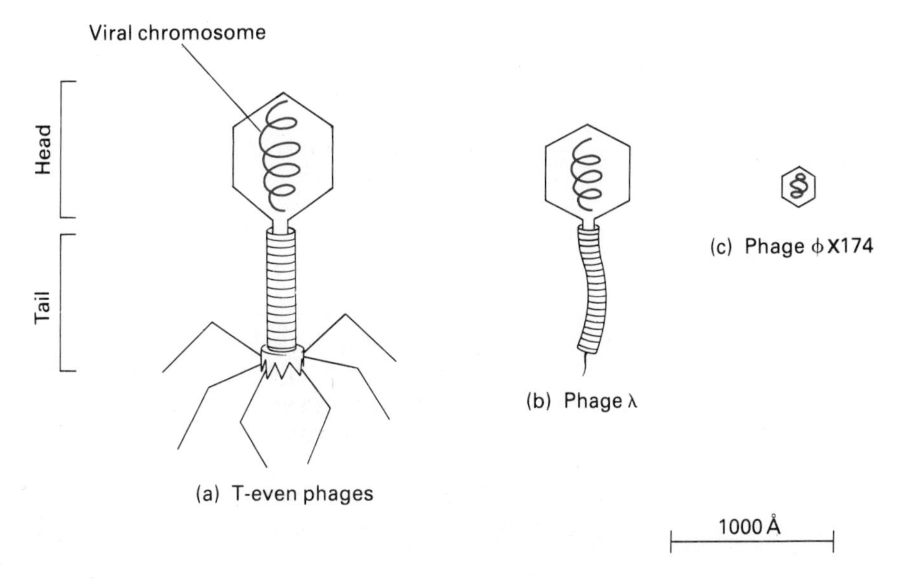

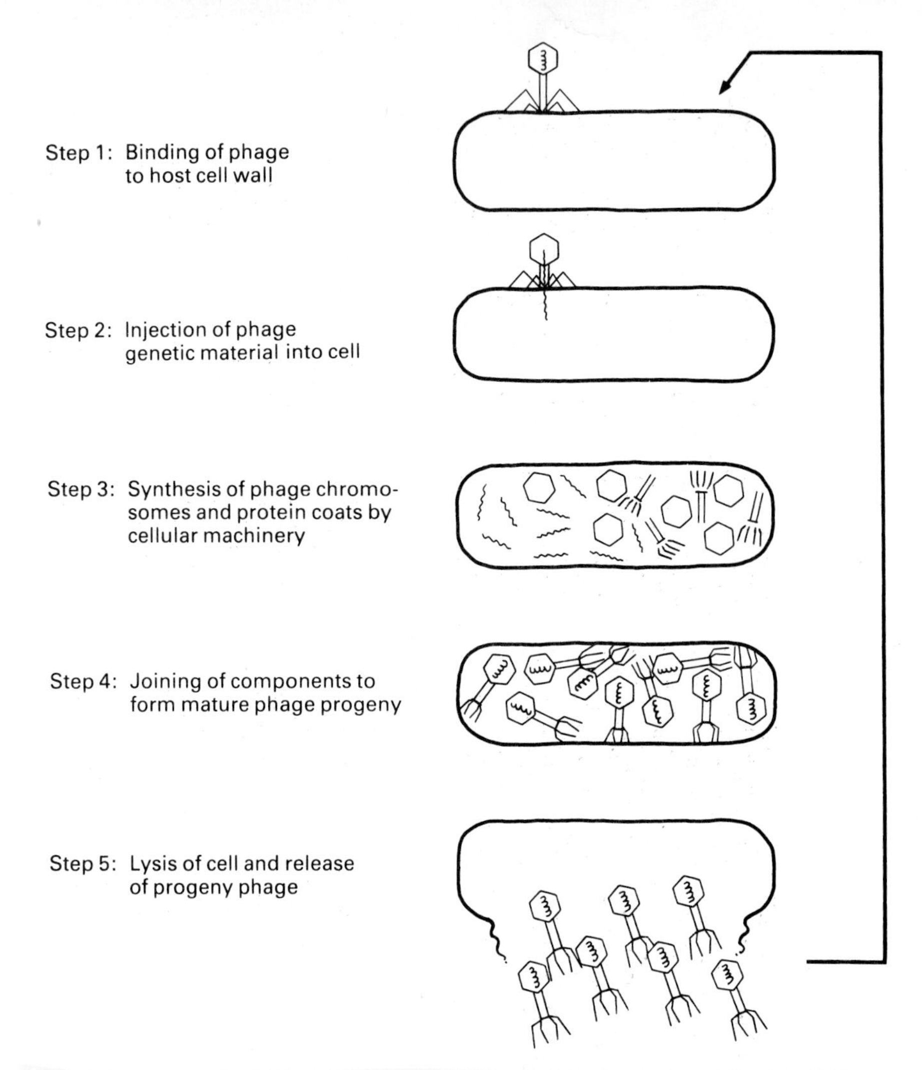
Step 1: Binding of phage to host cell wall
Step 2: Injection of phage genetic material into cell
Step 3: Synthesis of phage chromosomes and protein coats by cellular machinery
Step 4: Joining of components to form mature phage progeny
Step 5: Lysis of cell and release of progeny phage

Figure 3.7. Sequence of events during the growth cycle of phages. After one or more phages adsorb to receptor sites on the bacterial cell wall, the genetic material of the phage is injected into the cell. Phage genes direct the cell's synthetic machinery to produce many copies of the phage chromosome and protein head and tail. After the component parts join to produce mature phage progeny, lysis of the cell follows. Phages released through lysis are then free to initiate a new round of infection.

During reproduction, phage particles first bind to receptor sites located on the bacterial cell wall and inject their genetic material into the cell. The cellular functions of the bacterium are immediately diverted to the synthesis of progeny phages. Phage genes direct the replication of the viral chromosome and the synthesis of protein coats, making perhaps as many as 100 to 1000 progeny phages per cell in a time period as short as 15 minutes. The cell is finally lysed by enzymes produced by the developing phages, releasing mature phage particles to infect other cells, thus initiating a new round of phage reproduction (Fig. 3.8).

Bacterial viruses were first described and named in 1917 by F. D'Herelle, a French bacteriologist. It was not until the early 1950s that it became known that only the DNA of the phage, and not the protein coat, is injected into the host during the infection process. This finding, reported in 1952 by A. D. Hershey and M. Chase, further substantiated that DNA is the genetic material.

The Hershey-Chase Experiment. Hershey and Chase studied the reproduction of a particular class of bacterial viruses, known as T2, that infect the common intestinal bacterium *Escherichia coli*. In their experiment, Hershey and Chase used radioactive tracers to follow the transfer of genetic material from T2 phage particles into their host cells. Their experimental design made use of the differing chemical compositions of the DNA and protein coat of phages. DNA contains phosphorus as one of its basic ingredients, whereas most proteins do not. In contrast, DNA lacks sulfur, whereas proteins usually contain a substantial amount of this element. Hershey and Chase prepared two populations of phage particles. In one, they incorporated the radioactive isotope ^{32}P into the phage DNA; in the other, they incorporated the isotope ^{35}S into the phage protein. (Isotope incorporation is accomplished by allowing phages to reproduce in bacterial cells that are growing

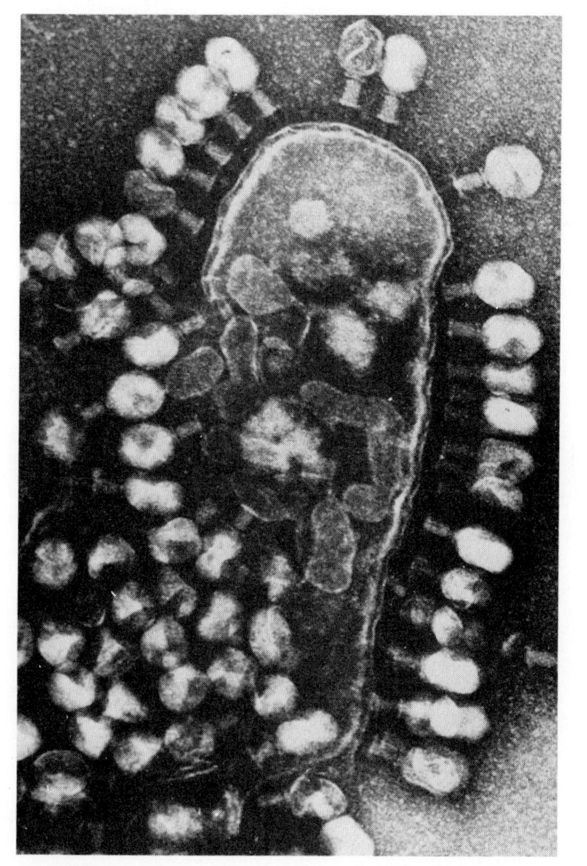

Figure 3.8. Electron micrograph of T4 phage in association with an *E. coli* cell. Numerous phage particles are seen attached by their tails to the cell wall. Progeny phage are being released from the lysed cell. *Source:* Courtesy of Lee D. Simon, Waksman Institute, Rutgers University, Piscataway, N.J.

in a chemically defined medium whose sole source of phosphorus and sulfur comprises radioactive isotopes of these elements.) Each culture of radioactive-labeled phages was allowed to infect a separate population of *E. coli*. After the investigators had allowed sufficient time for the genetic material of the phages to be injected into the cells, they agitated the two cultures separately in an electric food blender to shear off any phage parts that might still be attached to the cell walls. They then subjected the samples to low speed centrifugation, which forces the infected bacteria toward the bottom of the tube where they accumulate in the form of a pellet. Finally, the investigators used a radiation counter to determine the amount of radioactivity that remained in the supernatant and the amount that appeared in the pellet.

Hershey and Chase found that radiation emanating from the ^{32}P was localized in the pellet containing the infected cells, whereas most of the ^{35}S radioactivity remained in the supernatant (Fig. 3.9). In other words, shearing removes nearly all the ^{35}S from infected cells, but hardly any of the ^{32}P. From these results, Hershey

Figure 3.9. The Hershey-Chase experiment, which demonstrated that DNA is the genetic material of phage T2. (a) Phage DNA is labeled with ^{32}P; the radioactivity is subsequently found to be associated mainly with the cells. Very little radioactivity is found in the supernatant. (b) Phage protein coats are labeled with ^{35}S; the radioactivity is subsequently found in the supernatant containing the phage ghosts. These results indicate that only the DNA of the phages enters the cells. The protein coats serve merely to package the DNA prior to infection. On the basis of this experiment, Hershey and Chase concluded that DNA, and not protein, is the genetic material necessary for the production of progeny phage.

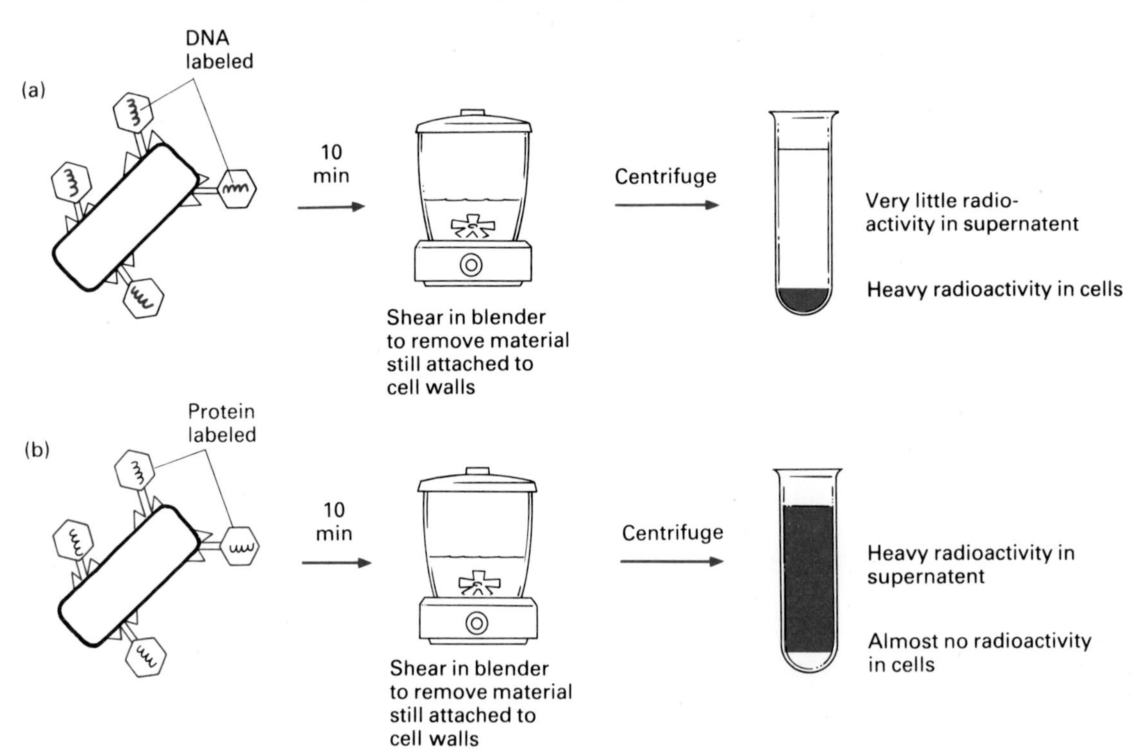

and Chase concluded that the DNA of the phage enters the bacterium, while the protein coat remains on the cell surface following infection. The shearing force generated by the blender breaks the tails by which the empty phage protein coats ("phage ghosts") are held to the cell wall, releasing them into the medium. These results clearly show that DNA is the material that enters the cells and directs the synthesis of phage progeny. DNA is thus the genetic material of the virus.

TO SUM UP

1. DNA is the genetic material in all prokaryotic and eukaryotic organisms that have been studied. A few viruses have RNA instead of DNA; RNA serves as the carrier of genetic information in these exceptions.
2. The transformation experiments of Avery, MacLeod, and McCarty, reported in 1944, were the first to show unequivocally that DNA is the genetic material. They demonstrated that naked fragments of DNA, isolated from a donor bacterium, could be taken up by a recipient bacterium, giving rise to a permanently inherited change in the genotype of the recipient cells.
3. The conclusion that genes are composed of DNA was reinforced by the radioactive labeling experiments of Hershey and Chase in 1952 on the T2 bacteriophage. Their work showed that the DNA of the phage, but not the protein coat, enters the bacterial cell to direct the synthesis of phage progeny.

Structure of DNA and RNA

In 1869, Friedrich Miescher, a Swiss physician, published a method for separating the nuclei of cells from the cytoplasm. Analysis of the chemical composition of the isolated nuclei revealed the presence of an acidic component with a high content of nitrogen and phosphorus. Miescher called this substance *nuclein.* Miescher's nuclein is now known as *nucleic acid*—either *DNA* (*deoxyribonucleic acid*) or *RNA* (*ribonucleic acid*). Nucleic acids are large polymers, whose molecular weight ranges from several hundred to several million. Because of their large size, nucleic acids are called *macromolecules.*

Primary Structure of Nucleic Acids

Both DNA and RNA are composed of long chains of **nucleotides,** the structural building blocks of the nucleic acid molecule. The different types of nucleotides alternate in an irregular fashion to make up a *polynucleotide chain.* Many different nucleotide sequences are possible, depending on how the different nucleotides happen to be arranged with respect to one another. The sequence of nucleotides that form a polynucleotide, which is the basic strand of nucleic acid, is known as the **primary structure** of the molecule. In this section, we will discuss the chemical structure of the different nucleotides and their linkage with one another to form a strand of DNA or RNA.

Structure of the Nucleotides. Each nucleotide is composed of three distinct parts: (1) a nitrogen-containing base, (2) a five-carbon sugar, and (3) a phosphate group. In the case of DNA, the nitrogenous base can be either *adenine* (A), *guanine*

(G), *cytosine* (C), or *thymine* (T). RNA also has adenine, guanine, and cytosine bases, but it contains the base *uracil* (U) in place of thymine. These different bases can be thought of as derivatives of two parent compounds, **purine** and **pyrimidine:**

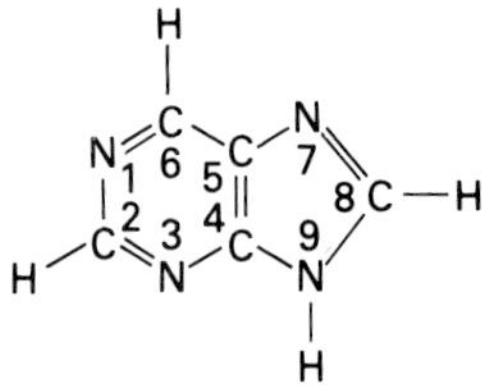

Purine base

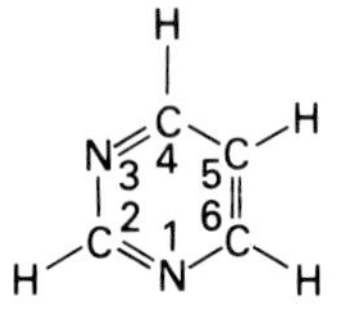

Pyrimidine base

Adenine and guanine are purine bases, whereas cytosine, thymine, and uracil are pyrimidine bases. The numbers provide a convenient means of referencing the atoms of each base. Thus, adenine is characterized by having an amino (—NH_2) group replacing hydrogen at position 6 in the purine ring, while guanine has an amino group at position 2 and an oxy (=O) group at position 6 (see Fig. 3.10). Similar substitutions at positions 2 and 4 in the pyrimidine ring differentiate the various pyrimidine bases, as well as the insertion of a methyl (—CH_3) group at position 5 in the case of thymine.

Figure 3.10. The principal purine and pyrimidine bases of nucleic acids. Adenine, guanine, cytosine, and thymine are present in DNA. Adenine, guanine, cytosine, and uracil are present in RNA.

(a) Purine bases

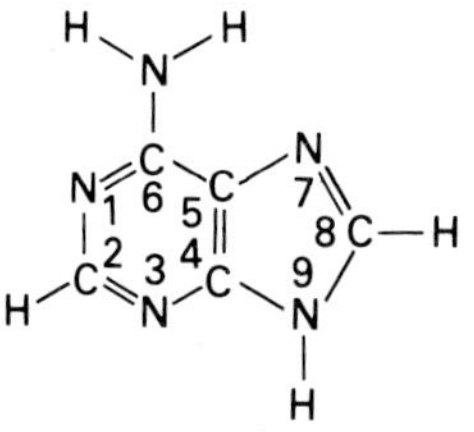

Adenine (A)

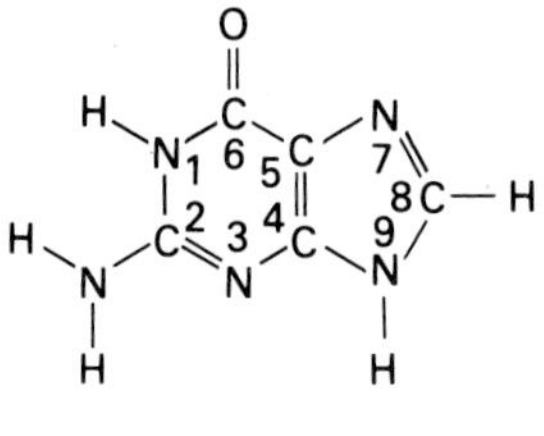

Guanine (G)

(b) Pyrimidine bases

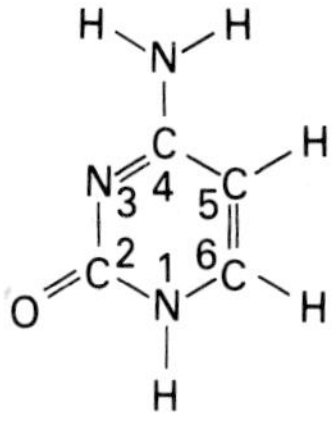

Cytosine (C)

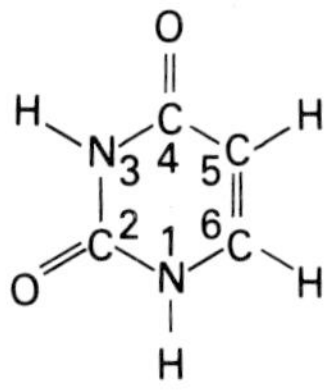

Uracil (U)

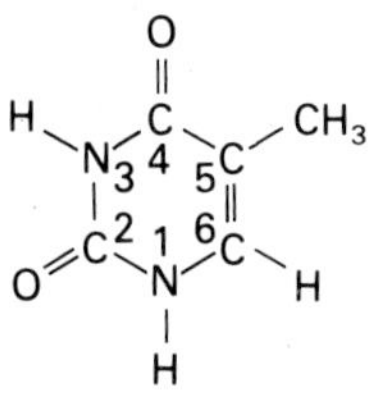

Thymine (T)

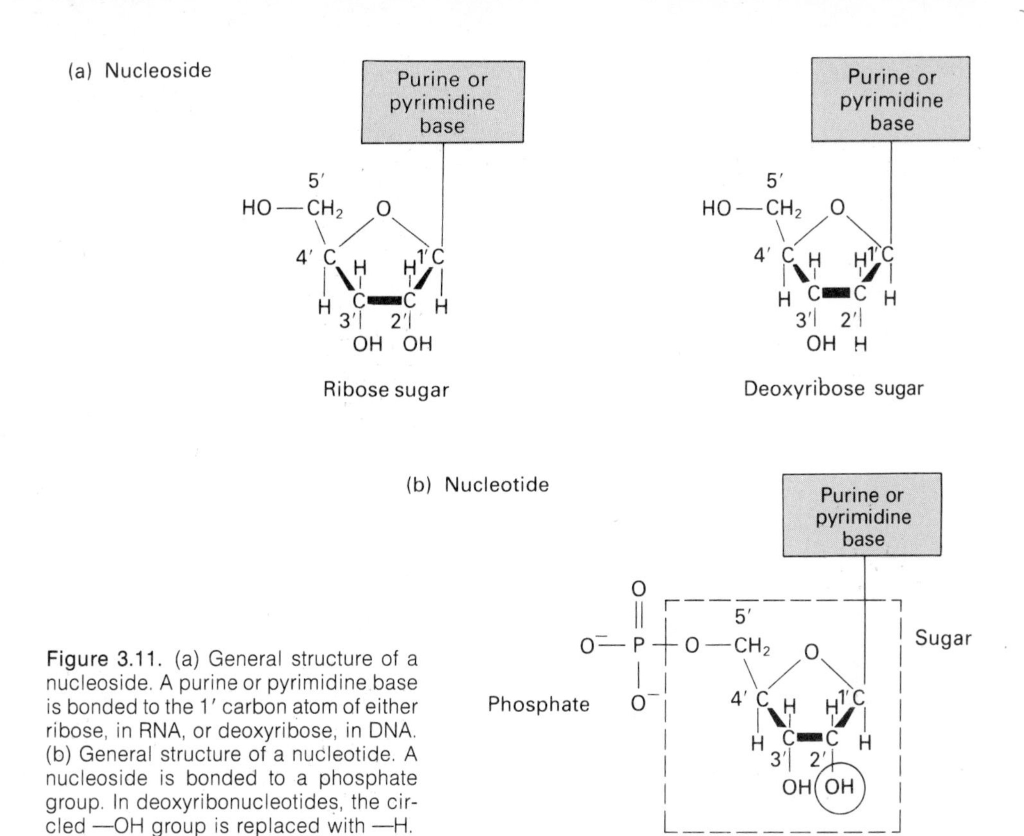

Figure 3.11. (a) General structure of a nucleoside. A purine or pyrimidine base is bonded to the 1′ carbon atom of either ribose, in RNA, or deoxyribose, in DNA. (b) General structure of a nucleotide. A nucleoside is bonded to a phosphate group. In deoxyribonucleotides, the circled —OH group is replaced with —H.

Position 9 of the purine ring and position 1 of the pyrimidine ring are involved in linking each base to the five-carbon sugar to form a **nucleoside** (Fig. 3.11a). The carbon atoms of the sugar are then numbered 1′, 2′, 3′, and so on, to distinguish them from the numbering of atoms in the base. In RNA, the five-carbon sugar is **ribose.** Ribose is a cyclic structure in which four carbon atoms and an oxygen atom are bonded together to form a ring, with the fifth carbon atom (the 5′ carbon) protruding from this ring. Note that both the 2′ and 3′ carbons carry hydroxyl (—OH) groups. In contrast, the DNA sugar, also shown in Fig. 3.11, lacks the hydroxyl group at the 2′ carbon, and is therefore called **deoxyribose.**

A nucleotide is formed when a phosphate group is bonded to the sugar portion of a nucleoside by way of an *ester* linkage (Fig. 3.11b). The nucleotides formed out of this process are named after their constituent bases and are usually described as monophosphates. The four nucleotides of DNA are thus deoxyadenosine monophosphate (dAMP), deoxyguanosine monophosphate (dGMP), deoxycytidine monophosphate (dCMP), and deoxythymidine monophosphate (dTMP). The nucleotides of RNA are designated in a similar manner, but without the prefix *deoxy.* To save space, we will usually denote the various nucleotides simply by their base symbols A, G, C, T, or U, making certain that it is clear whether we are speaking of DNA or RNA.

Organization of Nucleotides into Polynucleotide Chains. Nucleotides are linked to one another in a polynucleotide by **3′-5′ phosphodiester bonds.** To form a

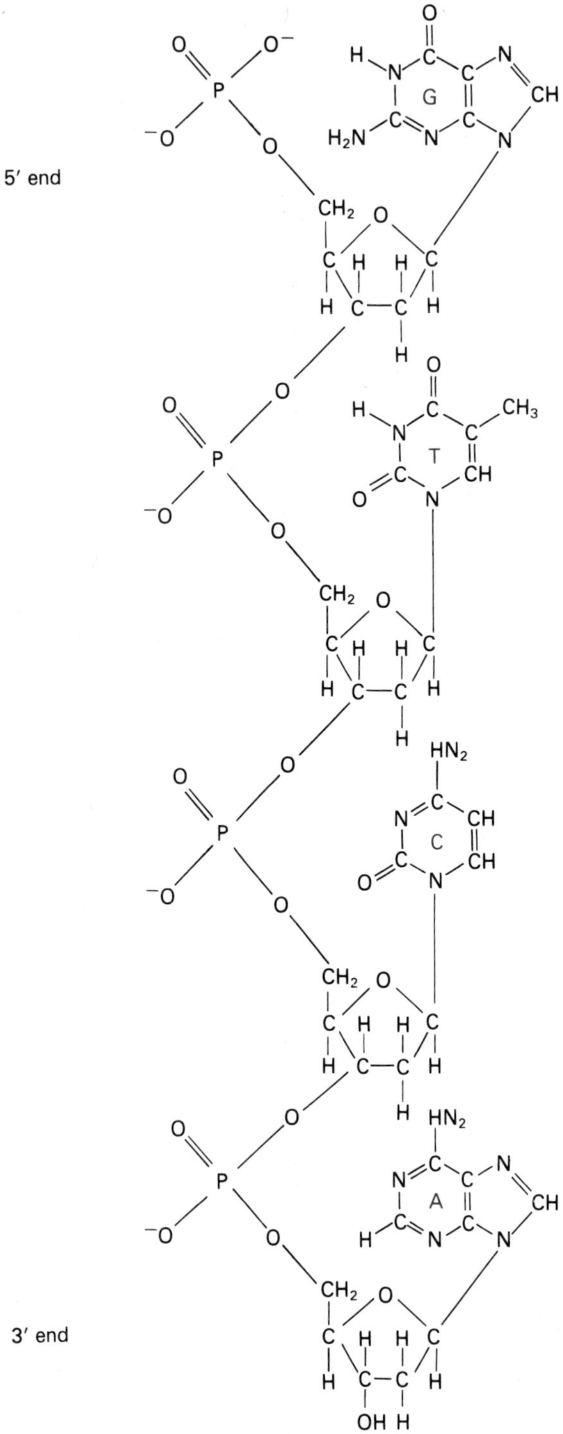

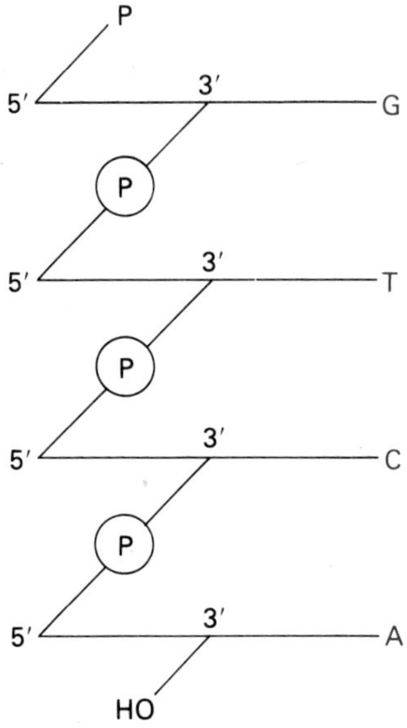

Figure 3.12. Representation of a DNA chain consisting of four nucleotides in length. Adjacent nucleotides are linked through phosphodiester bonds (P), joining the 3′ carbon of one sugar to the 5′ carbon of the next. In this example, the top of the chain is the 5′ end with a free phosphate group; the bottom of the chain is the 3′ end with a free hydroxyl group. The alternating 5′ and 3′ linkages give the backbone of the structure a 5′ → 3′ polarity, clearly shown in the schematic representation of the tetranucleotide on the right. DNA chains ordinarily consist of a large number of nucleotides.

polynucleotide, each phosphate group participates in two ester linkages, forming a phosphodiester bridge between the 3′ carbon of one nucleoside and the 5′ carbon of another. An example of a polynucleotide chain is shown in Fig. 3.12, along with an abbreviated version of the structure. Note that under the conditions normally present within a cell, the phosphate groups appear in their ionized form. The tendency of the phosphate groups to lose protons (hydrogen ions) and exist in an ionized state gives DNA and RNA their acidic characteristics.

In reading the order of nucleotides in a polynucleotide chain, we begin with the nucleotide whose 5′ phosphate group is free. This is the 5′ end of the chain (e.g., the guanine nucleotide in Fig. 3.12). We then identify the nucleotides in sequence, reading from the 5′ end toward the 3′ end (the end with a free 3′ hydroxyl group). For example, we would read the sequence of nucleotides in the chain shown in Fig. 3.12 as G-T-C-A.

Secondary Structure of DNA

The polynucleotide chain folds into a specific three-dimensional pattern, giving the nucleic acid molecule a particular spatial orientation that is crucial to determining its function. This three-dimensional pattern is its **secondary structure.** Because RNA includes a heterogeneous group of molecules with different functions and therefore with different secondary structures, we will consider the secondary and higher-level structures of RNA in Chapter 9, when we discuss the role of RNA in protein synthesis. Our present discussion is limited to DNA. The secondary structure of DNA is a double-stranded (double-chained) helix, the model for which was first proposed by James Watson and Francis Crick.

Nature of the Double Helix. In the early 1950s, investigators in laboratories over the world were seeking to determine the structure of DNA. An American biochemist, Erwin Chargaff, discovered a regular pattern in the chemical composition of DNA. He found that the amounts of purine and pyrimidine bases were equal to one another in the DNA of a wide variety of species. Stated in another way, the total amount of adenine and guanine equalled the total amount of cytosine and thymine. Moreover, Chargaff found that the amounts of adenine and thymine were equal, as were the amounts of guanine and cytosine. No consistent relationship appeared between the combined amounts of adenine and thymine [A + T] and those of guanine and cytosine [G + C], however. These patterns have since been named *Chargaff's rules,* summarized as follows:

1. Sum of purines = sum of pyrimidines
   [A + G] = [T + C]
2. Adenine = thymine; guanine = cytosine
   [A] = [T] [G] = [C]

At about the same time, the biophysicists Maurice Wilkins and Rosalind Franklin were studying DNA by a technique known as *X-ray diffraction.* They beamed X-rays at an oriented DNA fiber and then measured the diffraction of the rays from the DNA that appeared as spots on a film sensitive to X-rays. Such an

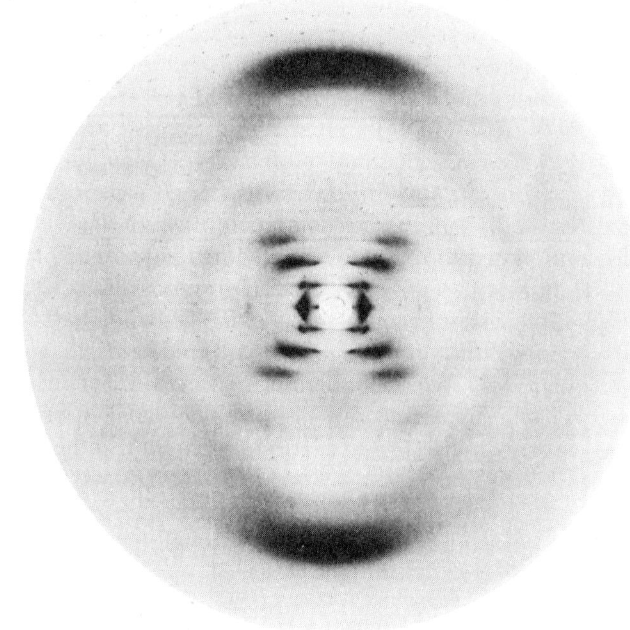

Figure 3.13. X-ray diffraction photograph of semi-crystalline DNA. *Source:* Courtesy of Professor M. H. F. Wilkins, Biophysics Department, King's College, London.

X-ray diffraction photograph of DNA is shown in Fig. 3.13. Only a trained specialist can interpret the pattern of the spots. Suffice it to say that this particular pattern suggests that a DNA molecule is long and slender, and that it is in the form of a helix, with repeating subunits spaced every 0.34 nm apart.

Watson and Crick combined the empirical rules of Chargaff and the X-ray diffraction data of Wilkins and Franklin in a model for the structure of DNA. They proposed that a molecule of DNA consists of two polynucleotide chains wound in a highly ordered helical, or spiral, configuration. Figure 3.14(a) shows the general form and dimensions of the double helix. The bases form the core of the double helix, while the alternating sugar and phosphate groups form the backbone on the periphery of the structure. The helical arrangement is stabilized by hydrogen bonding between the bases across from each other on the different strands, so that 10 base pairs occur along each turn of the molecule. (We will discuss this hydrogen bonding in further detail very shortly.) As illustrated by Figs. 3.14(b) and 3.14(c), the two bases of each pair lie flat and are stacked on top of the neighboring pair in a parallel array, perpendicular to the axis of the helix. This arrangement yields the "stacking energy" that is a major factor in maintaining the helical form. Because the bases are relatively insoluble in water, the stacked bases in the core exclude water from the interior of the molecule. This hydrophobic action further aids in maintaining the stability of the DNA structure in the aqueous environment that exists within a cell.

A final feature of the double-helix structure, illustrated in Fig. 3.15, is that the two strands of any one molecule show opposite polarity. That is, the 5′ phosphates of the two strands are at different ends of the molecule, so that the $5' \rightarrow 3'$ polarity

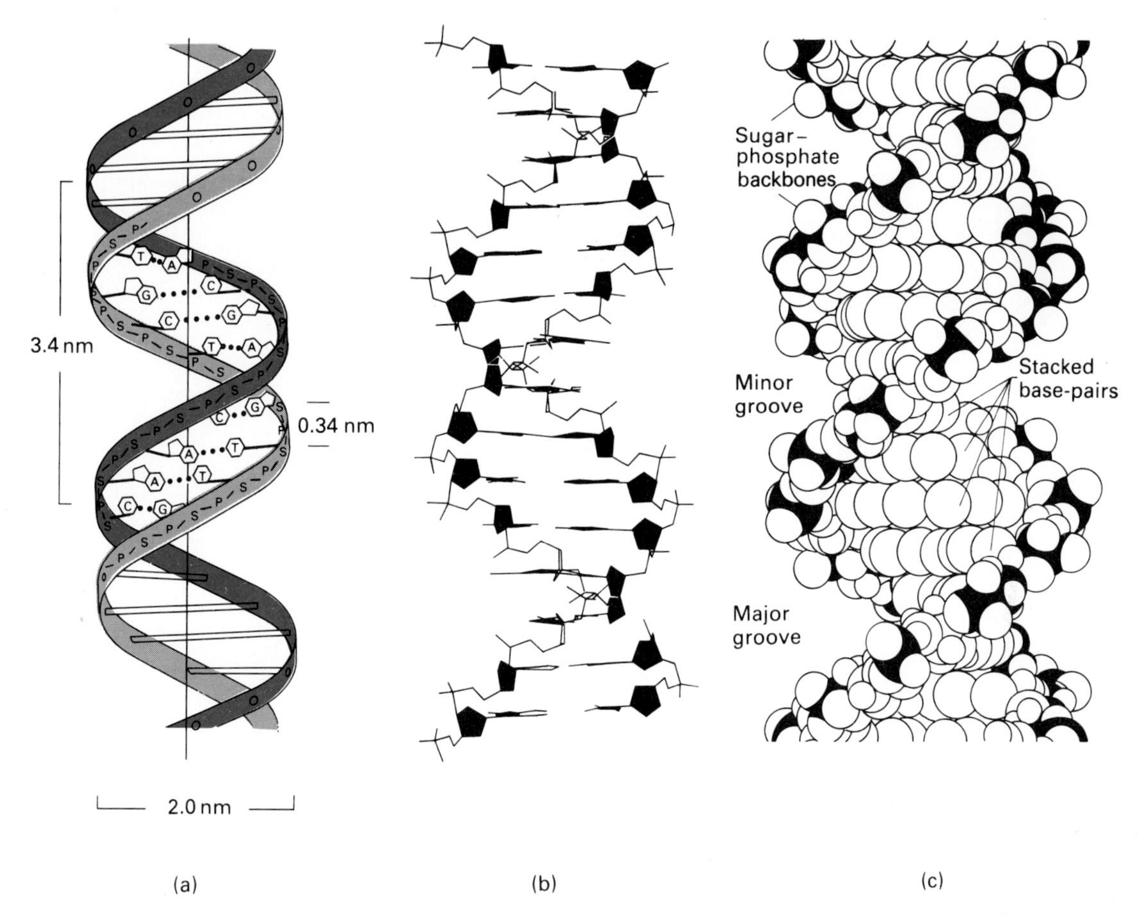

Figure 3.14. The structure of the DNA double helix. (a) The general form and dimensions of the molecule. The two "ribbons" represent the alternating sugar-phosphate backbone of the molecule, while the bases in horizontal pairs form the core. The bases at corresponding positions on the different strands form hydrogen bonds to each other across the helix (shown by the dots). The distance that is required to complete one full turn of the helix, known as its pitch, is 3.4 nm. Since there are ten base pairs in each turn of the helix, the distance between successive base pairs is 0.34 nm. (b) The molecule is shown in terms of the spatial orientation of its sugars and bases. The bases lie flat, perpendicular to the axis of the helix. (c) A space-filling model of DNA is shown. The stacking of hydrophobic base pairs in the interior of the molecule provides a stacking force that greatly contributes to maintaining the helical structure of the DNA molecule in an aqueous environment.

of the two chains runs in opposite directions. The chains are thus said to be **antiparallel;** as a consequence, rotation by 180° does not alter the appearance of the molecule.

So far, we see that the double-helix model fits well with the data of Wilkins and Franklin. Closer inspection shows that it also complies with Chargaff's rules. In constructing the model, Watson and Crick recognized that each base pair must consist of one purine and one pyrimidine if the width of the double helix is to be

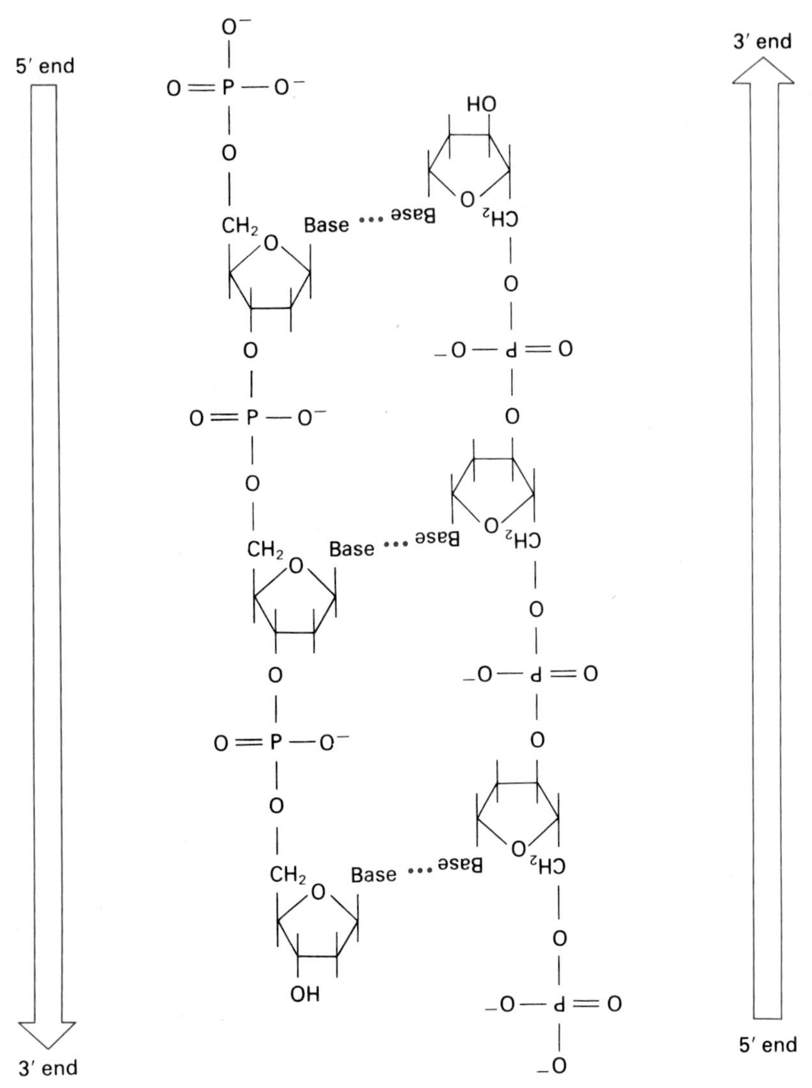

Figure 3.15. Chemical structure of the double helix (shown unwound) illustrating the antiparallel nature of the two strands. The strands run in opposite directions.

constant throughout its length. They also noticed that only certain purine–pyrimidine pairs could be formed through hydrogen bonding. A **hydrogen bond** is a weak linkage that forms between an electronegative (electron-seeking) atom (usually nitrogen or oxygen) and a hydrogen atom covalently bonded to another electronegative atom. An example of hydrogen bonding with oxygen as the acceptor is shown below:

$$\delta^- \; \mathrm{N} - \underset{\delta^+}{\mathrm{H}} \cdots \underset{\delta^-}{\mathrm{O}} = \mathrm{C} \; \delta^+$$

The hydrogen atom is linked covalently to a nitrogen atom, which has a tendency to attract an electron. The resulting partial separation of charge between the hydrogen and nitrogen atoms leaves the hydrogen with a partial positive (δ^+) charge. The oxygen and carbon atoms also undergo partial charge separation, leaving the oxygen with a slight negative (δ^-) charge. The juxtaposition of opposite charges creates an electrostatic force of attraction (• • •) called the hydrogen bond.

Under normal conditions, proper hydrogen bonding can occur only between adenine and thymine and between guanine and cytosine (Fig. 3.16). Pairing between adenine and cytosine and between guanine and thymine cannot take place, since the identical charges on both bases of each combination would cause them to repel one another. Thus, A normally pairs with T, and G with C. (Rare errors in base pairing do occur, and their consequences are discussed later in this chapter.) A double helix made up of only AT and GC base pairs thus contains the same amount of A as T, and the same amount of G as C, thereby satisfying Chargaff's rules. Note, however, that there is no restriction on the total proportions of AT vs. GC pairs in the molecule, so that the amount of A + T does not bear any set relationship to the amount of G + C.

Table 3.1 gives the base compositions of the DNA of various organisms. With the exception of the phage ϕX174, all of these organisms conform to Chargaff's

Figure 3.16. Base pairing in DNA. Only adenine with thymine and guanine with cytosine have the combination of positive and negative charges required to form hydrogen bonds. An AT pair is joined by two H bonds, a GC pair is joined by three H bonds.

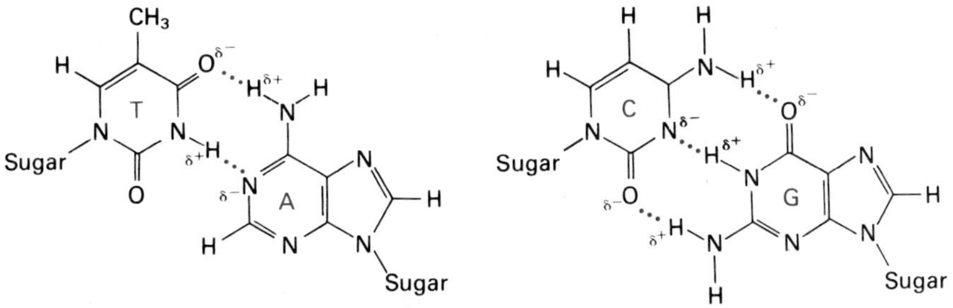

Hydrogen bonding in A-T and G-C base pairs

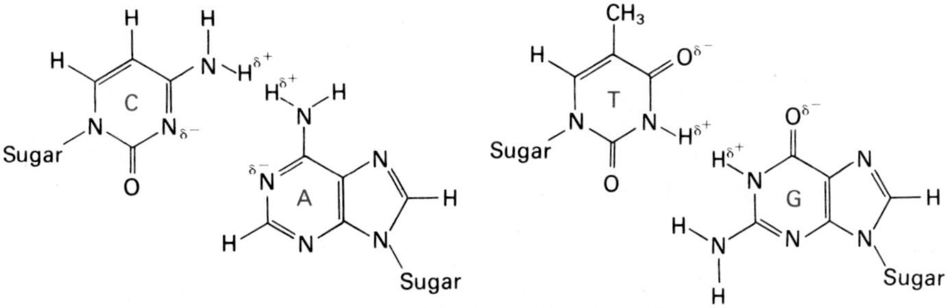

No hydrogen bonding of A and C or G and T

Table 3.1. DNA base compositions in a variety of organisms.

SOURCE	A	T	G	C	(A + T)/(G + C)	(A + G)/(C + T)
human	31.0	31.5	19.1	18.4	1.67	1.00
mouse	29.1	29.0	21.1	21.1	1.38	1.00
chicken	28.0	28.4	22.0	21.6	1.29	1.00
frog	26.3	26.4	23.5	23.8	1.11	0.99
fruit fly	27.3	27.6	22.5	22.5	1.22	0.99
corn	25.6	25.3	24.5	24.6	1.04	1.00
tobacco	29.7	30.4	19.8	20.0	1.51	0.98
bread mold	23.0	23.3	27.1	26.6	0.86	1.00
E. coli	24.6	24.3	25.5	25.6	0.96	1.00
pneumococcus	30.3	29.5	21.6	18.7	1.48	1.08
herpes simplex virus	13.8	12.8	37.7	35.6	0.36	1.06
phage lambda	26.0	25.8	23.8	24.3	1.08	0.99
phage ϕX174	24.7	32.7	24.1	18.5	1.35	0.95

rules, even though there is a wide range of variation in the ratio of A + T to G + C. The reason why phage ϕX174 appears to violate Chargaff's rule is that the DNA of this phage is single-stranded and does not normally exist in the form of a double helix.

Watson and Crick published their model for the structure of DNA in the magazine *Nature* in 1953. In 1962, Watson, Crick, and Wilkins were awarded the Nobel Prize for their work. Had Franklin not died in 1958 at the age of 37, she might have shared in the award.

EXAMPLE 3.1. Explain why the (A + G):(T + C) ratio in double-stranded DNA must equal 1, whereas the (A + T):(G + C) ratio can take on any value, depending on the species from which the DNA is derived. If the (A + G):(T + C) ratio is 1, can you conclude that the DNA in that organism is double-stranded, or would you need more information?

Solution. Since each adenine forms hydrogen bonds with a thymine, and each guanine forms hydrogen bonds with a cytosine, we can write a pair of relationships (Chargaff's rule 2):

$$[A] = [T]$$

and

$$[G] = [C].$$

Adding the terms on each side of the equal signs gives [A + G] = [T + C], so that a ratio of (A + G):(T + C) = 1.00. Just because this ratio is 1, however, does not imply that the DNA is double-stranded. We must know the proportions of each of the four bases in order to make this conclusion. For instance, a DNA molecule with a base composition of 20% A, 20% C, 30% G, and 30% T has an (A + G):(T + C) ratio of 1, but it is single-stranded since A $\neq$ T and G $\neq$ C.

The (A + T):(G + C) ratio can take on any value, because the double-helical structure imposes no restrictions on the proportions of AT and GC pairs. This ratio depends only on the relative amounts of AT and GC base pairs along the length of the DNA molecule.

Physicochemical Properties of the Double Helix. Because of the specific rules for base pairing, the sequence of bases in one strand of a double helix completely specifies the sequence of the other strand. The two polynucleotide chains are thus said to be **complementary** in base sequence. This complementarity has important implications in the replication of DNA, to be discussed shortly.

Another important property of DNA is the nature of the hydrogen bonding that holds the two strands together. Hydrogen bonds are weak enough to be broken under conditions that have no effect on covalent bonds. For instance, when a solution of DNA is exposed to high temperatures of about 80° to 100° C, a sufficient amount of thermal energy is supplied to disrupt the hydrogen bonds between the paired bases without breaking the phosphodiester bonds in the backbone of the helix. The double helix then unwinds, allowing the intact strands to separate. DNA that has been made single-stranded by such procedures is called **denatured DNA** or *melted DNA.* Theoretically, any agent that can break hydrogen bonds without disrupting covalent bonds is capable of converting a DNA double helix into two single polynucleotide chains.

The process of denaturation can be followed experimentally by measuring the increase in absorption of ultraviolet light by DNA that occurs during unwinding. Although DNA is not visible to the naked eye, it does absorb light in the ultraviolet region of the spectrum, with a maximum absorption at a wavelength of 260 nm. Furthermore, the ability to absorb light at this wavelength is enhanced when DNA becomes single-stranded. This increase in UV absorption that accompanies denaturation is shown by the melting curve in Fig. 3.17. Note that a sharp transition occurs over a comparatively narrow temperature range. The temperature that marks the midpoint of the transition from double-stranded to single-stranded DNA is called the **melting temperature,** T_m. The value of T_m depends directly on the GC content of DNA; the greater the proportion of GC base pairs in a given DNA specimen, the higher the melting temperature of the molecule. This relationship is a result of the greater stability provided by the three hydrogen bonds connecting the GC base pairs in the helix, as compared with only two hydrogen bonds for the AT pairs. It simply takes more thermal energy to break three hydrogen bonds than two. Careful measurements of melting temperatures, under fixed conditions of pH and ionic strength (salt concentration), can therefore be used to determine the base compositions of DNA from different organisms.

Another property of DNA that is dependent on GC content is its **buoyant density.** Those DNA molecules with a high proportion of GC base pairs have a slightly greater mass-to-volume ratio (density) than molecules with a high proportion of AT pairs.

The buoyant density of DNA is measured by a technique known as **density gradient equilibrium centrifugation.** This is an ultracentrifugal technique in

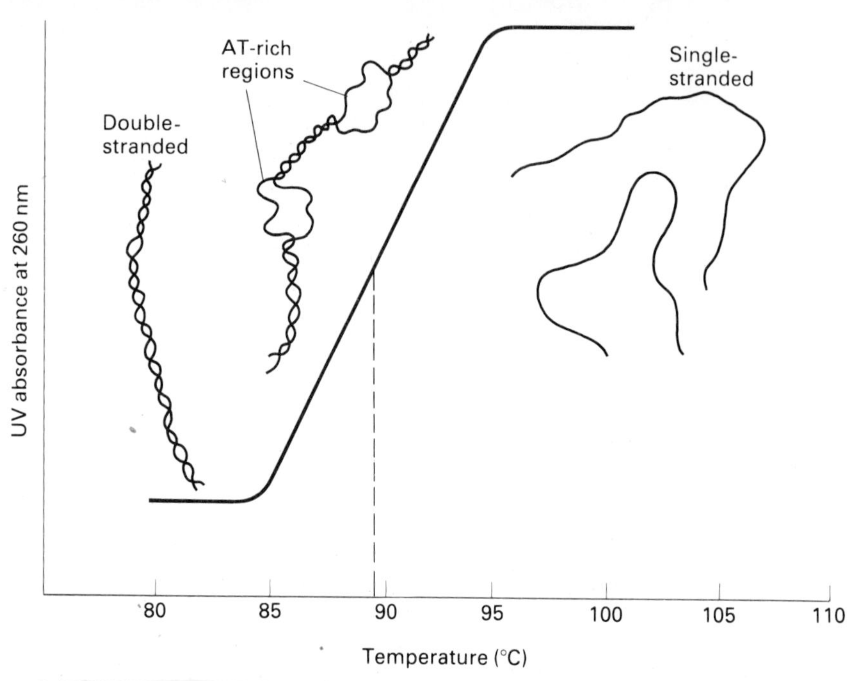

Figure 3.17. Melting curve of *E. coli* DNA, showing the sharp rise in absorption of ultraviolet light (hyperchromic shift) that accompanies thermal denaturation. Regions rich in AT base pairs denature first (shown as "bubbles"). The midpoint of the transition from a double helix to single strands is the melting temperature (T_m) of the DNA, which in this case is 89.8°C.

which a solution of CsCl (cesium chloride, a heavy salt) and the DNA whose density is to be measured is centrifuged at approximately 50,000 rpm for 50 to 60 hours (Fig. 3.18). When spun at such high speeds, the CsCl forms a density (and concentration) gradient, in which the density of the CsCl solution increases linearly from the top to the bottom of the centrifuge tube (Fig. 3.18b). Meanwhile, the DNA molecules, which are initially distributed throughout the tube, move to the position in the gradient that matches their own density (Fig. 3.18c). Those molecules that were initially in a region of CsCl with a density less than their own will sediment toward the bottom of the tube, and those in a region of density greater than their own will float toward the meniscus. The DNA therefore collects in a narrow band at the point in the gradient at which the densities of solute and solvent are equal. The band of DNA can then be detected by scanning the tube with ultraviolet light. The density of CsCl at the point in the gradient at which the band is positioned is taken as the buoyant density of the DNA.

DNA Hybridization. The denaturation of DNA is a reversible process. If a solution of denatured DNA is held at about 20°C below T_m, complementary strands of DNA will rewind spontaneously upon collision, forming an intact double helix. We can detect this **renaturation,** or **annealing,** of DNA experimentally as a drop in the absorption of ultraviolet light by DNA at 260 nm; the renaturation curve closely resembles a reversal of the melting curve.

The rate at which renaturation occurs depends on both the concentration of single strands in solution (which directly influences the probability of collision) and the chance that hydrogen bonds will form once complementary strands collide. The probability of hydrogen bonding depends, in turn, on the length of the DNA strands, since at least a dozen or so bases of the colliding strands must be in complete register in order to reestablish the helical form. The longer the strands, the more likely they are to collide out of register.

Single DNA strands from different sources, such as different strains or different but closely related species, can also join to form double helices. Sequence differences in these DNA strands lead to the formation of DNA duplex molecules, called *DNA-DNA hybrids*, with noncomplementary regions. The mismatched regions cannot hydrogen bond, and remain single stranded. Double helices can also be formed by the combination of complementary DNA and RNA strands, resulting in *DNA-RNA hybrids*. Such hybrids have a double-helical structure similar to DNA except that adenine in the DNA strand pairs with uracil instead of thymine. Nucleic acid hybridization has been extremely useful in genetic research, and several experiments using this technique are described in later chapters.

Figure 3.18. Density gradient equilibrium centrifugation. (a) A solution of CsCl and DNA is centrifuged at high speed for 50 to 60 hours. (b) The CsCl forms a linear density gradient, ranging from a salt concentration of about 1.55 g/cc at the meniscus to about 1.80 g/cc at the bottom of the tube. (c) The DNA moves to the position in the density gradient that corresponds to its own density, forming a band at that position.

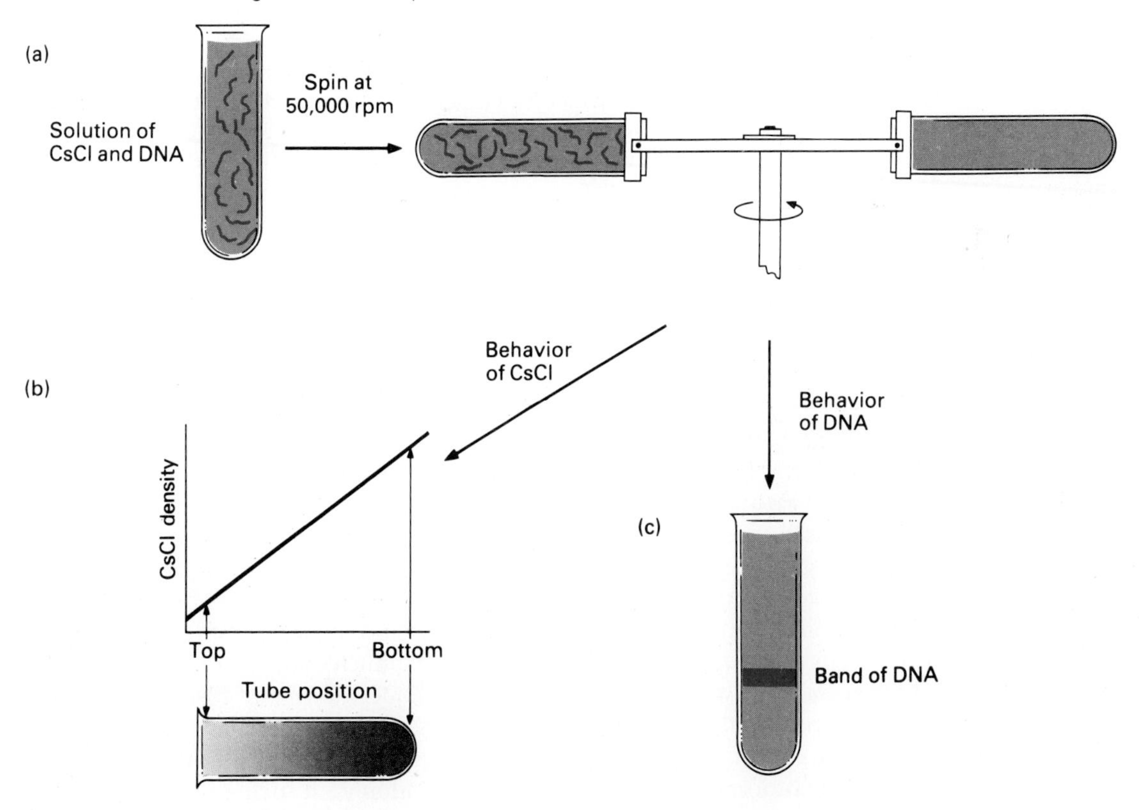

TO SUM UP

1. Nucleic acids are large polymeric molecules, in which the repeating monomeric units are nucleotides. The two types of nucleic acids are DNA (deoxyribonucleic acid) and RNA (ribonucleic acid).
2. Each nucleotide consists of a nitrogenous base (either a purine or pyrimidine), a five-carbon sugar, and an acidic phosphate group. In DNA, the purine bases are adenine (A) and guanine (G), the pyrimidine bases are cytosine (C) and thymine (T), and the sugar is deoxyribose. The nucleotides in RNA are the same as in DNA, except that the pyrimidine base uracil (U) occurs in place of thymine, and the sugar is ribose rather than deoxyribose.
3. Nucleotides are linked to one another by means of 3′-5′ phosphodiester bonds to form long polynucleotide chains. Most RNA molecules exist as single polynucleotide chains. In contrast, nearly all DNA molecules must be in the form of a double helix in order to be biologically active within a cell.
4. The DNA double helix consists of two polynucleotide chains wound around each other in a spiral configuration. The two DNA chains in a double helix are antiparallel, in that their $5' \rightarrow 3'$ polarity is reversed.
5. The two chains of a double helix are held together by hydrogen bonds between the bases across from each other on opposite strands. Normally, only adenine can form hydrogen bonds with thymine, and only guanine can bond with cytosine. These restrictions in base pairing are the explanation behind Chargaff's rules, which state that the amounts of purine and pyrimidine bases in a DNA molecule are equal, and that the amount of adenine equals that of thymine and the amount of guanine equals that of cytosine.
6. Exposing DNA to high temperatures causes denaturation or unfolding of the double helix. The input of excessive thermal energy disrupts the hydrogen bonds between the paired bases, allowing the two strands of a double helix to separate from each other. Renaturation, or rewinding, will occur spontaneously when temperatures are brought back into the biological range.
7. Both the melting (denaturation) temperature and the buoyant density of DNA are dependent on the GC content of the molecule, and can therefore be used to determine the base composition of the DNA being investigated. The buoyant density of DNA can be measured by a technique known as density gradient equilibrium centrifugation.

Functions of DNA

The chemical structure of DNA must allow it to meet certain basic requirements that are essential to function as a gene. As we pointed out earlier in this chapter, the genetic material must be capable of the storage and transfer of genetic information, replication, and mutation. A valuable feature of the Watson-Crick model is that it predicts the manner in which DNA can fulfill certain of these requirements. For instance, with the development of the model, it immediately became apparent how genetic information might be stored. Since no inherent restrictions are placed on the possible arrangements of bases in a polynucleotide chain, each DNA molecule could carry information in its own unique *sequence of base pairs*. Different arrangements of the four possible base pairs (AT, TA, GC, and CG) would then serve as a genetic code, just as different sequences of dots and dashes form the basis of the Morse code. The potential for variation with such a coding mechanism is almost limitless. For example, a region in DNA consisting of only 1000 nucleotide sites, with any one of the four base pairs at each site, could form $4 \times 4 \times 4 \times \ldots = 4^{1000}$ (or approximately 10^{600}) possible sequences. Even if only a small fraction of these sequences have biological meaning, this value is still many times greater than the number needed to encode the genetic information of even the most complex organism.

The Watson-Crick model also suggests mechanisms by which DNA can replicate and undergo mutation. In the sections that follow, we will briefly examine the topics of DNA replication and mutation. We will consider the molecular bases for these and other functions of DNA in greater detail in later chapters.

DNA Replication

In the same paper in which they proposed the double-helical model of DNA structure, Watson and Crick suggested a mechanism for its duplication. They recognized that the base pairing restrictions provide a built-in replication scheme, in which each strand in the double helix could serve as a **template** for the synthesis of a new complementary strand. Rather than assuming that the helix replicates intact, Watson and Crick suggested that the hydrogen bonds between the paired bases break, allowing the strands to separate. Figure 3.19 shows how such a replication process would proceed. According to the model, the helix unwinds simultaneous with the synthesis of new DNA strands. Each newly synthesized region immediately re-forms a double helix with its respective template strand. The resulting daughter helices are thus each composed of one old polynucleotide chain (the template chain) and one new chain. This arrangement of old and new chains in the daughter molecules is called *semiconservative replication.*

It is possible to conceive of two other arrangements of old and new material in the daughter molecules. One called *conservative replication* is diagrammed in Fig. 3.20, along with a third model, called *dispersive replication.* In conservative replication, the two newly synthesized chains form a separate helix, leaving the parental molecule intact; whereas, in dispersive replication, old and new material is interspersed along each strand of a newly formed double helix. (The mechanisms by which conservative and dispersive replication might be accomplished need not concern us here.) Experimentation was thus needed to establish which of the three models most accurately represents the replication process.

The Meselson-Stahl Experiment. The first conclusive experimental data pertaining to the mechanism of DNA replication was reported in 1958 by M. Meselson and F. Stahl, who studied replication in the bacterium *E. coli.* Meselson and Stahl realized that distinguishing among the three models required that the old parental strands be labeled in some way in order to differentiate them from the new strands synthesized during the replication process. Through labeling, they could then follow the fate of the parental strands upon replication. To label the parental strands, they made use of the fact that the purine and pyrimidine bases in DNA contain substantial amounts of nitrogen. The most common isotope of nitrogen occurring in the DNA is ^{14}N. But another isotope of nitrogen, ^{15}N, can be incorporated into DNA without harm to the cell. DNA constructed of ^{15}N (or heavy

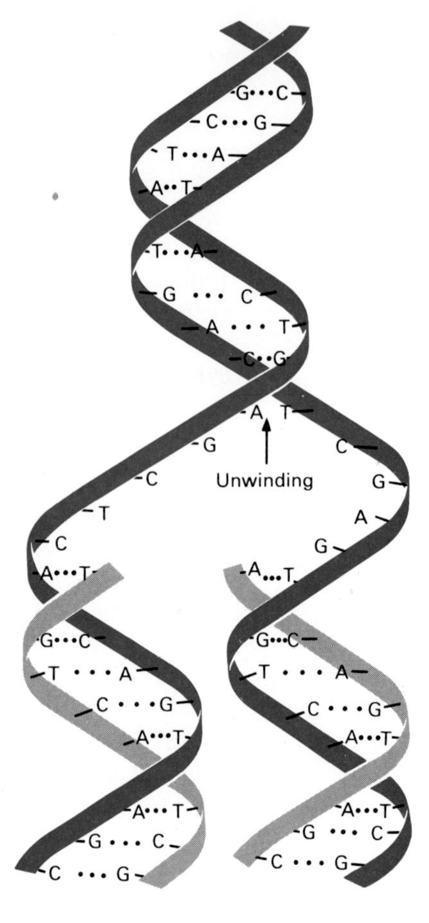

Figure 3.19. Replication of double helical DNA according to the scheme proposed by Watson and Crick. As the strands separate, each serves as a template for synthesis of a complementary new strand. The resulting helices each contain one old and one new strand, an arrangement termed semiconservative replication.

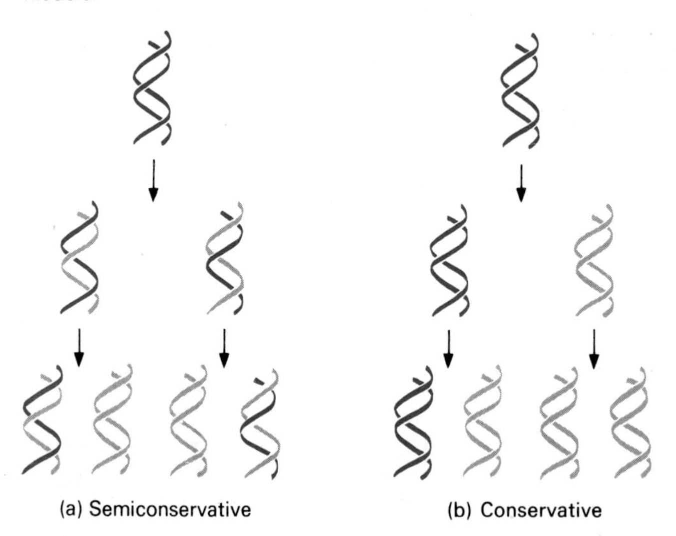

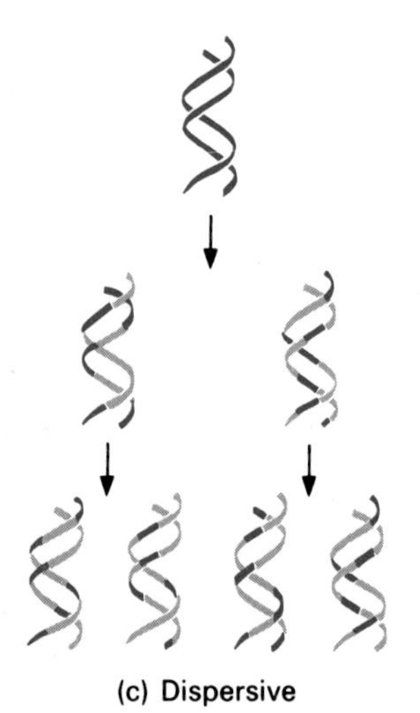

Figure 3.20. Three proposed models of DNA replication: (a) semiconservative, (b) conservative, and (c) dispersive. In each case, the parental strands act as templates for the synthesis of complementary new strands. It is the arrangement of old and new material into daughter helices that distinguishes the three models.

nitrogen, as it is called) is more dense ("heavier") than ordinary DNA containing ^{14}N (light DNA).

The difference in density between ^{14}N DNA and ^{15}N-labeled DNA can be detected by density gradient equilibrium centrifugation, which we discussed in the previous section of this chapter. When the DNA molecules in the centrifuge tube differ in density, more than one band forms in the CsCl gradient. Thus, in a mixture of ^{14}N DNA and ^{15}N-labeled DNA, the heavier ^{15}N-labeled molecules will band at a lower region of the tube than will the lighter ^{14}N molecules. Molecules that are composed of mixed ^{14}N and ^{15}N-labeled material will band at an intermediate position.

The Meselson-Stahl experiment was performed as follows: They grew *E. coli* cells for many generations in a medium containing ^{15}N as the sole nitrogen source, until virtually all the bacterial DNA molecules were labeled with the heavy isotope. This marks time zero, the time at which the parental molecules are fully labeled. The investigators then transferred the cells to a medium containing ^{14}N, rather than heavy nitrogen, and allowed the bacteria to grow for controlled periods of time. All the DNA strands synthesized in this medium will then contain only ^{14}N. The researchers withdrew and lysed samples of bacteria at successive generations of growth and extracted and centrifuged the DNA of the cells to determine its density.

The three replication models make very different predictions as to the density of the DNA following each generation of growth in ^{14}N medium (Fig. 3.21). According to the semiconservative scheme, one generation in ^{14}N medium yields DNA molecules that are all of intermediate density, with each molecule being composed of one heavy and one light strand (HL molecules). Only one band would form in the gradient, at a position of intermediate density. After two generations, however, two bands should form: one at the light density position, consisting of molecules with two light strands (LL molecules), and one at the intermediate position. The amount of DNA in each of these bands should be equal. In contrast, conservative replication would always yield two kinds of molecules: some light (LL molecules) and others heavy (HH molecules), representing the conserved parental helix. No DNA should band at an intermediate position. Finally, dispersive replication would always form molecules of intermediate density, composed of interspersed heavy and light material.

The actual results of the Meselson-Stahl experiment are shown in Fig. 3.22 in terms of ultraviolet scans, along with the interpretation of band positions in the centrifuge tube. After one generation of growth in ^{14}N medium, all of the DNA molecules band at the intermediate position; no heavy or light DNA is formed. Two generations yield equal proportions of molecules in each of the two bands: one at the intermediate position and the other at the light position. These results are clearly in agreement with the semiconservative mechanism proposed by Watson and Crick; they are inconsistent with the conservative and dispersive replication models.

Enzymatic Synthesis of DNA: The DNA Polymerases. The first major breakthrough in understanding the enzymatic mechanisms by which DNA is replicated came with the discovery by A. Kornberg of an enzyme that catalyzed the synthesis

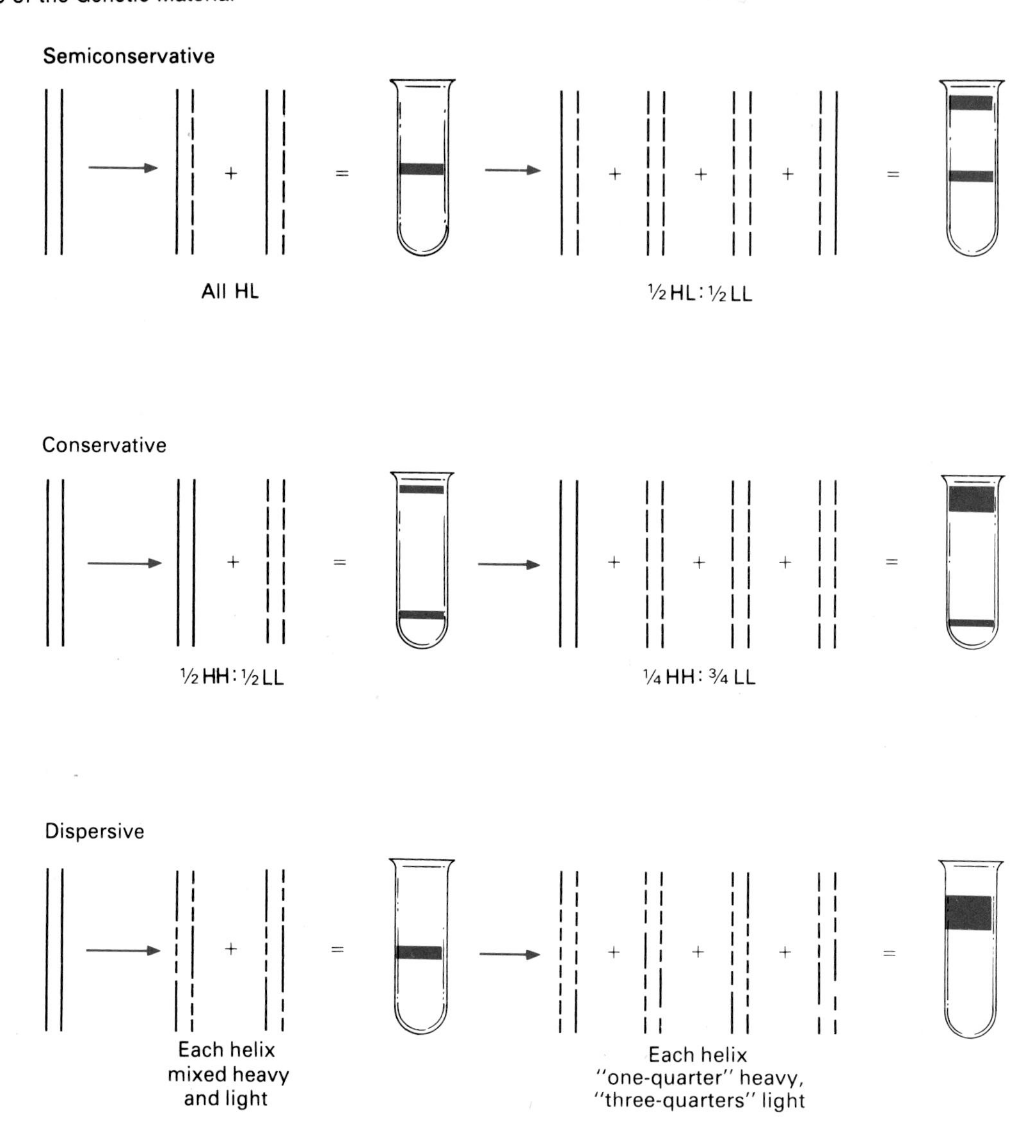

Figure 3.21. Predictions of the three models of DNA replication. Heavy (old) strands are shown as solid lines, light (new) strands are shown as dashed lines. After one replication in ^{14}N medium, the semiconservative model predicts that all helices will be of intermediate density, the conservative model predicts half heavy and half light helices, and the dispersive model predicts helices of intermediate density. Following two replications in ^{14}N medium, the semiconservative mode predicts that half the helices will be of intermediate density and half light, the conservative model predicts one-fourth heavy and three-fourths light helices, and the dispersive model predicts that all helices are composed of "one-quarter" heavy segments, giving them a density between the intermediate and light positions.

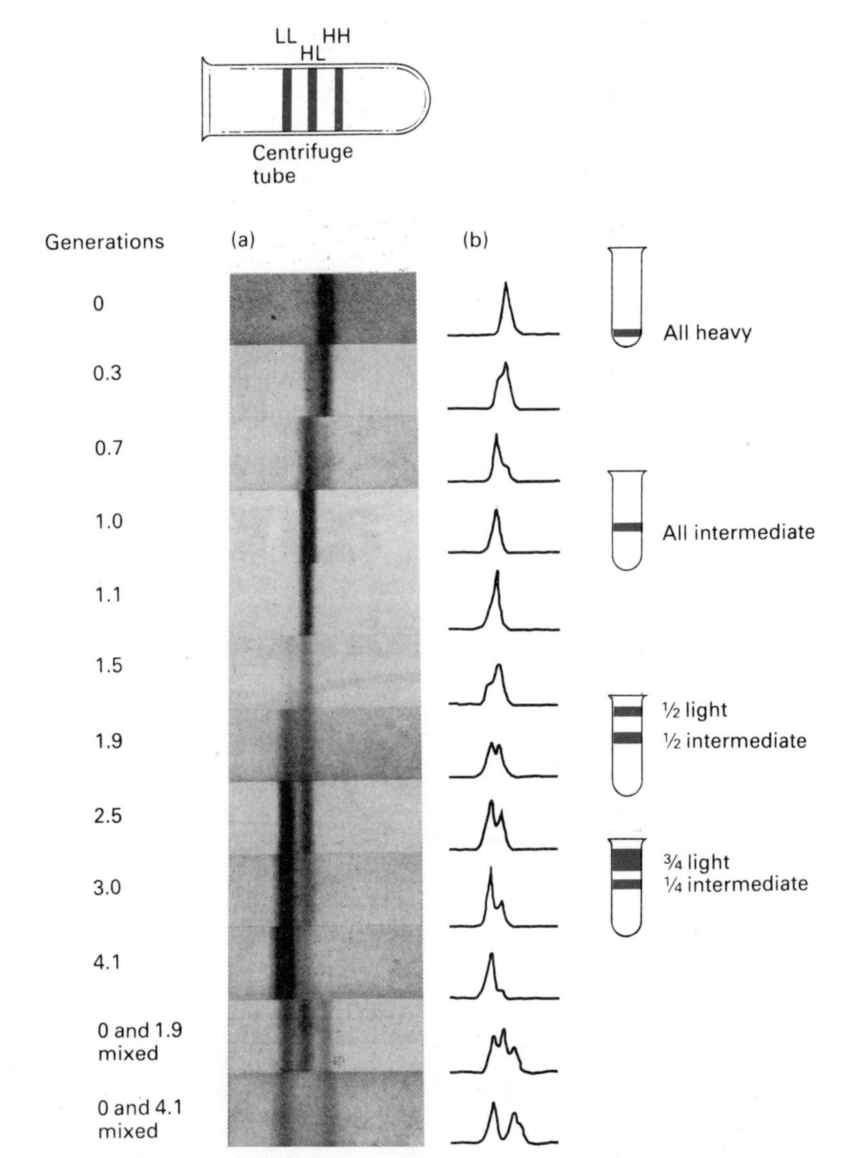

Figure 3.22. Results of the Meselson-Stahl experiment. (a) Photographs taken of the bands formed during centrifugation by scanning the centrifuge tubes with ultraviolet light. (b) Machine tracings of the photographs show peaks of UV absorption, with the height of a peak proportional to the amount of DNA. At time 0, all of the DNA is heavy, giving a single band and a single peak at the heavy position. After one generation in ^{14}N medium, a single peak is again observed, but at an intermediate density position. Two generations yield two peaks of equal height at the intermediate and light density positions, indicating equal proportions of DNA in these two bands. The results are clearly in agreement with the semiconservative model. *Source:* From M. Meselson and F. W. Stahl, *Proc. Natl. Acad. Sci., U.S.,* 44:671, 1958.

of DNA in a test tube (in vitro). The enzyme, now called **DNA polymerase I,** was isolated from *E. coli* and requires for its activity the presence of a DNA molecule, consisting of one partially synthesized **primer** strand and a longer template strand, and deoxyribonucleotides (present in the form of the four nucleoside triphosphates, dATP, dGTP, dCTP, and dTTP). The reaction catalyzed by the enzyme is as follows:

$$\underset{\text{primer strand}}{(\text{dNMP})_n} + \text{dNTP} \xrightarrow[\text{template strand}]{} \underset{\text{lengthened primer}}{(\text{dNMP})_{n+1}} + \text{PP}_i$$

where dNMP and dNTP denote deoxyribonucleoside monophosphate and triphosphate, respectively, and PP_i is inorganic pyrophosphate. The particular sequence of deoxyribonucleotides (dNTPs) used in the reaction depends on the base sequence of the template strand.

The synthesis of each new strand of DNA occurs by the mechanism shown in Fig. 3.23. The DNA polymerase enzyme cannot start synthesizing a new DNA molecule by itself; it can only add nucleotides on to a preexisting primer strand. The primer is lengthened nucleotide-by-nucleotide in a sequence that is complementary to the sequence of bases in the template strand. The nucleotides are inserted in accordance with the Watson-Crick base-pairing rules. Note that the nucleotides are added to the 3′ end of the growing strand, so that synthesis always proceeds in the 5′ → 3′ direction.

Since Kornberg's discovery of DNA polymerase I, two additional DNA polymerase enzymes have been identified that function in *E. coli*: DNA polymerase II and DNA polymerase III. The function of DNA polymerase II is not well understood, but much more is known about the other two polymerase enzymes. Both DNA polymerases I and III catalyze DNA strand elongation in the 5′ → 3′ direction by basically similar mechanisms. They also can function as **exonucleases.** An exonuclease is an enzyme that sequentially excises (hydrolyzes away) nucleotides from an exposed end of a polynucleotide chain. In the case of DNA polymerases I and III, strand digestion can occur in either direction, so that the enzymes possess both 5′ → 3′ and 3′ → 5′ exonuclease activity.

DNA polymerase III is the principle enzyme catalyzing DNA strand elongation during the replication of the *E. coli* chromosome. DNA polymerase I, the enzyme isolated by Kornberg, participates in strand elongation, but in a more specialized way. The Kornberg enzyme is a *repair enzyme* that excises damaged (or otherwise incorrect) polynucleotide segments and synthesizes new sections of DNA to replace the segments that have been removed (we will consider its role as a repair enzyme in Chapter 11).

Different DNA polymerases have also been discovered in eukaryotes. The eukaryotic enzymes synthesize DNA in much the same manner as the *E. coli* enzymes; however, the DNA polymerases of eukaryotic cells lack exonuclease activity.

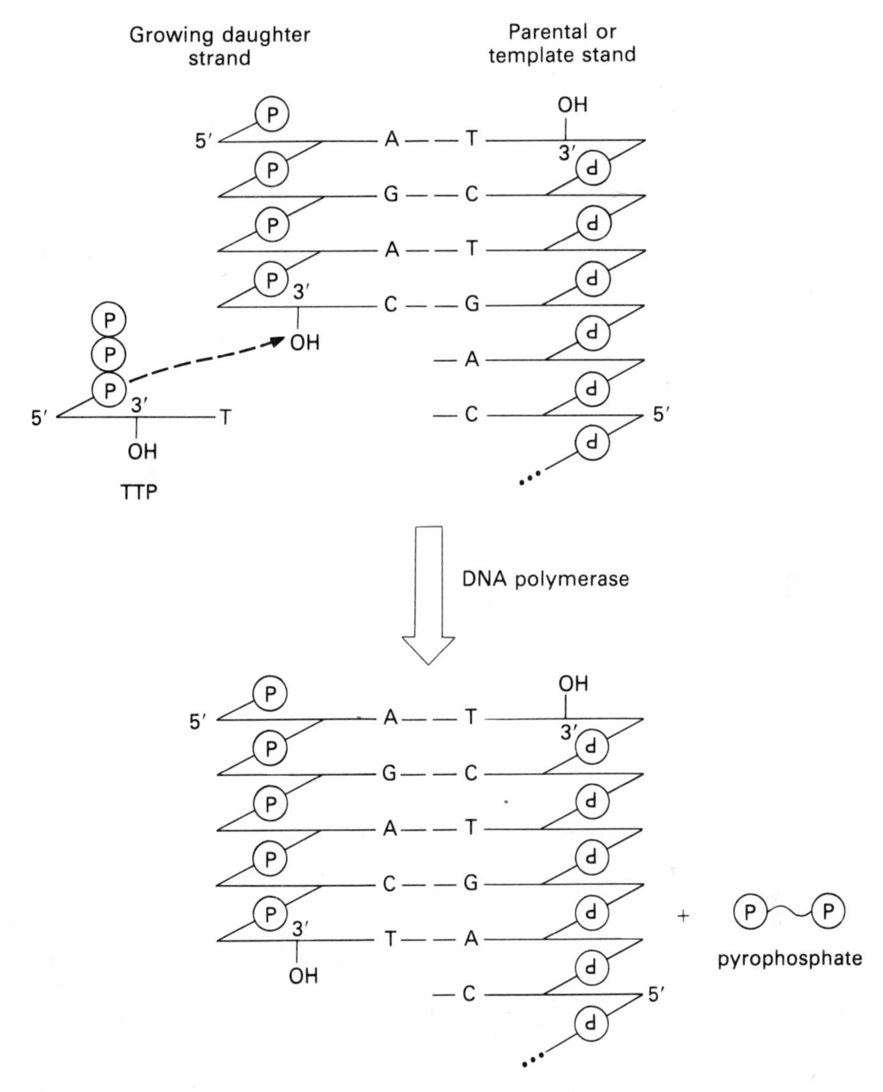

Figure 3.23. Addition of nucleotides to the 3′ hydroxyl end of the growing nucleotide strand by DNA polymerase. Only one strand of the parental double helix is shown here.

Discontinuous Replication of DNA. After the discovery of DNA polymerase I, it soon became apparent to geneticists that the DNA polymerases are not the only enzymes acting in DNA replication. We have seen that the DNA polymerase enzymes can catalyze strand elongation only in the 5′ → 3′ direction. Yet, studies on the *E. coli* chromosome (which we will describe in the following chapter) indicate that synthesis occurs simultaneously on both template strands at the point of unwinding (the replication fork). This would suggest that DNA is being lengthened in the 5′ → 3′ direction on one template strand and in the 3′ → 5′ direction on the other strand. Since no enzyme has been discovered that could catalyze 3′ → 5′ synthesis, how then does replication occur on both template strands at the replication fork? An answer to this question was provided in the late 1960s when R. Okazaki, a Japanese investigator, discovered that DNA is synthesized discontinuously on at least one of the template strands in the form of short pieces. These pieces, now popularly known as *Okazaki fragments,* are then joined together enzymatically to make progressively longer DNA segments, until the entire length of the molecule has been replicated. According to the current view of DNA replication, one growing strand, the *leading strand,* is synthesized *continuously* along its template from 5′ to 3′ while moving in the same direction as the replication fork (Fig. 3.24). The other growing strand, the *lagging strand,* is synthesized *discontinuously* in the form of Okazaki fragments. These fragments, having an average length of about 1000 nucleotides in prokaryotes and 200 nucleotides in eukaryotes, are also synthesized from 5′ to 3′, but in a direction opposite to the movement of the replication fork.

Another question arises regarding the requirement of the DNA polymerase enzymes for a primer. If the DNA polymerases cannot start synthesizing a new DNA strand on their own, what initiates replication of each Okazaki fragment? The question was answered when it was discovered that an Okazaki fragment is

Figure 3.24. Discontinuous synthesis of DNA. Short segments of new DNA, called Okazaki fragments, are synthesized (5′ → 3′) in a direction opposite to the movement of the replication fork. The ends of the Okazaki fragments are joined to form the lagging strand. The other strand, called the leading strand, is synthesized (5′ → 3′) in a continuous manner while moving in the same direction as the replication fork.

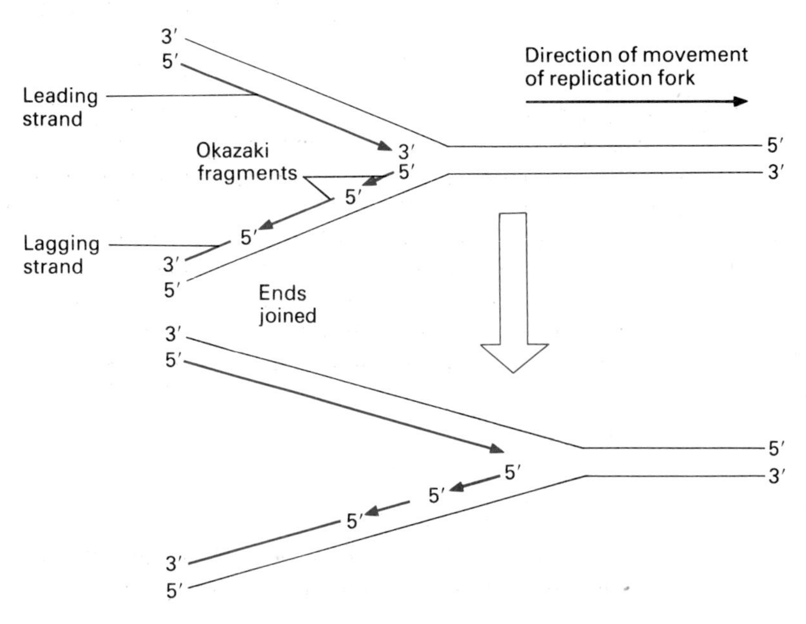

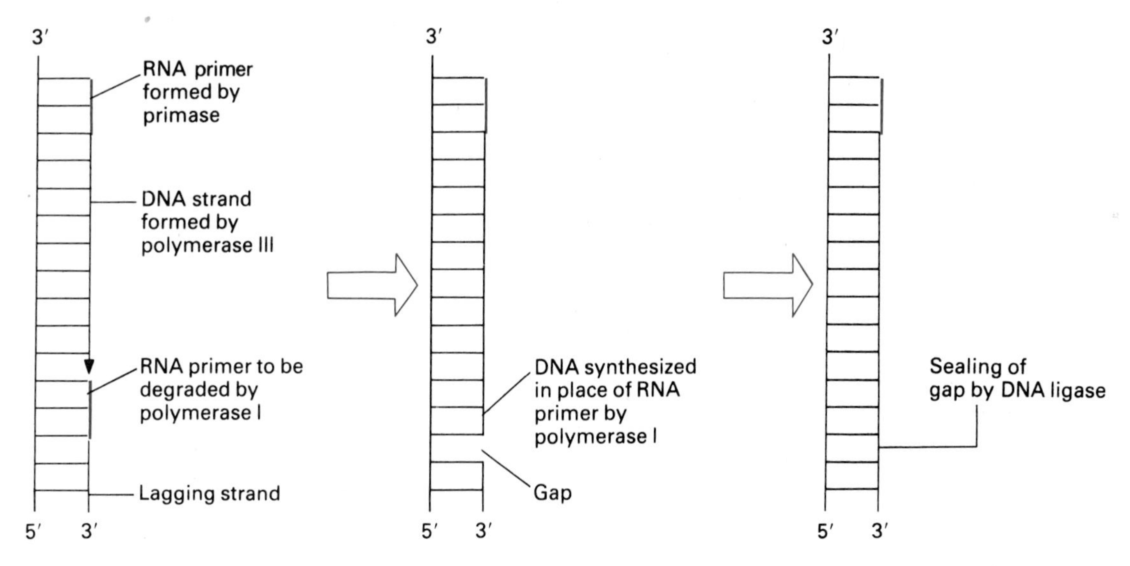

Figure 3.25. Steps involved in the synthesis of an Okazaki fragment in *E. coli*. Synthesis begins with the formation of an RNA primer. A DNA strand is then produced from the 3′ end of the primer by DNA polymerase III. The primer of the lagging strand is removed and replaced with DNA by DNA polymerase I. The fragment is subsequently joined to the remainder of the lagging strand by DNA ligase.

synthesized using a short *RNA primer* of just a few ribonucleotides in length. The RNA primer is synthesized on the DNA template by an enzyme known as **primase.** The action of primase is similar to that of the DNA polymerases except that primase requires the ribonucleotides ATP, GTP, CTP, and UTP as substrates and can initiate polynucleotide synthesis on its own.

The sequence of events involved in the discontinuous synthesis of DNA in *E. coli* is illustrated in Fig. 3.25. An RNA primer is first formed by the action of primase. Starting with the 3′ end of the RNA primer, deoxyribonucleotides are then added sequentially in the 5′ → 3′ direction by the action of DNA polymerase III. Elongation of the DNA strand continues until the 5′ end of the lagging strand is reached. DNA polymerase I then takes over: It removes the RNA primer of the lagging strand by means of its 5′ → 3′ exonuclease activity and replaces the excised primer with the corresponding DNA sequence. Finally, a splicing enzyme called **DNA ligase** joins the 3′-hydroxyl group of the fragment that has just been synthesized to the 5′-phosphate group of the lagging strand.

Ongoing research reveals that the replication of DNA is even more complex than we have so far indicated. A variety of proteins are required in addition to primase, DNA polymerases I and III, and DNA ligase in order to unwind the double helix and keep the two strands apart. Figure 3.26 diagrams the currently accepted scheme of DNA replication, showing the roles of these various proteins. Enzymes known as *helicases* act ahead of the replicating fork to unwind and separate the parental DNA strands. Energy for unwinding is provided by the

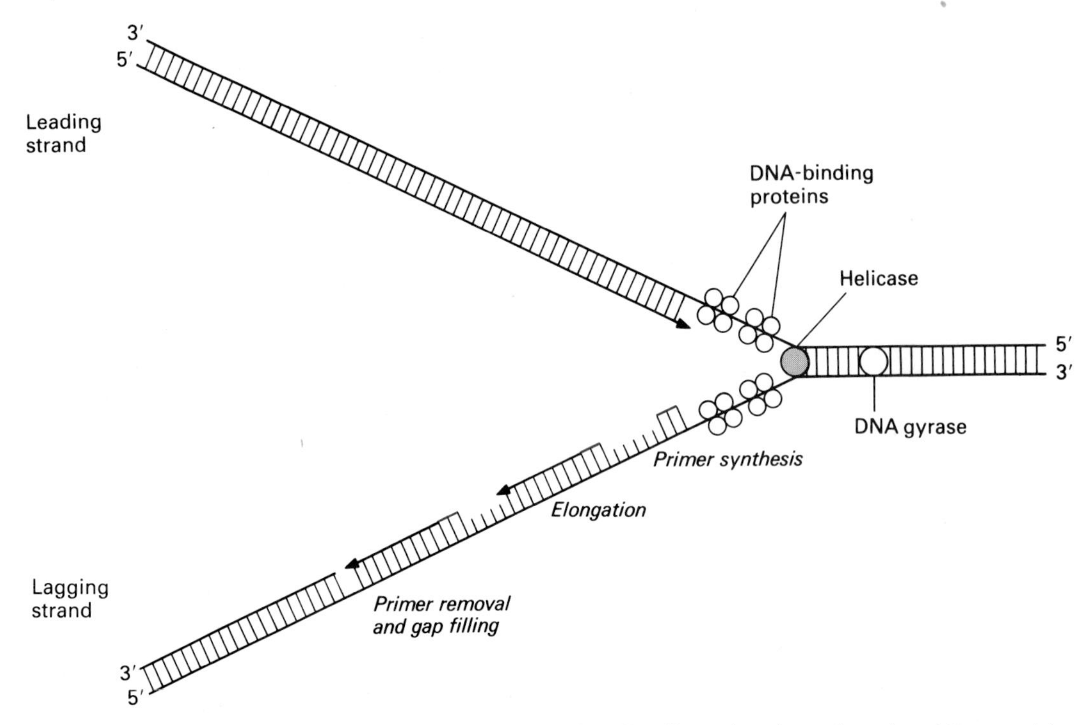

Figure 3.26. Current model of DNA replication. The illustration shows the roles of three proteins involved in the unwinding of DNA prior to replication: Helicase enzymes act to separate the parental DNA strands, DNA binding proteins prevent the strands from reforming the double helix, and DNA gyrase prevents twisting of the double helix ahead of the replication fork.

breakdown of ATP. *DNA binding proteins* (DBP) temporarily bind to the single-stranded regions that are exposed just prior to replication to prevent the parental helix from re-forming. As the unwinding proceeds, the double-helical region ahead of the replication fork is prevented from twisting by a *topoisomerase* enzyme, called *DNA gyrase*. The function of topoisomerases is considered further in the following chapter. Other proteins have also been implicated in the replication process. But much still needs to be done before we fully understand the biochemical basis of DNA replication.

Mutation of DNA

The Watson-Crick model of DNA also suggests a possible mechanism for mutation. A mutation could occur when a new base pair is erroneously substituted for one that is normally present at a given location. For example, an AT base pair at a particular site might be replaced with a GC pair, or vice versa. Such an event could give rise to a mutant gene that has a slightly different sequence of nucleotides than its normal allele. This process is shown in Fig. 3.27 where a gene is represented as a segment of DNA with a certain sequence of base pairs. Note that many

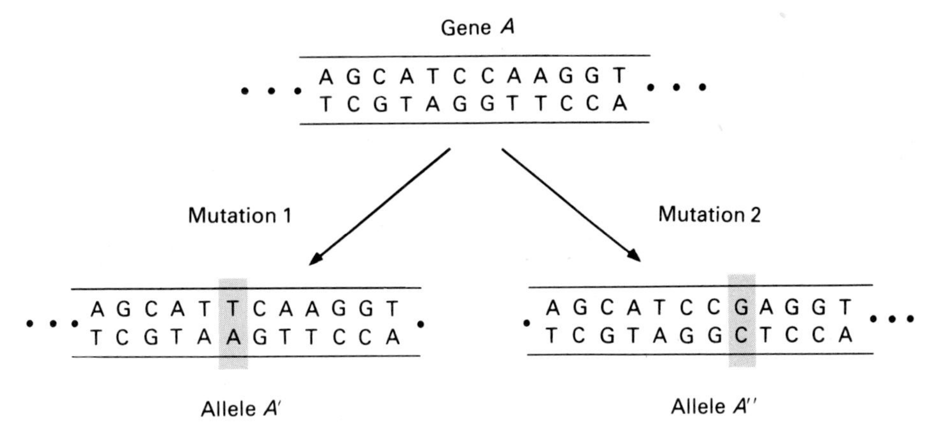

Figure 3.27. Origin of different allelic forms of a gene by mutation. A gene constitutes a segment of DNA that carries information in the sequence of its many base pairs, only 12 of which are shown in the diagram. Gene mutation occurs through the substitution of different base pairs, giving rise to different alleles.

different mutations are possible. Since every gene comprises many base pairs, the substitution of bases at various nucleotide sites could result in a whole family of mutant alleles.

Mutation as an Error in Base Pairing. When Watson and Crick proposed their model for the double-helical structure of DNA, they also suggested a means by which base substitutions might spontaneously occur. They recognized that the structures of the various purine and pyrimidine bases are not static but can chemically fluctuate. Changes in the distribution of hydrogen atoms are known to occur on rare occasions. These changes result from a shift in the location of a proton from one position to another in a purine or pyrimidine ring. Such a shift in proton position, called a **tautomeric shift,** can change the hydrogen bonding properties of the base in which it occurs. For example, adenine normally pairs with thymine. But if adenine undergoes a tautomeric shift involving a change in location of a hydrogen atom from its amino group at the number 6 position to its nitrogen at the number 1 position (Fig. 3.28), its hydrogen bonding properties are changed to those of guanine. Adenine in its rare form then pairs with cytosine. This change in base pairing properties is shown in Fig. 3.29(a). Note that the reverse situation is also true: A tautomeric shift in guanine can change its base pairing properties to those of adenine (Fig. 3.29b). All four bases are subject to these structural changes and can exist, at least temporarily, in their rare tautomeric state, with altered base pairing properties. In general, therefore, a purine in its rare state can pair with the "wrong" pyrimidine, and a pyrimidine in its rare state can pair with the "wrong" purine. As Watson and Crick pointed out, the existence of these rare tautomeric alternatives can lead to base substitutions as a result of errors in base pairing.

In order for a mutation to occur as a result of a base-pairing mistake, the base must appear in its rare tautomeric form at the moment of DNA replication. Only

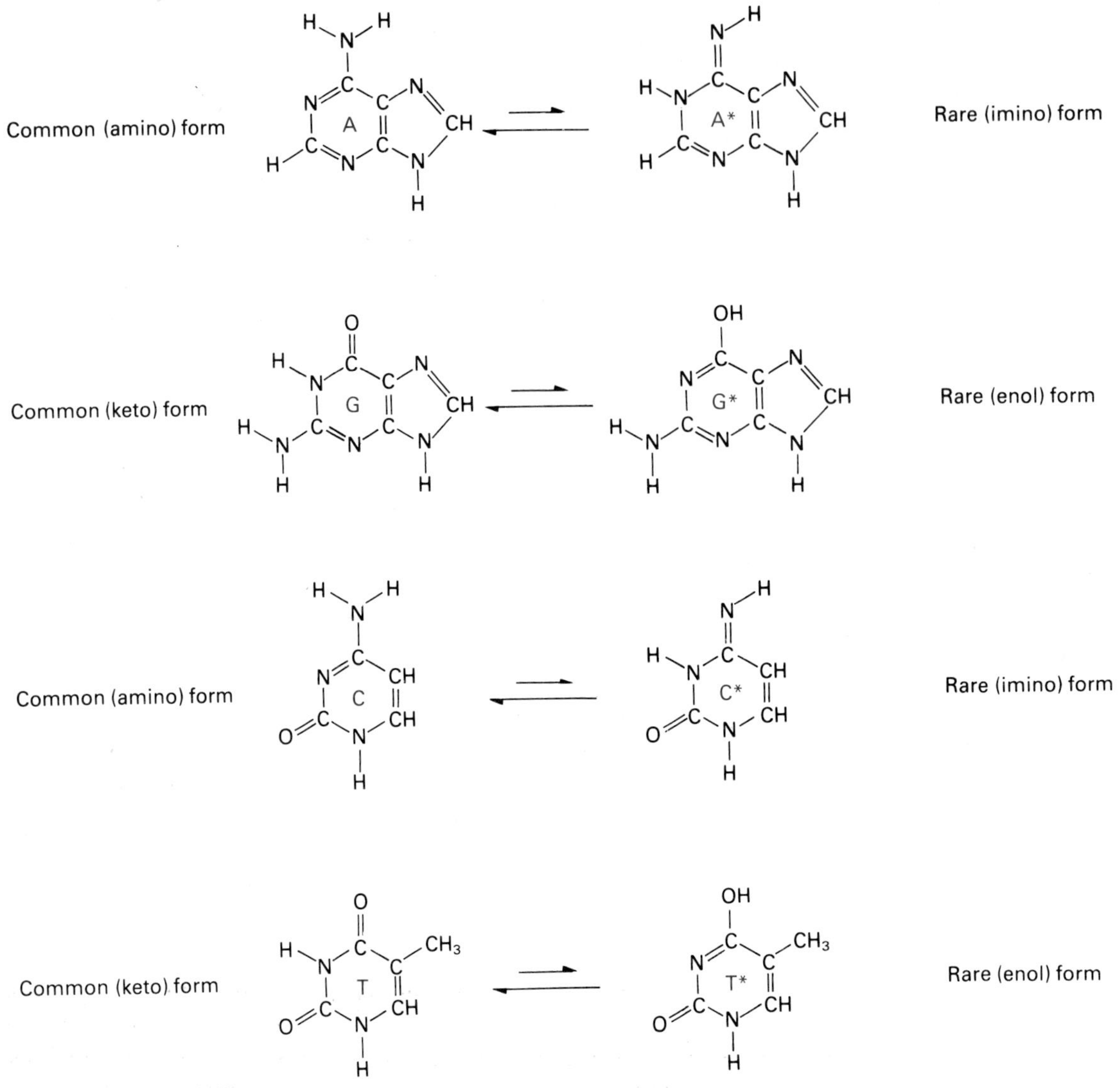

Common (amino) form — A ⇌ A* — Rare (imino) form

Common (keto) form — G ⇌ G* — Rare (enol) form

Common (amino) form — C ⇌ C* — Rare (imino) form

Common (keto) form — T ⇌ T* — Rare (enol) form

Figure 3.28. Rare forms of the four DNA bases, caused by a shift in the location of a proton from one position to another in a purine or pyrimidine ring. The usual forms of the bases are shown on the left, the rare tautomeric forms on the right.

then can a "mismatched" base pair be incorporated into a DNA molecule. This mechanism is illustrated in Fig. 3.30. Observe that there are two ways in which tautomeric shifts can induce base-pairing mistakes. One is through a *template error,* in which the rare tautomeric form of the base exists in the template strand during replication (Fig. 3.30a). The other is through an *incorporation error,* in which the

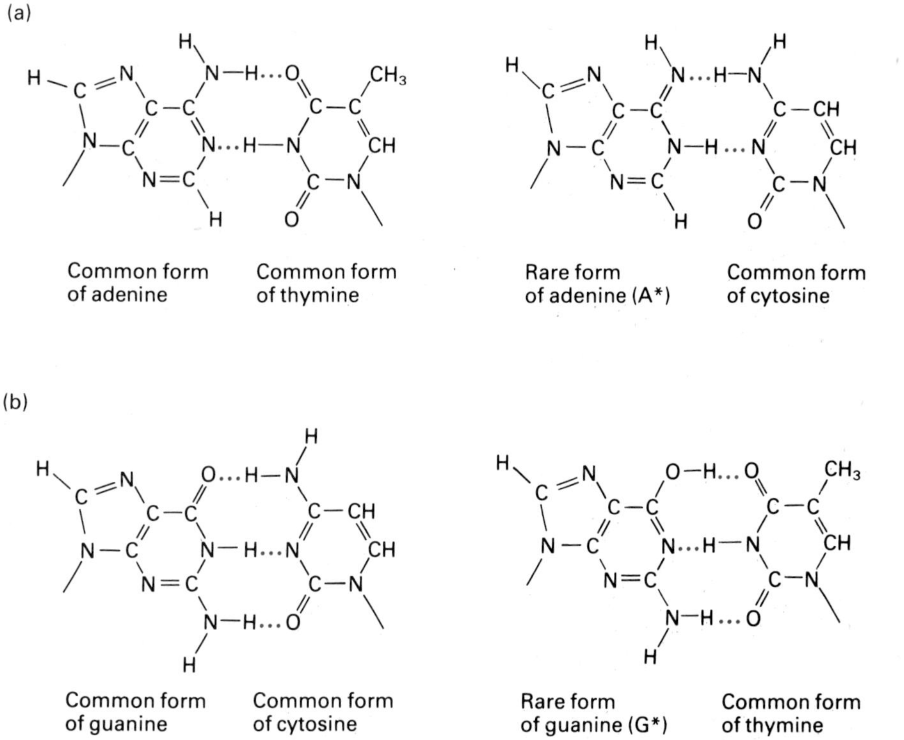

(a)

Common form of adenine — Common form of thymine

Rare form of adenine (A*) — Common form of cytosine

(b)

Common form of guanine — Common form of cytosine

Rare form of guanine (G*) — Common form of thymine

Figure 3.29. Errors in base pairing can occur through shifts in the positions of protons in the purine and pyrimidine bases. (a) The common form of adenine forms hydrogen bonds with thymine (left); in its rare tautomeric form, it forms hydrogen bonds to cytosine (right). (b) Guanine in its common form bonds to cytosine (left); in its rare tautomeric form, it bonds to thymine. All of the bases can exist in their rare tautomeric forms, with altered base pairing properties.

rare tautomeric form of the base exists in a deoxyribonucleotide precursor and is inserted into the growing chain (Fig. 3.30b). In either event, the rare base, which is now present in the molecule, is highly unstable and is expected to revert back to its normal state. The result is a DNA molecule, called a **heteroduplex,** with an incorrectly matched base pair. A second round of DNA replication is then required to segregate the mismatched base pair and produce a fully mutant double helix. Thus after two rounds of DNA replication, a base pairing mistake in the parental DNA molecule will result in the replacement of one base pair—either AT or GC—with a different base pair—either GC or AT, respectively—in one of the progeny DNA molecules.

The idea that mutation occurs through base-pairing mistakes has received considerable experimental support. In Chapter 11 we will consider this model further, along with a description of some other molecular mechanisms of mutation.

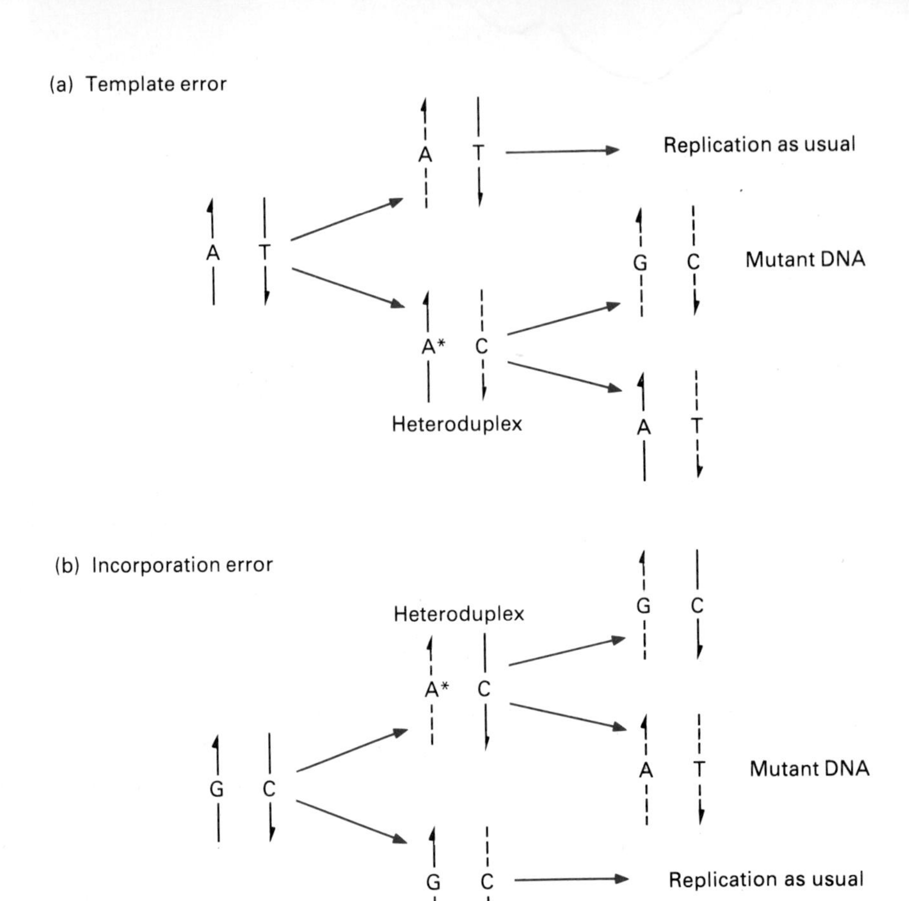

Figure 3.30. Spontaneous mutations involving base substitutions caused by the mispairing of adenine. (a) Tautomerism of adenine in a template strand leads to the base pair change AT → GC. (b) Incorporation of the tautomeric form of adenine into a daughter strand leads to the base pair change GC → AT.

TO SUM UP

1. Genetic information is carried by DNA in its sequence of base pairs. The information is organized into functional sequences, which we call genes.
2. The replication of DNA requires the unwinding and separation of its polynucleotide chains. Each strand of the helix serves as a template for the synthesis of a new complementary strand. This mode of synthesis, in which the daughter molecules of DNA have one old parental strand and one new complementary strand, is called semiconservative replication.
3. Replication of at least one of the strands of the double helix is discontinuous, proceeding in segments, called Okazaki fragments, that have a length of about 1000 nucleotides in prokaryotes and 200 nucleotides in eukaryotes. A DNA polymerase enzyme makes each segment separately, by extending the synthesis in a 5′ → 3′ direction from an RNA primer. Following excision of an RNA primer and repair synthesis by a DNA polymerase enzyme to fill in the gap, the segment is attached to the rest of the growing strand by DNA ligase.
4. A gene mutation can occur through the substitution of one base pair for another in

DNA as a result of an error in base pairing. A base-pairing mistake can result from a shift in the proton distribution of a base, called a tautomeric shift, which gives rise to a change in the hydrogen bonding properties of the affected purine or pyrimidine.

Questions and Problems

1. Define the following terms and distinguish between members of paired terms:
bacteriophage
buoyant density
Chargaff's rules
complementary strands–antiparallel strands
denaturation–renaturation
deoxyribose–ribose
DNA polymerase
genetic transformation
heteroduplex
hydrogen bond–3′-5′ phosphodiester bond
melting temperature
nucleotide–nucleoside
primary structure–secondary structure
purine–pyrimidine
semiconservative replication–conservative replication
tautomeric shift
template–primer
3′ end–5′ end

2. State whether each of the following statements is true or false. If it is false, explain why.

(a) Genetic transformation refers to the transfer of DNA from a donor cell to a recipient cell by a bacterial virus.

(b) The transformation of type R pneumococcus cells by an extract of dead S cells is not affected by RNase or protease.

(c) In the Hershey-Chase experiment, the finding that most of the ^{35}S remains in the supernatant indicates that DNA is the genetic material.

(d) Cytosine and guanine are purine bases.

(e) The phosphate group of a deoxyribonucleotide is bonded to the 2′ carbon of the sugar.

(f) All double-stranded DNA molecules, regardless of the organism from which they are isolated, have the same molecular dimensions.

(g) All double-stranded DNA has an (A + T):(G + C) ratio equal to 1.

(h) If the nucleotide sequence along one strand of a DNA molecule reads 5′-A-T-A-C-T-G-C-G-T-3′, then the sequence along the corresponding section of the opposite strand will be 5′-A-C-G-C-A-G-T-A-T-3′.

(i) If the DNA of species A denatures at a higher temperature than that of species B, then the DNA of species A must contain a higher proportion of AT base pairs.

(j) If the DNA of species A has a higher buoyant density than that of species B, it will band at a lower position in the centrifuge tube than will the DNA of species B.

(k) Even though one strand of DNA appears to be growing in the 3′ → 5′ direction, all known DNA polymerases synthesize DNA exclusively in the 5′ → 3′ direction on both template strands.

(l) Discontinuous synthesis of at least one of the two daughter strands of a replicating double helix is the rule in both prokaryotes and eukaryotes.

(m) Exonucleases degrade DNA by cleaving nucleotides from an exposed 5′ or 3′ end of a polynucleotide chain.

(n) A primer is a polynucleotide chain that is copied into a complementary strand by DNA polymerase.

(o) An inhibitor that blocks all RNA synthesis would have no effect upon DNA replication.

(p) DNA polymerase I is known as a repair enzyme, and it normally plays no role in DNA replication.

3. Fill in the blanks.

(a) The three bases common to both DNA and RNA are ______, ______, and ______.

(b) Both ______ in DNA and ______ in RNA form hydrogen bonds with adenine.

(c) The sequences of the two strands of a DNA double helix are ______.

(d) Complementarity of DNA strands requires

that the amount of ________ equal that of ________, and the amount of ________ equal that of ________.

(e) Bases across from each other on opposite DNA strands are held together by ________ bonds, whereas adjacent nucleotides along a DNA strand are connected by ________ bonds.

(f) Denaturation of DNA refers to the breakage of ________ bonds and the subsequent separation of the ________.

(g) The three functions of the genetic material that are met by DNA are ________, ________, and ________.

(h) Enzymes that catalyze the formation of phosphodiester bonds between adjacent nucleotides, with no actual synthesis of DNA, are called ________.

(i) The fragments that are intermediates in discontinuous DNA synthesis are about ________ nucleotides long in prokaryotes and ________ nucleotides long in eukaryotes.

(j) If adenine undergoes a tautomeric shift, it pairs with ________ instead of thymine.

(k) If thymine undergoes a tautomeric shift, it pairs with ________ instead of adenine.

4. Compare and contrast the structures of prokaryotic and eukaryotic cells. Include a list of organelles that are present in one type but not in the other, as well as a list of structures that are present in both types.

5. If 1000 cells per ml are used to start a bacterial population, and the number of cells doubles every 30 minutes on the average, **(a)** how many cells will be present after five hours? **(b)** After ten hours?

6. The base compositions of nucleic acids isolated from several different sources are given below. In each case, decide if the nucleic acid is DNA or RNA and whether it is single- or double-stranded.

Source 1: 20% A, 30% C, 30% G, 20% T
Source 2: 40% A, 10% C, 10% G, 40% T
Source 3: 30% A, 20% C, 20% G, 30% U
Source 4: 40% A, 10% C, 40% G, 10% U
Source 5: 30% A, 30% C, 20% G, 20% T

7. Studies show that the relationship between the melting temperature (T_m) and the percent GC content of DNA is expressed by the equation $T_m = 69.3 + 0.41(\%GC)$. Calculate the base composition of DNA from a human spleen cell, given that the melting temperature of the DNA is 86.5°C.

8. Studies show that the buoyant density of DNA is $1.660 + 0.00098(\%GC)$. Calculate the buoyant densities of two DNA samples: one having 45% GC base pairs, and the other having 55% GC pairs. (The units of density are grams per cubic centimeter.)

9. You have two strains of a certain species of bacteria. One strain is penicillin-sensitive (pen^s), and the other is penicillin-resistant (pen^r). You would like to determine whether or not this particular species is capable of undergoing genetic transformation. Describe an experiment involving the pen^s and pen^r strains that could be used to test this possibility.

10. Consider a "Meselson-Stahl" experiment in which bacterial cells are grown in medium containing ^{15}N until all of the DNA molecules are fully labeled. These cells are then transferred to a medium in which all of the nitrogen is ^{14}N and are permitted to duplicate further. **(a)** According to the semiconservative mechanism of replication, what proportion of the DNA molecules present after one round of duplication in ^{14}N medium will be of intermediate density? What proportion will be fully heavy? What proportion will be fully light? **(b)** What proportion of the DNA molecules present after three rounds of duplication in ^{14}N medium will be of each density class? **(c)** What result would Meselson and Stahl have obtained after three rounds of duplication if DNA replication were conservative? What result would they have obtained if replication were dispersive?

11. How many twists (complete turns) are there in a DNA molecule that consists of 2,000,000 base pairs?

12. A DNA molecule that is isolated from a particular virus is 130 μm in length. How many base pairs are contained in this molecule? What would be its molecular weight? (Assume an average weight of 660 per nucleotide pair.)

13. A double-stranded DNA molecule contains 30 percent adenine nucleotides. What are the proportions of the other three nucleotides?

14. One strand of a DNA molecule contains the nucleotide proportions 20% A, 30% C, 40% G, and 10% T. What are the proportions of the four base pairs expected in the double-stranded form of this DNA?

15. The bacterium *Escherichia coli* is approximately a cylinder 2 μm in length and 1 μm in diameter. A human liver cell is roughly a sphere 20 μm in diameter. How much greater is the volume of the liver cell than that of the bacterial cell? (The volume of a cylinder is $\pi r^2 h$, and that of a sphere is $(4/3)\pi r^3$, where r is the radius and h is the length. Note that π cancels when you divide one volume by the other.)

16. The capsid (head) of phage T4 contains a space of about 2×10^{-16} cm^3. The DNA of this phage is 2×10^5 nucleotide pairs in length. Calculate the volume of this DNA (assume a cylindrical shape) and compare it with the space inside the capsid into which the DNA is packed.

17. If heavy DNA (labeled with ^{15}N) that was isolated from bacteriophage λ is mixed with light λ DNA and is subjected to density gradient equilibrium centrifugation, two bands appear in the centrifuge tube. In contrast, if the DNA mixture is heated to 100°C and held at 65° prior to centrifugation, three bands appear in the centrifuge tube. Account for the difference in the results of centrifugation between the two experiments. What would be the position of the third band formed by the heated and slowly cooled material relative to the other two bands?

18. If the experiments of Problem 17 are repeated using a mixture of DNA from heavy λ and light T2 bacteriophages, only two bands appear following the heating and slow cooling. Furthermore, these bands are found at identical positions to the bands formed by the unheated mixture. How would you account for the absence of the third band in this case?

19. In an experiment, the DNA of a certain bacterium is broken into short double-stranded fragments and is then subjected to density gradient equilibrium centrifugation. The fragments form one band, at a density position that is identical to the band formed by the unfragmented DNA. What does this result tell you about the base pair composition along the DNA molecule of this organism?

20. The "Meselson-Stahl" experiment cannot distinguish between semiconservative and dispersive replication if the cells are allowed just one act of duplication in ^{14}N medium. Outline an additional experimental procedure that could be used to distinguish between the kinds of first-generation molecules that are predicted by the semiconservative and dispersive replication models.

21. The DNA of a newly discovered virus fails to yield an increase in absorption of ultraviolet light when heated to temperatures over 110°C. Postulate an explanation for this result.

22. A strain of pneumococcus that is sensitive to both the antibiotics penicillin (pen^s) and streptomycin (str^s) is mixed with DNA that is isolated from a doubly resistant ($pen^r\ str^r$) strain. Only 0.0004% of the recipient cells acquire both the donor pen^r and str^r characteristics. What percentage of the recipient cells are expected to have acquired only one of the genes for antibiotic resistance if the *pen* and *str* markers are so far apart on the bacterial chromosome as to be normally carried on different transforming DNA fragments? (Assume that several donor DNA fragments can potentially enter a recipient cell with equal efficiency.)

23. Figure 3.30 shows spontaneous mutation caused by the tautomerism of adenine and the possible base pair substitutions that result. Diagram the sequence of events that lead to mutation as a consequence of tautomerism in each of the other three DNA bases. Include in your diagram both template and incorporation errors.

24. A single-stranded DNA virus has an A:T base ratio of 2, a G:C ratio of ½, and a (A + T):(G + C) ratio of ¼. **(a)** What is the (A + G):(T + C) ratio in this molecule? **(b)** If this single-stranded DNA forms a complementary strand, what are these four ratios in the complementary strand? **(c)** What are these four ratios in the resulting DNA duplex (the original and complementary strands together)?

25. Remembering that nucleotide sequences are read in the $5' \rightarrow 3'$ direction, determine which of the following pairs of dinucleotides will be equal in frequency in a typical DNA molecule and which need not be: **(a)** G-G and C-C, **(b)** A-G and T-C, **(c)** A-G and C-T, **(d)** G-C and A-T, **(e)** C-T and A-G, **(f)** C-A and G-T.

Chapter 4

Structure and Replication of Chromosomes

In the preceding chapter, we concluded that, with the exception of certain viruses, genes are composed of DNA. We also determined that the hereditary information of each gene is carried by a specific sequence of base pairs and is replicated in a semiconservative fashion. We will now consider the manner in which the hereditary information is packaged within a cell or virus and is transmitted to the progeny produced when a cell divides or a virus undergoes intracellular multiplication.

Despite their nucleic acid compositions, genes seldom exist in the form of naked and extended molecules of nucleic acid; they are usually organized into chromosomes. From a purely structural standpoint, polynucleotide chains are extremely large polymeric molecules; in fact they are much too long to fit neatly into a virus or cell when fully extended or even when coiled into the Watson-Crick double helix. They must assume a higher level of structural organization, which involves extensive folding and supercoiling, to make them more compact. In many cases, the systematic condensation, or *packaging*, of the genetic material requires the complex association of the DNA molecule with specific kinds of proteins and various other molecules. In this chapter, we will discuss the organization of DNA (or RNA) into chromosomes in viruses, bacteria, and higher organisms. We will find both differences and similarities in chromosome structure existing among these different groups of organisms. We will also present the schemes that these different organisms employ to replicate their chromosomes.

Chromosomes of Viruses and Bacteria

Chromosomes of Viruses

There are three major groups of viruses, defined by the type of organism used as a host. *Animal viruses* are active only on animal cells, *plant viruses* infect only plant cells, and bacteriophages use only bacterial cells as hosts. Each particular type of virus within a group is further restricted in the range of specific hosts that it can successfully infect.

Viruses and their chromosomes exist in a wide variety of structures. Viral chromosomes are tightly packaged into protein coats, called capsids, which can be helical or polygonal in shape (Fig. 4.1). The negative charge of the nucleic acid molecule is neutralized by binding of the molecule with *polyamines,* a class of basic polymers that are found in association with viral and bacterial nucleic acid and with some eukaryotic DNA (e.g., the DNA of a sperm cell). Several animal viruses are also enclosed in an envelope composed of material from the cell membrane of the host; the envelope forms around the virus particle as it is released from the host cell.

Viruses have diverse nucleic acid structures. Many viral chromosomes exist in one form while in the viral capsid (the *mature state* or **virion**) and in another form while replicating in the host cell (the *vegetative state*). Some viruses also can exist in still a third form (the **provirus**), in which they lie essentially dormant but integrated within the chromosome of the host cell. We will consider each of these three states separately.

The Virion. The chromosome of a mature virus can be made up of RNA or DNA, can be linear (rod-shaped) or circular (without ends), and can exist in either a single-stranded or a double-stranded configuration. There seems to be no general rule that relates the specific way in which the nucleic acid is organized to the category of virus. For example, both plant and animal viruses are known that contain RNA rather than DNA, either in a single-stranded linear form (e.g., the tobacco mosaic virus), in a single-stranded circular form (e.g., the endomyocarditis virus), or as a double helix. Mammalian reovirus is an example of a virus that contains RNA in a double-helical configuration. Many types of RNA bacteriophages are also known, including Qβ, R17, and f2. The RNA of these viruses is single stranded, but it assumes a flowerlike structure consisting of hairpin loops. The loops form because of hydrogen bonding between complementary regions of the same strand.

A wide variety of chromosome structures is found among the DNA viruses as well. These include single-stranded linear forms (phage M13), single-stranded circular forms (φX174), double-stranded linear forms (herpes virus), and double-

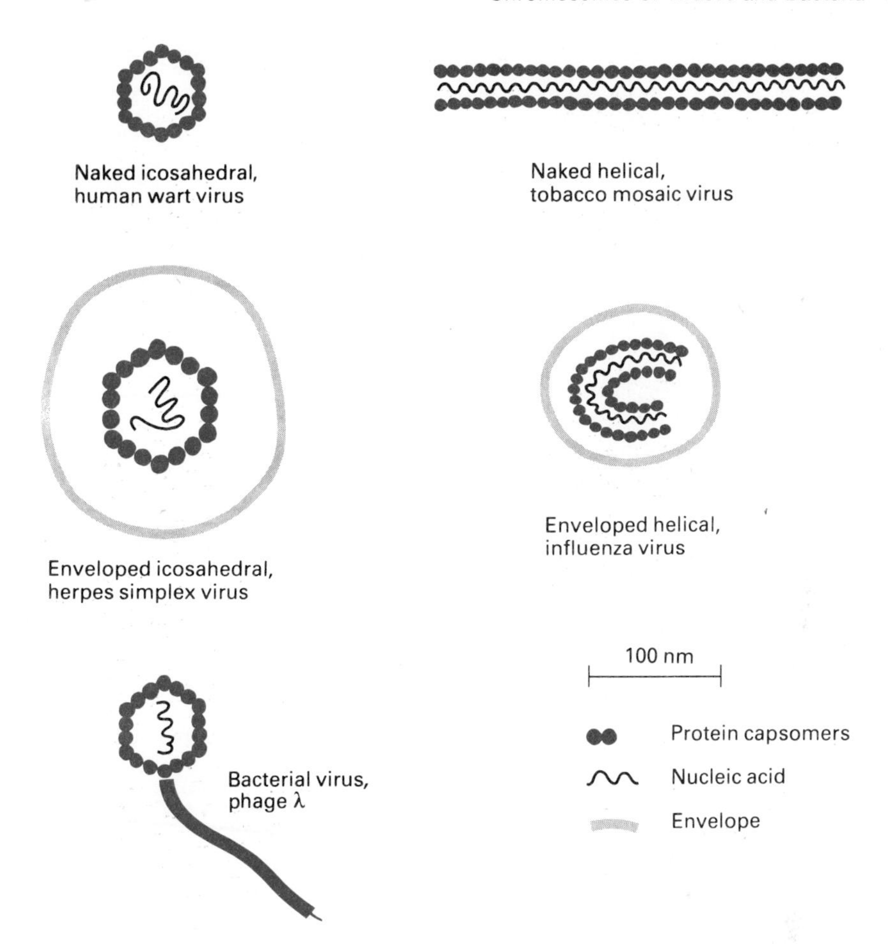

Figure 4.1. Representative animal, plant, and bacterial viruses. Viral protein coats are composed of subunits called capsomers; these coats, or capsids, can be polygonal or helical in shape. (An icosahedron is a polyhedron with 20 faces.) The viral nucleic acid is found inside the protein coat. Several animal viruses are also enclosed in an envelope derived from the cell membrane of the host.

stranded circular forms (polyoma tumor virus). Other variations also exist. For example, the hepatitis B viral chromosome is about half double- and half single-stranded.

Most of our discussion of viruses here centers on bacteriophages, since they are much better known genetically than their counterparts that infect eukaryotes. The chromosomal organization of viruses tends to resemble that of their hosts, so that the viruses of eukaryotes are more complex than those of bacteria.

The Vegetative Virus. When a virus infects its host by introducing its nucleic acid into a host cell, the chromosome of the virion is converted into a noninfectious replicative form, or vegetative virus. The replicative form of many viruses differs, often substantially, from the chromosome of the virion. As an example, the ϕX174 chromosome is single-stranded in the mature state, whereas the replicative form

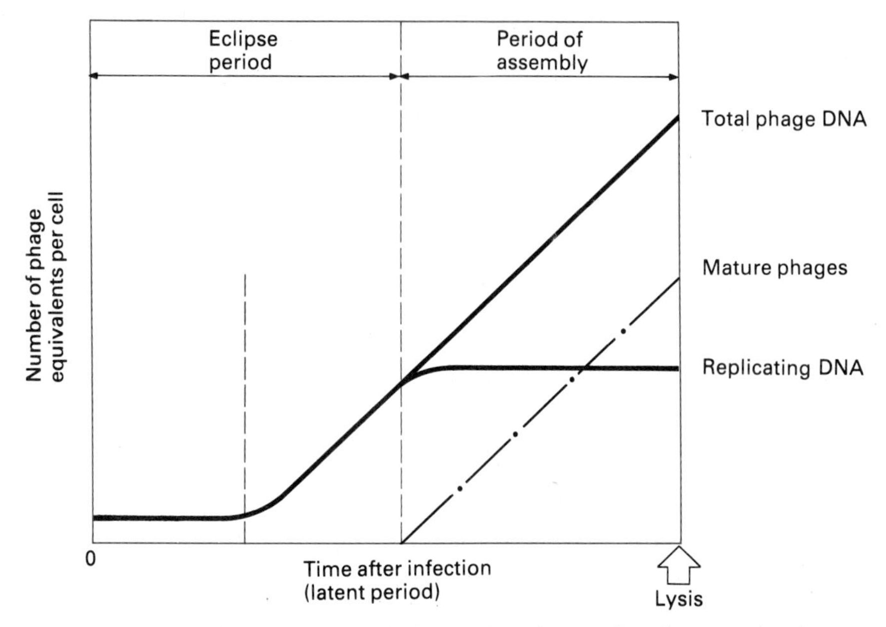

Figure 4.2. Idealized view of the changes in the number of vegetative phages and mature assembled phages during the stages of intracellular growth. The time required for intracellular growth, known as the latent period, is divided into the eclipse and assembly stages. The eclipse period, during which no mature phages are present, is characterized by an initial lag followed by a phase of replication of phage chromosomes. When assembly of mature phages begins, the amount of phage DNA in the vegetative form reaches a steady state, as phage chromosomes are withdrawn for maturation into infective viral particles as fast as they are produced by replication.

is double-stranded. Double-stranded virion chromosomes too may undergo a change in the vegetative state. The DNA of phage lambda (λ), for example, is linear in the mature state but is circular while replicating within the bacterial host.

The precise mechanisms by which viral chromosomes replicate vary. Even so, they all tend to have a similar growth pattern within the host cell. An idealized view of the stages of intracellular growth is shown in Fig. 4.2 in terms of the events that occur during infection by a DNA containing phage. Immediately following the injection of the phage genome into the bacterial host, there is a period known as the **eclipse** when no infective viral particles have yet been synthesized within the cell. The eclipse period includes an initial lag, followed by a phase of active chromosome replication. During the eclipse, the phage DNA directs the synthesis of specific proteins, including the structural components of the capsid and various enzymes that are involved in the intracellular growth process. Mature *phage assembly* then ensues, beginning about midway through the reproductive cycle. During assembly, the component parts are combined in an ordered sequence of steps, to form the intact virion. The steps involved in the assembly of phage T4 are shown in Fig. 4.3. The formation of the mature phage is a spontaneous (self-assembly) process. Each protein undergoes a conformational change upon assembly; this change exposes a binding site that is recognized by the next protein in

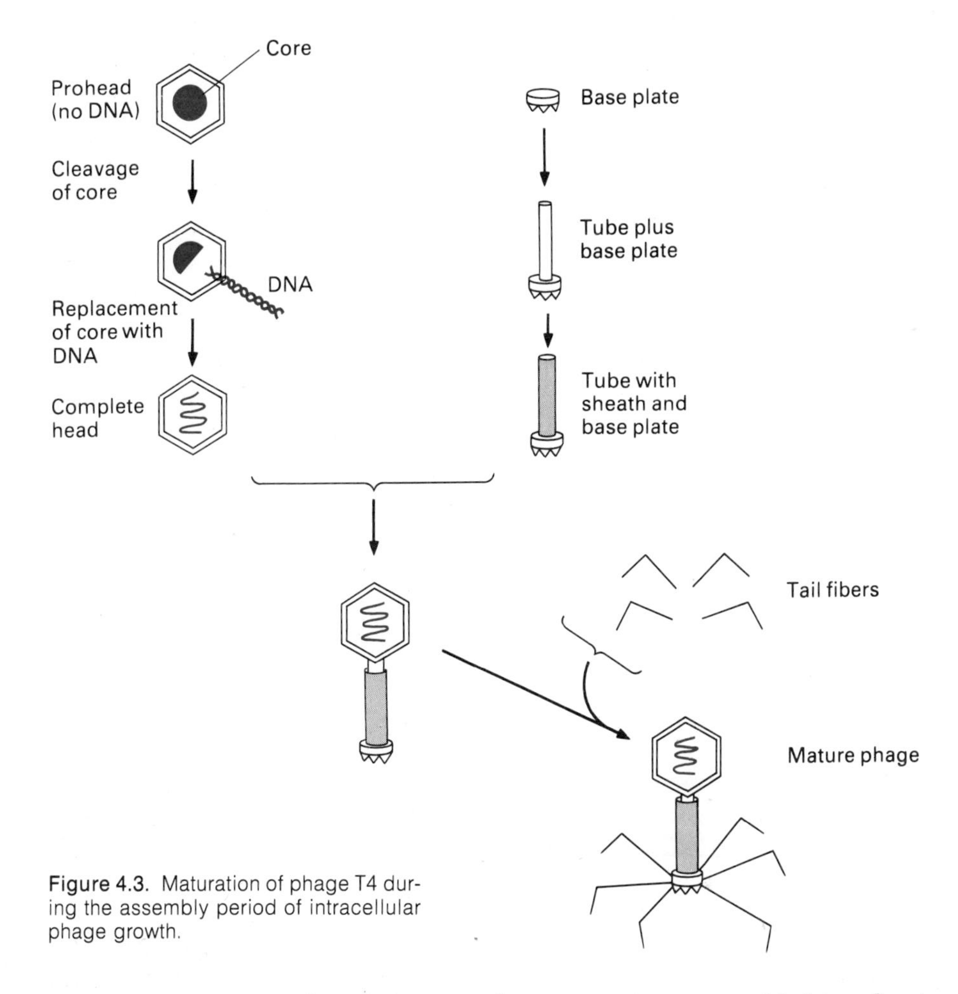

Figure 4.3. Maturation of phage T4 during the assembly period of intracellular phage growth.

the series, thus ensuring that the structural components are assembled together in the proper sequence. Phage assembly can be visualized as a steady state; in this state, the rate at which mature phage particles are formed keeps pace with the rate at which the phage chromosome is replicated, so that the number of vegetative phages in the replicating pool remains more or less constant. The period of intracellular phage growth (the **latent period**) culminates with the lysis of the bacterial cell. During lysis, the host cell bursts, and the newly synthesized viral particles are released.

The growth pattern depicted in Fig. 4.2 is accomplished by different replicative schemes in different phages. But since the vegetative stage of phage λ has been studied intensively, we will use this phage as an example of the viral replication process. Immediately following injection of the λ chromosome into the host cell, the λ DNA molecule assumes a circular form (Fig. 4.4). The formation of a circular molecule is made possible by base pairing between complementary single-stranded regions at the ends of the injected chromosome; these regions are known as **cohesive ends.** The circular chromosome then replicates in a semiconservative man-

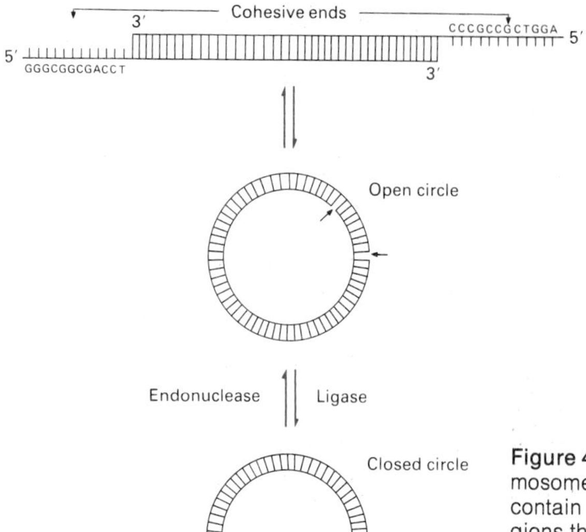

Figure 4.4. Formation of a circle by the λ chromosome. The ends of the linear λ chromosome contain complementary single-stranded regions that are 12 nucleotides in length. Hydrogen bonding between these complementary segments forms an open circle (arrows point to gap regions). The action of DNA ligase produces a covalently closed circle.

ner, beginning at the *ori* (origin of replication) site. At first, replication is *bidirectional,* with two replicating forks proceeding from the starting point around the circle in opposite directions (Fig. 4.5a). This mode of replication, known as **symmetric replication,** gives rise to a theta-shaped (θ) structure when a chromosome is caught midway through the act of duplicating. Theta structures can often be observed when replicating phage genomes are viewed with the aid of an electron microscope.

Partway through the latent period, the λ chromosome switches to a replication mechanism known as **asymmetric** (or *rolling-circle*) **replication** (Fig. 4.5b). In contrast with the symmetric scheme of replication, asymmetric replication is *unidirectional,* with only one replication fork per DNA molecule. This method produces a sigma-shaped (σ) replication intermediate. The process of asymmetric replication is initiated when one of the strands in the DNA double helix is nicked at a specific site by an **endonuclease** (a DNase enzyme that cleaves DNA internally). The unbroken circular strand acts as a template for the synthesis of a new complementary strand, with DNA polymerase using the exposed 3′ end of the nicked strand as a primer. Extension of synthesis in the 5′ → 3′ direction displaces the 5′ end of the nicked strand, so that it forms a "tail" on the circle while synthesizing its own complementary strand. (Notice the sigma shape of the replication intermediate at this step.) The linear daughter chromosomes that are generated by each succeeding "roll of the circle" remain attached end-to-end, forming a string of phage chromosomes that are linked together. These **concatamers** of chromosomes are then cut at their specific beginning and ending points into the linear DNA molecules of the mature phage. Each linear double helix is packaged into a

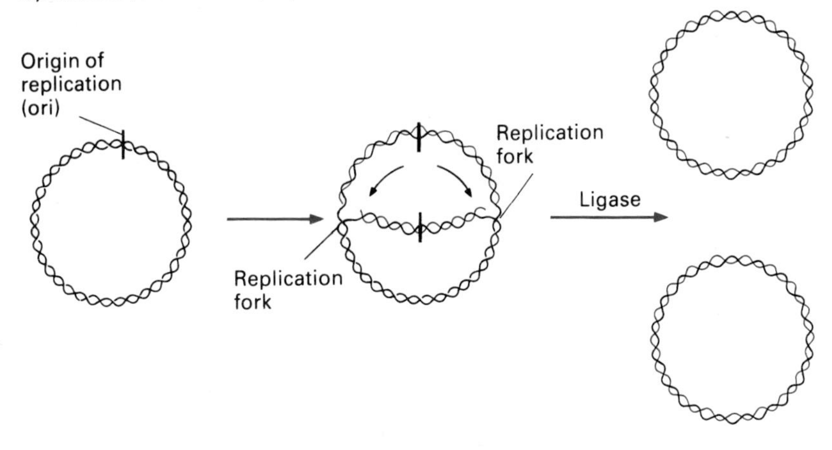

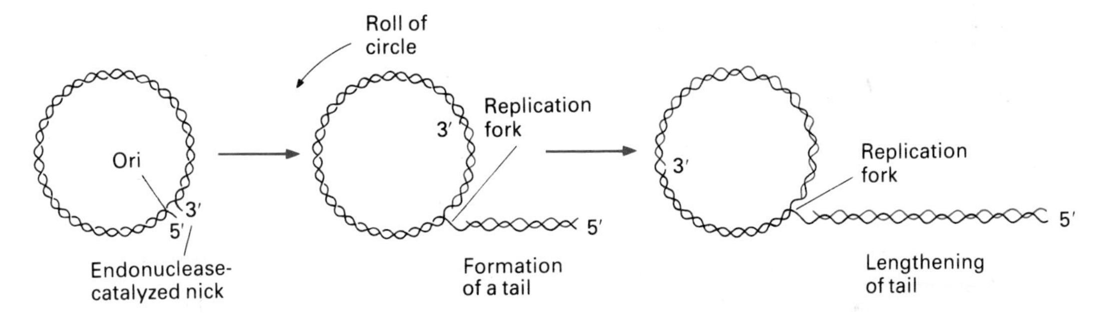

Figure 4.5. Two mechanisms of replication employed by circular chromosomes. (a) In symmetric replication, the parental double helix unwinds to expose the template strands for replication, which then proceeds in both directions from the replication origin. As a consequence, a theta-shaped (θ) replication intermediate is observed. (b) In contrast, asymmetric replication occurs as one strand of the parental molecule is progressively displaced from the circle, moving in only one direction from the replication origin. The result is a sigma-shaped (σ) replication intermediate. Replication forks mark the sites of the most recent unwinding and DNA synthesis. In both methods of replication, the direction of overall synthesis is shown; in each case, one daughter strand must be synthesized discontinuously.

protein capsid to form a virion. A summary of this rather complex replication system is shown in Fig. 4.6.

It is important to note that the two replication schemes—symmetric and asymmetric—contribute in different ways to the replicating DNA pool of the infecting phage. The switch from symmetric to asymmetric replication is marked by a change over time in the intracellular accumulation of phage DNA. Symmetric replication produces a geometric increase in the number of phage chromosomes, with each replicating DNA molecule forming two identical daughter molecules at each round of synthesis. This mode of replication is reponsible for the initial rise

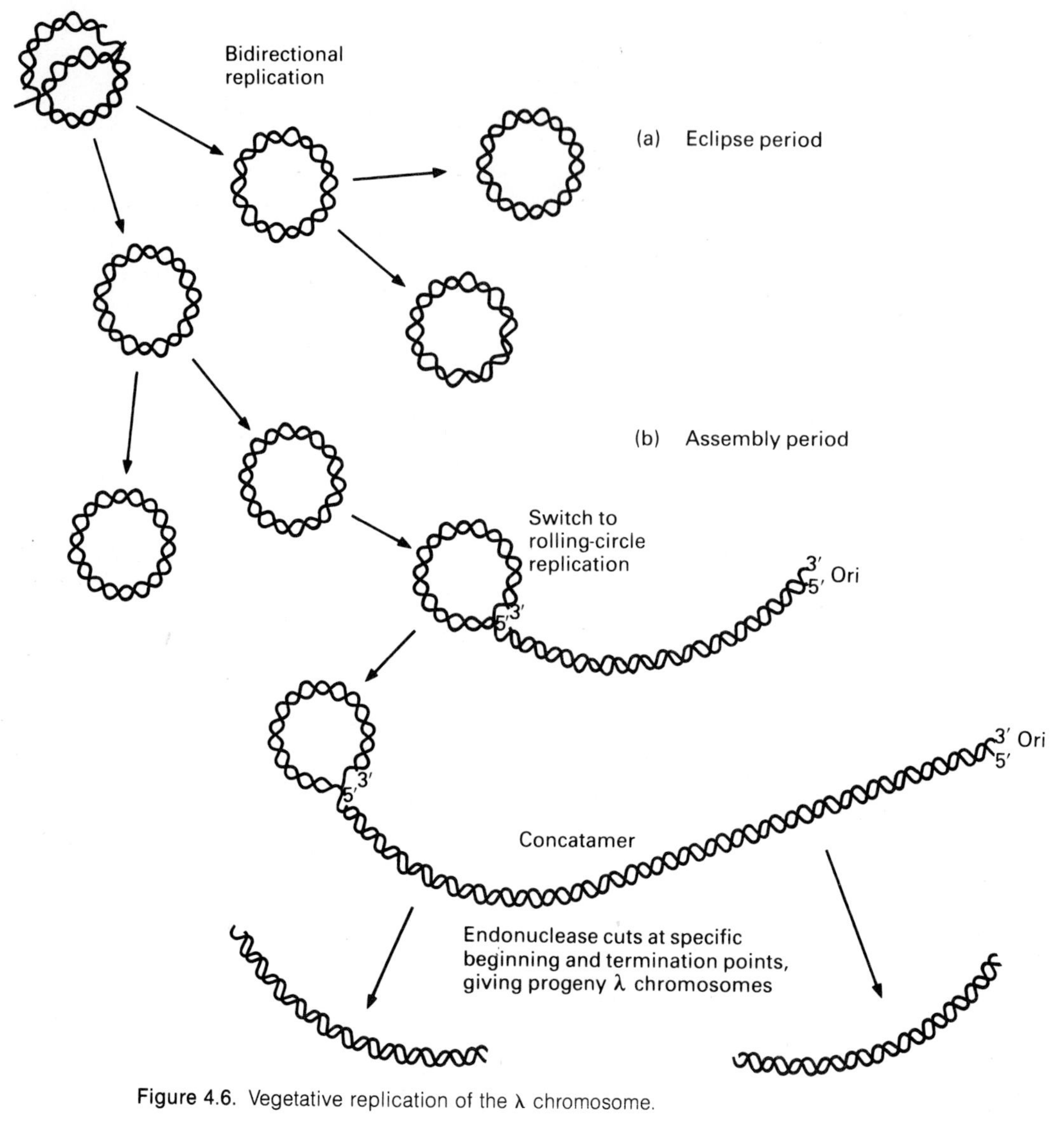

Figure 4.6. Vegetative replication of the λ chromosome.

in the number of phage chromosomes that occurs during the latter part of the eclipse period (refer to Fig. 4.2). Once phage assembly begins, DNA replication changes to the rolling-circle mechanism. During this final stage, only one of the two strands of a parental DNA molecule continues to serve as a template for successive rounds of replication. The other strand forms the distal tail of the sigma structure and does not further replicate. The number of replicating phage chromosomes will then remain constant, while the total amount of phage DNA will increase linearly over time.

Provirus. Certain strains of viruses are sometimes able to establish a long-term relationship with their host that does not usually harm the cell. In such cases, the viral DNA becomes integrated into a host chromosome, where it is known as a provirus. Bacteriophages that are capable of forming this integrated state are called **temperate phages.** Phage λ is an example of such a phage. The formation of a circle by the infecting λ chromosome is usually followed by its replication and the subsequent production of progeny viruses; through this process, the host cell (of species *E. coli*) is killed upon lysis. Because it leads to lysis of the host cell, this mode of phage reproduction is known as the **lytic cycle.** Occasionally, however, the circular λ chromosome follows an alternate route of replication, known as the **lysogenic cycle.** In this process, pairing occurs between specific sites on the λ and *E. coli* chromosomes, and a breakage and reunion event inserts the phage chromosome into that of the bacterium (Fig. 4.7). The resulting integrated phage chromosome, called a **prophage,** is a latent genetic structure that expresses none of the functions that would lead to autonomous multiplication of the phage and to lysis of the cell. The prophage replicates along with the host chromosome, as an extra section of that chromosome.

On rare occasions, the prophage is spontaneously released from the bacterial chromosome by a process that is essentially the reverse of that shown in Fig. 4.7. This deintegration process is known as **induction.** Induction occurs spontaneously

Figure 4.7. Integration of the lambda chromosome into the *E. coli* chromosome (shown as a circle). Following infection, the λ chromosome forms a circle. A short region of homology between the bacterial and phage chromosome provides for recognition and pairing. A single breakage and reunion of the two chromosomes serves to integrate the phage chromosome into that of its host.

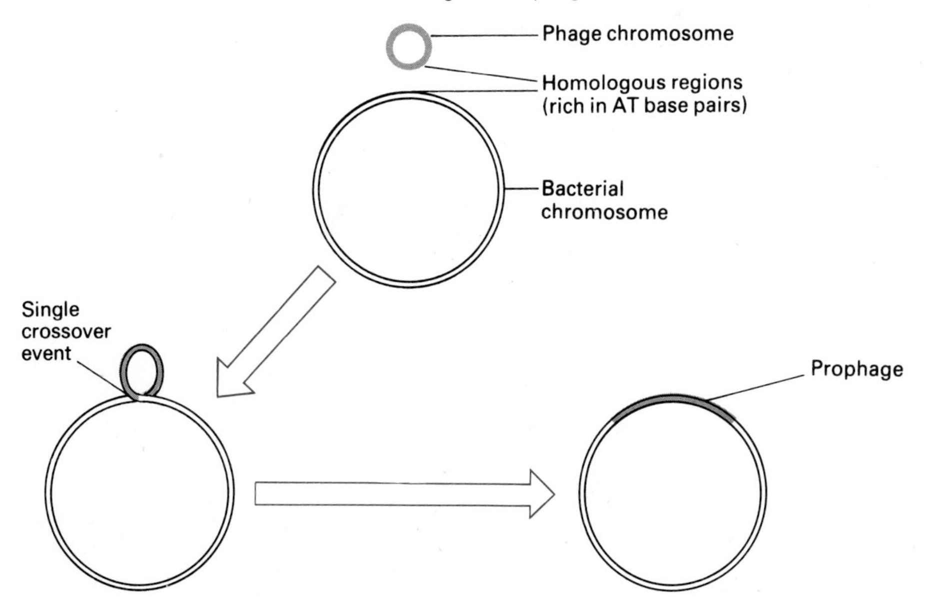

in only about one in every million cells, although ultraviolet light, a potent inducing agent, can be used experimentally to increase the chance of this event. Once free of its association with the host chromosome, the phage DNA replicates and produces progeny phage particles by way of the lytic cycle. Because the bacterial host of a prophage is always potentially subject to such a lysis, it is called a **lysogenic cell.**

There are several temperate phages that are used in genetic research. They include λ and P22, a phage of *Salmonella typhimurium,* as well as many others. In contrast, phages such as T2 and T4, which always lyse their hosts and are incapable of forming a prophage, are referred to as **virulent phages.** The life cycle of a temperate phage is summarized in Fig. 4.8.

Chromosomes of Bacteria

Bacteria do not exhibit the diversity in nucleic acid constitution that characterizes viruses. All bacterial chromosomes studied so far are circular and are composed of a single DNA double helix.

The first evidence for the circularity of bacterial chromosomes came from experiments conducted on *E. coli* by J. Cairns in the early 1960s. Cairns grew cells for varying periods of time in a medium containing ^3H-labeled thymidine, so that newly synthesized strands of DNA contained the radioactive isotope tritium (^3H). He lysed the cells very gently, in order to keep the chromosomes intact, and collected the chromosomes on membrane filters. He then subjected the chromosomes to a technique known as autoradiography. **Autoradiography** is a photographic procedure used to identify the distribution of a radioactive label. The

Figure 4.8. Life cycle of a temperate phage. Lysogenic bacteria replicate the prophage as a section of the bacterial chromosome (left). In induction (right), the prophage deintegrates from the host chromosome and replicates via the lytic cycle.

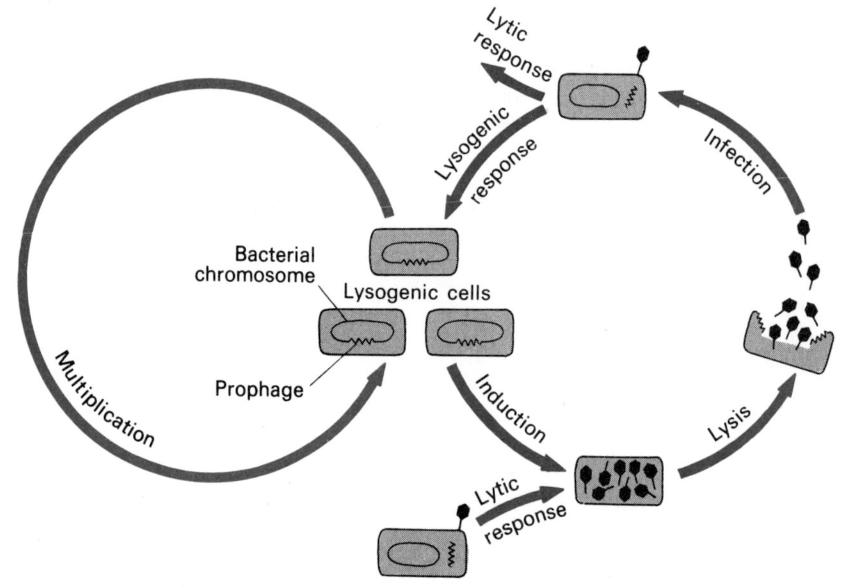

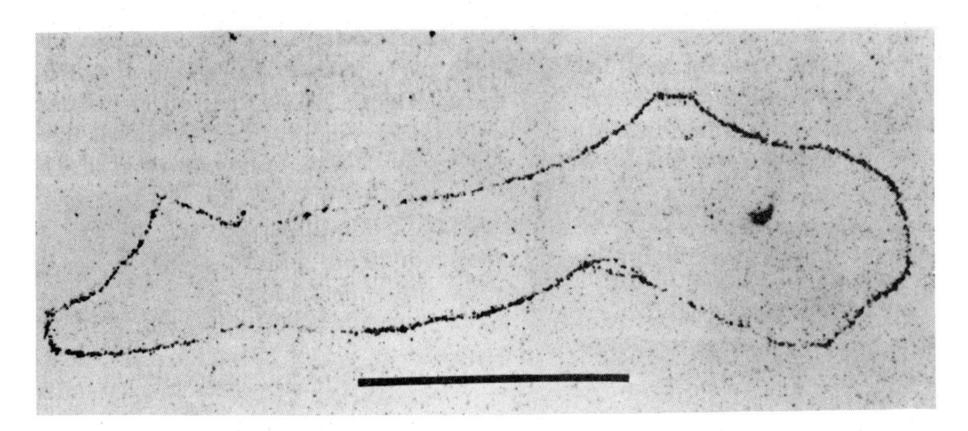

Figure 4.9. Autoradiogram of the *E. coli* chromosome replicating in ^{3}H-medium. The circular shape of the chromosome is apparent. The scale designates 100 μm. *Source:* Reprinted with permission from *Endeavour* 22:144, J. Cairns. Copyright 1963, Pergamon Press, Ltd. Photograph courtesy of J. Cairn.

isotope tritium decays by emitting high energy electrons, called β particles. If a photographic emulsion is placed over a chromosome spread that contains ^{3}H, a chemical reaction occurs wherever a β particle strikes the film, thus exposing the film. After several weeks to allow decay of a large portion of the tritium, the film is developed into a print, which shows the emission tracks of the β particles as black spots or grains. The distribution of radioactive atoms over the chromosomes can then be determined. Cairns found that after one round of DNA replication in the labeled medium, the decay of the tritium produced a faint circular outline of the bacterial chromosome on the autoradiogram (Fig. 4.9). The outline of emission tracks from the β particles is very narrow, because only one of the two strands of this double helix is labeled, in accordance with semiconservative replication. Thicker (more dense) outlines of β tracks were found when Cairns allowed two or more rounds of replication in the labeled medium, indicating the labeling of both strands of a newly synthesized DNA molecule.

The circular *E. coli* chromosome observed in autoradiograms has a contour length of 1360 μm. If we compare this length with the dimensions of an *E. coli* cell (about 1 μm by 2 μm), it becomes obvious that the chromosome cannot exist within the cell unless it is present in a more compact form. Electron micrographs show that the chromosome is indeed highly condensed; in fact, it exists in the form of a nucleoid (the functional chromosome) that constitutes only about 10 percent of the volume of the cell. Bacterial DNA is packaged into a compact state through the mediation of RNA, proteins, and polyamines (to neutralize surface charge) and by means of higher-level coiling.

When the *E. coli* chromosome is isolated by special procedures designed to maintain its highly compact in vivo state, the so-called *folded genome* structure is observed (Fig. 4.10). In its functional state, the *E. coli* chromosome is apparently arranged into 40 loops of DNA that are held together by RNA linkers. Researchers learned that the linkers are composed of RNA when they discovered that treating

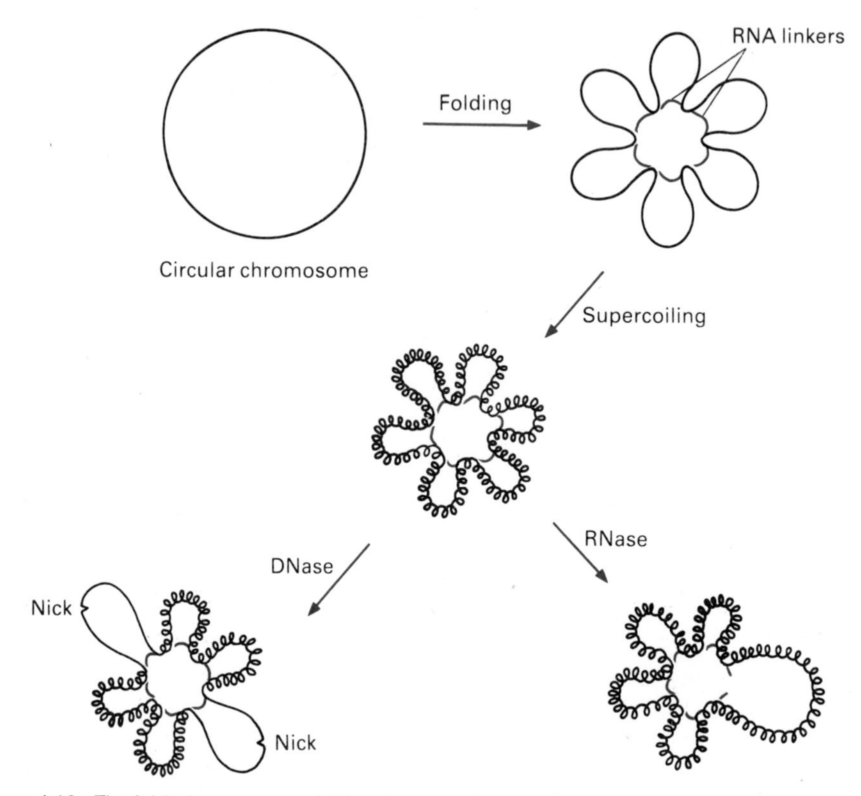

Figure 4.10. The folded genome model for chromosome structure in *E. coli*. Approximately 40 loops (only six of which are shown) are held together by RNA linkers. The DNA within each loop is supercoiled. RNase releases the loops, while DNase relaxes the supercoiling. *Source:* After D. E. Pettijohn and R. Hecht, *Cold Spring Harbor Symp. Quant. Biol.* 38:39, 1974.

the chromosome with the enzyme RNase releases the loops. The DNA double helix within each loop is further condensed by twisting, or supercoiling. The fact that this supercoiled structure is sensitive to the action of DNase, but not to RNase or proteases, indicates that the DNA forms supercoils spontaneously, without the aid of RNA or protein.

The term **supercoiling** refers to the twisting of a double helix to form a higher-level spiral structure (Fig. 4.11). Supercoiling is characteristic of any closed circular DNA molecule, including the replicating forms of many viruses; it occurs whenever a DNA molecule is underwound (with fewer helical turns than if it were completely relaxed and under no torsional strain) or overwound (with too many turns). The degree and direction of supercoiling is controlled by a class of enzymes known as topoisomerases. These enzymes (which include DNA gyrase) can relax supercoiled DNA or can introduce torsional strain into the molecule. It now appears that supercoiling is a widespread phenomenon that plays a crucial role in determining the function of DNA, as well as the three-dimensional structure of the chromosomes in various organisms.

Figure 4.11. Basic configurations of the double helix: (a) linear form, (b) open circle, and (c) closed circle. The open circle has incurred a single-stranded break, or nick (arrow), which relaxes it from the supercoiling of the unnicked (closed) circle shown in (c).

Plasmids. In addition to their main chromosome, bacteria often possess one to several very small chromosomes that replicate autonomously within the cytoplasm. These small autonomous structures, which are known as **plasmids,** consist of circular double-stranded molecules of DNA.

Many different kinds of plasmids are known. Some, such as the *F plasmid* (fertility factor), can replicate either as an autonomous unit or inserted like a prophage into the main chromosome of the bacterial cell. Plasmids with this dual replicative capability are known as **episomes.** The F factor is also an example of a *conjugative plasmid,* in that it is capable of being transmitted from one cell to another by means of bacterial conjugation. **Conjugation** is a mating process in bacteria that involves a one-way transfer of genetic material from one bacterium (the donor) to another bacterium (the recipient) during contact between the cells. A cell is a donor in the conjugation process if it contains one or more F plasmids; it is a recipient (designed F^-) if the F factor is absent. Donor cells possess special hairlike appendages called **sex pili** (singular, *pilus*) that protrude from their cell walls. Genes located on the F plasmid control the synthesis of these pili. When a donor cell randomly comes in contact with an F^- cell, some kind of recognition process occurs. A pilus physically connects the two cells and becomes modified into a hollow tube called the *conjugation tube*. The transfer of genetic material from donor to recipient can then take place through this tube.

Since the F plasmid is an episome and can replicate in one of two states, two types of donor cells exist. F^+ donor cells have the F plasmid present in the autonomous state, replicating independently of the main chromosome. Conjugative matings between F^+ and F^- cells result in the transfer of a replicated F plasmid from donor to recipient, leaving both cells F^+ (Fig. 4.12a). Alternatively, Hfr (high frequency recombination) donors have the F plasmid linearly inserted into their main chromosome. Hfr cells are formed from F^+ cells (and vice versa) through a process that integrates (or excises) the circular chromosome of the F plasmid in a manner similar to that involved in the integration (or excision) of a prophage (see Fig. 4.7). Unlike $F^+ \times F^-$ matings, in which only an F plasmid is transferred, part of the main chromosome of the donor is typically transferred along with part of the plasmid chromosome when an Hfr cell participates in conjugation with an F^- cell (Fig. 4.12b). We will consider the reason for this difference in the next section.

Figure 4.12. Two possible consequences of bacterial conjugation. (a) In $F^+ \times F^-$ matings, only a replicated copy of the F plasmid is transferred, leaving both cells F^+. (b) In Hfr $\times$ F^- matings, usually only part of a replicated copy of the Hfr chromosome is transferred, leaving the donor cell Hfr and the recipient cell F^-. Note that an F^+ cell forms an Hfr cell when the F plasmid becomes inserted into the main chromosome of the bacterium. Deintegration is also possible, converting an Hfr cell into an F^+ cell.

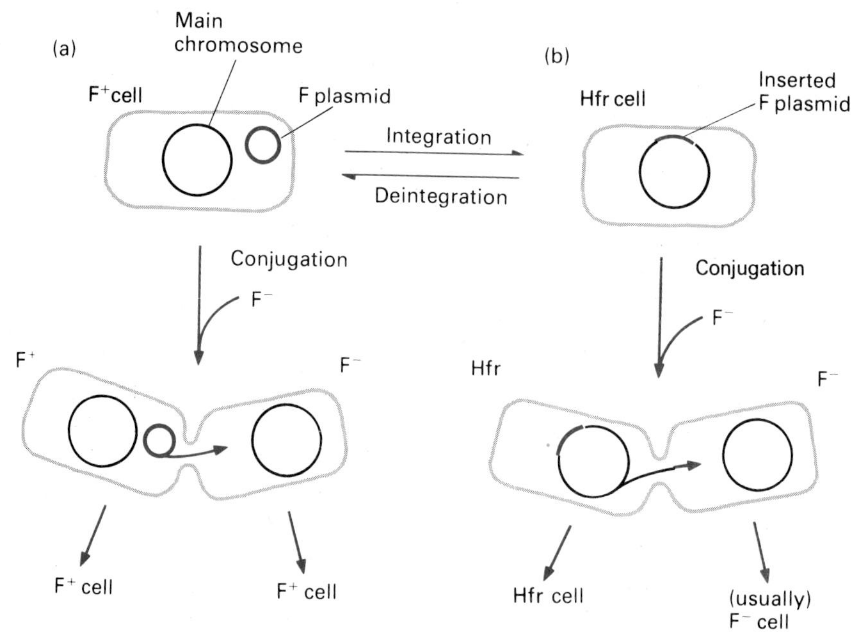

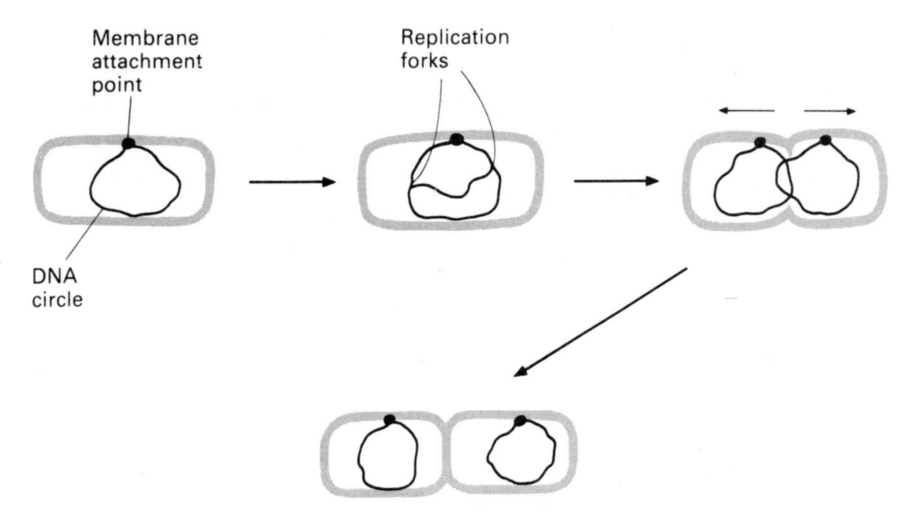

Figure 4.13. Model showing segregation of chromosomes into daughter cells in bacteria.

Other important types of plasmids include *colicinogens* (*col plasmids*) and *resistance* (*R*) *plasmids*. Col plasmids are carried by col$^+$ strains of bacteria. The col$^+$ cells secrete proteins called *colicins*, which are capable of killing bacteria of susceptible col$^-$ strains. The R plasmids transfer genes for antibiotic resistance from one bacterial cell to another. We will discuss the importance of R plasmids in the spread of antibiotic resistance in bacteria in Chapter 14.

Replication of Bacterial Chromosomes. DNA replication in actively dividing bacterial cells is *bidirectional*. DNA synthesis starts at a fixed point on the chromosome (the origin, O) and proceeds around the DNA circle in both directions. When replication is completed, it is believed that the two copies are transmitted to separate daughter cells while attached to the cell membrane (Fig. 4.13). The membrane between the two attachment points of the daughter chromosomes grows, causing the two DNA circles to move toward opposite ends of the cell. A lateral partition then develops between the two chromosomes, separating the cell into halves.

Chromosome replication also occurs during conjugation. In this case, however, DNA synthesis proceeds by way of the *rolling circle* mechanism (Fig. 4.14). In an Hfr × F$^-$ mating, the DNA double helix of the Hfr cell is nicked by an endonuclease at a point about in the center of the integrated F factor. The 3′ end of the nicked strand serves as a primer for DNA polymerase, which begins synthesizing a new complementary strand, using the unnicked strand as a template. The displaced 5′ end of the replicating circle then passes through the conjugation tube and enters the recipient cell. Because of the fragile nature of the conjugation tube, the passageway between the conjugating cells is often broken before the entire Hfr chromosome is transferred. The result is that usually only the O end of the F factor and a section of the main chromosome enter the recipient cell; the rest of

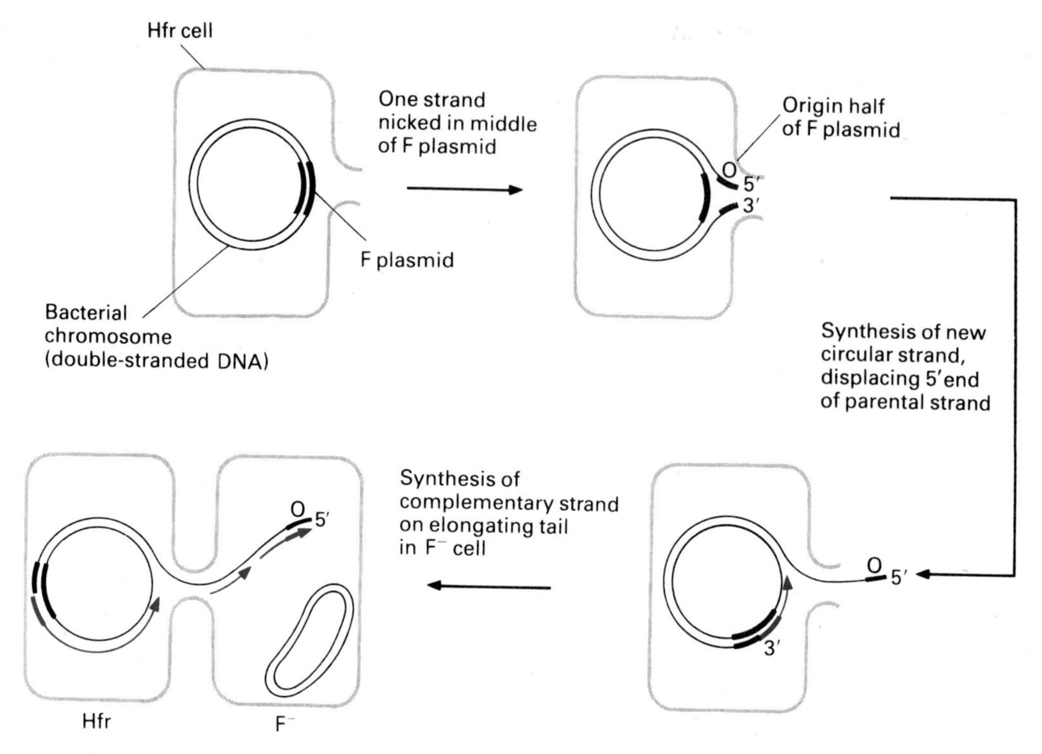

Figure 4.14. Transfer of the bacterial chromosome during conjugation, with accompanying rolling circle replication of the chromosome within the donor cell. For simplicity, the double helix of the donor chromosome is shown unwound. After one strand is nicked in the center of the F plasmid, DNA polymerase extends the synthesis of a new strand, using the exposed 3′ end as primer (new strands are shown in red). Continuing synthesis of the new complementary strand displaces the 5′ end, which migrates across the conjugation tube and is then itself replicated. Breakage of the conjugation tube usually limits transfer to just a section of the donor chromosome. The recipient cell thus remains F^-, except in those rare instances in which the conjugation tube remains intact long enough for the entire donor chromosome to be transferred. (Synthesis is discontinuous, at least on the transferred strand, to meet the polarity requirements of polymerase.)

the Hfr chromosome and the other half of the F factor remain within the donor cell. If the conjugation tube remains intact long enough, the remaining half of the F factor will be transferred last. Thus, the recipient cell usually remains F^-. The original Hfr cell retains its donor status, because replication accompanies the transfer.

Plasmids replicate in a manner similar to that of the main chromosome. Bidirectional replication is typical, except during conjugation. In $F^+ \times F^-$ matings, for example, the F plasmid replicates in a rolling-circle fashion during its transfer. The donor thus retains a copy of the F factor and remains F^+. The F^- recipient receives the transferred copy of the plasmid and is converted to the F^+ state in the process.

TO SUM UP

1. The structure of viral chromosomes varies greatly, depending on the particular virus being considered and the stage of the viral life cycle. In general, viral chromosomes can be either DNA or RNA, single- or double-stranded, and circular or linear.
2. The lytic cycle of phage growth is divided into two parts: the initial stage (called the eclipse), in which the phage DNA replicates and viral proteins accumulate, and a second stage (called phage assembly), in which the various components of the phage are assembled to form complete or mature infectious virions. The period of vegetative phage multiplication culminates in lysis of the cell.
3. Temperate phages are capable of two modes of replication: autonomously in the cytoplasm of the host (the lytic cycle) or in association with the bacterial chromosome (the lysogenic cycle). The lysogenic cycle is initiated when the chromosome of a temperate phage is inserted into the bacterial chromosome to become a prophage. Virulent phages, in contrast, are capable only of lytic growth.
4. All essential genes of a bacterium are carried on one main chromosome, which assumes a tightly folded configuration within the bacterial cell. The chromosome is composed of a supercoiled circular DNA molecule in a complex with varying amounts of RNA, proteins, and polyamines. In addition to the main chromosome, bacterial cells may also harbor one to several small circular DNA molecules, known as plasmids.
5. During cell division, replication of the bacterial chromosome is bidirectional, starting at the replication origin and proceeding around the circle in both directions. During conjugation, the bacterial chromosome undergoes rolling-circle replication, with the displaced strand moving from the donor through the conjugation tube into the recipient cell.
6. Conjugation is a form of gene exchange in bacteria that involves a one-way transfer of genetic material during cell-to-cell contact. Conjugation is mediated by a specific plasmid, the F plasmid. Donor cells in the conjugation process are either F^+ (with the F plasmid replicating autonomously within their cytoplasm) or Hfr (with the F plasmid inserted into their main chromosome). Recipient cells are F^- and lack an F plasmid.

Chromosomes of Eukaryotes

Unlike most bacteria and viruses, which carry their genes on a single main chromosome, eukaryotic cells typically contain several, comparatively large chromosomes within their nuclei. The number of chromosomes in each cell nucleus varies considerably from species to species as seen in Table 4.1. The numbers listed in the table are those of somatic (non-sex) cell nuclei. In most higher animals and many higher plants, somatic cells are **diploid,** containing two complete sets of chromosomes. Each of the sets carries the basic complement of genes of the organism, known as the **genome.** In sexually reproducing organisms, one set of chromosomes is derived from the maternal parent (the *maternal set*), and the other set is derived from the paternal parent (the *paternal set*). Chromosomes in the diploid state therefore exist in pairs, with each pair consisting of one maternally and one paternally derived chromosome. We call the matched members of each pair **homologous chromosomes,** or simply **homologs.**

Unlike the somatic cells of diploid organisms, the gametes (or sex cells) are **haploid.** Each gamete contains only half of the chromosomes present in the

Table 4.1. Chromosome numbers in the somatic cells of selected animals and plants.

ORGANISM	CHROMOSOME NUMBER	ORGANISM	CHROMOSOME NUMBER
Cat (*Felis domesticus*)	38	Barley (*Hordeum vulgare*)	14
Cattle (*Bos taurus*)	60	Bean (*Phaseolus vulgaris*)	22
Chicken (*Gallus domesticus*)	78	Bread wheat (*Triticum aestivum*)	42
Chimpanzee (*Pan troglodytes*)	48	Cabbage (*Brassica oleracea*)	18
Donkey (*Equus asinus*)	62	Corn (*Zea mays*)	20
Dog (*Canis familiaris*)	78	Cucumber (*Cucumis sativus*)	14
Fruit fly (*Drosophila melanogaster*)	8	Cultivated oat (*Avena sativa*)	42
Horse (*Equus caballus*)	64	Emmer wheat (*Triticum turgidum*)	28
Housefly (*Musca domestica*)	12	Green alga (*Acetabularia mediterranea*)	20
House mouse (*Mus musculus*)	40	Garden pea (*Pisum sativum*)	14
Human (*Homo sapiens*)	46	Potato (*Solanum tuberosum*)	48
Mosquito (*Culex pipiens*)	6	Radish (*Raphanus sativus*)	18
Platyfish (*Platypoecilus maculatus*)	48	Rice (*Oryza sativa*)	24
Rat (*Rattus norvegicus*)	42	Rye (*Secale cereale*)	14
Starfish (*Asterias forbesi*)	36	Tobacco (*Nicotiana tabacum*)	48
Turkey (*Meleagris gallopavo*)	82	Tomato (*Solanum lycopersicum*)	24

somatic cells of the same individual. For example, there are 23 chromosomes in a human egg or sperm, as compared to 46 in a human somatic cell. We symbolize the number of chromosomes in the gametes of an individual as n; so in humans, n (the gametic number) = 23 and $2n$ (the somatic number) = 46.

In humans and other diploid organisms, the gametic and somatic numbers of chromosomes are the haploid and diploid numbers of chromosomes of the species. This is not true of **polyploid** organisms, including many higher plants, in which there are several (*poly* = many) copies of the genome (see Chapter 12). In the case of polyploids, the monoploid (or genomic) number is symbolized as x so as to distinguish it from the gametic (n) number; the gametic number in polyploids includes more than a single genome.

The Cell Cycle

Nuclear chromosomes undergo periodic structural and biochemical changes in actively dividing cells. Each chromosome exists in a *dispersed* (loosely coiled) state during part of a cell's life cycle and in a *condensed* (tightly coiled) state during another part of the cycle. Similarly, each chromosome duplicates during one period of the cell cycle and its duplicated parts separate and move into daughter nuclei at a different and later period of the cycle.

The cell cycle can be divided into four fairly distinct phases: (1) a presynthetic period, designated G_1; (2) a period of DNA synthesis, designated S (for synthesis); (3) a postsynthetic period, designated G_2; and (4) **mitosis** (or M), the period of nuclear division. In a typical mammalian cell growing in culture, G_1 lasts about 11 hours, S takes about 8 hours, G_2 lasts for 4 hours, and M is completed in 1 hour, for an average cell cycle of 24 hours (Fig. 4.15).

Distinct nuclear chromosomes can be seen with a light microscope only during the M phase. For this reason, it is customary to refer to all the phases other than M collectively as **interphase.** During interphase, the chromosomes within a nucleus form a dispersed network of chromatin fibers. Each chromatin fiber consists of a thin strand of DNA in a complex with certain basic (positively charged) proteins called **histones** and a variety of relatively acidic (negatively charged) **nonhistone proteins.** Five different types of histones are found in each chromatin fiber: H1, H2A, H2B, H3, and H4, in the molar ratio of approximately 1 H1 : 2 H2A : 2 H2B : 2 H3 : 2 H4. The types differ in molecular weight and in their composition of amino acids (the primary structural components or "building blocks" of proteins). The positive charge of the histones at neutral pH is due to the presence of the amino acids arginine and lysine, which make up a large part of the amino acid composition of these proteins.

Figure 4.15. Phases of the life cycle of a dividing cell. The cell cycle is normally divided into four phases: G_1 (the period prior to DNA synthesis), S (the period of DNA synthesis), G_2 (the period between DNA synthesis and mitosis), and M (mitosis). The amount of time spent by a typical mammalian cell in culture in each different stage is given here for a 24-hour cycle time.

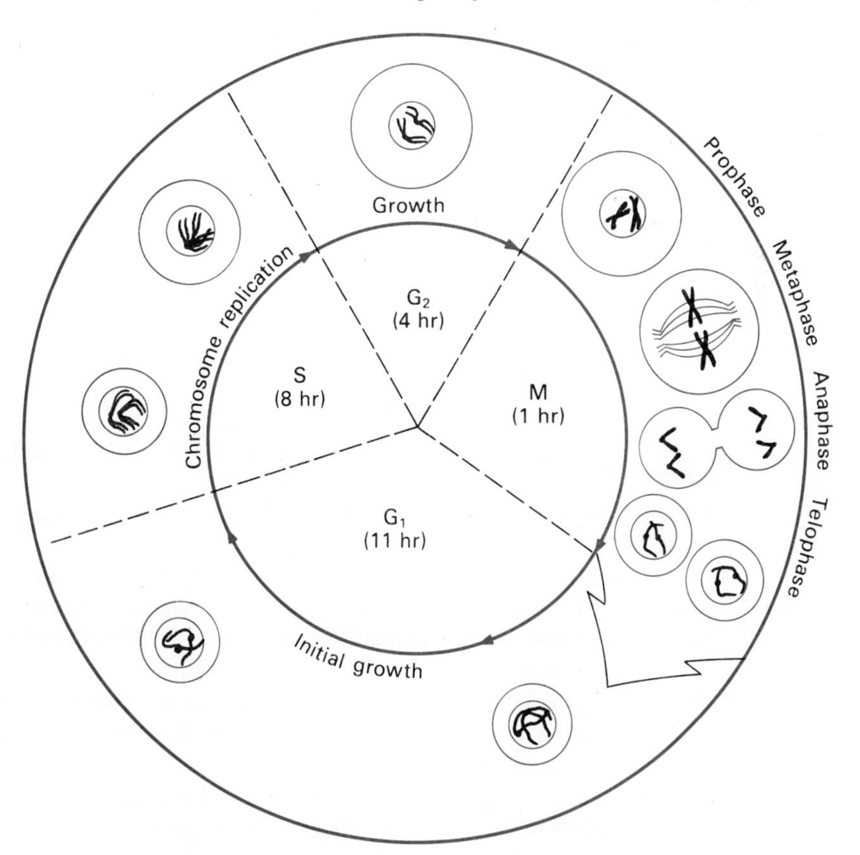

DNA Replication. The DNA of nuclear chromosomes is synthesized during the S period of interphase. Unlike the prokaryotic chromosome, which replicates as a single unit, a eukaryotic chromosome has multiple replication sites, called **replicons.** Each replicon consists of a length of DNA with a separate replication origin (Fig. 4.16). Replication is *bidirectional;* each replicating "bubble" has two replication forks that move from the origin in opposite directions. The existence of multiple replication sites in eukaryotic chromosomes is not surprising, because of the vastly greater amount of DNA that must be packaged by the higher organisms. For example, the human haploid genome consists of about 600 times more DNA than that of *E. coli.* If fully extended, this DNA would form a single bacteriumlike chromosome of almost a meter in length. Large numbers of replication sites are therefore needed to complete the duplication of chromosomes within the normal time frame allotted to the S phase.

EXAMPLE 4.1. A certain line of mammalian cells growing in culture has an S phase of 8 hours. During S, DNA synthesis within these cells occurs at a rate of 2000 base pairs per minute at each replicating fork. If the DNA in each cell nucleus has a total length of 1.36×10^9 nm, what is the minimum number of replicons that must be operating within each of these cells during chromosome duplication?

Solution. Each cell has a total of $(1.36 \times 10^9 \text{ nm})/(0.34 \text{ nm/base pair}) = 4 \times 10^9$ base pairs of DNA. Since there are two replicating forks per replicon, each replicon can synthesize $(2)(2000 \text{ base pairs/min})(8 \text{ hr})(60 \text{ min/hr}) = 1.92 \times 10^6$ base pairs of DNA during S. Therefore, the number of replicons needed to replicate all the DNA within a cell (assuming all operate simultaneously) is $4 \times 10^9/1.92 \times 10^6 = 2083$.

Mitosis

Mitosis is the last stage of the cell cycle and also the shortest, usually accounting for less than 10 percent of the entire cycle. Mitosis is a continuous process that proceeds without extensive interruptions. But for convenience, geneticists divide mitosis into four different stages in the following order: **prophase, metaphase, anaphase,** and **telophase** (Fig. 4.17).

Prophase. Prophase, the first stage of mitosis, is characterized by the condensation of chromosomes. While the chromosomes are condensing, the nuclear envelope gradually breaks down and the nucleolus disappears, as its component parts disperse throughout the cell.

The beginning of prophase is marked by the appearance of threadlike chromosomes, each consisting of two identical longitudinal halves that are held together at a narrowed region along their length called a **centromere.** The two longitudinal halves of each chromosome are the result of chromosome replication during interphase. These duplicated parts of a chromosome are called **sister chromatids** as long as they remain connected at the centromere.

The chromosomes continue to shorten and thicken as condensation progresses, so that by the end of prophase they have the appearance of doubled rods (Fig. 4.18). By this time, three morphological features are distinct enough to be used

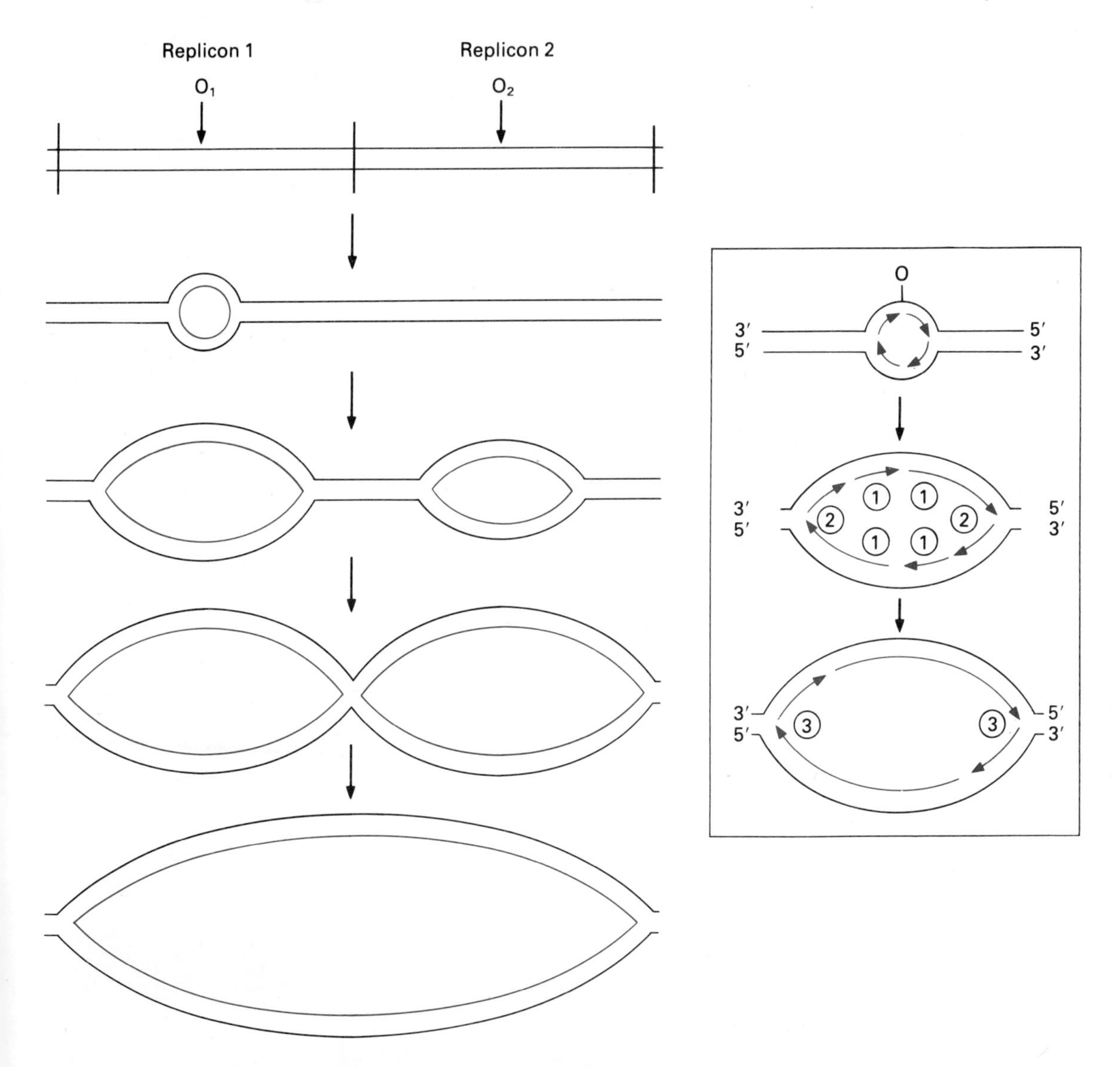

Figure 4.16. A replicating eukaryotic chromosome, showing a section that contains two replication origins. The separation of the DNA strands at these points allows synthesis of new DNA (red lines) to proceed in both directions from each origin, forming a replication "bubble." Each replication unit is called a replicon. Different replicons do not necessarily replicate at the same time, as illustrated here. The inset shows the discontinuous nature of DNA synthesis within each bubble; each Okazaki fragment (numbered in order of replication) is about 200 nucleotides long (see Chapter 3).

in identifying different chromosomes. These features are: (1) length, (2) position of the centromere (relative arm length), and (3) presence of small terminal knobs called **satellites.** Depending on the position of the centromere, chromosomes are classified at this time as **metacentric,** if the centromere is in a median (centrally

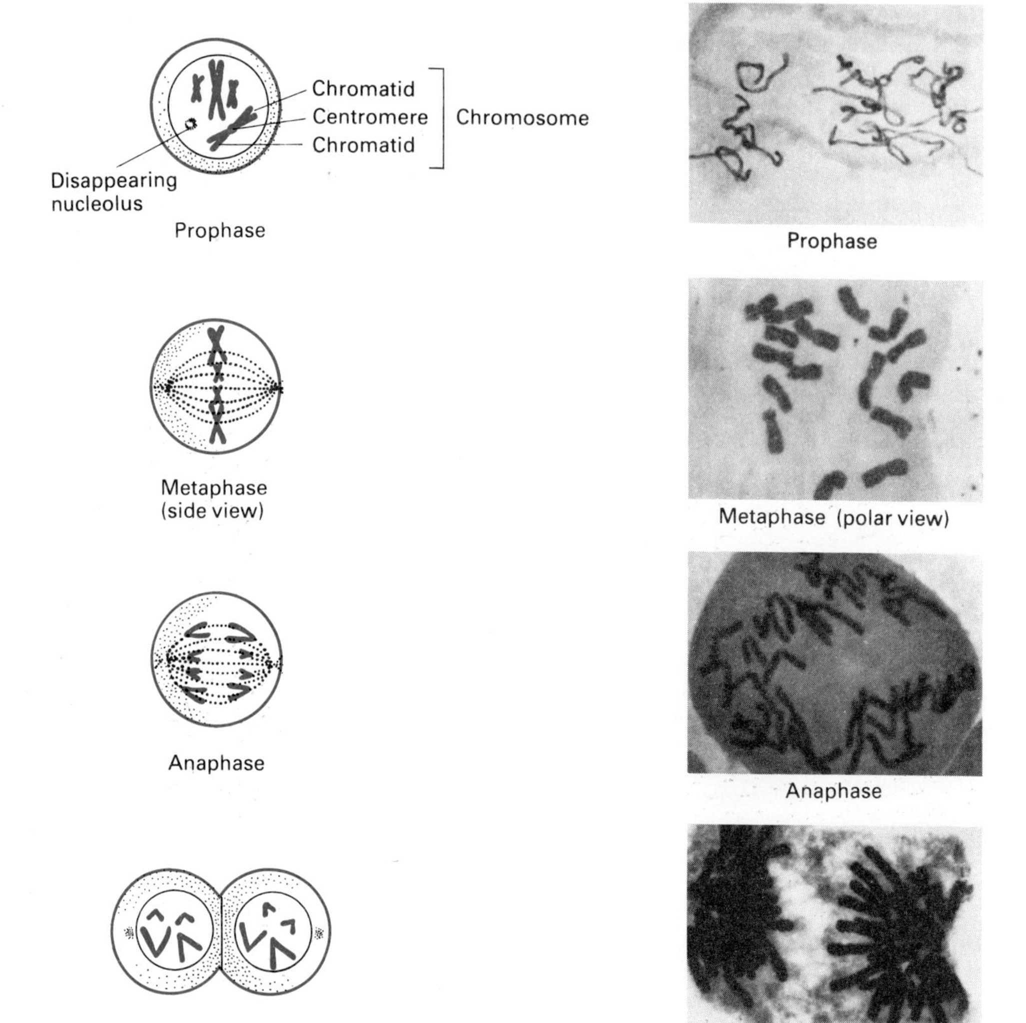

Figure 4.17. Stages of mitosis. (a) Diagrammatic representation of mitosis in an animal cell having two pairs of chromosomes. In prophase, the duplicated chromosomes condense, and the nuclear envelope and nucleolus disappear. The side view of a metaphase cell shows the chromosomes arranged at the equator of the newly formed spindle. In anaphase, the chromatids separate, and the daughter chromosomes move to the opposite poles of the spindle. In telophase, the chromosomes uncoil, and the nucleolus and nuclear envelope reappear. Cytokinesis occurs. (b) The actual stages of mitosis observed in a cell of an onion root tip having eight pairs of chromosomes. *Source:* Photographs courtesy of E. J. Gardner and W. S. Boyle.

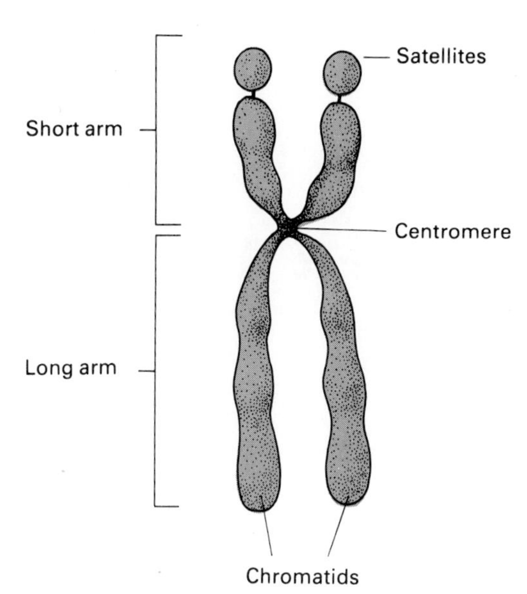

Figure 4.18. The basic structure of a nuclear chromosome when maximally condensed. Each chromosome is composed of identical sister chromatids held together at the centromere. Terminal knobs, called satellites, are seen on certain chromosomes.

located) position; **submetacentric,** if the centromere is in a submedian (off-center) position; **acrocentric,** if the centromere is very near one end; or **telocentric,** if the centromere is at a terminal location and only one arm is visible (Fig. 4.19).

The chromosome complement, or **karyotype,** of cells is usually studied following the completion of prophase when chromosomes are maximally condensed. The chromosomes are photographed, individually cut out of the photograph, and arranged in pairs in order of decreasing length. The result of one such study on

Figure 4.19. Different centromere locations in nuclear chromosomes. (a) Metacentric chromosomes have centromeres at a median position so that chromosome arms are about equal in length. (b) Submetacentric chromosomes have centromeres in a submedian position (offcenter). (c) Acrocentric chromosomes have centromeres close to one end. (d) Telocentric chromosomes have centromeres at one end so that only one arm of each chromatid is visible.

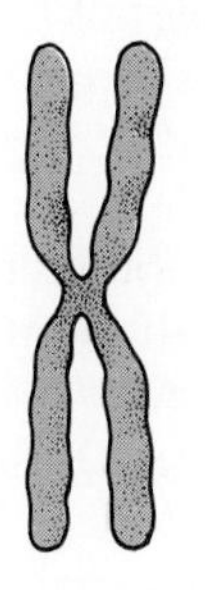
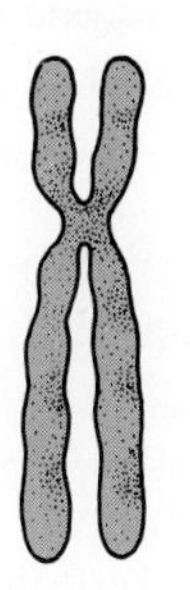
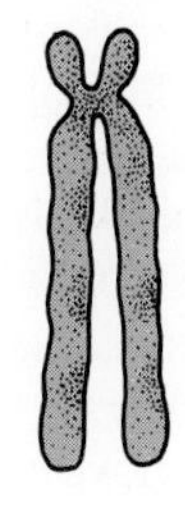

(a) Metacentric (b) Submetacentric (c) Acrocentric (d) Telocentric

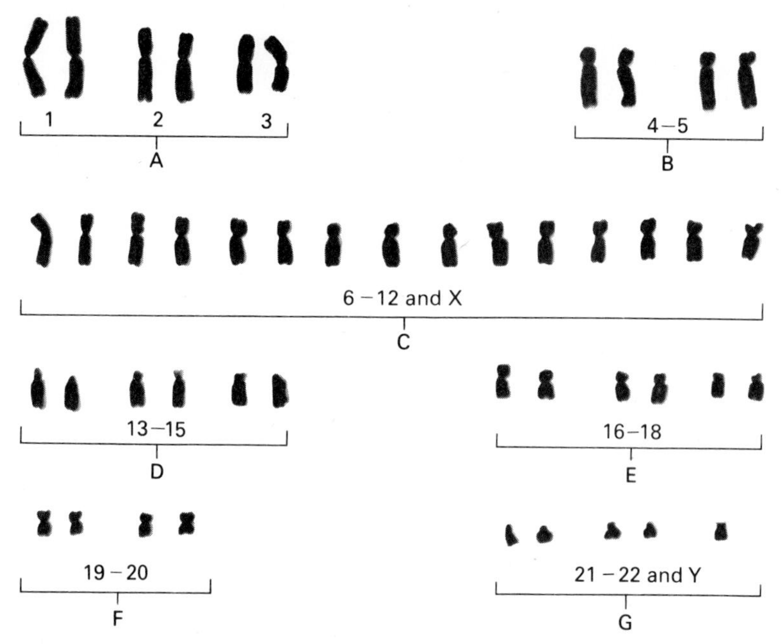

Figure 4.20. Human chromosomes arranged in their standard sequence. The chromosomes were photographed at metaphase, individually cut out of the photograph, and arranged in pairs on a standard karyotype analysis sheet in order of decreasing length. The chromosomes are assigned to seven lettered groups, A through G, on the basis of their relative length and the position of the centromere. The centromere constrictions of all chromosomes are placed at the same level to aid in correct assignment. This placement brings out more clearly any differences in arm lengths.

the normal human karyotype is shown in Fig. 4.20. Karyotype analyses similar to this are used for identifying certain chromosome abnormalities (see Chapter 12).

Metaphase. Metaphase, the second stage of mitosis, begins when the nuclear envelope disappears. The region formerly occupied by the nucleus is then replaced by a structure known as the **spindle.** The spindle is an organized system of hollow cylinders, called *microtubules,* that are assembled in the cytoplasm during prophase from smaller protein subunits. Located at the poles of the spindle in most animal cells and the cells of many lower plants are pairs of cylindrical organelles known as *centrioles,* which function in these cells in the development of the spindle microtubules.

The spindle is composed of two types of microtubules. The *interpolar microtubules* extend between the poles of the spindle without making connections to the chromosomes. By contrast, the *chromosomal microtubules* extend from opposite poles to the centromeric regions of chromosomes. During metaphase, the chromosomes move to align themselves on the equatorial plane of the spindle, each attached at its centromere to bundles of chromosomal microtubules (Fig. 4.21). Time-lapse photography reveals that the movement is somewhat erratic, with

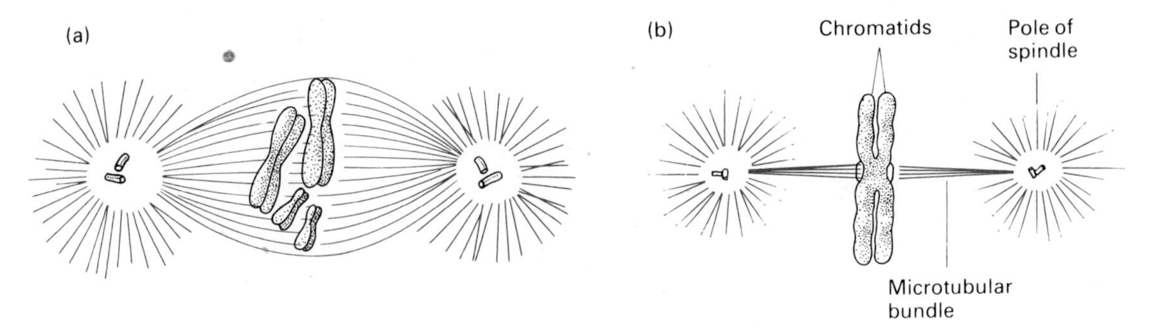

Figure 4.21. Chromosomes at metaphase. (a) Alignment of chromosomes at the equator of the spindle. (b) The sister chromatids are attached at the centromere to bundles of chromosomal microtubules leading to opposite poles.

chromosomes oscillating about the midpoint of the spindle as they maneuver into position. When the centromeres of the chromosomes are aligned on the equator of the spindle, the chromosomes are ready for the next phase.

Anaphase. Anaphase, the third stage of mitosis, begins with the separation of sister chromatids. Once the sister chromatids have separated, they are referred to as daughter chromosomes. These daughter chromosomes continue to move apart as they are pulled toward the opposite poles of the spindle (Fig. 4.22).

What supplies the force required for the poleward movement of the chromosomes? No one really knows. According to one hypothesis, chemical cross-bridges between the spindle microtubules are successively broken and reformed, to create a force that causes the microtubules to slide past one another. In this model, active sliding between the microtubules both pulls the chromosomes toward the poles and pushes the poles farther apart.

Telophase. Telophase, the fourth and final stage of mitosis, commences when the daughter chromosomes have completed their migration to the opposite poles. A nuclear envelope reassembles around the cluster of chromosomes at each pole,

Figure 4.22. Chromosomes at anaphase. (a) Disjunction of chromosomes on the mitotic spindle. (b) Once the sister chromatids separate, the duplicate structures are pulled toward opposite poles of the spindle by chromosomal microtubules.

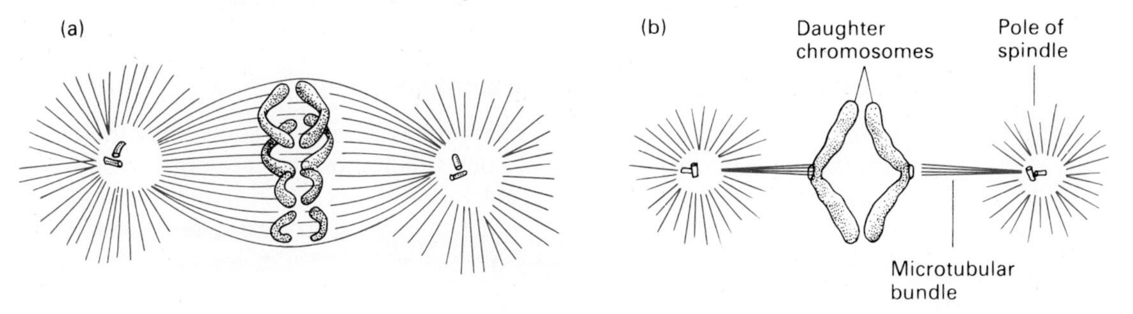

the spindle microtubules disappear, and the chromosomes gradually uncoil. With the reappearance of nucleoli, the daughter nuclei begin to assume their interphase morphology.

The process of mitosis has clearly evolved to provide for an exact disjunction (separation) of daughter chromosomes into separate nuclei. By the separation of duplicates of every chromosome, mitosis ensures that each daughter nucleus receives the same complement of chromosomes as the parent nucleus. **Cytokinesis,** or cytoplasmic division, occurs after chromosome disjunction is complete. During cytokinesis, the separate nuclei containing the two sets of chromosomes are incorporated into daughter cells.

EXAMPLE 4.2. The continuance of chromosomes in a duplicated state until after their connection to the spindle microtubules helps to ensure an orderly distribution of daughter chromosomes during cell division rather than the random assortment that might otherwise occur. Suppose that instead of the mechanism described, the following events were to occur: The sister chromatids separate at prophase. The daughter chromosomes that are formed align independently at the spindle equator, with half moving to one pole at anaphase and half moving to the other in random assortment. If a cell that has two pairs of chromosomes, A,A' and B,B', were to divide by this mechanism, what is the chance that the chromosome makeup of the daughter cells would be identical?

Solution. When the sister chromatids separate, there will be eight daughter chromosomes: 2 of A, 2 of A', 2 of B, and 2 of B'. To be identical in chromosome makeup, the daughter cells must receive one copy of each of these chromosomes. This event can occur in $2 \times 2 \times 2 \times 2 = 16$ different ways. Since any four chromosomes could move to one pole and the other four could move to the other pole in random assortment, there are $8!/4!4! = 70$ different ways of randomly selecting four of any type and four of another. The probability that both daughter cells are $AA'BB'$ is then

$$P(AA'BB') = \frac{\text{Ways of selecting } AA'BB'}{\text{Total ways}} = \frac{16}{70} = \frac{8}{35}$$

It is apparent that this hypothetical mechanism of chromosome disjunction would be extremely error prone, with less than ¼ of the daughter cells having the chromosomes makeup of the original cells from which they were derived. We should note that independent assortment of genes, which is an effective mechanism, is not the same as the separation of chromatids during cell division. Segregation and independent assortment of alleles occur during the reduction division of meiosis, not the cellular multiplication process of mitosis. We will discuss meiosis in detail in the next chapter.

Molecular Organization of Nuclear Chromosomes

Recently, a substantial body of information has developed concerning the way in which the molecular components of nuclear chromosomes are organized. Studies employing the electron microscope and various other physical and chemical techniques have shown that nuclear chromosomes are **uninemic;** that is, each chromosome has a single molecule of DNA running its entire length. For example, electron microscope studies of the DNA extracted from the chromosomes of yeast, *Saccharomyces cerevisiae,* have disclosed a range of molecule lengths that roughly correspond with the values estimated for the smallest and largest chromosomes in this species, based on the assumption that each chromosome contains a single uninterrupted duplex of DNA. Results that show even better agreement with the predicted values were obtained when the DNA lengths were measured by determining the rate at which the molecules sediment during high-speed centrifugation and by studying the elastic properties of the DNA when in solution (known as its viscoelasticity). The viscoelastic measurements of DNA lengths in the fruit fly (*Drosophila*) also indicate that each unduplicated chromosome comprises a single molecule of DNA.

The results of radioactive-labeling experiments are also consistent with a uninemic structure of unduplicated chromosomes. If eukaryotic cells are immersed in a solution containing tritiated (^{3}H-labeled) thymidine during the S phase of the cell cycle so that chromosomes replicate once in the presence of the label, both chromatids are observed to be radioactive in the following mitosis. If these same chromosomes are allowed to replicate again but in a nonlabeled medium, only one of the chromatids of each daughter chromosome remains radioactive at the following mitosis. The results of one such study by J. H. Taylor and his colleagues on root-tip cells of the broad bean, *Vicia faba,* are shown in Fig. 4.23. In this experiment, the location of radioactivity was detected by an autoradiographic procedure in which a photographic emulsion is layered over prepared slides of the chromosomes. Observe in Fig. 4.23(c) that the simplest way to interpret these results is to assume the presence of a single DNA molecule in each chromosome that replicates in a semiconservative manner.

Nucleosomes. Recent studies on isolated chromatin have revealed how chromosomal DNA interacts with proteins to form the chromatin fibers. These studies show that the chromatin fiber is a highly ordered structure, in which histone proteins are combined with the DNA to form repeating subunits called **nucleosomes.** The structure and organization of nucleosomes depend upon the histone composition of chromatin and on the ionic strength of the medium. At low ionic strength, chromatin that has been stripped of histone H1 appears in the electron microscope as a beaded filament; the beads represent the basic structural components of nucleosomes. Each bead, or *nucleosome core particle,* is formed from a protein core and a length of DNA consisting of 146 nucleotide pairs; the DNA is wrapped one and three-quarters times around the protein core in a helix (Fig. 4.24a). Each protein core is an octamer of histone that is composed of two molecules each of the core histones H2A, H2B, H3, and H4.

The addition of histone H1 to the complex yields what is called a *minimal nucleosome* (or *chromatosome*). In this structure, a single molecule of H1 is thought to associate with another 20 base pairs of filament DNA (10 pairs on each side of the 146 core pairs), so that two full turns of DNA (166 base pairs) are wrapped around the histone core (Fig. 4.24b). Adjacent nucleosomes are connected by approximately 34 base pairs of internucleosomal linker DNA (the number varies

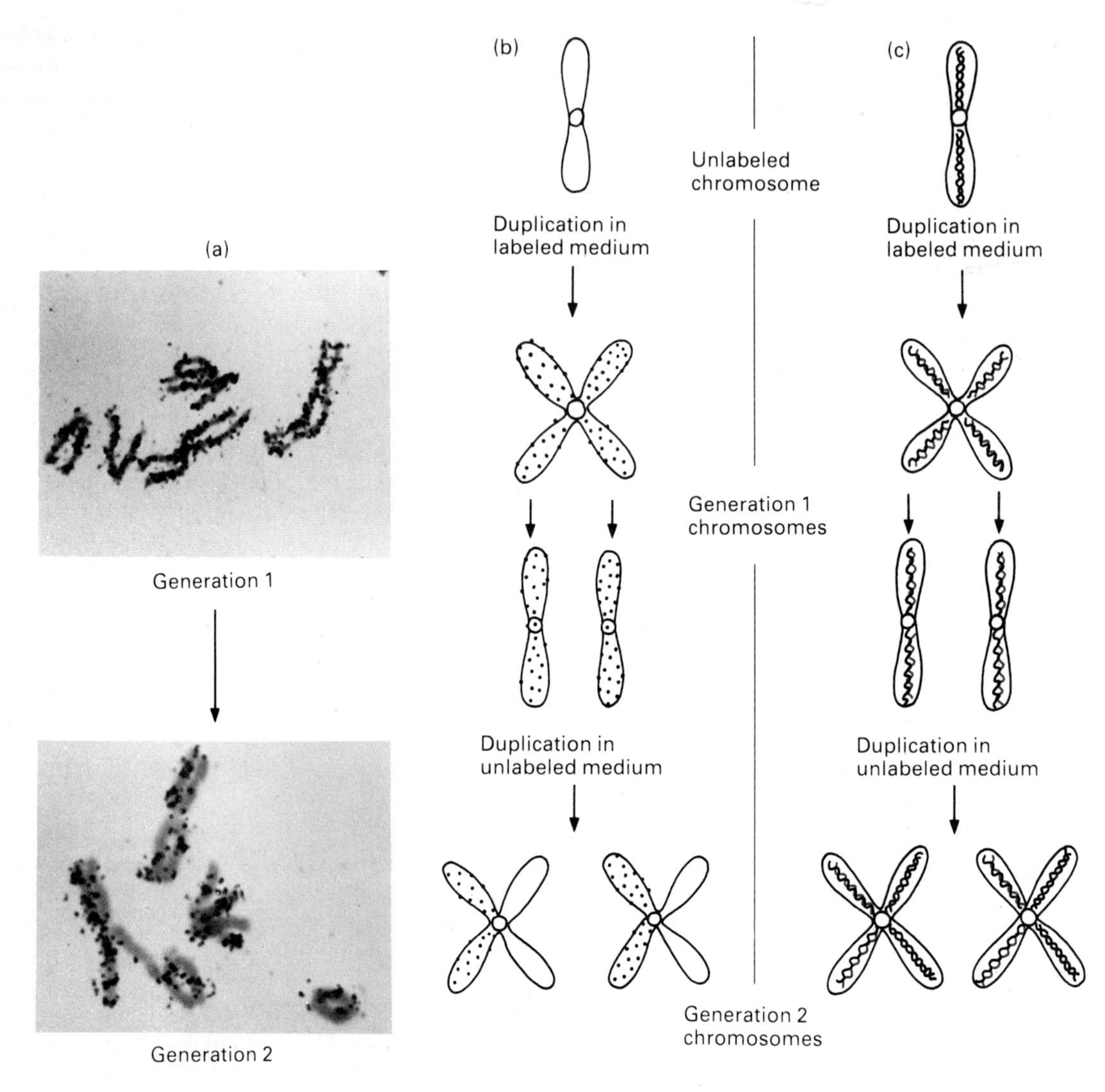

Figure 4.23. Results obtained by Taylor on chromosome replication in bean root tips. (a) Autoradiograms of chromosomes after one generation in a medium containing a tritium label (top), and after a second generation but in a medium without the label (bottom). (b) Schematic illustration of autoradiograms. Sister chromatids produced after one generation in the presence of the label are both labeled. Another generation, but in the absence of the label, produces sister chromatid pairs that contain one labeled and one unlabeled member (bottom). (c) The results are interpreted in terms of semiconservative replication of a single DNA double helix within each chromatid. Dark lines indicate radioactive strands. *Source:* From J. H. Taylor, *Molecular Genetics,* Part I, J. H. Taylor (ed.), Academic Press, Inc., 1963.

among species and among the tissues of a species). This nucleosome fiber, with a diameter of about 10 nm, has a zigzag appearance when viewed microscopically (Fig. 4.24c). The coiling of each 200-base-pair segment of DNA about a protein core of 10 nm in thickness reduces the length of the DNA double helix by approximately sevenfold (200 nucleotide pairs × 0.34 nm/pair = 68 nm, and 68 nm/10 nm = 6.8).

Further packing is needed to form the so-called unit chromatin fiber of about 25 nm thickness that can be seen in electron micrographs of eukaryotic chromosomes. Thoma and Koller have shown that as the ionic strength of the solvent is

Figure 4.24. Organization of DNA into nucleosomes. (a) Each core nucleosome consists of 146 base pairs of DNA wrapped in a helical fashion around a protein core that consists of eight histone molecules. (b) The association of a core particle with a molecule of histone H1 yields a minimal nucleosome, which contains a total of 166 base pairs of DNA. The 146 base pairs of the core particle are marked off by bars here, in order to show that the minimal nucleosome contains ten more base pairs at each end of the core DNA. Histone H1 is thought to bind to the DNA at the site at which the DNA enters and leaves the nucleosome. (c) Because the entry and exit points are on the same side of the nucleosome, the nucleosome fiber takes on a zigzag shape, with adjacent nucleosomes separated by approximately 34 base pairs of linker DNA. *Source:* After F. Thoma, Th. Koller, and A. Klug, *J. Cell Biol.*, 83:403–427, 1979, by copyright permission of The Rockefeller University Press.

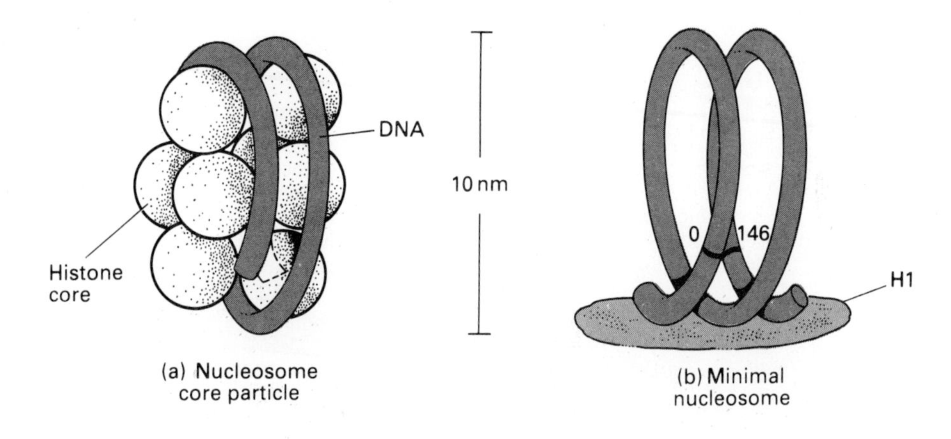

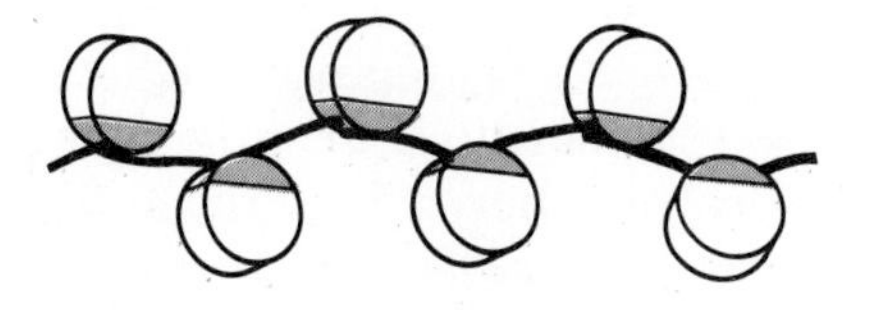

(c) Nucleosome fiber

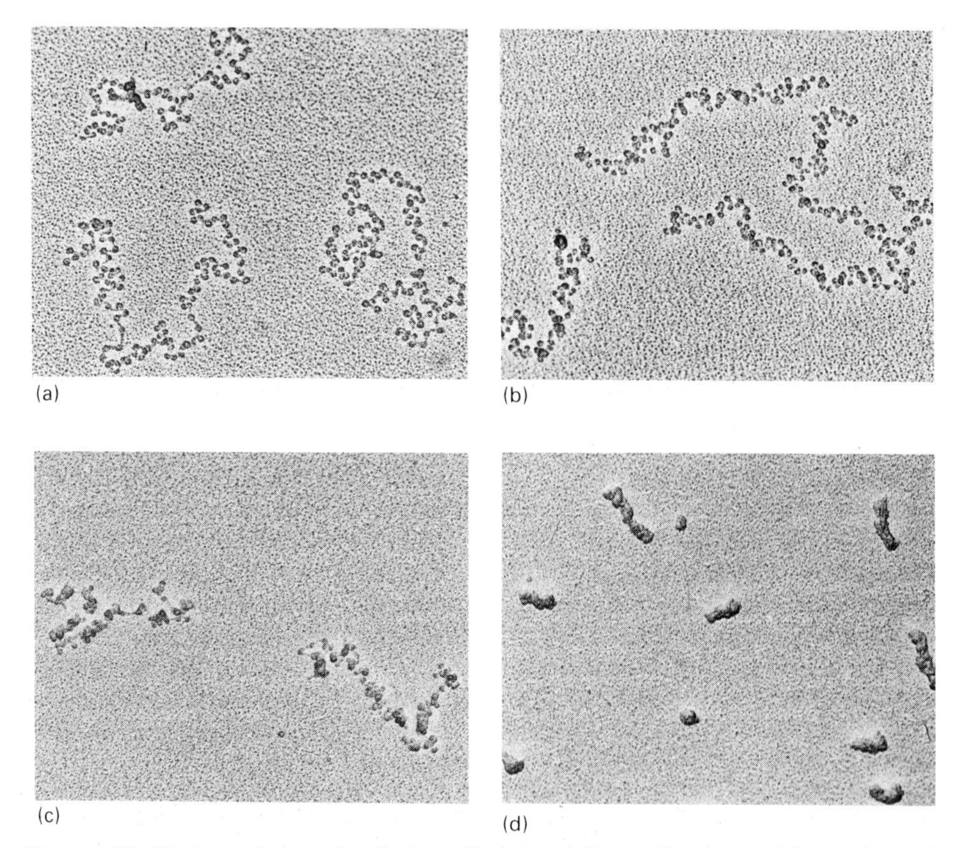

Figure 4.25. Electron micrographs of chromatin from rat liver cells, shown at increasing ionic strengths: (a) 1 mM NaCl, (b) 10 mM NaCl, (c) 40 mM NaCl, and (d) 100 mM NaCl. The nucleosome filaments that can be observed in (a) gradually condense to form closed, zigzag-shaped fibers in (b), then discontinuous, compact fibers in (c), and finally, continuous, compact fibers in (d). *Source:* Reprinted with permission from F. Thoma and Th. Koller, *J. Mol. Biol.*, 149:709–733, 1981. Copyright: Academic Press, Inc. (London) Ltd.

increased, a gradual condensation of the nucleosome fiber takes place (Fig. 4.25). At intermediate ionic strengths, increases in fiber diameter occur to where the filaments of DNA can no longer be seen, as the H1-binding regions of the nucleosomes approach each other more and more closely. High ionic strengths produce a cylindrical coil, or **solenoid,** that has a diameter of 25 nm, which is characteristic of the unit chromatin fiber. Using different experimental techniques, several investigators have shown that histone H1 is required for this condensation of chromatin into 25 nm fibers. Chromatin that is stripped of its H1 shows a less pronounced condensation and an irregular coiling pattern. A model of chromatin condensation is shown in Fig. 4.26.

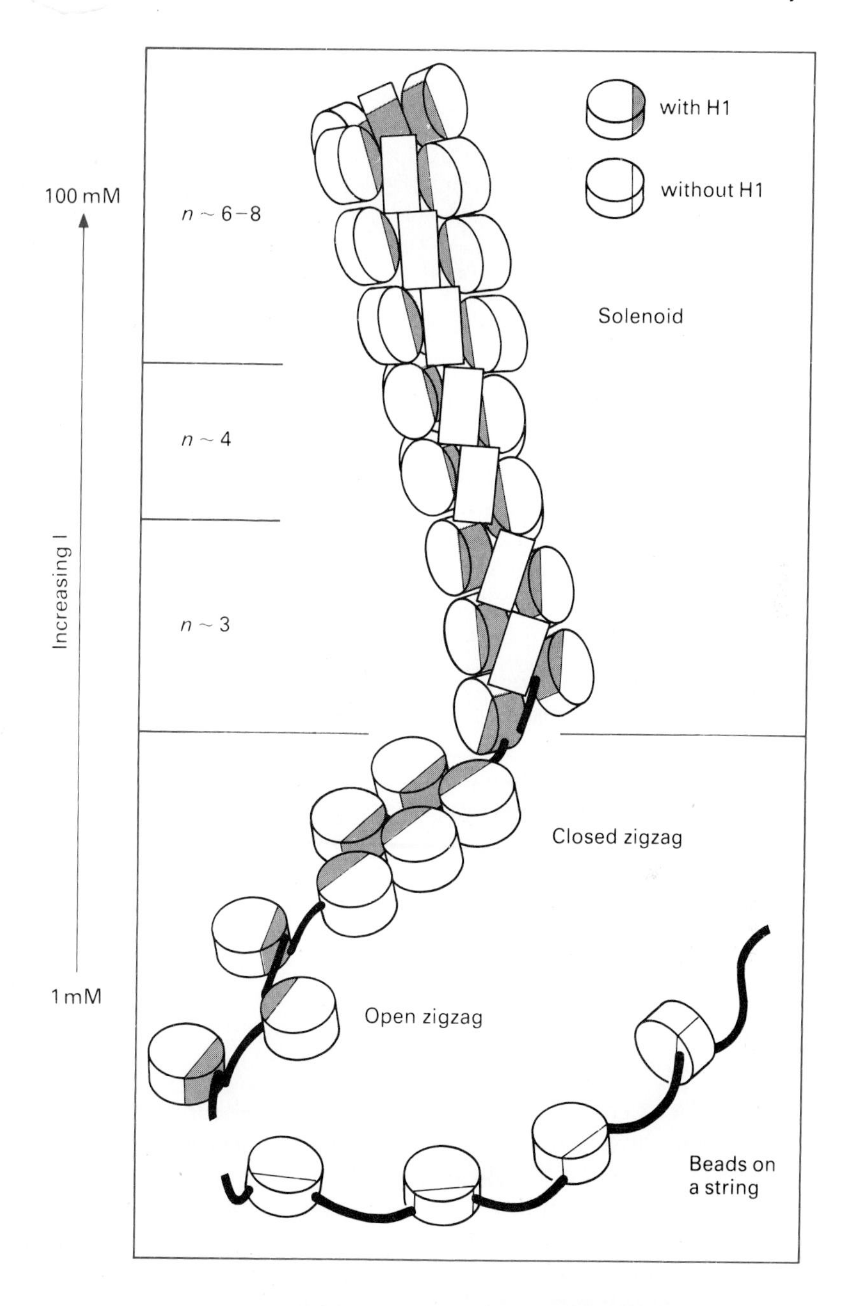

Figure 4.26. Model of the condensation of chromatin containing H1 with increasing ionic strength (I). The open zigzag fiber closes up to form helices with increasing numbers of nucleosomes per turn (n). When H1 is absent (bottom), no zigzags or higher-order levels of organization are found. *Source:* After F. Thoma, Th. Koller, and A. Klug, *J. Cell Biol.*, 83:403–427, 1979, by copyright permission of The Rockefeller University Press.

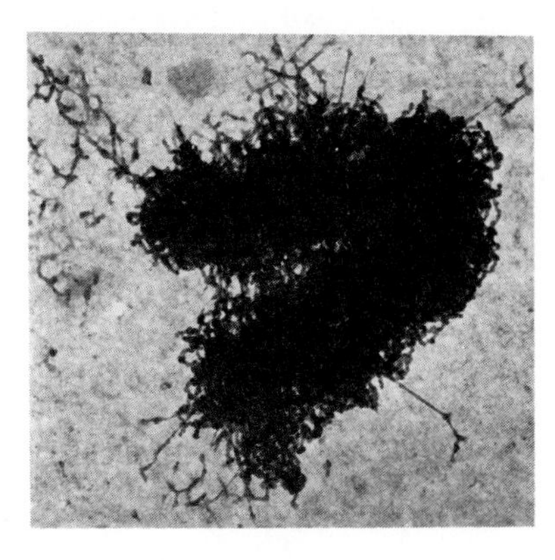

Figure 4.27. Electron micrograph of a bovine (*Bos taurus*) chromosome showing a highly coiled chromatin fiber making up each chromatid. The chromatin fibers are joined together at the centromere region located at the top of the chromosome. *Source:* Courtesy of Dr. Stephen L. Wolfe, University of California, Davis.

Much further coiling is needed to produce the highly compact state of mitotic chromosomes that is seen in electron micrographs (Fig. 4.27). The molecular basis for the condensation of chromosomes from the interphase to metaphase state is unknown. Various models have been proposed for the process. However, notions about coiling of chromatin above the level of the solenoid are still highly speculative.

Nonhistone Proteins. In addition to histones, chromatin contains various nonhistone proteins. The nonhistone components are believed to include two major classes of proteins: *enzymes* needed for the repair and replication of DNA and the transfer of genetic information, and *regulatory proteins,* which function in the control of certain genes. Some nonhistone proteins may also serve a purely structural role in chromosomes. Although nonhistone proteins do not appear to have a crucial influence on the nucleosome structure or on the development of the solenoid, they may provide some sort of supportive framework or "scaffold" in mitotic chromosomes. Evidence for this view has been provided by electron micrographs that reveal proteinaceous scaffolds left after histones were removed from mitotic chromosomes.

Organelles and Their Chromosomes

Not all eukaryotic DNA is located within the nucleus. Eukaryotic cells also possess DNA within special cytoplasmic organelles: mitochondria (found in both plant and animal cells) and chloroplasts (found only in plant cells).

Mitochondria are complex bodies, about the size of bacteria, that are bounded on the outside by a double membrane comprising two unit membranes separated by an intermembranous space (Fig. 4.28). The outer membrane of the mitochondrion appears smooth and continuous around the structure; in contrast, the inner membrane has folds, known as *cristae,* that extend into the interior, or matrix, of the organelle. The mitochondria are responsible for the bulk of the aerobic respiratory activities of a eukaryotic cell, which include the biochemical processes of the citric acid (or Krebs) cycle, electron transport, and oxidative phosphorylation. The enzymes that are involved in these metabolic activities are separated by the

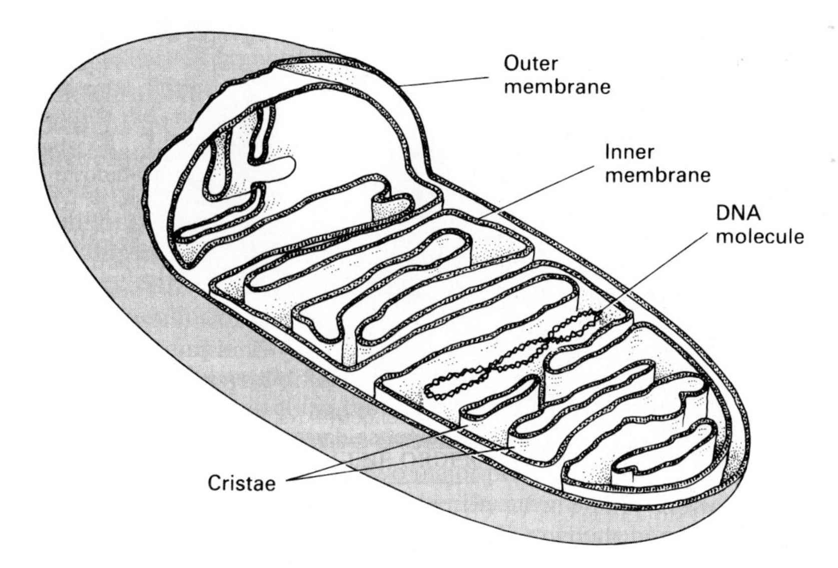

Figure 4.28. A mitochondrion showing the outer membrane that forms a continuous barrier around the organelle and the inner membrane that contains a series of tubular extensions or folds, the cristae. The inner membrane including the cristae contains the biochemical machinery that functions in aerobic respiration. The internal aqueous matrix of the mitochondrion contains various solutes as well as ribosomes and DNA.

membrane compartments of the organelle. For example, the enzymes that are involved in the Krebs cycle are found mainly within the matrix, whereas the enzymes that participate in the coupled processes of electron transport and oxidative phosphorylation are located in the inner membrane.

Chloroplasts are present only in eukaryotic cells that can carry out the vital function of photosynthesis. Chloroplasts vary in size and shape, but they are frequently disk-shaped and larger than a typical mitochondrion (Fig. 4.29). Like mitochondria, chloroplasts are bounded by a double membrane that encloses an interior matrix, or stroma. In chloroplasts, the inner membrane is intricately folded to form ordered stacks of disk-like structures that contain chlorophyll, called *grana*. The grana are formed during the development of mature chloroplasts from smaller colorless *proplastids*. The membranes of chloroplasts, like those of mitochondria, serve to compartmentalize functions. For example, the grana house the light-sensitive pigments and enzymes that are involved in the light-capturing phase of photosynthesis. The stroma contains the enzymes that are involved in the conversion of atmospheric carbon dioxide into sugars.

Chromosomes of Mitochondria and Chloroplasts. The chromosomes of mitochondria and chloroplasts in most organisms consist of circular duplex molecules of DNA. They vary in size, depending on the organism and the organelle. For example, DNA circles in mitochondria average around 5 μm in contour length in most animals and 20–30 μm in plants. The chromosomes of chloroplasts are consistently larger than those of mitochondria, averaging around 45 μm in cir-

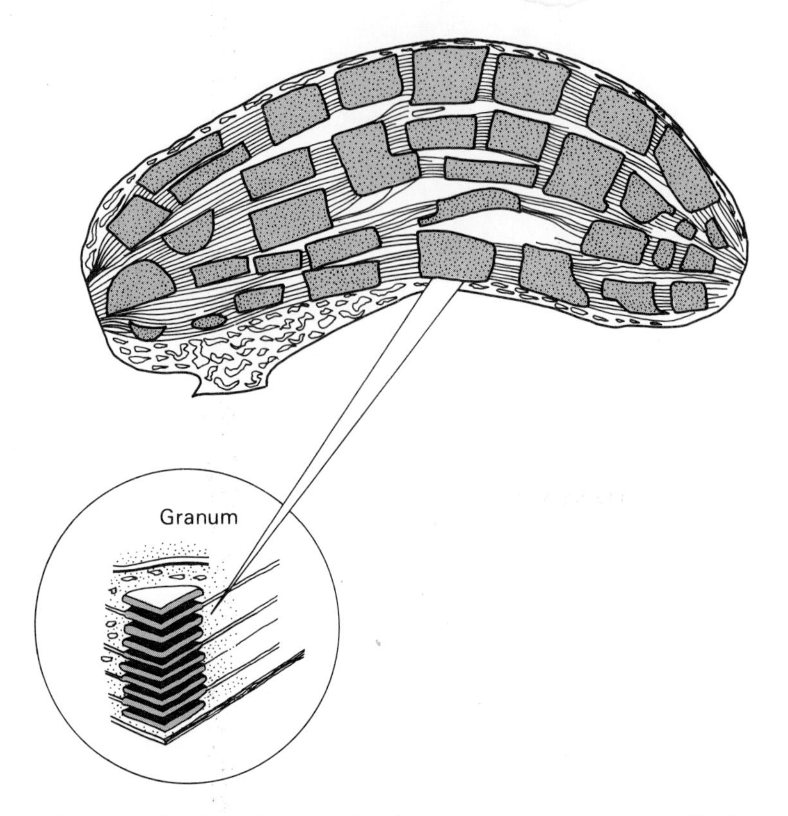

Figure 4.29. Drawing of a chloroplast, showing the outer membrane and the cylindrical stacks of flattened membraneous sacs known as the grana. DNA molecules are shown located between the membranous layers of the grana.

cumference in the higher plants. In general, organelle chromosomes are similar in size to bacterial plasmids.

Unlike nuclear chromosomes, mitochondrial and chloroplast chromosomes are not associated with histones. Neither are they surrounded by a membrane comparable to the nuclear envelope; they are organized, instead, into nucleoids within the matrix of each organelle. The DNA molecules of mitochondria and chloroplasts also differ in base composition from nuclear DNA, so that they can often be isolated in separate bands upon density gradient equilibrium centrifugation. The similarity of organelle chromosomes to prokaryotic rather than nuclear chromosomes (as well as many other similarities between organelles and prokaryotes) has prompted several investigators, led by Lynn Margulis, to propose that mitochondria and chloroplasts were once free-living bacteria that were engulfed by ancient eukaryotic cells and established a symbiotic relationship with them. Whether or not this theory correctly describes the origin of organelles within eukaryotic cells remains to be determined.

The genes carried in the chromosomes of mitochondria and chloroplasts code for some of the components of these organelles, while nuclear genes code for others. For example, some of the genes in the mitochondrial chromosomes are known to code for essential components of the cellular respiration process and of the protein synthesizing system in mitochondria. But the enzymes of the Krebs cycle and many of the proteins involved in electron transport are specified by nuclear genes. Organelles are thus interactive genetic systems, depending in large part on the nuclear genome for their structural and functional parts.

Mitochondria and chloroplasts (or proplastids) are self-replicating structures, which means that new organelles arise from existing ones rather than being assembled from scratch by the cell. The mechanism by which the organelles reproduce themselves is not fully understood; however, it is known that they divide during the process of cell division. The organelle chromosomes replicate prior to division by a process that resembles the one described for prokaryotes.

TO SUM UP

1. Nuclear chromosomes occur in pairs in the somatic cells of diploid organisms. The matched members of each pair are called homologs. One member of each pair of chromosomes is present in the gametes of a diploid organism, so that gametes are haploid with only half as many chromosomes as somatic cells.
2. A cell cycle consists of interphase followed by mitotic division. The comparatively long interphase is divided into a presynthetic period (G_1), a period of DNA synthesis (S), and a postsynthetic period (G_2). During interphase, nuclear chromosomes exist as dispersed chromatin fibers made up of DNA in a complex with histone and nonhistone proteins.
3. The synthesis of DNA in nuclear chromosomes occurs during the S period of interphase and proceeds simultaneously within multiple replication units called replicons. DNA synthesis within each replicon is bidirectional, proceeding in both directions from the origin.
4. Mitosis is the process of nuclear division that occurs in somatic cells. At the beginning of mitosis, chromosomes appear as threadlike structures that consist of two identical chromatids, joined by a centromere. The chromatids are formed by the duplication of each chromosome during interphase.
5. We divide mitosis into four separate stages. The chromosomes shorten and thicken during prophase while the nuclear envelope and nucleolus disappear. The chromosomes align on the equator of the spindle during metaphase, each connected at its centromere to the poles of the spindle by microtubules. The chromatids separate during anaphase, and the duplicated halves of each chromosome move to the opposite poles of the spindle. The chromosomes uncoil during telophase as nucleoli reappear and a nuclear envelope reforms around the chromosomes at each pole.
6. The basic structural unit of the nuclear chromosome is a molecule of DNA combined with histone proteins at periodic points, giving it the appearance of a string of beads. Each bead in the DNA-histone complex is a particle, called a nucleosome, that consists of a segment of DNA wrapped in helical fashion around a protein core of eight histone molecules. The 25 nm chromatin fiber observed in electron micrographs of mitotic chromosomes is formed through the higher-level coiling of the DNA-histone complex.
7. Mitochondria and chloroplasts contain their own chromosomes, which resemble those of prokaryotes in appearance. These chromosomes carry genes that specify some of the components needed to carry out the functions of the organelles.

Questions and Problems

1. Define the following terms:

autoradiography	karyotype
cell cycle	lysogenic cell
centromere	microtubules
concatamer	plasmid
conjugation	replicon
cytokinesis	satellite
episome	spindle
genome	solenoid
interphase	supercoiling

2. Distinguish between members of each of the following sets of terms:
   (a) virion, vegetative virus, provirus
   (b) eclipse period, latent period, period of assembly
   (c) asymmetric replication, symmetric replication
   (d) temperate phage, virulent phage
   (e) lysogenic cycle, lytic cycle
   (f) Hfr cell, F^+ cell, F^- cell
   (g) colicinogens, resistance plasmids
   (h) haploid number, diploid number
   (i) sister chromatids, homologous chromosomes
   (j) chromatid, chromosome arm
   (k) interpolar microtubules, chromosomal microtubules
   (l) cristae, grana
3. The following is a spread of several eukaryotic chromosomes at metaphase. Indicate whether each chromosome is metacentric, submetacentric, acrocentric, or telocentric. Which if any of these chromosomes have satellites?

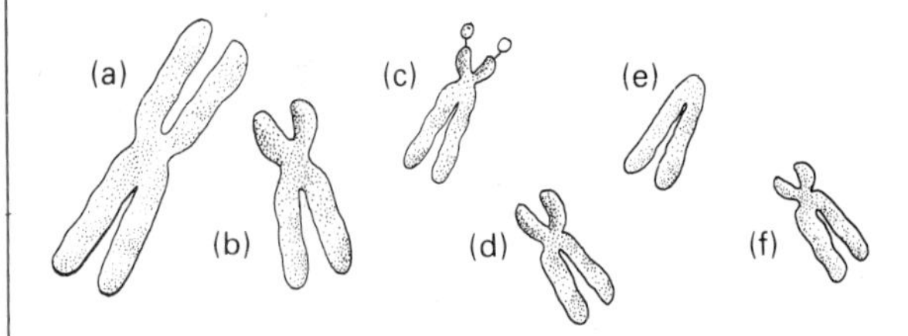

4. A certain plant cell has 20 chromosomes at G_1. How many chromosomes, centromeres, and chromatids are present within a single cell of this plant **(a)** at prophase? **(b)** At metaphase? **(c)** At anaphase? **(d)** At the close of telophase? (Assume that cytokinesis occurs at mid-telophase.)
5. The following are onion root-tip cells in the process of mitosis. In which stage of mitosis is each of these cells?

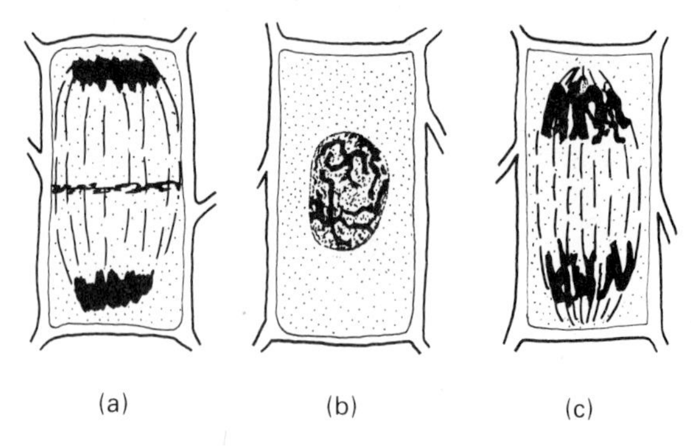

6. Give the specific stage in the cell cycle when each of the following events takes place: **(a)** Sister chromatids are first produced. **(b)** Sister chromatids are first visible with a light microscope. **(c)** Sister chromatids disjoin.
7. Suppose that a fertilized human egg undergoes cleavage (doubling of the number of cells) once every 48 hours throughout intrauterine development. Assume that at birth a baby has 10^{14} cells. **(a)** How long would it take to produce this number of cells under the assumed conditions? **(b)** Compare this time period with the actual gestation time. Why is there such a large difference?
8. The amount of DNA present at G_1 in the haploid set of chromosomes in humans is 2.8 picograms (2.8×10^{-12} g). **(a)** What is the DNA content of a human somatic cell nucleus at G_2? At anaphase? At the close of telophase? **(b)** Compute the average DNA content *per chromatid equivalent* in each of these phases. **(c)** Compute the average DNA content *per chromosome* in each of these phases and compare your answers with those obtained in part (b).
9. The circular *E. coli* chromosome observed in autoradiograms has a contour length of 1360 μm. Calculate the number of base pairs in the *E. coli* chromosome.
10. In *E. coli,* DNA biosynthesis occurs at a rate of 50,000 base pairs per minute per replication fork. How long does it take an *E. coli* cell to

duplicate its chromosome? Compare your answer with the generation time of *E. coli* (20 minutes under optimal growth conditions).

11. It takes 100 minutes for an Hfr donor cell of *E. coli* to transfer its entire chromosome during conjugation. What is the length of chromosome (in base pairs) that an Hfr cell transfers in 1 minute? Compare your answer with the rate of chromosome replication in *E. coli*.

12. The rate of DNA replication in *E. coli* is 25 times greater than the average eukaryotic rate of 2000 base pairs per minute. Give some possible reasons for the large difference in the rate of DNA replication in prokaryotic and eukaryotic cells.

13. About 6 minutes after infection of a bacterial cell by a phage T4 particle, replication of phage DNA begins. By 11 minutes from the time of infection, each infected cell contains approximately 32 phage equivalents of DNA, and mature phages begin to appear. The cells lyse at about 23 minutes. Assuming that once assembly begins the number of vegetative phages in the replicating DNA pool remains constant, determine the average number of mature phages released per infected cell (known as the "burst size" of the infected bacterial cells).

14. An average human chromosome at metaphase is made up of two chromatids, each 5.8 μm long by 0.8 μm thick. Assume that each chromatid consists of a chromatin fiber 825 μm long by 0.025 μm (25 nm) thick. **(a)** How much greater is the length of the fiber than that of the chromatid? **(b)** By what factor does the volume of the chromatid exceed that of its chromatin fiber? (*Hint:* Assume the shapes are cylindrical.)

15. When a sample of eukaryotic cells is grown in a culture, the cells tend to be randomly distributed with respect to stage of the cell cycle. At any given moment, some of the cells are in G_1, others are in S, and so on. The number of cells in each phase at any given moment depends solely on what fraction of the average cycle time is taken up by that stage. **(a)** Suppose that G_1, S, G_2, and M for a particular cell culture last for 11 hours, 8 hours, 4 hours, and 1 hour, respectively. What percentage of cells are expected to be in each of these phases at any given moment? **(b)** If four cells are selected at random from this same population, what is the chance that at least one of the cells is in some phase of mitosis? **(c)** When cells from this population are examined for mitotic activity, it is found that of the cells undergoing mitosis 40 percent are in prophase, 30 percent are in telophase, 20 percent are in metaphase, and 10 percent are in anaphase. How long (in minutes) is each of these mitotic stages in this cell population?

16. Growth of normal somatic cells in culture is limited to a fixed number of division cycles. The number depends on the species and the age of the individual from which the cells were obtained. For example, human embryonic cells can undergo only around 50 cell cycles before the cells begin to deteriorate and division stops. How many progeny cells can a human embryonic cell theoretically produce during its normal lifespan? Compare this figure with the 10^{14} to 10^{15} cells that are present in a human adult.

17. The limited capacity of cells to divide in culture (described in Problem 16) is not altered by artificially halting the growth. For example, when human embryonic cells are placed in a deep freeze to arrest cell growth, the only effect is to delay the inevitable deterioration of the culture. Freezing does not change the limit of 50 cell cycles that is characteristic of this cell type in culture. Suppose that a sample of 100 human embryonic cells from the twentieth cell cycle is frozen and later thawed. How many progeny cells can this sample theoretically produce by the end of its normal lifespan?

18. Consider again the hypothetical situation described in Example 4.2. In this problem, we assumed that daughter chromosomes separate in prophase and align independently on the metaphase spindle. Now suppose that instead of four daughter chromosomes always moving to one pole and four to the other, the direction of migration is strictly random, so that each chromosome has an equal chance of moving to either pole. **(a)** What is the probability that each daughter cell will receive exactly four chromosomes? (For now assume that any combination of four chromosomes will suffice.) **(b)** What is the chance that each daughter cell will receive one copy of each of the four chromosomes present in the parent cell, so that each has a chromosome makeup of *AA'BB'*?

Suggested Readings / Part II

Chapter 3

Avery, O., C. M. MacLeod, and M. McCarty. Studies on the chemical nature of the substance inducing transformation in pneumococcal types. *J. Expl. Med.* 79:137–158, 1944.

Crick, F. H. C. The structure of the hereditary material. *Scientific American* 191:54–61, 1954.

Hershey, A. D. and M. Chase. Independent functions of viral protein and nucleic acid in growth of bacteriophage. *J. Gen. Physiol.* 36:39–56, 1952.

Kornberg, A. Biologic synthesis of deoxyribonucleic acid. *Science* 131:1503–1508, 1960.

Kornberg, A. *DNA Replication,* Freeman, San Francisco, 1980. (Also *Supplement to DNA Replication,* 1982.)

Stent, G. S. and R. Calendar. *Molecular Genetics: An Introductory Narrative,* Freeman, San Francisco, 1978.

Swift, H. Quantitative aspects of nuclear nucleoprotein, *Int. Rev. Cytol.* 2:1–76, 1953.

Watson, J. D. *The Double Helix,* Antheneum, New York, 1968.

Watson, J. D. *Molecular Biology of the Gene,* Benjamin, Menlo Park, Calif., 1976.

Watson, J. D. and F. H. C. Crick. Molecular structure of nucleic acids. A structure for deoxyribose nucleic acid. *Nature* 171:737–738, 1953.

Watson, J. D. and F. H. C. Crick. Genetical implications of the structure of deoxyribonucleic acid. *Nature* 171:964–967, 1953.

Chapter 4

Borst, P. and L. A. Grivell. The mitochondrial genome of yeast. *Cell* 15:705–723, 1978.

Cairns, J. The bacterial chromosome. *Scientific American* 214:36–44, 1966.

Campbell, A. M. How viruses insert their DNA into the DNA of the host cell. *Scientific American* 235:103–113, 1976.

Gibor, A. and S. Granick. Plastids and mitochondria: inheritable systems. *Science* 145:890–897, 1964.

Grivell, L. A. Mitochondrial DNA. *Scientific American* 248:78–89, 1983.

Hayes, W. *The Genetics of Bacteria and their Viruses,* Wiley, New York, 1968.

Horne, R. W. The structure of viruses. *Scientific American* 208:48–56, 1963.

Kornberg, A. *DNA Replication,* Freeman, San Francisco, 1980. (Also *Supplement to DNA Replication,* 1982.)

Kornberg, A. and A. Klug. The nucleosome. *Scientific American* 244:52–79, 1981.

Margulis, L. Symbiosis and evolution. *Scientific American* 225:49–57, 1971.

Stent, G. S. and R. Calendar. *Molecular Genetics: An Introductory Narrative,* Freeman, San Francisco, 1978.

Swanson, C. P., T. Merz, and W. J. Young. *Cytogenetics,* Prentice-Hall, Englewood Cliffs, N.J., 1981.

Watson, J. D. *Molecular Biology of the Gene,* Benjamin, Menlo Park, Calif., 1976.

Wolfe, S. L. *Biology of the Cell,* Wadsworth, Belmont, Calif., 1981.

The Chromosomal Basis of Gene Transmission

Chapter 5

Chromosomes and Sexual Reproduction

Mitotic cell division is the mechanism responsible for the growth of multicellular organisms. Mitosis is also an asexual reproductive process that provides for the multiplication of unicellular eukaryotes as well as the vegetative propagation of many plants. Since mitosis preserves the chromosome complement from one generation to the next, its function in reproduction results in the establishment of clones, in which all individuals are genetically identical.

Most eukaryotes are capable of reproducing sexually through the formation and subsequent union of haploid gametes. This mechanism allows for the genetic material of two parents to mix. Through the production of new and different combinations of genes, sexual reproduction serves as the major source of genetic variability. This variability is essential for evolutionary progress, because it permits survival under diverse environmental conditions. Sexual reproduction is therefore adaptively advantageous to a species, and it is a common feature in the life cycles of most organisms.

In this chapter, we will discuss a wide range of topics related to sexual reproduction in eukaryotes. We will start by describing the origin of the haploid phase in multicellular organisms and will then examine the chromosomal basis of Mendel's laws. We will next look at differentiation in gametes and the formation of zygotes. We will also consider the mechanisms that determine the sex of an individual.

The Sexual Reproductive Process

Reproductive Cycles

The life cycle of many sexually reproducing eukaryotes is characterized by alternating haploid and diploid stages. In the higher animals, the diploid stage is by far the most prominent phase of the life cycle, since it includes the multicellular adult (Fig. 5.1a). The diploid adult is composed of millions of cells formed through repeated cell cycles, starting with the fertilized egg. For example, a human adult consists of about 10^{14} (or 2^{47}) total cells. Thus, about 47 division cycles are required for the complete development of the human diploid stage.

The haploid phase in animals is much less conspicuous than the diploid phase. The haploid phase consists solely of the short-lived egg and sperm cells. Because these cells are normally incapable of mitotic division, the haploid phase ends without further cell proliferation when the egg and sperm unite during fertilization.

By contrast, many plants undergo alternation of generations, in which a multicellular haploid stage is followed by a multicellular diploid stage, and so on. The haploid phase in many plants is therefore more complex than it is in animals. In land plants, the haploid phase starts with haploid cells known as **spores,** which are produced by the diploid plant generation, called the **sporophyte** (Fig. 5.1b). The spores divide mitotically to form the haploid plant generation, known as the **gametophyte.** The gametophyte matures to produce gametes, which unite during fertilization to restore the diploid condition. The mitotic development of the diploid zygote into the sporophyte generation then completes the cycle.

Meiosis

Fertilization marks the beginning of the diploid phase in both plants and animals that reproduce sexually. The haploid stage of the sexual cycle results from a nuclear division process known as **meiosis.** Meiosis occurs in certain specialized cells, called **meiocytes,** that are located within the reproductive tissues of an organism. Like mitosis, meiosis takes place after the G_1, S, and G_2 stages of interphase and provides for a precise distribution of chromosomes into daughter cells. Unlike mitosis, however, meiosis involves two successive division cycles (that occur *without an intervening synthesis of DNA*) and results in four daughter nuclei instead of two.

An overview of meiosis is given in Fig. 5.2. Note that the first division, called meiosis I, is a *reduction division.* Meiosis I converts the nucleus of a meiocyte, which contains $2n$ duplicated chromosomes, into two haploid daughter nuclei, each of which has n duplicated chromosomes. The chromosome number is reduced when the homologous pairs of chromosomes disjoin. As in mitosis, the disjoining proceeds by way of the spindle microtubules and is followed by cytokinesis.

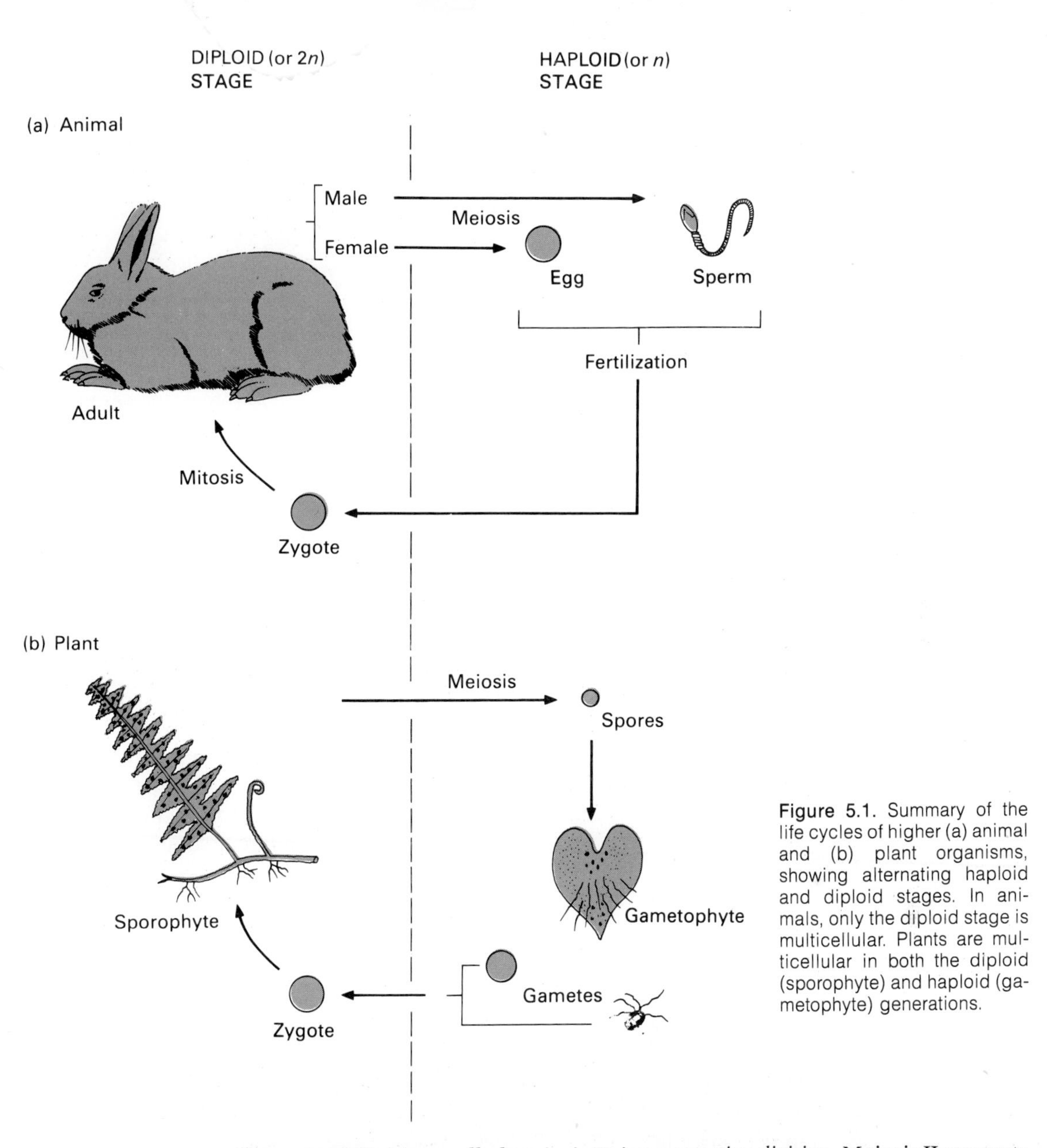

Figure 5.1. Summary of the life cycles of higher (a) animal and (b) plant organisms, showing alternating haploid and diploid stages. In animals, only the diploid stage is multicellular. Plants are multicellular in both the diploid (sporophyte) and haploid (gametophyte) generations.

The second division, called meiosis II, is an *equation division.* Meiosis II converts the two haploid products of the first meiotic division into four haploid nuclei, with n unduplicated chromosomes in each. This process is in many ways similar to mitosis, since its primary function is to provide for the exact disjunction of sister chromatids and the movement of daughter chromosomes into daughter nuclei.

The reason that two division cycles are needed instead of one is apparent when we consider the relative amounts of DNA in each cell nucleus at the various stages in the cell cycle. The meiocyte begins as a diploid cell. DNA synthesis in the S stage results in four times the amount of DNA as would be in a haploid cell

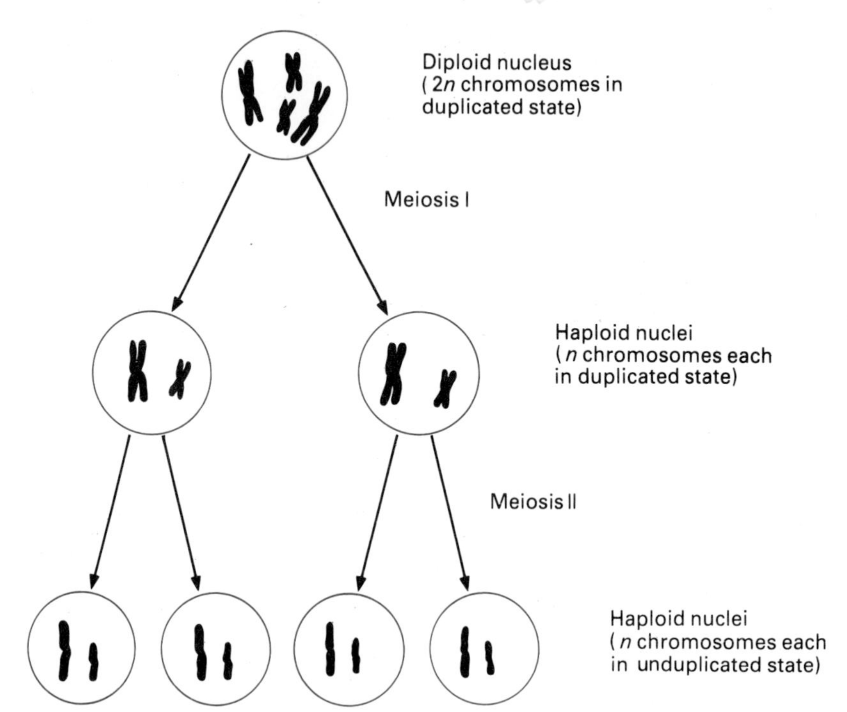

Figure 5.2. Summary of the nuclear events in meiosis. For the sake of simplicity, the haploid number of chromosomes here is equal to two. The first division is a reduction division, which serves to separate members of each homologous pair. The second division, like mitosis, separates the sister chromatids.

at G_1. Since a single cell division can only decrease the amount of DNA by half, two division cycles are needed to reduce the DNA content to a quarter of the amount the cell contains at the end of interphase. This reduction is shown as follows:

$$\underset{(4x \text{ DNA in each})}{1 \text{ nucleus}} \xrightarrow{\text{meiosis I}} \underset{(2x \text{ DNA in each})}{2 \text{ nuclei}} \xrightarrow{\text{meiosis II}} \underset{(1x \text{ DNA in each})}{4 \text{ nuclei}}$$

Fertilization then reestablishes the normal $2x$ DNA content of a diploid cell at G_1.

Stages of Meiosis. Each meiotic division cycle (meiosis I and meiosis II) is divided into the same four stages as mitosis: prophase, metaphase, anaphase, and telophase. The name of each stage is followed by a I or a II, depending on which division cycle is involved.

From the standpoint of genetics, *prophase I* is the most complex phase of meiosis. It is also the longest, taking several days, weeks, or even years to complete in different organisms. The major events that occur during prophase I are summarized here and diagrammed in Fig. 5.3: (1) In the leptotene stage (leptonema) the chromosomes condense to become visible as slender, threadlike structures. (2) In the zygotene stage (zygonema), the homologous chromosomes pair and

Figure 5.3. Stages of meiosis. Substages (a) through (e) are the major events of prophase I. The stages of meiosis are described in sequence in the text. Prophase II is not shown; it is a brief stage during which the spindle reforms. ▶

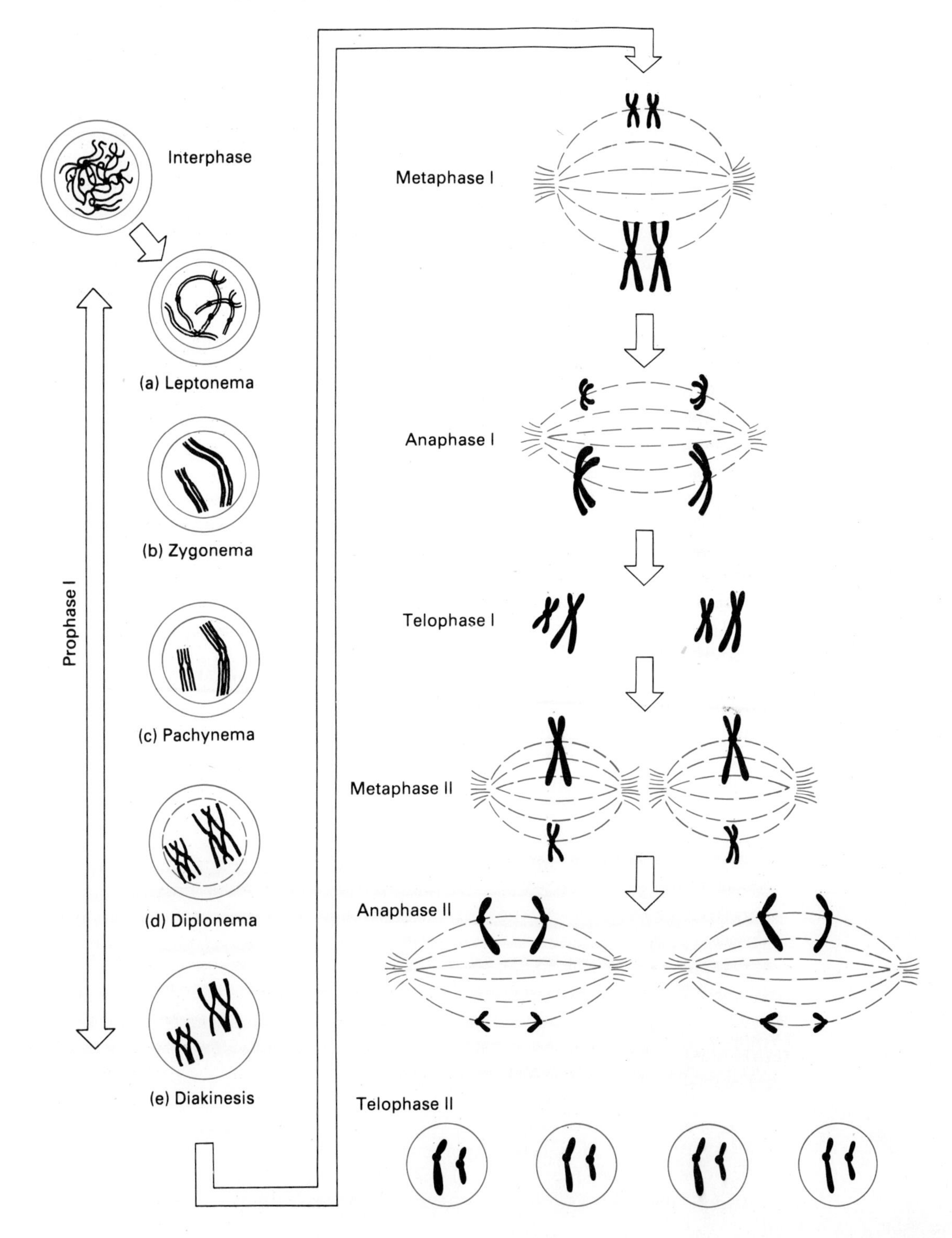
Interphase
Prophase I
(a) Leptonema
(b) Zygonema
(c) Pachynema
(d) Diplonema
(e) Diakinesis
Metaphase I
Anaphase I
Telophase I
Metaphase II
Anaphase II
Telophase II

begin to align lengthwise. The pairing process, called **synapsis,** results in a precise side-by-side alignment of homologous chromosomes. (3) In the pachytene stage (pachynema), the chromosomes continue to condense as the pairing process is completed. Each closely associated pair of homologs is now referred to as a **bivalent.** (4) In the diplotene stage (diplonema), the sister chromatids become clearly visible, as the pairing forces between the homologs relax. Paired chromatids of the two homologs, called nonsister chromatids, remain together at points of contact called **chiasmata** (singular: chiasma). Closer inspection reveals that nonsister chromatids cross over (exchange parts) at these points (see section on crossing over). (5) In diakinesis, the homologs coil up to become shorter and thicker structures. At this time, the chiasmata undergo a process of **terminalization,** in which the points of contact move toward the ends of the chromosome arms. The nucleolus and nuclear envelope have now disappeared, and the spindle is forming.

The lengthy prophase I is followed by a comparatively brief metaphase. In *metaphase I,* the bivalents align at the center of the spindle, with the homologous centromeres arranged on either side of the equatorial plane. Unlike mitotic metaphase, in which the centromeres are aligned at the midpoint of the spindle, the two centromeres of each bivalent in meiotic metaphase are directed toward opposite poles because they are attached to opposing chromosomal microtubules.

Anaphase I begins with the movement of homologous chromosomes toward the opposite poles of the spindle. In contrast to mitotic anaphase, only chromosomes of maternal and paternal origin disjoin. Sister chromatids do not separate at this time.

The arrival of homologs at the opposite poles of the spindle signals the start of *telophase I.* The extent of this phase is highly variable. In some species, such as humans, it is quite involved, with the chromosomes unwinding and the nucleoli and nuclear envelopes reforming. In these species, telophase I is usually followed by a brief interphase. In other species, telophase I may be highly abbreviated or entirely lacking, with the cells going immediately into the second meiotic division cycle. In either case, meiosis I has caused the homologs to separate and move to opposite sides of the cell, thus distributing the chromosomes evenly in preparation for the events of meiosis II.

The structural details of meiosis II are quite similar to those of mitosis. At *prophase II,* a new spindle forms about each of the haploid nuclei created in meiosis I. In *metaphase II,* the chromosomes align at the equator of their respective spindle, each attached at its centromere to chromosomal microtubules extending to both poles. During *anaphase II,* sister chromatids disjoin, and the duplicated halves of each chromosome migrate toward opposite poles. The arrival of the daughter chromosomes at the poles signals the start of *telophase II,* during which a nucleus forms around each of the four sets of chromosomes. Finally, cytokinesis occurs, and the meiotic process is completed.

EXAMPLE 5.1. How many chromosomes are present in a human reproductive cell at each of the following stages: **(a)** Anaphase I? **(b)** Metaphase II? **(c)** Anaphase II? (Assume that cytokinesis occurs at mid-telophase I and II.)

Solution. **(a)** At anaphase I, homologs are still in the process of separating. Chromatids do not disjoin at this stage. There are consequently $2n = 46$ duplicated chromosomes still present in each cell. **(b)** By metaphase II, homologs have already disjoined, so there are now $n = 23$ duplicated chromosomes aligned on the equator of each spindle. **(c)** Chromatids separate and daughter chromosomes disjoin at anaphase II. There are thus $2n = 46$ single (unduplicated) chromosomes within a single cell: 23 moving to one pole and 23 to the other.

The Meiotic Basis for Mendel's Laws. So far, we have stressed the function of the pairing and disjunction of homologs in developing the haploid state. These events also have a genetic function in that they are responsible for the basic Mendelian patterns of inheritance. For example, suppose that gene *A* is located on some chromosome of maternal origin and allele *a* is at the same position (or locus) on the homologous chromosome of paternal origin. This situation is illustrated in Fig. 5.4. Chromosome duplication has already occurred, so two copies of each allele exist, one on each of two sister chromatids. Since only one of the four chromosome strands of the bivalent can enter a single gamete, we can readily see that *the normal disjunction of homologs results in the segregation of alleles.*

Normal pairing and disjunction of chromosomes also accounts for the independent assortment of nonallelic genes. To illustrate, let us consider the dihybrid *A/a B/b,* where the slashes (/) are used to indicate that the two pairs of genes are

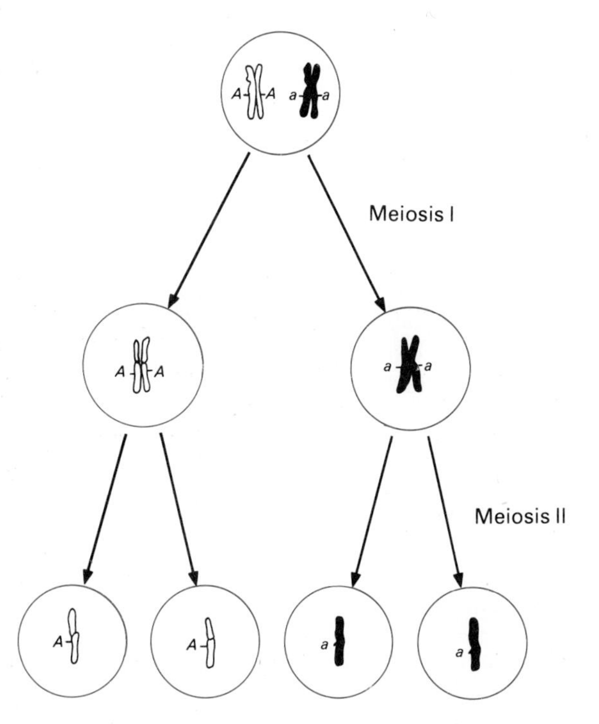

Figure 5.4. The chromosomal basis for the segregation of alleles. Maternally derived chromosomes are shown in white; paternally derived chromosomes in color. Because of chromosomal disjunction, only one member of each pair of alleles will enter a single gamete.

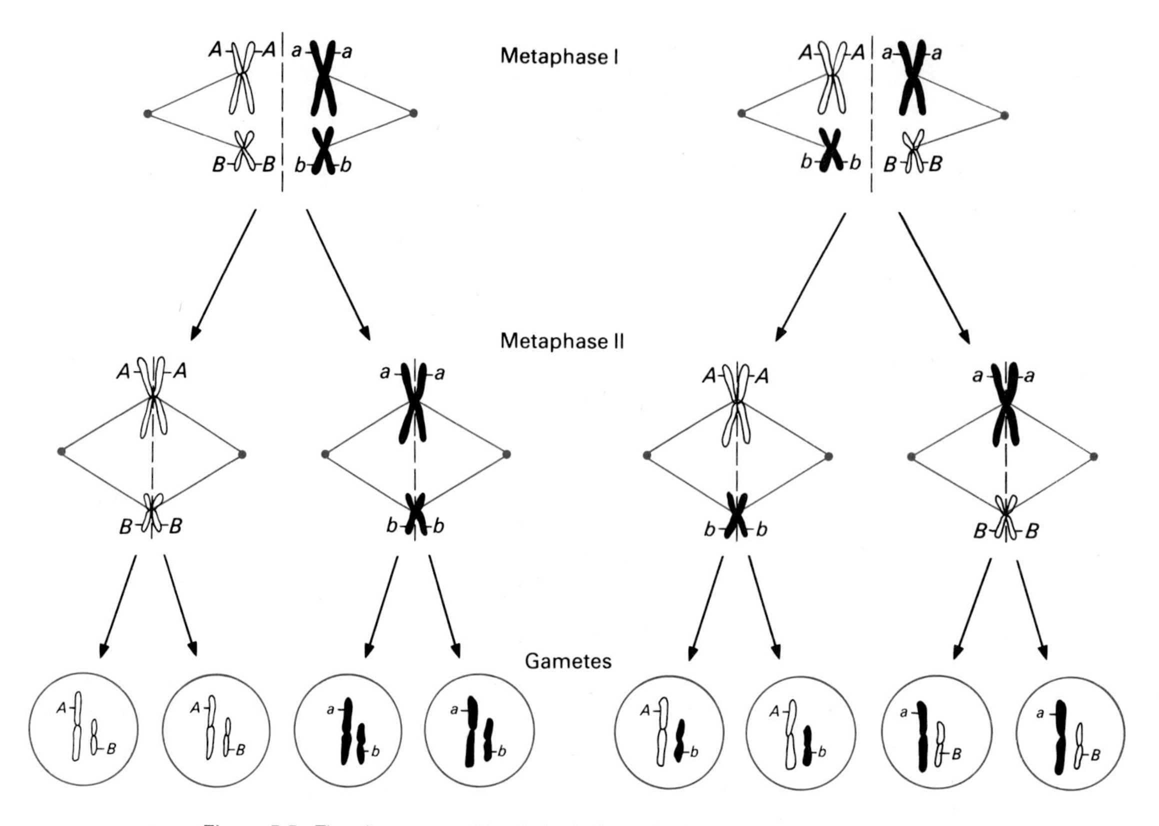

Figure 5.5. The chromosomal basis for independent assortment of nonallelic gene pairs. The figure shows two alternative alignments of nonhomologous chromosomes at metaphase I. Since the alignments are strictly random, all four gamete types ((AB), (Ab), (aB), and (ab)) are expected to occur in equal frequency.

located on different pairs of homologous chromosomes. Figure 5.5 shows the alignment of both pairs of chromosomes at metaphase I. At this stage of meiosis, bivalents align with the maternally and paternally derived centromeres arranged at random on either side of the equatorial plane. Two arrangements of nonallelic genes are therefore possible: one in which *A* and *B* go to one pole and *a* and *b* to the other, and another in which *A* and *b* go to one pole and *a* and *B* go to the other. Since both arrangements occur with equal likelihood, the haploid cells produced by meiosis will have the following genotypic ratio: $\frac{1}{4}$ (*AB*), $\frac{1}{4}$ (*Ab*), $\frac{1}{4}$ (*aB*), and $\frac{1}{4}$ (*ab*). Thus, *independent assortment is a characteristic of nonallelic genes that are located on different pairs of chromosomes.*

EXAMPLE 5.2. Concerning independently assorting genes that are present on different pairs of chromosomes: **(a)** How many arrangements are possible at the spindle equator for three gene pairs? **(b)** For four gene pairs? **(c)** For *n* gene pairs?

Solution. **(a)** If the first pair of alleles is aligned in a certain way, the second allele pair can align in two different ways in relation to the first, and the third pair of alleles can align in two different ways in relation to the second. There are thus $2 \times 2 = 4$ (or 2^2) different arrangements possible. They are $\frac{A\ B\ C}{a\ b\ c}$, $\frac{A\ B\ c}{a\ b\ C}$, $\frac{A\ b\ c}{a\ B\ C}$, and $\frac{A\ b\ C}{a\ B\ c}$. **(b)** With four pairs of alleles, there are $2 \times 2 \times 2 = 8$ (or 2^3) arrangements possible. **(c)** Extending this pattern to n gene pairs, there are then 2^{n-1} possible arrangements of genes.

Crossing Over. The normal behavior of chromosomes during meiosis is also responsible for the reshuffling, or *recombination,* of nonallelic genes located on the same pair of homologs. Genes that are carried by the same chromosome pair but at separate loci are said to be **linked.** Linked genes tend to remain together during meiosis, so that they are transmitted in the same combination in which they were inherited. On occasion, however, new combinations of nonallelic genes are formed by a mechanism known as **crossing over** (Fig. 5.6). Crossing over is a breakage and reunion process which results in the physical exchange of homologous segments between *nonsister* chromatids of a bivalent. As pictured in Fig. 5.6, crossing over takes place during prophase I after synapsis has occurred. Following the crossover event, the combination of nonallelic genes on each of the chromatids involved has changed.

The points at which crossing over occurs are the chiasmata that appear in late prophase I. These visible signs of chromatid exchange occur in many different places along the paired chromosomes, often averaging more than one per bivalent. When multiple crossovers occur along the same chromosome pair, they can involve various combinations of nonsister chromatids. Figure 5.7 illustrates the different types of double crossover events that are possible. Although double crossovers are the most common type of multiple exchange, other possibilities can occur, including triple crossovers or higher. The genetic consequences of crossover events will be discussed in Chapter 7.

EXAMPLE 5.3. Twenty percent of the gametes produced by meiosis in a particular organism show evidence of a crossover within a certain region of chromosome 1. Another 10 percent show evidence of a crossover within an adjacent region on this same chromosome. If we assume that these crossovers occur independently of each other, what percentage of the gametes will show evidence of crossing over in both regions?

Solution. Since crossovers are assumed to be independent in the two adjacent regions, the chance that crossovers will occur in both of them is then $0.2 \times 0.1 = 0.02$. Therefore, 2 percent of the gametes would be expected to be of this type.

Crossing over during meiosis is possible because of a structure known as the **synaptinemal complex.** This structure forms between the arms of homologous chromosomes during synapsis and disperses when they separate. The synaptinemal

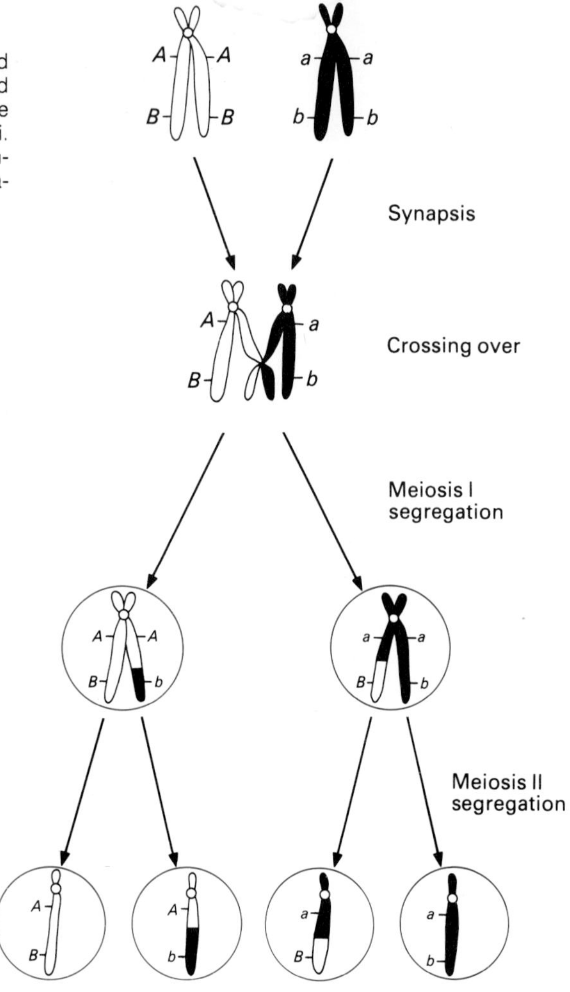

Figure 5.6. The recombination of linked genes through crossing over. The linked genes are assumed to be present on the same chromosome but at separate loci. Crossing over is the mechanism responsible for the formation of new combinations of linked genes.

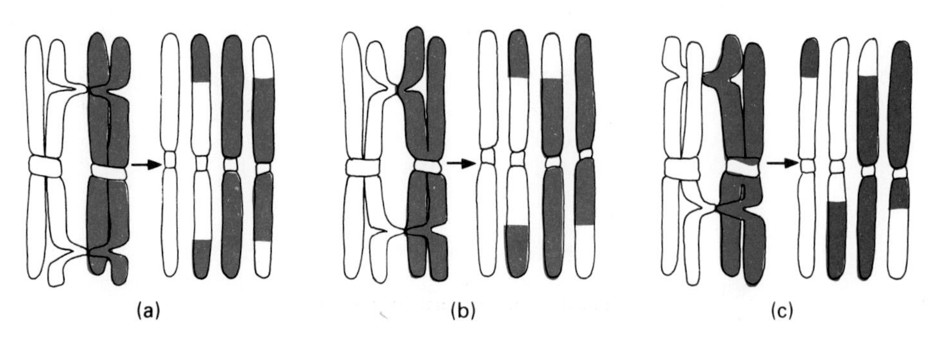

Figure 5.7. Different types of double crossover events. (a) Two-strand double crossover. (b) Three-strand double crossover. (c) Four-strand double crossover. Double crossovers are the most common type of multiple exchange. Triple crossovers, four crossovers, etc., can also occur, but at progressively lower frequencies.

(a)

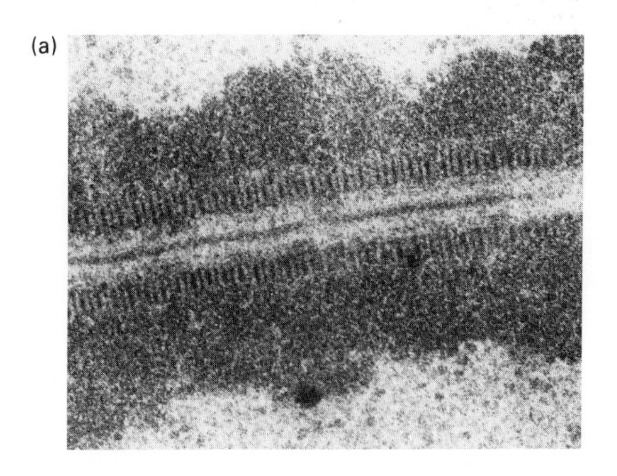

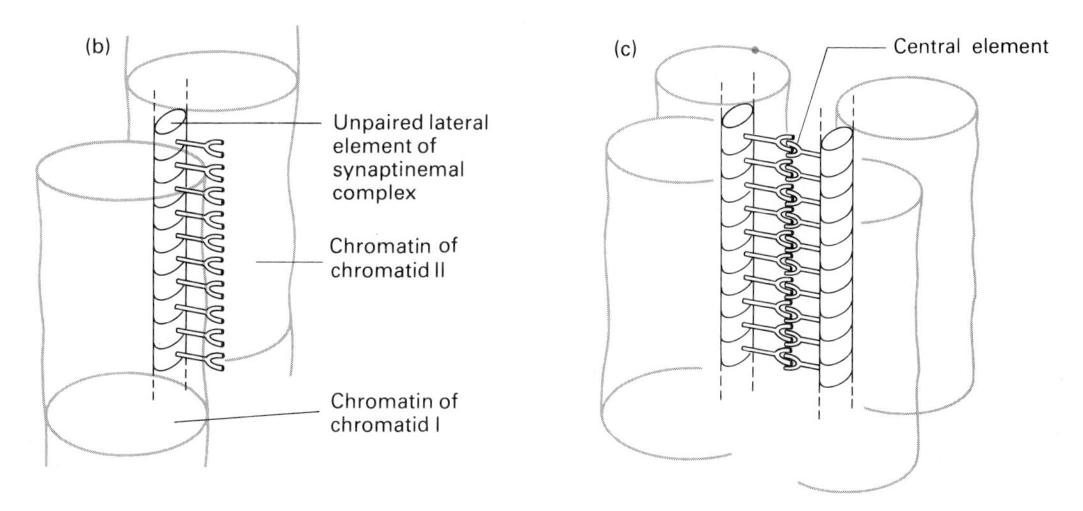

Figure 5.8. The synaptinemal complex. (a) Electron micrograph of the synaptinemal complex of the fungus *Neotiella* (× 70,000). This is a longitudinal section of a bivalent showing a dense central element present between the banded lateral elements that are associated with each of the two homologous chromosomes. (b) and (c) Model showing the development of the synaptinemal complex. Single lateral elements first appear between sister chromatids during early prophase I (b). The central element forms when homologs pair (c) through the joining of short fibrous projections that extend at right angles from each lateral element. The association of homologs is illustrated in a manner analogous to a zipper closing. *Source:* From *Biology of the Cell*, Second Edition by Stephen L. Wolfe, © 1981 by Wadsworth, Inc. Reprinted by permission of Wadsworth Publishing Company. (a) Micrograph courtesy of M. Westergaard and D. von Wettstein, *Ann. Rev. Genetics*, 6:74–110, 1972.

complex is about 100 μm wide at its peak. When viewed with an electron microscope, it appears to comprise three parallel bands of material (Fig. 5.8). The two outer bands, called *lateral elements,* are produced by the homologs early in prophase I. When the homologous chromosomes begin to pair, the inner band, or *central element,* is formed. Although the precise mechanism of action of the synaptinemal complex is not understood, it appears to be important in holding the paired homologs in close contact with each other, thus keeping them exactly aligned during crossing over.

While prolonged synapsis is needed for the nonsister chromatid exchange to occur during meiosis, it does not appear to be essential for all crossing over. Genes at linked loci can also be exchanged between homologs in somatic cells, in which extended pairing does not usually take place. *Somatic crossing over* is a rare event, however, occurring only once in every 100 or so mitotic cycles.

Gametogenesis

Spermatogenesis. The development and maturation of the male gamete, or sperm, in animals is known as **spermatogenesis.** The process of spermatogenesis has been studied most extensively in the testes of mammals. The testes are glands that have two complementary functions: (1) the production of sperm, and (2) the production of *androgens,* which are the male sex hormones. Sperm are produced in the germinal epithelial lining of the *seminiferous tubules* in the testes. The germinal epithelium layer is formed from sperm-producing cells that are arranged in an orderly sequence, beginning with the cells (called *spermatogonia*), which are located next to the basement membrane, and ending with the sperm, which are located in the lumen of the tubule (Fig. 5.9). The spermatogonia divide mitotically to maintain a ready supply of progenitor cells. Once they start to divide by meiosis, they are called *spermatocytes.* A spermatocyte, which is a meiocyte in the male, goes through the first and second meiotic divisions to produce four haploid cells known as *spermatids.* Each spermatid then undergoes a maturation process, during which its form and structure change dramatically. The spermatid loses nearly all the cytoplasm in its nuclear region, except for the Golgi apparatus (which forms a cap, or *acrosome,* around the nucleus), centrioles, and a few mitochondria. The spermatid also acquires a tail, which enables the mature sperm to be mobile. A mature sperm is shown in Fig. 5.10. Although the development of a meiocyte into mature sperm cells may take weeks or months to complete (it takes about 74 days in human males), the process is continuous in a sexually mature male, so a ready supply of sperm cells is always available.

Before the sperm cells are released from the body, various glands add secretions, which provide a source of energy and an alkaline environment needed by the sperm for motility. These secretions combined with the sperm constitute the *semen.* An average human ejaculation releases about 3 milliliters of this fluid; each milliliter contains around 10^8, or 100 million, sperm cells. Thus, an average human male may produce as many as 10^{12}, or a million million, gametes during his lifespan.

EXAMPLE 5.4. Cattle have 60 chromosomes in each somatic cell. What is the chance that one sperm cell of a certain bull contains only chromosomes derived from his mother?

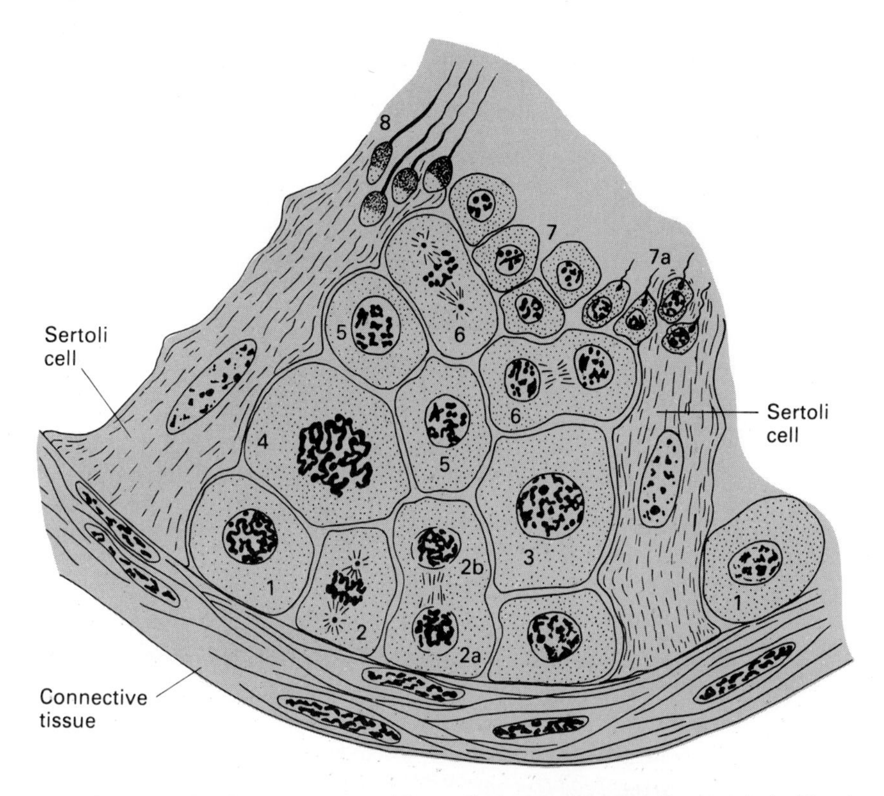

Figure 5.9. Diagram showing a segment of the wall of an active seminiferous tubule. The development of the sperm-producing cells proceeds in an orderly sequence from the basement membrane to the lumen of the tubule. A spermatogonium (1) undergoes mitosis (2) to form more spermatogonia (2a and 2b). A spermatogonium may develop into a primary spermatocyte (3), which will enter meiosis I (4) to become two secondary spermatocytes (5). These cells complete meiosis II (6) to yield four spermatids (7). The spermatids differentiate in the tip of a Sertoli cell (7a) to become mature sperm cells (8). *Source:* From B. M. Patten, *Human Embryology,* 3 ed., 1968, McGraw-Hill Book Co., Inc., New York.

Solution. One member of each of the 30 chromosome pairs in the bull is derived from his mother. Since the members of each pair disjoin at anaphase I, the maternally derived chromosome of any one particular pair has a one-half chance of being represented in the sperm cell of the bull. Therefore, the chance that all 30 chromosomes in the sperm cell are maternally derived is $(1/2)^{30}$.

Oogenesis. The formation of the female gamete, or egg, in animals is known as **oogenesis.** As with sperm development in the testes, the initiation of egg development has been studied most extensively in the ovaries of mammals. The ovaries, like the male gonads, have a dual function: (1) the production of eggs and (2) the production of hormones, including *estrogens* and *progesterone,* which are the female sex hormones.

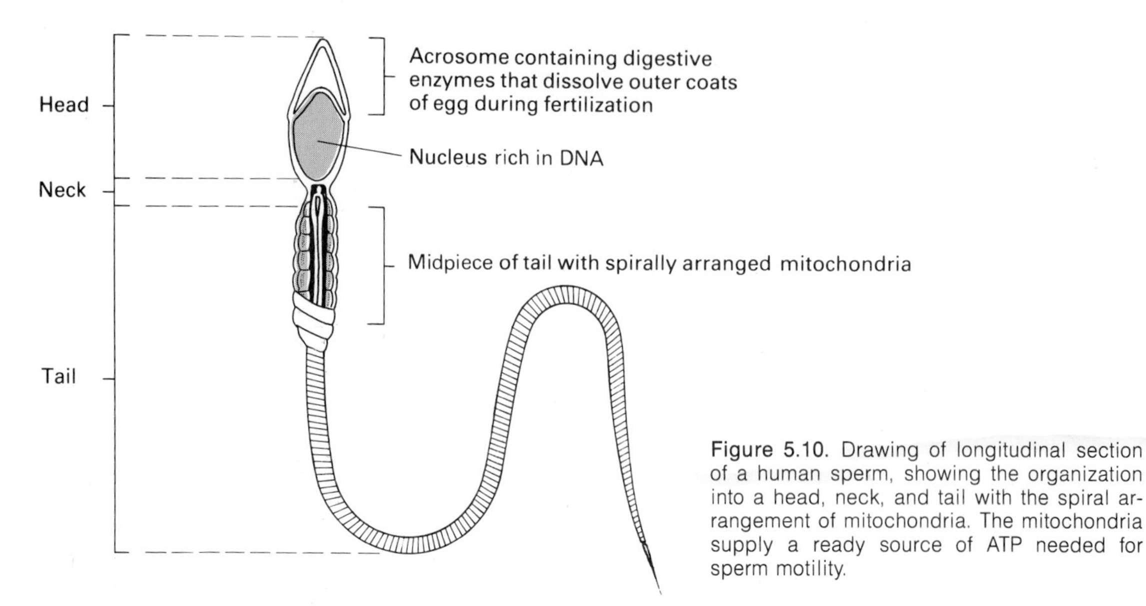

Figure 5.10. Drawing of longitudinal section of a human sperm, showing the organization into a head, neck, and tail with the spiral arrangement of mitochondria. The mitochondria supply a ready source of ATP needed for sperm motility.

Eggs are produced in mammals in separate patches of cells in the ovaries called *primary follicles* (Fig. 5.11). These follicles develop out of the germinal epithelium of the ovaries. A follicle consists of a large gamete-producing cell surrounded by a layer of follicle cells, which help to protect and nourish the maturing egg. The meiocytes within the follicles originate from cells called *oogonia* during embryonic development. Like the sequence in males, the oogonia enter meiosis to become *oocytes.* Unlike the process in males, however, the oocytes do not proceed directly through meiosis; they remain suspended in prophase I until the female has attained the reproductive age. After this age has been reached, one or more of the primary follicles mature during a short period of each reproductive (or estrous) cycle. During this period, the oocyte completes the first meiotic division. In the average lifespan of a human female, approximately 400 follicles, or only about 0.1 percent of the 400,000 or so estimated follicles present in the two ovaries at birth, mature and continue past the stage of prophase I.

As the follicle develops, the oocyte enlarges, the number of follicle cells increases, and a fluid-filled cavity forms around the oocyte (see Fig. 5.11). In most mammals, the primary oocyte completes the first meiotic division to give rise to a haploid secondary oocyte, which retains most of the cytoplasm, and a small **polar body,** which is also haploid but with very little cytoplasm. The polar body usually degenerates, but occasionally it may persist long enough to undergo a second meiotic division along with the remaining oocyte. The fully developed follicle bulges from the surface of the ovary and eventually ruptures, discharging the oocyte into the oviduct, a tube leading to the uterus. The release of the oocyte with its corona of follicle cells from the ovary into the oviduct is the process of *ovulation.*

Once the oocyte enters the oviduct, cilia and peristaltic movements of the oviduct propel it in the direction of the uterus. If a sperm cell penetrates the oocyte

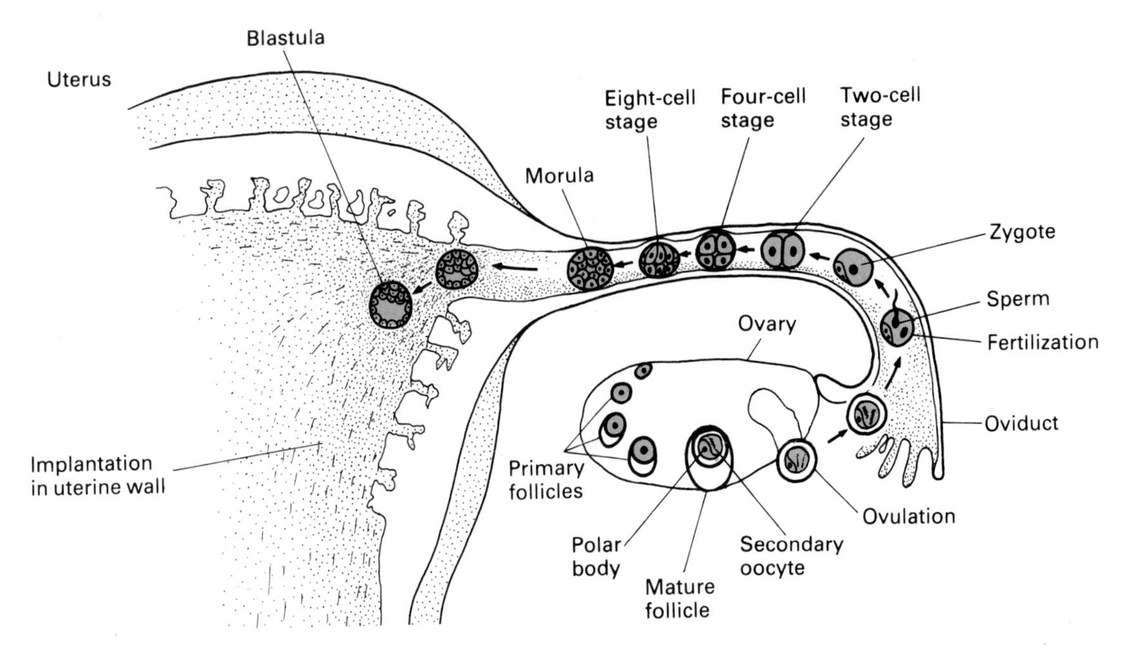

Figure 5.11. Maturation of the mammalian egg. Diagram of the ovary and associated oviduct showing the sequence of events involved in the origin, growth, and rupture of the ovarian follicle and the subsequent fertilization of the developing egg during its passage toward the uterus.

during its passage through the oviduct, the oocyte then completes the second meiotic division. As before, one cell retains most of the cytoplasm, becoming an *ootid,* while the other receives very little cytoplasm and becomes a nonfunctional polar body. The ootid then matures into the ovum, or egg. If sperm are not present within the oviduct, the oocyte degenerates without undergoing meiosis II.

Figure 5.12 summarizes the events involved in oogenesis and compares them with the process of spermatogenesis. The bulk of the differentiation that occurs in the developing egg is during meiosis I, while the greatest morphological changes that occur in sperm formation follow the completion of both meiotic divisions. Because the egg contains a greater accumulation of nutrient material than does the sperm, which has lost much of its excess cytoplasm, the egg is much larger than the sperm. The difference in size is directly related to the functional roles of the two gametes in fertilization and in the subsequent development of the embryo. The nutrient material accumulated in the egg is needed by the embryo for growth and differentiation, until it can produce its own nutrients or obtain them from the environment. The loss of excess cytoplasm by the sperm facilitates its motility.

Gametogenesis in Plants. Unlike gametogenesis in animals, meiosis and gamete development in plants are not directly linked. Meiosis in plants leads to the formation of haploid spores rather than sex cells. The spores then divide mitotically to form the gametophyte structure. The gametophyte is responsible for gamete formation. Figure 5.13 illustrates the process of spore and gamete formation in a flowering plant (or angiosperm).

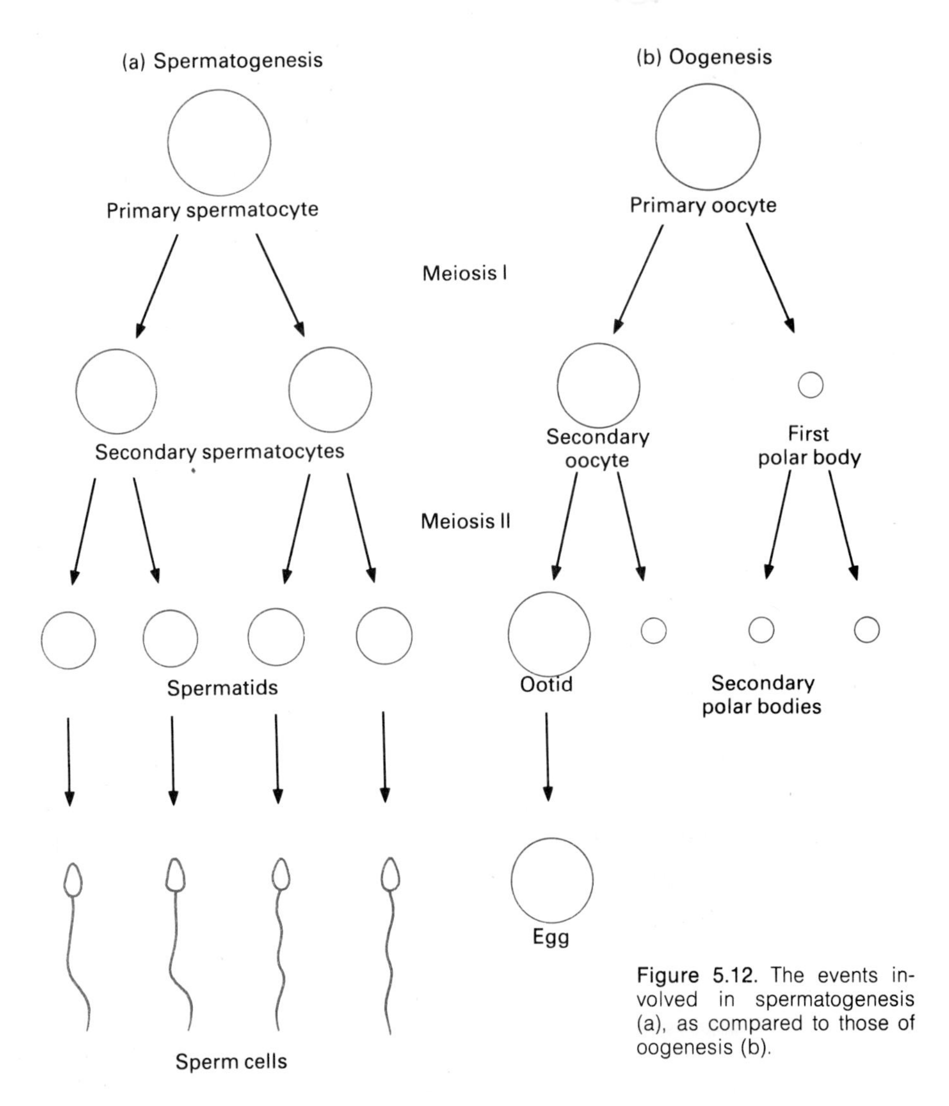

Figure 5.12. The events involved in spermatogenesis (a), as compared to those of oogenesis (b).

In the anther, which is the male organ of the flower, meiocytes (microspore mother cells) undergo meiosis to form haploid *microspores.* The microspores mature into *pollen* grains, which are the male gametophytes. During maturation, each microspore divides mitotically to give rise to a pollen grain with two haploid nuclei: a tube nucleus and a generative nucleus. When the pollen grain germinates to form a pollen tube, the generative nucleus divides mitotically once again to produce two male gametes, one of which will fertilize the egg.

Meanwhile, within the ovary of the flower, each meiocyte (megaspore mother cell) undergoes meiosis to form four haploid *megaspores.* Only one of these megaspores survives; it enlarges to become the *embryo sac,* which is the female gametophyte. The nucleus of the embryo sac undergoes three successive mitotic

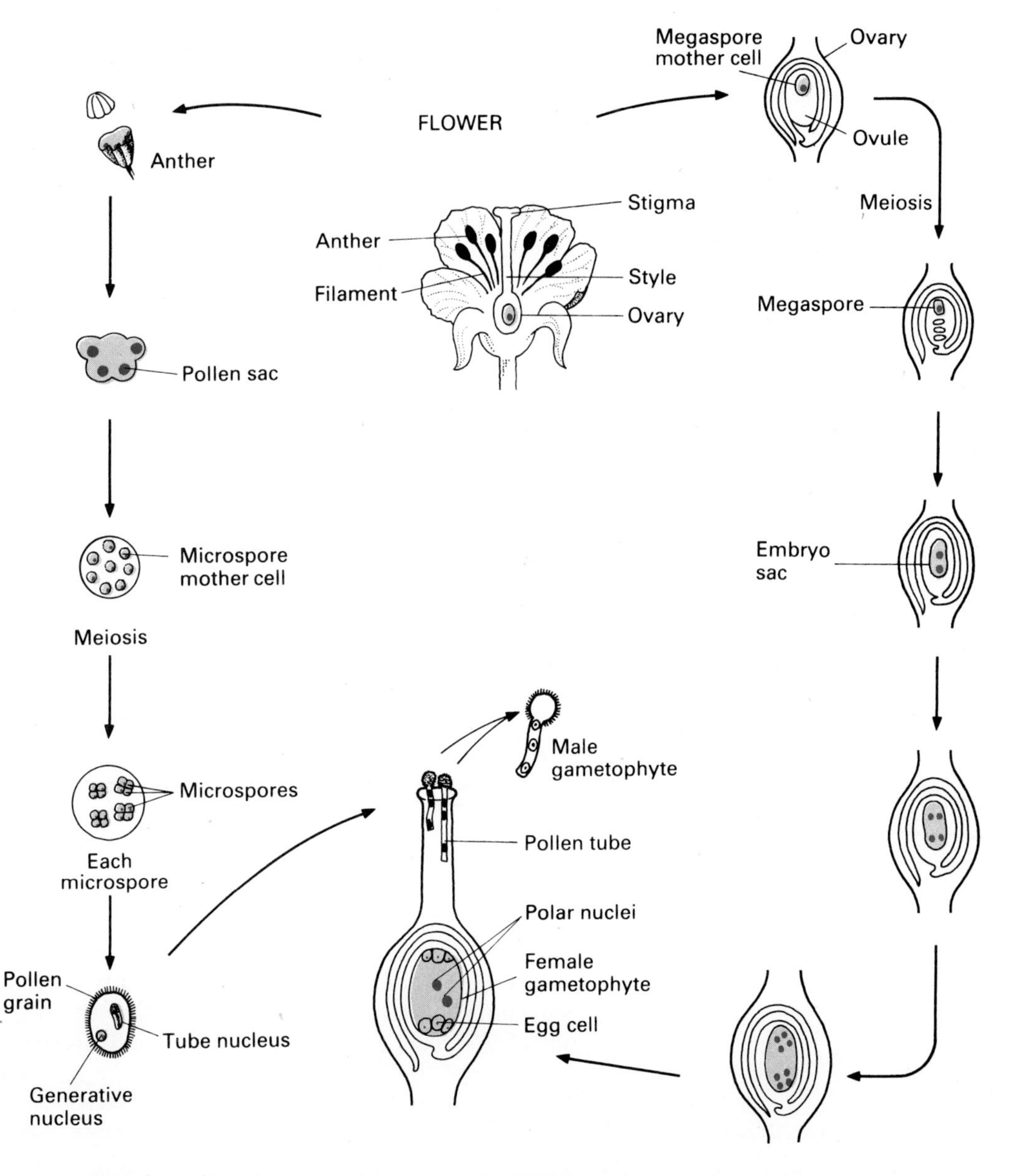

Figure 5.13. Spore and gamete formation in a flowering plant. Meiosis in the anther, the male organ of the flower, leads to the development of four haploid microspores from each microspore mother cell. During maturation, the nucleus of a microspore divides mitotically to produce a pollen grain with two haploid nuclei: a tube nucleus and a generative nucleus. When the pollen grain germinates to form a pollen tube, the generative nucleus divides mitotically once again to form two male gametes. Meiosis in the ovary of the flower forms four haploid megaspores from each megaspore mother cell. One megaspore survives and enlarges to become the embryo sac. Three successive mitotic divisions of the embryo sac produce eight nuclei. One nucleus forms the egg, and two others, known as polar nuclei, become the nuclei of the large central cell, and later combine to form the single diploid fusion nucleus.

cycles to produce eight nuclei. After the eight nuclei have migrated to various points within the embryo sac, one nucleus is enclosed in the egg cell of the embryo sac to become the *egg nucleus*. Two others, called polar nuclei, become the nuclei of the large central cell. They later combine to form a single diploid *fusion nucleus*.

Fertilization

Mammals. In animals, fertilization is the culmination of successful mating, and involves the penetration of a developing egg by a sperm and the subsequent union of their respective nuclei. We will discuss fertilization in humans here, which is fairly typical of the fertilization process in most mammals.

The human egg remains viable for only about a day after ovulation. Consequently, fertilization must take place within the oviduct of the female. In the oviduct, the oocyte is surrounded by a series of coats, or accessory layers, in addition to its plasma membrane. Outermost is a layer of follicle cells, called the *corona radiata*. This layer covers a thick envelope known as the *zona pellucida*, which is made up of a noncellular, protein-rich substance. Once contact is made with the developing egg, the sperm must penetrate both coats in order to achieve fertilization. The acrosome of the sperm releases lytic enzymes to aid in penetration. These enzymes can digest certain structural components of the layers surrounding the egg.

As soon as a sperm enters an oocyte, the developing egg becomes impenetrable to other sperm. When the head of the sperm has penetrated the two outer layers, it separates from the tail and becomes a *pronucleus*. The presence of the pronucleus causes the oocyte to undergo meiosis II and form its own pronucleus within the ootid. The two pronuclei fuse to form a single diploid structure, thus completing fertilization.

Flowering Plants. In flowering plants, fertilization is the result of pollination, an event that is usually accomplished in nature with the help of wind currents or insects. We will use the familiar corn plant to illustrate the process of fertilization in plants and its relationship to the development of seeds (Fig. 5.14). In corn, the female part of the flower extends from an embryo sac in the form of a long slender silk. When a pollen grain from the tassel alights on a silk of an ear, it germinates to produce a long pollen tube. Two male gametes (the sperm nuclei) are then carried within the tube to the embryo sac, where entrance is made. One sperm nucleus fuses with the egg nucleus to produce the zygote. The second sperm nucleus unites with the $2n$ fusion nucleus to form a triploid ($3n$) endosperm nucleus, containing a diploid set of maternal chromosomes and a haploid set of paternal chromosomes.

The development of the seed follows (Fig. 5.14b), as the $2n$ zygote divides mitotically to form the embryo of the kernel. The triploid nucleus also undergoes several mitotic divisions to form a multicellular tissue for food storage. This tissue, known as the *endosperm*, nourishes the embryo in the seed and the young seedling. The outermost cells of the endosperm form a specialized layer called the *aleurone* (important to us because of its contribution to kernel color). Surrounding the

(a) Fertilization

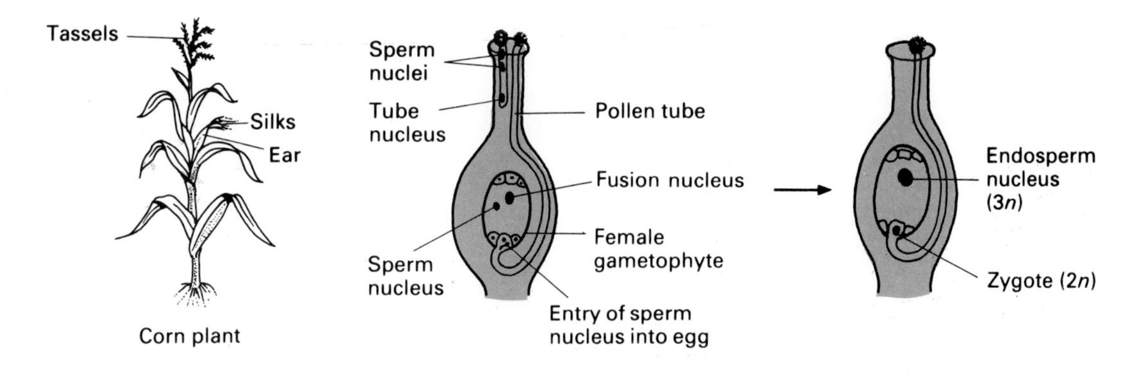

(b) Seed development

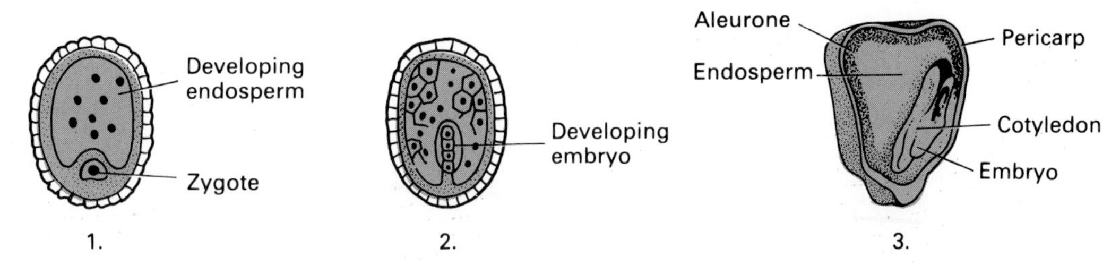

Figure 5.14. Fertilization (a) and seed development (b) in corn. Fertilization occurs when a pollen grain germinates on a silk to produce a pollen tube, through which sperm nuclei enter the embryo sac. One of the sperm nuclei fuses with the egg nucleus to form the diploid zygote, while the other unites with the $2n$ fusion nucleus to form the triploid endosperm nucleus. During development of the seed (b), the endosperm nucleus divides mitotically to produce many endosperm nuclei (1), which are then enclosed by cross walls (2). The zygote divides mitotically to form the embryo (2 and 3). The mature seed (3) contains the embryo with one cotyledon, surrounded by triploid cells of the endosperm tissue, the aleurone, and the seed coat, or pericarp.

endosperm is a seed coat, or *pericarp,* which is derived solely from $2n$ maternal cells.

One interesting aspect of kernel development is the $3n$ condition of the endosperm. In genetic terms, this means that a kernel with a heterozygous *Aa* embryo can have an endosperm genotype of either *AAa,* if it is derived from an *AA* fusion nucleus and an *a* male gamete, or *Aaa,* if its origins are reversed. The triploid nature of the endosperm normally does not affect the outcome of crosses involving endosperm characteristics, as long as one dose of the dominant gene is sufficient to mask completely the expression of a double dose of the recessive allele. One example where there is an effect on cross results, however, is in the inheritance of alleles for flinty (*F*) versus floury (*f*) endosperm. Flinty endosperm is the manifestation of an *FFF* or *FFf* endosperm genotype, whereas an *Fff* or *fff* genotype produces the floury phenotype. When pollen is taken from *FF* plants and is used to fertilize plants with an *ff* genotype, only floury (*Fff*) kernels develop. Conversely, only flinty (*FFf*) kernels are formed when the pollen and egg donors are reversed.

EXAMPLE 5.5. Consider the monohybrid cross *Ff* × *Ff* in corn, involving the alleles for flinty (*F*) and floury (*f*) endosperm. What ratio of endosperm phenotypes is expected among the kernels from this cross?

Solution. The cross can be diagrammed as follows. In the boxed area, the genotypes of the sperm and egg nuclei and the resulting embryo are written on the top in each case, and the genotypes of the second sperm and fusion nuclei and the resulting endosperm are written below.

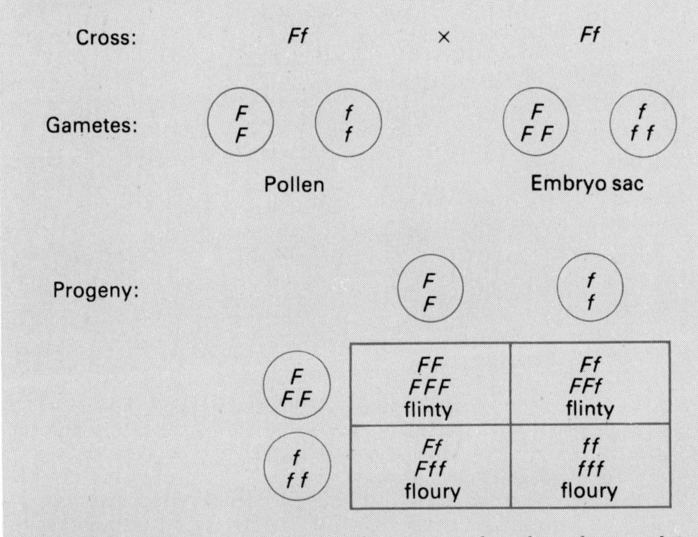

We get a 1:1 phenotypic ratio in this case, rather than the usual 3:1 ratio expected from a monohybrid cross. The 1:1 ratio results from the lack of total dominance by the *F* allele and the triploid (3*n*) nature of the endosperm.

TO SUM UP

1. The life cycles of sexually reproducing eukaryotes are characterized by alternating haploid and diploid stages. In higher animals, the haploid stage is very transitory, consisting of only the egg and sperm cells. In higher plants, by contrast, both the haploid (gametophyte) and the dominant diploid (sporophyte) generations are multicellular.
2. During gamete formation, the germ cells, or meiocytes, of an individual undergo meiosis. Meiosis consists of two successive nuclear divisions, each of which is composed of the following stages: prophase, metaphase, anaphase, and telophase. The first

meiotic division results in the separation of homologous chromosomes, while the second meiotic division is responsible for the separation of sister chromatids.

3. The separation of members of each homologous pair of chromosomes during meiosis results in the segregation of alleles. The independent alignment of each pair of homologs on the spindle followed by disjunction accounts for the independent assortment of nonallelic genes present on different chromosome pairs. Homologous chromosomes can also exchange segments during pairing through a process known as crossing over. The exchange of chromosome parts during crossing over permits linked genes to reassort. Linked genes are carried on the same chromosome pair but at separate loci.
4. In higher animals, there are two distinct types of gamete formation: spermatogenesis in males, and oogenesis in females. Spermatogenesis occurs continuously throughout the reproductive lifespan of the male, with all four meiotic products being functional, motile sperm cells. Oogenesis, on the other hand, begins during early embryonic development and proceeds through meiosis I around the time of ovulation. The second phase of meiotic division is only completed after fertilization has taken place. A single functional egg and at most three nonfunctional polar bodies are produced from each meiocyte in oogenesis.
5. Meiosis in flowering plants produces haploid microspores in the anther, which is the male organ, and haploid megaspores in the ovary of the flower. Each microspore matures through mitotic division to form a pollen grain (the male gametophyte), while one megaspore eventually develops into the embryo sac (the female gametophyte). During pollination, one male gamete from the pollen grain fuses with the egg nucleus, forming a zygote. The second male gamete combines with the two polar nuclei (or a single $2n$ fusion nucleus) within the embryo sac, giving rise to the endosperm (a $3n$, or triploid, tissue for food storage).

Sexual Reproduction in Haploid Eukaryotes

Certain algae and fungi spend most of their life cycle in the haploid phase. The diploid phase, when it is formed, is ordinarily of short duration. As in other sexually reproducing organisms, the diploid phase of these haploid eukaryotes starts with the fusion of haploid cells and ends in meiosis. But unlike higher plants and animals, the products formed by meiosis complete the sexual cycle by forming haploid adults, either directly, as in certain unicellular algae, or through repeated mitoses.

One example of sexual reproduction in a haploid eukaryote is shown by the life cycle of the pink bread mold, *Neurospora crassa.* This is a fungus that has been used extensively in genetic research. Figure 5.15 diagrams the life cycle of this organism, illustrating both the asexual and sexual modes of reproduction. The asexual (vegetative) cycle is initiated when a single reproductive cell, or *conidium,* germinates into a long threadlike structure called a *hypha.* The hypha branches repeatedly as it elongates to form a multinucleate mass of cytoplasm enclosed within a branched system of tubes, collectively referred to as the *mycelium.* The haploid mycelium releases masses of conidia, which, upon germination, repeat the vegetative reproductive cycle.

Sexual reproduction involves the union of cells of opposite mating types. There are two mating types in *Neurospora,* determined genetically by a single pair of

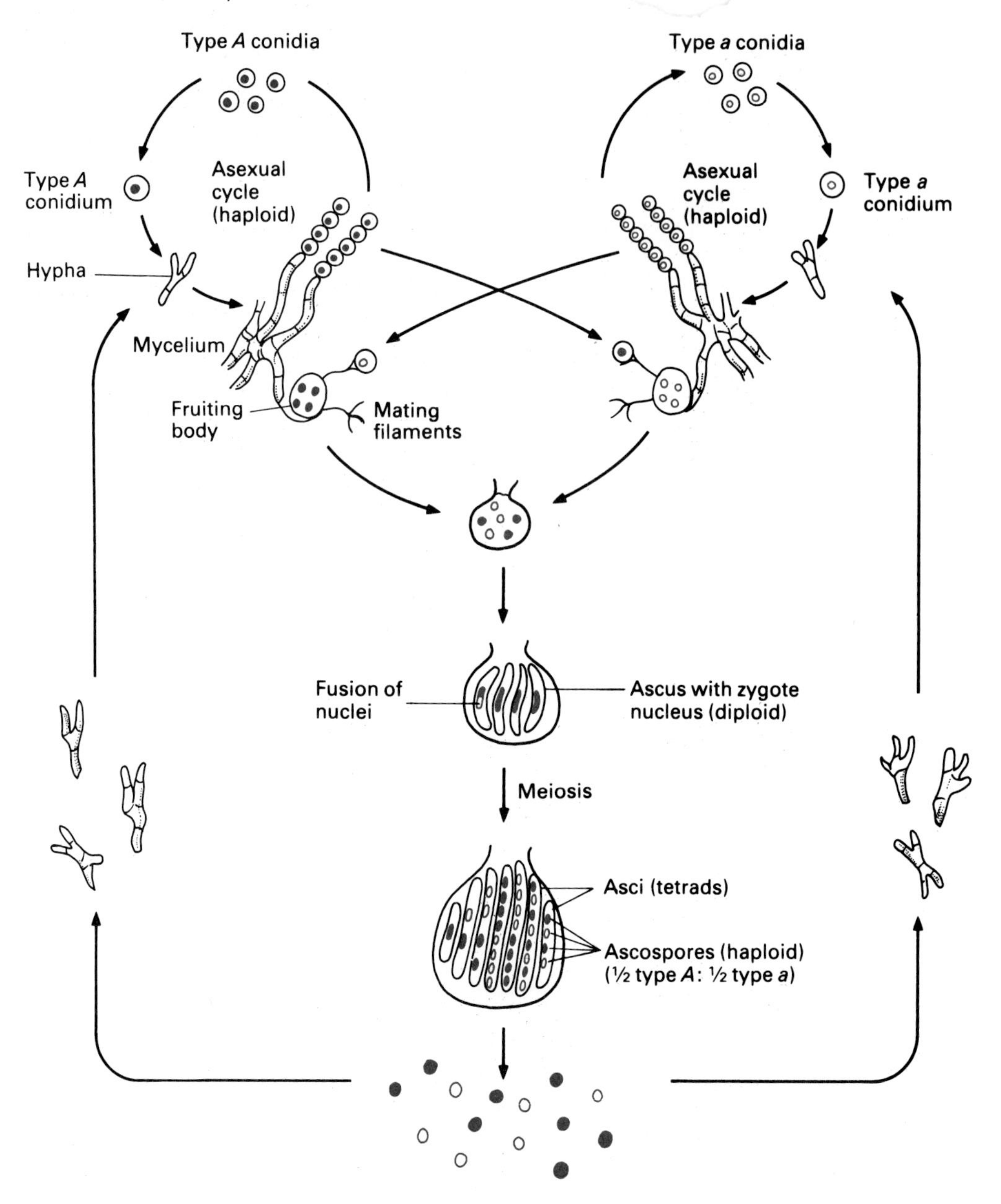

Figure 5.15. Asexual and sexual methods of reproduction of *Neurospora*. In the asexual cycle, individual spores (conidia) germinate into hyphal systems that release masses of conidia, each of which repeats the germination process. The more complicated sexual cycle of reproduction begins when a conidium of one mating type (*A* or *a*) encounters a fruiting body of the opposite mating type (*a* or *A*, respectively). The conidium enters the fruiting body as it divides by mitosis and fertilizes several of the maternal nuclei. Each resulting zygote is enclosed within a sac-like structure termed the ascus. Meiosis and a subsequent mitotic division yield mature asci, each containing four pairs of ascospores. Geneticists refer to the four pairs of ascospores as a tetrad. Release of ascospores allows their germination into hyphae and a repeat of either the sexual or asexual cycle.

alleles, *A* and *a*. Sexual reproduction is initiated when the mycelium produces fruiting bodies, each of which contains many haploid maternal nuclei. Specialized mating filaments that recognize the opposite mating type extend from each fruiting body. If a hypha or conidium of one mating type contacts a mating filament of the opposite mating type, it moves into the fruiting body by successive mitotic divisions and fertilizes several of the maternal nuclei. Each fertilization yields a diploid zygote nucleus, enclosed within an oval sac called an **ascus.** A mature fruiting body may contain a hundred or more such asci.

The diploid phase of *Neurospora* is of short duration. The zygote within each ascus immediately undergoes meiosis, yielding four haploid products, called **ascospores.** The four products of meiosis within a single ascus are collectively referred to as a **tetrad.** In the case of *Neurospora,* meiosis is followed by a single mitotic division, producing identical copies of each of the four ascospores. The mitotic division obviously does not alter the genetic composition of the tetrad.

Tetrad Analysis

In *Neurospora,* the ascus is so narrow that it prevents the meiotic and mitotic products from slipping past each other. The eight ascospores remain in a row in a fixed position relative to one another. By inspecting Fig. 5.16, we can see that the particular linear order of the ascospores within the ascus reflects the linear order in which the chromosomes separated during the first and second meiotic divisions. We observe not only a 2 *A* : 2 *a* segregation ratio (identical to the 1 : 1 Mendelian segregation ratio that occurs in the gametes of diploid organisms), but a linear order as well, reading either *AAaa* or *aaAA*. The precise linear order of the ascospores can be detected experimentally by dissecting the ascospores individually from their ascus and germinating them into hyphae (Fig. 5.17). Each ascospore can thus be analyzed genetically on the basis of its phenotype when grown in culture. Because the ascospores are arranged in the order in which they were produced by meiosis, *Neurospora* is said to produce **ordered tetrads.**

First- and Second-Division Segregation. A tetrad order of *AAaa* (or *aaAA*) directly demonstrates that the chromosome bearing the *A* allele segregated from its homolog bearing an *a* allele at the first meiotic division (refer to Fig. 5.16 to see why this is so). For this reason, the *AAaa* and *aaAA* arrangements are called **first-division segregation (FDS)** patterns. In addition to the two FDS patterns, four other tetrad arrangements are also often observed with regard to mating type. They are *AaAa, aAaA, aAAa,* and *AaaA*. Each of these tetrad arrangements can be explained by crossing over. Figure 5.18 shows that when the different pairs of nonsister chromatids participate in a single exchange between the locus that determines mating type and the centromere, one of these four other tetrad orders will result. In each case, the occurrence of a crossover brings the *A* and *a* alleles together on the same homolog and thereby delays the segregation of these alleles until the second meiotic division. Hence, these latter four arrangements are referred to as **second-division segregation (SDS)** patterns.

Genes on Different Chromosomes. Tetrad analysis also provides information concerning the assortment of two pairs of genes. Consider the cross *AB* × *ab* in

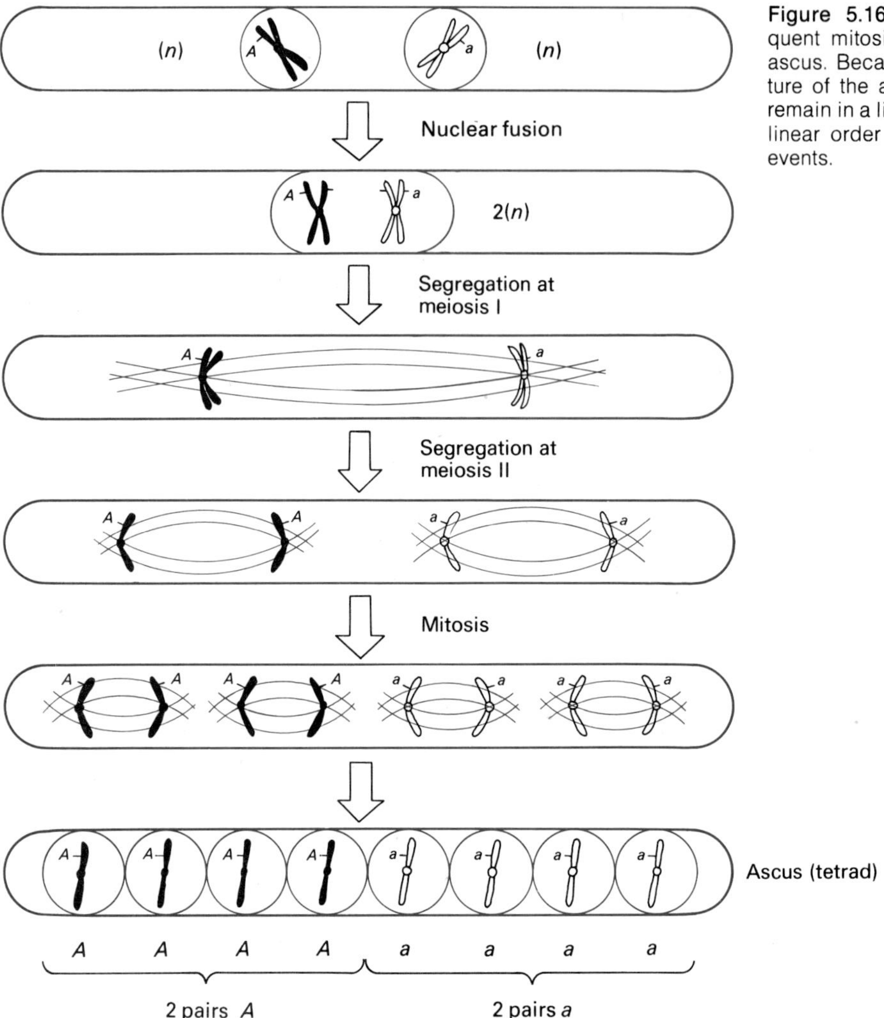

Figure 5.16. Meiosis, and subsequent mitosis, within a *Neurospora* ascus. Because of the narrow structure of the ascus, meiotic products remain in a linear order reflecting the linear order of meiotic segregation events.

Neurospora. The *A,a* alleles determine mating type, and the *B,b* alleles influence color, with *B* producing the normal pink color and *b* producing yellow. These two gene pairs are carried on different chromosomes, and must therefore assort independently from the two possible chromosome alignments shown in Fig. 5.19(a). The result is that two types of tetrads occur in *equal proportion*. One type, called a **parental ditype (PD)** tetrad, contains ascospores with only the parental genotypes (in this case, *AB* and *ab*). The other type, called a **nonparental ditype (NPD)** tetrad, contains ascospores with only nonparental gene combinations (in this case, *Ab* and *aB*). Thus, for independent assortment, the frequencies of PD and NPD tetrads are equal. It should be stressed that this result does not differ from earlier observations on independent assortment, concerning the equal occurrence of all genotypes among the meiotic products. If we were to remove the

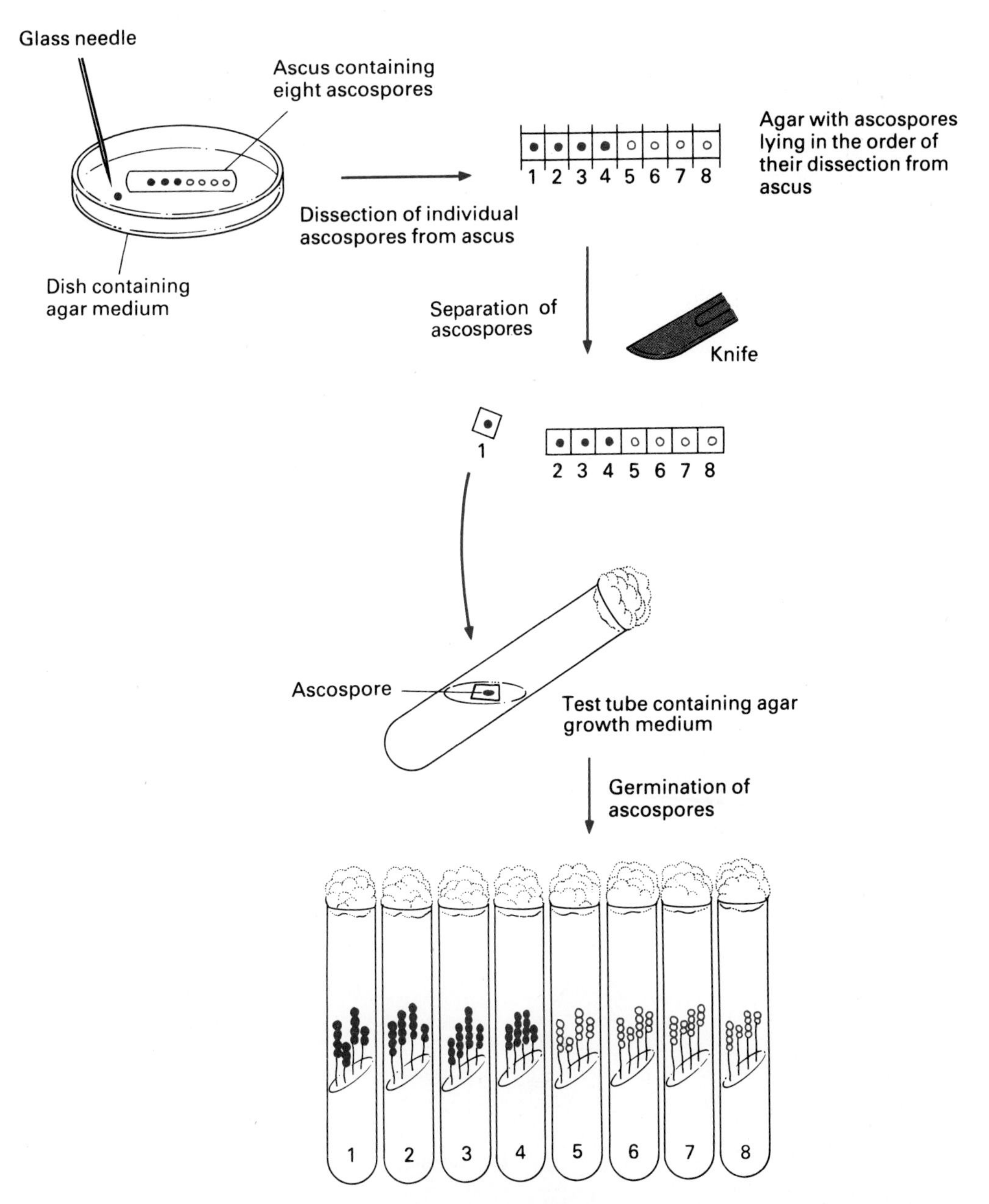

Figure 5.17. Technique used to isolate ascospores produced by a *Neurospora* cross.

ascospores from both kinds of asci and pool them, each of the four possible genotypes would make up one-fourth of the total number of ascospores present.

A third type of tetrad, called a **tetratype (T)** tetrad, can also be produced as a result of crossing over between a gene and the centromere of its chromosome

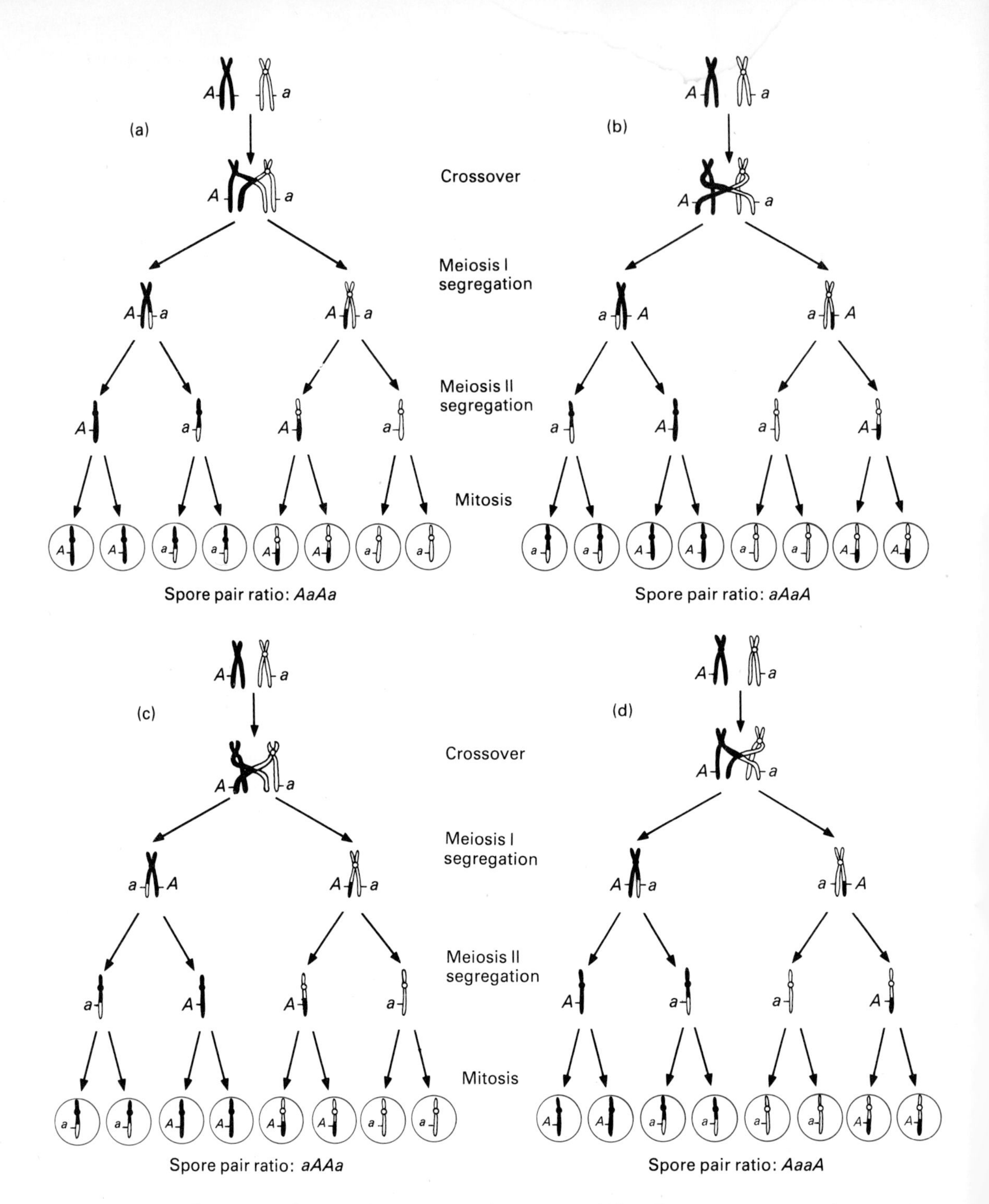

Figure 5.18. Second division segregation patterns within a *Neurospora* ascus, each caused by a single crossover between the centromere and the gene *a* locus. Since the crossover can involve any pair of nonsister chromatids, four tetrad arrangements are possible: *AaAa*, *aAaA*, *aAAa*, and *AaaA*.

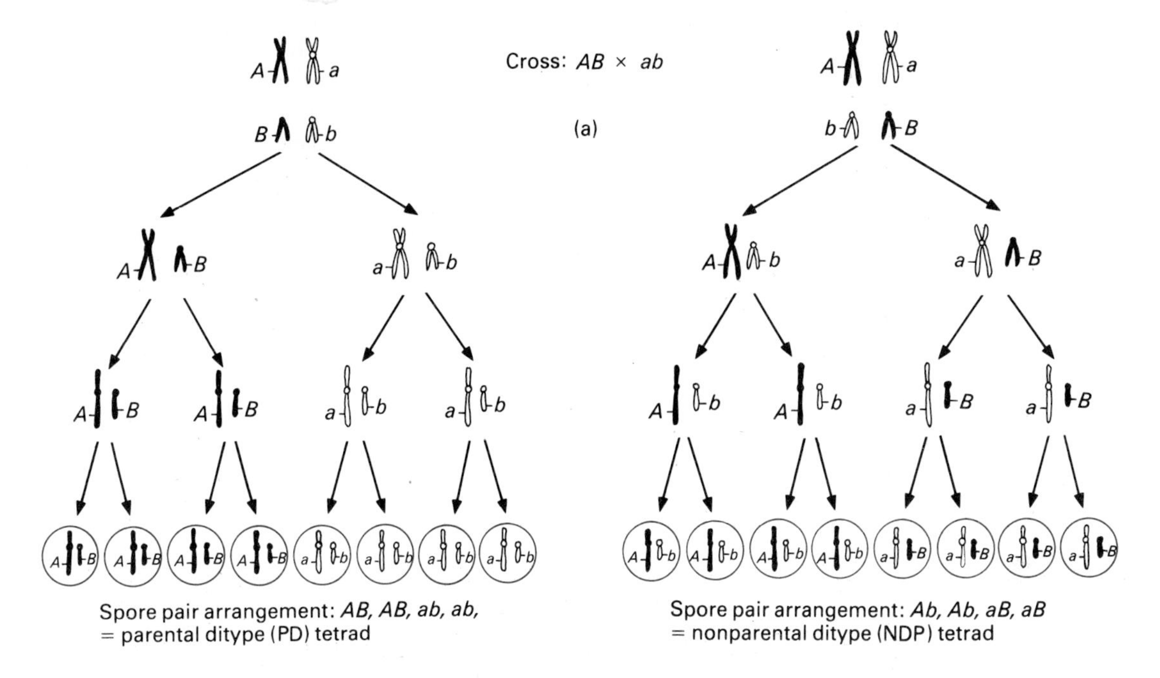

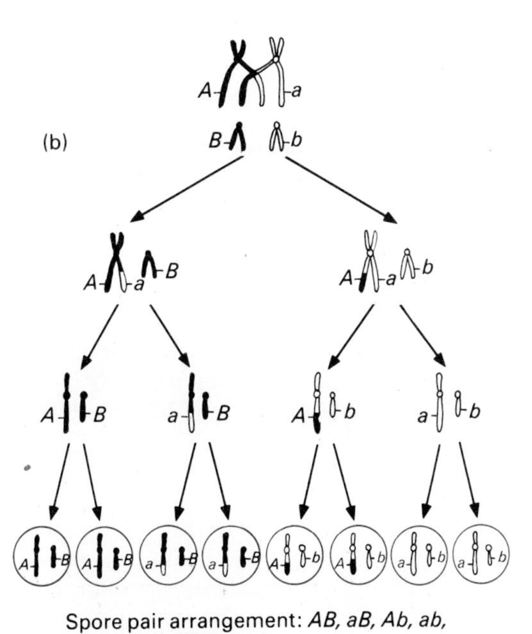

Figure 5.19. Origin of the three types of tetrads—PD, NPD, and T—for genes located on separate pairs of chromosomes. The cross is *AB* × *ab*. The PD and NPD tetrads occur with equal frequency. The T tetrads arise from crossing over between a gene and its centromere. Only one of several possible T arrangements is shown here.

(Fig. 5.19b). In a tetratype ascus, each pair of ascospores has a different genotype. Many different tetratype arrangements are possible in *Neurospora,* because exchanges between genes and centromeres can involve either or both pairs of chromosomes, and any two nonsister chromatids can participate in an exchange event. In Fig. 5.19(b), only one crossover has occurred, and it has taken place between the locus that determines mating type and the centromere. If we analyze the resulting tetratype tetrad one gene locus at a time, we see that the *A,a* alleles exhibit an SDS pattern, while the *B,b* alleles shown an FDS pattern, as expected. Of course, crossing over may involve only the pair of chromosomes on which the *B,b* alleles are located, or it may involve both pairs of chromosomes. If both pairs of chromosomes are involved in crossing over, then both pairs of alleles would exhibit an SDS pattern.

EXAMPLE 5.6. The cross *AB* × *ab* in *Neurospora* gives the following kinds and numbers of tetrads (spore pairs in each class of tetrads are listed vertically):

	(1)	(2)	(3)	(4)	
Spore pair 1:	*AB*	*Ab*	*AB*	*AB*	
Spore pair 2:	*AB*	*Ab*	*aB*	*Ab*	
Spore pair 3:	*ab*	*aB*	*Ab*	*aB*	
Spore pair 4:	*ab*	*aB*	*ab*	*ab*	
	135	135	90	40	= 400 total tetrads

(a) Identify each of the tetrad classes as PD, NPD, or T. **(b)** Show that the genes have assorted independently. **(c)** Calculate the percent SDS asci for each gene pair.

Solution. **(a)** All the spores in tetrad class 1 have the parental genotypes, making these tetrads PD. The class 2 tetrads consist of only spores with nonparental genotypes, so these are NPD tetrads. Classes 3 and 4 both consist of a mixture of spores with parental and nonparental genotypes, making these T classes. **(b)** Independent assortment of the nonallelic gene pairs is indicated by the equal frequency of PD (class 1) and NPD (class 2) tetrads. **(c)** Tetrad class 3 represents meioses in which crossing over between gene pair *A,a* and its centromere occurred, as illustrated by the SDS pattern for this pair of alleles. The percent SDS asci for the *A,a* allele pair is therefore $(90/400)100 = 22.4\%$. Tetrad class 4 represents crossing over between gene pair *B,b* and the centromere of its chromosome. The percent SDS asci for the *B,b* pair of alleles is $(40/400)100 = 10.0\%$.

TO SUM UP

1. Certain algae and fungi are haploid for most of their life cycle. Some, including *Neurospora,* have a brief diploid phase that starts by the fusion of haploid cells and ends in meiosis with the reestablishment of the haploid phase.
2. Haploid eukaryotes, such as *Neurospora,* have proved ideal for genetic studies, since the products of a single meiosis are kept together within a single structure, distinct from the products of other meiotic acts. Each meiotic product can be separated from the others and analyzed individually as to genotype. Since the meiotic products are haploid, there is no complication of dominance.
3. In *Neurospora,* the order of the four pairs of ascospores within the ascus forms an

ordered tetrad, in which the arrangement of the ascospores reflects the order of the segregation events during meiosis. Spore pair arrangements of the type *AAaa* and *aaAA* arise from first division segregation of alleles. Spore pair arrangements of the type *AaAa, aAaA, aAAa,* and *AaaA* result from second division segregation of alleles. Second-division segregation patterns are a consequence of a crossover between a gene and the centromere of its chromosome.

4. The combinations of nonallelic genes in the four pairs of ascospores within an ascus are the basis for classifying a tetrad as either parental ditype (PD), nonparental ditype (NPD), or tetratype (T). The frequencies of PD and NPD tetrads are the same for genes that assort independently.

Sex Determination

Individuals of certain species are bisexual and can produce both male and female gametes. Animals of this type are called **hermaphroditic,** whereas plants with both egg-bearing and pollen-bearing flowers on the same individual are called **monoecious.** Since all the cells of these individuals have the same genes, different sexual functions within the same organism must be the result of environmental differences during development, and not differences in the chromosome complement or genotype.

Other species have separate sexes, with zygotes developing into either of two functionally distinct types: one that produces only spermlike gametes, and the other that produces only egglike gametes. Such specialization of sexual function is characteristic of most higher animals and of **dioecious** plants, in which the egg-bearing and pollen-bearing flowers are on separate individuals. In such organisms, mechanisms have evolved to avoid the formation of intermediate types, such as intersexes. These mechanisms provide for the exact segregation of genes that determine sex, without the complications of crossing over and independent assortment. The precise mechanisms that have evolved vary with the particular group of organisms under study. Therefore, we will discuss only some of the most common and best understood forms of sex determination.

Chromosomal Basis of Sex Determination

Chromosome Dimorphism. Sex determination in species with separate sexes is often associated with chromosome dimorphism. In mammals, chromosome dimorphism occurs in the form of a pair of structurally different **sex chromosomes,** designated X and Y. In humans, for example, the X chromosome is of medium size and submetacentric, whereas the Y chromosome is quite small and acrocentric (Fig. 5.20). Female mammals have a pair of X chromosomes, and produce only X-carrying eggs. Male mammals, by contrast, have one X and one Y chromosome, so that half the sperm cells produced by a male will carry an X chromosome and half will carry a Y chromosome. Since females produce only one kind of gamete (with regard to the type of sex chromosome that it carries), whereas males produce two kinds of gametes, females are referred to as **homogametic** and males as **heterogametic.** To distinguish between the sex chromosomes and those chro-

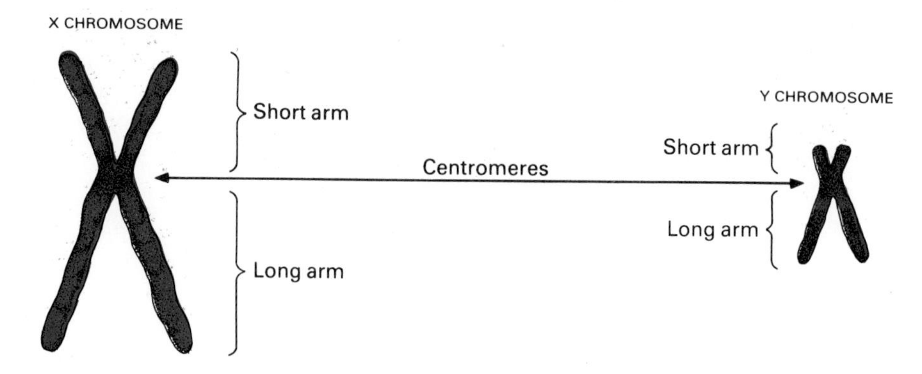

Figure 5.20. The sex chromosomes of humans: X chromosome (left) and Y chromosome (right).

mosomes not associated with the sex of the bearer, we refer to all chromosomes other than the sex chromosomes as **autosomes.** Unlike the sex chromosomes, autosomes occur in matching pairs in both sexes.

Investigators have encountered three major types of chromosome dimorphism in diploid organisms. One is the XX–XY system that we have just described, in which the female is homogametic (XX) and the male is heterogametic (XY). This system is characteristic of many different animals (including mammals) and some plants. A second type of chromosome dimorphism is the XX–XO system found in several kinds of insects, in which females are XX in genotype, with two sex chromosomes, and males are XO (pronounced X-oh), with only one sex chromosome. The third major type is the ZW system found in birds (including domestic fowl), moths, and some fishes. In species that possess this system, the male is homogametic (ZZ), with two identical sex chromosomes, and the female is heterogametic (ZW). These different types of chromosome dimorphism permit the grouping of the genetic determiners for one sex on one type of sex chromosome and for the opposite sex on either an alternative sex chromosome or on the autosomes. The recombination of genetic determiners is avoided as long as the members of a sex chromosome pair do not participate in crossing over. This restriction is made possible in mammals, for example, by the lack of structural similarity between the sex chromosomes.

The Genic Balance Mechanism. One mechanism for sex determination that has been extensively studied is that of the fruit fly, *Drosophila melanogaster.* As in mammals, the normal chromosomal makeup of the female is 2A + XX and that of the males is 2A + XY, where A refers to the haploid set of autosomes (*Drosophila* has a haploid set of three autosomes and one sex chromosome). Other arrangements of sex chromosomes are known to exist, however, in which a fly can have more than two X chromosomes and more than two haploid sets of autosomes.

In 1922, Bridges analyzed a number of these variants and obtained the results summarized in Table 5.1. Regardless of the absolute number of X chromosomes

(N_X) or haploid sets of autosomes (N_A), the N_X/N_A ratio indicates that a normal sex phenotype develops whenever N_X/N_A = 1.0 for females and 0.5 for males. Departures from these normal states of genic balance give rise to sexually abnormal individuals, such as metafemales when $N_X/N_A > 1.0$ and metamales when $N_X/N_A < 0.5$. These abnormal flies tend to show an exaggeration of the secondary sex characteristics and are invariably weak and sterile. When $0.5 < N_X/N_A < 1.0$, an intersex develops, displaying secondary sex characteristics that are between those of normal males and normal females.

In this case, the critical factor in sex determination seems to be the ratio of X chromosomes to haploid sets of autosomes, suggesting that the female determiners are carried on the X chromosome and the male determiners are carried on the autosomes. This theory has been further substantiated by the observation that the Y chromosome in *Drosophila* does not play a role in sex determination, although it is needed for sperm motility. Therefore, flies with a chromosome makeup of 2A + XO, where O indicates the absence of a second sex chromosome, are normal males in appearance, but they are sterile because they produce immotile sperm.

The Y Chromosome in Sex Determination. In contrast to *Drosophila*, in which the Y chromosome appears to lack sex determining genes, the Y chromosome in mammals plays an essential role in sex determination. In humans, for example, individuals with at least one Y chromosome are phenotypically male, regardless of the number of X chromosomes they possess. This result indicates that the Y chromosome must carry essential male-determining genes. The evidence indicates that the Y chromosome of humans, as in other mammals, determines sex by triggering the formation of testes during embryological development. Early in development, both XY and XX embryos form undifferentiated gonads, called *ovo-testes*, that have the potential of developing into either testes or ovaries. These undifferentiated gonads become testes only when the Y chromosome (or its male-determining-gene product) is present; they otherwise develop into ovaries.

The Y chromosome has also been shown to be important in sex determination in certain plants. One such case involves the wild campion, *Lychnis dioica* (formerly

Table 5.1. The relationship between chromosome constitution and sex phenotype in *Drosophila*.

NUMBER OF X CHROMOSOMES (N_X)	NUMBER OF SETS OF AUTOSOMES (N_A)	N_X/N_A RATIO	SEX DESIGNATION
3	2	1.5	Metafemale (sterile)
4	4	1.0	Normal female (4*n*)
3	3	1.0	Normal female (3*n*)
2	2	1.0	Normal female (2*n*)
2	3	0.67	Intersex (sterile)
1	2	0.5	Normal male (2*n*)
1	3	0.33	Metamale (sterile)

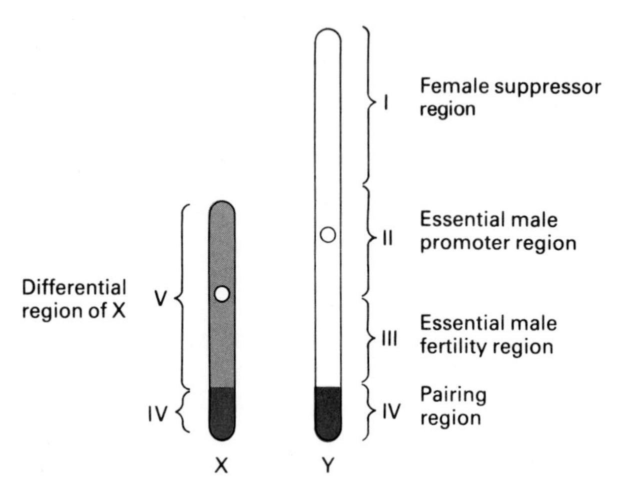

Figure 5.21. Regions of the X and Y chromosomes of *Lychnis* and the effects of each region on sex determination. Region I suppresses the development of female flower parts. Region II promotes the development of male flower parts. Region III is required for male fertility. Region IV is common to both the X and Y chromosomes and is needed for proper meiotic pairing. *Source:* After M. Westergaard, 1948. Courtesy of Hereditas.

of the genus *Melandrium*). The wild campion is a dioecious flowering plant that belongs to the pink family, a variety of garden flower. The egg-bearing (pistillate) plants of this species possess two X chromosomes, while the pollen-bearing (staminate) plants have an X and a larger Y chromosome. The Y chromosome in this plant plays three major functional roles in sex determination: It promotes the development of male flower parts, it suppresses the development of female flower parts, and it maintains male fertility. These functions were demonstrated by studies of plants from which portions of the X and Y chromosomes were deleted through mutation. Figure 5.21 summarizes the results of these studies diagrammatically. We can observe from the diagram that the Y chromosome contains four distinct regions (I–IV):

Region I suppresses the development of female flower parts. The deletion of this segment gives rise to monoecious plants.

Region II is essential for the development of male flower parts. Only pistillate plants develop when this segment is deleted.

Region III is required for male fertility. The loss of this segment produces sterile staminate plants.

Region IV is a region of homology needed for meiotic pairing and proper separation at anaphase. This region is common to both the X and Y chromosomes.

Genes Affecting Sex Determination

In some organisms, such as corn, a small number of genes can exert a major influence on the sex phenotype of an individual. Corn is normally a monoecious

(a) (b)

Figure 5.22. Separate sexes in corn. (a) Corn homozygous for the *ba* gene for barren stalk. They are unable to produce ears but have normal tassels. (b) Corn homozygous for the *ts* gene for tassel seed. They have ears at the top of the stem in place of tassels. *Source: The Ten Chromosomes of Maize,* DEKALB Ag Research, Inc., Dekalb, Ill.

plant, in which the tassels are pollen-bearing (staminate) flowers and the ears are egg-bearing (pistillate) flowers. But changes in genotype at two independent gene pairs, *Ba,ba* and *Ts,ts,* can produce a dioecious plant in which the egg-bearing and pollen-bearing flowers are on separate individuals (Fig. 5.22).

Genetic studies indicate that individuals with at least one of each of the dominant genes (*Ba/- Ts/-*) are bisexual. Individuals that are homozygous for the *ba* gene for barren stalk (*ba/ba Ts/-*) are staminate plants that are unable to produce ears but have normal tassels. Those that are homozygous for the *ts* gene for tassel seed (*Ba/- ts/ts*) are pistillate plants, because the recessive *ts* gene transforms the tassel into a pistillate, or egg-bearing structure. Plants that are homozygous for both recessive nonalleles (*ba/ba ts/ts*) are unable to produce lateral ears because of the lack of a *Ba* gene, but they are pistillate because a single ear develops at the top of the stalk in place of the tassel.

It is possible for such plants to evolve into a dioecious strain of corn merely by establishing the genotypes *ba/ba Ts/ts* (staminate) and *ba/ba ts/ts* (pistillate) within a population. The cross *ba/ba Ts/ts* × *ba/ba ts/ts* must then continue to produce staminate and pistillate plants in a 1:1 ratio.

EXAMPLE 5.7. Show that a mixture of *ba/ba Ts/ts* and *ba/ba ts/ts* plants constitutes a dioecious strain of corn that will continue to produce an equal number of pistillate and staminate offspring in each generation.

Solution. Plants with the *ba/ba Ts/ts* genotype are staminate, whereas those with the *ba/ba ts/ts* genotype are pistillate. Only one kind of cross is therefore possible, as shown below:

Parents: *ba/ba Ts/ts* × *ba/ba ts/ts*

Gametes: (*ba Ts*) (*ba ts*) (*ba ts*)

Progeny:

	ba Ts	*ba ts*
ba ts	*ba/ba Ts/ts*	*ba/ba ts/ts*

The parental genotypes are thus reformed in each succeeding generation in a ratio of 1:1.

Environmental Control of Sex Determination

Many organisms are extremely flexible in their ability to differentiate sexually, with each individual developing into either a male or female depending on the specific set of environmental conditions. When the sex of the individual is not genetically determined, every zygote must possess all the genes needed to produce both the male and female reproductive systems.

One example of such extreme flexibility in sexual development is that of the burrowing marine worm, *Bonellia viridis.* The female *Bonellia* has a body shaped like a walnut, with a slender proboscis about an inch long (Fig. 5.23). The male is only the size of a large protozoan and lives as a parasite within the reproductive tract of the female. During reproduction, fertilized eggs are released into the water and develop into free-swimming larvae. Those larvae in the vicinity of adult *Bonellia* are attracted to and settle on the proboscis of the nearest female. Each larva in contact with an adult female is eventually transformed into a male and migrates to the female's uterus. The remaining larvae, which failed to make contact with a free-living adult female, settle on the sand and develop into females. The finding that an extract made from the female proboscis acts just as effectively as actual contact in transforming larvae into males suggests that some hormonelike substance produced by adult females can initiate male sex determination in the young of this species.

Hormones and Sex Differentiation. Hormones are also important in regulating sex differentiation in higher organisms. This is seen in the development of primary sexual differences in humans. Prior to the sixth week of development, human embryos are sexually undifferentiated; they develop both *Müllerian ducts,* which

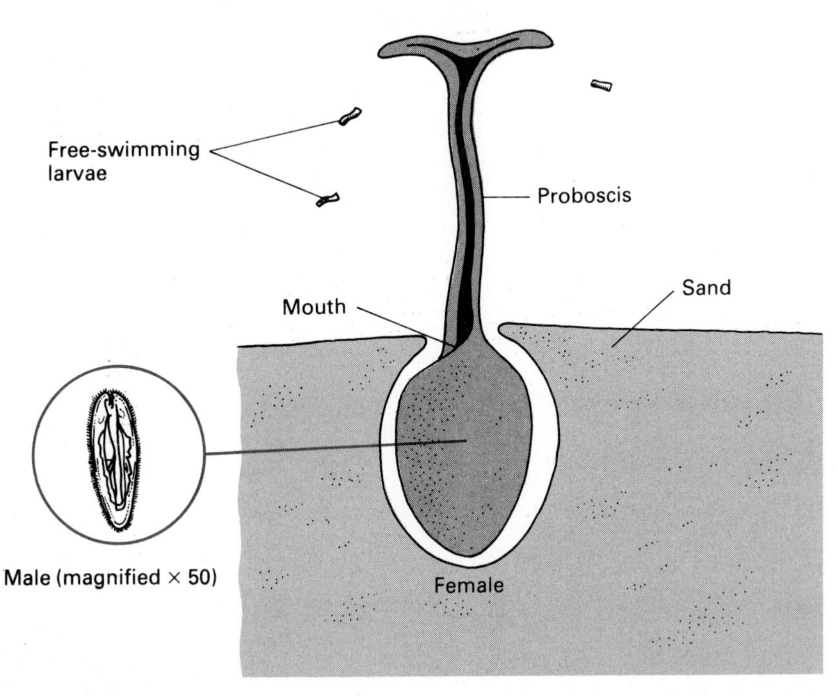

Figure 5.23. Male (left) and female (right) *Bonellia*, a burrowing marine worm that exhibits extreme flexibility in sexual development. The male lives as a parasite within the reproductive tract of the female.

are the precursors of the uterus and oviducts, and *Wolffian ducts,* which are the precursors of the prostate gland, seminal vesicles, and tubes of the male reproductive system (see Fig. 5.24). In addition, both XX and XY embryos develop a protuberance known as the *genital tubercle,* positioned above the *urogenital groove.* Once the development of the testes has been established in presumptive males, the testes secrete the male hormone testosterone, which induces the completion of the male genital-duct system from the Wolffian ducts and the formation of male external genitalia from the genital tubercle and its associated structures. The testes also produce a Müllerian inhibitor, which causes the Müllerian ducts to degenerate and prevents the appearance of female sex structures. In presumptive females, no Müllerian inhibitor is formed, so that Müllerian ducts are free to develop into the uterus and oviducts. External genitalia also develop spontaneously in females, with the genital tubercle forming the clitoris and the urogenital groove giving rise to the vagina.

The mechanism of sexual differentiation in mammals has been determined, in part, by experiments conducted on mammalian embryos. For example, when the gonads are surgically removed from a rabbit embryo before male or female differentiation has occurred, both XX and XY embryos follow the female course of development when they are returned to the uterus. This finding indicates that only the Müllerian ducts will develop in the absence of a signal from the gonads.

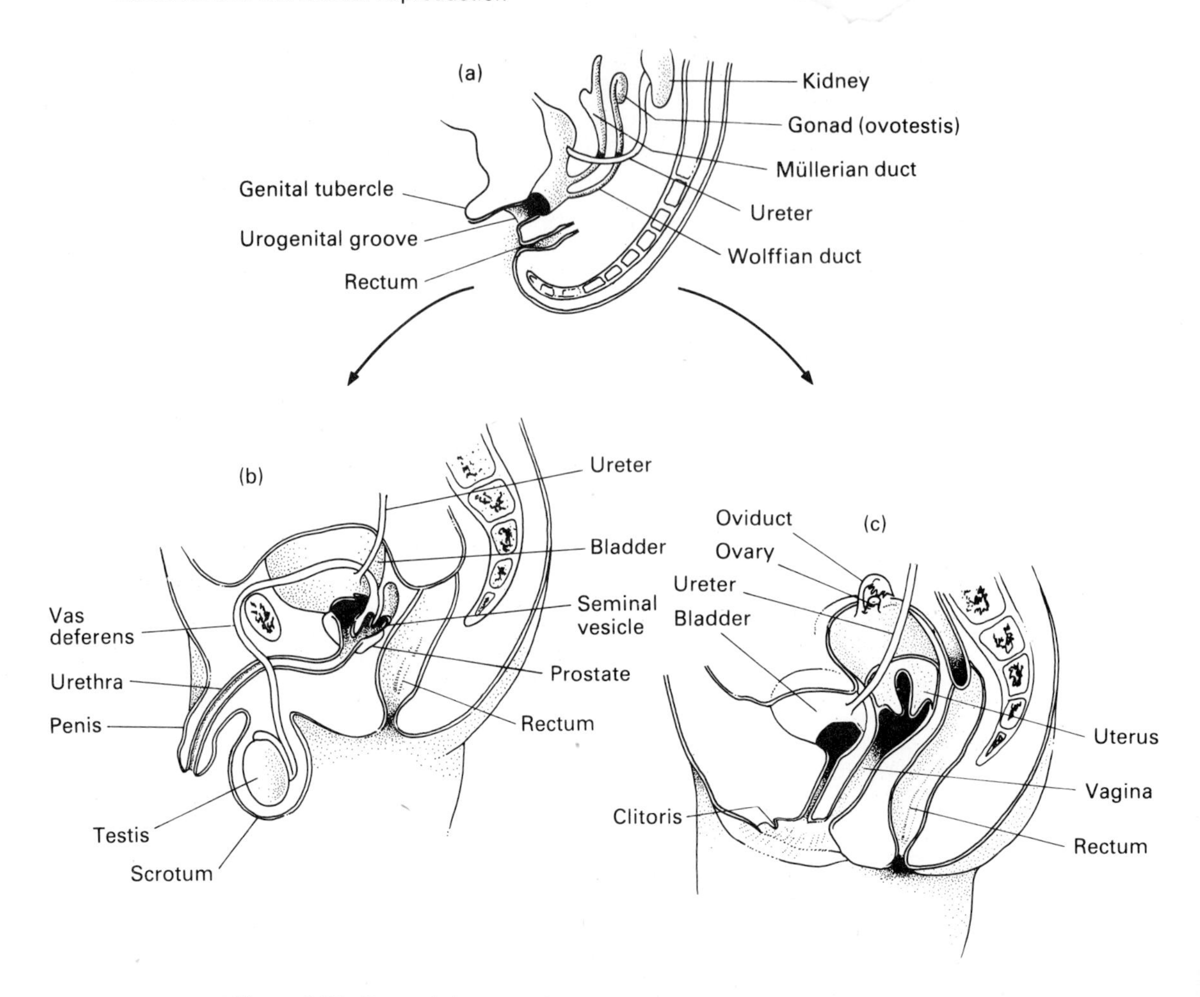

Figure 5.24. Prenatal development of internal sexual structures in XY and XX embryos. (a) In XY embryos, the Müllerian ducts degenerate and the Wolffian ducts develop into the prostate gland, seminal vesicles, and the tubes of the male reproductive system. The genital tubercle becomes the penis. In XX embryos, the Wolffian ducts regress and the Müllerian ducts form oviducts, uterus, and most of the vagina. A clitoris develops from the genital tubercle. (b) Sexual anatomy of a male at birth. (c) Sexual anatomy of a female at birth.

But when testosterone is administered to embryos whose gonads have been removed, normal male structures are formed, in addition to oviducts and the uterus. We can therefore conclude that an inhibitor of Müllerian duct development is needed in addition to testosterone in order for an embryo to develop into a completely normal male.

One example of a genetic defect that interrupts the normal sequence of sexual differentiation in humans is a condition known as *androgen insensitivity syndrome* (also called testicular feminization). Individuals affected with this condition have a male karyotype (XY) and possess normal, testosterone-secreting testes. The tissues of these individuals fail to respond to the action of the male hormone, however. As a result, female external genitalia develop, along with other sexual characteristics typical of normal females. Externally, these individuals are normal females. Internally, they lack the structures that would normally derive from the primitive sex ducts of the embryo. Since individuals with this condition are insensitive to testosterone stimulation, no derivatives of the Wolffian ducts are formed. Since the tissues do respond normally to Müllerian inhibitor, though, the oviducts and uterus are also absent. In this case, the karyotype and gonads are male, but the overt sexual appearance (distribution of hair, body proportions, breast development, etc.) and, to a certain extent, the genitalia, are female.

Sex Reversal. Sex reversal can also occur in higher organisms, indicating that a single karyotype is capable of producing both males and females. In female birds, for example, only one presumptive gonad develops into an ovary, while the other remains in an undifferentiated state. If the ovary is removed or destroyed by disease, the rudimentary gonad develops into a testis. Male hormone production by the testis then leads to a reversal of the secondary sex characteristics with the development of cock feathering and crowing ability. Birds that have undergone sex reversal can even father offspring. In a cross between a sex-reversed male and a normal female, both parents have the ZW sex chromosome makeup. This cross leads to a sex ratio of one male to two females, since WW zygotes are inviable.

EXAMPLE 5.8. A certain hen has a sex reversal to become a functional male and is then mated to a normal hen. Show that the offspring produced will have a sex ratio of 2 females : 1 male.

Solution. The cross is diagrammed as follows:

Parents: Female with sex reversal to become a functional male (ZW) × Normal female (ZW)

Gametes: Z W Z W

Progeny:

	Z	W
Z	ZZ male	ZW female
W	ZW female	WW lethal

The ratio is thus 2 females : 1 male among the surviving offspring.

Sex reversal is also possible in mammals to varying degrees. One example that is well known to cattle breeders is the development of a sterile intersex called a *freemartin*. The freemartin has an XX sex chromosome makeup and female external genitalia but shows some male development of its internal organs. It is the twin of a normal male calf; the freemartin arises when the placental membranes of the twins fuse to permit a mixing of fetal blood. Recent studies have shown that the gonads of the freemartin are sex-reversed and secrete male hormones. This finding suggests that the male twin produces some diffusible substance early in embryonic life that can induce male development, even in an embryo with a female karyotype.

TO SUM UP

1. Dimorphism of the sex chromosomes is common among species with separate sexes. The major types of dimorphism encountered in nature are the XX–XY (or XX–XO) system in which females are XX and males are XY (or XO), and the ZZ–ZW system in which males are ZZ and females are ZW.
2. Sex determination in *Drosophila* involves striking a proper balance between female-determining genes located on the X chromosome and male determiners present on the autosomes. Studies show that when the ratio of X chromosomes to haploid sets of autosomes is equal to 1.0, a normal female develops. When the ratio equals 0.5, a normal male is the result.
3. In certain organisms, such as mammals, the presence of a Y chromosome determines maleness, regardless of the number of X chromosomes present. The absence of a Y chromosome results in a female sex phenotype.
4. In some organisms, a small number of genes can exert a major influence on the sex phenotype. In corn, for example, changes at two gene loci can convert a plant that is normally bisexual into a strain in which the sexes are separate.
5. When both types of gametes or sex organs are produced by individuals of the same genotype, the determination of which type of gamete or which sex an individual will have depends primarily upon environmental differences. In the marine worm *Bonellia*, for example, the specific path of sex differentiation depends on the presence or absence of an external chemical stimulus. In higher animals, sex differentiation is influenced mainly by the internal environment of the animal, which includes the presence of internal inducers and the balance among the sex hormones.

Questions and Problems

1. Define the following terms and differentiate between paired terms:

   ascus
   bivalent
   chiasma
   crossing over
   dioecious–monoecious
   FDS—SDS segregation patterns
   freemartin
   gametophyte–sporophyte
   hermaphrodite
   linked genes
   meiocyte
   oogenesis–spermatogenesis
   ordered tetrad
   PD–NPD tetrads
   polar body
   synapsis
   T tetrad

2. Designate whether each of the following events occurs in mitosis, in meiosis, or in both kinds of nuclear division.
   (a) Chromosomes duplicate prior to the initial stage.
   (b) Homologs undergo pairing (synapsis).
   (c) Individual chromosomes align independently on the spindle equator.
   (d) Homologs separate and move toward opposite poles.
   (e) Diploid cells are formed in the process.
3. Why is meiosis I called a reduction division while meiosis II is called an equation division? Why are such terms misleading when meiosis is viewed from the standpoint of nuclear DNA content?
4. The cell of a diploid organism observed under a microscope appeared as follows:

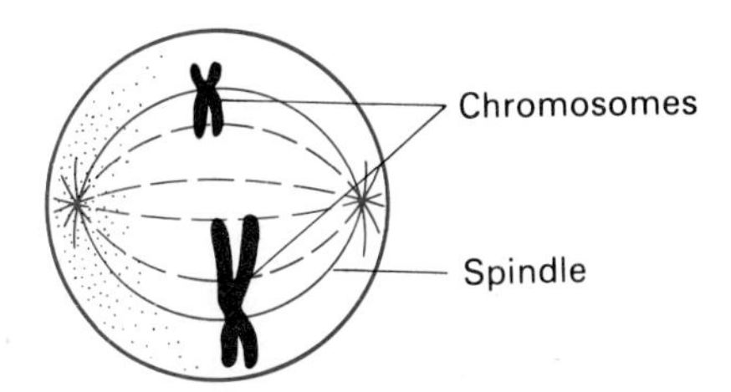

   Assuming no other chromosomes are present within the cell, identify the stage of meiosis and/or mitosis in this cell. Give reasons for your choice.
5. (a) How many functional sperm cells can be derived from 10 spermatogonia? (b) How many eggs can arise from 20 oogonia?
6. Evaluate the validity of the following statements about gametogenesis.
   (a) Half the autosomes present in a gamete of an animal are maternal in origin.
   (b) A sperm contains half the amount of DNA present in a spermatocyte of the same animal.
   (c) The presence of a chiasma reveals that the homologs involved have exchanged structural parts.
   (d) Meiosis II is the same as mitosis.
7. A certain plant has 20 chromosomes in each cell at G_1. How many chromosomes, centromeres, and chromatids are present in each cell at **(a)** prophase I? **(b)** Anaphase I? **(c)** The end of telophase I? **(d)** Anaphase II? **(e)** The end of telophase II? (Assume that cytokinesis occurs at mid-telophase in both division cycles.)
8. How many different combinations of chromosomes can be produced among the gametes of an individual with **(a)** two pairs of chromosomes? **(b)** Three pairs of chromosomes **(c)** n pairs of chromosomes? (*Hint:* Treat homologs in the same manner as alleles.)
9. The following diagrams are of an embryo sac and a germinating pollen grain in a flowering plant. The different nuclei are labeled A–K.

Embryo sac

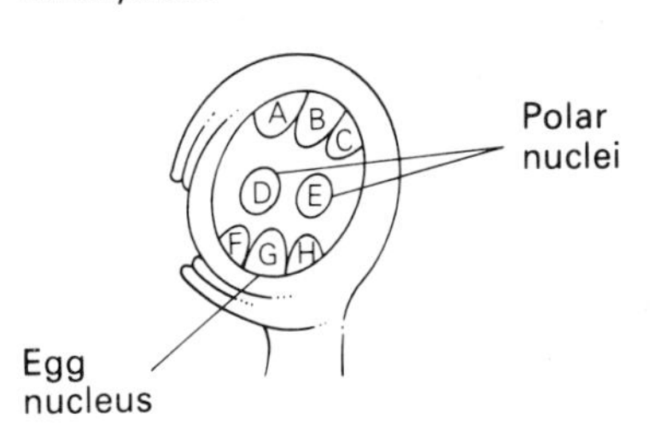

Pollen

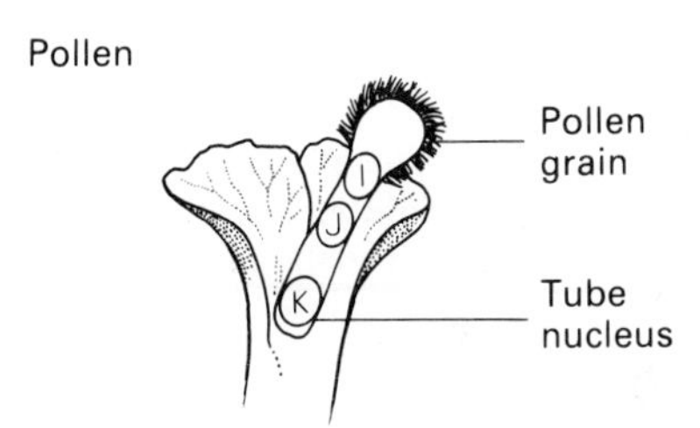

   (a) What nuclear combinations could give rise to the embryo? (Give the letters of each possible combination.)
   (b) What nuclear combinations could form the endosperm?
   (c) Which nuclei in the embryo sac contain chromosomes that are genetically identical? Which nuclei in the germinating pollen are genetically identical?
   (d) Which nuclei are not involved in a fertilization act?
10. Emmer wheat (*Triticum dicoccum*) has 28 chromosomes in each of its somatic cells, while rye (*Secale cereale*) has 14. Matings between these two cereal crops produce sterile hybrids with characteristics that are intermediate between wheat and rye. How many chromosomes do these hybrids possess?

11. An *Ff* corn plant that is heterozygous for the alleles for flinty (*F*) and floury (*f*) endosperm is fertilized by pollen from a plant that is *ff* in genotype. What ratio of endosperm phenotypes is expected among the offspring? What would be your answer if the genotypes of the egg and pollen donors were reversed?

12. Compare the relative duration of meiosis I in human males and females. What are the evolutionary advantages of a rapid rate of gametogenesis in males and a much slower rate in females?

13. Suppose that a cross involving the single pair of mating type alleles in *Neurospora* gives the following kinds and numbers of tetrads (spore pairs within each ascus are listed vertically):

(1)	(2)	(3)	(4)	(5)	(6)
A	*a*	*A*	*a*	*a*	*A*
A	*a*	*a*	*A*	*A*	*a*
a	*A*	*A*	*A*	*a*	*a*
a	*A*	*a*	*a*	*A*	*A*
82	77	10	12	9	10

Determine the percentages of FDS and SDS asci.

14. Gene pairs *A,a* and *D,d* are observed in a mating of *Neurospora* strains *Ad* and *aD*. The following classes of tetrads are obtained:

(1)	(2)	(3)	(4)	(5)	(6)	(7)
Ad	*AD*	*AD*	*Ad*	*Ad*	*AD*	*AD*
Ad	*AD*	*aD*	*AD*	*aD*	*ad*	*ad*
aD	*ad*	*Ad*	*ad*	*Ad*	*AD*	*Ad*
aD	*ad*	*ad*	*aD*	*aD*	*ad*	*aD*
13	12	41	70	30	33	45

(a) Identify each of the tetrad classes as PD, NPD, or T. **(b)** Demonstrate that the genes have assorted independently. **(c)** Calculate the percent SDS for each gene pair.

15. Assume a certain male animal has three pairs of chromosomes, X and Y, 1_m and 1_p, and 2_m and 2_p, where the subscripts m and p designate autosomes of maternal and paternal origin, respectively. **(a)** How many different combinations of chromosomes can this animal produce among its gametes? **(b)** What fraction of the gametes contains only chromosomes of paternal origin? **(c)** What fraction of the gametes are bearing a Y chromosome?

16. Assume that a human egg at fertilization weighs 1.5×10^{-6} g and that a sperm weighs 2×10^{-9} g. (Note: 1 g = 2.205×10^{-3} lb.) **(a)** Compare your present weight and your approximate weight at birth with the weight of the egg and sperm cell that gave rise to you (express all weights in the same units). **(b)** By what factor did your weight increase from the zygote stage to an infant at birth? By what factor did it increase from birth to your present age?

17. How might the long-term survival of a species be enhanced by a form of sex determination that assures approximately equal numbers of males and females?

18. A plant has three pairs of chromosomes, 1_m and 1_p, 2_m and 2_p, and 3_m and 3_p, where m and p designate different haploid sets. Each chromosome has a 30 percent chance of participating in one or more crossovers with its homolog. If the plant undergoes self-fertilization, **(a)** what is the chance that an offspring will receive chromosomes that are identical in all respects to those of the parental plant (i.e., that it receives all the members of both haploid sets and that no chromosome has exchanged parts through crossing over)? **(b)** What is the probability that there is no change in chromosome makeup in a particular plant derived from two generations of selfing?

19. A paternally derived chromosome in a certain individual carries the linear arrangement of genes *ABcd* along its length, with each gene controlling a different function. The homologous maternally derived chromosome carries the gene arrangement *AbCD*. **(a)** If crossing over can occur between any or all gene loci, how many kinds of gametes can such an individual produce? Identify where the crossing over occurred in each case. **(b)** Assume that the probability of crossing over between the *A,a* and *B,b* gene loci is 0.05, between the *B,b* and *C,c* genes is 0.15, and between the *C,c* and *D,d* genes is 0.10. What are the expected proportions of each possible gamete type?

20. Predict the sex of each of the following fruit flies, based on the number of X chromosomes and haploid sets of autosomes (each haploid set is designated as A): **(a)** AAAXXX. **(b)** AAX. **(c)** AAAAAXX.

21. Predict the sex of the following mammals: **(a)** AAXXX. **(b)** AAXXY. **(c)** AAXO.

22. Predict the sex of the following plants of the wild campion (*L. dioica*): **(a)** XY with a deletion of both regions I and II of the Y chromosome. **(b)** XY with a deletion of both regions I and III of the Y chromosome.

23. In a certain moth, the male possesses 62 chromosomes and the female possesses 61. Which is the heterogametic sex? Why?

24. Occasionally, human babies are born who are AAXO in chromosome makeup. The AAOY constitution has never been observed, either among live births or aborted fetuses. Why should we expect the complete absence of X chromosomes to be lethal, when the lack of a Y chromosome is not?

25. Among mammals, there appears to be a greater tendency for errors in sexual development to occur in males than in females. How might this difference be explained?

26. A few XY human babies are born who have both male and female internal organs, which are derived from the Müllerian and Wolffian duct systems. Otherwise, these babies have the appearance of normal males. How might such an error in sexual development occur?

27. In humans with a condition known as gonadal dysgenesis, neither ovaries nor testes are produced. Will the genitalia of such an individual be male or female? Explain.

28. The transformer gene *tra* in the fruit fly *Drosophila* transforms XX females into males when the gene is present in the homozygous condition. The transformed XX *tra/tra* males are similar in appearance to normal males, but they are sterile. The *tra* gene does not affect the sex phenotype or fertility of XY individuals. (a) Suppose that the cross *Tra/tra* × *Tra/tra* is performed. What sex ratio is expected among the progeny? (b) A cross involving the *tra* gene yields a sex ratio among the progeny of three males to one female. What are the genotypes of the male and female parents in this cross?

29. In the garden asparagus, *Asparagus officinalis*, sex is determined by a single pair of alleles. The plant is normally dioecious, with staminate individuals having the genotype *A*- and pistillate plants being *aa*. Show that continued matings in this plant will result in a 1:1 sex ratio consisting of half *Aa* (staminate) and half *aa* (pistillate) genotypes.

Chapter 6

Allelic Genes on Autosomes and Sex Chromosomes

We have seen that genes are located on chromosomes, with each kind of gene occupying its own particular site (or *locus*) on one of the chromosomes. The various genes of an organism are therefore located at different regions, either on the same chromosome or on different chromosomes. For example, the *a* gene occurs on a certain chromosome at a region called the *a* locus, the *b* gene occurs at the *b* locus, and so on. When two alleles of the same gene (such as *A* and *a*) are present, however, they occupy the same locus but on homologous chromosomes.

Paired genes occur in the strictest sense only on the autosomes. The autosomes themselves are present in matched pairs, so the genes they carry are also present in pairs. The paired genes may be in either the heterozygous or the homozygous condition, depending on whether two of the same or two different alleles are present. The disjunction of autosomes during meiosis will therefore produce a typical Mendelian pattern for the segregation of alleles, with each gamete receiving one member of every gene pair. This pattern is the same in both males and females. Thus, the hereditary roles of the two sexes are equal for genes that are carried on autosomes.

A somewhat different hereditary pattern exists for genes carried on the sex chromosomes. As we learned in Chapter 5, the X and Y (or Z and W) chromosomes are nonhomologous for most of their lengths. Consequently, many genes that are present on one type of sex chromosome are not present on the other. This lack of homology results in an inheritance pattern that is specifically related to sex. Traits inherited on the sex chromosomes will therefore have different probabilities for males and females.

This chapter provides an introduction to the chromosomal basis of gene transmission in diploid eukaryotes. We will consider inheritance patterns that can be interpreted in terms of alleles at a single gene locus. Some of these patterns concern dominant and recessive alleles carried on the autosomes. Other patterns show a discrepancy in phenotypic expression between males and females, indicating some form of sex-related inheritance. Still other patterns involve the production of several different phenotypes by a single locus with multiple (more than two) alleles. For each of these inheritance patterns, we will present distinguishing features to help you identify the pattern involved.

Gene Symbolism. Until now, we have used the same letter of the alphabet to represent the various alleles of a gene. We designated dominant alleles by capital letters, recessive alleles by lowercase letters, and codominant or incompletely dominant alleles by primed letters. We will continue to employ these symbols. But this chapter will also introduce the symbolism that is preferred by geneticists who use the fruit fly, *Drosophila,* as their experimental organism.

With *Drosophila,* the gene symbol is the first letter or an abbreviation of the name used to describe the phenotypic effect of a deviant or rare allele. The symbol is capitalized if the allele is dominant and lowercased if it is recessive. The wild-type, or normal, allele of the gene is the allele found most frequently in natural or standard laboratory populations. The wild-type allele is distinguished from all the alternatives by a + superscript. For example, black body color in *Drosophila* is recessive to the standard wild-type color. We therefore designate the alleles for black and wild-type color as b and b^+, respectively. When body color is the only trait being considered, we may also use the shortened version b and +.

The *Drosophila* symbolism places the emphasis on the most common and usually the most adapted form of a gene, regardless of its dominance relationship to other alleles, and uses this allele as the standard for comparison. Although plant and animal breeders do not generally use this symbolism, it has been adopted by most experimental geneticists, and students of genetics should be familiar with it.

Patterns of Transmission for Two Alleles

Pedigree Analysis

The study and recognition of patterns of inheritance are important in genetics. One of the most useful tools for this purpose is **pedigree analysis.** A pedigree is a diagram representing two or more generations of related individuals that shows the transmission pattern for a particular genetic characteristic. Individuals in the pedigree who express the characteristic are referred to as *affected* individuals; those who do not express the characteristic are referred to as *unaffected* individuals.

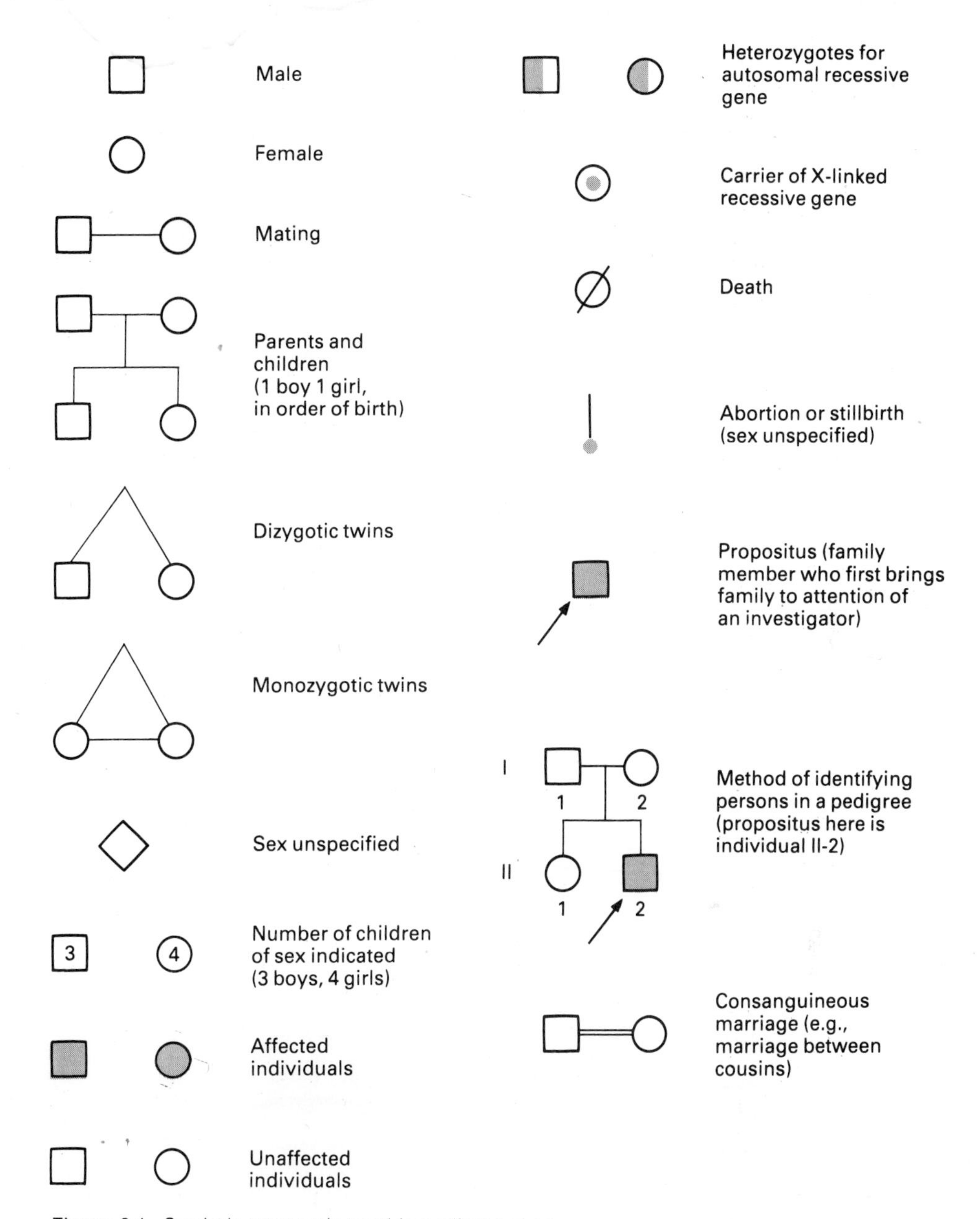

Figure 6.1. Symbols commonly used in pedigree charts.

Certain standard symbols are used in a pedigree or "family tree" (Fig. 6.1). A hypothetical human pedigree of four generations is shown in Fig. 6.2. This pedigree includes eight affected individuals (filled-in symbols). Females are represented by circles, males by squares. The generations are designated by Roman numerals (the first generation being generation I) and the individuals within each generation are designated by Arabic numbers. Individuals who marry into the family are shown and numbered in some cases (II-2, II-4, II-6, II-8, III-6, III-8), but not in others (mates of all members of generation III except for III-7). Individuals who marry into the family are assumed to be unaffected unless there is definite evidence to the contrary. For this reason, they are often omitted to save space in the diagram. Within each group of **sibs** (brothers and sisters), individuals are numbered in order of birth from oldest (lowest number) to youngest (higher number). Individuals IV-1, IV-2, and IV-3 have the same phenotype, as do individuals IV-10 and IV-11, and individuals IV-12, IV-13, and IV-14; they are grouped together under one symbol to conserve space. Sibs who are attached to the parental line at the same point are twins. Twins without a connecting line between them, such as III-3 and III-4, are fraternal (dizygotic) twins, arising from the fertilization of two eggs; twins with a connecting line between them, such as IV-16 and IV-17, are identical (monozygotic) twins, arising from a single egg.

Most pedigrees, particularly those involving humans, include a relatively small number of individuals. The number of offspring produced in a generation is therefore usually too small for the observed phenotypic ratio to be a reliable indicator of expected results. Substantial departures can arise simply owing to chance. However, if several different pedigrees for the same trait are examined, all but one inheritance pattern can often be excluded. Pedigree analysis can therefore provide an important means of study of inheritance patterns. Of course the inheritance of traits in experimental organisms such as corn, *Drosophila,* and peas is better studied through controlled crosses, rather than through pedigree analysis. However, most species are not well suited to the extensive genetic manipulation possible with these three organisms. In most cases then, pedigree analysis is the best practical method to elucidate inheritance mechanisms.

Autosomal Inheritance

The pedigrees of some simple traits show inheritance patterns that are consistent with the transmission of two alleles at a single locus on an autosome. In these pedigrees, father-to-son and mother-to-daughter transmission occurs as frequently, on the average, as father-to-daughter and mother-to-son transmission. In other words, the sex of the parent or the offspring does not influence the likelihood of either transmitting or inheriting the trait. This pattern of transmission is called autosomal inheritance, because the character being studied is specified by either a dominant or recessive gene located on an autosome.

Autosomal Dominant Characters. If a character is caused by a dominant gene, then inheritance of that gene from one or both parents will produce the characteristic. Thus, an individual affected with an autosomal dominant characteristic can be heterozygous or homozygous for the causative allele. Two indications that a characteristic is caused by a dominant gene are (1) *all affected individuals have at least one affected parent,* with no skipping of generations, and (2) *two unaffected parents have only unaffected offspring.* The basis for these criteria is fairly straightforward. An affected offspring must have received the causative allele from at least one of the parents. Since the allele is dominant, it would have been expressed in the parent(s) from which it was inherited. If neither parent exhibits the characteristic, then both are assumed to have the homozygous recessive genotype. These parents can then transmit only recessive alleles to their progeny, so that all the progeny will be unaffected.

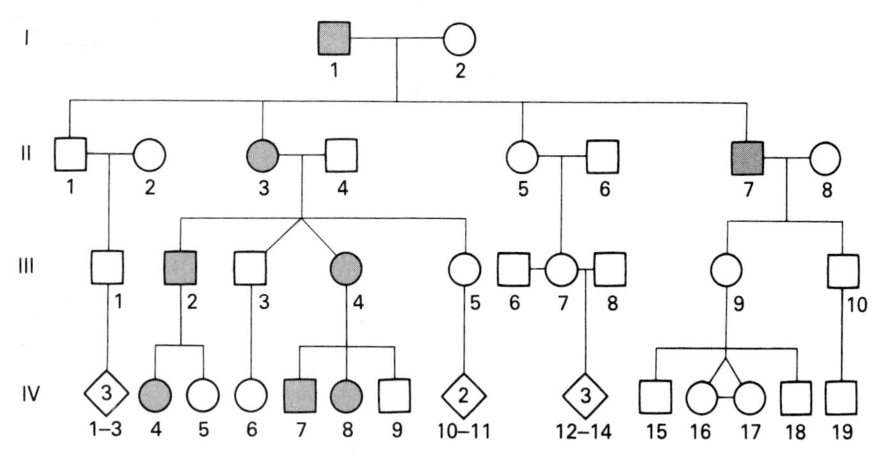

Figure 6.2. Hypothetical pedigree of an autosomal dominant characteristic, illustrating the symbols used in pedigree construction. The symbols are defined in Figure 6.1.

The pedigree in Fig. 6.2 was of an autosomal dominant characteristic. If we examine this pedigree, we see that both of the criteria for autosomal dominant inheritance are met. All seven individuals whose ancestry is known here have an affected parent. Furthermore, unaffected couples (which include family members II-1, II-5, III-1, III-3, III-5, III-7, III-9, and III-10) have only unaffected offspring.

After we have determined that a characteristic has an autosomal dominant basis, we can assign genotypes to the various individuals in the family. If we use *A* and *a* as allele symbols for the pedigree in Fig. 6.2, then unaffected individuals have genotype *aa*, and all affected individuals are *Aa*. (Convince yourself that there are no *AA* genotypes in this particular pedigree.) In each unaffected × affected mating in this pedigree, the offspring has a ½ chance of inheriting the characteristic.

One example of an autosomal dominant trait in humans is brachydactyly (Fig. 6.3). This relatively rare abnormality of the hands is caused by the shortening and

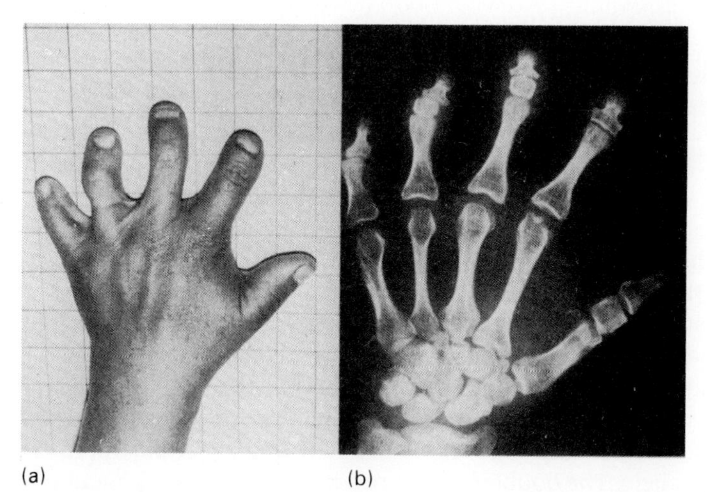

(a) (b)

Figure 6.3. A brachydactylous hand, with an X-ray photograph showing the shortening of the terminal joints. Brachydactyly is a common dominant abnormality of the hands, caused by the shortening and fusing of the terminal bones. *Source:* D. Hofnagel and D. S. Gerald, *Ann. Hum. Genet.* 29:377–382, 1966, Cambridge University Press.

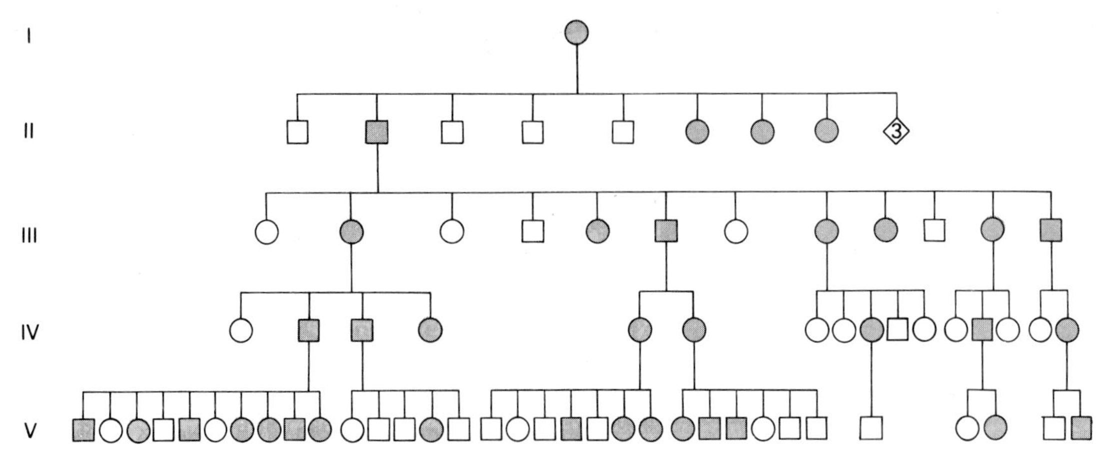

Figure 6.4. A pedigree showing the inheritance of brachydactyly. This pedigree, published in 1905, was the first pedigree of a simple dominant character in humans. Individuals marrying into this family were not affected with brachydactyly and are therefore not shown, to simplify the pedigree. *Source:* Reprinted from W. C. Farabee's *Inheritance of Digital Malformations in Man,* 1905, from the Peabody Museum Papers, vol. 3, no. 3, with the permission of the Peabody Museum of Archaeology and Ethnology.

fusing of the terminal bones. Figure 6.4 gives a pedigree showing the inheritance of this disorder.

It is sometimes difficult to prove with certainty that a disorder is caused by a dominant gene, since pedigree patterns that meet the criteria for dominant inheritance are sometimes consistent with the pattern for recessive inheritance as well. For example, if the characteristic depicted in Fig. 6.2 is not assumed to be a rare hereditary trait, we might alternatively assume that some of the individuals marrying into the family are carriers of a recessive gene for the trait. If the recessive allele is inherited from both parents, then recessive inheritance becomes a possible explanation. Thus, we must exercise caution when deciding on a mode of inheritance based on the analysis of a pedigree, unless we have sufficient evidence to rule out with certainty one or the other possibility.

Autosomal Recessive Characters. The major feature that distinguishes recessive from dominant inheritance is that *with recessive inheritance, unaffected parents can have affected offspring.* The unaffected parents of an affected offspring would both be heterozygous ($Aa \times Aa$). We would expect ¼ of the offspring of a mating between heterozygotes to be affected.

Let us consider the pedigree in Fig. 6.5(a), which shows the inheritance of phenylketonuria. In this disorder, mental retardation develops in infancy as a result of the accumulation of the amino acid phenylalanine. The disease is caused by the absence of the enzyme phenylalanine hydroxylase, which is responsible for the conversion of phenylalanine to tyrosine. As we can see from the pedigree, phenylketonuria is clearly caused by a recessive gene, since unaffected parents (IV-1 and IV-2) have affected offspring. Therefore, IV-1 and IV-2 must both be carriers of the recessive allele. The double line connecting the parents in generation IV means that they are related (in this case, second cousins).

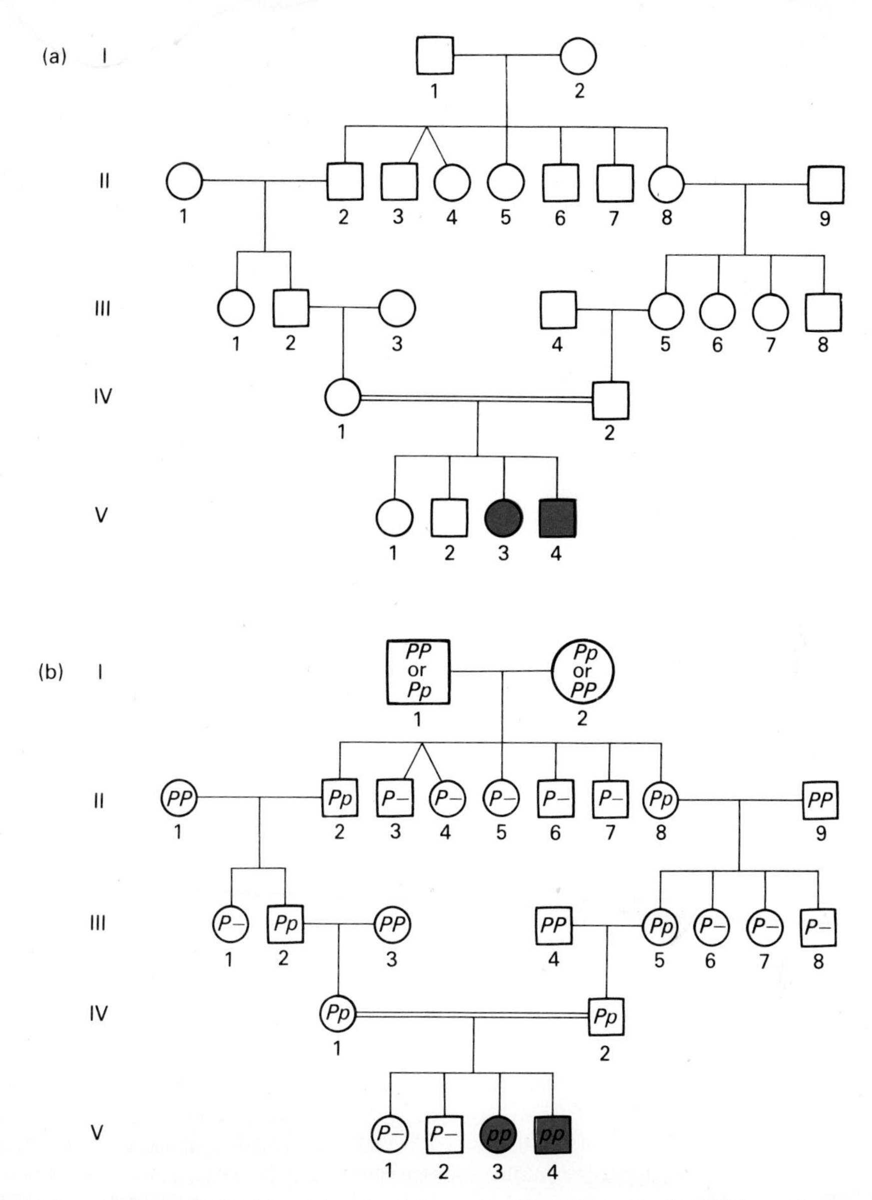

Figure 6.5. (a) A pedigree illustrating the inheritance of phenylketonuria (PKU), a biochemical disorder caused by a recessive gene. Affected individuals are unable to metabolize the amino acid phenylalanine, which accumulates in the body and causes mental retardation. The frequency of the disease is approximately 1 in 25,000. (b) The pedigree shown in (a) with genotypes of the individuals given in the pedigree symbols. Individuals marrying into the family are assumed to be homozygous for the dominant allele, since the recessive gene for phenylketonuria is rare in the general population.

While recessive genes are not uncommon, deleterious recessive genes are usually rare in a population. It is therefore unusual for two unrelated individuals both to be carriers of the same deleterious recessive gene. Very often, when both parents are heterozygous for the same deleterious recessive allele, family records show them to be related, suggesting the likelihood that the individuals inherited the allele from the same ancestor. In the absence of such data, we may still suspect that the parents have some common ancestry within the comparatively recent past if they are of similar geographic origin. As we will see in Chapter 16, recessive phenotypes are expressed much more frequently among the offspring of related parents than among offspring whose parents are not of common ancestry, especially if the frequency of the recessive gene in the population is low.

Figure 6.5(b) shows the pedigree given in Fig. 6.5(a) in terms of individual genotypes. Note that individuals II-2, II-8, III-2, III-5, IV-1, and IV-2 are all designated as heterozygotes, because this constitutes the most likely route of transmission of the recessive allele from generation I through generation IV. Individuals who marry into the family, including II-1, II-9, III-3, and III-4, are assumed to be homozygous for the dominant allele. Because the allele for phenylketonuria is relatively rare, any unaffected person is assumed to be homozygous for the dominant allele unless there is evidence to the contrary, based on inheritance patterns. The remaining family members could be either normal homozygotes or heterozygotes, the probability of which can often be determined. For example, what is the probability that individual V-1 is heterozygous? Since V-1 has affected sibs, her parents must both be carriers of the recessive allele ($Pp \times Pp$). The conditional probability that V-1 is heterozygous given that she is normal in phenotype (P-) is then ⅔. The number 3 is in the denominator rather than 4 since there are only 3 possible outcomes that produce the normal phenotype (2 Pp + 1 PP).

Sex-Linkage

Genes that are located on only one type of sex chromosome are said to be **sex-linked.** Some sex-linked genes are found exclusively on the X chromosome, without a corresponding locus on the Y chromosome. These genes are referred to as **X-linked** or X-located genes. Other sex-linked genes occur only on the Y chromosome, without a corresponding locus on the X chromosome. These genes are referred to as **Y-linked** or Y-located genes. Very few genes have been identified as being on the Y chromosome of any mammal (limited to genes for certain male-specific traits). By contrast, the X chromosome bears as many genes as would an autosome of similar size. Because nearly all sex-linked genes are X-located, the term sex-linkage almost always means X-linked rather than Y-linked inheritance. For this reason, traits that are determined by Y-linked genes are specifically referred to as **holandric** traits to avoid confusion.

Table 6.1 lists a few human traits that are caused by an X-located gene. Notice the wide variety of characteristics listed. The traits that are controlled by the vast majority of X-linked genes have nothing whatsoever to do with sexual function. The type of trait itself therefore gives us no clue as to whether it is X-linked or autosomally inherited. We must base our conclusions about the chromosomal location of a gene solely on the pattern of inheritance, not on the type of trait it produces.

Thomas Hunt Morgan discovered sex-linkage in 1910, when he was studying the inheritance of eye color in *Drosophila.* The standard or wild-type eye color is red. But many variant eye colors have been discovered, one of which is white (Fig. 6.6). When red-eyed females are crossed to white-eyed males, all the F_1 have

Table 6.1. Examples of X-linked characteristics in humans

Agammaglobulinemia (unusual proneness to bacterial infection)
*Color blindness
*Congenital deafness
*Diabetes
Glucose 6-phosphate dehydrogenase deficiency
Hemophilia
Lesch-Nyhan syndrome (mental retardation, "self-destructive" behavior)
*Ichthyosis (scaling of the skin)
*Immunodeficiency diseases
Juvenile (Duchenne) muscular dystrophy
*Night blindness
Ocular albinism
*Pituitary dwarfism
*Retinitis pigmentosa (progressive degeneration of the retina)
*Testicular feminization syndrome
Vitamin D-resistant rickets

* Disease has two or more different genetic causes, X-linkage being just one of the possibilities.

red eyes, which indicates that the allele for white eyes is recessive. Intercrosses among the F_1 give an F_2 ratio of three red-eyed flies to one white-eyed fly, as expected. All the white-eyed F_2 flies are males however; no white-eyed females are found. The F_2 ratio of eye color and sex is 2 red-eyed females : 1 red-eyed male : 1 white-eyed male.

The results of these crosses are explained in Fig. 6.7 by attributing eye color to an X-linked locus with two alleles: w = white eyes (recessive) and w^+ = red eyes. Since X-linked genes are found on the X but not on the Y chromosome, possible genotypes for males are $X^{w^+}Y$ and X^wY. Possible female genotypes are $X^{w^+}X^{w^+}$, $X^{w^+}X^w$, and X^wX^w. As we can see from this example, terms such as homozygous and heterozygous do not apply to males in the case of X-linked genes. The term **hemizygous** is used instead to refer to that genotype in the male that consists of only a single allele.

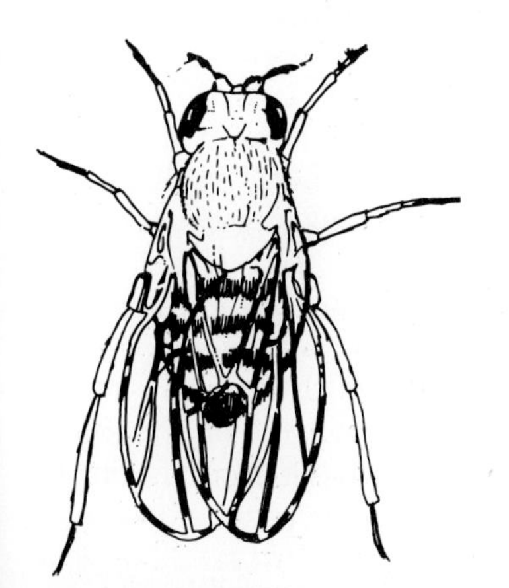

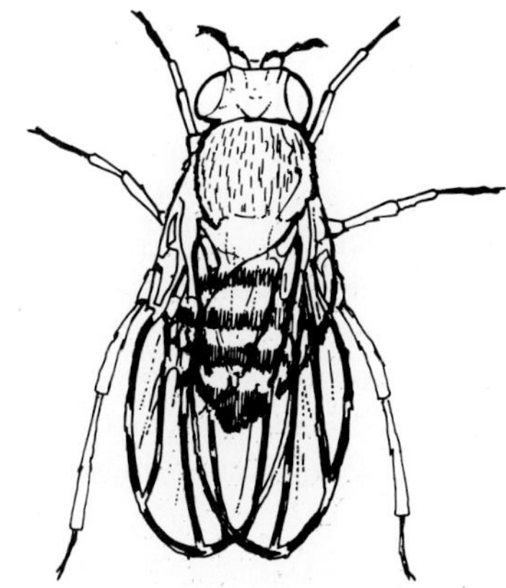

Figure 6.6. Wild-type *Drosophila* eye color is red (left). A sex-linked recessive mutant gene causes white eyes (right).

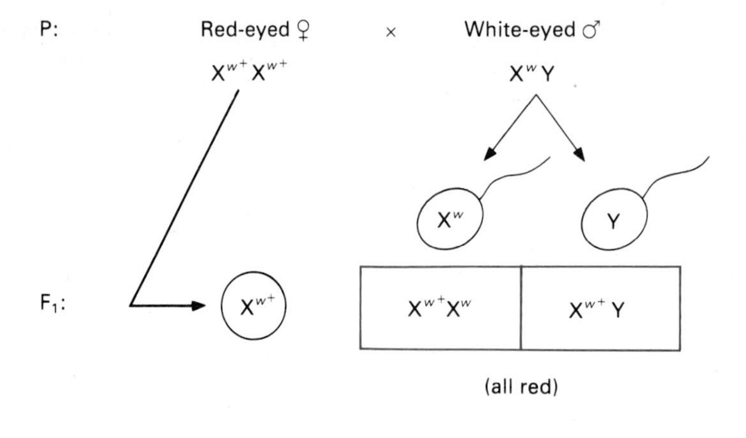

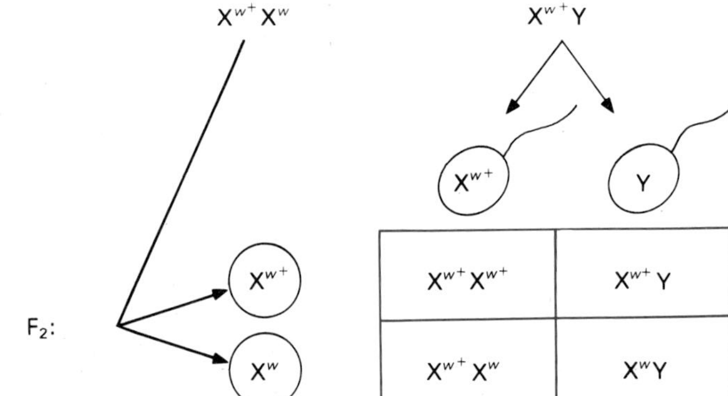

Figure 6.7. Chromosomal basis for eye color inheritance in *Drosophila*. The inheritance pattern is typical of an X-linked pair of alleles when the crossing sequence is started with a homozygous dominant female and a hemizygous recessive male.

The crossing sequence in Fig. 6.7 clearly shows that the mechanism for gamete formation and gene expression for X-linked genes follows Mendelian principles. The only difference between X-linkage and autosomal inheritance is that with X-linkage, one sex has only one copy of the gene rather than two. Therefore, the parents do not contribute equally to the genetic constitution of their offspring. As a result, X-linkage has three diagnostic features:

1. *A difference in phenotypic ratios between the sexes:* The proportions of males and females that express an X-linked character will usually differ. This difference reflects the fact that females carry two doses of an X-linked gene, whereas males have only one.
2. *Reciprocal crosses give different results:* When the progeny from reciprocal crosses are compared, different phenotypic ratios will be observed if the trait is X-linked. This difference is illustrated in Fig. 6.8, in which we compare the results of the cross $X^{w^+}X^{w^+} \times X^wY$ with the cross $X^wX^w \times X^{w^+}Y$ in *Drosophila*. In contrast, reciprocal crosses give the same result if the locus is on an autosome.

3. *Crisscross pattern of inheritance:* Since a male must receive his single X chromosome from the mother, X-linked genes are transmitted from mother to son. The inheritance pattern that results can be similar to that of the cross $X^wX^w \times X^{w^+}Y$ in Fig. 6.8. In this type of cross, the trait itself appears to pass from mother to son. An X-linked character cannot be passed from father to son, since a son inherits only a Y chromosome from its father.

The genetic results of sex-linkage—particularly the failure of reciprocal crosses to give the same progeny phenotypes—indicate that the parents do not contribute equally to the sex-linked traits of their offspring. Morgan recognized that these genetic results were consistent with the behavior of the X and Y chromosomes at meiosis, where one sex donates only an X chromosome to its gametes and the other sex donates either an X or a Y. This finding of parallel behavior in genes and chromosomes provided the first supporting evidence for the chromosome theory of inheritance.

X-Linked Recessive Characters. If an X-linked trait is recessive, it will occur more frequently among males than among females. This property stems from the fact that a single copy of the recessive gene is all that is needed for expression in males. A single copy of the recessive gene is masked by the dominant allele in females, in whom two doses of the recessive allele are needed for expression.

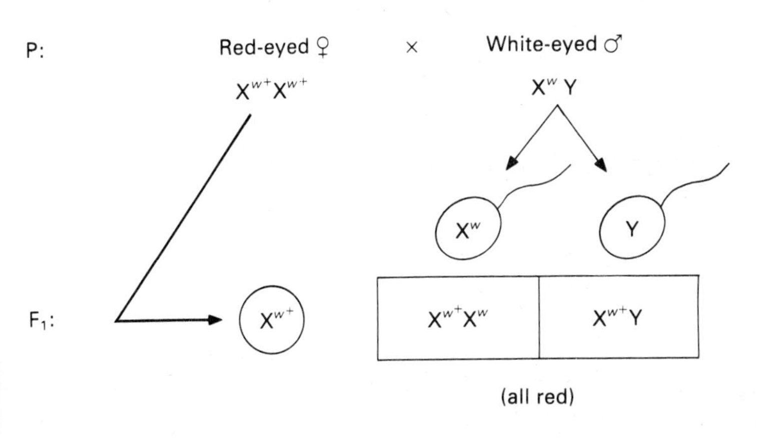

Figure 6.8. Failure of X-linked genes to give the same results in reciprocal crosses. In contrast, the results of reciprocal crosses are not expected to differ for autosomal genes.

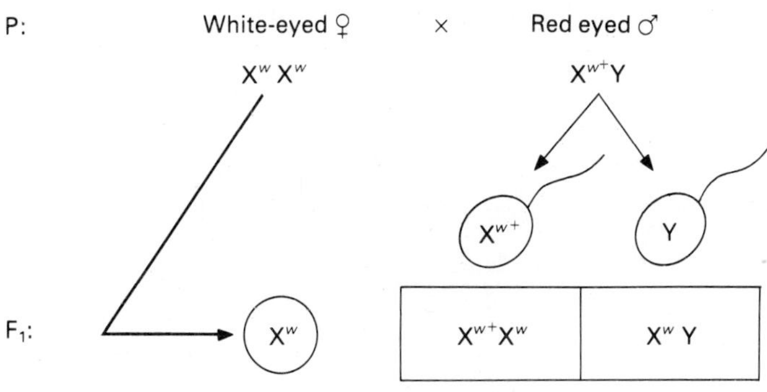

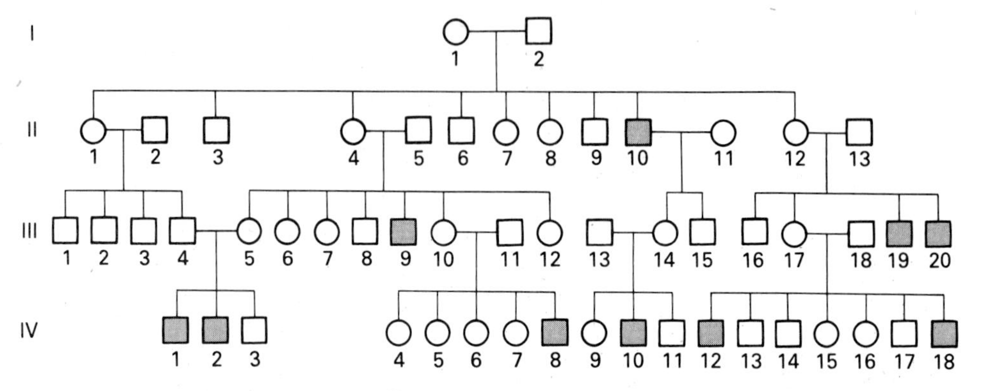

Figure 6.9. A pedigree of the family of Queen Victoria (I-1), showing the inheritance of hemophilia in the royal families of Europe. Some individuals of interest are Prince Albert (I-2), King Edward VII of England (II-3), Princess Alice of Hesse Darmstadt (II-4), Prince Leopold, Duke of Albany (II-10), Princess Beatrice of Battenberg (II-12), Princess Irene of Hessen (III-6), Princess Alix, wife of Czar Nicholas of Russia (III-10), Princess Alice, wife of Alexander, Prince of Teck (III-14), Princess Victoria, wife of King Alfonso of Spain (III-17), Czarevitch Alexis of Russia (IV-8), Lord Trematon (IV-10), Prince Alfonso (IV-12), and Prince Gonzalo (IV-18). The last three died after automobile accidents. Queen Victoria passed the hemophilia gene to her daughters Princesses Alice and Beatrice, and to her son Prince Leopold. It then spread throughout their children and grandchildren.

Since it is less likely for an individual to receive two copies of the same allele than just one (e.g., $(\frac{1}{2})^2 < \frac{1}{2}$), males will tend to be affected more often than females.

Another feature that is commonly observed in the pedigrees of rare X-linked recessive characteristics is their tendency to appear in alternate generations. This property of X-linked recessive traits originates from the crisscross pattern of gene transmission. A recessive allele is transmitted from the father, in whom it is expressed, to his heterozygous daughters, in whom it is not expressed, and then to his grandsons, in whom the gene is again expressed.

Well-known examples of X-linked recessive characteristics in humans include red-green color blindness and hemophilia. Individuals with red-green color blindness have reduced amounts of the pigment for either red color vision or green color vision in the cone cells of their retinas. In hemophilia, there is a deficiency of certain components of the blood that are required for clotting. In both disorders, males are affected more often than females. The greater prevalence of affected males is shown in a pedigree of the descendants of Queen Victoria of England (Fig. 6.9). Queen Victoria was heterozygous for the hemophilia gene, which subsequently spread throughout the royal families of Europe (see Fig. 6.10).

EXAMPLE 6.1. A certain couple have a daughter who is color blind and a son with normal vision. What are the genotypes of the parents in this cross?

Solution. If we designate the alleles for color blindness and normal color vision as c and C, respectively, the genotypes of the daughter and son are: X^cX^c and X^CY. Since the daughter must have received an allele for color blindness from each of her parents, and the son must have received only the C allele from his mother, the mother must be heterozygous normal, X^CX^c, and the father must be color blind, X^cY.

Figure 6.10. Queen Victoria (a) and some of her family, as they posed for a photograph in the late 1800s. Her daughter Princess Beatrice (b) transmitted the gene for hemophilia to the Spanish royal family. Granddaughter Princess Alix (c) transmitted the gene to the royal family of Russia. Granddaughters Princess Irene (d) and Princess Alice (e) also inherited the hemophilia gene from Queen Victoria. *Source:* From the Photography Collection, Humanities Research Center, University of Texas.

Some X-linked recessive traits invariably have a lethal effect on the individuals who express them. Since males are affected more often than females, these traits produce a modified sex ratio among the surviving offspring. Duchenne muscular dystrophy in humans is an example of such a trait. Individuals affected with this disorder appear normal in early childhood but then exhibit progressive wasting away of the muscles until they die, usually in their teens (Fig. 6.11). The causative gene, *d*, is carried by the mother and transmitted to half her sons on the average:

$$X^DX^d \times X^DY$$
$$\downarrow$$
$$\tfrac{1}{4}\ X^DX^D : \tfrac{1}{4}\ X^DX^d : \tfrac{1}{4}\ X^DY : \tfrac{1}{4}\ X^dY$$

Normal daughters (first two classes); Normal sons; Affected sons

The 2 : 2 ratio of female to male offspring at birth becomes modified to a 2 : 1 ratio of female to male offspring by adolescence. The finding at birth or sometime thereafter of a 2 : 1 overall sex ratio is strong evidence for the presence of a sex-linked lethal gene. One of the first genetic studies done on the survivors of the atomic explosions over Hiroshima and Nagasaki was to measure the sex ratio among the offspring of people exposed to the radiation to see if the radiation had caused any recessive lethal mutations on the X chromosome.

Figure 6.12 shows the pedigree of a family affected with Duchenne muscular dystrophy. The mother-to-son inheritance pattern is evident, as is the failure of

Figure 6.11. The Gowers sign, the characteristic "climbing up himself" maneuver by which a child affected with Duchenne muscular dystrophy rises from the prone position.

affected males to reproduce. Because affected males die before they reach reproductive age, a female child could never inherit the allele from both parents. For this reason, Duchenne muscular dystrophy is expected to occur only in male children, never in females.

X-Linked Dominant Characters. Most known X-linked disorders are caused by recessive genes. But a few X-linked dominant characteristics have been documented. Unlike their recessive counterparts, X-linked dominant traits tend to occur more frequently among females than among males. One example of an X-linked dominant trait in humans is hypophosphatemia, better known as vitamin D-resistant rickets. Individuals affected with this disorder have a form of rickets that does not respond to normal therapeutic doses of vitamin D. Among the offspring of crosses between normal females and affected males ($X^rX^r \times X^RY$), all

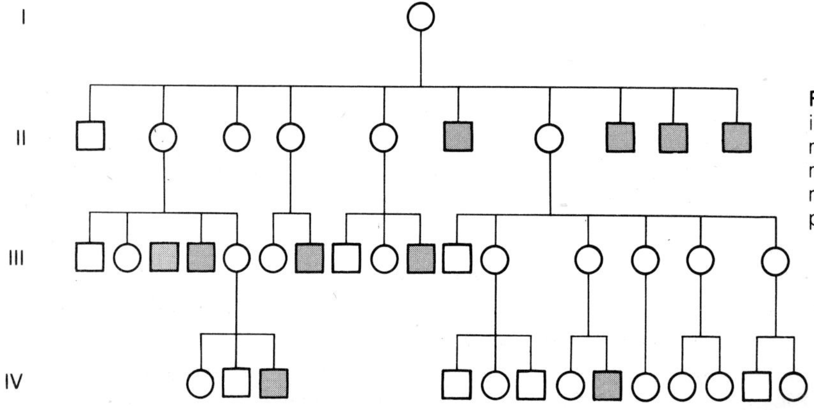

Figure 6.12. Pedigree of a family affected with Duchenne muscular dystrophy. Notice the mother-to-son pattern of transmission. Affected males die prior to reproductive age.

the daughters are affected, with the genotype X^Rx^r, and all the sons are normal, with the genotype X^rY. The major feature of X-linked dominant inheritance is that affected males transmit the character to all of their daughters and to none of their sons, as we have seen by this example. The reverse is not true, since affected females may transmit the character to both their daughters and their sons. Consequently, we cannot distinguish between autosomal and X-linked inheritance of a dominant allele merely by observing its transmission to the progeny of affected females.

EXAMPLE 6.2. Which of the following patterns of inheritance is consistent with the pedigree shown below? **(a)** Autosomal dominant gene. **(b)** Autosomal recessive gene. **(c)** X-linked dominant gene. **(d)** X-linked recessive gene. (Assume that the trait represented by the filled-in symbols is rare in the general population.)

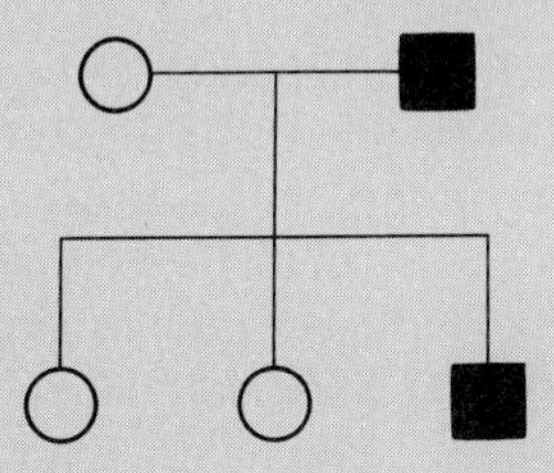

Solution. **(a)** Autosomal dominant gene. Since the trait is rare, the female in the cross is most likely not heterozygous for the causative allele. This assumption, coupled with the fact that an offspring is affected, would rule out autosomal recessive inheritance. (Recall that the offspring would have to receive a copy of the same allele from each of the parents if the trait were recessive.) The fact that the male transmits the gene to his son rules out X-linkage.

Z-Linked Characters. The general features of X-linkage discussed so far pertain to any organism in which the female is the homogametic sex, with two X chromosomes, and the male is the heterogametic sex, with one X and one Y chromosome. As we pointed out in Chapter 5, however, females are not always the homogametic sex. In birds, for example, in which the situation is reversed, more females than males express a sex-linked recessive phenotype. A pair of alleles that control feather growth in chickens illustrates this point. Normal feather growth is determined by dominant gene *K*, and slow feather growth is determined by recessive allele *k*. When normal females are mated with slow males in the cross $Z^KW \times Z^kZ^k$, the male offspring are all normal, Z^KZ^k, and the female offspring are all slow, Z^kW. This trait has practical value as a means of identifying the sex of chicks, since feathers begin to grow very soon after hatching.

Holandric Traits. Holandric, or Y-linked, genes are located exclusively on the Y chromosome. They are expressed in all males that carry them and by all their male descendants, but they are never expressed or transmitted by females. The only indisputable examples of holandric genes in humans are the genes involved in sex determination, such as the *TDF* (testis-determining factor) gene. Yet, there is some evidence to indicate that the gene responsible for the trait "hairy ears" may also be on the Y chromosome. This gene is very common in India and is expressed only in males. Not all male descendants of an affected male are themselves affected, however, and the age of onset of the character varies. These complications, along with some poor sampling data, make it difficult to establish the genetic basis of this trait with any high degree of certainty.

Sex-Limited and Sex-Influenced Traits

Sex hormones are an important environmental influence on the action of certain genes. In the extreme case, a given genotype is so dependent on the presence of these hormones that its expression is limited exclusively to one sex. The result is a **sex-limited trait.** Sex-limited traits are usually determined by autosomal genes, even though they are expressed only in males or only in females.

Sex-limited traits primarily concern the secondary sex characteristics. Beard development in males and milk production in females are examples. Both males and females possess the autosomal genes for these characteristics, but the expression of these genes is limited to one sex by the sex hormones. Thus, both parents contribute equally to the genetic basis of these traits in their offspring. Breeders of dairy cattle are therefore just as concerned with the quality of the milk production genes carried by the bull as with those of the cow. The bull does not express these genes himself, but he transmits them to his daughters, who do express them.

The differences in plumage exhibited by males and females in many species of birds demonstrate that sex-limited traits depend on the proper combination of genotype and hormonal background. In all breeds of chickens, females are "hen-feathered," with comparatively short, straight feathers on the neck and tail. In Leghorns and most other breeds, males are "cock-feathered," with a more showy plumage that has longer, curved feathers (Fig. 6.13). In a few breeds such as the Hamburgs and Wyandottes, males can be either hen- or cock-feathered,

Figure 6.13. Hen-feathering (left) and cock-feathering (right) in chickens. The feathering trait is controlled by a sex-limited gene.

while in others such as the Sebright bantams, all males are hen-feathered like the females.

The feathering type is determined by a single autosomal pair of alleles, *H* and *h*, which show complete dominance. The *H* allele produces hen-feathering, and *h* produces cock-feathering. But the recessive phenotype is expressed only in males, as the following chart illustrates:

GENOTYPE	FEMALES	MALES
HH	hen-feathered	hen-feathered
Hh	hen-feathered	hen-feathered
hh	hen-feathered	cock-feathered

Thus, of the various breeds of chickens, Leghorns all have the genotype *hh*, Wyandottes and Hamburgs can be either *H-* or *hh*, and Sebright bantams are all *HH*.

The effects of sex hormones on the alleles for male and female plumage have been studied in birds whose ovaries and testes have been removed surgically or destroyed by disease. All such birds become cock-feathered at first molt, regardless of their genotype. Apparently, the *H* allele inhibits cock-feathering in the presence of either male or female sex hormones. In birds without testes or ovaries, the sex hormone is absent, so the inhibiting effect is not produced. The *h* allele, on the other hand, permits cock-feathering to be expressed in *hh* homozygotes, but only in the absence of the female sex hormone. Thus, cock-feathering would normally be restricted to males.

Sex limitation is an extreme example of how the expression of a gene can be controlled by the hormones. In other, less extreme cases of sex-controlled characteristics, only the dominance relationship of the two alleles is affected. Characteristics of this type are known as **sex-influenced traits.**

Pattern baldness in humans is an example of a sex-influenced trait. This trait is characterized by the premature loss of hair from the front and top of the head. It is more common in males than in females. Women who have the genotype for pattern baldness typically show only thinning of the hair rather than complete loss. A pair of alleles, B_1 and B_2, on an autosome produce the trait. The presence of at least one B_1 allele results in baldness in males. The trait is sex-influenced in that the B_1 allele behaves as the dominant allele in males but as the recessive allele in females:

GENOTYPE	MALES	FEMALES
B_1B_1	baldness	baldness
B_1B_2	baldness	no baldness
B_2B_2	no baldness	no baldness

Examples of sex-influenced traits in other organisms include horns in sheep (dominant in males, recessive in females), beards in goats (dominant in males, recessive in females), and spotting in cattle (mahogany-and-white is dominant in males, red-and-white is dominant in females).

We must emphasize that sex-limited or sex-influenced expression of a gene is not the same as sex-linkage. Because of their association with the sex of the individual, sex-limited and sex-influenced traits are sometimes mistakenly thought to be sex-linked. Although a few genes of this sort are located on the X chromosome, most are autosomal. In the majority of cases, therefore, the inheritance pattern of a sex-limited or sex-influenced character is distinct from that of a sex-linked trait.

EXAMPLE 6.3. In the pedigree presented below, the trait represented by the filled-in symbols is a fairly common human hereditary characteristic. Which of the following patterns of transmission are consistent with the pedigree? **(a)** Autosomal dominant gene. **(b)** Autosomal recessive gene. **(c)** X-linked dominant gene. **(d)** X-linked recessive gene. **(e)** Y-linked gene. **(f)** Sex-limited autosomal gene. **(g)** Sex-influenced autosomal gene.

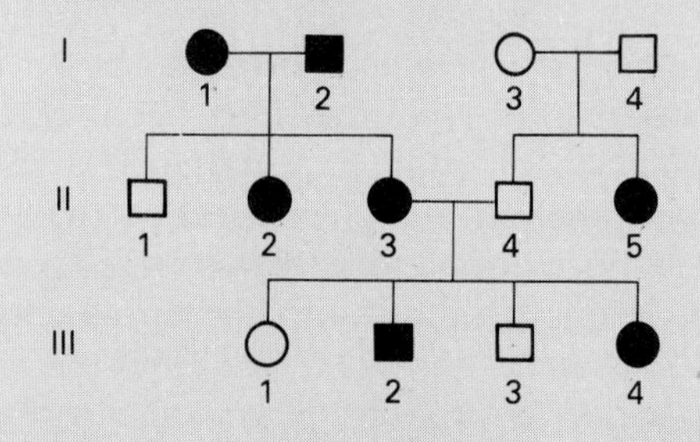

Solution. **(g)** A sex-influenced autosomal gene that is dominant in females, recessive in males. The cross involving I-1 and I-2 rules out simple autosomal recessive or X-linked recessive inheritance. Similarly, the cross involving I-3 and I-4 rules out simple autosomal dominant or X-linked dominant inheritance. They fact that both males and females show the trait rules out the possibility that it is a Y-linked or sex-limited characteristic. But, if the gene (call it A_1) is dominant in females and recessive in males, the hereditary pattern could then be accounted for as follows:

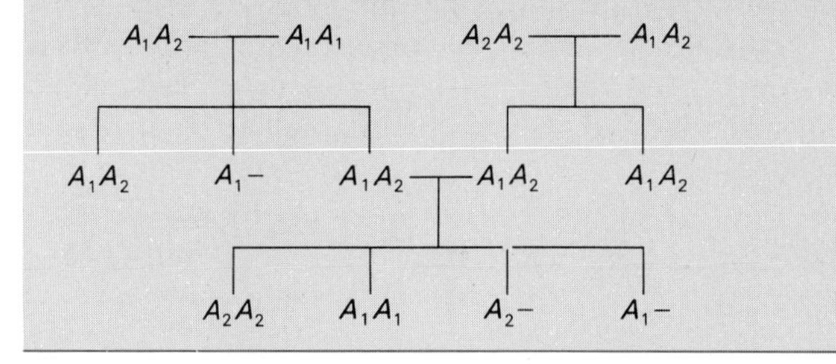

Penetrance and Expressivity

The sex hormones are only one of the many genetic and environmental factors to which a gene might respond. A myriad of other genetic and environmental variables can modify the expression of genes. In Chapter 1 we observed that the expression of a gene can be suppressed or otherwise modified by the action (or

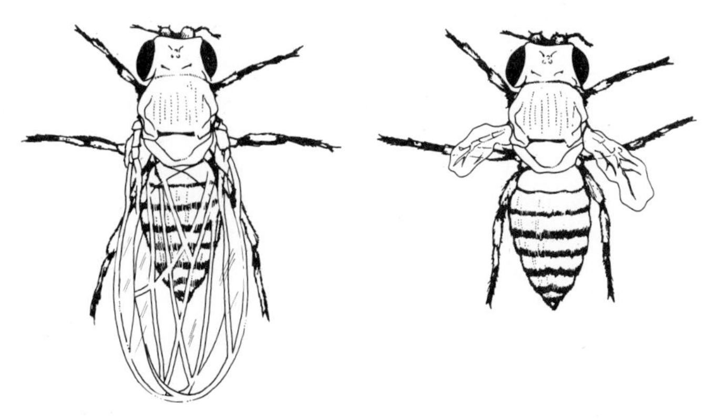

Figure 6.14. *Drosophila melanogaster*, the common fruit fly. An individual with wild-type wings is shown on the left, one with vestigial wings is shown on the right. Vestigial is caused by a recessive mutant gene in homozygous condition.

inaction) of other genes. These epistatic interactions depend upon the genotypes present at various gene loci.

The external environment can also alter the expression of genes. In *Drosophila*, for example, the normal wild-type wings are full and extend beyond the end of the abdomen (Fig. 6.14). A recessive mutant gene, *vg*, causes the wings of *vg/vg* homozygotes to develop improperly at ordinary temperatures, so that only rudimentary, or vestigial, wings are formed. But if the zygotes from *vg/vg* × *vg/vg* crosses are raised at abnormally high temperatures, such as temperatures above 30°C (86°F), many of the *vg/vg* progeny fail to express the expected vestigial-winged trait and instead develop normal, wild-type wings. The environmental stress has caused an extreme modification of gene expression. We can readily establish that the condition is environmentally induced, because the wild-type wing is not passed on to the next generation when environmental conditions are returned to normal. If the seemingly wild-type, but genetically mutant flies are interbred at ordinary temperatures, their *vg/vg* offspring will once again develop vestigial wings. For a variety of different reasons then, a genotype may fail to express its expected phenotype. When this occurs, the gene is said to lack **penetrance.**

The penetrance of a gene is measured quantitatively as the percentage of individuals of a certain genotype that show the expected phenotype. For example, if 60 out of 100 *vg/vg* flies raised at high temperatures show vestigial wings and the other 40 have the wild-type phenotype, the penetrance of the recessive allele in the population is 60 percent. Penetrance depends greatly on both environmental conditions and the influence of other genes. In another group of *vg/vg* flies, the gene might show 20 percent, or even 90 percent, penetrance. Because the penetrance of a gene can vary so much from one population to the next, we use the catchall term *incomplete penetrance* to describe any gene that expresses the expected phenotype less than 100 percent of the time.

Incomplete penetrance often complicates pedigree analysis. In the case of a dominant trait, for example, an incompletely penetrant gene can cause an apparent skipping of generations, which violates the criteria for dominant inheritance. This problem occurs in the pedigree of brachydactyly shown in Fig. 6.15. The daughter in generation II apparently has the dominant gene but does not express it. Her son inherits the dominant gene from her and does exhibit the phenotypic characteristics of the disorder.

Even if a gene is 100 percent penetrant, its degree of expression, or **expressivity,** may vary among individuals. For example, *vg/vg* flies raised at 70°F all have vestigial wings, but the precise wing size varies. In such a case, the gene is said to exhibit *variable expressivity.* As with incomplete penetrance, the reasons for such variation in phenotypic expression can be genetic, environmental, or a combination of the two, and are usually unknown.

A human characteristic that shows variable expressivity in terms of the age of onset is Huntington's chorea, discussed in Chapter 1. The symptoms of this disease usually first appear in middle age, but certain individuals have developed symptoms as early as 30 years and as late as 60 years of age (Fig. 6.16). This variable expressivity can make it appear as if the genotype is incompletely penetrant. Since individuals who carry the dominant gene sometimes die from other causes before manifesting the symptoms of chorea, they are mistakenly recorded in the pedigree as being unaffected. Such a pedigree might then not meet the criteria for dominant inheritance.

EXAMPLE 6.4. The failure of II-2 in the pedigree below to express the trait designated by the filled-in symbols violates the pattern expected for simple dominant inheritance. Show that this apparent case of incomplete penetrance can be explained on the basis of two complementary gene pairs that interact to produce a single trait.

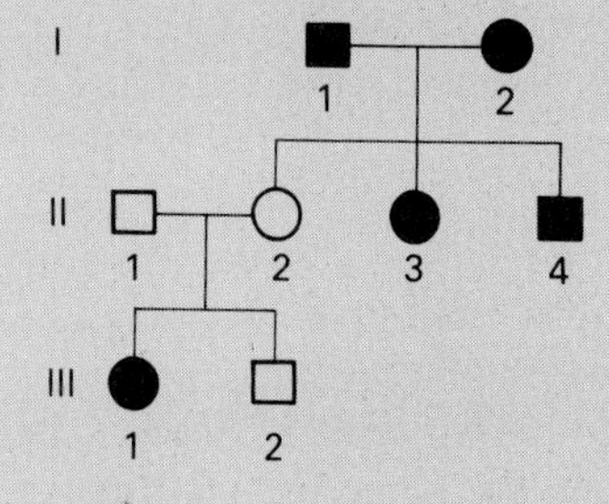

Solution. Assume that individuals with at least one dominant allele for each of two gene pairs (*A-B-*) show the trait in question. All other genotypes (*A-bb, aaB-,* and *aabb*) do not. One of many possible combinations of genotypes that could explain the pedigree is as follows:

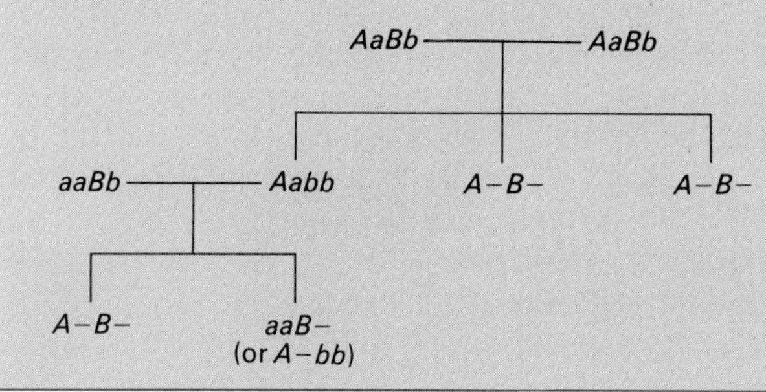

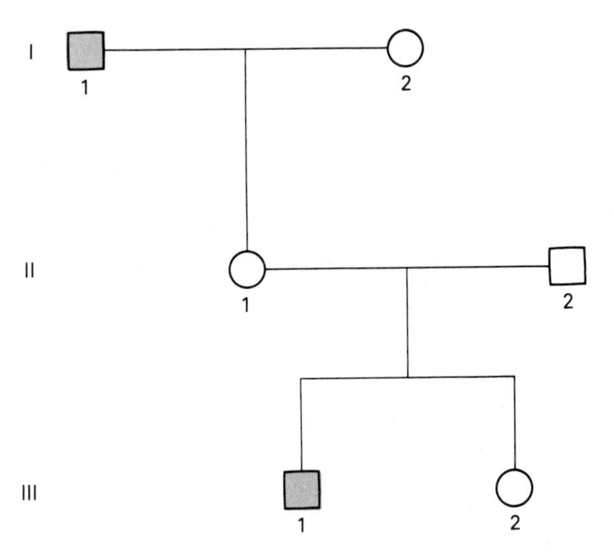

Figure 6.15. Part of the pedigree of a family that expresses the dominant character brachydactyly. Individual II-1 is heterozygous but, for unknown reasons, does not express the condition. The brachydactyly gene is said to exhibit incomplete penetrance.

Figure 6.16. Graph showing the distribution of the age of onset of Huntington's chorea. The vertical axis gives the proportion (as percentage) of heterozygotes (*Hh*) who show symptoms of the disease. *Source:* From data of T. E. Reed and J. H. Chandler. *Am. J. Hum. Gen.* 10:201–225, Copyright 1958 by The University of Chicago.

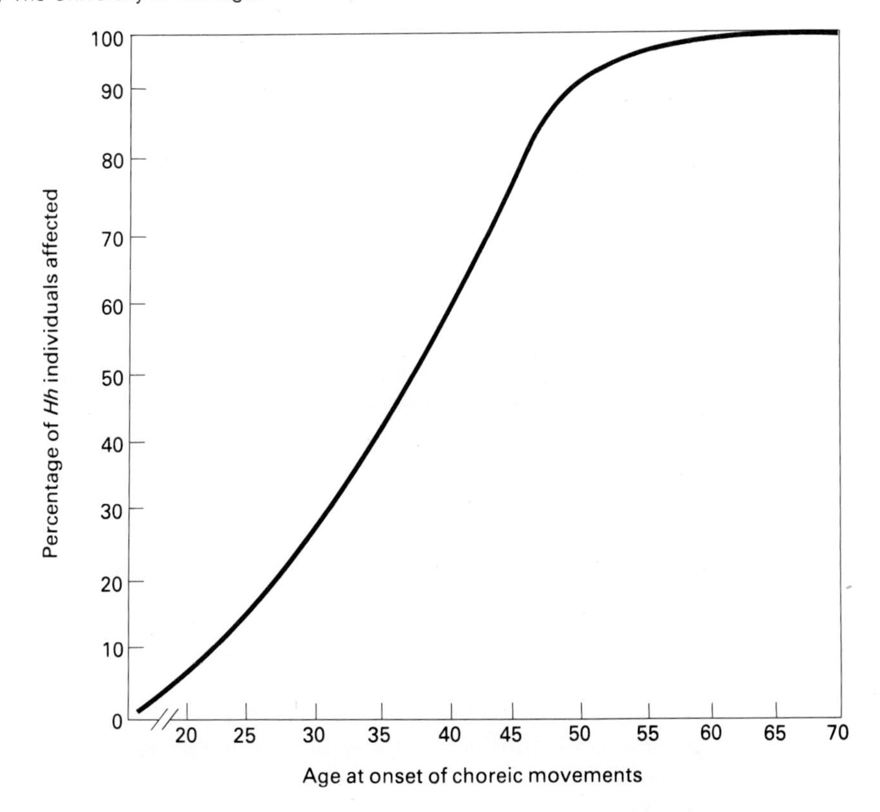

TO SUM UP

1. A pedigree is a way of classifying data on a family to show the inheritance pattern of a genetic characteristic. In many cases, this pattern has features that suggest an autosomal dominant or recessive basis for the characteristic.
2. When a gene is located on an autosome, it is transmitted independently of the sex of the individual. The inheritance pattern of a dominant gene shows that (a) affected offspring have at least one affected parent and (b) two unaffected individuals do not have affected progeny. In recessive inheritance, unaffected individuals may have one or more affected offspring.
3. In X-linked inheritance, pedigrees show differences in phenotypic ratios between the sexes. If the male is the heterogametic sex, then more males than females express the recessive phenotype, and more females than males express the dominant phenotype. An X-linked gene also gives different results for reciprocal crosses.
4. A crisscross inheritance pattern is another indication of X-linkage. Father-to-grandson and mother-to-son patterns are frequently observed for X-linked recessive characteristics. A father-to-daughter pattern is typical of an X-linked dominant gene.
5. Holandric, or Y-linked, genes have an inheritance pattern in which all male descendants of an affected male express the trait. These genes are not transmitted to females.
6. Sex-limited characters are expressed only by one sex. These characteristics are usually controlled by autosomal genes whose action depends on the presence of certain sex hormones.
7. Sex-influenced traits are determined by genes whose dominance relationship is controlled by the sex of the individual. The allele that is dominant in one sex is recessive in the other.
8. Inheritance patterns are sometimes complicated by environmental and genetic factors that influence the expression of genes. Genes that are incompletely penetrant are not always expressed by the individuals who carry them. Incomplete penetrance poses a particular problem in identifying a dominant inheritance pattern, since the criteria for dominant inheritance may be violated. Genotypes can also exhibit variable expressivity, so that individuals of the same genotype differ from one another in their degree of phenotypic expression.

Multiple Alleles and Genetic Variability

All the inheritance patterns we have discussed so far involve a single gene locus with two alleles. But as we have pointed out in Chapter 3, the number of allelic forms of a gene is not limited to two. A new allele arises as a mutation of a wild-type gene or of another allele. These mutations occur many times during the history of a species. Therefore, several allelic forms of a gene can exist in a population, even though a single diploid individual can possess at most two of these alleles (if heterozygous) and may only possess one (if homozygous). When three or more allelic forms of a gene exist in a population, they are referred to as **multiple alleles.**

Combining Alleles into Genotypes

As the number of alleles in a multiple allelic series increases, the number of diploid genotypes that can be constructed from this series also increases. Let us consider a locus with only three possible alleles: A_1, A_2, and A_3. These alleles yield three homozygous genotypes—A_1A_1, A_2A_2, and A_3A_3—and three heterozygous genotypes—A_1A_2, A_1A_3, and A_2A_3. A total of six different genotypes can thus be found in a population. By the addition of just one allele beyond the traditional Mendelian number of two, we have doubled the number of possible genotypes, increasing them from three to six.

The increase in the number of genotypes becomes even greater with a larger number of alleles. We will take the white-eye locus in *Drosophila* as an example. Besides the red (wild-type) and white alleles previously discussed, many other alleles for this locus are known to exist. Fourteen of the many alleles in this series are w^+ = red, w^{cf} = coffee, w^{col} = colored, w^{bf} = buff, w^{sat} = satsuma, w^a = apricot, w^{ec} = ecru, w^{dp} = deep purple, w^{co} = coral, w^{ch} = cherry, w^h = honey, w^e = eosin, w^{sp} = spotted, and w = white. These alleles yield 14 different possible homozygous genotypes, one for each allele. A large number of heterozygous genotypes are also possible. The number of heterozygous genotypes can be calculated using the combination formula we learned in Chapter 2. Let the symbol N represent the number of alleles in a multiple allelic series (in this particular case, $N = 14$). Any two of the 14 different alleles can be combined to produce a heterozygote. If we call the two alleles that are chosen successes, then the remaining 12 are called failures. The number of combinations that gives two successes and 12 failures is $14!/2!12! = 91$. So 91 different heterozygotes can be formed from 14 alleles. When we add in the 14 homozygotes, the total number of possible genotypes becomes $91 + 14 = 105$.

We can easily generalize this formula to allow us to calculate the total number of different genotypes for any value of N. The number of different homozygous genotypes is simply N. The number of different heterozygous genotypes is then $N!/2!(N - 2)!$. Adding together the homozygotes and heterozygotes, the total number of genotypes is then $N + N!/2!(N - 2)!$. If we simplify this formula algebraically to eliminate the factorials, it is equal to the quantity $N(N + 1)/2$. Thus we can write the formula as follows:

	Number of homozygotes	+	Number of heterozygotes	=	Total number of genotypes
For N alleles:	N	+	$\frac{N!}{2!(N-2)!}$	=	$\frac{N(N+1)}{2}$

Dominance Relationships and Number of Phenotypes

When all the alleles in a multiple allelic series show either incomplete dominance or codominance, each genotype has its own distinct phenotype. In this case, the number of possible phenotypes equals the number of genotypes. This type of dominance relationship yields the maximum number of possible phenotypes for any specified number of alleles. At the other extreme, the alleles in a series may show a simple hierarchy of complete dominance (e.g., A_1 is dominant to A_2 is dominant to A_3). With simple dominance, each allele in the series is dominant to all the remaining alleles in the series. In this case, the number of possible phenotypes is the minimum number, which is equal to the number of alleles.

A classical example of a hierarchy of dominance with multiple alleles was discovered several years ago in rabbits (Fig. 6.17). The wild-type coat color in

Figure 6.17. Coat color in rabbits. Wild-type or agouti color (a) is the result of a black or brown tip on each hair, succeeded by a yellow band, with the portion of the shaft nearest the skin being gray. Chinchilla (b) rabbits lack the yellow band and so appear silver-gray. Himalayan color (c) is white with black extremities, and albino (d) rabbits totally lack pigmentation.

rabbits is agouti, which is determined by gene c^+. Several alternative colors exist, including chinchilla, Himalayan, and albino. Chinchilla rabbits are gray, as a result of a mixed pattern of a color and white. The gene for chinchilla is an allele of c^+, designated c^{ch}. Himalayan rabbits are white with black extremities (ears, nose, and tail). The Himalayan gene is also allelic to c^+, and is denoted c^h. The differential coloration that is characteristic of Himalayan rabbits is due to the effect of the environment on the expression of the c^h gene. At the lower temperatures found in the extremities of the body, the enzyme that is specified by the Himalayan allele is active, yielding black pigmentation. However, under the influence of warmer body temperatures, the enzyme specified by the c^h allele is inactivated, and no pigment forms, yielding white fur. Some rabbits are totally white; this coloration is determined by a fourth allele in the series, designated c.

Controlled matings between rabbits of the various colors show that c^+ is dominant to all the other alleles, c^{ch} is dominant to c^h and c, and c^h is dominant to c. The dominance hierarchy can therefore be summarized as $c^+ > c^{ch} > c^h > c$. These alleles yield four phenotypes, which is the same as the number of alleles.

The relationship between the genotypes and phenotypes is as follows:

GENOTYPE	PHENOTYPE
c^+c^+, c^+c^{ch}, c^+c^h, c^+c	Agouti
$c^{ch}c^{ch}$, $c^{ch}c^h$, $c^{ch}c$	Chinchilla
c^hc^h, c^hc	Himalayan
cc	Albino

Several other series of multiple alleles are known to have a simple hierarchy of dominance. But many series of multiple alleles exhibit a complex mixture of dominance, with some alleles showing complete dominance and others being incompletely dominant or codominant. We cannot formulate a general rule to relate the number of phenotypes to the number of alleles in the series, since each pair of alleles in a series will have its own particular dominance relationship. We will next consider the ABO and Rh blood group systems, two examples of allelic series that show a complex pattern of dominance relationships. Each of these systems has its own special mix of completely dominant and codominant alleles.

EXAMPLE 6.5. One multiple allelic trait that is of interest to animal breeders is the pattern of white spotting that characterizes various breeds of cattle. Four spotting patterns are known. They have the following order of dominance: Dutch belt > white-faced or Hereford-type spotting > solid color > Holstein- and Guernsey-type spotting. How many different genotypes (in total) are involved in the determination of spotting pattern?

Solution. There are four phenotypes in this simple hierarchy of complete dominance. There must then be four alleles responsible for these four phenotypes. If we let $N = 4$ and apply the formula $N(N + 1)/2$, we calculate the number of genotypes to be $(4)(5)/2 = 10$.

EXAMPLE 6.6. In corn, different varieties produce kernels that turn either red, orange, or pink when exposed to sunlight. In contrast, normal kernels remain yellow. Crosses are made with the following results:

P	F_1	F_2
red × orange	all red	¾ red : ¼ orange
red × pink	all red	¾ red : ¼ pink
orange × pink	all orange	¾ orange : ¼ pink

Determine the number of alleles involved and arrange the phenotypes in order of dominance.

Solution. Each series of crosses yields the results expected for a simple pair of alleles in which one allele is completely dominant to the other. The number of alleles involved is three, which is equal to the number of phenotypes. The dominance hierarchy is red dominant to orange dominant to pink (red > orange > pink).

The ABO Blood Group. The allelic series responsible for the ABO blood group consists of three major alleles, designated I^A, I^B, and i. The I^A and I^B alleles in the series are codominant with each other and are completely dominant to i (we symbolize this relationship as $I^A = I^B > i$). This dominance relationship leads to the following phenotypes and genotypes:

Phenotype (blood group):	A	B	AB	O
Genotypes:	I^AI^A and I^Ai	I^BI^B and I^Bi	I^AI^B	ii

The type of blood a person has is based on the kind of *antigen* present on the surface of the red blood cells. Antigens are complex molecules, usually proteins or protein derivatives. If the body recognizes these molecules as being foreign to it, *antibodies* will be formed, thereby initiating an immune response. Antibodies are protein molecules produced by an individual in response to a foreign antigen. In the ABO system, the antigens are large glycoprotein (sugar-protein complex) molecules which make up part of the membrane of the red blood cell. The two types of antigens—type A and type B—differ from one another in the precise nature of the sugar portion of the molecule. Both antigen types arise from a precursor or core molecule known as the H substance, which is normally located on all red blood cells (Fig. 6.18). An enzyme coded for by the I^A allele catalyzes the conversion of the H substance to A antigen, whereas an enzyme coded for by

Figure 6.18. Biosynthesis of the A and B antigens on the red blood cell. The A and B antigens are formed from H substance. By the action of specific enzymes, different sugars (N-acetylgalactosamine for antigen A and galactose for antigen B) are added to the carbohydrate portion of the glycoprotein molecule. The core molecule, H substance, is also formed by a complex series of reactions under the direction of different genes.

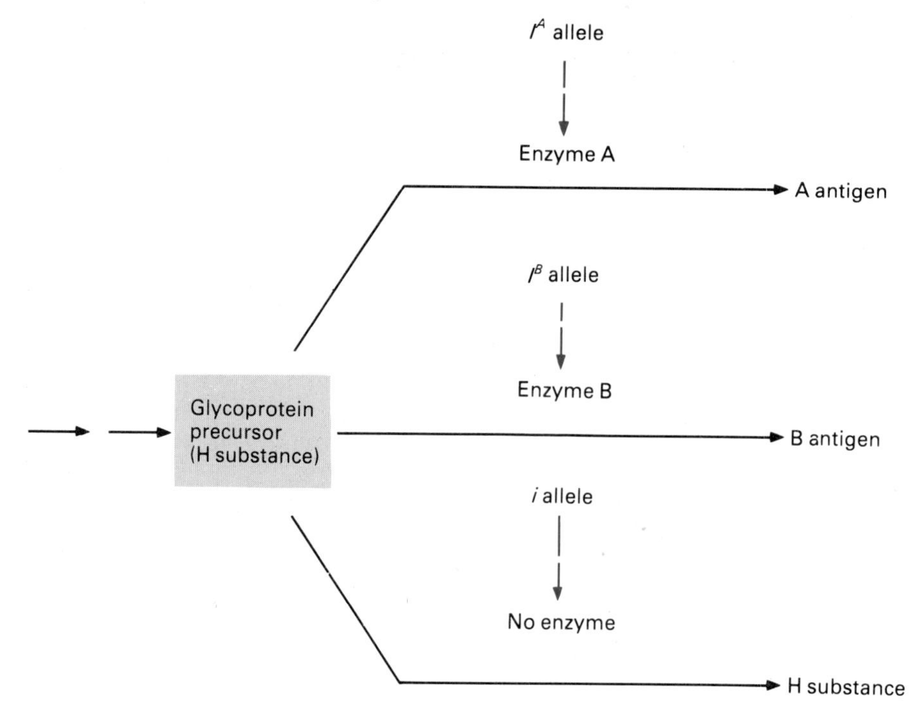

the I^B allele changes the H substance to B antigen. The i allele codes for neither enzyme, so substance H remains on the red blood cells of ii individuals in its unconverted form. Thus, individuals with the I^AI^A or I^Ai genotype have only type A antigen, those with the I^BI^B or I^Bi genotype have only type B antigen, I^AI^B heterozygotes have both antigens, and ii individuals have neither antigen.

EXAMPLE 6.7. A family consists of four children, one type A, one type B, one type AB, and one type O. What are the genotypes of the parents?

Solution. The type O offspring, being ii in genotype, must have received an i allele from each of its parents. The AB offspring, being I^AI^B in genotype, must have received an I^A allele from one parent and an I^B allele from the other. Thus, one parent must be I^Ai (type A), and the other must be I^Bi (type B).

The classical test for ABO blood type is based on an agglutination, or clumping, reaction that occurs when red blood cells that contain one or both antigens are mixed with a serum containing a specific antibody. (Serum is the fluid portion of the blood, with the blood cells removed.) While each person is capable of producing a tremendous variety of different antibodies, each type of antibody can recognize and combine with only one particular kind of antigen. The reaction between antigens and antibodies is therefore highly specific.

In 1900, Karl Landsteiner, working in Vienna, observed that red blood cells collected from one person would sometimes coalesce into visible clumps when mixed with serum from another person. When clumping occurs, the antigens located on the red blood cells of the first individual have reacted with antibodies in the serum of the second person. This reaction causes the red blood cells to adhere to one another so that they precipitate out of the suspension of blood, forming a clumped mass of cells (Fig. 6.19). Because of the specificity of antigen-antibody recognition, the A antigen reacts only with anti-A antibody, and the B antigen reacts only with anti-B antibody. Therefore, red blood cells from type A individuals will clump only when exposed to anti-A serum, those from type B individuals will clump only when exposed to anti-B serum, those from type AB individuals will clump when exposed to either antiserum, and those from type O individuals will react with neither antiserum (Fig. 6.20). These antigen-antibody reactions form the basis of testing for blood type. Landsteiner received a Nobel prize in 1930 for his investigations of these reactions.

EXAMPLE 6.8. Using lectins (plant extracts that agglutinate red blood cells), it is possible to distinguish different subgroups of type A blood. The two major divisions are A_1 and A_2; approximately 80 percent of type A individuals are A_1. The alleles that govern the two major subgroups of type A blood show the following dominance relationships to each other and to the other alleles in the ABO series: $(I^{A1} > I^{A2}) = I^B > i$. **(a)** If we include the two major subgroups of type A, how many different genotypes are possible in the ABO system? **(b)** How many different blood types do these genotypes produce?

Solution. **(a)** There are now four alleles in the series. Letting $N = 4$, we obtain $(4)(5)/2 = 10$ different genotypes. **(b)** There are six distinguishable blood types, distributed among the genotypes as follows:

Blood types:	A_1	A_2	B	A_1B	A_2B	O
Genotypes:	$I^{A1}I^{A1}$	$I^{A2}I^{A2}$	I^BI^B	$I^{A1}I^B$	$I^{A2}I^B$	ii
	$I^{A1}I^{A2}$	$I^{A2}i$	I^Bi			
	$I^{A1}i$					

A remarkable feature of the ABO system is that each person automatically has antibodies in his or her blood against whichever antigen is not present on his or her red blood cells. In other words, prior contact with a foreign A or B antigen is not a prerequisite for antibody production. Each blood type therefore contains the

Figure 6.19. The reaction between red blood cell antigens and antibodies against them. (a) The appearance of red blood cells under the microscope (about 500× magnification). In (b), these cells have been clumped by the addition of antibody against the cell surface antigens. Clumps form as cells associate with each other through cell-to-cell linkage, caused by combination of antibody with antigen. A molecular view of this reaction is shown in (c) as a lock and key type based on complementary shapes between antigen and antibody molecules.

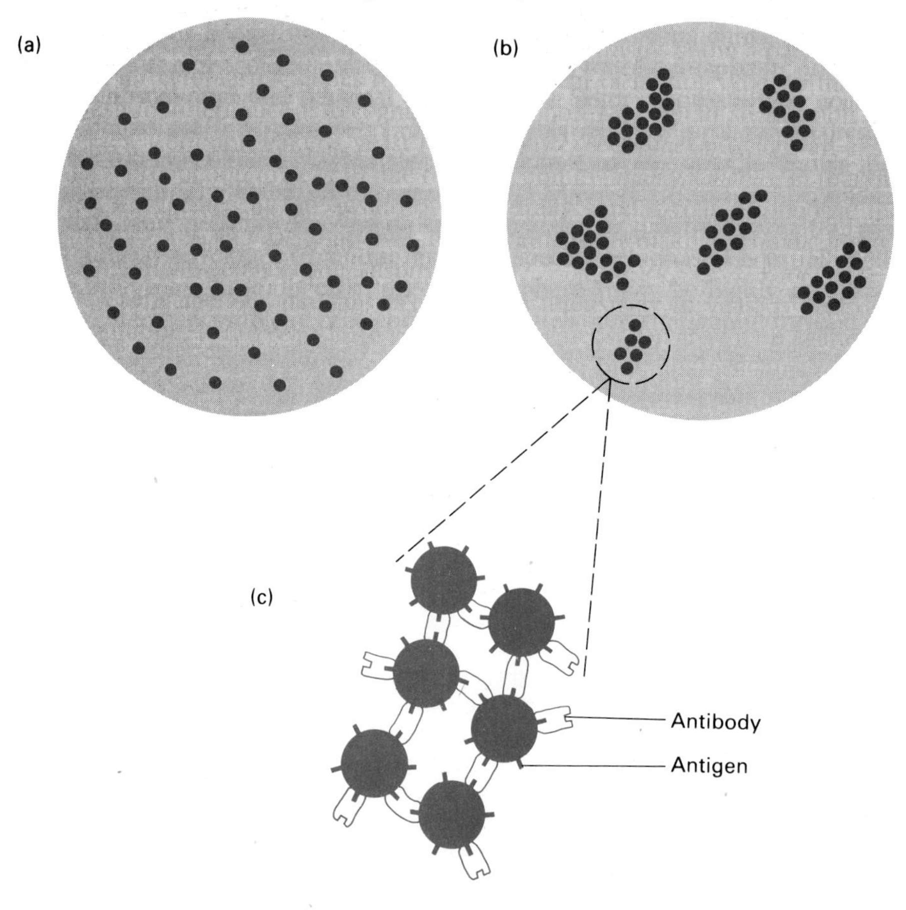

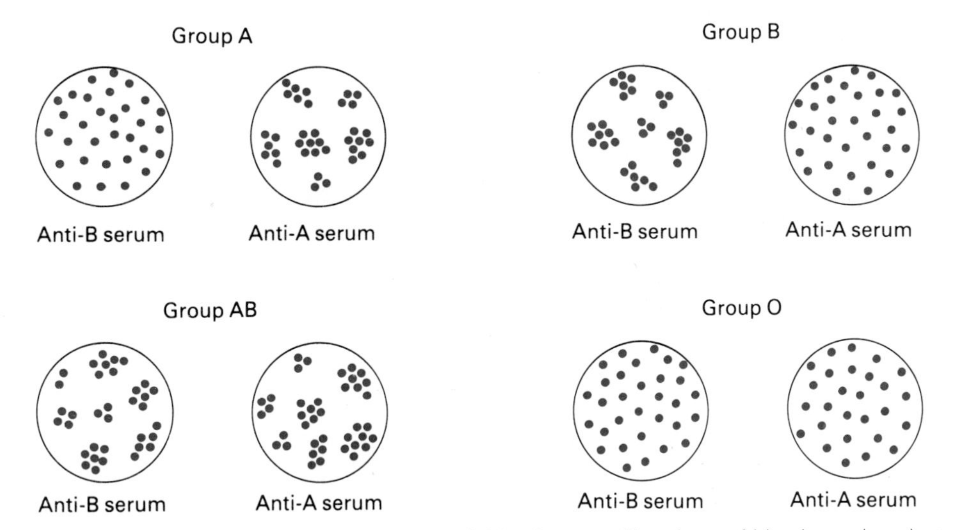

Figure 6.20. Agglutination reactions of the ABO blood groups. Two drops of blood are placed on a slide. To the first drop is added serum containing antibody against the B antigen. To the second drop is added antibody against the A antigen. Blood type A clumps only when exposed to anti-A serum; blood type B clumps only when exposed to anti-B serum; blood type AB clumps when exposed to either antiserum; and blood type O will not clump when exposed to either antiserum.

following antigens and antibodies:

BLOOD TYPE	ANTIGEN ON CELLS	ANTIBODY IN SERUM
A	type A only	anti-B only
B	type B only	anti-A only
AB	types A and B	neither anti-A nor anti-B
O	neither A nor B	both anti-A and anti-B

In most of the other immune systems of the body, actual contact with a foreign antigen is required before the specific antibodies are produced. The automatic presence of anti-A and anti-B in certain blood types explains Landsteiner's findings of clumping when the blood of two persons of different blood types is mixed. Knowledge of this system gives us information on which combinations are incompatible in blood transfusions. Clumping in the bloodstream of the recipient of a transfusion could be fatal. An incompatible combination occurs when the red blood cells of the donor carry an antigen that corresponds to the antibodies in the serum of the recipient. For example, a type A donor and type B recipient would be an incompatible combination, since the serum of a type B person contains anti-A antibodies.

The presence of antibody is also taken into account in the cross-matching test used to determine compatibility for transfusions. Type O individuals, who have neither antigen, are referred to as "universal donors." Type O blood does contain both antibodies, however, so that in practice it is transfused to persons whose blood type is other than O only in true emergencies. In such situations, the blood is transfused very slowly, to allow the antibodies present in the donor serum to

Table 6.2. The compatibility of ABO blood types for transfusions.

Recipient blood type	Donor blood type A	B	AB	O
A	C	I	I	E
B	I	C	I	E
AB	E	E	C	E
O	I	I	I	C

C = compatible; transfusion permitted as long as all other blood group systems are also compatible
I = incompatible; transfusion not permitted because of reaction of antigen on donor cells with antibody in recipient's plasma
E = transfusion permitted under emergency circumstances; antibodies in donor serum react with antigens of recipient

dilute rapidly in the blood of the recipient. Table 6.2 lists the ABO combinations that are permitted and prohibited in transfusions.

Even though as many as 500,000 A or B antigenic sites may be located on a single red blood cell, other blood group antigens are also known to be present. At least 20 other blood group systems have been identified, stemming from a minimum of 20 different gene loci that are involved in antigen determination. These other blood group systems include MN, Duffy, Diego, Kell, Lewis, Lutheran, and Rh, just to name a few. Because the antigen-antibody reactions of most of these systems do not have the severe effects caused by ABO or Rh incompatibility, they are not routinely tested for in cross-matching for blood types. The MN blood group, for example, consists of two very weak antigens, type M and type N, specified by the alleles L^M and L^N, respectively. These alleles are codominant, yielding three blood group phenotypes: M (L^ML^M), MN (L^ML^N), and N (L^NL^N). Most individuals fail to react (or react very weakly) to the M and N antigens, so they are of no importance in transfusions.

In addition to their role in the immune response, blood group antigens have other important applications. Blood tests involving the ABO series, as well as other blood groups, are used in medical and legal situations. In cases of disputed parentage, the results of these tests can be of considerable value in determining whether or not a person is the parent of a particular child. Evidence from blood tests is only conclusive in excluding the possibility of parentage, however. These tests cannot prove that a person is the parent of a child. Thus, in cases of illegitimacy or abandonment, most courts will admit this genetic evidence only when it provides information that rules out parentage by the person in question.

The Rh Blood Group. The Rh blood group system is well known for the incompatibility reaction it can cause in the fetus or newborn infant. As in the ABO system, the responsible antigen is located on the surface of red blood cells in individuals of certain genotypes. Only 10,000 to 20,000 Rh antigen sites occur on the surface of a red blood cell, so they are less numerous than the sites for the A and B antigens. The Rh antigen is commonly known as Rh factor; individuals who have Rh factor are described as being Rh-positive. Approximately 85 percent of Americans are Rh-positive. The smaller percentage of people who do not have the Rh factor are described as being Rh-negative.

The genetic basis of the Rh system is not clearly understood. We know that at least ten distinct classes of Rh antigen exist. Two major hypotheses have been advanced to explain the situation, one by Weiner and another by Fisher and Race. According to Weiner, a single gene locus with multiple alleles, similar to the ABO system, is responsible for the Rh trait. Weiner identifies at least ten alleles at this locus, the most common being r, r', r'', R^0, R^1, R^{1w}, R^2, and R^z. Each of these alleles determines one or more variant types of the Rh antigen, the most common being Rh_0, hr″, hr′, rh′, rh, hr, and rh″. Table 6.3 shows the types of antigen specified by each allele. The type that causes the major problem in transfusions and in pregnancies is the one symbolized Rh_0; this antigen is produced by alleles R^0, R^1, R^{1w}, R^2, and R^z, but not by any of the small-letter alleles. Therefore, we designate the phenotype of any genotype with one or more capital-letter alleles as Rh-positive and the phenotype produced by any combination of small-letter alleles as Rh-negative. Although we all produce some kind of Rh antigen, which makes us all therefore technically Rh-positive, only those individuals who produce the potent Rh_0 antigen are given the positive designation. If we use the symbol R to represent any one of the capital-letter alleles and the symbol r to represent any one of the small-letter alleles, we can approximate the genetic basis for the Rh trait as a two-allele system. Any R- genotype would be Rh-positive and the rr genotypes would be Rh-negative.

In contrast, Fisher and Race hypothesize that at least three loci are involved in the production of antigen, rather than just one locus with several alleles. These three loci are designated d, c, and e, and they are assumed to be located very close to one another on the same chromosome. Each locus has at least two alleles (D and d; C, C^w, and c; E and e), yielding many possible genotypes, the most common

Table 6.3. Two hypotheses for the Rh gene system and its antigens.

GENES PRESENT			ANTIGENS							
			Rh_0	hr″	hr′	rh′	rh	hr	rh″	(WEINER)
WEINER	FISHER-RACE		D	e	c	C	Ce	ce	E	(FISHER-RACE)
R^0	Dce	Rh-positive	+	+	+	−	−	+	−	
R^1	DCe	Rh-positive	+	+	−	+	+	−	−	
R^{1w}	DC^we	Rh-positive	+	+	−	+	+	−	−	
R^2	DcE	Rh-positive	+	−	+	−	−	−	+	
R^z	DCE	Rh-positive	+	−	−	+	−	−	+	
r	dce	Rh-negative	−	+	+	−	−	+	−	
r'	dCe	Rh-negative	−	+	−	+	+	−	−	
r''	dcE	Rh-negative	−	−	+	−	−	−	+	

of which are listed in the second column of Table 6.3. The major antigen, which was symbolized Rh_0 in the Weiner system, is called the D antigen by Fisher and Race. All the genotypes that have at least one *D* allele produce this antigen and are Rh-positive. Those genotypes that include the homozygous *dd* condition do not produce this class of antigen and are Rh-negative.

The presence or absence of the Rh_0 (or D) antigen is routinely determined for blood donors and recipients. This procedure is done to avoid complications that would arise if Rh-positive blood were transfused into an Rh-negative person, causing the recipient to develop anti-Rh antibodies. As in most immune systems (the ABO system is an exception), antibody production occurs only when an Rh-negative individual is exposed to the antigen. The exposed individual is then said to have been sensitized or immunized.

Very serious problems can develop when an Rh-negative woman is sensitized and subsequently becomes pregnant with an Rh-positive child. During the third trimester of pregnancy, some of the mother's antibodies are transmitted through the placenta into the fetal bloodstream. These antibodies provide the infant with immunity to various diseases for three to six months after birth. But the transmission of anti-Rh antibodies can have harmful effects. The red blood cells of an Rh-positive fetus may react with anti-Rh antibodies obtained from the mother, causing severe hemolytic anemia (lysis of the fetal red blood cells). Immature red blood cells (erythroblasts) enter the fetal circulation to compensate for the loss of mature red blood cells. This action creates the condition known as *erythroblastosis fetalis.* The severity of the condition varies greatly. Over 10 percent of the affected fetuses are stillborn prior to full term. A live-born infant may be premature; if carried to full term, it may require an immediate mass transfusion of blood, known as a perfusion. In the mildest cases, the infant may suffer only from mild jaundice, so that a transfusion is not necessary.

An Rh-negative woman is usually sensitized not through transfusion but through an accidental leakage of blood from an Rh-positive fetus into her bloodstream during a previous pregnancy. The leakage often occurs through the placental barrier shortly before or during delivery. This leakage causes no harm to the child, but it may sensitize the mother. Once sensitized, an Rh-negative woman responds very strongly to further contact with the Rh_0(D) antigen. A strong antibody response to an Rh-positive fetus in a subsequent pregnancy (Fig. 6.21a) can cause destruction of fetal red blood cells. For this reason, Rh incompatibility is usually not a problem with the firstborn Rh-positive child; instead it affects one or more later pregnancies. However, even in cases where the husband of the Rh-negative woman is homozygous *RR,* making all fetuses Rh-positive, the complications of Rh incompatibility arise in only about 12 percent of pregnancies. This low frequency apparently reflects variations in the degree of fetal red blood cell leakage into the mother during subsequent pregnancies, in the antibody response of the mother, or in the susceptibility of the fetus.

A technique was recently developed for transfusing blood into the fetus during the third trimester of pregnancy. This technique has saved many of the "Rh babies" that otherwise would have died. An even more recent procedure involves the use of Rh_0(D) antibody that has been isolated from human plasma. This anti-Rh_0(D) antibody is injected into the Rh-negative mother within 72 hours of delivery of

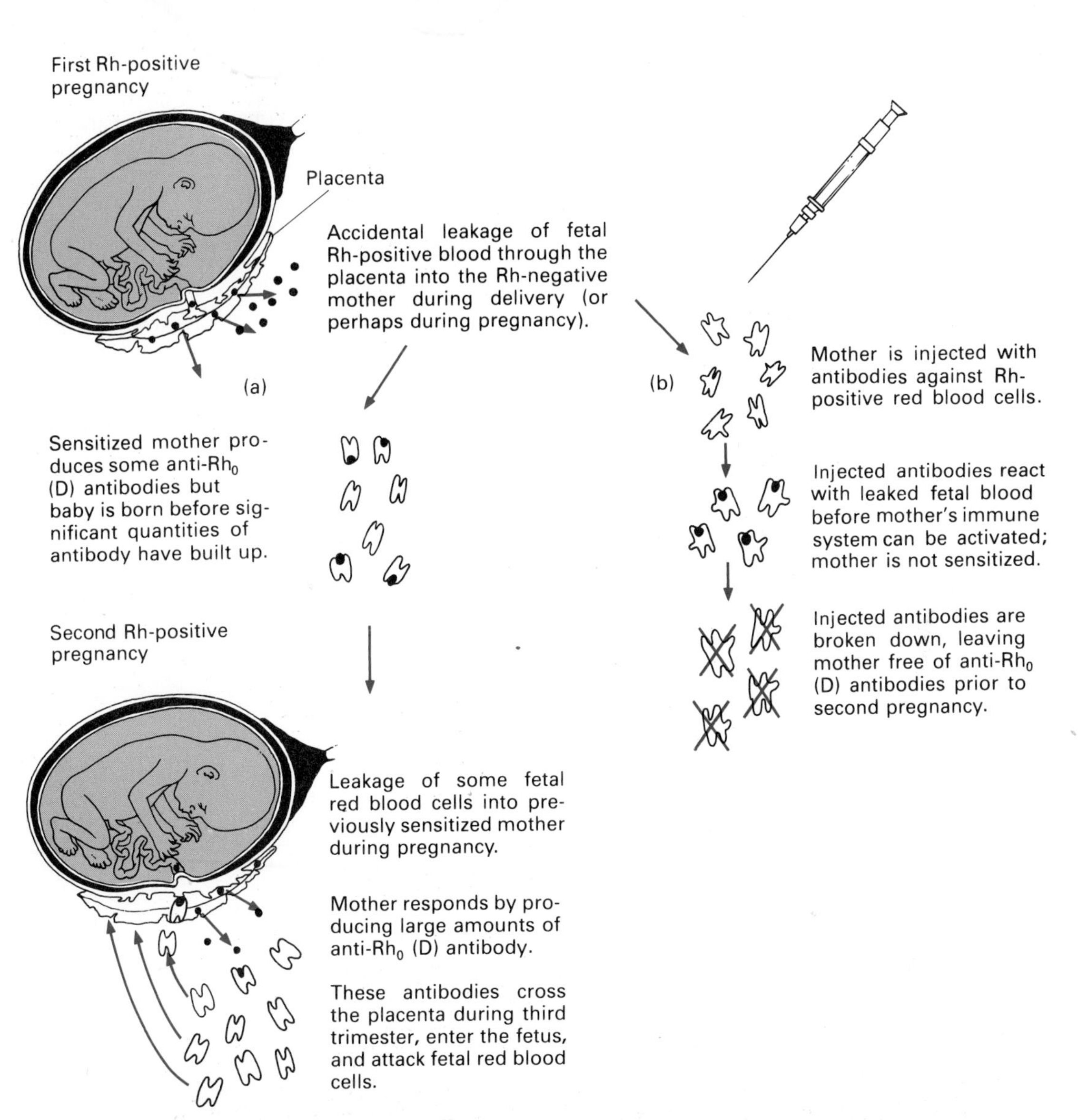
First Rh-positive pregnancy
Placenta
Accidental leakage of fetal Rh-positive blood through the placenta into the Rh-negative mother during delivery (or perhaps during pregnancy).
(a)
(b)
Mother is injected with antibodies against Rh-positive red blood cells.
Sensitized mother produces some anti-Rh_0 (D) antibodies but baby is born before significant quantities of antibody have built up.
Injected antibodies react with leaked fetal blood before mother's immune system can be activated; mother is not sensitized.
Second Rh-positive pregnancy
Injected antibodies are broken down, leaving mother free of anti-Rh_0 (D) antibodies prior to second pregnancy.
Leakage of some fetal red blood cells into previously sensitized mother during pregnancy.
Mother responds by producing large amounts of anti-Rh_0 (D) antibody.
These antibodies cross the placenta during third trimester, enter the fetus, and attack fetal red blood cells.

Figure 6.21. (a) The course of events leading to hemolytic disease of the newborn (erythroblastosis fetalis). An Rh-negative woman is exposed to the Rh factor during pregnancy with an Rh-positive fetus. Sensitization of the mother leads to a strong antibody response upon subsequent exposure to the Rh antigen in succeeding pregnancies. (b) Protection against hemolytic disease of the newborn. The Rh-negative mother is injected with anti-Rh_0(D) antibody shortly after each pregnancy with an Rh-positive fetus (following live birth, stillbirth, or abortion). The injected antibody destroys any fetal red blood cells that have leaked into the mother, eliminating the need for her to produce her own antibodies. She is therefore not sensitized, so that subsequent pregnancies are in no danger from Rh-incompatibility.

an Rh-positive child. If any blood from the fetus has accidentally entered the mother's bloodstream, the injected antibody will combine with and destroy the Rh antigens on these cells so that the mother's immune system is not stimulated to produce her own antibody molecules. The injected antibody therefore, in effect, suppresses antibody response by the mother, so that she is not sensitized. The injected antibody dissipates within a few months, presenting no danger to the fetus during subsequent pregnancies (Fig. 6.21b).

Tests for Allelism

Alternative phenotypes for a single trait can be the result of two or more alleles at a single locus or the expression of genes at two or more separate loci. When a new phenotype appears in a population, therefore, geneticists usually begin by attempting to determine whether the gene for the variant phenotype is allelic or nonallelic to any other gene known to affect the trait in question.

For example, let us suppose that in a certain species of plants that normally produce purple flowers, we discover two true-breeding variants: one with red flowers and another with yellow flowers. When each variant is mated back to the standard type, only purple-flowered offspring are produced. Are these differences in flower color caused by different alleles at the same locus, or are they a result of mutational changes at separate loci? If we hypothesize different alleles at the same locus, we might choose to designate three alleles: A (purple), a_1 (red), and a_2 (yellow), with A dominant to the others. The genotypes of true-breeding purple, red, and yellow varieties would then be AA, a_1a_1, and a_2a_2, respectively. If we hypothesize different gene loci, we might consider the complementary interaction of two gene pairs: A,a and B,b. We would then designate $aaBB$ as red, $AAbb$ as yellow, and $AABB$ as purple. In either case, crossing the variant types back to the purple variety would produce only purple-flowered offspring:

Case 1: $a_1a_1 \times AA \longrightarrow Aa_1$ and $a_2a_2 \times AA \longrightarrow Aa_2$
Case 2: $aaBB \times AABB \longrightarrow AaBB$ and $AAbb \times AABB \longrightarrow AABb$

One test that can be used to determine whether variant types are allelic is the **complementation test.** In this test, we cross the newly discovered true-breeding variants and observe the phenotype of the F_1. If the genes for the contrasting phenotypes are allelic, then all the offspring will exhibit a variant phenotype. For example, if case 1 is the correct hypothesis, then

$$a_1a_1 \times a_2a_2 \longrightarrow a_1a_2 \quad \text{(mutant)}.$$

The exact nature of the F_1 phenotype will be determined by the dominance relationship between the alleles. The phenotype produced by the a_1a_2 genotype might be the same as one of the parental phenotypes (e.g., red or yellow) or intermediate to them. But if the genes for the alternative phenotypes are nonallelic, then the wild-type forms of both gene pairs should complement each other in the heterozygous state. In other words, each gamete would contribute the wild-type gene missing in the other gamete. In this case, the presence of at least one wild-type gene at each locus is necessary for the completion of the standard phenotype. When both wild-type nonalleles are present, the F_1 will exhibit the standard (wild-type) character. Therefore, if case 2 is the correct hypothesis, then

$$aaBB \times AAbb \longrightarrow AaBb \quad \text{(wild-type)}.$$

Another test for allelic vs. nonallelic genes is the *analysis of segregation ratios* produced upon intercrossing the F_1. If the genes are allelic, we would predict a monohybrid ratio of 3:1 or 1:2:1 in the F_2, depending on the dominance relationship present. For example, if the three alleles in the example described above have a dominance hierarchy of $A > a_1 > a_2$, the results in the F_2 would be:

$$Aa_1 \times Aa_1 \longrightarrow \tfrac{3}{4}\ A\text{-} : \tfrac{1}{4}\ a_1a_1 \text{ (3 purple : 1 red)}$$
$$Aa_2 \times Aa_2 \longrightarrow \tfrac{3}{4}\ A\text{-} : \tfrac{1}{4}\ a_2a_2 \text{ (3 purple : 1 yellow)}$$
$$a_1a_2 \times a_1a_2 \longrightarrow \tfrac{3}{4}\ a_1\text{-} : \tfrac{1}{4}\ a_2a_2 \text{ (3 red : 1 yellow)}$$

These results contrast with the outcome expected for nonallelic genes in the case of the *AaBb* dihybrids. If the hypothesis of nonallelic genes is the correct one, intercrossing these F_1 would give a dihybrid ratio of 9:3:3:1 or 9:3:4, depending on the expression of the *aabb* genotype. Therefore, case 2 would yield the following results in the F_2 produced by *AaBb* F_1 intercrosses:

$$AaBb \times AaBb \longrightarrow \tfrac{9}{16}\ A\text{-}B\text{-} : \tfrac{3}{16}\ A\text{-}bb : \tfrac{3}{16}\ aaB\text{-} : \tfrac{1}{16}\ aabb$$
(9 purple : 3 yellow : 3 red : 1 *aabb* phenotype)

TO SUM UP

1. When more than two allelic forms for a gene locus exist, that locus is said to have multiple alleles. The existence of multiple alleles increases the number of possible genotypes. If N represents the number of alleles in the series, then the total number of possible genotypes is calculated as

$$\underbrace{N}_{\text{number of homozygotes}} + \underbrace{N!/2!(N-2)!}_{\text{number of heterozygotes}} = N(N+1)/2.$$

2. Each pair of alleles in a multiple allelic series has its own dominance relationship. When all alleles in a series show incomplete dominance or codominance, the number of possible phenotypes for the locus is equal to the number of possible genotypes.
3. The minimum possible number of phenotypes for a multiple allelic series occurs when the alleles of the series show a simple dominance hierarchy, in which each allele is completely dominant to all the remaining alleles in the series, when alleles are arranged in order of dominance. The potential number of phenotypes then equals the number of alleles.
4. The ABO blood group is an example of a multiple allelic series in which the alleles exhibit both codominance and complete dominance to other alleles in the series. The ABO system consists of three alleles, I^A, I^B, and i, which determine four blood groups, designated A, B, AB, and O. Alleles I^A and I^B are codominant with each other, and both are completely dominant to i. The blood types in this system are characterized by the presence (or absence) of the A and B antigens on the red blood cells and by the ability (or inability) to produce certain antibodies. The ABO system is of importance in blood transfusions in determining the compatibility of donors and recipients.
5. One test to determine whether two variant recessive genes that affect the same trait are alleles involves a cross between individuals that are homozygous for the variant phenotypes. If the F_1 have one or the other variant phenotype or a mixture of the two, then the genes are allelic. If the F_1 are wild-type in character, then the variants are caused by complementary genes at separate loci that happen to specify the same trait.
6. Another test to determine if genes are allelic is to perform an $F_1 \times F_1$ intercross and observe the phenotypic ratio in the F_2. If the variant genes are allelic, a monohybrid ratio of 3:1 or 1:2:1 should appear in the F_2. If the genes are nonallelic, a dihybrid ratio of 9:3:3:1 or some modification of this ratio should appear in the F_2.

Questions and Problems

1. Define the following terms and differentiate between paired terms:

antigen–antibody	multiple alleles
complementation test	sex-influenced gene
gene locus	sex-limited gene
hemizygous	sex-linked gene
holandric gene	variable expressivity
incomplete penetrance	

2. A notch in the tip of the ear is the expression of an autosomal gene in Ayrshire cattle. In the following pedigree, the filled-in symbols represent individuals with notched ears, and the open symbols represent individuals with normal ears. **(a)** Is the gene for notched ears dominant or recessive to its allele for normal ears? **(b)** Assume that III-2 and III-6 are mated. What is the probability that an offspring from this mating will have notched ears?

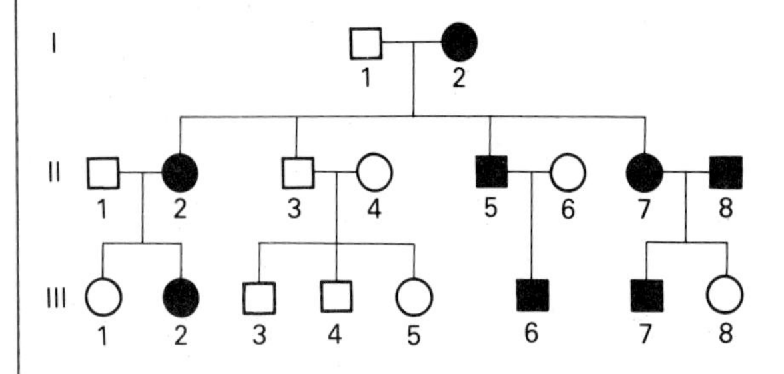

3. In the following pedigree, the filled-in symbols represent individuals with a hereditary form of deafness. Determine whether the inheritance pattern is consistent with a single pair of autosomal genes or requires the involvement of two or more complementary gene pairs.

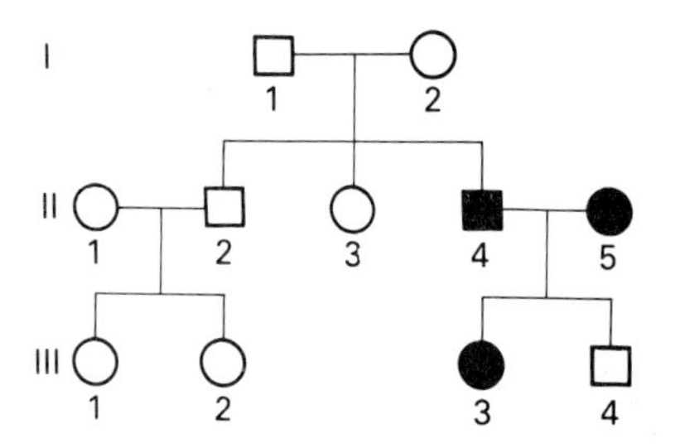

4. Show that in a family of four children, the probability of getting the expected ratio of 3 *A-* : 1 *aa* from the monohybrid cross *Aa* × *Aa* is less than ½.

5. Suppose that a comparatively rare trait in humans may express itself in either an autosomal dominant, autosomal recessive, X-linked dominant, or X-linked recessive mode of inheritance. Which of the following patterns would you *not* expect to find for each mode of inheritance? **(a)** Males and females affected equally. **(b)** Father and son both affected. **(c)** Excess of affected males. **(d)** Excess of affected females. **(e)** Both parents affected. **(f)** Parents closely related (e.g., first cousins).

6. Normal parents have a son affected with muscular dystrophy. What are the chances that any subsequent children will be affected if the form of muscular dystrophy is inherited as **(a)** an autosomal recessive trait? **(b)** An X-linked recessive trait?

7. Which of the following is the most likely mode of inheritance for each of the pedigrees shown below? **(a)** Autosomal dominant. **(b)** Autosomal recessive. **(c)** X-linked dominant. **(d)** X-linked recessive. Assume that the traits designated by the filled-in symbols in pedigrees 1 and 2 are rare in the general population.

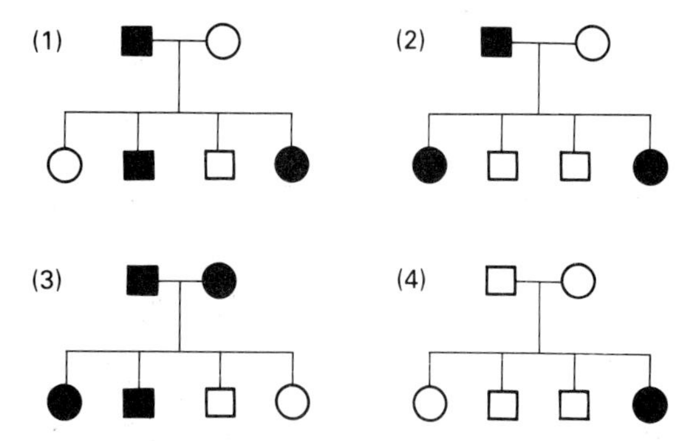

8. In the house cat (*Felix domesticus*), coat color among females can be yellow, black, or tortoiseshell (a mixture of black and yellow fur); males are either yellow or black. These coat colors are determined by an X-linked pair of alleles. **(a)** Using the first letter of the alphabet, give the genotypes that produce these different phenotypes for coat color in male and female cats. **(b)** Suppose that a mating produces a litter of three kittens, consisting of two black males and one yellow female. What are the colors of the male and female parents in this cross?

(c) A black cat gives a litter of two tortoiseshell and two black kittens. What is the sex of the black kittens?

9. In this chapter, we learned that eye color in *Drosophila* is determined by a series of alleles at a locus on the X chromosome. A particular mating yields a ratio in the progeny of 2 red-eyed females : 1 eosin-eyed male : 1 white-eyed male. Give the genotypes of the parents in this cross.

10. In another cross involving *Drosophila,* a female with short thoracic bristles is mated to a male with normal (long) thoracic bristles. The progeny produced by this cross occur in a ratio of 1 long-bristled female : 1 short-bristled female : 1 long-bristled male. Provide a genetic explanation for these results.

11. In the Barred Plymouth Rock breed of chickens, the barred pigmentation pattern consists of white bars on black feathers. This pattern is determined by a dominant Z-linked gene (*B*). The recessive allele of this gene (*b*) produces a nonbarred pattern in homozygous males and hemizygous females. **(a)** What phenotypic ratio is expected among the progeny of a cross between a barred female and a nonbarred male? **(b)** Another mating, in which both the parents are barred, produces a nonbarred female offspring. What are the genotypes of the parents in this cross? **(c)** A barred female loses her ovary through disease and is then sex-reversed to a functional male. This bird is mated to a normal nonbarred female. What phenotypic ratio is expected in the F_1?

12. Red-green color blindness is an X-linked recessive trait. A woman with normal color vision whose father was color blind has five sons. We know nothing about the color vision of her husband. **(a)** What is the probability that her oldest son is color blind? **(b)** What is the probability that three of her sons are color blind and two have normal vision?

13. An Rh-negative woman with normal vision marries an Rh-positive red-green–blind man whose sister had erythroblastosis fetalis. Of their two children, the oldest is color blind. What is the probability that the youngest child is **(a)** Rh-positive? **(b)** Red-green color blind? **(c)** An Rh-positive, normal-visioned girl?

14. In humans, the gene for pattern baldness is dominant in males and recessive in females. A nonbald man marries a nonbald woman whose mother is bald. What is the chance that their first child is a male with the genetic predisposition for baldness?

15. The presence of horns in sheep is a sex-influenced trait that is dominant in males and recessive in females. **(a)** The cross of a horned ewe (female) and a hornless ram produces a horned offspring. What is the sex of this offspring? **(b)** A series of crosses are performed in which hornless rams are mated to horned ewes and their offspring (the F_1) are then crossed among themselves. What phenotypic proportions are expected in the F_2?

16. Mahogany-and-white spotting in Ayrshire cattle is dominant in males and recessive in females. The reverse is true of red-and-white spotting. A farmer who has a herd of mahogany-and-white cattle purchases a red-and-white spotted bull. Describe the breeding procedure the farmer must use to establish a herd that is pure breeding for red-and-white spotting.

17. Hen-feathering is a sex-limited trait in poultry that is dominant in males. **(a)** A mating between a cock-feathered male and a hen-feathered female produces both cock-feathered and hen-feathered male offspring. What are the genotypes of the parents in this cross? **(b)** Suppose that the female in part (a), whose feathers are also barred (a dominant Z-linked trait), is mated to a hen-feathered nonbarred male. They produce several offspring; the first offspring to hatch becomes a cock-feathered, barred male. What phenotypic proportions are expected among their other offspring?

18. Tay-Sachs disease is a lethal and untreatable abnormality that is inherited as an autosomal recessive trait. A husband and wife both have siblings who died of this disorder. **(a)** What is the probability that their first-born child will have Tay-Sachs disease? **(b)** It is now possible to perform blood tests on suspected carriers to determine if the Tay-Sachs allele is present. What would your answer in part (a) be if the blood tests of this same husband and wife showed that both are carriers of this allele?

19. Suppose there are ten allelic forms of a gene at a given locus. Calculate **(a)** the largest number and **(b)** the smallest number of possible different phenotypes for this multiple allelic series. Indicate the dominance relationships of the alleles in each case.

20. In a certain species of plant, three alleles (A_1, A_2, A_3) have their locus on chromosome 1 and five alleles of another series (B_1, B_2, B_3, B_4, B_5) have their locus on chromosome 2. **(a)** How many different genotypes are theoretically possible for each allelic series considered separately? **(b)** How many different genotypes are theoretically possible for both allelic series considered together? **(c)** Studies conducted on the three allelic forms of gene *A* show the following dominance relationships: $A_1 = A_2 > A_3$. How many different phenotypes can be produced by this allelic series alone? **(d)** How many different phenotypes can be produced by the cross $A_1A_2B_1B_2 \times A_1A_3B_3B_4$, if the *B* alleles show a simple hierarchy of dominance ($B_1 > B_2 > B_3 > B_4 > B_5$)?

21. Show that by increasing the number of alleles (N) at a given locus the ratio of heterozygotes to homozygotes will increase as $(N - 1)/2$, while for a simple hierarchy of complete dominance, the ratio of genotypes to phenotypes will increase as $(N + 1)/2$.

22. Three of the many known coat colors in mink (*Mustela vison*) are the standard wild-type (black-brown) color, a blue-gray color known as silverblu (or platinum), and a darker blue-gray color known as steelblu. Matings between wild-type mink can produce all wild-type, or 3 wild-type : 1 silverblu, or 3 wild-type : 1 steelblu. Crosses of steelblu × steelblu sometimes produce all steelblu and sometimes produce 3 steelblu : 1 silverblu. Crosses of silverblu × silverblu produce only silverblu. **(a)** Arrange the coat colors in order of dominance. **(b)** How many alleles are responsible for the inheritance of these coat colors in mink? **(c)** How many different genotypes are responsible for steelblu color? For silverblu? **(d)** A particular mating yields progeny in a ratio of 1 steelblu : 2 wild-type : 1 silverblu. What are the coat colors of the parents in this cross?

23. Suppose that blood serum is taken from each of five students in a genetics laboratory and tested in different paired combinations with red blood cells obtained from each of the same individuals. Agglutination of red blood cells occurs in some sera (designated by +) but not in others (designated by −). The following table gives the results that were observed:

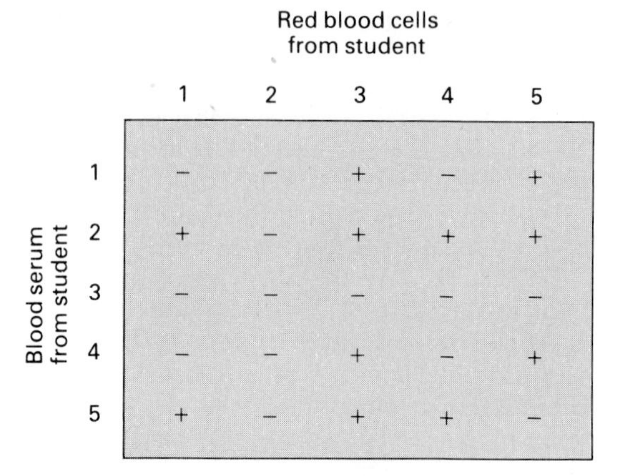

Blood serum from student	Red blood cells from student: 1	2	3	4	5
1	−	−	+	−	+
2	+	−	+	+	+
3	−	−	−	−	−
4	−	−	+	−	+
5	+	−	+	+	−

If the blood type of student 1 is known to be type A, what are the blood types of the four remaining students?

24. If we consider only the three ABO blood-type alleles I^A, I^B, and i, then give the genotypes of the parents in a cross that produces offspring with **(a)** 4 blood types; **(b)** 3 blood types, of which type B is most common; **(c)** 3 blood types, of which type AB is most common; **(d)** 2 blood types, of which type A is most common.

25. The ABO blood group system is valuable in helping to settle cases of disputed parentage. The following table lists the blood types of various mother–child combinations. In each case, list the blood types that can be excluded as possibilities in the father.

Blood type of child	Blood type of mother	Blood types that father cannot have
O	O	
O	B	
A	B	
B	O	
AB	A	
AB	B	

26. Three babies are born in a hospital on the same night. The blood types of the babies are A, B, and AB. If we know that the blood types of the three pairs of parents are **(a)** A and B, **(b)** AB and O, and **(c)** B and B, assign each baby to its proper parents.

27. On rare occasions, a newborn in a hospital is assigned to the wrong parents. Suppose that Mr. and Mrs. X believe that the baby they brought home from the hospital may not really be theirs. Blood tests reveal that Mr. X is type

A M, Mrs. X is AB MN, and the baby is O MN. Does the baby belong to Mr. and Mrs. X? Why or why not?

28. In the following pedigree, the phenotypes given below each symbol designate the presence (+) or absence (−) of the A antigen, B antigen, and $Rh_0(D)$ antigen, in that order. Determine the ABO and Rh genotypes of each person in the pedigree.

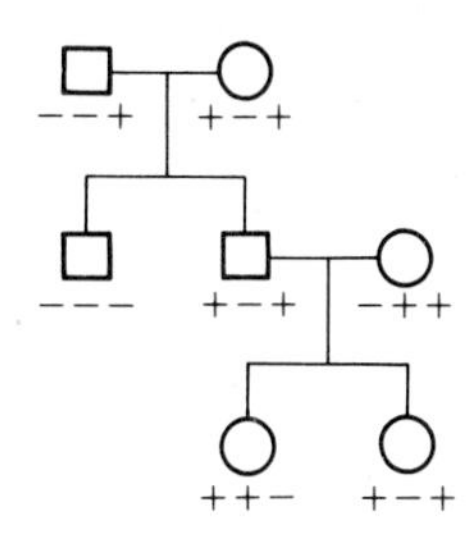

29. Each of the following pedigrees involves a different hereditary trait in humans. Pedigree 1 shows the transmission of a fairly common characteristic, whereas pedigree 2 shows the transmission of a trait that is rare in the general population. Affected individuals are designated by filled-in symbols. Which of the following patterns of transmission are consistent with each pedigree? Give reasons for your answers. **(a)** Dominant X-linked gene. **(b)** Recessive X-linked gene. **(c)** Y-linked gene. **(d)** Sex-influenced gene. **(e)** Sex-limited gene.

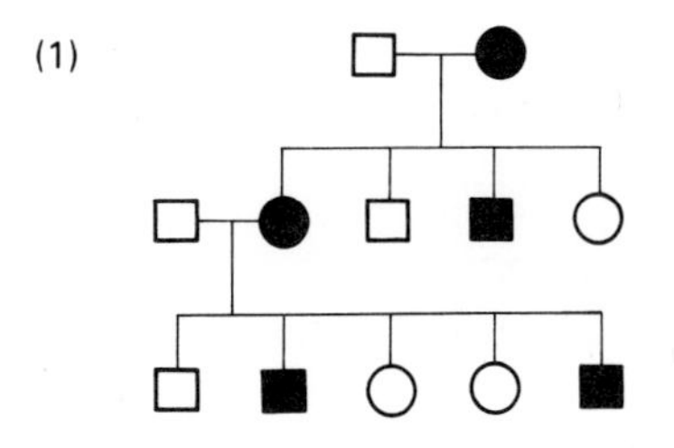

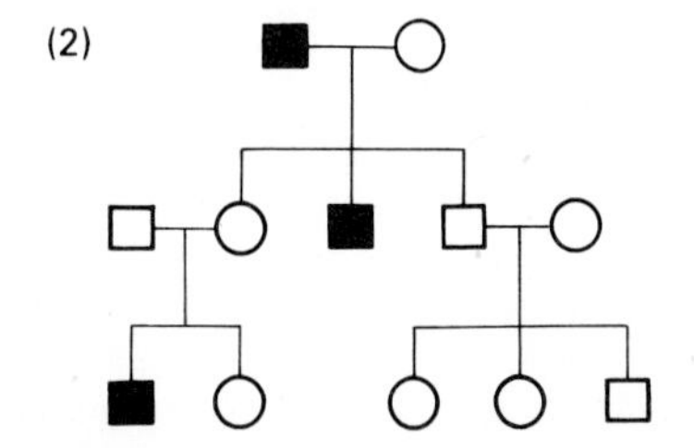

30. Quite often, information is needed from more than one pedigree to establish correctly the mode of inheritance for a particular trait. This situation is shown below by means of two pedigrees for the same hereditary trait. Affected individuals are designated by shaded symbols. Determine which of the following patterns of transmission are consistent with the trait when each pedigree is considered separately and then when both pedigrees are considered jointly: **(a)** autosomal dominant gene, **(b)** autosomal recessive gene, **(c)** X-linked dominant gene, **(d)** X-linked recessive gene, **(e)** sex-influenced gene.

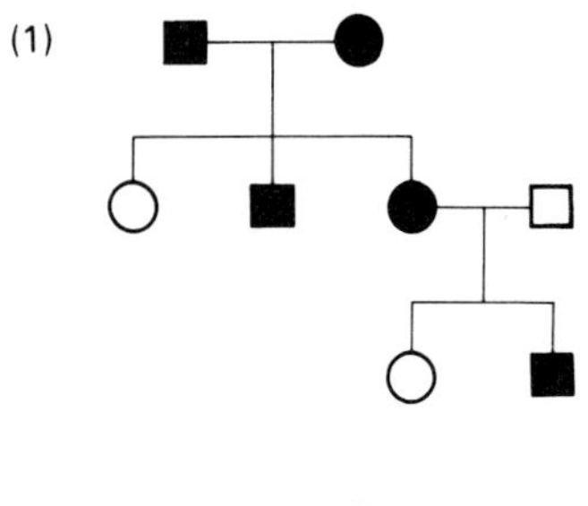

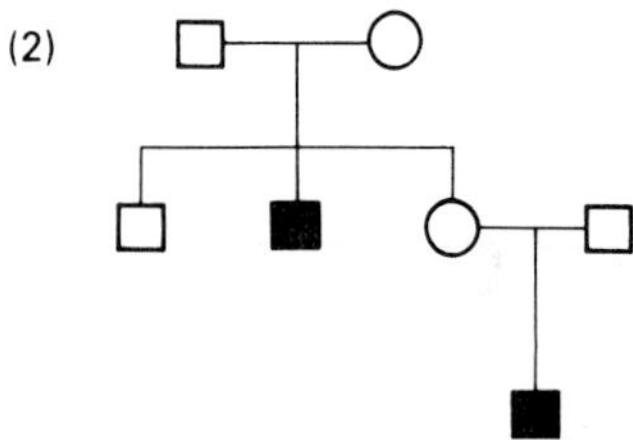

31. The filled-in, shaded, and open symbols in the following pedigree designate three different coat colors in a certain species of mammal. Determine whether the inheritance pattern is consistent with **(a)** a single pair of incompletely dominant or codominant alleles, **(b)** three alleles at a locus that shows a simple hierarchy of dominance, or **(c)** two different pairs of alleles with a complementary form of interaction.

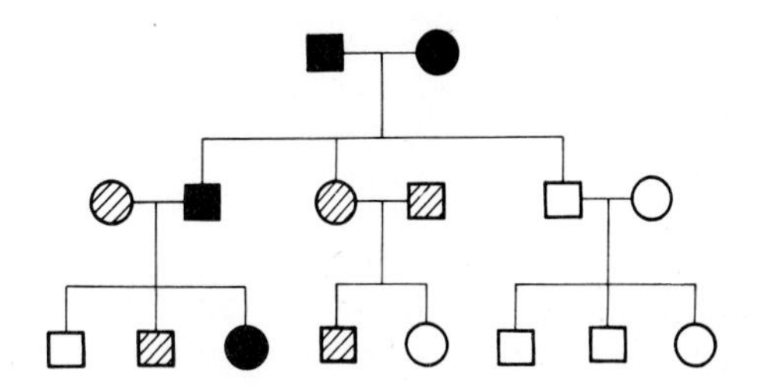

Chapter 7

Linked Genes and Chromosome Mapping in Eukaryotes

In Chapter 6, we limited our discussion to patterns of inheritance that characterize a single gene locus. It was observed that a gene may exist in several allelic forms, each form having arisen by mutation in a more ancestral allele. Since each gene pair resides on a single pair of chromosomes, we can predict the basic patterns for the transmission of alleles based on our knowledge of the behavior of chromosomes during meiosis. The information that can be derived from studying the inheritance of alleles at a single locus is limited, however. This knowledge is largely restricted to the nature of dominance relationships, the effects of the environment, and the type of chromosome on which the gene is located (sex chromosome or autosome).

When we consider two or more gene loci jointly, additional principles emerge. Two general patterns of joint transmission are possible: (1) **independent assortment** and (2) **linkage.** Independent assortment is typically associated with genes located on different chromosomes. The independent alignment of each pair of chromosomes on the meiotic spindle gives rise to random combinations of nonallelic genes in the gametes. Thus, when genes are located on different chromosomes, the dihybrid *AaBb* will produce $2^2 = 4$ gamete types ((*AB*), (*Ab*), (*aB*), and (*ab*)) in a 1:1:1:1 ratio; the trihybrid

AaBbCc will produce $2^3 = 8$ gamete types, also in equal proportions; and so on. For genes that assort independently, as for genes at a single locus, we can predict genotypic ratios a priori, employing only the basic rules of probability.

Linkage, which is the second general pattern, involves genes that are carried on the same chromosome, but at different loci. These nonallelic genes, known as **linked genes,** tend to remain together in the same combination in which they were inherited and do not assort independently. Thus, if the dihybrid *AaBb* is formed from the fusion of (*AB*) and (*ab*) gametes and the genes are linked, the gametes that the dihybrid produces are also likely to be *AB* and *ab* in genotype. The results are similar for the dihybrid *AaBb* formed from the fusion of (*Ab*) and (*aB*) gametes; in this case, the nonallelic genes tend to remain together in the *Ab* and *aB* combinations. But linked genes do not remain together 100 percent of the time. As we learned in Chapter 5, the linkage can be broken through a process known as *crossing over.* Crossing over causes a dihybrid to produce all possible genotypes, although the two inherited gamete genotypes will predominate for linked genes. Linkage is therefore detected, not by the inability of dihybrids to produce certain gamete types, but by their failure to produce the 1:1:1:1 ratio characteristic of independent assortment.

In this chapter, we will discuss the transmission of genes carried on the same chromosome. We begin with the detection of linkage and the interpretation of linkage data. We then consider some of the methods used in constructing maps of genes on chromosomes.

Recombination of Linked Genes

Linked genes remain together and are transmitted in the same combination in which they were inherited unless crossing over occurs. Crossing over, which is the exchange of parts between homologous nonsister chromatids, results in new gene combinations. Figure 7.1 illustrates crossing over in terms of a breakage and reunion process that occurs at identical points on the two chromosomes of a pair. The new gene combinations that are formed by this process are referred to as **recombinant types,** while gametes that retain the gene arrangements of the parents are called **parental types.** When genes are linked, recombination through crossing over is the only way in which nonparental gene combinations can arise. Crossing over is thus the major source of variability for genes carried on the same chromosome.

Identification and Measurement of Linkage

Because crossing over occurs, two linked genes will produce the same number of gametic arrangements in dihybrids as two unlinked (independently assorting) genes. The difference lies in the frequencies with which the various gamete types are formed. The different gamete types are equal in frequency when nonallelic genes assort independently. But when genes are linked, recombinant types are less frequent than parental types. For linked genes, the proportions of recombinant-type and parental-type gametes differ because crossing over occurs with less than absolute certainty between most genes on a chromosome. The probability of crossing over between linked genes can actually be quite small when the genes lie close together. This means that the majority of meiocytes will produce gametes

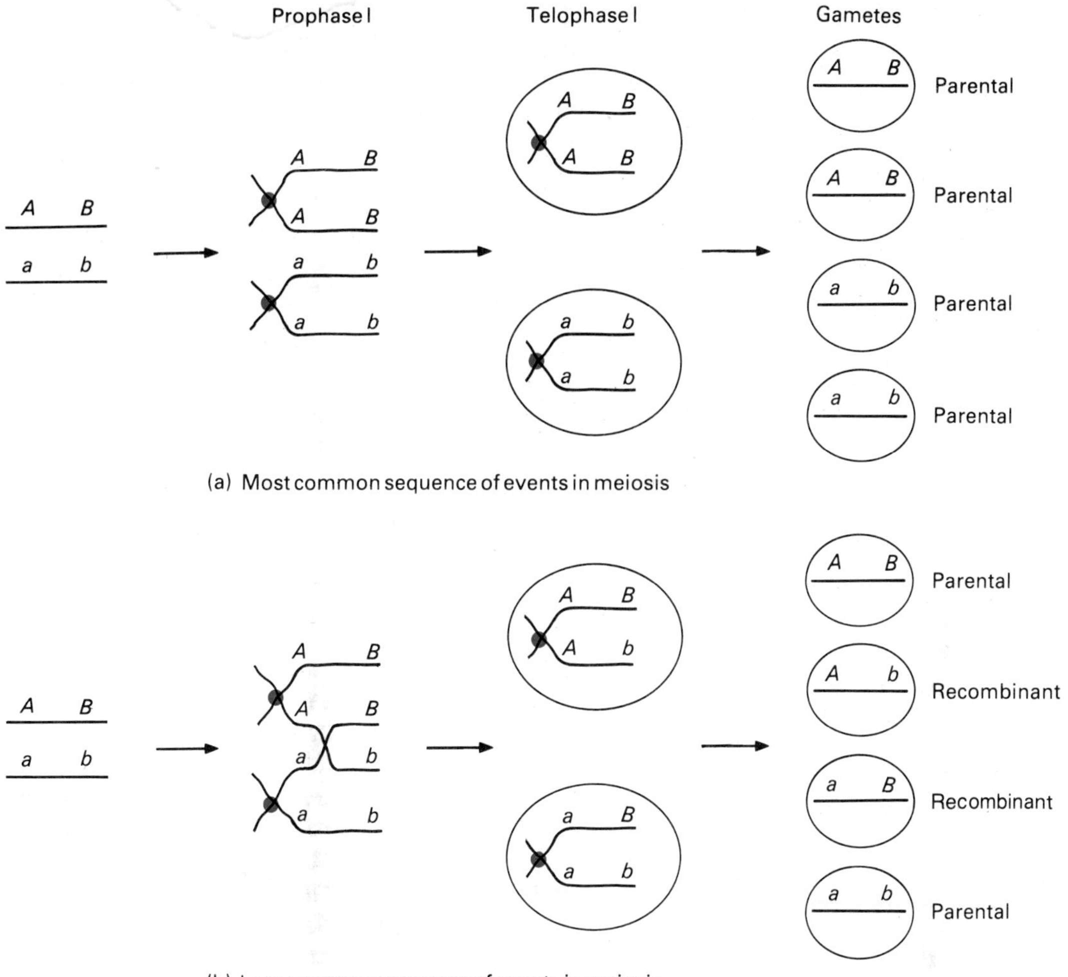

Figure 7.1. Production of gametes (a) without and (b) with crossing over. If loci *a* and *b* lie close together on the same chromosome, then crossing over, shown in sequence (b), occurs in fewer meiocytes than the events in sequence (a). As a result, fewer recombinant gametes than parental ones are present in the gamete pool of the heterozygote. For example, if 25 percent of the meiocytes undergo crossing over as in sequence (b), the pool of gametes would contain only 12.5 percent recombinant types.

with the parental-type gene arrangement. A smaller number of meiocytes will experience a crossover in the interval between the loci and will therefore yield recombinant types. The net result is an excess of parental types in the total pool of gametes formed by a dihybrid parent, with a smaller amount of recombinant gametes present (see Fig. 7.1). This outcome gives us one means of identifying linkage: Linked genes yield greater than 50 percent parental-type gametes and less than 50 percent recombinant-type gametes. The exact proportion of each type varies, depending on the interval between the particular genes being studied (we will elaborate on this point later in the chapter).

Detecting Linkage from Testcross Results. The most straightforward way to demonstrate linkage is to perform a testcross. Consider a testcross of the dihybrid shown in Fig. 7.1. The cross could be written simply as $AaBb \times aabb$. But since we know the genes are linked, the cross should ideally be expressed in some manner to reflect this linkage, such as

$$\frac{A\ B}{a\ b} \times \frac{a\ b}{a\ b}.$$

Each pair of parallel lines represents the chromosome pair on which the genes are located. If the genes were actually on different chromosome pairs, then all the classes of gametes produced by the heterozygous parent would occur in equal frequency. With linkage, however, the ($\underline{A\ B}$) and ($\underline{a\ b}$) gamete types will be present in the highest frequency since they have the parental-type gene arrangement. Thus, the criterion for linkage in this testcross is as follows:

Testcross: $\frac{A\ B}{a\ b} \times \frac{a\ b}{a\ b}$

	Gametes:				Testcross progeny:	
> 50%	($\underline{A\ B}$)	+	($\underline{a\ b}$)	→	$\frac{A\ B}{a\ b}$	> 50%
	($\underline{a\ b}$)	+	($\underline{a\ b}$)	→	$\frac{a\ b}{a\ b}$	
< 50%	($\underline{A\ b}$)	+	($\underline{a\ b}$)	→	$\frac{A\ b}{a\ b}$	< 50%
	($\underline{a\ B}$)	+	($\underline{a\ b}$)	→	$\frac{a\ B}{a\ b}$	

The offspring from the testcross will consist of greater than 50 percent $\frac{A\ B}{a\ b}$ + $\frac{a\ b}{a\ b}$ genotypes and less than 50 percent $\frac{A\ b}{a\ b}$ + $\frac{a\ B}{a\ b}$ genotypes. In other words, the offspring will consist of *two large frequency classes and two smaller frequency classes.* We recognize the presence of linkage from the lack of agreement with the 1:1:1:1 ratio expected with independent assortment.

As an actual example, consider two traits of tomatoes: fruit shape and flower arrangement (inflorescence). Fruit shape can be round (dominant gene *O*) or elongate (allele *o*), and flower pattern can be simple (dominant gene *S*) or compound (allele *s*). True-breeding round, simple plants are crossed with elongate, compound plants, and a testcross is then performed using the resulting F_1. The complete crossing sequence, including the observed results and interpretation of the data, is shown in Table 7.1. This testcross produces progeny consisting of 83 round, simple : 85 elongate, compound : 23 round, compound : 19 elongate, simple. The apparent deviation from a 1:1:1:1 ratio in the testcross progeny clearly demonstrates that the *o* locus is linked to the *s* locus. The largest classes of offspring

Table 7.1. Observed and expected results for a crossing sequence involving fruit shape (round vs. elongate) and inflorescence pattern (simple vs. compound) in tomatoes.

A. Observed Results

P: round, simple × elongate, compound

F_1: round, simple

Testcross: F_1 × elongate, compound

Testcross progeny:

round, simple = 83   round, compound = 23

elongate, compound = 85   elongate, simple = 19

B. Interpretation of Results: Idealized Case

P: $\frac{O\ S}{O\ S} \times \frac{o\ s}{o\ s}$

Gametes: $(\underline{O\ S})$ $(\underline{o\ s})$

F_1: $\frac{O\ S}{o\ s}$

Testcross: $\frac{O\ S}{o\ s} \times \frac{o\ s}{o\ s}$

Gametes: $(\underline{O\ S})$ 40% $(\underline{o\ s})$ 100%

$(\underline{o\ s})$ 40%

$(\underline{O\ s})$ 10%

$(\underline{o\ S})$ 10%

Testcross progeny (numbers are those expected out of 210 offspring):

$\frac{O\ S}{o\ s}$ = 84 (40%) $\frac{O\ s}{o\ s}$ = 21 (10%)

$\frac{o\ s}{o\ s}$ = 84 (40%) $\frac{o\ S}{o\ s}$ = 21 (10%)

---

result from parental-type gametes produced by the heterozygous F_1 parent. Thus, $(\underline{O\ S})$ and $(\underline{o\ s})$ are the parental-type gametes; they occur at a combined frequency of (83 + 85)/210 = 0.8. The two offspring classes that occur less frequently are formed from recombinant-type gametes $(\underline{o\ S})$ and $(\underline{O\ s})$. The total frequency of recombinants is then (23 + 19)/210 = 0.2. Observe that when a large number of offspring are examined, complementary gamete types are expected to occur with equal frequency. The complementary types $(\underline{O\ S})$ and $(\underline{o\ s})$ should be equal in frequency, as should $(\underline{O\ s})$ and $(\underline{o\ S})$.

Now let us suppose that a somewhat different crossing sequence is performed, starting with true-breeding round, compound and elongate, simple plants:

$$\frac{O\ s}{O\ s} \times \frac{o\ S}{o\ S}.$$

The F_1 will be $\frac{O\ s}{o\ S}$ in genotype, having received an $\underline{O\ s}$ chromosome from one parent and an $\underline{o\ S}$ chromosome from the other. The parental-type gametes produced by this F_1 are $(\underline{O\ s})$ and $(\underline{o\ S})$. By contrast, $(\underline{O\ S})$ and $(\underline{o\ s})$ gametes are the recombinant types, formed as a result of crossing over. In the offspring from a testcross of the F_1, the proportions of the genotypes will be reversed from those produced by the first crossing sequence. We now expect 80 percent $\frac{O\ s}{o\ s} + \frac{o\ S}{o\ s}$ types, with only 20 percent $\frac{O\ S}{o\ s} + \frac{o\ s}{o\ s}$ classes.

These two examples show that when we are dealing with linked genes, it is very important to note the arrangement of the genes in the heterozygous parent. When both dominant genes (in this case, O and S) are located on one member of the chromosome pair, with two recessives (o and s) on the other, the genes are said to be in a **cis** (or **coupling**) configuration:

$$\frac{O\ S}{o\ s} \quad \text{cis (or coupling) arrangement of genes}$$

Cis-type offspring will then be the most common among the testcross progeny. Conversely, if the dominant allele of one gene pair and the recessive allele of the other pair are located together on one chromosome (e.g., O with s and o with S), then the genes are said to be in a **trans** (or **repulsion**) configuration:

$$\frac{O\ s}{o\ S} \quad \text{trans (or repulsion) arrangement of genes}$$

Trans-type offspring will then be the most frequent. Figure 7.2 compares crossing sequences involving linked genes in the cis and trans arrangements.

EXAMPLE 7.1. The pedigree presented below for three generations of a family shows the joint inheritance of two linked pairs of alleles. One pair is for myotonic dystrophy (D) vs. normal (d), and the other pair is for secretor (S) vs. nonsecretor (s) (a secretor is an individual whose body secretions have certain blood group antigens). Individuals with myotonic dystrophy are represented by the filled-in symbols. The secretor status is given below the symbol for each individual.

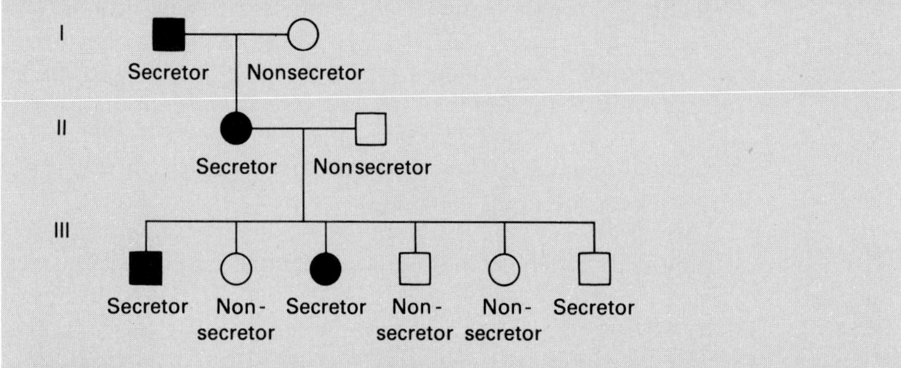

Determine the linkage phase (cis or trans) of the heterozygote in generation II and identify which of the progeny in generation III received a recombinant gamete from the heterozygote.

Solution. The heterozygous female in generation II has a cis arrangement of genes, since she received a $\underline{D\ S}$ chromosome from her father and a $\underline{d\ s}$ chromosome from her mother. The mating in generation II is a testcross of the type $\frac{D\ S}{d\ s} \times \frac{d\ s}{d\ s}$. Only the normal, secretor male in generation III (the last symbol on the right) is recombinant, having received a $\underline{d\ S}$ chromosome from his mother and a $\underline{d\ s}$ chromosome from his father. All the other offspring in generation III received a parental-type arrangement, either $\underline{D\ S}$ or $\underline{d\ s}$, from their mother.

Figure 7.2. Comparison of results expected from testcrosses involving (a) cis and (b) trans arrangements of linked genes. Linkage is revealed as a non-1:1:1:1 ratio in the offspring of the testcross. The arrangement of genes in the gametes that gave rise to the F_1 determines the parental arrangement of gametes produced by the F_1. This gene arrangement, in turn, dictates which offspring classes will appear in higher frequencies.

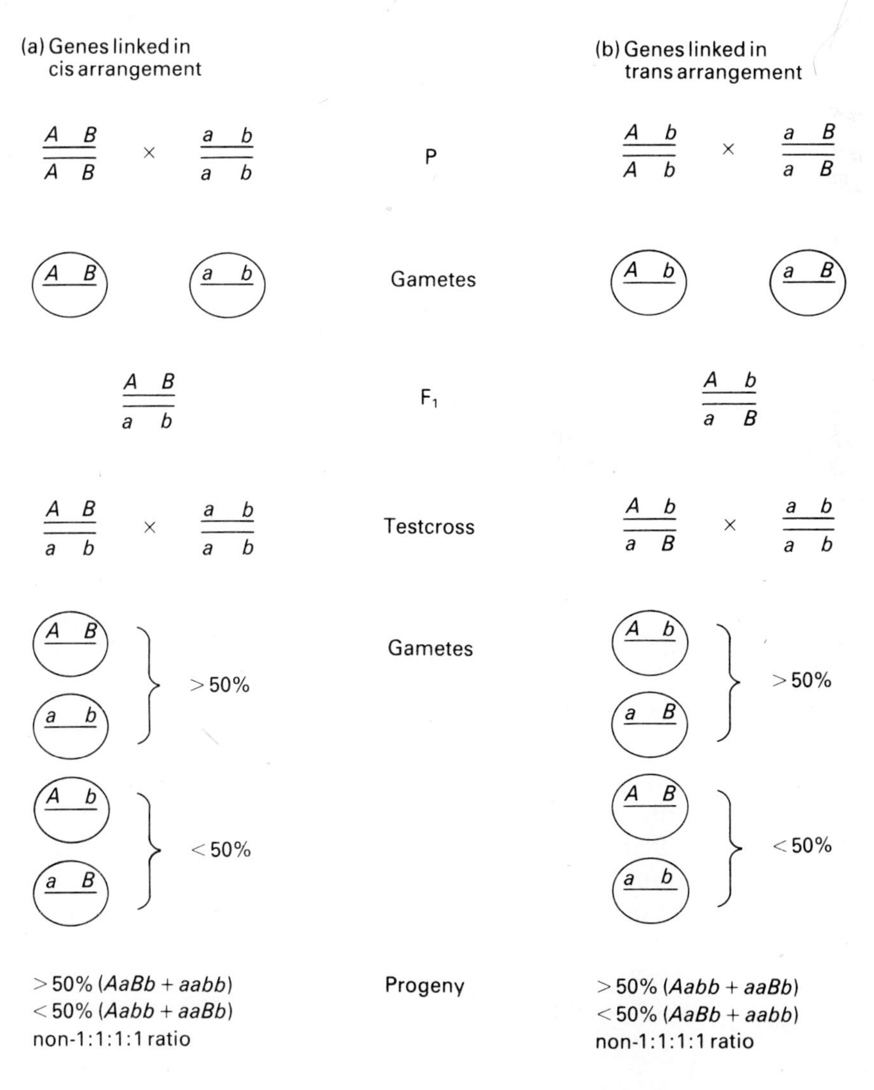

Maximum Frequency of Recombination. There are numerous points along a chromosome at which crossing over can occur. As the length of the interval between linked genes increases, the number of opportunities for crossing over therefore also increases. For this reason, crossing over occurs only rarely between closely linked loci, whereas it occurs more frequently between genes that are farther apart. If the genes are spaced very far apart, then all bivalents will experience crossing over in the interval separating the loci. The proportion of recombinants will not be 100 percent in the resulting gametes, however. Let us suppose, for example, that every bivalent undergoes a single crossover event involving two of the four chromatids, as illustrated in Fig. 7.1. The pool of gametes will then consist of 50 percent recombinant and 50 percent parental types. The upper limit of recombination would thus be 50 percent, no matter how far apart the two genes may be.

This upper limit of 50 percent recombinant types holds even when the bivalents undergo multiple exchanges; that is, two or even three or more crossover events along the same interval. The multiple exchanges can occur along the same chromatid pair, or they can involve various combinations of nonsister chromatids. We will first consider multiple crossovers along the same chromatid pair, as diagrammed in Fig. 7.3 for a cis arrangement. In Case 1, the genes are separated by a very short interval so that crossing over occurs only rarely, and multiple crossover events do not occur at all. The result is a few recombinant gametes within the gamete pool, in a combined proportion much less than 50 percent. In Case 2, the genes are farther apart, so that an occasional bivalent undergoes a double crossover. Notice that double crossovers re-create the parental gene arrangement; no recombinant-type gametes are produced by a double crossover. The second crossover nullifies the effect of the first. Case 3 shows two loci even farther apart, so that an occasional triple crossover is possible. Two recombinant gametes are produced by a triple crossover—a result that is no different from that of a single crossover. Case 4 shows that four crossover events between the two chromatids re-create the parental arrangement for genes bordering the crossover interval.

By now, you should begin to see the emergence of a general pattern. Any odd number of crossovers (1, 3, 5, . . .) between a pair of chromatids yields two recombinant gametes. Any even number of crossover events (2, 4, 6, . . .) yields two parental gametes. Two loci spaced far apart will undergo various types of multiple exchange events in different meiocytes. Furthermore, both pairs of nonsister chromatids can participate in crossing over. Because as many even-numbered as odd-numbered exchanges are expected to occur overall, the recombination frequency cannot exceed a maximum of 50 percent. The additional crossing over allowed by longer distance re-creates the parental types as often as it yields recombinants.

If different combinations of nonsister chromatids are involved in multiple crossing over, the immediate results are somewhat different in terms of the gamete types produced by one meiotic event. But the average proportion of recombinant gametes is still not expected to be greater than 50 percent. We will illustrate this outcome using double crossovers, although the same general results hold for all higher order events. Figure 7.4 shows the different double crossover events that are possible involving the various pairs of nonsister chromatids. If crossing over is

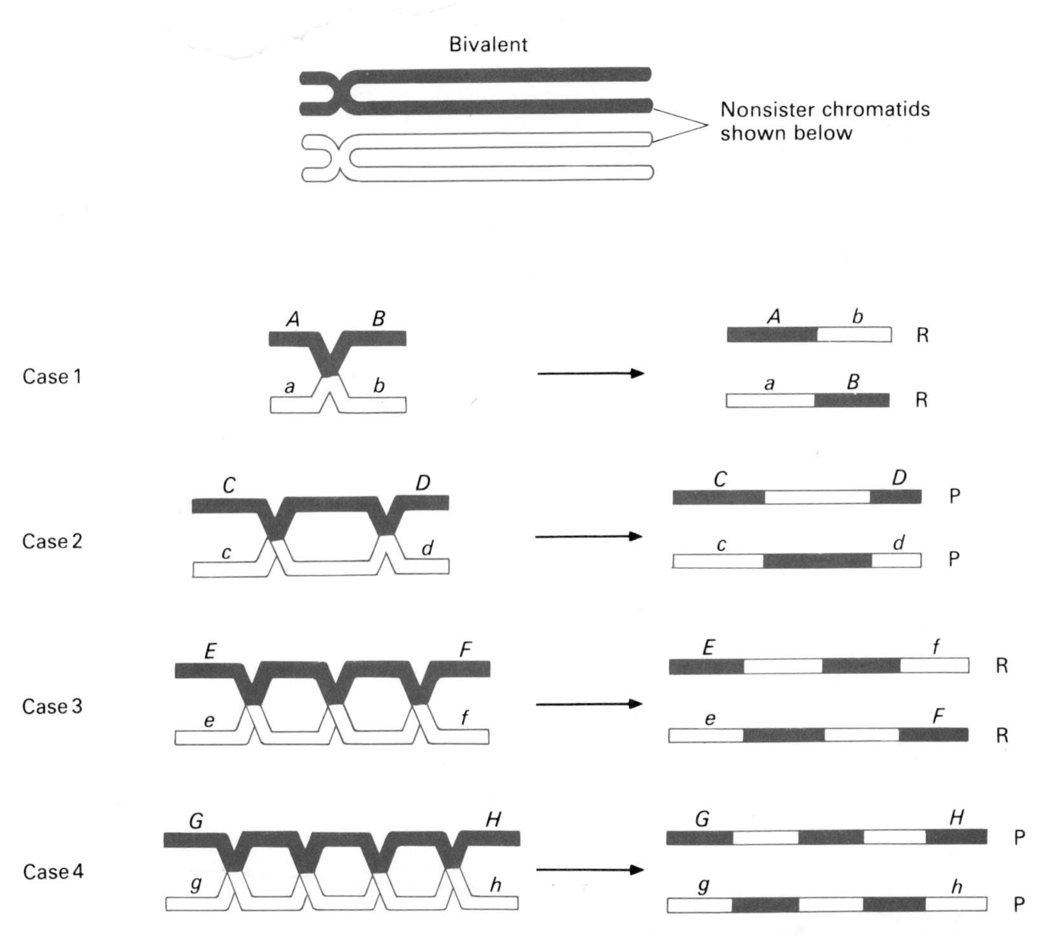

Figure 7.3. The effects of increased frequency of crossing over on the frequency of observed recombination, when all crossovers involve the same two nonsister chromatids within the bivalent. As the length of the interval separating the two loci increases, more crossing over takes place. However, the additional exchange events are not registered as additional recombination. Only odd-numbered events generate recombinants, and no matter how many events take place, only two recombinant types are formed. Even-numbered crossovers yield no recombinants at all.

random, four combinations of nonsister chromatids should be equally likely to occur in double exchanges. These types are (1) two-strand double exchanges, (2 and 3) two types of three-strand double exchanges, and (4) four-strand double exchanges. Case 1 yields four parental-type gametes. Cases 2 and 3 each yield two parental and two recombinant gametes. One parental type is derived from a chromatid that did not participate in crossing over; the other is derived from the recreation of the parental arrangement by the second crossover event. Case 4 yields four recombinant gametes. The net result in the gamete pool is 4 + 2 + 2 = 8 parental types and 2 + 2 + 4 = 8 recombinant types, for a recombination frequency of 50 percent.

At this point, we can conclude that even though the frequency of recombination will tend to increase as the distance between linked genes increases, it will never exceed a maximum value of 50 percent. Therefore, genes spaced very far

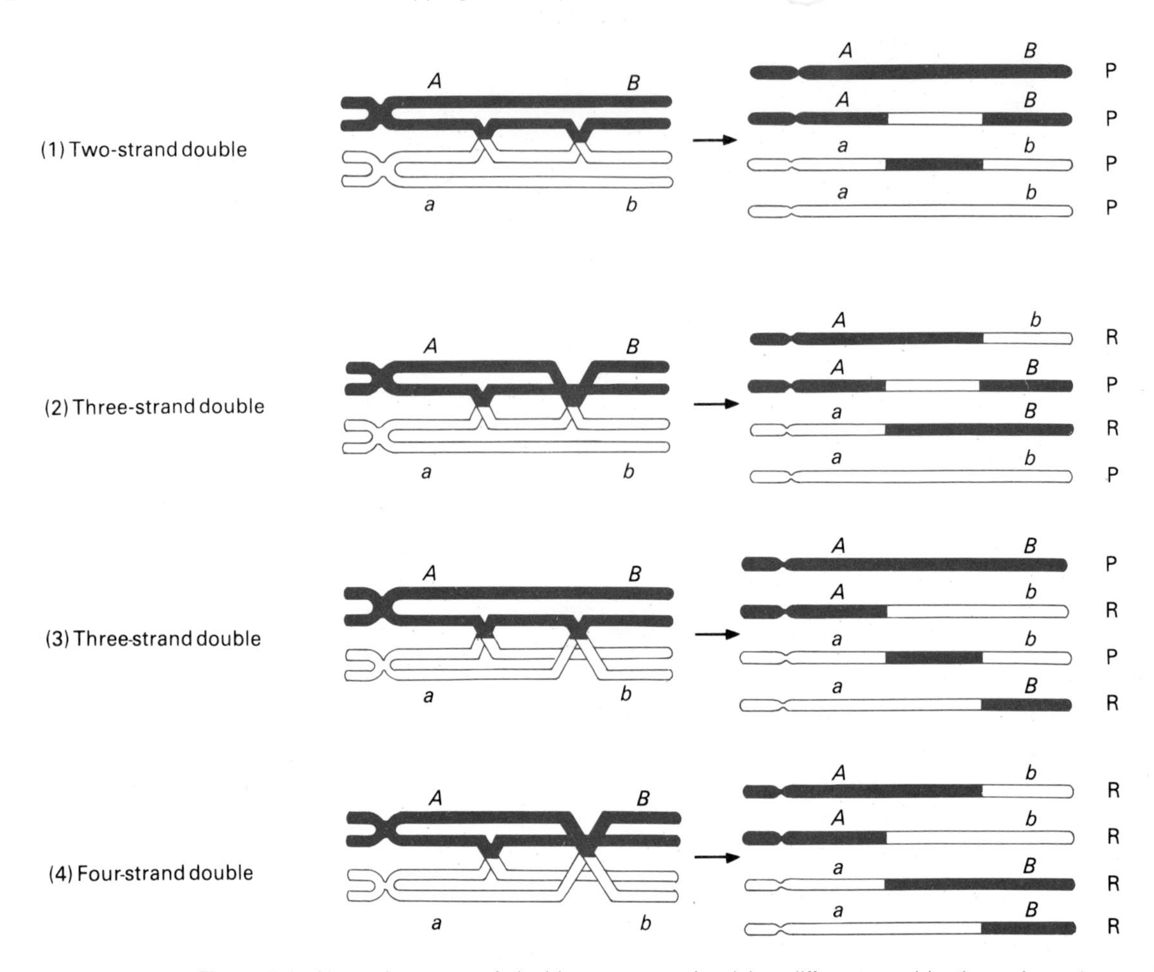

Figure 7.4. Alternative types of double crossovers involving different combinations of nonsister chromatids. The overall result is 50 percent parental and 50 percent recombinant gametes.

apart on the same chromosome will appear to exhibit 50 percent recombination. In this case, the four gamete classes produced by a dihybrid—whether in the cis or trans arrangement—will occur in a proportion of 25 percent each. This result is exactly the same as is expected of genes located on different chromosomes that assort independently in meiosis. Thus, genes may be located far enough apart on the same chromosome that the linkage is not shown in the testcross results. In one sense, such loci can be said to be *physically* linked, in that they do lie on the same chromosome. They do not appear to be *genetically* linked, though; that is, the data from genetic crosses do not show ratios characteristic of linkage. Instead, such genes appear to assort independently, and we could easily make the mistake of assigning them to different chromosomes. How can we determine that they are actually located on the same chromosome? The linkage of such genes can be discovered by finding that both show less than 50 percent recombination with some third gene. For example, suppose genes *a* and *b* yield 50 percent recombinants, but *a* and *c* yield 30 percent recombinants, and *b* and *c* yield 35 percent recombinants. Since both *a* and *b* are linked to the same common gene *c*, all three must be located on the same chromosome.

As we can see from this example, the genetic interpretations of the chromosomal locations of genes do not always coincide with the actual physical locations. It is therefore often useful to use another term to distinguish between genes located on different chromosomes and those located on the same chromosome. **Syntenic genes** are defined as those located on the same chromosome, regardless of whether or not the genetic data show their linkage. Table 7.2 summarizes some of the characteristics of the different classes of syntenic genes, comparing them with genes on different chromosomes.

Detecting Linkage from Dihybrid Cross Results. We have seen that the detection of linkage from testcross data is straightforward. A significant difference from a 1:1:1:1 ratio will appear if the genes are linked. Moreover, the percent of recombinant-type offspring in the cross is equal to the frequency of recombinant-type gametes produced by the dihybrid parent.

If a mating procedure other than a testcross is used, the effect of linkage can be considerably more complex. To illustrate, let us consider a dihybrid cross in tomatoes, involving once again the inheritance of fruit shape and inflorescence pattern. Assume that two cis-dihybrids, $\frac{O\ S}{o\ s}$, are crossed. *Each* parent now produces four kinds of gametes in the ratio of 4 (O S) : 4 (o s) : 1 (O s) : 1 (o S). Figure 7.5 works out the cross by means of a Punnett square using decimal fractions. A

Table 7.2. The possible kinds and proportions of gametes formed by double heterozygotes.

LOCATION OF GENES	PERCENTAGE OF GAMETES *AB*	*ab*	*Ab*	*aB*	PERCENTAGE RECOMBIN-ATION	NOTATION OF DOUBLE HETEROZYGOTE
syntenic; cis arrangement; close together*	$\frac{100 - \%R}{2}$	$\frac{100 - \%R}{2}$	$\frac{\%R}{2}$	$\frac{\%R}{2}$	$0 < \%R < 50$ (R = number between 0 and 0.5)	$\frac{A\ \ B}{a\ \ b}$
syntenic; trans arrangement; close together*	$\frac{\%R}{2}$	$\frac{\%R}{2}$	$\frac{100 - \%R}{2}$	$\frac{100 - \%R}{2}$	$0 < \%R < 50$	$\frac{A\ \ b}{a\ \ B}$
syntenic; cis arrangement; very far apart	25	25	25	25	$\%R = 50$	$\frac{A\qquad\qquad B}{a\qquad\qquad b}$
syntenic; trans arrangement; very far apart	25	25	25	25	$\%R = 50$	$\frac{A\qquad\qquad b}{a\qquad\qquad B}$
on nonhomologous chromosomes	25	25	25	25	$\%R = 50$	$\frac{A}{a}\ \frac{B}{b}$

* Only in these cases do we say we have genetic linkage of nonallelic genes.

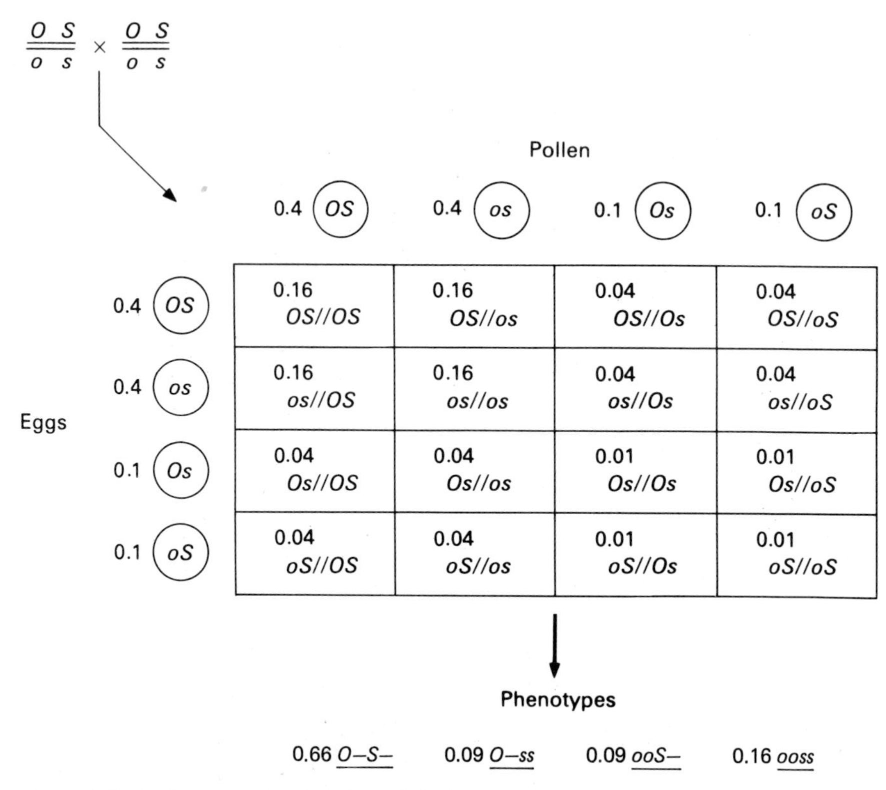

Figure 7.5. A dihybrid cross involving linked genes. The expected proportion of each zygote is calculated by multiplication of appropriate gametic frequencies. The phenotypic ratio is determined by summation of the relevant genotypic classes.

summary of the phenotypes in this square gives

Round, simple (*O-S-*) = 0.66
Round, compound (*O-ss*) = 0.09
Elongate, simple (*ooS-*) = 0.09
Elongate, compound (*ooss*) = 0.16

The results do not conform to a 9:3:3:1 Mendelian ratio ($9/16$ = 0.5625, $3/16$ = 0.1875, $1/16$ = 0.0625). Since a 9:3:3:1 ratio is expected only when genes assort independently, linkage is thus detectable by a significant departure from this ratio. Linkage is also indicated by the fact that a cross between cis-dihybrids does not give the same proportions in the offspring as a cross between trans-dihybrids. By doing the calculations, you can convince yourself that the mating $\frac{O\ s}{o\ S} \times \frac{O\ s}{o\ S}$ gives a phenotypic ratio of 51:24:24:1, rather than the 66:9:9:16 ratio obtained for dihybrids in the cis configuration.

An important observation about both F_2 results is that when each trait is considered separately, the ratio is the familiar 3 dominant : 1 recessive. For example, the cross of the cis-dihybrids gave an overall ratio of 66:9:9:16. Factoring out the round vs. elongate trait, we get 66 + 9 = 75 round and 9 + 16 = 25 elongate, for a ratio of 75:25, or 3:1. A comparison of simple versus compound also gives a 75:25 = 3:1 ratio. Similarly, the 51:24:24:1 ratio obtained when the trans-dihybrids are crossed can also be factored into proportions of 3 round : 1 elongate (51 + 24 = 75, 24 + 1 = 25) and 3 simple : 1 compound. That the ratio for each trait alone is 3:1 shows that Mendel's principle of segregation of alleles still applies to the members of each allele pair, regardless of whether the different pairs exhibit linkage or independent assortment. The fact that we are dealing with linked genes does make Mendel's second principle invalid, however. In mathematical terms, the product rule for computing the results of a cross no longer holds, since linked genes do not behave independently. In other words, we cannot simply multiply (3 + 1) × (3 + 1) to get 9:3:3:1 as we did for genes on separate chromosomes; rather, we must consider all possible gamete combinations separately, by use of a Punnett square or some other acceptable procedure.

EXAMPLE 7.2. When conducting linkage studies on a naturally self-fertilizing species of plant, it is usually more convenient to allow dihybrids to self-fertilize and to analyze the dihybrid offspring than to make a separate testcross. Suppose that you are interested in determining whether the genes for normal leaves (*M*) versus mottle (*m*) and smooth fruit skin (*P*) versus peach (*p*) in tomatoes are linked. A dihybrid *MmPp* plant is selfed and produces four phenotypic classes in the offspring, of which 0.16 percent are doubly recessive *mmpp*. Are the genes *M,m* and *P,p* linked, and if so, what is the linkage phase (cis or trans) of the heterozygous parent?

Solution. The genes appear to be linked, since the frequency of doubly recessive offspring is significantly less than 1/16 (6.25 percent), the proportion expected for a dihybrid cross with independent assortment. The linkage phase of the heterozygous parent is trans: $\frac{M\ p}{m\ P}$, since the $\underline{m\ p}$ gamete type is recombinant. We deduce that it is recombinant by noting that double recessives, $\frac{m\ p}{m\ p}$, are formed by the combination of two $\underline{m\ p}$ gametes. The expected frequency of this genotype is then

$$\mathrm{f}\left(\frac{m\ p}{m\ p}\right) = \mathrm{f}(\underline{m\ p}) \times \mathrm{f}(\underline{m\ p})$$

or

$$0.0016 = 0.04 \times 0.04.$$

Thus, the $\underline{m\ p}$ gene arrangement occurs in only 4 percent of the gamete types produced by the heterozygous parent. Since this proportion is less than the 1/4 or 25 percent proportion expected for independent assortment, gametes that carry the $\underline{m\ p}$ chromosome must be recombinant and must have been formed through crossing over. This result could occur only if the genes in the heterozygote are normally linked in the trans phase.

Experimental Verification of Crossing Over

It is difficult to demonstrate directly that an exchange of chromosomal segments occurs during recombination of linked genes. The members of a chromosome pair are not visibly distinguishable under ordinary circumstances. Only in those rare cases in which the homologs are structurally different can an exchange be followed. Two classical experiments in genetics have exploited unusual differences between homologs to confirm directly that recombination does result from a physical interchange between homologous chromosomes. These experiments were performed in 1931 by Stern with *Drosophila* and by Creighton and McClintock working with corn. Both experiments showed a direct correlation between the behavior of two genes and two cytological markers on chromosomes. This correlation is clearly evident in the results of the Creighton-McClintock experiment on corn. Creighton and McClintock used an aberrant chromosome 9 that has a dark-staining knob at one end and an added segment from another chromosome at the other end. The two gene pairs they selected were *C* (for colored kernels) and *c* (colorless) and *Wx* (for starchy kernels) and *wx* (waxy). Plants that were heterozygous for chromosome 9 and for the *c* and *wx* loci were crossed to $\frac{c\ Wx}{c\ wx}$ plants with normal chromosomes. The cross is illustrated as

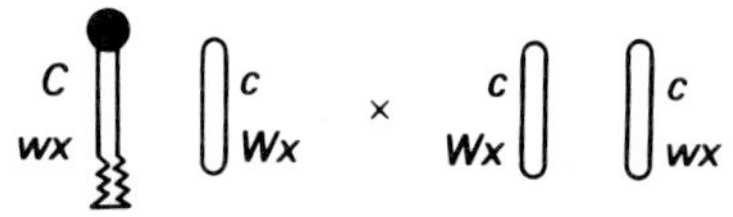

C wx   c Wx   ×   c Wx   c wx

If crossing over occurs in the doubly heterozygous parent, the recombinant gene arrangements are $\underline{C\ Wx}$ and $\underline{c\ wx}$. Since the parent with both normal chromosomes can only produce gametes with the $\overline{c\ Wx}$ and $\overline{c\ wx}$ arrangements, the recombinant classes of progeny are then $\frac{C\ Wx}{c\ Wx}$, $\frac{C\ Wx}{c\ wx}$, $\frac{c\ wx}{c\ Wx}$, and $\frac{c\ wx}{c\ wx}$. When recombinant plants were selected and examined cytologically, they were found in every case to have received an aberrant chromosome from their doubly heterozygous parent. Each aberrant chromosome had only one identifiable segment, either a terminal knob still associated with the *C* gene, or the added segment still associated with the *wx* gene:

C Wx   c wx

Therefore, progeny plants that were genetically recombinant in this cross were also recombinant for the chromosome markers. This analysis revealed that recombination of the *c* and *wx* loci in the doubly heterozygous parent was always accompanied by a physical exchange of chromosome segments.

Recombination at the Molecular Level

Crossing over occurs by a physical interchange of homologous chromosome segments. It must therefore involve a breakage and reunion of DNA molecules. Many models of the breakage and reunion process have been advanced (and abandoned) over the years. No one model has gained universal acceptance. Present evidence suggests that a crossover involves staggered cuts in the DNA duplex of each participating chromatid, producing a region of hybrid overlap in the strands of the recombinant products. In this section, we will discuss a model for the origin of such hybrid (heteroduplex) recombinant DNA as a means of explaining the events that appear to occur during recombination.

The Hybrid DNA Model of Breakage and Reunion. The model shown in Fig. 7.6 is patterned after the one proposed by Holliday in 1964 to account for the process of recombination as it is known to occur in eukaryotes. The model has as its centerpiece a hybrid DNA intermediate called the *Holliday intermediate,* or *chi* (χ) *structure,* which consists of two chromosomes held together at a region of homology by two of the four strands of the recombining DNA molecules. The Holliday intermediate is shown in Fig. 7.6(e) and (f) by two different planar views of the three-dimensional structure. Several different mechanisms have been proposed for the origin of the chi structure; only two are shown in Fig.7.6. One mechanism (Fig. 7.6b–d) is initiated by an enzymatic cleavage (b) of two DNA strands with

Figure 7.6. The hybrid DNA model of recombination. Homologous chromosomes are shown synapsed in (a). Each is composed of a double helix with strands of opposite polarity (+ and −). Strand exchange by either one of the two mechanisms shown (b-d or b′-d′) yields the Holliday intermediate, or chi form (e and f). Strand breakage and subsequent ligation (g-i) yields linear heteroduplex molecules. Of the four possible molecules so produced, two are parental and two are recombinant for genes flanking the region of overlap. The pathway is based on the model of R. Holliday with modifications by H. Potter and D. Dressler.

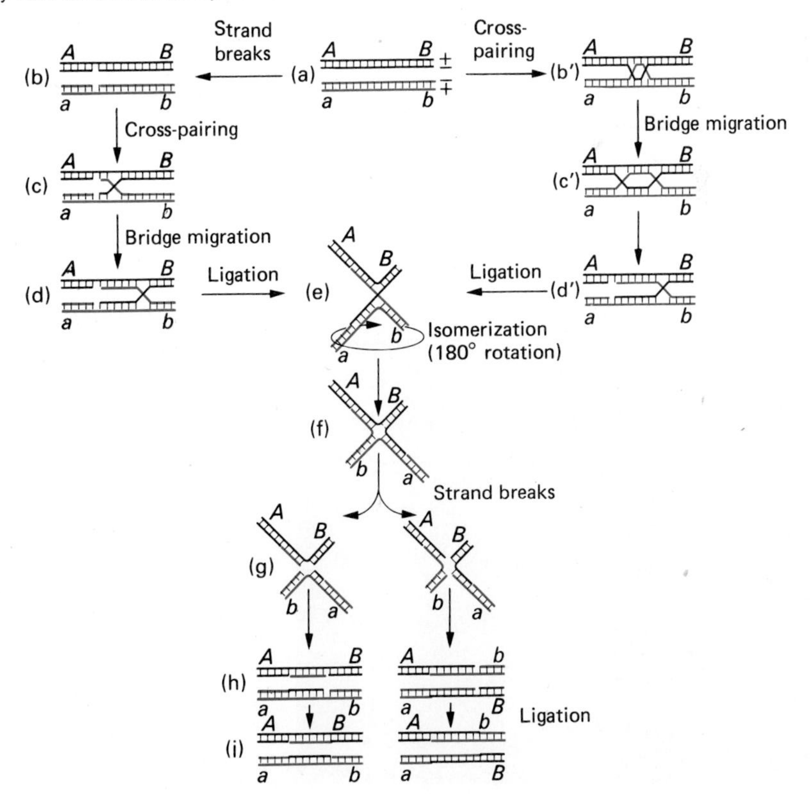

the same polarity from nonsister chromatids. (The breaks do not have to occur at equivalent sites, since gaps can be filled and strands can be tailored by repair enzymes to reform intact DNA molecules.) Strand breakage is then followed by an exchange of single strands of DNA between nonsister chromatids. The result is a cross-bridged structure (c), in which the DNA of each chromatid contains a hybrid, or heteroduplex, region (one strand rust, the other strand black). This heteroduplex region often includes minor differences in base sequences between homologous chromosomes. Migration of the cross-bridge (d) can result in an even longer region of hybrid overlap. Joining of the cleaved strands by DNA ligase would then produce a chi structure in which the homologs are held together by covalent bridges.

Another mechanism for the formation of a Holliday intermediate (Fig. 7.6b′–d′) is initiated by the pairing of DNA strands of nonsister chromatids (b′) in a localized region of denaturation; in this case, pairing precedes strand breakage. The region of pairing is then extended in either or both directions, resulting in a heteroduplex structure with two cross-bridges (c′). (Because of the constraints placed on the pairing process by the structure of DNA, a nicking-closing, or topoisomerase, enzyme is needed in this mechanism to introduce transient nicks into the structure and allow the rotation necessary for helix formation.) The introduction of single-strand breaks (d′) and the subsequent joining of the DNA strands by DNA ligase can then produce the covalently-bonded chi structure.

Once formed, the Holliday intermediate undergoes maturation through enzymatic cleavage of the cross-bridge by either horizontal or vertical cuts (Fig. 7.6g and h). Subsequent action by DNA ligase restores the intact linear molecules (i): One arrangement, called a patch-type heteroduplex, is parental for the genes that flank the region of hybrid overlap (left), and the other arrangement, called a splice-type heteroduplex, is recombinant for those genes (right).

Cytological support for the formation of a χ-shaped intermediate in recombination has come from studies on the small DNA molecules of viruses and plasmids. For example, electron micrographs of the DNA in a plasmid of *E. coli* show a structure resembling the cross-bridged Holliday intermediate (Fig. 7.7).

The original model proposed by Holliday has undergone several modifications over the years as a consequence of both theoretical and experimental advances.

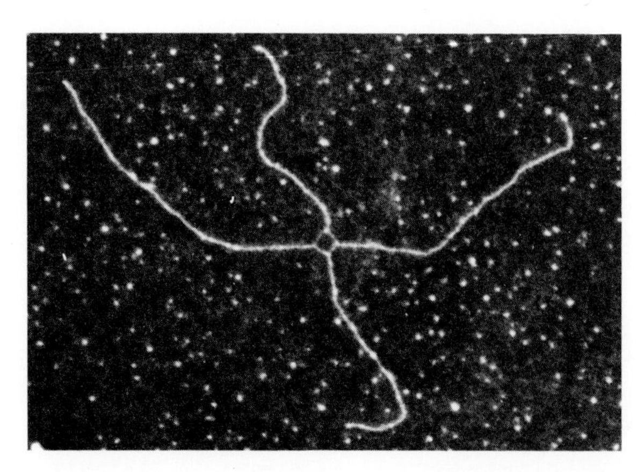

Figure 7.7. Electron micrograph of *E. coli* plasmid DNA, showing a striking resemblance to the "Holliday intermediate." Note in particular the single-stranded connections at the intersection of the four "arms." (Normally, plasmid DNA is circular; this DNA has been cleaved by an endonuclease.) *Source:* From H. Potter and D. Dressler, *Proc. Natl. Acad. Sci.,* U.S., 73:3000, 1976.

Most of these advances relate to the biochemical steps involved in the formation of the Holliday intermediate. Two enzymes have been identified in *E. coli* that participate in this process. One is the *recBC protein*, which is the product of the *recB* and *recC* genes in *E. coli*. The recBC protein is a multifunctional enzyme that acts as both an endonuclease and an exonuclease. But it is possible that its primary function is to unwind the DNA duplex to expose single-stranded regions for the recombination act. DNA binding proteins aid in the process by keeping the exposed regions from re-forming the double helix. The second enzyme, which is specified by the *recA* gene and is called the *recA protein*, then binds to the exposed single strands in the donor DNA molecule and mediates pairing with a homologous region in a recipient molecule. Single-strand nicks that may occur at this time would be catalyzed by an endonuclease (possibly the recBC protein). We will consider more about the function of the recA protein in Chapter 11.

TO SUM UP

1. Linked genes are nonallelic genes that are carried on the same chromosome. They are characterized by their tendency to remain together in the same combination as they were inherited, and they do not assort independently.
2. Linked genes form new (recombinant) gene arrangements through the process of crossing over, which is an exchange of segments between homologous chromosomes. The actual frequency of crossing over between two gene loci depends on the distance separating them. Genes close together form very few recombinant-type gametes through crossing over, whereas genes farther apart form a larger proportion of these gamete types.
3. The most straightforward way to detect linkage experimentally is to examine the progeny of a testcross in which the dominant parent is dihybrid for the genes being considered. If the genes are linked, the offspring will be distributed in two high-frequency and two low-frequency classes, rather than in the 1:1:1:1 ratio expected for genes that assort independently. The lower-frequency classes are those that have the recombinant gene arrangements.
4. The observed frequency of recombinant gametes cannot exceed 50 percent, regardless of the distance between linked genes. This limit to the recombination frequency results from the occurrence of multiple crossovers when distances are large. These multiple exchanges re-create the parental-type gene arrangement as often as they yield recombinants.
5. Genes that are spaced far apart on the same chromosome may not show the characteristics of linkage in testcross data. We can conclude that such genes are actually physically linked on the same chromosome by a finding that each shows linkage with some third gene.
6. Linkage is detected from the results of a dihybrid cross by a significant departure (in a particular direction) of the phenotypic ratio from the 9:3:3:1 ratio that is expected when genes assort independently. In the results of a dihybrid cross with linked genes, there is no one-to-one correspondence between the frequencies of the phenotypic classes in the offspring and the frequencies of the different gamete types produced by each heterozygous parent.
7. Homologous segments of chromosomes exchange parts during the recombination of linked genes. The recombination act proceeds by means of the breakage of the paired chromosomes, followed by the physical exchange of their parts.

8. Current studies on the molecular basis of recombination indicate that crossing over proceeds in a stepwise manner, forming an intermediate χ-shaped structure, called the Holliday intermediate, in which the two recombining DNA molecules are connected by two of their four polynucleotide chains. The Holliday intermediate is then processed in either of two related ways to produce two kinds of heteroduplex product DNA molecules: Some exhibit a parental arrangement of genes that flank the heteroduplex region, and others exhibit a recombinant arrangement of flanking genes.

Construction of Genetic Maps

We can construct a **genetic** (or **linkage**) **map** of each chromosome by studying the linkage relationships of various combinations of genes. A genetic map gives the relative locations of genes along the chromosome and the distances between successive loci. Since genes are microscopically indistinguishable, true physical distances between them cannot be measured. Instead of physical distance, geneticists use the concept of **genetic distance.** In genetic distances, the length of the interval separating two gene loci is defined as being equal to the frequency of crossing over (in percent) that occurs between them. This equation assumes that the probability of a crossover is directly proportional to the physical length of the interval. Therefore, two genes that are far apart on the chromosome should have a proportionally higher chance of experiencing a crossover between them than two genes that are closer together.

Recombination Frequency as a Measure of Map Distance

Crossover events cannot be accurately counted microscopically; they are usually inferred from the presence of recombinant-type gametes. Therefore, we assume that the frequency of crossing over can be measured by the frequency of recombinants among the gametes of a heterozygote. A unit of genetic distance, known as a **map unit,** is thus defined as being equal to a 1 percent recombination frequency.

Map distance can be evaluated directly from testcross data by calculating the proportion of testcross progeny with recombinant gene arrangements. For a testcross then, the recombination frequency (R) is given by

$$R = \frac{\text{Number of recombinant testcross progeny}}{\text{Total number of testcross progeny}}.$$

Using this formula, the total map distance is $100R$, and 1 map unit $= 1\%R$. For example, testcross results indicated that R was 20 percent for the genes involved in fruit shape and inflorescence pattern in tomatoes. The *o* and *s* loci are therefore separated by a genetic distance of 20 map units. A linkage map showing the relative locations of these genes is given as follows:

```
o        20        s
+------------------+
```

Once the distance between two genes is known, they can be mapped in relation to a third gene, a fourth, and so on within the linkage group. For a hypothetical example, let us suppose that testcross data show genes *a* and *b* to be separated by 5 map units. A third locus, *c*, is found to be 7 units from *a* and 12 units from *b*. Since *c* and *b* are farthest apart, the linkage map becomes

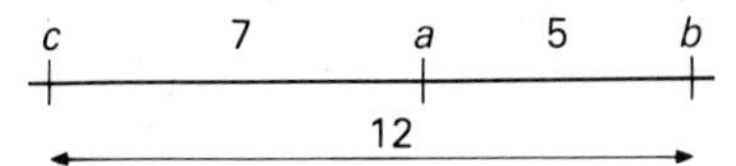

Other genes can be added to this map, until all the genes within the linkage group have been assigned positions. The gene lying closest to one end of the linkage group is assigned the map position 0.0, and all other loci are then given map position values according to their distance from the first gene. In the hypothetical example just given, let us suppose that gene *d* is found to lie 2 units to the left of the *c* locus and that no gene is found to map closer to the left end of the chromosome than *d*. Given in terms of genetic distances, the map is

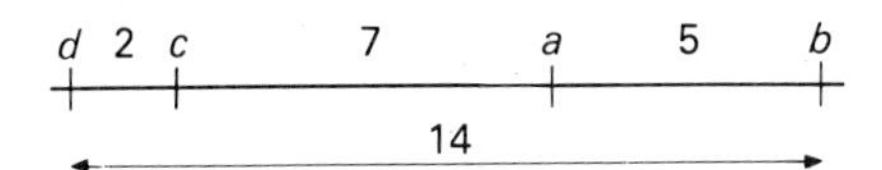

These relationships are more conveniently represented as

d	*c*	*a*	*b*
0.0	2.0	9.0	14.0

The crossover data we used in the preceding example yielded a precise linear order of genes along the chromosome. Linear linkage maps are in fact the rule in a wide variety of plants and animals.

EXAMPLE 7.3. Two of the many known pairs of traits in *Drosophila* are gray body (b^+) vs. black body (b) and long wings (vg^+) vs. vestigial wings (vg). The genes responsible for body color and wing length are linked. Suppose that a testcross $b^+b\ vg^+vg \times bb\ vgvg$ yields offspring in the ratio 985 gray, long : 962 black, vestigial : 236 black, long : 217 gray, vestigial. What is the map distance between the *b* and *vg* loci?

Solution. To compute the map distance, we must determine *R*, the frequency of recombination. The black, long and the gray, vestigial classes must be the recombinant types, since they occur in the lowest frequencies among the testcross progeny. Thus,

$$R = \frac{236 + 217}{2400 \text{ total}} = 0.189, \text{ or } 18.9 \text{ percent}$$

The map for the *b* and *vg* loci would then be

b —— 18.9 —— *vg*

Effects of Multiple Crossovers. As a rule, more crossing over occurs between widely separated loci than is registered as recombinant gametes. The recombination frequency thus becomes an underestimate of the true distance between widely separated loci. This discrepancy arises because multiple crossover events occur in longer intervals between genes. Observe in Fig. 7.3 that any even number of exchanges (2, 4, 6, . . .) will not be detected as a recombinant. Moreover, any odd number of crossovers, regardless of how many occur, will yield no more recombinants than would a single exchange. Thus, the increasing frequency of crossing over in longer intervals is simply not matched by a proportional gain in recombinants.

The precise form of the relationship between the recombination frequency and the true genetic distance is shown in Fig. 7.8. Notice that the recombination frequency is linearly related to the crossover frequency only over distances up to about 10 to 15 units. A discrepancy develops at longer distances, with the recombination frequency approaching a maximum value of 50 percent despite continued increases in the crossover frequency.

One approach that we can take to avoid the inaccuracies that may develop at longer distances is to subdivide the linkage map into a series of shorter intervals, each of which is no longer than 10 to 15 map units. To subdivide the map, we make use of one or more loci known to be present between the genes in question but that have not been studied in previous crosses. For example, suppose that we found genes *e* and *f* to be 16 map units apart on the basis of recombination frequency computed from testcross data. The preliminary map is then

e 16 *f*

Because of multiple crossovers, however, this distance probably represents an underestimate of the true distance. To obtain a more accurate value, we redo the mapping study, making use of gene *h* located between *e* and *f*. (The *h* locus has always been present; we just did not score for its phenotype in the previous

Figure 7.8. The frequency of recombination as a function of the true genetic distance (based on the percentage of crossing over). The recombination frequency is linearly related to the frequency of crossing over only at short distances. Recombination values greater than 10 to 15 percent are less than the actual frequency of crossing over. Notice that the maximum recombination frequency is 50 percent. (Various models have been developed by geneticists to provide a mathematical basis for this graphical relationship. One of the simplest of these models, called the Haldane mapping function, is given by the formula $R = \frac{1}{2}(1 - e^{-2d})$, in which R is the frequency of recombination, d is the frequency of crossing over, and e is the base of natural logarithms. In this model, crossovers are assumed to occur at random along the length of a chromosome pair.)

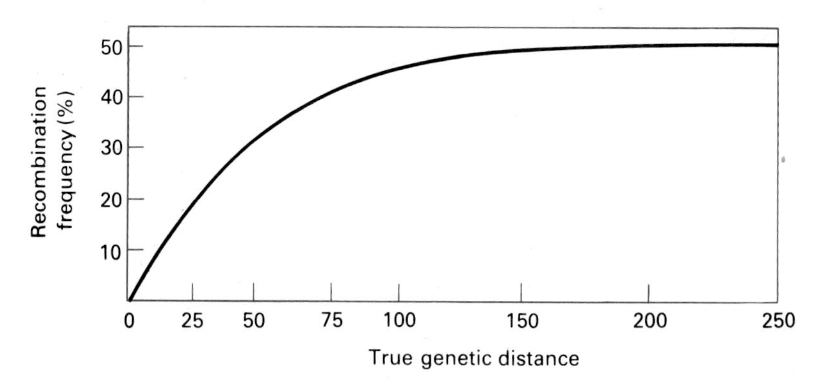

experiment.) We now make two testcrosses, one *EeHh* × *eehh* and the other *HhFf* × *hhff*. Assume that the first cross yields 8 percent recombinant offspring and the second cross yields 9 percent recombinant offspring. The improved map then becomes

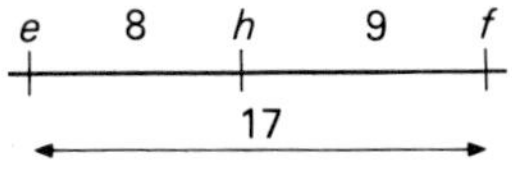

$$\underset{e}{+} \xrightarrow{8} \underset{h}{+} \xrightarrow{9} \underset{f}{+} \qquad \overleftrightarrow{\;17\;}$$

The distance between *e* and *f* is now greater than had been estimated earlier; it is equal to the sum of the two smaller values: 8 + 9 = 17 map units.

We could extend this procedure to include the entire range of nonallelic genes known to be carried on a chromosome (referred to collectively as a **linkage group**). By doing so, we could construct a genetic map in which the total map distance (estimated by summing the intervals within it) is much greater than the maximum theoretical distance of 50 map units given by the recombination frequency for any two widely separated loci. Examples of this outcome are illustrated in Figs. 7.9 and 7.10 by the linkage groups established for two well-known organisms in genetic research, *Zea mays* (corn) and *Drosophila*. The larger map position values have been determined by summing the shorter intervals.

The Three-Point Testcross Method for Mapping Genes

In the preceding discussion, we implicitly assumed that map distances are determined from the results of *two-point* (or *two-factor*) *testcrosses*, in which the dominant parent is heterozygous for only two gene pairs. There are two main disadvantages to mapping linked genes by this approach. First, by following the joint transmission of only two gene pairs at a time, we require a separate two-factor cross to detect linkage between each combination of two gene pairs. For example, we would have to identify and count the recombinant offspring of three separate testcrosses, *AaBb* × *aabb*, *BbCc* × *bbcc*, and *AaCc* × *aacc*, in order to arrange unambiguously the three gene loci *a*, *b*, and *c*. Second, we cannot directly measure the genetic effects of multiple crossovers by using two-point crosses; we can only infer the occurrence of double and higher-order crossovers from the failure of different matings to yield additive values of the recombination frequency.

Because of the drawbacks of obtaining linkage data from two-point crosses, a *three-point testcross* is the preferred method for mapping linked genes. In a three-point testcross, one parent is trihybrid, so that three genetic markers can be followed in a single mating. Moreover, the presence of three linked loci, spanning two adjacent crossover intervals, permits us to identify recombinants that arise from double-exchange events. The cross

$$\frac{A\ B\ C}{a\ b\ c} \times \frac{a\ b\ c}{a\ b\ c}$$

provides an example in which all genes are in the cis arrangement in the trihybrid parent. (Any gene arrangement can be used; we are using a cis-type arrangement to illustrate the cross for convenience only. Other possible trihybrid parents for the testcross are $\frac{A\ B\ c}{a\ b\ C}$, $\frac{A\ b\ C}{a\ B\ c}$, and $\frac{A\ b\ c}{a\ B\ C}$.) The trihybrid in the cross we have selected

Figure 7.9. Linkage map of the ten chromosomes of corn. *Source:* Reproduced from *Corn and Corn Improvement*, ASA Monograph No. 18, 1977, pp. 111–223 by permission of the American Society of Agronomy.

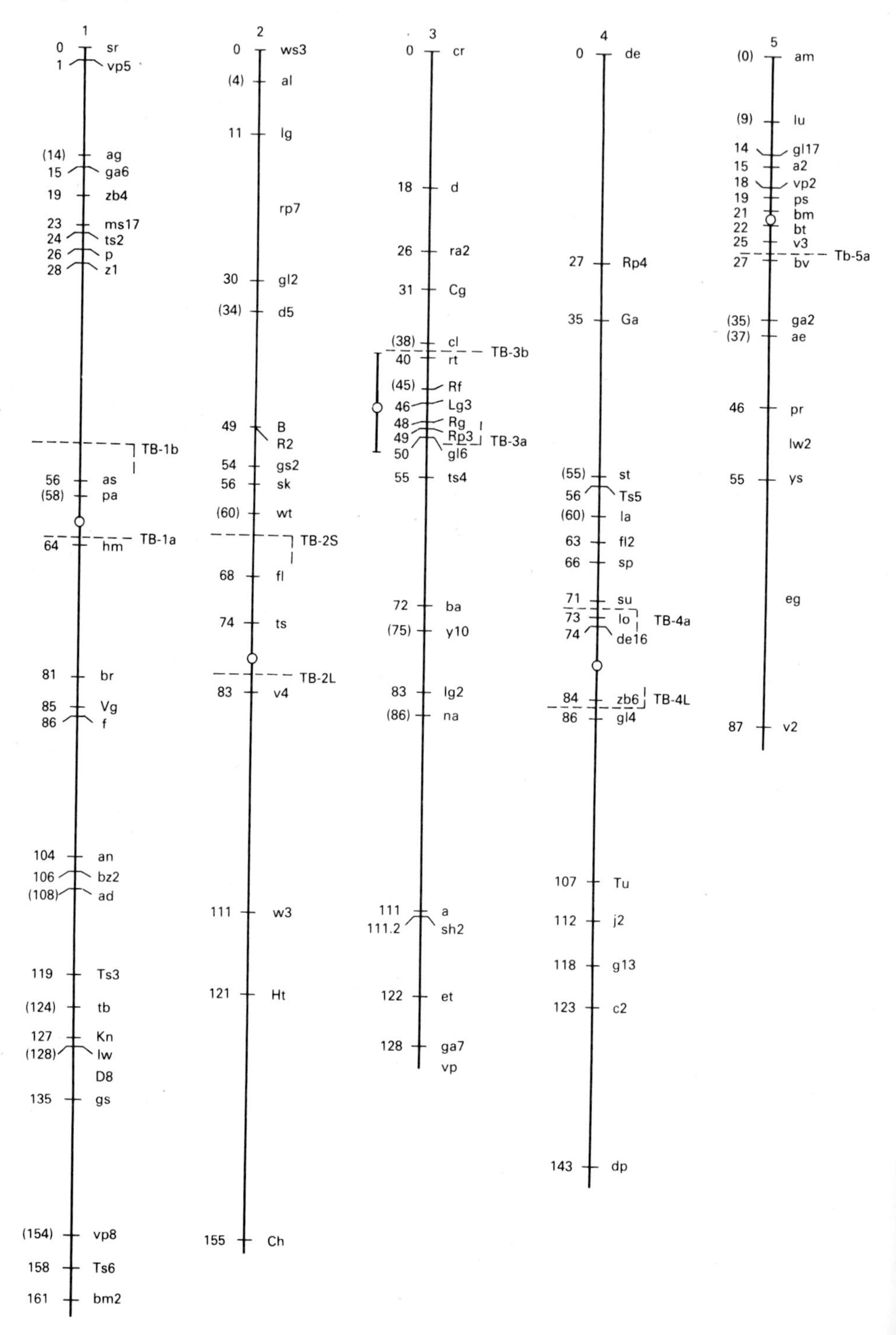

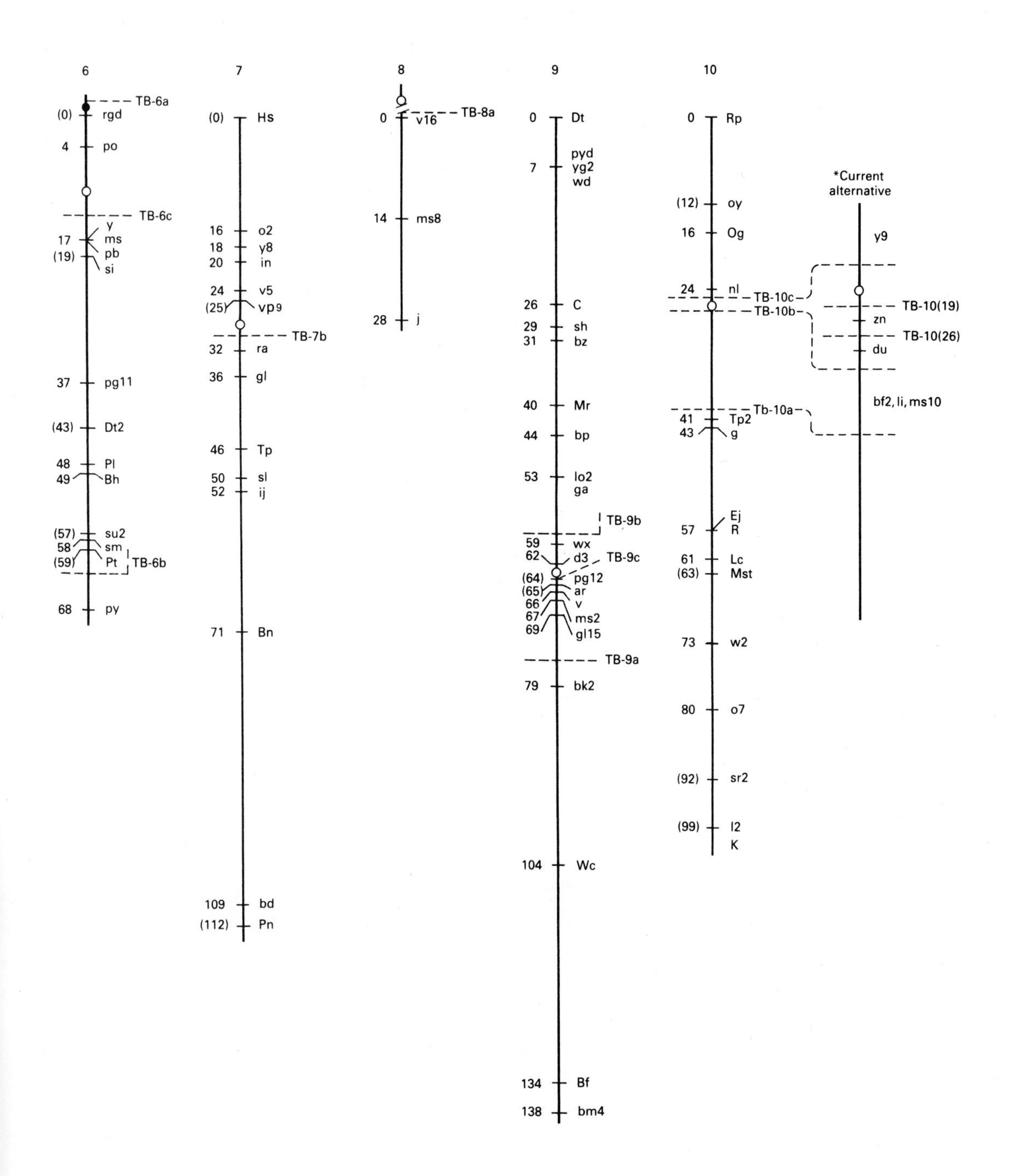
6
TB-6a
(0) rgd
4 po
TB-6c
y
17 ms
(19) pb
si
37 pg11
(43) Dt2
48 Pl
49 Bh
(57) su2
58 sm
(59) Pt
TB-6b
68 py
7
(0) Hs
16 o2
18 y8
20 in
24 v5
(25) vp9
TB-7b
32 ra
36 gl
46 Tp
50 sl
52 ij
71 Bn
109 bd
(112) Pn
8
0 v16
TB-8a
14 ms8
28 j
9
0 Dt
pyd
7 yg2
wd
26 C
29 sh
31 bz
40 Mr
44 bp
53 lo2
ga
TB-9b
59 wx
62 d3
TB-9c
(64) pg12
(65) ar
66 v
67 ms2
69 gl15
TB-9a
79 bk2
104 Wc
134 Bf
138 bm4
10
0 Rp
(12) oy
16 Og
24 nl
TB-10c
TB-10b
Tb-10a
41 Tp2
43 g
57 Ej
R
61 Lc
(63) Mst
73 w2
80 o7
(92) sr2
(99) l2
K
*Current alternative
y9
TB-10(19)
zn
TB-10(26)
du
bf2, li, ms10

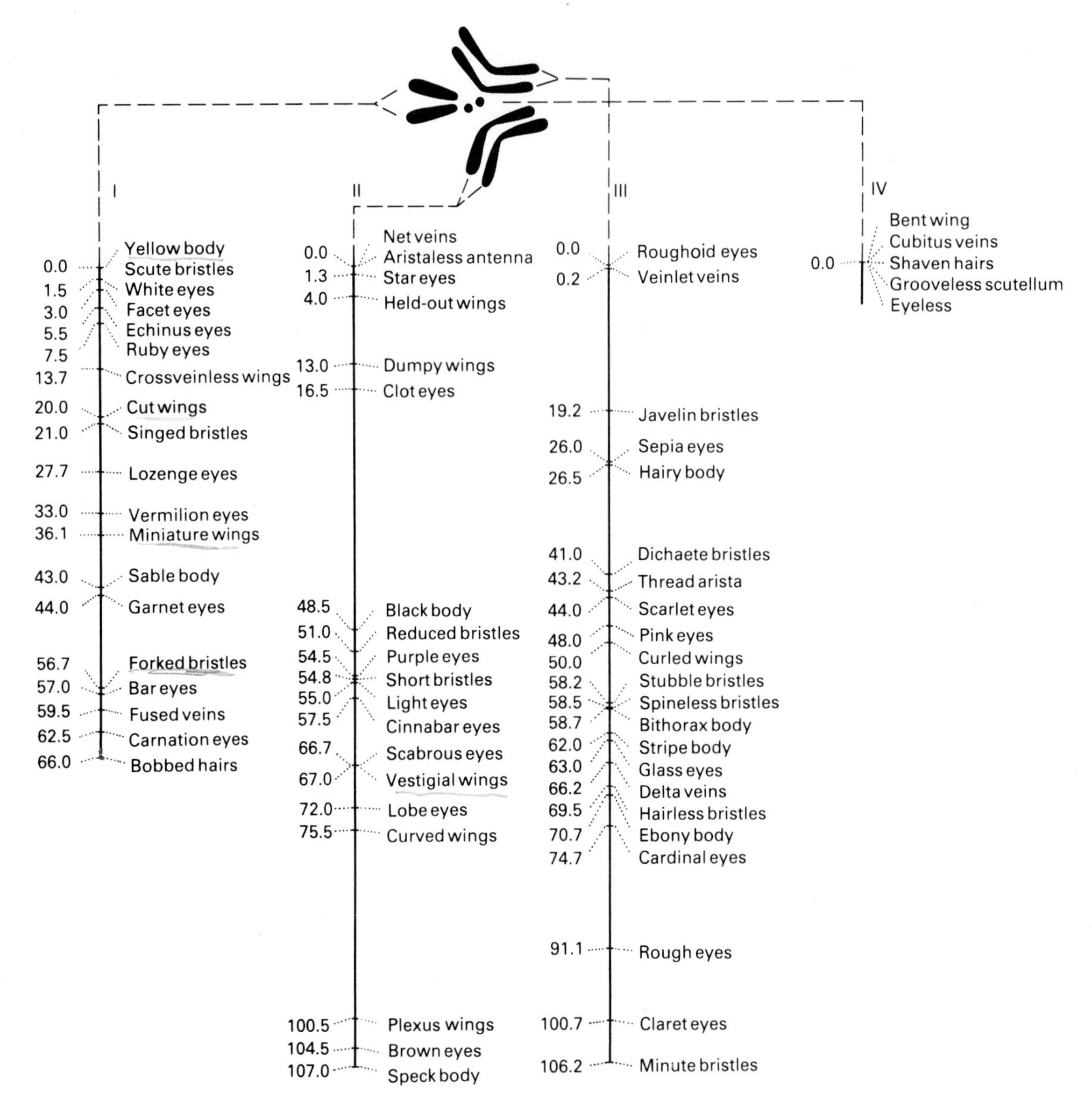

Figure 7.10. Linkage map of *Drosophila*, showing the chromosomes corresponding to each linkage group.

can produce eight kinds of gametes, which are listed in four classes as follows:

(1)	(2)	(3)	(4)
ABC	*Abc*	*ABc*	*AbC*
abc	*aBC*	*abC*	*aBc*

The simplest possible origin of each of these gamete classes is shown in Fig. 7.11 in terms of the crossover events that produced them. Class 1 includes parental-type gametes. For three genes that are linked, these gamete types will occur in the highest frequencies. Class 2 and Class 3 each have gamete types that result from a single recombination act, Class 2 involving the *a* and *b* loci and Class 3 involving the *b* and *c* loci. Class 4 contains the double-recombinant gamete types, involving recombination events for both sets of markers. Since the probability that two crossovers will occur simultaneously is less than the chance that only one crossover event will occur, double recombinant gametes are the gamete types present in the lowest frequency.

Determining Map Distance. We can use the frequencies of the various recombinant gamete classes given in the preceding example to compute map distances for the different sets of loci. To illustrate, we use *f*( ) to mean "frequency of" in this and all discussions that follow. We will also use R_{a-b}, R_{b-c}, and R_{a-c} to denote the recombination frequencies for the designated loci, such as we might have determined from three separate two-point testcrosses. Since all three genes are

Figure 7.11. Diagrams depicting the simplest possible origins of the various gamete classes produced by a trihybrid *A B C*//*a b c*. More complex events are also possible, since any odd number of exchanges in an interval can give rise to a recombinant arrangement of genes, while no crossover or any even number of crossovers can produce a parental arrangement.

(1) No crossing over

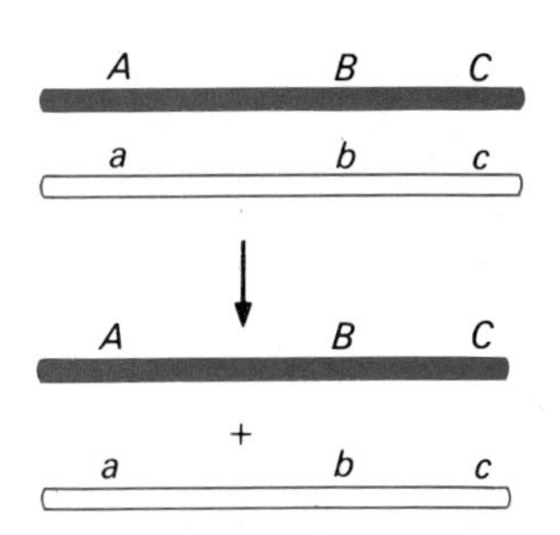

Parental for *a-b*; parental for *b-c*

(2) Crossing over *a-b*; no crossing over *b-c*

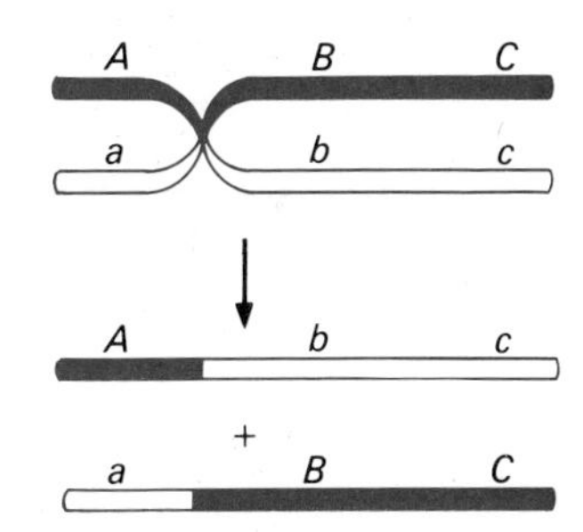

Recombinant for *a-b*; parental for *b-c*

(3) No crossing over *a-b*; crossing over *b-c*

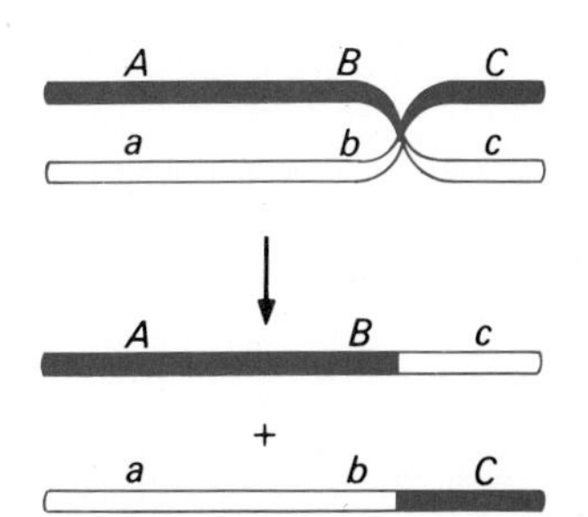

Parental for *a-b*; recombinant for *b-c*

(4) Crossing over *a-b*; crossing over *b-c*

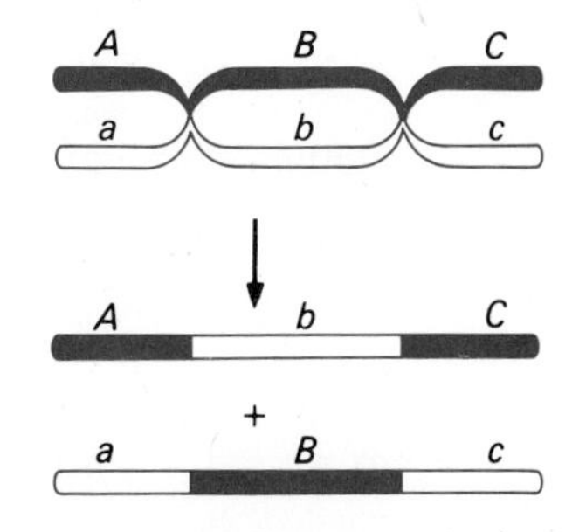

Recombinant for *a-b*; recombinant for *b-c*

linked in the cis arrangement in this example, the recombinant-type arrangements for each pair of loci are *Ab* and *aB*, *Bc* and *bC*, and *Ac* and *aC*, respectively. For this example then, $R_{a-b} = f(Ab) + f(aB)$, $R_{b-c} = f(Bc) + f(bC)$, and $R_{a-c} = f(Ac) + f(aC)$, where $f(Ab)$, $f(aB)$, and so on are the frequencies of the designated gene arrangements. If we analyze each pair of loci separately (indicated by making a slash through the locus not under consideration), then we can calculate the recombination frequencies from the frequencies of the trihybrid gamete classes—$f(Abc)$, $f(aBC)$, and so on—in the following manner:

$$R_{a-b} = f(Ab\not{c}) + f(aB\not{C}) + f(Ab\not{C}) + f(aBc)$$
$$R_{b-c} = f(\not{A}Bc) + f(\not{a}bC) + f(\not{A}bC) + f(\not{a}Bc)$$
$$R_{a-c} = f(A\not{b}c) + f(a\not{B}C) + f(A\not{B}c) + f(a\not{b}C)$$

To calculate R_{a-b}, we ignore the *c* locus and add the frequencies of all gametes with recombinant gene arrangements for the *a* and *b* loci. In like manner, we determine the *R* values for *b–c* and *a–c* from the frequencies of their recombinant gene arrangements, ignoring *a* while calculating R_{b-c} and ignoring *b* while calculating R_{a-c}. By using this procedure, we are analyzing the data from a single three-point testcross as though we had conducted three separate two-point testcrosses, each involving a different pair of gene loci.

We are now ready to calculate the distances between loci from the results of an actual three-point testcross. Corn is an excellent subject for the evaluation of testcross data. Since each ear of corn contains several hundred seeds, the results of an entire three-point cross can be obtained by simply recording the kernel phenotypes of some 10 to 20 ears. In our three-point cross, we will consider traits produced by the following three gene pairs:

Aleurone color: colored kernels (*C*) vs. colorless (*c*)
Amount of endosperm: plump kernels (*Sh*) vs. shrunken (*sh*)
Appearance of endosperm: starchy kernels (*Wx*) vs. waxy (*wx*)

For the time being, we will denote the trihybrid for these gene pairs as *Cc Shsh Wxwx*. We have listed the genes in alphabetical order, since the correct gene sequence is unknown to us, as is the arrangement of the nonallelic genes (cis or trans).

The results of a testcross from an experiment of Hutchison (1922) are given in Table 7.3. The phenotypes of the offspring are listed in order of frequency, along with the genotype of the gamete received from the trihybrid parent. The lack of a 1:1:1:1:1:1:1:1 ratio of phenotypes in the progeny clearly shows the linkage of genes. Furthermore, the parental types can be readily identified as those with the highest frequencies. The finding that *C sh Wx* and *c Sh wx* are the parental gene combinations means that the genotype of the trihybrid parent is $\frac{C\ sh\ Wx}{c\ Sh\ wx}$. The designated sequence is only tentative, however, since we still do not know if this is the correct gene order.

We can use the testcross results given in Table 7.3 to calculate genetic distances for the intervals between the different sets of genes. If we consider only two genes at a time, recombinant-type arrangements for the different sets of loci are *C Sh*

Table 7.3. Results of a trihybrid testcross in corn involving alleles for colored (*C*) vs. colorless (*c*) aleurone, plump (*Sh*) vs. shrunken (*sh*) endosperm, and starchy (*Wx*) vs. waxy (*wx*) endosperm.

KERNEL PHENOTYPE	GAMETE RECEIVED FROM HETEROZYGOUS PARENT	NUMBER
colorless, plump, waxy	*c Sh wx*	2708
colored, shrunken, starchy	*C sh Wx*	2538
colorless, plump, starchy	*c Sh Wx*	626
colored, shrunken, waxy	*C sh wx*	601
colorless, shrunken, starchy	*c sh Wx*	116
colored, plump, waxy	*C Sh wx*	113
colored, plump, starchy	*C Sh Wx*	4
colorless, shrunken, waxy	*c sh wx*	2
		6708

and *c sh, Sh Wx* and *sh wx,* and *C wx* and *c Wx*. The map distances are then

$$100R_{c\text{–}sh} = 100[f(C\ Sh\ \cancel{wx}) + f(c\ sh\ \cancel{Wx}) + f(C\ Sh\ \cancel{Wx}) + f(c\ sh\ \cancel{wx})]$$
$$= 100\frac{(113 + 116 + 4 + 2)}{6708 \text{ total}} = 3.5,$$

$$100R_{sh\text{–}wx} = 100[f(\cancel{c}\ Sh\ Wx) + f(\cancel{C}\ sh\ wx) + f(\cancel{C}\ Sh\ Wx) + f(\cancel{c}\ sh\ wx)]$$
$$= 100\frac{(626 + 601 + 4 + 2)}{6708 \text{ total}} = 18.4,$$

$$100R_{c\text{–}wx} = 100[f(C\ \cancel{Sh}\ wx) + f(c\ \cancel{sh}\ Wx) + f(c\ \cancel{Sh}\ Wx) + f(C\ \cancel{sh}\ wx)]$$
$$= 100\frac{(113 + 116 + 626 + 601)}{6708 \text{ total}} = 21.7.$$

Determining Gene Order. Another step involved in mapping genes by the three-point testcross method is to determine the proper gene order. The correct gene order may not be known to the investigator before a three-point cross is performed and it must be determined from the testcross results. One approach is to calculate the various map distances, as we did in the preceding analysis, and then to determine the proper gene order from the relative magnitudes of these calculated values. For example, the preceding calculations gave map distances for the *c–sh*, *sh–wx*, and *c–wx* intervals of 3.5, 18.4, and 21.7 map units, respectively. Since 21.7 is the longest of the three distances, we conclude that *c* and *wx* are the outside markers, with *sh* between them. The map is then

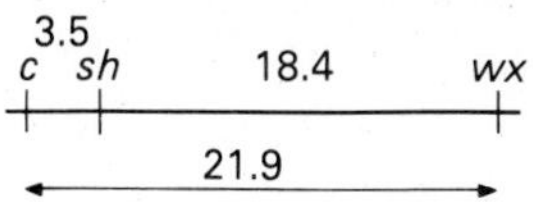

3.5
c sh 18.4 *wx*
21.9

The sum 3.5 + 18.4 = 21.9 is a more accurate measure of total distance than 21.7, since it compensates for the nullifying effects of an even number of exchanges between the outside markers made up of an odd number of exchanges in each of the included intervals (see Fig. 7.12).

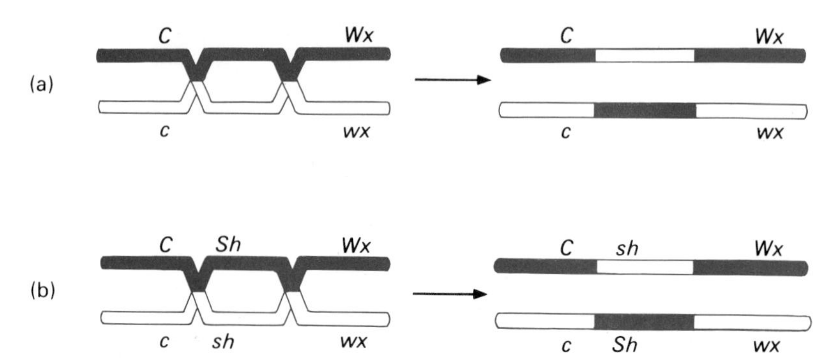

Figure 7.12. Double crossing over along the interval separating the *c* and *wx* loci. In (a), there is no way to detect the crossover event, because the outside markers are back in a parental arrangement. The middle section of the chromosome is recombinant, but we have no way of knowing this. In (b), the use of a gene to mark the middle section of the chromosome allows detection of double crossovers as recombinants of the middle gene with each outside gene.

It is also possible to determine the proper gene order without having to calculate the map distances first. All we need to do in this second approach is to deduce the sequence of genes in the parental chromosomes that would produce the genotypes of the double-recombinant class by a double crossover event. The double-recombinant class is produced by a single crossover event (more precisely, an odd-numbered event) in each of the included intervals. Therefore, only the middle pair of alleles will be affected by the double exchange. For example, suppose that among the eight types of gametes produced by the trihybrid *AaBbCc* in a three-point testcross, *AbC* and *aBc* appear in the highest frequency, whereas *Abc* and *aBC* occur in the lowest frequency. What is the correct order of the *a*, *b*, and *c* gene loci on the chromosome? It is apparent that gametes *AbC* and *aBc*, which occur in the highest frequency, are the parental types and that gametes *Abc* and *aBC*, which occur in the lowest frequency, are the double-recombinant types. Note that *AbC* differs from *Abc* and that *aBc* differs from *aBC* only by the nature of the allele at the *c* locus. Thus, double-recombinant gametes are formed from the parental gene arrangement by an exchange of chromosome parts involving only the *C,c* gene pair. Since a minimum of two crossovers are required to achieve this result, the *c* locus must be in the middle:

A C b / *a c B* → *A c b* / *a C B*

EXAMPLE 7.4. In *Drosophila*, the three gene pairs for normal bristles (f^+) vs. forked (bent or split bristles, *f*), red eye color (g^+) vs. garnet (pinkish eye color, *g*), and long wings (m^+) vs. miniature (*m*) are known to be present on the same chromosome. A trihybrid $f^+f\,g^+g\,m^+m$ female is crossed to a wild-type (normal-bristled, red-eyed, long-winged) male and produces the following offspring (in each case, only those genes that are known to be present from the phenotypes of the flies are listed):

Females (1000): All wild type

Males (1000): $f^+g^+m^+$ 408, $f^+g\ m$ 64, $f\ g\ m$ 399, $f\ g^+m^+$ 59
f^+g^+m 30, $f^+g\ m^+$ 4, $f\ g\ m^+$ 33, $f\ g^+m$ 3

(a) To which chromosome do these genes belong? How can you tell?
(b) Construct a linkage map of the *f, g,* and *m* loci.

Solution. **(a)** The three gene loci are present on the X chromosome. The female offspring all express the dominant wild-type genes that they received on the paternally derived X chromosome. The males received an X chromosome from their maternal parent only. Since the males are hemizygous, they will express all the X-linked genes inherited from their mother. Thus, the parental-type and recombinant-type gametes formed by the female parent can be observed directly in the male offspring. **(b)** First note that $f^+g^+m^+$ and $f\ g\ m$ are parental-type arrangements since they occur in the highest frequencies. Therefore, $f^+g\ m$ and $f\ g^+m^+$ are single-recombinant types for *f-g;* f^+g^+m and $f\ g\ m^+$ are single-recombinant types for *g-m;* and $f^+g\ m^+$ and $f\ g^+m$ are the double-recombinant types. The map distances can then be calculated as

$$100R_{f\text{-}g} = \frac{100(64 + 59 + 4 + 3)}{1000} = 13.0;$$

$$100R_{g\text{-}m} = \frac{100(30 + 33 + 4 + 3)}{1000} = 7.0;$$

and

$$100R_{f\text{-}m} = \frac{100(64 + 59 + 33 + 30)}{1000} = 18.6.$$

The linkage map becomes:

```
f        13        g     7     m
+------------------+-----------+
```

Thus, mapping data for X-linked genes can be derived directly from the male offspring of heterozygous females without the necessity of a testcross.

Interference and Coincidence. The previous analysis on the *c, sh,* and *wx* loci in corn gave recombination frequencies of $R_{c\text{-}sh} = 0.035$ and $R_{sh\text{-}wx} = 0.184$ for the two included intervals in the map. These numbers are also the probabilities of an odd number of crossovers in the designated regions. If crossovers in the *c–sh* region were independent of crossovers in the adjacent *sh–wx* region, then the expected proportion of double-recombinant gametes would be $0.035 \times 0.184 = 0.00644$, or $0.00644 \times 6708 = 43$ double recombinants out of the total number of offspring from the testcross. But note that the actual data contain only six (4 *C Sh Wx* and 2 *c sh wx*) double recombinants. This outcome suggests that crossovers that occur in adjacent intervals are not independent events but tend to have an inhibiting effect on each other.

The overall effect of a crossover in one region on the chance that another crossover will occur in an adjacent region is known as **interference.** The degree of interference is measured by the **coefficient of coincidence,** which is simply the ratio of the observed frequency of double recombinants to the frequency expected if the crossovers were truly independent:

$$\text{Coefficient of coincidence} = \frac{\text{Observed number of double recombinants}}{\text{Expected number of double recombinants}}.$$

For the preceding data, the numerical value for the coefficient of coincidence is $6/43 = 0.14$. When the coefficient of coincidence is between 0 and 1, the interference is called **positive interference.** Positive interference is characteristic of crossing over in many higher organisms. Positive interference means that one crossover decreases the likelihood of a second crossover nearby.

In the case of very closely linked genes, particularly in some microorganisms, (for example, the bacterial viruses), the occurrence of one crossover can actually increase the chance of additional crossing over. Double-recombinant types are then more frequent than would be expected for independent events—a phenomenon known as **negative interference.** In this case, the coefficient of coincidence is greater than 1.

EXAMPLE 7.5. Suppose that three loci gave the following linkage map:

a —— 8 —— b —— 9 —— c

Assuming no interference between crossovers in adjacent intervals, predict the frequencies of the recombinant classes produced by the three-point testcross $\frac{A\ B\ C}{a\ b\ c} \times \frac{a\ b\ c}{a\ b\ c}$.

Solution. The recombination frequencies R_{a-b} and R_{b-c} are the probabilities of obtaining a recombinant gene arrangement for each of the two pairs of loci. Thus, the probability of getting an Ab or an aB arrangement is $R_{a-b} = 0.08$, and the probability of getting a Bc or a bC arrangement is $R_{b-c} = 0.09$. The probabilities of obtaining parental-type arrangements are then $1 - R_{a-b}$ and $1 - R_{b-c}$, respectively. Also note that without interference, crossovers that occur in adjacent intervals are independent events. Applying the multiplication rule of probability, the frequencies of the three recombinant classes can be calculated as follows:

RECOMBINANT CLASS	FREQUENCY	TYPE
$\frac{A\ b\ c}{a\ b\ c}$, $\frac{a\ B\ C}{a\ b\ c}$	$R_{a-b}(1 - R_{b-c}) = 0.08 \times 0.91 = 0.0728$	Single, *a–b*
$\frac{A\ B\ c}{a\ b\ c}$, $\frac{a\ b\ C}{a\ b\ c}$	$(1 - R_{a-b})R_{b-c} = 0.92 \times 0.09 = 0.0828$	Single, *b–c*
$\frac{A\ b\ C}{a\ b\ c}$, $\frac{a\ B\ c}{a\ b\ c}$	$R_{a-b} \times R_{b-c} = 0.08 \times 0.09 = 0.0072$	Double, *a–b* and *b–c*

Mapping by Tetrad Analysis

The scope of our discussion so far in this chapter has been limited to linkage analysis in diploid eukaryotes. Linkage analysis in these organisms constitutes what is known as a random progeny analysis, since inferences made about crossing over are based on the progeny derived from a random sampling of all the meiotic products of a heterozygote. Linkage studies can also be conducted on haploid eukaryotes, such as *Neurospora*, by means of tetrad analysis. In a tetrad analysis, all four products of a single meiosis are recovered and tested individually to determine their genotypes. Tetrad analysis thus provides information about details of chromosome behavior during meiosis that are not ordinarily attainable from studies on higher organisms.

Mapping Two Linked Genes by Tetrad Analysis. Recall from Chapter 5 that if two gene pairs assort independently, they give equal frequencies of parental ditype (PD) and nonparental ditype (NPD) tetrads. In contrast, linkage of two genes results in a large excess of PD over NPD tetrads. Consider the cross *AC* × *ac* in *Neurospora* in which the genes are linked. In this cross, *A* and *a* are the mating type alleles and *C* and *c* are alleles that influence coloration, with *c* being colorless (albino). The most likely origin of each of the three major tetrad classes in this cross are shown in Fig. 7.13. The PD tetrads result primarily from meioses in

Figure 7.13. Main source of the three types of tetrads in the case of linked genes. The majority of the PD tetrads are nonrecombinant (a), whereas a single crossover is responsible for most of the T tetrads (b), and a four-strand double crossover is required to produce NPD tetrads (c). Linkage is thus manifested as an excess of PD over NPD tetrads. (For simplicity, only one member of each spore pair is shown.)

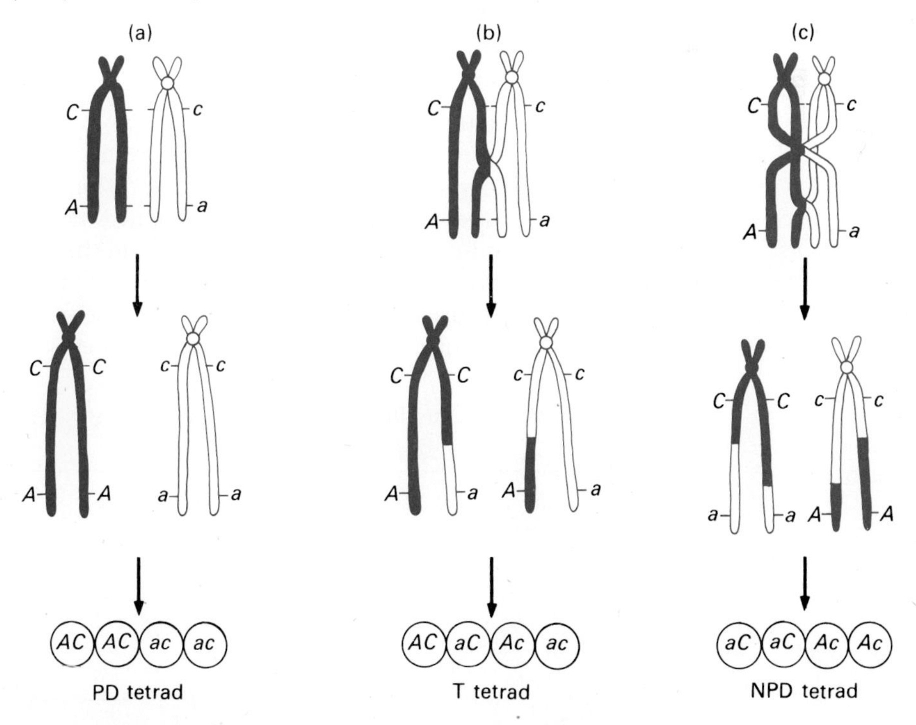

which recombination did not take place (or, less often, from two-strand double exchanges). The NPD tetrads result from meioses in which a four-strand double crossover took place; NPD tetrads for linked genes are therefore expected to be rare in comparison to PD tetrads. Thus, the major criterion of linkage is an *excess of PD over NPD tetrads.* The tetratype (T) tetrads result primarily from meioses in which a single crossover occurred (or, less often, from three-strand double crossovers or even higher degrees of crossing over). The frequency of T tetrads with linked genes is expected to be much higher than the frequency of the NPD class.

Observe in Fig. 7.13 that all four of the chromatids in NPD tetrads are recombinant, while only half of the chromatids in T tetrads are recombinant. We can therefore calculate the recombination frequency between the genes as

$$R = f(\text{NPD}) + (1/2)f(\text{T}) \tag{7.1}$$

where f(NPD) and f(T) are the frequencies of the NPD and T tetrad classes, respectively.

As an example of mapping by tetrad analysis, suppose that the cross $AC \times ac$ yields the classes and numbers of tetrads given in Table 7.4. We see that classes 1 and 5, which contain only parental type ascospores, are PD tetrads; classes 2 and 6, which contain only recombinant type ascospores, are NPD tetrads; and classes 3, 4, and 7, in which half the ascospores are parental type and half are recombinant type, are T tetrads. Linkage is immediately evident in this example from the excess of PD tetrads (175 + 7 = 182) over NPD tetrads (5 + 6 = 11). The map distance between the two genes is therefore

$$100R_{a-c} = 100[f(\text{NPD}) + (1/2)f(\text{T})]$$

$$= 100\left[\frac{5 + 6 + (1/2)(80 + 120 + 7)}{400 \text{ total}}\right] = 28.6 \text{ map units.}$$

Determining Gene-Centromere Distance in *Neurospora.* The ability to recover and analyze ordered tetrads in certain fungi, such as *Neurospora,* permits an investigator to detect crossing over between a gene and the centromere of its chromosome. We have shown in Chapter 5 that crossing over in the interval between a gene locus and the centromere locus produces the spore pair arrangement known as a second-division segregation (SDS) pattern. Since an SDS pattern is a result of crossing over, the total frequency of SDS asci produced by a cross in *Neurospora* can then be used to estimate the distance between a gene and the centromere. Recall that only two of the four chromatids cross over in producing

Table 7.4. The classes and numbers of tetrads produced by the *Neurospora* cross $AC \times ac$.

	(1)	(2)	(3)	(4)	(5)	(6)	(7)	
Spore pair 1:	*AC*	*A c*	*AC*	*AC*	*AC*	*A c*	*AC*	
Spore pair 2:	*AC*	*A c*	*a C*	*A c*	*a c*	*a C*	*a c*	
Spore pair 3:	*a c*	*a C*	*A c*	*a C*	*AC*	*A c*	*A c*	
Spore pair 4:	*a c*	*a C*	*a c*	*a c*	*a c*	*a C*	*a C*	
	175	5	80	120	7	6	7	= 400 total

Tetrad class		Segregation pattern A,a	C,c	Origin
1	*AC* *AC* *ac* *ac*	FDS	FDS	No exchange
2	*Ac* *Ac* *aC* *aC*	FDS	FDS	Four-strand double exchange — both in I or both in II
3	*AC* *aC* *Ac* *ac*	SDS	FDS	Two-strand single exchange — in I
4	*AC* *Ac* *aC* *ac*	FDS	SDS	Two-strand single exchange — in II
5	*AC* *ac* *AC* *ac*	SDS	SDS	Two-strand double exchange — one in I and one in II
6	*Ac* *aC* *Ac* *aC*	SDS	SDS	Four-strand double exchange — one in I and one in II
7	*AC* *ac* *Ac* *aC*	SDS	SDS	Three-strand double exchange — one in I and one in II

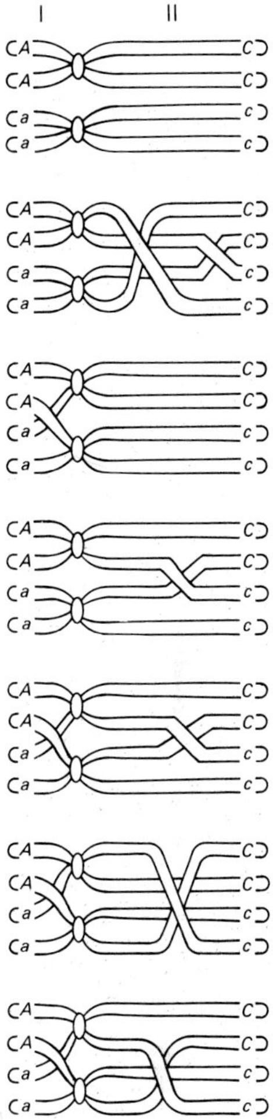

Figure 7.14. The origin of the seven classes of ordered tetrads arising from a *Neurospora* cross that generated the data given in Table 7.4. Classes 1 and 5 are PD tetrads, with class 1 representing the majority of this type. Classes 2 and 6 are NPD tetrads. The remaining classes—3, 4, and 7—are T tetrads, with class 7 being the minority class in this group. The recombination frequency between the genes is estimated as $R = f\,(\text{NPD}) + (1/2)f(\text{T})$. This formula assumes that the vast majority of NPD tetrads have undergone a double exchange, whereas nearly all the T tetrads result from a single exchange. The formula underestimates the distance somewhat, because the occasional triple-exchange NPD tetrads and some double-exchange T tetrads are not accounted for, and also because a few PD tetrads (class 5) have undergone crossing over but are not included in the calculation of the recombination frequency.

an SDS tetrad. The gene–centromere distance is therefore equal to one-half the percent of SDS asci.

The cross results in Table 7.4 provide an example of the kind of data that can be used for determining gene-centromere distance. The segregation patterns of both genes in the various tetrads are given in Fig. 7.14, along with the simplest origin of each tetrad class. The procedure used to determine the segregation pattern is the same as that described in Chapter 5; consider one gene pair at a time by looking only at the spore-pair arrangements of the gene of interest. By combining the SDS asci for each gene pair, the distance between the genes and their centromeres can be calculated as follows:

$$\text{Centromere to locus } a \text{ distance} = \tfrac{1}{2}(\%\text{SDS}_{A,a}) = \frac{1}{2}\left[\frac{(80 + 7 + 6 + 7)(100)}{400 \text{ total}}\right] = 12.5$$

$$\text{Centromere to locus } c \text{ distance} = \tfrac{1}{2}(\%\text{SDS}_{C,c}) = \frac{1}{2}\left[\frac{(120 + 7 + 6 + 7)(100)}{400 \text{ total}}\right] = 17.5$$

We can conclude that the genes lie on opposite arms of their chromosome from the fact that the distance between the *a* and *c* loci (determined earlier as 28.6) is compatible with the sum of the gene-centromere distances, rather than with their difference. The map is thus constructed as follows:

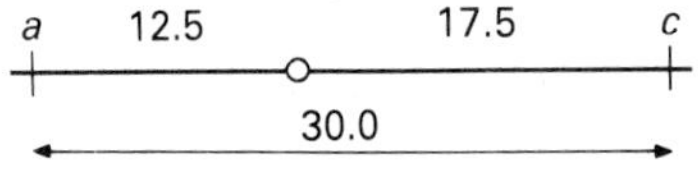

Unordered Tetrads. Unlike *Neurospora*, many organisms such as yeast and certain algae produce **unordered tetrads** from sexual reproduction. In unordered tetrads, the four meiotic products occur together in a jumbled arrangement, rather than in a precise linear array. Therefore the order in which the meiotic products are released is purely arbitrary.

The analysis of unordered tetrads is basically the same as that previously described for *Neurospora*, except that FDS and SDS patterns cannot be (directly) determined and used to map the centromere. Since the tetrads need not be ordered for us to classify them as PD, NPD, or T, however, unordered tetrads can be used to estimate linkage between genes. Map distance is then calculated in the usual manner, applying Eq. (7.1).

An example of the analysis of unordered tetrads is given in Table 7.5 by the results of a three-point cross $a\ b^+c^+ \times a^+b\ c$ in *Neurospora*. The results were obtained by allowing the asci to mature and release their ascospores themselves, thus avoiding the tedious task of dissecting individual asci. Although the ascospores of each individual ascus can be recovered and analyzed genetically, the precise order of the spores within an ascus cannot be established. Table 7.5 shows 191 tetrads of nine different types. (In general, many more than nine classes are possible, depending on the distances between the genes and the extent of crossing over.) Because of the wide variety of exchange events that are possible in the intervals that separate the loci, simultaneous analysis of all three gene loci in the tetrad is meaningless. Instead, we analyze the loci in pairs, beginning with the *a–b* region. Tetrad classes 1, 3, and 8 are PD with respect to the gene pairs a^+,a and b^+,b (ignoring the *c* locus); these three classes contain only $a\ b^+$ and a^+b ascospores, which are the parental arrangements. Class 9 is the only NPD tetrad with respect to these two loci. All the other classes are tetratype with respect to the combination of a^+,a and b^+,b alleles in their ascospores. Linkage of the *a* and *b* loci is demonstrated by the excess of PD and T tetrads over the number of NPD

Table 7.5. The classes and numbers of unordered tetrads produced by the three-point *Neurospora* cross $a\ b^+c^+ \times a^+b\ c$. Since the tetrads are unordered, the order in which the spores are listed within each tetrad is purely arbitrary.

(1)	(2)	(3)	(4)	(5)	(6)	(7)	(8)	(9)
$a\ b^+c^+$	$a\ b^+c^+$	$a\ b^+c^+$	$a\ b^+c^+$	$a\ b^+c^+$	$a\ b^+c$	$a\ b^+c$	$a\ b^+c$	$a\ b\ c^+$
$a\ b^+c^+$	$a\ b\ c$	$a\ b^+c$	$a\ b\ c^+$	$a\ b\ c$	$a\ b\ c^+$	$a\ b\ c$	$a\ b^+c$	$a\ b\ c$
$a^+b\ c$	$a^+b^+c^+$	$a^+b\ c^+$	a^+b^+c	a^+b^+c	$a^+b^+c^+$	$a^+b^+c^+$	$a^+b\ c^+$	$a^+b^+c^+$
$a^+b\ c$	$a^+b\ c$	$a^+b\ c$	$a^+b\ c$	$a^+b\ c^+$	$a^+b\ c$	$a^+b\ c^+$	$a^+b\ c^+$	a^+b^+c
14	54	103	2	6	3	6	2	1

tetrads. We find the distance separating the loci by using Eq. (7.1):

$$100R = 100\left[\frac{1 + \tfrac{1}{2}(54 + 2 + 6 + 3 + 6)}{191}\right] = 0.5 + \frac{1}{2}(37.2) = 19.1.$$

The other two intervals (a to c and b to c) can be analyzed in a similar fashion, giving the results shown in Table 7.6.

Gene Conversion. **Gene conversion** is a phenomenon observed in studies of fungi where there is an apparent conversion of one allele into another during recombination. Gene conversion is detected in tetrad analysis by a non-Mendelian segregation ratio among tetrads. The phenomenon is illustrated by the different spore-pair arrangements for the m gene and its wild-type allele in Fig. 7.15. In the absence of crossing over between a gene and its centromere, we would ordinarily expect to see a 4:4 segregation pattern for alleles (Fig. 7.15a). However, in about 1 percent (or less) of the asci, non-Mendelian ratios of the form 6:2 (or 2:6) or 5:3 (or 3:5) are observed. The 6:2 ratio is the result of *chromatid conversion,* which occurs when an allele on one chromatid is somehow converted into the

Table 7.6. Evaluation of map distances from the three-point tetrad data given in Table 7.5.

	INTERVALS		
	a–b	a–c	b–c
%PD	62.4	8.4	35.6
%NPD	0.5	4.2	1.0
%T	37.2	87.4	63.4

Map: a 19.1 b 32.7 c
51.8

$$100R_{a-b} = 0.5 + \tfrac{1}{2}(37.2) = 19.1$$
$$100R_{a-c} = 4.2 + \tfrac{1}{2}(87.4) = 47.9$$
$$100R_{b-c} = 1.0 + \tfrac{1}{2}(63.4) = 32.7$$

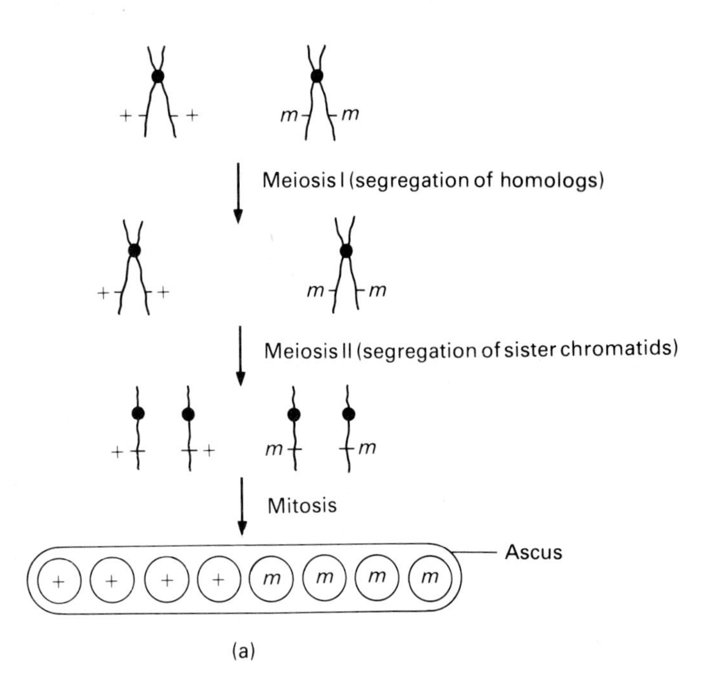

Figure 7.15. The phenomenon of gene conversion. The 4:4 spore ratio in the tetrad on (a) shows Mendelian segregation of the m^+, m allele pair during meiosis. The tetrad in (b) shows chromatid conversion (conversion of an m allele to m^+) and the resulting 6:2 (or 2:6) ratio of ascospores. The tetrad in (c) shows half-chromatid conversion and the resulting 5:3 (or 3:5) spore ratio.

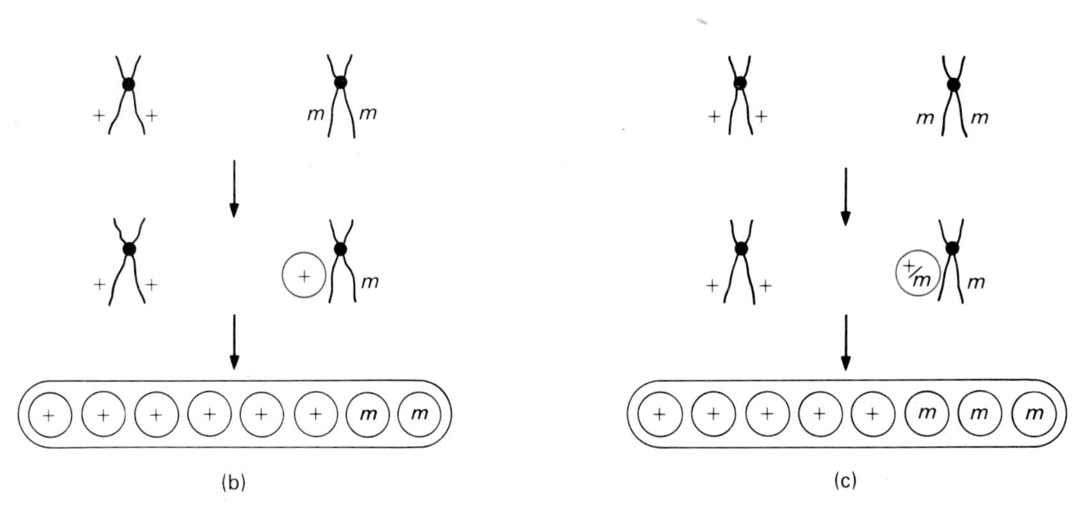

Figure 7.16. Explanation of gene conversion in terms of an excision and repair process that operates on the heteroduplex region of the hybrid DNA model. The lack of proper base pairing in the heteroduplex region causes a distortion of the double helix (middle, left). This distortion is recognized by repair enzymes, which first excise the mismatched region in one of the two strands in each chromatid (the m-bearing strand in this case). The enzymes then repair that strand, observing the correct base pairing rules (middle, right). The result is a 6:2 ratio of m^+ : m alleles, with half of the converted chromatids having a parental arrangement of flanking genes (bottom, left), while the other half have a recombinant arrangement of flanking genes (bottom, right).

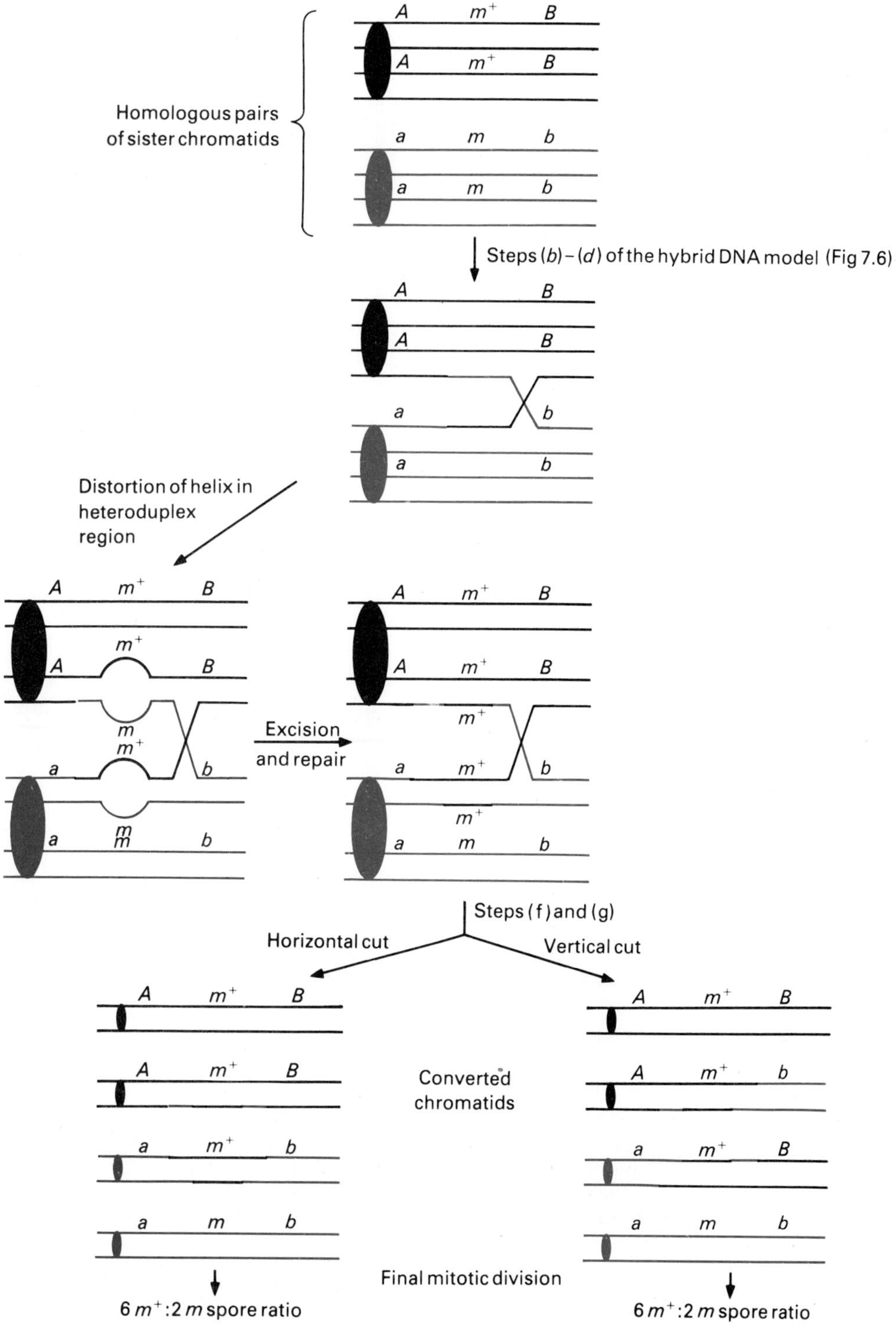

A m^+ B

A m^+ B

Homologous pairs of sister chromatids

a m b

a m b

Steps (b)–(d) of the hybrid DNA model (Fig 7.6)

A B

A B

a b

a b

Distortion of helix in heteroduplex region

A m^+ B

m^+

A B

m

m^+

a b

a m m b

Excision and repair

A m^+ B

A m^+ B

m^+

a m^+ b

m^+

a m b

Steps (f) and (g)

Horizontal cut

Vertical cut

A m^+ B

A m^+ B

a m^+ b

a m b

Converted chromatids

A m^+ B

A m^+ b

a m^+ B

a m b

Final mitotic division

6 m^+:2 m spore ratio

6 m^+:2 m spore ratio

alternative allele present on the homologous chromosome (Fig. 7.15b). Both members of a pair of spores will then carry copies of the converted allele. The 5:3 ratio is the result of *half-chromatid conversion,* in which only half the chromatid seems to be affected (Fig. 7.15c). In this case, only one member of a pair of spores carries the converted allele. Both chromatid and half-chromatid conversion occur at too high a frequency to be accounted for by mutation. An indication that they are somehow part of a recombination process is the finding that about 50 percent of the time, they are associated with the recombination of gene loci on either side of the converted allele.

It now appears that gene conversion is the result of a repair of mismatched bases that occurs in the heteroduplex region formed in DNA during recombination. A model for gene conversion is shown in Fig. 7.16, in which the locus of the converted allele is within the region of overlap. The lack of recognition between noncomplementary bases in the heteroduplex regions causes a distortion in each double helix. Repair enzymes recognize these base mismatches and excise the mismatched region on one of the two strands of each involved chromatid. A repair DNA polymerase then fills in the resulting gap in accordance with the Watson-Crick base-pairing rules. In the repair shown in Fig. 7.16, the regions that contain the m allele are excised and replaced, thereby "converting" the m allele to m^+ on each chromatid. If we compare the process shown in Fig. 7.16 with the sequence of steps involved in the Holliday model (see Fig. 7.6), it becomes apparent that we would obtain a 6:2 ratio of m^+: m alleles; we also note that half of the converted chromatids are recombinant for the genes that flank the converted region, while the other half are parental for these genes. The 5:3 ratio can occur when only one of the two mismatches shown in Fig. 7.16 is repaired.

TO SUM UP

1. Genetic maps are topological representations of chromosomes, showing the relative locations of genes and the distances between them. Geneticists have defined a unit of map distance as being equivalent to a 1 percent crossover frequency. In practice, map distances are based on the percentage of recombination observed in testcross progeny.
2. The recombination frequency underestimates the true genetic distance when the intervals between the genes under consideration are long. The discrepancy between crossover and recombination frequencies arises from the occurrence of multiple crossover events that either yield no recombinants or no more than the number produced by a single exchange.
3. If two genes yield more than 10 to 15 percent recombinant gametes, a better estimate of the true distance can be obtained by subdividing the interval between the genes into shorter regions. This segmentation can be accomplished by making use of other genes known to be located between the original two loci. The long distance is then best estimated as the sum of the two or more shorter distances.
4. The three-point testcross is a convenient method for mapping three genes in one experiment and for subdividing a long interval on a chromosome map. This technique also permits us to detect the effects of interference, in which one crossover alters the chance that another crossover will occur in an adjacent interval. Interference is usually measured by the coefficient of coincidence, defined as the ratio of the observed to the expected frequency of double recombinants.

5. Tetrad analysis provides a useful method of mapping genes in haploid eukaryotes. Linkage is then determined from the relative frequencies of parental ditype (PD), nonparental ditype (NPD), and tetratype (T) tetrads. The frequencies of PD and NPD tetrads are the same for unlinked genes, whereas linkage results in more PD than NPD tetrads.
6. The recombination frequency between genes is calculated in a tetrad analysis as $R = f(\text{NPD}) + (\frac{1}{2})f(\text{T})$, where $f(\text{NPD})$ and $f(\text{T})$ are the frequencies of nonparental ditype and tetratype tetrads, respectively. This relationship holds for both ordered and unordered tetrads. The distance between a gene locus and the centromere locus can also be determined in ordered tetrad analysis as one-half the percentage of second-division segregation (SDS) tetrads.
7. Tetrad analysis sometimes reveals non-Mendelian segregation patterns, a phenomenon known as gene conversion. Gene conversion is apparently part of a recombination process, where the locus of the converted allele is within a heteroduplex region that is formed in DNA during recombination and is subsequently excised and repaired.

Somatic Cell Genetics

Multicellular organisms such as humans have rather long life spans and small numbers of progeny. In comparison, their somatic cells have rapid reproductive rates when cultured outside the body and can undergo several division cycles in a short period of time. The large reproductive potential of somatic cells has made them extremely useful tools in genetic research, enabling a geneticist to examine several generations consisting of large numbers of cells within a relatively brief time interval. Moreover, the ability to culture one or just a few cell types under carefully controlled conditions has provided a refinement of genetic analysis that would be impossible with the intact organism. The many advantages of using somatic cells for a variety of genetic investigations, including the identification of genetic defects in cultured cells, have led to the development of a new area of cell research known as *somatic cell genetics.* What follows is a brief introduction to the methods used in this area, with special emphasis given to the procedures that are employed to detect linkage and assign genes to particular chromosomes. We will discuss several other applications of these methods in later chapters.

Somatic Cell Hybridization

The traditional methods used for the detection of linkage and the measurement of map distances apply in theory to all organisms, including humans. The opportunities to apply them in practice to humans have been rather limited, however, because of the long reproductive cycle and the obvious impossibility of making experimental crosses. Until recently, geneticists have had to rely on family pedigrees in order to deduce the linkage of genes for various traits. Such data are useful for genes located on the X chromosome, since sex-linkage is relatively easy to demonstrate. By contrast, pedigree information is usually inadequate for assigning autosomal genes to particular chromosomes. It has only been in the past couple of decades that comparatively rapid methods for localizing genes on human chromosomes have become available. One of the most useful of these methods is a technique called **somatic cell hybridization** that was developed in the early

1960s. Scientists use this technique to form hybrid cells from the somatic cells of different species. Although the technique was not purposefully developed for the study of linkage, it has been adapted by human geneticists for use in determining the chromosomal location of genes.

Formation of Hybrid Cells. When somatic cells, such as fibroblasts, are grown together in a culture medium, they are capable of undergoing fusion. Under normal circumstances, the fusion process occurs quite rarely, but the incidence can be greatly enhanced by adding certain chemicals or inactivated viruses to the suspension of cells. In plant cells, for instance, protoplasts whose cell walls have been removed by enzymatic digestion can be induced to fuse when treated with polyethylene glycol. Researchers can obtain similar results with certain animal cells, either after treating them with chemicals or after exposing them to Sendai virus particles that have first been killed by treatment with ultraviolet light. Although the inactivated viruses are incapable of infection, they can still bind to the surface of cells and render them more susceptible to fusion once contact is made with another cell.

To begin the procedure for somatic cell fusion, researchers mix two different lines of cells in the presence of a fusing agent, such as inactivated Sendai virus (Fig. 7.17). Fusion of the cytoplasm occurs first, during which a **heterokaryon** forms. A heterokaryon is a single fused cell that has two separate nuclei, one from each of the parental lines. In the process of cell division, chromosomes from both cell lines are organized into a single nuclear envelope, giving rise to a **hybrid cell** with a single nucleus. Hybrid cells, which express some genes of both parental lines, are capable of functioning and reproducing themselves. If the nuclei that fuse are from separate species, however, the hybrids tend to lose the chromosomes from one of the species during mitosis, eventually returning to a diploid state.

We do not yet know the reason why chromosomes are lost. Interestingly, only one of the sets of chromosomes in the interspecific hybrid cell is usually lost. For example, in hybrid cells made from mouse and human cells, the human chromosomes are lost preferentially until most of the cells in the line eventually contain only mouse chromosomes.

Detection of Hybrid Cells. Once cell fusion occurs, the geneticist must be able to distinguish the hybrid cells from all the other cells in the medium and to isolate selectively those hybrids whose genotypes are desired for study. Two readily identifiable cell characteristics are commonly used for this purpose: (1) the ability of certain cells to grow in the absence of specific nutrients and (2) the resistance of certain cells to the actions of various drugs. Suppose, for example, that two cell types, labeled B^+D^s and B^-D^r are mixed for the purpose of cell fusion. The symbol B^+ stands for the ability to synthesize substance B (a vitamin or amino acid needed for growth), and D^r designates resistance to the lethal effects of drug D. The symbols B^- and D^s would then signify the inability to produce nutrient B and sensitivity to drug D, respectively. Thus, B^+D^s cells can grow in the absence of nutrient B (since they can synthesize the compound from other substances provided in the medium), and B^-D^r cells can grow in the presence of drug D (because they are resistant to the lethal action of the drug), yet neither cell type can do both. By

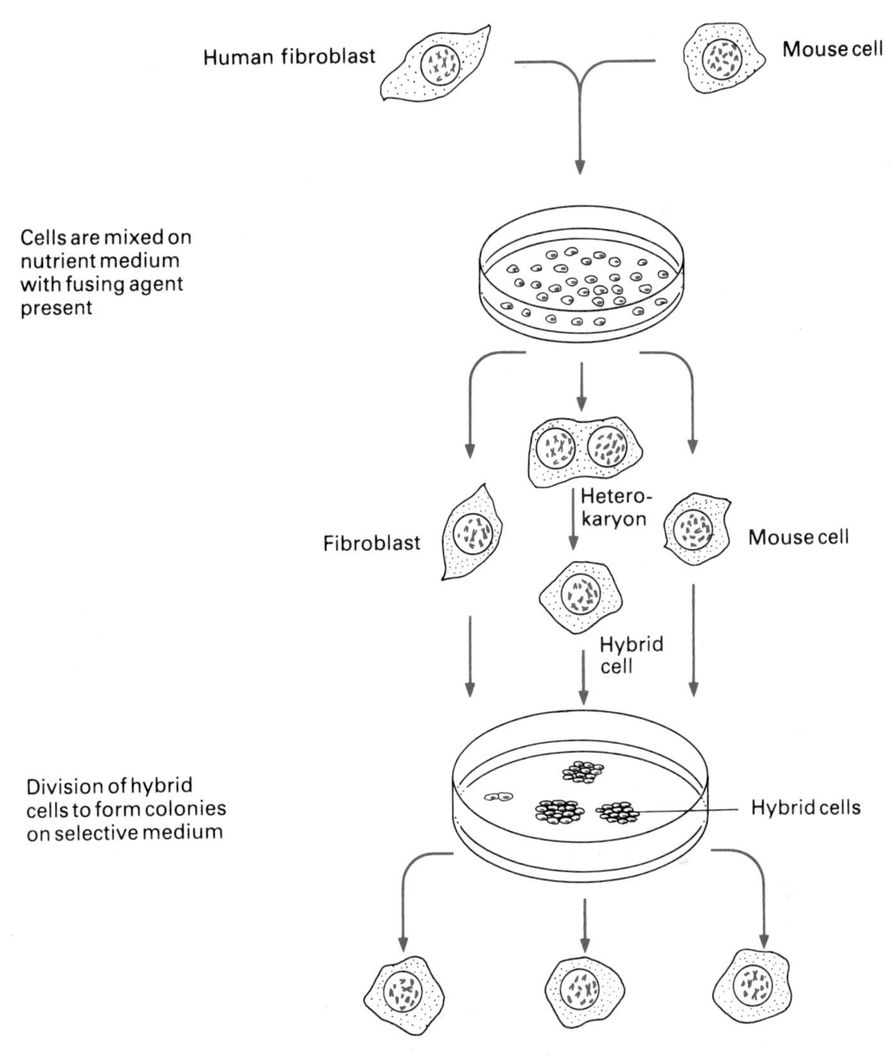

Figure 7.17. Cell fusion techniques applied to human and mouse cells. Cells are fused on a medium containing a fusing agent, such as inactivated Sendai virus. Cytoplasmic fusion first occurs, producing a heterokaryon with a nucleus from each parental cell line. Subsequent fusion of the nuclei gives rise to hybrid cells, which are isolated on a selective medium.

contrast, hybrid cells will have both capabilities, provided that B^+ and D^r are dominant traits. In this hypothetical case then, a medium that lacks nutrient B but contains drug D would serve as a selective system for the purpose of isolating the hybrid cells from a background of other cell types.

An actual example of a selective method used for the isolation of hybrid cells is the HAT technique. The medium for this technique contains hypoxanthine and thymidine (H and T), which are normal precursors of DNA, and aminopterin (A), a drug that specifically blocks one pathway leading to DNA synthesis. Normal cells can synthesize DNA by way of two metabolic pathways: the major pathway, which

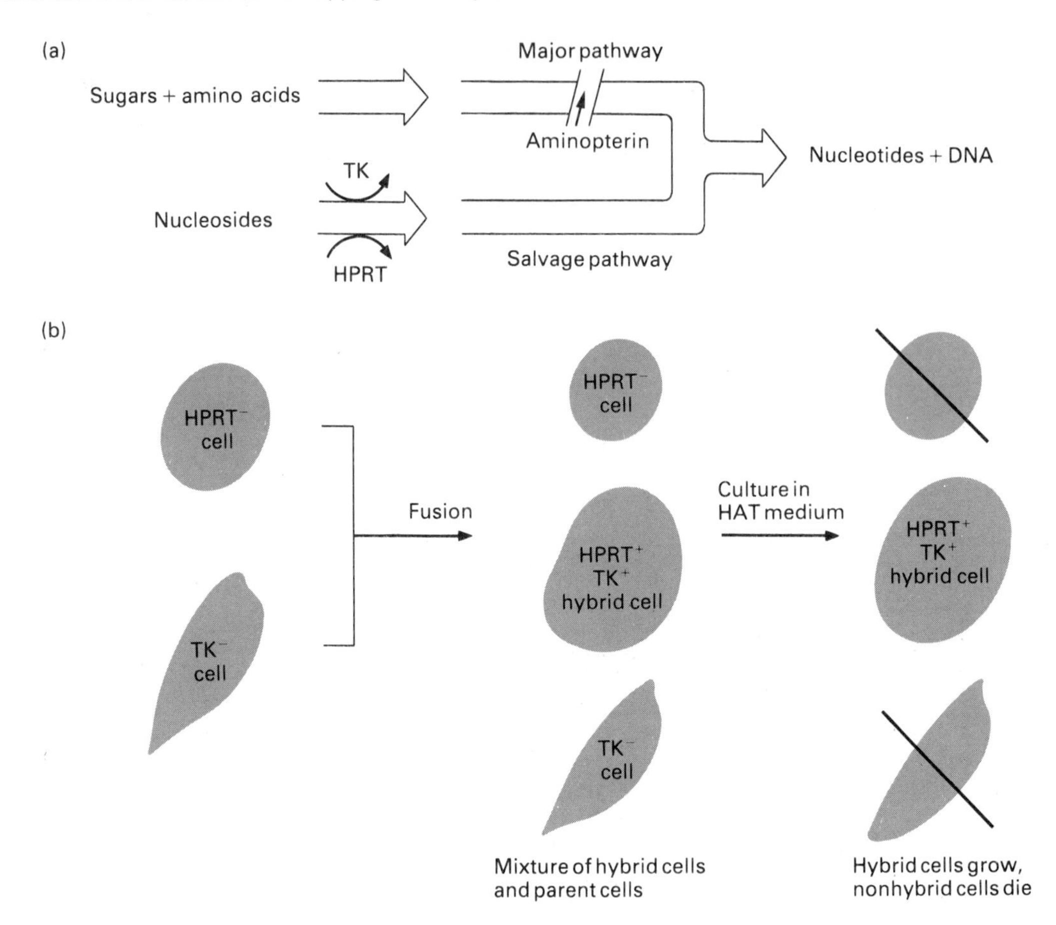

Figure 7.18. (a) Two metabolic pathways used in the synthesis of DNA: the major pathway, which is specifically blocked by aminopterin, and a salvage pathway. Utilization of the salvage pathway requires the enzymes thymidine kinase (TK) and hypoxanthine-guanine phosphoribosyl transferase (HPRT). (b) Cells incapable of producing the enzymes TK or HPRT will die when placed in a medium containing aminopterin (HAT medium). Fused hybrid cells that possess both capabilities will survive and grow.

is blocked by aminopterin, and a salvage pathway (Fig. 7.18a). The major pathway produces DNA enzymatically from scratch, starting with simple sugars and amino acids. The salvage pathway synthesizes DNA from more complex precursors (nucleosides), supplied by the breakdown of RNA and from dietary sources. This alternative (salvage) pathway thus enables cells to utilize starting materials that are already supplied in the diet, without having to break them down completely and resynthesize them from scratch.

Two functioning enzymes are required for completion of the salvage pathway: thymidine kinase (TK) and hypoxanthine-guanine phosphoribosyl transferase (HPRT). Cells that are unable to produce either TK (TK^- cells) or HPRT ($HPRT^-$ cells) cannot convert precursors into DNA along the salvage pathway and must depend on the major pathway for DNA synthesis. Thus, if the major pathway is blocked by the presence of aminopterin in the medium, such cells would be unable to grow. The application of this principle is shown in Fig. 7.18(b). Note that when TK^- mouse cells are fused with $HPRT^-$ human cells and plated on a medium containing aminopterin, both parental strains die. Only hybrid cells that have both the mouse $HPRT^+$ character and the human TK^+ trait will survive, since they are now capable of producing the functional HPRT and TK enzymes that are needed for the synthesis of DNA along the salvage pathway.

Assigning Genes to Chromosomes. Once hybrid cells have been isolated, they can be used to determine the location of genes on chromosomes. In the case of mouse–human hybrid cells, for example, the preferential loss of human chromosomes provides the basis for this procedure. Hybrid cells only gradually eliminate the human chromosomes as they divide, so that the order and timing of the losses are not regular (Fig. 7.19). We can follow this process microscopically, since mouse and human chromosomes are fairly easy to distinguish.

If the loss of a particular human chromosome from the hybrid cell line is always accompanied by the loss of a certain gene product (e.g., an enzyme or some other protein), then we can conclude that the gene responsible for that product is located on the chromosome that has been lost. For example, let us suppose that we find the enzyme β-galactosidase (an enzyme that breaks down the milk sugar lactose) in the hybrid cells only as long as at least one copy of human chromosome 22 is retained in the hybrid cell nucleus. We therefore conclude that the gene that codes for β-galactosidase is located somewhere on chro-

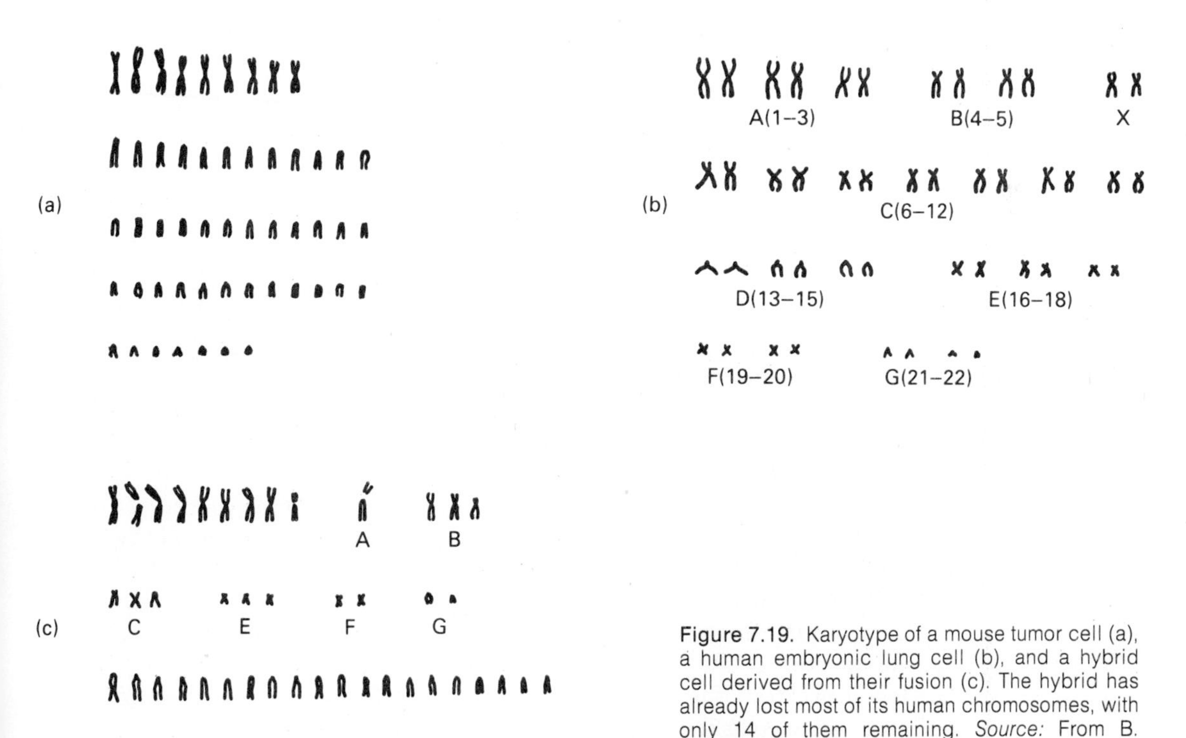

Figure 7.19. Karyotype of a mouse tumor cell (a), a human embryonic lung cell (b), and a hybrid cell derived from their fusion (c). The hybrid has already lost most of its human chromosomes, with only 14 of them remaining. *Source:* From B. Ephrussi and M. C. Weiss. *Scientific American* 220:26–35, 1969.

mosome 22. This procedure depends on having a reliable biochemical test for identifying the gene product; it is thus limited to those gene products for which such a test has been developed.

Another approach is to use a selective medium on which only those hybrid cells that have retained a certain chromosome grow. Let us take as an example the gene responsible for the enzyme thymidine kinase (TK). Only those hybrid cells that have retained the human TK^+ gene can survive on HAT medium. Now let us suppose that after extensive loss of chromosomes, we perform a karyotype analysis in order to identify the human chromosomes remaining in the surviving cells. The results we obtain from three representative cells are as follows:

	TK^+ CELL LINES		
	(a)	(b)	(c)
Human chromosomes remaining:	(1, 4, 7, 17, 19)	(2, 8, 17)	(5, 9, 11, 17, 21, X)

The only human chromosomes that all three cell lines still have in common is chromosome 17. We can therefore conclude that the *TK* gene locus in humans is on chromosome 17.

We will discuss additional methods for localizing genes in the following section and in later chapters. As these kinds of techniques become more refined, more progress will be made in mapping the genes of organisms that do not lend themselves well to the standard crossing procedures. Figure 7.20 will give you an indication of our state of knowledge at the present time about the chromosomal location of human genes.

Gene Transfer by Individual Chromosomes and Purified DNA

The techniques just described can only assign genes to certain chromosomes. They do not give the precise location of any gene on a chromosome map, nor do they suggest the relative order and degree of linkage of the different loci. A more detailed analysis is necessary to pinpoint the location of genes along chromosomes.

One promising new technique for the mapping of genes on chromosomes is *chromosome-mediated gene transfer.* In this method, genes are transferred from one cell to another by way of chromosomes isolated at metaphase. The experimental procedure that was used in the first demonstration of this mechanism is shown in Fig. 7.21. The investigators isolated free chromosomes at metaphase from $HPRT^+$ Chinese hamster cells; they subsequently incubated these chromosomes with $HPRT^-$ mouse cells at a ratio of one cell-equivalent of donor chromosomes to one recipient cell. After a period of incubation, the cells were placed on a nutrient medium for growth. A few transformed $HPRT^+$ mouse cells were then isolated using the HAT selection technique.

Chromosome-mediated gene transfer is initiated when the recipient cell takes up one or more of the donor chromosomes through phagocytosis (engulfment). The ingested chromatin is then transferred within the intracellular vacuoles to the nucleus, where usually only a small portion of the transferred material becomes associated with the chromosomes of the recipient cell (Fig. 7.22). During uptake and incorporation, the donor chromosomes are subjected to a certain amount of

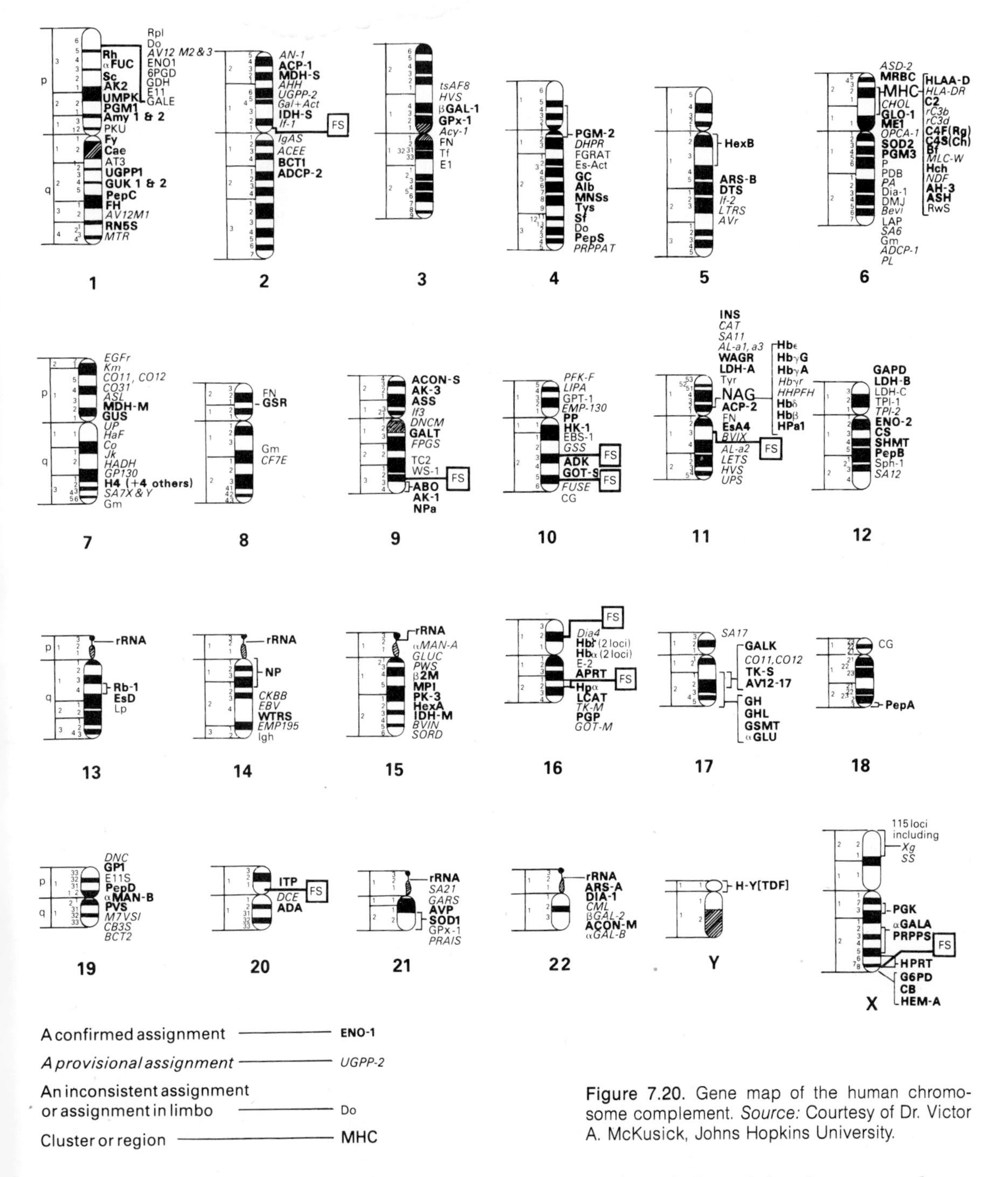

Figure 7.20. Gene map of the human chromosome complement. *Source:* Courtesy of Dr. Victor A. McKusick, Johns Hopkins University.

breakage and degradation. This enzymatic degradation of the chromosomal material tends to break the DNA into shorter fragments so that only genes that are closely linked have a good chance of being incorporated together (*cotransferred*) into a recipient chromosome. The probability of cotransfer declines as the distance between the linked genes increases, because of the greater likelihood of a breakage event. This inverse relationship between the probability of cotransfer of genes and the distance between linked genes on a chromosome enables geneticists to calculate map distances. Using this technique, we subtract the percentage of cotransfer frequency from 100 percent (the value expected if the genes mapped at the same locus). For example, in a certain experiment, the *TK* (thymidine kinase) gene was transferred with the *GALK* (galactokinase) gene in 19 percent of the cell lines and with the *PCI* (procollagen type I) gene in 74 percent of the cell lines. We can then calculate the map distances as follows: 100 − 19 = 81 map units for the *TK–GALK* interval, and 100 − 74 = 26 map units for the *TK–PCI* interval. The actual arrangement of genes could not be precisely determined from this experiment. In one transformed cell, however, the *TK* and *GALK* genes were transferred together without the *PCI* gene, suggesting an order of *PCI TK GALK*.

The genetic transformation of eukaryotic cells is not restricted to the transfer of genes by whole chromosomes or chromosome fragments, but, as in the case of

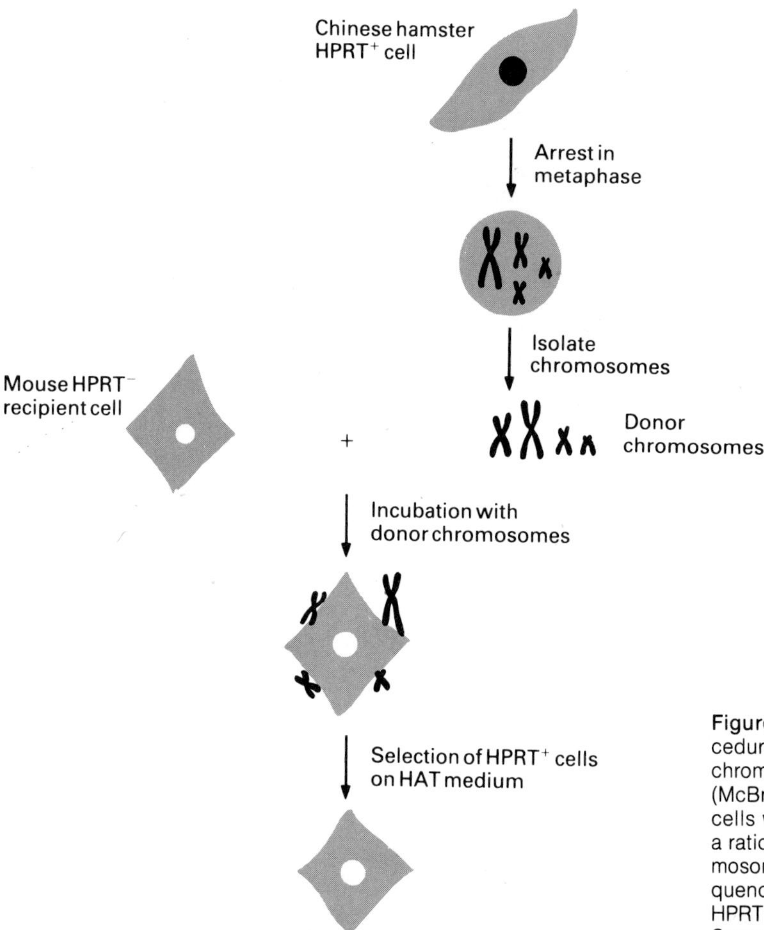

Figure 7.21. Outline of experimental procedure used for the first demonstration of chromosome mediated gene transfer (McBride and Ozer, 1973). The recipient cells were incubated with chromosomes at a ratio of one cell equivalent of donor chromosomes per recipient cell. The initial frequencies of transfer were very low, near 1 $HPRT^+$ transformant per 10^7 recipient cells. *Source:* Reproduced, with permission, from the Annual Review of Biochemistry, Volume 50. © 1981 by Annual Reviews Inc.

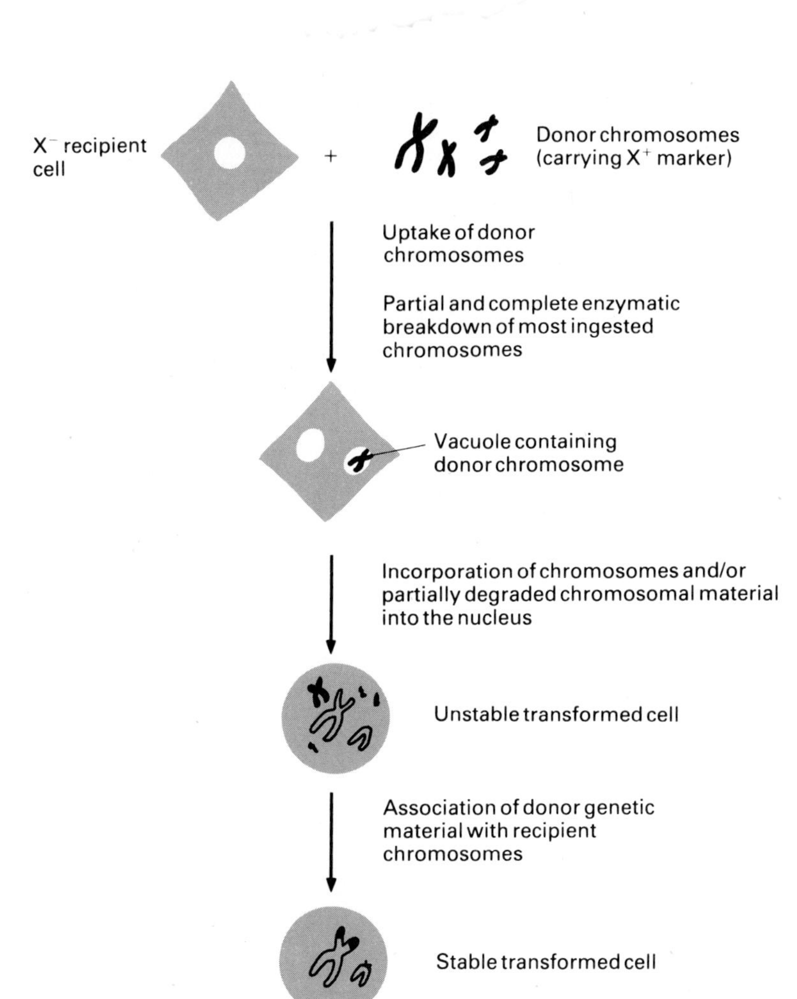

Figure 7.22. Sequence of steps that is thought to occur in the process of chromosome-mediated gene transfer. Donor chromosomes are represented by solid figures, recipient chromosomes by open figures. *Source:* Reproduced, with permission, from the Annual Review of Biochemistry, Volume 50. © 1981 by Annual Reviews Inc.

prokaryotes, it can also be accomplished by a process known as *DNA-mediated gene transfer,* in which the transferred genes are carried in the form of purified DNA. Yeast cells, for example, can be transformed with high efficiency after their outer walls have been degraded with enzymes and the resulting protoplasts have been exposed to DNA from a different strain of yeast in the presence of polyethylene glycol. The donor DNA that is taken up by the cells becomes stably integrated into the recipient genome, presumably at a homologous chromosomal location. Transformation can also occur naturally in yeast cells, but at a much lower efficiency than when it occurs by artificial means.

Researchers have recently developed a successful technique for transforming mammalian cells grown in culture. The procedure (see Fig. 7.23) involves exposing recipient cells that are deficient in some detectable biochemical function to a calcium phosphate precipitate of DNA fragments carrying the wild-type genes. After the competent cells have taken up transforming DNA, they are exposed to a selective medium that permits only the transformants (and a few spontaneously

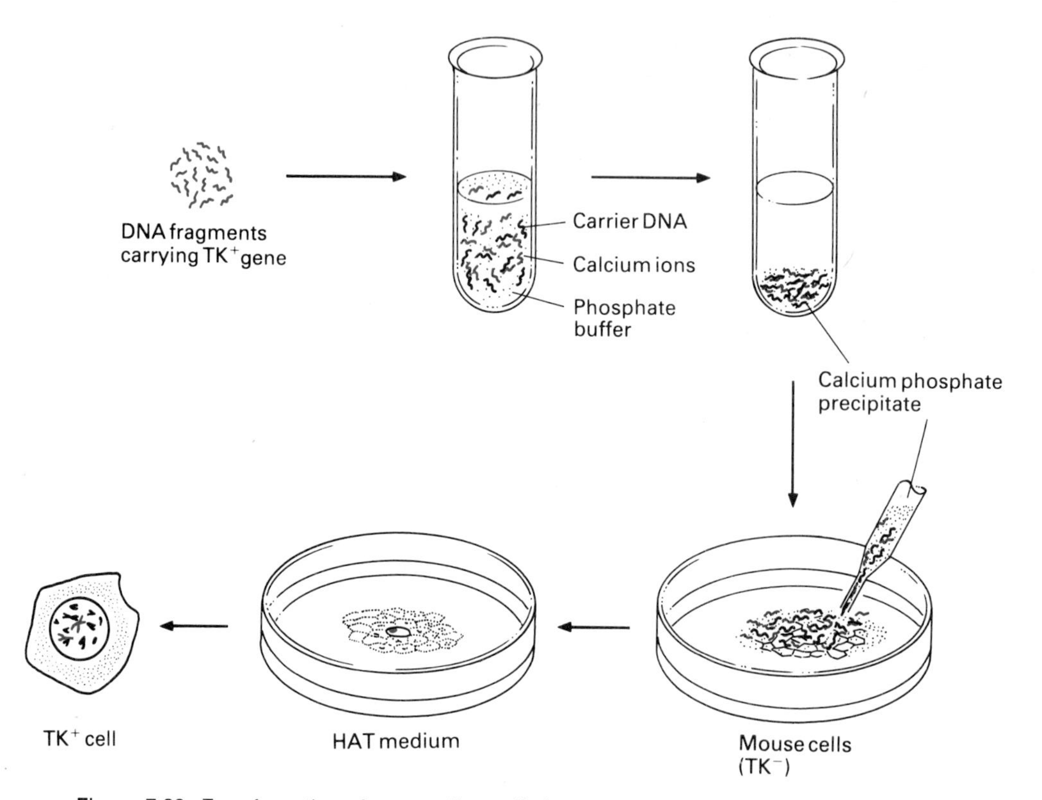

Figure 7.23. Transformation of mammalian cells by means of DNA-mediated gene transfer. DNA fragments carrying the TK^+ gene are mixed with a carrier DNA in a phosphate buffer solution. Addition of calcium ions forms a precipitate of DNA. The precipitate is collected and added to a culture of TK^- cells. Only cells that take up the TK^+ gene and are transformed to the TK^+ condition survive in HAT medium.

mutated cells) with the desired biochemical properties to survive and replicate. For example, transformants for the thymidine kinase (*TK*) gene can be selected for on HAT medium on which TK-deficient cells die. Similarly, transformants for dominant-acting markers for drug resistance, such as the marker for methotrexate resistance, can be isolated on a medium that contains high concentrations of the particular drug.

Studies reveal that only a very small fraction of the cell population (from $1/10^6$ to $1/10^5$) becomes stably transformed for the genetic marker selected. In these competent cells, cellular enzymes act on the incorporated genetic fragments to form large DNA molecules, called *concatamers,* which contain several copies of each gene in the original fragment. A concatamer may then recombine with a particular site on a recipient chromosome and become inserted into the genome of the transformed cell.

TO SUM UP

1. The genes on human chromosomes present a problem to the traditional methods of genetic mapping, in that controlled testcrosses cannot be performed. Researchers must therefore employ alternative methods, such as somatic cell hybridization and chro-

mosome-mediated gene transfer, to determine the chromosomal location of human genes.

2. Somatic cell hybridization has been particularly useful for assigning human genes to chromosomes. The usual procedure is to form mouse-human hybrid cells through cell fusion, and then to isolate the hybrid cells on a selective medium. The hybrid cells preferentially lose the human chromosomes. We can make the proper chromosome assignments by correlating the loss of certain traits with the loss of specific chromosomes.
3. We can determine the chromosomal arrangement of some genes by following their transfer from one cell to another on isolated metaphase chromosomes. Closely linked genes tend to be transferred together; in contrast, enzymatic breakage of donor chromosomes in the recipient cells often disrupts the linkage of loci that are more distantly separated.

Questions and Problems

1. Define the following terms:
   chromosome-mediated gene transfer
   genetic distance
   heterokaryon
   hybrid cell
   interference (positive vs. negative)
   linkage group
   linkage map
   linked genes
   ordered tetrad
   somatic cell hybridization
   syntenic genes
   unordered tetrad
2. Suppose that all the gene loci in a particular organism map into seven separate linkage groups. How many chromosomes are present in the somatic cells of this species?
3. Criticize the following statements:
   (a) Crossing over results in the formation of recombinant types.
   (b) Genes that are present on the same chromosome fail to assort independently.
   (c) A genetic map shows the dimensional relationships of genes on a chromosome.
   (d) The farther apart two genes are on a chromosome, the greater is their frequency of recombination.
4. Consider the following linkage data:

Gene loci:	*a–b*	*b–c*	*c–d*	*d–e*	*c–e*
Map distance:	8	6	2	4	6

   Construct a chromosome map of these loci.
5. Albinism in mice can arise as the homozygous expression of one or the other (or both) of two recessive genes, *c* and *d*. Individuals must have the dominant alleles of both genes (*C-D-*) to express normal color. Suppose that matings between a dihybrid male (*CcDd*) and females of a doubly recessive albino strain (*ccdd*) produce 148 albino and 52 colored offspring. Are the two loci linked?
6. In tomatoes, the gene for round fruit shape (*O*) is dominant to its allele for elongate shape (*o*), and the gene for smooth fruit skin (*P*) is dominant to its allele for peach (*p*). Two series of crosses involving these genes produced the following results:

   Cross 1: P: round, smooth × elongate, peach
   F_1: all round, smooth

   Cross 2: P: round, peach × elongate, smooth
   F_1: all round, smooth

   When the F_1 of Cross 1 were testcrossed, they produced offspring in the following proportions: 46% round, smooth; 4% round, peach; 4% elongate, smooth; and 46% elongate, peach. In contrast, testcrosses involving the F_1 of Cross 2 gave the following results: 4% round, smooth; 46% round, peach; 46% elongate, smooth; and 4% elongate, peach. Explain these results, designating the genotypes of all individuals concerned.
7. The following data were obtained from a series of two-factor testcrosses. Explain these results in terms of the relative locations of the different genes on a linkage map.

Gene loci:	*a–b*	*b–c*	*a–c*	*b–d*	*a–e*	*d–e*	*a–d*
Percent recombination:	42	44	50	48	50	50	50

8. In rabbits, genes at two loci, *b* and *c*, interact to produce coat color; *B-C-* individuals are

black, *bbC-* individuals are brown, and *--cc* (*B-cc* or *bbcc*) individuals are albino. Certain crossing data suggest that these loci are linked. In one series of testcrosses, matings between a dihybrid male and females of the doubly recessive albino strain resulted in 65 black, 34 brown, and 101 albino offspring. **(a)** Are the genes linked in the cis or trans arrangement in the dihybrid parent? **(b)** Calculate the map distance between these loci. **(c)** The dihybrid male in this series of testcrosses was the product of a mating between members of two homozygous strains. What were the genotypes of these homozygous strains?

9. Nail-patella syndrome in humans is characterized by congenital abnormalities of the fingernails (and sometimes toenails) and of the patellae (kneecaps). The gene for this disorder is dominant and is located on chromosome 9 about 10 map units away from the *ABO* locus. Suppose that a man with nail-patella syndrome and type A blood marries a normal woman with B blood. The mothers of both the husband and wife are normal and have blood type O. **(a)** The husband and wife have two children, one with type A blood and the other with type B blood. What is the probability that both children are normal (that is, do not have nail-patella syndrome)? **(b)** The couple are now expecting another child. What is the chance that the child will have nail-patella syndrome and O blood? **(c)** An amniocentesis is performed, which reveals that the fetus has AB blood. What is the chance that this fetus has nail-patella syndrome?

10. Elliptocytosis is a hereditary trait in humans that is characterized by the presence of oval-shaped red blood cells, rather than the normal biconcave-shaped cells. The gene for elliptocytosis is dominant to its normal allele and is located on chromosome 1, about 20 map units away from the *Rh* locus. The following pedigree shows the joint inheritance of the genes for Rh factor and elliptocytosis in three generations of a family. The filled-in symbols represent individuals with elliptocytosis, while the presence or absence of Rh factor is shown below each symbol. **(a)** Determine the genotypes of the individuals in the pedigree. **(b)** What is the linkage phase (cis or trans) of individual II-2? **(c)** Which individuals in generation III are recombinant types? **(d)** Calculate the recombination frequency for the progeny in generation III. How does your estimated map distance compare with the value given earlier in the problem?

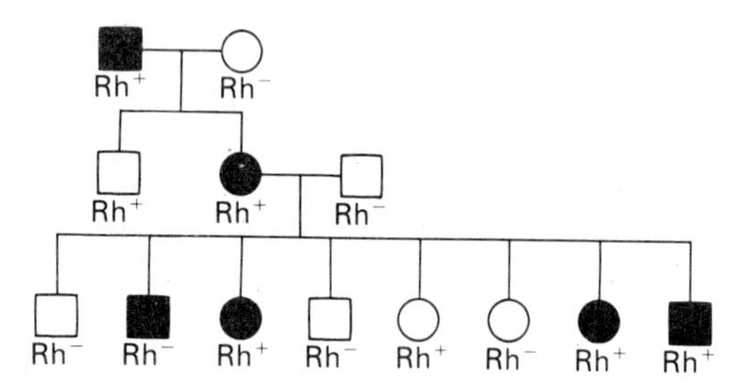

11. The recessive genes for hemophilia A and red-green color blindness in humans are located about 10 map units apart on the X chromosome. The *ABO* locus is on chromosome 9. A normal man with type AB blood marries a normal, type AB woman whose mother was color blind and whose father was hemophilic. They have several children. **(a)** What is the chance that their first child is color blind? **(b)** What is the chance that their second child is a boy with type A blood who is both hemophilic and color blind? **(c)** What is the chance that the third child will, like the parents, be normal in color vision and clotting time and AB in blood type?

12. In corn, the genes for colored kernels (*C*) vs. colorless (*c*) and plump kernels (*Sh*) vs. shrunken (*sh*) are about 4 map units apart on chromosome 9. The dihybrid cross $\frac{C\ sh}{c\ Sh} \times \frac{C\ sh}{c\ Sh}$ is made. **(a)** Compute the genotypic ratio in the gametes produced by each of the parents in the cross. **(b)** What percentage of the offspring will be $\frac{c\ sh}{c\ sh}$ in genotype? **(c)** What percentage of the offspring will have the colored, shrunken phenotype? **(d)** What percentage of the offspring will be true-breeding?

13. In *Drosophila,* crossing over occurs in females but not in males. Suppose that the dihybrid cross $\frac{pr^+\ b}{pr\ \ b^+} \times \frac{pr^+\ b}{pr\ \ b^+}$ is made, in which the genes for the dominant red eyes (pr^+) vs. purple (pr) and for the dominant gray body (b^+) vs. black (b) are in the trans arrangement. The genes for these traits are located about 6 map units apart on chromosome II. **(a)** Predict the phenotypic ratio among the progeny of this cross. **(b)** Does the ratio you predicted indicate anything about how far apart the *pr* and *b* loci are on chromosome II? Explain.

14. Three plants that are dihybrid for different pairs of gene loci are selfed. The frequency of double-recessive offspring produced by each mating is given below. In each case, determine whether or not the genes are linked. If the genes are linked, evaluate the linkage phase (cis or trans) in the dihybrid parent, and calculate the map distance between the genes involved.

Genotype of self-pollinated plant:	*AaBb*	*BbCc*	*CcDd*
Frequency of double-recessive offspring:	1.00% (*aabb*)	6.25% (*bbcc*)	9.00% (*ccdd*)

15. A testcross was made using the trihybrid *PpRrSs* as the dominant parent. The testcross results revealed that the trihybrid produced the following gametes:

PRS	11	*pRS*	15
PRs	238	*pRs*	240
PrS	242	*prS*	230
Prs	11	*prs*	13

(a) Which loci are linked, and which assort independently? **(b)** Determine the arrangement (cis or trans) of the linked genes in the trihybrid parent. **(c)** Calculate the map distances between all linked loci.

16. In *Drosophila*, the three gene pairs for red eyes (cn^+) vs. cinnabar (*cn*), normal bristle number (rd^+) vs. reduced (*rd*), and long wings (vg^+) vs. vestigial (*vg*) are known to have their loci on chromosome II. Suppose that a three-point testcross yields the following offspring:

cinnabar, reduced, vestigial	406
cinnabar, reduced, long	46
cinnabar, normal, vestigial	28
cinnabar, normal, long	3
red, normal, long	438
red, normal, vestigial	45
red, reduced, long	33
red, reduced, vestigial	1

(a) Which progeny classes are the parental types? Which are the single-recombinant types? Which are the double-recombinant types? **(b)** Calculate the map distances between the genes, and construct a linkage map of these loci. **(c)** Identify the most probable origin of each of the recombinant classes, with regard to the number and location of crossovers. **(d)** Determine the coefficient of coincidence for this set of loci.

17. In the seedling stage, a completely recessive corn plant was glossy (leaves have a shiny appearance), virescent (poor in chlorophyll), and liguleless (lacking certain appendages at the base of the leaves). This plant is crossed to a trihybrid that as a seedling had dull leaves, normal chlorophyll content, and ligules. Of the many seedlings produced in the next generation, a random sample of 1000 had the following characteristics:

dull, normal, with ligules	28
dull, normal, liguleless	179
dull, virescent, with ligules	69
dull, virescent, liguleless	250
glossy, normal, with ligules	198
glossy, normal, liguleless	70
glossy, virescent, with ligules	183
glossy, virescent, liguleless	23

(a) Calculate the map distances between the genes, and construct a linkage map of these loci (glossy = *gl*, virescent = *v*, and liguleless = *lg*). **(b)** Give the genotype of the trihybrid parent, designating the proper gene arrangement on the pair of chromosomes.

18. Assume the following linkage map:

a —— 10 —— b —— 6 —— c

(a) Calculate the frequencies of the parental, single-recombinant, and double-recombinant classes expected among the progeny of the three-point testcross $\frac{A\ B\ C}{a\ b\ c} \times \frac{a\ b\ c}{a\ b\ c}$, assuming that crossing over occurs without interference. **(b)** Repeat your calculations, but now assume interference, with a coefficient of coincidence of 0.5.

19. Three two-point testcrosses gave the following recombination frequencies:

$$R_{p\text{-}q} = 12\%, \quad R_{q\text{-}r} = 8\%, \quad R_{p\text{-}r} = 5\%.$$

(a) Construct a linkage map of the *p*, *q*, and *r* loci. Why are the distances not additive?

(b) Calculate the frequencies of the parental, single-recombinant, and double-recombinant classes expected among the progeny of the three-point testcross *PpQqRr* × *ppqqrr*, based on the values of R_{p-q}, R_{q-r}, and R_{p-r} given above. **(c)** Determine the coefficient of coincidence for this set of loci.

20. Consider the four-point testcross $\frac{A\ B\ C\ D}{a\ \ b\ \ c\ \ d} \times \frac{a\ b\ c\ d}{a\ b\ c\ d}$. **(a)** How many recombinant classes can we possibly detect in the progeny of this cross? **(b)** Identify the most probable origin of each recombinant class, with regard to the number and location of crossovers. **(c)** Suppose that data from previous two-point crosses indicate that the members of each of the pairs of gene loci (*a-b*, *b-c*, and *c-d*) are 2 map units apart. How many offspring must we examine from this four-point cross in order to observe about eight individuals in the lowest frequency class?

21. You discover a recessive mutant gene in *Drosophila* that is not reported in the scientific literature, and you want to determine on which chromosome this gene is located. Assume that you have access to two indicator strains of flies, one homozygous for a gene on chromosome II and another for a gene on chromosome III. Describe an experimental breeding procedure you could use that would permit you to assign this mutant gene to one of the four chromosomes in this species.

22. In a series of experiments with *Neurospora*, the following numbers of FDS and SDS asci were observed for each gene pair. Determine the distance between the gene and its centromere in each case.

GENE PAIR	NUMBER OF TETRADS FDS	SDS
(a) f^+ vs. f (fluffy) growth	82	118
(b) c^+ vs. c (colonial) growth	78	16
(c) r^+ vs. r (ropy) growth	67	42

23. *Neurospora* of opposite mating types with the genotypes *CD* and *cd* are crossed. Analysis of the 389 ordered tetrads produced from this cross yields the following data:

(1)	(2)	(3)	(4)	(5)	(6)	(7)
C D	*C d*	*C D*	*C D*	*C D*	*C d*	*C D*
C D	*C d*	*c D*	*C d*	*c d*	*c D*	*c d*
c d	*c D*	*C d*	*c D*	*C D*	*C d*	*C d*
c d	*c D*	*c d*	*c d*	*c d*	*c D*	*c D*
259	14	29	65	7	7	8

Determine whether or not the genes are linked. If they are linked, calculate the distances between the genes. What are the gene-centromere distances?

24. Gene pairs c^+,*c* (*c* = compact growth) and leu^+, *leu* (*leu* = requirement for the amino acid leucine) are observed in a mating of *Neurospora* strains $c^+\ leu^+ \times c\ leu$. The following classes of ordered tetrads are obtained:

(1)	(2)	(3)	(4)
+ +	+ *leu*	+ +	+ +
+ +	+ *leu*	*c* +	+ *leu*
c leu	*c* +	+ *leu*	*c* +
c leu	*c* +	*c leu*	*c leu*
367	4	11	50

(5)	(6)	(7)
+ +	+ *leu*	+ +
c leu	*c* +	*c leu*
+ +	+ *leu*	+ *leu*
c leu	*c* +	*c* +
60	2	6

Determine linkage relationships and the map distances between the genes.

25. Unordered tetrad analysis in yeast gives the following data:

(1)	(2)	(3)
+ *b*	+ +	+ *b*
+ *b*	+ +	+ +
a +	*a b*	*a b*
a +	*a b*	*a* +
363	3	234

Evaluate these data, and determine the distance between genes if they are linked.

26. Explain the origin of each of the tetrad classes

in problems 23 through 25 by diagramming the events of meiosis (assortment, crossing over) that were responsible for their production.

27. Unordered tetrad analysis of a cross between $a^+b^+c^+$ and $a\ b\ c$ strains in yeast yields the following data. Determine the map distances (genes are linked in the order given).

(1)	(2)	(3)	(4)	(5)
+ + +	+ + +	+ + +	+ + +	+ + +
+ + +	+ *b* *c*	+ + *c*	+ *b* +	+ *b* *c*
a *b* *c*	*a* + +	*a* *b* +	*a* + *c*	*a* + *c*
a *b* *c*	*a* *b* *c*	*a* *b* *c*	*a* *b* *c*	*a* *b* +
412	24	48	3	2

(6)	(7)	(8)	(9)
+ + *c*	+ + *c*	+ *b* *c*	+ + *c*
+ *b* +	+ *b* *c*	+ *b* *c*	+ + *c*
a + +	*a* + +	*a* + +	*a* *b* +
a *b* *c*	*a* *b* +	*a* + +	*a* *b* +
3	2	1	5

28. Unordered tetrad analysis of a cross between $d^+e^+f^+$ and $d\ e\ f$ strains in yeast gives the following data (genes are linked but not necessarily in the order given). Determine the correct gene order and genetic distances.

(1)	(2)	(3)	(4)	(5)
+ + +	+ + +	+ + +	+ + +	+ + +
+ + +	*d* + *f*	+ + *f*	*d* + +	*d* + *f*
d *e* *f*	+ *e* +	*d* *e* +	+ *e* *f*	+ *e* *f*
d *e* *f*	*d* *e* *f*	*d* *e* *f*	*d* *e* *f*	*d* *e* +
299	42	90	10	11

(6)	(7)	(8)	(9)	(10)
+ + *f*	+ + *f*	*d* + *f*	+ + *f*	*d* + +
d + +	*d* + *f*	*d* + *f*	+ + *f*	*d* + *f*
+ *e* +	+ *e* +	+ *e* +	*d* *e* +	+ *e* +
d *e* *f*	*d* *e* +	+ *e* +	*d* *e* +	+ *e* *f*
10	12	5	20	1

29. Suppose that in a study involving chromosome-mediated gene transfer, human metaphase chromosomes are used to transform $HPRT^-$ mouse cells to the $HPRT^+$ condition. Of the transformed $HPRT^+$ cells, 12 percent also received the human gene that codes for phosphoglycerate kinase (PGK) and 25 percent received the human gene that codes for glucose 6-phosphate dehydrogenase (G6PD). In a few cases, the $HPRT^+$ transformants received the PGK gene but not the G6PD gene. Using these data on cotransfer frequencies, construct a linkage map of the three loci.

30. Human cells are fused with mouse cells, and the resulting hybrid cells are placed into HAT medium. After extensive chromosome loss, six different cell lines are selected and tested for the presence of human chromosomes as well as for the presence of four human enzymes, one of which is thymidine kinase (TK). The results are tabulated below. (The presence of a chromosome or enzyme is indicated by +, the absence by −.)

		CELL LINES					
		(a)	(b)	(c)	(d)	(e)	(f)
Human Enzymes	E_1	+	+	−	−	−	−
	E_2	−	+	+	−	+	+
	E_3	+	+	−	−	−	+
	E_4	+	+	+	+	+	+
Human Chromosomes	1	+	+	−	−	−	+
	11	+	+	−	−	−	−
	17	+	+	+	+	+	+
	X	−	+	+	−	+	+

(a) Identify the chromosome that carries the gene for each enzyme. **(b)** Which of the enzymes (E_1, E_2, E_3, or E_4) is thymidine kinase (TK)? How can you tell? **(c)** The other enzymes tested for were lactate dehydrogenase (LDH), 6-phosphogluconate dehydrogenase (6PGD), and phosphoglycerate kinase (PGK). Using Fig. 7.20, identify the three remaining enzymes.

Chapter 8

Extensions of Gene Transmission Analysis

We have limited our discussion of gene transmission in the three preceding chapters to the nuclear genes of eukaryotes. In this chapter, we will consider the patterns of gene transmission that occur in three simpler genetic systems: viruses, bacteria, and cytoplasmic organelles. The genetic material of these systems is similar to that in nuclear chromosomes, but it is packaged differently and is subject to different rules for transmission. None of these systems undergoes the meiotic process that occurs in the nuclei of eukaryotic cells. However, all of them can be experimentally manipulated in such a way that a form of diploidy, followed by synapsis and crossing over, can be made to occur. Recombination can then be studied and used to evaluate the relative positions of genes along the chromosome.

Genetic Analysis of Bacteria and Their Viruses

Because of their ease of handling and rapid reproductive rates, bacteria and viruses have been used extensively in genetic research. You will recall that it was the study of bacteria and their viruses that led to the identification of the genetic material. The genetic analysis of bacteria and viruses is in many ways similar to that of eukaryotic organisms. However, the unique features of their life cycles require special mapping techniques. We will consider some of these techniques in the following sections.

Mapping Phage Chromosomes

Viral chromosomes can undergo recombination while in the intracellular vegetative pool. Because of this feature, viruses can be studied genetically, in a manner patterned after that of higher organisms. In fact, the comparatively simple structure of the viral chromosome makes viruses ideal experimental subjects to use in testing the correspondence between the genetic map and the arrangement of genes on the DNA molecule.

In this section, we will examine the genetic maps of lambda and the T-even phages (specifically, T2 and T4), comparing each genetic map to the molecular structure of its respective chromosome. We will find that there is a direct correspondence between the genetic map and DNA of phage λ, whereas the linkage maps of the T-even phages do not correspond with their chromosomal structure. But before we can go into this topic more deeply, we must first learn some related genetic concepts.

A Phage Cross. The methods for genetic mapping in phages superficially resemble those used for other organisms. The progeny that are produced by phages of contrasting genotypes are analyzed for recombinants, and the frequency of recombinant types is then used as a measure of the distance between linked genes. A phage cross, however, is not a cross in the traditional sense but is a *mixed infection.* A bacterium is infected by a mixture of phage particles of two (or more) different genotypes. These infecting phages serve as parents in the cross. The concentration of each parental phage is kept high enough to ensure that most bacterial cells are simultaneously infected by both phage types. A sufficiently high ratio usually means a ratio of about five or so phage particles of each genotype per bacterium.

The progeny of phage crosses are analyzed by means of a **plaque assay.** In the first step of this procedure, a small sample from a dilute suspension of phage particles is spread over the surface of an agar medium that contains an opaque "lawn" of bacterial growth. This lawn is the result of the formation of colonies by a large number of bacterial cells; the adjacent colonies overlap each other to produce confluent growth. Because the suspension of phage particles is diluted, each phage particle is likely to infect a different bacterial cell, lysing that cell and releasing several more phages. These phages, in turn, infect other cells in the immediate vicinity, lysing them and releasing another generation of infective phage particles. This recurring infection–lysis process continues until clear areas, or **plaques,** appear on the bacterial lawn (Fig. 8.1). Many genetically determined traits such as plaque size and clearness can then be assessed by visual inspection. Since each plaque originates from an infection by a single phage particle, the phenotype of a plaque can be taken as an indication of the genotype of the original phage.

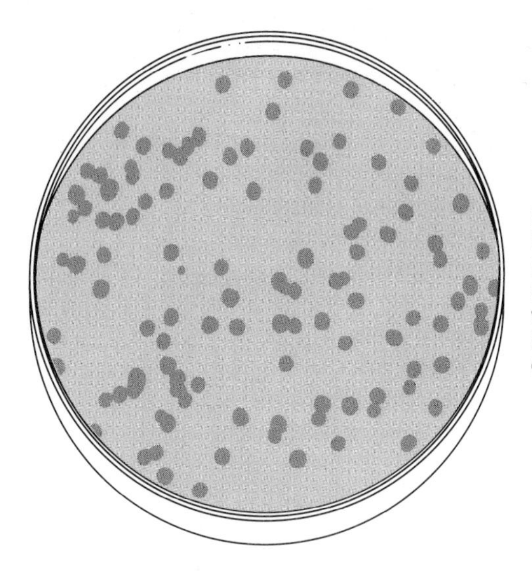

Figure 8.1. The appearance of phage plaques. A small volume of liquid phage suspension is spread over the surface of a bacterial lawn. Each phage particle infects a bacterial cell and lyses it, releasing progeny that infect and lyse other cells in the immediate vicinity. The result is a cleared area, called a plaque, on the bacterial lawn. The phenotype of a plaque identifies the genetic character of its single progenitor phage.

To illustrate the nature of a phage cross and the methods used in the analysis of phage recombinants, let us consider the allelic pairs for two traits that are exhibited by bacteriophage T2:

Plaque morphology: small plaques (r^+) vs. large plaques (r)

Host range: ability to infect *E. coli* strain B but not B/2 (h^+) vs. ability to infect both strains B and B/2 (h)

Since phages have a haploid genotype, a cross between a mutant for plaque morphology and a mutant for host range can be written as $h^+r \times h\ r^+$. To accomplish this cross, *E. coli* strain B is infected with equal numbers of both h^+r and $h\ r^+$ parental phages. The phage replication that follows produces multiple copies of each parental chromosome in the DNA pool; some of these copies participate in something akin to synapsis (called a mating by phage geneticists) and crossing over, yielding recombinant types:

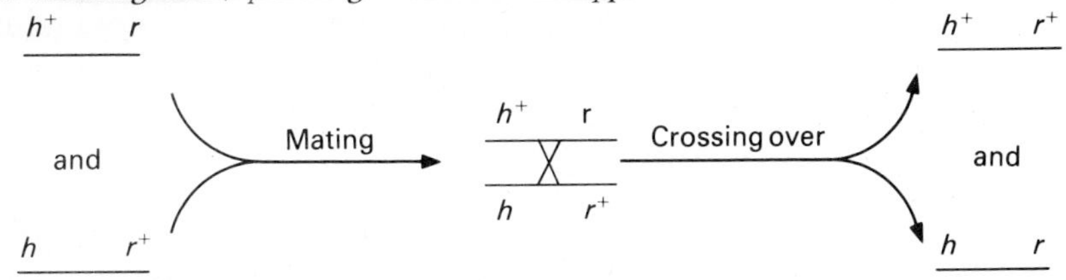

After mature progeny phages have been produced and the host cell has been lysed, these progeny are analyzed according to phenotype by spreading them on a lawn of mixed bacterial strains B and B/2. Phages that carry the h^+ allele produce cloudy plaques since they are unable to infect the B/2 bacteria in the mixture. Phages with the h allele infect both strains, yielding clear plaques. The four possible phenotypes among the progeny of the cross are shown in Fig. 8.2, and their

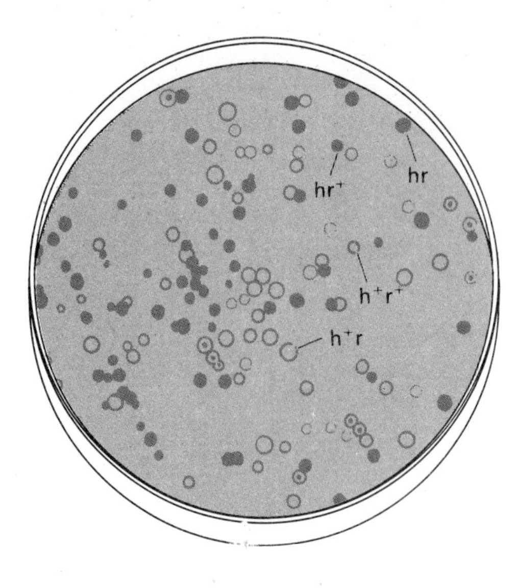

Figure 8.2. The appearance of plaques formed by $h\ r^+$, h^+r, h^+r^+, and $h\ r$ genotypes in bacteriophage T2 when plated on a lawn of mixed *E. coli* strains B and B/2. The *r* phages produce larger plaques than do r^+ phages. Phages carrying the *h* allele make clearer plaques on this type of bacterial lawn than do h^+ phages.

inferred genotypes are listed below:

APPEARANCE OF PLAQUE	GENOTYPE	
Cloudy and large	h^+r	Parental types
Clear and small	$h\ r^+$	
Clear and large	$h\ r$	Recombinant types
Cloudy and small	h^+r^+	

The recombination frequency is then calculated as

$$R = \frac{\text{Number of } h\ r \text{ plus number of } h^+r^+}{\text{Total number of plaques}}.$$

Although the analysis so far seems straightforward, chromosome mapping in bacteriophages is subject to several complications. First, each parental phage is duplicated several times early after infection. Therefore, multiple copies of the chromosomes of both parental phages will exist in a single cell. Thus, paired matings will not be restricted to contrasting genotypes, but can occur between two chromosomes of the same genotype as well. In our example, matings $\frac{h^+r}{h^+r}$ and $\frac{h\ r^+}{h\ r^+}$ will tend to occur as often as $\frac{h^+r}{h\ \ r^+}$. Because crossing over between phages of the same genotype would not yield recombinants, the recombination frequency is lowered, so that it might seriously underestimate the true genetic distance. The situation is further complicated in that a single chromosome can participate in several acts of mating while it replicates in the DNA pool. Even though these multiple matings might initially add to the frequency of recombinants, they can later decrease this frequency, because recombinants that are formed by one act of mating and crossing over can be lost through a subsequent pairing and exchange event. Researchers have developed mathematical models to convert the recombination frequency into a true genetic distance, based on these complexities. These models are, however, beyond the scope of our present discussion.

Another unusual feature of phage crosses is that the genetic maps of different kinds of phages (such as T2 and lambda, for example) are not comparable. During its replication cycle, each kind of phage undergoes its own characteristic average number of rounds of mating and crossing over. Thus, a recombination frequency in T2 as compared with λ will reflect an inherently different probability of events, so that similar recombination frequencies in these viruses may actually represent different lengths along their DNA molecules. We should also note that a unit of recombination in a bacterial virus does not correspond to the same physical chromosomal length as would a unit of recombination in a higher organism.

EXAMPLE 8.1. Three two-point crosses were made using different strains of bacteriophage T2. The following results are obtained:

Cross 1: $a^+b \times a\,b^+ \rightarrow$ 1773 a^+b, 1747 $a\,b^+$, 104 a^+b^+, 96 $a\,b$
Cross 2: $b^+c \times b\,c^+ \rightarrow$ 1348 b^+c, 1312 $b\,c^+$, 124 b^+c^+, 108 $b\,c$
Cross 3: $a^+c \times a\,c^+ \rightarrow$ 1443 a^+c, 1483 $a\,c^+$, 51 a^+c^+, 55 $a\,c$

(a) Construct a linkage map of these three loci. **(b)** Calculate the coefficient of coincidence from these results. What type of interference appears to be operating in this case?

Solution. **(a)** The recombination frequencies can be calculated as follows:

$$R_{a-b} = \frac{104 + 96}{3720} = 0.054,\ R_{b-c} = \frac{124 + 108}{2892} = 0.080, \text{ and } R_{a-c} = \frac{51 + 55}{3032} = 0.035.$$

Thus, the linkage map becomes:

b 5.4 *a* 3.5 *c*

(b) To determine the coefficient of coincidence, first note that when calculating the map distances for the adjacent intervals *a–b* and *a–c*, we not only included single recombinants in our calculations, but also double recombinants that, by chance, experienced an odd number of crossovers in each region. We can therefore express R_{a-b} and R_{a-c} as the sum of two frequencies as follows: R_{a-b} = *f*(single recombinants, *a–b* only) + *f*(double recombinants, *a–b* and *a–c*) and R_{a-c} = *f*(single recombinants, *a–c* only) + *f*(double recombinants, *a–b* and *a–c*). In contrast, recombinants for the outside markers *b–c* could have only undergone a single recombination act, which occurred in either one of the two included intervals. Thus, R_{b-c} = *f*(single recombinants, *a–b* only) + *f*(single recombinants, *a–c* only). These expressions can be solved simultaneously to yield *f* (double recombinants, *a–b* and *a–c*) = $\frac{1}{2}(R_{a-b} + R_{a-c} - R_{b-c})$. Since the frequency of double recombinants in the absence of interference has been defined earlier as $R_{a-b} \times R_{a-c}$, the coefficient of coincidence can be expressed as

$$\frac{\frac{1}{2}(R_{a-b} + R_{a-c} - R_{b-c})}{R_{a-b} \times R_{a-c}} = \frac{\frac{1}{2}(0.054 + 0.035 - 0.080)}{0.054 \times 0.035} = 2.38.$$

This is a case of negative interference, since the value is greater than one.

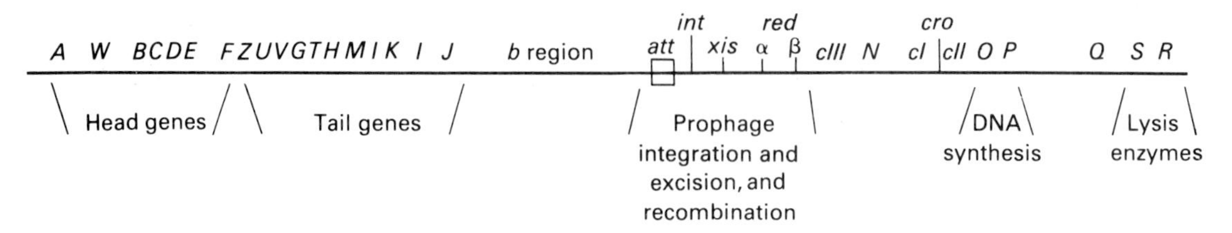

Figure 8.3. Genetic map of the λ chromosome. The genes that code for subunits of the phage protein head are grouped together, as are the genes coding for tail protein. The *att* region is the site of homology with the *E. coli* chromosome; the synapsis that occurs at this site leads to the integration of the λ prophage. Genes *int* and *xis* code for integration and excision enzymes, respectively, and the *red* genes code for proteins that function in recombination. The *cI*, *cII*, and *cIII* genes function in the establishment and maintenance of lysogeny. Genes *O* and *P* control DNA replication during the lytic cycle, and *S* and *R* code for lysozymes. Genes *N* and *Q* are positive regulatory genes, whose function we will discuss later.

Linear Map of the Lambda Chromosome. The linkage map of phage λ is presented in Fig. 8.3. The map is linear, reflecting the linear nature of the chromosome that is found in the λ virion. We may expect a linear map for any phage that has a unique starting and termination point to its chromosome. All phage particles in the population then share the same sequence of genetic markers. As we shall see in the next section, this feature is not common to all types of phages.

Studies of the structure of DNA molecules in phage λ that carry physical aberrations show that the genetic and physical maps are colinear. One useful aberration in this regard is a deletion called the *b*2 mutation, in which a segment of about 5000 base pairs is missing from the DNA molecule. This deletion has been found to be located near the center of the λ chromosome through the use of an approach known as **heteroduplex mapping.** In this procedure, DNA molecules from wild-type and *b*2 phages are denatured, and the single strands are isolated by density gradient equilibrium centrifugation. When a wild-type strand is renatured with its complementary *b*2 strand, a heteroduplex molecule results in which the double helix is interrupted at the mutant site by a single-stranded loop in the wild-type strand that corresponds in length to the deletion (Fig. 8.4). Electron microscopy reveals that this deletion is located 44.3 percent of the distance from one end along the λ chromosome. The physical position of the *b*2 mutation agrees precisely with its position on the genetic map. Similar studies with a variety of other aberrations confirm the correspondence between the genetic map and the chromosome structure of the virion.

Circularity of the T2 and T4 Linkage Maps. Unlike phage λ, whose genetic markers have a linear arrangement, the T-even phages evidence a circular linkage map (Fig. 8.5). Yet paradoxical as it may seem, the chromosomes of T2 and T4 are linear. How does a linear chromosome produce a circular linkage map? The answer in this case involves the **terminal redundancy** of the T2 and T4 chromosomes. Each chromosome ends in the same gene sequence as the one in which it began. An example would be a phage genome consisting of the sequence of gene markers $\underline{a\ b\ c\ d\ e\ f\ g\ a\ b}$, where the $\underline{a\ b}$ sequence is repeated at both ends. After the linear phage genome has infected the host cell and subsequently repli-

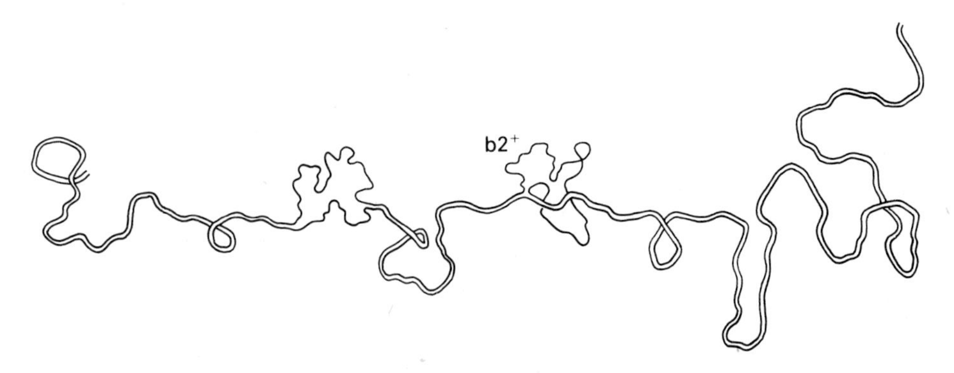

Figure 8.4. An interpretive drawing based on an electron micrograph of heteroduplex lambda DNA, in which one strand carries the *b2* deletion. The molecule appears double-stranded except in two regions. The single-stranded loop corresponds to the *b2*+ region that makes up the central section of the λ strand opposite the deletion. The second discontinuity in the molecule is an unpaired region where the two strands are nonhomologous. *Source:* From B. C. Westmoreland, W. Szybalski, and H. R. Ris, *Science* 163:1345, 1969. Copyright 1969 by the American Association for the Advancement of Science.

cated, multiple copies of the chromosome appear to join end-to-end, possibly by pairing and crossing over in their regions of terminal redundancy:

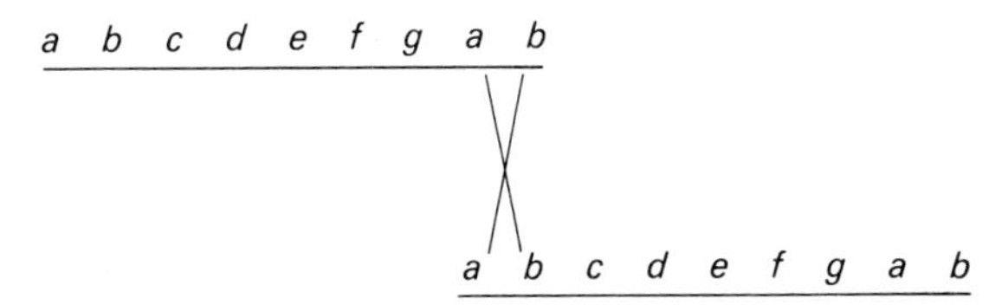

The end result is a longer chromosome that contains a repeat of the basic genome: *a b c d e f g a b c d e f g a b*. These concatamers are regularly found when cells infected with T2 or T4 are prematurely lysed to study the intermediate stages of phage production. Many are several chromosome lengths in size.

During the maturation of phage progeny, the concatamers of T2 and T4 are processed by what is known as the *headful mechanism*. The concatamers are cut into chromosome lengths of just the size that can be incorporated into the heads of the developing progeny. This enzymatic process is not restricted to a single genetic site, but can occur between any pair of markers. (Compare this process with the one that cuts λ chromosomes out of the concatamers generated by rolling-circle replication.) Let us assume, for example, that processing according to the headful mechanism means a cut in the concatamer every nine genes, as shown below:

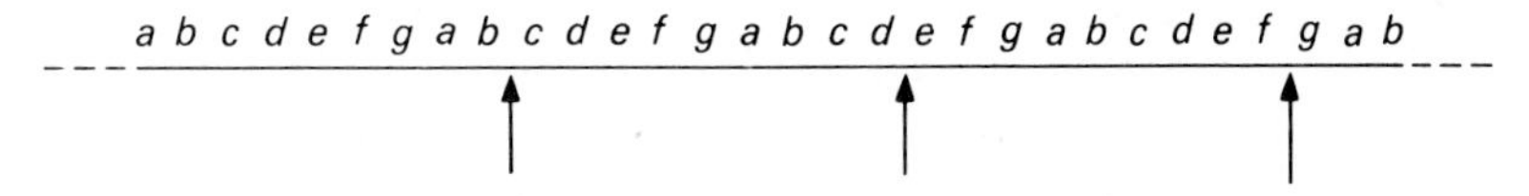

The result of these regularly spaced sites of enzymatic cleavage is a *circularly permuted* collection of phage chromosomes, with some reading *a b c d e f g a b*, others reading *c d e f g a b c d*, still others reading *e f g a b c d e f*, and so on. Be-

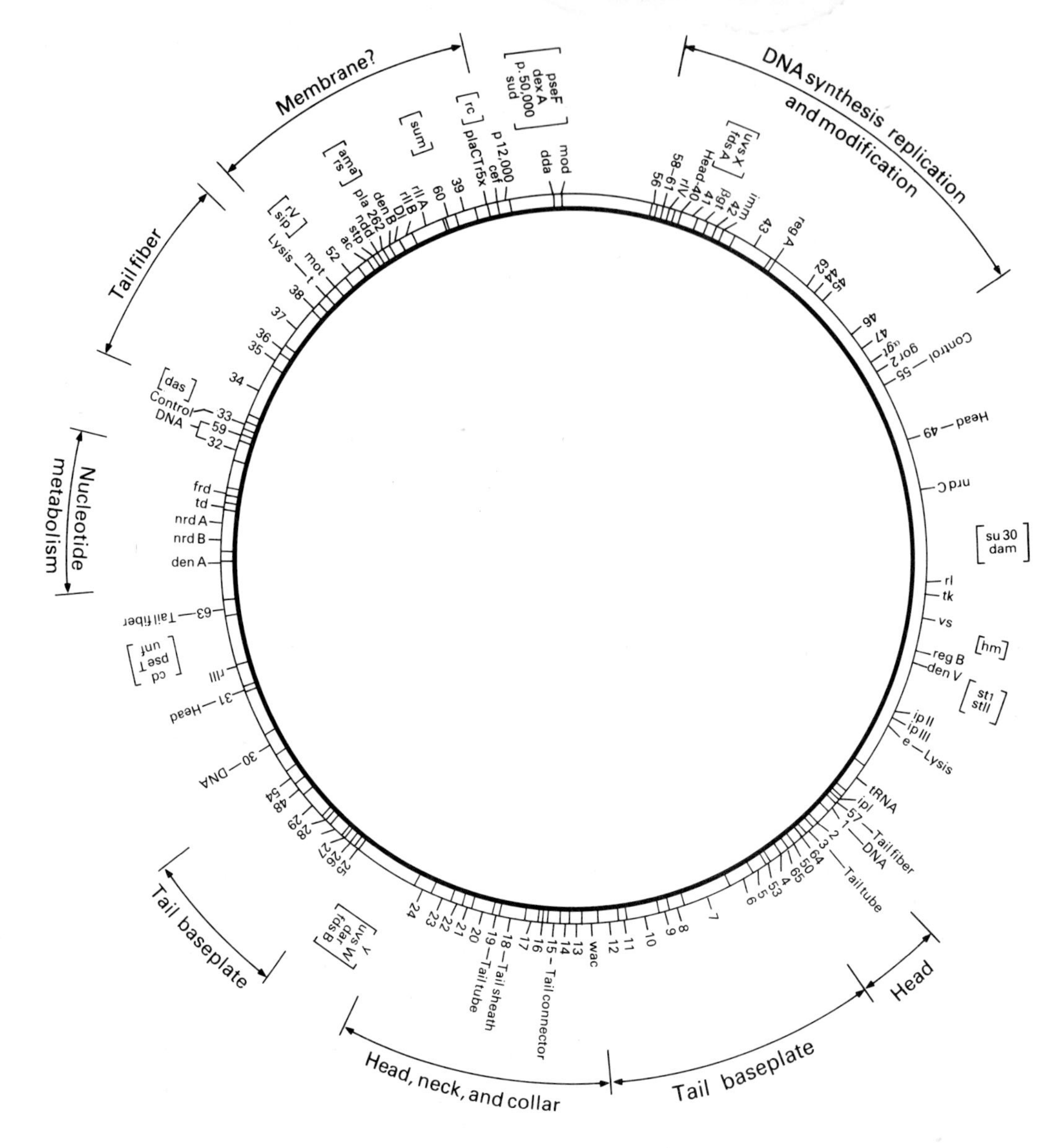

Figure 8.5. The linkage map of phage T4. Genes whose exact order is not known are enclosed in brackets in arbitrary order. Notice the clustering of related functions. *Source:* From W. B. Wood and H. R. Revel. *Bact. Rev.* 40:847–868, 1976.

cause of this cleavage system, recombination analysis of these progeny would reveal that gene *a* is closely linked not only to *b* but also to *g*, the end marker in the sequence. Similarly, gene *b* is closely linked to both *a* and *c*, gene *c* is closely linked to both *b* and *d*, and so on. These relationships give rise to a linkage map that can best be represented in the form of a circle. Thus, the circular linkage map of the T-even phages is derived from a circularly permuted collection of linear progeny chromosomes. In general, a circular map is expected for any phage that uses the headful mechanism to package its genetic material.

EXAMPLE 8.2. Four gene loci were discovered in a certain bacteriophage. We will designate these loci as *a, b, c,* and *d.* A series of two-point phage crosses gave the following unique relationships for the recombination frequencies:

$$R_{a-c} = R_{b-d} > R_{a-b} = R_{b-c} = R_{c-d} = R_{a-d}.$$

Explain these results in terms of a suitable linkage map.

Solution. If we were to ignore R_{a-d} for the moment, the relationships $R_{a-c} = R_{b-d} > R_{a-b} = R_{b-c} = R_{c-d}$ are consistent with a simple linear map in which the markers are arranged in the order *a–b–c–d* and are equidistant from one another. But the fact that *a* is also closely linked to *d* is best explained by a circular map, as follows:

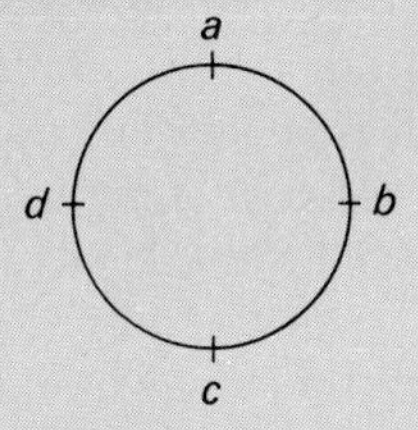

Fine-Structure Mapping. Until the 1940s, a gene was regarded as a unit of chromosome structure that was not subdivisible by crossing over. Crossing over was thought to occur only between genes, never within the genes themselves. According to this view, different mutation sites within a gene could never recombine and would always map at the same point. This view of the gene as an indivisible particle has since been shown to be completely wrong. We now know that crossing over occurs within the boundaries of genes as well as between them. Furthermore, different mutable sites within a gene are separable by crossing over. The ability to detect intragenic crossing over has allowed geneticists to construct **fine-structure maps** of gene loci, which show the relative positions of mutation sites within genes. Fine-structure mapping is difficult to perform in higher organisms, because of the extremely short distances separating base pairs that undergo a detectable mutational change. The shortness of these lengths makes it necessary to analyze 100,000 or more progeny in order to obtain accurate recombination frequencies. Such large population sizes are readily obtainable with phages, however, as well as with bacteria and certain lower eukaryotes, such as yeast.

The procedures used in fine-structure mapping are basically similar to those employed in mapping the chromosomal locations of different genes. Two-point crosses are conducted between different (independently isolated) mutant strains that have undergone mutation in the same gene. Since the different mutants may be indistinguishable in phenotype, crossing over between the strains is detected by the appearance of wild-type recombinants:

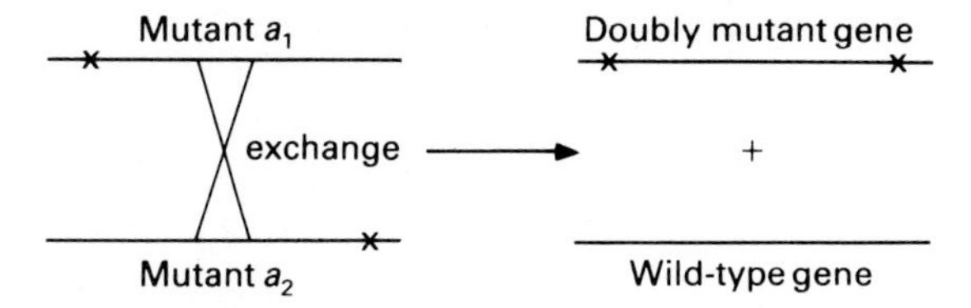

Since only half of the recombinant progeny are expected to have wild-type markers at both sites, the map distance is equal, in this case, to twice the frequency of wild-type recombinants.

Completely mapping all mutation sites within a gene is a formidable task. For example, to construct a map with 1000 mutant strains (each presumably differing from the others at a single base pair) requires a total of 1000!/2!998!, or about 500,000, two-point crosses! Various modifications of this basic procedure are therefore used to make things easier. One useful approach is to cross the different mutant strains that we have isolated with *deletion mutants* for the same gene. Unlike single-site mutants, a deletion mutant is missing a segment of DNA; the deletion can be of variable length and it includes multiple bases. When a single-site mutant is crossed with a deletion mutant, or when two deletion mutants are crossed, they will recombine and produce wild-type progeny only if their mutations *do not* share a common region. The theoretical basis for this genetic behavior is shown in Fig. 8.6 in terms of a hypothetical example involving two deletion mutants and a single-site mutant. Note that wild-type offspring are only possible when the mutable site (base pair) affected in the single-site mutant is not one of those missing in the deletion mutant. Since the two deletions in Fig. 8.6 overlap, we have, in effect, established a **deletion map:**

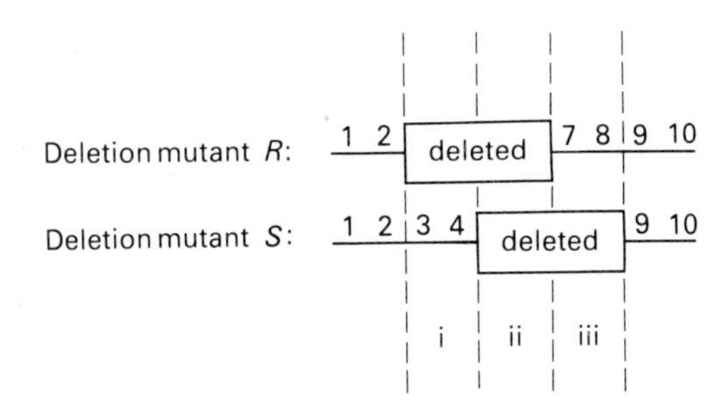

The overlap of the different deletions defines three regions within the gene: i, ii, and iii. Any single-site mutant can now be tested for its rough location. The possibilities are as follows:

1. Only crosses between the single-site mutant and deletion mutant *R* give wild-type recombinants, which means that the single-site mutation lies in region iii.
2. Only crosses between the single-site mutant and deletion mutant *S* give wild-type recombinants, which means that the single-site mutation lies in region i.
3. The single-site mutant gives no wild-type recombinants when crossed to either *R* or *S*, which means that the single-site mutation lies in region ii.

Once the rough location of the single-site mutation has been established, its precise location can be pinned down more closely by crossing the mutant to a

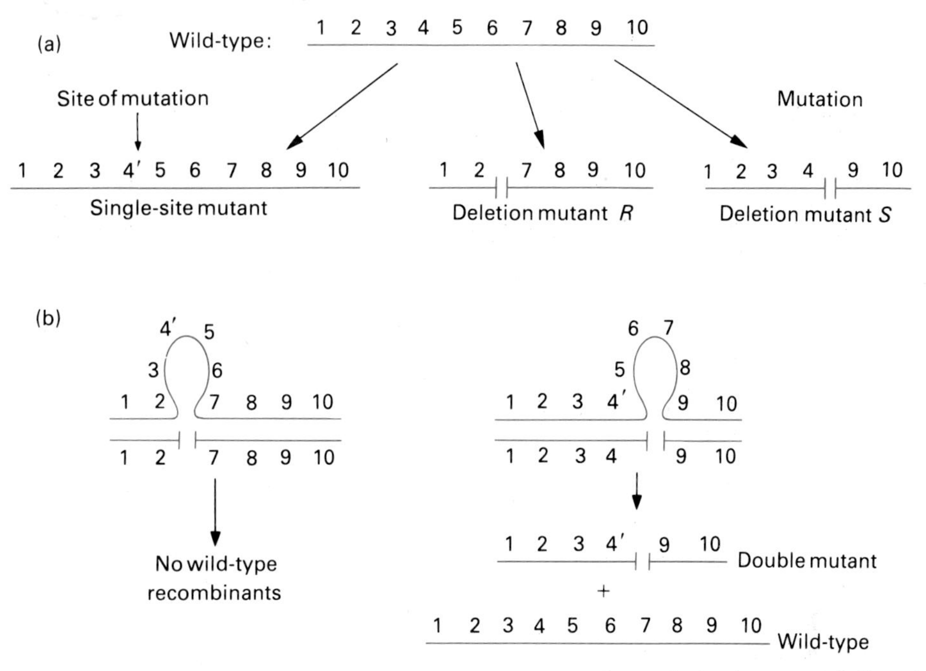

Figure 8.6. (a) The origin of three hypothetical mutants, including a single-site mutant (left) and two deletion mutants, *R* (left) and *S* (right). (b) The effects of crossing over between the single-site mutant and each of the deletion mutants. Wild-type recombinants are produced only by the crossover involving deletion mutant *S* (right), since the *S* deletion does not overlap the point of the single-site mutation, as does deletion *R*.

series of other known deletion strains that are missing progressively smaller amounts of DNA in the general region containing the mutation site (Fig. 8.7). The distances between the different mutation sites that have been localized in this manner can then be determined from two factor crosses, as described earlier.

EXAMPLE 8.3. Deletion mutants can be intercrossed and their deletions mapped just like single-site mutations. As an example of how a deletion map is constructed, let us consider the following problem. Four deletion mutants are crossed in paired combinations to test for their ability to produce wild-type recombinants. The results are given in the table below, where + indicates that recombinants were observed. Draw the deletion map that corresponds to this group of mutations, and label the regions into which the deletion map subdivides the locus.

		DELETION MUTANTS			
		1	2	3	4
DELETION MUTANTS	1	−	−	+	+
	2		−	−	+
	3			−	−
	4				−

Solution. Deletions 1 and 3, deletions 1 and 4, and deletions 2 and 4 do not overlap, as judged by their ability to give wild-type recombinants when crossed. However, deletion 3 overlaps 4, and deletion 2 overlaps both 1 and 3. A topological representation of these results is as follows:

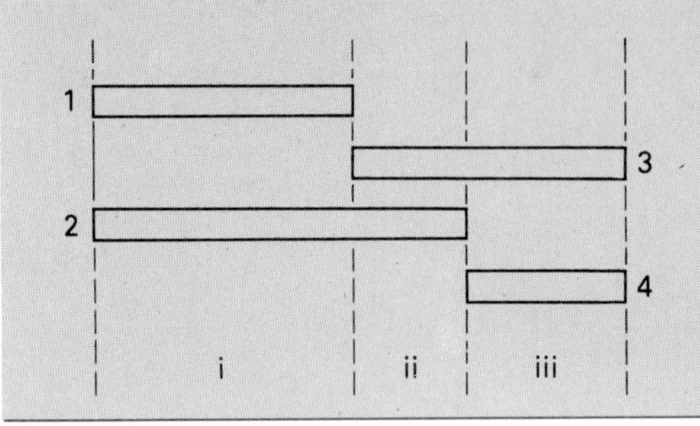

Figure 8.7. Localization of the point of a single-site mutation by using a series of known deletions (*R*, *S*, *R*1, *S*1, *R*1*a*, and *S*1*a*). The deletion mutants are missing progressively smaller amounts of DNA in the chromosomal region that contains the mutation site. This progressively reduces the chromosomal region within which the single-site mutation is known to lie.

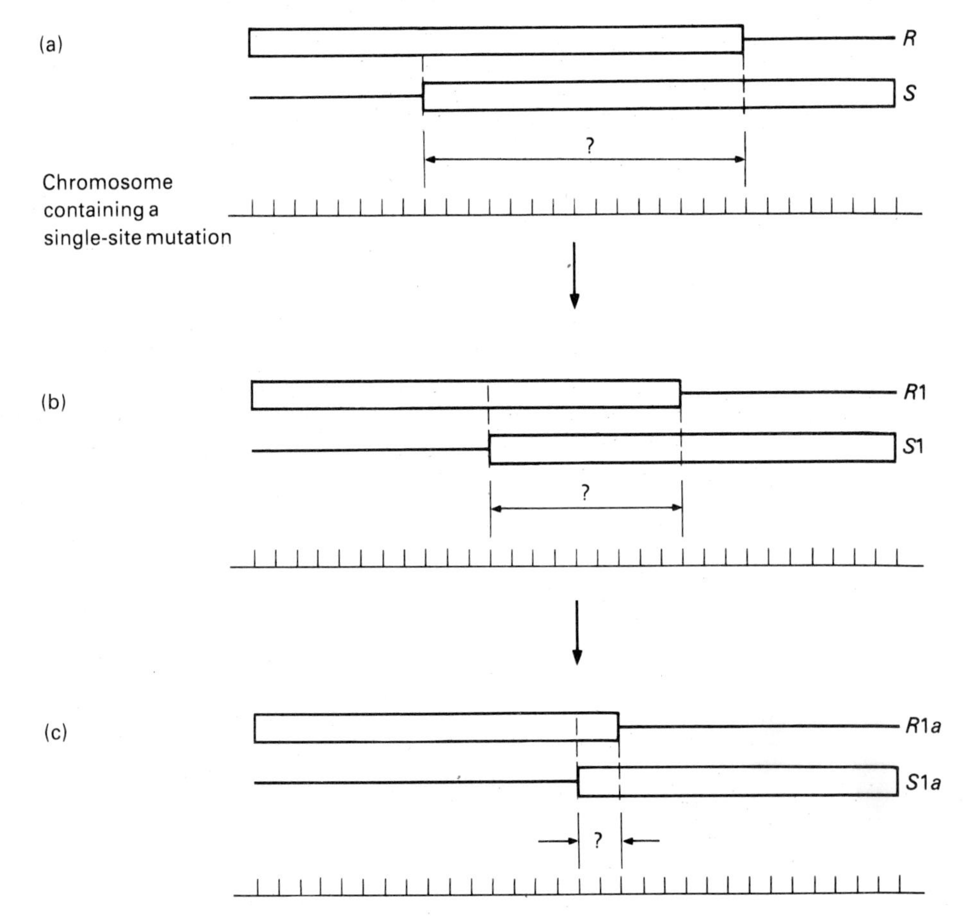

Many of the methods used in fine-structure mapping were developed by Seymour Benzer in the 1950s. Benzer mapped the *r*II locus of phage T4. The *r*II locus consists of two genes, *A* and *B*, that influence plaque size and host specificity. Wild-type (r^+) phages form small plaques on both *E. coli* strains B and K(λ). Mutant *r*II phages produce large plaques on strain B but are unable to infect strain K(λ). Benzer began his experiments with a group of about 3000 independently isolated *r*II mutants. By employing procedures similar to those we have just described, Benzer was able to subdivide the *r*II locus into 47 smaller regions on the basis of deletion mutants. He then used his deletion map to identify over 300 mutation sites that were separable by recombination. While 300 sites might seem large when considered in the context of a genetic map, they comprise only a small fraction of the total of 2650 or so base pairs that make up the *r*II region. Thus, many potentially mutable sites were not detected. As we shall see in Chapter 11, one likely reason for Benzer's failure to identify all potentially mutable sites is the inability of many base-pair changes to yield a mutant phenotype.

Gene Mapping in Bacteria

As in the case of viruses, the genetic analysis of bacteria provides a means for comparing the order of genes in a linkage map with their arrangement on a DNA molecule. Because bacteria do not undergo the meiotic process characteristic of eukaryotes, geneticists must rely on other mechanisms of gene exchange to study recombination in this group of organisms. Of the different mechanisms for genetic transfer in bacteria, **conjugation** has proved to be the most useful in mapping studies. A relatively large segment of the bacterial chromosome is ordinarily transferred during conjugation, enabling geneticists to map genes that are separated by large distances. It is possible to obtain a rough map of the entire bacterial chromosome by combining data from several crosses, involving many different genes. As we shall see in the sections that follow, these studies reveal that the bacterial linkage map, like the structure of the bacterial chromosome, is circular.

Under normal circumstances, conjugation does not provide the resolving power needed to map genetic markers that are very closely spaced. Furthermore, it occurs in only a few bacterial species. These limitations necessitate the use of additional mechanisms of gene exchange for chromosome mapping. Two mechanisms that have been important in this regard are **transformation,** in which the genes are transferred directly in the form of naked DNA (see Chapter 3), and **transduction,** which involves the transfer of bacterial genes by a temperate virus. Only very short lengths of chromosomal material are normally transferred by these mechanisms, thus limiting their usefulness to the mapping of genes that lie very close together. In both cases, the recipient cell becomes a "partial diploid," having its own chromosome plus a chromosome fragment from the donor cell. Synapsis and crossing over between the recipient chromosome and the donor fragment can then yield recombinants, in proportions that relate to the distance separating the genes involved.

Interrupted Conjugation. Conjugation was first used for the purpose of chromosome mapping in the late 1950s by Wollman and Jacob. Their method is referred to as the *interrupted mating technique.* Rather than allowing conjugation to proceed until the conjugation tube randomly breaks, the interrupted mating procedure controls the precise length of time allowed for the transfer of the donor chromosome. It then measures the proportion of recombinants formed for each gene as a function of the transfer time. To control the time allowed for mating, the researchers withdraw samples from a population of conjugating cells at regular time intervals. The samples are immediately placed in an electric food blender and are subjected to agitation in order to separate the conjugating pairs. The shearing breaks the conjugation tubes, but does not harm the cells themselves.

To illustrate the method, let us consider the cross of Hfr $str^s a^+ b^+ c^+ d^+ \times$ F^- $str^r a^- b^- c^- d^-$ shown in Fig. 8.8. The *str* locus determines the reaction of the bacterium to the antibiotic streptomycin. Hfr cells are sensitive to streptomycin (str^s), while F^- cells are resistant (str^r). If the cells are placed in a growth medium that contains streptomycin immediately after conjugation is terminated, only F^- cells will survive. This is a convenient way to eliminate the Hfr cells from the population after they have donated their genes, so that we can selectively study only the F^- recipient cells.

At defined time intervals, then, samples of conjugating cells are removed from the original mixture of Hfr $\times$ F^- cells and placed into a food blender to terminate gene transfer. The cells are next placed on a medium containing streptomycin to kill the Hfr cells. The surviving F^- colonies are individually tested for the presence of allelic characteristics from the donor strain. For example, the investigators may

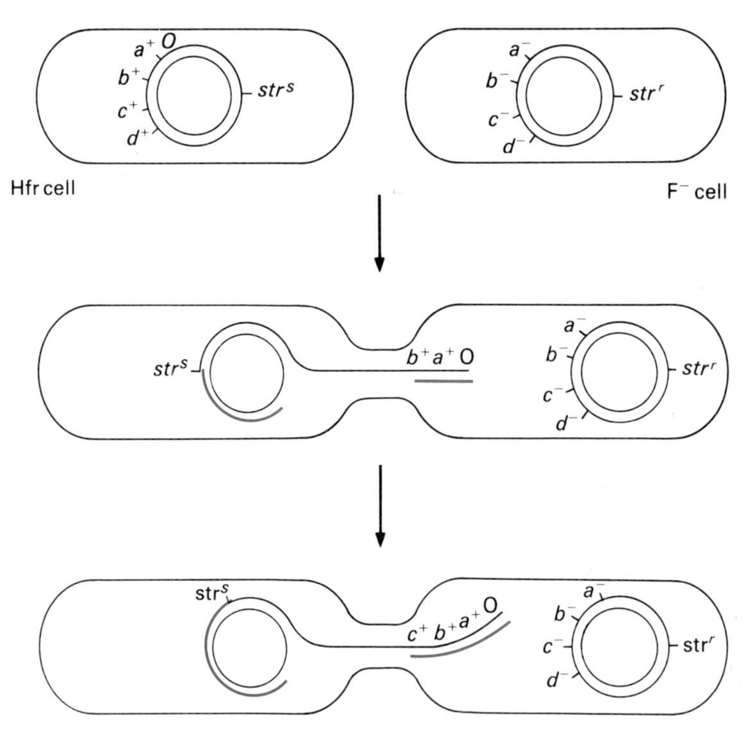

Figure 8.8. The linear transfer of the Hfr chromosome into the F^- recipient during conjugation. The O marks the origin half of the integrated F factor, which leads chromosome transfer. The Hfr chromosome replicates itself as transfer proceeds (newly replicated regions are shown in red). See Figure 4.14 for a detailed description of conjugation.

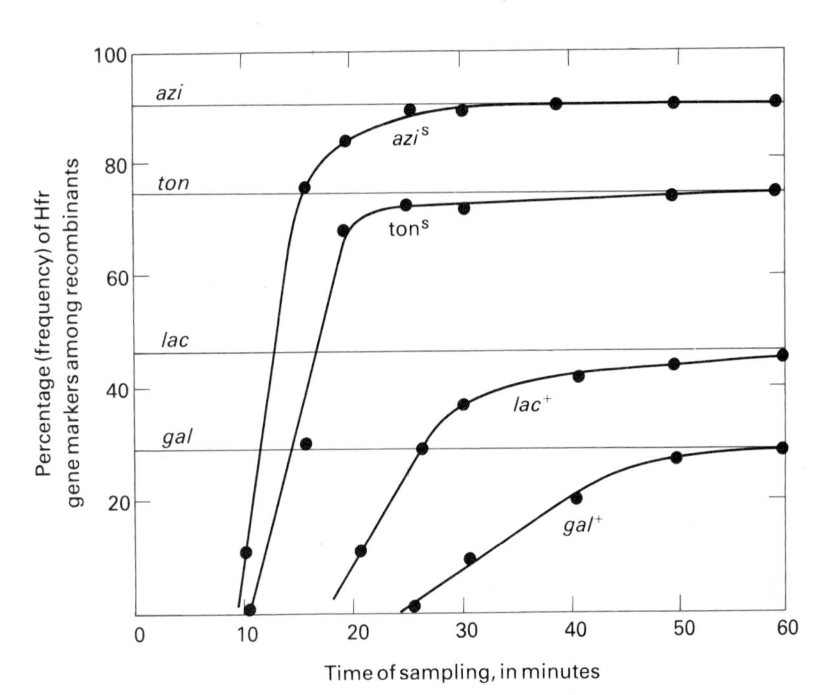

Figure 8.9. Interrupted mating experiment of Wollman and Jacob. F^- *str^r azi^r lac^- gal^-* cells were allowed to take up genetic material from Hfr *str^s azi^s ton^s lac^+ gal^+* cells. At specific time intervals, samples were withdrawn and the mating process was terminated by disruption in a blender. Cells were then placed in medium containing streptomycin, to kill the Hfr. Surviving F^- cells were then tested to determine their genotype and the results were plotted as percent of each Hfr characteristic among the surviving cells, as a function of the time allowed for mating. *Source:* From E. L. Wollman, and F. Jacob, *C.R. Acad. Sci.*, Paris, 240:2449, 1955.

transfer samples of each colony to various media to determine the acquisition of such donor traits as the ability (or inability) to utilize certain compounds for growth, the ability (or inability) to grow in the absence of certain nutrients, or the resistance (or sensitivity) of the cells to different drugs and viruses.

Figure 8.9 is a graph of the results published by Wollman and Jacob. The genes *azi^s* (azide sensitivity), *ton^s* (sensitivity to T1 virus), *lac^+* (ability to metabolize the sugar lactose), and *gal^+* (ability to metabolize the sugar galactose) correspond to the a^+, b^+, c^+, and d^+ markers used in our generalized description of the experiment. Notice in the results that each donor gene first appears in the F^- recipients at a specific time after mating begins. For instance, the *azi^s* marker first makes its appearance at about 9 minutes after the start of mating, whereas the *ton^s* marker does not first appear until 11 minutes. Furthermore, genes that appear later yield lower maximal levels of F^- recombinants. These results are expected if the Hfr chromosome is transferred to the F^- cell in a linear fashion, beginning at a specific point (the O end of the broken F plasmid). Gene loci near point O are likely to be transferred even during brief periods of conjugation. Genes farther away from point O have an increasingly smaller chance of entering the recipient cell, since the conjugation tube often breaks by itself even before the sample is

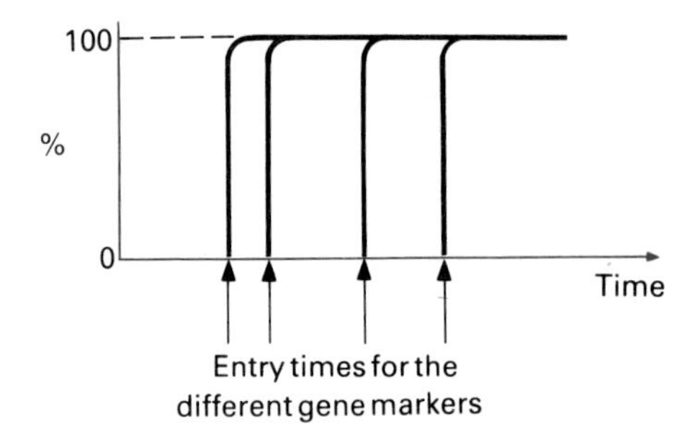

(a) Idealized results: All cells begin conjugation at the same time and continue transfer of their chromosomes until mating is artificially interrupted.

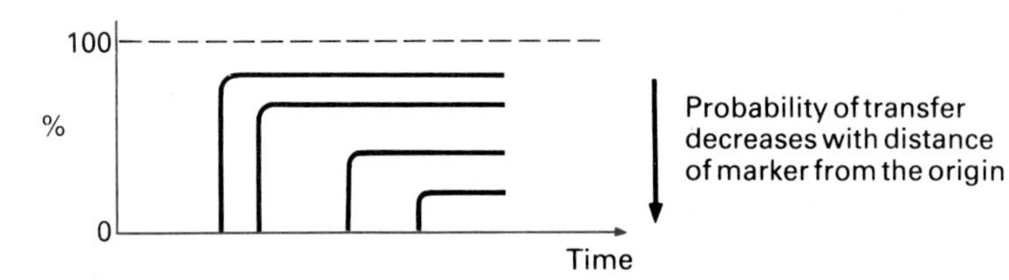

(b) Less idealized results: Mated cells separate spontaneously prior to and during artificial rupture.

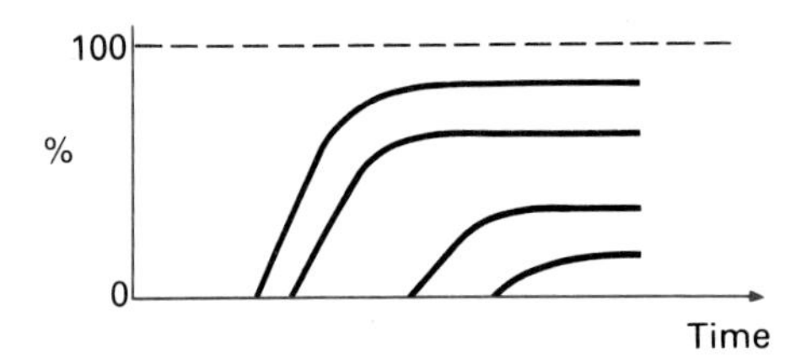

Actual results: Variability in the time of pair formation, as well as spontaneous separation of donor and recipient cells.

Figure 8.10. Explanation of the graphical results shown in Figure 8.9. The shape of each curve is a consequence of three main effects: artificial rupture of mated cells, which delineates the time that each marker first enters the recipient; spontaneous rupture of mated cells, which decreases the probability that markers distal to the origin will eventually be received by the recipient; and variation in the time of pair formation.

subjected to mechanical shearing. Thus, the later in time that a gene is transferred, the less likely it is to enter a recipient cell and to participate in recombination. Figure 8.10 further analyzes these relationships.

If we assume that the transfer of the chromosome occurs at a constant rate, we can use the length of time that it takes each donor gene to first appear among the F^- cells as a measure of the distance of the gene from the origin. The unit of distance commonly used for this purpose is one minute of transfer time. A map based on the results of Wollman and Jacob is shown below.

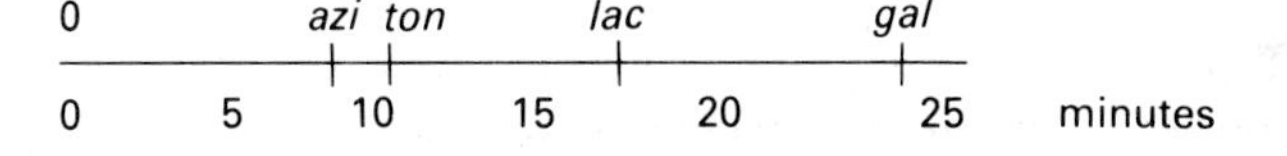

Circularity of the Linkage Map. The circular nature of the bacterial chromosome was first suggested when mapping studies revealed that the linkage map of *E. coli* is circular. This circular linkage map is generated from the combined results of interrupted mating experiments that employ different Hfr strains. These different strains are formed by the insertion of the F factor at different sites in the bacterial chromosome (Fig. 8.11). There are apparently several different possible sites of integration for the F plasmid during the conversion of an F^+ cell to an Hfr strain. Moreover, the F factor has polarity and can apparently orient itself in either of two directions during its integration. The end result is that different Hfr strains will transfer genes during conjugation from different starting points and in different directions. For example, interrupted mating experiments using six different Hfr strains may give rise to six different maps, based on the order of transfer for the three hypothetical genes *a, b,* and *c*:

HFR STRAIN	RELATIVE GENE ORDER
1	O–*a*–*b*–*c*
2	O–*c*–*b*–*a*
3	O–*b*–*c*–*a*
4	O–*a*–*c*–*b*
5	O–*c*–*a*–*b*
6	O–*b*–*a*–*c*

The various maps produced using these strains show the genes differing in their relationship to the point of origin (O) or in their polarity of transfer. As we can see from Fig. 8.11, the six gene orders are circular permutations of one another, produced by cutting a single circular map at different points and transferring it in different directions.

As we learned previously, the entire Hfr chromosome is not usually transferred in any one mating. But different sections along the chromosome can be mapped by using a variety of Hfr strains. These maps are then pieced together to yield a composite map of the entire chromosome. One such map of the *E. coli* chromosome is shown in Fig. 8.12. The entire length of this map is 100 minutes, which is the time needed for the entire donor chromosome to be transferred to the recipient cell during the conjugation process. Observe that the map is circular, corresponding in overall shape to the bacterial chromosome itself.

Mapping by interrupted mating is accurate in dealing with genes that are separated from one another by at least two minutes of transfer time. When the genes are closer than this distance, the map length and gene order can be correctly established only by using other recombination techniques such as transduction or transformation.

Transduction. Transduction is the transfer of genes from one bacterial cell to another by way of a temperate phage. During phage multiplication within the donor cell, a small piece of donor bacterial DNA is sometimes accidentally packaged into a phage protein coat in place of all or part of the phage chromosome. A phage particle that contains bacterial DNA is called a *transducing phage.* If this phage subsequently infects a recipient cell, the transferred block of donor bacterial genes may then be added to or exchanged with a corresponding segment of the recipient chromosome by crossing over. The transduced cell can express the donor genes if it survives further viral infection by becoming lysogenic for the temperate phage.

Two mechanisms of transduction are known: *specialized transduction,* in which genetic transfer is restricted to bacterial genes that are located adjacent to the site of prophage attachment, and *generalized transduction,* in which virtually any locus in the donor chromosome can be transferred with equal efficiency. We can illustrate the mechanism of specialized transduction by observing the course of events that leads to the formation of λ transducing phages. The λ prophage is inserted next to the galactose (*gal*) region on the bacterial chromosome. Deintegration of the prophage usually occurs without error. But occasionally the prophage does not loop out completely from the host chromosome; instead, an altered loop is formed

Figure 8.11. Origin of different Hfr strains. The insertion of an F plasmid at one of many potential integration sites in the bacterial chromosome leads to the conversion of an F⁺ cell into a specific Hfr strain that transfers its chromosome during conjugation from a unique starting point. The direction of the arrow shows the direction with which the Hfr chromosome is transferred. The F plasmid has polarity and can thus orient itself in either of two directions during integration.

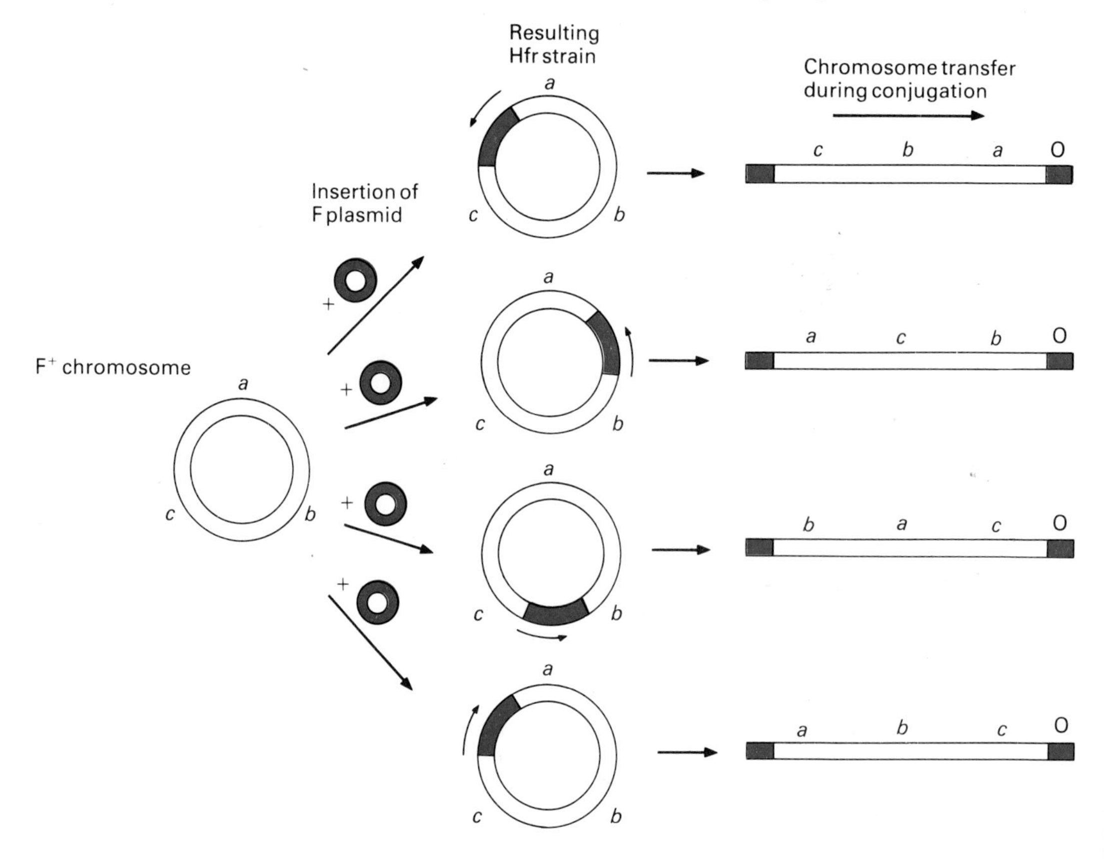

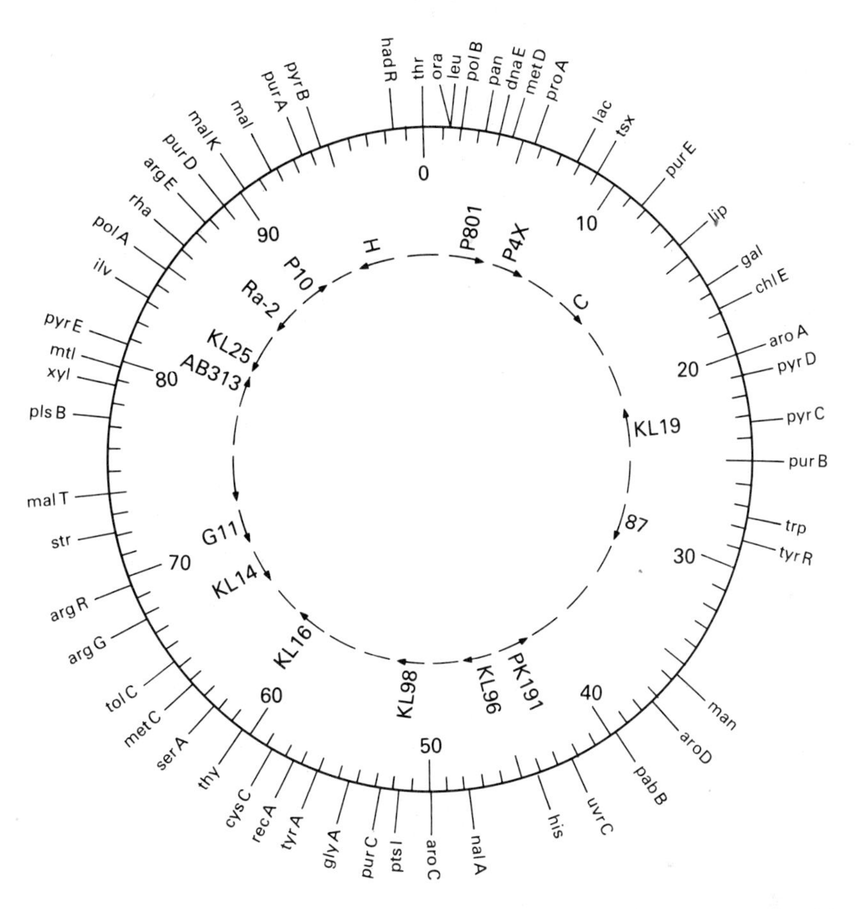

Figure 8.12. Map of the *E. coli* chromosome, showing the locations of some of its genes. The total map length is 100 minutes. The arrows on the dashed inner circle indicate the point of origin and the polarity of a few well-known Hfr strains. Some of the better-known genes are *leu* = leucine synthesis, *pol B* = DNA polymerase II, *dna E* (*pol C*) = DNA polymerase III, *met* = methionine synthesis, *pro* = proline synthesis, *lac* = lactose operon, *tsx* = T6 resistance, *pur* = purine synthesis, *pyr* = pyrimidine synthesis, *gal* = galactose operon, *trp* = tryptophan synthesis, *tyr* = tyrosine synthesis, *uvr C* = ultraviolet sensitivity (repair), *his* = histidine synthesis, *rec A* = recombination and repair, *cys* = cysteine synthesis, *str* = streptomycin sensitivity, *pol A* = DNA polymerase I. *Source:* From B. J. Bachmann, K. B. Low and A. L. Taylor, *Bact. Rev.* 40:116–167, 1976.

that contains the bacterial *gal* locus and about half of the phage chromosome (Fig. 8.13). The resulting hybrid phage chromosome is defective, since it lacks essential phage genes, and is incapable of producing progeny by the lytic cycle. Normal function is provided when the bacterial cell is subsequently infected with normal λ phages. The normal phage chromosomes act as "helpers" by supplying the defective phage with the missing phage genes that it requires to synthesize progeny. Lysis of the cell then releases a mixture of normal λ phages and λ-transducing phages.

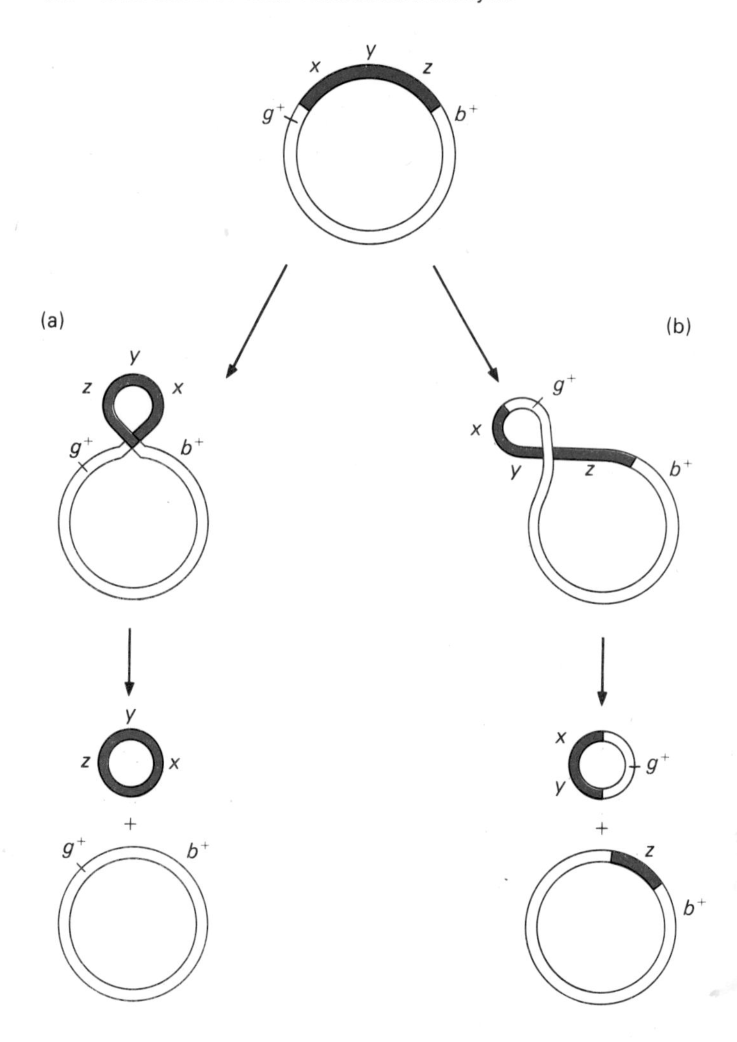

Figure 8.13. Induction of the λ prophage. Usually, deintegration is the exact reversal of integration, as shown in (a). Occasionally, however, the prophage loops out incorrectly (b), resulting in a chromosome that consists of approximately half of the phage genes, together with a short section of bacterial chromosome on which the *gal* (g^+) gene is located.

The transducing phages produced in this way are known as λ*dgal* phages (d meaning defective). After a recipient bacterium is infected by λdgal, the transducing chromosome can be integrated into the host chromosome in a manner similar to the way in which normal λ becomes a prophage (Fig. 8.14). The host now has two copies of the *gal* region on its chromosome, a condition that is referred to as partial diploidy. If the recipient cell is *gal*$^-$ (unable to metabolize the sugar galactose) and the donor cell is *gal*$^+$ (able to metabolize galactose), the researchers can detect the transduction experimentally by observing the conversion of the recipient cell to the gal$^+$ phenotype.

In contrast with phage λ, in which transduction is restricted to genes that are next to the prophage site, almost any bacterial gene can be transferred by phage P22 in *Salmonella*. This system is an example of generalized transduction. Generalized transduction appears to involve transducing phages that carry very little or no phage DNA; instead, they consist of host chromosomal material, surrounded

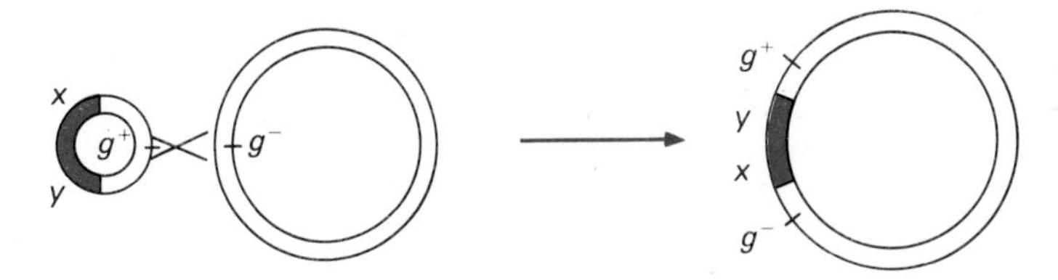

Figure 8.14. Integration of a λ*dgal* chromosome into a recipient bacterial chromosome. The *gal* loci provide a region of homology where crossing over can occur. The recipient chromosome then contains two copies of the *gal* locus. Phenotypically, it is a gal$^+$ transductant.

by a phage protein coat. The exact mechanism by which P22 transducing phages are formed is not known. It is apparently quite different from the mechanism that produces λ transducing phages. After induction in a cell that is lysogenic for P22, the bacterial chromosome becomes fragmented. One of these fragments is accidentally incorporated into a phage head in place of phage DNA (Fig. 8.15). Upon

Figure 8.15. Generalized transduction by phage P22. Following phage infection of a donor cell, the phage replicates progeny chromosomes and protein coats, and the bacterial chromosome is fragmented. Occasionally, a piece of bacterial chromosome is accidentally included in a phage head, to produce a transducing phage. When a recipient cell is infected by one of these transducing phages and a double crossover event occurs, the cell is transduced, in this case to a^-.

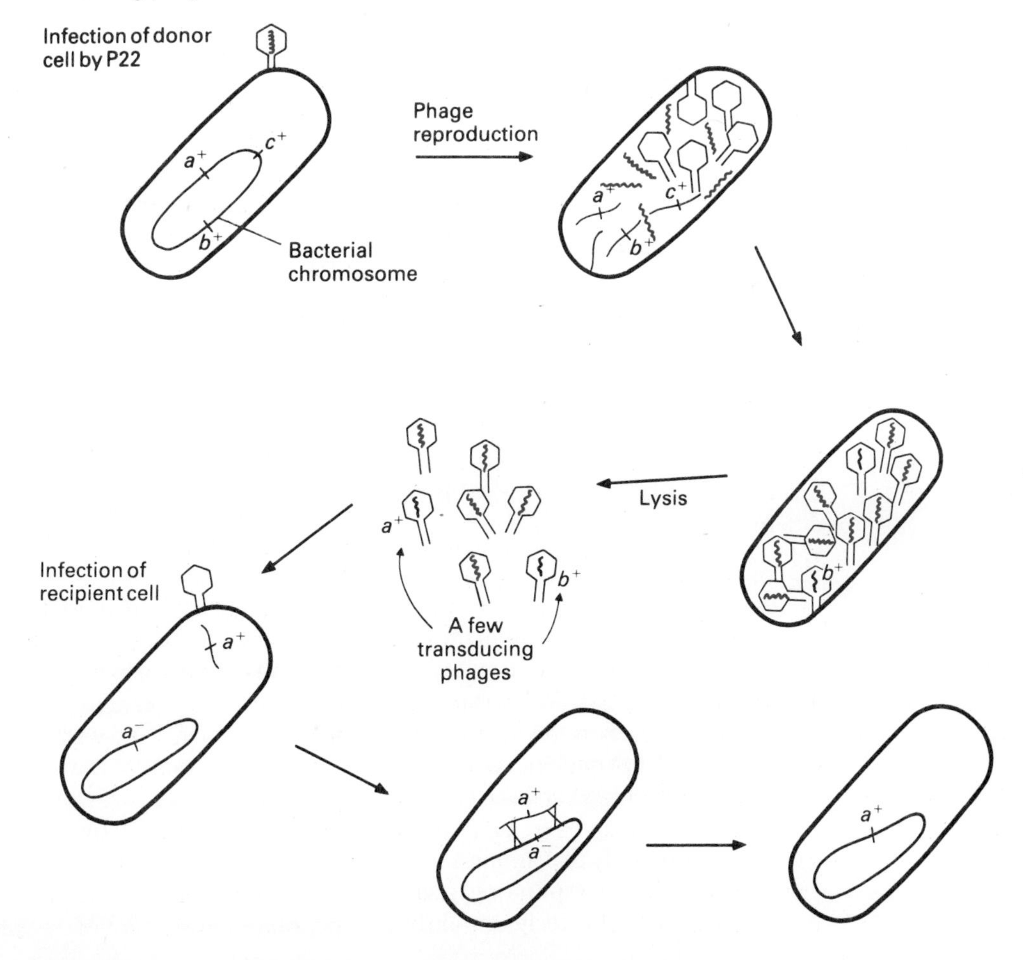

Table 8.1. A comparison of specialized and generalized transduction.

	SPECIALIZED TRANSDUCTION	GENERALIZED TRANSDUCTION
(A) Origin of the transducing phage.	Bacterial genes are acquired as an error in the deintegration of the prophage during induction.	Bacterial genes are acquired as an error in the assembly of the mature phage during lytic multiplication.
(B) Nature of the transducing phage.	The phage particle contains a chromosome in which some phage genes are missing and replaced by genes of the bacterial host.	The phage particle contains mainly (or all) bacterial genes on a chromosome fragment incorporated into the capsid by mistake.
(C) Nature of the donor genes transferred.	Restricted to bacterial genes that are normally located close to the site of prophage attachment.	Any set of closely linked bacterial genes may be transferred without restriction as to chromosomal location.
(D) Incorporation of the donor genes into the recipient chromosome.	Donor genes are incorporated along with phage genes when the chromosome of the transducing phage is inserted as a prophage in the recipient genome.	Donor genes are incorporated by crossing over between the donor fragment and a homologous region on the recipient chromosome.

entering the recipient cell, this donor fragment is then inserted into the recipient chromosome in place of allelic genes, through homologous pairing and recombination. Generalized transduction thus occurs through an exchange of genetic material in the recipient cell, in contrast with specialized transduction, which involves the addition of extra genetic material to the bacterial chromosome. The differences between specialized and generalized transduction are summarized in Table 8.1.

Recombination Mapping. In principle, the process by which a recipient cell is converted to a stable recombinant is the same in conjugation, transformation, and generalized transduction. In every case, the recipient of chromosome transfer is initially a partial diploid, or **merozygote,** having its own chromosome plus a chromosome fragment from the donor cell. The segment of donor chromosome within the recipient cell is called the **exogenote,** which may now synapse with the homologous section of the recipient chromosome, known as the **endogenote** (Fig. 8.16). Crossing over then yields recombinant bacterial genotypes. Although a double crossover is the simplest mechanism by which part or all of the exogenote can replace the corresponding segment on the endogenote and thus be stably incorporated into the recipient chromosome, any *even number* of exchanges will

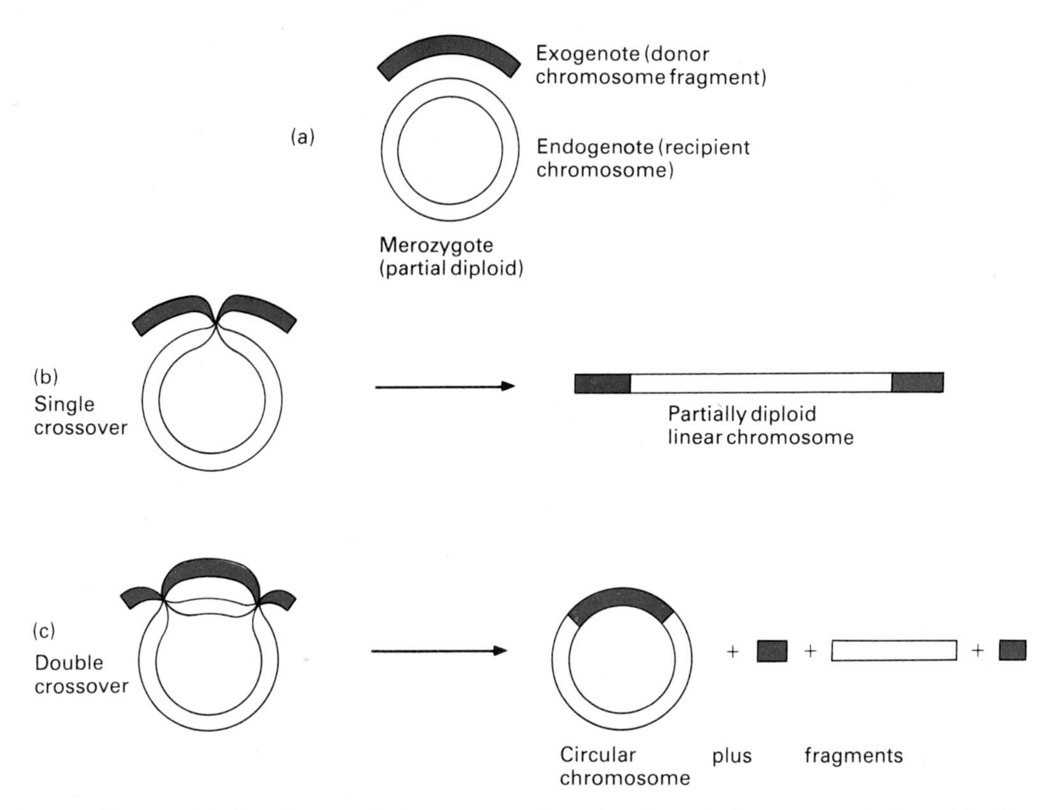

Figure 8.16. Crossing over between exogenote and endogenote in a merozygote formed by the transfer of part of the donor chromosome during conjugation, transformation, or transduction. Synapsis between exo- and endogenotes occurs in (a). In (b), a single crossover leads to a partially diploid linear chromosome, which would be lethal. (c) A double (or other even-numbered) crossover event must occur in order to keep the circular chromosome intact. This recipient is now recombinant, having exchanged a section of its chromosome for the homologous section of the donor chromosome. The remaining chromosome fragments are lost from the cell upon division.

suffice. A single or other odd-numbered exchange would break open the recipient chromosome to produce a linear chromosome with extra donor sections at both ends. Such a chromosome would be lethal.

Only bacterial genes that are closely linked can be carried on the same chromosome fragment during transduction or transformation. Physical proximity does not guarantee cotransfer, however. In transduction, the linkage between genes can be broken during the packaging of a head-sized piece of bacterial DNA into the protein coat of the transducing phage. In transformation, linkage can be broken if the transforming DNA is fragmented during its release from a lysed donor cell. Closely linked genes may also be separated by crossing over during the incorporation of the donor fragment into the recipient chromosome. For example, suppose that homologous pairing of the donor fragment with the recipient chromosome yields the following temporary diploid condition:

a^+ b^+ c^+ Exogenote

(1) (2) (3) (4)

... a b c ... Endogenote

The numbers indicate potential crossover intervals. If only two crossovers occur, one in region 1 and the other in region 3, the linkage would be broken between *c* and the other two loci, since only the a^+b^+ segment would then be incorporated into the genome:

a^+ b^+ c^+ → a b c^+

a b c → a^+ b^+ c

Similarly, a crossover in region 2 and another in region 4 would separate gene a^+ from the others:

Of course, these examples illustrate only two of many possible outcomes, since any even-numbered set of crossovers can act to insert the donor genes.

Despite the many ways in which linkage might be broken, we can reason that the closer two genes are on the bacterial chromosome, the greater the likelihood that they will both be present on the same donor fragment and will be incorporated together into the recipient genome. Map distance is therefore inversely related to the frequency with which genes are jointly transferred; it can be calculated by subtracting the percentage of the cotransfer frequency from 100 percent, which is the value we would expect if the genes mapped at the same locus.

To illustrate how we may establish a linkage map from recombination data, let us consider a hypothetical case of generalized transduction. Assume that P22 phages are grown on a wild-type $a^+b^+c^+$ donor bacterium and are then used to transduce the mutant bacterial strain *a b c*. Suppose that among a^+ transductants, 90 percent were also transduced for b^+ and 70 percent were also transduced for c^+. In addition, suppose that 80 percent of the b^+ transductants were also transduced for c^+. Map distances in this case are therefore $100 - 90 = 10$ map units for *a–b*, $100 - 70 = 30$ map units for *a–c*, and $100 - 80 = 20$ map units for *b–c*. Thus, the linkage map would be

a —10— b —20— c

We can relate these map distances to the time distances in interrupted conjugation by using the empirically derived relationship of approximately 20 percent recombination units to 1 minute of transfer time.

The level of resolution that is obtainable with recombination analysis is extremely high, enabling geneticists to do fine-structure mapping in bacteria in a manner similar to that described for bacteriophages. The results of some fine-structure studies are described in later chapters.

TO SUM UP

1. A phage cross involves the joint infection of a bacterium with parental phages of contrasting genotypes. The frequency of recombinant-type phages among the progeny is used as a measure of the distance between linked genes.

2. Some phages, such as lambda, have linear maps, while others, such as T2 and T4, have circular maps. Circular maps in bacterial viruses do not normally indicate that the chromosome itself is circular; rather, they are a result of a circularly permuted gene arrangement.
3. Phages are used for the detailed mapping of mutation sites within a gene locus. Maps of this sort are called fine-structure maps and are derived from the results of crosses between phages with mutational changes at different base pairs within the same gene.
4. To facilitate the construction of a fine-structure map, geneticists subdivide a gene into smaller intervals based on the location, length, and overlap of the DNA segments missing in deletion mutants. Single-site mutations can then be quickly mapped into defined intervals by crossing the mutant strains with each of the overlapping deletion mutants.
5. The interrupted mating technique is a method used to control the duration of mating in bacteria by artificially separating conjugating cells. The data obtained with this procedure yield a linkage map that shows the minimum time of entry of each gene into the recipient cell. Linkage data obtained in this way from several different Hfr strains give a circular composite map of the bacterial chromosome.
6. Transduction is the transfer of bacterial genes from donor to recipient cells by means of a temperate phage. In specialized transduction, the genes that are transferred are restricted to a region in the vicinity of the prophage; in generalized transduction, the genes may be from any single area throughout the bacterial chromosome.

Analysis of Extranuclear Genes in Eukaryotes

In Chapter 4 we learned that DNA is present within the mitochondria and chloroplasts of eukaryotic cells. The DNA of these organelles is packaged into chromosomes whose size and structure vary from those of chromosomes within the nucleus. Organelle chromosomes are also transmitted into the daughter cells at cell division in a way that is different from nuclear chromosomes. As a consequence, the genes on the organelle chromosomes show an inheritance pattern that differs from the simple Mendelian pattern of nuclear genes. The pattern of inheritance that is shown by organelle genes is known as **cytoplasmic inheritance,** in order to distinguish it from inheritance patterns that we associate with the nucleus. In addition to organelles, cytoplasmic inheritance is characteristic of certain intracellular parasites and symbionts that have their own genetic material.

Cytoplasmic Inheritance

The existence of extranuclear genes clearly complicates the analysis of genetic characteristics. Not only must we establish that a trait is genetically caused, but we must also determine whether the trait, if inherited, is the expression of nuclear or cytoplasmic genes. When attempting to demonstrate a cytoplasmic basis for inheritance, geneticists usually start by ruling out the possible role of nuclear genes. For example, **nuclear transplantation** can be performed with the eggs of certain animals such as frogs and toads. By using a fine needle, the scientist can remove the nucleus from an egg and replace it with one from another egg. When the egg with the nuclear transplant develops, the researchers can then determine the differential effects of the cytoplasm and nucleus on the trait in question.

Table 8.2. Differences between extranuclear and nuclear inheritance patterns.

	NUCLEAR GENES	EXTRANUCLEAR GENES
(A) Relative genetic contribution of the parents.	Male and female parents make equal genetic contributions to their offspring.	The female makes a greater genetic contribution to the offspring than the male parent.
(B) Reciprocal cross results.	With the exception of sex linkage, reciprocal crosses give the same progeny results.	Reciprocal crosses yield different progenies.
(C) Segregation ratios.	Mendelian segregation ratios are observed, based on the behavior of chromosomes during meiosis.	Non-Mendelian segregation ratios are observed.
(D) Linkage and mappability.	A gene can be mapped to a nuclear chromosome at a particular position relative to other nuclear genes.	A gene shows no linkage to known nuclear genes, but can often be mapped to an organelle chromosome.

Another technique for establishing a cytoplasmic basis for inheritance is the **heterokaryon test,** in which cells of different genotypes are fused to form heterokaryons. This test is particularly useful in studies on the filamentous fungi, such as *Neurospora*. The technique involves fusing a cell with a known nuclear genetic marker with a cell that has a genetic marker that is believed to be carried in the cytoplasm. The investigators then select uninucleate progeny cells (or spores) derived from the heterokaryon and test each spore for the presence of both genetic markers. You may recall that in heterokaryons, the two cytoplasms mix but the nuclei remain separate, so that different nuclear genes cannot recombine. Therefore, the only way a uninucleate cell can acquire and express both markers is if the gene whose location is unknown is indeed present in the cytoplasm.

But usually a geneticist must rely on less direct methods, such as genetic crosses, for identifying extranuclear genes. When this is the case, the criteria listed in Table 8.2 are then useful in distinguishing between nuclear and cytoplasmic inheritance. We will now briefly discuss the kinds of observations that suggest a cytoplasmic basis for transmission.

Maternal Inheritance. In many kinds of plants, chloroplasts are transmitted from generation to generation by the egg and not by the pollen. These plants will tend to show an inheritance pattern for chloroplast genes that is known as **maternal inheritance.** In maternal inheritance, the trait is inherited exclusively through the female parent, with no contribution by the male. For example, certain plants, including many common house plants, exhibit a characteristic known as variegation. Variegated plants, such as the ornamental four-o'clock (*Mirabilis jalapa*) shown in Fig. 8.17, have the appearance of yellow or white patches or streaks on an otherwise green leaf or stem. The green coloration comes from the

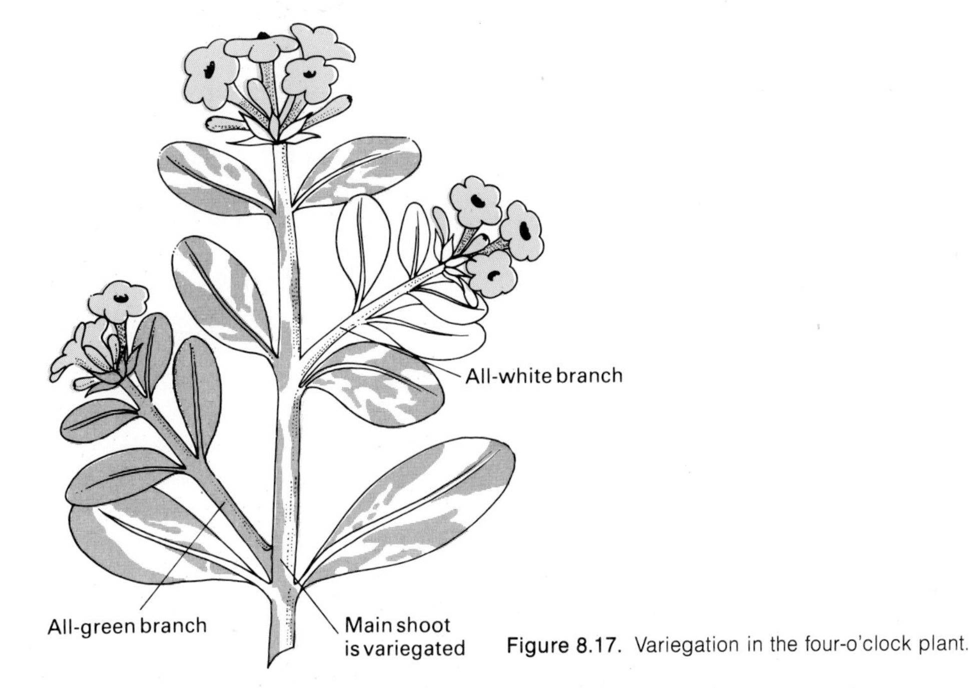

Figure 8.17. Variegation in the four-o'clock plant.

presence of normal chloroplasts that contain chlorophyll. The cells within a white region contain only mutant proplastids that lack chlorophyll; these cells would die were it not for the transfer of nutrients from the green regions of the plant that do contain chlorophyll. If a plant embryo contains a mixture of normal chloroplasts and mutant proplastids, some of the cells that are produced by division as the plant grows will receive normal chloroplasts and will produce green branches, while others will receive only proplastids and will produce white branches, and still others will inherit both types of plastids, giving rise to variegated branches.

One of the pioneering studies on the inheritance of variegation was conducted on the four-o'clock by C. Correns in the early 1900s. The results of his studies are

Table 8.3. Progeny resulting from crosses between different phenotypes of the four-o'clock plant.

POLLEN FROM BRANCH OF TYPE	POLLINATED FLOWERS ON BRANCH OF TYPE	PROGENY GROWN FROM SEED
white	white	white
	green	green
	variegated	white, green, variegated
green	white	white
	green	green
	variegated	white, green, variegated
variegated	white	white
	green	green
	variegated	white, green, variegated

summarized in Table 8.3. Notice that the phenotype of the progeny depends only on the phenotype of the female (or seed-bearing) parent and not on the male (or pollen-bearing) plant. Thus, ovules derived from green portions of the plant produce only green progeny, ovules derived from white branches yield only white offspring, and ovules derived from variegated branches give rise to all three progeny types, regardless of the source of the pollen.

We should point out, however, that an inheritance pattern that shows a strong maternal influence does not always mean that the causative gene is located in the cytoplasm. Since the female contributes much more cytoplasm to the zygote than does the male, maternal substances of nuclear origin that are already present in the egg at fertilization, such as RNA molecules, can influence the phenotype of the offspring. In such instances, the maternal effect would depend on the nuclear genotype of the mother and not on the cytoplasmic genes. We should also note that traits that show cytoplasmic inheritance in one case may demonstrate nuclear inheritance in another. Variegation provides an example of this difference. The recessive gene *j* on chromosome 8 in corn produces a green and white striped appearance in homozygous *jj* plants. Unlike variegation in the four-o'clock, maternal inheritance is not the pattern in this instance, since the cross *JJ* × *jj* yields only green offspring, regardless of whether the variegated *jj* parent is the egg or the pollen donor. Moreover, the cross *Jj* × *Jj* results in the classical 3 green (*J-* : 1 stripped (*jj*) ratio that is expected for the segregation of nuclear genes.

EXAMPLE 8.4. The freshwater snail *Limnaea peregra* is a polymorphic species. Some individuals exhibit right-handed (dextral) coiling of the shell, while others have left-handed (sinistral) coiling. This characteristic is completely determined by the nuclear genotype of the mother. The mother's genotype establishes the orientation of the spindle that develops in metaphase during the first cleavage division following the formation of the zygote. The spindle orientation, in turn, influences further cell division and determines the coiling pattern. Since only the mother's genotype, and not her phenotype or any cytoplasmic factor, is involved in controlling the coiling pattern in the offspring, this trait is said to show a **maternal effect,** as opposed to maternal inheritance. It is not an example of extranuclear inheritance, even though some of the characteristics of the phenotypic ratios in the progeny mimic those obtained with cytoplasmic transmission. A single pair of alleles is involved in the transmission of this trait: s^+ = dextral and s = sinistral coiling, with s^+ dominant to s. If females from a true-breeding dextral line are crossed with males from a true-breeding sinistral line, **(a)** what are the expected genotypic and phenotypic ratios in the F_1? In the F_2? **(b)** Because these snails are hermaphroditic, they are capable of both cross- and self-fertilization. If each genotypic class in the F_2 is inbred, what will be the resulting genotypic and phenotypic ratios in the F_3?

Solution. **(a)** The starting cross is s^+s^+ female × ss male. The F_1 are all genotypically s^+s and all exhibit dextral coiling, like their mothers. F_1 intercrosses yield a genotypic ratio of 1 s^+s^+ : 2 s^+s : 1 ss in the F_2; they have a phenotypic ratio of all dextral, based on their mother's genotype. **(b)** Inbreeding of the s^+s^+ and s^+s F_2 classes yields only dextral progeny, whereas inbreeding of the ss class yields all sinistral F_3.

Differences in Reciprocal Cross Results. In maternal inheritance, we observed that the maternal parent has the deciding influence on the phenotypes of the offspring. Thus for two alternative traits, say *A* and *B*, we might expect to see cross results of the following type: *A* female × *B* male → all *A* progeny, and *B* female × *A* male → all *B* progeny. Maternal inheritance is therefore characterized by a difference in the results of reciprocal crosses. This characteristic of cytoplasmic inheritance also extends to organisms in which there are no observable physical differences between mating types. The single-celled alga *Chlamydomonas reinhardi* provides an example (see Fig. 8.18). *Chlamydomonas* does not have sexes; instead, there are two mating types that appear to be identical under the light microscope.

Figure 8.18. The life cycle of the green alga *Chlamydomonas reinhardi*.

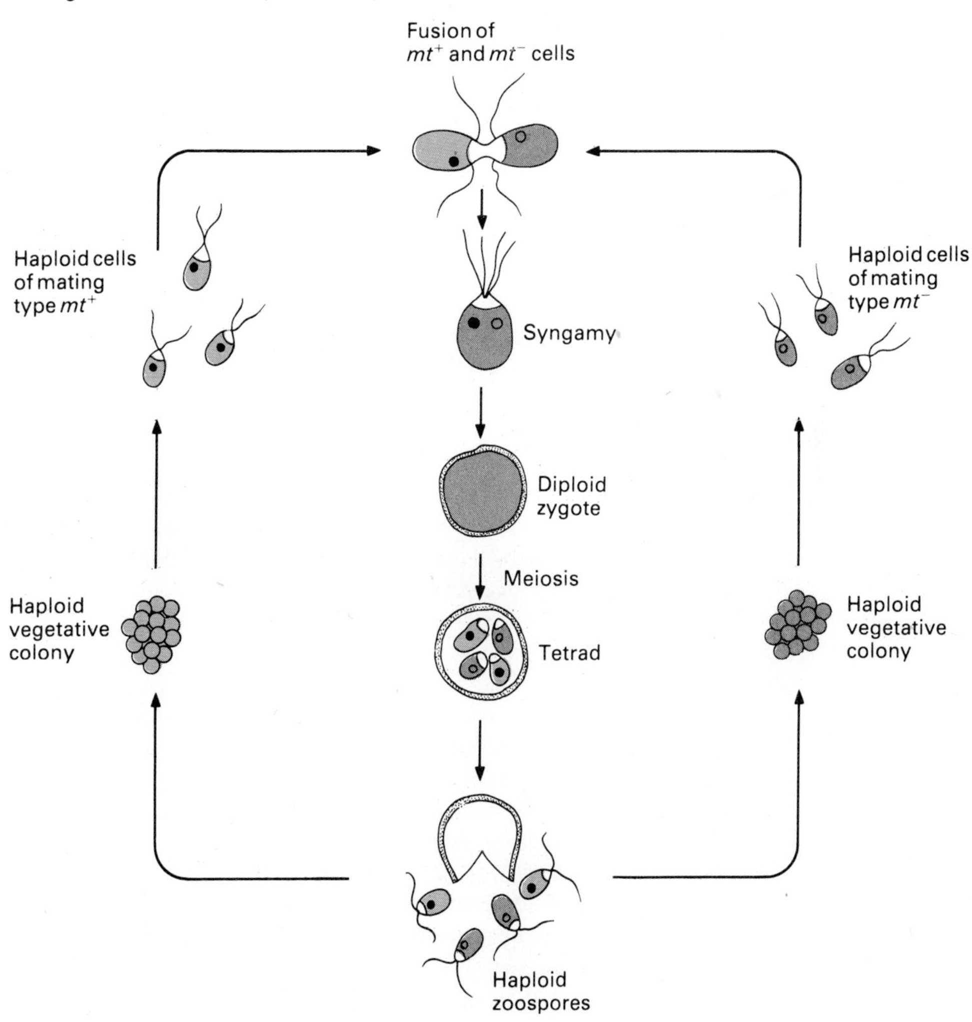

When two haploid cells of opposite mating type fuse, they produce a diploid zygote. The zygote then matures and undergoes meiosis to produce four haploid cells, completing the life cycle. The two mating types of *Chlamydomonas* are determined by a pair of alleles designated mt^+ and mt^-. The alleles are present on nuclear chromosomes, so that the cross $mt^+ \times mt^-$ yields a 1:1 ratio of the two mating types in the progeny.

Various kinds of antibiotic resistance in *Chlamydomonas* result from effects on the chloroplast genome. One such case involves the alleles for streptomycin resistance (sm^r) and sensitivity (sm^s). When crosses are made between sensitive and resistant strains, we observe the following pattern of uniparental inheritance:

$$sm^s\ mt^+ \times sm^r\ mt^- \longrightarrow \text{streptomycin-sensitive progeny,}$$

$$sm^r\ mt^+ \times sm^s\ mt^- \longrightarrow \text{streptomycin-resistant progeny.}$$

The mt^+ mating type is thus behaving in a manner analogous to the female parent in cases of maternal inheritance. Studies have shown that this inheritance pattern is due to the exclusion of mt^- chloroplast genes from the progeny because the mt^- chloroplast DNA undergoes enzymatic breakdown after the two cells mate. Only the mt^+ chloroplast genes are then expressed in the offspring.

Non-Mendelian Segregation Ratios. Certain organisms show a pattern of transmission for cytoplasmic genes that is independent of mating type. In these cases, cytoplasmic inheritance is recognized by the non-Mendelian segregation pattern for alleles. Common baker's yeast (*Saccharomyces cerevisiae*) provides an example. Sexual reproduction in yeast is like reproduction in *Chlamydomonas,* in that diploid zygotes are formed by the fusion of structurally similar cells of opposite mating types (Fig. 8.19). In yeast, however, both mating types make an equal contribution to the organelle genes of the offspring.

One characteristic that has been studied intensively in yeast is the so-called *petite* mutation. Petite yeast form very small colonies when grown on agar medium, as compared to wild-type or *grande* yeast. Petite mutants lack functional mitochondria; as a consequence, they are deficient in respiratory function, because they are unable to produce ATP through oxidative phosphorylation. Some ATP is provided by cellular fermentation, but not enough to produce the larger colonies that are characteristic of the wild-type cells. Genetic studies have succeeded in distinguishing three types of petites, two of which show cytoplasmic inheritance, while the other shows a nuclear inheritance pattern. The first two types are referred to as *cytoplasmic petites,* to distinguish them from the nuclear or *segregational petites.* The two types of cytoplasmic petites are designated *neutral petites* and *suppressive petites.* Neutral petites have lost all mitochondrial DNA and thus retain no mitochondrial genetic markers. Suppressive petites, in contrast, have mitochondrial DNA but this DNA has large deletions of random parts of the genome.

The two types of cytoplasmic petites can be distinguished from segregational petites and from each other on the basis of cross results (see Fig. 8.20). When crossed to a wild-type or grande cell, the segregational petite produces a tetrad of haploid ascospores with a ratio of 2 grande : 2 petite. The cross of a cytoplasmic petite to a grande cell produces all grande ascospores in the case of neutral petite

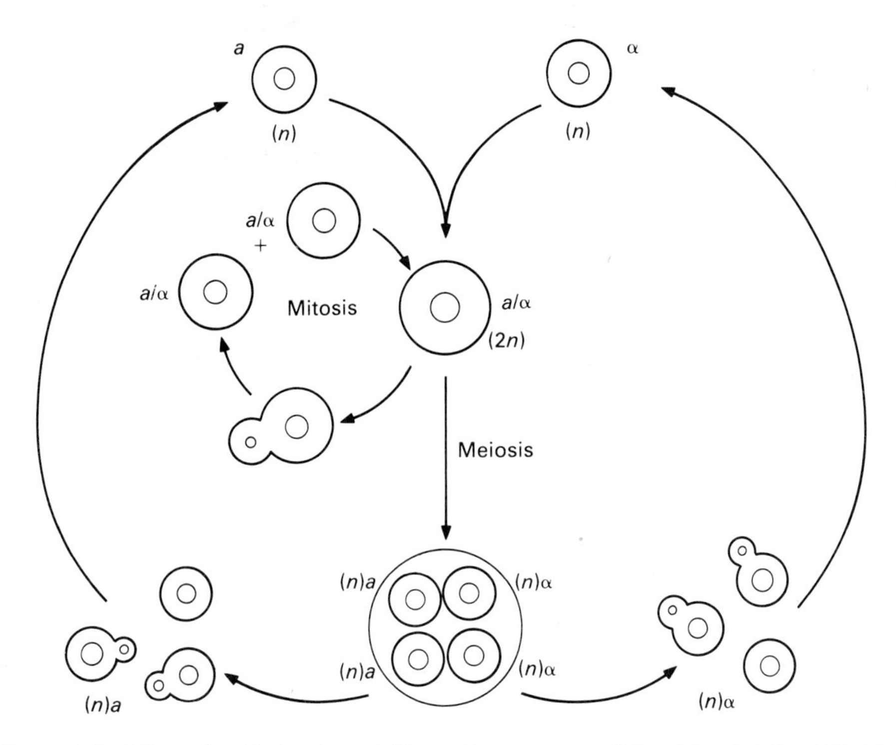

Figure 8.19. Life cycle of bakers' yeast. The nuclear a and α alleles determine the mating type. Fusion between cells of opposite mating types produces a diploid cell, which contains equal proportions of cytoplasm from each parental cell. Diploid cells normally reproduce by mitosis (budding). By plating on special growth medium, however, the fusion nucleus can be induced to undergo meiosis instead, producing four haploid products.

Figure 8.20. Results expected from crosses of grande (wild-type) yeast to various petite mutants. The cross involving segregational petite yields a Mendelian ratio of 2 grande : 2 petite offspring, which is a reflection of nuclear inheritance. In contrast, the crosses of grande × suppressive and grande × neutral petites yield progeny ratios that are more characteristic of cytoplasmic inheritance.

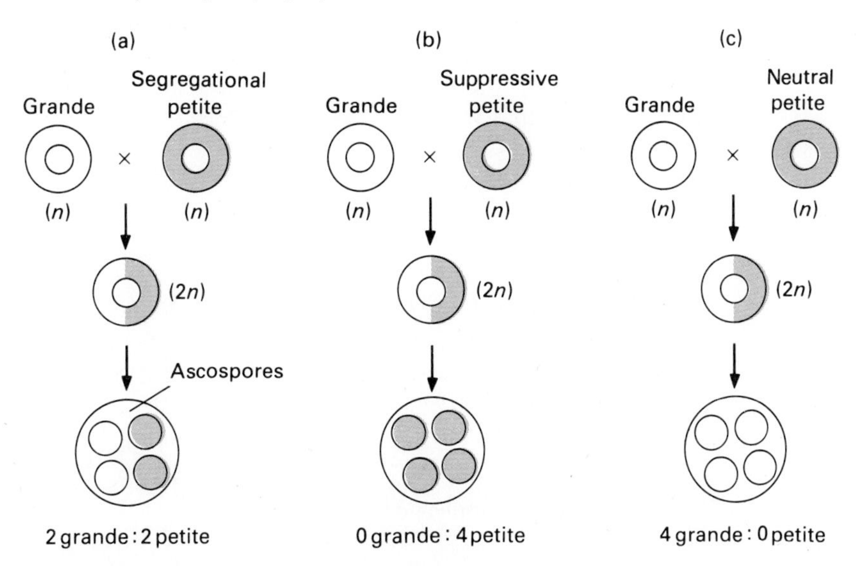

× grande, and all petite ascospores in the case of suppressive petite × grande. When cytoplasmic factors are involved in the cross then, the result is a 4:0 (or 0:4) tetrad ratio.

Nuclear-Cytoplasmic Interactions

As we mentioned earlier, some of the components of cytoplasmic organelles are coded for by cytoplasmic genes, and others are coded for by nuclear genes. We would therefore expect to find some inheritance patterns that involve an interaction between nuclear and cytoplasmic effects. One example of such an interaction is the recessive gene *iojap* (*ij*), which is located on chromosome 7 in corn. Corn plants that are homozygous for *ij* are variegated, with a green and white striped appearance that is similar to the one produced by the *j* gene described earlier. When a plant of this genotype serves as the male parent (or pollen donor) in the cross *Ij*/*Ij* female × *ij*/*ij* male, all of the offspring are green (Fig. 8.21). But when the reciprocal cross *Ij*/*Ij* male × *ij*/*ij* female is performed, using *ij*/*ij* plants as the female parent (or egg donor), three offspring types are produced: green, white, and variegated. The three types appear despite the presence of the dominant *Ij* allele in the heterozygous offspring. In fact, once the iojap trait is established, the *ij* allele is not even required for the continued expression of variegation, as seen by the F_2 results in Fig. 8.21. Even though the *ij* allele is needed initially to induce

Figure 8.21. Results of reciprocal crosses between homozygous green and variegated plants. (a) When the male is variegated, the cross gives the results that are expected for nuclear inheritance. (b) When the female is variegated, however, the results are those that are expected for cytoplasmic inheritance. The conclusion is that the iojap trait is the result of an interaction between nuclear and cytoplasmic determiners.

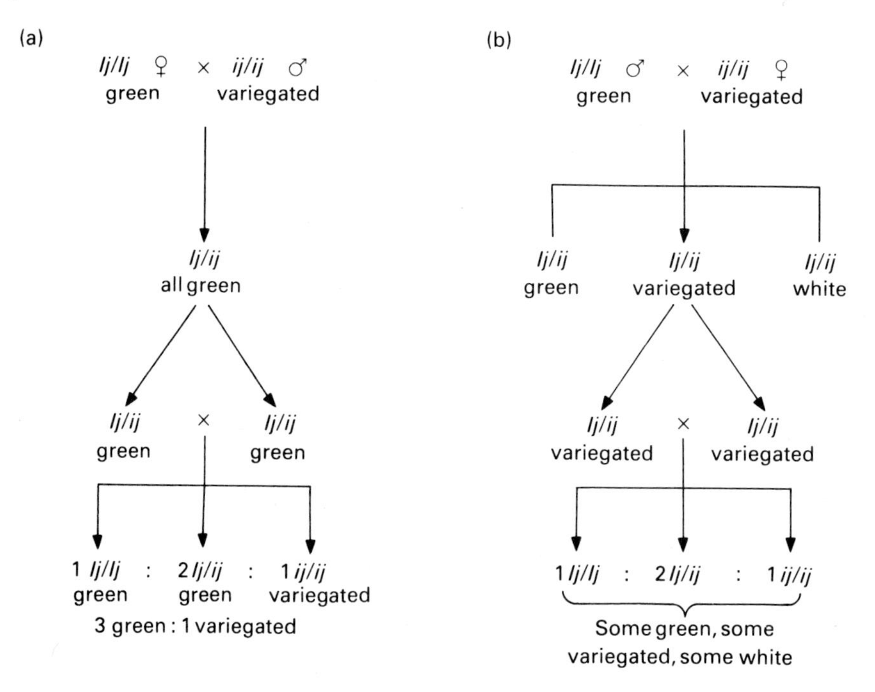

the conversion of some of the normal chloroplasts into mutant proplastids, the effect is permanent, so that the altered proplastids remain colorless. Therefore, once the altered proplastids are formed, they are no longer dependent on the genotype at the *ij* locus and exhibit the maternal inheritance that is expected for an organelle.

Another example of a nuclear-cytoplasmic interaction is **cytoplasmic sterility** in plants, in which a nuclear gene masks the effect of a cytoplasmic factor. Various nuclear genes can cause male sterility in plants by preventing the production of normal pollen. But in corn, onions, sorghum, and certain other plants, male sterility can also be caused by cytoplasmic factors. Cytoplasmic male sterility, as opposed to the sterility produced by nuclear genes, shows maternal inheritance and is transmitted by the female gamete.

The cytoplasmic factors that are involved in cytoplasmic male sterility do not act alone; rather, they interact with certain nuclear genes. For example, the dominant nuclear gene *R* in corn can restore fertility in plants with cytoplasmic male sterility. Corn with the *S* (male sterile) cytoplasm and the *rr* nuclear genotype (*Srr*) produce poorly developed anthers without pollen, whereas plants with an *F* (male fertile) cytoplasm or an *R-* nuclear genotype or both (*FR-*, *SR-*, or *Frr*) produce normal pollen. Plant breeders have used this nuclear-cytoplasmic interaction extensively in the commercial production of hybrid seed. In corn, highly vigorous hybrids are produced commercially by crossing less vigorous homozygous varieties. To ensure cross pollination and to avoid selfing, the variety chosen as female parent in the cross must either be detasseled or have cytoplasmic male sterility. If one of the homozygous varieties is male sterile (*Srr*) and the other variety is fertile (e.g., *FRR*), hybrid *SRr* seed can be obtained that will develop into fertile plants. An example of the procedure used by plant breeders to produce hybrid corn commercially is shown in Fig. 8.22. We will consider this crossing method further in Chapter 16.

Mapping Mitochondrial Genes in Yeast

Mitochondrial genes have been mapped in yeast, taking advantage of the mitochondrial chromosome deletions that are present in suppressive petites. Various genetic markers on the mitochondrial chromosome have been used; these markers include resistance to a number of different antibiotics and certain point mutations that result in the loss of specific respiratory functions (designated mit^-).

One mapping technique that has proved useful in yeast is *mapping by marker retention*. In this procedure, the investigators use a mutagen such as ethidium bromide to induce random deletions in the mitochondrial DNA of a multiple drug-resistant strain of yeast, so as to produce various suppressive petites. Each petite strain produced in this way is then tested to determine which resistant markers are retained (Fig. 8.23). Since the deletions are randomly induced, it is therefore likely that different drug-resistant genes will be lost in different strains. The deletions can be quite extensive, so that the simultaneous loss of more than one resistance marker is possible. Thus, a genetic analysis of these strains might yield results similar to those given in Fig. 8.24. Note that by combining the data on

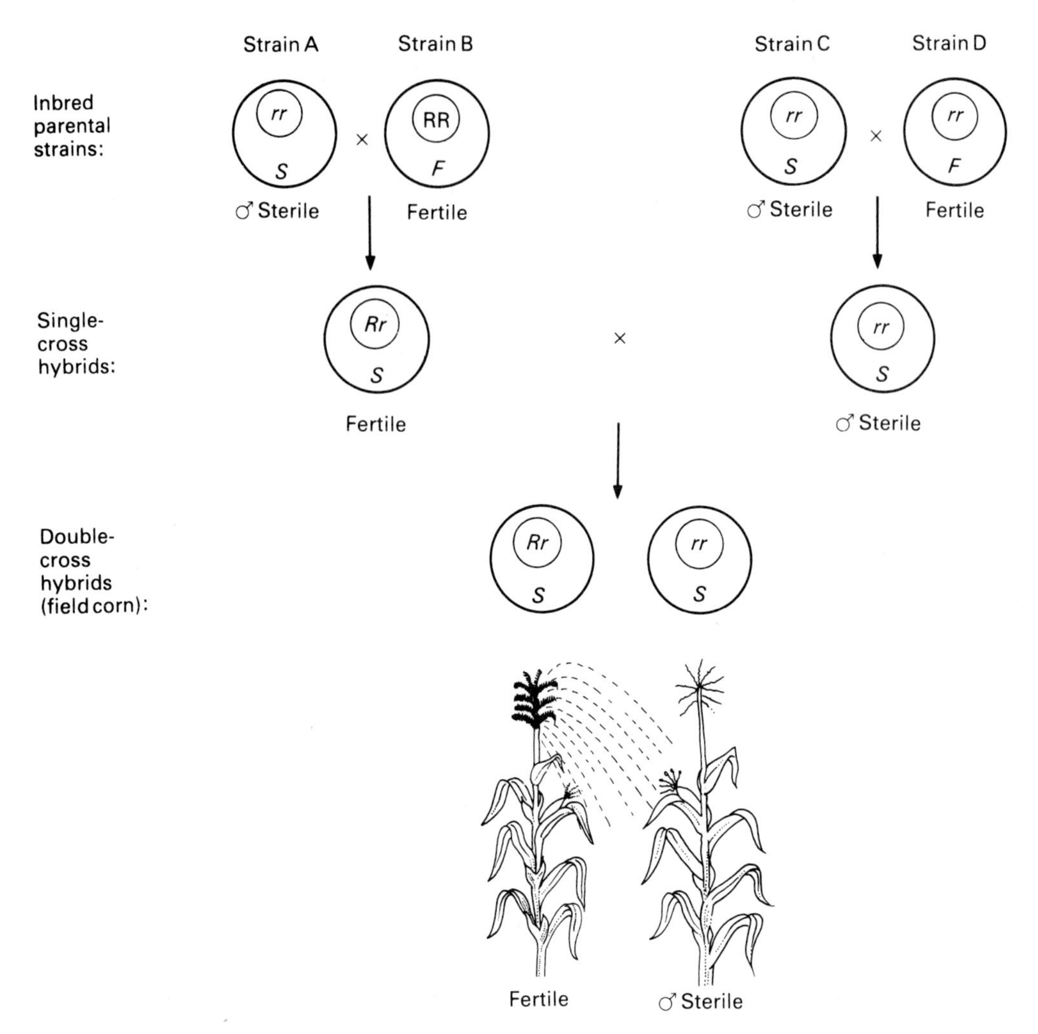

Figure 8.22. Procedure used in the commercial production of hybrid seed for corn. Type *S* cytoplasm yields cytoplasmic male sterility; the nuclear gene *R* restores fertility in type *S* plants. The fertile single-cross hybrid (*SRr*) is sold to the farmer as seed. In the past, prior to the development of many of the modern farming methods and the associated high corn yields, the double-cross hybrid seed was planted by the farmer. It is produced by crossing the single-cross hybrids. The double-cross plants produce sufficient pollen to fertilize all plants, even though some are male-sterile. Further information related to this crossing method will be presented in Chapter 16.

marker retention from these different strains, we get a series of overlapping segments that can be arranged to form a circular map. Maps of mitochondrial chromosomes in yeast and humans will be given in Chapter 17. These maps were obtained by various experimental methods (including techniques similar to those just described and newer molecular techniques that will be discussed in Chapter 17) and are circular, like the structure of the mitochondrial chromosome.

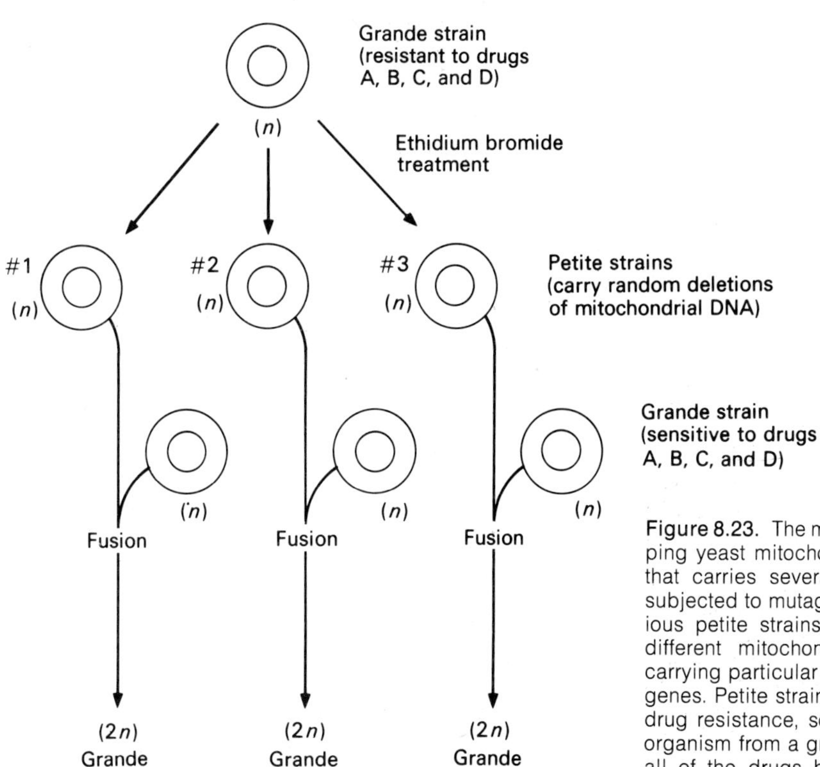

Figure 8.23. The marker retention method of mapping yeast mitochondrial genes. A grande strain that carries several drug-resistance markers is subjected to mutagen treatment. The result is various petite strains, each of which has retained different mitochondrial chromosome segments carrying particular combinations of drug-resistant genes. Petite strains cannot be directly tested for drug resistance, so each petite is fused with an organism from a grande strain that is sensitive to all of the drugs being tested. The retention of resistance to a certain drug by the petite is then measured in terms of the resistance acquired by the diploid fusion product as a result of recombination between the sensitive grande and the resistant petite.

Figure 8.24. (a) Hypothetical data resulting from a marker retention analysis. Each petite strain has retained a particular combination of resistance markers. (b) By combining the data on marker resistance from these different strains, we obtain a group of overlapping chromosome segments that fit together into a circular map.

(a)

\+ marker retained
− marker lost

Petite strain	Mitochondrial genotype (markers retained)				
	A	*B*	*C*	*D*	*E*
1	+	+	+	−	−
2	+	+	−	−	+
3	−	−	−	+	+
4	−	−	+	+	−

(b) Map:

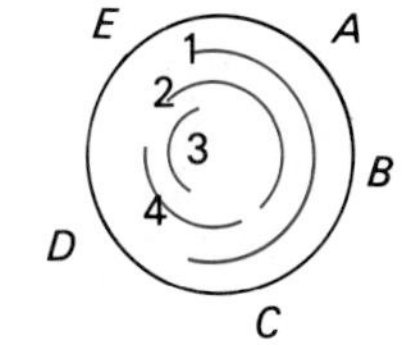

TO SUM UP

1. Mitochondria and chloroplast genes exhibit the characteristics of extranuclear inheritance, including a difference in the results of reciprocal crosses and non-Mendelian ratios. Both of these phenomena are due to the inheritance of the genes through the cytoplasm, rather than through the nucleus.
2. The genes of cytoplasmic organelles sometimes exhibit complex inheritance patterns, in which the phenotypic effects are influenced by genes in the nucleus. This interaction between cytoplasmic and nuclear genes is largely a result of the inability of organelles to code for all of their structural and functional proteins. The organelles must rely on the nucleus for many of their component parts.
3. Organelle genes recombine and segregate in a manner that shows linkage, thus enabling geneticists to construct chromosome maps of the organelle genomes. The linkage map of mitochondria is circular, which agrees with the known structure of the mitochondrial chromosome.

Questions and Problems

1. Define the following terms, and differentiate between members of paired terms:
cytoplasmic inheritance–nuclear inheritance
cytoplasmic male sterility
cytoplasmic petites–segregational petites
deletion mutant
endogenote–exogenote
fine-structure mapping
interrupted mating technique
maternal inheritance
merozygote
mixed infection
neutral petites–suppressive petites
plaque
terminal redundancy
transduction

2. Bacteriophage particles are diluted to a concentration of 10^3 per ml. A 0.2-ml sample is added to 0.2 ml of bacteria sensitive to that phage, taken from a culture at a concentration of 10^8 bacteria per ml. The suspension is mixed, incubated until essentially all the phages have infected bacterial cells, and then spread over the surface of a nutrient agar medium in a petri dish. **(a)** How many phage particles were most probably mixed with the 0.2 ml sample of bacteria? **(b)** How many plaques are expected to appear on the agar surface after overnight incubation?

3. Assume two mutant strains of bacteriophage T2. The one designated *h* is unable to synthesize head protein, and the other, designated *t*, is unable to form tail protein. Mature forms of both mutant strains can inject their chromosomes into *E. coli*, but neither strain by itself can complete the reproductive cycle and produce lysis. When phages of both strains infect the same cell, lysis occurs with the release of the same number of progeny viruses as would normally be produced by wild-type T2. Most of the progeny of such mixed infections are genetically mutant, but about 6 percent are wild type, having the ability to complete the lytic cycle by themselves to produce progeny. **(a)** Explain these results in terms of the life cycle of bacterial viruses. **(b)** From the data given, estimate the distance between the *h* and *t* loci.

4. The chromosome of a temperate phage with characteristics similar to phage lambda has the gene arrangement $\underline{a\ b\ c\ d\ e}$ when replicating in the lytic cycle. The corresponding gene arrangement in the prophage of this temperate virus is $\underline{c\ b\ a\ e\ d}$. Explain how these gene arrangements are related, indicating the manner in which the prophage map is derived from that of the vegetative phage.

5. Four different Hfr strains of *E. coli* transfer their genes during conjugation in different sequences. After each strain mates with F^- cells, the following results are obtained:

HFR STRAIN	GENE ORDER
1	*P D X A L*
2	*A L U R T*
3	*P W Q C M*
4	*Q C M T R*

Determine the sequence of genes on the chromosome of the F^+ strain from which these Hfr strains were derived.

6. Four different Hfr strains of *E. coli* transfer their genes in the order given below:

HFR STRAIN					
1	Gene markers:	*A*	*B*	*C*	*F*
	Time (minutes):	4	54	72	79
2	Gene markers:	*D*	*E*	*F*	*G*
	Time (minutes):	7	16	67	80
3	Gene markers:	*E*	*D*	*A*	*G*
	Time (minutes):	11	20	35	47
4	Gene markers:	*F*	*B*	*E*	
	Time (minutes):	2	27	53	

Construct a linkage map of the *E. coli* chromosome from these results, using the entry times to calculate the distances in transfer time between the adjacent pairs of loci.

7. There is a particular type of slow-growing mutant in *Neurospora* called *poky*. Crosses between wild-type females and poky males yield all wild-type offspring, whereas crosses between poky females and wild-type males yield all poky progeny. **(a)** What seems to be the inheritance mechanism of the poky characteristic? **(b)** Following the formation of a heterokaryon between arg^+ *poky* and arg^- $poky^+$ cells, about one-fourth of all the uninucleate spores from the heterokaryon are totally wild-type (arg^+ $poky^+$). Is this result consistent with your answer to the previous question? Explain.

8. In barley, virescent leaves are the result of either a cytoplasmic factor (L_1 = normal leaves, L_2 = virescent leaves) or the recessive nuclear gene *v* (*vv* = virescent leaves). What genotypic and phenotypic results would you expect to obtain from each of the following crosses? **(a)** Pure-breeding normal female × L_1vv male. **(b)** L_1vv female × pure-breeding normal male. **(c)** Pure-breeding normal female × L_2vv male. **(d)** L_2vv female × pure-breeding normal male. **(e)** Female F_1 from cross (a) × male F_1 from cross (d). **(f)** Male F_1 from cross (a) × female F_1 from cross (d).

9. In onions, male sterility is due to an interaction between a chromosomal gene pair (*ms/ms*) and a male-sterile cytoplasm (*S*). Only *S ms/ms* plants are male sterile. Starting with true-breeding lines, outline a method of producing hybrid F_1 seed for the commercial crop.

10. Consider the interaction between the dominant nuclear restorer gene *R* and cytoplasmic male sterility in corn. **(a)** If a cytoplasmically sterile plant that is homozygous for nuclear gene *R* is crossed to a male-sterile plant, what genotypic and phenotypic ratios would be expected in the F_1 progeny? **(b)** If these F_1 plants are selfed, what genotypic and phenotypic ratios would be expected in the F_2?

11. In a certain strain of barley, foliage can be green or yellow, and ears can be long-awned or short-awned. When a true-breeding green, long-awned plant is fertilized by pollen derived from a true-breeding yellow, short-awned plant, all of the offspring are green and long-awned. Intercrosses among these F_1 yield F_2 consisting of 624 green, long-awned plants and 290 green, short-awned plants. In contrast, when yellow, short-awned plants are pollinated by green, long-awned plants, the F_1 are all yellow and long-awned, and the F_2 consist of 648 yellow, long-awned plants and 310 yellow, short-awned plants. Explain the genetic bases for these color and ear traits.

12. How many different two-factor phage crosses are needed to map unambiguously **(a)** four mutant sites? **(b)** Five mutant sites? **(c)** *N* mutant sites? (*Hint:* Consider the different combinations of genetic markers, taken two at a time.)

13. You have discovered a form of gene transfer between two strains of bacteria. Assume that you have access to an enzyme that can catalyze the degradation of isolated (naked) bacterial DNA and you also have a device that allows you physically to separate the different bacterial strains without interrupting the free flow of smaller materials (e.g., viruses, chromosome fragments) between them. Explain how you could differentiate between conjugation, transduction, and transformation as the possible basis for this transfer of genetic material.

14. Generalized transducing phages obtained from $a^+b^+c^+$ donor bacteria were used to infect recipient cells of genotype $a^- b^- c^-$. Transductants for a^+ were selected and later tested for the cotransfer of the other genes. Of a total of 500 a^+ transductants examined, 265 were b^+c^+, 165 were b^-c^+, 5 were b^+c^-, and 65 were b^-c^-. **(a)** Calculate the cotransduction frequencies for the markers a^+ and b^+ and for a^+ and c^+. **(b)** Determine the order of the three loci on the bacterial chromosome.

15. When Hfr cells conjugate with F^- cells that are lysogenic for lambda, the recipient cells usually survive. But when Hfr cells that are lysogenic for λ conjugate with F^- nonlysogens, the recipient cells lyse after a short period of time. This phenomenon is known as zygotic induction. **(a)** Explain the basis of zygotic induction. **(b)** How could you determine the chromosomal locus of the integrated prophage?

16. In pneumococci, strain 1 has resistance to drugs A, B, C, and D, because of genes *a, b, c,* and *d,* respectively. Strain 2 is sensitive to all of these drugs, because it carries the wild-type alleles of the genes. DNA from strain 1 is used to transform strain 2, and the recipient cells are then plated on a nutrient agar medium containing various combinations of the four drugs. The following results are obtained:

DRUGS ADDED TO GROWTH MEDIUM	NUMBER OF COLONIES
none	10,000
A	1,177
B	1,152
C	1,196
D	1,183
A, B	420
A, C	31
A, D	710
B, C	39
B, D	580
C, D	26
A, B, C	22
A, B, D	410
A, C, D	18
B, C, D	16
A, B, C, D	20

(a) Which three of the four genetic markers are so closely linked as to be normally present on the same DNA fragment? **(b)** Give the sequence of these loci and calculate the map distances, based on the frequency of cotransfer.

17. If female snails from a sinistral line are crossed with males from a true-breeding dextral line (the reciprocal cross of the one in Example 8.4), **(a)** What are the expected genotypic and phenotypic ratios in the F_1? **(b)** What are the expected genotypic and phenotypic ratios in the F_2? **(c)** If each F_2 genotypic class is inbred, what genotypic and phenotypic ratios are expected in the F_3?

18. There are three genetic mechanisms that can cause reciprocal crosses to give different results: **(a)** sex linkage, **(b)** extranuclear inheritance, and **(c)** maternal effects. How could you distinguish experimentally among these three alternatives?

19. In his fine-structure mapping studies, Benzer observed a minimum recombination frequency of 0.01% between mutant sites. Assuming 2000 nucleotides as the length of the *r*II locus (which has a genetic distance of 8 map units), calculate the minimum number of nucleotides in a recon, which is defined as the smallest unit of recombination.

20. Three point mutants were tested against the deletion mutants described in Example 8.3 through paired crosses. The results of these crosses are given below. As in Example 8.3, a + indicates that wild-type recombinants were observed. What is the relative order of the point mutants?

		DELETION MUTANTS			
		1	2	3	4
POINT MUTANTS	*a*	+	−	−	+
	b	+	+	−	−
	c	−	−	+	+

21. Seven deletion mutants of bacteriophage T4 were crossed in paired combinations, to test for their ability to produce wild-type recombinants. The results are given in the table below, where a + indicates that recombinants were formed. Construct a deletion map based on these findings.

		DELETION MUTANTS						
		1	2	3	4	5	6	7
DELETION MUTANTS	1	−	+	−	−	−	−	−
	2		−	+	−	+	−	−
	3			−	+	−	+	−
	4				−	−	−	−
	5					−	+	−
	6						−	−
	7							−

22. Five point mutants were tested in paired crosses with the deletion mutants listed in problem 21 for their ability to yield recombinants. The results of these crosses are given below. What is the relative order of the point mutants?

		DELETION MUTANTS						
		1	2	3	4	5	6	7
POINT MUTANTS	*a*	+	−	+	−	+	−	−
	b	−	+	−	+	−	+	−
	c	−	+	+	−	−	+	−
	d	−	+	+	−	+	−	−
	e	−	+	−	+	−	+	+

Suggested Readings / Part III

Chapter 5

Bridges, C. B. Sex in relation to chromosomes and genes. *Am. Nat.* 59:127–137, 1925.

DuPraw, E. J. *Cell and Molecular Biology,* Academic Press, New York, 1968.

Gordon, J. W. and F. H. Ruddle. Mammalian gonadal determination and gametogenesis. *Science* 211:1265–1271, 1981.

Lewis, K. R. and B. John. The chromosomal basis of sex determination. *Int. Rev. Cytol.* 23:277–379, 1968.

Mittwoch, U. *Genetics and Sex Differentiation,* Academic Press, New York, 1973.

Swanson, C. P., T. Merz, and W. J. Young. *Cytogenetics,* Prentice-Hall, Englewood Cliffs, N.J., 1981.

Swanson, C. P. and P. Webster. *The Cell,* Prentice-Hall, Englewood Cliffs, N.J., 1977.

Wachtel, S. S. H-Y antigen and the genetics of sex determination. *Science* 198:797–799, 1977.

Westergaard, M. The mechanism of sex determination in dioecious flowering plants. *Adv. Genet.* 9:217–281, 1958.

Wolfe, S. L. *Biology of the Cell,* Wadsworth, Belmont, Calif., 1981.

Chapter 6

Grant, V. *Genetics of Flowering Plants,* Columbia University Press, New York, 1975.

Harris, H. *The Principles of Human Biochemical Genetics,* Elsevier Biomedical Press, Amsterdam, 1975.

Hutt, F. B. *Animal Genetics,* Ronald Press, New York, 1964.

Landsteiner, K. and A. S. Weiner. An agglutinable factor in human blood recognized by immune sera for rhesus blood. *Proc. Soc. Exp. Biol. Med.* 43:223, 1940.

Mange, A. P. and E. J. Mange. *Genetics: Human Aspects,* Saunders, Philadelphia, 1980.

McKusick, V. A. *Mendelian Inheritance in Man,* Johns Hopkins University Press, Baltimore, 1983.

Morgan, T. H. Sex limited inheritance in *Drosophila. Science* 32:120–122, 1910.

Thompson, J. S. and M. W. Thompson. *Genetics in Medicine,* Saunders, Philadelphia, 1980.

Todd, N. B. Cats and commerce. *Scientific American* 237:100–107, 1977.

Weiner, A. S. (ed.) *Advances in Blood Grouping, III,* Grune and Stratton, New York, 1970.

Chapter 7

Burdette, W. J. (ed.) *Methodology in Basic Genetics,* Holden-Day, San Francisco, 1963.

Creighton, H. B. and B. McClintock. A correlation of cytological and genetical crossing over in *Zea mays. Proc. Natl. Acad. Sci., U.S.* 17:492–497, 1931.

Fincham, J. R. S., P. R. Day, and A. Radford. *Fungal Genetics,* Blackwell, Oxford, 1978.

King, R. C. (ed.) *Handbook of Genetics,* Plenum, New York, 1974.

Lindsley, D. L. and E. H. Grell. *Genetic Variations of Drosophila melanogaster,* Carnegie Inst. Wash. Publ. No. 627, 1967.

McKusick, V. A. and F. H. Ruddle. The status of the gene map of the human chromosome. *Science* 196:390–405, 1977.

O'Brien, S. J. and W. J. Nash. Genetic mapping in mammals: chromosome map of domestic cat. *Science* 216:257–265, 1982.

Rick, C. M. The tomato. *Scientific American* 239:76–89, 1978.

Ruddle, F. H. and R. S. Kucherlapati. Hybrid cells and human genes. *Scientific American* 231:36–49, 1974.

Stern, C. Zytologisch-genetische untersuchungen als beweise für die Morgansche theorie des faktorenaustauchs. *Biol. Zentralbl.* 51:547–587, 1931.

Chapter 8

Beckett, J. B. Classification of male-sterile cytoplasms in maize (*Zea mays* L.). *Crop Sci.* 11:724–727, 1971.

Benzer, S. The fine structure of the gene. *Scientific American* 206:70–87, 1962.

DNA: Replication and recombination, Cold Spring Harbor Sympos. Quant. Biol. 43, 1979.

Edwardson, J. R. Cytoplasmic male sterility. *Bot. Rev.* 36:341–420, 1970.

Grivell, L. A. Mitochondrial DNA. *Scientific American* 248:78–89, 1983.

Hayes, W. *The Genetics of Bacteria and their Viruses,* Wiley, New York, 1968.

Jacob, F. and E. L. Wollman. *Sexuality and the Genetics of Bacteria,* Academic Press, New York, 1961.

Meselson, M. On the mechanism of genetic recombination between DNA molecules. *J. Mol. Biol.* 9:734–745, 1964.

Nass, M. M. K. Mitochondrial DNA, in R. C. King (ed.) *Handbook of Genetics,* Vol. 5, Plenum, New York, 1976, pp. 477–533.

Sager, R. and G. Schlanger. Chloroplast DNA: physical and genetic studies, in R. C. King (ed.) *Handbook of Genetics,* Vol. 5, Plenum, New York, 1976, pp. 371–423.

Stahl, F. W. *Genetic Recombination: Thinking About it in Phage and Fungi,* Freeman, San Francisco, 1979.

Stent, G. S. and R. Calendar. *Molecular Genetics: An Introductory Narrative,* Freeman, San Francisco, 1978.

Part IV

The Molecular Basis of Gene Expression

Chapter 9

The Transfer of Genetic Information

So far, we have considered the structure and replication of the genetic material and have been specifically concerned with the manner in which the genetic information in DNA is transmitted from one cell to its daughter cells and from one generation to the next. We are now ready to discuss the way in which this information is transmitted to different parts of the same cell or organism and is translated into the phenotype of its bearer.

The genes themselves do not participate directly in producing an individual's phenotype. Rather, each gene directs the synthesis of a "messenger" molecule; the messenger then moves into the cytoplasm, carrying instructions for the production of a specific protein. It is this protein, interacting with many other proteins within the organism, that is responsible for the expression of an inherited trait.

An organism may possess thousands of genes and can produce about as many different kinds of proteins. While each kind of protein has its own particular function, all are synthesized by the same overall process. In this chapter, we will discuss the general characteristics of proteins, including their functional roles in cellular activities. We will then learn how the function of a protein is dependent on its structure and how this structure, in turn, is related to the information contained in the gene. Finally, we will consider the actual process of protein synthesis.

Structure and Function of Proteins

Proteins, like nucleic acids, are complex polymers that are composed of repeating monomeric units. While the nucleic acids serve mainly to store and transfer information, the proteins play a more direct role in the architecture and functioning of living matter. Some proteins serve as structural components; others serve as oxygen carriers (e.g., hemoglobin), hormones (e.g., insulin), or antibodies. Many act as enzymes—a class of proteins that has been of particular importance to geneticists in unraveling the mystery of how genes function. Despite their diverse activities, all proteins have several structural features in common. We therefore begin with a description of the structural characteristics shared by proteins as a group.

Protein Structure

A protein molecule consists of one or more **polypeptides,** folded into a specific three-dimensional (often globular) shape. Each polypeptide chain is a polymer composed of a linear sequence of **amino acids.** There are 20 different amino acids that occur naturally in proteins, as shown in Fig. 9.1. The chemical formulas of some of the amino acids (such as proline, tryptophan, and histidine) are rather complicated. But note that all 20 amino acids have the same basic structure:

$$\begin{array}{c} NH_2\text{—}\underset{\displaystyle R}{\underset{|}{CH}}\text{—}COOH \end{array}$$

where R stands for the particular side chain of the amino acid. There are 20 possible side chains, ranging from the simple H atom of glycine to the complex ring structures of tryptophan, phenylalanine, and tyrosine. The side chains are critical factors in determining the size, shape, and charge of the amino acid—qualities that affect the folding patterns of polypeptides.

By inspecting the basic structure, we can see that in general, amino acids are characterized by the presence of an *amino group* (NH_2—) and a *carboxyl group* (—COOH). These groups give the amino acids both basic (hydrogen acceptor) and acidic (hydrogen donor) properties. The NH_2 group can accept an H^+ ion, making it positively charged (NH_3^+—), whereas the COOH group can lose an H^+ ion, making it negatively charged (—COO^-). In a polypeptide chain, the amino group of one amino acid is joined to the carboxyl group of the adjacent amino acid by a **peptide bond,** which forms by the elimination of water (Fig. 9.2). When only two amino acids are linked together in this manner, the result is a *dipeptide;* three amino acids linked together yield a *tripeptide,* and so on. The end result is a polypeptide consisting of many (up to a thousand or more) amino acid residues, linked together by the NH_2 and COOH groups that they all possess. The R groups (side chains)

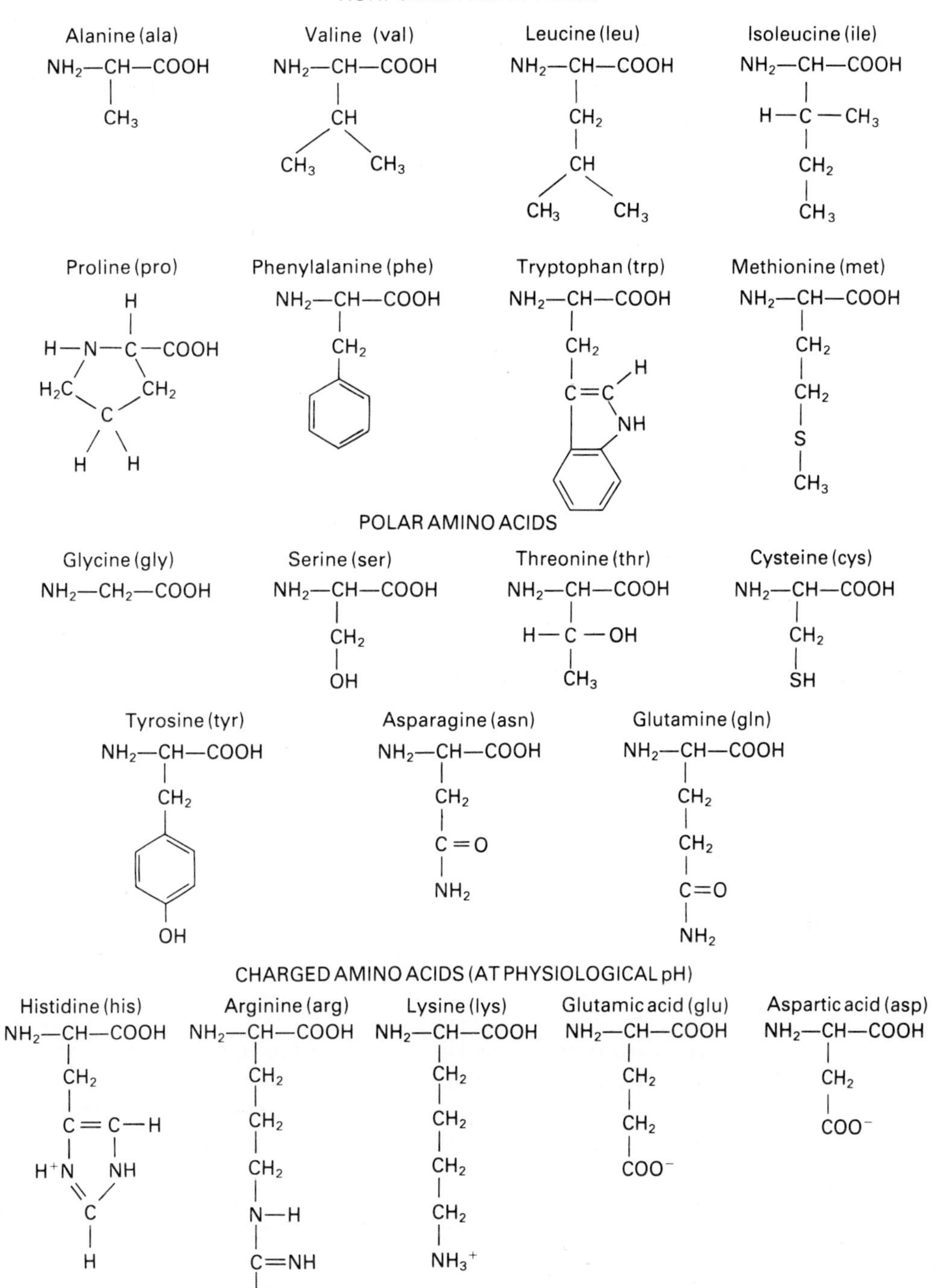

Figure 9.1. Structures of the 20 amino acids that occur naturally, arranged in groups having similar chemical properties. The NH_2—CH—COOH portion is common to all amino acids with the exception of proline.

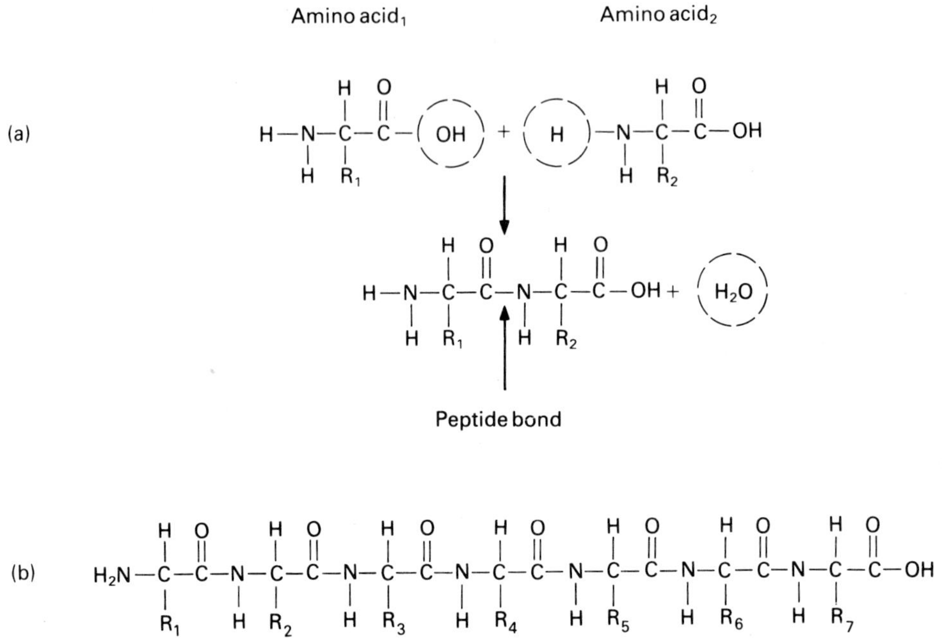

Amino $acid_1$ Amino $acid_2$

(a)

Peptide bond

(b)

Figure 9.2. (a) Formation of a peptide bond between two amino acids, yielding a dipeptide. (b) The general structure of a polypeptide—in this case, one composed of seven amino acids.

are not involved in linking the amino acids together in a polypeptide chain; they are thus free to interact chemically with one another to form higher levels of structural organization.

As with DNA, we may visualize the structure of proteins as existing at different levels of complexity. The simplest level, or primary structure, is the sequence of amino acids within the polypeptide. The peptide bond is responsible for maintaining this level of structure. The next higher level, or secondary structure, is the three-dimensional conformation of a polypeptide chain, maintained exclusively by hydrogen bonds that form between the amino and carboxyl groups of different amino acids. One common form of secondary structure is a helical conformation known as the alpha helix, shown in Fig. 9.3. But proteins do not usually appear as the long coiled threads suggested by the structure of the alpha helix; rather, they tend to be more spherical or globular than threadlike. This globular, three-dimensional shape results from the additional folding of the chain into the tertiary structure of the protein. The tertiary structure is exemplified by the folding pattern of myoglobin (a protein that carries oxygen) in the sperm whale, shown in Fig. 9.4. The tertiary structure is maintained by various bonds (hydrogen bonds, ionic bonds, and others) between the side groups of different amino acids. Many proteins also adopt a quaternary structure when two or more polypeptides combine to form a single aggregate, called an *oligomeric protein.* The bonds that maintain this aggregate are basically the same as those involved in the tertiary structure; at the quaternary level, however, these bonds occur between the R groups of amino acids in different polypeptide chains. The **disulfide bond** is an important type of covalent bond that acts to hold together polypeptide chains. It is formed when the sulfhydryl (—SH) side group of a cysteine amino acid bonds chemically with the sulfhydryl group of another cysteine, forming an S–S linkage.

A critical factor in determining the three-dimensional shape of proteins is the polarity of the amino acid side chains. Nonpolar amino acids (such as alanine, valine, leucine, and isoleucine, among others) are hydrophobic ("fear of water"). Their mutual repulsion of water causes them to associate together in the interior of a protein molecule, protected from hydration. The association of these compounds through their mutual repulsion of water is known as **hydrophobic bonding.** (Recall that hydrophobic interactions are also important in stabilizing the DNA double helix.) In contrast, polar amino acids are hydrophilic ("love water") and are usually found on the surface of the protein, accessible to hydration. The folding of a polypeptide so as to protect its hydrophobic amino acids and expose its hydrophilic ones often imparts a globular shape to the protein molecule.

In conclusion, we should state two major principles concerning protein structure, function, and assembly: (1) *The activity of a protein depends on its three-*

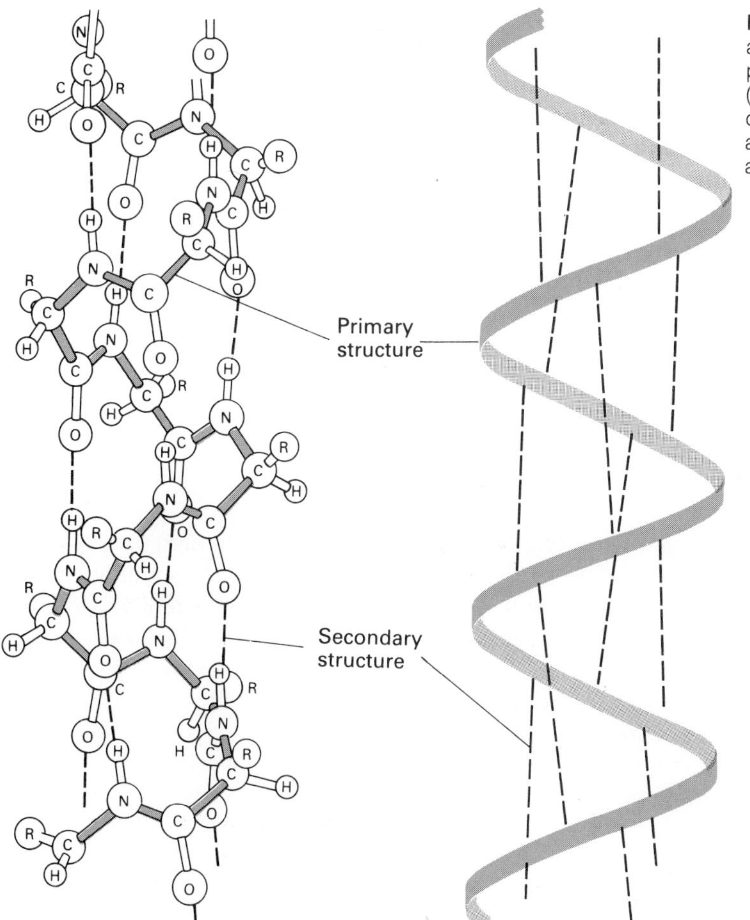

Figure 9.3. The alpha-helical secondary structure that is often assumed by a polypeptide chain. The hydrogen bonds (dashed lines) between the amino and carboxyl groups of different amino acids along the chain maintain the helical arrangement.

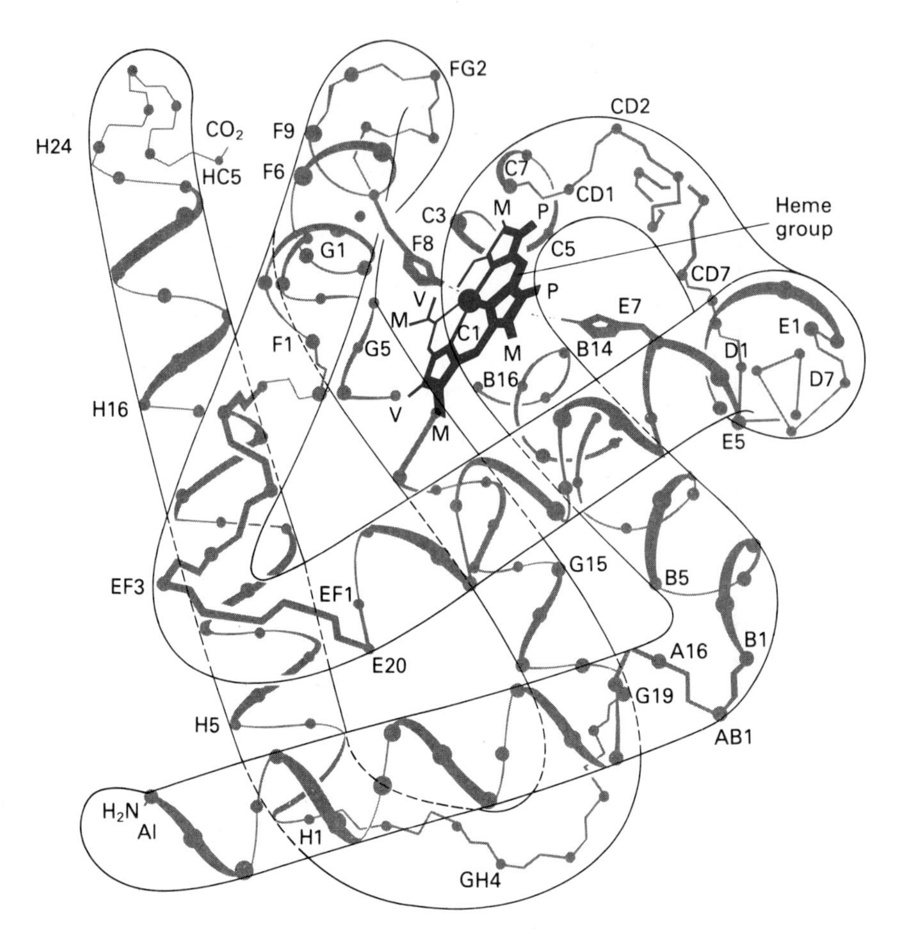

Figure 9.4. The tertiary structure of a myoglobin polypeptide chain from the sperm whale. The folding pattern is maintained by hydrogen and other noncovalent bonds that form between the amino acid side groups within the chain. Each dot represents the carbon atom attached to an amino acid side group; the symbols refer to regions of the chain. *Source:* From R. E. Dickerson, in *The Proteins*, Vol. II, H. Neurath (ed.), Academic Press Inc., 1964, pp. 603–778.

dimensional conformation. If a protein is denatured by exposure to high temperatures or extremes in pH, its secondary, tertiary, and quaternary structures will be greatly altered. This change in conformation is associated with a concurrent loss of biological activity. (2) *The three-dimensional conformation of a protein is assumed spontaneously and is determined entirely by the specific amino acid sequence of each polypeptide chain.* The sequence of amino acids yields the unique folding pattern and, therefore, the unique activity of a protein.

Function of Proteins as Enzymes

Nowhere is the relationship between the structure and function of proteins better illustrated than in the activity of proteins as enzymes. Enzymes are biological catalysts; like other catalysts, they act to increase the rate of chemical reactions

Precursor compound —Enzyme 1 / Step 1→ Intermediate compound 1 —Enzyme 2 / Step 2→ Intermediate compound 2 —Enzyme 3 / Step 3→ End product

Figure 9.5. Hypothetical metabolic pathway illustrating the step-by-step conversion of a precursor substance to its end product. Each step of the pathway represents a chemical reaction that is catalyzed by a particular enzyme.

without being used up in the process. The catalytic activity of enzymes enables biochemically important reactions to proceed at rapid rates at temperatures as low as 37°C and lower.

Enzymes manifest a high degree of specificity for the reactions they catalyze. Within living organisms, these chemical reactions occur in complex and interconnected reaction sequences, called *metabolic pathways* (Fig. 9.5). These pathways may be *biosynthetic,* involving the step-by-step transformation of simpler substances into more complex substances, or they may be *degradative,* involving the breakdown of materials and the attendant production of ATP. Different enzymes catalyze the different steps in a pathway, so that each reaction occurs only in the presence of its own particular protein catalyst. Thus, if an enzyme is rendered nonfunctional as a result of mutation, the reaction that it would otherwise catalyze is effectively blocked, and further transformation of the chemical intermediate cannot occur, at least along its normal sequence of steps. A blocked step in a chemical pathway is called a **metabolic block.**

The Active Site. The ability of an enzyme to recognize the specific substrate that it normally acts upon results from the three-dimensional folding of the enzyme molecule to form what is known as its **active site.** The active site is a localized region, or "pocket," on the surface of the enzyme that can interact in a complementary fashion with the substrate molecule (Fig. 9.6). During catalysis, the enzyme first combines with the substrate to form an intermediate enzyme–substrate complex. Something akin to a "lock-and-key" fit is achieved through the complementarity of shape between the substrate and the active site. During the binding of the two molecules, functional groups within the side chains of the amino acids at the active site form (often weak) chemical bonds with complementary groups on the substrate. This combination of reactant and catalyst promotes a chemical change by increasing the probability of reaction. The enzyme may participate in the transfer, addition, or removal of groups in the substrate, or it may simply bring different reactants into close proximity and into the proper orientation for a favorable interaction. Following this conversion, the enzyme molecule is released and is then free to be used again as a catalyst.

If the shape of an enzyme is changed in such a way that the active site is affected, the enzyme's ability to carry out its normal function will be impaired. For example, the replacement by mutation of a charged or polar amino acid with a hydrophobic one could distort the shape of the enzyme and alter the distribution of functional groups at the active site. Even just one such change could be enough to destroy the normal function of an enzyme completely, especially if such a change involved an essential amino acid at the active site. It is important to note,

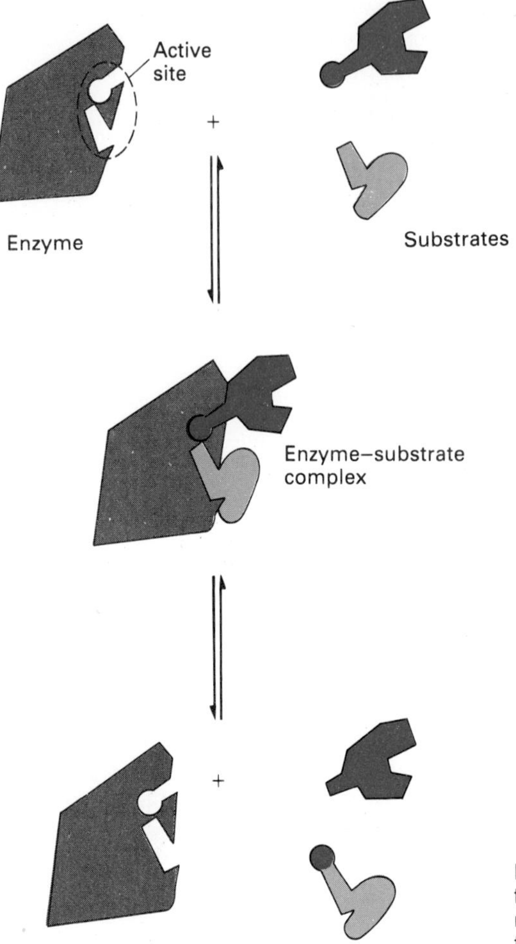

Figure 9.6. Recognition between the active site of an enzyme and its substrate, by means of their complementary shapes. The formation of an enzyme-substrate complex is followed by catalysis.

however, that the side chains of several amino acids are involved in the proper folding of a protein molecule, so that changes at any of these amino acid residue sites may lead to a loss or alteration of enzyme activity. Because of the irregular folding of a polypeptide chain, such changes might appear in widely separated and seemingly unrelated locations. As we shall see in Chapter 11, many well-known human diseases result from amino acid changes that have rendered various enzymes nonfunctional.

The Allosteric Site. Many enzymes possess not only a binding, or *catalytic site* for the substrate but also a second site, called an **allosteric site,** which is involved in the regulation of enzyme activity. The allosteric site is separate from the active site on the surface of the enzyme; it functions by specifically binding certain substances of comparatively low molecular weight, called *effectors.* The binding of an effector molecule to the allosteric site has the effect of either *increasing* the activity of the enzyme or *inhibiting* it. In the first instance, the allosteric effector would act as an activator, while in the second, it would behave as an inhibitor. The allosteric effector achieves this regulation during its reversible binding to the enzyme by altering the three-dimensional structure of the enzyme. The interaction of the effector with the allosteric site produces a change in protein conformation that carries through to the active site, even though the allosteric site may be far (in molecular terms) from the site of substrate binding. If the effector is an activator, the affinity of the altered active site for its substrate is then enhanced; if it is an inhibitor, the affinity is reduced.

A well-known example of how an effector molecule controls the activity of an enzyme is that of **feedback inhibition.** In many cases that have been studied, feedback inhibition involves the inhibition of the enzyme that catalyzes the first step (the so-called committed step) of a metabolic sequence by the end product of that pathway. Figure 9.7 illustrates one example of this form of regulation in terms of the pathway involved in the synthesis of the amino acid histidine. In this case, histidine itself acts as an allosteric effector; when it is present in excess, histidine combines with and inhibits the enzyme that catalyzes the first step in the reaction sequence. In this manner, an end product of a biosynthetic pathway can, when it accumulates, temporarily turn off the enzymes needed for its own formation.

The histidine pathway is a negative feedback system, in that it inhibits enzyme activity. Instances of positive feedback, or activation of enzyme function, are also known. One example is the synthesis of the pyrimidine nucleotides in bacteria, diagrammed in Fig. 9.8. The first step of the metabolic pathway that produces CMP, UMP, and TMP is subject to feedback inhibition by CMP and to feedback activation by the purine nucleoside triphosphate ATP. Thus, pyrimidines are synthesized only when the level of ATP is high and the levels of pyrimidines are low.

Figure 9.7. The metabolic pathway for histidine synthesis in *Salmonella typhimurium.* Excess histidine binds allosterically to the enzyme that catalyzes the first step of the pathway. This binding inactivates the enzyme and shuts off further synthesis of histidine temporarily.

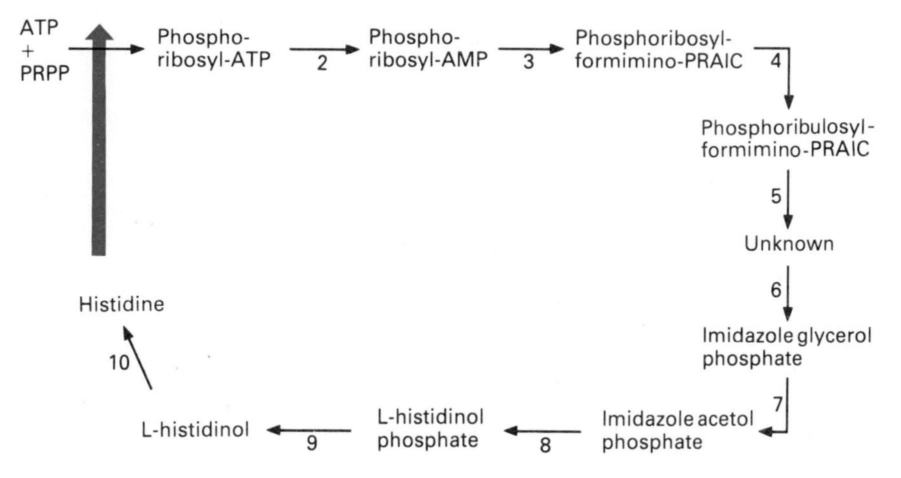

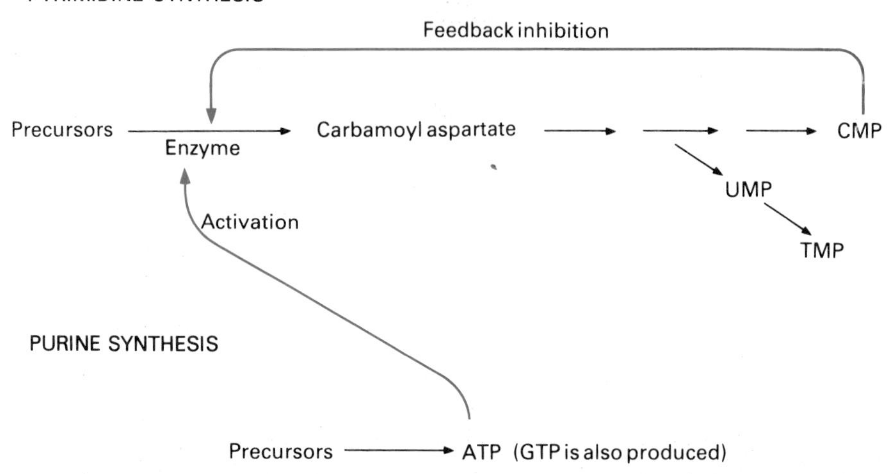

Figure 9.8. Feedback activation of the pyrimidine pathway. ATP binds allosterically to the enzyme responsible for producing carbamoyl aspartate, an intermediate in the pathway, serving to activate the enzyme. High levels of ATP thus ensure that pyrimidine synthesis will keep pace with purine synthesis.

This dual feedback system ensures a balance in the amounts of purine and pyrimidine nucleotides available for the synthesis of nucleic acids.

Regulation by Covalent Modification. In addition to the reversible interaction of enzymes with effector molecules, catalytic activity can also be regulated through covalent modification of the structure of the enzyme itself. A second enzyme is often required to catalyze a covalent structural modification in the first enzyme, thus setting up a hierarchical system. The covalent binding of a chemical residue to an enzyme usually results in an alteration that is reversible. For example, phosphorylation, methylation, and adenylation (the covalent addition of AMP) are involved in the conversion of certain enzymes from their inactive to their active forms. Some covalent changes are irreversible, such as the conversion of the pancreatic zymogens (inactive precursor proteins) into their active enzyme forms. In one such conversion process, the pancreas secretes an inactive precursor of chymotrypsin, called chymotrypsinogen, which is a single polypeptide chain 245 amino acids in length (Fig. 9.9). The precursor is converted into active chymotrypsin by enzymatic hydrolysis, which cleaves the zymogen into three polypeptides, held together by two disulfide bonds. The catalytically active amino acid residues then lie in separate chains, but the folding of the three polypeptide chains is thought to bring these sites together to form the active site of chymotrypsin.

The Vertebrate Hemoglobins

Hemoglobins provide an excellent example of proteins that are composed of several polypeptide chains, thus exhibiting quaternary structure. Hemoglobin is the oxy-

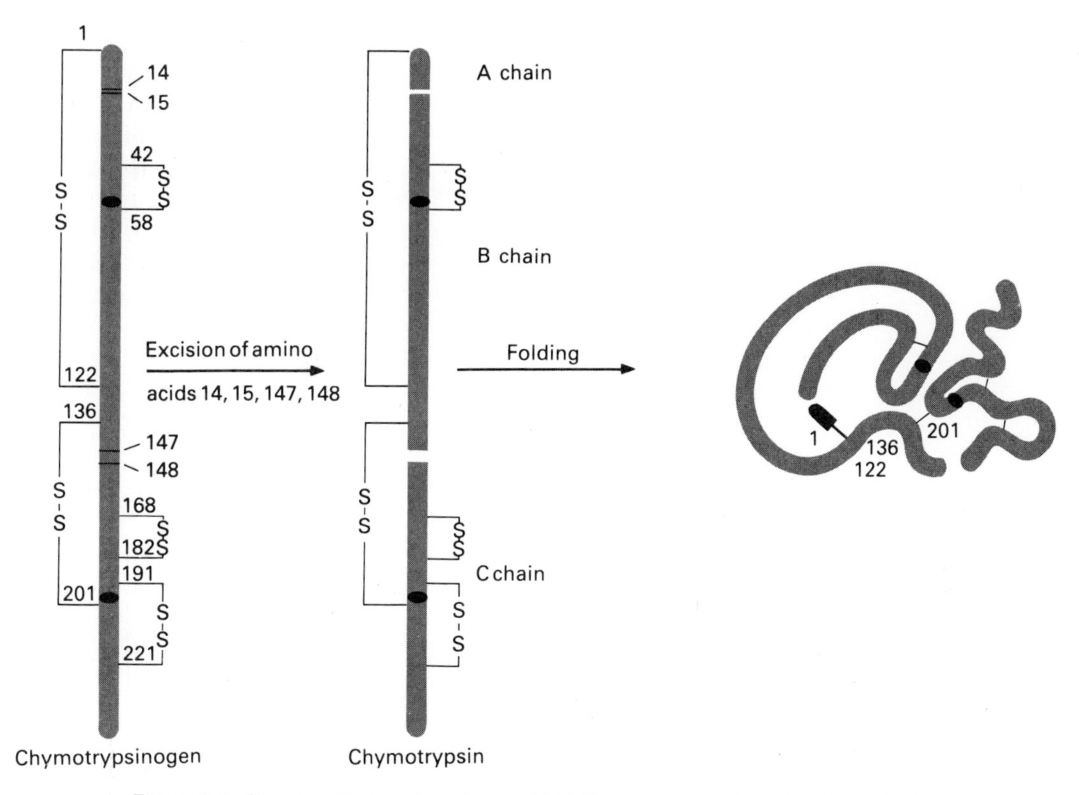

Figure 9.9. Structure of chymotrypsinogen (left), the precursor polypeptide from which chymotrypsin (right) is cut. After excision of two dipeptides (residues 14–15 and 147–148), the resulting three polypeptide chains (A, B, and C) are connected only by two disulfide bridges. Folding brings together the two catalytically active amino acid residues (color) to form the active site.

gen-carrier molecule found in red blood cells. This oxygen-carrying function is a result of its ability to combine reversibly with four molecules of oxygen in the following manner:

$$1\ \text{Hb} + 4\ O_2 \underset{\text{In capillaries of other tissues}}{\overset{\text{In lung capillaries}}{\rightleftharpoons}} 1\ \text{Hb}\cdot(O_2)_4$$

The hemoglobin molecule is composed of a protein portion, known as globin, which is associated with four iron-containing heme groups that are responsible for the property of the molecule to combine with oxygen. The globin portion of the molecule consists of four polypeptides—two identical alpha (α) chains and two identical beta (β) chains—which are folded together in a very precise manner (Fig. 9.10). Hemoglobin is one of an increasing number of proteins whose exact amino acid sequence has been elucidated. Each α chain consists of 141 amino acids, and each β chain consists of 146 amino acids. The primary structure of hemoglobin is shown in Fig. 9.11.

Perhaps the best-known genetic example involving hemoglobin is the disease sickle-cell anemia (see Chapter 1). This disorder is caused by a mutational alteration in the globin portion of the molecule. Normal hemoglobin (symbolized as HbA) contains the amino acid *glutamic acid* at the sixth position of the β chain, as shown in Fig. 9.11. In sickle-cell hemoglobin (HbS), the glutamic acid residue at this position is replaced by another amino acid, *valine*. This single amino acid substitution is the only difference between the primary structures of HbA and HbS. Researchers now believe that the large nonpolar side chain of valine affects the quaternary structure of the hemoglobin molecule when it is deoxygenated (that is, when it is not carrying an oxygen molecule). When the oxygen tension falls within the blood of an individual with this condition, as many as 16 HbS molecules in a red blood cell polymerize, apparently through intermolecular hydrophobic bonding. The growing aggregate distorts the shape of the red blood cell, yielding the characteristic sickled appearance.

Hemoglobin A is not the only form of hemoglobin that can transport oxygen. Adults also have a small amount of hemoglobin that consists of two α chains and

Figure 9.10. The three-dimensional structure of a normal hemoglobin molecule. The two α polypeptides are shown in light gray, the two β polypeptides in dark gray. The disks represent heme groups. *Source:* From M. F. Perutz, The hemoglobin molecule, *Scientific American*, 211:64–67, 1964. Copyright © 1964 by Scientific American, Inc. All rights reserved.

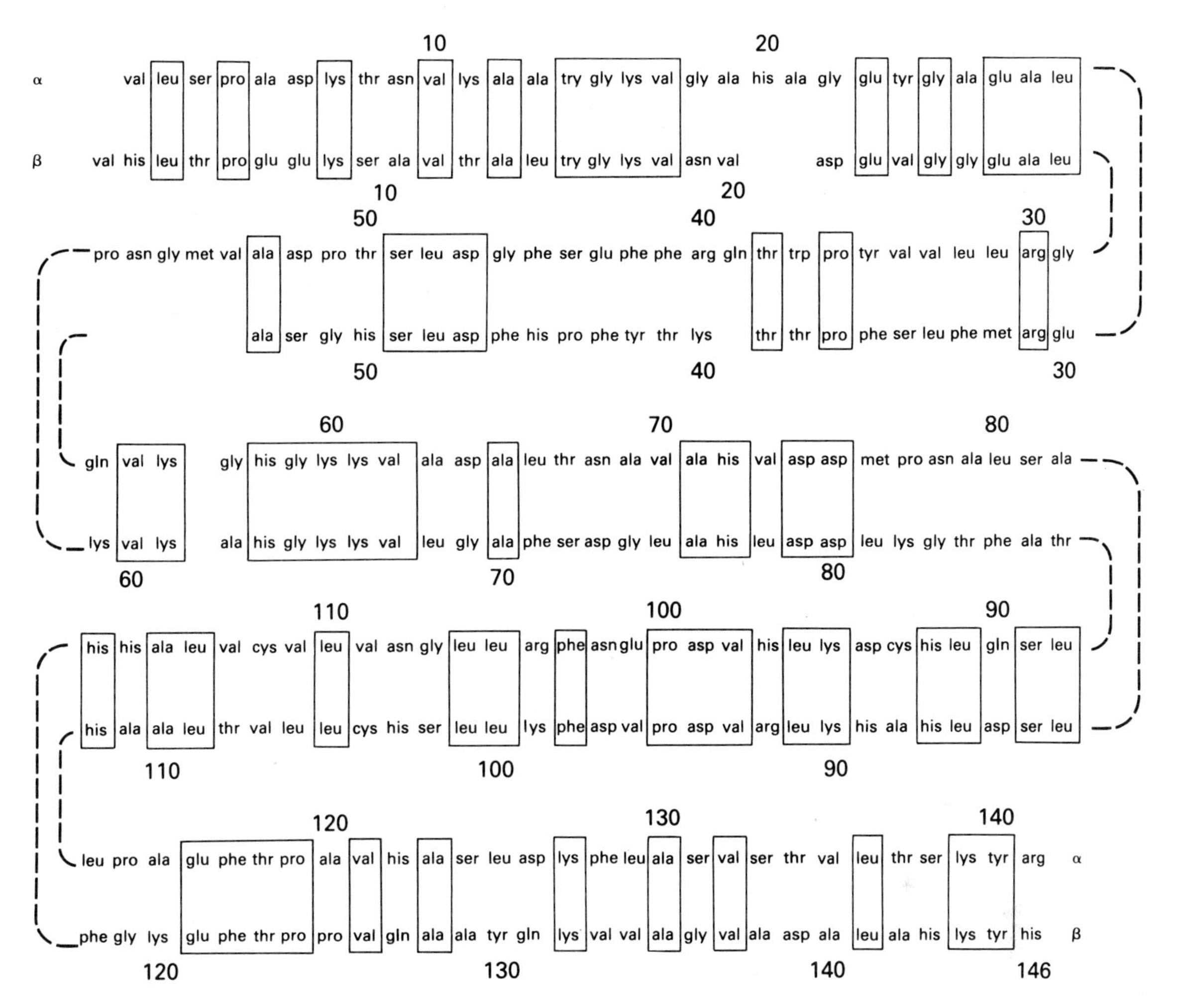

Figure 9.11. Primary structure of the normal α and β chains of human hemoglobin. The numbers indicate the positions of the amino acids. Boxes enclose amino acids that occupy the same relative positions in the two chains. *Source:* From V. M. Ingram, *The Hemoglobins in Genetics and Evolution,* Columbia University Press, New York, 1963.

two delta (δ) chains, which take the place of the β chains in HbA. Furthermore, the human fetus contains no HbA but instead has a form known as fetal hemoglobin (HbF). Fetal hemoglobin is composed of two α chains plus two gamma (γ) chains. The γ chain is a β-like chain that is synthesized at a relatively high rate until the time of birth. The level of HbF then falls to less than 1% by six months of age, as β polypeptides gradually replace the γ chains. Individuals with one form of the disease thalassemia are unable to produce β chains and continue to form HbF in adult life. The switch from fetal to adult hemoglobin indicates that the expression of the genes that code for β and γ peptides is regulated. The β gene is "turned on" and the γ gene is "turned off" during fetal development.

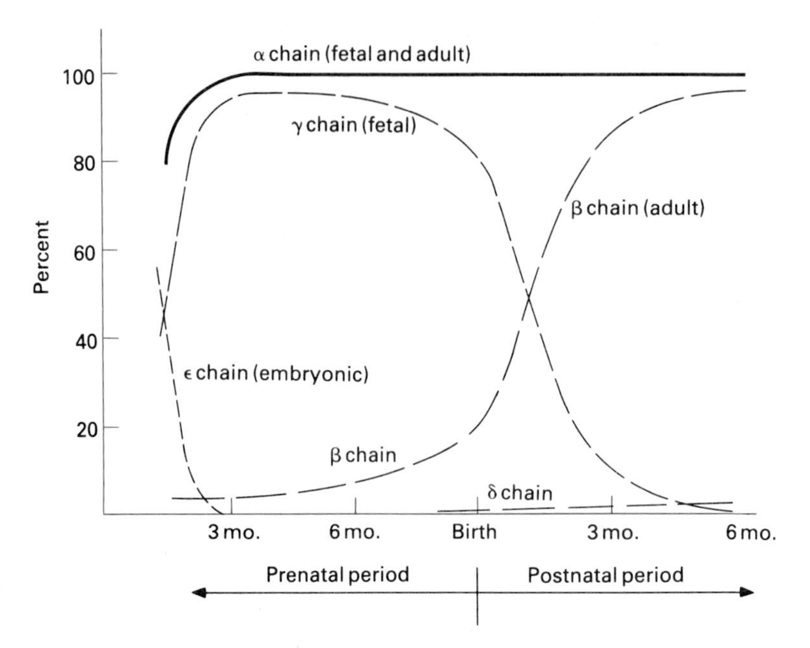

Figure 9.12. Changes in the kinds and levels of human hemoglobin chains produced from the embryonic period to the adult stage. The hemoglobin molecule is a tetramer, with two α chains. The other two chains can be either ε, γ, or β, depending on the stage in the life cycle. The δ chain is a minor hemoglobin component that replaces the β chain in the adult hemoglobin of some individuals. *Source:* From E. R. Huehns, N. Dance, G. H. Beaven, F. Hecht, and A. G. Motulsky, *Cold Spring Harbor Sympos. Quant. Biol.* 29:327–331, 1964.

TO SUM UP

1. Proteins are large molecules that are composed of one or more polypeptide chains. Each polypeptide chain is a linear polymer of amino acids that are linked to each other by peptide bonds.
2. We may consider the structure of proteins as existing at different levels of complexity: (1) the primary structure comprises the amino acid sequence of the polymer chain, (2) the secondary structure comprises the regular three-dimensional array (often a helix) that is maintained by hydrogen bonds, (3) the tertiary structure comprises the three-dimensional (often globular) folded structure that is maintained by the bonding between amino acid side chains, and (4) the quaternary structure comprises the aggregation of two or more polypeptide chains.
3. The activity of a protein depends on its three-dimensional conformation; this conformation, which is assumed spontaneously, is determined entirely by the primary structure of each polypeptide chain. Changes in the amino acid sequence, caused by mutation, can alter the spatial conformation of a protein and can thereby modify or destroy its biological activity.
4. Proteins exhibit a wide variety of functions. Two examples that we have considered here are (1) enzymes, a highly specific class of catalysts, and (2) vertebrate hemoglobins, which function in the transport of oxygen to the tissues.

Genes and Proteins

The association between genes and proteins was first recognized in 1909 by an English physician, Archibald Garrod. His studies of the causes of human congenital diseases led Garrod to suggest that many heritable disorders are the result of metabolic blocks, which he referred to as *inborn errors of metabolism.*

Garrod worked extensively with the disease alkaptonuria, a rare disorder with comparatively minor consequences. Individuals affected with this disorder experience some hardening and darkening of the cartilage, associated with arthritis to varying degrees. The disease is diagnosed by testing the urine, which turns black upon exposure to oxygen. This darkening of the urine and cartilage is caused by the accumulation of homogentisic acid (also called alkapton) in the body. Homogentisic acid is an intermediate in a metabolic pathway that is concerned with the utilization of dietary protein. Garrod postulated that a missing enzyme led to a block in this pathway at the step in the pathway at which homogentisic acid is degraded. Since this compound could no longer be converted to the next metabolic intermediate in the pathway, it accumulated within the body and was subsequently excreted in large amounts in the urine. From studies of family pedigrees and consultations with early geneticists, Garrod was able to show that the disease followed the pattern of inheritance that is expected for an autosomal recessive gene. Garrod thus established the first connection between a mutant gene and the deficiency of a particular enzyme.

Relationship between Genes and Proteins

Garrod's work gave birth to the idea that genes specify enzymes and that a missing or nonfunctional enzyme is the basis of a mutant phenotype. Out of this idea came an important principle of genetics as a result of the experiments begun around 1940 by G. Beadle and E. Tatum.

Beadle and Tatum worked with the common bread mold, *Neurospora crassa.* Even though *Neurospora* is a eukaryotic organism, it can be easily manipulated in the laboratory and can be grown on solid medium in petri dishes in much the same manner as *E. coli.* These properties have made *Neurospora* a useful organism for genetic studies. Wild-type *Neurospora,* like *E. coli,* is able to synthesize all of the complex organic compounds it needs for growth, such as amino acids and various purine and pyrimidine bases. Organisms with this ability to synthesize their own essential organic compounds are said to be **prototrophic** in their nutritional requirements. They are able to grow on a simple aqueous medium (known as *minimal medium*); in the case of *Neurospora,* this medium consists of only inorganic salts, sucrose (as a carbon and energy source), and the plant growth factor biotin. But a variety of mutant strains of *Neurospora* exist that have lost the ability to synthesize one or more of these organic growth requirements. Such mutants are called **auxotrophs;** they can grow only if the medium is supplemented with whatever compound they cannot synthesize for themselves. As an analogy, humans and rabbits are auxotrophs for vitamin C; we are unable to synthesize this vitamin and must therefore include it in our diet. Beadle and Tatum worked with a class of *Neurospora* mutants known as *arginine auxotrophs* (arg^-).

None of these mutants can grow on minimal medium, but all will grow if the amino acid arginine is added to the medium as a supplement. Extensive genetic investigations revealed that the arginine mutants could be divided into three classes: *arg-1* mutants, *arg-2* mutants, and *arg-3* mutants. Each class carried a mutation in a different gene. In other words, there are (at least) three separate genes concerned with the synthesis of arginine, and a mutation in any one of them will cause an arg$^-$ auxotrophic state.

Beadle and Tatum carefully tested each class of arg$^-$ mutants for their growth response to various compounds involved in the biosynthesis of arginine. The arg-1 mutants had the least restrictive growth requirements; they grew equally well if either arginine, citrulline (a compound whose structure is similar to arginine), or another related compound, ornithine, was added to the medium. In contrast, arg-2 mutants grew only if the medium was supplemented with either arginine or citrulline. The arg-3 mutants were even more restrictive in their growth requirements; they grew only if arginine was added to the medium. The growth responses of each mutant strain and the structures of the three required compounds are shown in Fig. 9.13.

Figure 9.13. (a) The structures of the three compounds utilized by the various arg auxotrophic mutants of *Neurospora*; the segment of the arginine biosynthesis pathway that involves these compounds is also shown. (b) Growth data that led Beadle and Tatum to deduce the order of the compounds in the pathway (+ = growth, − = no growth).

(a)

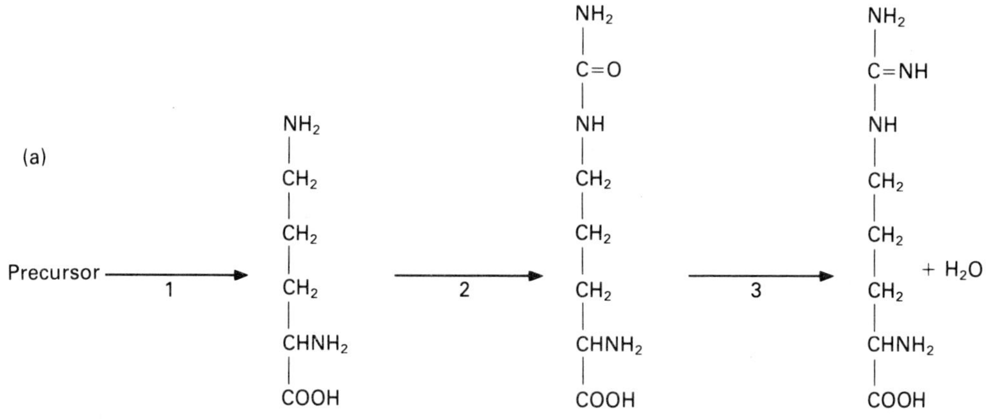

Precursor $\xrightarrow{1}$ Ornithine $\xrightarrow{2}$ Citrulline $\xrightarrow{3}$ Arginine + H_2O

Ornithine: NH_2—CH_2—CH_2—CH_2—$CHNH_2$—COOH

Citrulline: NH_2—C=O—NH—CH_2—CH_2—CH_2—$CHNH_2$—COOH

Arginine: NH_2—C=NH—NH—CH_2—CH_2—CH_2—$CHNH_2$—COOH

(b)

Growth on:

MUTANT STRAIN	Minimal medium	Ornithine	Citrulline	Arginine	REACTION BLOCKED
1	−	+	+	+	1
2	−	−	+	+	2
3	−	−	−	+	3

To explain these results, Beadle and Tatum postulated that each of the various *arg* genes codes for a different enzyme in the pathway leading to arginine:

$$\text{Precursor} \xrightarrow[\text{Enzyme 1}]{\text{Step 1}} \text{Ornithine} \xrightarrow[\text{Enzyme 2}]{\text{Step 2}} \text{Citrulline} \xrightarrow[\text{Enzyme 3}]{\text{Step 3}} \text{Arginine}$$

They recognized that in certain mutant strains, the required end product can be formed only from compounds that occur in the pathway after the blocked step. Only these precursor substances, or the end product itself, will promote growth. For example, arg-1 mutants respond to all three compounds, which indicates that the block in their pathway occurs prior to the formation of any of these substances—that is, at step 1. Arg-2 mutants respond to arginine and citrulline but not to ornithine. The pathway in these mutants must therefore be blocked at a step prior to citrulline and arginine but after ornithine—that is, at step 2. Finally, arg-3 mutants grow only on arginine; the addition of citrulline or ornithine to the medium does nothing to stimulate their growth. The pathway for arginine synthesis in arg-3 mutants must therefore be blocked at a step prior to arginine but after citrulline and ornithine—that is, at step 3. Thus, arg-1 mutants are deficient in enzyme 1 (their complete genotype is $arg\text{-}1^{-}\ arg\text{-}2^{+}\ arg\text{-}3^{+}$), arg-2 mutants are deficient in enzyme 2, and arg-3 mutants are deficient in enzyme 3. The concept that each gene specifies one particular enzyme became known as the *one gene : one enzyme* hypothesis.

Subsequent experimentation verified the one gene : one enzyme hypothesis in all cases in which the enzyme is composed of just one type of polypeptide chain. As we have already discussed, however, many enzymes are made up of two or more different polypeptides. Moreover, a large number of proteins are not enzymes but function in other capacities instead. It is also possible for a single polypeptide to possess more than one enzymatic activity. For instance, in animal tissues, the seven enzymatic steps involved in fatty acid biosynthesis are catalyzed by seven different active sites on a single protein. In this case, the different active sites are formed by the three-dimensional folding of a single polypeptide chain. To account for these exceptions, the one gene : one enzyme hypothesis was modified to what is now known as the *one gene : one polypeptide* hypothesis. This more general relationship states that a gene codes for a unique polypeptide, which alone or folded together with other polypeptides makes up a functional protein. For example, the hemoglobin molecule is coded for by two separate genes: one for the α chains and another for the β chains.

Genetic Complementation

The one gene : one polypeptide hypothesis provides geneticists with the basis for determining whether two mutants with similar phenotypes have mutations in the same gene or in different genes. The procedure that is used by geneticists for this purpose is the complementation test. We encountered this test in Chapter 6, where it is described as a means for distinguishing between allelic and nonallelic genes. The test involves introducing separate copies of two independent recessive mutations into the same cell or organism. If the resulting heterozygote is normal (wild type) in phenotype, the mutant chromosomes are said to *complement* each other and the mutations are judged to be in different genes. If the heterozygote is mutant in phenotype, the mutant chromosomes fail to complement each other and the mutations are judged to be in the same gene.

The theoretical basis for the complementation test is shown in Fig. 9.14 in terms of two closely linked mutation sites. We see in Fig. 9.14(a) that when the mutations are in the same gene, neither allele can produce a functional polypeptide as long as the mutation sites are in the *trans* configuration. But when the sites are in the *cis* configuration, one of the alleles can still produce a fully functional polypeptide. Thus, the phenotype of the trans configuration is mutant and the phenotype of the cis configuration is wild type. Figure 9.14(b) shows what happens when the mutations are in different genes. A normal allele of each gene is then present in the heterozygote to supply the missing function of its mutant allele. The functional polypeptides of both genes are then formed, so that a normal phenotype is produced in both the cis and trans configurations.

The main technical difficulty in performing the complementation test is introducing separate mutations into a common cytoplasm. In haploid organisms, such as *E. coli* and *Neurospora,* the test has to be accompanied by the formation of a diploid or pseudodiploid condition. The manner in which this condition is achieved varies with different organisms and their modes of reproduction. For example, partial diploids are formed in *E. coli* by specialized transduction (see Chapter 8). Each transduced bacterium that is formed by this method contains two copies of a chromosomal region: its own copy and the copy carried by the transducing phage. A complementation test can also be accomplished in *Neurospora* by making heterokaryons of two different mutants. For instance, heterokaryons are formed between the arg-1 and arg-2 mutants in *Neurospora* (described in the previous section) when the mutant hyphae fuse and both types of nuclei enter a common cytoplasm. Each type of nucleus then supplies the missing function in the other, so that arginine is produced in the shared cytoplasm.

We see that complementation depends only on the interaction of gene products within a common cytoplasm; recombination, or any other direct interaction of chromosomes, is not involved. The complementation test thus provides us with a means for identifying *functional alleles* based on the phenotype of the trans heterozygote. S. Benzer, using the terms cis and trans, coined the word **cistron** to mean the unit of function that is operationally defined on the basis of the complementation test. Usually a cistron and a gene are synonomous. But, as we shall presently see, there are cases in which a gene and a cistron do not correspond.

Intragenic Complementation. Complementation has been observed in a number of cases between mutations that lead to different alterations in the same polypeptide. According to the one gene : one polypeptide hypothesis, these mutations are in the same gene. Yet, if we were to base our judgment solely on the results of a complementation test and disregard all else, we would classify these mutations into different cistrons. How can we account for this exception to the rule that complementation can occur only between mutations in different genes? According to one hypothesis, intragenic complementation can result when a gene codes for a *multimeric protein,* such as a dimer or tetramer, of the same kind of polypeptide.

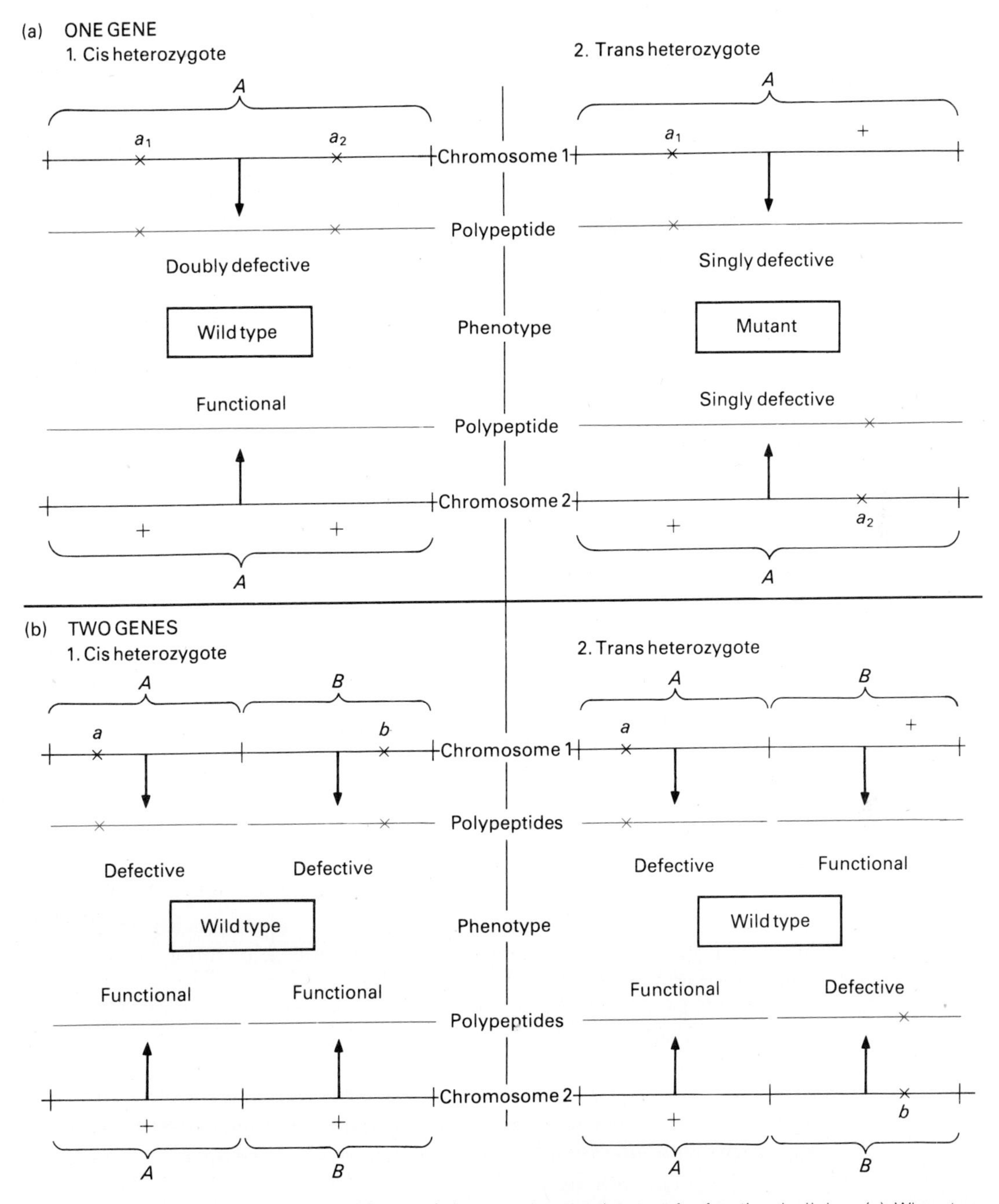

Figure 9.14. Theoretical basis of the complementation test for functional alleles. (a) When two mutation sites, which affect the same gene, are present together in a heterozygote, a functional polypeptide can be produced only in the cis configuration. The cis configuration therefore yields a wild-type phenotype, whereas the trans configuration produces a mutant phenotype. (b) When the mutation sites affect different genes, normal alleles of each gene are present to produce a wild-type phenotypic effect in both the cis and trans configuration.

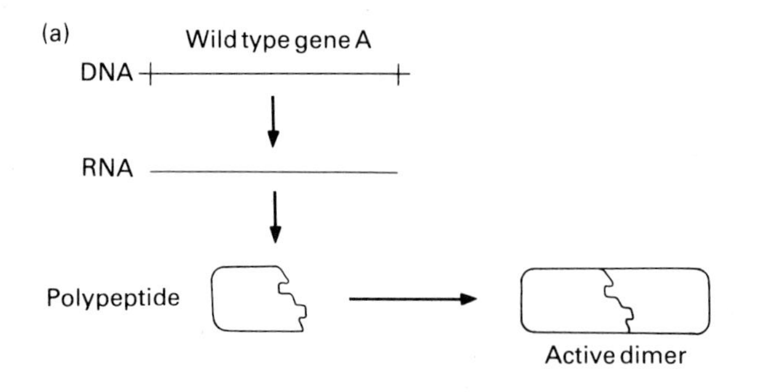

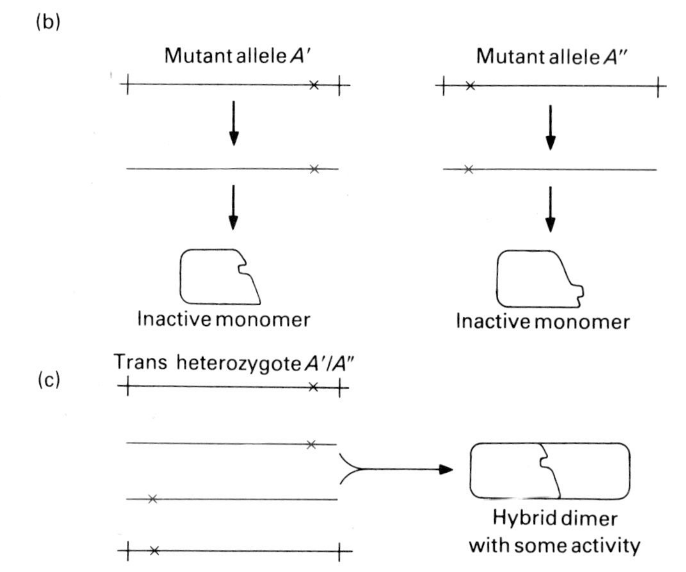

Figure 9.15. Intragenic complementation involving a dimeric protein. (a) An active dimer is formed by the combination of two wild-type polypeptides. (b) Different mutations within the gene can lead to the production of altered polypeptide chains. Each altered polypeptide is unable to join with another of the same type to form an active dimer. (c) Intragenic complementation can occur in a trans heterozygote when two different mutant polypeptides interact to form a hybrid dimer with some wild-type activity.

Different alleles of a gene might then code for polypeptide chains that are nonfunctional by themselves, but which interact when present together to form a functional protein product. One possible mechanism for intragenic complementation is shown in Fig. 9.15. In this example, the mutational changes in these different alleles are assumed to affect the interface between the polypeptide subunits of a dimeric protein. Both mutations (A' and A'') alter the structure of the polypeptide in such a way that it is unable to join with another of the same type to form the functional protein. However, when both types of polypeptides are present within the same cell, one mutational alteration compensates for the second so that the different mutant polypeptide chains can combine and at least partially restore enzymatic activity.

The mixed multimer explanation for intragenic complementation is strongly supported by demonstrations that mixtures of polypeptide chains from different inactive multimeric proteins can combine spontaneously within a test tube to form hybrid multimers with some biological activity. Studies also reveal that complementation can occur between alleles that code for proteins with more than one enzymatic activity. As we have seen, multifunctional enzymes can have several active sites contained within a single polypeptide chain. Thus, if one active site were nonfunctional in one mutant strain and another active site were nonfunctional in a different strain, both enzymatic activities would be present in the hybrid that produced a mixture of these two mutant enzymes.

Colinearity of Gene and Polypeptide

The information that specifies a protein is built into the nucleotide sequence of a particular gene. Numerous experiments with bacteria and viruses have shown that there is a 1:1 correspondence between the coding sequence of the gene and the sequence of amino acids that make up the polypeptide for which it codes. In other words, there is a linear correlation between the coding sequence and the amino acid sequence—a coding relationship known as **colinearity.** The first code word of the gene specifies the first amino acid of the polypeptide, the second code word specifies the second amino acid, and so on.

The first strong evidence for colinearity came from studies by C. Yanofsky on the *trpA* gene and its polypeptide product in *E. coli.* The *trpA* gene codes for one of the polypeptides (peptide α) that make up the *E. coli* enzyme tryptophan synthetase. This enzyme catalyzes the last step in the metabolic pathway that is concerned with the biosynthesis of tryptophan. Yanofsky and his colleagues worked with a collection of tryptophan synthetase mutants that produced an inactive form of this enzyme. Mapping studies revealed that these mutants differed from one another in the precise position of the mutation sites within the *trpA* gene. Furthermore, sequence analysis of each mutant α peptide chain revealed that the different mutants had amino acid changes at different residue locations. The order of these mutation sites within the gene corresponded perfectly with the position of the amino acid substitutions in the polypeptide, as shown in Fig. 9.16. Eleven different mutation sites are indicated, which correspond to the eleven different amino acid changes. The mutation site farthest to the left on the gene causes a change in the 15th amino acid position of the polypeptide. The next mutation site on the gene causes a change in amino acid number 22. The third

Figure 9.16. Colinearity between the α peptide of tryptophan synthetase in *E. coli* and the gene that codes for this enzyme. There is a linear correlation between the mutation sites mapped along the gene and the relative position of the amino acid changes in the peptide. *Source:* After C. Yanofsky, *Scientific American* 216:80–94, 1967.

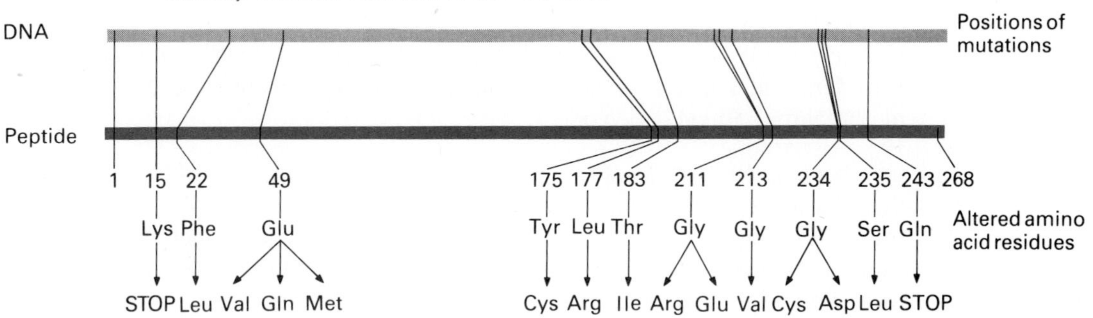

mutation site causes a change in amino acid number 49, and so on. Besides the correlation between the order of the mutation sites and the order of the changed amino acid sites, Yanofsky observed that the farther apart two mutation sites are on the map, the more amino acids there are between the corresponding positions in the polypeptide, thus lending additional support to the concept of colinearity.

The Triplet Code. Because the two strands of a DNA molecule are complementary, they carry different information. Thus, only one strand for any given gene is copied when information is transferred and translated into a polypeptide chain; the strand that is copied is known as the **sense strand.** Although each gene always uses the same strand as the sense strand, different genes, even adjacent ones, may use a different strand. Exactly how the sense strand of a gene is determined is not fully understood.

The results of many experiments conducted during the past two decades demonstrate conclusively that the information for the primary structure of a polypeptide chain is carried in the sense strand of a gene in the form of a regular sequence of code words, called **codons.** Each codon in the gene comprises a specific sequence of three nucleotides; the sequence is translated by the cell's metabolic machinery to call for a particular amino acid in a polypeptide. (We will discuss the details of the translation process later in this chapter.) Such a code, in which three nucleotides compose a codon, is called a *triplet code.*

Researchers initially deduced the triplet nature of each codon by comparing the number of naturally occurring amino acids with the number of sequences possible for different numbers of nucleotides. There are 20 different kinds of amino acids but only four different bases. It is thus apparent that a codon cannot consist of a single nucleotide. A doublet code, in which a group of two nucleotides specifies one amino acid, would also be inadequate, since this arrangement would yield only $4 \times 4 = 16$ possible sequences. A triplet code, on the other hand, provides more than enough code words to specify all 20 amino acids. If nucleotides are read in groups of three, there would be $4 \times 4 \times 4 = 64$ possible codons. Investigators working on the code therefore expected that it would be triplet. Moreover, some elegant genetic experiments by Crick's group (see Example 11.2) firmly established that the codons were indeed triplet.

In terms of colinearity, we can now view a gene as a linear sequence of triplet code words, read consecutively without punctuation and translated into a linear sequence of amino acids. For example, if the sense strand of a gene has the sequence of nucleotides TACTCTAGA . . ., TAC would code for the first amino acid in the polypeptide chain, TCT would code for the second amino acid, and so on. Precisely which of the 64 possible codons designates which amino acid will be considered later in this chapter, along with the experiments that were important in "breaking" the genetic code.

Overlapping Genes. From the mid-1960s to the late 1970s, genes were thought to be nonoverlapping, such that each segment of DNA contained the information of only one gene. Recent discoveries made on certain viruses and human mitochondrial DNA have challenged this established idea about the structure of genes and have caused geneticists to modify their views.

Phage ϕX174 has been a favorite subject of experiments in molecular genetics, owing to its small size, the ease with which it can be manipulated, and the fact that its DNA can be replicated in a test tube. Based on the molecular weight of the ϕX174 chromosome, we can estimate that the DNA molecule contains about 5400 (5375) nucleotides. Assuming that the average gene in this phage is 1000 nucleotides long, then the chromosome has enough information to code for five or six genes. But genetic studies show that ϕX174 has nine genes. This discrepancy is even more apparent when we consider the nine proteins coded for by these nine genes; if we use the combined molecular weight of these proteins to estimate the total number of amino acids they contain, we can arrive at the total number of code words needed. The actual number is considerably greater than $5400/3$, which is the minimum value possible for a nonoverlapping code. In brief, the ϕX174 chromosome appears to contain an insufficient amount of genetic material for all of its genes!

The puzzle was finally resolved in 1977 when the complete nucleotide sequence of the ϕX174 chromosome was determined. When the nucleotide sequence of the viral chromosome was compared with the amino acid sequences of the nine proteins, a surprising result was obtained. The nine genes did not each occur at a unique position on the chromosome. Instead, two cases of **overlapping genes** were found: genes *A* and *B* overlap, as do genes *D* and *E* (Fig. 9.17a). Each pair of overlapping genes is able to produce two different polypeptides because the genes begin and end at different places, as illustrated in Fig. 9.17(b) for the *D* and *E* genes.

Figure 9.17. (a) Arrangement of genes on the chromosome of ϕX174. Two pairs of overlapping genes occur. (b) The nucleotide and amino acid sequences of the *D* and *E* genes and their respective polypeptides. The nucleotides are numbered from the start of the *D* gene. Gene *E* begins at nucleotide 179, with its triplet code word sequence offset from that of gene *D*. The *D* and *E* genes thus code for different amino acid sequences. *Source:* After J. C. Fiddes, *Scientific American* 237:54–67, 1977.

(a)

A C D J F G H

B E

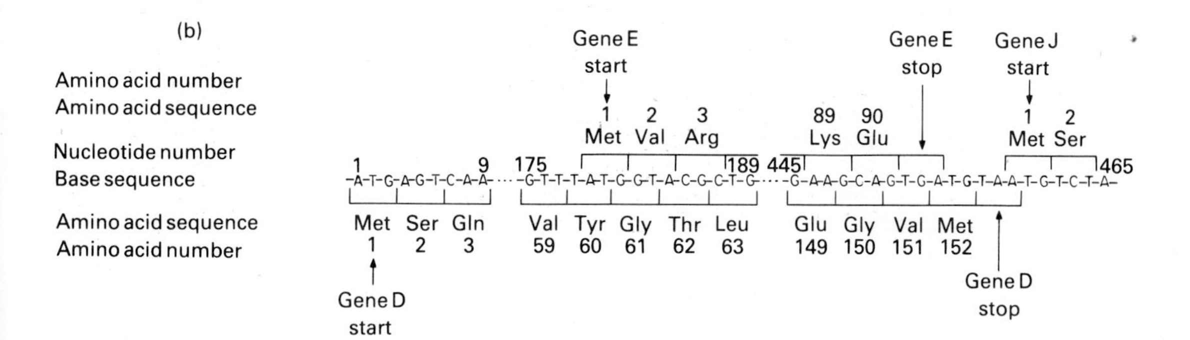

Gene overlap raises an interesting evolutionary question. A nucleotide change at a single point on a chromosome within an overlapping sequence causes both genes in the sequence to mutate. If both mutant genes are advantageous or both are deleterious, this change poses no problem. But if one is beneficial and the other is harmful, how will selection proceed?

Split Genes. In prokaryotes, essentially all nucleotides in the sense strand of a gene code for amino acids. Since an average-size protein molecule contains about 330 amino acids, the gene that codes for an average protein in prokaryotic cells must therefore consist of approximately 1000 (3 × 330) nucleotide pairs. This is not generally the case in eukaryotic cells, however. Most of the genes in eukaryotes are *discontinuous,* or *split.* In these organisms, the DNA that codes for a single polypeptide is interrupted at various points by noncoding sequences. Researchers discovered the existence of **split genes** through the use of complicated molecular techniques that were designed to isolate and purify the gene for ovalbumin production in chickens (the ovalbumin gene is responsible for producing the major egg-white protein). Instead of a gene-sized fragment of DNA, the purification process yielded several smaller fragments; when spliced together, these fragments constituted the entire gene. Investigators have since obtained similar results for nearly all eukaryotic genes studied (the histone and interferon genes are notable exceptions).

The discontinuous nature of the ovalbumin gene is shown in Fig. 9.18. This chromosomal locus is composed of eight sections of coding information for amino acids, separated by seven noncoding segments. The coding regions are called **exons,** and the noncoding regions are called **introns** (short for intervening sequences). The entire locus is 7700 base pairs long but only 1872 of these pairs code for the ovalbumin protein. Some eukaryotic genes are split even more than the ovalbumin gene, being separated into as many as 20 (or more) coding segments. It is not uncommon for a coding sequence that is 1000 nucleotides long to be separated into exons that are distributed over a length of DNA ten times that size.

The split gene arrangement helps to explain why eukaryotes contain much more DNA than that which is required to code for all their proteins. The chro-

Figure 9.18. Split gene arrangement of the ovalbumin gene. The gene is 7700 base pairs long and consists of eight coding sections called exons (shaded), separated by seven noncoding regions called introns (unshaded). The size of the introns varies from 251 base pairs (intron *B*) to about 1600 base pairs (intron G). The total length of the DNA that codes for the ovalbumin peptide is the sum of just the exons, or 1872 base pairs. *Source:* From P. Chambon, Split genes, *Scientific American* 244:60–71, 1981, based on work by A. Dugaiczyk *et al., Nature* 274:328, 1980. Copyright © 1981 by Scientific American, Inc. All rights reserved.

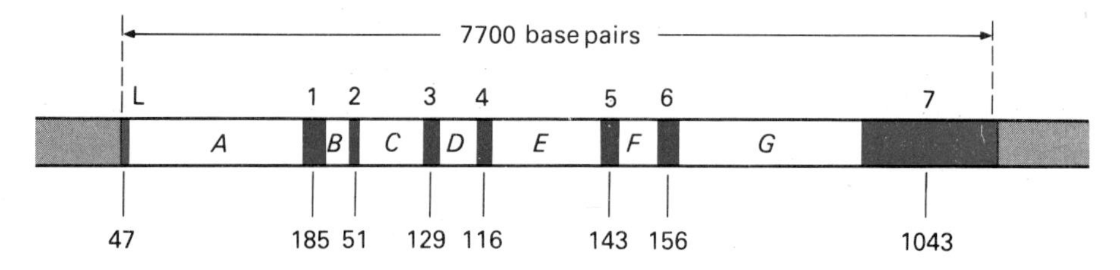

mosomes of mammals consist of approximately four billion nucleotide pairs of DNA—enough to code for over three million proteins. Yet all evidence suggests that maybe 150,000 different proteins, at the very most, are synthesized by any one organism. Some of the excess DNA probably serves in a structural capacity within the chromosomes (see Chapter 10). Additional amounts occur in spacer regions between genes. However, a major portion of this excess DNA may simply represent noncoding intervening sequences.

The function of the split gene organization is unknown. It may serve some role in regulating the activity of genes, such as controlling the rate at which the protein is coded for by the gene. Certain investigators have suggested that protein variation might arise if the exons and introns are occasionally switched. The most attractive proposal concerns the potential benefit of the split gene arrangement to evolutionary change. It is becoming increasingly clear that DNA is subject to a surprising degree of movement within the genome, with small fragments of DNA being excised from one location and inserted into another (see Chapters 10 and 11). Such rearrangements would be harmful if they disturbed an existing gene whose products are needed, but they could be of benefit if a new combination of exons were formed, creating a *new* gene with a different function. Geneticists are only beginning to understand the implications of these ideas.

TO SUM UP

1. The one gene : one polypeptide hypothesis states that each gene codes for a unique polypeptide chain. This hypothesis is a generalized extension of the one gene : one enzyme hypothesis, which was formally advanced by Beadle and Tatum, since it also applies to oligomeric proteins and to proteins that do not function as enzymes.
2. The complementation test is used by geneticists to determine whether two recessive mutations are in the same gene or in different genes. The test involves introducing two independent mutations into the same cell or organism in a heterozygous condition. If the mutations are in separate genes, the mutant chromosomes complement each other; the normal allele of each mutation is then present to produce a functional polypeptide, giving the heterozygote a wild-type phenotype. If the mutations are in the same gene, neither allele can produce a functional polypeptide, which gives the heterozygote a mutant phenotype.
3. Intragenic complementation is said to occur when two different mutations within the same gene give a wild-type phenotype in a complementation test. Intragenic complementation is usually associated with genes that code for multimeric proteins or for proteins with more than one enzymatic activity.
4. The information of a gene is read as a sequence of triplet code words, called codons. Each codon specifies a particular amino acid at a residue site in the polypeptide. The relationship between a gene and the polypeptide for which it codes is colinear; that is, a linear sequence of code words in the gene corresponds to the linear sequence of amino acids in its polypeptide product.
5. Most genes in eukaryotes, unlike prokaryotes, are split into alternating coding and noncoding regions. A gene in a eukaryote therefore extends over a length of DNA that can be many times longer than the length needed for coding. Overlapping genes, which have been discovered in viruses and mitochondria, represent another variation in the way genetic information is arranged in DNA.

Information Flow from DNA

We have seen that the code or information for all biological processes is stored in DNA. In order to produce a phenotype, this information must be transferred from DNA to other macromolecules. The flow of information in biological systems is outlined by a scheme that Crick called the *Central Dogma* of molecular biology:

The solid arrows in the diagram trace the major directions of information flow, whereas the dashed arrows represent allowable but rare transfers, which are accomplished only by certain viruses that contain RNA in place of DNA. If we ignore these rare transfers, the overall scheme of information flow in both prokaryotes and eukaryotes is

$$\text{DNA} \xrightarrow{\text{transcription}} \text{RNA} \xrightarrow{\text{translation}} \text{protein}$$

The circular arrow represents the transfer of genetic information from one generation to the next, through the semiconservative replication of DNA. The straight arrows represent the flow of information that occurs when genes express themselves phenotypically. The DNA of the gene is first *transcribed* into RNA through the formation of a complementary strand of RNA. Some of this RNA is then *translated* by the cellular machinery into protein. The process of information transfer is thus divided into two stages: **transcription** and **translation.**

Transcription

Individual genes and groups of genes are selectively transcribed in cells to produce the three major types of cellular RNA: (1) **messenger RNA** (abbreviated **mRNA**), which is coded for by the **structural** (or protein-coding) **genes,** (2) **ribosomal RNA** (or **rRNA**), and (3) **transfer RNA** (or **tRNA**). In cells that are actively synthesizing protein, the regions of DNA molecules that contain these kinds of genes will undergo a separation of strands in a localized area, in order to expose the polynucleotide chains of DNA in that area of transcription (Fig. 9.19). A *complementary* single-stranded chain of RNA is then synthesized from its nucleotide precursors (ATP, GTP, CTP, and UTP), using only the *sense strand* of DNA as the template. The reaction equation is as follows:

$$\underset{\substack{\text{RNA}\\\text{strand}}}{(\text{NMP})_n} + \text{NTP} \xrightarrow[\substack{\text{DNA sense}\\\text{strand}}]{\qquad\qquad} \underset{\substack{\text{Lengthened}\\\text{RNA}}}{(\text{NMP})_{n+1}} + \text{PP}_i$$

where NMP and NTP are the ribonucleoside monophosphate and triphosphate, respectively, and PP_i is inorganic pyrophosphate. Note from Fig. 9.19 that RNA, like DNA, is always synthesized in the $5' \longrightarrow 3'$ direction.

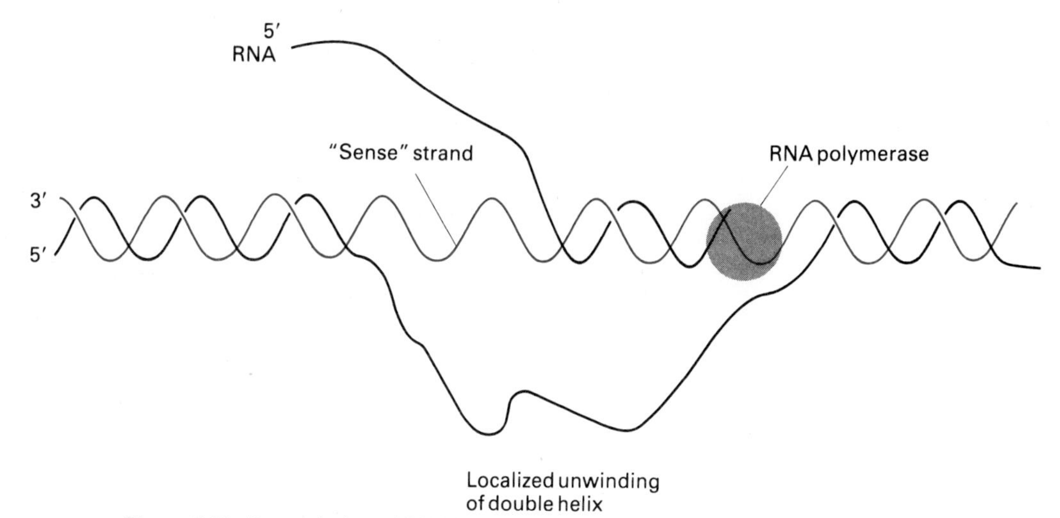

Figure 9.19. Transcription of RNA from a DNA template. A short region of the DNA is unwound. Within this region, the "sense" strand serves as a template for the synthesis of a complementary RNA molecule. The direction of RNA synthesis is always 5′ → 3′. Transcription is catalyzed by the enzyme RNA polymerase.

Transcription is catalyzed by an enzyme called **RNA polymerase.** This enzyme selects the proper DNA template strand and combines with DNA at a particular site, called a **promoter.** The RNA polymerase recognizes its binding site on the DNA molecule by the highly specific sequence of nucleotides in the promoter region (Fig. 9.20).

The RNA polymerase molecule in prokaryotes is composed of an oligomeric *core protein,* which can combine with yet another polypeptide called the *sigma* (σ)

Figure 9.20. The sequence of nucleotides in several bacterial and viral DNA template strands, showing the initial position of binding by RNA polymerase and the first base to be transcribed (in color), λ, T7, ϕX174, and SV40 are viruses; *lac* is an *E. coli* gene.

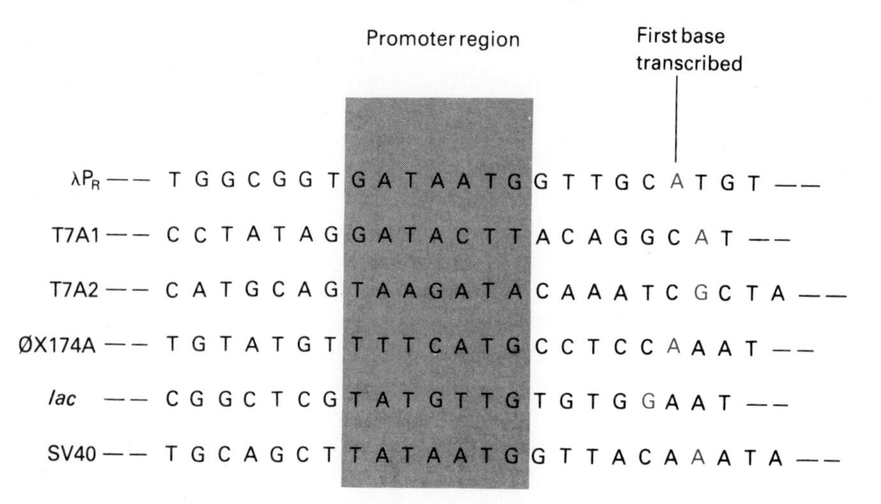

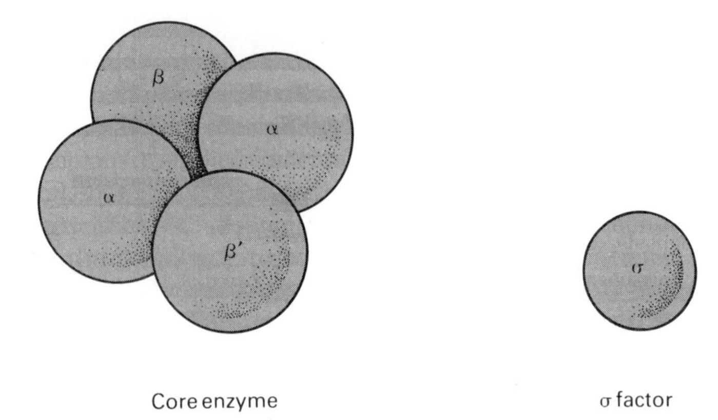

Figure 9.21. The subunit structure of the DNA polymerase molecule in *E. coli*. The core enzyme is composed of two identical α polypeptides, a β polypeptide, and a β′ polypeptide. The core enzyme forms a complex with a fifth polypeptide, σ, to yield the holoenzyme.

factor (Fig. 9.21). The core protein in combination with the σ factor gives rise to the complete enzyme, or *holoenzyme*. In vitro studies of the transcription process in *E. coli* reveal that the presence of the σ factor allows the enzyme to recognize specific promoter regions. If the σ factor is missing, the polymerase core enzyme will begin its transcription at a randomly selected initiation point. Once RNA synthesis has begun, the σ factor dissociates from the core enzyme. Synthesis then proceeds until a termination site is reached. The enzyme and completed RNA chain are then released through a process that often requires the participation of a protein termination factor, called *rho* (ρ). We can therefore regard transcription as occurring in three phases: *initiation, chain elongation,* and *termination*. The events occurring in these different phases of transcription in *E. coli* are summarized in Fig. 9.22.

In eukaryotes, the three classes of RNA are synthesized by three different RNA polymerase enzymes. Polymerase I transcribes rRNA genes into ribosomal RNA; RNA polymerase II transcribes structural genes into mRNA; and RNA polymerase III synthesizes tRNA and 5S RNA (a component of ribosomes). We will now discuss each class of RNA in turn and the role each plays in the transfer of genetic information from DNA.

Nature and Modification of mRNA. In prokaryotes, some mRNA contains information transcribed from only a single gene. Other mRNA molecules are multigenic, each representing the transcripts of two or more adjacent genes. In either case, once the mRNA is formed, it is immediately used to synthesize proteins and is then degraded. Prokaryotes do not ordinarily modify their mRNA after transcription, nor do they store it for use at a later time.

The situation is much more complex in the cells of eukaryotes. The original transcript undergoes a series of modifications to produce the eukaryotic mRNA. While all the details are not yet clear, the overall scheme of messenger RNA production appears to be as follows: An RNA transcript is synthesized, using a single structural gene in the DNA as a template. Since most eukaryotic genes are split, transcription normally produces a long RNA molecule, called **heterogeneous nuclear RNA** (or **hnRNA**). The hnRNA molecule contains all the alternating coding and noncoding segments of the original template DNA strand, but in a complementary base sequence. This hnRNA must then be *processed* into mRNA, which contains only the exons, or coding sequences. Among the first steps involved in the processing of hnRNA are the addition of a methylated guanine "cap" to the 5′ end of the molecule and a string of adenine nucleotides to the 3′ end of the molecule to give it a poly-A tail. The 5′ cap facilitates the binding of mRNA to ribosomes; it may also function with the poly-A tail to help protect eukaryotic mRNA from enzymatic destruction. Following capping and the addition of a poly-A tail, the modified transcript is then *spliced* to give the final mRNA. The **splicing** reactions involve the removal of intron regions from hnRNA by nuclease activity and the end-to-end joining of exons in the proper arrangement to form the completed mRNA molecule. When RNA has been experimentally extracted from nuclei at different times, the RNA molecules appear to change in length, indicating that the splicing is done in a step-by-step manner. A clue to the mechanism of splicing has come from the study of another class of RNA molecules in the nucleus, called *small nuclear RNA* (or *snRNA*). These small RNAs, each about 100 nucleotides in length, contain base sequences that exhibit complementarity to the sequences at the ends of each intron. One of the functions of these small RNAs may be to form hydrogen bonds with complementary bases in the splice junctions of the hnRNAs and align the ends of adjacent exons together to be spliced at the correct

Figure 9.22. Action of RNA polymerase. When the core enzyme binds with the σ factor, the polymerase is able to recognize the promoter region. Once RNA synthesis has begun, the σ factor dissociates from the core enzyme. Transcription then proceeds until a termination site is reached. The ρ factor associates with the core enzyme at the terminator site to release the polymerase molecule and the completed RNA strand from the DNA template.

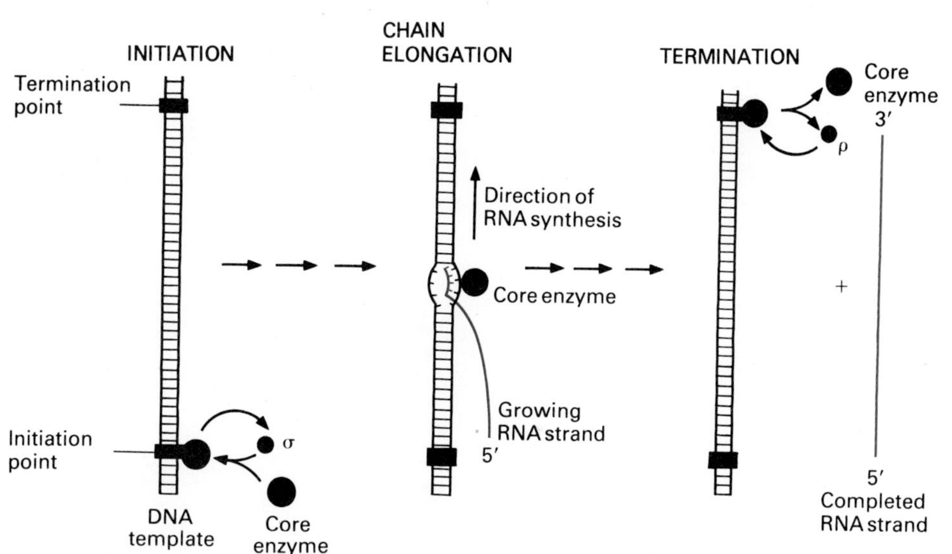

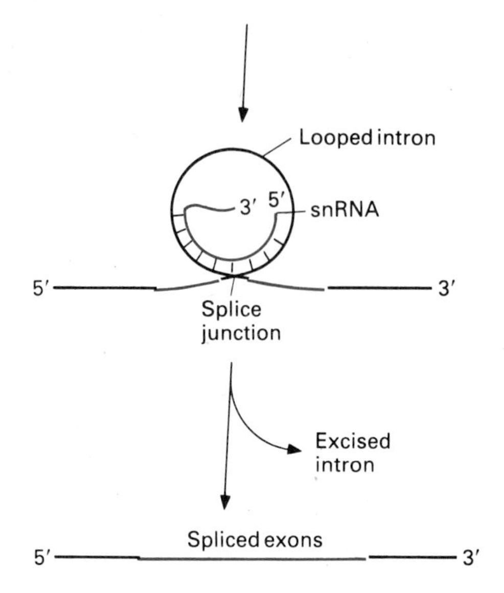

Figure 9.23. A proposed model for splicing of hnRNA involving small nuclear RNA (snRNA). The snRNA has a base sequence that is complementary in part to the ends of each intron, and can align the ends of adjacent exons together to be spliced at the correct point.

point (Fig. 9.23). A summary of the major steps involved in the processing of the RNA transcript of the ovalbumin gene into ovalbumin mRNA is given in Fig. 9.24.

All of the reactions involved in the processing of hnRNA occur in the nucleus of the cell; the RNA molecule is thus in the form of mature mRNA before it enters

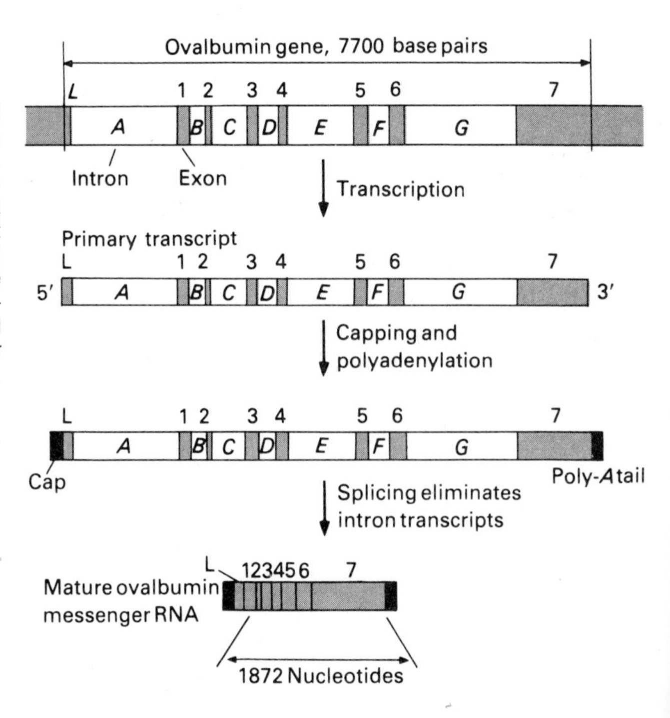

Figure 9.24. Steps involved in the production of mRNA from the ovalbumin gene. First, the entire gene is transcribed. This precursor RNA transcript is then capped at the 5′ end, and a poly-A tail is added to the 3′ end. The intron regions are then excised from the transcript, and the exons are spliced together to produce the mature mRNA.

the cytoplasm where proteins are synthesized. Upon leaving the nucleus, the completed mRNA molecule combines with special proteins, which appear to guide the mRNA through the pores of the nuclear envelope. The RNA fragments that are "spliced out" during processing remain in the nucleus to be degraded by enzymes into ribonucleotides. These ribonucleotides can then be used again in the synthesis of more RNA.

Characteristics of rRNA and Ribosomes. Ribosomes are small spherical particles that serve as the cellular sites where mRNA is translated into proteins. Electron micrographs reveal that each ribosome consists of two subunits, one about half the size of the other (Fig. 9.25). The size of the ribosome is often expressed in terms of its sedimentation coefficient, which measures the rate at which a particle will sediment in a unit centrifugal field of force. In general, the larger the size of the particle, the faster it will move in response to high-speed centrifugation. Thus, in the case of the *E. coli* ribosome, the complete functional particle has a sedimentation coefficient of 70S, while that of the smaller subunit is 30S and that of the larger subunit is 50S. The subunit values are not additive (30S + 50S ≠ 70S) because the velocity of sedimentation depends not only on size but also on shape,

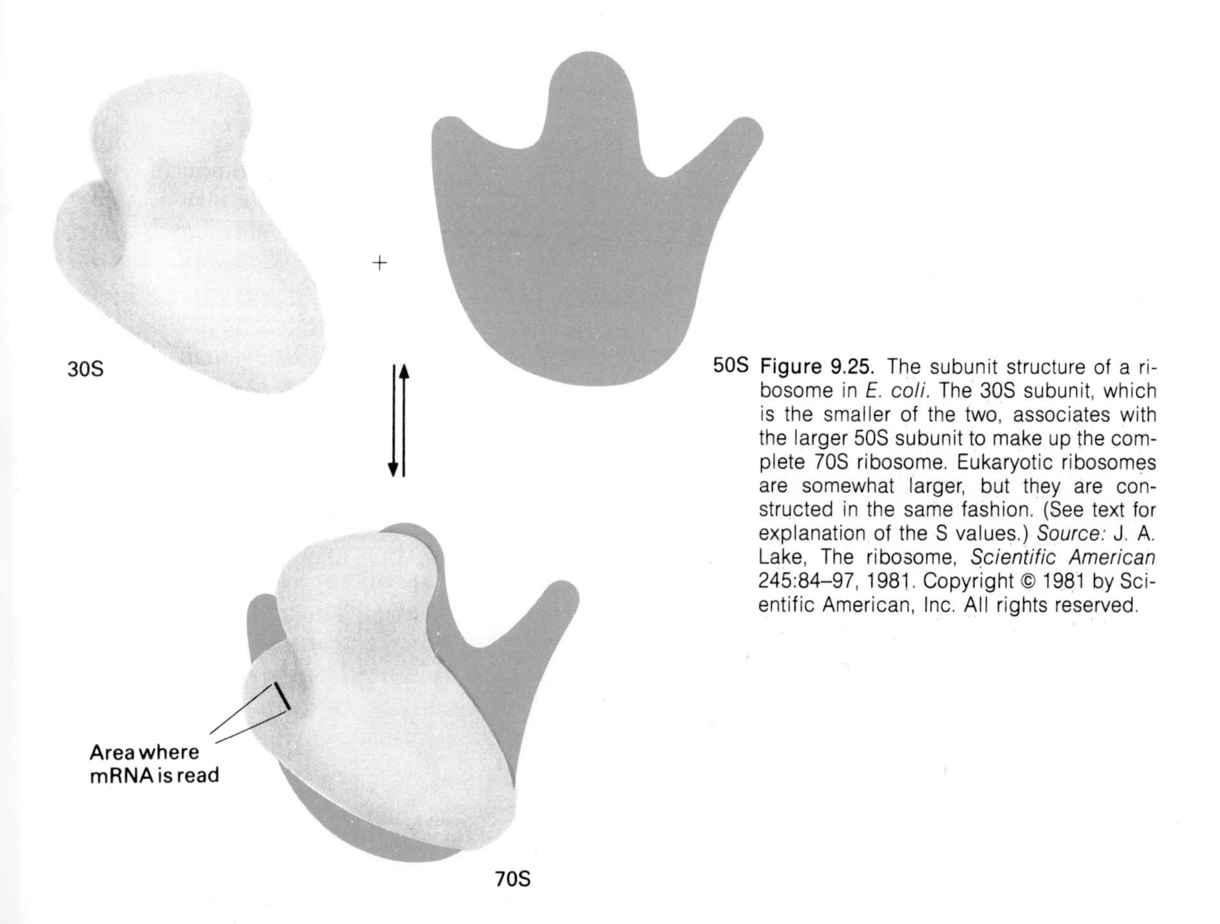

Figure 9.25. The subunit structure of a ribosome in *E. coli*. The 30S subunit, which is the smaller of the two, associates with the larger 50S subunit to make up the complete 70S ribosome. Eukaryotic ribosomes are somewhat larger, but they are constructed in the same fashion. (See text for explanation of the S values.) *Source:* J. A. Lake, The ribosome, *Scientific American* 245:84–97, 1981. Copyright © 1981 by Scientific American, Inc. All rights reserved.

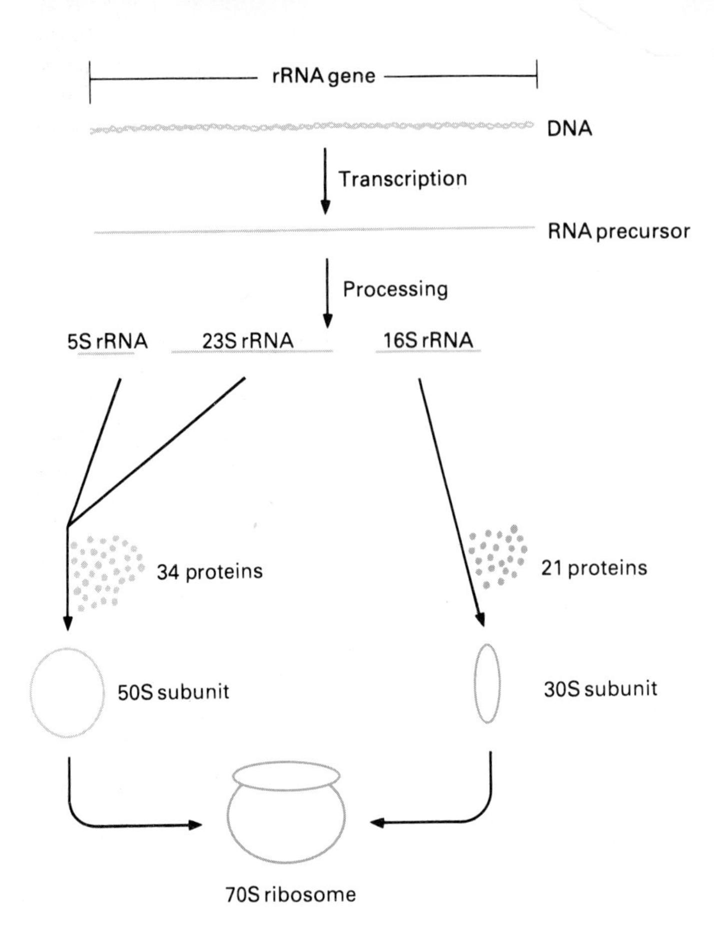

Figure 9.26. Production of a ribosome in *E. coli*. The bacterial chromosome contains 5 to 10 copies of the rRNA gene, one of which is shown above. Transcription of this gene yields a 30S precursor RNA molecule, which is cleaved into 5S, 23S, and 16S rRNA molecules. The 23S and 5S rRNA form a complex with 34 ribosomal proteins to yield the 50S subunit. The 16S rRNA molecule associates with 21 ribosomal proteins to yield the 30S subunit.

which differs in the three components. The slightly larger eukaryotic ribosomes vary in size in different species, but on the average are composed of 40S and 60S subunits, with a sedimentation coefficient of 80S for the complete structure.

Ribosomes are composed of roughly equal amounts of RNA and protein. There are 5 to 10 copies of the rRNA gene in *E. coli,* with at least one copy at each of three regions on the bacterial chromosome. Transcription of each rRNA gene yields a 30S precursor RNA molecule, which is then processed into three classes of rRNA: 16S (about 1700 nucleotides), 23S (about 3700 nucleotides), and 5S (about 120 nucleotides), as shown in Fig. 9.26. A 16S molecule combines with 21 different ribosomal proteins to yield the 30S ribosomal subunit. A 23S and a 5S molecule combine with 34 different ribosomal proteins to yield the 50S subunit.

Not as many details are known about the development and structure of eukaryotic ribosomes. The rRNA genes of higher organisms are present in hundreds to thousands of copies, mostly located in tandem in the so-called **nucleolar organizer regions** of chromosomes. These represent constricted areas (secondary constrictions) on the chromosomes that give rise to nucleoli. In humans, nucleolar organizers are found on the short arms of chromosomes 13, 14, 15, 21, and 22. In *Drosophila* and corn, on the other hand, nucleolar organizers are present on a single pair of chromosomes. The transcription of an rRNA gene yields a large precursor molecule, which is later processed into 18S, 28S, and 5.8S rRNA. Transcription from scattered sites other than the nucleolar organizer regions produces the 5S rRNA. Union of the 18S molecule with ribosomal proteins forms the 40S subunit of the ribosome. The 28S, 5.8S, and 5S molecules unite with proteins to yield the 60S subunit.

Transfer RNA. The function of tRNA is to transfer amino acids to the ribosome in the sequence dictated by the mRNA. There are one or more kinds of tRNA for each particular amino acid. Despite their differences in amino acid specificity, all of these tRNAs have a number of features in common. (1) All tRNAs are very small molecules, about 80 nucleotides in length, that are transcribed from clusters of repeated tRNA genes scattered throughout the genome. As with rRNA, transcription yields a large precursor molecule that is processed into the functional RNA. (2) All tRNAs have similar secondary and tertiary structures, despite the differences in their primary structure, or sequence of nucleotides. Studies of tRNA show a large amount of complementarity between different parts of each tRNA strand, which leads to hydrogen bonding and causes the strand to fold back upon itself into a characteristic "cloverleaf" two-dimensional arrangement and an "L-shaped" three-dimensional structure (Fig. 9.27). (3) All tRNAs contain several

Figure 9.27. The structure of tRNA. (a) The nucleotide sequence and the cloverleaf secondary structure of the tRNA for alanine in yeast. The structure contains three loops, which are produced by hydrogen bonding within the chain. The P and OH refer to the phosphate and hydroxyl groups at the 5′ and 3′ ends, respectively. (b) The tertiary structure of this molecule. The numbers indicate the positions of nucleotides, beginning with the 5′ end. The region labeled anticodon is responsible for recognizing the mRNA code words. *Source:* (a) From R. W. Holley *et al.*, *Science* 147:1462–1465, 1965; (b) From S. H. Kim *et al.*, *Science* 185:435–440, 1974. Copyright 1965 and 1974 by the American Association for the Advancement of Science.

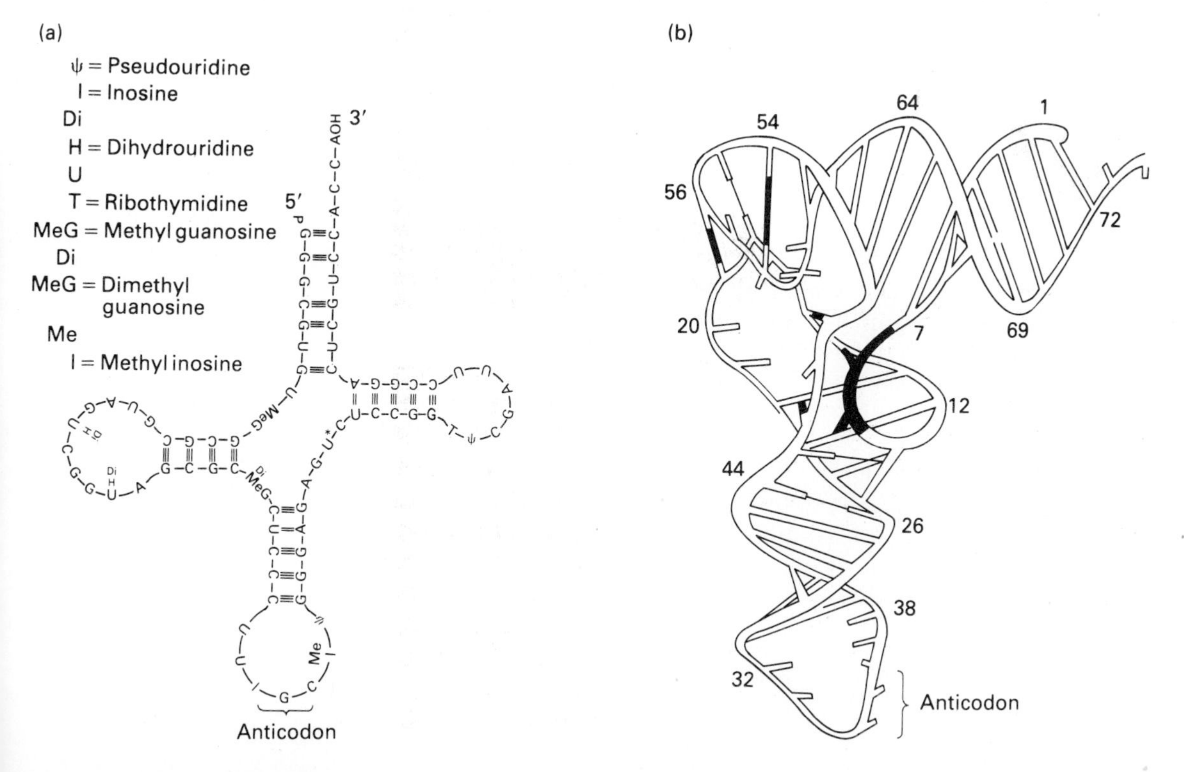

unusual or modified nucleosides. The post-transcriptional modification of the tRNA transcripts alters many of the usual nucleosides, to form residues that are not commonly found in other types of RNA. These unusual nucleosides include inosine, which contains the purine hypoxanthine, ribothymidine, in which ribose is bonded to thymine, and pseudouridine, which has an unusual carbon-to-carbon bond between the base and the sugar.

Each tRNA molecule must exhibit two types of specificity in order to perform its role in protein synthesis. It must be able (1) to recognize a specific amino acid and then (2) to recognize the specific codon for that amino acid in the mRNA. The *amino acid recognition* is needed so that the tRNA will bond chemically to only one type of amino acid and will transfer that particular amino acid to the ribosome. The *codon recognition* is needed so that the tRNA will bind to the surface of the ribosome under the direction of mRNA, thus ensuring the insertion of the amino acid at its proper position within the polypeptide chain. These specificity requirements are illustrated in Fig. 9.28.

Amino acid recognition is accomplished by a class of enzymes known as **aminoacyl-tRNA synthetases.** There are 20 of these enzymes, each of which is specific for a different amino acid. The enzyme catalyzes the bonding of the amino acid that it recognizes to the 3′ end of a particular tRNA; the energy for this reaction is provided by the hydrolysis of ATP. This combining process of the amino acid with tRNA occurs in a two-step reaction. In the first step, the amino acid is *activated* by forming a complex with AMP and the enzyme. This step is illustrated below for amino acid type 1 and synthetase enzyme 1 (E_1):

$$\text{Amino acid}_1 + \text{ATP} + E_1 \longrightarrow \text{Amino acid}_1\text{-AMP-}E_1 + PP_i$$

where PP_i represents inorganic pyrophosphate. The amino acid-AMP-enzyme complex then undergoes the second step in the reaction, which results in the formation of a bond between the amino acid and a tRNA molecule of the proper type:

$$\text{Amino acid}_1\text{-AMP-}E_1 + \text{tRNA}_1 \longrightarrow \text{Amino acid}_1\text{-tRNA}_1 + E_1 + \text{AMP}$$

A tRNA molecule that is combined with its amino acid is referred to as either an *aminoacyl-tRNA* or a *charged tRNA*.

The second form of specificity, that of recognizing a particular codon in the mRNA strand, is accomplished by way of a specific sequence of three nucleotides, called the **anticodon,** in the tRNA molecule (see Fig. 9.27). The anticodon in each different tRNA is complementary in sequence to one (or, in certain cases, more than one) codon in the mRNA. This complementarity allows an aminoacyl-tRNA to hydrogen bond specifically to an mRNA codon, and thereby bring its attached amino acid to the ribosome surface when dictated by the genetic instructions in the mRNA. Transfer RNA thus serves as an adaptor molecule to mediate the transfer of information from the mRNA to the polypeptide.

Translation

During protein synthesis, the codon sequence in mRNA is translated into a corresponding sequence of amino acids in the polypeptide chain. It is a complex chemical process that requires the participation of ribosomes, charged tRNAs,

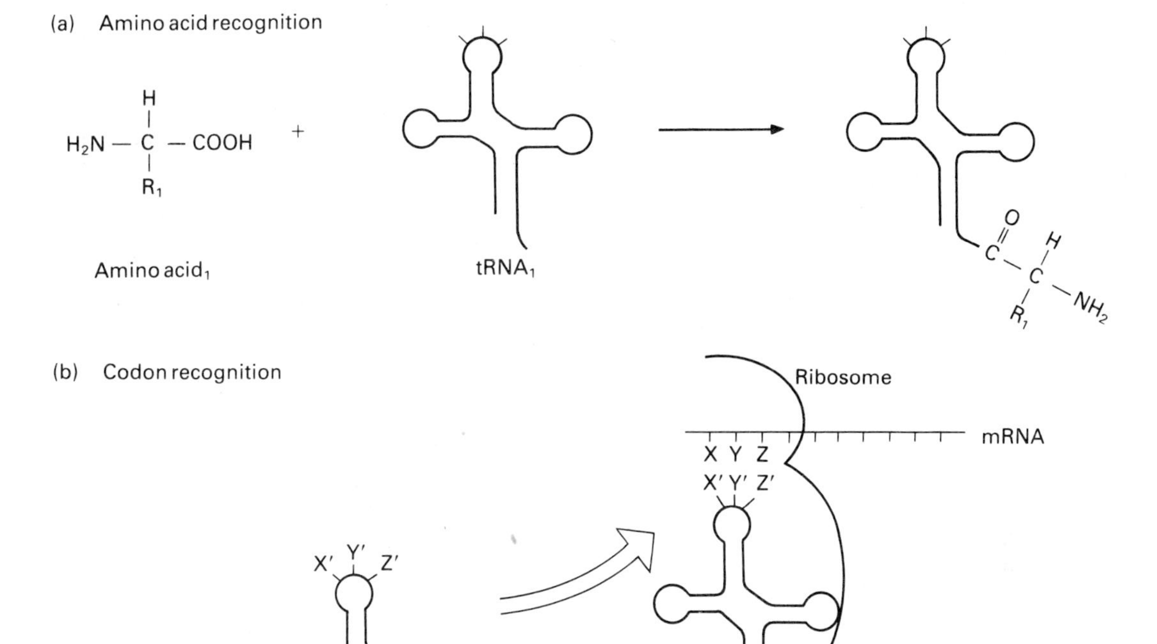

Figure 9.28. The two specifications shown by a tRNA molecule: amino acid recognition and codon recognition. (a) Each tRNA molecule combines with a specific amino acid under the direction of the enzyme aminoacyl-tRNA synthetase. (b) Once bonded to its amino acid, the tRNA hydrogen-bonds to the specific codon for that amino acid in the mRNA on the surface of the ribosome. X, Y, and Z represent the three bases in a particular codon, while X′, Y′, and Z′ represent the bases in the anticodon that enable the tRNA molecule to recognize the codon.

several different protein factors (enzymes), and GTP as an energy source. The translation of the gene message begins when a ribosome becomes attached to the 5′ end of the mRNA. The mRNA molecule then acts as a tape of instructions; the instructions are read consecutively, codon by codon, as the mRNA moves along the ribosome. While only one polypeptide can be formed during the process on any single ribosome, many identical copies of the coded polypeptide can be synthesized in a short time. This is accomplished through simultaneous translation of the mRNA molecule by several ribosomes, spaced about 90 nucleotides apart (Fig. 9.29). The complex of ribosomes plus mRNA is referred to as a **polysome.** Electron micrographs of cells that are actively engaged in protein synthesis show extensive polysome complexes, with the length of the polypeptide chains increasing on the ribosomes that are located toward the 3′ end of the mRNA molecule. In bacteria,

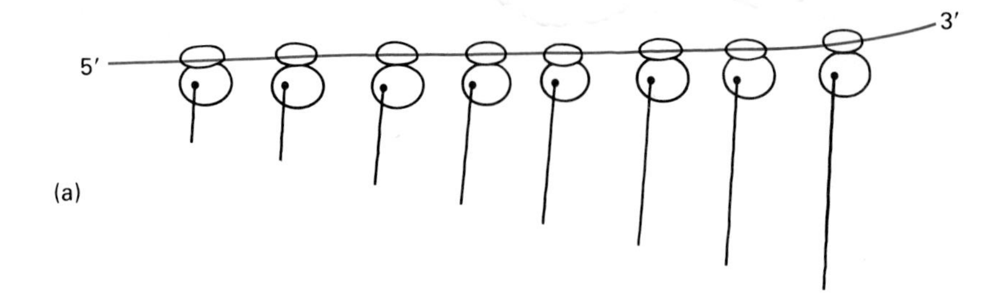

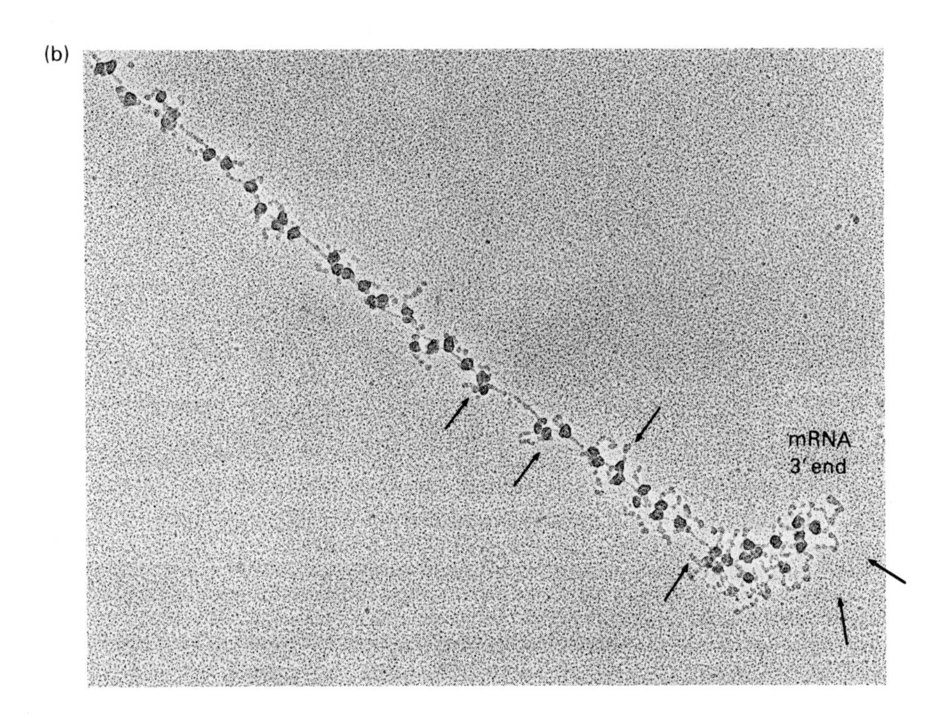

Figure 9.29. (a) A single mRNA molecule associates with many ribosomes to yield a polysome. Each ribosome is in a different stage of polypeptide synthesis, so that the lengths of the polypeptide being synthesized form a gradation, from the shortest at the 5′ end of the mRNA molecule to the longest at the 3′ end. (b) An electron micrograph, showing the translation of the mRNA for the fibroin protein in the silkworm, *Bombyx mori*. In the micrograph, we can actually see the gradation in length of the nascent polypeptide chains. The arrows point to polypeptides; the ribosomes are the dark, spherical bodies. *Source:* Photograph courtesy of S. L. McKnight and O. L. Miller, Jr., Department of Biology, University of Virginia.

polysomes are formed at the same time that mRNA is being synthesized (Fig. 9.30). Such coupling of transcription and translation is possible only in prokaryotes, in which there is no nuclear envelope to separate the DNA from the protein-synthetizing apparatus.

In addition to combining with mRNA, each ribosome can also bind with charged tRNAs. Every ribosome has two binding sites that are capable of combining with tRNA (Fig. 9.31a). One binding site is called the *peptidyl* or *P site,* and the other is called the *aminoacyl* or *A site.* During translation, the most recently added

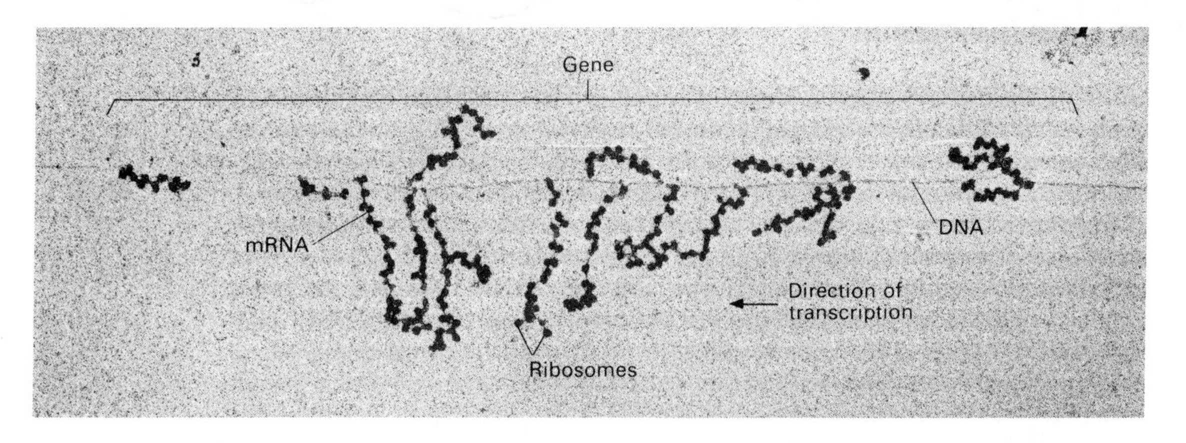

Figure 9.30. An *E. coli* gene being simultaneously transcribed and translated. The DNA, mRNA, and ribosomes are visible, but the polypeptides and tRNA are not. *Source:* From O. L. Miller, Jr., B. A. Hamkalo, and C. A. Thomas, Jr., *Science* 169:392–395, 1970. Copyright 1970 by the American Association for the Advancement of Science.

charged tRNA is always bound at the A site. After a peptide bond has been formed, linking its amino acid to the previous one added to the chain, the tRNA moves to the P site. The partially completed polypeptide remains attached to the 3′ end of the tRNA whose amino acid has just been added to the chain. Thus, when both sites are properly filled for further elongation of the chain, a tRNA with a partially completed polypeptide (peptidyl-tRNA) will be bound to the P site and a newly attached aminoacyl-tRNA will be present at the A site (Fig. 9.31b).

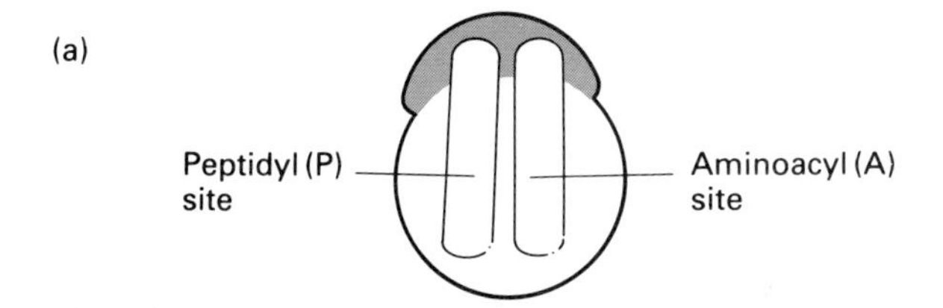

Figure 9.31. (a) A ribosome has two binding sites capable of combining with tRNA: the peptidyl or P site and the aminoacyl or A site. (b) During polypeptide synthesis, a tRNA attached to a partially completed polypeptide is bound to the P site, and an aminoacyl-tRNA is bound to the A site. A ribosome in this state is in the proper form for further elongation of the polypeptide chain.

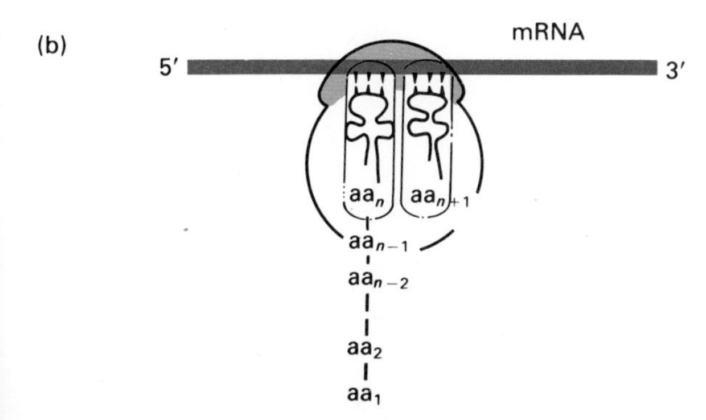

Initiation of Polypeptide Synthesis. Translation is initiated when the 5′ end of the mRNA molecule combines with the smaller subunit of the ribosome and with the *initiator tRNA,* which carries the first amino acid to start polypeptide chain growth. The process requires an energy source (mainly GTP) and several protein *initiation factors*—three, called IF-1, IF-2, and IF-3, in prokaryotes and as many as eight or more, labeled eIF-1, eIF-2, eIF-3, etc., in eukaryotes. The initiator tRNA that is involved in this process is charged with the amino acid *methionine* in eukaryotes. In prokaryotes, methionine in the initiator tRNA is formylated to produce *formylmethionyl-tRNA* (*fmet-tRNA*). Therefore, methionine will be the initial amino acid of a newly synthesized polypeptide in eukaryotes, and formylmethionine will be the initial amino acid in prokaryotes. The formyl group is removed before the protein becomes functional, however, and the initial methionine residue may be modified or removed as well. For this reason, functional proteins in prokaryotes and eukaryotes often do not contain methionine as their initial amino acid.

The initiation of translation in prokaryotes is illustrated in Fig. 9.32. We see that in prokaryotes initiation proceeds in three steps. The first step involves the binding of the 5′ end of the mRNA to the 30S ribosomal subunit. The dissociation of the 70S ribosome in the presence of IF-1 and IF-3 and the association of IF-3 with the 30S subunit are essential prerequisites for this initial binding step.

The second step in the initiation process involves the binding of the initiator tRNA, along with GTP and IF-2, to the 30S complex. Binding of the initiator tRNA to the 30S complex occurs with the formation of hydrogen bonds between the anticodon of the initiator tRNA (UAC) and the complementary mRNA codon (AUG). The codon AUG normally specifies the amino acid methionine when present in the interior of the gene message, where it is specific for methionyl-$tRNA^{met}$. However, when AUG occurs at the beginning of the gene message in prokaryotes, it binds only to fmet-$tRNA^{fmet}$. Since AUG is the first mRNA codon to be recognized during translation, it is known as the *initiator codon.*

The 50S subunit of the ribosome then combines with the 30S complex in the third step of initiation. This step requires the breakdown of GTP and results in the release of the initiation factors. The complete ribosomal structure, called the 70S initiation complex, is then formed, in which the charged initiator tRNA is bound to the P site. The ribosomal complex is now ready to carry out the process of protein synthesis.

Elongation of the Polypeptide Chain. We can view the elongation of the polypeptide chain as a cyclic process that occurs through the repetition of three successive steps: (1) *aminoacyl-tRNA binding,* (2) *peptide bond synthesis,* and (3) *translocation* (Fig. 9.33). In aminoacyl-tRNA binding, an aminoacyl-tRNA associates with the vacant A site of the ribosome. This step is mediated by specific pairing between the codon and anticodon and occurs with the breakdown of GTP and in the presence of certain protein *elongation factors,* symbolized EF-Tu and EF-Ts in bacteria.

Once the incoming charged tRNA is stabilized at the A site, a peptide bond forms between the carboxyl group of the first amino acid to be carried to the

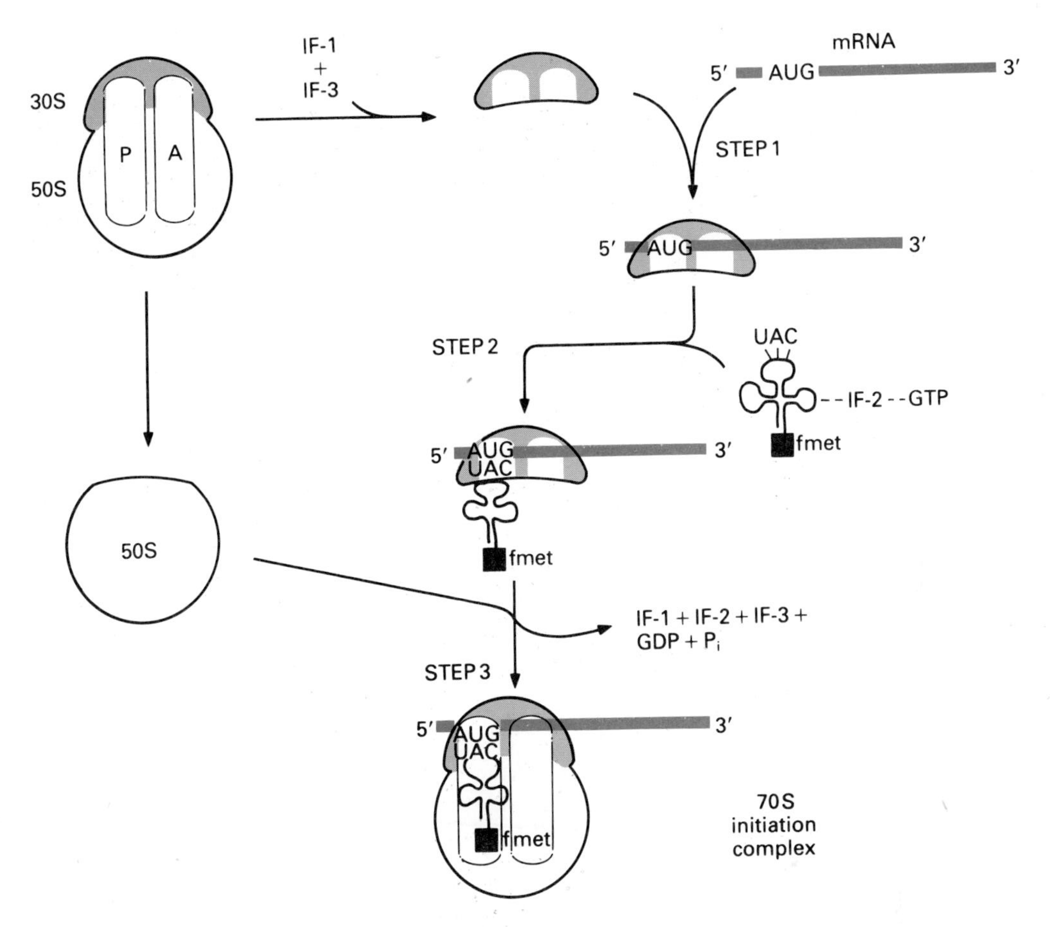

Figure 9.32. Initiation of translation. The smaller ribosomal subunit forms a complex with various initiation factors (IFs), GTP, and the 5′ end of the mRNA molecule. In response to the initiator codon AUG on the mRNA, a special initiator tRNA that is charged with the amino acid formylmethionine (methionine in eukaryotes) binds to the P site of the ribosomal subunit. The specificity for binding is provided by the complementary nature of the initiator codon (AUG) and the tRNA anticodon (UAC). The larger ribosomal subunit then associates with the developing structure to form the complete 70S initiation complex.

ribosome and the amino group of the second. The formation of the peptide bond is catalyzed by an enzyme known as *peptidyl transferase,* which is one of the proteins that make up the larger ribosomal subunit.

The final step in the cycle is **translocation,** which is needed to move the peptidyl-tRNA that is now at the A site to the P site, thus vacating the A site for further aminoacyl-tRNA binding. During the process, the tRNA formerly bound to the P site is released for reuse. The transfer of the peptidyl-tRNA from the A site to the P site is accompanied by the movement of the mRNA along the ribosome by one codon and requires elongation factor G (also called *translocase*) and the breakdown of another molecule of GTP. Translocation is believed to involve a change in the three-dimensional conformation of the entire ribosome in order to

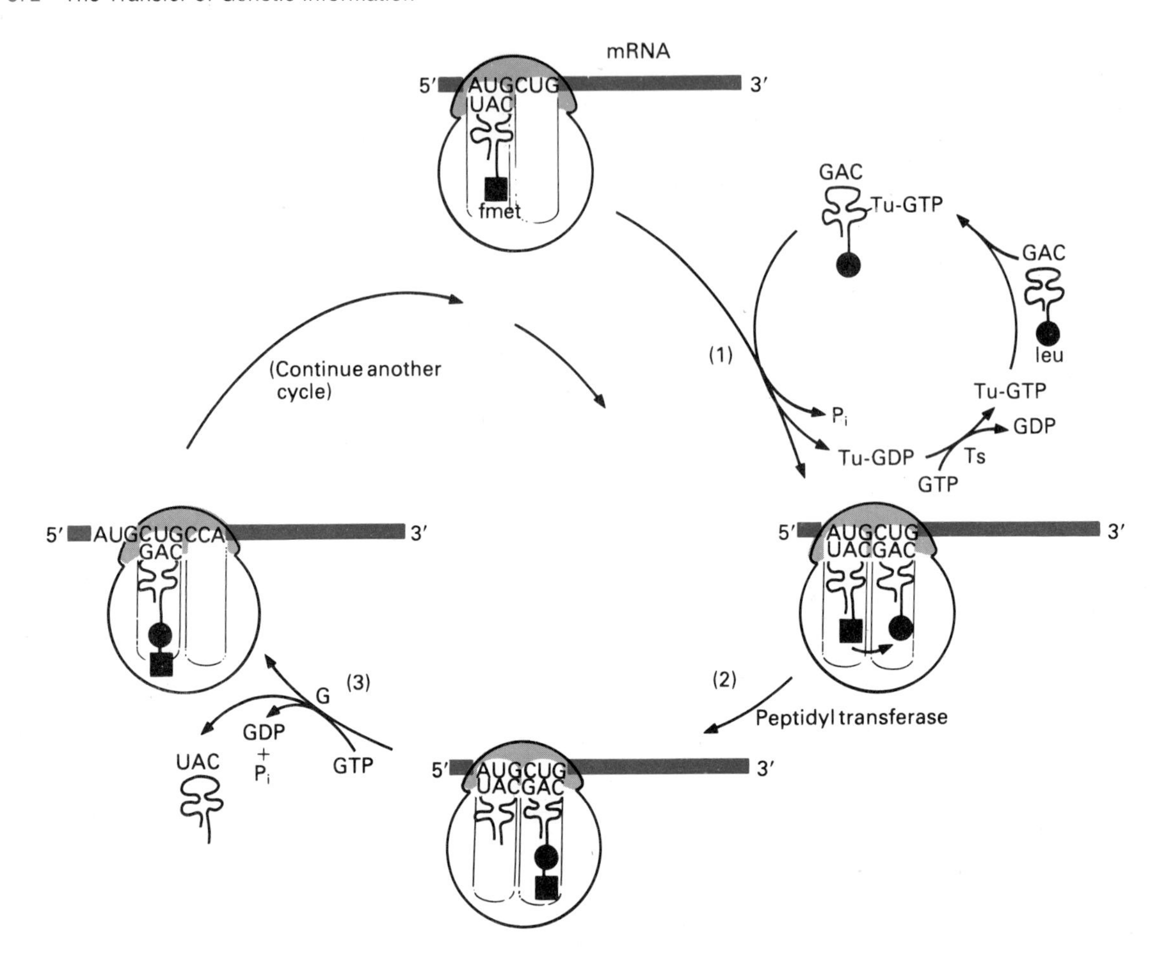

Figure 9.33. Elongation of the polypeptide chain. In response to the second codon (CUG), a leucyl-tRNA binds to the A site (step 1). Binding requires the presence of elongation factors EF–Tu and EF–Ts (Tu and Ts in the diagram) and the hydrolysis of GTP. Specificity for binding is provided by the recognition between the codon and anticodon. The enzyme peptidyl transferase then catalyzes the formation of a peptide bond between the carboxyl group of fmet and the amino group of leu (step 2). The final step of elongation (step 3) is translocation, during which the mRNA moves up by one codon and the peptidyl-tRNA simultaneously moves from the A to the P site. Translocation requires an elongation factor G and the hydrolysis of GTP. The elongation cycle repeats itself as each amino acid is added to the growing polypeptide chain.

produce the one-codon shift in the tRNA and mRNA. Once translocation is complete, the ribosome is now ready for another elongation cycle.

Termination. The elongation cycle of protein synthesis continues until all of the amino acids are incorporated into the completed polypeptide chain. Termination then occurs through the process shown in Fig. 9.34. Termination requires the presence of certain *termination* (or *release*) *factors,* which can recognize the *terminator* (or *stop*) *codons* UAA, UAG, and UGA in mRNA. In *E. coli,* for example, one release factor (RF-1) recognizes UAA and UAG, while another (RF-2) recognizes UGA and UAA. Once a terminator codon arrives opposite the A site on

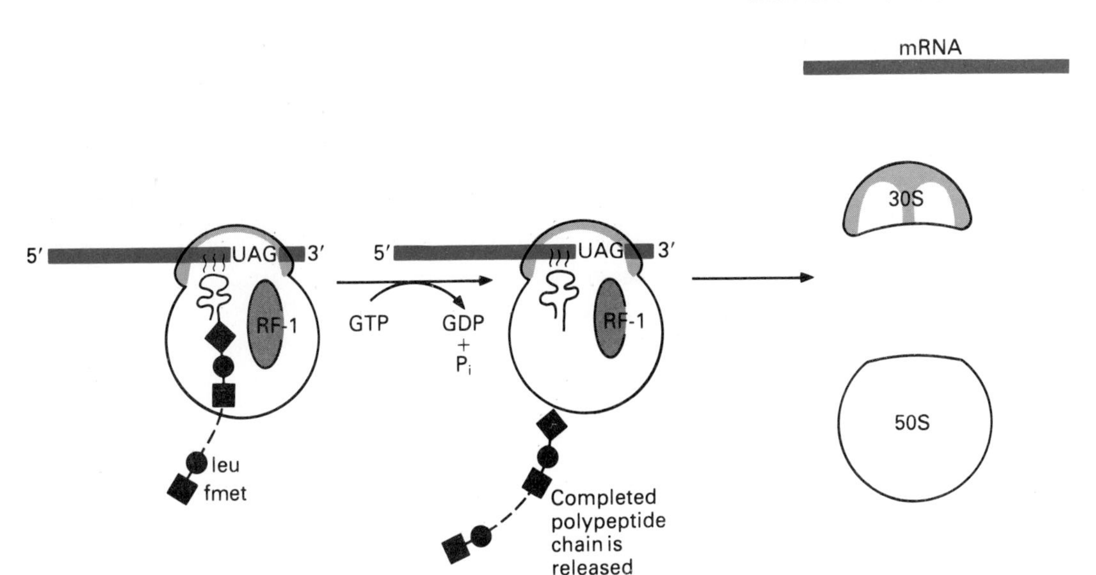

Figure 9.34. Termination of polypeptide synthesis. A terminator codon, in this case UAG, occupies the A site. A termination or release factor (RF-1 in this case) combines to the A site. The ribosome then releases the completed polypeptide and the mRNA. The ribosome dissociates into its subunits, which are used again in the synthesis of other polypeptides.

the ribosome, it is recognized by the appropriate release factor, which then blocks further chain elongation. In the presence of another release factor (RF-3), the completed polypeptide chain dissociates from the tRNA. The ribosome then releases the mRNA and the uncharged tRNA and is free to engage in another round of protein synthesis.

The Genetic Code

We have seen that the genetic code is a triplet code, which consists of three-letter code words, called codons, that serve to specify the sequence of amino acids in a polypeptide chain. We will now consider the genetic code in more detail with special emphasis on how the genetic code was solved.

Deciphering the Code. The genetic code was deciphered through work in several laboratories in the early 1960s after an in vitro system was developed that permitted researchers to synthesize polypeptides with the use of synthetic mRNAs of known base composition. In the first experiments designed to crack the code, the synthetic mRNAs were synthesized with the enzyme polynucleotide phosphorylase. This enzyme catalyzes the formation of single-stranded RNA molecules from nucleoside diphosphates. The enzyme has the interesting property of not using a template to specify the nucleotide sequence. The proportion of each base in the RNA molecules formed in the reaction therefore depends solely on its proportion in the reaction mixture of nucleoside diphosphates. Thus, if ADP is the only nucleoside diphosphate present in the reaction mixture, the RNA made by polynucleotide phosphorylase will be a homopolymer composed only of A, whereas if the reaction mixture has two different nucleoside diphosphates, consisting shall we say of 75 percent ADP and 25 percent CDP, the RNA that is formed will be a heteropolymer that contains three-fourths A and one-fourth C. Notice that although we can predict the overall base composition of the RNA that is produced in this manner, we have no way of knowing the specific sequence of nucleotides.

The first RNA codon to be identified with the use of synthetic mRNAs was the code word for phenylalanine. This identification was made by M. Nirenberg and H. Matthaei in 1961 when they discovered that the synthetic RNA polyuridylic acid (poly U) promoted the synthesis of only the polypeptide polyphenylalanine, despite the presence of all 20 amino acids in the reaction mixture. Nirenberg and Matthaei correctly interpreted the result of this experiment to mean that the RNA codon for phenylalanine must be UUU. Shortly after this discovery, investigators found that poly A and poly C yield polylysine and polyproline, respectively. Thus, AAA must be the codon for lysine and CCC must be the codon for proline.

Codons with more than one kind of nucleotide were identified by using mRNAs that contain two types of bases. The nucleotide sequences in these copolymers are essentially random, so that it is possible to predict the frequencies with which different codons appear in the synthetic mRNA. By relating the frequencies of the various amino acids in the polypeptides that the RNA codes for to the expected frequencies of the various triplet sequences in the random RNA copolymer, it is possible to deduce the base composition of the codons for all 20 amino acids. For example, a reaction mixture consisting of $\frac{3}{4}$ ADP and $\frac{1}{4}$ CDP should produce a random copolymer in which the codons AAC, ACA and CAA will each occur at a frequency of $(\frac{3}{4})(\frac{3}{4})(\frac{1}{4}) = \frac{9}{64}$. The finding of asparagine at about this frequency in the polypeptide would indicate that a codon consisting of 2 A and 1 C codes for asparagine.

Experiments with randomly ordered synthetic RNAs provided information about the base composition of the various codons; they did not establish the actual *sequence* of the nucleotides within any codon, except for the trivial cases of AAA, CCC, GGG, and UUU. Two other experimental procedures were later developed that led to the assignment of base sequences within codons. One procedure was the synthesis of polypeptides using artificial mRNAs of defined rather than random nucleotide sequence. For example, H. Khorana synthesized the RNA copolymer poly-UG with the alternating codon sequence –UGU–GUG–UGU–GUG–. Use of this copolymer as an mRNA gave rise to a polypeptide chain with an alternating amino acid sequence –cysteine–valine–cysteine–valine–. Since it was already known that the triplet UG_2 codes for valine but not for cysteine, the codons for cysteine and valine must then be UGU and GUG, respectively. The second experimental procedure was developed by Nirenberg and P. Leder when they discovered that RNA triplets of known sequence can attach to ribosomes in a manner similar to mRNAs and promote the specific binding of aminoacyl-tRNAs that carry the complementary anticodons. For example, when the triplet UUG was added to ribosomes, it promoted the specific binding of leucyl-tRNA to the ribosome surface. Thus, UUG must code for the amino acid leucine. With these procedures, investigators soon established the code words for each of the 20 amino acids.

EXAMPLE 9.1. Several of the first experiments designed to crack the genetic code were conducted in the laboratory of S. Ochoa. In one experiment reported by Ochoa's group in 1963, an artificial mRNA was synthesized using a reaction mixture consisting of 5/6 ADP and 1/6 CDP and polynucleotide phosphorylase. This mRNA, when used in a cell-free extract, coded for the synthesis of a polypeptide with the following amino acid composition:

AMINO ACID	OBSERVED FREQUENCY RELATIVE TO LYSINE
lysine	100%
threonine	26%
asparagine	24%
glutamine	24%
proline	7%
histidine	6%

From these data, postulate codon assignments for each of the amino acids listed.

Solution. Since AAA is the most frequently occurring codon in the artificial mRNA, we can express the expected codon frequencies relative to AAA in the following manner:

POSSIBLE CODONS IN mRNA		EXPECTED CODON FREQUENCY	CODON FREQUENCY RELATIVE TO AAA
AAA		$(5/6)^3 = 125/216$	100%
AAC	each	$(5/6)^2(1/6) = 25/216$	20%
ACA			20%
CAA			20%
ACC	each	$(5/6)(1/6)^2 = 5/216$	4%
CAC			4%
CCA			4%
CCC		$(1/6)^3 = 1/216$	0.8%

The codon AAA specifies lysine. This codon assignment was known before the experiment was conducted and prompted the investigators to express their data relative to lysine. Threonine, at a relative frequency of 26%, would appear to be coded for by one of the triplets that is made up of 2 A's + 1 C (either AAC or ACA or CAA) and by one of the triplets that is composed of 1 A + 2 C's (either ACC or CAC or CCA). Its expected relative frequency would then be 20% + 4% = 24%. Asparagine would then be coded for by another of the 2 A + 1 C triplets, with an expected relative frequency of 20%. The final 2 A + 1 C triplet would be assigned to glutamine, also with a frequency of 20%. We would postulate that proline is coded for by another of the 1 A + 2 C triplets and by the CCC triplet, for an expected relative frequency of 4.8%. Finally, we would assign histidine the remaining 1 A + 2 C triplet, for an expected frequency relative to the codon for lysine of 4%. Note that some of the actual frequencies must be inferred by taking the group of amino acids produced as a whole. For example, we might guess that asparagine is coded for by one 2 A + 1 C triplet and one 1 A + 2 C triplet, to give an expected frequency of 24%—the same as the observed frequency. If this were the case, however, we would be short one arrangement of triplets to account for the observed frequencies of proline and histidine.

Table 9.1. The mRNA codon assignments.

FIRST (5′) LETTER	SECOND LETTER U		SECOND LETTER C		SECOND LETTER A		SECOND LETTER G		THIRD (3′) LETTER
U	UUU	Phe	UCU	Ser	UAU	Tyr	UGU	Cys	U
	UUC		UCC		UAC		UGC		C
	UUA	Leu	UCA		UAA	terminator	UGA	terminator	A
	UUG		UCG		UAG	terminator	UGG	Trp	G
C	CUU	Leu	CCU	Pro	CAU	His	CGU	Arg	U
	CUC		CCC		CAC		CGC		C
	CUA		CCA		CAA	Gln	CGA		A
	CUG		CCG		CAG		CGG		G
A	AUU	Ile	ACU	Thr	AAU	Asn	AGU	Ser	U
	AUC		ACC		AAC		AGC		C
	AUA		ACA		AAA	Lys	AGA	Arg	A
	AUG	Met	ACG		AAG		AGG		G
G	GUU	Val	GCU	Ala	GAU	Asp	GGU	Gly	U
	GUC		GCC		GAC		GGC		C
	GUA		GCA		GAA	Glu	GGA		A
	GUG		GCG		GAG		GGG		G

Characteristics of the Code. The genetic code dictionary for RNA is given in Table 9.1. It has several important features that deserve mention. First, of the 64 possible triplet codons in RNA, all but three (UAA, UAG, and UGA) code for specific amino acids. As we have seen in an earlier section, these three code words function as stop or terminator codons. The three are usually found either singly or in combination at the end of a gene where they signal the completion of the gene message. Second, there are 61 codons remaining that code for the 20 amino acids; the average number of codons per amino acid is thus approximately three ($61/20 \simeq 3$). We therefore say that the genetic code is **degenerate,** which means that more than one codon can specify the same amino acid. For example, six different code words specify leucine; there are also six codons for serine and arginine. Several amino acids have four codons each; one (isoleucine) has three codons; and several others have two. Only methionine and tryptophan are each specified by a single codon.

A third major point is that extensive study with a wide variety of organisms and in vitro systems suggests that, with the exception of certain cytoplasmic organelles, the genetic code is *universal*; that is, the same code is used in viruses and prokaryotes and by the nuclear genes of eukaryotes. Some organelles appear to employ a somewhat different codon dictionary. For example, in human mitochondria, AUA, which is normally the codon for isoleucine, codes for methionine, while UGA, which is normally a terminator codon, codes for tryptophan. The basis for these and other differences in the genetic code of mitochondria are still unknown.

The degeneracy of the genetic code raises some interesting questions concerning its evolution. The genetic code seems to have evolved in a way that provides a buffer against the effects of mutation. For instance, a nucleotide change in the third position of the code word GUU would not alter the amino acid sequence of the polypeptide chain, because all codons that start with GU code for the same amino acid. The same pattern is true of a number of other amino acid codons. Furthermore, if we compare the genetic code with the amino acid structures given in Fig. 9.1, we can see that amino acids with similar chemical properties often have code words that differ by only one nucleotide. Thus, even if a nucleotide change results in an amino acid change, the newly substituted amino acid may be similar enough to the original so as not to destroy protein function.

Wobble Hypothesis. Since there are 61 mRNA codons that specify amino acids, we might expect to find at least 61 different tRNA molecules (that is, tRNA molecules with 61 different anticodons) in a cell. The evidence now indicates that because of the degeneracy in the genetic code, fewer tRNAs are actually needed. Crick has suggested that fewer types of tRNA than mRNA codons are needed because of a lack of pairing specificity in the third base of many anticodons. He proposed that the first and second bases of the codon and anticodon pair according to strict base-pairing rules (i.e., A with U, and G with C), but that a degree of flexibility, or wobble, is allowed in the pairing of bases at the third position. This flexibility results in a less stringent set of rules for third-position pairing, given in Table 9.2. For example, an anticodon that consists of the nucleotide sequence AGG will pair not only with the exact complement UCC in mRNA, but with a certain degree of wobble, it can also recognize and bond to the mRNA codon UCU. The first two bases of codons therefore confer most of the codon-anticodon specificity, since the wobble rules dictate that the mRNA codons ending in either C or U, or even A in some cases, can specify the amino acid carried by a single type of tRNA.

Table 9.2. Pairing at the third position of the mRNA codon and tRNA anticodon.

NUCLEOSIDE ON ANTICODON	BASES RECOGNIZED ON CODON
U	A, G
C	G
A	U
G	U, C
I*	U, C, A

* I = inosine (the purine hypoxanthine bonded to ribose).

TO SUM UP

1. Information flows from DNA to RNA in a process known as transcription. A single strand of DNA, called the sense strand, serves as the template for synthesizing a complementary strand of RNA. The transcription process is catalyzed by the enzyme RNA polymerase.
2. Transcription yields three classes of RNA. They are mRNA, which carries the message of structural genes; rRNA, which is a constituent of ribosomes, the sites of protein synthesis; and tRNA, which brings amino acids to the ribosomes in the sequence dictated by the messenger RNA.
3. All three classes of RNA must undergo modification after transcription in order to be converted to their active forms. Eukaryotic mRNA and the rRNAs and tRNAs of both eukaryotes and prokaryotes are cut out of larger precursor transcripts. A cap and a poly-A tail are also added to the eukaryotic mRNA. The rRNA forms a complex with certain proteins to make up a ribosome, and the tRNA becomes active when it binds chemically to an amino acid.
4. Translation is the process in which the message in the mRNA is decoded into an amino acid sequence. Translation occurs on the ribósomes and is mediated by various protein factors. Polypeptide synthesis is initiated by a special initiator tRNA, which brings in a methionine (formylmethionine in prokaryotes) in response to the mRNA initiator codon AUG. Translation is terminated when the ribosome encounters any of the three terminator codons UAA, UAG, and UGA in mRNA.
5. The elongation cycle of protein synthesis consists of three steps: aminoacyl-tRNA binding, in which the anticodon of an incoming charged tRNA is paired with a complementary codon in mRNA; peptide bond synthesis, which is catalyzed by the enzyme peptidyl transferase; and translocation, which involves the transfer of a peptidyl-tRNA from the A site to the P site of the ribosome, accompanied by the movement of the mRNA along the ribosome by one codon.
6. The genetic code was deciphered by three types of experiments: (1) the use of polyribonucleotides with random base sequences as mRNAs in cell-free protein synthesizing systems, (2) the use of synthetic mRNAs of defined base sequences, and (3) the binding of trinucleotides of known sequence and their complementary aminoacyl-tRNAs to ribosomes. These experiments established that the code is degenerate, with usually more than one codon specifying the same amino acid, and (nearly) universal, applying to bacteria and viruses and to all nuclear genes of eukaryotes.

Questions and Problems

1. Define the following terms, and distinguish between the members of paired terms:
   active site–allosteric site
   amino acid
   aminoacyl-tRNA synthetase
   cistron
   codon–anticodon
   colinearity
   enzyme
   feedback inhibition
   hnRNA
   metabolic block
   mRNA
   peptide bond
   polypeptide
   polysome
   promoter
   prototrophic–auxotrophic
   ribosomes
   RNA polymerase
   rRNA
   split genes–overlapping genes
   transcription–translation
   tRNA

2. State whether each of the following statements is true or false, and, if false, explain why.

(a) There are 20 different amino acids that occur naturally in DNA.

(b) Disulfide bonds are important in the three dimensional folding of many polypeptides.

(c) All proteins are enzymes.

(d) The activity of a protein is ultimately determined by the specific sequence of amino acids in its polypeptide chains.

(e) Hydrophobic bonding in proteins refers to the association of nonpolar side chains of amino acids in such a way as to protect them from hydration.

(f) Only the amino acid changes that occur within the active site of an enzyme can destroy its catalytic activity.

(g) The two strands of a DNA molecule contain the same genetic information.

(h) Within one gene, the same codon always specifies the same amino acid, but in a different gene, that codon may specify a different amino acid.

(i) There are four amino acids that have six codons each.

(j) Only two amino acids, methionine and tryptophan, are specified by just one codon each.

(k) Some organisms contain ten (or more) times the amount of DNA than is needed to code for their different proteins.

3. Fill in the blanks.

(a) A protein that is an aggregate of two or more polypeptide chains is said to be a(n) ______________ protein.

(b) The alpha helix is one possible ________ structure of proteins.

(c) Amino acids are linked together into a chain by ______________ bonds.

(d) The activity of many proteins is regulated by the binding of effector molecules to the ______________ site. This binding alters the structure of the protein, which in turn affects the ______________ site.

(e) Heritable disorders resulting from blocks in metabolic pathways were named ______________ by Garrod.

(f) The ______________ of the genetic code refers to the fact that most amino acids are specified by more than one codon.

(g) Most eukaryotic genes are split into coding regions called ______________, separated by noncoding regions called ______________.

(h) The transfer of genetic information from RNA to proteins is called ______________.

(i) The regions of the eukaryotic chromosome(s) that contain the rRNA genes are referred to as ______________ regions.

(j) Amino acids are carried to the ribosomes by ______________ RNA.

(k) A tRNA molecule charged with its specific amino acid is called a(n) ______________.

(l) A tRNA molecule must possess the specificity to recognize both a(n) ______________ and the corresponding ______________.

(m) The unusual nucleoside ______________ is found in tRNA, in which it can participate in codon-anticodon recognition.

4. Referring to the list of amino acids and their chemical natures given in Fig. 9.1, hypothesize what effect each of the following mutations might have on the conformation of the protein:

RESIDUE IN WILD-TYPE PROTEIN	RESIDUE IN MUTANT PROTEIN
arginine	leucine
cysteine	glutamine
serine	threonine
isoleucine	serine
arginine	aspartic acid
proline	histidine

5. A short section of a particular gene includes the following sequence of nucleotide pairs:

antisense strand
↓
(3′) –G–T–C–T–T–A–C–G–C–T–A–G– (5′)

(5′) –C–A–G–A–A–T–G–C–G–A–T–C– (3′)
↑
sense strand

(a) What will be the base sequence of the mRNA? **(b)** List the anticodons on the tRNA molecules that recognize each mRNA codon (assume no wobble in this case). **(c)** Give the amino acid sequence of the polypeptide made by translation.

6. If an organism synthesizes 150,000 different proteins but contains 4×10^9 nucleotide pairs of DNA, what fraction of the DNA is actually coding for protein? (Assume 1000 coding base pairs per gene.)
7. Consider the enzyme that catalyzes the first step in the histidine biosynthetic pathway diagrammed in Fig. 9.7. Compare the probable effect on enzyme activity and histidine synthesis of a mutation at the active site of this enzyme with the consequences of a mutation at the allosteric site.
8. Refer to the genetic code in Table 9.1. **(a)** In how many cases would you be able to identify the amino acid specified by a particular codon if you knew only the first two nucleotides of that codon? **(b)** Conversely, in how many instances would you be able to identify the first two nucleotides of a codon if you knew the amino acid that it specifies?
9. The amino acid sequence of a certain polypeptide is leu-tyr-arg-trp-ser. **(a)** How many nucleotides are necessary in the DNA to code for this pentapeptide? **(b)** How many different nucleotide sequences in the DNA could possibly code for this pentapeptide?
10. Assuming no restriction on the relative frequencies of the 20 kinds of amino acids, calculate the number of possible sequences that can conceivably exist in a polypeptide consisting of **(a)** 2, **(b)** 3, **(c)** 4, and **(d)** *n* amino acids in length.
11. Two auxotrophic mutant strains of *Neurospora* lack the ability to synthesize the amino acid leucine. Strain 1 can grow if the medium is supplemented with either α-ketoisocaproic acid or leucine. Strain 2 can grow on leucine but not on α-ketoisocaproic acid. Diagram the steps involved in the synthesis of leucine, and indicate which step in the pathway is blocked in each of the mutant strains.
12. Four auxotrophic mutant strains of *E. coli* (1, 2, 3, and 4) are each blocked at a different step of the same metabolic pathway. This pathway involves compounds A, B, C, D, and E. Each of these compounds is tested for its ability to support the growth of mutant strains, with the results given below (+ = growth, − = no growth):

		COMPOUND USED AS SUPPLEMENT				
		A	B	C	D	E
MUTANT STRAIN	1	+	−	+	−	+
	2	−	−	+	−	−
	3	+	−	+	+	+
	4	−	−	+	−	+

Give the order of the compounds in the metabolic pathway, and indicate the specific step that is blocked in each mutant strain.

13. Five mutant strains of *Neurospora* give the following results in complementation tests (+ = complementation, − = no complementation, blank = not tested):

		MUTANT STRAIN					
		1	2	3	4	5	
MUTANT STRAIN	1	−	+		+	−	Results of different pairwise combinations
	2		−	+			
	3			−	−		
	4				−	+	
	5					−	

Arrange the mutant loci into cistrons.

14. The following table provides just enough information about a section of a particular gene to allow you to determine **(a)** the sequence of base pairs along the DNA, **(b)** which DNA strand is the template for mRNA transcription, **(c)** the mRNA nucleotide sequence, **(d)** the tRNA anticodons, and **(e)** the amino acid sequence of the polypeptide. Complete the table.

<table>
<tr><td></td><td></td><td></td><td>G</td><td></td><td></td><td></td><td></td><td>T</td><td></td><td></td><td></td><td></td><td></td><td></td></tr>
<tr><td></td><td></td><td></td><td></td><td></td><td></td><td>T</td><td>T</td><td></td><td></td><td></td><td></td><td>T</td><td>G</td><td>A</td></tr>
<tr><td></td><td></td><td></td><td></td><td>U</td><td>C</td><td></td><td>U</td><td></td><td></td><td></td><td></td><td></td><td></td><td></td></tr>
<tr><td></td><td></td><td></td><td></td><td></td><td></td><td></td><td></td><td></td><td>A</td><td>U</td><td>G</td><td></td><td></td><td></td></tr>
<tr><td colspan="3">met</td><td colspan="3"></td><td colspan="3"></td><td colspan="3"></td><td colspan="3"></td></tr>
</table>

15. Because of wobble, many tRNA molecules can often recognize two (or more) mRNA codons. Refer to the mRNA codon assignments given in Table 9.1. What would be the minimum number of different tRNA molecules required to recognize the leucine codons?
16. A certain DNA molecule contains 5250 nucleotide pairs. If the genes in this organism are nonoverlapping sequences of nucleotide pairs along the DNA, how many proteins of molec-

ular weight 30,000 could be coded for by this DNA? (Take the average residue weight of an amino acid to be 120.)

17. In tabular form, summarize the distinguishing features of the origin, structure, and function of the three classes of RNA that participate in translation. Compare prokaryotic and eukaryotic systems where possible.

18. Describe the events involved in the three phases of translation—initiation, elongation, and termination—as they are known to occur in *E. coli.*

19. Suppose you use polynucleotide phosphorylase and a reaction mixture consisting of GDP and UDP to synthesize an artificial mRNA. If the RNA is produced from a mixture that contains an excess of UDP, more valine than glycine is incorporated into protein. But when GDP is in excess, more glycine than valine is incorporated. Without looking at Table 9.1, list the codons containing G and U that could possibly be coding for val and gly.

20. Determine the relative frequencies of all possible triplet codons in the RNA produced by a reaction mixture consisting of 60% UDP, 40% CDP, and polynucleotide phosphorylase.

21. Suppose that the RNA of the preceding problem is used as a messenger in a protein synthesizing system. Using Table 9.1, determine what percentage of the various amino acids are expected to be incorporated into polypeptides under the direction of this RNA. Express each percentage relative to the amount of phenylalanine.

22. Using Table 9.1, give the amino acid sequences in the polypeptides produced in an in vitro protein synthesizing system under the direction of artificial mRNAs with the following defined sequences: **(a)** AUAUAUAUAUAU . . . , **(b)** GAUGAUGAU . . . , **(c)** UAAUAAUAA. . . . Assume that under the conditions used in the in vitro system, translation is initiated at randomly selected codons in the artificial mRNAs. How would your results differ if the genetic code were actually a doublet code?

Chapter 10

Organization and Regulation of Genes in Chromosomes

Chromosomes are by no means inert vehicles of inheritance, carrying genes in some random assortment along their lengths, but possess extensive functional as well as structural organization. This organization is particularly evident in prokaryotes, in which the clustering of functionally related genes plays an important role in regulating gene expression. In this chapter, we will look at how genes are organized into chromosomes so as to function as integrated units. We will also consider different mechanisms of gene control, as well as some potential consequences of the loss of normal gene regulation. We shall find, for example, that the clustering of related genes is not typical of eukaryotes, which indicates a fundamental difference between prokaryotes and eukaryotes with respect to their manner of gene control.

Functionally Related Genes in Bacteria and Viruses

Genetic studies on bacteria and their viruses reveal that genes with related functions tend to be clustered together in the same general region of the linkage map. In bacteriophage T4, for example, the genome is roughly divided into two parts: one containing the genes that are necessary for replication (the so-called *early genes*), and the other containing the genes for protein heads and tails and for the enzyme lysozyme, which digests the wall of the host cell (called the *late genes*). Investigators have also observed that groups of related genes in a cluster are often regulated as a unit, in that they are transcribed by RNA polymerase as a single polygenic strand of mRNA. Such groups of related structural genes that are under joint transcriptional control are known as **operons.**

Operon Structure and Control: General Considerations

Figure 10.1 shows an idealized operon consisting of three adjacent structural genes labeled S_1, S_2, and S_3. Also present are a promoter (p), for initiating transcription, and one or more terminator (t) sites (two are shown in the diagram), for stopping the growth of the RNA chain. Not shown in Fig. 10.1 are various regions that serve strictly in a regulatory capacity. These regions include supplemental binding sites that interact with the products of certain genes that control transcription of the operon. Genes that exert control over an operon are referred to, in general, as *regulatory genes*. The relative locations of regulatory genes vary; they may be closely linked to or located some distance away from the operon(s) that they affect. The precise functions of the proteins produced by these regulatory genes also vary, depending on the type of control they exert. Some of the better-known mechanisms for the transcriptional control of operons are shown in Fig. 10.1 and are listed in the following outline:

A. *Positive control,* in which the product of the regulatory gene stimulates transcription by:
   1. Enhanced binding of RNA polymerase to the promoter.
   2. Preventing early termination of RNA chain growth at a terminator site.

B. *Negative control,* in which the product of the regulatory gene inhibits transcription by:
   1. Interfering with the binding of RNA polymerase to the promoter.

The negative control mechanism was first discovered in 1961 by Jacob and Monod in the bacterium *E. coli*; it has since been shown to be important in other bacteria and in viruses as well. In this case, the product of the regulatory gene is a **repressor protein,** which blocks the binding of RNA polymerase to the promoter region. The repressor interferes with this binding by combining with an **operator** (o) **site** (see Fig. 10.2). The operator locus is typically adjacent to, if not overlapping, the promoter region, so that RNA polymerase cannot transcribe the structural genes when the repressor is bound at this location.

Functional Organization of Bacterial Chromosomes

The clustering of functionally related genes into operons is an important characteristic of the bacterial genome. Most bacterial operons control the transcription of genes that code for metabolic enzymes; in this capacity, they tend to have certain features in common. First, the structural genes within a bacterial operon ordinarily code for enzymes that are involved in the same metabolic pathway. For example, the genes of the *gal* operon code for enzymes that are necessary for the

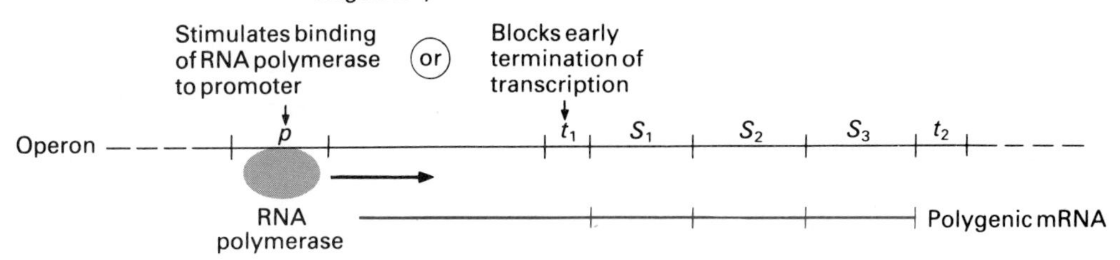

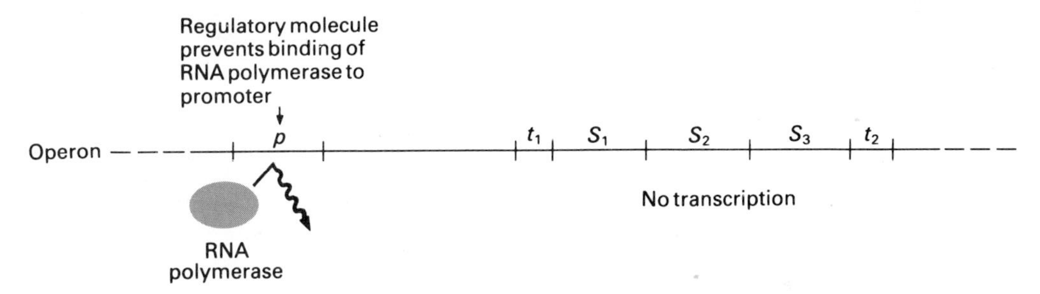

Figure 10.1. Basic components of operons subject to (a) positive or (b) negative control. Each of the operons shown here consists of three structural genes (S_1, S_2, and S_3), which are transcribed onto a single molecule of mRNA, a promoter region (p), and two transcription termination sites (t_1 and t_2). The t_1 site is an early terminator, whereas t_2 signifies the end of transcription of the operon. Transcription is controlled by the product of a regulatory gene (not shown), which can either directly stimulate transcription (in the case of operons under positive control) or inhibit it (in the case of operons under negative control). In positive control systems, either the product of the regulatory gene enhances binding of RNA polymerase to the promoter region, or it overcomes early termination of transcription. In contrast, in negative control systems, the regulatory gene product interferes with the access to the promoter region by RNA polymerase.

degradation of galactose, those of the *trp* operon code for enzymes that are used in the biosynthesis of tryptophan, and so on. Second, the structural genes within an operon are frequently under more than one type of transcriptional control. It is not uncommon for an operon to be affected by two different regulatory gene products, one of which exerts negative control while the other exerts positive control. Third, the regulatory gene products tend to be allosteric proteins; in addition to attaching to a specific regulatory region in the DNA by their primary binding (active) site, they also interact with certain key metabolites at their allosteric site. The key metabolite is often the first reactant in a degradative pathway or the end product of a biosynthetic pathway. These metabolites act as *effector molecules* by promoting changes in the conformation of the regulatory gene proteins, thereby altering their capacity to bind to the DNA.

Inducible and Repressible Operons. Bacterial operons fall into two major categories, based on the actions the effector molecule has on the regulatory gene product. If the effector molecule alters a regulatory gene product in such a way that transcription of the structural genes is "turned on," the operon is said to be *inducible.* Conversely, if the effector substance modifies the regulatory product so that transcription of the structural genes under its control is "turned off," the operon is said to be *repressible.*

Inducible operons are usually concerned with the synthesis of enzymes involved in degradative pathways, such as those that function in the breakdown of certain sugars. In this case, the effector molecule, called an **inducer,** is often the initial substrate in the degradative pathway. For example, galactose serves as the inducer of the *gal* operon. When galactose is absent from the growth medium, the enzymes required for the breakdown of galactose occur only in very small quantities within each cell. When galactose is present, these same enzymes are produced in large amounts. The molecular basis for the action of inducers such as galactose is shown by the general model in Fig. 10.2. Note that in the absence of the inducer, the operon is under the negative control of a regulatory gene. The regulatory gene product, or repressor, binds to the operator site and prevents transcription. When the inducer is present, however, it combines with the allosteric site on the repressor, causing the repressor protein to undergo a change in conformation. Transcription and subsequent enzyme synthesis are enhanced, since the ability of the modified repressor to bind to the operator is much reduced. We can thus view induction as resulting from the allosteric inhibition of the repressor protein.

Repressible operons, in contrast, are mainly associated with metabolic pathways that are involved in the synthesis of required substances, such as amino acids. The effector molecule, now called the **corepressor,** is often the end product of the biosynthetic pathway. For example, the amino acid tryptophan serves as the corepressor of the *trp* operon. The enzymes that are involved in tryptophan synthesis are produced by the *trp* operon when the concentration of tryptophan within the cell is low. The synthesis of enzymes by this operon is repressed when the concentration of tryptophan is high. The general model in Fig. 10.3 shows the molecular basis for the action of corepressors. Again, we see that the effector molecule exerts its influence by combining with the allosteric site of a regulatory protein. In this case, however, the unaltered product of the regulatory gene is an inactive **aporepressor,** which is unable to combine with the operator locus. The structural genes of a repressible operon are thus transcribed in the absence of the corepressor. When the corepressor is present, it exerts its effect by combining with the aporepressor, converting it to an active repressor. The repressor can then bind to the operator, shutting off enzyme synthesis.

Inducible and repressible operons are adaptations that exemplify the general economy of protein synthesis. The synthesis of proteins requires the expenditure of large amounts of cellular energy. It is thus clearly advantageous for a cell to produce proteins only when they are definitely needed.

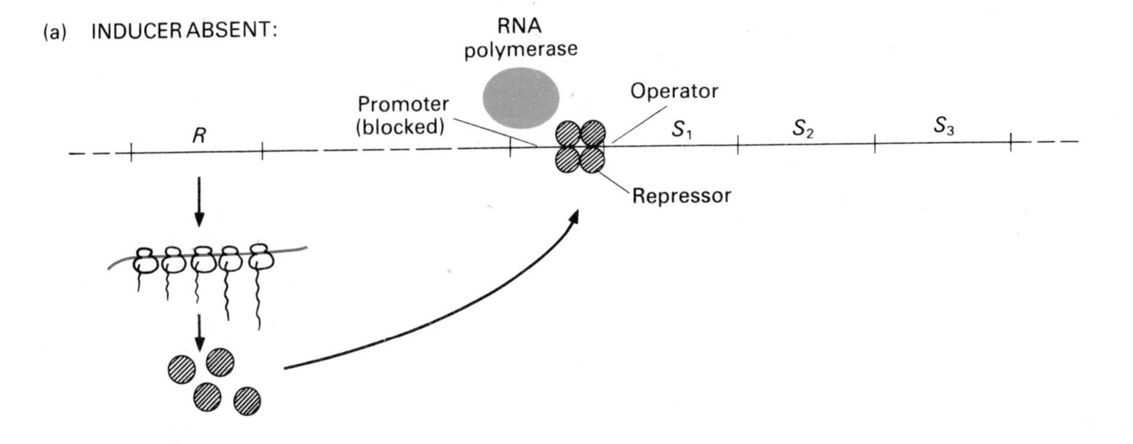

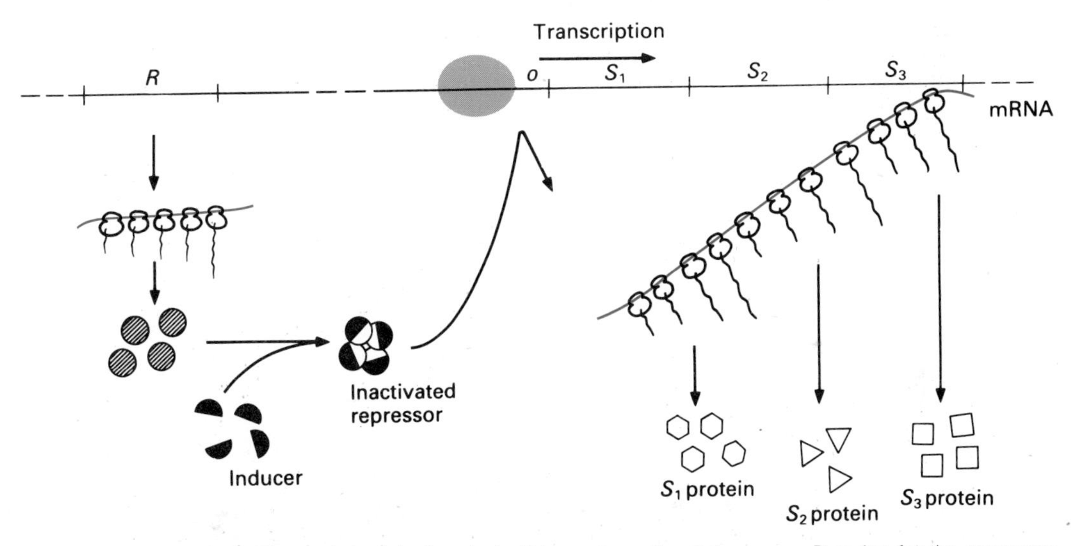

Figure 10.2. General model of an inducible operon. Regulatory gene *R* codes for the repressor protein. (a) In the absence of the inducer, the repressor binds to the operator region, thereby blocking the binding of RNA polymerase to the adjacent promoter site. No transcription of structural genes can then take place. (b) When the inducer is present, however, it allosterically inactivates the repressor, leaving it unable to bind to the operator. The RNA polymerase molecules can now initiate transcription at the promoter and can transcribe the three structural genes into mRNA.

Structural genes that are subject to regulation are in contrast to *constitutive genes,* which are continually being expressed. Constitutive genes code for proteins that are essential for growth and are required by the cell in relatively constant amounts, regardless of the composition of the growth medium.

The Lactose Genes: An Inducible Operon. The *lac* operon of *E. coli* is possibly the most thoroughly studied operon in bacteria. It was first analyzed by the French

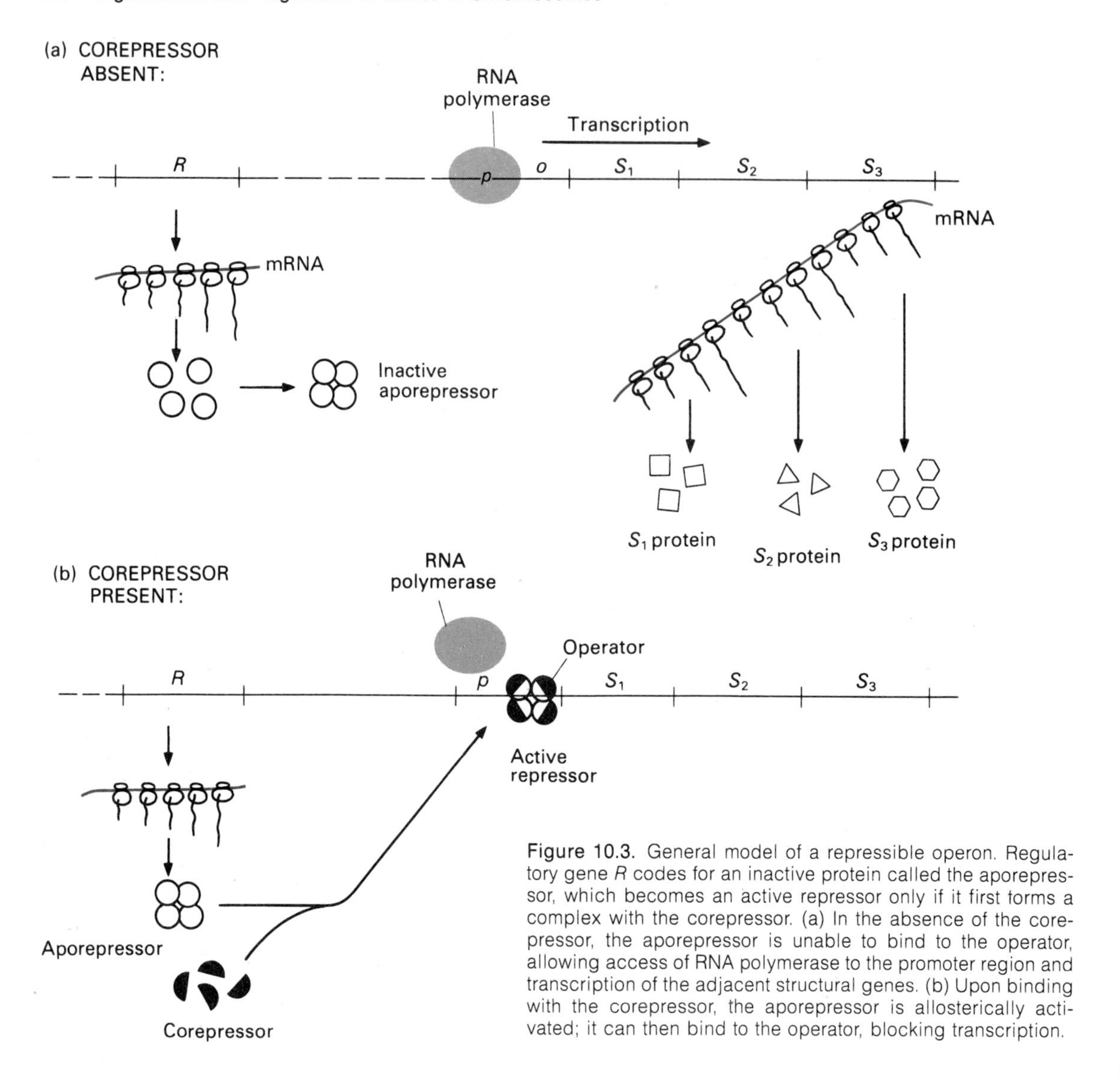

Figure 10.3. General model of a repressible operon. Regulatory gene R codes for an inactive protein called the aporepressor, which becomes an active repressor only if it first forms a complex with the corepressor. (a) In the absence of the corepressor, the aporepressor is unable to bind to the operator, allowing access of RNA polymerase to the promoter region and transcription of the adjacent structural genes. (b) Upon binding with the corepressor, the aporepressor is allosterically activated; it can then bind to the operator, blocking transcription.

microbiologists F. Jacob and J. Monod and served as the prototype for the operon model, which these investigators first described in 1961. The *lac* operon is an inducible operon with three structural genes that code for enzymes that are involved in the metabolism of lactose. The *z* gene codes for the enzyme β-galactosidase, which catalyzes the conversion of lactose to galactose and glucose:

$$\text{Lactose} + H_2O \xrightarrow{\text{Galactosidase}} \text{Galactose} + \text{Glucose}.$$

Gene *y* codes for galactoside permease, a cell protein located in the bacterial cell membrane that aids in the transport of lactose molecules into the cell. Gene *a* codes for the enzyme thiogalactoside transacetylase, whose function is still unclear.

Genetic studies have shown that the order of the three structural genes on the genetic map is *z-y-a* (see Fig. 10.4). These structural genes share one promoter and one operator site and are under negative control by regulatory gene *i*. Gene *i* codes for a repressor protein that exists in its active form as a tetramer made up of four identical polypeptide chains. Lactose (actually the isomeric form, allolactose) acts as an inducer in this system, combining with the repressor and distorting its shape so that the repressor can no longer bind to the *lac* operator site.

Jacob and Monod formulated the operon model based on the properties of various *lac* mutations that resulted in altered patterns of lactose metabolism. The major classes of these mutations are the following:

A. *Structural gene mutations* affect the structure of the enzymes produced.
B. *Operon control mutations* affect only the regulation of enzyme synthesis.
   1. *Constitutive mutants* always produce enzymes, even in the absence of inducer.
   2. *Super-repressed mutants* never produce enzymes, even in the presence of inducer.

Constitutive mutations can be either regulatory (i^c) or operator (o^c) in origin. In addition to mapping at distinctly different sites, o^c and i^c mutations can be differentiated by their properties when present in a partially diploid condition. (Recall from Chapter 8 that partial diploids in bacteria are usually constructed using either conjugation or transduction.) The constitutive expression of the o^c mutation involves only those structural genes that are located on the same chromosome as the mutant o^c locus. Thus, a partial diploid of genotype $o^+z^-y^-a^-/o^cz^+y^+a^+$ would show constitutive synthesis of all three enzymes, making it appear that o^c is dominant to o^+. When the genes are arranged differently, however, as in a cell with genotype $o^+z^+y^+a^+/o^cz^-y^-a^-$, the three enzymes are synthesized only in the presence of the inducer. The o^c mutation is thus said to have a **cis-dominant** effect. In comparison, i^c mutations do not show a cis-dominant effect;

Figure 10.4. The genes that make up the *lac* operon in *E. coli*. Structural genes z, y, and a code for enzymes that are involved in the utilization of lactose by the cell. The coordinate transcription of these genes is controlled by regulatory gene *i*, which codes for a repressor protein. The binding of the repressor molecule to the operator site blocks transcription of the structural genes.

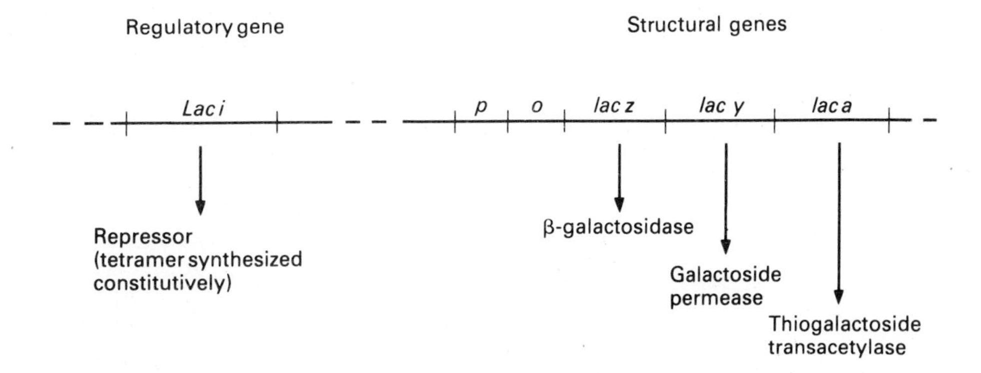

these mutations are recessive to i^+ regardless of whether the arrangement is cis or trans in relation to the alleles at the structural gene loci.

Mutations of the super-repressed variety can occur in either the regulatory gene or the promoter region of the operon. The regulatory super-repressed (i^s) mutation is especially interesting, since it shows complete dominance to i^+ in both the cis and trans positions. The dominant effect of this mutation is attributable to the production of an altered, but diffusible repressor that can attach to either operator site in the partial diploid, thus having the potential of turning off transcription of the structural genes on both homologs.

Figure 10.5 shows the simplest origin of these mutations. Observe that constitutive mutations eliminate the binding of the repressor to the operator site. This inability to bind may result from an altered operator locus, as in the o^c mutation, or from an altered repressor molecule, as in the i^c mutation. The super-repressed expression of the i^s mutation results from a defect in the binding site for the inducer on the repressor molecule (the allosteric site). The defective repressor is still able to bind to the operator locus, turning off gene activity, but it does not respond to the effects of the inducer.

EXAMPLE 10.1. For each of the following situations involving operons and the effector molecules that influence them, determine whether or not transcription of the adjacent structural genes occurs. (R = the regulatory gene of the operon.)

INDUCIBLE OPERON			REPRESSIBLE OPERON		
(a)	$R^+p^+o^+$,	inducer absent	**(d)**	$R^+p^+o^+$,	corepressor absent
(b)	$R^-p^+o^+$,	inducer absent	**(e)**	$R^+p^+o^c$,	corepressor present
(c)	$R^+p^+o^c$,	inducer absent			

Solution. In an inducible system like the *lac* operon discussed in the text, the R^+ gene specifies an active repressor. Without an inducer to inactivate this repressor, transcription of the structural genes of the operon remains turned off **(a)**. In **(b)**, however, there is no active repressor made, because of the mutant condition of the *R* gene. Thus, transcription occurs. Transcription occurs in **(c)** as well, even though the *R* gene is wild-type and the inducer is absent, because the mutant *o* site prevents the repressor from binding to the DNA to block transcription.

In a repressible system, the corepressor is needed to activate the aporepressor (the product of gene *R*), in order to turn off transcription. Hence, transcription does occur in **(d)**. Transcription also occurs in **(e)** because of the mutant condition of the *o* region.

Positive Control of the *Lac* Operon. The *lac* operon, as well as certain other inducible operons, is also under a form of positive control by a regulatory gene product known as *catabolite activator protein* (CAP). CAP exerts its effect on the *lac* operon by binding to a site within the promoter region, immediately adjacent to the binding site for RNA polymerase. For unknown reasons, CAP must be bound to the *lac* promoter in order for transcription to occur. We might speculate that CAP influences the conformation of the rest of the promoter region, making it suitable for binding by RNA polymerase.

(a) OPERATOR CONSTITUTIVE MUTATION
(repressor cannot bind to altered operator site):

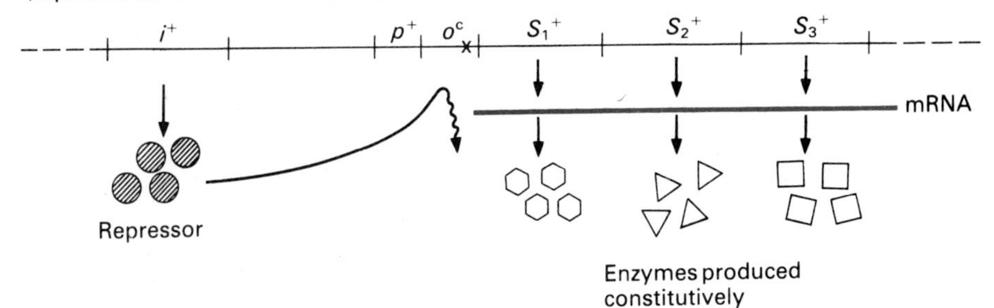

(b) REGULATORY CONSTITUTIVE MUTATION
(altered repressor cannot bind to operator):

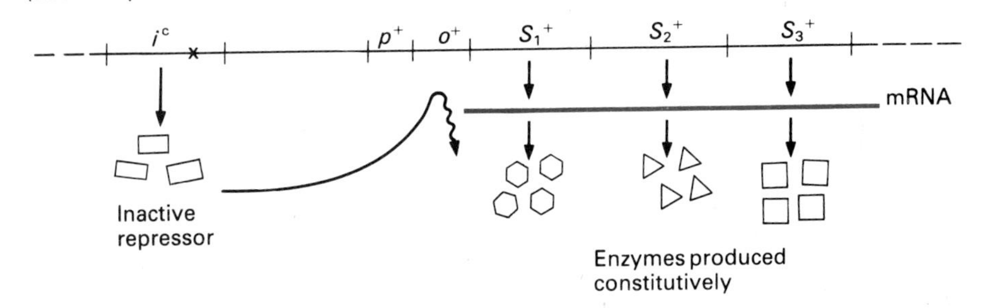

(c) SUPER-REPRESSED REGULATORY MUTATION
(altered repressor cannot bind with inducer):

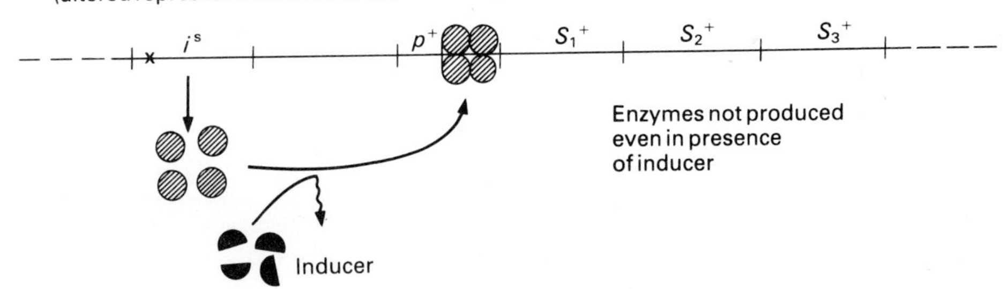

Figure 10.5. Three classes of mutations that affect the control of the inducible *lac* operon. (a) A mutation in the operator region destroys its ability to bind the repressor, resulting in constitutive transcription and enzyme synthesis. (b) A mutation in the regulatory gene *i* affects the site on the repressor protein that normally binds to the operator region. The result is again constitutive transcription of the operon. (c) A different mutation in the regulatory gene *i* affects the site on the repressor that normally recognizes the inducer. As a result, the repressor cannot be allosterically inactivated by the inducer, and transcription of the operon is blocked by the continued binding of the repressor to the operator.

Before binding to the promoter, CAP must first be converted to an active form by combining with an effector known as cyclic adenosine monophosphate, or cAMP (Fig. 10.6). Cyclic AMP plays an important role in the function of cells; it exists within cells at a concentration that is inversely related to the concentration

(a)

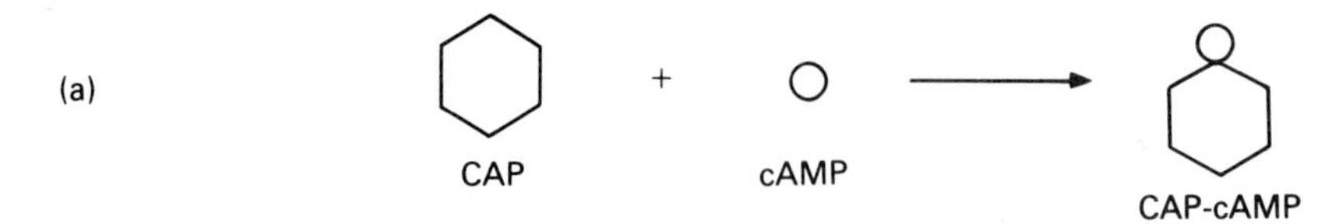

(b)

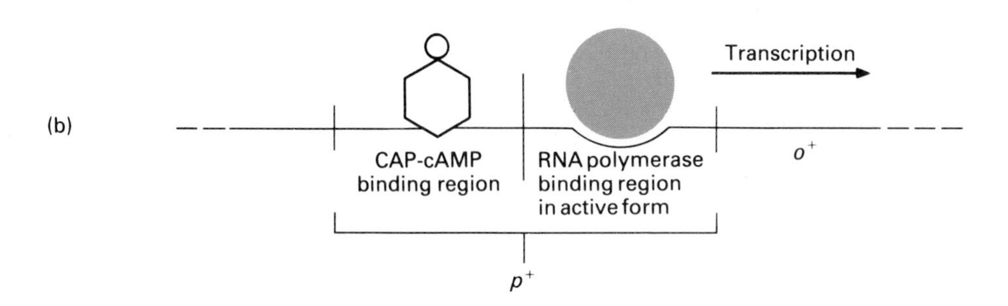

(c)

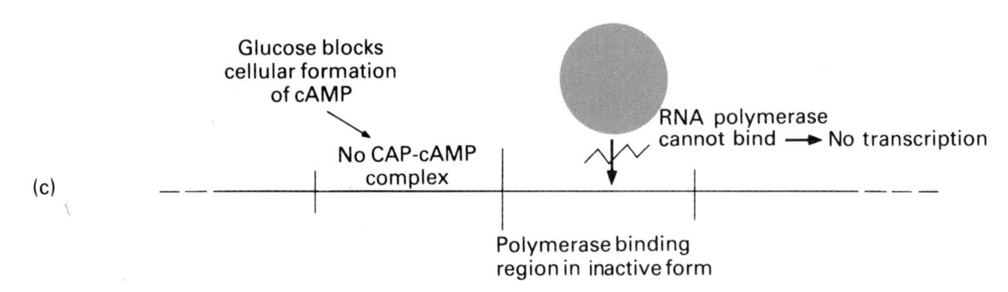

Figure 10.6. The effect of the complex of catabolite activator protein (CAP) and cyclic AMP (cAMP) on the transcription of the *lac* operon. (a) CAP is inactive until it binds with cAMP. (b) The CAP-cAMP complex binds to a section of the promoter region, stimulating the binding of RNA polymerase to the adjacent section of the promoter. (c) Without the CAP-cAMP complex, the polymerase molecule is unable to bind to the promoter region. The presence of glucose blocks production of cAMP by the cell and thereby prevents transcription of the operon.

of glucose present. An as yet unidentified breakdown product of glucose has been found to interfere with the cellular conversion of ATP to cAMP. Thus, when glucose levels are high, cAMP concentrations will be low, and vice versa. If cAMP levels are low, as they are in the presence of glucose, the *lac* operon cannot be induced, regardless of how much lactose is present. A cell that is provided with both glucose and lactose will therefore preferentially utilize glucose as a carbon and energy source. This apparent inhibition of the *lac* operon, as well as certain other inducible operons, by the presence of glucose is known as **catabolite repression.**

Functional Organization of Bacteriophage Chromosomes

In bacteriophages, operons provide a mechanism to ensure that different sets of functionally related genes are expressed in an orderly temporal sequence during phage development. For example, when phages such as lambda, T4 and T7, and

SPO1 (a *B. subtilis* phage) infect a bacterium, one or more products of the first phage operon in the sequence act to turn on the genes of the second operon, one or more products of the second operon then turn on the genes of the third operon, and so on, until all pertinent sets of functionally related genes have been transcribed. The life cycle of a bacteriophage thus follows a programmed developmental sequence that is controlled at the level of transcription.

Operon Control of Phage Lambda. Phage λ is the best-known example of a genome organized into operons. The genes of phage λ are grouped into three operons, as shown in Fig. 10.7. There are two groups of early genes, reflecting the fact that an infecting λ chromosome can follow either of two growth cycles. Those genes contained in the *early right operon* are concerned with DNA replication during the lytic cycle. Those genes contained in the *early left operon* are involved in lysogenic growth. The genes that code for the head, the tail, and lysozyme are found in the *late operon*. Each operon is transcribed from its own promoter site, designated P_L (early left), P_R (early right), and $P_{R'}$ (late).

Six genes, *N, Q, cI, cII, cIII,* and *cro,* act in the regulation of λ phage development. Genes *N* and *Q* are regulatory genes that exert positive transcriptional control. The protein coded for by the *N* gene, which is one of the first to be produced after infection, allows transcription of the early operons by preventing early termination of RNA synthesis at the terminator sites marked t_L and t_R (Fig. 10.8a). The protein coded for by the *Q* gene, which also prevents the termination

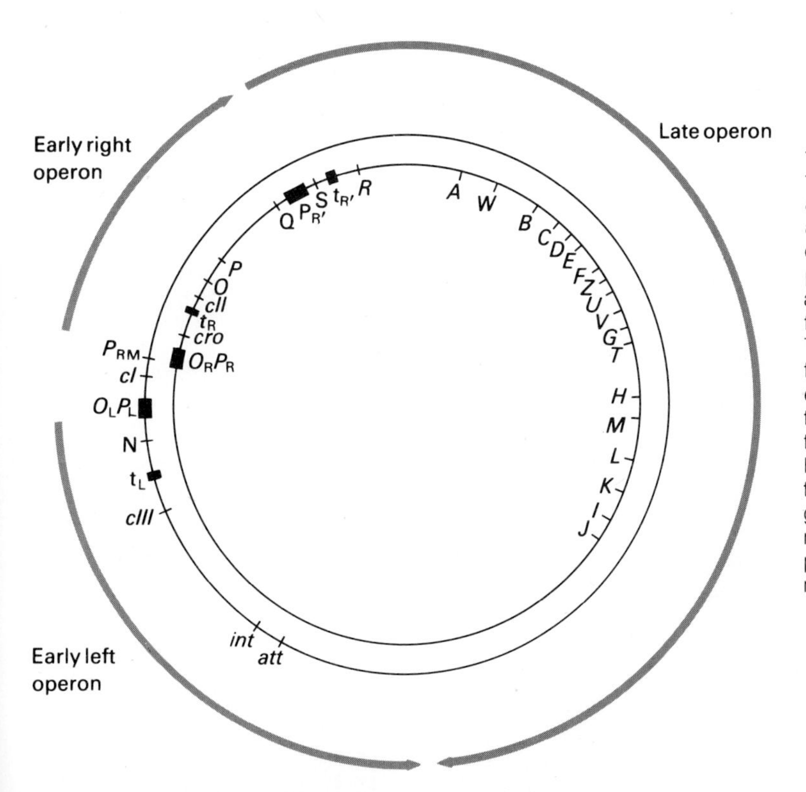

Figure 10.7. Functional organization and transcription of the vegetative λ chromosome. Three operons are shown: two concerned with early and one with late functions. The early operons are each controlled by a promoter-operator complex (O_RP_R and O_LP_L) and are transcribed in different directions (shaded arrows). The early right operon is transcribed from the inner DNA strand, while the early left operon is transcribed from the outer strand. The late operon is transcribed from the inner strand, beginning with promoter $P_{R'}$. Notice that the region that includes the *cI* gene is not part of either early operon. It is transcribed separately from promoter P_{RM} during the maintenance of the lysogenic state.

(a) EARLY EVENTS: Transcription of early left and early right operons. N protein prevents early terminations; cro protein prevents transcription of *cI* gene by blocking promoter P_{RM}.

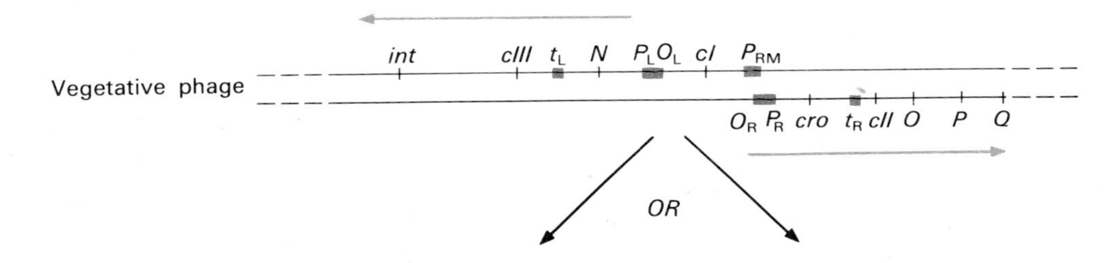

(b) LYSOGENIC CYCLE: Formation of prophage. The cII-cIII protein complex activates the transcription of *cI* (repressor) and *int* (integration protein) from their individual promoters; following prophage formation, only *cI* is transcribed.

(c) LYTIC CYCLE: Transcription of operons needed for replication and assembly. Q protein prevents early termination of late operon transcription; cro protein blocks *cI* and early left transcriptions.

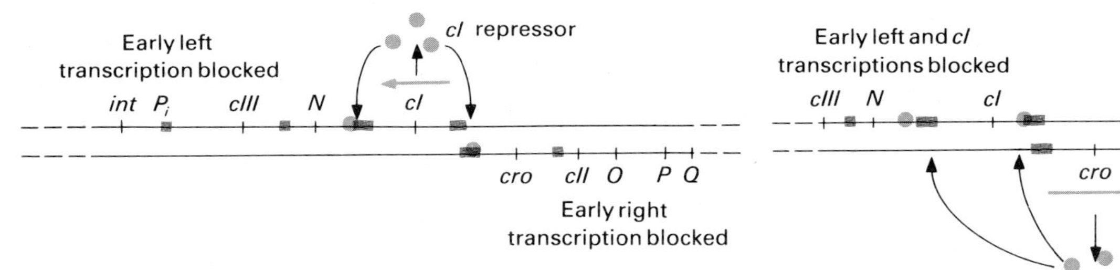

PROPHAGE

VEGETATIVE PHAGE

Figure 10.8. Sequence of transcriptional events that occur during (a) the initial stages of lambda growth and during (b) lysogeny or (c) lytic growth. Immediate early transcription is terminated at t_L and t_R; the product of gene *N* overcomes these terminations, allowing transcription of the early left and early right operons. The product of gene *cro* blocks transcription of the *cI* gene during the early stages of vegetative growth. But the lysogenic cycle can be established if large amounts of cII and cIII proteins are synthesized. The cII-cIII protein complex stimulates the transcription of *cI* and *int*. The cI repressor blocks transcription of the early left and early right operons by binding to their respective operators. The int protein promotes integration of the phage chromosome into the bacterial chromosome, to form a prophage. Following integration, *cI* is the only gene expressed by the prophage. The cI repressor keeps all of the vegetative functions of the phage shut down. In contrast, the lytic cycle occurs if the repression of *cI* transcription caused by the cro protein predominates. Build-up of the product of the *Q* gene overcomes early termination at the $t_{R'}$ site of transcription of the late operon, allowing the genes for the protein components and lysozymes of the phage to be expressed.

of transcription, is produced after the early right operon is transcribed, and acts at the $t_{R'}$ site to prevent termination of transcription of the late operon.

Genes *cI, cII, CIII,* and *cro* are involved in the "choice" between the lysogenic and lytic growth cycles made by an infecting phage chromosome (Fig. 10.8b and 10.8c). The conditions within the bacterial host cell heavily influence the decision. The lysogenic response is favored in starved cells and in cells that have been infected by a high ratio of phage to bacteria. Under either of these conditions, the concentration of the cII protein is high. In a relationship that is complex and incompletely understood, the cII protein acts together with the product of the *cIII* gene to inhibit lytic development and to activate the transcription of the *cI* gene. The activity of the *cII* gene is thought to be the determining factor in the choice of growth pathway. Simply put, if the activity of *cII* is high, resulting in a high concentration of the cII protein, then the lysogenic pathway ensues; if *cII* activity is low, then lytic growth is followed. In the lysogenic cycle, the product of the *cI* gene (a repressor protein) blocks the transcription of both early operons. In the lytic cycle, the product of the *cro* gene (also a repressor) blocks the transcription of *cI*.

The activity of the *cI* gene is critical for the lysogenic state. The *cI* gene specifies a repressor protein that binds to both operators O_L and O_R, which are adjacent to the promoters for the early left and early right operons, respectively. The binding of the repressor prevents transcription of both early operons and blocks the synthesis of lytic proteins. In its active form, bound to the operator sites on lambda DNA, the cI repressor is a dimer. Each monomer unit of the dimer is a single polypeptide chain that folds into two globular domains (Fig. 10.9a), giving a repressor molecule that looks like a pair of dumbbells (Fig. 10.9b). The two domains have different functions. The domain that includes the amino (NH_2) terminus of the polypeptide chain binds to operator DNA (Fig. 10.9c). The carboxyl-domain, on the other hand, has no direct role in binding to the DNA. Instead, the carboxyl-domain is responsible for the association of two polypeptide chains into a dimer. This dimerization is required for the efficient binding of the repressor molecule to lambda DNA.

The prophage must continue to produce the cI repressor in order to maintain the lysogenic state. It has recently been discovered that in addition to the negative control that *cI* exerts over most other lambda genes, the *cI* gene is also a positive regulator of gene expression, in that the cI repressor stimulates the transcription of its own gene (a phenomenon known as *autoregulation*). At low levels, the cI protein acts as a positive regulator on a promoter near *cI* called P_{RM}, which stimulates the transcription. Thus, once the prophage state has been established, *cI* is the only gene expressed by the phage.

The cI repressor not only prevents the transcription of prophage genes, but it also blocks the transcription of chromosomes from other λ phages that might later infect the cell. For this reason, a lysogenic cell is said to be immune to infection by another phage of the same type as the one that is carried as a prophage. Immunity is not the same as phage resistance. In immunity, the infecting phage chromosome enters the cell but is unable to replicate. By contrast, phage resistance typically involves the loss of functional receptor sites on the bacterial cell wall, so that a phage particle is unable to attach itself to the resistant cell.

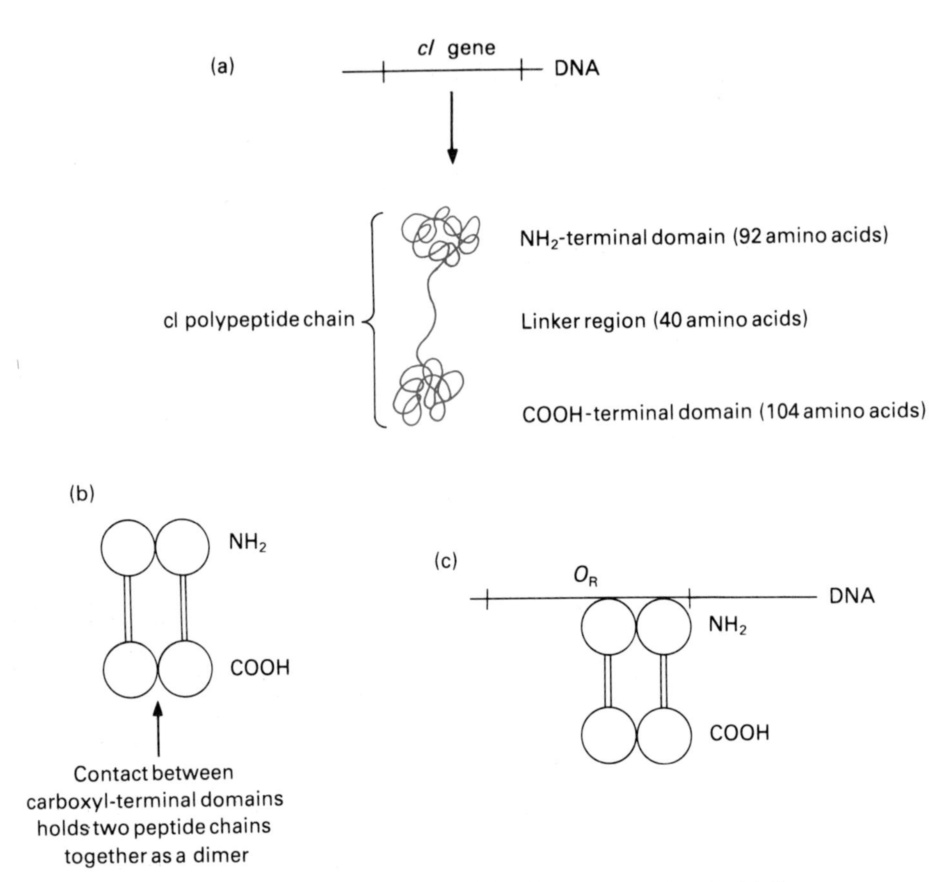

Figure 10.9. The structure and action of the lambda cI repressor molecule. (a) The monomer unit of the repressor is a polypeptide chain that is coded for by gene *cI*. The polypeptide folds into two domains, linked together by a region of the chain that is susceptible to cleavage by proteases. (b) A functional repressor molecule is a dimer. The two polypeptide chains are held together by contact between their carboxyl-terminal domains. (c) A repressor molecule binds to DNA at its amino-terminal domains. Only dimers bind efficiently to the DNA. Binding of repressor to the O_R and O_L operators blocks the transcription of the early right and early left operons, thereby preventing the expression of the genes that are necessary for lytic growth.

While the *cI* gene is required for lysogeny, the *cro* gene is essential to lytic growth. Like *cI,* the *cro* gene produces a repressor that binds to both operators O_L and O_R, but with different results. The cro repressor preferentially binds to a site in O_R that blocks *cI* transcription from the P_{RM} promoter. Thus, *cI* and *cro* are directly antagonistic in function. The cI protein represses the activity of *cro* during lysogeny, and the cro protein inhibits transcription of *cI* during the lytic cycle. The cro protein is much smaller than the cI repressor, but like the repressor, cro binds to the DNA as a dimer. Each monomer unit folds into a single domain, which mediates both binding to the DNA and dimerization.

The question as to how two regulatory proteins can bind to the same region of DNA and yet have opposite effects on gene expression can be answered by first describing the organization of the O_R control region (Fig. 10.10a). The O_R region contains 80 base pairs of DNA, bracketed by the *cI* gene immediately to the left and the *cro* gene immediately to the right. There are three distinct recognition sites within the O_R region, designated O_R1, O_R2, and O_R3. Each site is 17 base pairs long. The nucleotide sequences of the three sites are not identical, but they are similar enough that either cI repressor or cro protein can bind to any of the three sites. However, cI repressor and cro protein differ in their affinities for the three

Figure 10.10. (a) Organization of the O_R region and the adjacent *cI* and *cro* promoters and genes on the DNA of phage λ. The O_R region is divided into three recognition sites: O_R1, O_R2, and O_R3. The *cI* promoter overlaps O_R3 and O_R2, and the *cro* promoter overlaps O_R1 and O_R2. (b) Transcription of the prophage during the lysogenic state. The cI repressor preferentially binds to O_R1 and O_R2, blocking the transcription of *cro* and other genes of the early right operon. RNA polymerase is able to bind to the *cI* promoter region, however, so that transcription of *cI* occurs. The repressor bound to O_R2 is thought to make contact with the RNA polymerase molecule and this contact is hypothesized to aid the binding of polymerase to the *cI* promoter region. In this manner, *cI* stimulates its own synthesis. (c) Transcription of the λ genome during early lytic growth. The lack of functional cI repressor molecules leaves O_R1 and O_R2 open; RNA polymerase binds to the rightward promoter, transcribing *cro* and other early lytic genes. The *cro* protein binds to O_R3, blocking further transcription of *cI*.

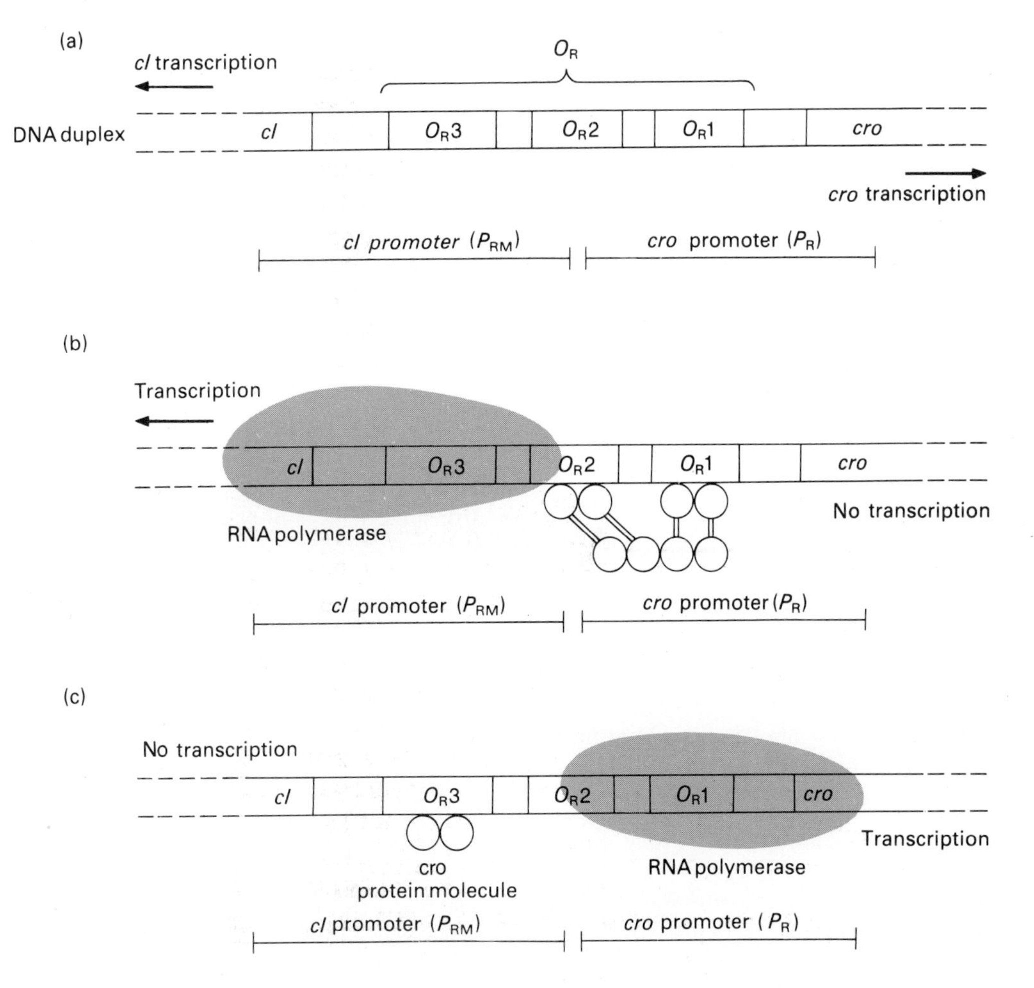

sites, and this difference is largely responsible for their opposite effects on transcription.

The O_R region is overlapped on each side by promoter regions. One of the promoters (P_{RM}) directs RNA polymerase to read in a leftward direction, transcribing the *cI* gene. The other promoter (P_R) is oppositely oriented, and it directs RNA polymerase to read in a rightward direction, transcribing *cro* and the other genes of the early right operon, all of which are concerned with the early stages of lytic growth.

In a lysogenic cell, the cI repressor is attracted most strongly to O_R1. Once a repressor dimer is bound to O_R1, site O_R2 is made more attractive to repressor, and a second dimer binds there. The carboxy-terminal domain of the dimer bound at O_R1 is thought to make contact with the carboxy-terminal domain of the second dimer, thereby aiding the binding at O_R2. The O_R3 site ordinarily remains free of repressor (unless there is an abnormally high concentration of repressor within the cell). The binding of cI repressor molecules at O_R1 and O_R2 blocks the access of RNA polymerase to the rightward promoter (P_R) and thereby prevents the transcription of *cro* and other early right genes (Fig. 10.10b). At the same time, the binding of repressor stimulates the synthesis of more repressor (autoregulation). The amino-terminal domains of the repressor bound at O_R2 apparently come into contact with the RNA polymerase molecule and help it to bind to the DNA at P_{RM}.

As we pointed out in Chapter 4, induction of the λ prophage occurs spontaneously at a low rate. Exposure of lysogenic cells to certain agents, such as ultraviolet light, effectively induces all of the cells. The switch from the lysogenic to the lytic cycle occurs when the cI repressor is inactivated. The inactivation is the result of the cleavage of repressor molecules between the two domains by a bacterial protein, recA. Ordinarily, the recA protein functions in recombination (see Chapter 7). However, under certain circumstances, the recA protein can assume a protease (protein degrading) activity. (The change in recA activity to a protease is thought to be mediated by some factor that is produced as a result of the damage to DNA caused by inducing agents such as ultraviolet light.) The loss of the carboxyl-domains from the repressor molecules greatly reduces the efficiency of dimerization, which, in turn, affects the ability of the repressor to bind to the DNA. Thus, the concentration of active repressor in the cell falls, and the O_R1 and O_R2 sites are left open. As a result, the rate of synthesis of repressor decreases (because polymerase is no longer stimulated to bind to P_{RM} through autoregulation), and, in addition, the RNA polymerase molecule now has access to the rightward promoter (P_R). The *cro* and other early right genes are then transcribed. The cro protein binds to the O_R3 site, which blocks the P_{RM} promoter and thus abolishes the synthesis of cI repressor (Fig. 10.10c).

TO SUM UP

1. Many of the genes of bacteria are organized into functionally related groups known as operons. The genes within each operon are regulated as a group and are transcribed jointly. Most bacterial operons control the transcription of genes that code for metabolic enzymes and are either inducible or repressible.

2. Inducible operons respond to an effector substance that is an inducer by turning on the transcription of the structural genes. Inducible operons are usually concerned with the synthesis of enzymes that function in degradative pathways.
3. Repressible operons respond to an effector substance that is a corepressor by turning off the transcription of the structural genes. Repressible operons are usually involved with the synthesis of enzymes that function in biosynthetic pathways.
4. Functionally related genes in bacteriophages, as in bacteria, tend to be organized into operons. The genes that specify the replication of the phage genome are typically under separate control from the genes that specify later functions, such as head and tail synthesis and the production of lysozymes. The chromosomes of temperate phages also include a third group of genes—those involved in the lysogenic growth cycle.

Organization and Regulation of Genes in Eukaryotes

Certain genes (for example, those that code for rRNA, tRNA, and other components of the translational machinery) are required to function in all cells of a eukaryotic organism. Most genes, however, are active in some cells but not in others. This differential activity of genes is necessary in order for groups of cells to differentiate into tissues with specialized functions. All of the cells of an organism possess the same genetic complement—the one that was present in the fertilized egg or zygote. Differentiation must therefore occur not through the gain or loss of genetic material but rather through the regulatory mechanisms that turn particular genes on and off in different cells at specific times during development. While many of the mechanisms of gene regulation have been studied intensively in prokaryotes, those that operate in eukaryotes are still poorly understood. It is becoming increasingly clear, however, that the functional organization and regulation of genes in eukaryotes are different from those in prokaryotes. But before we consider the control mechanisms that are unique to eukaryotic cells, we first need to discuss how eukaryotic DNA differs from prokaryotic DNA in sequence organization.

Organization of DNA Sequences in Chromatin

One feature that is unique to the chromosomes of eukaryotes is the presence in the DNA of repeated nucleotide sequences, known as *repetitive DNA*. Some DNA sequences are repeated many times throughout the genome of the organism. In *Drosophila*, for example, numerous repeats of DNA sequences such as $\begin{matrix}\overline{\text{AATAT}}\\ \underline{\text{TTATA}}\end{matrix}$ and $\begin{matrix}\overline{\text{AATATAT}}\\ \underline{\text{TTATATA}}\end{matrix}$ are known to occur. In contrast, prokaryotic DNA lacks substantial amounts of repetition; it exists predominantly in the form of *unique-copy* (or *single-copy*) sequences.

The first evidence for repetitive DNA came from studies in which eukaryotic DNA was subjected to centrifugation in a CsCl gradient (see Chapter 3). You may recall that when bacterial DNA that contains only normal isotopes of carbon and nitrogen is centrifuged in this manner, it forms a single band in the region of the centrifuge tube where its buoyant density equals the density of the gradient. Even when the DNA is highly fragmented into smaller pieces, we observe only a single band, which indicates that all portions of the bacterial chromosome have the same average density. In contrast, when fragmented eukaryotic DNA is centrifuged in CsCl, it yields one large band, called the **mainband DNA,** plus one or more smaller bands, composed of **satellite DNA** (Fig. 10.11). This technique shows that eukaryotic DNA is nonuniform with respect to density distribution.

As we learned in Chapter 3, the density of DNA depends directly on the proportion of guanine–cytosine (GC) base pairs. Therefore, the higher the percentage of GC pairs in the DNA, the greater its buoyant density. Satellite DNA that bands at a high density region of the tube (that is, below the main band) thus has a greater content of GC pairs than mainband DNA. By contrast, satellite DNA of comparatively low density is rich in AT base pairs. When satellite DNA of either density is isolated and analyzed for base sequence, most of it is found to be *repetitive,* consisting of short nucleotide sequences (e.g., $\frac{\overline{\text{AATAT}}}{\underline{\text{TTATA}}}$) repeated some 10^5 to 10^7 times within the genome. This is the class of satellite DNA that we are concerned with in our present discussion. Other satellite DNA bands are composed of the DNA from cellular organelles, such as the mitochondria and chloroplasts.

A technique known as *in situ hybridization* has been used to determine the location of highly repetitive DNA in eukaryotic chromosomes (Fig. 10.12). In this procedure, cells are grown in a medium containing ^{3}H until all the chromosomal material is labeled. The chromosomal DNA is then extracted and fragmented, and the satellite DNA fraction is isolated by centrifugation. This is followed by dena-

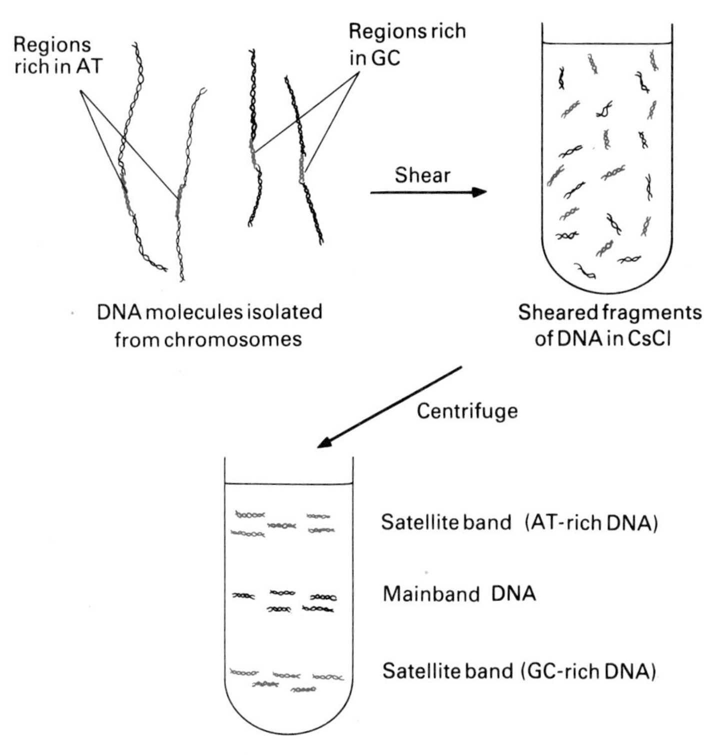

Figure 10.11. Heterogeneity of density in eukaryotic DNA. Density gradient equilibrium centrifugation of sheared eukaryotic DNA typically yields mainband DNA plus at least one smaller satellite band. Satellite bands are formed by DNA fragments that are rich in either AT or GC base pairs. See text for details.

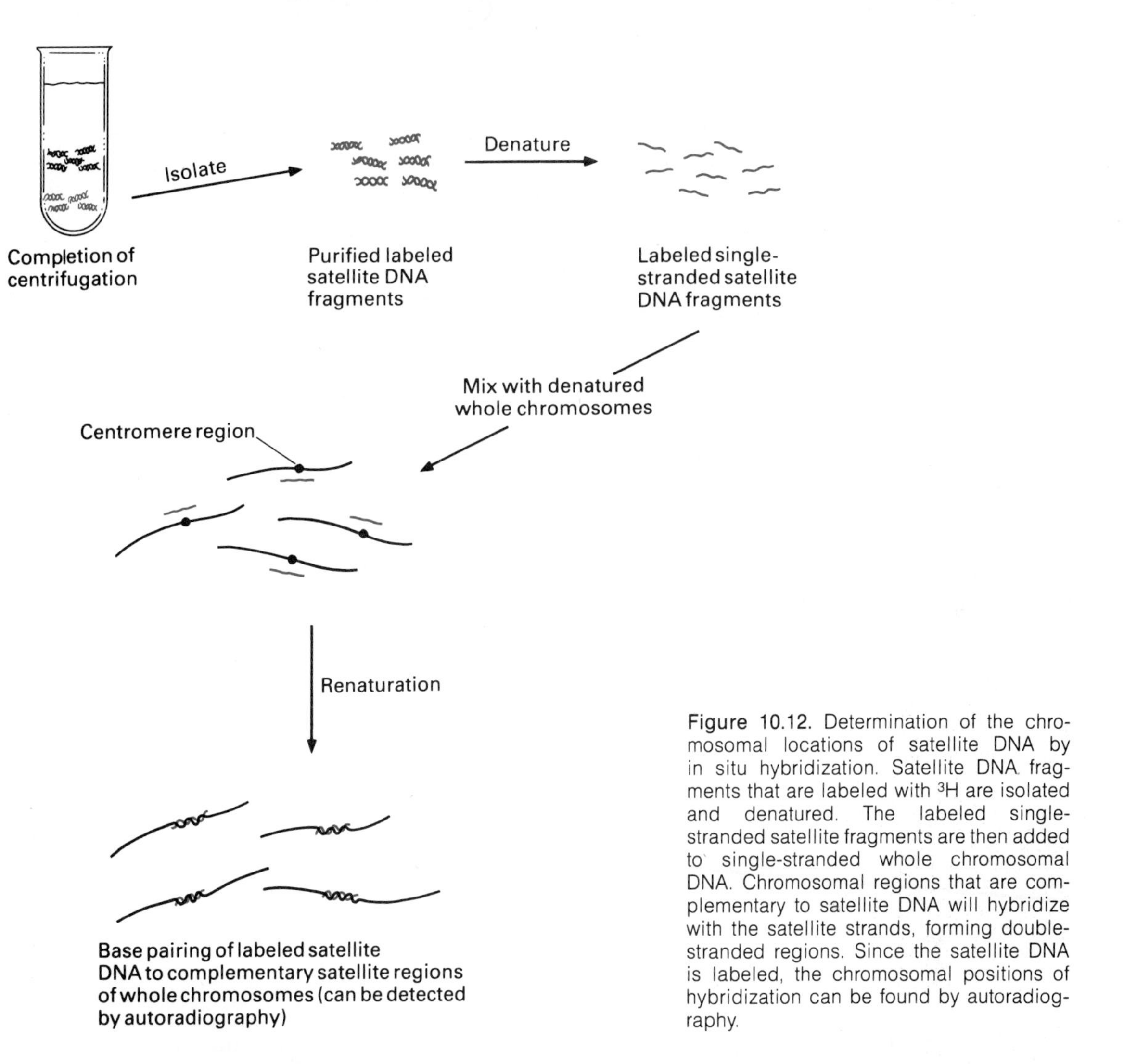

Figure 10.12. Determination of the chromosomal locations of satellite DNA by in situ hybridization. Satellite DNA fragments that are labeled with ^{3}H are isolated and denatured. The labeled single-stranded satellite fragments are then added to single-stranded whole chromosomal DNA. Chromosomal regions that are complementary to satellite DNA will hybridize with the satellite strands, forming double-stranded regions. Since the satellite DNA is labeled, the chromosomal positions of hybridization can be found by autoradiography.

turation of the satellite DNA, converting it to single-stranded molecules. This labeled single-stranded satellite DNA is then added to unlabeled whole chromosomes that have also been denatured. The investigators reverse the conditions used in denaturing the DNA to allow the satellite DNA to renature (hybridize) with complementary regions of the whole chromosomes. Autoradiography is then used to determine the chromosomal locations of the double-helical hybrid regions, made up of one labeled and one unlabeled strand. The autoradiogram reveals that highly repetitive satellite DNA is localized mainly within the centromere regions of the chromosomes. It is currently believed that highly repetitive satellite DNA has a structural role at the centromere regions. This DNA has no known genetic function and is regarded to be genetically inert.

Characterization of Repetitive DNA. Different types of repetitive DNA have been found in the chromosomes of eukaryotic organisms. These classes of DNA have been discovered by studying the renaturation of small fragments of single-stranded

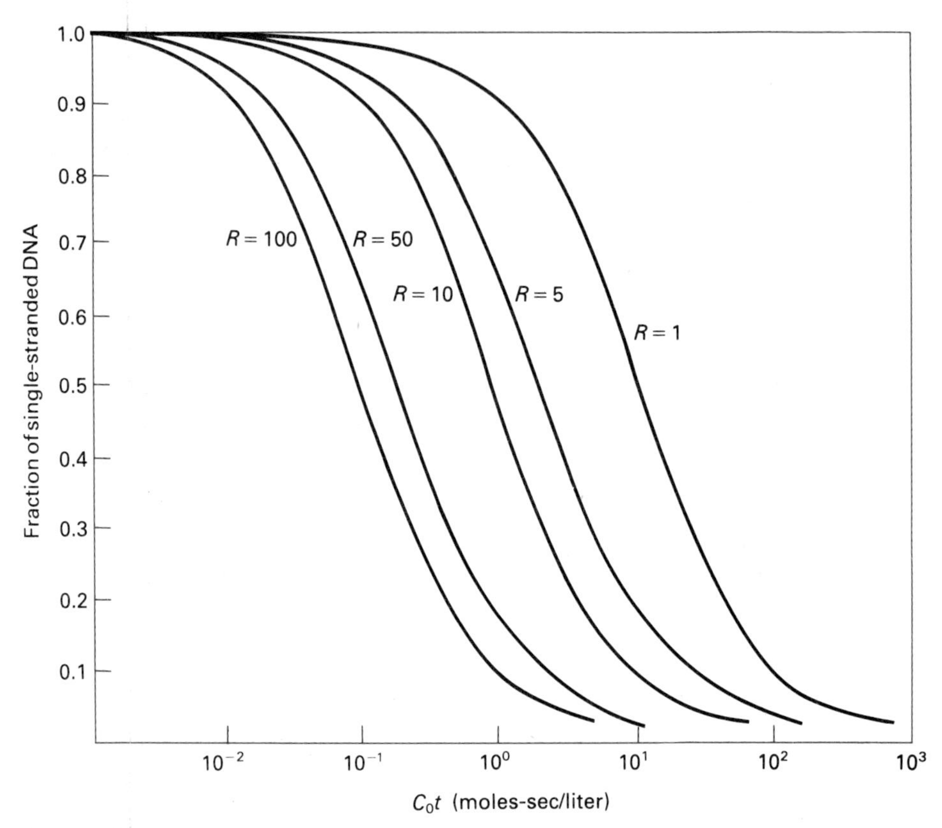

Figure 10.13. Graphs of Eq. (10.1) for different values of R. The plots are commonly known as Cot curves and are shown for a genome length equivalent to that of *E. coli*. For longer total genome lengths, such as found in eukaryotes, the curves will be shifted to the right on the C_0t axis, since it will then be less likely that each single-stranded fragemnt of size 400 nucleotides will collide with its complement. $R = 1$ for *E. coli* DNA, which lacks significant repetition of base sequences.

DNA. The procedure commonly used is to break DNA into uniform fragments of about 400 nucleotide pairs, denature the fragments, and then allow the single-stranded fragments to renature into double-helices under the conditions of ionic strength, DNA concentration, and temperature that will produce a suitable rate of reassociation. Under these conditions, the probability of renaturation depends upon three factors: (1) the initial concentration of single DNA strands (C_0), (2) the amount of time that the DNA is allowed to renature (t), and (3) the number of repeats of a given nucleotide sequence in each haploid equivalent of DNA (R, the repetition number). First, we should note that the total number of collisions between molecules in a solution will increase with time and with the concentration of molecules. Therefore, the chance that a particular DNA chain collides with another of opposite polarity will vary directly with the product $C_0 \cdot t$, or C_0t. We should also note that the chance that the colliding strands are complementary in base sequence will increase with the number of repeats of each given nucleotide sequence in the mixture. Thus, the probability that two DNA strands will collide in time t *and* remain joined together by hydrogen bonds is directly proportional to the product RC_0t.

We can incorporate the effects of R, C_0, and t into a single useful equation by noting that since they all directly influence the probability of renaturation, *the fraction of DNA chains that do not undergo renaturation in time t* will be inversely related to the product of these factors. More precisely, it can be shown that

$$\text{Fraction of single strands remaining at time } t = \frac{C}{C_0} = \frac{1}{1 + KRC_0t} \tag{10.1}$$

where C is the concentration of single strands remaining at time t and K is a proportionality constant that incorporates the effects of all potential variables other than R, C_0, and t, such as temperature and ionic strength. Observe that when $t = 0$, $C/C_0 = 1$, which means that all the DNA is still single-stranded. When t gets large, C/C_0 approaches zero, which means that most of the DNA is now present as double helices.

Figure 10.13 shows plots of Eq. (10.1) for different values of R. We refer to such curves as **Cot curves,** using the colloquial expression "Cot" to describe the progress of reassociation in terms of the product C_0t. The curve in Fig. 10.13 for $R = 1$ is typical of bacterial DNA, which lacks repetition of base sequences that can be detected by this form of analysis. DNA which contains repeated base sequences will renature faster than bacterial DNA, as shown by the Cot curves for R values greater than one in Fig. 10.13. Since the DNA isolated from a eukaryotic cell contains a mixture of different repetitive base sequences, renaturation studies on eukaryotic DNA reveal a much more complex pattern, as illustrated by the Cot curve in Fig. 10.14. Curves of this format usually indicate that the DNA has

Figure 10.14. The renaturation curve of human DNA. The complex renaturation pattern indicates the presence of more than one renaturing class, based on the degree of repetitiveness in the DNA. *Source:* Data from C. W. Schmid and P. O. Deininger, *Cell* 6:345–358, 1975. Copyright © by The MIT Press.

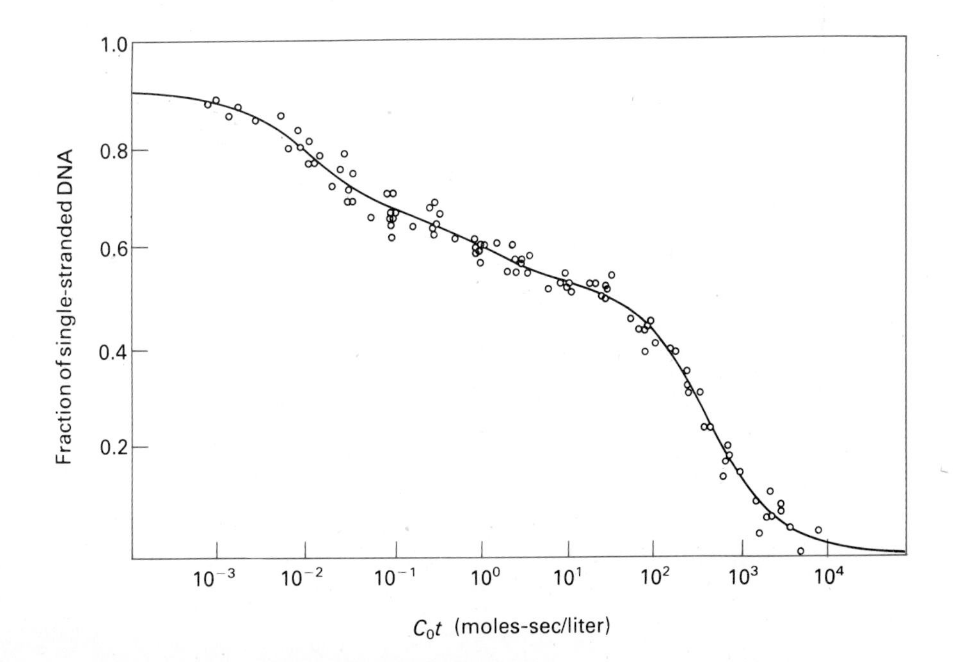

different component parts that are renaturing at different times. The simplest interpretation of these results is shown in Fig. 10.15, in which we distinguish three renaturing classes of DNA: (1) a rapid-renaturing class composed of **highly repetitive DNA** for which R is very large, (2) a slow-renaturing class made up of **unique-sequence DNA,** for which $R = 1$, and (3) a heterogeneous middle-renaturing class consisting of **moderately repetitive DNA,** for which the values of R are intermediate. We can thus view the overall curve as the sum of the different Cot curves for the three renaturing components within eukaryotic DNA.

Moderately repetitive DNA consists of sequences that are repeated anywhere from 10 to 10^5 times, whereas highly repetitive DNA is characterized by repetition numbers greater than 10^5. Different organisms exhibit a wide range of variation in the proportion of their DNA that falls into each of the three repetition classes. Anywhere from 10 to 80 percent of the DNA can be unique, 10 to 50 percent can be moderately repetitive, and 0 to 50 percent can be highly repetitive in base sequence. (The DNA from the highly repetitive class will also be distinguishable as satellite DNA if its base composition differs substantially from that of the other two classes.) The importance of the varying proportions of the three repetition classes of DNA among even closely related organisms is not known; this variation is an interesting aspect of current evolutionary studies. Table 10.1 summarizes the characteristics of the three renaturing classes of eukaryotic DNA.

Figure 10.15. Simplest interpretation of the renaturation pattern for an idealized Cot curve based on human DNA. The main curve (solid line) is the sum of the separate components (dashed lines) adjusted to their relative contributions. F, I, and S are the curves of the fast (highly repetitive), intermediate (moderately repetitive), and slow (unique sequence) renaturing components taken separately. Extrapolated intercepts give the percentage of each component class in the total DNA. [In humans, as in many eukaryotes, the Cot curve shows a fourth class of DNA; this class renatures faster than the fastest renaturation time (or Cot value) that can be detected experimentally. This extremely rapidly renaturing component consists of DNA fragments that have complementary regions within the strand, allowing them to fold back upon themselves. This class comprises about 9% of human DNA.]

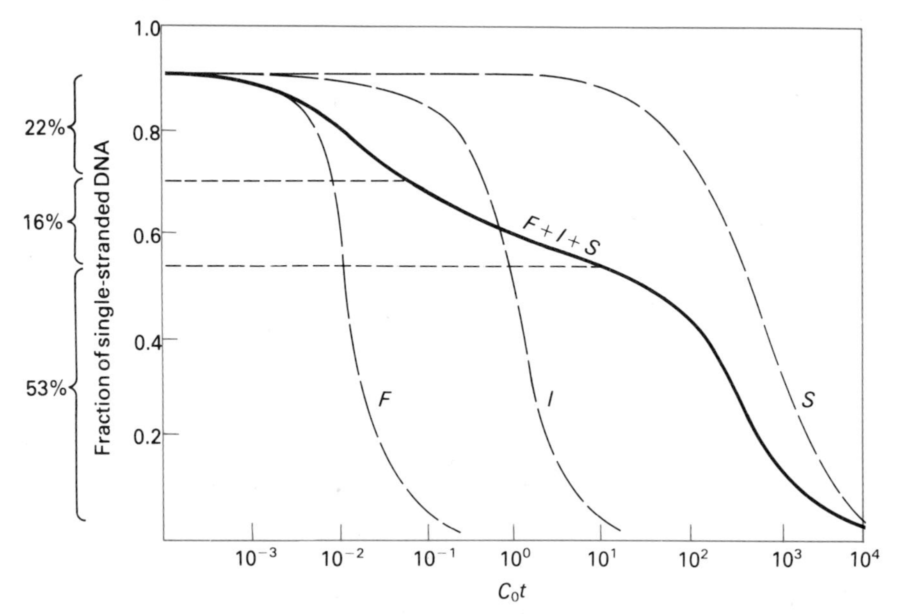

Table 10.1. Characteristics of the three major classes of eukaryotic DNA that can be distinguished on the basis of renaturation studies.

CLASS	PERCENTAGE OF GENOME	R VALUE*	GENERAL CHARACTERISTICS	LOCATION IN GENOME AND FUNCTION
Highly repetitive	0–50	$> 10^5$	5–300 base pairs per repeating unit; distinguishable as satellite DNA if its base composition differs from that of the rest of the DNA	DNA sequences at centromere region; thought to have no coding function
Moderately repetitive	10–50	10^1–10^5	10^2–10^7 base pairs per repeating unit; heterogeneous class, with distinct subclasses having different R values being found in some organisms.	**(1)** Intragenic and intergenic spacer regions, function unknown **(2)** Repeated genes such as those coding for rRNA, tRNA, histones, and antibodies
Unique sequence	10–80	1	$>10^8$ base pairs per repeating unit	Most structural genes

*The R value is the average repetition frequency of a variety of individual repetition sequences dispersed about the genome, each having a different length and degree of repetition.

EXAMPLE 10.2. Suppose that the fast renaturing component of DNA makes up 10 percent of the total DNA complement of 10^8 base pairs in a given organism and has a repetition number of 10^6. **(a)** Calculate the size (in number of base pairs) of the base-pair sequence that is repeated in this particular component of DNA. **(b)** Show that the concentration of this repetitive DNA component that is still single stranded after t minutes of renaturation can be expressed as

$$\frac{(0.1)C_0}{1 + (10^5)KC_0t}.$$

Solution. **(a)** Since the fast renaturing component makes up 10 percent of the organism's complement of DNA, the total number of base pairs in the fast renaturing form is $(0.1)(10^8) = 10^7$. Furthermore, since $R = 10^6$, we know that there are 10^6 copies of the repeated base-pair sequence. Thus, the number of base pairs in each repeated sequence is

$$\frac{\text{base pairs in fast renaturing DNA}}{\text{copies of the repeated sequence}} = \frac{10^7}{10^6} = 10.$$

(We have assumed that the fast renaturing component is made up of just one kind of base pair sequence.) **(b)** The concentration of the fast renaturing component at $t = 0$ is $(0.1)C_0$. Since $R = 10^6$, then after t minutes of incubation, the concentration of the single-stranded form of this component can be calculated from Eq. (10.1) as

$$\frac{(0.1)C_0}{1 + K(10^6)(0.1)C_0t} = \frac{(0.1)C_0}{1 + (10^5)KC_0t}.$$

Interspersion of Repetitive and Nonrepetitive Sequences. As much as 80 percent of the DNA of such diverse organisms as sea urchins, *Xenopus*, and humans exhibits a pattern of *dispersed repeats*. The regions of the chromosomes that contain the genes are primarily unique-sequence DNA that is interspersed with sections of moderately repetitive DNA. This arrangement is known as the *short-period interspersion pattern*, which can be depicted as follows:

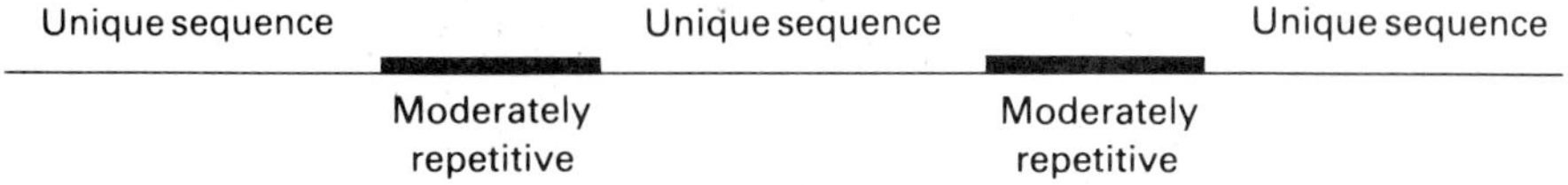

Unique sequence — Moderately repetitive — Unique sequence — Moderately repetitive — Unique sequence

In the DNA of many eukaryotes, the unique sequences are segments that are 1000 to 2000 base pairs long, alternating with moderately repetitive sequences that are 100 to 300 base pairs long. The presence of the short-period interspersion pattern in the DNA of an organism does not preclude the presence of other sequence organizations within the same DNA. These other patterns can include clustered repeats that are not interspersed with unique sequences, long unique sequences that are not interspersed with short repeated sequences, and repeated sequences that are much longer than 300 base pairs.

Organisms that display the short-period interspersion pattern differ from one another in their moderately repetitive regions in at least one major respect. In some organisms (e.g., primates and rodents), many of the short, interspersed, moderately repetitive regions are similar enough in sequence to belong to a single "family." A sequence family is defined as including those regions that are similar enough in base sequence that they can undergo renaturation (hybridize with one another) under standard conditions; members of different families do not hybridize. The most prominent interspersed repetitive sequence family in primates and rodents is known as the *Alu* family. Its more than 300,000 members may make up as much as half of the interspersed repeats of human DNA and at least 3 to 6 percent of the total human genomic DNA. Human *Alu* family members differ from one another by no more than 10 percent of their nucleotide pairs. In addition, *Alu* sequences in rodents and primates show considerable homology to each other, as if the *Alu* sequences of both groups are descendants of a common ancestral sequence. In contrast, the short, interspersed, moderately repetitive sequences in sea urchins belong to many unrelated families, with no single sequence family being the most abundant.

Not all eukaryotic species exhibit the short-period interspersion pattern. A notable exception is *Drosophila*. The DNA of *Drosophila* is characterized by a *long-period interspersion pattern*, in which moderately repetitive sequences that are 5000 to 7000 base pairs long are interspersed among unique sequences that are as long as 35,000 base pairs. Between 15 and 20 percent of *Drosophila* DNA is thought to be made up of long interspersed repetitive sequences that can be grouped into approximately 70 different sequence families whose members range in occurrence between 3 and 100 copies per haploid genome. Although the interspersion pattern of the DNA of most eukaryotes is either of the long-period or the short-period pattern, the DNA of some organisms contains an intermediate-period dispersion pattern. For example, chicken DNA has an interspersion pattern consisting of 4500 nucleotide pairs of unique sequence DNA bracketed by moderately repetitive sequences that are approximately 2000 base pairs in length.

The ubiquity of interspersed repetitive sequences in eukaryotic DNA raises questions about the mechanism by which these repeats have been dispersed throughout eukaryotic genomes. It is now known that certain genetic elements found in eukaryotic cells are mobile; that is, they are able to move from one chromosomal location to another. Some mobile genetic elements are also able to effect the movement of DNA sequences that are located near the site of entry of an element into a chromosome. Mobile genetic elements may be involved in the dispersion of the moderately repetitive regions. Supporting this idea is the finding that dispersed repeats in *Drosophila* DNA are not always present at the same chromosomal locations. We will discuss this idea further in Chapter 11.

The finding that some form of dispersed repetitive sequence organization characterizes all eukaryotic DNA raises the interesting possibility that such a pattern serves some common function. Several investigators have suggested that the *Alu* sequences—along with dispersed repetitive sequences in general—function in the control of gene activity, although there is no conclusive experimental evidence to support this idea at the present time. It is known that the majority of these sequences are transcribed into hnRNA, and most are then removed during processing of the RNA. Their removal suggests that they do not serve a function in mature mRNA molecules. Many workers feel that these DNA sequences constitute "extra" DNA that confers no functional benefit to the organism, being potentially important only in evolutionary processes.

An interesting recent finding that may shed some light on the function of dispersed repetitive sequences is that some *Alu* sequences appear to be transcribed as discrete short RNA molecules. Discrete low–molecular-weight RNA molecules of sizes 4.5S and 7S have been isolated from mouse, hamster, and rat cells, and these molecules exhibit considerable homology with *Alu* sequences. The function of these RNA molecules is presently unknown, but two possibilities have been suggested. The homology of the small RNA molecules to the *Alu* regions suggests that the RNA molecules might base pair with the *Alu* sequences in the hnRNA and somehow function in the processing of hnRNA. [See Chapter 9 for a discussion of other small nuclear RNA molecules (not homologous to *Alu* sequences) that may be involved in the processing of RNA.] On the other hand, the small RNA molecules might play a role in the dispersal of the *Alu* sequences throughout mammalian DNA. This latter possibility could conceivably occur by the RNA molecules serving as templates for the synthesis of complementary DNA (see discussion of RNA tumor virus for the enzymatic basis of reverse transcription).

Regulation of Gene Expression

Experiments have shown that the genes that affect a particular trait or a particular stage of development in eukaryotes are not usually clustered in one region of a chromosome. The histone and hemoglobin genes are among the few instances in which functionally related loci are known to be clustered in eukaryotes. In the case of hemoglobin, the α-type and β-type polypeptides are coded for by two corresponding groups of genes that are arranged in both clusters according to their sequence of expression during development (see Fig. 9.12). Within the β cluster, the ε gene is expressed first, followed in order by the γ gene (for which there are two variant forms), the δ gene, and the β gene. The ε gene is preceded by a **pseudogene,** a β-type gene from which no corresponding polypeptide is synthesized. The different genes in the cluster are widely separated by long spacer regions, so that of 50,000 base pairs that make up the entire cluster region, only 10,000 constitute genes. Of these 10,000, 3000 actually code for amino acids; the other 7000 are introns.

We do not know whether or not the clustered arrangement of the hemoglobin genes reflects some aspect of their control. No operon systems have been discovered in eukaryotes. The histone genes are perhaps more a candidate for this type of control than are the hemoglobin genes, since the individual histone genes are much closer together within their cluster. The five histone genes (*H1, H2A, H2B, H3,* and *H4*) occur as a unit of DNA that is 6000 base pairs long and contains one copy of each gene; this unit is repeated anywhere from 10 to 1000 times within the genome, depending on the species. Still, the five histone genes within each unit appear to be transcribed as separate mRNA molecules, rather than as a single polygenic mRNA. The importance of operon control is therefore in doubt, at least in higher eukaryotes.

In addition to lacking the extensive clustering of functionally related genes that is so characteristic of bacteria and viruses, eukaryotes also differ from prokaryotes in one other fundamental aspect of gene control. The regulation of transcription in bacteria is mainly geared toward allowing the expression of genes to adjust rapidly in response to sudden changes in the environment. This form of regulation is both *short-term* and *reversible.* Eukaryotic genes, too, can respond in this manner. We know, for example, that sudden changes in certain hormone levels within the body can lead to reversible alterations in gene expression. Unlike in bacteria, however, many of the changes in gene activity in the higher eukaryotes are *long-term* and are largely *irreversible* in character. These more-or-less permanent changes are associated with the process of cellular differentiation that occurs during the development of multicellular organisms. Much of the study of gene regulation in eukaryotes is therefore tied to the studies of embryology and developmental biology. *Developmental genetics* is currently emerging as an active area of research.

Many hypotheses have been advanced to explain how genes are regulated in eukaryotes. Some of these hypotheses have received more experimental support than others; yet none has anywhere near the experimental backing that has been given to the operon model in bacteria. We will briefly consider here some of the more important models that have been proposed.

Chromatin Condensation. One mechanism that is believed to regulate transcription in eukaryotes is the extent to which chromatin is coiled. The chromatin fiber of a chromosome typically remains tightly coiled, or in a condensed state, during cell division and becomes loosely coiled, or in a dispersed state, during interphase. Variation does occur, however, in both the degree and pattern of coiling. Chromatin fibers show a greater degree of coiling in certain parts of chromosomes, especially in regions adjacent to the centromere and at the tips, than they do in other areas. The areas where coiling is the greatest tend to stain more heavily with dyes that bind to DNA. These areas of tightly packed, dark-staining chromatin are referred to as **heterochromatin,** as opposed to the more loosely packed and less intensely staining chromatin found in other areas, called **euchromatin.** Unlike euchromatin, which becomes dispersed at the end of cell division, heterochromatin remains in its condensed state throughout interphase as well as division.

Geneticists recognize two different types of heterochromatin: constitutive and facultative. *Constitutive heterochromatin* remains in a permanently condensed, highly stainable condition at all stages of the life cycle of the organism. Constitutive heterochromatin commonly appears in the region of the centromere and at the tips of chromosomes. In contrast, *facultative heterochromatin* may appear as heterochromatin at some stages of the life cycle of the organism and as euchromatin at others. Facultative heterochromatin may involve whole chromosomes or entire chromosome sets. For example, in the male mealybug, the entire paternally derived set of chromosomes becomes heterochromatic during early embryonic development and remains heterochromatic throughout the developmental process. Because of its changeability, facultative heterochromatin is considered to be an induced rather than a permanent feature of chromosomes, which is formed in response to certain physiological or developmental conditions.

Heterochromatin appears to be, for the most part, genetically inactive. This inactivity has been demonstrated in the mealybug, in which genetic tests reveal that the entire heterochromatic set of paternally derived chromosomes is inactive in the male. A negative correlation appears to exist between the degree of coiling of the chromatin fiber and the extent of genetic activity, with the highly condensed heterochromatic regions of chromosomes containing few if any functioning genes. By contrast, transcriptionally active chromatin is less tightly coiled and has a more open nucleosome structure, making the DNA more accessible to RNA polymerase. There is now reason to believe that the transition between transcriptionally active chromatin and condensed, inactive chromatin is mediated by the chemical modification of certain constituents of chromatin. Different modifications have been implicated in the process, including phosphorylation, acetylation, and methylation of histones as well as methylation of DNA.

X-Chromosome Inactivation. Somatic cells of a female mammal contain only one completely functioning X chromosome. The other X chromosome becomes heterochromatic and remains condensed and, for the most part, genetically inactive throughout the cell cycle. The condensed X chromosome, which consists of facultative heterochromatin, forms a dark-staining structure referred to as the **sex chromatin body** or **Barr body** after its discoverer, M. Barr. Since the number of X chromosomes differs between the cells of males and females, the inactivation of one of the X chromosomes in females serves to equalize the dosage of functional X chromatin in the sexes.

The inactivation of an X chromosome is referred to as **Lyonization,** after M. Lyon, an early investigator of the phenomenon. Although it is still incompletely understood, X chromosome inactivation is known to occur during early embryological development. Studies reveal that both X chromosomes are euchromatic in the early development of the female embryo. At a certain point in the development of the embryo, one of the X chromosomes in each embryonic cell becomes condensed. In all of the descendants of the cell, that X chromosome will show up as the sex chromatin body.

Which of the two X chromosomes becomes inactivated in each cell is largely a random occurrence. Thus, in about half of the cells of the female embryo, the maternally derived X chromosome becomes condensed and the paternally derived X chromosome remains active. The reverse situation occurs in the remaining cells. As a consequence, all female mammals are mosaics, with some patches of tissue derived from a Lyonized cell in which the maternally derived X chromosome remained active and others derived from a cell in which the paternally derived X chromosome remained active.

The mosaic nature of female tissue has been demonstrated using cell cultures derived from females who are heterozygous for genes carried on the X chromosome. In one study, skin cells were cultured from a female who was heterozygous for the gene *Gd*, which controls the production of the enzyme G6PD (glucose 6-phosphate dehydrogenase), an enzyme active in glucose metabolism. The recessive allele *gd* produces an alternative form of G6PD, which shows a much lower level of activity than the normal enzyme. When individual *Gd/gd* cells were grown into separate colonies (or clones), half of these clones produced only cells whose functional X chromosome bore the *Gd* allele; these cells showed normal G6PD activity. The remaining clones contained only cells with the defective form of the enzyme, the condition that is expected when the only X chromosome that is functionally active is the one that bears the *gd* allele.

Researchers have developed staining techniques that enable geneticists to detect the presence of X chromatin in cell nuclei at interphase. These procedures are extremely valuable for identifying various abnormalities in the number of X chromosomes. In persons with varying numbers of X chromosomes, all X chromosomes except one become inactivated; such persons have one less Barr body than they have X chromosomes per cell. The presence of more than one Barr body in a female, for example, or of a Barr body in a male would thus indicate an abnormal number of X chromosomes. A Barr body analysis is performed on cells from a buccal smear, a simple procedure in which epithelial cells are scraped from the lining of the mouth. The cells are spread on a microscope slide, fixed in ethanol, and stained with a basic dye. The Barr body appears as a small, compact structure (1 μm or less in diameter), located close to or touching the inner surface of the nuclear envelope (Fig. 10.16).

Hormonal Regulation. Certain hormones are known that control the transcriptional activity of genes by affecting the level of coiling of chromatin. For example, studies of *Drosophila* and other dipterans reveal that ecdysone, the molting hormone in insects, induces specific puffing patterns of the chromosomes in the larval stage. **Chromosome puffs** are chromosomal regions that are enlarged or ballooned (Fig. 10.17). The puffed areas are actively engaged in transcription, so that the puffed appearance represents a loosening of the coiling of the chromosome in

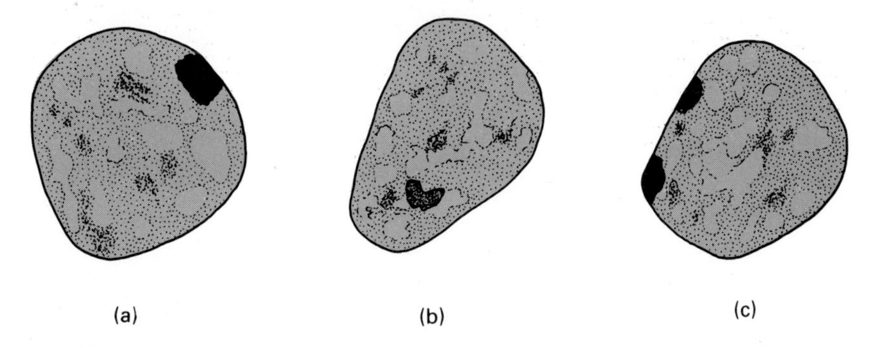

Figure 10.16. (a) The X chromatin (or Barr) body in the nucleus of an epithelial cell from a human female. When epithelial cells from the buccal mucosa are spread on a microscope slide, fixed in ethanol, and stained, the X chromatin body appears as a dark-stained structure closely associated with the inner surface of the nuclear envelope. (b) The nucleus of an epithelial cell from a human male. This cell does not have an X chromatin body. (c) An abnormal XXX female nucleus, with two X chromatin bodies.

transcriptionally active regions. Experiments reveal that certain puffs, called "early" puffs, develop rapidly in response to the molting hormone. These early puffs form even when protein synthesis is inhibited by different drugs. Several hours later, the so-called "late" puffs appear, but only if protein synthesis is allowed to occur. The need for protein synthesis has been interpreted to mean that the proteins coded for by the early puffs somehow induce transcription at the sites at which the late puffs develop. The genetic regions that are included in the early puffs thus behave as regulatory genes.

Ecdysone is a steroid hormone that, like the mammalian steroid hormones (e.g., estrogen), is important in short-term regulation. The steroid hormones appear

Figure 10.17. (a) Photomicrograph and (b) autoradiograph of a polytene chromosome of the fly *Chironomus tentans*. Three giant puffs characterize this chromosome. Puffs are also known as Balbiani rings, and those shown here are symbolized BR1, BR2, and BR3. *Source:* From B. Daneholt, S. T. Case, J. Hyde, L. Nelson, and L. Weislander, *Progress in Nucleic Acid Research and Molecular Biology* 19:319–334, 1976.

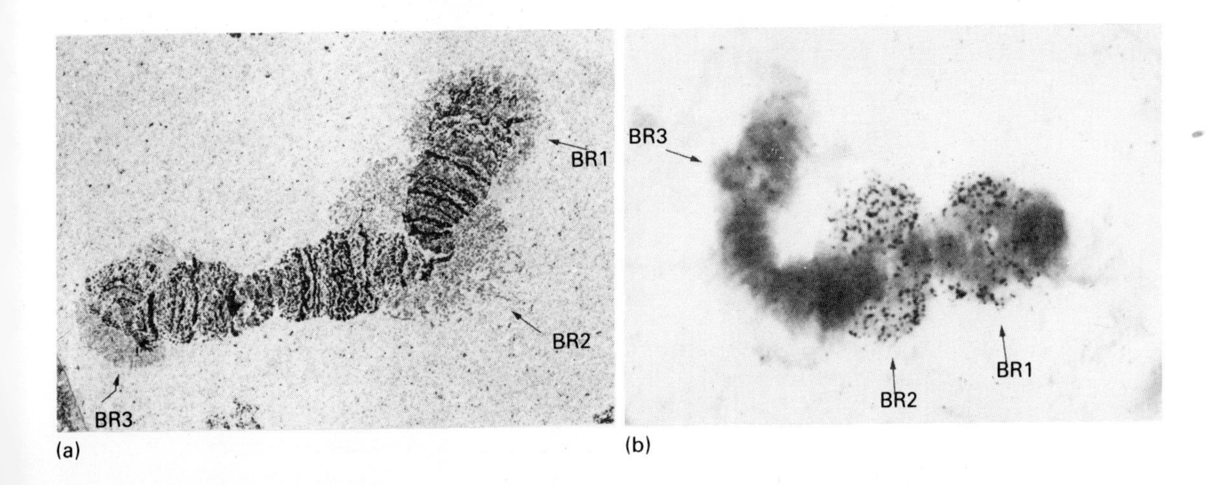

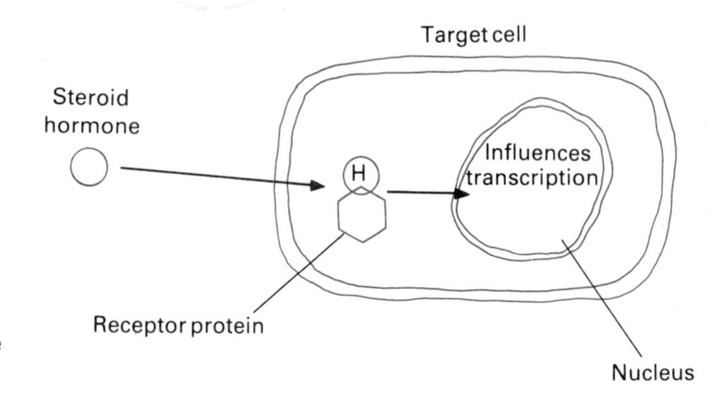

Figure 10.18. Model of the action of a steroid hormone.

to serve as effector molecules that induce transcription in certain target cells (Fig. 10.18). The hormone enters the cytoplasm of the cell and then combines with and alters the structure of a specific *receptor protein.* The hormone is subsequently carried to the nucleus, where the steroid-receptor protein complex binds to the chromatin and induces the transcription of specific genes.

In contrast to the steroid hormones, which regulate gene expression at the level of transcription, the polypeptide hormones affect gene expression by the posttranslational modification of proteins. The polypeptide hormones include insulin and glucagon, which are involved in the control of glucose metabolism, and somatotropin (growth hormone), which affects the rate of skeletal growth and the gain in body weight. These hormones act by way of an intermediate, or "second messenger": cyclic AMP (Fig. 10.19). The primary effect of a polypeptide hormone (glucagon, for example) is initiated when it binds to a receptor site on its target

Figure 10.19. Action of a polypeptide hormone in exerting post-translational control over the activity of a protein kinase. The hormone binds to a protein receptor on the target cell. The hormone-receptor complex activates a membrane-bound enzyme, adenyl cyclase, which is responsible for the conversion of ATP to cAMP. The cAMP so produced then serves as a second messenger by allosterically interacting with a protein kinase. These kinases are composed of a regulatory subunit (R) and an enzymatically active subunit (E). Interaction with cAMP frees the enzymatically active subunit, allowing it to catalyze the phosphorylation of various cellular proteins. The phosphorylation is a means of controlling the activity of these proteins.

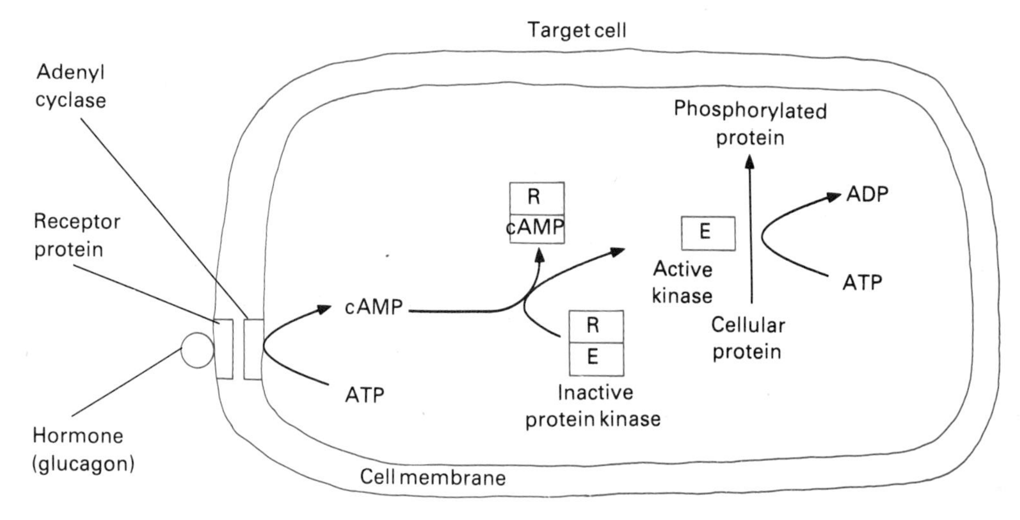

cell. The hormone and receptor form a complex that activates the enzyme adenyl cyclase, located within the cell membrane. The activated enzyme then converts ATP to cAMP. The major role of cAMP is to activate allosterically certain enzymes known as *protein kinases.* The protein kinases are regulatory enzymes that alter the activity of various proteins within the cell. They do so by a reaction known as phosphorylation, in which they catalyze the attachment of phosphate groups to certain amino acid components of the proteins. Several different proteins are affected by these phosphorylation reactions, including the enzymes glycogen synthetase and phosphorylase, which are involved in glycogen metabolism. The proteins of membranes, microtubules, and ribosomes are also affected, as are the histones. In its role as an allosteric effector, cAMP combines with a regulatory subunit (R) within the protein kinase. The cAMP-R complex then disassociates from the rest of the enzyme, leaving the active form of the enzyme (designated E) free to participate in the phosphorylation reactions. The actions of the polypeptide hormones thus regulate the expression of genes by controlling the activity of their protein products.

Cancer: A Loss of Normal Gene Regulation

Many cells of the body, including skin cells and blood cells, must continue to divide throughout the individual's life in order to replace the cells that are lost or destroyed. Since the body can maintain a constant number of cells only by balancing the rate of cell production with the rate of cell death or loss, cell division within the organism must be under strict control. On occasion, however, abnormal cells appear that can escape the usual constraints and proceed to multiply at the expense of the neighboring cells. When these abnormal cells remain localized in their place of origin and do not invade the surrounding tissues, the resulting cell mass is called a *benign tumor.* The abnormal growths can usually be removed surgically without danger of recurrence. If, on the other hand, the abnormal cells spread (metastasize) to the surrounding tissues or to distant sites by way of the blood stream, the abnormal growth is then called a *malignant tumor* or *cancer.* Thus, in addition to their ability to proliferate without normal regulation, the cells of a malignant tumor are also able to spread to and produce secondary growths in other parts of the body.

Malignant tumors are invasive, in part because of changes that occur in the surface properties of the cancer cells as compared to their normal counterparts. These surface alterations make the cancer cells unresponsive to the growth restraints of contact between cells, known as **contact inhibition** (Fig. 10.20). The alterations also cause the cells to be unresponsive to the regulatory actions of serum factors (e.g., polypeptide hormones) that normally exert their effects by combining to sites on the plasma membrane. In addition, the modifications in the surface of the cell result in the loss of normal affinities between cells. Normal cells tend to show affinity for and produce aggregates with cells of the same kind. For example, normal kidney cells tend to stick to other kidney cells, and liver cells stick to other liver cells. Cancer cells, on the other hand, show aberrant interactions between cells and form aggregates in which the arrangements of cells are disorganized.

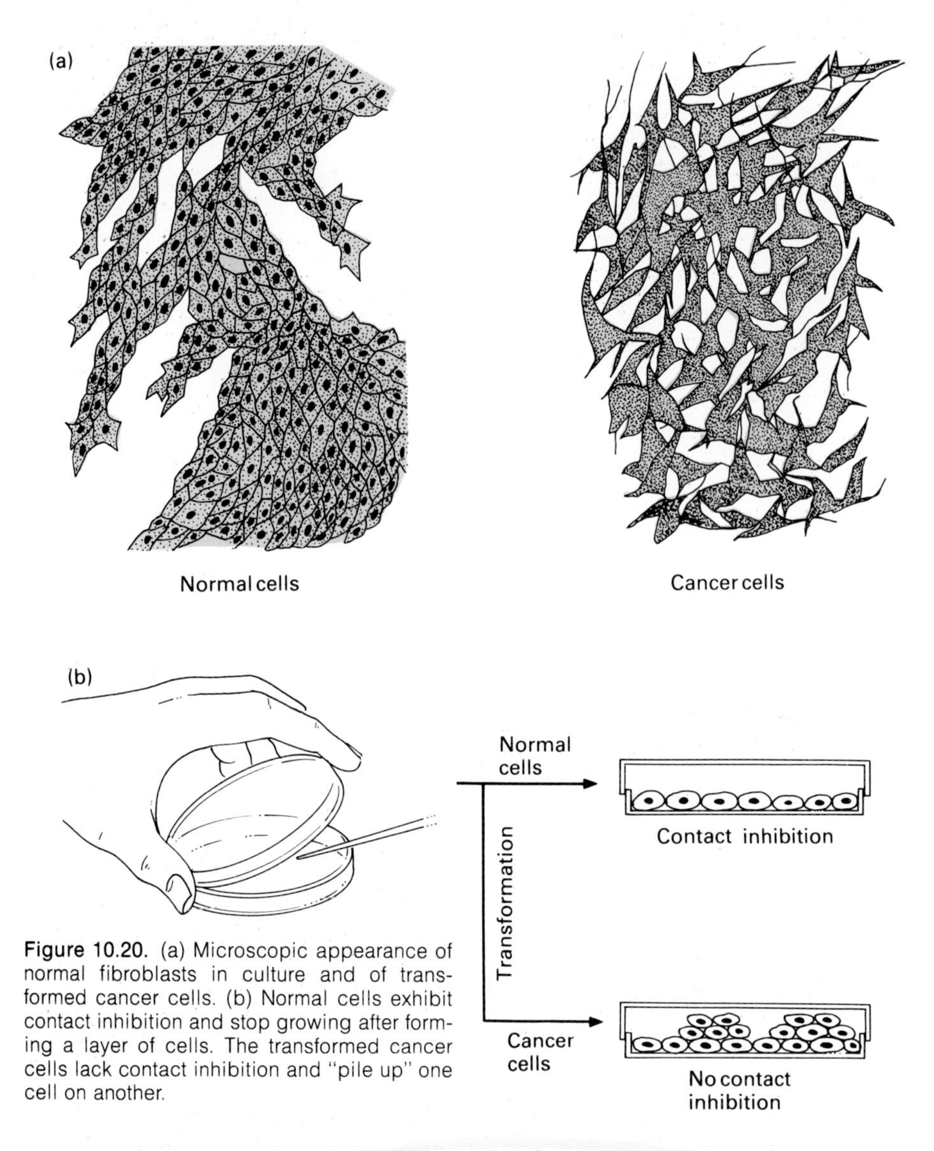

Figure 10.20. (a) Microscopic appearance of normal fibroblasts in culture and of transformed cancer cells. (b) Normal cells exhibit contact inhibition and stop growing after forming a layer of cells. The transformed cancer cells lack contact inhibition and "pile up" one cell on another.

Approximately 200 distinct varieties of human cancer are recognized. Most of these types belong to three major groups. The *carcinomas,* which include about 90 percent of all types of cancer, constitute the largest group. These tumors originate from the epithelial tissues (e.g., skin and the linings of the respiratory and digestive tracts). Cancers of the lung, large intestine, and breast make up a large percentage of this group (40 to 50 percent). Most of the remaining types of cancer are either *sarcomas,* which are derived from the supporting tissues of the body (bone, cartilage, muscle, and fat), or *leukemias,* which originate in the progenitors of the white blood cells. In humans, sarcomas and leukemias make up about 8 percent of the types of cancer; in other mammals that have been studied, however, these two major groups are not nearly so rare.

Studies indicate that the cells derived from a tumor tend to differ from the norm and from one another in chromosome number, often averaging three to four times the haploid number found in normal cells. For example, in the HeLa cell line, which is derived from a human cervical carcinoma, the chromosome number differs from cell to cell, ranging in most cells from about 70 to 90. Because of their variable karyotype, such cell populations are said to be *heteroploid* and are described by the most frequently occurring, or modal, chromosome number.

Despite differences in karyotypes among the cells, each tumor is basically a clone, originating from a single abnormal cell. This is what we would expect if the initial tumor cell arose as a consequence of one or a few relatively rare random events. Since these events are likely to involve hereditary changes in the genome of the cell, we might be interested in determining whether the malignant condition behaves as a dominant or a recessive trait. One study of this type has been performed on a highly malignant mouse cell line (Ehrlich cell line), which produces a fatal malignancy when injected into mice. As illustrated in Fig. 10.21, the

Figure 10.21. Results of fusing normal and cancer cells. When malignant and nonmalignant mouse cells are fused, and the hybrid cells are injected into mice, no malignancy develops. The cancerous state, in this case, acts as a recessive character. But genetic supression of the malignant state is relaxed over time with the loss of chromosomes from the hybrid cells.

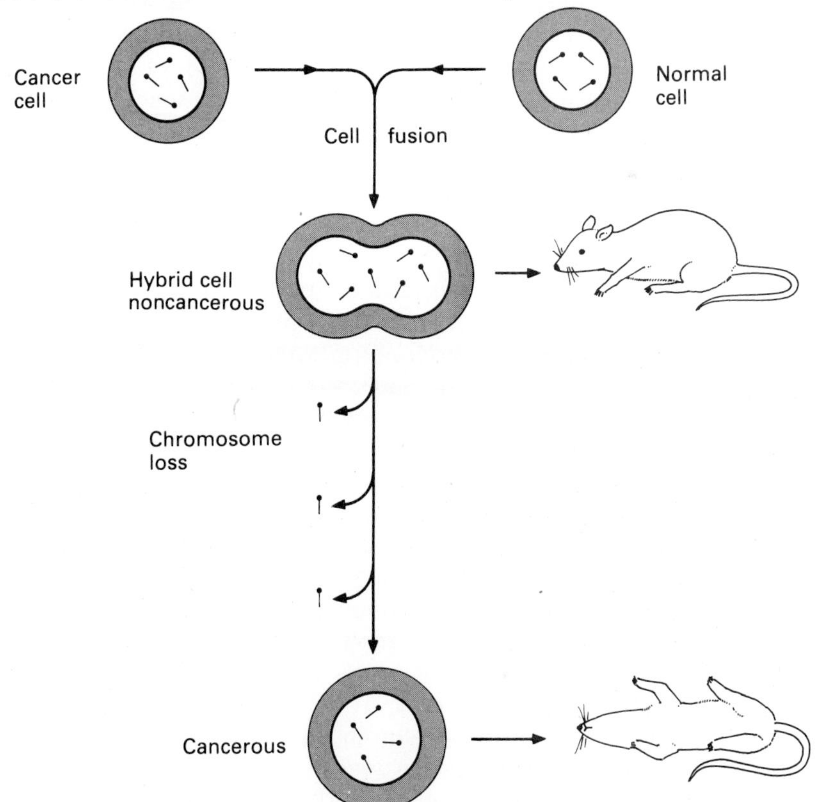

investigators first fused the malignant cells to normal cells and then injected the resulting Ehrlich/normal hybrid cells into mice. No malignancy developed as long as the normal chromosome numbers were maintained in the hybrid cells, which indicates that the malignant state acts as a recessive character in this particular case. But if an Ehrlich/normal hybrid cell lost many of its chromosomes during prolonged culture, the malignant condition would reappear. This finding suggests that the chromosomes of normal cells carry specific suppressors of the malignant state. We should keep in mind, however, that cancer is expressed as a dominant condition in some cell fusion experiments.

Oncogenic Viruses. There may be several different changes in the normal biochemistry of a cell that can lead to malignancy, even in tumors of the same type, as well as many different factors that can cause these changes. For instance, tumors may develop spontaneously from mutations in somatic cells or may be caused by **carcinogens** (cancer-producing agents). Among the important carcinogens are the **oncogenic viruses.** Oncogenic, or tumor-producing, viruses are divided into two general classes, the DNA tumor viruses and the RNA tumor viruses, based on the nature of their genetic material.

The life cycle of SV40, a DNA tumor virus, is illustrated in Fig. 10.22. In general, infection with a DNA tumor virus can have one of two possible consequences: (1) *productive infection,* in which progeny virus particles are formed and released and eventually destroy the host cell, or (2) *cell transformation,* in which the malignant state develops with little or no virus production. In cell transformation, the viral DNA is inserted into the host chromosome. While in the integrated state, the viral DNA replicates in synchrony with the host DNA and transforms the cell into a cancerous one by some as yet unknown mechanism. The presence of viral genes in transformed cells can be detected by means of cell fusion. In one experiment, for example, heterokaryons were formed when malignant cells transformed by SV40 were fused with normal cells that were known not to contain SV40. The heterokaryons released mature SV40 particles, even though the virally transformed line had lost this ability. Thus, transformed cells carry all the genes needed for the production of the intact virus.

The RNA tumor viruses belong to a family known as **retroviruses.** The retroviruses derive their name from their ability to form a DNA copy of their RNA chromosome within an infected cell by using an enzyme called **reverse transcriptase.** The life cycle of a retrovirus known as Rous sarcoma virus is shown in Fig. 10.23. Transformation by this virus occurs when the DNA copy of the viral RNA is inserted into the host chromosome, forming a provirus. RNA polymerase can then transcribe the provirus, producing both viral mRNA, which codes for the viral proteins, and progeny RNA, which is encapsulated to form new tumor viruses. As opposed to the DNA tumor viruses, the cell infected by an RNA tumor virus is usually not killed but continues to produce progeny virus particles while in the transformed state.

Because viral cancers are known to exist in animals, and because cultured human cells can be transformed by viruses to the malignant state in the laboratory, scientists were moved to search for viruses that can cause cancer in humans. Despite the vast amount of work that has been done in this area in recent years,

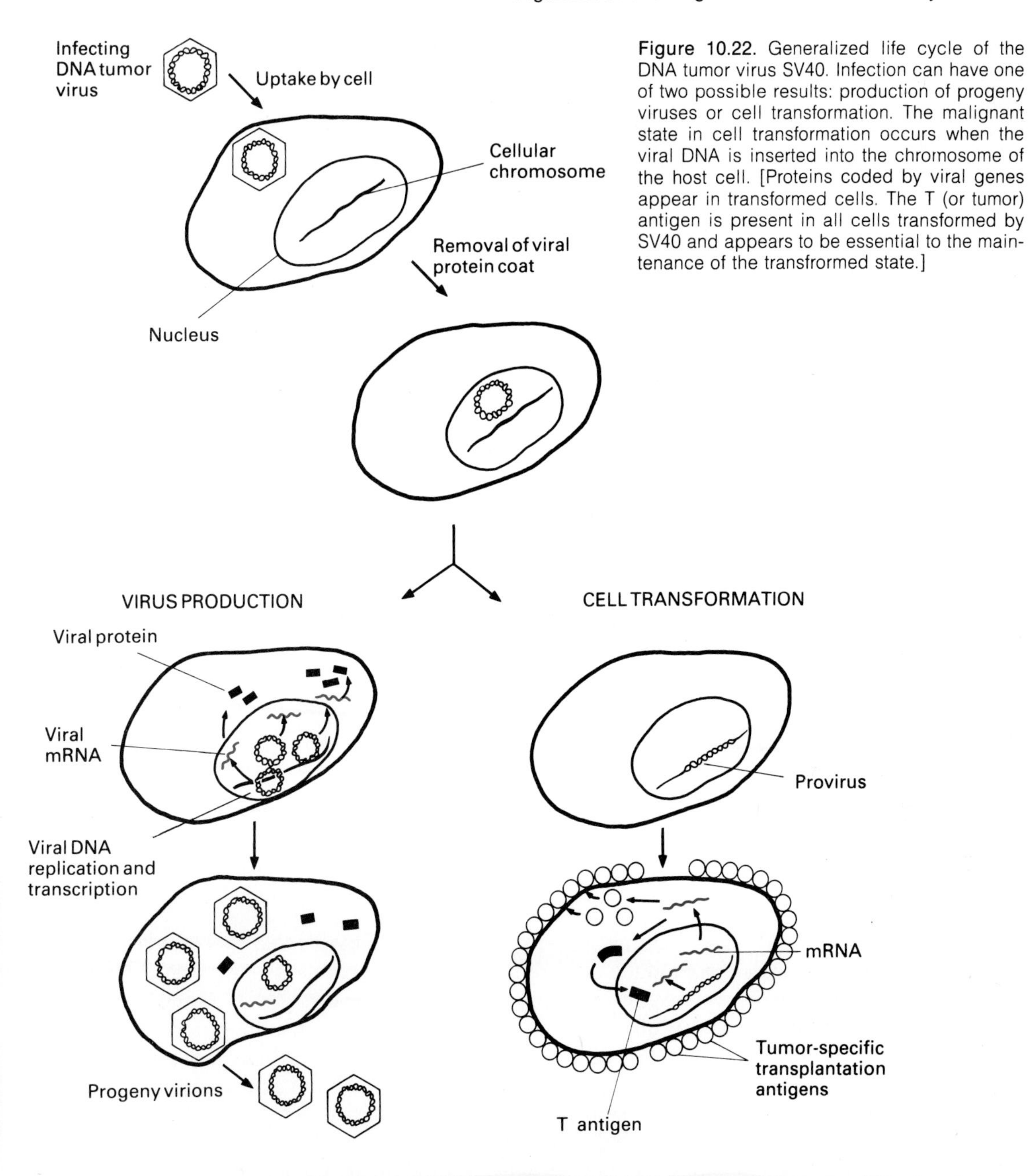

Figure 10.22. Generalized life cycle of the DNA tumor virus SV40. Infection can have one of two possible results: production of progeny viruses or cell transformation. The malignant state in cell transformation occurs when the viral DNA is inserted into the chromosome of the host cell. [Proteins coded by viral genes appear in transformed cells. The T (or tumor) antigen is present in all cells transformed by SV40 and appears to be essential to the maintenance of the transfrormed state.]

the search for human cancer viruses has been largely unsuccessful. This failure has led some experts in the field to doubt that viruses will ever prove to be a major cause of human cancer, other than perhaps leukemias. Nevertheless, the study of oncogenic viruses has turned out to be immensely important in determining the factors that are responsible for cancerous growth. One such factor is the oncogene, which we will consider next.

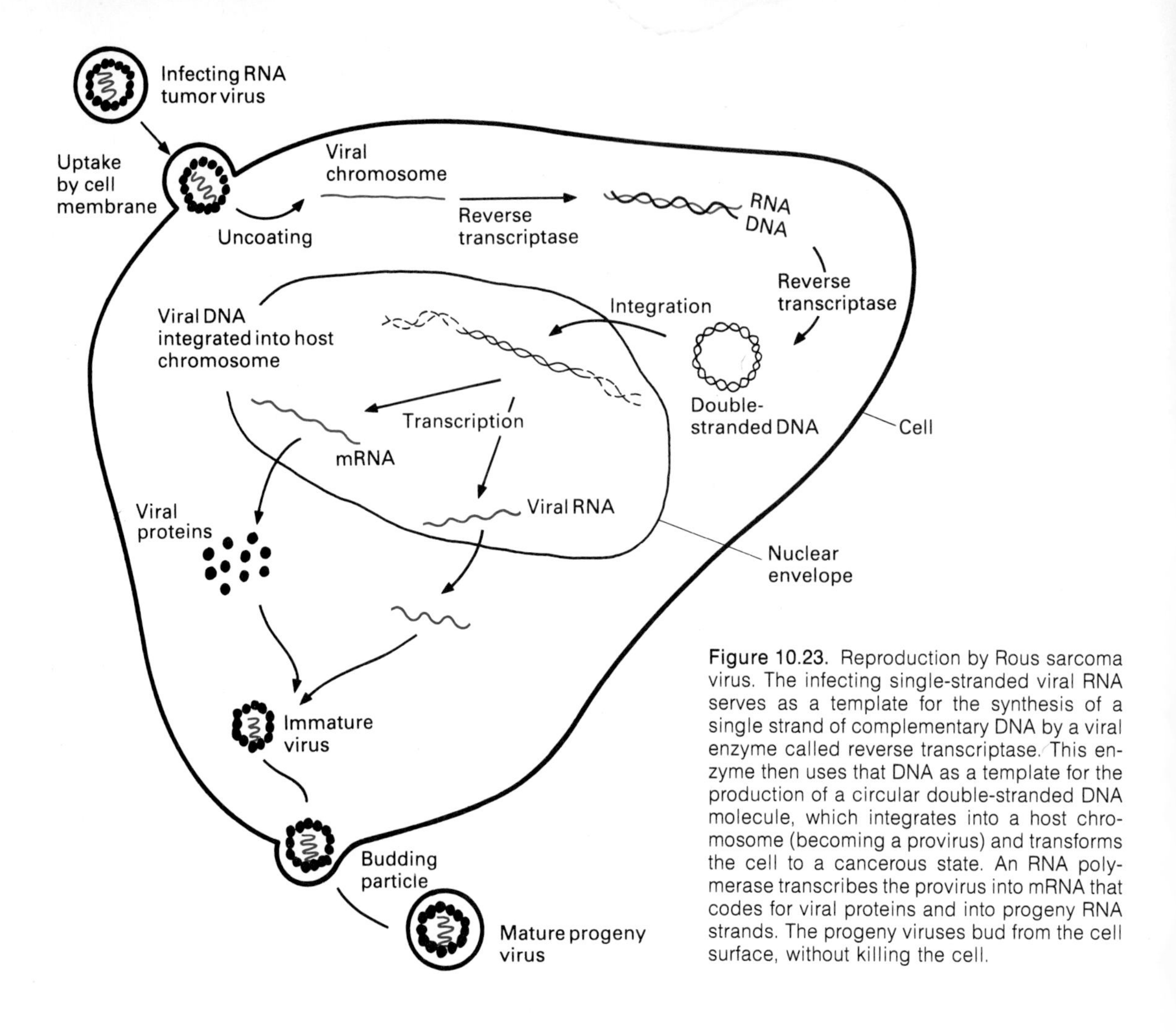

Figure 10.23. Reproduction by Rous sarcoma virus. The infecting single-stranded viral RNA serves as a template for the synthesis of a single strand of complementary DNA by a viral enzyme called reverse transcriptase. This enzyme then uses that DNA as a template for the production of a circular double-stranded DNA molecule, which integrates into a host chromosome (becoming a provirus) and transforms the cell to a cancerous state. An RNA polymerase transcribes the provirus into mRNA that codes for viral proteins and into progeny RNA strands. The progeny viruses bud from the cell surface, without killing the cell.

Oncogenes. Studies on retroviruses have revealed that the ability of many of these viruses to induce cell transformation depends on the presence of a gene known as an **oncogene.** Put simply, an oncogene is a cancer-producing gene. An oncogene was first discovered in the Rous sarcoma virus; the gene was designated *src,* for sarcoma. The *src* gene codes for a specific protein kinase that phosphorylates proteins by attaching a phosphate group to the amino acid tyrosine. This ability to phosphorylate tyrosine is shared with certain other protein kinases that are known to stimulate cell division. Although the mechanism is not fully understood, the protein kinase of the *src* gene is thought to trigger cell transformation by phosphorylating certain proteins in the plasma membrane.

Since their initial discovery, oncogenes similar to *src* have been found in a variety of retroviruses. Even more surprising was the discovery that such genes are also present in the normal cells of a number of different birds and mammals. This finding has led scientists to believe that oncogenes similar to those found in retroviruses are normal constituents of the genome of a cell.

But how can such a gene, which is known to cause cancer when it is brought into a cell by a virus, function as a normal part of the cellular DNA? One possible explanation for this paradox is given by the model in Fig. 10.24. According to this model, the normal expression of the oncogene is required for normal cell growth and development. But if the expression of this gene is augmented by a tumor virus, which carries another copy of the gene, or is altered in some way by some other type of carcinogen, its expression may lead to cancerous growth. The cell thus carries the seeds of its own destruction.

The model shown in Fig. 10.24 provides a common basis for the actions of different carcinogens in terms of a single oncogene. Recent evidence indicates, however, that there may be several different cancer genes, each of which is active in a different type of tumor. Clearly, much remains to be done in defining the genetic basis of cancer.

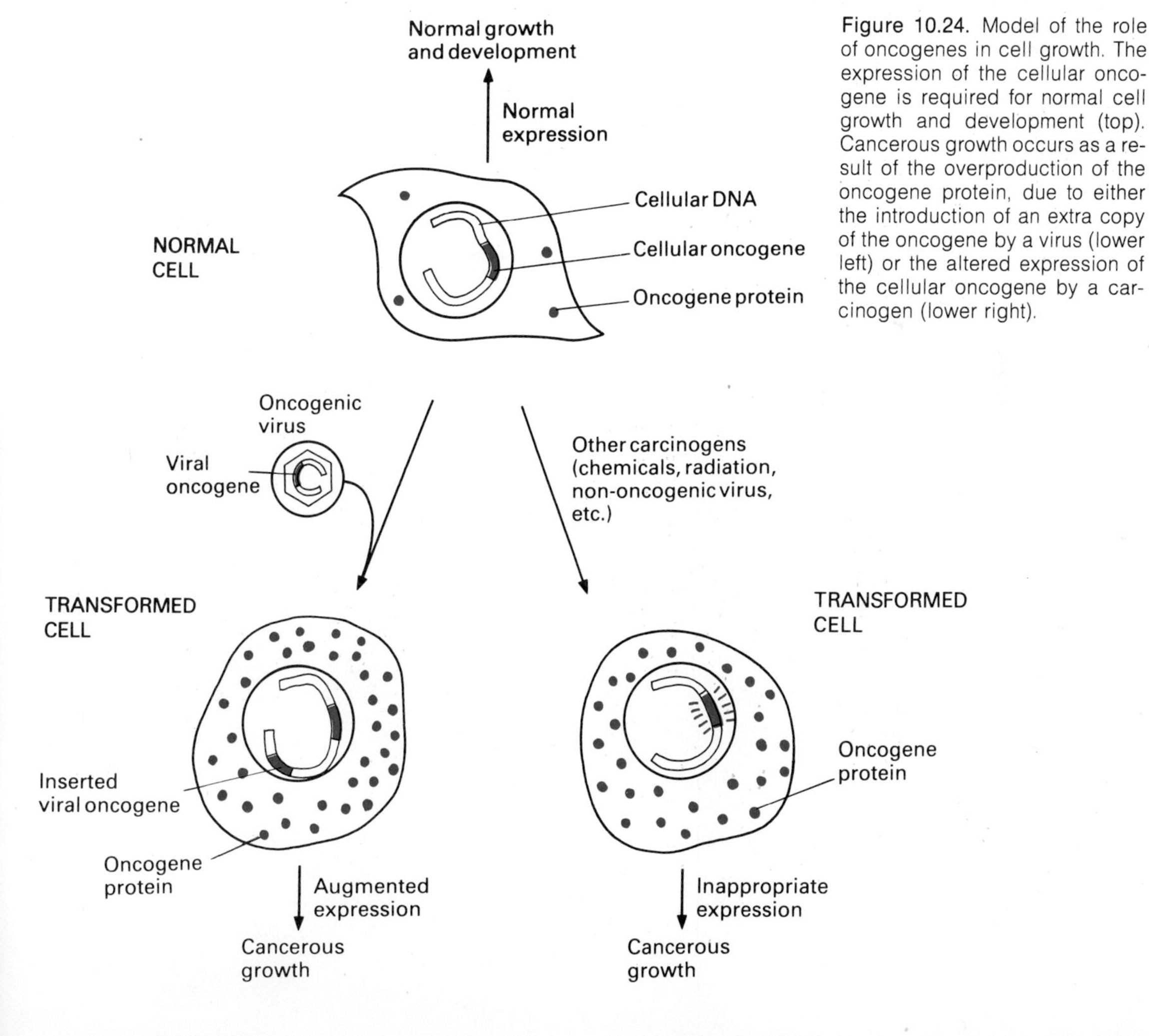

Figure 10.24. Model of the role of oncogenes in cell growth. The expression of the cellular oncogene is required for normal cell growth and development (top). Cancerous growth occurs as a result of the overproduction of the oncogene protein, due to either the introduction of an extra copy of the oncogene by a virus (lower left) or the altered expression of the cellular oncogene by a carcinogen (lower right).

TO SUM UP

1. Eukaryotic chromosomes contain DNA that is composed of repeated nucleotide sequences; some sequences (called highly repetitive DNA) are repeated many times throughout the genome of an organism. Renaturation studies reveal that the nucleotide sequences fall into three frequency classes: highly repetitive, moderately repetitive, and nonrepetitive or unique.
2. The chromosomal locations and functions of the repetitive and nonrepetitive classes of DNA differ. Nonrepetitive DNA is localized within the genes of an organism. Moderately repetitive DNA includes a few repeated gene sequences (gene clusters) and certain sequences that are interspersed with nonrepetitive DNA. Highly repetitive DNA is localized in the heterochromatic non-genic regions of chromosomes, such as the centromere, where it might serve a structural role.
3. Much less is known about the mechanisms of gene control in eukaryotes than in prokaryotes. In contrast with prokaryotes, functionally related genes of eukaryotic chromosomes are not organized into operons. The differences between prokaryotes and eukaryotes with respect to the nucleotide sequences of their chromosomes, gene structure, and processing of RNA may account for the differences in regulatory systems between the two groups.
4. Regulation of eukaryotic gene expression may occur at the level of transcription through the differential coiling of chromatin. Highly condensed chromatin (heterochromatin) appears to be genetically inactive. In contrast, transcriptionally active chromatin is less tightly coiled during interphase, becoming condensed only during cell division. Chromatin that is loosely packed and less intensely staining is termed euchromatin.
5. One of the X chromosomes in each cell of a female mammal becomes heterochromatic early in embryological development and remains, for the most part, genetically inactive in all descendant cells, at all stages of the cell life cycle. Heterochromatic X chromosomes can be microscopically detected in interphase cells as darkly staining Barr bodies.
6. Steroid hormones also appear to play an important role in the regulation of transcription in eukaryotes. In contrast, the polypeptide hormones function in the posttranslational control of protein activity.
7. Abnormal tissue growths in which the cells are capable of spreading to neighboring tissues or to distant sites by way of the blood stream are known as malignant tumors or cancers. Cancer cells have altered surface properties with the result that they are unresponsive to the growth restraints of contact inhibition and to the regulatory actions of serum factors, and they lack the normal affinities between like cells.
8. Oncogenic viruses are those that are capable of causing cancer; they include the DNA and RNA tumor viruses. The malignant state is induced by both types of tumor viruses when the viral DNA or the DNA copy of the viral RNA is inserted into the chromosome of a host cell. Both an RNA tumor virus and the cell that it infects possess a gene called an oncogene, which is responsible for the tumor-producing character of the virus.

Questions and Problems

1. Define the following terms, and distinguish between members of paired terms:
   benign tumor–malignant tumor
   Barr body
   chromosome puff
   contact inhibition
   Cot curve
   euchromatin–heterochromatin
   inducible–repressible operon
   oncogene

oncogenic virus
operon
regulator gene–operator gene
repetitive DNA–unique-sequence DNA
repressor–corepressor
satellite band
short-period interspersion pattern

2. Distinguish between the DNA of prokaryotes and eukaryotes with respect to each of the following properties:
  (a) Total amount of DNA per cell.
  (b) Variability of GC content from region to region within a single genome.
  (c) Presence of repetitive DNA.

3. In the bacterium *Salmonella typhimurium,* the genes that code for the enzymes of the biosynthetic pathway for histidine are organized into a repressible operon, with histidine acting as the corepressor. Diagram a model of the repressible *his* operon, showing the action of histidine as corepressor. [Recall that histidine synthesis is also subject to feedback inhibition by excess histidine (see Fig. 9.6).]

4. One method for solving Eq. (10.1) for K is by using the value of C_0t on the Cot curve that occurs when half of the renaturing DNA is still in its single-stranded form (i.e., when $C = \frac{1}{2}C_0$). The value of C_0t at this point is customarily designated $C_0t_{1/2}$. Show that with nonrepetitive DNA, K will equal $1/C_0t_{1/2}$.

5. Several different haploid and diploid genotypes for the *lac* operon are given below. For each genotype, determine whether or not β-galactosidase and galactoside permease will be produced under conditions of noninduction (no lactose present) and induction (lactose present). Fill in the table below with a + where enzymes are produced and a − where they are not.

GENOTYPE	β-GALACTOSIDASE NO LACTOSE	β-GALACTOSIDASE LACTOSE	GALACTOSIDE PERMEASE NO LACTOSE	GALACTOSIDE PERMEASE LACTOSE
(a) $i^+p^+o^+z^+y^+$	−	+	−	+
(b) $i^cp^+o^+z^+y^+$				
(c) $i^+p^-o^+z^+y^+$				
(d) $i^+p^+o^cz^+y^-$				
(e) $i^+p^+o^+z^-y^-/i^cp^+o^+z^+y^+$				
(f) $i^+p^+o^cz^+y^-/i^+p^+o^+z^-y^+$				
(g) $i^+p^+o^cz^-y^+/i^cp^+o^+z^+y^-$				
(h) $i^sp^+o^+z^+y^-/i^cp^+o^+z^-y^+$				
(i) $i^+p^-o^cz^+y^+/i^+p^+o^+z^-y^-$				

6. There are three linked loci in a bacterial species: a, b, and c. These loci are concerned with the synthesis of a group of jointly controlled enzymes. In the table below, + means that the enzymes are synthesized, and − means that they are not. One of these genes is a regulator, one is an operator, and one is a structural gene. Which is which?

GENOTYPE	ENZYME SYNTHESIS INDUCER ABSENT	ENZYME SYNTHESIS INDUCER PRESENT
$a^+b^+c^+$	−	+
$a^+b^+c^-$	+	+
$a^-b^+c^+$	−	−
$a^+b^-c^+$	+	+
$a^+b^-c^+/a^-b^+c^-$	+	+
$a^-b^+c^+/a^+b^-c^-$	+	+
$a^+b^+c^-/a^-b^-c^+$	−	+

7. The DNA of a particular eukaryote has a mixture of three renaturing components, as seen in the following Cot curve:

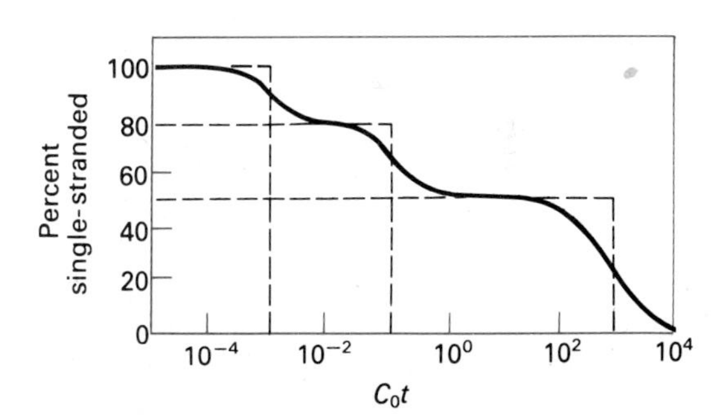

(a) Calculate the fraction of each renaturing component in the genome of this eukaryote.
(b) Estimate the value of $C_0t_{1/2}$ for each renaturing component in the mixture. (Recall from Problem 6 that $C_0t_{1/2}$ is the value of C_0t when half of the component has undergone renaturation.)
(c) Suppose that each renaturing component of DNA is isolated in pure form. What $C_0t_{1/2}$ values would you expect to obtain for the different components when each is renatured separately? (*Note*: C_0 is now the initial concentration of a pure form of DNA rather than the initial concentration of DNA in the mixture.)
(d) Calculate the repetition number for each renaturing component. Assume that K equals 2×10^{-3}.
(e) If there are 10^9 base pairs in the total genome of this eukaryote, how many base pairs would be present in each repeating unit in the most highly repetitive component of the DNA?

8. Two models have been proposed in order to account for the molecular basis of transcriptional control in an inducible operon:

I. The regulatory gene (R) produces a repressor protein that in its active form combines with the operator-promoter site (o) of the operon to block transcription of the adjacent structural gene (S). The inducer acts by allosterically inactivating the repressor protein.
II. The regulatory gene (R) produces an expressor protein that in its active form combines with the operator-promoter site (o) to promote transcription of the structural gene (S). The inducer acts by allosterically activating the expressor protein.

Suppose that you have access to three kinds of mutations: R^i = a mutation that destroys the ability of the regulatory gene product to combine with the inducer; R^o = a mutation that destroys the ability of the regulator gene product to combine with the operator-promoter site; and o^- = a mutation that destroys the ability of the operator-promoter site to combine with the regulator gene product. For each of the following genotypes, indicate whether enzyme synthesis according to hypotheses I and II will be **(a)** constitutive and independent of the concentration of inducer, **(b)** inducible and occurring in substantial amounts only in the presence of the inducer, or **(c)** repressed, with enzyme synthesis lacking or occurring in only trace amounts regardless of the level of inducer.

GENOTYPE	PHENOTYPE EXPECTED FOR I	II
$R^i o^+ S^+$	repressed (c)	repressed (c)
$R^o o^+ S^+$		
$R^+ o^- S^+$		
$R^i o^+ S^+/R^+ o^+ S^+$		
$R^o o^+ S^+/R^+ o^+ S^+$		
$R^+ o^- S^+/R^+ o^+ S^-$		

9. It is currently believed that tumors develop from a single abnormal cell (that is, have a clonal origin), rather than from a group of altered cells. Suppose that tumors are removed from females who are heterozygous (*Gd/gd*) for alleles carried at the G6PD locus on their X chromosomes. Tests are then performed to determine whether the cells of each tumor are only *Gd* or *gd* in genotype, or a mixture of both *Gd* and *gd* types. What result is expected if the clonal origin of tumors is correct?

10. According to the Lyon hypothesis, X chromosome inactivation is strictly random, so that each cell in an early female embryo has an equal chance of having the maternally derived or the paternally derived X chromosome become heterochromatic. Suppose that X inactivation occurs in a certain mammal at the eight-cell stage of embryonic development. **(a)** What is the probability that in exactly half of the cells the maternally derived X chromosome is inactivated? **(b)** What is the chance that the female that develops from this embryo is not mosaic but has either all of its maternally derived or all of its paternally derived X chromosomes inactivated? **(c)** Suggest a way to determine the embryonic stage when X inactivation occurs in a certain species of mammal.

Suggested Readings / Part IV

Chapter 9

Beadle, G. W. The genes of men and molds. *Scientific American* 179:30–39, 1948.

Beadle, G. W. and E. L. Tatum. Genetic control of biochemical reactions in *Neurospora. Proc. Natl. Acad. Sci., U.S.* 27:499–506, 1941.

Chambon, P. Split genes. *Scientific American* 244:60–71, 1981.

Crick, F. H. C. The genetic code. *Scientific American* 207:66–74, 1962.

Crick, F. H. C. The genetic code: III. *Scientific American* 215:55–62, 1966.

Doty, P. Proteins. *Scientific American* 197:173–184, 1957.

The Genetic Code. Cold Spring Harbor Sympos. Quant. Biol. 31, 1966.

Miller, O. L. The visualization of genes in action. *Scientific American* 228:34–42, 1973.

Nirenberg, M. W. The genetic code: II. *Scientific American* 208:80–94, 1963.

Phillips, D. C. The three-dimensional structure of an enzyme molecule. *Scientific American* 215:78–90, 1966.

Rich, A. and S. H. Kim. The three-dimensional structure of transfer RNA. *Scientific American* 238:52–73, 1978.

Watson, J. D. *Molecular Biology of the Gene,* Benjamin, Menlo Park, Calif., 1976.

Yanofsky, C. Gene structure and protein structure. *Scientific American* 216:80–94, 1967.

Chapter 10

Barr, M. L. Sexual dimorphism in interphase nuclei. *Am. J. Hum. Genet.* 12:118–127, 1960.

Beckwith, J. and D. Zipser (eds.). *The Lactose Operon,* Cold Spring Harbor Laboratory, Cold Spring Harbor, New York, 1970.

Bishop, J. M. Oncogenes. *Scientific American* 246:80–93, 1982.

Britten, R. J. and E. H. Davidson. Gene regulation for higher cells: a theory. *Science* 165:349–357, 1969.

Croce, C. M. and H. Koprowski. The genetics of human cancer. *Scientific American* 238:117–125, 1978.

Davidson, E. and R. Britten. Organization, transcription and regulation in the animal genome. *Quart. Rev. Biol.* 48:565–613, 1973.

DeRobertis, E. M. and J. B. Gurdon. Gene transplantation and analysis of development. *Scientific American* 241:74–93, 1979.

DuPraw, E. J. *DNA and Chromosomes,* Holt, Rinehart & Winston, New York, 1970.

Gurdon, J. B. Transplanted nuclei and cell differentiation. *Scientific American* 219:24–35, 1968.

Hood, L. E., J. H. Wilson and W. B. Wood. *Molecular Biology of Eucaryotic Cells,* Benjamin, Menlo Park, Calif., 1975.

Jacob, F. and J. Monod. Genetic regulatory mechanisms in the synthesis of proteins. *J. Mol. Biol.* 3:318–356, 1961.

Lewin, B. *Gene Expression, Vol. 2: Eucaryotic Chromosomes,* Wiley-Interscience, New York, 1980.

Lyon, M. F. Sex chromatin and gene action in the mammalian X-chromosome. *Am. J. Hum. Genet.* 14:135–148, 1962.

Ptashne, M., A. D. Johnson, and C. O. Pabo. Regulatory proteins interact with DNA to turn genes off and on. *Scientific American* 247:128–141, 1982.

Part V

Mutation

Chapter 11

Mechanisms of Mutation and Repair

In previous chapters, we considered the mechanisms by which DNA stores information, replicates, and recombines, and the way in which the genetic information is translated into a biologically meaningful phenotype. We are now ready to discuss the molecular mechanisms of another function of DNA, namely, mutation. We will also be introducing the concept of repair of damage to DNA, an idea whose importance is only now becoming understood. Although knowledge of both of these processes is incomplete, enough is known to allow us to appreciate how vastly complex are the molecular functions of DNA. We begin our discussion with gene mutation.

Nature of Gene Mutation

Mutations are generally classified as **somatic mutations,** when they occur in the body cells of an organism, and as **germinal mutations,** when they occur in tissue that will ultimately form gametes. Because germinal mutations can be passed on to future generations, we can speak of their effects on the entire organism, not just on a localized group of cells. For this reason, the term mutation is usually taken to mean a germinal mutation, unless otherwise indicated.

Mutations are also classified as either *gene mutations* or *chromosomal mutations,* according to the size of the alteration. Chromosomal mutations are changes in the structure and number of chromosomes, whereas gene mutation refers to a change at the level of a single gene locus. We will thoroughly discuss chromosome mutations in Chapter 12 and will limit our discussion here to changes that occur within a single gene. Broadly speaking, a gene mutation is any heritable change in the genetic material. As such, it constitutes the mechanism by which the alleles of a gene are formed and thereby serves as the original source of all genetic variability.

The very nature of the structure of DNA enables a gene to undergo mutational alteration. Under natural conditions, mutations occur spontaneously at a very low rate. Some mutations give rise to changes in phenotype that make the organism ill-suited to survive and reproduce in its environment. We tend to be most familiar with this type of aberration. In fact, the term mutation often suggests an abnormality, such as a physical or mental defect. Other mutations may yield no recognizable change in phenotype, or produce a phenotype that is advantageous to the individual. Such mutations are ordinarily expressed in all the environments to which the organism is exposed. In contrast, still other mutations, termed **conditional mutations**, are expressed only under a particular environmental condition (called the *restrictive* condition); the wild-type phenotype is expressed under other circumstances (called *permissive* conditions). Temperature is often the environmental factor that controls the phenotypic expression of a conditional mutant gene. For instance, dominant heat-sensitive lethal genes exhibit the wild-type phenotype at low temperatures but are lethal when the temperature is raised. Conditional mutants have been extremely useful in genetic studies because they provide a means to maintain deleterious mutant genes in laboratory populations under permissive conditions. A shift to the restrictive condition then allows expression of the mutation.

Rates of Mutation

Gene mutations are rare events; newly mutated alleles at a given locus generally occur in less than 1 out of every 1000 gametes. The probability that a mutation will occur is usually expressed in the form of a **mutation rate.** For our purposes, we will define the mutation rate as the probability that a mutation will occur at a given locus per gamete per generation (or per cell per generation in the case of bacteria and fungi). Mutation rates vary with the gene and with the organism, ranging from a low rate of about 10^{-9} to a high rate of about 10^{-4}. Some examples of mutation rates that have been observed in various organisms are given in Table 11.1. At first glance, the rates shown for the selected *E. coli* and *Neurospora* genes seem much lower than those for the eukaryotic genes. But most of this discrepancy is a consequence of the different units used to express mutation rates. The buildup of a pool of meiocytes and gametes in eukaryotes requires many cellular generations. Mutation rate expressed on a per gamete basis is therefore actually a measure of the accumulated effects of the mutations that occurred at each of the acts of cell division that formed the line of meiocytes leading to the gamete. Moreover,

Table 11.1. Mutation rates at specific loci for various organisms.

ORGANISM	TRAIT	MUTATION RATE (PER GAMETE PER GENERATION)*
Escherichia coli	Phage T1 resistance	3×10^{-8}
	Arginine independence	4×10^{-9}
Neurospora crassa	Adenine independence	4×10^{-8}
	Inositol independence	8×10^{-8}
Zea mays	Shrunken (Sh)	1×10^{-6}
	Purple (Pr)	1×10^{-5}
Drosophila melanogaster	Yellow body color	1×10^{-4}
	Ebony body color	2×10^{-5}
	White eye color	3×10^{-5}
Mus musculus	Brown coat	4×10^{-6}
	Albino coat	1×10^{-5}
Homo sapiens	Achondroplasia	7×10^{-5}
	Color blindness (total)	3×10^{-5}
	Huntington's chorea	1×10^{-6}
	Retinoblastoma	1×10^{-5}

* Mutation rates for *E. coli* and *N. crassa* were measured on a per cell per generation basis.

mutations that occur early in a meiocyte line are replicated many times over and may contribute to forming several mutant gametes, thereby compounding the extent of mutation.

The mutation rate for a particular gene locus is not always constant. It can be substantially altered by both genetic and environmental factors. For example, radiation and chemical substances can act as **mutagens** (agents that increase the likelihood of mutation). It is also known that organisms possess extensive *repair systems,* which under normal conditions efficiently repair both spontaneous and induced mutational damage to DNA. Defective repair and replication mechanisms in bacteria can increase the mutation rate by as much as 100,000-fold. If we consider the repair mechanisms, we can think of mutation either as an escape from the normal repair process or as a misrepair. Since the enzymes that are involved in repairing damaged DNA are coded for by the genes of an organism, we can conclude that the mutation rates are under the genetic control of the organism itself.

Besides being rare, mutations are also *reversible.* A mutant allele can undergo secondary mutation back to its original state. The change to a mutant form ($a^+ \rightarrow a$) is referred to as a **forward mutation,** whereas the change from a mutant allele back to the wild-type ($a \rightarrow a^+$) is called a **back mutation** (or **reversion**). True reversion is the exact reversal of the base pair alteration in the DNA that produced the original mutation. A second process, known as **suppression,** mimics the phenotypic consequences of reversion. Suppression occurs when a second mutation at a different site on the DNA somehow compensates for the effects of the first. This compensation restores the wild-type phenotype but leaves the genotype in a mutant condition.

Molecular Mechanisms of Mutation

Most gene mutations arise from changes in a single base pair, either through a base substitution or through the addition or deletion of a nucleotide pair. Gene mutation is thus often called **point mutation,** the "point" referring to the location of a single base pair along the length of DNA. We will now turn our attention to the molecular chain of events that is thought to occur during the mutation process. While these events can be induced in the laboratory, many can also happen spontaneously, without being mediated by any external factor. For the time being, we will concern ourselves with these spontaneous changes in the genetic material. We will consider induced mutations in a later section of this chapter.

Base Substitutions. As we pointed out in Chapter 3, base substitutions are thought to be the consequence of rare base pairing errors that take place during DNA replication. Tautomeric shifts cause the hydrogen bonding properties of a purine or pyrimidine base to change, and these changes, in turn, lead to the formation of a mismatched base pair. In the next round of replication, a base pair substitution occurs in one of the daughter helices.

Twelve different base pair changes are possible. Four are **transitions,** in which a purine in one strand of DNA is replaced with the other purine and the pyrimidine in the complementary strand is replaced with the other pyrimidine. The four possible transitions are thus AT → GC, TA → CG, GC → AT, and CG → TA. If we disregard which strand of DNA is designated first, we can use the general notation AT ⇌ GC to represent the different transition events. In reference to this notation, there are two general ways in which transitions can occur. An AT pair (meaning AT or TA) can be replaced by a GC pair (meaning GC or CG, respectively), or a GC pair can be replaced by an AT pair. In contrast, base substitutions in which a purine replaces a pyrimidine and a pyrimidine replaces a purine are called **transversions.** Eight different transversions are possible: AT ⇌ TA, AT ⇌ CG, CG ⇌ GC, and GC ⇌ TA. Transitions and transversions are compared in Fig. 11.1.

While tautomerism can explain the origin of transitions, it is not at all clear how transversions arise. Some investigators have postulated that rare purine–purine base pairing errors occur as intermediates in the formation of transversions. Because transversions are regularly observed, investigators will continue trying to determine the mechanism by which they arise. The mutant allele that codes for sickle-cell hemoglobin, for example, appears to be the result of the transversion of AT → TA, as shown in Fig. 11.2.

Effect of Base Substitution on Protein Structure. A base pair substitution changes a single codon in the DNA. The mutant codon may or may not then specify an amino acid change along the encoded polypeptide chain. Its effect on the amino acid sequence can take one of three possible forms: silent mutation, missense mutation, or nonsense mutation. The three possible consequences that a base pair

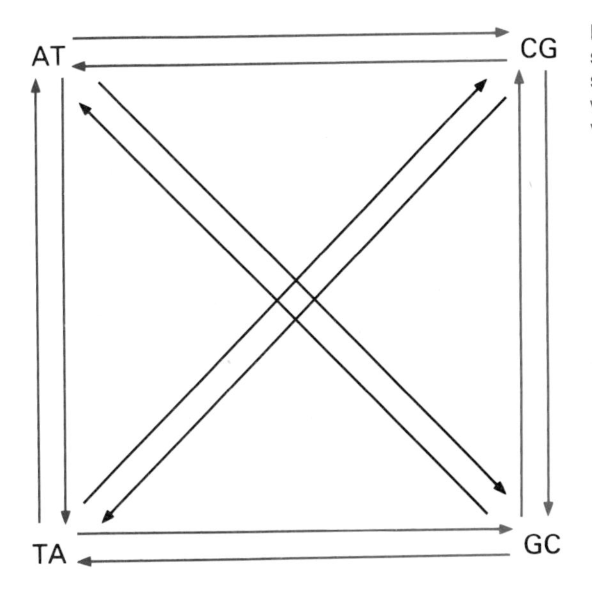

Figure 11.1. The possible base pair substitutions in DNA. The substitutions shown on the diagonals are transitions, while those on the periphery are transversions.

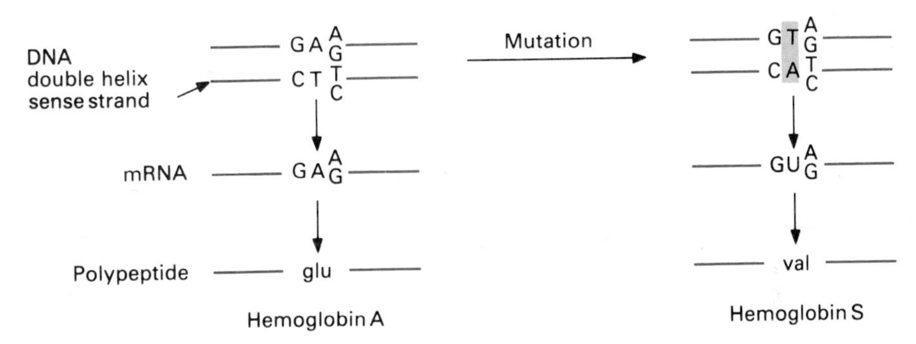

Figure 11.2. Nature of the base substitution thought to give rise to sickle-cell hemoglobin (HbS). Normal hemoglobin (HbA) is known to carry the amino acid glutamic acid (glu) in the sixth position of the β polypeptide, whereas HbS carries valine (val) at that position. From the genetic codon assignments given in Table 9.1, we find that the DNA code word for glu that allows the change to val by way of a single base substitution could be either CTT or CTC (denoted CT^{T}_{C}). The base pair substitution changing the normal gene to its mutant allele would be AT → TA, a transversion.

substitution can have on the amino acid sequence are compared in Fig. 11.3. In the case of a **silent mutation,** the base pair substitution does not alter the amino acid sequence of the polypeptide. As a result of the degeneracy of the genetic code, many base substitutions do not change the amino acid assignment of the affected codon. This outcome shows the difficulty we have in attaching a strict definition to the term mutation. A DNA code word is altered in a silent mutation, but no mutation occurs at the level of the protein coded for by the mutant gene. In discussing mutation, we therefore must specify the biological level (DNA, amino acid sequence, protein function, phenotype) that we are considering. This distinc-

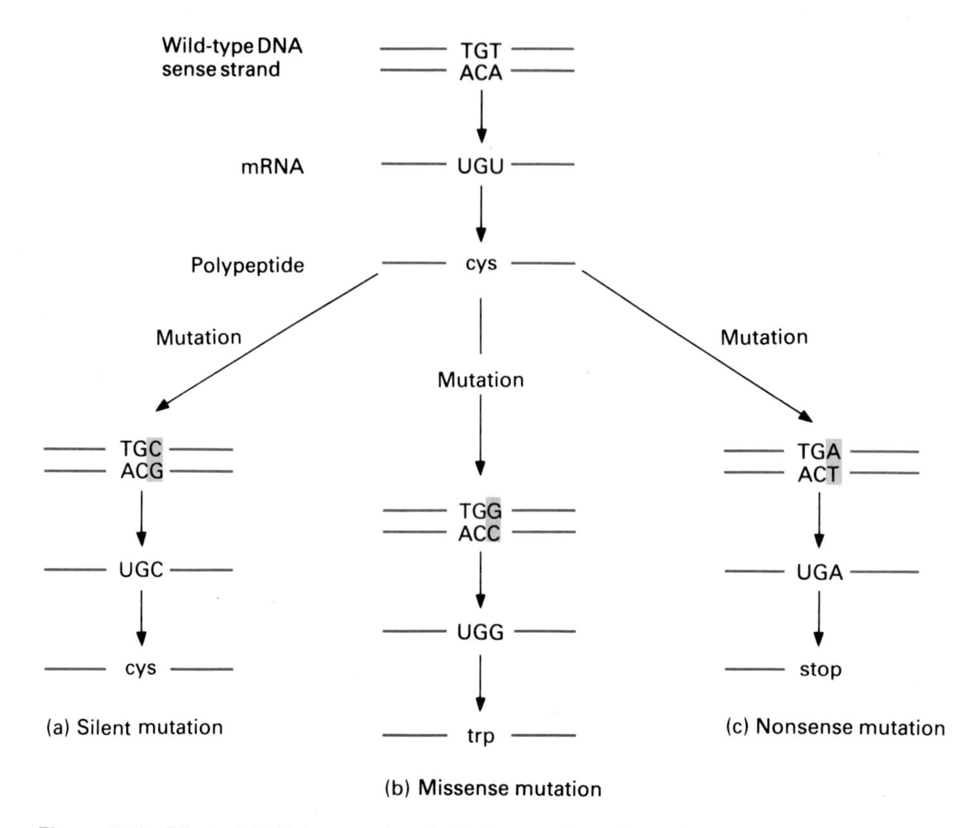

Figure 11.3. Effect of DNA base pair substitution on the amino acid sequence of the encoded polypeptide. The bottom DNA strand serves as the template for transcription in all four cases. Base substitution at the third position of this particular DNA code word can give rise to (a) a silent mutation, (b) a missense mutation, and (c) a nonsense mutation.

tion will become more apparent as we proceed through the chapter. A mutation at one level may or may not have mutant consequences at the next level.

A **missense mutation** is defined as an amino acid substitution in the polypeptide chain. The term *mis*sense refers to a codon that makes the "wrong kind of sense" to the ribosomal apparatus for translation; that is, the codon codes for a different amino acid than the one originally intended. The consequences of a missense mutation may or may not be severe. If we inspect the genetic codon assignments given in Table 9.1, we see that amino acids with similar structures are more often than not coded for by similar codons. A missense mutation will therefore quite often involve amino acids with similar chemical properties. Recall that it is the amino acid sequence of the polypeptide chain that determines its folded structure and, consequently, its activity. If the substituted amino acid can participate in intrachain and interchain bonding, the normal folding pattern and protein function will be maintained. It is thus quite possible that a mutation at the level of the amino acid sequence will not manifest itself by altering the activity of the protein. A missense mutation that does not change the function of the protein is sometimes referred to as a **neutral mutation.**

On the other hand, some missense mutations severely affect protein function. A mutation that affects protein function usually involves substitution by an amino acid whose chemical properties differ from those of the amino acid that normally occupies that position in the polypeptide chain. In this event, the intrachain and interchain interactions that are involved in the folding pattern of the polypeptide are so altered that a functional folding pattern cannot be attained. A missense mutation of this type is the molecular basis of sickle-cell anemia, as shown by Fig. 11.2. We will return to a discussion of hemoglobin mutations shortly.

We have seen that the effects of a missense mutation on protein function depend on the *type of amino acid change* involved. The consequences of an amino acid substitution also depend on the *amino acid position* that is affected. Chemical studies of proteins have revealed that some amino acid positions along a polypepide chain are much more critical than others in establishing the normal protein structure. An obvious example concerns the amino acids that, through the folding pattern, make up the active site of an enzyme molecule. Any change in these amino acids would be very likely to render the enzyme nonfunctional. Only a few proteins have been analyzed in enough detail to determine which amino acid positions are most critical to function, partly because so few proteins have been sequenced. One protein whose amino acid sequence we know is hemoglobin. Researchers have identified more than 100 variant hemoglobins in the human population. The majority of these variants differ from HbA in just a single amino acid in either the α or β chain. Some of the better-known hemoglobin mutations are shown in Fig. 11.4. The sixth position on the β polypeptide chain appears to be critical to protein function, because both HbS and HbC cause sickling of the red blood cells. Some of the other variants have deleterious effects that are not as severe as those of HbS. Still others transport oxygen as efficiently as HbA.

The final type of mutational effect on the amino acid sequence is a **nonsense mutation.** A nonsense mutation involves the change of any one of the 61 sense codons into a triplet that does not code for an amino acid (Fig. 11.3c). Polypeptide synthesis is prematurely halted, resulting in a shortened, and almost certainly nonfunctional, protein. Nonsense mutations usually have a very severe effect on protein function.

Frameshift Mutations. Although we would expect that the majority of spontaneous gene mutations would be the consequence of base substitutions, a surprising proportion of the mutations studied in bacteria and phages are caused by the addition or deletion of a single nucleotide pair. The addition or deletion causes the reading frame of the triplet code words along the DNA sense strand to be shifted by one nucleotide, as illustrated by Fig. 11.5. For this reason, nucleotide additions and deletions are known as **frameshift mutations.** The shift in the reading frame originates at the point where the nucleotide has been added or deleted, and it changes the entire remaining sequence of codons. Some of the new codons are silent mutations, others are missense mutations, and still others are nonsense mutations. Since the amino acid sequence of the encoded polypeptide will be drastically altered, a frameshift mutation usually has severe effects. The mechanism by which DNA spontaneously gains or loses nucleotide pairs is not known.

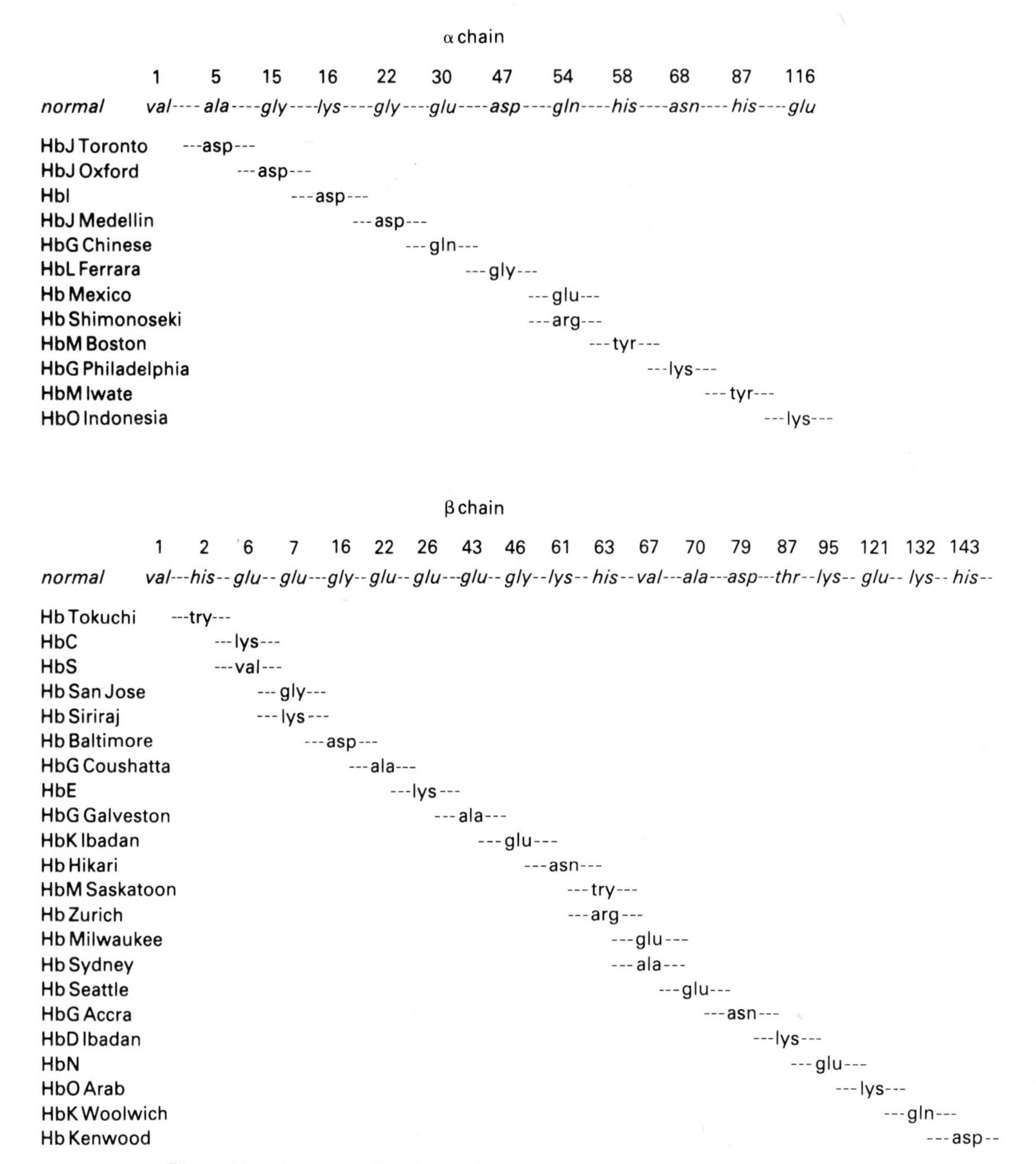

Figure 11.4. A sample of the known hemoglobin mutations. The numbers above the normal amino acid sequences indicate the specific amino acid that has been changed in each mutant type. *Source:* From L. T. Hunt, M. R. Souchard and M. O. Dayhoff, In *Atlas of Protein Sequence and Structure,* Vol. 5, M. Dayhoff (ed.), National Biomed. Res. Found., Washington, D.C., pp. 67–87, 1972.

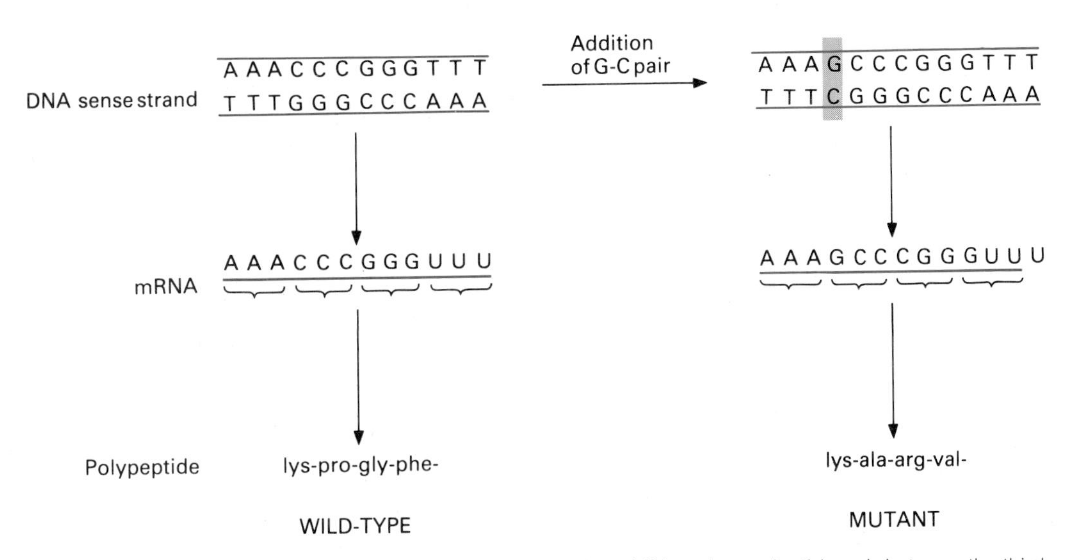

Figure 11.5. Frameshift mutation resulting from the addition of a nucleotide pair between the third and fourth base pairs of a wild-type gene. The reading frame of the mutant gene is shifted by one base pair, beginning at the point of the addition. This shift, in turn, changes the amino acid sequence of the polypeptide chain.

Secondary Consequences of Mutation

While an amino acid change is the immediate effect of a gene mutation, the resulting secondary consequence to protein function is what actually causes the mutant phenotype. In this section, we will discover how the impairment of protein function from a gene mutation leads to a detectable change at the phenotypic level.

The transition from genotype to phenotype encompasses several levels of biological organization. Consider a comparatively simple trait such as eye color as an example. The eye color that we can observe is determined by the presence of certain pigments called *melanin* in the iris of the eye. Melanin is produced by special cells called *melanocytes,* which synthesize the pigments in specialized membranous structures known as *melanosomes.* Melanin is formed in the melanosomes through a complex sequence of chemical steps, each of which is catalyzed by a different enzyme. Each enzyme, in turn, is coded for by one or more genes. Thus, in order to relate a complex phenotypic change to a particular mutant gene, we need to understand not only the genetics of the situation but other biological and chemical aspects of the situation as well.

Metabolic Basis of Inherited Disease. As early as the turn of this century, geneticists had made the association between mutant genes and malfunctioning metabolic pathways. In Chapter 9, we discussed the ideas of Archibald Garrod, in which he related a mutation to a block in a metabolic pathway. The block occurs when a mutant gene is unable to code for the functional enzyme that would normally catalyze the blocked reaction. Garrod worked with several disorders in humans that he proposed were consequences of metabolic blocks, but he was particularly interested in alkaptonuria. The dark substance that is found in the urine of individuals who are affected with this disease is alkapton, which is produced from the oxidation of homogentisic acid. The chemical structure of homogentisic acid resembles that of the amino acids phenylalanine and tyrosine. Recognizing this similarity, Garrod proposed that homogentisic acid is merely a product of the breakdown of these amino acids. He suggested that the phenylalanine and tyrosine of ingested foods are converted to homogentisic acid, which is then normally broken down into simpler substances. Individuals with alkaptonuria must therefore have a block in this degradative pathway after the step at which homogentisic acid is formed. Lacking the necessary digestive enzyme, such individuals would accumulate homogentisic acid and excrete it in their urine.

The complete metabolic pathway used by cells for the breakdown of phenylalanine and tyrosine is shown in Fig. 11.6. In this metabolic sequence, protein that has been taken into the body in the diet is broken down into its constituent amino acids, two of which are phenylalanine and tyrosine. These amino acids are then converted, via a series of metabolic reactions, to various end products. Besides alkaptonuria, blocks in other steps of this pathway give rise to other well-known genetic disorders. One of these conditions is *phenylketonuria* (*PKU*), which results from a block at the step that converts phenylalanine to tyrosine. The recessive PKU gene causes cells to fail to produce a liver enzyme called phenylalanine hydroxylase. Then phenylaline and its derivatives, including phenylpyruvic acid, attain abnormally high concentrations in the blood and urine and accumulate in the cells. The accumulation of the phenylalanine derivatives, particularly phenylpyruvic acid, within the cells is toxic to the central nervous system, causing irreversible brain damage during the first few months of postnatal life.

There is no cure for PKU. If the condition is diagnosed shortly after birth, however, the infant can be placed on a controlled diet that minimizes the amount of protein that contains phenylalanine. (Because phenylalanine is one of the essential amino acids for humans, which means that it cannot be produced by the body's own metabolic machinery, some phenylalanine must be included in the diet so that protein for body structures can be synthesized.) Most individuals with PKU who are placed on this diet within the first month of life have been able to discontinue it by the age of six years, with no serious effects on brain function. Of the 50 states, 48 now require hospitals to screen all newborns for phenylketonuria, using the Guthrie test. In his procedure, a drop of blood that is obtained from the heel of the newborn is placed on a disk of absorbant paper. This disk is then placed on the surface of an agar-based bacterial growth medium. Over the growth medium are spread bacteria from a strain that is incapable of multiplying unless a large amount of phenylalanine is added as a supplement to the growth medium. If the infant's blood contains large amounts of phenylalanine, enough of the amino acid will diffuse from the disk into the medium to support bacterial growth. Thus, a halo of bacterial colonies surrounding the disk is diagnostic of the disease. Prior to testing and use of dietary therapy for PKU, about 1 percent of patients in institutions for the mentally retarded suffered from this condition. Phenylketonuria is the most common inherited metabolic disorder that affects brain development.

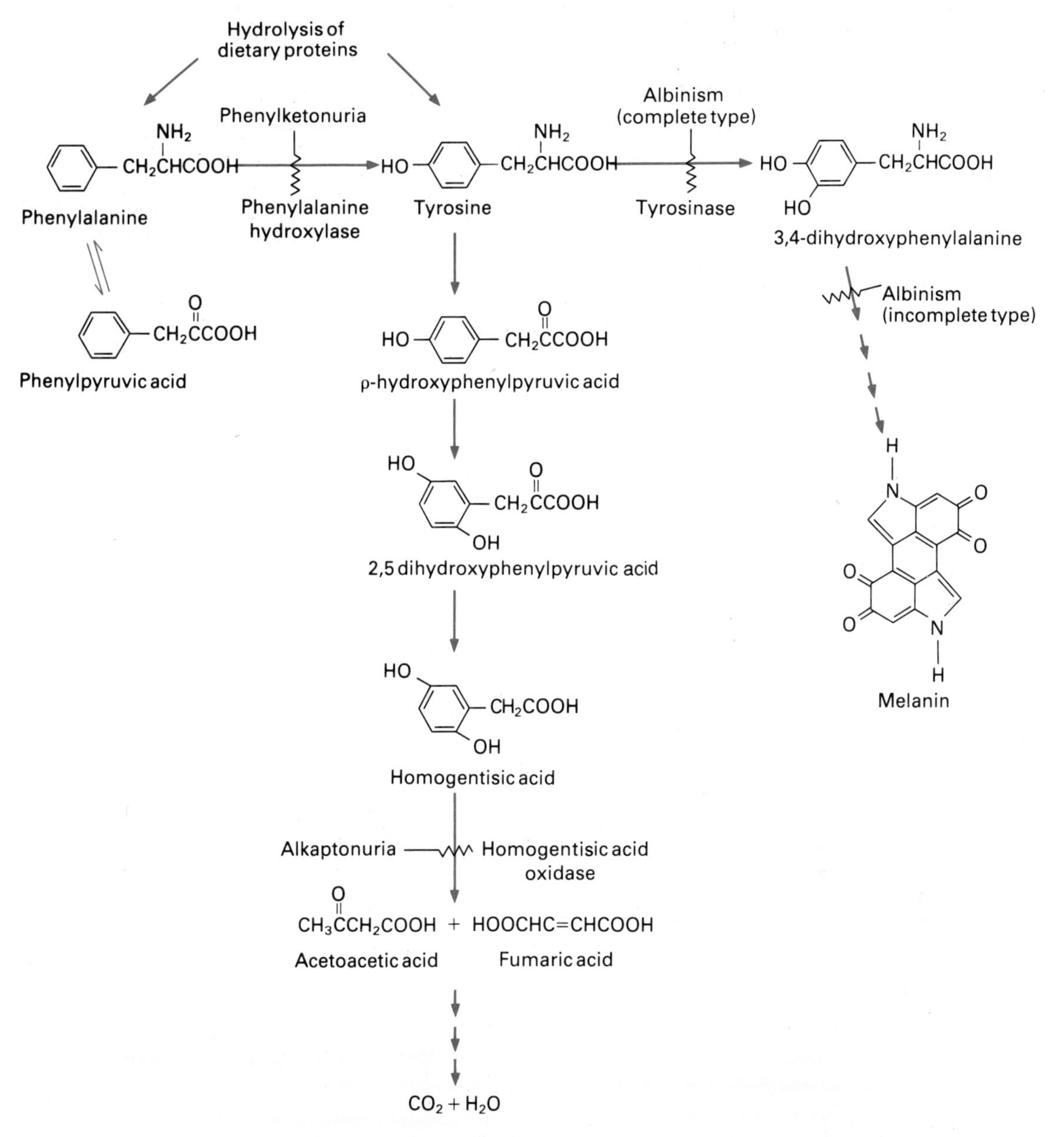

Figure 11.6. Metabolic pathway responsible for the degradation of phenylalanine and tyrosine. Homogentisic acid is the intermediate that accumulates in individuals who are affected with the disease alkaptonuria. Two other diseases that are caused by blocks at steps in this pathway are phenylketonuria and albinism.

Another condition caused by the lack of an enzyme that is normally operating in this same metabolic pathway is albinism. In this case, the block in the pathway results in failure to produce the pigment melanin, rather than an accumulation of a toxic product. Albinism can be caused by a blockage at either of two steps leading from tyrosine to melanin. *Complete* (or *classical*) *albinism* is caused by a deficiency of the functional tyrosinase enzyme, owing to a recessive mutant gene at a single locus. In this type of albinism, pigment is totally lacking in the skin, hair, and eyes. *Incomplete albinism,* which is characterized by the development of some pigmentation by adulthood, is caused by a block at a step that converts dihydroxyphenylalanine to melanin.

Phenylketonuria and albinism illustrate the two major ways in which a metabolic block can cause disease. Phenylketonuria results from the *accumulation of a substance that is toxic to cells at high concentrations,* while albinism results from the *lack of the end product of a metabolic pathway.* Most metabolic diseases can be attributed to one or the other of these causes, although affected individuals might display some of the features of both. Individuals with phenylketonuria, for example, often are fair-skinned, with blond hair and blue eyes, owing to their lower-than-normal levels of melanin.

Many heritable human diseases are the result of a single gene mutation that leads to a nonfunctional enzyme and a metabolic block. Victor McKusick, a well-known human geneticist, lists about 1500 such diseases (along with their symptoms and the affected enzyme) in his catalog of human genetic disorders, entitled *Mendelian Disorders in Man.* Some of the more widely known examples include cystic fibrosis, galactosemia, Hurler syndrome, Hunter syndrome, Lesch-Nyhan syndrome, maple syrup urine disease, porphyria, and Tay-Sachs disease. Table 11.2 lists some of these metabolic diseases for which the enzyme defect has been identified. Observe that diet or drug therapy is available in many cases to alleviate certain symptoms of the disease.

Many metabolic disorders can now be detected in the fetus, using a method known as **amniocentesis.** This procedure involves inserting a needle through the mother's abdomen into the fluid-filled space that surrounds the developing fetus, called the *amniotic cavity.* A small sample of the amniotic fluid is then removed by means of a syringe (Fig. 11.7). Cells of the fetus that have been shed into the amniotic fluid are then grown in culture and used for chromosome and enzyme analysis. Although amniocentesis is currently the standard prenatal test for birth defects, a newer method, called a *chorion biopsy,* may soon be generally available. In a chorion biopsy, a catheter is inserted into the uterus to remove a small tissue sample from the outer fetal membrane, or chorion, that is attached to the placenta. A chorion biopsy has two main advantages over amniocentesis: the results are available overnight as compared with up to four weeks for amniocentesis, and the method can be used as early as the seventh week of pregnancy, while the current test cannot be done until the fourteenth to sixteenth week.

One metabolic disorder for which amniocentesis is currently employed for prenatal diagnosis is the Lesch Nyhan syndrome. This disease, inherited as an X-linked recessive trait (passed from a carrier mother to her sons about 50 percent of the time, was first identified and described by Lesch and Nyhan in 1964. Symptoms of the disorder develop gradually, beginning several months after birth. They include extremely high concentrations of uric acid in the urine and blood (uremia), cerebral palsy (involuntary spasms of the limbs, neck, and trunk), mental retardation, and compulsive self-mutilation by biting the fingers, lips, and inside of the mouth. Drug therapy is successful in reducing the uric acid level but has no effect on the behavioral aspects of this disorder. The degree of self-mutilation

Table 11.2. Some metabolic disorders in humans for which the enzyme defect has been identified.

DISORDER (AND GENETIC BASIS)	BRIEF DESCRIPTION	EARLIEST POSSIBLE DIAGNOSIS	THERAPY
Favism (G6PD deficiency) (X-linked recessive)	Deficiency of the enzyme glucose 6-phosphate dehydrogenase. Characterized by severe hemolytic anemia when exposed to fava beans and certain drugs.	Prenatal, by amniocentesis.	Avoid fava beans and specific drugs.
Galactosemia (autosomal recessive)	Deficiency of an enzyme needed to break down galactose. Characterized by excess galactose, enlarged liver, eye defects, mental deterioration, and early death.	Prenatal, by amniocentesis.	Galactose-free diet.
Hurler syndrome (autosomal recessive)	Deficiency of an enzyme needed to break down complex mucopoly-saccharides found in connective tissue. Characterized by stiff joints, growth retardation, and death in childhood.	Prenatal, by amniocentesis.	—
Lesch-Nyhan syndrome (X-linked recessive)	Deficiency of an enzyme needed in purine metabolism. Characterized by mental retardation, high uric acid levels, muscle spasms, and compulsive self-mutilation.	Prenatal, by amniocentesis.	Drugs, for uric acid. None for behavioral problems.
Maple syrup urine disease (autosomal recessive)	Deficiency of an enzyme needed to break down branched-chain amino acids. Characterized by maple syrup odor of urine, progressive degeneration, and early death.	Prenatal, by amniocentesis.	Diet low in the amino acids valine, leucine, and isoleucine.
Phenylketonuria (autosomal recessive)	Deficiency of an enzyme needed to convert phenylalanine to tyrosine. Characterized by excess phenyl-alanine in blood and severe mental deterioration.	Newborns, by screening blood samples.	Diet low in phenylalanine.
Tay-Sachs disease (autosomal recessive)	Deficiency of an enzyme needed to break down a ganglioside lipid. Characterized by progressive deterioration and death in infancy.	Prenatal, by amniocentesis.	—

varies extensively from one patient to the next, as well as in the same patient under different environmental conditions. In some cases, stressful situations appear to trigger episodes of self-mutilation. Death usually occurs before age 20, from pneumonia, uremia, or kidney failure.

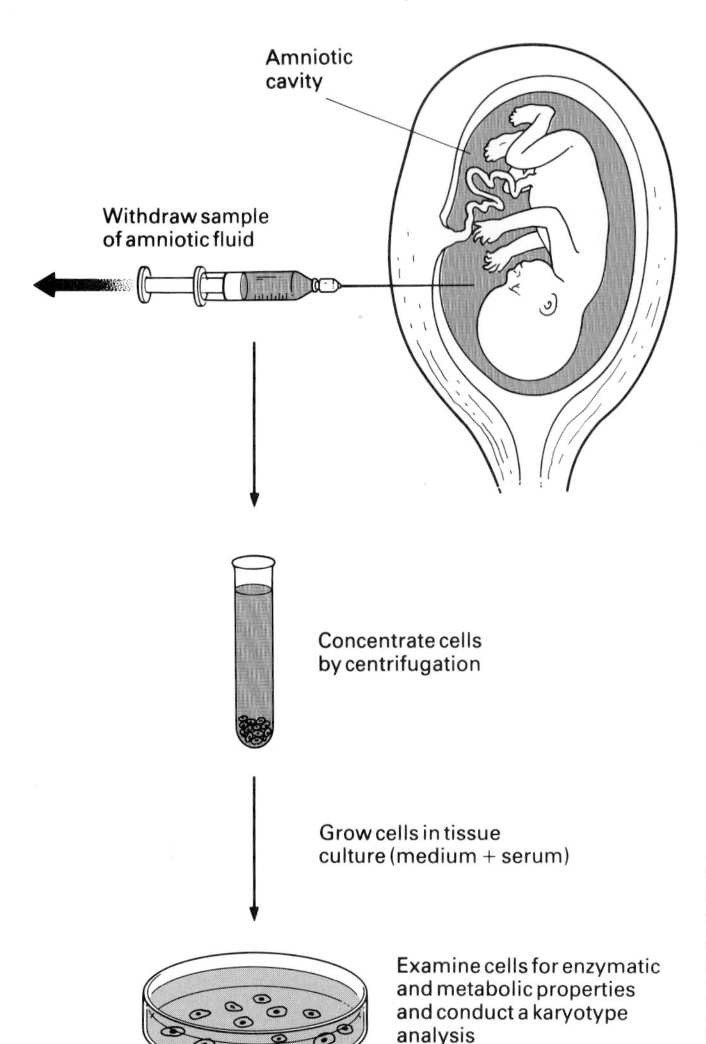
Amniotic cavity
Withdraw sample of amniotic fluid
Concentrate cells by centrifugation
Grow cells in tissue culture (medium + serum)
Examine cells for enzymatic and metabolic properties and conduct a karyotype analysis

Figure 11.7. Prenatal diagnosis using amniocentesis. The procedure, which is performed sometime between the fourteenth and sixteenth week of pregnancy, involves removing a small sample of the amniotic fluid by means of a syringe. Cells of the fetus that have been shed into the amniotic fluid can then be grown in culture and used for biochemical studies and to conduct a karyotype analysis.

The complicated variety of problems associated with Lesch-Nyhan syndrome arise from a block in purine metabolism, resulting from a total lack of functional HPRT enzyme. Recall from Chapter 7 that HPRT recycles purine nucleotides from degraded nucleic acids by way of the salvage pathway (see Fig. 7.18). When the salvage pathway fails to function, excess purines resulting from the breakdown of nucleic acids are degraded to uric acid, which then attains abnormally high levels in the blood and urine. Since the HPRT enzyme can be detected in cultures of fetal cells derived from the amniotic fluid, prenatal diagnosis by amniocentesis can detect an affected fetus carried by a woman who is heterozygous for the causative allele. Because of the dosage compensation, a carrier female shows mosaicism for the amounts of HPRT in the cells; this finding can thus be used to test whether a woman is heterozygous if her family history leads her to suspect that she carries the defective allele.

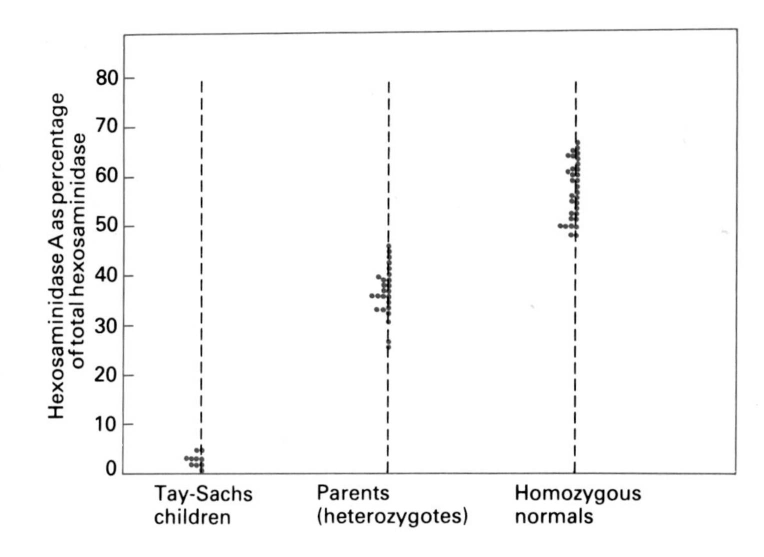

Figure 11.8. The detection of heterozygotes for the recessive allele that causes Tay-Sachs disease. The serum concentration of the enzyme hexosaminidase A in parents of affected children is about 50% of the level in homozygous normal individuals. *Source:* Adapted by the authors from data appearing in the *New England Journal of Medicine*, 283:15–20, 1970.

In the metabolic defects that we have just discussed, the dominant or recessive expression of the gene takes the form of the presence or absence of the functional enzyme. Complete dominance results when one dose of the normal gene is all that is needed to produce a physiologically adequate amount of the enzyme. Both heterozygotes and dominant homozygotes are expected to be normal in phenotype. In contrast to outward appearance, however, dominance at the level of enzyme concentration is usually incomplete. Heterozygotes tend to have an intermediate amount of the enzyme as compared to homozygous normal individuals and persons affected with the disease. This lack of complete dominance at the enzymatic level has been used to detect carriers, such as women who carry the allele for HPRT deficiency and individuals who carry the autosomal gene for Tay-Sachs disease (Fig. 11.8).

Of course, not all diseases caused by metabolic blockage are particularly severe. Some of the factors that act to alleviate the severity of metabolic blocks are summarized in Fig. 11.9. In many cases, the excess amount of a metabolic inter-

Figure 11.9. Factors that reduce the severity of a metabolic block. A hypothetical pathway is blocked at the step that normally converts compound X to compound Y. The severity of the effects that result from the accumulation of compound X are reduced by the excretion of excess compound X and by utilizing compound X in an alternative pathway. These factors are particularly important in degradative pathways. Problems associated with the lack of an end product can be compensated for by an alternative pathway that synthesizes the same end product or by a direct dietary source of the end product.

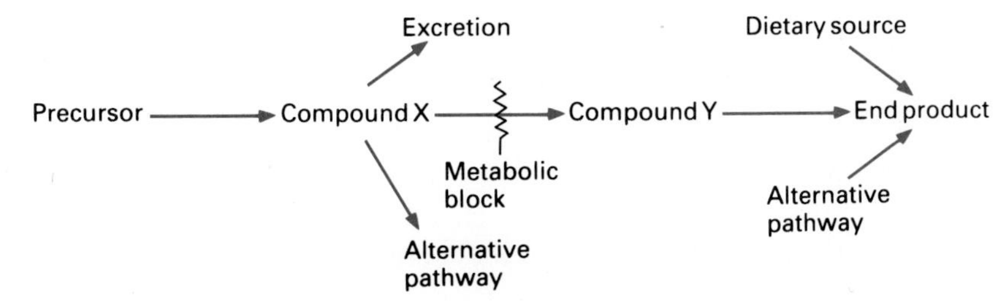

mediate can be excreted in the urine, and the cells can use an alternative pathway to compensate for the blocked one, thereby preventing a shortage of the end product (alkaptonuria is an example of this type of compensation). In other instances, the affected pathway is used so little under normal dietary conditions that the blockage has no severe effect unless an abnormally high amount of a particular dietary substance or an unusual substance is ingested. Certain drug reactions are now known to fall into this category. Some individuals lack an enzyme that is required for the normal metabolism of a particular drug. Theraputic doses of the drug, which would normally be metabolized with no ill effects, can result in severe reactions in individuals who lack one of these enzymes. The relationship between certain genotypes and responses to drugs forms the basis of a relatively new area of genetic study, known as *pharmacogenetics*. Researchers have found genetic variation in the ability of humans to metabolize many types of drugs, including alcohol.

TO SUM UP

1. The ability to mutate is an intrinsic property of the structure of DNA, giving rise to spontaneous mutations at a very low rate per gene per generation.
2. The phenotypic effects of mutant genes are varied. These effects include recognizable alterations in morphology, metabolism, survival, and resistance to chemicals and viruses that are normally harmful to an organism. An organism with a conditional mutation expresses the altered phenotype only under certain (restrictive) environmental conditions.
3. Many gene (or point) mutations arise from a single base pair substitution. Several types of substitutions are possible; they can be classified as either transitions or transversions. The changed DNA code word may not specify an amino acid change in the encoded polypeptide (a silent mutation), or it may result in an amino acid substitution (a missense mutation) or in a termination codon (a nonsense mutation). The effect of a missense mutation on protein function depends on the nature of the amino acid change and on the position of the amino acid affected.
4. Frameshift mutations occur when a nucleotide pair is added or deleted. The reading frame of triplet code words along the DNA is shifted, causing a severe alteration of protein structure.
5. The phenotypic change that is associated with a mutant gene results from the loss of function of the protein that is coded for by the wild-type allele of that gene. A lack of functional enzyme creates a block in a metabolic pathway. The consequences of this blockage depend on the degree to which the metabolic intermediates accumulate and the extent to which the end product is deficient.
6. Dominance can be thought of as the presence of enzyme function, while recessiveness is its absence. Dominance at the level of enzyme concentration in the cells is usually incomplete, with heterozygotes having an intermediate amount of the enzyme as compared to both homozygous types.

Mutagens and Their Effects

Any substance that increases the rate of mutation is known as a mutagen. Today's concern for the effects of chemicals on our environment and on human health has focused attention on the mutagenic properties of such substances as cosmetics, drugs, food additives, pesticides, and synthetic industrial compounds like plastics and other petrocarbon materials. Hundreds of chemicals that we encounter in our daily lives are now known to have mutagenic effects, with the effects ranging from very slight to very potent. Besides chemicals, radiation must also be added to our list of mutagenic agents. In this category, the radiation that occurs naturally contributes much more to the average total radiation to which we are exposed over a lifetime than all human-made sources combined. In addition to environmental mutagens, such as chemicals and radiation, geneticists have recently recognized that certain factors within the cell can also influence the mutation rate. We now know, for example, that some genes exist whose mutation can be brought about by transposable genetic elements. These elements are segments of DNA that move around within the chromosomes of a bacterial, plant, or animal cell. Mutation is presumably caused when such an element inserts itself into a gene locus. Geneticists are only now determining the importance of these transposable genetic elements and the way in which they influence mutation.

The mode of action of most mutagens is not fully understood. Although mutation has long been a favorite topic of genetic research, only recently have geneticists achieved some understanding of the molecular aspects of mutation. The mutagenic actions that are best known are those of certain chemicals (such as base analogs, alkylating agents, and acridine dyes) and of radiation (including X-rays and ultraviolet light). The damage that can be done to DNA by such mutagens is illustrated by Fig. 11.10; the types of damage include chemical modification of bases, purine loss (apurination), single- and double-strand breakage, intra- and interstrand crosslinkage, and intercalation. All mutagens share one important property, regardless of the type of damage they inflict upon DNA. They induce mutation at randomly affected loci; no specific gene is preferentially mutated by any particular mutagen.

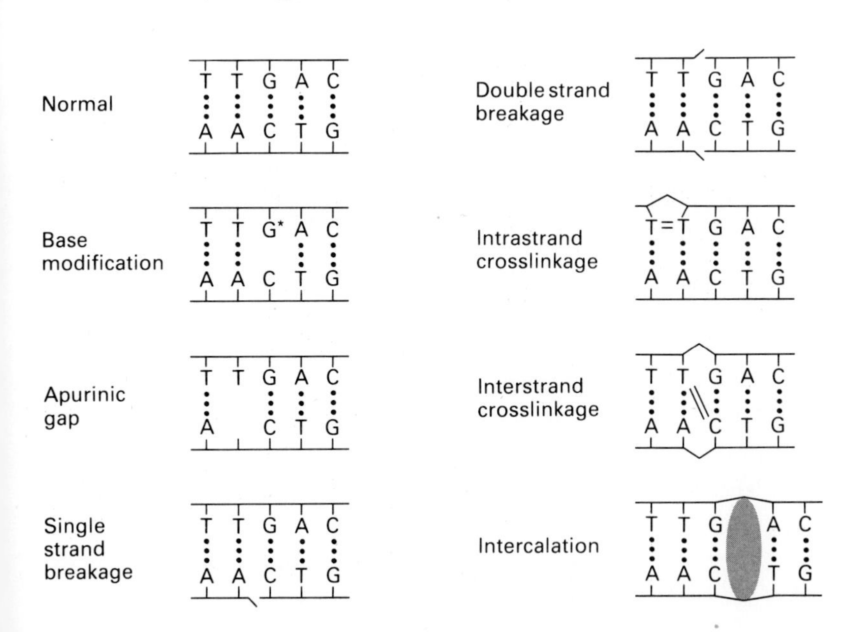

Figure 11.10. Various ways of damaging DNA that can subsequently give rise to mutation.

Some mutagens, such as X-rays, are extremely diverse in their effects and can induce various types of both gene and chromosomal mutations. Chromosomal mutations can have particularly serious consequences on a pregnant female. Because a rearranged, partially duplicated, or incomplete genetic complement upsets the normal balance of genes, chromosomal mutations nearly always affect the process of development of the embryo and fetus. Mutagens that cause chromosomal mutations in the developing embryo or fetus will therefore usually result in severe congenital malformations. If the affected infant survives to reproductive age, these defects are, of course, hereditary.

It is important to note that congenital malformations can also be the result of nongenetic factors. A **teratogen** is any environmental agent that interferes with development. A teratogen is distinguished from a mutagen in that a teratogen does not affect the genetic material; as a consequence, abnormalities caused by teratogens are not heritable. Examples of teratogenic agents and their effects on humans are given in Table 11.3. Rubella virus and thalidomide are perhaps the best known of these agents. Alcohol and anticonvulsant drugs differ in their effects from the rest of the teratogens listed, in that not all fetuses of exposed mothers develop abnormally. Apparently, the genotypes of both the embryo and the mother are important in determining the teratogenic effects of these two agents. In contrast, all mothers who had taken thalidomide at a specific early stage of pregnancy delivered babies who had some degree of abnormality.

The study of mutagens is closely tied to the study of substances thought to be *carcinogenic* (cancer causing). Nearly all known carcinogens have been shown to be mutagenic as well; this finding strongly suggests that most forms of cancer are due, at least in part, to changes in DNA.

In addition to the obvious practical benefits of understanding the action of mutagenic agents, studies with mutagens have also been invaluable in probing the nature of the spontaneous mutation process. For example, consider a mutagen that is known specifically to cause GC $\rightarrow$ AT base pair substitutions in DNA. Suppose that we find this mutagen able to revert a spontaneous mutant allele

Table 11.3. Examples of teratogenic agents and their effects on human development.

AGENT	PHENOTYPIC EFFECT
Rubella virus	Congenital heart defects, cataracts, deafness
Cytomegalovirus	Mental retardation
Radiation	Microcephaly (abnormally small head)
Progestin	Masculinization of female fetus
Aminopterin	Abnormal cranial structure, broad nasal bridge, low-set ears, upswept scalp hair
Thalidomide	Phocomelia (limbs extremely shortened so that feet or hands arise close to trunk)
Alcohol	Growth retardation, mental retardation, microcephaly
Anticonvulsants	Retardation of physical and mental development

back to its wild-type condition. We can then conclude that the nature of the original spontaneous mutation event was an AT → GC base pair change.

Chemical Mutagens

Much of our knowledge concerning the molecular action of mutagens is based on tests of the mutagenic effects of various chemical substances, using viral DNA (or RNA), purified transforming DNA, seed DNA, or *Drosophila* as the test system. Except in *Drosophila,* no metabolic activation (or deactivation) of the mutagen can occur in such a system. It is now well known that many chemical substances are altered during the course of their metabolism by bacterial and animal cells, being transformed into substances with entirely different mutagenic potential than that of their parent compounds. In addition, with in vitro systems, such as purified viral or bacterial DNA, there is no possibility for the toxicity of a mutagen to be a result of damage to cellular components other than DNA. Mutagenesis in these in vitro systems can often be related to a specific chemical modification of a base pair in DNA. This simple explanation of mutation as a function of a single base pair change cannot always be extended to in vivo systems, however. In in vivo systems, the mutagen must first be able to enter the cell. It may then be metabolically altered, and any damage that it does to DNA may be repaired by a pathway that itself may be mutagenic. Hence, in natural systems, the measurement of mutagenesis is greatly affected by the particular test system being employed, and the effects of a mutagen on DNA are often not well understood.

The chemical mutagens that we will discuss can be divided into two groups: those that are effective only on DNA that is actively replicating (base analogs) and those that do not require concurrent DNA replication for their primary action (hydroxylamine, alkylating agents, and acridines). The base analog mutagens and alkylating agents cause two-way transitions (AT ⇌ GC). Alkylating agents can also cause transversions and frameshift mutations. Hydroxylamine is the most specific mutagen of this group, in that it induces only GC → AT transitions. Finally, acridines cause frameshift mutations. The chemical structures of some of these mutagens are shown in Fig. 11.11.

Base Analogs. A **base analog** is a substance that is similar enough in structure to one of the four bases to be incorporated into a nucleotide and then into a strand of DNA in place of the base it resembles. Two of the most widely cited base analogs are *5-bromouracil* (*BU*) and *2-aminopurine* (*AP*). The former is a pyrimidine analog of thymine, while the latter is a purine analog of adenine. Incorporation of a base analog into a strand of DNA does no direct harm to the DNA. Because the hydrogen bonding properties of BU and AP resemble those of T and A, respectively, BU will pair with adenine as does thymine, and AP will pair with thymine as does adenine. The mutagenic effect of base analogs is derived from their relatively high rate of tautomerism. Base analogs undergo tautomeric shifts much more frequently than do the bases that are normally found in DNA. Recall from Chapter 3 that tautomerism changes the hydrogen bonding properties of a base. The tautomeric form of a base analog has the same altered hydrogen bonding properties as the tautomeric form of the base it resembles. Thus, the tautomeric form of BU (BU*) pairs with guanine, just as does the tautomeric form of thymine (Fig. 11.12).

BASE ANALOGS

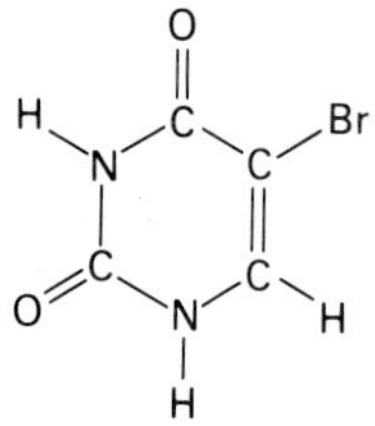

5-bromouracil (BU)

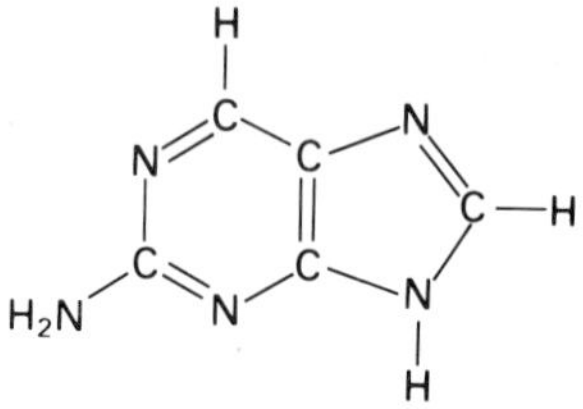

2-aminopurine (AP)

Figure 11.11. Structures of some of the better known chemical mutagens.

ALKYLATING AGENTS

$CH_3–CH_2–O–SO_2–CH_3$
Ethylmethane sulfonate (EMS)

$CH_3–CH_2–O–SO_2–CH_2–CH_3$
Ethylethane sulfonate (EES)

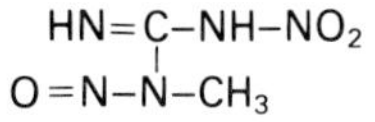

$HN=C–NH–NO_2$
$O=N–N–CH_3$
Nitrosoguanidine (NG)

$Cl–CH_2–CH_2–S–CH_2–CH_2–Cl$
Sulfur mustard (mustard gas)

ACRIDINES

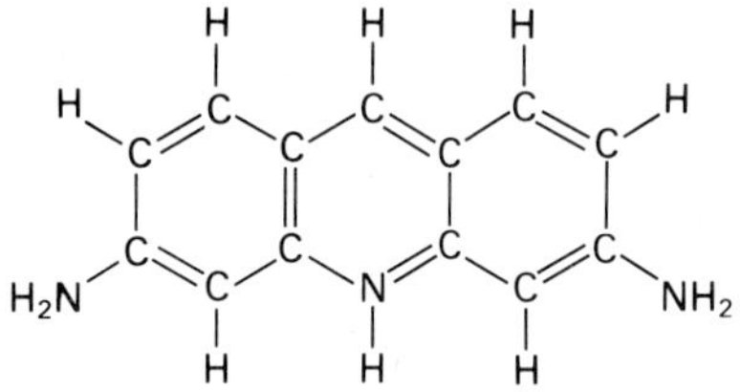

Proflavin

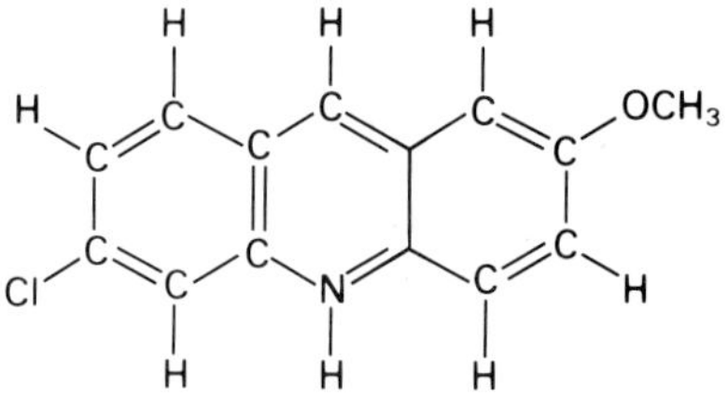

Quinacrine

MISCELLANEOUS

HNO_2
Nitrous acid (NA)

NH_2OH
Hydroxylamine (HA)

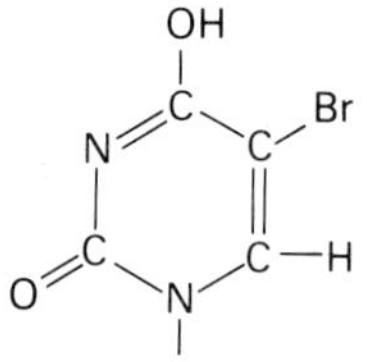

Tautomeric form of BU

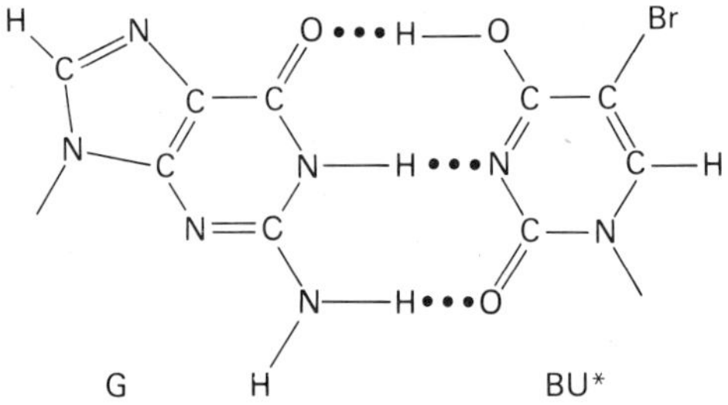

G BU*

Pairing of BU* with G

Figure 11.12. The tautomeric form of bromouracil (BU*) has the base pairing properties of cytosine, and thus it forms hydrogen bonds with guanine.

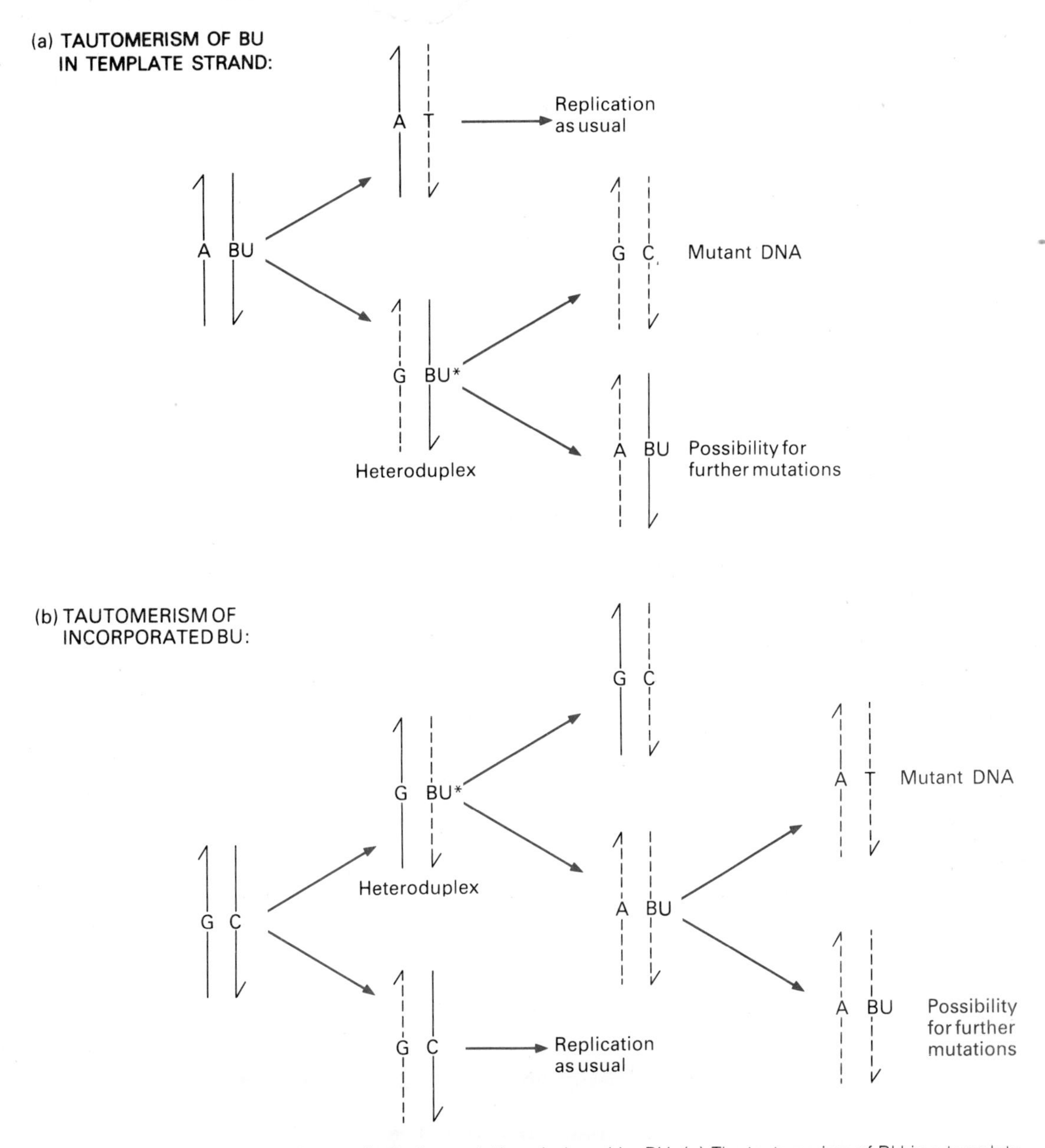

Figure 11.13. Base substitution mutations induced by BU. (a) The tautomerism of BU in a template strand leads to the transition AT → GC. (b) Incorporation of the tautomeric form of BU into a daughter strand gives the transition GC → AT.

As is the case for a spontaneous mutation caused by tautomerism, a mutation induced by a base analog can occur through either an incorporation or a template error during the replication of DNA. Figure 11.13 shows the base substitutions that can be induced by BU. An error in the template strand gives rise to an AT → GC transition, whereas an error in incorporation yields a GC → AT transition.

Studies of the mechanism of action for base analogs have been a major tool used by geneticists to decipher the molecular basis of the spontaneous mutation

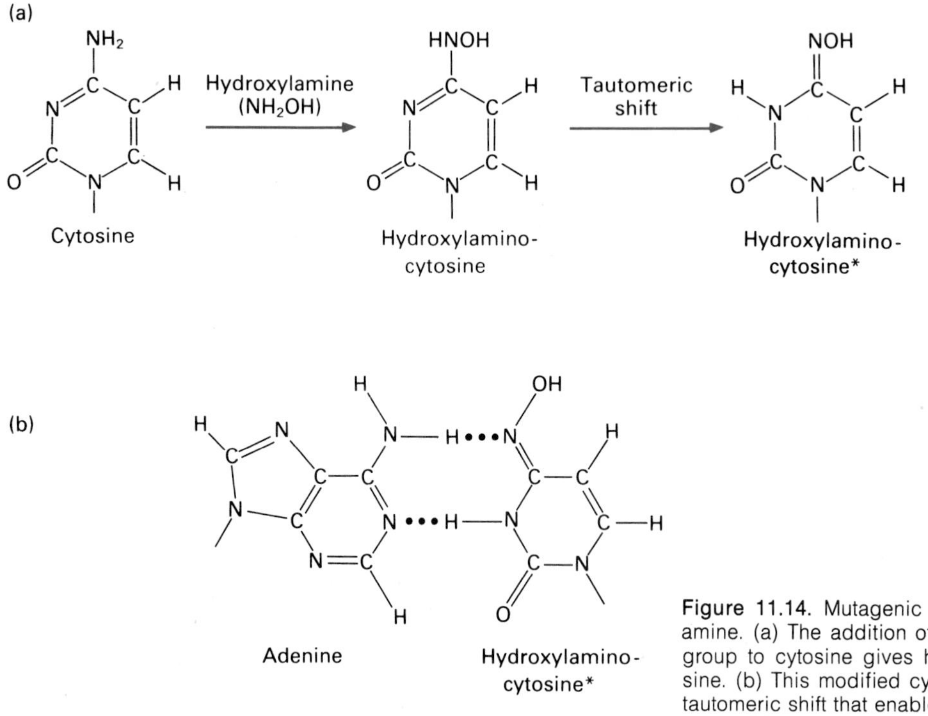

(a)

Cytosine — Hydroxylamine (NH_2OH) → Hydroxylamino-cytosine — Tautomeric shift → Hydroxylamino-cytosine*

(b)

Adenine — Hydroxylamino-cytosine*

(c) GC → AC* → AT

Figure 11.14. Mutagenic action of hydroxylamine. (a) The addition of a hydroxyl (—OH) group to cytosine gives hydroxylamino cytosine. (b) This modified cytosine undergoes a tautomeric shift that enables it to pair with adenine. (c) The result is the substitution of AT for GC.

process. The fact that an increased incidence of tautomerism can explain the mutagenic effects of base analogs lends strong evidence to the theory that spontaneous mutation occurs through tautomerism and errors in base pairing.

DNA Modifiers. Hydroxylamine, alkylating agents, and acridines act by chemically modifying the DNA. For example, Fig. 11.14 shows that *hydroxylamine (HA)* (which is thought to be an intermediate in nitrate reduction in vivo) reacts specifically with cytosine, adding a hydroxyl group (—OH) to yield hydroxylamino-cytosine. The modified cytosine is then able to undergo a tautomeric shift to C*, which pairs with adenine. The overall result is the base pair change GC → AT (GC → AC* → AT).

Alkylating agents react with DNA in a variety of ways; they therefore have diverse mutagenic effects. These effects include transitions, transversions, frameshift mutations, and chromosome mutations. Alkylating agents include ethylmethane sulfonate (EMS), ethylethane sulfonate (EES), nitrosoguanidine (NG) and other *N*-nitroso compounds (which are also highly carcinogenic), and nitrogen and sulfur mustards (see Fig. 11.11). These compounds act by transferring an alkyl group (CH_3-, CH_3-CH_2-, etc.) to a base in a DNA strand. Sulfur mustard, commonly known as mustard gas, was the first chemical mutagen to be discovered. Its use as a poison gas during World War II prompted investigations into its mutagenic properties.

The alkylating agent whose characteristics are best known is EMS. This compound is known to cause mutation in at least two direct ways, as shown by Fig. 11.15. One major mechanism of mutagenesis involves alkylation of the oxy group

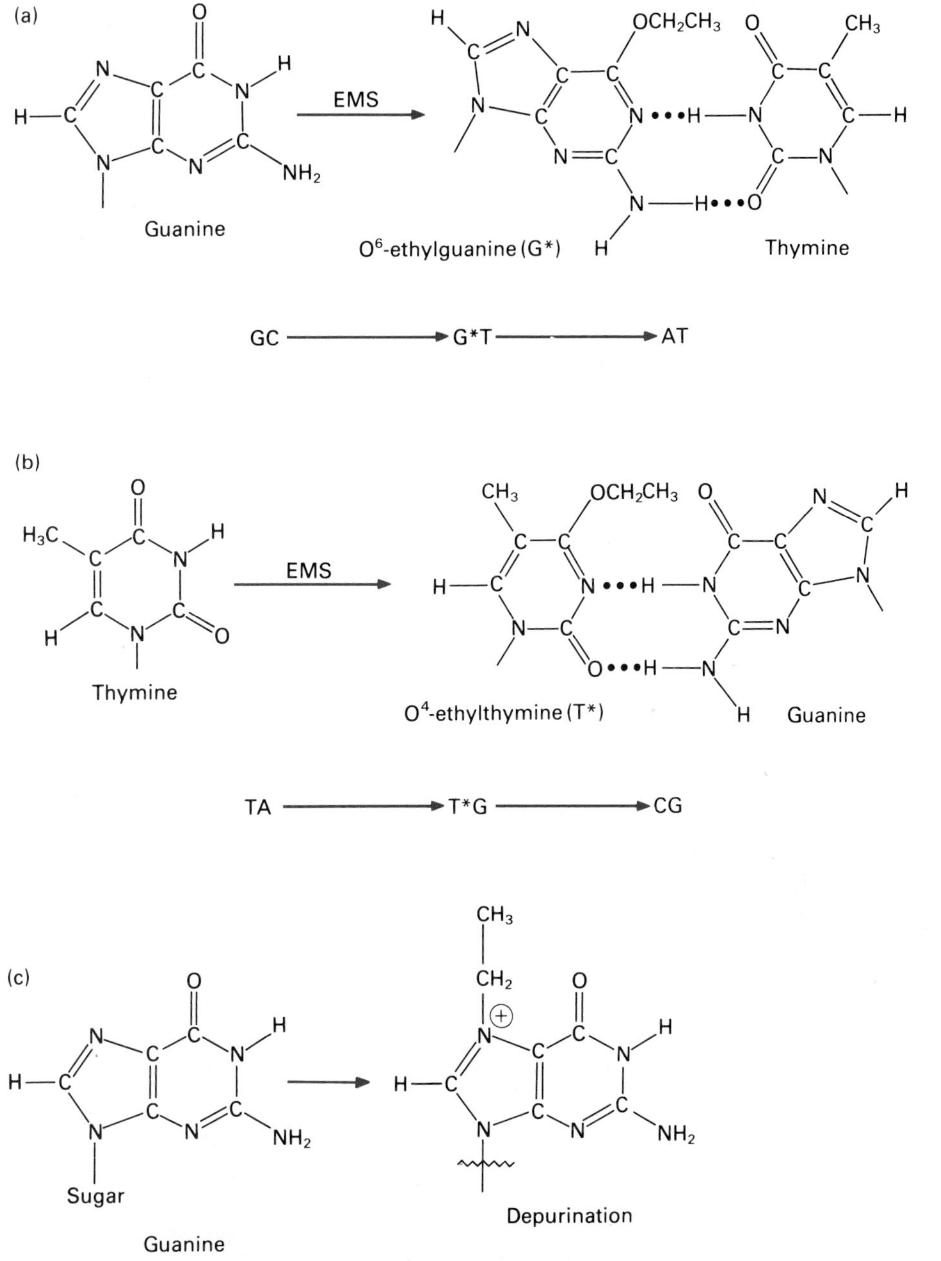

(a)

Guanine → EMS → O^6-ethylguanine (G*) · Thymine

GC ⟶ G*T ⟶ AT

(b)

Thymine → EMS → O^4-ethylthymine (T*) · Guanine

TA ⟶ T*G ⟶ CG

(c)

Guanine (Sugar) ⟶ Depurination

Figure 11.15. Direct mutagenic actions of the alkylating agent EMS. (a) The alkylation of the oxygen in the sixth position in guanine yields a guanine derivative (G*), which pairs with thymine. The transition GC → AT results. (b) The alkylation of thymine yields a thymine derivative (T*), which pairs with guanine. The result is the transition TA → CG. (c) The alkylation of the nitrogen in the seventh position of guanine can weaken the bond to deoxyribose, leading to the loss of the guanine base from the nucleotide. If the resulting gap is present during replication, any nucleotide could be inserted into the daughter strand, possibly yielding a transition or a transversion. If the gap is skipped during replication, a frameshift mutation results.

in either guanine or thymine (Fig. 11.15a and b), which changes their hydrogen bonding properties to those of adenine or cytosine, respectively. The alkylation of guanine results in the transition GC → AT (GC → G*T → AT), whereas the reaction with thymine produces the transition TA → CG (TA → T*G → CG). The second mechanism of action of the alkylating agents also involves a reaction with guanine; in this case, the alkyl group bonds with a nitrogen in the seventh position, resulting in depurination (Fig. 11.15c). In depurination, a purine base separates from the sugar-phosphate backbone, leaving a gap in the DNA. If the apurinic gap is present during replication, either a base substitution or a frameshift mutation is possible. A base substitution could arise because any nucleotide could conceivably be inserted into the complementary strand opposite the gap; a transition or transversion might result. On the other hand, the gap could simply be skipped by the replication enzymes, resulting in a nucleotide deletion and a frameshift mutation.

In addition to their direct mutagenic effects, it is believed that alkylating agents also act as mutagens indirectly, by causing errors in the cellular systems that are designed to repair DNA damage. The bulky alkyl groups apparently distort the helical structure of the DNA molecule. Nitrogen and sulfur mustards produce such distortions by causing intra- and interchain crosslinkage. The repair processes apparently recognize such lesions and are usually efficient enough to restore the damaged section of DNA to its original condition. But if the repair systems are unable to keep up with the magnitude of damage done by a mutagen, or if for some other reason a mistake is made in repair (that is, a misrepair occurs), then gene mutation might result. We will discuss the idea that mutation can result from the misrepair of DNA damage later in this chapter.

Acridines (for example, proflavin and acridine orange) have large, flat hydrophobic surfaces, which allow them to insert between the base pairs of a DNA double helix. The dimensions of an acridine approximate those of a base pair, making possible the insertion into DNA, known as intercalation (Fig. 11.16). Acridines are known to cause the addition or deletion of a single nucleotide pair; they are most potent when present during times of DNA replication or recombi-

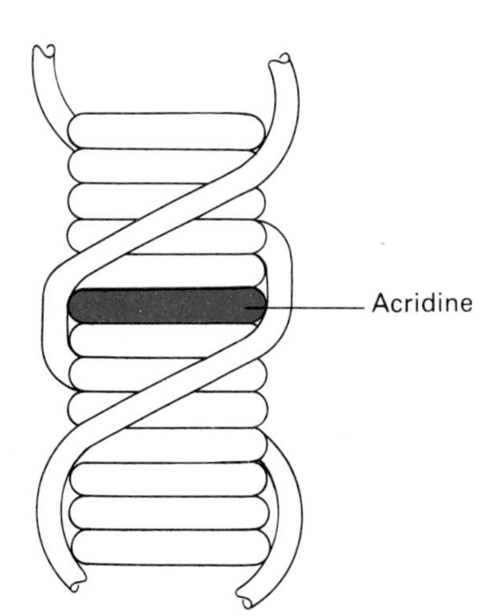

Figure 11.16. Model of intercalation of an acridine molecule into the DNA double helix, between stacked base pairs. *Source:* After L. S. Lerman, *J. Cellular and Comparative Physiology,* 64(Suppl. 1):1, 1964.

Table 11.4. Summary of the action of selected chemical mutagens.

MUTAGEN	DAMAGE TO DNA	MUTAGENIC EFFECT
Base analogs (BU and AP)	Base-pairing errors through tautomerism	Two-way transitions (AT ⇌ GC)
Hydroxylamine (HA)	Base-pairing error through modification of C	One-way transition (GC → AT)
Alkylating agents (EMS, EES, NG, mustards)	Base-pairing errors through modifications of G and T Depurination Nucleotide deletion Structural distortion of double helix	Two-way transitions (AT ⇌ GC) Transversions Frameshift mutations
Acridines (proflavin, acridine orange)	Intercalation into double helix	Frameshift mutations

nation. The mechanism by which the mutation occurs is not yet understood, although it may involve breakage by repair enzymes of the DNA strand near the point of intercalation.

The subject of mutagenesis is a complex one. The diversity of mutagenic chemicals, the fact that each may produce more than one type of lesion, the fact that the mechanism of action of a mutagen may differ depending on the test system employed, and our incomplete understanding of the action of most mutagens, make it difficult to organize the topic in a way that is easily comprehensible. To aid you in remembering the actions of the various chemical mutagens we have discussed here, a summary of the mutagens along with the major types of mutations that each induces is given in Table 11.4.

EXAMPLE 11.1. A spontaneous *rII* mutant in the bacteriophage T4 is exposed to a series of mutagens, including HA, BU, EMS, and proflavin, in a test to determine which mutagen(s) is/are able to cause reversion back to the wild-type phenotype. Only EMS causes the mutant to revert back to wild-type. What is the probable nature of the original spontaneous mutation?

Solution. Alkylating agents such as EMS can cause transitions, transversions, and frameshift mutations. But since HA and BU, which also cause transitions, were unable to cause the mutant to revert, the original mutation was not a transition. Neither was it a frameshift mutation, since proflavin was unable to cause a reversion. This leaves transversion as the nature of the original mutation that produced the *rII* allele.

Radiation

Electromagnetic radiations of wavelengths that are too short to be visible to us are potent mutagens. These forms of radiation include emissions from radioactive sources; human-made X-rays, protons, and neutrons; cosmic rays; and ultraviolet light. All of these forms are classified as *ionizing radiation* except for ultraviolet light, which is a *nonionizing* form of radiation. Emissions from radioactive sources and X-rays are found in the high frequency (or short wavelength) portion of the radiant energy spectrum. They are classified as ionizing radiation because they are energetic enough to dislodge electrons from their orbits, so as to convert neutral atoms into electrically charged particles, or ions.

Ionizing radiation is perhaps best known for its ability to cause chromosome (DNA) breakage, with subsequent chromosomal mutations. All of us are normally exposed to the ionizing radiation of cosmic rays from outer space and the radioactive sources in rocks and soil. Many people also receive substantial doses of radiation from the diagnostic and therapeutic use of X-rays. These sources undoubtedly contribute to the "spontaneous" origin of chromosomal mutations.

X-rays were the first mutagens to be discovered. Besides their ability to induce chromosome breaks and resulting chromosomal mutations, X-rays can also induce point mutations or at least make changes in DNA so small as to mimic the phenotypic consequences of point mutations. In 1927, H. J. Muller (who was awarded the Nobel Prize in 1949) demonstrated that exposure to X-rays significantly increases the frequency of sex-linked lethal mutations in *Drosophila*. The number of gene mutations induced depends on the radiation dose, which is measured in *roentgens* (r), named after the discoverer of X-rays. One r is equivalent to the amount of radiation that produces 2×10^9 ion pairs (one electrostatic unit of charge) per cubic centimeter of air. Since the ionizing effects of X-rays depend on their interaction with living tissue, however, the radiation dose is often measured in terms of the energy absorbed by irradiated matter, in either *rads* or *rems*, instead of roentgens. One rad is the amount of radiation that releases 100 ergs of energy per gram of irradiated matter; it is slightly larger than a roentgen. The rem, which stands for roentgen-equivalent-man, expresses the biological effect of one rad on human tissue.

The frequency of point mutations induced by X-rays is proportional to the radiation dose. In the case of irradiated sperm from *Drosophila*, this relationship yields a straight line (Fig. 11.17). The lack of a threshold below which no mutational damage is done means that there is no lower limit beneath which X-rays are "safe" in *Drosophila*. The effect of a particular dose of X-rays varies from one organism to the next, however. *Drosophila* flies sustain the same amount of mutational damage from a given dose of X-rays, regardless of whether the dose was administered as low-intensity radiation over a long period of time (chronic exposure) or high-intensity radiation over a short period (acute exposure). In contrast, the effects of a given dose of radiation on mice differ with the way in which it is administered. Chronic exposure induces significantly fewer mutations than does the same dose administered through acute exposure. The different dosage responses in the two organisms probably reflect their different capacities to repair damage done to DNA. If humans respond to ionizing radiation in a manner similar to mice, we would expect that repeated exposure to low-intensity radiation would cause fewer mutations than would be the case if the same total dose were to be administered all at one time.

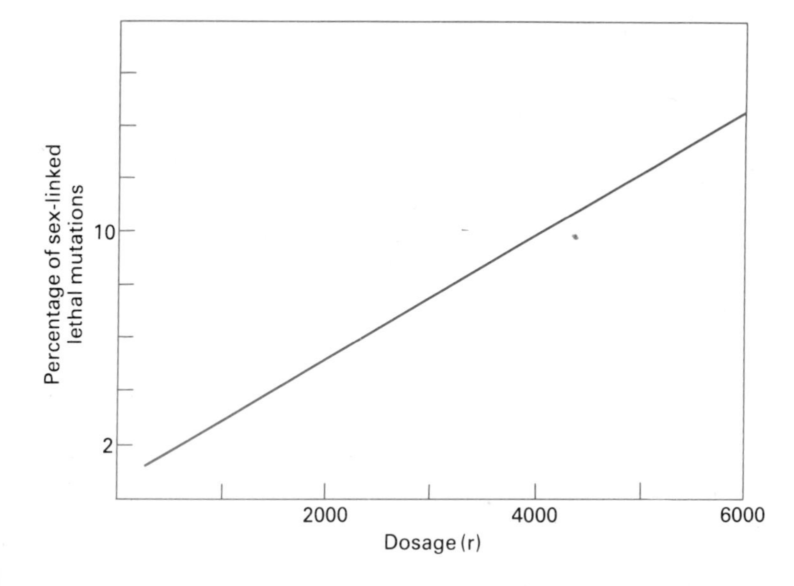

Figure 11.17. Direct proportionality between the frequency of sex-linked lethals in *Drosophila* and X-ray dose, measured in roentgens.

Natural, or background, radiation is chronic radiation, whereas diagnostic and therapeutic X-rays and nuclear explosions are acute forms. Background radiation includes cosmic rays and the radioactive elements in the earth's crust, which emit alpha and beta particles and gamma rays. Exposure to background radiation varies with geographical location; exposure to cosmic rays also varies with altitude. Doses of radiation from the earth's crust vary nearly 100-fold from one location to another around the world. There is no evidence, though, of a significant correlation between variations in background radiation and mutation rates. The average total exposure of a human to background radiation per generation (30 years) is estimated at about 4 to 10 rems. An X-ray machine delivers about 0.2 rem per exposure.

In contrast to ionizing radiation, nonionizing radiation, such as ultraviolet (UV) light, does not cause the formation of ions. In fact, nonionizing radiations are of too low an energy to penetrate living tissue. The outer layer of skin is enough to protect organisms from the penetration of the sun's ultraviolet rays. Ultraviolet light is a powerful germicidal agent to bacteria, however, if administered at high doses. At lower doses, it is an effective bacterial mutagen.

Both lethal and mutagenic effects of UV light on bacteria are the result of its being directly absorbed by the DNA, specifically by the purine and pyrimidine bases. The absorbed ultraviolet rays dissipate their energy to the atoms of the bases, thereby raising them to a higher energy level, or "excited" state. This state, in turn, increases the chemical reactivity of the bases. Many different kinds of chemical modifications have been found in DNA treated with UV light. The most prevalent modifications are **pyrimidine dimers,** which result from the covalent bonding of adjacent pyrimidines on the same strand of DNA (a form of intrachain crosslinkage). The most common pyrimidine dimers are **thymine dimers** (Fig. 11.18). The presence of a thymine dimer in the DNA distorts the double helical

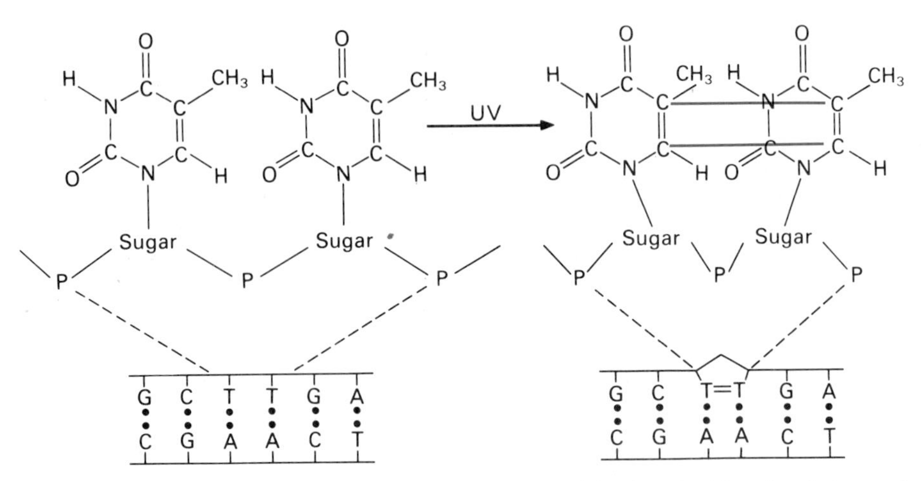

Figure 11.18. Formation of a thymine dimer by ultraviolet light. Covalent bonding between the number 4 carbons and between the number 5 carbons of adjacent thymine bases along a DNA strand produces a thymine–thymine dimer. Dimerization causes a bulge in the double helix.

structure, which interferes with the ability of the thymines to form hydrogen bonds with the adenines on the opposite strand. The failure to form hydrogen bonds, in turn, interferes with DNA replication and transcription; it is ultimately lethal unless repaired. The mutagenic effects of UV light are attributable to a misrepair of the area in which the dimer is located.

Screening Tests for Mutagens

About 63,000 different chemicals are now in use in this country, and more than 500 new chemicals are introduced each year. A number of different systems have been devised to test these chemicals for mutagenic properties. Some tests employ experimental animals such as rats and mice, while others make use of microorganisms. In the *dominant lethal test,* male rats or mice are treated with a chemical and are mated with untreated virgin females. The females are killed during gestation; the presence of an unusually high number of resorbed embryos and abnormal fetuses is taken as an indication that the sperm derived from the treated male carried at least one dominant lethal gene. The obvious drawback to this test is that it can identify only one specific class of mutant genes: dominant lethals. This problem can be overcome through the use of the *Drosophila test,* which looks for mutants in multiple generations derived from treated ancestors.

Short-term mutagenicity tests have become vital to the screening process because of the length of time required to conduct whole-animal studies, as well as their cost. The characteristics of these short-term assays are described in Table 11.5. In vitro *cytogenetic tests* apply the chemical being tested to human cells that are growing in culture. Microscopic examination of these cells reveals any increased incidence of chromosomal mutation. *E. coli growth inhibition* uses the inhibition of growth cells of bacterial strains unable to repair DNA lesions as an indication of the mutagenic properties of the chemical being tested. *Eukaryotic DNA replication inhibition* is based on the induction of damage to DNA, which prevents its replication as the basis for detection of mutagenicity. Other tests evaluate mutagenicity by observing the occurrence of *unscheduled DNA synthesis* in treated cells. One drawback to these kinds of tests is that specific kinds of point mutations will not be revealed. Another drawback is that a chemical that is mutagenic to a whole organism may not be mutagenic to cells growing in culture or to bacterial cells, since cultured cells and bacterial systems do not mimic the metabolic system of the whole eukaryotic organism. For example, cultured cells will fail to respond to substances that are not mutagenic in their natural form and become mutagenic only after they have been chemically modified by certain metabolic enzymes. An example of a group of mutagens that are metabolically activated are the nitrates and nitrites, which have been used as food preservatives. These substances are harmless until they are metabolically converted by the liver into highly mutagenic and carcinogenic nitrosamines.

The tests we have just described evaluate the mutagenic potential of a substance by its ability to break or otherwise damage DNA. In contrast, several more direct tests have been devised that use the induction of point mutations as a criterion of the mutagenic activity of a substance. The most commonly used of these test systems at present is the *Ames test,* which utilizes the bacterium *Salmonella typhimurium* as the test organism. The Ames test combines the ease of handling and large population size of microorganisms with the eukaryotic metabolic system. Rat liver enzymes are isolated by sedimenting the extract of liver cells into what is called the microsomal fraction. The liver enzymes are added to the medium of the bacteria, along with the chemical being tested for mutagenic activity. The enzymes convert the chemical to the same types of metabolic intermediate compounds as would the metabolic system of the whole organism, thereby metabolically activating a potentially mutagenic substance.

Table 11.5. Characteristics of commonly-used short-term assays for mutagenicity.

	PRINCIPLE UPON WHICH MUTAGENICITY IS EVALUATED	
ASSAY	GENE MUTATION	DNA DAMAGE
Cytogenetics (chromosomal aberration, sister chromatid exchange, etc.)		+
E. coli growth inhibition (using repair deficient strains)		+
Eukaryotic DNA replication inhibition		+
Unscheduled DNA synthesis		+
Eukaryotic microorganisms (*S. cerevisiae, N. crassa, A. nidulans*)	+	
Cell culture and mutagenesis	+	
Ames/*Salmonella*	+	

Source: Courtesy of Dr. Ralph J. Rascati, Department of Biological Sciences, Illinois State University, Normal, Il.

The bacteria used in the Ames test are histidine auxotrophic mutants (*his*$^-$), so they cannot grow on minimal medium. But when placed on an agar-based minimal medium, the bacterial cells do not die; they remain dormant unless they happen to mutate back to the wild-type state:

$$\underset{\text{No growth}}{his^-} \xrightarrow{\text{Reverse mutation}} \underset{\text{Growth}}{his^+}$$

The Ames test is conducted by spreading *his*$^-$ cells over the surface of a petri dish containing minimal medium and the rat liver enzyme extract. A disk saturated with the test chemical is placed in the center of the plate (Fig. 11.19). The chemical diffuses outward from the disk, forming a concentration gradient. The number of bacterial colonies at various distances from the disk measures the effectiveness of the chemical in inducing back mutation. Because the strain of *Salmonella* used in the Ames test is very sensitive to mutagens, very weak mutagenic agents can be detected. The Ames test can even distinguish between a mutagen that causes base substitutions and one that induces frameshift mutations. Discrimination at this level is possible because various *his*$^-$ strains whose DNA lesions are known are available for use as indicator bacteria. For example, if a substance is found to cause reversion in his^-_{bs} bacteria (in which the mutation is of the base substitution type) but not in his^-_{fs} bacteria (in which it is of the frameshift type), then we can conclude that the substance in question specifically induces base substitution mutations. As we have already discussed, some mutagens require DNA replication for their action. We can accommodate this requirement in an experiment with *his*$^-$ mutants by adding a trace of histidine to the minimal medium, just enough to allow a few generations of growth, but not enough to allow the formation of colonies.

Ames and his colleagues, who have now tested hundreds of chemicals using this procedure, have observed a strong correlation between mutagenicity and carcinogenicity. Not only have roughly 90% of the known carcinogens proved to be mutagenic in this test, but more than 90% of substances that are not thought to be carcinogenic have proved negative in the test. The Ames screening procedure has therefore become highly regarded as a test for carcinogenicity. It is especially valuable as a fast, simple, sensitive, and economical means of doing a preliminary screening on a wide variety of chemicals, and is one of the four mutagenicity tests that the Federal Drug Administration currently allows as a substitute for preliminary cancer screening tests for drugs intended for use in food animals. Combinations of various short-term tests may be allowed in the near future to serve as preliminary screening tests for additives intended for use in human foods.

Transposable Genetic Elements

Geneticists have just recently given recognition to a mutational process discovered and described over 30 years ago. In this process, the insertion of a short length of DNA into a gene causes the gene to mutate. Both bacterial and eukaryotic cells have been found to contain certain structurally and genetically discrete segments of DNA that have the ability to move around among the chromosomes, inserting and excising themselves from various chromosomal localities. These segments have

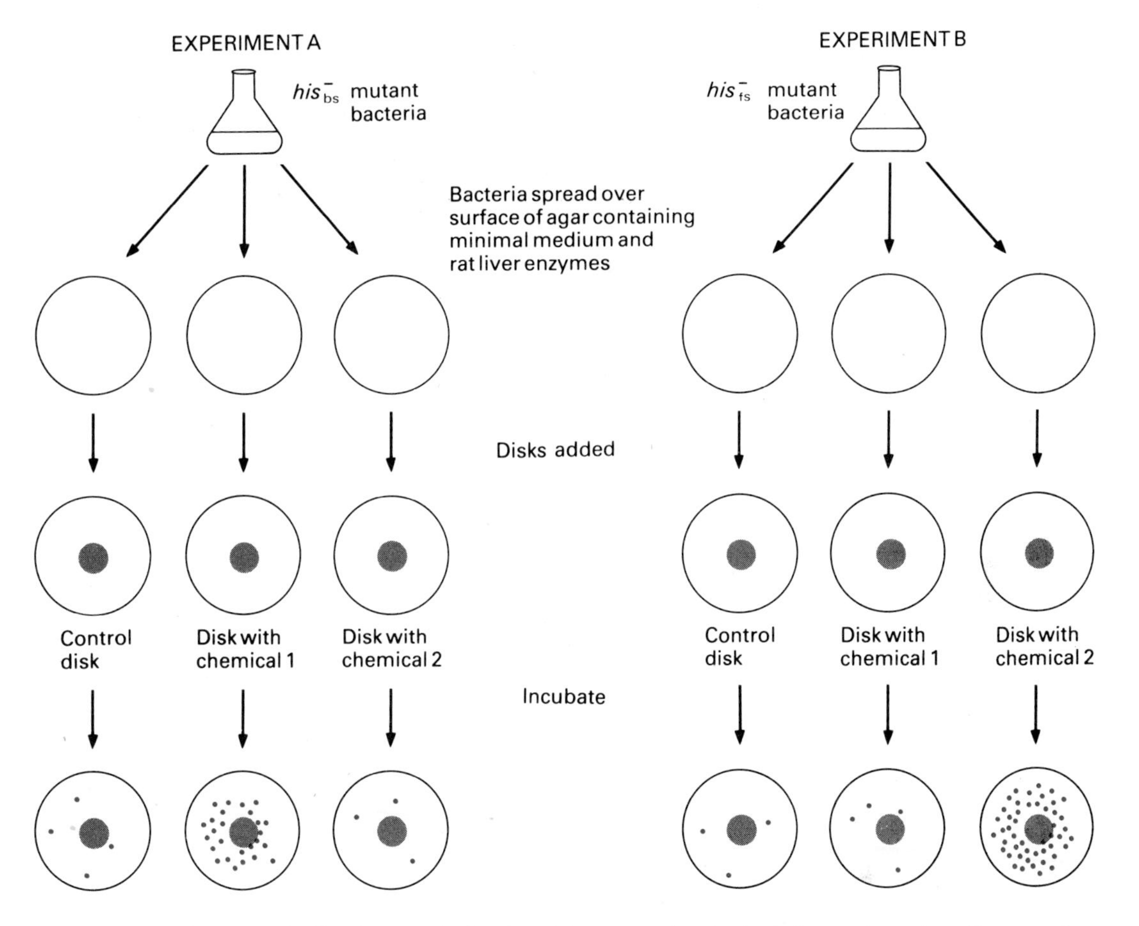

Figure 11.19. The Ames test, developed to screen chemical substances for mutagenic properties. Two tests are shown: the first (a) utilizes a *his*⁻ base substitution mutant strain (his^-_{bs}) as the indicator, while the second (b) utilizes a *his*⁻ frameshift mutant strain (his^-_{fs}) as the indicator. A suspension containing the bacteria is spread over the surface of an agar-based minimal medium containing rat liver enzymes. The *his*⁻ mutants are unable to grow on minimal medium. Disks saturated with the chemicals being tested are added to the plates. Control plates receive disks that do not contain any chemical. After incubation for a short time, the plates are scored for revertant (his^+) colonies. The control plates show the background level of spontaneous reversion. In the experiments diagrammed here, it is clear that chemical 1 induces base substitution mutations but not frameshift mutations, whereas chemical 2 causes frameshift mutations but not base substitutions.

been referred to by a variety of names, including jumping genes, roving genes, mobile genes, and transposons. Because these segments range in size to include anywhere from a single gene to several genes, the term **transposable genetic elements** (or **transposons**) seems best suited to describe them as a group.

Transposable genetic elements were first described in the late 1940s by B. McClintock, who has studied the inheritance of kernel color in corn. She discovered the genetic element *Ds*, which causes chromosomal breakage near its location on the chromosome. This breakage can be readily detected by cytological examination. The most unusual feature of *Ds* that McClintock noticed was that it moves around from one chromosomal location to another, rather than staying in one place. In its wanderings, *Ds* occasionally inserts itself into or near a gene locus that determines the coloration of kernels, causing the gene to mutate. At this locus, the dominant gene for color, *C*, causes the kernel to be solidly pigmented; spontaneous mutation of *C* to *c* results in colorless kernels. If, however, the mutation of *C* is caused by the insertion of a *Ds* element, some of the kernels are not colorless but instead have a mottled or dotted appearance, owing to the presence of pigment spots on an otherwise colorless background. One explanation for the spots is that the mutation caused by *Ds* has a high frequency of reversion back to *C*. Revertant cells would produce pigment, whereas those cells that were still mutant would produce no pigment. According to this explanation, the high reversion frequency could be the result of the high frequency with which the *Ds* element is excised from the *C* gene. This explanation has indeed been found to be the case. Excision of *Ds* is under the control of another gene, *Ac* (activator of reversion to the colored phenotype). In the presence of *Ac*, the colorless mutation caused by the insertion of *Ds* into the *C* gene is highly unstable. In fact, the mutant gene produced in this manner is denoted as c^u. If the *Ac* gene is not present (genotype Ac^+Ac^+), then the colorless mutation caused by the insertion of *Ds* is stable. The *Ac* element also moves from site to site among the chromosomes. The relationship between genotype and phenotype in this system is summarized in Fig. 11.20.

McClintock concluded from her studies that there are genetic factors in corn that move around the genome and influence the expression of other genes. She called these factors *controlling elements*. Researchers have now identified five such systems in corn populations. Each is composed of three parts: (1) a *target gene* (in this case *C*) that can be inactivated by (2) a specific *receptor* element (in this case *Ds*), which inserts itself into or near the target gene and exerts a cis effect on the function of the target gene, and (3) a *regulatory* gene (in this case *Ac*), which causes the mutation of the target gene to be unstable by its effect on the receptor element. The receptor and regulatory genes are both transposable genetic elements and are highly specific in their interaction.

Mutable and Mutator Genes. Genes that exhibit a high rate of mutation or that are highly unstable when they are mutant (such as the c^u gene) are collectively referred to as **mutable genes.** Examples of mutable genes have been discovered

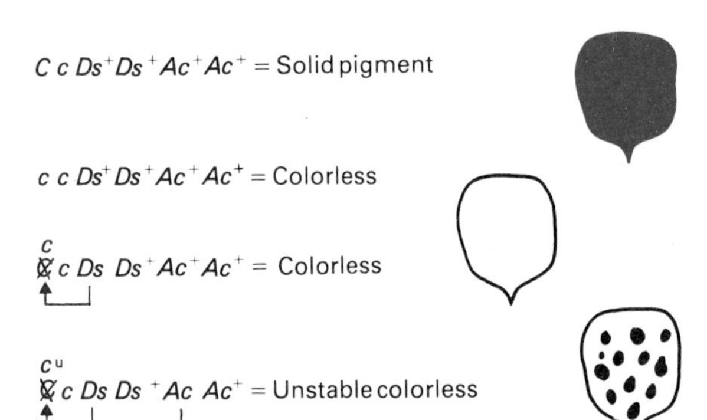

Figure 11.20. Relationship between genotype and phenotype of three loci concerned with kernel coloration in corn. *C* = color, *c* = no color; *Ds* = presence of *Ds* element, Ds^+ = lack of *Ds* element; *Ac* = activator of reversion of *c*, Ac^+ = no activator of reversion.

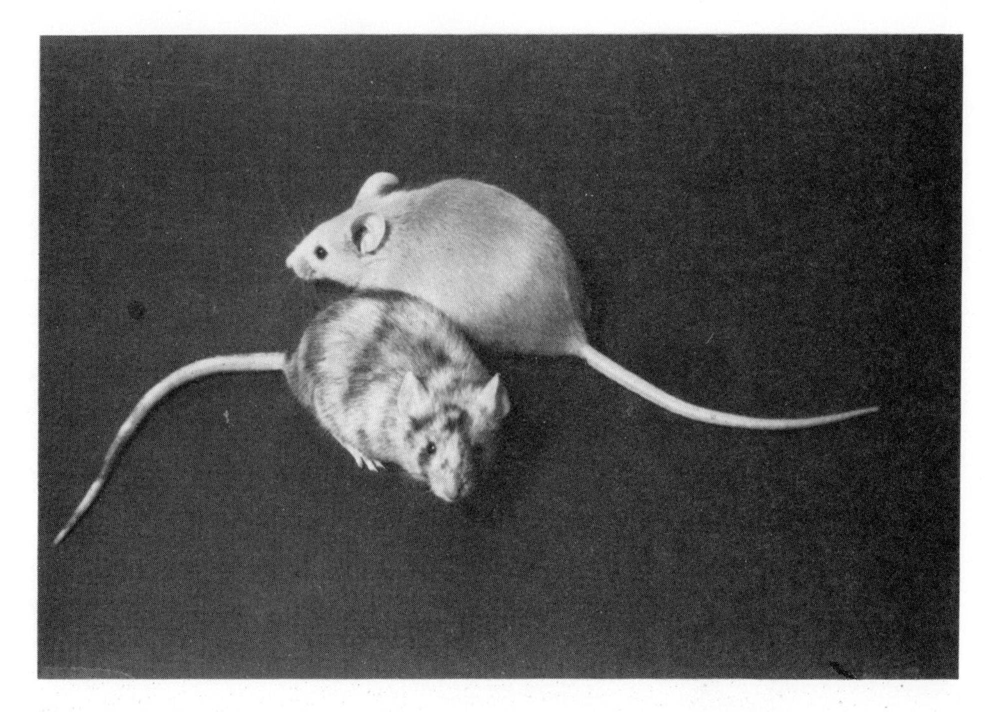

Figure 11.21. Two mice of the genotype $p^{un}p^{un}$. The pale color is typical of mice with a mutation at the locus for pink-eyed dilution. The mouse in the lower portion of the photograph shows extensive areas of revertant (black) pigmentation against the pale-colored background. It is proposed that the pink-eyed unstable allele p^{un} arises when a transposable genetic element is integrated into or near the wild-type p^+ locus. This event mutates the color locus, producing the pink-eyed dilution phenotype, which resembles the stable phenotype produced by the *p* allele. The black spots on $p^{un}p^{un}$ mice may result when the transposable element is excised, thus restoring the affected gene to its normal state. *Source:* From M. L. Lamoreux and J. B. Whitney, *J. Heredity,* 73:12–18, 1982. Copyright 1982 by the American Genetic Association. Photograph courtesy of M. L. Lamoreux and J. B. Whitney III.

in bacteria, animals, and plants. The c^u gene expresses its mutable characteristic phenotypically as the mottling of kernels. Many other mutable genes in plants and animals are also recognized by color variegations. For example, a certain mutant locus for eye color in *Drosophila* exhibits a high rate of mutation back to wild-type and to other mutant alleles. McClintock was the first to point out that a mutable gene and a target gene may be one and the same. She concluded that mutable genes may be unstable because of their association with transposable genetic elements. As support for this idea, transposable elements have recently been hypothesized as a control factor for genetic instabilities in mammals (Fig. 11.21).

Those genes that influence the stability of other genes are called **mutator genes.** This category includes *Ds* and *Ac,* along with many other transposable elements that integrate into or near gene loci, creating mutable genes or highly unstable mutant genes. Mutator genes other than transposable elements are also known. Examples include the genes that code for mutant DNA polymerases. Any malfunction of these polymerase enzymes will result in a high frequency of error in the incorporation of nucleotides during DNA synthesis. The overall effect is to increase the mutation rate for all of the genes that make up the genetic complement of the organism.

Characteristics of Transposable Genetic Elements. McClintock's postulate that eukaryotic cells contain movable genetic elements that can mutate or otherwise affect the expression of target genes was not well accepted by most geneticists. It was not until nearly 20 years later, when similar phenomena were discovered in prokaryotes, that the validity of her ideas was recognized. Only recently has she received the credit due her important work, being awarded a Nobel Prize in 1983.

In the mid-1960's, transposable genetic elements were discovered in bacteria. DNA segments, up to 2000 nucleotide pairs in length, were found that could insert themselves within any of a variety of genes, thereby turning off the expression of the gene. These elements have been named **insertion sequences,** or **IS elements.** We now know that the wild-type *E. coli* chromosome contains at least 13 IS elements as part of its normal genetic makeup. These elements cause mutation only when they accidentally end up in abnormal positions, within a gene locus.

Shortly after the discovery of IS elements, certain genes responsible for the resistance of bacteria to antibiotics were found to act as transposable genetic elements. The resistance of bacteria to antibiotics is a problem of growing medical concern. Resistance is frequently caused by the presence in the bacterial cell of *R plasmids*. Recall from Chapter 4 that plasmids are small circular double helices of DNA that replicate autonomously within the bacterial cytoplasm. In effect, they constitute small chromosomes that are independent of the bacterial chromosome with respect to both replication and function. Researchers have observed that R plasmids are larger than other plasmids that do not carry genes for antibiotic resistance. This finding indicates that R plasmids have acquired the genes for resistance through the addition of genetic material, rather than through point mutation. As we would expect, it turns out that each incorporated gene for antibiotic resistance is included within a transposon, which makes the gene a transposable genetic element. In addition to the transposable elements that carry genes for antibiotic resistance, transposons have been discovered that encode a wide variety of other functions. We now know that one bacteriophage, Mu, is actually a transposable element.

Originally, a distinction in nomenclature was made between small transposable elements that encode only transposition functions (that is, their only message is to insert themselves into the chromosome), and larger elements that encode transposition as well as additional information (such as antibiotic resistance). More recently, however, both kinds of elements have been grouped together under the general classification of transposable elements. In general, then, all transposable genetic elements encode (and perhaps even regulate) the expression of their own transposition function. They may or may not carry accessory functions.

The mechanism by which a transposable element enters and leaves a chromosome is only partially understood. All bacterial transposons studied so far have ends that consist of **inverted repeat (IR) sequences.** That is, the nucleotide sequences of the two ends of each DNA strand of the transposable segment are complementary to each other if one is read in reverse order. In the example given

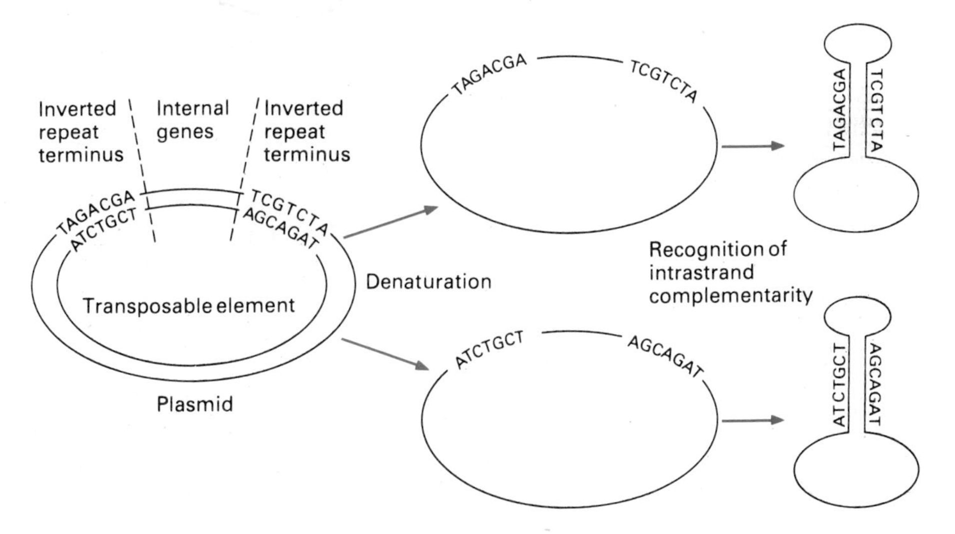

Figure 11.22. Plasmid DNA that contains a transposable genetic element is characterized by the inverted repeat sequences at its termini. The TAGACGA sequence at one end of the outer strand is complementary to the reverse order of the nucleotide sequence at the other end of the same strand. Similar inverted complementarity holds for the inner strand. Upon strand separation, the inverted repeat sequences cause each strand to fold back upon itself into a "stem-and-loop" configuration.

in Fig. 11.22, the nucleotide sequence at the left end of the transposon in the outer strand (TAGACGA) is complementary to the nucleotide sequence at the right end in the same strand, if the latter is read in the reverse order as ATCTGCT. A similar complementarity is observed on the inner strand. If DNA that contains inverted repeat sequences is denatured so that the two strands separate, the inverted complementary sequences in each strand are then free to pair and each molecule will fold back upon itself. This action generates a characteristic "stem-and-loop" structure, demonstrated by the electron micrograph of plasmid pSC105 shown in Fig. 11.23. There apparently exists some highly specific mechanism, coded for by the transposable element itself, that recognizes the inverted repeat termini and cleaves the DNA at these locations, allowing it to leave the chromosome.

A recent finding in the analysis of nucleotide sequences relates to the question of how transposable elements are inserted. In all cases studied so far, the base sequences of the chromosomal nucleotide pairs that bracket the inserted transposon are exact repeats (Fig. 11.24). The length of the repeating sequences ranges from less than 10 to several hundred base pairs, depending on the particular transposable element being studied. One way to account for this finding is to postulate that the recipient chromosome is enzymatically cleaved at staggered sites on opposite strands of DNA. The transposable element is then inserted and repair enzymes duplicate the short single-stranded stretches on either side, in each case using the complementary strand as a template. This process would always yield a duplicate nucleotide sequence on either side of an inserted element.

Transposable elements have been found to promote a number of different rearrangements of DNA in addition to simple transposition, including deletions (loss of a chromosome segment), duplications (repeated segment along a chromosome), and the reversal of the gene order along a segment of chromosome. Rearrangement of the genetic material is an important source of genetic diversity for evolution. Some scientists have suggested that transposable genetic elements may be "nature's tools for genetic engineering." In addition, transposable elements may play a role in moving genes into and out of regulatory positions. Recent evidence suggests that any gene can be transposed if IS elements are inserted into the DNA on either side of its locus. It has even been speculated that transposable elements are responsible for the rearrangements of genes that occur during development in some cell lines, such as during the maturation of an antibody-producing cell. In addition, there is now firm evidence that genes have been transferred from mitochondria and chloroplasts to the nucleus during the course of the evolution of eukaryotic cells. This finding adds support to the symbiotic theory of the origin of eukaryotic cells. As discussed in Chapter 4, this theory proposes that mitochondria and chloroplasts arose from free-living prokaryotes that entered and established a symbiotic relationship with a progenitor of the eukaryotic cell. Since the genomes of these organelles now serve only a small proportion of their genetic needs, with the remainder provided by the nucleus, the theory has to allow for the transfer of genes from the organelles to the nucleus.

The experimental method of nucleotide sequence analysis has recently revealed that the *Alu* sequences (discussed in Chapter 10) are flanked on either side

Figure 11.23. The "stem-and-loop" structure exhibited by plasmid pSC105. Enlargement is 230,000 diameters. *Source:* From S. N. Cohen and J. A. Shapiro, Transposable genetic elements, *Scientific American,* 242:40–49, 1980. Courtesy of Professor S. N. Cohen, Department of Genetics, Stanford University.

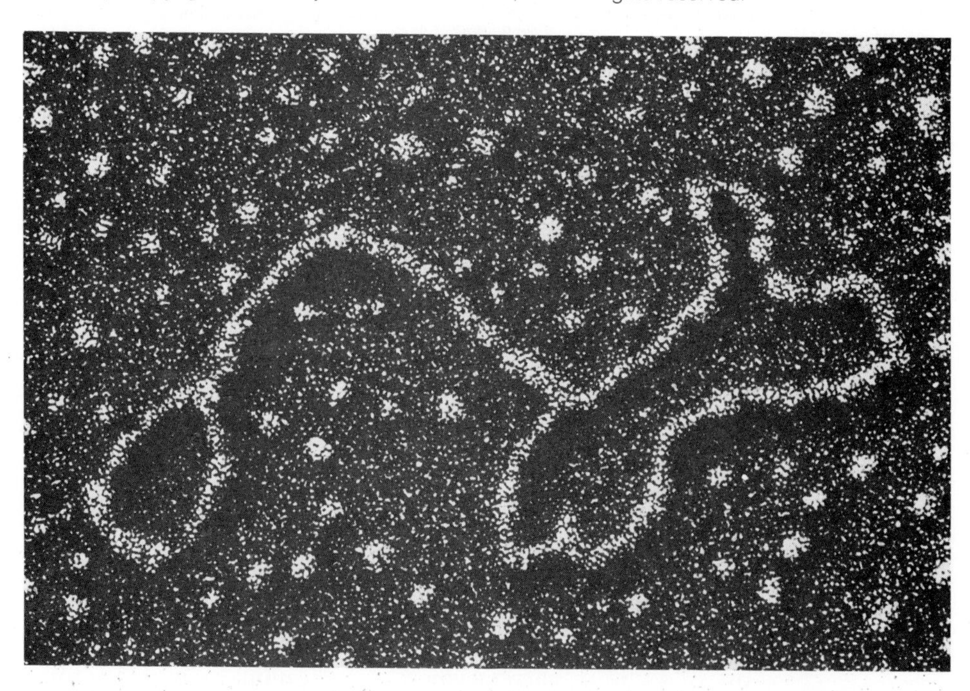

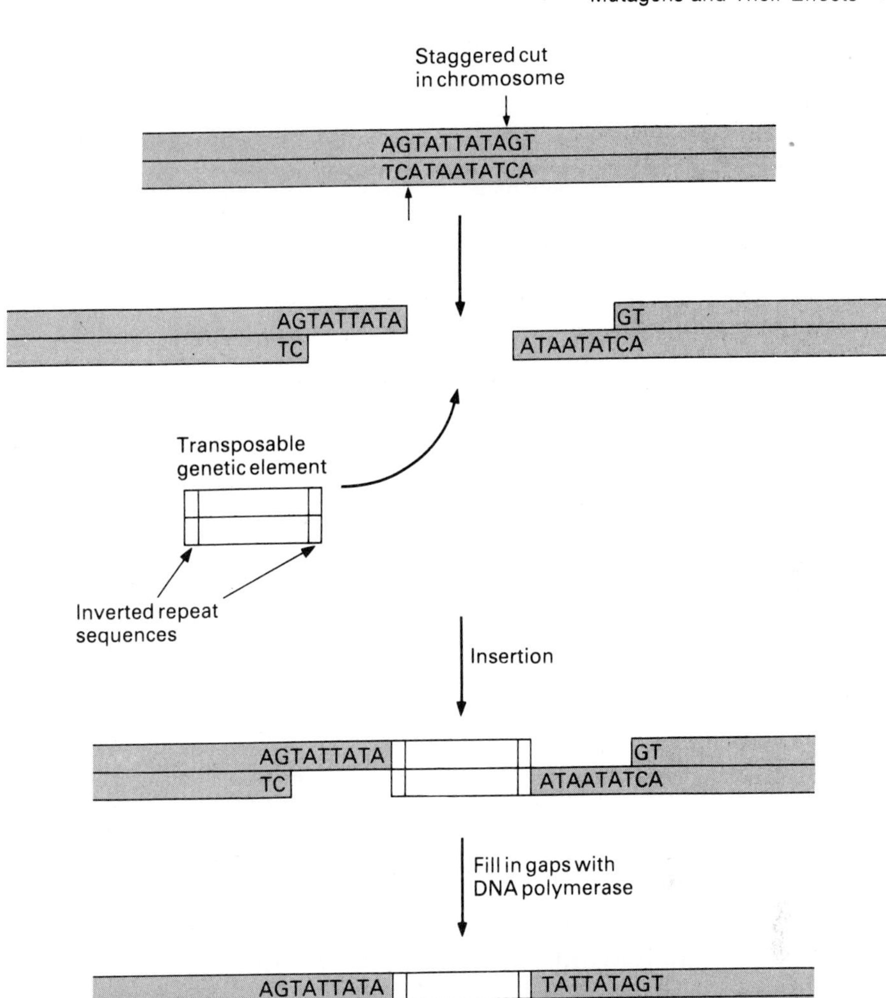

Figure 11.24. Duplication of nucleotide pairs in the recipient DNA is associated with insertion of a transposable element. This figure shows how that duplication may come about. The two strands of recipient DNA are cleaved at staggered sites seven nucleotides apart. (In reality, these cleavage sites are 300–500 nucleotides apart.) The transposable element then inserts itself, and single-stranded regions on either side of the inserted element are filled in, to produce the chromosome shown at bottom.

by nucleotide sequences that are direct repeats. This finding suggests that the *Alu* family members are transposable elements within the DNA. A similar finding has been made for the long-period dispersed repeated sequences found in *Drosophila* DNA. The *Alu* sequences do not possess inverted repeat sequences at their termini, as do the long-period repeated sequences of *Drosophila* and bacterial transposons, however. This difference suggests that the short-period and long-period repeated sequences disperse throughout chromosomal DNA by different mechanisms.

TO SUM UP

1. Mutagens are chemical substances, radiation, or genetic elements that increase the likelihood of mutation. Most chemical mutagens probably act by increasing the frequency of base pairing errors; the tautomerism or chemical modification of DNA bases caused by these mutagens leads to a base pair substitution. Other chemical substances cause mutation by DNA crosslinkage or by intercalation into the double helix.
2. Both ionizing radiation (e.g., X-rays) and nonionizing radiation (e.g., ultraviolet light) can induce point mutations. The precise mechanism by which this process occurs is not yet fully understood, although it is thought to involve misrepair of the DNA that has been damaged by the radiation.
3. Nonionizing radiation is mutagenic to bacteria, in which it is directly absorbed by the DNA. Ultraviolet light excites the atoms of the nucleotide bases, which increases their reactivity; the most common result is the dimerization of adjacent thymine bases along a strand of DNA.
4. Chemical substances can be screened for mutagenic activity by a variety of tests. The best known of these tests at the present time is the Ames test. This procedure combines the metabolic enzymes of a eukaryotic system with a bacterial test organism. Because of the high correlation between mutagenicity and carcinogenicity, the Ames test is widely regarded as a screening test for carcinogens.
5. The mutation of some genes is mediated by the insertion of a mutator (or receptor) element into or near the gene locus. Such mutations are typically unstable, however, because of the influence of a regulator element that promotes the excision of the receptor element from the target gene.
6. Transposable genetic elements are DNA sequences that move from one chromosomal location to another. They are best known in higher plants and in bacteria, but they have been discovered in animal cells as well. Their exact function remains unclear.

Mechanisms of Repair

All DNA, whether in a bacterial cell or in the millions of cells of a higher organism, is vulnerable to damage by radiation and by chemical agents in the environment. Any lesion that interferes with proper base pairing will disrupt the fundamental functions of DNA, since the recognition of base-sequence complementary is necessary for the normal replication, recombination, and transcription of DNA. Unless the damage is quickly repaired, it can cause the cell to die. The stability of DNA is maintained by enzymatic systems that repair the genetic lesions caused by mutagens.

Even in the absence of mutagens, spontaneous errors in base pairing continually occur through the process of tautomerism. It is thought that spontaneous errors in base pairing are normally detected by the same system that replicates the DNA. During replication, the repair enzymes "edit" the nucleotide sequence that is being laid down to form a new strand. These enzymes usually recognize the mispaired bases, excise them from the strand, and insert the correct bases.

The continual repair of induced and spontaneous damage to DNA preserves its information content. Because extensive repair systems operate in both prokaryotic and eukaryotic cells, the expressed incidence of mutation is much less than the actual amount of damage incurred in the DNA. Mutations can then be viewed either as an escape from the normal cellular repair processes (such as might occur if the damage from a mutagen were extensive, thus swamping the repair systems) or as a mistake made during the repair process (under this view, mutation becomes equivalent to misrepair).

We will now discuss the molecular basis of DNA repair and misrepair. Our discussion centers around *E. coli*, the organism in which the majority of experimental studies of repair have been carried out. All of the kinds of mutagen-induced damage to DNA that were considered in the previous section are repairable; these types include deleted bases and nucleotides, mispaired bases, chemically altered bases, crosslinkages, and single- and double-stranded breakages. But by far the best-studied example is the repair of the thymine dimer.

Repair of Damage Caused by Ultraviolet Light

The well-defined genetic system of *E. coli* has promoted an understanding of the enzymatic functions that are used in the repair process, particularly in response to damage induced by ultraviolet light. Over 65 *E. coli* genes play a role in one or more repair processes. Many of these genes are also required for the normal replication and recombination of DNA. Thymine dimers induced by UV light can be repaired by at least four different systems: (1) photoreactivation, (2) excision-repair, (3) post-replication daughter strand gap repair, and (4) trans-dimer repair. Photoreactivation and excision-repair appear to be the major systems employed by *E. coli* to repair UV damage accurately.

Photoreactivation. The simplest mechanism for repair of thymine dimers is the direct reversal of the dimer-forming process by photoreactivation. The photoreactivating enzyme binds to the region on the DNA that contains the dimer. When visible light is absorbed by the enzyme-dimer complex, the enzyme is activated and it splits the bonds holding the thymines together, restoring each to its monomer form (Fig. 11.25). Photoreactivation involves no excision or replacement of nucleotides, so there is no possibility for mutation to occur during the repair process.

Excision-Repair. In contrast to photoreactivation, excision-repair uses a complex of several enzyme activities to repair the dimerized area, and visible light is not required. Figure 11.26 illustrates the four major enzymes involved in the excision-repair process: DNA polymerase I, endonuclease, exonuclease, and ligase. We discussed all of these enzymes in Chapter 3, so we will only review their activities briefly here. DNA polymerases catalyze the addition of nucleotides to the 3′ end of a growing DNA strand, using the parental strand as a template (refer also to Fig. 3.23). Endonucleases split a single phosphodiester bond within a DNA strand, in a process referred to as *nicking*. Exonucleases sequentially chop nucleotides from an exposed single-stranded end of DNA. Some exonucleases cleave nucleotides exclusively in the 5′ → 3′ direction, while others work only in the reverse direction. The action of a ligase is to catalyze the formation of a phosphodiester bond between adjacent nucleotides that for some reason were not previously linked together.

As the term excision-repair implies, the area that includes the thymine dimer is first excised from the DNA strand; the resulting gap is then filled in with the

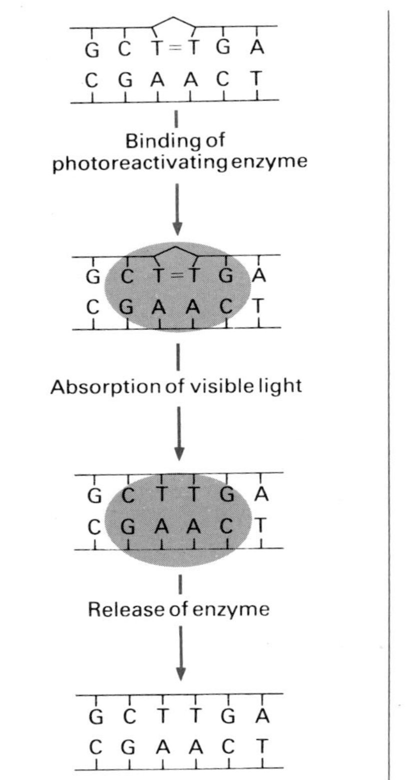

Figure 11.25. Photoreactivation, a process that repairs the pyrimidine dimers induced by ultraviolet irradiation. The photoreactivating enzyme recognizes the dimer by the associated distortion in the double helix. After positioning itself on the DNA at the damaged region, the enzyme is activated by absorption of visible light. The activated enzyme splits the covalent bonds responsible for dimerization, returning the DNA to its normal condition.

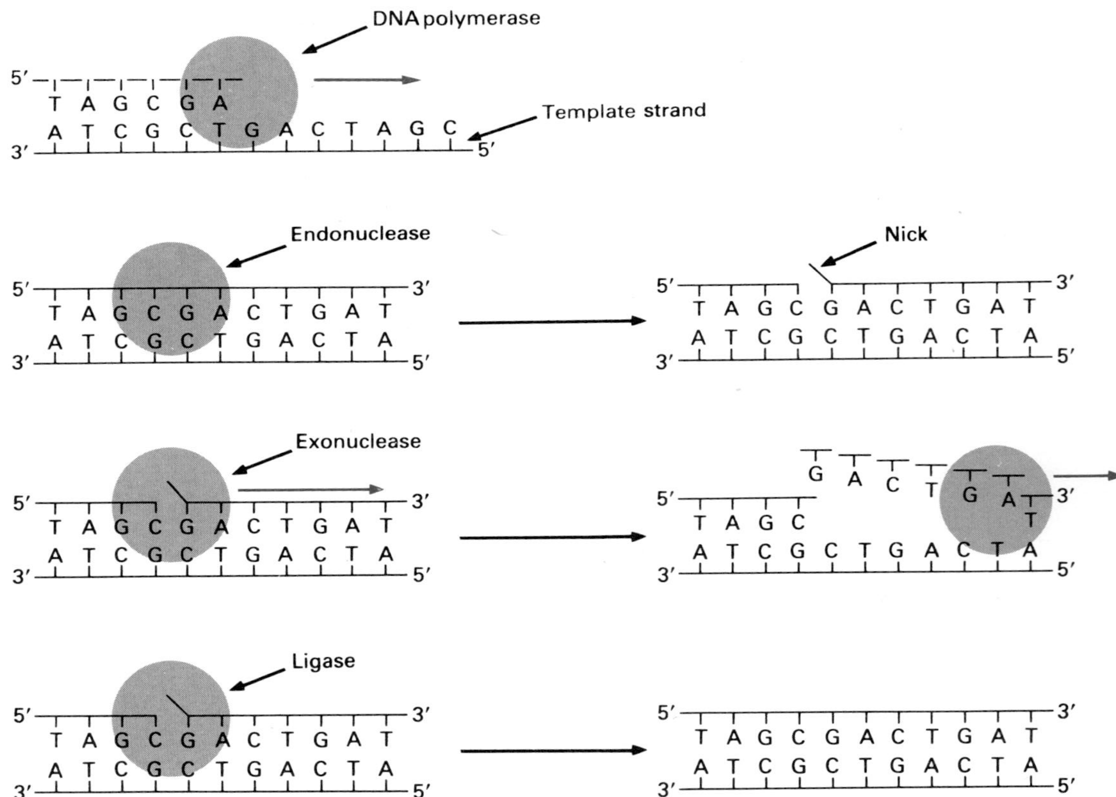

Figure 11.26. Four major enzyme activities involved in excision-repair. (a) DNA polymerase catalyzes the synthesis of single-stranded DNA in the 5′ → 3′ direction, using one of the strands of the parental double helix as a template. (b) Endonucleases split phosphodiester bonds, producing one or more nicks in a DNA strand, but they do not excise any nucleotides. (c) Exonucleases sequentially chop nucleotides from an exposed single-stranded end. The exonuclease shown in the diagram is operating in the 5′ → 3′ direction, but 3′ → 5′ exonucleases are also known. (d) Ligase catalyzes the formation of a phosphodiester bond between adjacent nucleotides that were not previously linked together.

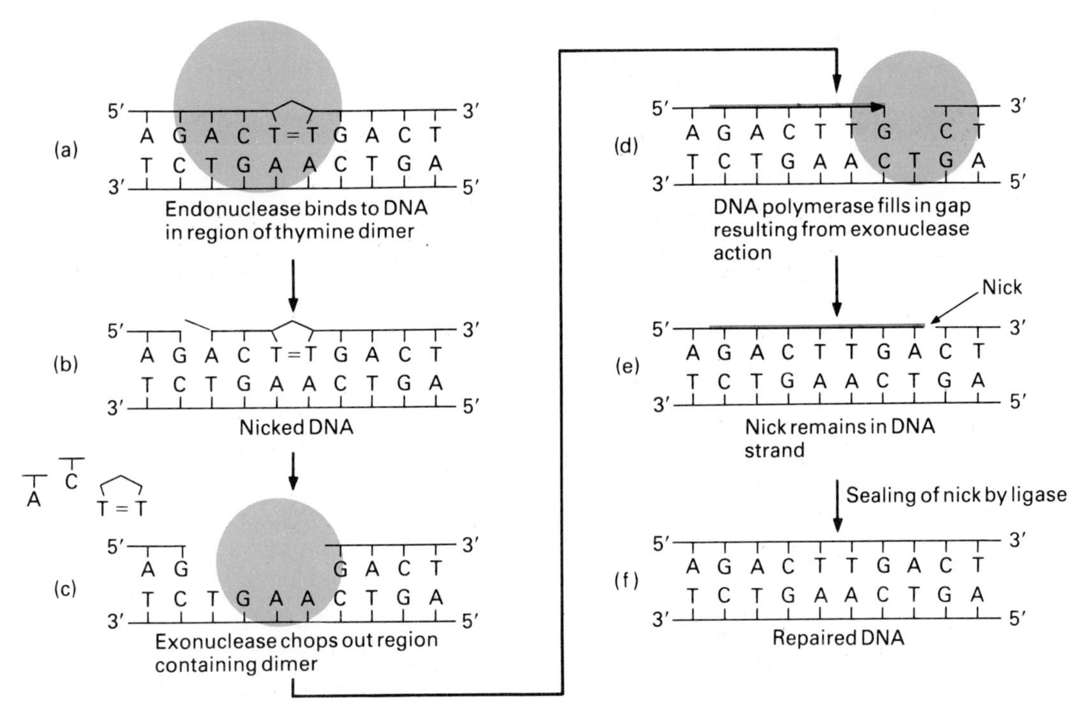

Figure 11.27. Excision-repair of DNA. A dimerized region is recognized by an endonuclease (a), which nicks the damaged strand upstream from the dimer (b). Exonuclease activity chops out the region containing the dimer (c), and DNA polymerase fills in the resulting gap, synthesizing new DNA in the 5′ → 3′ direction (d). At the remaining nick in the DNA (e), the 3′ end of the newly synthesized region is joined to the parental strand by ligase, completing the repair process (f).

proper nucleotide sequence by DNA polymerase I. Figure 11.27 diagrams the excision-repair process. The repair system apparently recognizes the lesion by the associated distortion of the double helix. An endonuclease makes a nick near the dimer and the strands unwind to expose a single-stranded DNA end at the nick. An exonuclease can then sequentially cleave nucleotides from this strand, chopping out the dimer as a single unit. A region of approximately 10 to 12 nucleotides surrounding the dimer is excised. DNA polymerase I then fills in the gap, using the undamaged strand as a template to ensure that the proper sequence of nucleotides is synthesized. After the last nucleotide is filled into the gap, a nick remains (see Fig. 11.27e). This nick represents the need for a final phosphodiester bond that cannot be catalyzed by DNA polymerase. A ligase is thus required to seal the nick, completing the repair process.

Several excision and repair systems are thought to exist in *E. coli,* as well as in mammalian cells. The excision activities include *glycosylases,* which cleave the base-sugar bond of a nonconventional nucleotide residue, as well as endonucleases and exonucleases. Repair processes of the excision and repair type possibly comprise the largest group of enzymes acting on DNA in the cell. The spontaneous decay of DNA, which is thought to occur mainly through the loss of bases due to depurination and depyrimidination and through deamination of cytosine and adenine residues is apparently greater than is generally recognized. However such lesions are repaired so efficiently that they are seldom the basis of spontaneous mutation.

Post-Replication Repair and Misrepair. When the exposure to ultraviolet light is moderate, photoreactivation and excision-repair are probably able to repair most damage efficiently prior to the next round of DNA replication. The subsequent duplication of the DNA molecule then proceeds unimpeded, and the damage is not passed on to the daughter helices. If for some reason, however, a dimer is not repaired before that area of DNA is replicated, different repair systems are called into play. These systems are the post-replication processes. They are most easily studied in bacteria that carry mutations for the genes that code for photoreactivating and excision-repair enzymes, since such cells are forced to utilize the post-replication repair processes.

Figure 11.28 shows a replicating double helix about to encounter a thymine dimer on one of the template strands. The dimer interferes with the reading of the template by DNA polymerase, thus blocking the elongation of the daughter strand. It is thought that in many instances of this type, the polymerase simply skips over the damaged region of the template and resumes reading the template on the other side of the dimer. Post-replication *daughter strand gap repair* can then act to repair this gap.

In the first step of this procedure, specific proteins bind to the sister double helices. These proteins mediate homologous pairing between the region of one helix that contains the gap and the corresponding region of the sister helix. Endonucleases make two cuts spaced up to 1000 nucleotides apart on the undamaged parental strand. The undamaged section is released by the cuts and fitted into the post-replication gap; the ends are sealed with ligase. Of course, a new gap is now left on the originally undamaged parental strand. Unlike the first gap, however, this new gap is opposite an undamaged region that can be used as a template by DNA polymerase I. The polymerase fills in this second gap, and a ligase seals it into place, completing the repair. The dimer is still present on one strand of one of the daughter helices, so that, in a sense, this system is not really a repair. Rather, we can view it as a means of "stalling for time" until a photoreactivating enzyme or the enzymes of the excision-repair system can correct the damage.

Daughter strand gap repair is referred to by some investigators as recombinational repair, because part of a strand is transferred in the process and because this type of repair uses some of the same enzymes that are involved in recombination. For example, mutants for the *recA* gene in *E. coli* exhibit a deficiency of recombination and an increased sensitivity to radiation; irradiated DNA that is isolated from these mutants contains an unusually large number of gaps. These findings implicate the involvement of the *recA* gene in both repair and recombination. Daughter strand gap repair also requires some biochemical functions that are not needed for genetic recombination, though, so that it is not completely analogous to a recombination process.

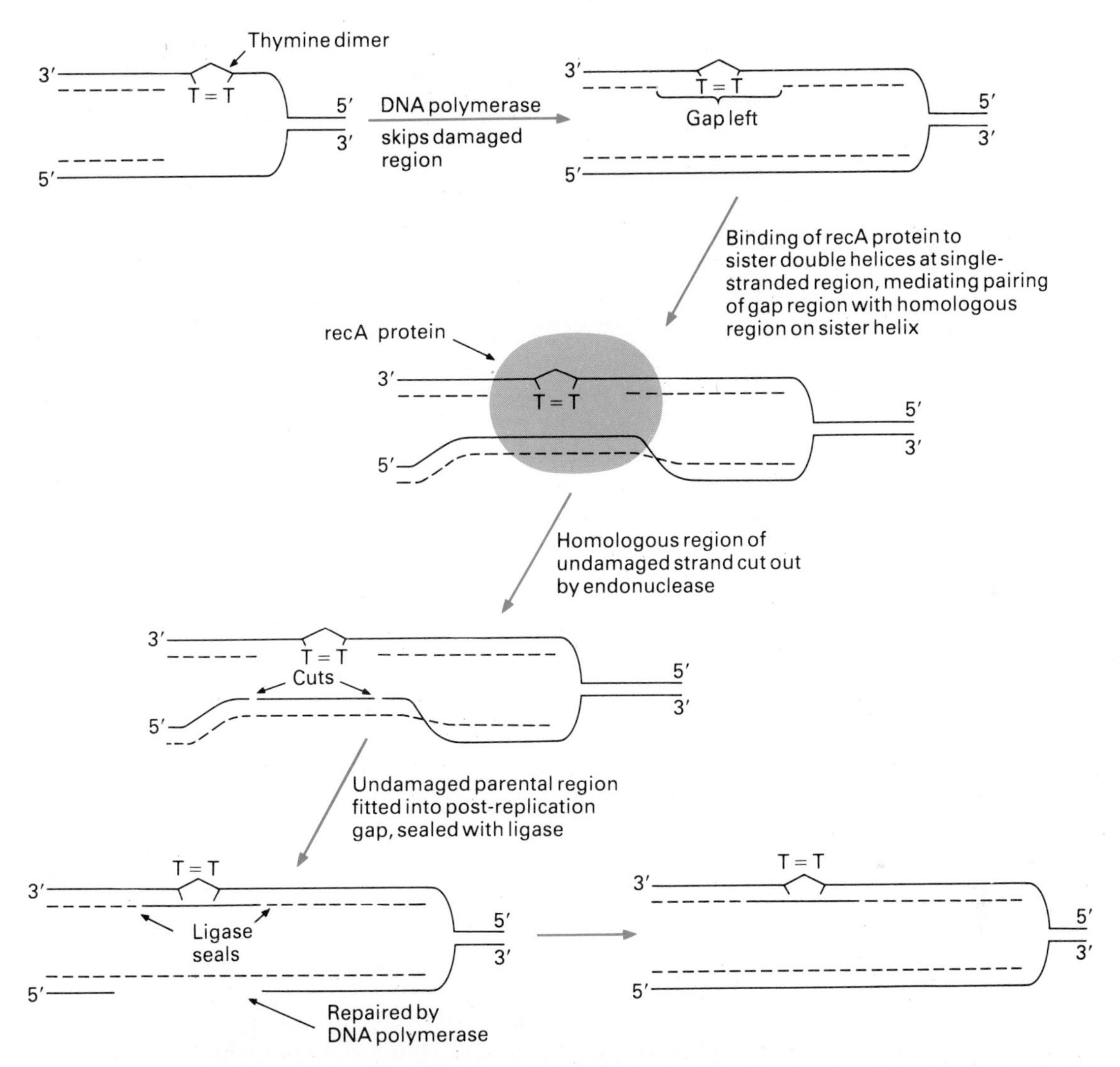

Figure 11.28. Post-replication daughter strand gap repair. The thymine dimer interferes with the reading of the template by DNA polymerase during replication. The polymerase skips over the damaged region, leaving a post-replication gap (top right). The recA protein binds to the sister double helices at the single-stranded region containing the dimer and mediates homologous pairing of the gap region with the undamaged parental strand of the sister helix. The region of the undamaged parental strand that is homologous to the damaged area is cut out by an endonuclease and is filled into the post-replication gap. It is sealed into place with a ligase (bottom left). Using the daughter strand as a template, DNA polymerase fills in the gap that was left on the undamaged parental strand, and the repair is complete.

Inducible Repair. The enzymes and other proteins that act in the three major repair systems just described are thought to be present continually in bacterial cells in low amounts, so that they are immediately available to repair moderate levels of mutagen-induced or spontaneous DNA damage. Recently, an additional level of *inducible repair response,* also known as the "SOS" response, has been discovered in *E. coli.* The various repair activities of the "SOS" response are coordinately regulated by a control molecule called the *lexA protein.* This protein, coded for by the *lexA* gene, binds to the DNA just ahead of several genes that code for repair enzymes, to the operator regions of these genes. Fig. 11.29(a) shows the binding of lexA protein to the operator regions of the *uvrA, uvrB, uvrC,* and *recA* genes, which code for some of the enzymes involved in excision-repair and post-replication daughter strand gap repair. The binding of lexA protein to these operator sites represses transcription of the adjacent genes. In this case, however, the binding of the lexA repressor protein does not completely prevent transcription, so that the repair enzymes are constantly available in the cells in the low levels needed to repair occasional spontaneous damage to DNA.

Inactivation of the lexA repressor (Fig. 11.29b) allows enhanced transcription from the *uvr* and *rec* genes, in order to make large amounts of repair enzymes available in response to mutagen exposure. Inactivation of the lexA repressor is thought to occur when the recA protein, which normally has repair and recombination activities, is changed into a protease, with protein-degrading capacity. The protease activity of the recA protein cleaves the lexA repressor molecules. We do not know what mechanism changes the function of the recA protein into a protease. Some kind of interaction may occur between the recA protein and certain short pieces of DNA that are released from the replicating chromosome when it is stalled at the site of a DNA lesion.

The "SOS" response is thought to involve a wide range of repair proteins in addition to the ones just described, all of which are under the control of the lexA repressor protein. One model constructed to explain the increased mutagenesis that follows large doses of ultraviolet light proposes that an (unknown) inducible protein under the control of the lexA repressor protein modifies one or more of the normal DNA polymerases. The modified polymerase is then able to continue replicating past the DNA lesions, in a process known as *transdimer synthesis.* The dimerized region affects the accurate reading of the template by this polymerase, however, leading to error-prone synthesis and gene mutation.

Mutagenesis from ultraviolet light in *E. coli* is probably mediated by several pathways, all of which involve errors that were made during repair processes. Mistakes made during the excision-repair process account for the majority of mutations induced by UV light. But when we consider that excision-repair is the major repair system normally operating in *E. coli,* we realize that proportionally, the frequency of errors in relationship to the amount of DNA synthesized during repair is low. The inducible repair response is apparently much more error prone than the non-induced response. Daughter strand gap repair has been found to be considerably more error-prone than excision-repair. Transdimer synthesis is very error prone, but it is not yet known how frequently or under what circumstances this misrepair system operates in wild-type bacterial cells.

Proofreading Function of Prokaryotic DNA Polymerases

The DNA polymerases of *E. coli* also act in a form of repair that operates directly at the level of DNA synthesis. In this repair process, the polymerases correct their own errors.

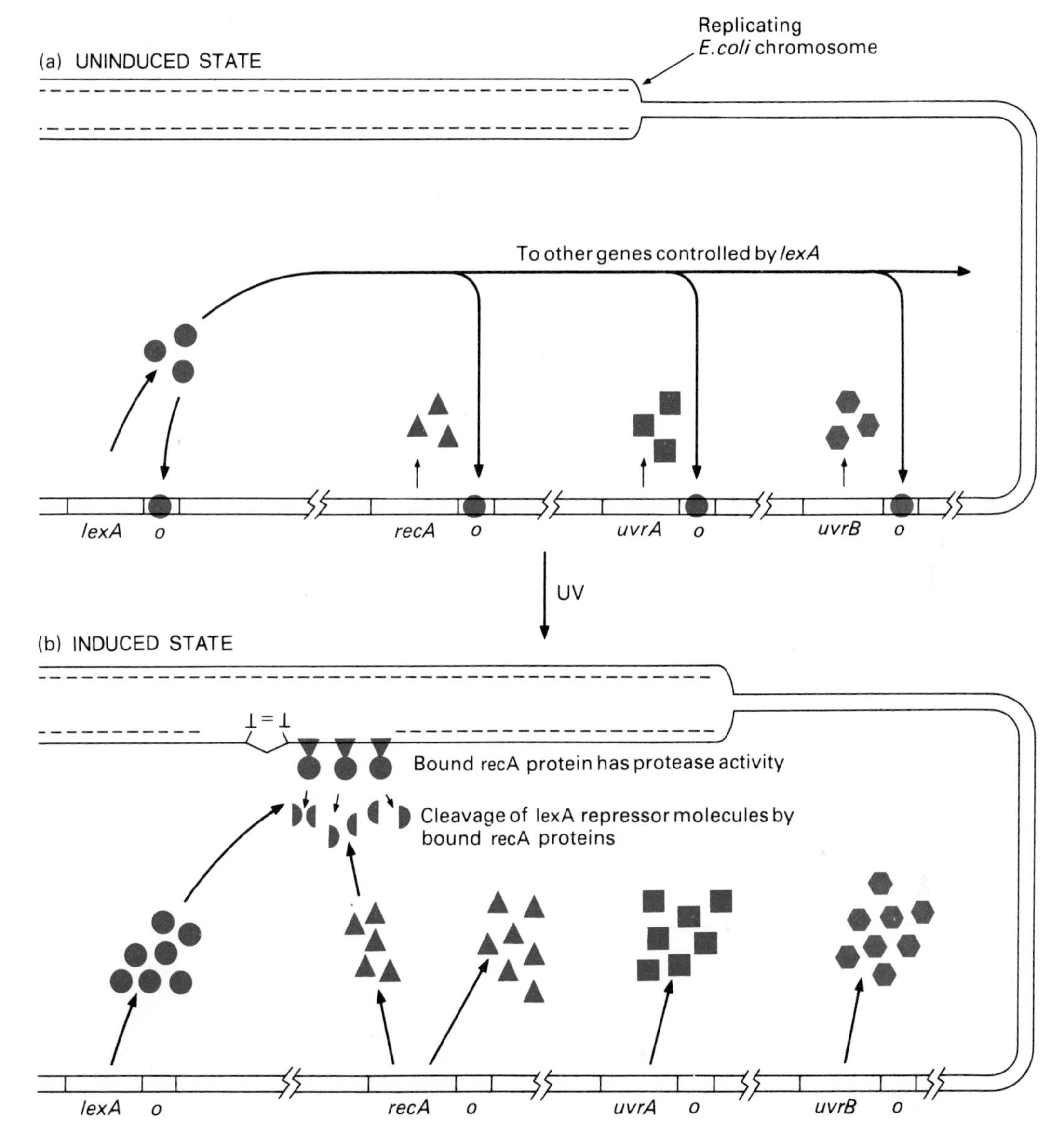

Figure 11.29. Model of an inducible repair response, in which the lexA protein controls the expression of the genes that code for the repair enzymes. (a) In the uninduced state, the lexA repressor blocks the operator sites of several repair genes, including *recA* (involved in daughter strand gap repair) and the *uvr* genes (involved in excision-repair). Blockage by the repressor is not complete, however, so that synthesis of repair enzymes still occurs in amounts sufficient to repair spontaneous damage to DNA. Since the lexA repressor blocks its own operator as well, its levels are kept low enough to ensure that the other operators are not completely blocked. (b) Replication following exposure to a mutagen leaves post-replication gaps. The binding of the recA protein to the damaged area of DNA on the template strand opposite a gap allosterically changes its activity to that of a protease. The protease inactivates the lexA repressor molecule, allowing the *recA* and *uvr* genes to be fully expressed. *Source:* After P. Howard-Flanders, *Scientific American* 245:72–103, 1981.

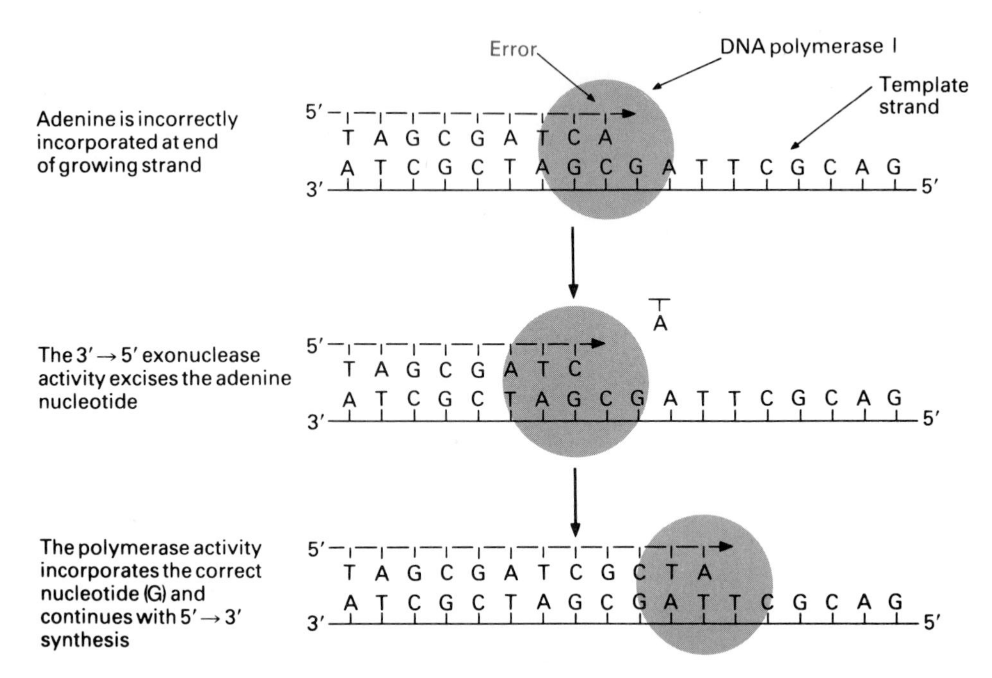

Figure 11.30. Proofreading function of DNA polymerase I. The 3′ → 5′ exonuclease activity carried out by the enzyme excises the incorrect nucleotide addition (such as the A nucleotide in the top diagram). The polymerase activity of the enzyme then guides the addition of the correct nucleotide in the 5′ → 3′ direction, using the complementary strand as a template.

In 1968, Kornberg found that DNA polymerase I possesses dual enzyme functions that allow it to edit the DNA as it is being synthesized during repair. Both polymerization and exonuclease activities are carried out by this one enzyme. Because of these dual activities, polymerase I can proofread the newly synthesized DNA and can immediately excise from the daughter strand bases that are incorrectly paired (Fig. 11.30).

The accuracy of the repair synthesis thus depends on two factors: (1) the success of the polymerase activity in selecting the proper base and (2) the ability of the enzyme to recognize and excise its own mistakes. It is now known that DNA polymerase III also possesses both polymerase and exonuclease activities. Thus, much of the spontaneous and induced mutation that occurs may indirectly result from the occasional failure of the DNA polymerases to carry out their editing function properly, either during replication itself or during repair synthesis. Support for this idea comes from in vitro studies of phage T4, in which the mutator polymerase is associated with an increased ratio of polymerase activity to exonuclease activity as compared to the wild-type polymerase. In contrast, antimutator polymerases, whose overall rate of gene mutation is lower than normal, are associated with a decreased polymerase-to-exonuclease ratio. As attractive as these ideas are, however, more work is necessary, particularly with in vivo systems, before they can be accepted as explanations for mutagenesis.

Eukaryotic Repair Systems

Various repair systems in bacteria have been known for years, yet only recently were the molecular details of their mechanisms elucidated. Repair systems other than the ones we have discussed are already known and others will undoubtedly be discovered. Even less is known about the repair processes that operate in higher organisms, although it is currently thought that eukaryotic cells employ most of the same kinds of repair systems as prokaryotes. In fact, the major pathways of DNA repair seem to be universally distributed in living cells, as would be expected if repair enzymes appeared very early during biological evolution.

It appears that the same kinds of repair enzymes that evolved to avoid mutation to auxotrophy or lethality may also be involved in repairing DNA lesions that would otherwise lead to the transformation of the cell to the cancerous state, since no additional repair pathways for this purpose have been detected in higher cells. The study of repair of eukaryotic DNA is especially important in light of the increasing evidence that damage to DNA and its repair are factors in human cancer and in the aging process. Investigators are currently studying several human disorders that may be related to deficiencies in the repair of DNA. A connection between human skin cancer and DNA repair has already been firmly established in the recessive hereditary disease *xeroderma pigmentosum* (Fig. 11.31). This disease

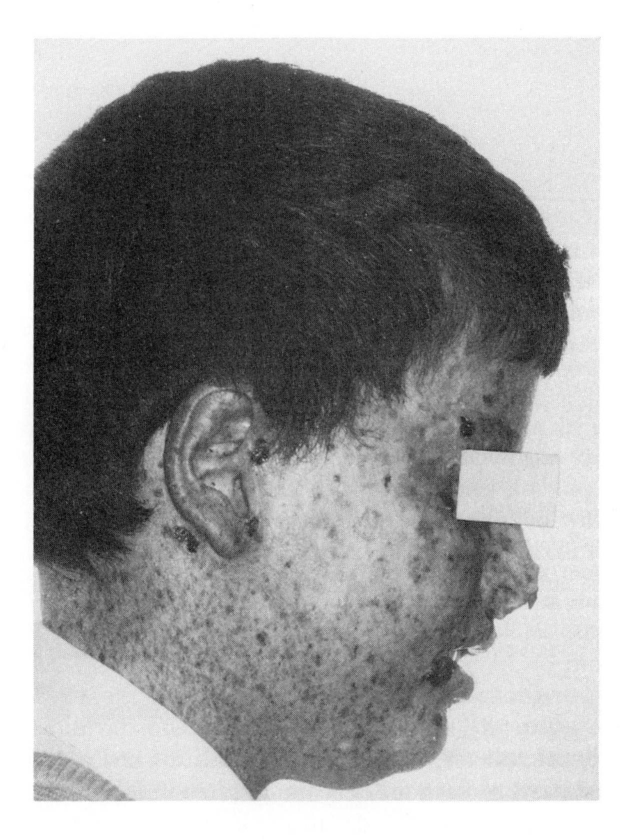

Figure 11.31. Xeroderma pigmentosum, a disease caused by a defect in DNA repair synthesis. Affected individuals exhibit extensive skin tumors in areas of skin that are exposed to ultraviolet light. *Source:* Photograph courtesy of P. E. Polani, Guy's Hospital, London.

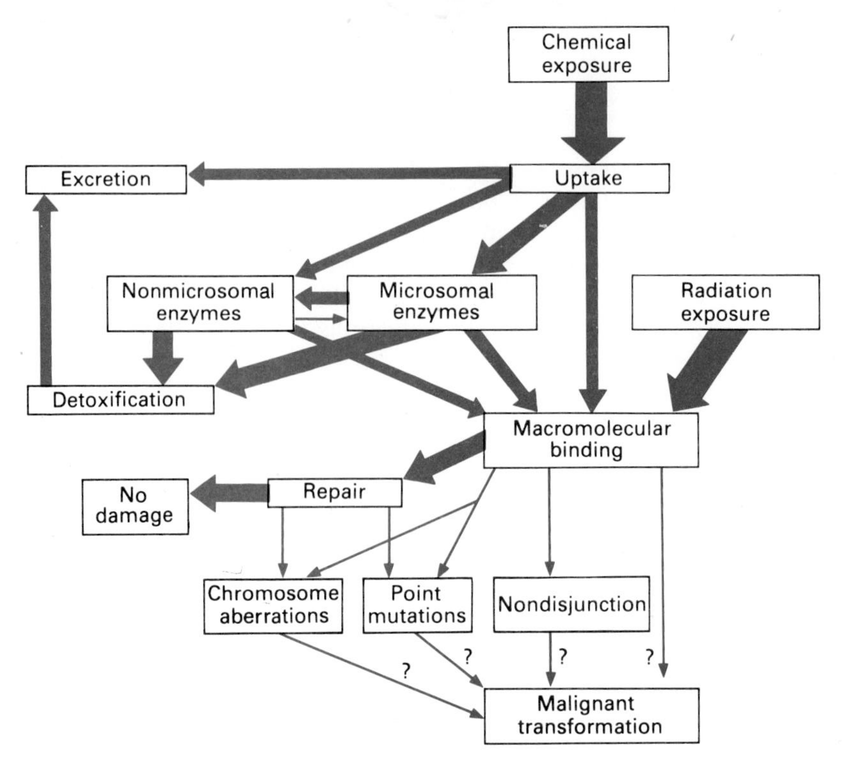

Figure 11.32. Possible sequence of reactions in a cell exposed to a foreign chemical or radiation. The heavier arrows indicate the most likely reactions. In this model, transformation to the cancerous state results from the relatively infrequent escape from repair or from misrepair. *Source:* After J. K. Selkirk and M. C. MacLeod, *BioScience*, 32:601–605. Copyright 1982 by the American Institute of Biological Sciences.

is characterized by very heavy freckling with open sores and cancers in areas of the skin that are exposed to ultraviolet light. Death almost always occurs before the individual reaches adulthood. While cultured cells that are derived from normal individuals can repair damage done by ultraviolet light to DNA, cells from individuals with xeroderma pigmentosum have a much reduced repair capacity. In some cases, the reduced capacity appears to result from a lack of functional endonuclease, thus blocking the excision-repair process. The disease is inherited as an autosomal recessive trait. It is conceivable that a mutation in any of the genes that code for repair enzymes could cause this disease.

Figure 11.32 summarizes the possible chain of events that might occur in a cell during the time between its exposure to a chemical or to radiation and its transformation to the cancerous state. In this scheme, the repair processes are an important intervening barrier that prohibits the cell from being transformed. Carcinogenesis is viewed as the result of incomplete or faulty repair.

Protective Mechanisms

Repair systems are not the only means by which organisms can defend themselves against the consequences of damage to DNA. Defense against mutation is an intrinsic property of the genetic code. Because of degeneracy, many single base substitutions do not result in an amino acid change. In fact, 24 percent of all possible base substitutions in the three positions of the genetic code are silent mutations. Furthermore, since similar amino acids tend to have similar codons, a missense mutation does not always affect the function of the polypeptide (a result known as neutral mutation). The genetic code is organized in such a way as to form a buffer against the effects of base substitutions.

Defense against the effects of mutation can also take the form of suppression. A second mutation (which might occur spontaneously or because of a mutagen) might occur that is able to suppress the mutant phenotype (not genotype) caused by a previous mutation. Both *intragenic* and *intergenic* suppression are known to exist. In an intragenic suppression, both the original and the suppressor mutation occur within the same gene locus. We can most easily illustrate this type of suppression using frameshift mutations, as shown by Fig. 11.33. In this example, the original mutation is the addition of a base pair, which shifts the reading frame one nucleotide out of phase. If a second frameshift mutation in the form of a deletion of one nucleotide should occur near the point of the first mutation, the reading frame may be restored before severe mutational damage is done. If intra-

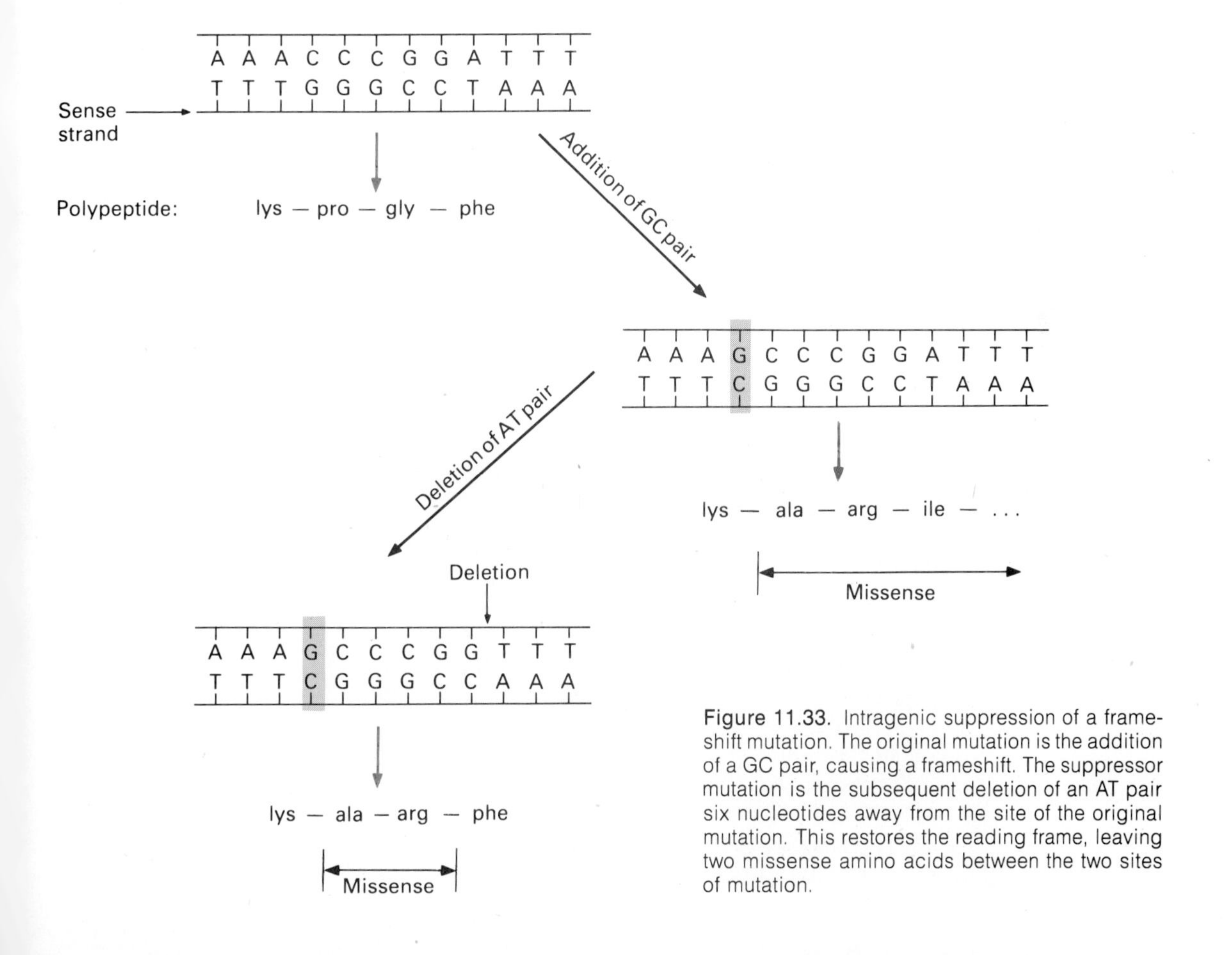

Figure 11.33. Intragenic suppression of a frameshift mutation. The original mutation is the addition of a GC pair, causing a frameshift. The suppressor mutation is the subsequent deletion of an AT pair six nucleotides away from the site of the original mutation. This restores the reading frame, leaving two missense amino acids between the two sites of mutation.

genic suppression has occurred, only those codons that are between the points of addition and deletion are mutated. We cannot predict in general how severe the effect of this stretch of code words will be. But this effect is certain to be less severe than it would have been if suppression had not taken place.

Intergenic suppression involves a second mutation that occurs in a different gene than the first. A well-known example of intergenic suppression involves *nonsense suppressors.* Nonsense suppressors are mutant tRNA genes that code for tRNA molecules with altered anticodons. The altered anticodons are able to recognize the three terminator code words within a strand of mRNA. The mutant tRNA is thus able to read a nonsense mRNA code word at the ribosome as an amino acid, so as to avoid the serious consequences of premature polypeptide termination.

EXAMPLE 11.2. Crick and his associates first suggested that the genetic code is triplet on the basis of experiments with frameshift mutations in the *rII* gene of bacteriophage T4. The mutant *rII* manifests itself phenotypically by shortening the time between the phage infection of a cell and the lysis of the cell. Thus, phages with mutant *rII* genes are referred to as rapid lysis mutants. Crick's group found that many spontaneous mutations of the *rII* locus, as well as those induced by proflavin (a particular type of acridine mutagen) involve the addition or deletion of a single base pair. These mutations can therefore be classified as either (+) or (−), for addition or deletion, respectively. Researchers can also use proflavin to induce such a mutation to revert to the wild-type phenotype. This reversion is made possible by causing a mutation at a second site (a suppressor mutation). The double mutant produced in this way contains a second mutation within the *rII* locus. A double mutant that reverts to wild-type is always of the variety (+)(−), never (+)(+) or (−)(−). Triple and higher-order mutants of various types that revert to wild-type can also be constructed. Given the following mutants and their phenotypes, explain how Crick's group arrived at the conclusion that the genetic code is read in groups of three nucleotides.

MUTANT	PHENOTYPE
(+)	rapid lysis
(−)	rapid lysis
(+)(+)	rapid lysis
(−)(−)	rapid lysis
(+)(−)	wild-type
(+)(−)(+)	rapid lysis
(−)(−)(+)	rapid lysis
(+)(+)(+)	wild-type
(−)(−)(−)	wild-type
(+)(+)(+)(+)	rapid lysis
(−)(−)(−)(−)	rapid lysis
(+)(+)(+)(+)(−)	wild-type
(−)(−)(−)(−)(+)	wild-type
(+)(+)(−)(−)(−)	rapid lysis

Solution. Each addition or deletion shifts the reading frame of the DNA so that the rest of the message is misread. The fact that a (+)(−) mutant produces the wild-type phenotype indicates that it is possible to restore the reading frame while leaving in the polypeptide a mutant section of protein that is short enough so as not to affect protein activity severely. The data also show that any combination of mutations resulting in three +'s or three −'s restores the reading frame; other combinations do not. This finding suggests that the code is read in groups of three's rather than two's, four's, or some other multiple. An example of how the various frameshift mutations in the *rII* locus might affect the reading frame is as follows:

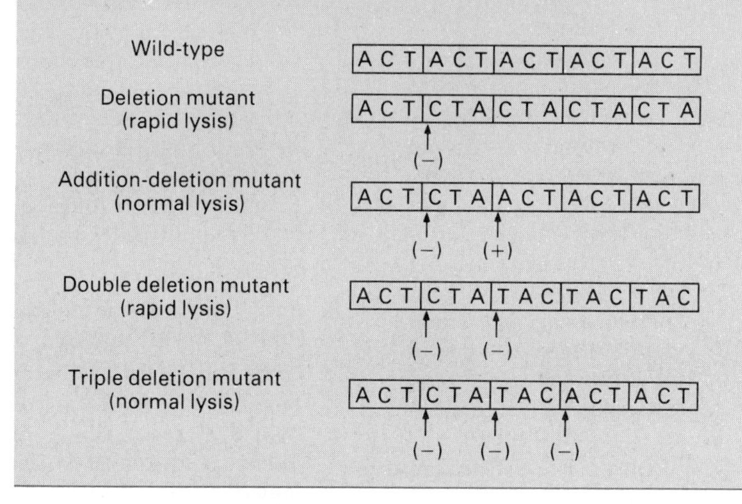

TO SUM UP

1. Both spontaneous and induced damage to DNA can usually be repaired by the cellular enzymatic systems of repair. Mutation occurs when the repair processes fail to recognize a damaged area or when they repair it incorrectly.

2. *Escherichia coli* that has been damaged by ultraviolet light is repaired by any of several enzymatic systems, including photoreactivation, excision-repair, and post-replication repair. The enzymes involved in these repair processes are the photoreactivating enzyme, DNA polymerases, exonuclease, endonuclease, and ligase, as well as other enzymes.

3. Errors in base pairing that occur during replication or repair synthesis are usually recognized by the editing function of DNA polymerase. In *E. coli,* DNA polymerase immediately cuts out a mispaired nucleotide and then inserts the correct base.

4. The degeneracy of the genetic code, the grouping of similar amino acids with respect to their codon assignments, and suppression are additional ways in which organisms protect themselves against the consequences of mutational damage to DNA.

Questions and Problems

1. Define the following terms:

Ames test	neutral mutation
base analog	nonsense mutation
carcinogen	reverse mutation
conditional mutant	silent mutation
forward mutation	suppressor mutation
frameshift mutation	teratogen
missense mutation	thymine dimer
mutagen	transitions
mutation rate	transversions

2. Explain why all base pair substitutions in DNA do not give rise to an amino acid substitution. Explain why amino acid substitutions do not always produce a protein with reduced activity.

3. If a transition mutation occurs in a codon that specifies trp, what amino acids could be substituted in place of trp? Answer this same question if a transversion occurs.

4. Hemoglobin C was the next hemoglobin variant to be discovered after HbS. It carries lysine in place of glutamic acid at the sixth position of its β chains. Postulate the mutational change in the DNA that gave rise to the HbC allele.

5. Which of the following amino acid changes can result from a single base pair substitution? **(a)** arg → leu, **(b)** cys → glu, **(c)** ser → thr, **(d)** ile → ser, **(e)** arg → asp, **(f)** pro → his.

6. List the mRNA codons that, if altered by a single base change, could give rise to chain termination during protein synthesis.

7. Categorize each of the following base substitution mutations in the DNA sense strand in terms of its effect on the amino acid sequence of the encoded polypeptide: **(a)** CCT → CCA, **(b)** CCT → CAT, **(c)** CCT → ACT, **(d)** CTA → CTC, **(e)** AAC → GAC, **(f)** AGT → GGT, **(g)** ACC → ATC.

8. Aniridia, a form of blindness caused by the absence of the iris, is an autosomal dominant trait that shows full penetrance. Of 4,664,799 persons reported born in the state of Michigan during the period 1919–1959, 41 were aniridic offspring of normal parents. What is the mutation rate to aniridia? Could the mutation rate of a recessive mutant gene be estimated by this same kind of study? Explain.

9. The navel orange and the Delicious apple both arose as the result of somatic mutation. In each case, a spontaneous mutation occurred in a single cell that, through successive mitotic divisions, produced an entire branch with the characteristics of the mutant type. Both the navel orange and the Delicious apple now enjoy widespread popularity. Why has it been possible to take advantage of the products of somatic mutation in plants, while such products in animals are uncommon?

10. Show that the base analog 2-aminopurine causes AT ⇌ GC transitions.

11. Diagram the mutational consequences of rare AG pairing. Which of the transversion events have we not been able to explain by purine-purine pairing?

12. Suppose you are studying a protein, part of whose amino acid sequence is

trp-lys-ala-arg-thr-val

Several mutants for the gene that specifies this protein are isolated and the amino acid sequences of their proteins are determined:

mutant 1: arg-lys-ala-arg-thr-val
mutant 2: trp-met-ala-arg-thr-val
mutant 3: trp-lys-val-arg-thr-val
mutant 4: trp-lys-ala
mutant 5: trp-lys-cys-ser-asn-gly
mutant 6: trp-lys-cys-ser-thr-val

(a) Determine the mutation event at the level of DNA that gave rise to each mutant allele. **(b)** Give the most probable nucleotide sequence along the sense strand of the wild-type gene. **(c)** How many base substitution mutations are possible in the DNA that codes for this polypeptide?

13. Among a group of men who work in a small town at a chemical factory, two have sons that are affected with juvenile muscular dystrophy. In both cases, the disease has not previously occurred in either the man's or his wife's known ancestry. They feel that two cases of this same rare disease is more than just a coincidence and initiate action to sue the company for damages. Do you think their suits are justified? Explain.

14. An investigator irradiates *Drosophila* gametes with different doses of X-rays. A dose of 2000 r produces sex-linked lethal mutations in 6 percent of the treated gametes. Based on your knowledge of the relationship between mutation rate and radiation dose in *Drosophila,* give the expected frequency of sex-linked lethals induced by each of the following X-ray doses: **(a)** 1000 r, **(b)** 4000 r, **(c)** 5000 r.

15. The Ames test can be used either to determine the mutagenicity of a chemical substance or to analyze a spontaneous mutation for a particular type of mutational lesion. Consider its use in characterizing spontaneous mutations. An investigator exposes each of four spontaneously mutant bacterial strains (1, 2, 3, and 4) to a series of mutagens, to see if reversion will occur. The results are given below, where a + indicates reversion and a − indicates no reversion:

		MUTAGEN			
		HA	BU	EMS	ACRIDINE
BACTERIAL STRAIN	1	+	+	+	−
	2	−	−	+	−
	3	−	−	−	+
	4	−	+	+	−

Considering the known modes of action of these mutagens, describe the nature of each original spontaneous mutational event. Be as specific as possible.

16. Summarize the distinguishing features of the three major repair processes that occur in bacteria, paying special attention to the specific enzyme activities involved in each process.

17. Several instances have been recorded in which all of the offspring of two albino parents are normally pigmented. Explain how this occurrence is possible.

18. Of the more than 140 mutational forms of hemoglobin besides HbS that have been discovered so far, two of these forms are hemoglobin C and hemoglobin Hopkins-2 (Ho-2). When an individual who carries both the *S* and *C* genes marries a homozygous normal individual, each of their offspring will have either the sickle cell or the hemoglobin C trait (recall that having the trait as opposed to the disease refers to the heterozygous condition). By contrast, when an individual who carries both the sickle cell and Hopkins-2 genes marries a homozygous normal individual, four different types of offspring can be produced: some are homozygous normal, others have either the sickle cell or the Hopkins-2 trait, and still others have both traits. Explain the genetic bases of the HbC and Ho-2 variants in relation to the sickle cell gene.

19. A certain species of extraterrestrial being normally has a red glow to its body. Two variant types also occur, however, which are caused by recessive mutations in nonallelic genes located on different chromosomes. Mutant gene *s* gives a bright scarlet glow, whereas mutant gene *b* imparts a brown glow to the organism. The metabolic pathway for synthesis of glow pigment has recently been discovered:

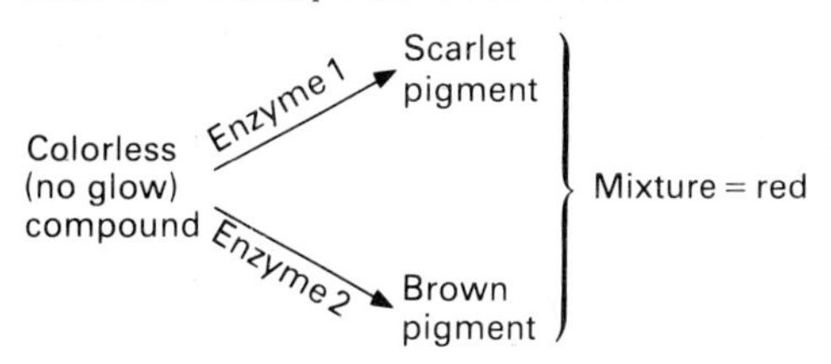

(a) Which enzyme is nonfunctional in scarlet individuals? Which is nonfunctional in brown individuals? **(b)** One particular mating between a scarlet and a brown individual gives only red offspring. Give the genotypes of the scarlet and brown parents and of the red progeny. **(c)** If the red offspring from the mating described in part (b) intercross, what phenotypic classes and proportions are expected among their progeny?

20. In a certain plant, flower color is usually purple. Both blue and white mutant variants are known, with the metabolic pathway for pigment production as follows:

$$\text{white} \xrightarrow{\text{enzyme A}} \text{blue} \xrightarrow{\text{enzyme B}} \text{purple}$$

(a) What color is expected of a pure-breeding plant that is unable to catalyze the first reaction? **(b)** What color is expected of a pure-breeding plant that is unable to catalyze the second reaction? **(c)** If the plant described in part (a) is crossed to the plant described in part (b), what will be the flower color of the F_1? (d) From intercrosses of these F_1, what phenotypic ratio is expected in the F_2? (e) Explain why flower color and the white clover trait discussed in Chapter 1 give different phenotypic ratios in the F_2.

21. At two of the many loci that affect eye color in *Drosophila* are the autosomal recessive genes *pr* (purple) and *st* (scarlet). These genes are located on nonhomologous chromosomes. Wild-type flies (of genotype $pr^+/-\ st^+/-$) have red eyes, whereas flies that are homozygous for both mutant genes (of genotype *pr/pr st/st*) have white eyes. Construct a metabolic pathway for pigment synthesis that accounts for these phenotypes, and indicate which gene locus specifies the enzyme that catalyzes each step of the pathway.

22. Gene *A* codes for an enzyme that is needed to catalyze the formation of a flower pigment in a particular diploid plant. Its mutant allele, *a*, codes for an inactive enzyme that can no longer combine with its substrate. Gene *A* is said to be completely dominant, since the phenotype of *AA* homozygotes is indistinguishable from that expressed by *Aa* heterozygotes. When studies were conducted to determine the enzyme levels present in *AA* and *Aa* plants, it was discovered that under normal conditions, the amount of active enzyme (estimated by the quantity of substrate converted to product per minute by tissue extracts) was the same in both genotypes. When large amounts of a particular effector molecule, which is required by this enzyme for activity, were added to enzyme samples taken from *AA* and *Aa* individuals, the enzyme activity increased in both samples; but the amount of substrate converted by the tissue extracts from *AA* plants was twice that converted by corresponding extracts from *Aa* plants. **(a)** Explain the possible molecular basis for complete dominance in this case. **(b)** A mutation in a different gene is known that converts allele *A* to an incompletely dominant gene. Describe the probable molecular basis of this mutation in light of your previous explanation for complete dominance.

23. Enzyme X is a protein molecule whose activity depends on the presence of an activator substance. Only when the activator is present in the cell is protein X functional as an enzyme. Protein X and the activator substance are coded for by loci that assort independently and show complete dominance:

gene *A* = activator present

allele *a* = no activator

gene *B* = protein X

allele *b* = no protein X

If enzyme X catalyzes the formation of a pigment from a colorless precursor compound, describe the phenotypic ratio that would result from this gene interaction system in the dihybrid cross *AaBb* × *AaBb*.

24. Enzyme Y, which catalyzes the synthesis of a different pigment from a colorless precursor compound, is active only in the absence of a repressor substance. The independently assorting genes that are responsible for the repressor and for protein Y are as follows:

gene *C* = repressor present

allele *c* = no repressor

gene *D* = protein Y

allele *d* = no protein Y

What phenotypic ratio will result among the offspring of a dihybrid cross in this case?

25. A certain pigment is the metabolic end product of either of two separate reactions, as follows:

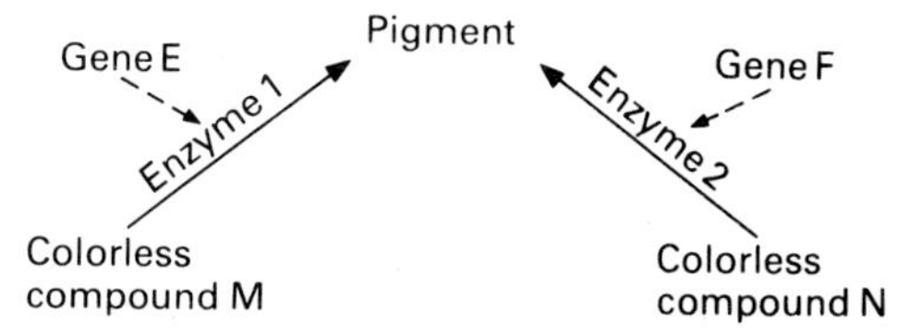

(a) If enzymes 1 and 2 each catalyze the production of exactly one half the amount of pigment that the cell can use, what phenotypic ratio will result from a dihybrid cross? (Assume that the intensity of the coloration is directly proportional to the amount of pigment produced.) **(b)** If enzymes 1 and 2 each catalyze the production of a saturating amount of pigment (meaning that only one of the genes need be active to obtain full amounts of end product), what phenotypic ratio will result from a dihybrid cross?

26. Of the three gene pairs that influence color in a certain organism (*A,a*; *B,b*; and *C,c*), one is a target gene (the locus for color), one is a dominant receptor element that mutates the target gene, and one is a dominant regulatory gene that causes mutational instability of the target gene. From the information given below, match each gene with its particular function.

GENOTYPE	PHENOTYPE
Aabbcc	no color
AaBbcc	no color
AaBbCc	spotted
aaBbcc	solid color
aaBbCc	solid color
aabbCc	no color

Chapter 12

Chromosome Abnormalities

Thus far we have implicitly assumed that chromosomes behave normally during mitosis and meiosis, so that the karyotype of a species remains the same from one generation to the next. Irregularities do occur, however, and they can lead to changes in the number and arrangements of genes, chromosomes, or even whole genomes. In this chapter, we will look at these microscopically visible deviations from the norm, which are referred to as *chromosome mutations.* These abnormalities may occur spontaneously as rare aberrations, or they may be induced by human intervention.

Geneticists generally recognize two major types of chromosome mutations: (1) changes in the number of chromosomes and (2) changes in the structure of chromosomes. Changes in the number of chromosomes can be further subdivided into **aneuploidy,** which is the loss or addition of less than a complete set of chromosomes, and **euploidy,** in which the number of chromosomes varies by a multiple of the whole genome. While each specific type of aberration within these categories is relatively rare, chromosome mutations taken as a group are quite common. For example, it is possible that 10 percent or more of all human zygotes have some type of chromosome defect. Although most chromosome mutations in higher organisms are probably lethal, many of the aberrations that are compatible with life lead to alterations in phenotype. These mutations are therefore an important source of variation and have undoubtedly played an important role in evolution.

Aneuploidy

Nature and Production of Aneuploids

Aneuploid organisms have a chromosome complement in the somatic cells that is not an exact multiple of the haploid number. Several possibilities exist. Two of the most common are individuals with one additional chromosome ($2n + 1$) or with one less chromosome ($2n - 1$) than normal. Individuals with an additional chromosome, called **trisomics,** have three copies of one of their chromosomes, while individuals missing a chromosome, called **monosomics,** have only one copy of a certain chromosome and two copies of all the remaining chromosomes. These variations are compared with the normal $2n$ condition, which serves as the standard of reference.

With regard to chromosome number, trisomic and monosomic individuals tend to produce more than one kind of gamete. In trisomics, for example, the chromosomes that are present in triplicate arrange themselves on the equator of the meiotic spindle so that two chromosomes will move to one pole, and one will move to the other pole (Fig. 12.1). Segregation results in two kinds of gametes: one kind consisting of n chromosomes, and the other kind consisting of $n + 1$ chromosomes. Thus, if both gamete types are equally viable and fertile, a mating between a trisomic and a normal individual is expected to produce equal numbers of trisomic ($(n + 1) + (n) \rightarrow 2n + 1$) and normal ($(n) + (n) \rightarrow 2n$) offspring.

The chromosome abnormalities that are characteristic of aneuploidy can arise as a result of a phenomenon known as **meiotic nondisjunction.** Meiotic nondisjunction is the failure of chromosomes to separate during meiosis. Figure 12.2 illustrates the consequences of nondisjunction in a single pair of chromosomes. Note that nondisjunction can occur at either the first or the second meiotic division. If it occurs at anaphase I, all four products of meiosis are abnormal: two have $n + 1$ and two have $n - 1$ chromosomes. If these abnormal gametes unite with normal gametes during fertilization, both trisomic and monosomic offspring will result. On the other hand, if nondisjunction occurs in anaphase II, only two of

Figure 12.1. A 2:1 chromosome segregation pattern in a trisomic. During meiosis I, two of the homologs migrate to one pole of the spindle, and one homolog migrates to the other pole.

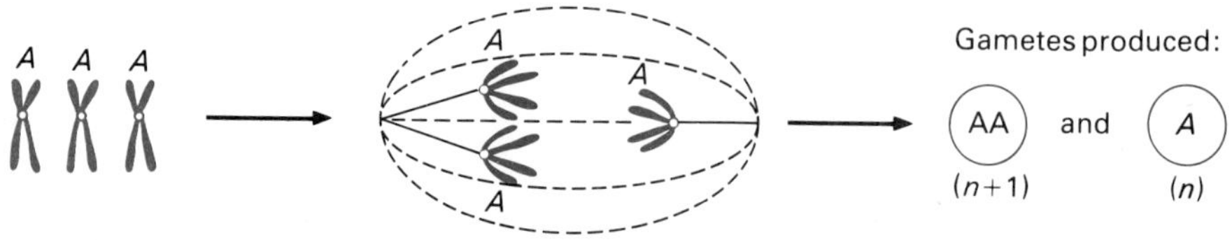

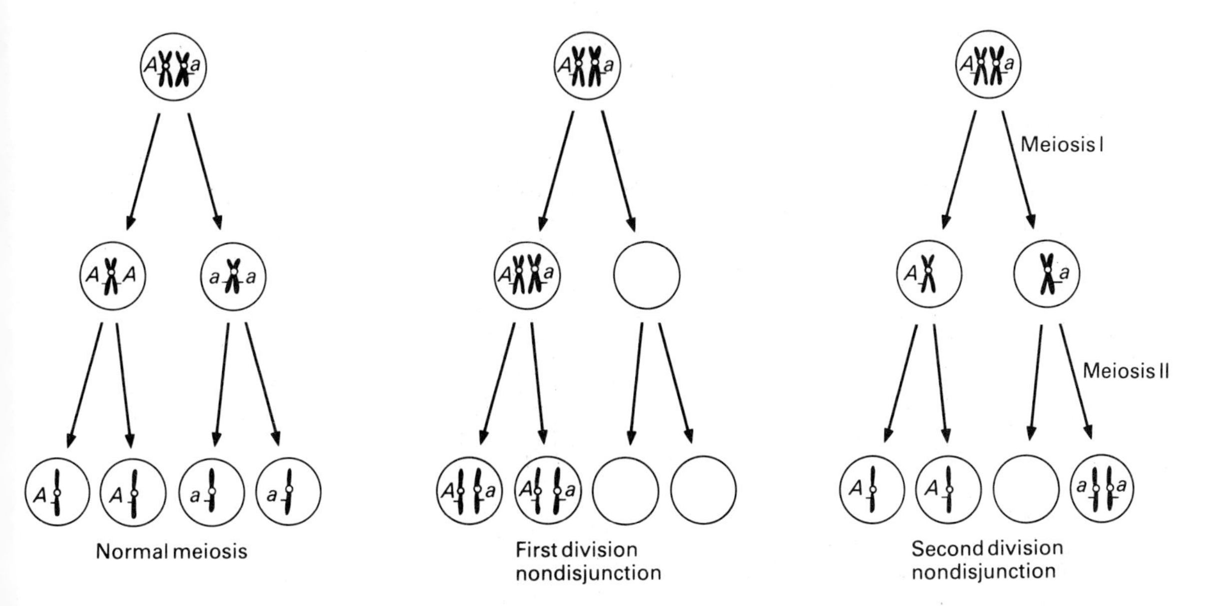

Figure 12.2. The results of chromosomal nondisjunction at the first and second meiotic divisions. All other chromosomes are assumed to segregate normally. First division nondisjunction results in all four gametes being abnormal in chromosome makeup. Two are disomic ($n + 1$) and two are nullisomic ($n - 1$). In second division nondisjunction, only two of the four potential gametes are abnormal, one being disomic and another being nullisomic.

the meiotic products will be abnormal. The two abnormal products will lead to both $n + 1$ and $n - 1$ gamete types, once again.

Genetic Analysis of Trisomics. Geneticists have studied aneuploidy quite extensively in a number of flowering plants, such as Jimson weed, corn, tomatoes, and wheat, and in the fruit fly, *Drosophila*. In the Jimson weed, *Datura stramonium*, researchers have identified 12 different trisomics, which correspond to the 12 different chromosomes that make up the haploid set in this organism. Each of the 12 trisomics has its own characteristic phenotype. For example, an extra dose of a certain chromosome produces large fruit capsules with long spines, while an extra dose of another chromosome yields small, depressed capsules, and so on. The fruit capsules of the normal diploid strain and of the 12 possible trisomics are illustrated in Fig. 12.3.

Genetic studies on trisomic strains of *Datura* reveal that crosses involving trisomic plants do not result in the standard Mendelian ratios. Consider, for example, the purple locus on chromosome 9. At this locus, gene *P* (for purple color) is completely dominant to its recessive allele *p* (for white). Crosses that involved only purple-flowered plants with the *PPp* genotype produced offspring at a ratio of 17 purple : 1 white. How might such a ratio be explained?

Recall that in trisomics, two homologs normally move to one pole of the spindle at anaphase I and one moves to the other pole. Four genotypes are therefore possible in the gametes produced by this cross: (*PP*), (*Pp*), (*P*), and (*p*), with the $n + 1$ and n gamete types occurring in equal numbers. Since two of the three homologs carry the *P* allele, the ratio of (*P*) : (*p*), and hence of (*Pp*) : (*PP*), should be 2:1. The overall ratio in the gametes should therefore be 1 (PP) : 2 (Pp) : 2 (P) : 1 (p).

This theoretical ratio does occur among the eggs in *Datura*. But the only microspores in this organism that are functional are the haploid ones. Therefore, only haploid pollen is produced, at a ratio of 2 (P) : 1 (p). A *PPp* × *PPp* cross in *Datura* is thus expected to yield white-flowered (*pp*) and purple-flowered (P-- and

Figure 12.3. Fruit capsules of the normal diploid strain of *Datura stramonium* and of the 12 possible trisomics. The numbers 1 through 12 correspond to the twelve different chromosomes making up the haploid set of *Datura*. *Source:* From A. F. Blakeslee, *J. Hered.*, 25:87, 1934.

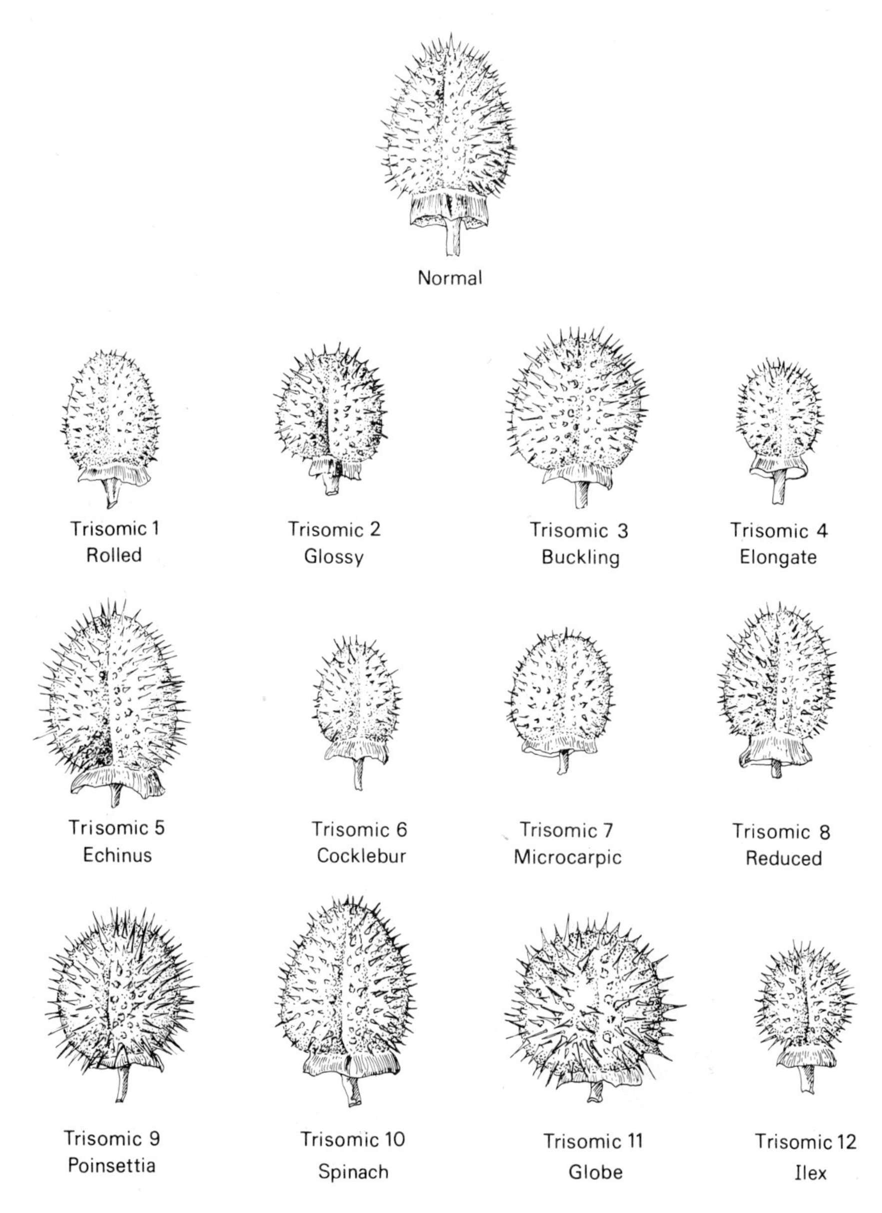

Cross: Purple PPp × Purple PPp

	eggs (1/6) (*PP*)	(2/6) (*Pp*)	(2/6) (*P*)	(1/6) (*p*)
Pollen (2/3) (*P*)	(2/18) *PPP*	(4/18) *PPp*	(4/18) *PP*	(2/18) *Pp*
Pollen (1/3) (*p*)	(1/18) *PPp*	(2/18) *Ppp*	(2/18) *Pp*	(1/18) *pp*

Genotypic ratio: 2 *PPP* : 5 *PPp* : 2 *Ppp* : 4 *PP* : 4 *Pp* : 1 *pp*

Phenotypic ratio: 17 Purple (*P*-- and *P*-): 1 white (*pp*)

Figure 12.4. Diagram of cross between two purple-flowered trisomic strains of *Datura* with the genotype *PPp*. The gamete ratio among eggs conforms to that expected if $n + 1$ and n gamete types occur in equal numbers. Among pollen, however, only the n gametes are functional.

P-) offspring in the proportions of $\frac{1}{6}\ p$ (in eggs) $\times\ \frac{1}{3}\ p$ (in pollen) $= \frac{1}{18}\ pp$, and $1 - \frac{1}{18} = \frac{17}{18}$ *P*-- and *P*-. The results of this cross are analyzed further in Fig. 12.4. Other crosses, yielding different phenotypic ratios, are considered in Example 12.1 and in the problems at the end of the chapter.

EXAMPLE 12.1. Predict the ratio of purple-flowered (*P*-- and *P*-) and white-flowered (*ppp* and *pp*) *Datura* plants from the cross *Ppp* × *Ppp*.

Solution. Since two of the three homologs in these trisomic plants carry the *p* allele, the gamete ratio among the pollen is 2 (*p*) : 1 (*P*). The gamete ratio among the eggs is 1 (*pp*) : 2 (*Pp*) : 2 (*p*) : 1 (*P*). We can then predict the results of the cross as follows:

	1 (*P*)	2 (*Pp*)	2 (*p*)	1 (*pp*)
1 (*P*)	1 *PP*	2 *PPp*	2 *Pp*	1 *Ppp*
2 (*p*)	2 *Pp*	4 *Ppp*	4 *pp*	2 *ppp*

These results yield $\frac{12}{18}$ (*P*-- and *P*-) : $\frac{6}{18}$ (*ppp* and *pp*), which reduces to 2 purple : 1 white.

Associating Linkage Groups with Particular Chromosomes. Aneuploids have been used in genetic mapping studies to associate linkage groups with particular chromosomes. Chromosome assignments can often be made in plant species that are normally diploid by using trisomic or monosomic varieties. In this case, the assignment is made by comparing the observed phenotypic ratio with the expected ratio for trisomic or monosomic inheritance. For example, supppose that the recessive *r* allele is located on one of the chromosomes that appear in triplicate in a trisomic individual that has the genotype *RRr*. This plant can then produce four genotypes among the eggs, in a ratio of 1 (*RR*) : 2 (*Rr*) : 2 (*R*) : 1 (*r*). If we cross this plant with pollen donated by an *rr* individual (*RRr* × *rr*), then the progeny would occur in a phenotypic ratio of 5 dominant (*R*-- and *R*-) : 1 recessive (*rr*). If the female plant instead had only two alleles at the *r* locus, the testcross would be *Rr* × *rr*, producing a ratio in the offspring of 1 dominant (*Rr*) : 1 recessive (*rr*). Thus, if various trisomic plants, each of which has a different extra chromosome and all of which are heterozygous for a single *r* allele, are crossed with pollen donated by an *rr* individual, we can make a chromosome assignment on the basis of which trisomic variety gives a phenotypic ratio nearest 5:1 in the offspring.

Aneuploidy in Humans

Most severe alterations of the normal human karyotype result in a condition that is incompatible with life. Comparisons between the observed number of abnormalities and the expected number bear out this relationship. For example, 24 kinds of trisomics and 24 kinds of monosomics are possible in humans, one for each of the 22 different autosomes and one for each sex chromosome, X and Y. Of these 48 possibilities, only seven have been observed with any regularity among liveborn infants (Table 12.1). The chromosomally defective fetuses that survive to term are predominately trisomic for a sex chromosome or for one of the smaller autosomes. Individuals with such a defect are usually born with some collection of phenotypic abnormalities, known as a *syndrome.* Trisomy and monosomy of the larger autosomes do occur. But these other chromosomal abnormalities create such severe developmental abnormalities that they result in spontaneous abortion, usually long before the time of normal birth.

Table 12.2 lists the frequencies in spontaneous abortions of trisomies for the seven different chromosome groups of humans. The expected frequencies are based on the percentage of chromosomes that are in each group. In contrast to trisomy, most monosomies are rarely seen in aborted fetuses. It is possible that their effects are so severe that most $2n - 1$ zygotes are lost even before they can be detected as miscarriages.

Autosomal Aneuploids. *Trisomy 21* (*Down syndrome*) is the most common of the autosomal aneuploids. It is also the only form of autosomal aneuploidy in which the individual survives through infancy with any regularity. Down syndrome produces certain characteristic features such as short stature, a rounded

Table 12.1. Human aneuploids found in live births.

CHROMOSOME ABNORMALITY	SYNDROME	ESTIMATED FREQUENCY
Trisomy 21	Down syndrome	1 in 700 births
Trisomy 18	Edwards syndrome	1 in 4000 births
Trisomy 13	Patau syndrome	1 in 5000 births
47, XXX	Triple-X syndrome	1 in 1000 female births
47, XXY	Klinefelter syndrome	1 in 1000 male births
47, XYY	—	1 in 1000 male births
45, XO	Turner syndrome	1 in 5000 female births

Table 12.2. Observed frequencies of trisomies for the different chromosome groups in spontaneous abortions.

CHROMOSOME GROUP*	OBSERVED FREQUENCY (% OF TOTAL CASES)	EXPECTED FREQUENCY (% OF TOTAL CASES)†
A (chromosome pairs 1–3)	3	14
B (chromosome pairs 4–5)	1	9
C (chromosome pairs 6–12 and X)	16	31
D (chromosome pairs 13–15)	22	14
E (chromosome pairs 16–18)	37	14
F (chromosome pairs 19–20)	3	9
G (chromosome pairs 21–22 and Y)	18	9

* Chromosome groups are listed in order of decreasing size, from A (largest) through G (smallest).
† Expected frequencies are based on the total number of chromosomes in each group.

face and broad head, and epicanthal folds (folds of skin on the upper eyelid) (Fig. 12.5). Individuals with Down syndrome are also likely to have various physical abnormalities, including heart disease and an increased susceptibility to respiratory infections. Most individuals with this disorder have cheerful, affectionate dispositions. Unfortunately, many suffer from severe mental retardation. A few affected females have produced offspring; half of the offspring were affected with Down syndrome, and half were normal. This result is exactly what we would expect when equal numbers of eggs with 23 and 24 chromosomes are produced. Down syndrome is not restricted to humans; it has also been observed in a chimpanzee, in whom the phenotypic effects were similar to those of affected humans.

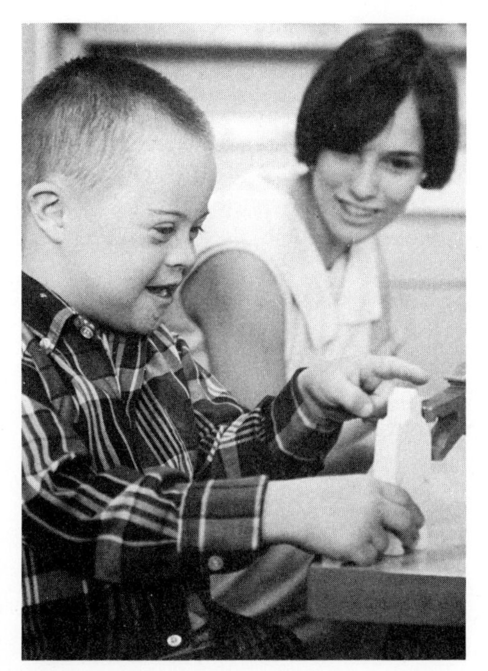

Figure 12.5. (a) A young boy with Down syndrome. (b) G-banded chromosomes of a female with Down syndrome. (See page 520 for discussion of G bands.)

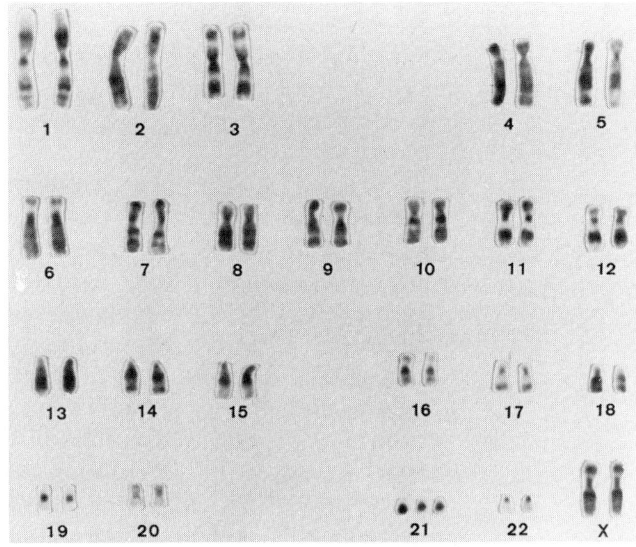

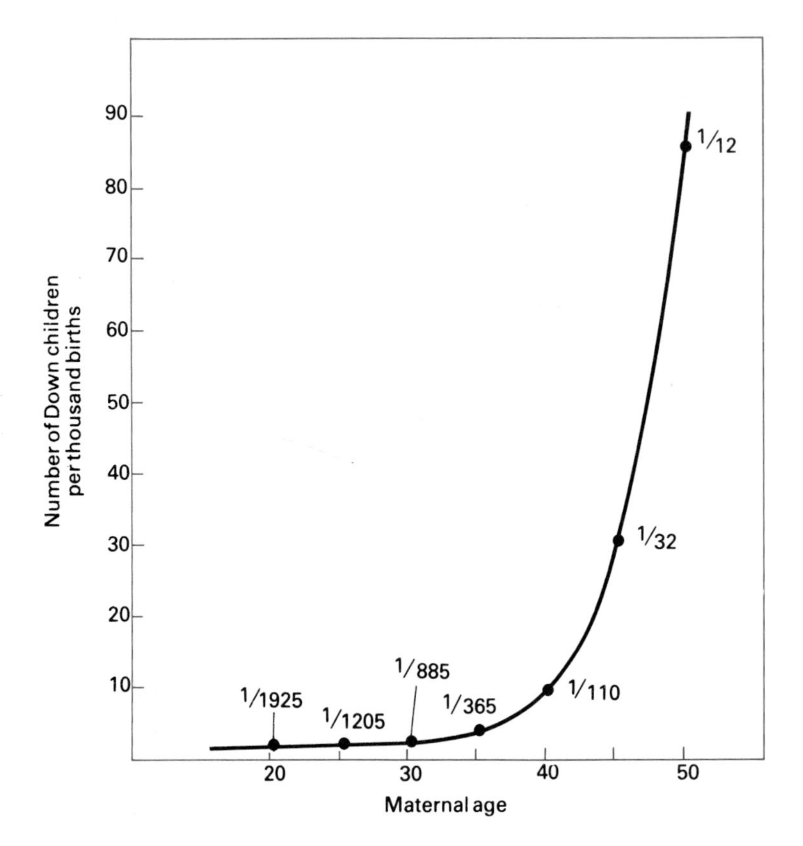

Figure 12.6. Estimated incidence of Down syndrome for different maternal ages. The risk of having a child with Down syndrome increases dramatically in older mothers. *Source:* Data from E. B. Hook and G. M. Chambers. Estimated rates of Down syndrome in live births by one year maternal age intervals for mothers aged 20-49 in a New York State study—Implications of the risk figures for genetic counseling and cost-benefit analysis of prenatal diagnosis programs. In D. Bergsma and R. B. Lowry (eds.): "Numerical Taxonomy of Birth Defects and Polygenic Disorders." New York: Alan R. Liss for The National Foundation-March of Dimes, BD:OAS XIII (3A):123-141, 1977.

The two other forms of autosomal aneuploidy listed in Table 12.1, *trisomy 18* (Edward syndrome) and *trisomy 13* (Patau syndrome), result in severe abnormalities that have a number of features in common. Children born with these disorders are severely retarded in mental and physical development and have multiple malformations, including low set malformed ears, defective eyes, flexion deformities such as clenched fists, and heart defects. Survival of individuals with these syndromes is very limited; most of the affected individuals die by the age of six months.

Researchers have shown that the incidence of these autosomal aneuploidies increases with the age of the mother. This correlation is illustrated in Fig. 12.6 with data on Down syndrome. As we can see from this figure, the relationship to maternal age is curvilinear, rising sharply at later ages.

We do not yet know why nondisjunction is correlated with maternal age. Some geneticists believe that the relationship develops from the fact that a human female at birth has all the potential eggs that she will ever produce, arrested in prophase I. The oocytes remain in this stage until she reaches puberty; from then on, one oocyte completes meiosis I during each succeeding menstrual cycle. Thus, the oocytes that are released during ovulation in a 40-year-old woman will have remained in a "dormant" state for almost twice as long as those released in a 20-year-old woman. Because the time delay before ovulation is dependent on age, the cumulative chance of damage to chromosomal microtubules by viruses, radiation, and other environmental agents is expected to rise with age. The exposure to these damaging agents presumably contributes to a higher rate of nondisjunction in the oocytes of older women.

An increase in the incidence of nondisjunction with paternal age has also been recorded. The factors that contribute to the correlation with paternal age are unknown.

Sex Chromosome Aneuploids. Aneuploids for the X chromosome generally have less severe phenotypic effects than those for the autosomes. The lesser effect is believed to be a consequence of the process of Lyonization, in which all X chromosomes except for one become inactivated, as we have seen. Nevertheless, inactivation of the additional X chromosomes is not complete, and all aneuploids for the X chromosome suffer physical defects and mental retardation to varying degrees.

One of the first abnormalities involving the X chromosome to be described was *monosomy X* (*Turner syndrome*), with a karyotype formula of 45,XO. Although individuals with Turner syndrome are phenotypically female, they have only one X chromosome and they can therefore be diagnosed by the absence of a Barr body in their cell nuclei. The physical symptoms of this abnormality include short stature, infantile genitalia, sparse pubic hair, wide-spaced and underdeveloped breasts, and webbing of the neck. Individuals with Turner syndrome are usually sterile because their ovaries have failed to develop properly, and they do not menstruate. The low incidence of this disorder (1 in 5000 female births) is a consequence of its high rate of fetal mortality. Over 90 percent of XO fetuses spontaneously abort.

There are two conditions that arise as a result of there being one too many X chromosomes: one has a karyotype formula of 47,XXY (*Klinefelter syndrome*), and the other has a karyotype formula of 47,XXX (*triple-X syndrome*). Like autosomal trisomies, the risk of giving birth to both XXY and XXX children increases with the age of the mother. Individuals with Klinefelter syndrome are male in that they have a Y chromosome, but they are also sex-chromatin positive in that they have a Barr body. These individuals typically have longer than average limbs, sparse body hair, and some degree of breast development. Many are mentally retarded and are sterile because their testes are underdeveloped.

Triple-X individuals are female, but with two Barr bodies. The behavioral and physical expressions of this disorder are quite variable. Some triple-X females are of normal intelligence and fertility, which indicates that an additional X chromosome does not necessarily lead to an abnormal condition.

In addition to abnormalities in the number of X chromosomes, individuals may also possess an abnormal number of Y chromosomes. Researchers have recently focused on individuals with 47 chromosomes and an XYY sex chromosome condition. These individuals are males who tend to be taller than their counterparts with a normal chromosome number, and they often have a history of severe facial acne. Several authors have linked the presence of an additional Y chromosome with criminal behavior, because some studies claim to have shown that the frequency of XYY persons in mental and penal institutions is greater than

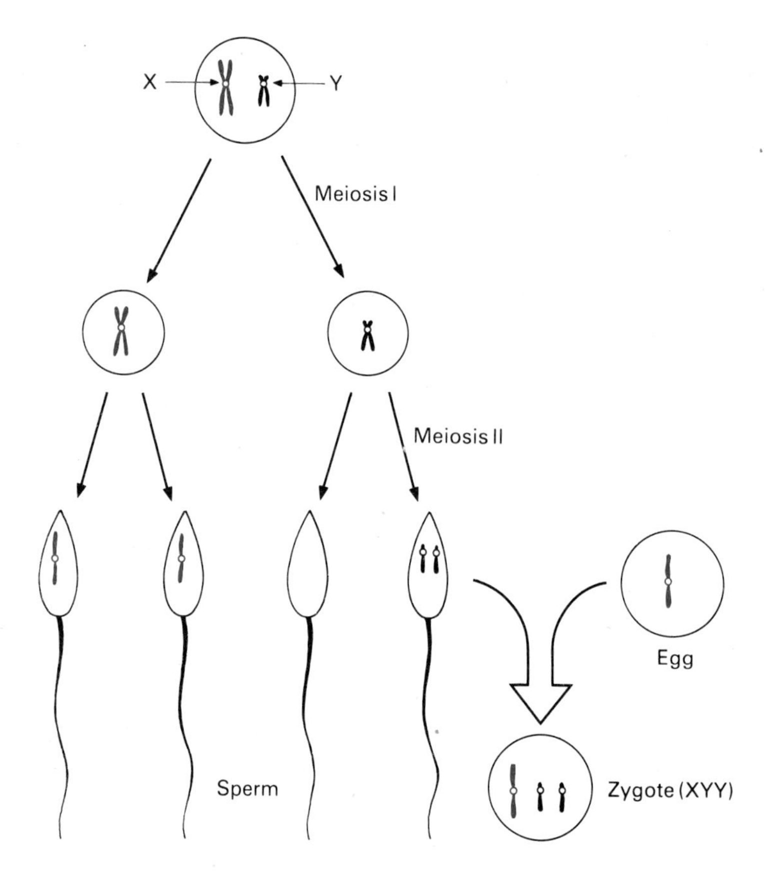

Figure 12.7. Origin of XYY male through second division nondisjunction of the Y chromosome.

in the general population. The actual relationship, if any, between an extra Y chromosome and antisocial behavior is still not known, however. Many, if not most, XYY males appear to lead normal and productive lives. Zygotes of the XYY type most likely form when a normal egg fuses with a sperm bearing two Y chromosomes, resulting from nondisjunction at anaphase II in a chromosomally normal male (Fig. 12.7).

Chromosome Mosaics

Aneuploid chromosome conditions can also be detected in individuals who have a mixture of aneuploid and normal cell lines. These individuals, known as **chromosome mosaics,** frequently possess abnormal phenotypes. In *Drosophila,* for example, sexual mosaics (called **gynandromorphs**) can occur that are part male and part female in sex phenotype. Bilateral gynandromorphs, such as the one shown in Fig. 12.8, develop when an XX zygote loses one of the X chromosomes during the first cleavage division, so that one cell in the two-cell stage is XX and the other is XO. Since one of the two cells divides to produce the tissues of the right half of the body and the other produces the tissues of the left half, the XO/XX mosaic that results is half male (XO) and half female (XX) in appearance.

Chromosome mosaics may develop in one of two ways. First, they may result from the failure of chromosomes to separate during mitosis. Such **mitotic non-**

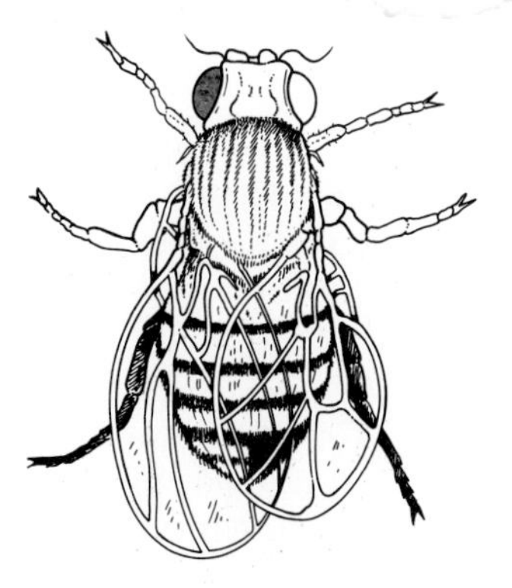

Figure 12.8. A bilateral gynandromorph of *Drosophila melanogaster.* Such individuals are female on one side of the body and male on the other. The female side of the bilateral mosaic (recognizable in this case by the red compound eye) consists of normal XX cells, whereas the male side (recognizable by the white eye) is derived from an XO cell line. Note the dark colored male tip of the abdomen (right).

disjunction tends to give rise to three cell types: one normal, a second deficient, and a third with an extra chromosome (Fig. 12.9). This process can result in a large amount of abnormal tissue if it occurs at an early stage of embryological development. A second mechanism for the formation of chromosome mosaics is *anaphase lag.* In anaphase lag, a chromosome fails to migrate properly during mitotic anaphase and is not incorporated into a daughter nucleus. In contrast to nondisjunction, anaphase lag will produce only two cell types, one of which has the normal chromosome constitution and the other of which is monosomic for the lost chromosome.

Investigators have reported several interesting examples of mosaics with two and three cell lines in humans. Mosaics involving the sex chromosomes in humans include the forms XX/XO, XX/XO/XXX, XY/XXY, and XXY/XX/XXYY. Individuals with one of the first two forms tend to show mild symptoms of Turner syndrome,

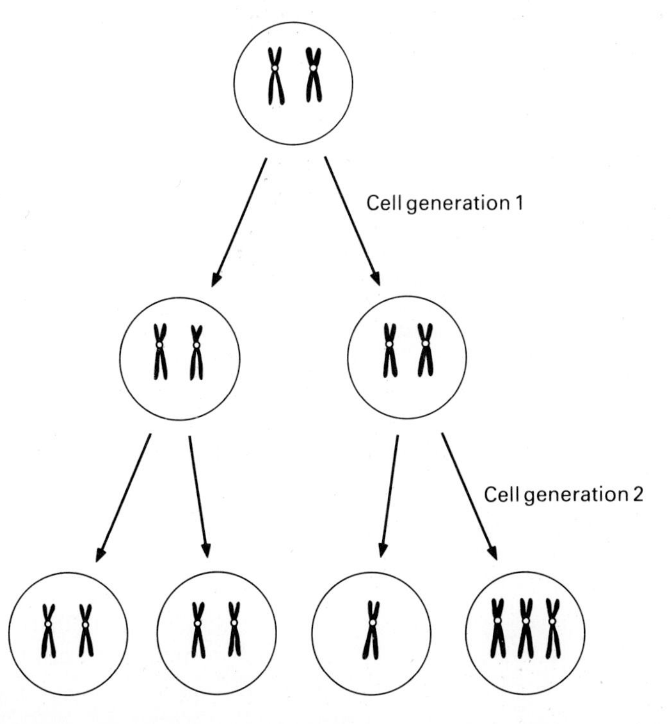

Figure 12.9. Mitotic nondisjunction. Three cell types are formed: one normal, a second missing a chromosome, and a third with an extra chromosome.

while individuals with one of the latter two forms tend to have the appearance of Klinefelter syndrome. Mosaics with the XX/XY constitution have been reported in some hermaphrodites, which accounts for the development of both ovarian and testicular tissues. Hermaphrodites of this type are rare in humans; they tend to have ambiguous external genitalia and are ordinarily sterile, because of rudimentary ovotestes.

Autosomal mosaics have also been reported, including several individuals who had a mixture of normal and trisomy 21 cells. Individuals who are mosaics for trisomy 21 usually show only mild symptoms of Down syndrome.

TO SUM UP

1. Aneuploid organisms have a somatic chromosome complement that is not an exact multiple of the haploid number. Two of the most common types of aneuploids are trisomics (with one additional chromosome) and monosomics (with one less chromosome than normal).
2. Nondisjunction is the failure of chromosomes to separate during meiosis or mitosis. Meiotic nondisjunction can occur in either the first or the second meiotic division. It leads to the production of gametes with one more or one less chromosome than normal. If the abnormal gamete unites with a normal gamete during fertilization, an aneuploid state can result in all the cells of the offspring.
3. In some plants, such as the Jimson weed, different trisomic strains are viable and can be used in genetic crosses. Since trisomics have three copies of each gene carried on the affected chromosome, the segregation of alleles in heterozygotes produces results that differ from the standard Mendelian ratios.
4. Aneuploids have been used for associating a gene with a particular chromosome. The altered inheritance pattern of the gene is correlated with a particular monosomy or trisomy condition, allowing assignment of that gene to the chromosome involved in the aneuploidy.
5. Only three autosomal aneuploids occur with any reasonable frequency among liveborn human infants. These conditions are trisomy 21 (Down syndrome), trisomy 18 (Edwards syndrome), and trisomy 13 (Patau syndrome). Researchers have shown that the incidence of all three conditions increases with the mother's age.
6. Unlike most autosomal aneuploids, aneuploids for the sex chromosomes in humans are usually viable. The four most common types of sex chromosome abnormalities are the monosomic condition 45,XO (Turner syndrome), and three trisomies: 47,XXY (Klinefelter syndrome), 47,XXX (triple-X syndrome), and 47,XYY.
7. Individuals known as chromosome mosaics consist of mixtures of aneuploid and normal cell lines. Chromosome mosaics are formed by mitotic nondisjunction or by anaphase lag, which involves the failure of a chromosome to migrate properly during mitosis and to become incorporated into a daughter nucleus.

Euploidy

Euploid organisms have a somatic chromosome complement that is an exact multiple of the basic set, or genome. Thus, if we use the symbol x to represent the genomic number, the euploid series comprises the chromosome states of monoploid (x), diploid ($2x$), triploid ($3x$), tetraploid ($4x$), and so on. In plants of the rose genus, for example, the basic set consists of 7 chromosomes, but various species with 14, 21, 28, 35, 42, and 56 chromosomes are known. In this case, 14 is the diploid number, 21 is the triploid number, 28 is the tetraploid number, and so on, each constituting a basic multiple of 7 in the euploid series.

Changes in the euploid state from the normal condition arise when complete sets of chromosomes are lost or added. These changes can occur spontaneously as rare aberrations, or they can be induced by artificial means. In this section, we will discuss the nature of these changes and their applications to humans.

Changes in the Euploid State

Monoploids. Monoploids are individuals that possess only the haploid set of chromosomes in a species that is normally diploid. Monoploids can arise from unfertilized gametes (as do the monoploid males of bees, wasps, and ants) and are ordinarily smaller and less vigorous than their diploid counterparts. Monoploid organisms are typically sterile. (There are exceptions; meiosis can be bypassed in some monoploid organisms, such as male honeybees, which produce gametes by mitotic division.) The sterility of monoploids is due to the production of gametes that contain less than the haploid number of chromosomes. Such gametes are nonfunctional. Chromosome pairing is impossible in monoploids, since each chromosome occurs singly at meiosis. Chromosome migration at anaphase I is therefore disorganized. Thus when the chromosome number is large, the chance that a normal haploid spore or gamete will be produced is extremely small, and most of the products of meiosis are missing one or more chromosomes.

Monoploids are of special interest to plant geneticists, who derive them by artificial means. The technique of anther culture, which is summarized in Fig. 12.10, has been used for this purpose. In this procedure, haploid cells produced by meiosis in the anthers of a plant are induced to grow in culture in the presence of certain plant hormones. This growth results in multicellular structures called **embryoids,** which resemble normal stages in the embryonic development of the plants. With the proper treatment and care, the embryoids develop into small

Figure 12.10. Anther culture, an artificial means of generating a monoploid plant. Haploid cells derived from meiosis in the anthers of a plant are induced by hormones to grow into multicellular haploid embryoids. These embryoids can then be stimulated to grow into small plantlets and whole plants.

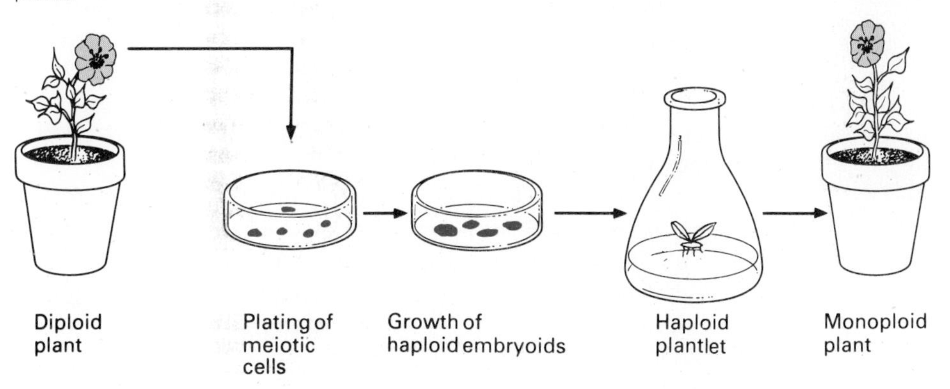

plantlets, which can then be used to form the mature monoploid plant with roots, stems, leaves, and flowers.

The plants produced by this method are sterile. But geneticists may wish to reproduce certain of these plants that have a desirable combination of genes. Monoploid plants can be induced to form diploid tissue when their growing parts are treated with the alkaloid **colchicine** (a drug extracted from the autumn crocus plant) to arrest cell division at the metaphase stage. This drug interferes with the formation of spindle fibers. The chromosomes of cells treated with colchicine are unaffected during interphase and will duplicate properly. But the cytoplasmic phase of cell division fails to occur, resulting in a doubling of the chromosome number in the growing cells. When tissue treated in this way is propagated vegetatively (that is, by asexual means), fertile diploid plants can form that are homozygous for all the genes of the monoploid parent. This procedure can therefore be used to establish stable, true-breeding lines in a very short time. We will consider this and other applications of anther culture more fully in Chapter 17.

Autopolyploids. Euploid organisms that possess three or more complete sets of chromosomes are referred to as **polyploids.** Polyploids are usually classified according to the origin and nature of their multiple sets of chromosomes. They are called **autopolyploids** when their sets of chromosomes are multiples of the same genome (i.e., homologous sets).

The number of homologous sets increases from the duplication of the basic chromosome complement of a single species. This duplication can come about in different ways. Irregularities in meiosis can lead to the production of unreduced diploid gametes. Let us suppose, for example, that in a mating between diploid parents ($2x \times 2x$), one parent produces a diploid ($2x$) gamete, owing to complete nondisjunction, while the other produces a gamete that has the normal haploid (x) complement of chromosomes. The result is a triploid ($3x$) offspring. If meiotic irregularities occur in both parents, the offspring produced could have $4x$ chromosomes or an even higher level of ploidy.

Autotetraploids in plants may also result from the process of **somatic doubling.** In somatic doubling, mitotic division is aborted, so that there is a doubling of the number of chromosomes without a corresponding cell division. Somatic doubling is usually accompanied by some form of asexual reproduction (such as runners or stolens) which enables the tetraploid somatic tissue to form an entirely new plant. Researchers can also induce tetraploidy through somatic doubling artificially by applying colchicine to diploid plant tissue. Colchicine interferes with the formation of the mitotic spindle so that cells with four chromosome sets are produced. The resulting tetraploid tissue is then propagated vegetatively by means of slips or grafts.

There are three important genetic consequences of autopolyploidy. First, an increase in the number of chromosomes is often accompanied by an *increase in cell size.* This increase in turn leads to an increase in the size of plant parts that are commercially important (flowers, fruit, leaves, etc.). Consequently, many familiar food and crop plants that are marketed are autopolyploids or at least have some polyploid varieties.

A second consequence of autopolyploidy is a *reduction in fertility* relative to diploid organisms of the same species. The reduction in fertility is particularly evident in triploids and other odd-numbered types ($5x$, $7x$, etc.). Because of inherent reproductive difficulties, such polyploid varieties are normally restricted to organisms that reproduce asexually. They usually encounter no difficulty during mitosis, since each chromosome behaves independently on the mitotic spindle. The problem that occurs during meiosis results from an unequal distribution of the extra sets of chromosomes during anaphase I. These odd-numbered polyploids tend to produce nonfunctional spores or gametes that contain unbalanced sets of chromosomes, with either an excess or a deficiency of chromosomes.

Let us consider meiosis in triploids. Three homologs are present in triploids for each kind of chromosome. Two homologs of a kind move to one pole of the meiotic spindle and one homolog moves to the other pole. This mechanism is the same as in trisomy, except that in trisomy only one type of chromosome is involved (see Fig. 12.1), whereas in a triploid all the chromosome triplets are involved. Each product of meiosis in a triploid organism thus has a ½ chance of receiving only one copy of a particular chromosome. Since each triplet of homologs undergoes disjunction independently of the other triplets, the chance that a fully balanced haploid (or diploid) spore or gamete is formed is equal to $(1/2)^x$ where x is the number of chromosomes in the haploid complement. Observe that when x is large, $(1/2)^x$ is a very small value. Therefore, most of the spores or gametes produced by a triploid organism are genetically unbalanced. These unbalanced products are usually nonfunctional and fail to participate in reproduction. Even if they do participate, the development of the zygote is usually impaired. In either case, the triploid organism is effectively sterile. Plant breeders have used to advantage the sterile nature of triploid plants by developing seedless varieties such as the common banana and the seedless watermelon (Fig. 12.11).

Figure 12.11. A diploid variety of watermelon (Sugar Bush VP), left, compared with a triploid seedless variety, right. *Source:* Courtesy of the W. Atlee Burpee Co.

(a)

(b)

EXAMPLE 12.2. The diploid number of chromosomes in the watermelon is 22. The seedless variety of watermelon is a triploid, with 33 chromosomes. Assuming that functional spores must contain a haploid or diploid complement of chromosomes, what is the probability that the triploid variety will produce a functional seed?

Solution. The chance that a spore of the triploid watermelon will receive a single copy of a chromosome is ½ for each kind of chromosome in its complement. The chance of receiving two copies of a chromosome is also ½ for each kind of chromosome. Hence, the chance that a spore is haploid, having received one copy of all chromosomes is $(1/2)^x$. This is also the chance that the spore is diploid, having received two copies of all x chromosomes. The overall probability of a spore's receiving a balanced complement of chromosomes is therefore

$$
\begin{aligned}
P(\text{balanced complement}) &= P(\text{haploid complement}) + P(\text{diploid complement}) \\
&= (1/2)^x + (1/2)^x \\
&= 2(1/2)^x \\
&= (1/2)^{x-1}
\end{aligned}
$$

The haploid number of chromosomes in the watermelon is $x = 1/2(22) = 11$. The probability that the plant produces a spore with a balanced complement of chromosomes (and thus a functional seed) is then $(1/2)^{10} = 1/1024$.

In contrast to varieties with an odd number of chromosome sets, organisms with an even number of sets have a much better chance of being fertile. In certain tetraploids, for example, meiotic pairing gives rise to a 2:2 segregation pattern, in which two homologs of a particular chromosome move to one pole of the spindle and two move to the other pole. These tetraploids are interfertile, in that their mating produces fertile tetraploid offspring (Fig. 12.12). They are also reproductively isolated from their diploid ancestors, since matings between a tetraploid and a diploid result in sterile triploid progeny. These matings have a practical use, however, as they are a ready source of the commercially desirable triploid varieties. For example, seeds from a tetraploid × diploid cross can be used to grow seedless watermelons.

Figure 12.12. Consequences of 2:2 chromosome segregation in autotetraploids. Tetraploids will produce diploid gametes and are interfertile, whereas matings between tetraploids and diploids tend to yield triploid offspring, which are sterile.

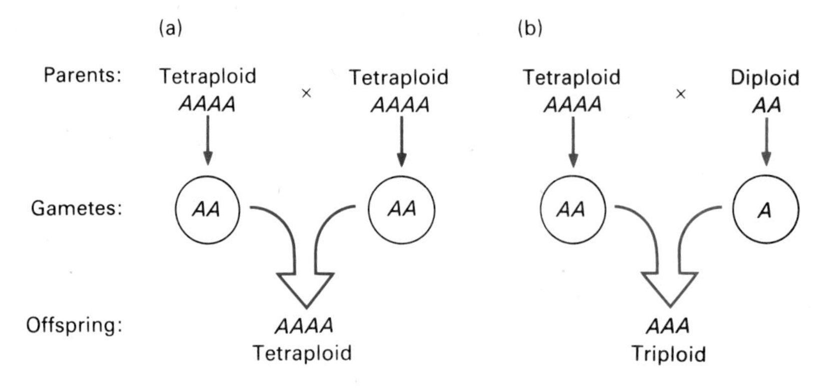

A third important consequence of polyploidy is a *decline in the frequency of recessive phenotypes,* by comparison to the frequency in the diploid varieties. Polyploidy thus reduces the chance that deleterious recessive genes are expressed in a population. To illustrate, let us consider the pair of alleles *A,a* that are present in heterozygous form in the tetraploid *AAaa.* Let us further assume a perfect 2:2 segregation pattern, so that the alleles segregate in pairs in random assortment. What then is the probability that a gamete of this tetraploid is a fully recessive (*aa*)? To follow the alleles better in this case, we will label them $A_1A_2a_3a_4$. Note that there are 4!/2!2! = 6 different combinations of alleles that can arise from disjunction: (A_1A_2), (A_1a_3), (A_1a_4), (A_2a_3), (A_2a_4), and (a_3a_4). Of these combinations, only (a_3a_4) includes both recessive alleles. The probability that a gamete of this tetraploid is fully recessive is therefore 1/6. In the mating *AAaa* × *AAaa,* we would expect the frequencies of recessive (*aaaa*) and dominant (*A---*) offspring to be:

$$P(aaaa) = P(aa)P(aa) = (1/6)(1/6) = 1/36$$
$$P(A---) = 1 - P(aaaa) = 1 - 1/6 = 35/36$$

Thus, given a perfectly random 2:2 segregation pattern of alleles and complete dominance, a monohybrid cross in tetraploids will yield a phenotypic ratio of 35:1. We can compare this ratio with the 3:1 ratio expected for the corresponding *Aa* × *Aa* cross in diploids. Since homozygotes among tetraploids are comparatively rare, we can see that the additional sets of chromosomes would act to hide the expression of any deleterious recessive genes. This masking effect is even further enhanced in organisms of higher ploidy. Overall it permits the load of deleterious recessive genes in the population to build up, since these genes are so rarely expressed.

EXAMPLE 12.3. Consider the pair of alleles *A* and *A'*, which are lacking in dominance, so that each genotype expresses a different phenotype. Determine the number of phenotypic classes produced by a cross between the tetraploids *AAA'A'* × *AAA'A'*, and predict the expected frequency of each class. (Assume a perfectly random 2:2 segregation pattern of alleles.)

Solution. The cross will yield five different phenotypic classes, corresponding to the five different genotypes: *AAAA, AAAA', AAA'A', AA'A'A'*, and *A'A'A'A'*. We can deduce the relative proportion of each class by means of a Punnett square. First, recall that each tetraploid of this genotype can produce six combinations of alleles. These combinations form three genotypes in the gametes, in the ratio 1 (*AA*) : 4 (*AA'*) : 1 (*A'A'*). Random fertilization produces the following possible zygotic combinations:

	1 (*AA*)	4 (*AA'*)	1 (*A'A'*)
1 (*AA*)	1 *AAAA*	4 *AAAA'*	1 *AAA'A'*
4 (*AA'*)	4 *AAAA'*	16 *AAA'A'*	4 *AA'A'A'*
1 (*A'A'*)	1 *AAA'A'*	4 *AA'A'A'*	1 *A'A'A'A'*

These results yield a genotypic, and thus a phenotypic, ratio of 1 *AAAA* : 8 *AAAA'* : 18 *AAA'A'* : 8 *AA'A'A'* : 1 *A'A'A'A'*. Note that with incomplete dominance, polyploids can maintain a greater amount of phenotypic variation than can diploids.

While relatively common in plants, autotetraploids are quite rare among animal species. In animals, most naturally occurring polyploids are either hermaphroditic (such as earthworms) or can reproduce parthenogenetically (that is, they can produce progeny without fertilization).

The polyploid state tends to be lethal in humans and other mammals. Almost all recorded cases of polyploidy in humans have resulted in spontaneous abortions. The few known live-born infants that seemed to be polyploid actually turned out in most cases to be mosaics, with both normal and polyploid cell lines.

One possible explanation for the deleterious effect of polyploidy in animals is the imbalance that it creates between the autosomes and the sex chromosomes (recall that in mammals, all but one sex chromosome is genetically inactivated early in embryonic development). The imbalance disturbs the normal ratio of genes that determine the sex of the individual, resulting in intersexes and sterile individuals.

Allopolyploids. The second major type of increase in the number of genomes is through the formation of **allopolyploids.** When the multiple sets of chromosomes in a polyploid are initially derived from different species (that is, they are nonhomologous sets), the resulting organism is known as an allopolyploid.

Many plant species are capable of producing hybrids from the mating of two different species. These hybrids are often sterile, however, because they lack the genetic homology needed for accurate meiotic pairing, which in turn provides an equal distribution of chromosomes to the gametes. Fertile allopolyploids can be produced from these sterile hybrids by means of chromosome doubling. This mechanism results in an allotetraploid (also called an **amphidiploid**), which has a diploid number of chromosomes from both parental species. Each chromosome then has a partner for normal pairing, and meiosis can proceed in the regular way.

Amphidiploids are altogether separate species from their diploid parents. They are fertile among themselves, but are reproductively isolated from all other species, including those that gave rise to them initially. Figure 12.13 summarizes the sequence of events that can produce an amphidiploid.

An interesting example of the development of an amphidiploid is the origin of the plant *Raphanobrassica.* This plant was initially derived from hybrids formed by cross-pollinating radishes (*Raphanus sativus*) and cabbages (*Brassica oleracea*). Hybrids produced by this mating, which have 18 chromosomes (9 from the radish and 9 from the cabbage), are normally sterile. But the breeders discovered a hybrid that was not sterile. This interfertile hybrid, which was called *Raphanobrassica* after its two parent species, turned out to be an amphidiploid, with two sets of radish chromosomes and two sets of cabbage chromosomes. The species is regarded as somewhat of a novelty, but unfortunately it has no direct economic importance, since it has the root of a cabbage and the leaves of a radish.

The distinction between allopolyploids and autopolyploids is not always as clearcut as the preceding discussion might imply. Studies reveal that chromosome sets are not always strictly homologous (with identical pairing affinities at meiosis) or nonhomologous (with no pairing affinity). Varying degrees of similarity exist, depending upon the closeness of the relationship between the parent species. If the species are closely related in their evolution, the chromosomes they contribute

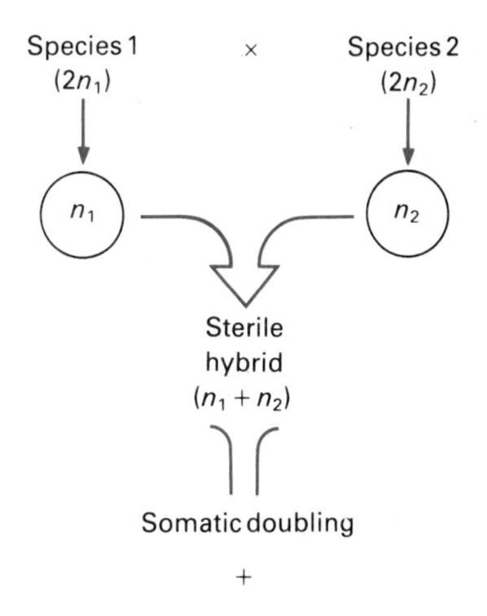

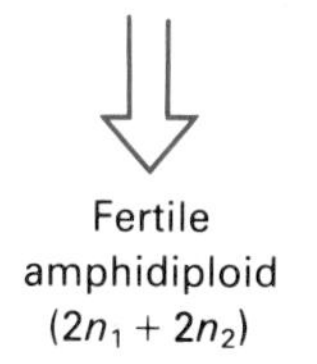

Figure 12.13. The possible origin of an amphidiploid in a plant species.

to the allopolyploid will not have had sufficient time to diverge completely in character and will be likely to show some amount of similarity. Chromosomes that are derived from different species but still retain some partial similarity are called **homeologous** (meaning similar) as opposed to *homologous* (meaning identical).

Genetic similarities in different sets of chromosomes can lead to pairing irregularities in allopolyploids during meiosis, with homeologs of different sets competing with homologs for pairing partners. These allopolyploids may then be less fertile than expected, as a result of producing spores or gametes with unbalanced chromosome numbers, making them nonfunctional. Allopolyploids of this sort could be mistakenly identified as autopolyploids.

One example of an allopolyploid species in which the extra sets of chromosomes exhibit some degree of similarity is the common bread wheat (*Triticum aestivum*). Common wheat is a hexaploid of complex origin, which exists in over 20,000 cultivated varieties. The total chromosome complement in this organism consists of two sets of chromosomes in each of three genomes, designated A, B, and D. The formula for the genomes of common wheat is then AABBDD. Since each genome is made up of $x = 7$ chromosomes, the somatic cells of this species contain 14 (for A) + 14 (for B) + 14 (for D) = 42 chromosomes.

EXAMPLE 12.4. Short oat (*Avena brevis*) has 14 chromosomes, Abyssinian oat (*Avena abyssinica*) has 28 chromosomes, and the common cultivated oat (*Avena sativa*) has 42 chromsomes. **(a)** How many chromosomes appear to make up the basic genomic set in the oat series? **(b)** How many genomic sets occur in the common oat?

Solution. **(a)** Short oat appears to be a diploid in this series, with $2x = 14$; its chromosome number is exactly divisible by 2 but not by any larger integer other than 7. Thus, $x = 7$ is the number of chromosomes in the basic genomic set. **(b)** Common oat is a hexaploid, with $6 \times 7 = 42$ chromosomes.

Studies show that the A, B, and D genomes of wheat form seven homeologous groups of three chromosomes each. Chromosomes 1A, 1B, and 1D, for example, are all genetically similar. Chromosomes 2A, 2B, and 2D are also homeologous, and so on. Despite the close relationship between the three genomes, chromosome pairing in wheat is like that of a diploid organism, with synapsis occurring only between true homologs. Homeologous pairing is suppressed as a result of the action of a gene named *Ph* (for "pairing homeologous"), which is present on the long arm of chromosome 5B. When this gene is present, only homologs pair during meiosis. When it is absent, homeologs also pair, but less frequently.

Researchers have demonstrated that the A, B, and D genomes are homeologous by developing viable **nullisomics** in wheat. Nullisomics are individuals that have lost both members of a pair of chromosomes. In diploids, this condition (designated $2n - 2$) is ordinarily lethal, because vital gene functions are lost. In the hexaploid bread wheat, however, homeologous pairs of chromosomes can compensate for the loss to some extent, because they have similar genes. Although the nullisomics are typically less fertile and vigorous than normal plants, their survival is at least assured by the existence of genes in the partially duplicated genomes that are similar to the missing ones.

Experimental Production of Polyploids

In the decades ahead, we will need an expanding food supply to accommodate the world's growing population. A major part of any increase in food production will most likely come from improvements we are able to make by traditional breeding techniques in our current species of crop plants. The use of these techniques has already resulted in impressive gains in yield and in nutritional quality during recent years. There is every indication that these techniques will lead to further advances in the future.

But the approach to crop improvement that concerns us here is the creation of entirely new and possibly more productive genetic varieties. This area of research includes the synthesis of allopolyploids by experimental techniques.

Induction of Polyploidy. Geneticists now routinely produce allopolyploids by using colchicine to induce the doubling of chromosome number in interspecific hybrids. This procedure has been successful in the production of the new genus *Triticale* from a hybrid of wheat (genus *Triticum*) and rye (genus *Secale*). First-generation seeds from such a cross are sterile, even though some germinate. If the tips of the growing sprouts from these seeds are treated with colchicine, however, doubling of chromosome number occurs, and the flowers that develop produce the fertile amphidiploid seeds of the *Triticale* plant.

The procedures that plant breeders used in the initial development of *Triticale* are illustrated in Fig. 12.14. Notice that *Triticale* is a hexaploid like the common

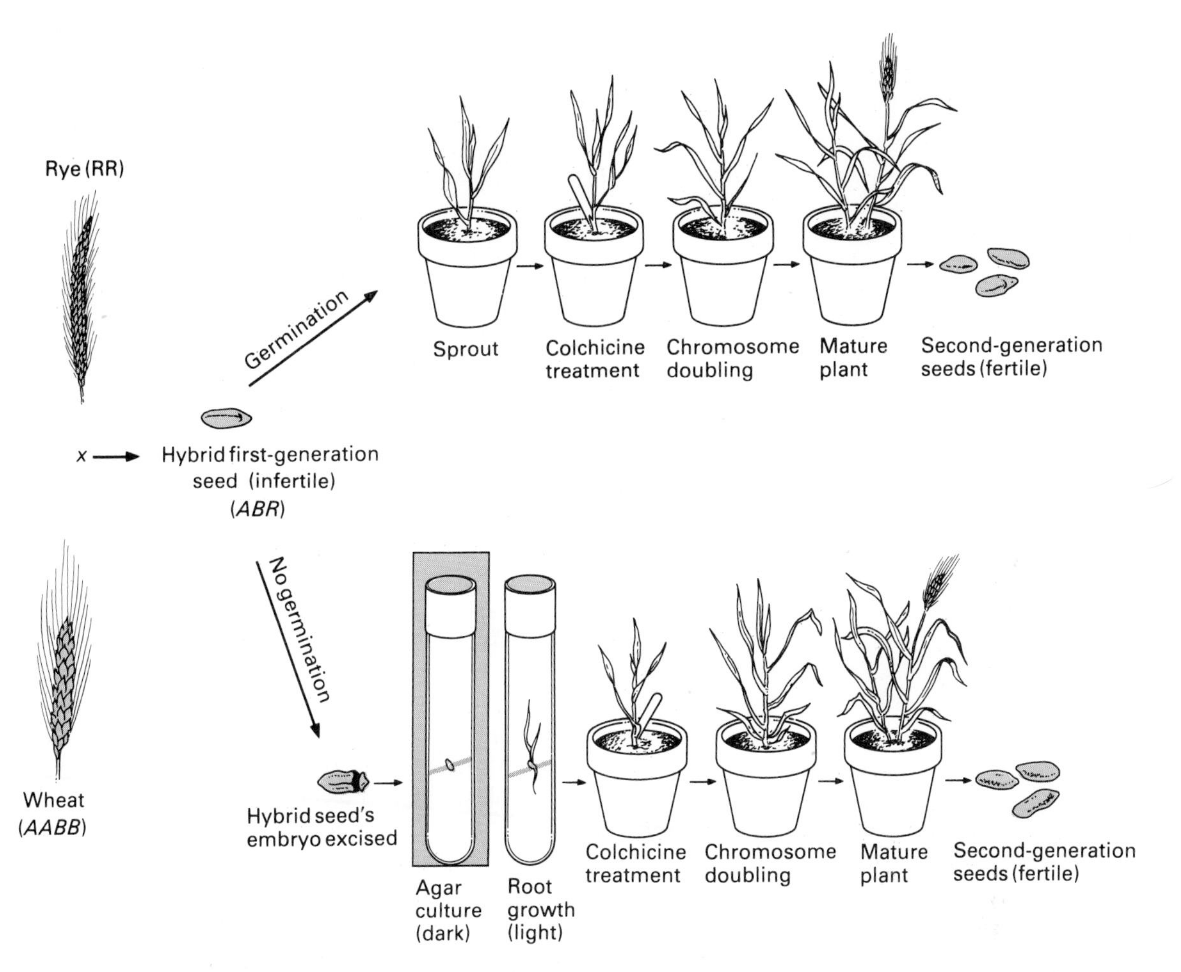

Figure 12.14. Production of the amphidiploid *Triticale*. The cross of rye with wheat yields first-generation hybrid seeds that are usually sterile, although some may germinate (top sequence of events). Sterility in that case is overcome by treatment of the tips of growing sprouts with colchicine. This doubles the chromosome number, so that the mature plant (with the genomic formula AABBRR) flowers produce fertile second-generation seeds. In contrast, if the first-generation hybrid seed does not germinate (bottom sequence of events), embryo culture is used to produce the first-generation seedlings. Embryos are excised from immature seeds and cultured in the dark until the embryo sprouts. *Source:* After J. H. Hulse and D. Spurgeon, *Triticale, Scientific American* 231:72–80, 1974.

bread wheat, but with a genomic formula of AABBRR. The A and B genomes are inherited from a tetraploid species of wheat (*T. turgidum*) and the R genome is inherited from rye.

The overall performance of *Triticale* was at first disappointing. Breeders have since produced improved genetic strains by selecting for traits such as high fertility and early maturity, greater insensitivity to day length, and greater resistance to lodging (collapse of the plants, resulting from a weak straw and a heavy head). Disease resistance has also been enhanced. Certain strains of *Triticale* appear to be more resistant to rust than wheat is. (Rust, a disease produced by a fungus, has limited wheat production in many parts of the world.)

The improved varieties of *Triticale* combine certain desirable features of both parent species, such as the high yield of wheat and the ruggedness of rye. In some instances they also combine the high protein content of wheat and the high lysine content of rye. Lysine is an essential amino acid that is present in only small amounts in the proteins of many cereals. A high lysine content therefore improves the quality of the protein.

Polyploids from Somatic Cell Fusion. The technique of *somatic cell fusion* has recently provided geneticists with a method of producing amphidiploids artificially. This procedure is based on the observation that plant cells whose cell walls have been removed (called **protoplasts**) are capable of fusing with similarly treated protoplasts from the same plant or from different plant species, to form tetraploid hybrids. The hybrid protoplasts contain a complete diploid set of chromosomes from both parent cell types. With the proper treatment and care, they can develop into mature tetraploid plants. This technique does not require any form of sexual reproduction. It therefore has the potential of producing hybrids between even distantly related plant species that are incapable of cross pollination.

In Chapter 7 we discussed the application of somatic cell fusion to animal cell cultures. The procedure as applied to plant cells is summarized in Fig. 12.15. Investigators begin by taking plant cells, usually from leaf tissue, and treating them with enzymes that catalyze the degradation of cell wall material, in order to convert the cells into protoplasts. Next, the researchers mix suspensions of protoplasts from different species and combine them with the chemical polyethylene glycol, which stimulates cell fusion. They subsequently place the protoplasts on an agar medium that permits the hybrid cells to grow. Each hybrid cell forms a clump of cells known as a **callus.** The investigators can induce a callus to generate shoots, leaves, and roots by treating it with various plant hormones. In this way, the callus gives rise to the mature plant.

One example of this procedure is the use of cell fusion to develop a fertile hybrid formed between two species of *Nicotiana* (tobacco): *N. langsdorffi* × *N. glauca.* These species can also produce amphidiploids by cross pollination and chromosome doubling. Researchers found that the amphidiploid plants that were derived by means of cell fusion were indistinguishable from those formed by standard sexual means in the several characteristics for which they tested.

Evolution Through Polyploidy

Polyploidy has been an important mechanism in the evolution of higher plants. Most estimates suggest that more than a third of the flowering plants were derived in this manner. The most impressive feature of polyploid evolution is the sudden emergence of a new species. This emergence can occur in just one or two generations, without a long history of genetic changes.

Despite the suddenness of the process of polyploid evolution, polyploidy may serve largely as a conservative force, because it limits the rate at which changes occur in response to the demands of the environment. It does so through the masking effect of duplicate genomes on the expression of recessive variations. As we have seen, polyploids tend to express a lower proportion of recessive pheno-

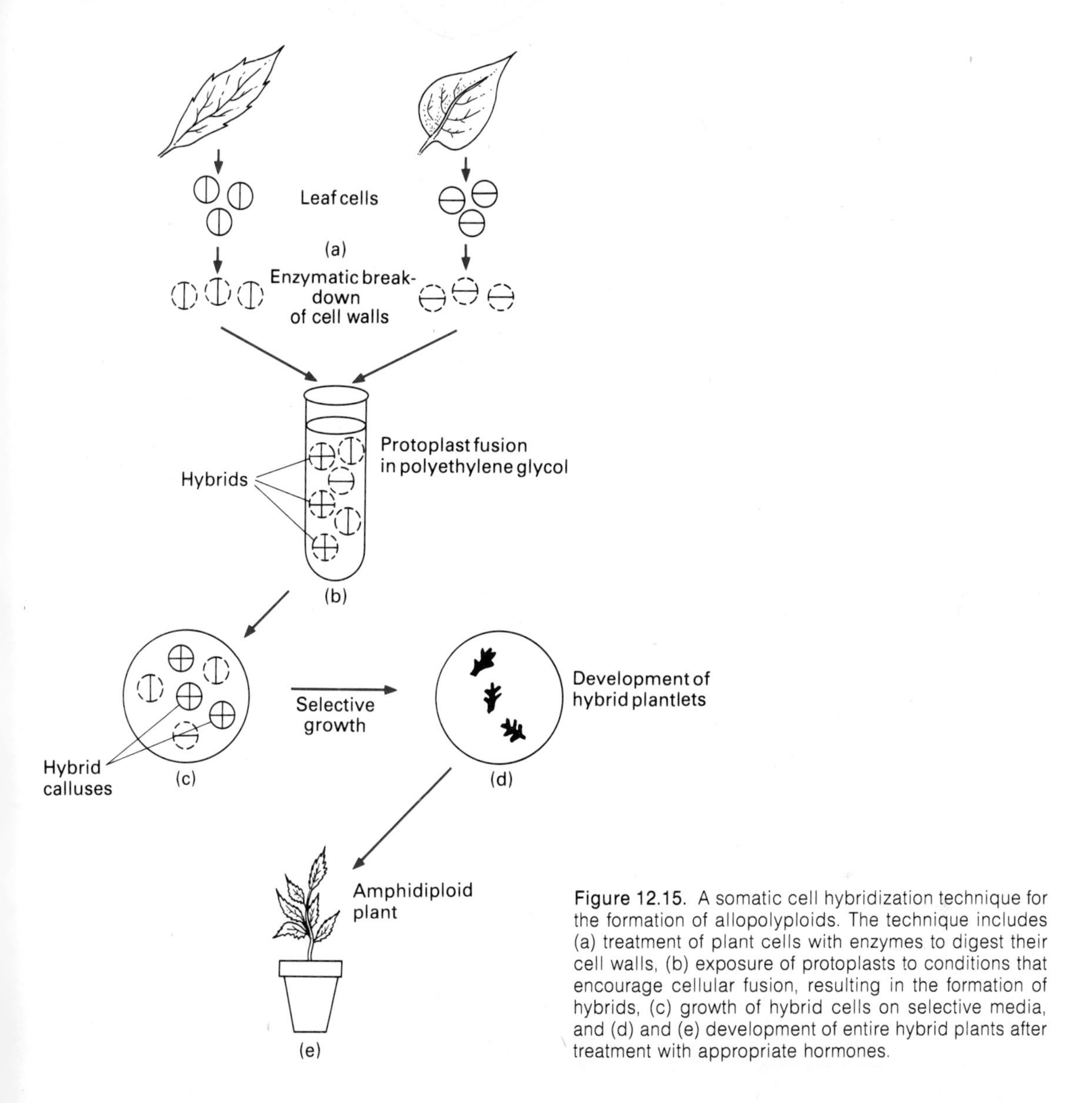

Figure 12.15. A somatic cell hybridization technique for the formation of allopolyploids. The technique includes (a) treatment of plant cells with enzymes to digest their cell walls, (b) exposure of protoplasts to conditions that encourage cellular fusion, resulting in the formation of hybrids, (c) growth of hybrid cells on selective media, and (d) and (e) development of entire hybrid plants after treatment with appropriate hormones.

types than do diploids and thus express fewer recessive variations to be selected for or against by the environment.

Evolution of Wheat. One example of polyploid evolution has occurred in the development of the present-day species of bread wheat. The genus *Triticum* contains a number of different species, which can be subdivided into three groups according to chromosome number: those that have 14 chromosomes, 28 chromosomes, and 42 chromosomes. The group with 14 chromosomes contains, among others, the wild and cultivated forms of einkorn wheat, *T. monococcum* (genomic formula AA). The group with 28 chromosomes includes the wild and cultivated forms of emmer wheat, *T. turgidum* (genomic formula AABB). The group with 42 chromosomes includes the different varieties of common bread wheat, *T.*

aestivum (genomic formula AABBDD). Geneticists currently believe that each group gave rise to the progenitors of the next higher group in the series through the process of allopolyploidy.

The theorized sequence of changes involved in the development of the common bread wheat is illustrated in Fig. 12.16. Two hybridization events have occurred, each of which is thought to involve a wild grasslike species with a chromosome number of $2x = 14$. The first accidental cross-pollination may have taken place between einkorn wheat and the wild grass *T. searsii,* an inhabitant of wheat fields in southwestern Asia. The subsequent doubling of chromosomes in the F_1 then produced an amphidiploid species of emmer wheat with $4x = 28$ chromosomes.

The second interspecies cross involved emmer wheat and the wild grass *T. tauschii,* which inhabits wheat fields in the Mediterranean region. After this second cross, the doubling of chromosomes in the F_1 gave rise to the progenitor of bread wheat, with $6x = 42$ chromosomes. Thus, it appears that the AABBDD genomic constitution of bread wheat consists of chromosome sets derived from three diploid species: the AA of *T. monococcum,* the BB of *T. searsii,* and the DD of *T. tauschii.*

Two American scientists, McFadden and Sears, provided evidence to support this theory when they succeeded in hybridizing emmer wheat (AABB) with *T.*

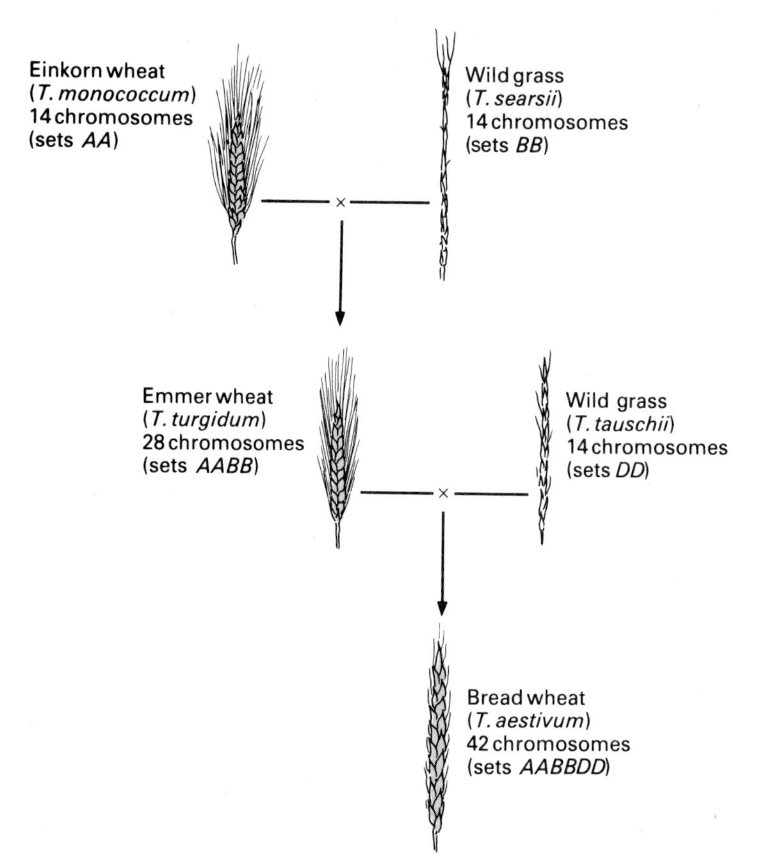

Figure 12.16. Evolution of wheat (genus *Triticum*). Sequence of hybridization events believed to be involved in the evolutionary development of common bread wheat. As described in the text, each hybridization is followed by chromosome doubling, to form fertile amphidiploids.

tauschii (DD). Colchicine was then used to artificially induce doubling of chromosomes in the F_1. The 42-chromosome amphidiploid (AABBDD) that resulted had many characteristics in common with bread wheat. Moreover, it could readily cross with bread wheat to form fertile hybrids.

TO SUM UP

1. Euploid organisms have a chromosome complement in their somatic cells that is an exact multiple of the basic set, or genome. Changes in the euploid state from the normal diploid condition come about when complete sets of chromosomes are added or lost.
2. Monoploid individuals possess only the haploid set of chromosomes of a normally diploid species. They tend to be smaller and less vigorous than diploids and are usually sterile. Investigators can produce monoploid plants in the laboratory by using a technique known as anther culture.
3. Organisms with three or more complete sets of chromosomes are polyploid. Autopolyploids possess multiple chromosome sets that are derived from homologous genomes of the same species. They are formed from zygotes produced by unreduced, diploid gametes or by the doubling of chromosomes in the somatic cells and the subsequent vegetative reproduction of the resulting tetraploid tissue.
4. Many autopolyploid plants have great commercial value because of their increased size and vigor. The inability of many autopolyploid plants to produce balanced gametes frequently results in infertility, however, so that many commercially useful polyploids must be propagated by asexual means.
5. Allopolyploids possess multiple chromosome sets that are derived from members of different species. Fertile amphidiploids, which are formed through hybridization and the subsequent doubling of the chromosome number, have the diploid set of chromosomes from each of the two parental species. They constitute an important source of new species in higher plants.
6. Researchers can produce amphidiploids in the laboratory by using the drug colchicine to induce the doubling of chromosome number in sterile interspecific hybrids. Amphidiploids between species that cannot cross pollinate can also be formed by the technique of somatic cell hybridization.

Structural Alterations of Chromosomes

Chromosome mutations also involve changes in the structure of chromosomes. Many of these structural abnormalities are large enough to be seen with an ordinary light microscope and can be studied within the cells. These microscopically visible aberrations, called *macrolesions,* occur when whole chromosomes break and fairly large chromosomal segments become rearranged.

Types of Aberrations

There are four basic types of structural changes: (1) deficiencies, (2) duplications, (3) inversions, and (4) translocations. The major classes and the possible origin of each class are shown in Fig. 12.17. We can usually distinguish the different types of macrolesions by observing the pairing arrangements of homologs during synapsis in an organism that is heterozygous for an aberrant chromosome. Since

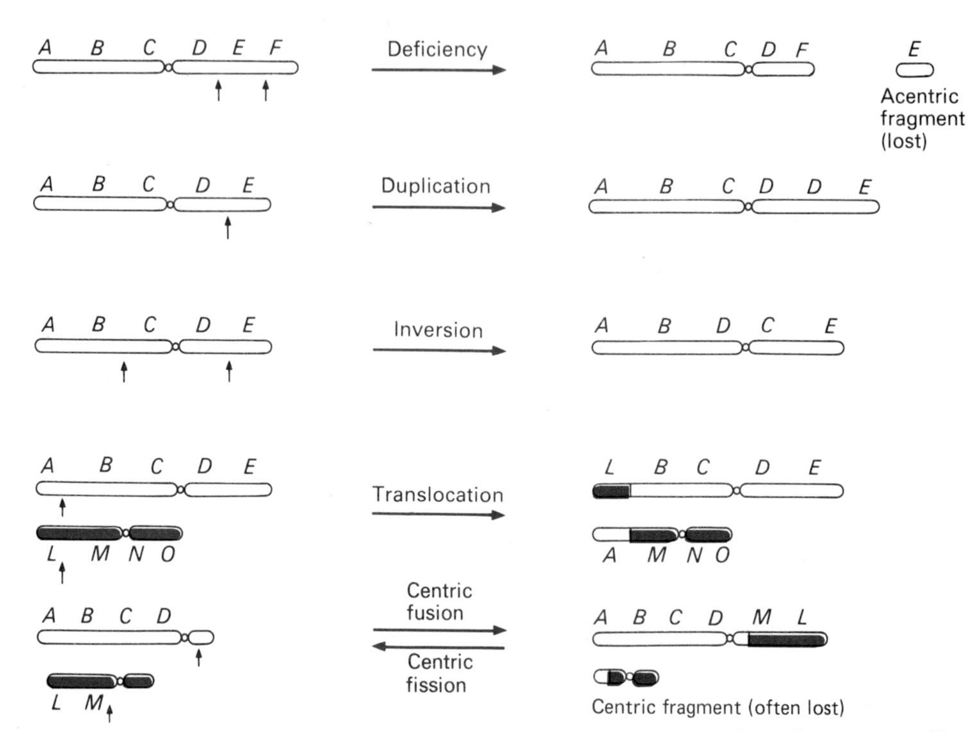

Figure 12.17. The major classes of chromosome aberrations and possible origin of each. The arrows indicate points of breakage and alteration.

homologs must pair in a precise manner, bivalents that include a structurally modified chromosome take on configurations that are unique for each type of aberration. These aberrations can therefore be identified in meiocytes during the first meiotic prophase.

The different structural aberrations have been studied in the finest detail in the large chromosomes of certain secretory tissues (such as salivary glands) in the dipteran insects (flies, mosquitoes, etc.). These so-called **polytene** (i.e., many stranded) **chromosomes** have enlarged through several successive doublings of their genetic material without undergoing nuclear division. The many duplicate strands in each chromosome coil differentially along their lengths, resulting in a unique banding pattern that becomes evident upon staining (Fig. 12.18). Moreover, the homologs of each chromosome remain tightly paired in a continuous state of somatic synapsis. These three features of polytene chromosomes—their large size, differential banding pattern, and somatic pairing—enable geneticists to recognize each type of aberration and to identify its precise location on the chromosome pair.

Deficiencies. A deficiency (or deletion) occurs when a segment is lost from a chromosome. The deficiency is *terminal* if an end of the chromosome is lost; it is *interstitial* if an internal segment is missing. Thus, if $\underline{A\ B\ C\ D}$ represents the normal arrangement of genes within a chromosome, $\underline{A\ B\ C}$ would constitute a terminal deficiency and $\underline{A\ C\ D}$ would constitute an interstitial deficiency.

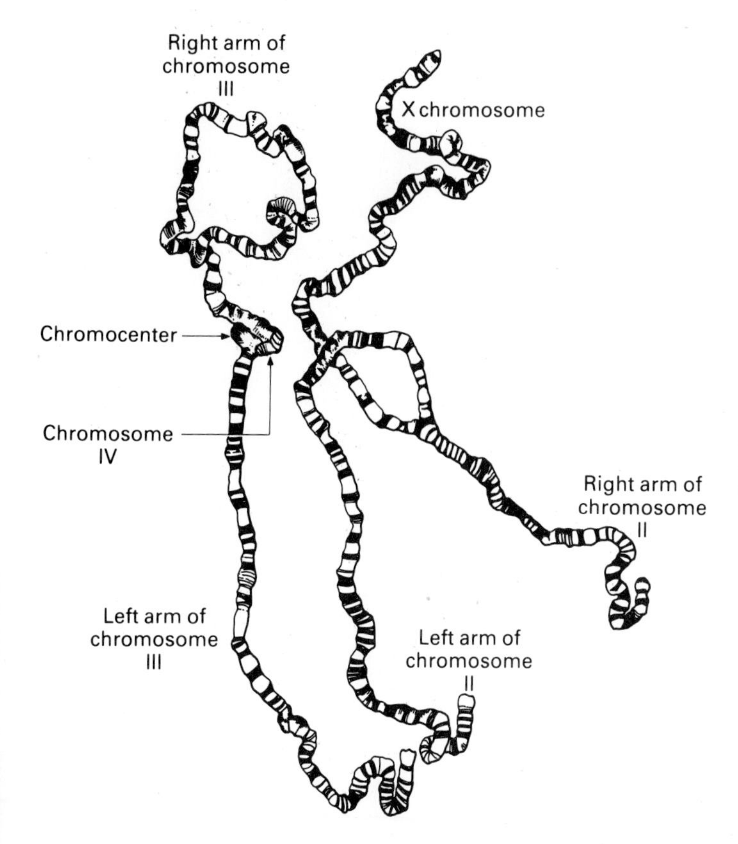

Figure 12.18. Salivary gland chromosomes of *Drosophila melanogaster.* The chromosomes are enlarged through successive doublings of their genetic material without cell division. They are further enlarged by the close pairing of homologs (somatic pairing). Differential coiling of chromosomal material results in precise banding patterns that become evident upon staining.

Deficiencies that are microscopically detectable normally involve a substantial loss of genetic material. Most homozygous deletions are therefore lethal. Their phenotypic effect in the heterozygous state depends in part on the length of the missing segment. The smaller deficiencies may have little or no effect and may behave in inheritance as recessive lethals. But larger deletions may have the effect of a monosomy and may be deleterious or even lethal in the heterozygous state.

We can detect deficiencies in the heterozygous state cytologically by observing a characteristic loop in the chromosomes during synapsis (Fig. 12.19a). The normal

Figure 12.19. Pairing arrangements of synapsed homologs in heterozygotes for aberrant chromosomes. (a) Deficiency. (b) Duplication.

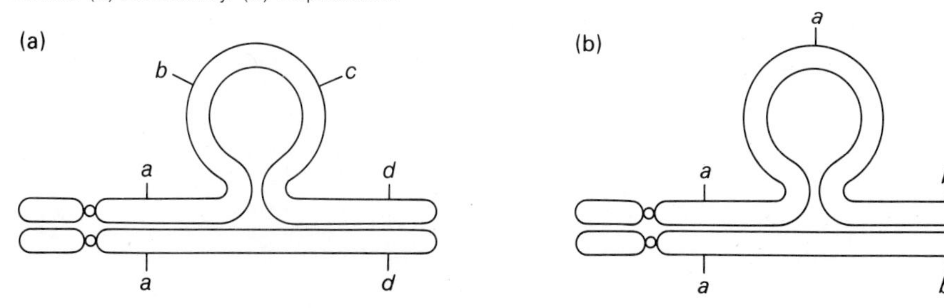

chromosome loops out in the region that corresponds to the missing segment, in order to achieve proper pairing with its aberrant homolog.

Duplications. A duplication occurs when a segment in a chromosome is repeated. For example, if $\underline{A\ B\ C\ D}$ represents the normal arrangement of genes, then $\underline{A\ B\ B\ C\ D}$ would constitute a tandem duplication of the *B* locus. Like deficiencies, duplications can be detected cytologically by the characteristic single loop-out exhibited by bivalents that are heterozygous for the duplication (Fig. 12.19b). Both deficiencies and duplications alter the genetic balance of a cell by changing the dosage of genes involved in the abnormality. They differ from one another in that duplications behave as partial trisomies, whereas deficiencies behave as partial monosomies. The phenotypic consequences of duplications are therefore less severe.

An example of a duplication with a distinct phenotypic effect is the *Bar* duplication in *Drosophila*. The *Bar* duplication is a tandem repeat of the segment of the X chromosome that is designated 16A in the banding pattern of chromosomes in the salivary glands of *Drosophila* (Fig. 12.20). The duplication acts to narrow the compound eyes by reducing the number of eye facets from an average of 770 per eye in normal flies to about 358 in duplication heterozygotes, and still further, to about 68 in duplication homozygotes.

Once a chromosome contains a tandem repeat of the same region, or of one that is similar in genetic sequence, additional duplications can arise when asym-

Figure 12.20. The Bar mutation in *Drosophila*. (a) Comparison of the eye phenotypes of flies with 1 (wild type), 2 (bar), and 3 (double-bar) doses of region 16A of the X chromosome. (b) Comparison of altered regions in the salivary gland chromosomes of bar and double-bar mutants with that of wild type flies. *Source:* After Bridges.

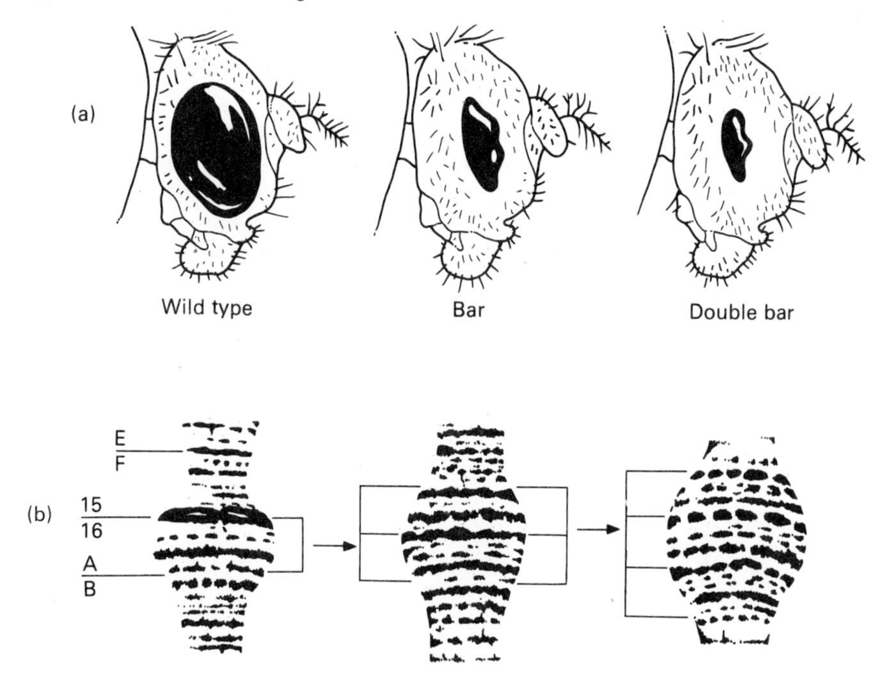

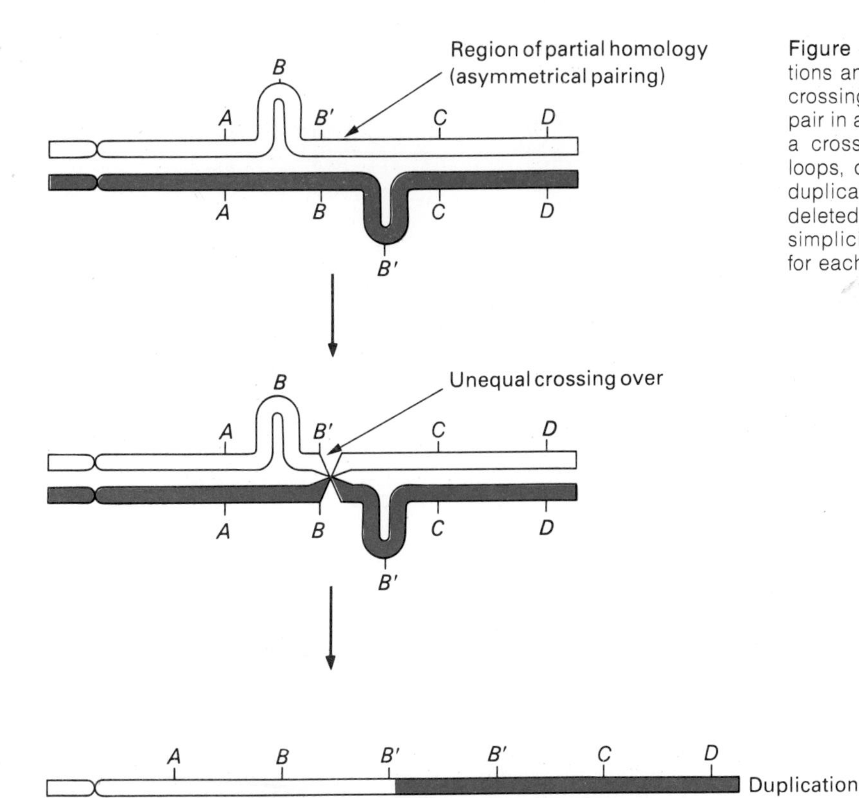

Figure 12.21. Production of duplications and deficiencies through unequal crossing over. If homologs mistakenly pair in a region of partial homology, and a crossover then occurs between the loops, one of the products will carry a duplication of the region. That region is deleted from the other homolog. (For simplicity, only one chromatid is shown for each chromosome.)

metrically paired homologs cross over during prophase I (Fig. 12.21). Such *unequal crossing over* produces one homolog with a tandem repeat of an identical genetic segment and the other homolog with a deletion of the same region. For example, unequal crossing over occurs on occasion in homozygotes for the *Bar* duplication in *Drosophila,* giving rise to so-called double-bar offspring. Region 16A is present in triplicate in the X chromosome of these offspring (see Fig. 12.20).

Inversions. The remaining types of structural alterations—inversions and translocations—can be thought of as *segmental rearrangements,* since no gain or loss of material is involved, only a change in gene arrangement. An inversion occurs when a segment in a chromosome is reversed. The rearrangement is a **pericentric inversion** if the inverted segment includes the centromere; it is a **paracentric inversion** if it does not include the centromere. For example, if *A*–o–*B C D* represents the normal arrangement of genes, where –o– designates the location of the centromere, then *A*–o–*C B D* would be a paracentric inversion and *A C B*–o–*D* would constitute a pericentric inversion. The microscopic appearance of a chromosome with a pericentric inversion may be quite different from that of its homolog, because of the change in the location of the centromere.

We can often detect a chromosome with an inverted segment by observing the *double-loop* pairing arrangement that occurs in an inversion heterozygote (Fig.

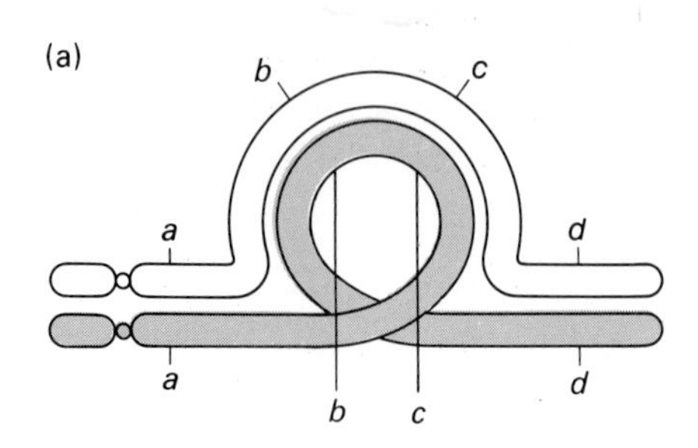

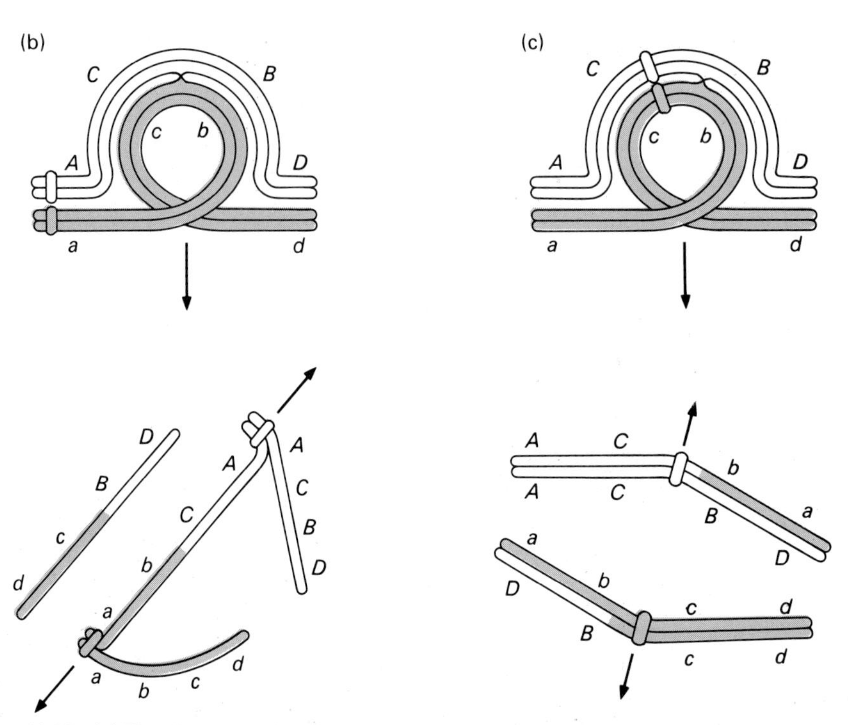

Figure 12.22. (a) The double-loop pairing arrangement of synapsed homologs in a chromosome pair that is heterozygous for an inversion. The lower diagrams show the results of crossing over within an inversion loop in heterozygotes for a (b) paracentric and (c) pericentric inversion. Crossing over in a paracentric inversion loop gives rise to a fragment that lacks a centromere and to a chromosome with two centromeres that forms a bridge during meiosis. Crossing over in a pericentric inversion loop yields two chromosomes with large deletions and duplications.

12.22a). This unique pairing configuration causes inversions to act as *crossover suppressors.* This term was originally proposed to describe the apparent inhibitory effect of inversions on the recombination of nonallelic genes. Geneticists at that time thought that the anomaly interfered with crossing over within the inverted part of the chromosome. We now know that inversions do not necessarily suppress crossing over; rather, inversion heterozygotes form nonfunctional meiotic products when such crossovers do occur (Fig. 12.22b and c). When a crossover occurs within the loop of a paracentric inversion (Fig. 12.22b), it results in a fragment that lacks a centromere (an **acentric** fragment) and a chromosome with two centromeres (a **dicentric** chromosome). The acentric fragment fails to migrate to one pole or the other on the meiotic spindle, whereas the dicentric chromosome forms a bridge during anaphase that subsequently breaks or that interferes with the completion of meiosis. By contrast, a crossover within the loop of a pericentric inversion (Fig. 12.22c) produces two chromosomes that contain large deletions and duplications. Either type of inversion has the effect of suppressing crossing over. Spores or gametes that have received a crossover chromosome are never recovered, because they are nonfunctional or fail to produce viable zygotes. Thus, the block of genes in an inverted segment of a chromosome seemingly tends to retain its original sequence, and the genes behave like a single unit, or "super gene," in their pattern of inheritance.

Translocations. Translocation is the second major type of segmental rearrangement. A translocation occurs when a segment is transferred to a new location in the same chromosome or in a completely different chromosome. The transfers are frequently reciprocal, resulting in the exchange of parts between nonhomologous chromosomes. For example, if $\underline{A\ B\ C\ D}$ and $\underline{E\ F\ G\ H}$ are the arrangements of genes in different chromosomes, the products of a reciprocal translocation might appear as $\underline{A\ B\ G\ H}$ and $\underline{E\ F\ C\ D}$.

Reciprocal translocations in the heterozygous state are fairly easy to detect cytologically. The chromosomes will synapse in a cross configuration that includes two aberrant and two normal chromosomes (Fig. 12.23a (on p. 513)). This synaptic arrangement is responsible for the reduction in fertility that often occurs in translocation heterozygotes. Disjunction from such a configuration tends to produce meiotic products that are unbalanced, such as spores or gametes that have large deficiencies and duplications (Fig. 12.23b). The only way that functional products can be produced is when the translocation arrangement in synapsis forms a zigzag or figure-eight pattern during disjunction.

EXAMPLE 12.5. A barley plant that is heterozygous for a reciprocal translocation undergoes self-fertilization. Show that half of the offspring of this barley plant are also expected to be heterozygous for the translocation.

Solution. We can illustrate the results of this mating by designating the synaptic arrangement at metaphase I in the following manner:

A | A
B | B
D C | G H
D C | G H
F | F
E | E

Only two gamete types are viable: $\frac{A\ B\ C\ D}{E\ F\ G\ H}$ and $\frac{A\ B\ G\ H}{D\ C\ F\ E}$

Upon self-fertilization of the plant, these gametes will yield the following zygote types:

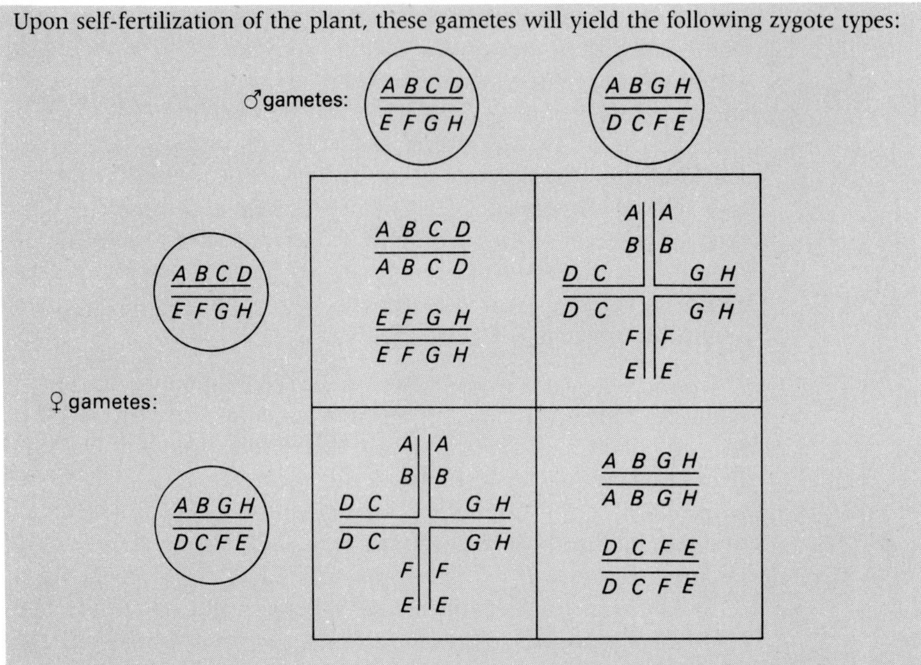

The progeny occur in the ratio of 1 standard homozygote : 2 translocation heterozygotes : 1 translocation homozygote.

Use of Aberrations in Gene Mapping

Geneticists have used chromosome aberrations successfully to localize genes on chromosomes. We can infer the presence of a gene on a particular chromosome or in a particular region of a chromosome by noting when an aberration is consistently associated with a change in the expression of the gene or with an alteration in its pattern of inheritance. For instance, certain aberrations are characterized by their ability to affect the results of crosses by changing the genotypic (and phenotypic) ratios for genes that are carried on abnormal chromosomes. The aberration thus serves as a marker for the chromosomal location of the affected linkage group, simplifying the task of associating a gene with a particular chromosome.

Translocations have been used extensively as genetic markers, especially in plants, in which heterozygotes for a translocation typically have reduced fertility (see Fig. 12.23). In corn, for example, about half the meiotic products of translocation heterozygotes are nonfunctional. Consequently, half the pollen grains produced by these carriers are aberrant (because they are abnormally small and deficient in starch content), and the ears produced have only about half the normal number of kernels. Of the remaining functional gametes, half must carry both translocation chromosomes (see Fig. 12.23). Therefore, half of the progeny of a testcross between a translocation heterozygote and a plant with the normal chromosome constitution will have reduced fertility. A testcross of this type is shown

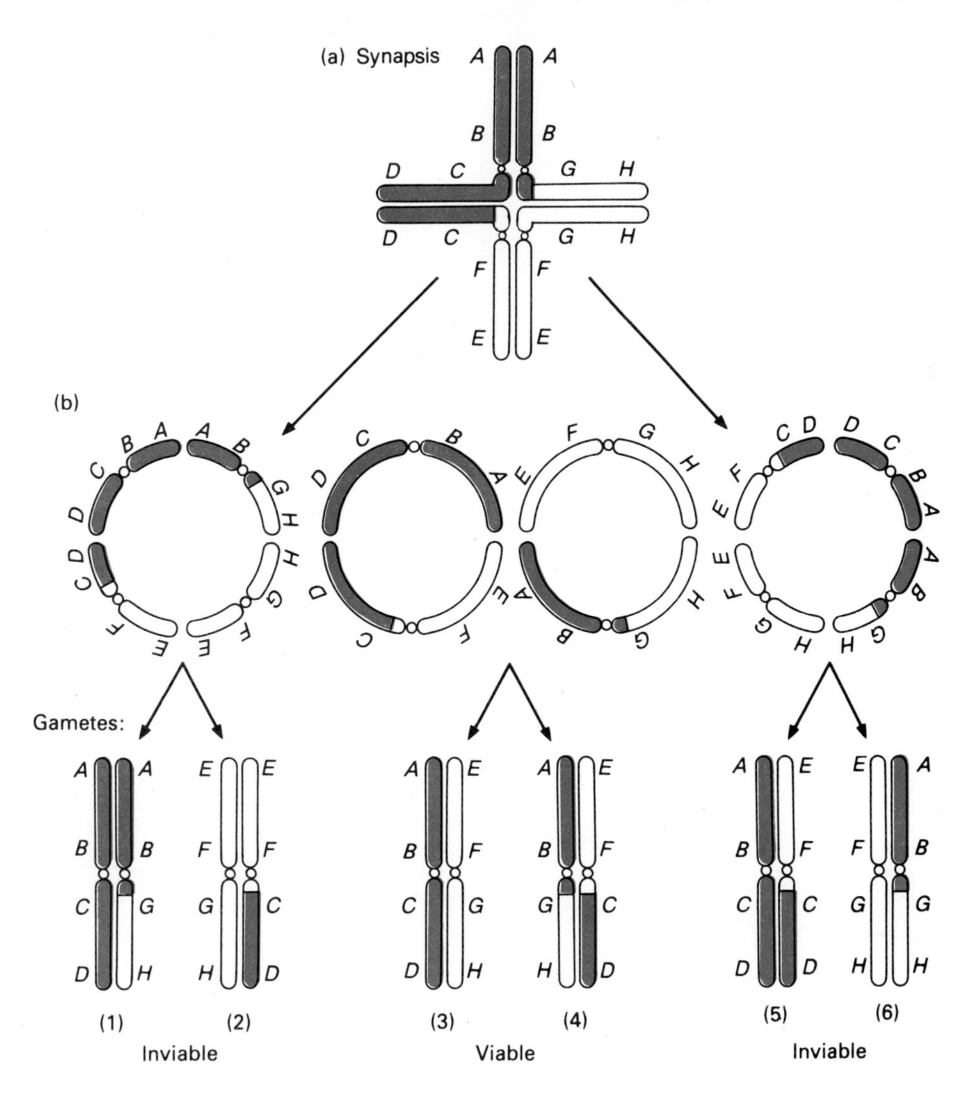

Figure 12.23. Chromosome behavior in translocation heterozygotes. (a) The pairing arrangement of synapsed chromosomes forms a cross configuration in which two of the chromosomes are normal and two have undergone a reciprocal translocation. (b) Meioses in translocation heterozygotes show three segregation patterns at anaphase I. Gametes with a balanced complement of genes can only be produced by the meiotic translocation configuration that forms a figure-eight during disjunction. (The middle segregation pattern, where the two normal chromosomes segregate together, as do the two translocated ones, is called *alternate segregation*. The left-hand and right-hand segregation patterns, in which each normal chromosome segregates with one of the translocated chromosomes, are called *adjacent segregation*. Both types of adjacent segregation produce unbalanced gametes, containing duplications and deficiencies, that usually have lethal effects.)

in Fig. 12.24. Note that any gene that is linked to the translocation break will also show linkage to the condition of reduced fertility. Thus, a gene will be assigned to one of the translocation chromosomes if it fails to recombine at random with the condition of reduced fertility.

The presence of a gene in a particular region can be inferred from the procedure known as deletion mapping (discussed in Chapter 8 in reference to fine-structure

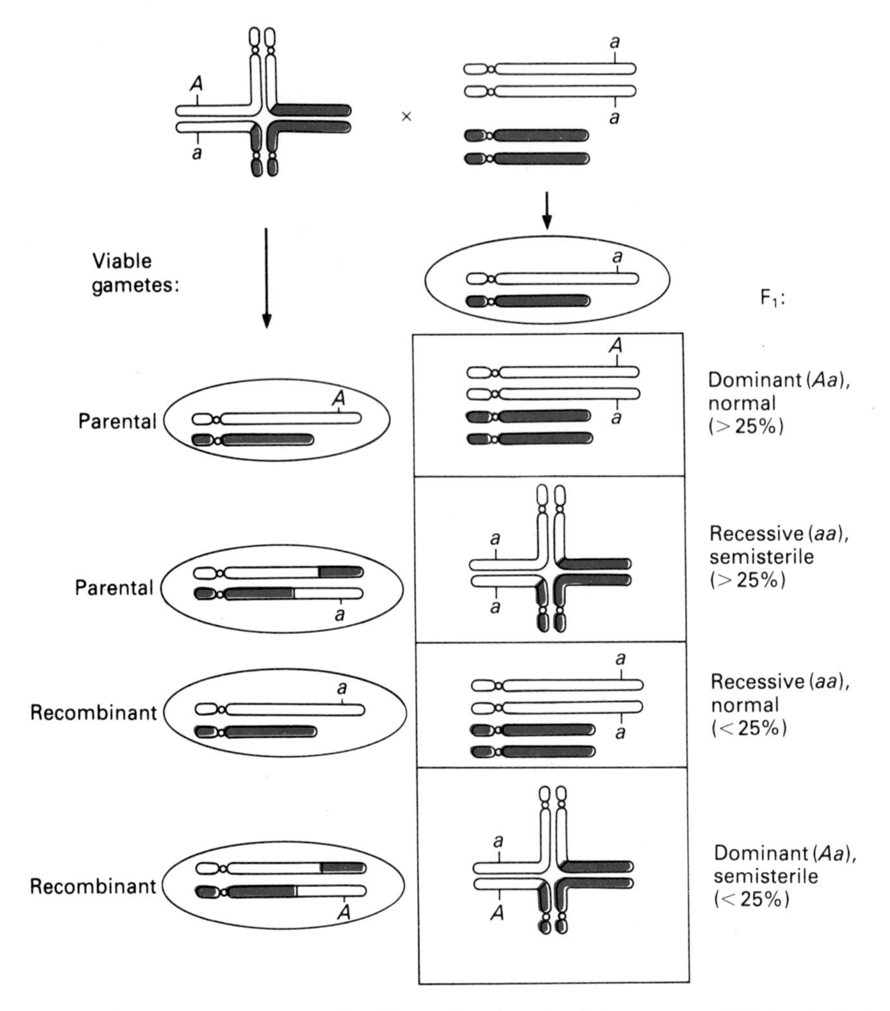

Figure 12.24. Diagram of a testcross involving a translocation heterozygote which is also heterozygous for the *A,a* alleles. Linkage between the gene locus and the translocation break can be detected by the failure of the allelic pair to recombine at random with the condition of reduced fertility (the semisterile condition). The recombinant types are formed as a result of crossing over between the gene locus and the translocation break.

mapping). This technique, which makes use of overlapping deficiencies, has been particularly helpful in generating physical maps of chromosomes, permitting geneticists to identify the position of a gene locus in relation to some visible landmark, such as the centromere or a particular band. In this procedure, individuals with a certain recessive gene are mated to heterozygous carriers for various deletions, in order to bring the gene into heterozygous combination with the various deficiencies of known length. If the homologous wild-type segment of the gene is missing in the deletion chromosome in question, the recessive allele will be expressed since it is present in a hemizygous state. We can therefore determine the location of the gene by correlating the occurrence of recessive offspring with the segment that is missing in the chromosome received from the carrier for the deletion.

To illustrate the method of deletion mapping, let us assume that we would like to know the location of a newly discovered recessive gene called *r*, which we suspect is situated near one end of a particular chromosome. This chromosome is designated $\underline{1\ 2\ 3\ 4\ 5}$, where the numbers represent identifiable landmarks, such as cross bands. Let us also assume that we have available three carriers of overlapping deletions: (a) $\frac{2\ 3\ 4\ 5}{1\ 2\ 3\ 4\ 5}$, (b) $\frac{3\ 4\ 5}{1\ 2\ 3\ 4\ 5}$, and (c) $\frac{4\ 5}{1\ 2\ 3\ 4\ 5}$. These carriers differ from one another in the length of the segment lost from the end of the chromosome that carries the wild-type *R* allele. We begin crossing each deletion heterozygote (of the *R* phenotype) with an *rr* homozygote of normal chromosome construction. Some progeny will have the *r* phenotype because they inherited a chromosome from the deletion parent in which the missing segment includes the *R* locus. These progeny, which are hemizygous for the *r* allele (e.g., $\underset{r}{\text{—+==}}$), will express the recessive trait. Thus, if our experiment shows that only crosses involving the deletion carrier (c) give rise to recessive progeny, then the *r* locus must be in region (band) 3 on the chromosome, since only deletion (c) is missing this region. If both (b) and (c) produce recessive offspring, however, we would conclude that the *r* locus is in region 2, which is absent in both (b) and (c) but not in (a). If all three deletion carriers produce recessive offspring, the *r* locus must be in region 1, the only segment that is missing in all three deletions.

The expression of a single copy of a recessive gene due to loss of its dominant allele is known as **pseudodominance.** Figure 12.25 illustrates the use of overlapping deletions of the X chromosome and pseudodominance to physically map two sex-linked genes of *Drosophila*.

Structural Aberrations and Evolution

Macrolesions have undoubtedly played an important role in the evolutionary development of higher plants and animals. Duplications and deletions have provided a mechanism for the gain and loss of genetic material. In addition, duplications are responsible for the development of a wide variety of related gene functions. For example, the genes that code for the pancreatic zymogens trypsinogen and chymotrypsinogen (the inactive precursors of the proteolytic enzymes trypsin and chymotrypsin) may have evolved after having been duplicated from a common ancestral gene. The inactive regions of these precursor proteins are quite similar to one another in amino acid sequence. There are many other examples of related genes that may have developed through successive duplication, such as the genes for the different hemoglobins and histones, discussed in Chapter 10. Once a particular gene has been duplicated, one copy can maintain some semblance of its original function, leaving the other copy free to change in various ways without resulting in a harmful loss to the organism.

Inversions and translocations have been important in evolution by providing a means for genes to rearrange themselves. In addition, translocations that involve

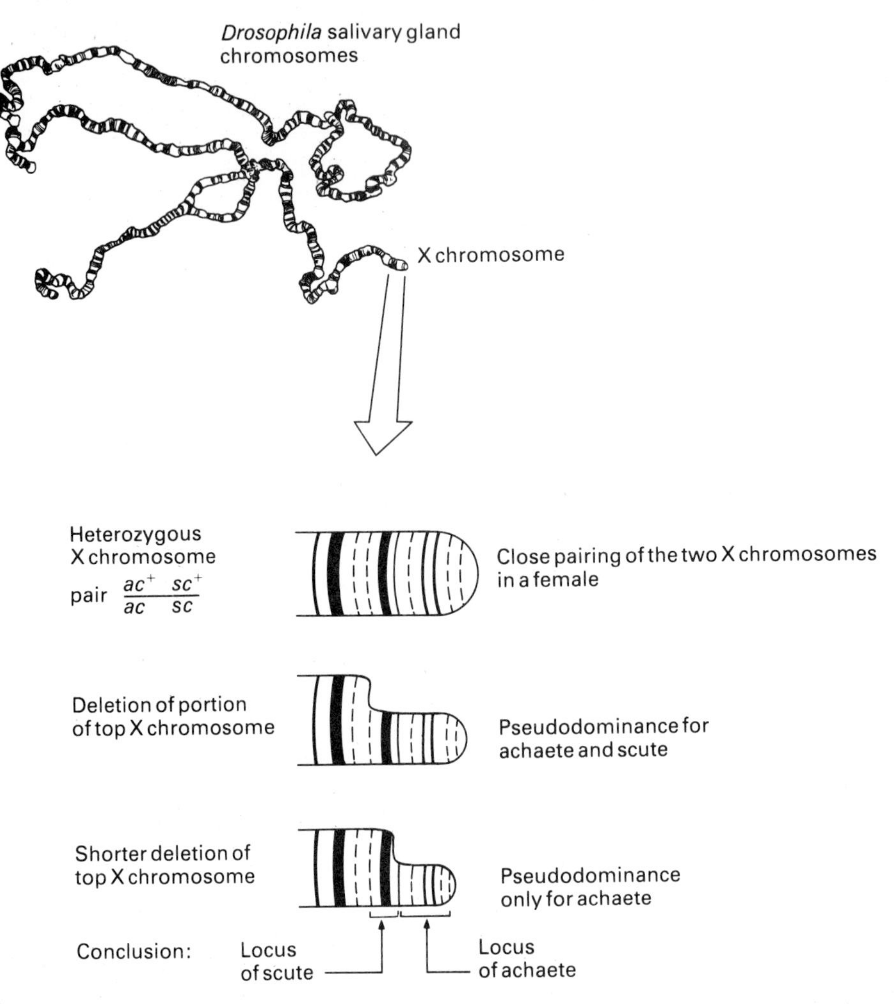

Figure 12.25. Localizing genes in the salivary gland chromosomes of *Drosophila* utilizing overlapping deficiencies. The X chromosome tip shown on top is heterozygous for the *ac* (achaete) and *sc* (scute) loci. The fly expresses the wild-type phenotype for both genes. A deletion of the tip of one of the X chromosomes (middle) is easily visible in the giant chromosomes, and is found to be associated with expression of the achaete and scute phenotypes, a phenomenon known as pseudodominance (expression of a single copy of a recessive gene due to loss of its dominant allele). Other individuals carrying a deletion of one of the X chromosomes exhibit pseudodominance only for achaete (bottom). The location of the scute gene is thus placed within the band or bands that differentiate the overlapping deletions.

the fusion of the entire long arms of two acrocentric chromosomes, called **centric** (or **Robertsonian**) **fusions**, have been responsible for the reduction in chromosome number in certain species. A comparison of the banding patterns in the salivary gland chromosomes of *Drosophila* species shows that evolution has proceeded in these chromosomes by centric fusion, decreasing the haploid set from a primitive number of six to five, four, and even three in some species (Fig. 12.26).

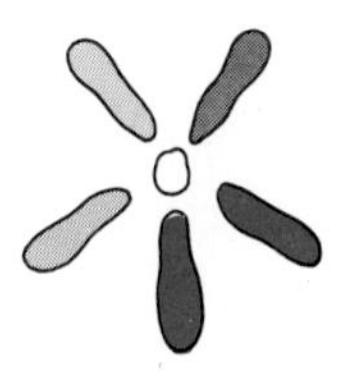

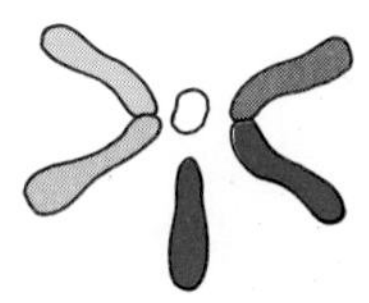

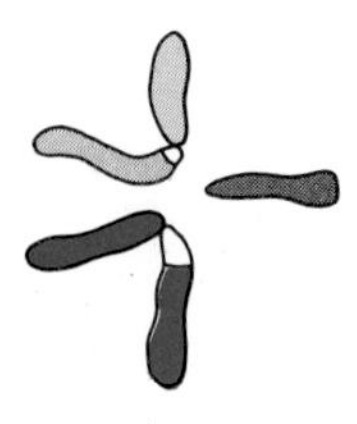

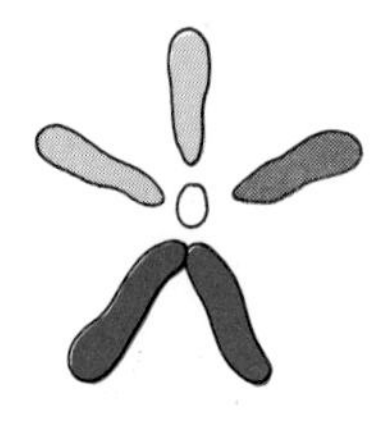

D. ananassae

Figure 12.26. The haploid chromosome complements of five *Drosophila* species. Shading identifies homologous chromosome arms. The ancestral karyotype for this genus is thought to involve five pairs of acrocentric chromosomes and one pair of dot-like chromosomes, as in *D. subobscura*. Centric fusions and inversions could have produced the other karyotypes shown here.

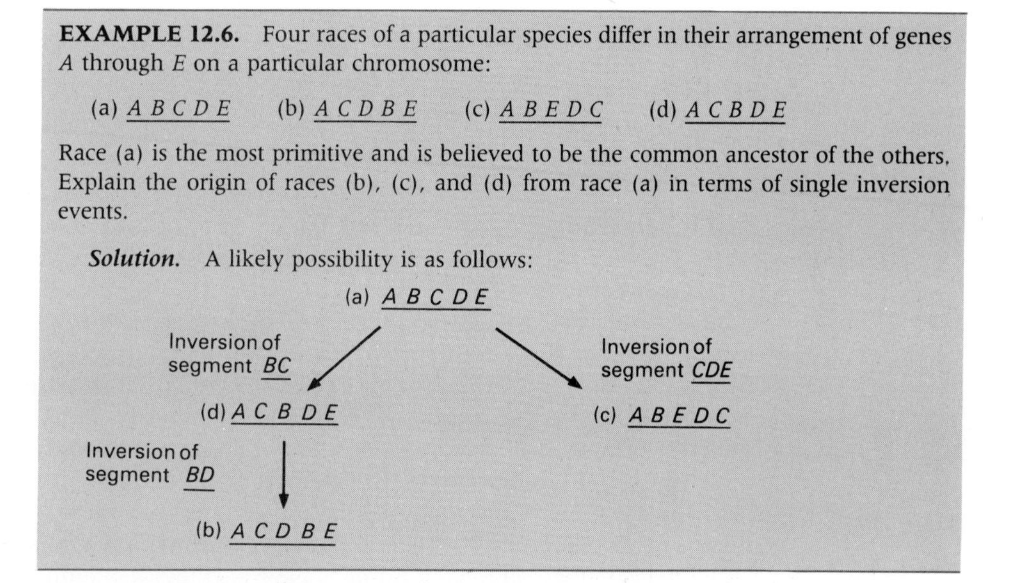

EXAMPLE 12.6. Four races of a particular species differ in their arrangement of genes *A* through *E* on a particular chromosome:

(a) *A B C D E* (b) *A C D B E* (c) *A B E D C* (d) *A C B D E*

Race (a) is the most primitive and is believed to be the common ancestor of the others. Explain the origin of races (b), (c), and (d) from race (a) in terms of single inversion events.

Solution. A likely possibility is as follows:

Structural Aberrations in Humans

Types of Aberrations. Several different chromosome aberrations have been detected in humans. Deficiencies and translocations have the most striking effects. The best-known example of a deficiency in humans is a deletion in the short arm of chromosome 5, which gives rise to *cri-du-chat* (or cat's cry) *syndrome.* The symptoms of this disorder include mental retardation, microcephaly (a small head),

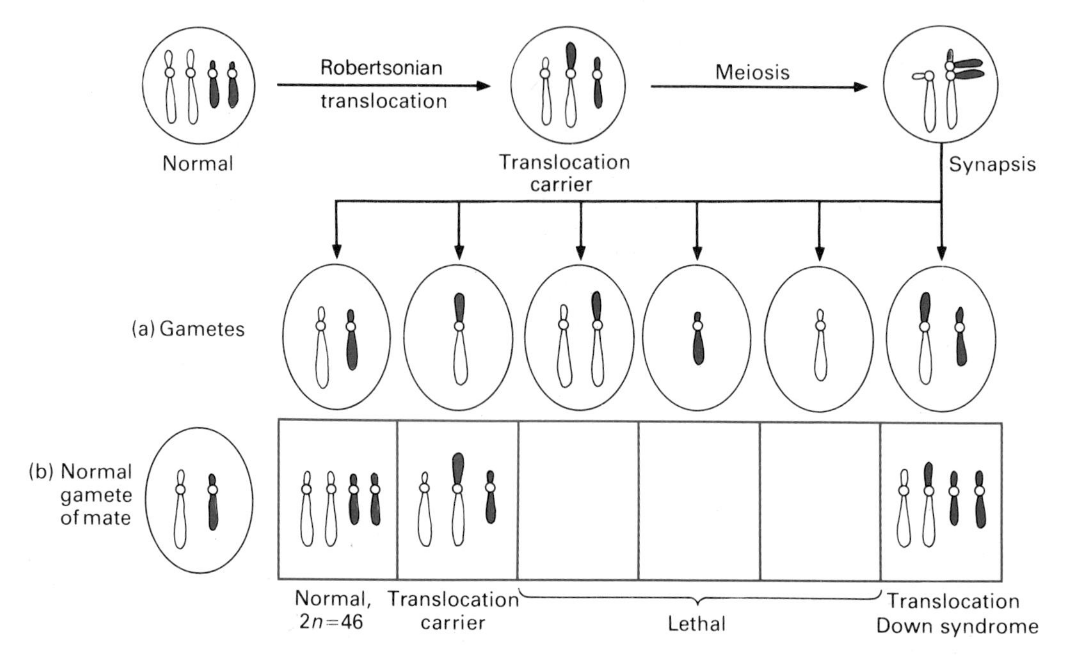

Figure 12.27. Chromosome segregation in a 14-21 translocation carrier. (a) The six types of gametes that can be produced by the carrier. (b) Types of offspring produced when a carrier mates with an individual with a normal karyotype.

a broad "moon" face, and a cry in infancy that sounds like a cat's cry. This syndrome is not necessarily lethal in infancy; a few affected individuals have survived to adulthood.

Other nonlethal deficiencies have also been reported. For example, partial deletions of either the long or the short arm of chromosome 18 are associated with relatively good survival rates of affected individuals. Ring chromosomes from every chromosome group have also been detected in live births. A *ring chromosome* is caused when terminal deletions occur at both ends of the normal chromosome, and the broken ends of the centric (centromere containing) fragment subsequently fuse to form a ring. While many of these deletions are not associated with any particular syndrome, they characteristically produce mental retardation.

Cytological investigations in humans have also revealed a number of different types of translocations. The most extensively studied translocation in humans is that associated with a certain type of Down syndrome. This chromosomal abnormality is responsible for 4 percent of the reported cases of Down syndrome. It differs from the common form of trisomy 21 in that individuals with translocation Down syndrome have 46 chromosomes rather than 47. They also possess one rather long chromosome, consisting of a copy of chromosome 14 (or another D group chromosome) to which chromosome 21 has been translocated. In almost every instance of this type, one of the parents of the individual with Down syndrome is a translocation carrier, with only 45 chromosomes, among which there is only one copy of chromosome 21, one copy of chromosome 14, and a combined 14-21 chromosome produced by centric fusion.

Down translocation carriers have a normal phenotype but are capable of producing six different gametes, shown in Fig. 12.27. Carriers not only have a high risk of producing offspring with Down syndrome, but they can also give birth to normal carriers like themselves. Thus, while the common form of trisomy 21 is usually not passed on to offspring, the translocation abnormality can be passed on by seemingly normal parents. Since translocation Down syndrome does not involve nondisjunction, its frequency does not increase with the age of the mother.

Analysis of Aberrant Chromosomes. During the past 20 years, new staining methods have been discovered that have revolutionized the study of human chromosomes. These methods, called chromosome banding, enable an investigator to stain chromosome regions differentially, permitting in some cases the exact identification of chromosomes involved in structural rearrangements. Figure 12.28 summarizes the staining procedures used and the results of some of these tech-

Figure 12.28. Four chromosome banding techniques commonly used in genetics. Only highlights of the basic procedures are given.

BANDING TECHNIQUE | GENERAL APPEARANCE OF CHROMOSOMES

G banding — Treat metaphase spreads with enzyme that digests part of chromosomal protein. Stain with Giemsa stain. Observe banding pattern with light microscope.

Q banding — Treat metaphase spreads with the chemical quinacrine mustard. Observe fluorescent banding pattern with a special ultraviolet light microscope.

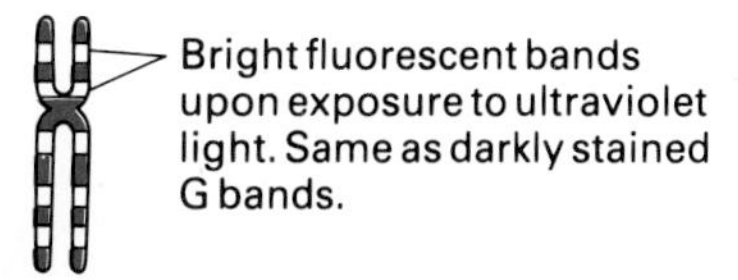

Bright fluorescent bands upon exposure to ultraviolet light. Same as darkly stained G bands.

R banding — Heat metaphase spreads at high temperatures to achieve partial denaturation of DNA. Stain with Giemsa stain. Observe microscopically.

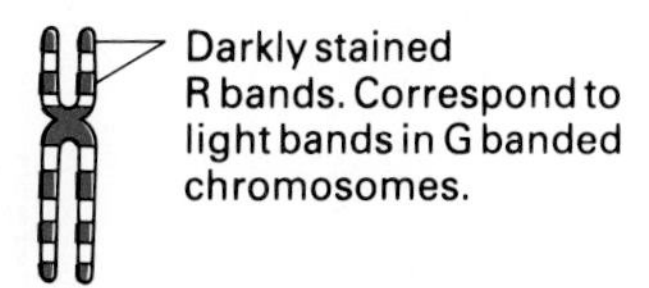

Darkly stained R bands. Correspond to light bands in G banded chromosomes.

C banding — Chemically treat metaphase spreads to extract DNA except from centromeric regions of chromosomes. Stain with Giemsa stain and observe microscopically.

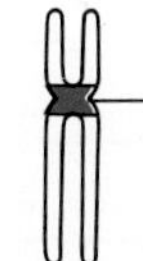

Darkly stained C band in centromeric region of the chromosome. Corresponds to region of constitutive hetero-chromatin.

niques. One of the more frequently used of these procedures is G banding. This technique involves treating the chromosomes with Giemsa stain (a mixture of basic dyes) after they have been exposed to an enzyme that digests part of the chromosomal protein. Dark-stained bands called *G bands* (G for Giemsa) are produced that can be studied under a light microscope. A diagrammatic representation of the G banding pattern for human chromosomes is given in Fig. 7.20.

Cytogeneticists have adopted a convenient method for identifying the bands produced by the G staining technique. The method is illustrated in Fig. 12.29. According to this procedure, each chromosome is separated into two segments beginning at the centromere, with the letter p (for petite) used to designate the short arm and q used for the long arm. Each segment, or arm, is divided into regions that are numbered consecutively from the centromere. The end of each region is defined on the basis of a landmark that is a consistent feature, such as a major band, that is useful in identifying the chromosome. Both the light and dark bands within each region are numbered consecutively, moving outward along the arm. A band bisected by the centromere is regarded as two bands, each one starting the first region of its designated segment. A landmark band is not bisected, however; rather, it constitutes the first band of the region distal to the designated boundary line. It is thus possible to identify a particular band by using four designations: the chromosome's number, the segment (p or q), the region number, and the band number within the region. For example, band 3q25 refers to band 5 in region 2 of the long arm of chromosome 3.

Geneticists have attempted to standardize chromosome terminology to cope better with the rapidly increasing discoveries of human chromosome anomalies.

Figure 12.29. Illustrations showing the method used to identify bands produced by the Q- and G-banding techniques. The dark regions on each chromosome represent darkly stained G bands or fluorescent Q bands. A particular band is identified by designating four items: the chromosome number, the segment (*p* or *q*), the region number, and the band number within the region.

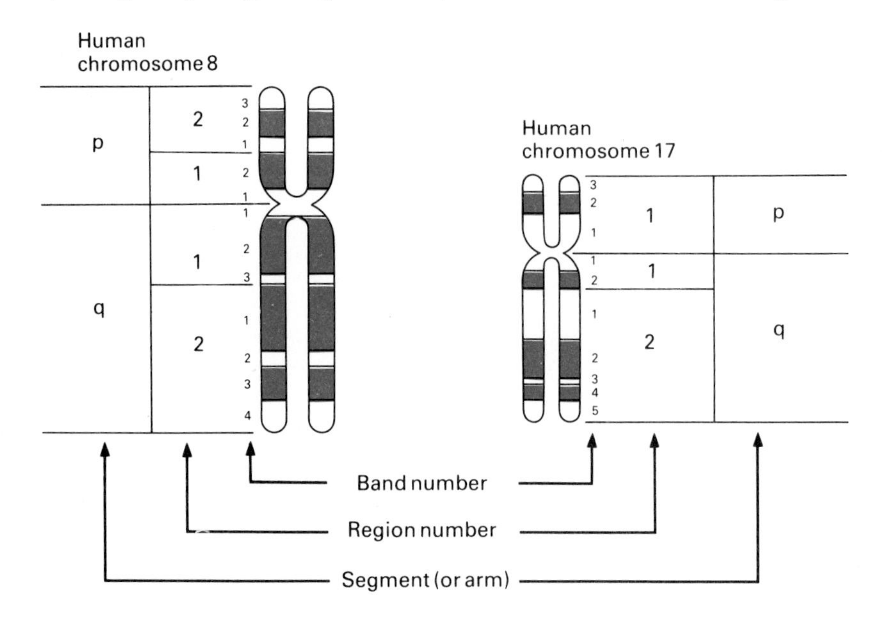

The nomenclature in use today describes a karyotype by indicating (1) the number of chromosomes, (2) the status of the sex chromosomes, (3) the presence or absence of a particular chromosome (designated by a + or − placed before the number assigned to the chromosome), and (4) the nature of any structural rearrangement. In addition, a diagonal (/) is used to separate cell lines in describing mosaicism.

Several symbols are used to indicate specific structural rearrangements, such as *del* for deletion, *dup* for duplication, *inv* for inversion, and *t* for translocation. For example, a male with cri-du-chat syndrome would be designated as 46,XY,del(5p), showing that the deletion is in the short arm of chromosome 5. At times it is also possible to determine precise break points by chromosome banding and to designate them in terms of specific band numbers. Whenever break points are known, they are given within parentheses following the chromosome designation, along with the segment (p or q). For example, if a male has a segment missing from chromosome 2 between band q24 and band q32, his karyotype designation is given as 46,XY,del(2)(q24q32).

TO SUM UP

1. Macrolesions involve the breakage and structural rearrangement of whole chromosomes. The four basic types of macrolesions are deficiencies (loss of a segment), duplications (repeat of a segment), inversions (reversal of a segment), and translocations (transfer of a segment to a different region on the same or a different chromosome).

2. We can recognize a macrolesion cytologically by observing the pairing arrangements of chromosomes during meiosis in heterozygotes for the aberrant chromosome. A chromosome with a deficiency or duplication will pair with its normal homolog by way of a single loop arrangement. An inversion chromosome pairs with its homolog through a double, inverted loop arrangement. In a reciprocal translocation, two aberrant chromosomes and their normal homologs pair in a cross configuration.

3. Macrolesions often result in a reduction in fertility and viability. Large deficiencies are usually lethal when homozygous. Reciprocal translocations are not lethal, but they decrease the fertility of heterozygotes by producing some nonfunctional spores or gametes. Inversion heterozygotes also form some nonfunctional products; in this case, they are formed when crossovers occur within the inverted segment of an aberrant chromosome. This effect has led geneticists to describe inversions as crossover suppressors, since the results of such crossovers are never detected.

4. Geneticists often make use of chromosome aberrations to localize genes on chromosomes. One common procedure is deletion mapping, in which they use overlapping deficiencies to identify the position of a gene in relation to some visible landmark on the chromosome. The procedure is based on the principle that a recessive gene will be expressed when it is placed in heterozygous combination with a homologous chromosome that has lost the corresponding dominant allele as part of a deficiency. Translocations can also be used to associate a gene with a particular chromosome. This aberration alters the inheritance pattern of the genes in the translocated region, thus serving as a marker for the chromosomal location of the affected linkage group.

5. The macrolesions with the most striking effects in humans are deficiencies and translocations. The best-known examples are cri-du-chat syndrome, which involves the deletion of the short arm of chromosome 5, and translocation Down syndrome, which usually involves the fusion of the long arms of chromosomes 14 and 21. Unlike the usual trisomy 21, translocation Down syndrome is transmitted by phenotypically normal carriers of the translocation chromosome. Since nondisjunction is not a factor, translocation Down syndrome does not increase in frequency with the maternal age.

6. Different staining methods have been developed that produce bands in metaphase chromosomes. These techniques include the use of modified Giemsa staining (G bands), which permit each chromosome to be unequivocally identified.

Questions and Problems

1. Define the following terms and differentiate between paired or among grouped terms:
   allopolyploid–autopolyploid
   amphidiploid
   aneuploidy–euploidy
   centric fusion
   chromosome banding
   homeologous
   inversion
   monosomic–trisomic–nullisomic
   mosaic
   nondisjunction
   somatic doubling
   translocation
2. The haploid number of chromosomes in a certain organism is 12. How many chromosomes would be present in a monosomic of this species?
3. How many different kinds of trisomics are possible in an organism with 10 pairs of chromosomes in its somatic cells?
4. A certain woman has unusually prominent satellites on one of her copies of chromosome 21. Her mother also carries this unusual chromosome. **(a)** If the woman has a baby, what is the chance that her child will inherit this chromosome? **(b)** The woman has a child with Down syndrome; the child has one normal copy of chromosome 21 and two copies of the unusual chromosome 21. At what stage of oogenesis did nondisjunction occur?
5. A child with Down syndrome is found to have 46 chromosomes. The child has a normal 46,XY brother and an apparently normal sister with 45 chromosomes. The father of these children has a normal complement of chromosomes. How many chromosomes does the children's mother possess? Explain your answer.
6. Two white blood cells from a normal human have abnormal chromosome numbers. One has 52 chromosomes, while the other has 69 chromosomes. One is aneuploid and the other is polyploid. Which is which?
7. Three different somatic cells of an animal were examined and were found to contain 59, 62, and 93 chromosomes. One of the cells is normal, one is polyploid, and the third is aneuploid (not necessarily in the order given). What is the haploid number of chromosomes in this species?
8. In corn, red aleurone is determined by the completely dominant gene *R*, and colorless aleurone is determined by the recessive allele *r*. **(a)** The pollen from a trisomic plant of genotype *Rrr* is used to fertilize a diploid plant of genotype *rr*. What phenotypic ratio is expected in the progeny, assuming that $n + 1$ pollen are nonfunctional? **(b)** How would you answer part (a) of this question if the genotypes of the egg and pollen donors were reversed?
9. A pair of twins born to normal parents appear to be identical in every way (same blood types, can exchange skin grafts, and so on) except that one twin is a normal male and the other is a female with Turner syndrome. What is the likely explanation for this result?
10. Criticize the following statements:
    (a) Inversions suppress crossing over.
    (b) Multiple sets of chromosomes are homologous in autopolyploids and nonhomologous in allopolyploids.
    (c) Polyploidy is responsible for rapid evolutionary changes in some groups of organisms.
11. Various species of raspberries have 14, 21, 28, 42, and 49 chromosomes. **(a)** How should the chromosome complements of these species be designated in relation to the basic number of chromosomes, *x*? **(b)** How does evolution appear to be taking place in this plant group?

12. An amphidiploid has 46 chromosomes, comprising chromosome sets derived from species 1 and species 2. If the haploid number of chromosomes in species 1 is 7, how many chromosomes make up the haploid set of species 2?

13. Emmer wheat has 28 chromosomes, comprising 14 chromosomes of genome A and 14 chromosomes of genome B. How many different nullisomics are possible in this species?

14. Defective tooth enamel is a dominant X-linked human trait. A person with Klinefelter syndrome whose parents have defective teeth has normal tooth enamel. In what parent and at what stage of meiosis did nondisjunction most probably occur?

15. Consider the testcross *AAaa* × *aaaa*, in which the alleles show complete dominance and a perfect 2:2 segregation pattern. What phenotypic ratio is expected among the offspring?

16. Three pairs of duplicate genes are responsible for kernel color in wheat: R_1 and r_1, R_2 and r_2, and R_3 and r_3. A genotype of R_1----- (or --R_2--- or ----R_3-) produces red kernels, and a genotype of $r_1r_1r_2r_2r_3r_3$ produces white kernels. The three gene pairs are carried on homeologous sets of chromosomes and assort independently of each other. Suppose that varieties which breed true for red seeds and white seeds are crossed, and their offspring are intercrossed to give an F_2 generation. What phenotypic ratio is expected in the F_2?

17. Suppose that an autotetraploid that is heterozygous for two gene pairs (*AAaaBBbb*) is selfed. Assuming complete dominance and a 2:2 segregation pattern for each gene pair, what fraction of the offspring is expected to have the doubly recessive (*aaaabbbb*) genotype?

18. The following diagram represents part of a pair of polytene chromosomes in the salivary glands of *Drosophila melanogaster.* The bands on the homologs have been arbitrarily designated by the letters a through f.

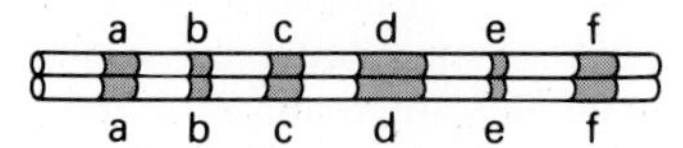

Draw a diagram showing the appearance of the chromosome pair when one member has **(a)** a deletion of band c, **(b)** a tandem duplication of band d, and **(c)** an inversion of the segment that includes bands c and d.

19. Identify the band designated 4q27 on a G-banded preparation of human chromosomes. Is the band lightly or darkly stained? (See Fig. 7.20 for G bands.)

20. Several human females have been reported who are 46, XY in chromosome makeup but are missing part of the short arm of the Y chromosome. These females have poorly developed (streak) gonads and are sterile. Individuals who are 46,XY and missing part of the long arm of the Y are also known, but these individuals appear to be normal males. What does this finding suggest concerning the location of the gene (or genes) for the factor that determines maleness?

21. Five races of the fruit fly can be distinguished by the sequence of bands A through F on one of their polytene chromosomes. It is believed that these races were separated by single inversion differences, beginning with the most primitive banding sequence, ABCDEF. Propose an evolutionary pathway that relates the banding patterns of the following five races: **(a)** AD CEBF, **(b)** ABCDEF, **(c)** BCDAEF, **(d)** ADCFBE, **(e)** ADCBEF.

22. Certain groups of closely related species have chromosome numbers that are not simple multiples of some basic integer but that vary almost continuously over a wide range of values. What mechanisms of chromosome mutation might be responsible for such variation?

23. Account for the following changes in karyotype:

(a) Seven pairs of small acrocentric chromosomes → five pairs of small acrocentric chromosomes plus one pair of large metacentric chromosomes.

(b) Seven pairs of small acrocentric chromosomes → six pairs of small acrocentric chromosomes plus one pair of small metacentric chromosomes.

24. Suppose that a plant that is heterozygous for a reciprocal translocation is crossed to one with the normal karyotype. **(a)** What will be the ratio of normal : translocation heterozygote : translocation homozygote among the offspring? **(b)** Translocation homozygotes are often phenotypically indistinguishable from the normal stock. What specific crosses could be used to determine which is which?

25. Suppose that by using X-rays to irradiate a plant that is homozygous for the dominant *A* gene, you induce a deficiency of the chromosomal segment that includes the *A* gene on one of the homologs. Assuming that you also have access to *aa* plants, outline a procedure that you could use to determine cytologically the location of the *A* gene locus on the chromosome.

26. Down syndrome occurs in approximately 1 out of 700 live births in the general population. **(a)** The incidence of Down syndrome is about 22 times greater among children of 45-year-old women than in the population at large. What is the chance that a child with Down syndrome will be born to a woman in this age group? **(b)** A 45-year-old woman is about 60 times more likely to give birth to a child with Down syndrome than is a woman of 20. What is the incidence of Down syndrome among the offspring of 20-year-olds? **(c)** If women over 35 years of age give birth to 60 percent of the children with Down syndrome but produce only 15 percent of all children, what is the frequency of children with Down syndrome among the children born to women in this age group?

27. While humans have 46 chromosomes, chimpanzees (*Pan troglodytes*), gorillas (*Gorilla gorilla*), and orangutans (*Pongo pygmalus*) have 48. These different apes possess two pairs of acrocentric chromosomes that are not present in humans but that show similarities in G-banding patterns to the long and short arms of the large submetacentric human chromosome 2. Suggest a mechanism that could account for the difference in chromosome number between humans and apes.

28. A team of geneticists uses X-rays to induce five deletions in corn, each spanning a different region of chromosome 9. Plants that are heterozygous for these deletions are then crossed to individuals of the normal karyotype, each of which is homozygous for a different recessive gene, labeled *a* through *f*. The results are given below (+ indicates the presence of recessive progeny, – designates their absence). Construct a linkage map of the various gene loci in these crosses.

Deletion heterozygotes	Homozygotes for gene *a*	*b*	*c*	*d*	*e*	*f*
1	+	+	–	+	–	–
2	+	–	+	–	+	–
3	+	+	–	+	+	–
4	–	+	–	–	–	+
5	–	–	+	–	+	–

29. The gene for starchy endosperm in corn (*Wx*) is dominant to its allele for waxy endosperm (*wx*). A plant that is heterozygous for a translocation involving chromosomes 8 and 9 and that developed from a starchy kernel is crossed to one that is homozygous for *wx*. The cross gave the following offspring: 554 semisterile, *Wx;* 194 semisterile, *wx;* 166 normal fertility, *Wx;* and 586 normal fertility, *wx*. How far is the *wx* locus from the translocation break?

30. A 2:2 segregation pattern of alleles is possible in tetraploids only when the gene locus is closely linked to the centromere, so that crossing over in the region between the gene and the centromere is negligible. At the other extreme are genes that are separated from the centromere by a very large distance. When the distance is large, crossing over occurs so frequently that all eight chromatids of the four homologs assort independently in the region of the gene locus. Any two of the eight allelic genes can thus enter a gamete at random, resulting in $8!/2!6! = 28$ possible allelic combinations. **(a)** Suppose that a dominant gene *A* and its recessive allele *a* are located far from the centromere on their chromosome. What ratio of *AA*:*Aa*:*aa* will a tetraploid of genotype *AAaa* produce in the gametes? **(b)** If the *AAaa* tetraploid is selfed, what ratio of dominant : recessive phenotypes is expected among the progeny? (*Hint:* In working part (a) of this problem, designate the eight allelic genes as A_1, A_1', A_2, A_2', a_3, a_3', a_4, a_4', where primes are used to distinguish alleles on sister chromatids. Using these designations, you should be able to determine the allelic combinations required to produce the different gamete types.)

Suggested Readings / Part V

Chapter 11

Auerbach, C. *Mutation Research,* Chapman & Hall, London, 1976.

Devoret, R. Bacterial tests for potential carcinogens. *Scientific American* 241:40–49, 1979.

Drake, J. W. *The Molecular Basis of Mutation,* Holden-Day, San Francisco, 1970.

Fuchs, F. Genetic amniocentesis. *Scientific American* 242:47–53, 1980.

Neel, J. V. Frequency of spontaneous and induced "point" mutations in higher eukaryotes. *J. Hered.* 74:2–15, 1983.

Oppenheimer, S. B. Prevention of cancer. *Am. Lab.* 15:66–72, 1983.

Schull, W. J., M. Otake and J. V. Neel. Genetic effects of the atomic bombs: a reappraisal. *Science* 213:1220–1227, 1981.

Upton, A. C. The biological effects of low-level ionizing radiation. *Scientific American* 246:41–49, 1982.

Chapter 12

Blakeslee, A. F. New Jimson weeds from old chromosomes. *J. Hered.* 25:80–108, 1934.

Feldman, M. and E. R. Sears. The wild gene resources of wheat. *Scientific American* 244:102–113, 1981.

Simmonds, N. W. *Evolution of Crop Plants,* Longman, New York, 1976.

Swanson, C. P., T. Merz and W. J. Young. *Cytogenetics,* Prentice-Hall, Englewood Cliffs, N.J., 1981.

Thompson, J. S. and M. W. Thompson. *Genetics in Medicine,* Saunders, Philadelphia, 1980.

Part VI

Population and Quantitative Genetics

Chapter 13

Genetics of Populations

So far in our discussion of the nature of genes, we have focused our attention on the individual, such as a single cell, organism, or strain, and have limited our consideration of gene transmission to the results of specific matings, such as those encountered in the analysis of a pedigree or breeding experiment. We will now view genes from a somewhat different perspective and consider the fate of genes in a large collection of individuals under the less restrictive conditions that prevail in nature. In order to study gene transmission in nature (i.e., in the field), we must extend our investigations to the level of the population and must then analyze the genetic characteristics of whole breeding groups. These characteristics, which are collectively known as the genetic structure of a population, include the nature and frequencies of the alleles and genotypes in the population and the pattern in which the individuals of the population select their mates.

The study of the genetic mechanisms of interbreeding populations constitutes the field of population genetics. Population geneticists seek to understand the means by which genetic variation is derived and maintained and to describe the nature of the processes involved in adaptive change. Geneticists make extensive use of mathematical models in these theoretical investigations, in order to gain insight into the basic mechanisms of evolution. Population genetics is also a field that has practical value, in that it has many applications to other scientific areas, such as medicine, anthropology, and plant and animal breeding. For example, population genetics has been used to account for the origin and persistence of genetic defects in human populations and to supplement pedigree analysis in determining the mode of inheritance of certain traits.

This chapter is an introduction to the concepts of population genetics. In the discussion that follows, we will begin by describing the genetic characteristics of sexually reproducing populations, including in our discussion some mathematical methods that are employed in analyzing these populations. This section is followed by a quantitative treatment of the genetic consequences of random mating.

Allelic Variation Within Populations

A sexually reproducing population is a group of interbreeding individuals that occupies a particular space at a particular time. The individuals within such a group are typically variable in genetic makeup, so that they differ in their allelic and genotypic compositions. To understand better the extent of these differences, we will begin by considering the nature and measurement of allelic variations in populations.

Genotype and Allele Frequencies

Populations usually differ from one another in terms of the number of individuals that fall into each genotypic class. But populations also differ from one another in terms of total size. It is therefore desirable for the investigator to eliminate population size as a variable when comparing groups; this factor can be eliminated by expressing the genotypic composition of a group in terms of the relative frequency of each genotype instead of the actual number of individuals in each genotypic class. The number of individuals in a class can be converted to a relative frequency by simply dividing the number of individuals with that genotype by the total number of individuals in the entire group. For example, suppose that among 500 individuals in a population, 250 are *AA* in genotype, 100 are *Aa*, and 150 are *aa*. The corresponding genotype frequencies are then $f(AA) = 250/500 = 0.5$, $f(Aa) = 100/500 = 0.2$, and $f(aa) = 150/500 = 0.3$. Note that throughout this chapter, the expression "frequency of" is shortened to $f(\ )$.

We should emphasize that unlike the progeny of a particular mating, the genotype frequencies in a natural population are in theory not limited to a specific set of values. These frequencies may instead take on any numerical distribution; the only restriction is that their sum must equal one. Thus, while the frequencies of the genotypes *AA*, *Aa*, and *aa* in the preceding example population are 0.5, 0.2, and 0.3, they could be 0.2, 0.3, and 0.5 (or any other combination of values that add to one) in another group. You should also note that since the frequencies are simply fractions of the total number, they can be treated mathematically in much the same way as probabilities.

While the genotype frequencies describe the genetic composition of the diploid phase of a particular generation, the **allele** (or **gene**) **frequencies** describe the genetic composition of the haploid phase.The frequency of an allele in a population is defined as

$$\text{Allele (or gene) frequency} = \frac{\text{Number of copies of a given allele}}{\text{Total copies of all alleles at the locus}}.$$

This value measures the average contribution to each allele made at any moment by the various diploid genotypes in the population. As long as the various genotypes are equally viable and fertile, the allele frequency also represents the relative proportion of all gametes of a population that carry the allele in question. For example, if 30 percent of the gametes carry allele A and 70 percent carry a, then the frequencies of these alleles in the hypothetical population are $f(A) = 0.3$ and $f(a) = 0.7$, assuming that all genotypes (AA, Aa, and aa) are equally viable and fertile.

If the genotypes in a population differ from one another in terms of the average number of surviving offspring they produce, then allelic compositions of the haploid and diploid phases will not be the same, and they will tend to change from one generation to the next. This change in the genetic makeup of a population owing to reproductive inequalities among the genotypes constitutes the process of selection, which we will discuss in greater detail in the next chapter.

Calculating Allele Frequencies from Genotype Frequencies. An allele frequency is statistically a numerical average that is computed by multiplying the frequency of each diploid genotype by its fractional contribution to the allele in question. We will use the following population to show how to determine allele frequency:

	Alleles		Genotypes		
	A	a	AA	Aa	aa
Frequencies:	$f(A)$	$f(a)$	$f(AA)$	$f(Aa)$	$f(aa)$

Observe that all of the alleles in AA individuals and half of the alleles in Aa individuals are of type A. The average contribution of the individuals in this population to the frequency of the A allele is then $(1)f(AA) + (\frac{1}{2})f(Aa) + (0)f(aa)$. Similarly, all of the alleles in aa homozygotes and half of the alleles in Aa heterozygotes are of the alternative allelic form a. Thus, the average contribution of these genotypes to the frequency of the a allele is $(1)f(aa) + (\frac{1}{2})f(Aa) + (0)f(AA)$. By reducing these equations to their simplest forms, the relationships between the allele and genotype frequencies can be written as follows:

$$\text{Frequency of } A \text{ allele:} \quad f(A) = f(AA) + \tfrac{1}{2}f(Aa)$$
$$\text{Frequency of } a \text{ allele:} \quad f(a) = f(aa) + \tfrac{1}{2}f(Aa)$$

where $f(A) + f(a) = 1$.

To use a numerical example, let us suppose that we wish to determine the frequencies of the R and R' alleles in a population of Shorthorn cattle that consists of 50 red (RR), 300 roan (RR'), and 150 white ($R'R'$). Since the total population consists of 500 individuals, we first convert the observed numbers to genotype frequencies by dividing each value by 500. The genotype frequencies are thus $f(RR) = {}^{50}/_{500} = 0.1$, $f(RR') = {}^{300}/_{500} = 0.6$, and $f(R'R') = {}^{150}/_{500} = 0.3$. We can then compute the allele frequencies as

$$f(R) = 0.1 + \tfrac{1}{2}(0.6) = 0.4 \quad \text{and} \quad f(R') = 0.3 + \tfrac{1}{2}(0.6) = 0.6.$$

Since the sum of both allele frequencies must equal 1, we can arrive at the frequency of either allele by simply subtracting the frequency of the other allele from one. Thus, if the frequency of an allele is 0.4, as in this example, we can calculate the frequency of the alternative form of this gene as $1 - 0.4 = 0.6$.

We can extend this procedure to multiple alleles as well. To illustrate, let us assume N total alleles for a locus in a certain population, with the alleles designated as $A_1, A_2, \ldots, A_N$. The genotypes in the population are then A_1A_1, A_1A_2, A_2A_2, $\ldots$, A_NA_N. Each allele is carried in duplicate by one homozygous genotype and occurs in a single dose in $N - 1$ heterozygous types. When it is expressed as an average contribution of the different genotypes, therefore, the frequency of the ith allele becomes

$$f(A_i) = f(A_iA_i) + \tfrac{1}{2}f(A_iA_1) + \tfrac{1}{2}f(A_iA_2) + \ldots + \tfrac{1}{2}f(A_iA_N)$$

If we generalize this procedure for any number of alleles, *the frequency of an allelic gene is equal to the frequency of the homozygous genotype for this allele plus one-half the sum of all the frequencies of the heterozygous carriers of that allele.*

EXAMPLE 13.1. In a certain population, the genotypes for the three alleles A_1, A_2, and A_3 occur in the following ratio: $1\ A_1A_1{:}2\ A_2A_2{:}1\ A_3A_3{:}\ 2\ A_1A_2{:}4\ A_2A_3{:}2\ A_1A_3$. What are the frequencies of the three alleles in this population?

Solution. We observe that the sum of the terms in the ratio gives $1 + 2 + 1 + 2 + 4 + 2 = 12$. The frequencies of ($A_1A_1$, A_2A_2, A_3A_3, A_1A_2, A_2A_3, A_1A_3) are therefore ($^1/_{12}$, $^2/_{12}$, $^1/_{12}$, $^2/_{12}$, $^4/_{12}$, $^2/_{12}$). We can then calculate the frequencies of the alleles as follows:

$$f(A_1) = f(A_1A_1) + \tfrac{1}{2}f(A_1A_2) + \tfrac{1}{2}f(A_1A_3) = \tfrac{1}{12} + \tfrac{1}{2}(\tfrac{2}{12}) + \tfrac{1}{2}(\tfrac{2}{12}) = \tfrac{3}{12}$$
$$f(A_2) = f(A_2A_2) + \tfrac{1}{2}f(A_1A_2) + \tfrac{1}{2}f(A_2A_3) = \tfrac{2}{12} + \tfrac{1}{2}(\tfrac{2}{12}) + \tfrac{1}{2}(\tfrac{4}{12}) = \tfrac{5}{12}$$
$$f(A_3) = f(A_3A_3) + \tfrac{1}{2}f(A_1A_3) + \tfrac{1}{2}f(A_2A_3) = \tfrac{1}{12} + \tfrac{1}{2}(\tfrac{2}{12}) + \tfrac{1}{2}(\tfrac{4}{12}) = \tfrac{4}{12}$$

Detecting Allelic Variability

Geneticists ordinarily determine the extent of allelic variability within a population by estimating the number of alleles that are present and their respective frequencies. To make this determination, the geneticists must first be able to identify different genotypes. The procedure is relatively straightforward for traits that show incomplete dominance or codominance. In this case, each genotype is expressed as a distinguishable phenotype, so that the allele frequencies can be determined directly from the phenotype frequencies.

One example of a codominant trait that has been studied extensively in human populations is the MN blood group. The different blood group antigens in this system are detected by using purified immune serum that contains anti-M or anti-N antibodies. Because the alleles for these antigens are codominant, this

Table 13.1. Frequencies of genotypes and alleles for the MN blood group locus among various human populations.

		GENOTYPE FREQUENCY			ALLELE FREQUENCY	
POPULATION	LOCATION	L^ML^M	L^ML^N	L^NL^N	L^M	L^N
Aborigines	S. Australia	0.024	0.304	0.672	0.176	0.824
Bengali	India	0.354	0.508	0.138	0.608	0.392
Eskimo	Baffin Island	0.662	0.310	0.028	0.817	0.183
German	Berlin	0.284	0.499	0.217	0.533	0.467
Japanese	Tokyo	0.285	0.510	0.205	0.540	0.460
Polynesian	Hawaii	0.125	0.417	0.458	0.333	0.667

procedure in effect also tests for the occurrence of the corresponding blood group genotypes.

The results of a few studies on the MN blood group are given in Table 13.1. Two basic patterns emerge from these results. First, it is apparent that *differences exist among the individuals within each population.* Most populations are genetically variable for this trait, with both blood group alleles present at significant frequencies. The MN trait and other traits for which two or more forms commonly occur within a population are said to be **polymorphic.** A **monomorphic** character, by comparison, describes a population in which virtually all the individuals are of one form for that trait. Second, we can see from the MN data in Table 13.1, *the allele frequencies tend to differ among the populations.* These differences are particularly evident in populations that are separated by wide geographic barriers. Thus, more than 60 percent of the Eskimos tested on Baffin Island were type M, and less than 3 percent were type N. The reverse is true among the Aborigines in Australia, where type N is the most common.

Unlike the MN blood group system, many alleles show complete dominance at the level of their expression by the individual organism. In the case of complete dominance, heterozygotes appear phenotypically the same as dominant homozygotes, so that it is difficult to determine the allele frequencies directly from the phenotype frequencies, as in codominance, by using the standard methods. Crossing techniques have proven to be useful in detecting the presence of certain rare recessive alleles. Heterozygote frequencies are then measured indirectly, providing a means for calculating allele frequencies. Cytological techniques have also been used in certain instances in which an allele is associated with an identifiable chromosome rearrangement (e.g., an inversion). In recent years, geneticists have measured genetic variation at the molecular level. It is this biochemical approach to the study of variation that is considered in the following section.

Detecting Variation in Protein Structure. Recent experimental developments in molecular genetics have made it possible to test for the occurrence of alleles that are masked in the heterozygous state by identifying amino acid changes in their protein products. A direct relationship usually exists between allelic variation in a structural gene and variation in the amino acid sequence of the polypeptide for

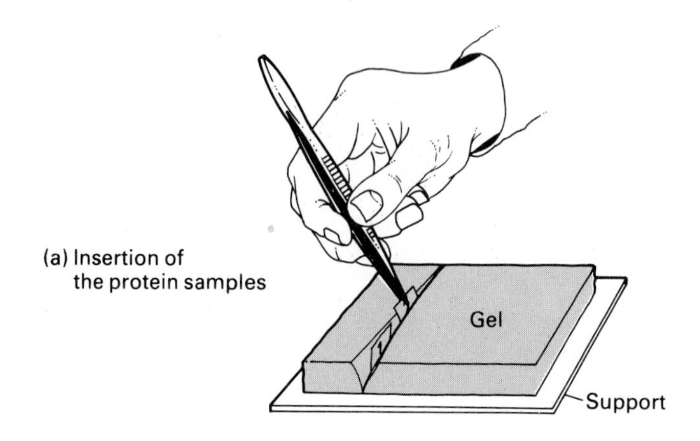

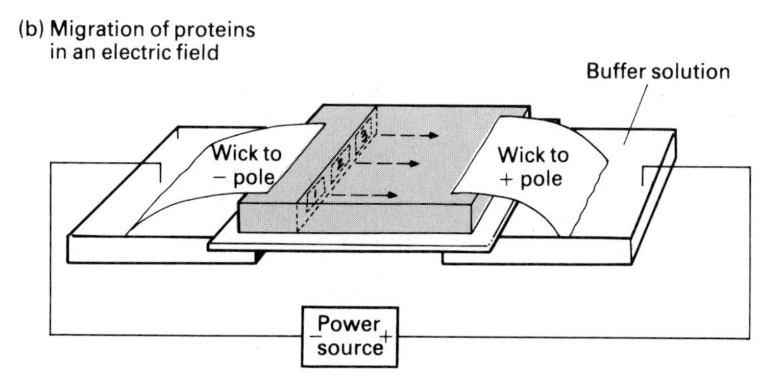

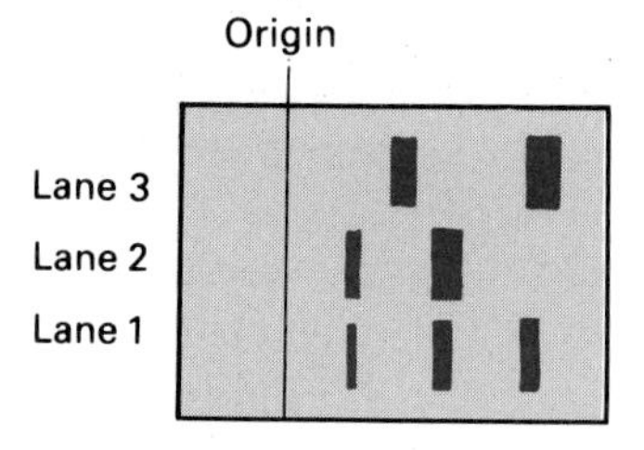

Figure 13.1. Identification of variation among proteins by the procedure of gel electrophoresis. (a) Solutions that contain the various proteins are adsorbed onto small paper squares, which are then inserted into a slit in the gel. (b) An electric current is applied across the gel, causing the charged proteins to migrate through the gel. (In this case, the buffer pH has been adjusted so that all of the proteins bear a net negative charge and therefore migrate toward the positive pole.) The rate of migration depends on the magnitude of the charge carried by a protein and on its conformation (the more highly charged, smaller, spherical molecules move faster). (c) Staining of the gel reveals the final positions of the protein variants. The protein bands can also be cut from the gel for chemical analysis.

which it codes. For example, a base substitution that converts the codon GAA to AAA in the mRNA would cause lysine to replace glutamic acid in the corresponding polypeptide chain. Using chemical techniques, we can detect amino acid changes of this type in heterozygotes regardless of whether or not these changes produce an observable difference in external appearance. Thus, by evaluating the phenotype closer to the level of the gene itself, it is possible to bypass many of the problems that are associated with dominance and to measure allele frequencies directly.

The ideal approach to take in identifying protein differences is to determine the complete amino acid sequences in the normal and variant polypeptide chains. While this has been accomplished in certain cases, complete amino acid sequencing is normally too expensive and time-consuming to be performed routinely in practice. We usually have to be satisfied with swifter and less costly methods that give only partial information concerning the amino acid changes.

One substitute procedure that is commonly used is **electrophoresis** (Fig. 13.1). In the gel electrophoresis method, tissue samples are homogenized to release the cell proteins. For each sample, a small amount of liquid containing the proteins is then placed on a layer of gel. A direct current is passed through the gel, causing the charged proteins to migrate in response to the electric field. Molecular variants of a given protein that differ in charge (i.e., in the mix of negatively charged and positively charged amino acids) and conformation (i.e., in molecular size and shape) tend to migrate at different rates and to become separated from one another on the gel. After a certain amount of time, the gel is removed from the electric field and is stained for proteins. The result is a series of dark-stained spots that designate the final positions of the different protein forms (Fig. 13.2).

Gel electrophoresis can only distinguish between proteins that differ sufficiently in charge (and, to a certain extent, in size and shape) so that they become separated in an electric field. For instance, the replacement of a negatively charged amino acid with one that is positive or neutral in charge (e.g., the replacement of glutamic acid with lysine or glycine) would probably alter the rate of migration of the protein sufficiently to be detected. But the replacement of an amino acid with one of the same charge (e.g., the replacement of glycine with alanine) may not. Only about one-third of all amino acid replacements result in a change in charge. Consequently, gel electrophoresis tends to underestimate the amount of genetic variability at a given gene locus.

The results of electrophoretic studies on various human enzymes are given in Table 13.2. It is apparent that substantial variation exists at these structural gene loci, with at least two alleles appearing for every trait (three for red-cell acid phosphatase and peptidase D). The allele frequencies in Table 13.2. were determined from the frequencies of electrophoretically detectable variants for each enzyme. These molecular variants, called **allozymes,** are alternative enzyme forms that are coded for by different alleles at the same locus.

Studies on allozyme variation have been conducted on a wide variety of species, including some species that had not been previously analyzed by standard genetic techniques. For most species, the results indicate that about one-third of

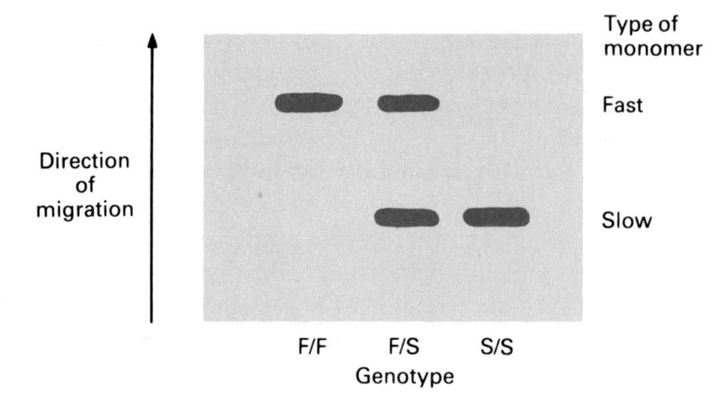

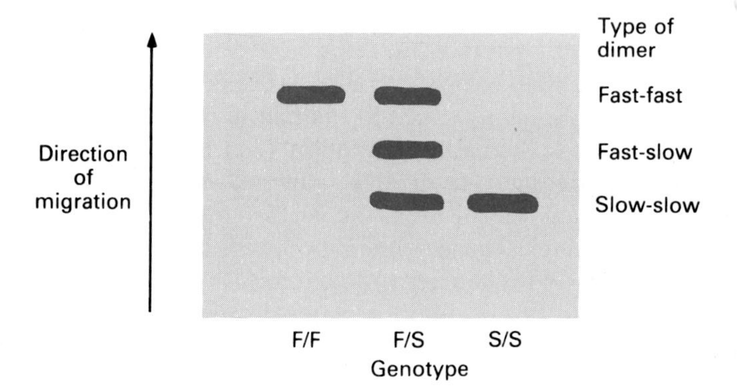

Figure 13.2. Variation in electrophoretic patterns of proteins that can be explained on the basis of two alleles. (a) A monomeric protein, which consists of a single polypeptide, will show only two forms in heterozygotes: a fast-migrating form associated with allele F and a slow-migrating form associated with allele S. (b) A dimeric protein, which consists of two polypeptides, can show three forms in heterozygotes: A slowly migrating homodimer (slow-slow), a rapidly migrating homodimer (fast-fast), and a heterodimer (fast-slow) with an intermediate electrophoretic mobility.

Table 13.2. Allele frequencies determined from an electrophoretic survey of certain enzymes in European and African populations.

	EUROPEANS			AFRICANS		
		Allele			Allele	
ENZYME	(1)	(2)	(3)	(1)	(2)	(3)
Adenosine deaminase	0.94	0.06	—*	0.97	0.03	—
Adenylate kinase	0.95	0.05	—	1.00	—	—
Peptidase A	1.00	—	—	0.90	0.10	—
Peptidase D	0.99	0.01	—	0.95	0.03	0.02
Phosphoglucomutase						
Locus PGM_1	0.77	0.23	—	0.79	0.21	—
Locus PGM_3	0.74	0.26	—	0.37	0.63	—
Red-cell acid phosphatase	0.36	0.60	0.04	0.17	0.83	—

* A dash (—) indicates that the allele occurs with a frequency of less than 0.01.
Source: Data originally published in *The Principles of Human Biochemical Genetics,* by H. Harris, 1970. Reproduced with permission from Elsevier Biomedical Press, Amsterdam.

all structural gene loci are polymorphic, on the average. In other words, about 1 out of every 3 loci tested has two or more alleles segregating in the population. The precise number of alleles in each instance and the frequencies at which they occur vary with the locus and species studied.

EXAMPLE 13.2. A particular enzyme in humans is known to be a tetramer that consists of four polypeptide chains coded for at a single locus. Studies reveal that the enzyme isolated from most individuals produces only a single band when subjected to gel electrophoresis, but that this enzyme band varies in electrophoretic mobility, with a fast form of the enzyme occurring in some individuals and a slow form occurring in others. In contrast, the enzyme isolated from a few individuals within the population produces five bands upon electrophoresis. These bands range in mobility from the fast form to the slow form and occur in the relative amounts of 1:4:6:4:1. Give a genetic explanation of these results, indicating the number of alleles that are most probably responsible for protein variation in this case.

Solution. The data can best be explained on the basis of two alleles, A and A', which produce polypeptides that we shall designate α and α'. The fast and slow enzyme forms are then α_4 and α_4', which are produced by the AA and $A'A'$ homozygotes in the population. The AA' heterozygotes would then produce both α and α' polypeptide chains, presumably in equal frequency. If these polypeptides combine at random to form the tetramer, they should yield five allozymes—α_4, $\alpha_3\alpha'$, $\alpha_2\alpha_2'$, $\alpha\alpha_3'$, and α_4'—in the proportions given by the binomial expansion $(1/2 + 1/2)^4 = 1/16 + 4/16 + 6/16 + 4/16 + 1/16$. (Recall from Chapter 2 that the value of each term in this expansion can be calculated by the binomial formula.)

TO SUM UP

1. The study of the genetic mechanisms of interbreeding populations is called population genetics. Population geneticists attempt to explain the origin and maintenance of genetic variation in natural populations and to describe the genetic processes involved in evolution.
2. Genotype frequencies and allele (or gene) frequencies, both of which are expressed as proportions of the total, are used to characterize the genetic structure of a population. The allele frequencies can be computed from the genotype frequencies by adding the frequency of the homozygote for the allele in question to one-half the sum of the frequencies of heterozygous carriers for the allele.
3. Gel electrophoresis provides information on the replacement of one amino acid in a protein by another. Geneticists use this technique to detect allelic variation in structural genes based on changes in the electrophoretic mobility of the protein products of the genes. Electrophoretic studies reveal that most populations maintain considerable genetic variation, with about one-third of all loci studied having two or more alleles.

Organization of Genetic Diversity

Genes are comparatively stable entities that are transmitted intact from one generation to the next. Genotypes, in contrast, have only a temporary existence. Since alleles are separated during meiosis and reshuffled into new combinations when mating occurs, genotypes must be formed anew in each generation. The frequencies of the genotypes produced therefore depend on the kinds and frequencies of matings in the parental population.

Populations usually consist of more than a single genotype. The two or more genotypes in the population typically result in several different kinds of matings, each of which produces its own characteristic genotypic ratio. Therefore, the genotype frequencies in a given generation are statistical averages of the genotype frequencies produced by each different parental mating. For example, if two-thirds of all matings in a population are *aa* × *aa* and one-third are *AA* × *AA*, and both produce on average the same number of offspring, then two-thirds of the progeny will be *aa* in genotype and one-third will be *AA*.

Random Mating Frequencies

While genotype frequencies can be computed as statistical averages from the mating frequencies, mating frequencies cannot be calculated from the genotype frequencies unless precise information is known about the mating system of the population. The mating system is the pattern by which mates are selected by members of the parental generation. The simplest mating system to describe—and the one that serves as the primary basis for comparison—is **random mating.** Mating is said to be random when mates are selected without regard to genotype. In other words, random mating occurs when the genotypes of prospective mates are statistically uncorrelated, so that there is no tendency for individuals to select mates that are similar to or different from themselves in genotype.

Natural populations very nearly have random mating for many traits that do not play an important role in the mating procedure. For example, human populations tend to have random mating with regard to blood type. On the other hand, mating patterns for such traits as height and intelligence show significant departures from randomness.

When mating is random, the genotypes of each pair of mates are statistically independent. Our ability to assume statistical independence greatly simplifies the numerical calculations, since the frequency of a particular mating can then be expressed as the product of the frequencies of the combined genotypes. Suppose, for example, that $f(AA) = 1/2$ and mating is random, so that any individual in the population has a 50 percent chance of selecting an *AA* mate. The frequency of *AA* × *AA* matings will then be $(1/2)(1/2) = 1/4$.

To extend this example further, suppose that all other individuals in the population are *aa* in genotype, so that the population consists of the two genotypes *AA* and *aa* in equal proportions. Four kinds of matings are then possible: *AA* male × *AA* female, *AA* male × *aa* female, *aa* male × *AA* female, and *aa* male × *aa* female. Each kind occurs with a frequency of 1/4. If all four kinds of matings produce the same number of offspring on the average, the collective progeny formed by these matings will occur in a ratio of 1 *AA* : 2 *Aa* : 1 *aa* (see Fig. 13.3). Note that this is the same result as we would expect from the cross *Aa* male × *Aa* female. As we shall see later, it is also the progeny ratio produced by numerous other mixtures of parental genotypes under random mating, such as 1 *AA* : 2 *Aa* :

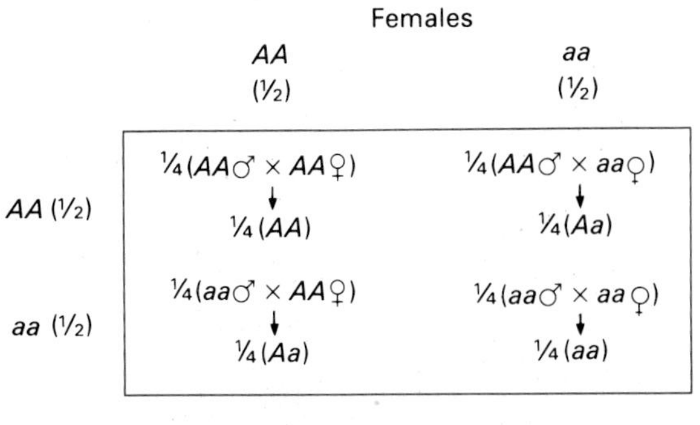

Figure 13.3. Results of random mating in a population that consists of a mixture of *AA* and *aa* genotypes in a ratio of 1:1. When mating is random, the frequency of each kind of mating equals the product of the frequencies of the combined genotypes. It is assumed that all females produce the same average number of offspring, so that the proportion of all offspring produced in each case equals the mating frequency.

1 *aa* and 2 *AA* : 1 *Aa* : 2 *aa*, or any other combination in which the ratio of alleles among the parents is 1 *A* : 1 *a*.

To avoid giving the impression that random mating must always result in a 1 : 2 : 1 ratio, Fig. 13.4 shows the same example, but with a starting ratio of p *AA* : q *aa*. In this case, the progeny genotypes appear in frequencies of p^2 *AA*, $2pq$ *Aa*, and q^2 *aa*. As we shall see, this general result is expected for any combination of parental genotypes in which the ratio of alleles is p *A* : q *a*.

The preceding examples illustrate two important consequences of the mating process. First, *the genotypic composition of the progeny generation depends on the allele frequencies of the parents.* The allele frequencies of the parents are critical factors in determining the genetic composition of a population; in contrast, the parental genotype frequencies are not critical to this determination. For example, a random mating population with an allele ratio of 1 *A* : 1 *a* will produce a progeny population of 1 *AA* : 2 *Aa* : 1 *aa*, regardless of whether the genotypes among the parents are 1 *AA* : 1 *aa*, 1 *AA* : 2 *Aa* : 1 *aa*, or 2 *AA* : 1 *Aa* : 2 *aa*.

Second, unlike the case in specific genetic crosses, *we cannot predict the genotypic composition of the parental generation by knowing only the genotype frequencies of their offspring.* For example, if we know that the progeny occur in a ratio of 1 *AA* : 2 *Aa* : 1 *aa*, we cannot determine the genotypic ratio among the parents; we can only deduce that the allele ratio must have been 1 *A* : 1 *a*. It is therefore impossible to reconstruct the past genetic history of a population using only the genotypic proportions that exist in a given generation.

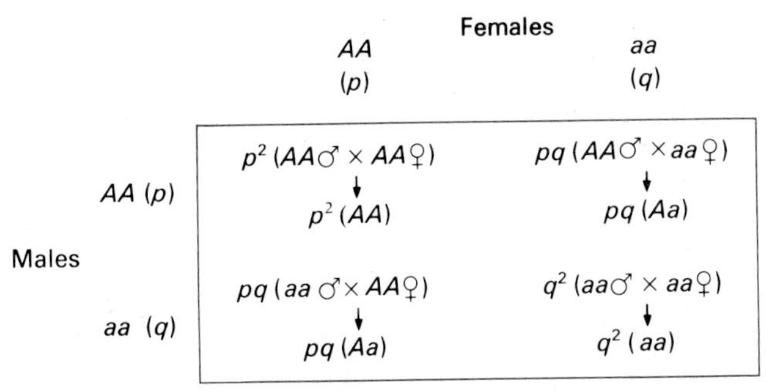

Females
AA
(p)
aa
(q)
Males
AA (p)
aa (q)
p² (AA♂ × AA♀)
p² (AA)
pq (AA♂ × aa♀)
pq (Aa)
pq (aa ♂× AA♀)
pq (Aa)
q² (aa♂ × aa♀)
q² (aa)
Total offspring = p² AA + 2pq Aa + q² aa

Figure 13.4. Results of random mating in a population that consists of a mixture of *AA* and *aa* genotypes in a ratio of *p*:*q*. Again, it is assumed that all females produce the same average number of offspring, so that the proportion of all offspring produced in each case equals the mating frequency.

EXAMPLE 13.3. In a certain population, the genotypes *AA*, *Aa*, and *aa* occur at frequencies 0.6, 0.4, and 0, respectively. If this population mates at random, what proportion of all matings should occur between mates that are of different genotypes?

Solution. Four matings are possible in this population: *AA* male × *AA* female, *AA* male × *Aa* female, *Aa* male × *AA* female, and *Aa* male × *Aa* female. Of these matings, only *AA* male × *Aa* female and its reciprocal, *Aa* male × *AA* female, involve different genotypes. Since mating is random, their combined frequency will be

$$f(AA \text{ male} \times Aa \text{ female}) + f(Aa \text{ male} \times AA \text{ female}) = (0.6)(0.4) + (0.4)(0.6) = 2(0.24) = 0.48.$$

Random Union of Gametes. One useful feature of the random mating process is that the offspring are derived in a manner that is equivalent to the random union of gametes in the parental generation. Each parent has $f(AA)$, $f(Aa)$, and $f(aa)$ chance of mating with an individual of genotype *AA*, *Aa*, or *aa*, respectively. If we assume allele frequencies of p *A* and q *a*, a sperm (or egg) of this individual thus has a probability of $f(AA) + \frac{1}{2}f(Aa) = p$ of combining with a gamete that carries the *A* allele and a probability of $f(aa) + \frac{1}{2}f(Aa) = q$ of combining with a gamete that carries the *a* allele. When we consider all of the mating combinations, the genotype frequencies that we would expect among the progeny of this population will be (see Fig. 13.5):

$$\text{Frequency of } AA \text{ progeny} = p(A \text{ in sperm}) \times p(A \text{ in eggs}) = p^2$$
$$\text{Frequency of } Aa \text{ progeny} = [p(A \text{ in sperm}) \times q(a \text{ in eggs})] + [q(a \text{ in sperm}) \times p(A \text{ in eggs})] = 2pq$$
$$\text{Frequency of } aa \text{ progeny} = q(a \text{ in sperm}) \times q(a \text{ in eggs}) = q^2$$

The relationships that exist between genotype and allele frequencies are depicted graphically in Figure 13.6. We should note that these are general relationships that apply to all random mating populations, regardless of the genotype frequencies of the parents. Thus, different parental populations can give rise to the same genotypic distribution among their offspring, as long as their allele frequen-

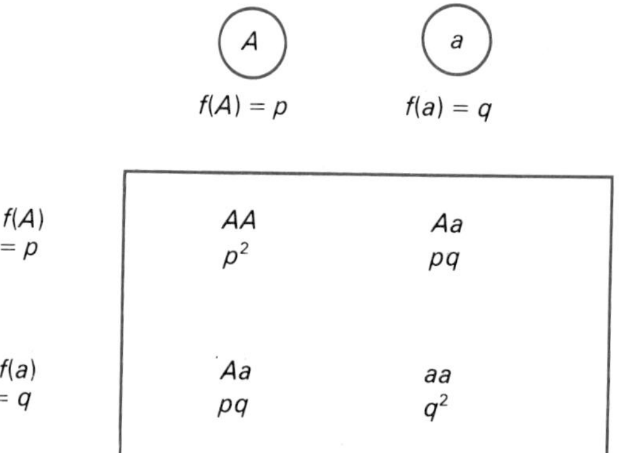

Figure 13.5. The distribution of genotype frequencies formed from the random combination of gametes during random mating. The combined frequencies p^2 (*AA*) + $2pq$(*Aa*) + q^2(*aa*) correspond to the terms of the binomial expansion $(p + q)^2$.

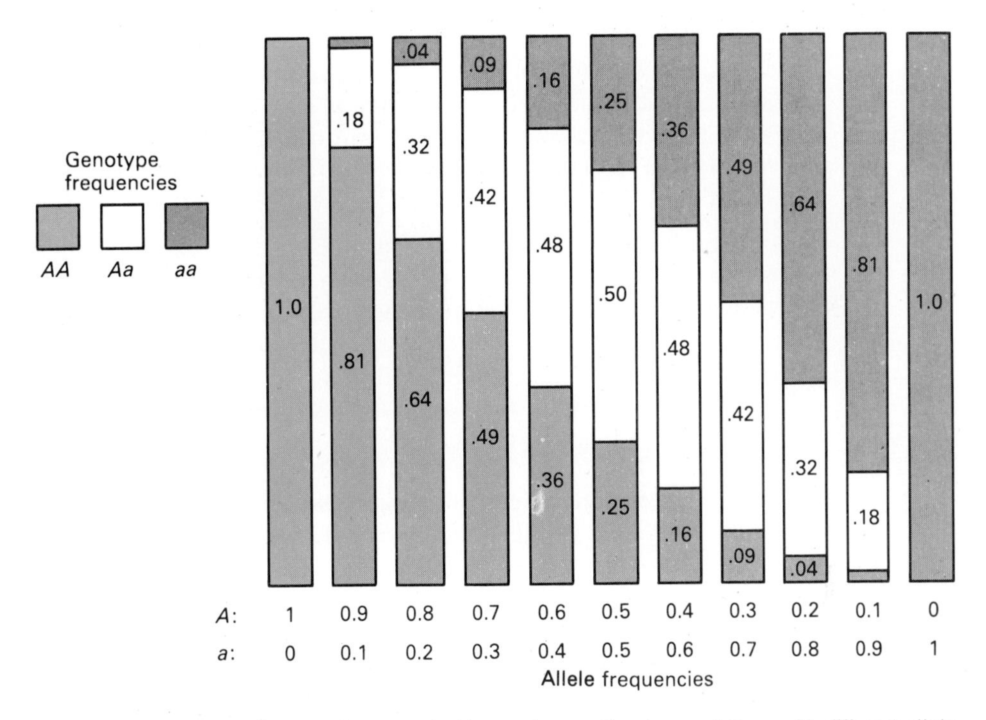

Figure 13.6. Genotype frequencies expected for random mating in populations with different allele frequencies. To obtain the phenotype frequencies expected for conditions of complete dominance, sum the frequencies of *AA* and *Aa* in each column.

cies correspond in value. For instance, parental populations as divergent as 0.2 *AA* : 0.1 *Aa* : 0.7 *aa* and 0.05 *AA* : 0.4 *Aa* : 0.55 *aa* yield identical progeny distributions upon random mating, since both have the same allele frequencies of $p = 0.2 + \frac{1}{2}(0.1) = 0.05 + \frac{1}{2}(0.4) = 0.25$ and $q = 0.75$.

We should also observe that while the genotype frequencies adjust to fit the values of p^2, $2pq$, and q^2, the allele frequencies retain their initial values of $f(A) = f(AA) + \frac{1}{2}f(Aa) = p^2 + \frac{1}{2}(2pq) = p(p + q) = p$ and $f(a) = 1 - p = q$. Thus, the allele frequencies remain unaffected by the mating process and are not expected to change. For example, if a population starts with genotype frequencies of $f(AA) = 0.7$, $f(Aa) = 0.2$, and $f(aa) = 0.1$, the allele frequencies are $f(A) = 0.7 + \frac{1}{2}(0.2) = 0.8$ and $f(a) = 0.2$. The genotype frequencies in this population will be altered after random mating to become $f(AA) = (0.8)^2 = 0.64$, $f(Aa) = 2(0.8)(0.2) = 0.32$, and $f(aa) = (0.2)^2 = 0.04$. But the values of $f(A)$ and $f(a)$ are still $0.64 + \frac{1}{2}(0.32) = 0.8$ and 0.2, respectively; these values are identical to what they were in the initial population.

Random Mating Equilibrium: The Hardy-Weinberg Law

In 1908, G. H. Hardy, a British mathematician, and W. Weinberg, a German physician, independently pointed out that genotype frequencies in populations tend to stabilize after one generation of random mating and remain constant in all subsequent generations. Since the genotype frequencies stay constant from generation to generation, we say that an equilibrium is established in the population. The equilibrium is determined by internal biological properties of the group such as mating behavior, rather than by any outside force; it is expected to occur in large random mating populations in which all genotypes are equally viable and fertile (i.e., there is no selection), migration is negligible, and mutation can be ignored.

Equilibrium at a Single Autosomal Locus. The equilibrium condition for an autosomal locus with two alleles is characterized by a binomial distribution of genotype frequencies. The frequency of each genotype is expressed as the product of the frequencies of its respective alleles. If we use A and a to symbolize the alleles and p and q to symbolize their respective frequencies, we can express the genotype proportions as the terms of the binomial expansion:

$$[p(A) + q(a)]^2 = p^2(AA) + 2pq(Aa) + q^2(aa).$$

As we have already learned, this relationship describes the chance distribution that is expected whenever the alleles are combined in a random fashion through the process of random mating.

The preceding binomial array of genotype frequencies is only one of a potentially large number of equilibrium states that can theoretically occur. In general, a genetic equilibrium is formed whenever the effects of the matings that act to increase the frequency of each genotypic class are exactly counterbalanced by the matings that act to reduce these same frequencies. In the case of two alleles, only three types of matings can act to change the genotype frequencies; they are AA male × aa female, aa male × AA female, and Aa male × Aa female. The first two types of matings serve to increase the frequency of heterozygotes in relation to homozygotes by producing only Aa offspring. The third type of mating has the opposite effect, producing homozygous ($\frac{1}{4}$ AA + $\frac{1}{4}$ aa) as well as heterozygous ($\frac{1}{2}$ Aa) progeny. Since only half of the offspring from Aa male × Aa female matings have genotypes that differ from their parents, these matings are only half as effective as the other two types in altering the genotype frequencies. In order to create a balance, therefore, Aa male × Aa female matings must occur at twice the frequency and thus produce twice the total offspring of AA male × aa female and aa male × AA female matings. Stated in symbolic terms, an equilibrium is established when

$$f(Aa \text{ male} \times Aa \text{ female}) = 2[f(AA \text{ male} \times aa \text{ female}) + f(aa \text{ male} \times AA \text{ female})].$$

When this relationship occurs, a balance is produced in the types of progeny that result, which leads to stability of the genotype frequencies.

In the case of random mating, the mating frequencies depend on the products of the genotype frequencies. Therefore, $f(Aa$ male × Aa female$) = (2pq)(2pq) = 4p^2q^2$, and $f(AA$ male × aa female$) = f(aa$ male × AA female$) = (p^2)(q^2)$. Substituting these values into the equation given above for the equilibrium state, we have the equality $4p^2q^2 = 4p^2q^2$, which is in agreement with the general condition for genetic equilibrium.

We must emphasize that an equilibrium that is established through random mating not only depends on the distributions of genotype and mating frequencies but also requires that the allele frequencies remain constant. Once a genetic equilibrium is established, the genotype frequencies will retain specific values of p^2, $2pq$, and q^2 only as long as the allele frequencies stay the same. If the allele frequencies should change, say from p and q to p' and q', the genotypes will attain a new equilibrium state in the next generation with frequencies of p'^2, $2p'q'$, and q'^2. This new distribution will remain until the allele frequencies change once again; the genotype frequencies will not tend to return to the old set of values.

EXAMPLE 13.4. Show that crosses other than $AA \times aa$ and $Aa \times Aa$ can never lead to changes in genotype frequencies among the collective progeny of a population, as long as all types of matings produce the same average number of offspring.

Solution. Four other crosses are possible in addition to those that are listed above. These crosses are $AA \times AA$, $AA \times Aa$, $Aa \times aa$, and $aa \times aa$. To demonstrate the inability of these crosses to alter the genotype frequencies, it is sufficient to show that the offspring produced by these matings have the same genotypes and occur in the same proportions as their parents. Since the mating types are assumed to be equally fertile, the one-to-one relationship between parents and offspring is thus apparent: $AA \times AA \rightarrow$ all AA, $AA \times Aa \rightarrow \frac{1}{2}AA + \frac{1}{2}Aa$, $Aa \times aa \rightarrow \frac{1}{2}Aa + \frac{1}{2}aa$, and $aa \times aa \rightarrow$ all aa.

Changes in the allele frequencies are caused by factors that are excluded from the model we are considering here. These factors include small population size, reproductive inequalities among the genotypes, and the introduction of alleles into the population through mutation and migration. *Allele frequencies are not altered by the mating scheme.* Neither are they affected by the degree of dominance of the allele in question. *The frequency of an allele does not change simply because the allele is dominant.* Dominance refers to the ability of one allele to mask its recessive counterpart in the heterozygote, not to its ability to increase numerically in the population.

In summary, there are three main parts to the Hardy-Weinberg law:

1. The allele frequencies at an autosomal locus will not change in a large random mating population in the absence of selection, mutation, and migration.
2. The genotype frequencies will attain an equilibrium distribution that conforms to the terms of the binomial expansion $[p(A) + q(a)]^2 = p^2(AA) + 2pq(Aa) + q^2(aa)$.
3. The equilibrium distribution for genotype frequencies is established after one generation of random mating.

This last characteristic is important to populations because it provides resilience to change. If for any reason a population should depart from its binomial array of genotype frequencies, the equilibrium distribution will be reestablished in only a single generation of random mating.

EXAMPLE 13.5. One assumption that is embodied in the Hardy-Weinberg law is that the allele frequencies are the same for both males and females. If they are not the same, then equilibrium genotype frequencies are not established in a population until after two generations of random mating. Show that this one-generation delay in attaining equilibrium is true for a population that consists of an equal mixture of *AA* males and *aa* females.

Solution. A population of *AA* males and *aa* females will produce offspring that consist of *Aa* heterozygotes of both sexes. The offspring of this first generation do not conform to the Hardy-Weinberg frequencies. But after still another generation of random mating, they will produce an equilibrium population of ¼ *AA*, ½ *Aa*, and ¼ *aa*. In this example, we see that it takes one generation of random mating just to achieve an equality of allele frequencies between the sexes. The next generation and all subsequent generations of random mating will then correspond to the Hardy-Weinberg principle presented in the text.

Applications of the Hardy-Weinberg Law

From a purist standpoint, the assumptions that are embodied in the Hardy-Weinberg law (e.g., no selection, mutation, or migration) are too restrictive to apply exactly to any real population. Nevertheless, many populations in nature do approximate a random mating equilibrium over a limited period of time. In these breeding groups, departures from the ideal conditions that are assumed in the model are small enough to be ignored, at least during the period of study. When these departures can be ignored, we can use the Hardy-Weinberg law to provide us with information about the genetic structure of natural populations. Three important applications of the Hardy-Weinberg law are described in this section.

Confirmation of Random Mating. One useful application of the Hardy-Weinberg model is to test for conditions of random mating in sample populations. For codominant and incompletely dominant traits, experiments of this type are relatively straightforward. They usually involve comparing the observed numbers in the various genotypic classes with those expected on the basis of random mating. If the observed numbers conform to the expected numbers, which are calculated on the basis of the Hardy-Weinberg formula, the population is considered to be in genetic equilibrium, since the data indicate that the genotypes were derived from random mating in preceding generations.

To test for agreement with the Hardy-Weinberg expectations, we usually begin by calculating the allele frequencies from the observed genotype frequencies. The expected numbers in the different genotypic classes are then derived by substituting the allele frequencies into the binomial formula $p^2 + 2pq + q^2$. A chi-square test is subsequently performed to obtain a measure of goodness-of-fit.

For a numerical example of the above procedure, let us consider the inheritance of the MN blood group alleles L^M and L^N. The results of one study are shown in Table 13.3. Are these data consistent with the hypothesis of random mating? To answer this question, we start by obtaining estimates of the L^M and L^N allele

Table 13.3. Chi-square test of goodness-of-fit to the Hardy-Weinberg law of blood group frequencies at the MN locus in a sample of 140 Pueblo Indians.

	M ($L^M L^M$)	MN ($L^M L^N$)	N ($L^N L^N$)	TOTAL
Observed numbers	83	46	11	140
Expected proportions	p^2	$2pq$	q^2	1.0
	(0.573)	(0.368)	(0.059)	1.0
Expected numbers	80.2	51.5	8.3	140
$(O - E)^2/E$	0.098	0.587	0.878	$\chi^2 = 1.56$

Note: The expected proportions were derived using $p = 0.757$ and $q = 0.243$. See the text for further details.

frequencies, as follows:

$$f(L^M) = p = 0.593 + \tfrac{1}{2}(0.328) = 0.757 \quad \text{and}$$
$$f(L^N) = q = 1 - 0.757 = 0.243.$$

Using the allele frequencies, we can then calculate the expected frequencies for the M, MN, and N blood types, assuming random mating: $f(\text{M}) = p^2 = (0.757)^2 = 0.573$, $f(\text{MN}) = 2pq = 2(0.757)(0.243) = 0.368$, and $f(\text{N}) = q^2 = (0.243)^2 = 0.059$. Since the expected and observed results are in close agreement, these data obviously conform to the Hardy-Weinberg law. This conclusion is verified by the chi-square value calculated in Table 13.3. The value of $\chi^2 = 1.56$ corresponds to a probability between 0.2 and 0.3, which indicates that the observed and expected values are not significantly different. Note that there is only one degree of freedom in this case. The allele frequencies in natural populations, unlike those in controlled genetic crosses, do not have values that are theoretically predictable; the values must be calculated from the observations themselves. Even though the example has three phenotypic classes, we lose an additional degree of freedom by estimating p from the experimental data. As a general rule in these situations, the number of degrees of freedom equals the number of phenotypic classes minus the number of alleles.

Unlike the preceding example, natural populations do not always agree so closely to the Hardy-Weinberg expectations. A lack of agreement between the observed and expected genotype frequencies can be caused by departures from the ideal conditions that are assumed in the model, including nonrandom mating and unequal viability and fertility of genotypes. The Hardy-Weinberg law is not very sensitive to certain of these departures, however, and the mere fact that a population conforms to the equilibrium law does not mean that all of the assumed conditions are met.

Allele-Frequency Analysis When Dominance Is Involved. Another useful application of the Hardy-Weinberg law is in computing the allele and genotype frequencies when dominant/recessive inheritance is involved. The equilibrium principle greatly simplifies the calculations under these circumstances, since it permits us to predict the genetic composition of a population if we know only the frequency of the recessive homozygotes.

Suppose that we determine that 16 percent of a random mating population has the recessive genotype *aa*. What are the corresponding frequencies of *AA* and *Aa* individuals? Since mating is random, we can make the following conclusions:

1. The frequency of recessive homozygotes is $q^2 = 0.16$.
2. The frequency of the *a* allele is the square root of 0.16, or $q = 0.4$.
3. The frequency of the *A* allele is $p = 1 - 0.4 = 0.6$.
4. The frequency of the *AA* genotype is therefore $p^2 = (0.6)^2 = 0.36$, and that of the *Aa* genotype is $2pq = 2(0.6)(0.4) = 0.48$.

One interesting human characteristic that has been analyzed successfully with this procedure is the ability to taste the compound phenylthiocarbamide ($C_7H_8N_2S$), otherwise known as PTC. The ability to taste PTC is determined by the dominant allele *T*. The test for this tasting ability is quite simple. It entails placing a strip of filter paper that has been previously impregnated with PTC on the tip of the tongue. Tasters of PTC (*T-*) experience a bitter taste, whereas nontasters (*tt*) find the compound virtually tasteless. If you were to test your classmates for this ability, you would probably find that 70 percent are tasters and 30 percent are nontasters. Since the ability to taste PTC obviously has nothing to do with how an individual selects a mate, we can reasonably assume a random mating equilibrium. The allele frequencies for PTC tasting are therefore calculated to be

$$q = \sqrt{0.3} = 0.55 \quad \text{and} \quad p = 0.45.$$

We then use the allele frequencies to calculate the expected genotype frequencies:

$$f(TT) = (0.45)^2 = 0.2; \quad f(Tt) = 2(0.45)(0.55) = 0.5; \quad f(tt) = 0.3.$$

This procedure can also be used to estimate the frequency of certain rare recessive alleles that have a detrimental effect when they are in the homozygous state. One example involves the recessive disorder phenylketonuria (PKU). One child in 10,000 live births has PKU. Although persons with PKU do not compose a random mating group, the disorder is so rare that the affected individuals make a negligible contribution to the allele frequencies in subsequent generations. We can thus reasonably assume that the genotype frequencies among the progeny of the population are determined almost exclusively by the unaffected genotypes *AA* and *Aa*, and that the pairing of individuals with these genotypes is essentially random. We can therefore equate 1/10,000 PKU offspring to q^2, so that $q = 0.01$ and $p = 1 - q = 0.99$. Using these values, we can estimate the frequency of heterozygous carriers of the PKU allele as $2pq = 2(0.99)(0.01) = 0.0198$. Therefore, about 2 in 100 individuals are carriers of this recessive allele.

Observe that when the frequency of a recessive allele is low, most copies of the allele are carried in a concealed state by heterozygotes. We can demonstrate this relationship more clearly by writing the ratio of heterozygotes to recessive homozygotes at equilibrium as

$$f(Aa)/f(aa) = 2pq/q^2 = 2p/q.$$

In the case of a rare recessive disorder such as PKU, q may be 0.01 or less. In this instance, heterozygotes outnumber the recessive homozygotes by a factor of at least 2(0.99)/0.01, or approximately 200-fold.

Testing for Dominant/Recessive Inheritance. The Hardy-Weinberg law can also be used in conjunction with pedigree analysis to verify whether the inheritance pattern for a particular character is dominant or recessive. For example, suppose that pedigree analysis gives us reason to believe that a trait is inherited as a simple recessive character. As a second source of evidence to help substantiate the recessive mode of inheritance, we can calculate the frequency of suspected recessive offspring from each of two kinds of matings: (1) crosses in which one parent is dominant and the other is recessive ($A\text{-} \times aa$) and (2) crosses in which both parents have the supposed dominant phenotype ($A\text{-} \times A\text{-}$). An offspring must receive a copy of a recessive allele from any recessive parent in a cross; but the offspring is less than certain to receive a copy of the recessive allele from a dominant parent. The expected frequency of recessive offspring in the first cross ($A\text{-} \times aa$) must therefore equal the probability of receiving an a allele from the dominant A- parent, which is designated as $P(a \text{ from } A\text{-})$. Both parents are dominant in the second cross ($A\text{-} \times A\text{-}$), so that the expected frequency of recessive offspring from this cross will equal $[P(a \text{ from } A\text{-})]^2$.

Since heterozygotes and dominant homozygotes are indistinguishable in phenotype, the results of these crosses can only be interpreted by using a genetic model such as the Hardy-Weinberg principle, which allows us to predict the frequencies of the genotypes in the population. If mating is random, the genotypes will appear in the population as a whole in the frequencies p^2, $2pq$, and q^2. The probability that an offspring will receive a recessive allele from a dominant parent will then be

$$P(a \text{ from } A\text{-}) = \frac{\text{Frequency of } a \text{ among } A\text{- parents}}{\text{Total frequency of } A \text{ and } a \text{ among } A\text{- parents}} = \frac{pq}{p^2 + 2pq} = \frac{q}{1+q}$$

where $p^2 + 2pq = p(p + 2q) = p(1 - q + 2q) = p(1 + q)$. Therefore, the fraction of recessive homozygotes produced by dominant $\times$ recessive matings can be calculated as $q/(1 + q)$, and the fraction of recessive homozygotes produced by dominant $\times$ dominant matings becomes $[q/(1 + q)]^2$. We can calculate both expected fractions from the frequency of the recessive allele in the population and can compare these values with the observed results. If the expected values agree with those observed, a simple dominant/recessive inheritance is indicated.

A numerical example of this procedure is provided by the Rh blood-group alleles. One study involving the Rh phenotype of 375 offspring yielded the results shown in Table 13.4. Are these data consistent with a simple dominant/recessive mode of inheritance for the Rh^+/Rh^- alternatives? Certainly the fact that no Rh^+ progeny are produced by $Rh^- \times Rh^-$ mating would support the hypothesis that

Table 13.4. Inheritance of the Rh phenotype among 375 total offspring.

	NUMBER OF OFFSPRING		
MATING	Rh^+ (Rh-positive)	Rh^- (Rh-negative)	TOTAL
$Rh^+ \times Rh^+$	240	25	265
$Rh^+ \times Rh^-$	54	23	77
$Rh^- \times Rh^-$	0	33	33

Rh^- is a recessive trait. To test this possibility further, we first calculate the frequency of the Rh^- phenotype in the sample:

$$f(Rh^-) = (25 + 23 + 33)/375 = 81/375 = 0.216.$$

If Rh^- is recessive, $f(Rh^-) = q^2$, assuming random mating, so that $q = \sqrt{0.216} = 0.46$. Moreover, the expected frequency of recessive homozygotes from dominant × recessive matings is $q/(1 + q) = 0.46/1.46 = 0.315$, and the expected frequency from dominant × dominant matings is $[q/(1 + q)]^2 = 0.099$. The expected numbers of Rh^- offspring from these matings can then be calculated as follows:

$$\text{Expected number of } Rh^- \text{ from } Rh^+ \times Rh^+ = (0.099)(265) = 26.2,$$
$$\text{Expected number of } Rh^- \text{ from } Rh^+ \times Rh^- = (0.315)(77) = 24.2.$$

If we compare these expected values with the observed values (25 and 23), it is obvious that no statistical test is needed in this case, since the expected and observed results are in very close agreement. These data are therefore consistent with the notion that the Rh^- phenotype is inherited as a simple Mendelian recessive characteristic.

EXAMPLE 13.6. In a random mating population, four times as many recessive offspring are produced by dominant × recessive matings as are produced by dominant × dominant mating. **(a)** What is the frequency of the recessive allele in this population? **(b)** What is the frequency of the recessive phenotype in this population? **(c)** What proportion of the offspring from dominant × dominant matings show the dominant phenotype?

Solution. **(a)** The frequency of recessive offspring from dominant × recessive matings is $q/(1 + q)$, and the frequency from dominant × dominant matings is $[q/(1 + q)]^2$. For the population given, we can therefore write $\frac{q/(1 + q)}{[q/(1 + q)]^2} = 4$. Solving for q, we obtain $q/(1 + q) = 1/4$, which gives $q = 1/4 + q/4$ or $3q/4 = 1/4$. Thus, $q = 1/3$. **(b)** The frequency of the recessive phenotype is $q^2 = 1/9$. **(c)** Since the proportion of recessive offspring from dominant × dominant matings is $[q/(1 + q)]^2 = (1/4)^2 = 1/16$, the proportion of dominant offspring from such matings is $1 - 1/16 = 15/16$.

TO SUM UP

1. The genotype frequencies of a population can be computed from mating frequencies as the statistical averages of the proportions of each genotype produced by the different parental matings. We can also calculate the mating frequencies from the genotype frequencies when the precise mating pattern in the population is known. If mating is random, so that mates are selected without regard to genotype, the frequency of each type of mating can be expressed as the product of the frequencies of the genotypes involved.
2. The Hardy-Weinberg law states that after one generation of random mating, a large population will attain an equilibrium in which the genotype frequencies remain constant. The equilibrium is characterized by a binomial distribution of genotype frequen-

cies, in which the genotypes occur in proportions that can be expressed as the products of the frequencies of their respective alleles. The equilibrium distribution of the genotype frequencies is a chance distribution that is formed from the random combination of gametes during random mating.

3. An equilibrium of genotype frequencies requires that the allele frequencies remain constant. The frequency of an allele is not changed by random mating. Neither is it altered by whether the allele is dominant or recessive. The allele frequencies are affected by factors that are not considered in the Hardy-Weinberg model, such as small population size, selection, mutation, and migration.
4. The Hardy-Weinberg law provides a test for random mating that involves comparing the observed genotype frequencies with those expected under the conditions of a random-mating equilibrium. The Hardy-Weinberg law can also be used to help substantiate a recessive mode of inheritance for a trait by comparing the frequencies of recessive offspring produced by certain crosses in a population with the frequencies expected for random mating.

Extensions of the Hardy-Weinberg Law

The Hardy-Weinberg law is a general principle that covers more than just the genetic effects of random mating on two alleles at a single autosomal locus. In this section, we will consider three important extensions of this principle that are frequently encountered in studies on population genetics.

Multiple Alleles

As we learned in Chapter 6, multiple alleles can greatly increase the number of genotypes within a population. Despite this increased potential for variation, random mating will still produce a genetic equilibrium among the progeny after a single generation. Figure 13.7 shows the genetic makeup of a population at

Female gametes \ Male gametes	A_1 (p)	A_2 (q)	A_3 (r)
A_1 (p)	A_1A_1 p^2	A_1A_2 pq	A_1A_3 pr
A_2 (q)	A_1A_2 pq	A_2A_2 q^2	A_2A_3 qr
A_3 (r)	A_1A_3 pr	A_2A_3 qr	A_3A_3 r^2

Combined frequencies: A_1A_1, p^2; A_1A_2, $2pq$; A_1A_3, $2pr$; A_2A_2, q^2; A_2A_3, $2qr$; A_3A_3, r^2

Figure 13.7. Genotype frequencies produced with three alleles after one generation of random mating. Allele frequencies are assumed to equal p, q, and r in the parental generation.

equilibrium for a locus with three alleles, A_1, A_2, and A_3, which have frequencies p, q, and r, respectively. Note that the genotype frequencies are given by the terms of the binomial expansion $(p + q + r)^2 = p^2 + 2pq + q^2 + 2pr + 2qr + r^2$, so that their equilibrium values are again simple products of the allele frequencies.

Allele Frequencies at the ABO Locus. When more than two alleles occur within a population, our evaluation of the allele frequencies is often complicated by the particular dominance relationships that exist among the alleles. A case in point is the ABO blood group system, which has been extensively studied in various human populations. Recall that three alleles (I^A, I^B, and i) contribute to this system, with I^A codominant with I^B, and both dominant to i. If we define the allele frequencies in this system as $p = f(I^A)$, $q = f(I^B)$, and $r = f(i)$, the genotype and blood group frequencies at equilibrium then become

Blood group:	A	B	AB	O
Genotype:	I^AI^A and I^Ai	I^BI^B and I^Bi	I^AI^B	ii
Frequency:	$p^2 + 2pr$	$q^2 + 2qr$	$2pq$	r^2

By assuming a random mating equilibrium, we can then calculate the allele frequencies. Since $f(\text{O}) = r^2$, $f(i)$ can be evaluated as

$$r = \sqrt{f(\text{O})}.$$

To solve for $f(I^B)$, note that $f(\text{A}) + f(\text{O}) = f(I^AI^A) + f(I^Ai) + f(ii) = p^2 + 2pr + r^2 = (p + r)^2$. Thus, $p + r = \sqrt{f(\text{A}) + f(\text{O})}$. Since $p + q + r = 1$ for a system with three alleles, $q = 1 - (p + r)$. Therefore,

$$q = 1 - \sqrt{f(\text{A}) + f(\text{O})}.$$

Similarly, $q + r = \sqrt{f(\text{B}) + f(\text{O})}$. Hence,

$$p = 1 - \sqrt{f(\text{B}) + f(\text{O})}.$$

The following distribution of blood group frequencies observed in a genetics laboratory provides a numerical example of this procedure:

Blood group:	A	B	AB	O	Total
Observed number:	31	16	5	48	100

The allele frequencies for this sample are

$$r = \sqrt{48/100} = 0.69$$

$$q = 1 - \sqrt{(31 + 48)/100} = 0.11$$

$$p = 1 - \sqrt{(16 + 48)/100} = 0.20$$

EXAMPLE 13.7. If 36 percent of all persons within a random mating population are of blood type O and 45 percent are of type A, what percentages are expected to be of type B and type AB?

Solution. Since the frequency of type O persons is $r^2 = 0.36$, the frequency of the i allele must be $r = 0.6$. We can also write that $p + r = \sqrt{f(\text{A}) + f(\text{O})}$. Therefore, $p = \sqrt{0.45 + 0.36} - 0.6 = \sqrt{0.81} - 0.6 = 0.3$. From these values, we can calculate that $q = 1 - p - r = 1 - 0.3 - 0.6 = 0.1$. The frequencies of types B and AB then become

$$f(\text{B}) = q^2 + 2qr = (0.1)^2 + 2(0.1)(0.6) = 0.13, \text{ or 13 percent, and}$$
$$f(\text{AB}) = 2pq = 2(0.3)(0.1) = 0.06, \text{ or 6 percent.}$$

Sex-Linked Genes

The Hardy-Weinberg law also applies to genes on the X chromosomes, although not without a few modifications. Since the homogametic sex (which we shall assume to be female in this case) has two X chromosomes, its X-linked genes will follow the same general pattern as is observed for genes on the autosomes. In other words, with two alleles A and a, females should attain a random mating equilibrium in which the genotype frequencies are

Female genotype:	AA	Aa	aa
Equilibrium frequency:	p^2	$2pq$	q^2

The primary difference between the equilibrium established at an X-linked locus and the equilibrium for an autosomal gene involves the males. Since males carry only a single dose of each X-linked gene, their genotype frequencies are the same as the allele frequencies and can be expressed as follows:

Male genotype:	AY	aY
Equilibrium frequency:	p	q

The result of a single generation of random mating involving X-linked genes is shown in Fig. 13.8. In this case, the random union of sperm and eggs yields five different genotypes rather than three, because of the presence of the Y-bearing sperm.

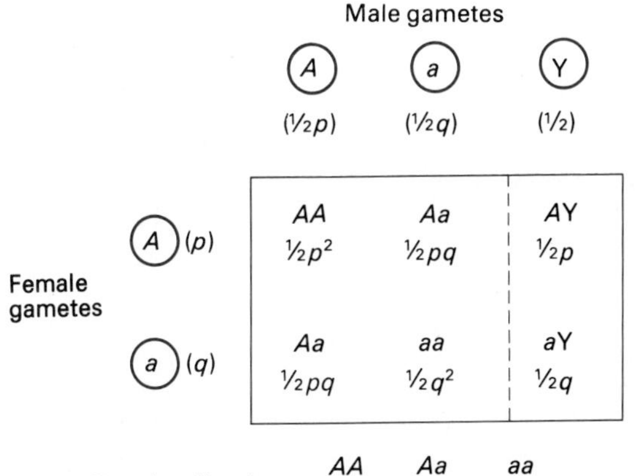

Figure 13.8. Genotype frequencies for a sex-linked pair of alleles produced among male and female offspring after one generation of random mating. The genotype frequencies among females correspond to the frequencies expected at an autosomal locus. The genotype frequencies among males are equal to the allele frequencies.

It is a relatively simple matter to evaluate p and q for sex-linked loci, since in the heterogametic sex, the allele frequencies are the same as the genotype frequencies. Thus, when the male sex is heterogametic, the incidence of an X-linked recessive phenotype is q among males. Its corresponding frequency is q^2 among females, if we assume that the allele frequencies are the same in both sexes. The ratio of females to males exhibiting a recessive trait is then $q^2/q = q$. The ratio illustrates why rare X-linked recessive characteristics appear more frequently among males than among females. Let us take hemophilia as an example. About 1 male in 10,000 has hemophilia. The frequency of the recessive allele for hemophilia is thus $q = 1/10{,}000 = 0.0001$. We would therefore expect to observe only 1 hemophilic female in the population for every 10,000 similarly afflicted males.

The reverse situation holds for rare dominant traits. In this case, the ratio of females to males exhibiting the trait in question will be $(p^2 + 2pq)/p = p + 2q = 1 + q$. When the dominant trait is rare in the population, $q = 1 - p$ approaches 1, so that $1 + q$ approaches 2. About twice as many females as males are therefore expected to have the trait.

Equilibrium at Two Loci

The equilibrium situation is considerably more complex when two (or more) gene loci are considered simultaneously. The added complications are due in part to the fact that when two (or more) gene pairs are considered, the allele and gamete frequencies need no longer be the same. To illustrate, assume two unlinked gene loci with two alleles at each locus. Let the frequencies of each pair of alleles be

$$\text{locus 1:} \quad f(A) = p_1 \qquad f(a) = q_1,$$
$$\text{locus 2:} \quad f(B) = p_2 \qquad f(b) = q_2,$$

where $p_1 + q_1 = 1$ and $p_2 + q_2 = 1$.

At any given time, four gamete types exist in the population: (AB), (Ab), (aB), and (ab). By restricting our attention to only those gametes that carry each specific allele, we can see that the allele frequencies are related to the gamete frequencies in the following manner:

$$\text{locus 1:} \quad p_1 = f(AB) + f(Ab) \qquad q_1 = f(aB) + f(ab),$$
$$\text{locus 2:} \quad p_2 = f(AB) + f(aB) \qquad q_2 = f(Ab) + f(ab).$$

From these relationships, it is apparent that the gamete frequencies can take on a variety of numerical values for any fixed set of allele frequencies. For example, a population that consists of only *AaBb* individuals would produce all gamete types at frequencies of ¼ each, assuming that the genes are unlinked. The allele frequencies would then equal ½ at both loci. These same allele frequencies would also characterize a population that consists of an equal mixture of *AABB* and *aabb* genotypes, in which only (AB) and (ab) gametes are formed. We can see from this example that only a loose association exists between the allele and gamete frequencies, so that it is possible for the allele frequencies to be at their expected equilibrium levels even though the gamete frequencies are not.

Table 13.5. Genotype frequencies in a population at random mating equilibrium for two pairs of alleles A,a and B,b at frequencies of p_1, q_1 and p_2, q_2, respectively.

				Sum
	$AABB$ $p_1^2p_2^2$	$AABb$ $2p_1^2p_2q_2$	$AAbb$ $p_1^2q_2^2$	p_1^2
	$AaBB$ $2p_1q_1p_2^2$	$AaBb$ $4p_1q_1p_2q_2$	$Aabb$ $2p_1q_1q_2^2$	$2p_1q_1$
	$aaBB$ $q_1^2p_2^2$	$aaBb$ $2q_1^2p_2q_2$	$aabb$ $q_1^2q_2^2$	q_1^2
Sum:	p_2^2	$2p_2q_2$	q_2^2	

As in the case of a single autosomal locus, the joint equilibrium for two gene loci is established when the genotypes of the zygotes occur in proportions that are products of the frequencies of their respective genes. The equilibrium distribution of the genotype frequencies for two gene loci with two alleles each is summarized in Table 13.5. Note that the random union of gametes will give rise to an equilibrium distribution among the zygotes that equals the product of the equilibrium distributions at each separate locus: $(p_1^2 + 2p_1q_1 + q_1^2)(p_2^2 + 2p_2q_2 + q_2^2)$. Because the nonallelic genes are associated randomly in a population at equilibrium, the gamete types will also occur in the proportions determined by the products of their respective genes (see Fig. 13.9). The apparent statistical independence of nonallelic genes at equilibrium applies to linked genes as well as unlinked genes. For this reason, a population is said to be in **linkage equilibrium** when the alleles at different loci are randomly distributed in the gametes; and the population is said to be in *linkage disequilibrium* when they are not. It is impossible to detect linkage from the observed genetic characteristics of a population in linkage equilibrium, since both the gamete and genotype frequencies then depend strictly on the allele frequencies and not on the location of the genes on chromosomes.

Figure 13.9. Random association of nonallelic genes within the gametes of a population at linkage equilibrium. At equilibrium, the gamete types occur in the proportions determined by the products of their respective gene frequencies.

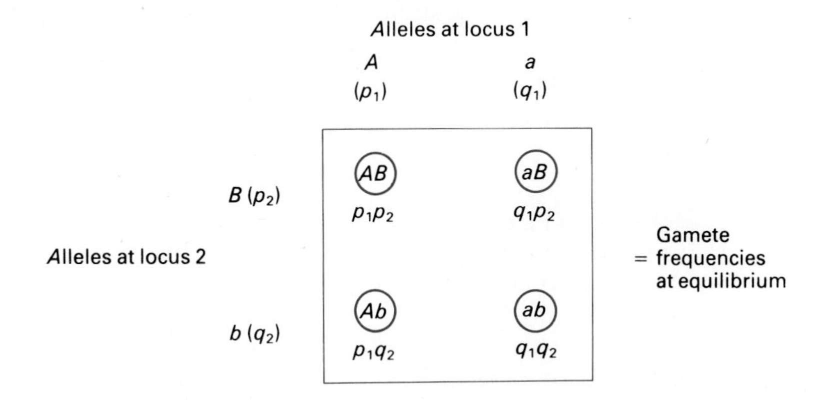

Table 13.6. Gradual approach to random-mating equilibrium for two unlinked genes in a population started with equal numbers of *AABB* and *aabb* genotypes.

GENERATION	GAMETE FREQUENCIES			
	AB	*Ab*	*aB*	*ab*
0	1/2	0	0	1/2
1	3/8	1/8	1/8	3/8
2	5/16	3/16	3/16	5/16
3	9/32	7/32	7/32	9/32
4	17/64	15/64	15/64	17/64
5	33/128	31/128	31/128	33/128
.	.	.	.	.
.	.	.	.	.
.	.	.	.	.
limit	1/4	1/4	1/4	1/4

Gradual Approach to Equilibrium. Unlike the single autosomal locus, which achieves equilibrium in a single generation, two loci considered jointly approach their combined equilibrium state in a gradual manner over several generations. The extra time for two loci as opposed to one is required in order for the nonallelic genes to become randomly associated within the gametes of the population.

To give an example, let us consider a population that consists of equal numbers of *AABB* and *aabb* individuals of both sexes. Assume that the two gene pairs are unlinked. All of the allele frequencies in this starting population are equal to 1/2, since $f(A) = f(a)$ and $f(B) = f(b)$. Thus, proportions for the gametes at equilibrium are expected to be 1/4 (*AB*), 1/4 (*Ab*), 1/4 (*aB*), and 1/4 (*ab*). Initially, only two types of gametes are formed (1/2 (*AB*) and 1/2 (*ab*)); these types unite in the next generation (generation 1) to produce three diploid genotypes (*AABB*, *AaBb*, and *aabb*) in a ratio of 1:2:1. Of these genotypes, 1/4 *AABB* yields 1/4 (*AB*) gametes, 1/4 *aabb* yields 1/4 (*ab*) gametes, and 1/2 *AaBb* yields 1/2(1/4 (*AB*), 1/4 (*Ab*), 1/4 (*aB*), and 1/4 (*ab*)) gamete types, so that the genotypes collectively form gametes in the proportions 3/8 (*AB*), 1/8 (*Ab*), 1/8 (*aB*), and 3/8 (*ab*). If we extend our calculations over several generations, we get the results shown in Table 13.6. Note that a gradual trend emerges, with an overall increase in the frequencies of the (*Ab*) and (*aB*) gametes and a corresponding reduction in the frequencies of the (*AB*) and (*ab*) types. Eventually, an equilibrium develops at which all of the gamete frequencies are equal in this particular case.

EXAMPLE 13.8. Calculate the allele frequencies and gamete frequencies for a population consisting of *AABB*, *AaBb*, and *aabb* at a ratio of 7:8:1. Assume that the *A,a* and *B,b* gene pairs are unlinked.

Solution. Since 7 + 8 + 1 = 16, the genotype frequencies are $f(AABB)$ = 7/16, $f(AaBb)$ = 8/16, and $f(aabb)$ = 1/16. The gamete frequencies can now be calculated from the genotype frequencies as follows:

$$f(AB) = f(AABB) + \tfrac{1}{4}f(AaBb) = \tfrac{7}{16} + \tfrac{1}{4}(\tfrac{8}{16}) = \tfrac{9}{16},$$
$$f(Ab) = \tfrac{1}{4}f(AaBb) = \tfrac{1}{4}(\tfrac{8}{16}) = \tfrac{2}{16},$$
$$f(aB) = \tfrac{1}{4}f(AaBb) = \tfrac{1}{4}(\tfrac{8}{16}) = \tfrac{2}{16},$$
$$f(ab) = f(aabb) + \tfrac{1}{4}f(AaBb) = \tfrac{1}{16} + \tfrac{1}{4}(\tfrac{8}{16}) = \tfrac{3}{16}.$$

The allele frequencies can then be computed from the gamete frequencies as follows:

$$f(A) = p_1 = f(AB) + f(Ab) = \tfrac{11}{16}, \qquad f(a) = q_1 = f(ab) + f(aB) = \tfrac{5}{16},$$
$$f(B) = p_2 = f(AB) + f(aB) = \tfrac{11}{16}, \qquad f(b) = q_2 = f(ab) + f(Ab) = \tfrac{5}{16}.$$

Even though this example differs from the single-gene model in terms of the time it takes for the gametes to attain their equilibrium frequencies, once these frequencies are reached, the genotype frequencies will be in equilibrium in the next generation, as in the case for a single locus. We can clarify this point by considering a population of all *AaBb* individuals. With only dihybrids in the population, the genotype frequencies are obviously not in equilibrium. As long as the genes are unlinked, however, the gamete frequencies are in equilibrium, since they are the products of their respective gene frequencies (i.e., $\tfrac{1}{2} \times \tfrac{1}{2} = \tfrac{1}{4}$). An equilibrium for genotype frequencies is then immediately established among the zygotes in the next generation, with a genotypic ratio of 1:2:1:2:4:2:1:2:1.

It is important to note that much the same pattern would result if the genes were linked, except that with linked genes, the time it takes to reach an equilibrium is longer. To illustrate this point numerically, let us again assume equal numbers of *AB/AB* and *ab/ab* genotypes, but with the two gene loci positioned 25 map units apart on the same chromosome. Figure 13.10 depicts the frequencies of each type

Figure 13.10. Beginning of the approach toward equilibrium by two gene loci that are positioned 25 map units apart on the same chromosome. Only two generations are shown, starting with a population that consists of equal numbers of *AABB* and *aabb* genotypes. Once linkage equilibrium is established, the alleles at the *A,a* and *B,b* loci will be randomly distributed in the gametes.

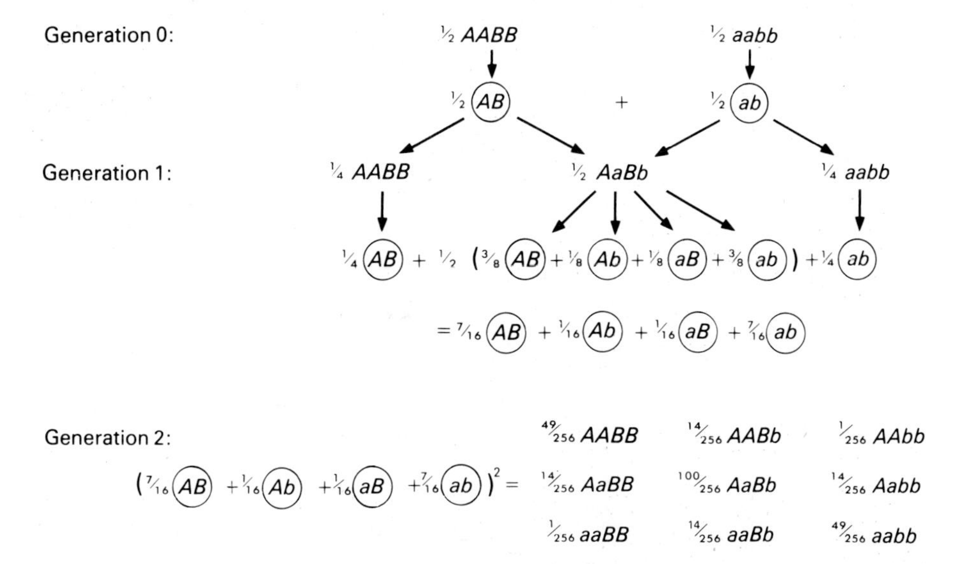

of gamete and zygote for two generations of this hypothetical starting population. As in the case of unlinked genes, all four gamete types appear by the first progeny generation, although the (*Ab*) and (*aB*) gametes make up a smaller fraction of the total gamete pool in this particular example. The state of linkage disequilibrium is only temporary, however; it gradually diminishes in succeeding generations, but at a slower rate than is expected for unlinked genes. This pattern is demonstrated in Fig. 13.11, in which linkage is seen to prolong the time needed to achieve an equilibrium state. In general, the length of time it takes two linked loci to reach linkage equilibrium is inversely proportional to the distance between the loci; the closer the two loci are on a chromosome, the smaller the chance of recombination, and consequently, the greater the number of generations required for the alleles at these different loci to become randomly distributed in the gametes.

Most pairs of gene loci that have been studied so far in natural populations show little or no evidence of linkage disequilibrium, which suggests that these loci are in or very near equilibrium with one another. There are several interesting exceptions, however, including the close associations that have been observed between the MN and Ss blood group loci and between certain loci within the HLA (human lymphocyte A) complex in humans.

Three explanations have been given for the failure of certain loci to be in linkage equilibrium. First, the different loci may be very closely linked and may not have had the time to reach the equilibrium state. Second, a recent admixture of populations with different allele frequencies may have upset the equilibrium that previously existed. Third, certain combinations of nonallelic genes may confer on individual carriers an increase (or decrease) in the survival or reproductive potential, so that these combinations are maintained at higher (or lower) frequencies than would be predicted for random mating equilibrium. Unlike the first two

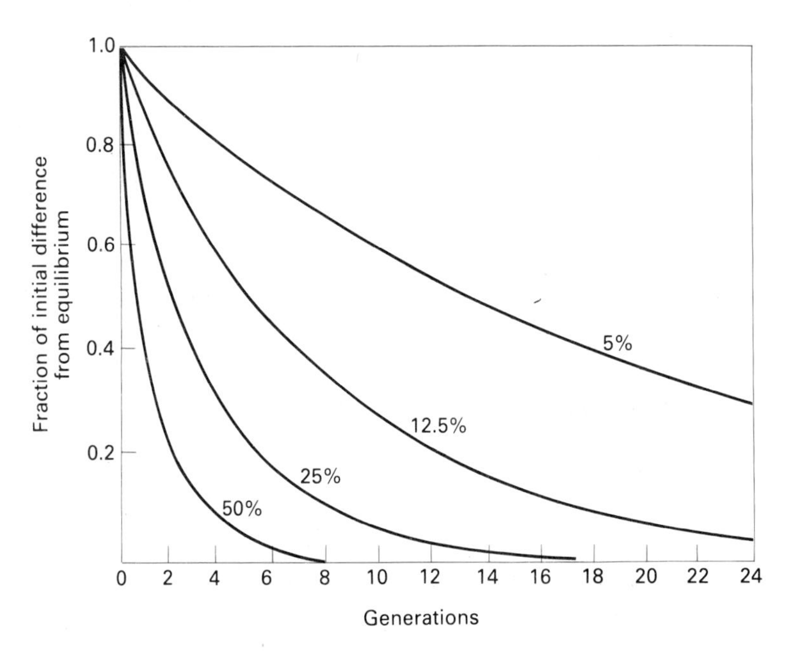

Figure 13.11. Gradual approach to the equilibrium state by two loci considered jointly. The four curves are plotted for different percentages of recombination between the loci. The most rapid approach to equilibrium occurs with unlinked genes (50% recombination). As we can see by the other curves, linkage prolongs the time needed to achieve the equilibrium state.

possibilities, which are merely temporary departures from the equilibrium state, the third proposed mechanism can lead to a condition of permanent linkage disequilibrium. An extreme example of this third possibility would occur if only genotypes whose genes are linked in the cis position (*AB*/*AB*, *AB*/*ab*, and *ab*/*ab*) are viable. Complete linkage disequilibrium would occur in these cases, since the *Ab* and *aB* combinations produced by recombination in some heterozygotes would be eliminated in each generation. A state of permanent linkage disequilibrium can also be maintained for genes that are involved in a chromosome inversion (see Chapter 12). Recall that inversions behave as crossover suppressors in heterozygotes for the inversion, so that the genes within an inverted segment of a chromosome are always transmitted together intact in the form of a single unit, or "supergene." Inversions thus serve as an important mechanism by which certain combinations of genes can be assured of joint inheritance and can avoid being split up through crossing over into new and possibly less adaptive genotypes.

TO SUM UP

1. The Hardy-Weinberg principle can be readily extended to include multiple alleles. In general, multiple alleles with frequencies $p, q, r, \ldots$ produce equilibrium genotype frequencies that conform to the terms of the expansion $(p + q + r + \ldots)^2$, in which $p^2, q^2, r^2, \ldots$ are the frequencies of the homozygotes and $2pq, 2pr, 2qr, \ldots$ are the frequencies of the heterozygotes. As in the case for a single locus with two alleles, the equilibrium is established after a single generation of random mating.
2. The random mating equilibrium that is established for a sex-linked gene differs from the equilibrium at an autosomal locus, because the heterogametic sex has only a single dose of the gene in question. In the case of an X-linked gene, the equilibrium genotype frequencies in the heterogametic sex will correspond to the allele frequencies. The equilibrium genotype frequencies in the homogametic sex, on the other hand, will show the same distribution as we would expect for an autosomal locus.
3. The joint equilibrium for two gene loci is established when the genotypes of both the zygotes and gametes occur in proportions that are the products of the frequencies of their respective genes. At this point, the alleles at the different loci are associated randomly in the gametes of the population.
4. Whereas a single autosomal locus attains a random mating equilibrium in one generation, two autosomal loci when considered jointly gradually approach their combined equilibrium over many generations. The additional time is needed for the gamete frequencies to attain their equilibrium values. Once these levels have been reached, the equilibrium genotype frequencies among the zygotes are established in the very next generation, like those in the single-locus case.

Questions and Problems

1. Distinguish among the terms allele frequency, genotype frequency, and mating frequency. Show how the three are related in a random mating population.
2. Criticize the following statements:
   - **(a)** Since fertilization is random, the frequency of the recessive genotype in a population must equal the square of the frequency of the recessive allele.
   - **(b)** Because of the masking effect exerted by a dominant allele in heterozygotes, the frequency of a dominant trait in a population will always exceed that of the recessive counterpart.

(c) Because of the equal segregation of alleles, an equilibrium will tend to be established in a population in which three-fourths of the individuals show the dominant trait.
(d) The statistical independence of nonallelic genes at equilibrium in a random mating population is merely a consequence of the independent assortment of the different gene pairs during meiosis.

3. A sample of 1400 persons living in New York City disclosed 408 persons of blood type M, 694 of type MN, and 298 of type N. Determine the frequencies of the M, MN, and N blood types in this sample and the frequencies of the L^M and L^N alleles.

4. A large population consists of the genotypes *AA, Aa,* and *aa* at frequencies of 0.1, 0.6, and 0.3, respectively. **(a)** What are the allele frequencies in this population? **(b)** Calculate the expected allele and genotype frequencies after one generation of random mating.

5. Of a random mating population that shows 1 percent recessive homozygotes, what percentage is heterozygous?

6. In a certain population, 80 percent of all persons are Rh^+, and 20 percent are Rh^-. Assume that mating is random for the Rh factor. **(a)** Of all marriages in this population, what percentage is expected to be an Rh^+ husband and an Rh^- wife? **(b)** In what percentage of all marriages is one person Rh^+ and the other Rh^-?

7. A farmer planted a large field of corn from seed that consisted of a mixture of purple (dominant) and yellow (recessive) kernels. The mature plants that developed were allowed to pollinate at random. Of the seeds produced in the next generation, 91 percent were purple and 9 percent were yellow. **(a)** What are the allele frequencies in this population? **(b)** Can the farmer deduce from these results the phenotypic ratio among the seeds that were initially planted? Why or why not? **(c)** Suppose that the seeds planted by the farmer were also the result of random pollination. Given this additional information, compute the expected ratio of purple to yellow kernels among the seeds that the farmer initially planted.

8. Assume that spotted and nonspotted (black) wings in a certain species of beetle are expressions of alleles at a single locus. One study revealed that 36 percent of a large population of beetles were spotted and 64 percent were black. The geneticist doing the study concluded that the frequency of the allele for spotting is equal to 0.6. **(a)** Is the geneticist justified in drawing this conclusion from the data given? Why or why not? **(b)** Previous studies have shown that beetles mate at random for the spotted and nonspotted phenotypes. Does this added information change your answer in part (a)? Explain. **(c)** The results of another study revealed that among matings between spotted beetles, about 20 percent of the offspring were black, whereas black × black matings produced only black progeny. Based on this information, what are your estimates of the allele frequencies in the beetle population? Justify any difference between the allele frequency for spotting that you obtain and the value calculated by the geneticist.

9. Determine the frequency of heterozygotes in a random mating population in which dominant individuals outnumber recessive individuals by a factor of 8 to 1.

10. Calculate the frequency of the recessive allele in a random mating population in which the frequency of heterozygotes is four times greater than the frequency of recessive homozygotes.

11. Given $p + q = 1$, verify the following equalities:
(a) $p + 2q = 1 + q$
(b) $p(1 + q) = 1 - q^2$
(c) $q^3 + pq^2 = q^2$
(d) $p^2q + pq^2 = pq$

12. The following are frequencies of genotypes *AA, Aa,* and *aa* in four different populations: Population 1: 0.64, 0.20, and 0.16; Population 2: 0.40, 0, and 0.60; Population 3: 0.01, 0.18, and 0.81; and Population 4: ⅓, ⅓, and ⅓. **(a)** Calculate the allele frequencies for each of the four populations. **(b)** Determine which of the four populations are not in random mating equilibrium, and calculate the genotype frequencies that these populations would have after one generation of random mating.

13. Suppose that studies on coat colors in two large herds of Shorthorn cattle give the following data:

NUMBER OF CATTLE

	Red	Roan	White	Total
HERD 1:	112	56	32	200
HERD 2:	98	84	18	200

Which herd conforms and which herd does not conform to a random mating population with respect to coat color frequencies?

14. Show that for a rare autosomal recessive disorder, the frequency of heterozygous carriers is approximately equal to twice the frequency of the recessive allele.

15. Cystic fibrosis is an autosomal recessive disorder, in which the mucus-secreting tissues of the affected individual are abnormal. Intestinal obstruction is an early symptom of the disease, and clogged respiratory passages often occur. The incidence of affected persons in the United States white population has been estimated to be 1 in 2500. **(a)** Determine the expected frequency of the allele for cystic fibrosis and the frequency of carriers of this disorder in the designated population. **(b)** What will be the frequency of marriages that involve two carriers?

16. Colorblindness is the result of a recessive X-linked gene. Forty males in a sample of 1000 males are found to be colorblind. **(a)** What percentage of females is expected to be colorblind in this population? **(b)** What percentage of females is expected to be heterozygous? **(c)** What percentage of marriages in this population can produce only normal offspring?

17. Two disorders, A and B, are known to be determined by different X-linked genes. Disorder A occurs 100 times more frequently among males than females. Disorder B, in contrast, is expressed in about 1.98 times as many females as males. **(a)** Which trait is dominant and which is recessive? **(b)** Calculate the allele frequencies for both disorders in the population.

18. Suppose that an enzyme is a dimer, consisting of two polypeptide chains that are coded for at a single locus. If the polypeptides combine at random during the formation of the enzyme, how many different allozymes are theoretically possible in a population in which there are two alleles that segregate at this locus? Three alleles? *N* alleles?

19. If a trait is known to show a simple dominant/recessive pattern of inheritance, the expected proportions of recessive offspring from dominant × dominant and dominant × recessive matings can then be used to test whether a population is mating at random for this trait. For example, Snyder (1934) measured the frequency of the dominant allele for PTC tasting to be $p = 0.463$. Among 761 children produced by taster × nontaster matings in this population, 483 were found to be tasters and 278 were nontasters. Does mating appear to be at random in this population for PTC tasting ability? Give reasons for your answer. (*Note:* If necessary, you can determine the goodness-of-fit between the observed and expected values by means of a chi-square test with 1 degree of freedom in this case.)

20. Suppose that two communities, each with different frequencies of M, MN, and N blood types, combine to form one freely intermarrying population. Blood tests reveal that in 4 percent of all marriages in this amalgamated group, both husband and wife have blood type M, in 16 percent both are type MN, and in 16 percent one is type M and the other is type N. **(a)** From these results, calculate the expected frequency of the M blood type among the progeny of this population. Do the same for types MN and N, assuming random mating. **(b)** Given that mating is random for blood type, calculate the expected frequencies of the M, MN, and N blood types in the two communities just after amalgamation.

21. The following are the observed ABO blood group phenotypes among a group of 600 American Indian students: 200 A, 196 B, 104 AB, and 100 O. **(a)** Calculate the frequencies of the I^A, I^B, and i alleles in this group, assuming random mating. **(b)** Determine the expected numbers of the blood group phenotypes if the students were selected from a random mating population.

22. Suppose that a person of unknown blood type is involved in an accident and is given a transfusion with type A blood. If the frequencies of the I^A, I^B, and i alleles are 0.6, 0.1, and 0.3, respectively, and if mating is random in the population for this trait, what is the chance that the individual will possess antibodies against the donated blood cells? (Consider only the ABO system.)

23. The results of ABO blood tests reveal that the frequencies of the I^A, I^B, and i alleles in a particular random mating population are 0.2, 0.1, and 0.7, respectively. **(a)** Of the matings in this population, what proportion is expected to be between two persons with type A blood? **(b)** What is the expected proportion of type O children from matings between type A persons in this population?

24. A random mating population is in linkage equilibrium for the *A,a* and *B,b* loci. The allele frequencies in this population have been determined to be $f(A) = 0.6$, $f(a) = 0.4$, $f(B) = 0.3$, and $f(b) = 0.7$. **(a)** Calculate the frequencies of the gamete types produced by this population. **(b)** Suppose that an individual is selected at random from this population. What is the probability that the genotype of the selected individual is *Aabb*? **(c)** Does your answer in part (b) indicate anything about the chromosomal location of the two loci relative to each other? Explain.

25. Geneticists sometimes find a direct association between the occurrence of a specific disease and the presence of an identifiable allelic marker. Two explanations for such an association are (1) direct causation, in which the allele in question causes the disease, and (2) linkage disequilibrium, in which the allele in question is closely linked to the gene that causes the disease and tends to segregate with it. Suppose that you discover an association between a genetic marker and a specific disease. How might you determine which of these two mechanisms is operating in your particular case?

26. Assume that a random mating population consists of equal numbers of three genotypes, *AABB*, *AaBb*, and *aabb*, and that the *A,a* and *B,b* loci are carried on different chromosomes. **(a)** Calculate the frequencies of the gamete types produced by this population. **(b)** What are the allele frequencies in this population? **(c)** Calculate the genotype frequencies among the offspring of this population. **(d)** Are the progeny genotypes in equilibrium for the *A,a* and *B,b* loci when both are considered jointly? When each locus is considered separately? **(e)** Predict the genotype frequencies for this population at equilibrium.

27. Suppose that the *A,a* and *B,b* loci in Problem 26 are linked and that there is a 25 percent chance of recombination between them. Calculate the allele and gamete frequencies for a population that consists of equal numbers of *AB/AB*, *AB/ab*, and *ab/ab* genotypes.

28. A random mating population was observed to have the following genotype frequencies for alleles at a sex-linked locus:

FEMALES			MALES	
AA	*Aa*	*aa*	*A*	*a*
0.04	0.32	0.64	0.6	0.4

(a) Calculate the genotype frequencies in the next generation.

(b) What genotype frequencies will appear in the population once an equilibrium is established?

Chapter 14

Genetic Processes of Evolution

The Hardy-Weinberg law describes a static situation in which the types and frequencies of genes and their distribution among the offspring remain constant from generation to generation. It clearly describes a state in which evolution is not occurring. In order to evolve, a population must undergo genetic change, which can only come about through alterations in gene frequency. We can therefore conclude that the factors that are responsible for evolutionary change are the very factors that were assumed not to operate in a population at random-mating equilibrium.

There are four basic evolutionary forces that can modify the frequencies of genes. These forces are: (1) selection, which results from genotypes that are unequally viable and fertile, (2) mutation, (3) migration, and (4) random genetic drift, which consists of random fluctuations in gene frequency that occur mainly in small populations. In this chapter, we will consider each of these forces in some detail, describing the mechanisms by which they act on the genetic structure of a population.

Natural Selection

Selection is the differential reproduction of genotypes. In other words, when individuals of a specific genotype produce more (or fewer) surviving offspring, on the average, than individuals of other genotypes, we can say that selection is occurring. In nature, these genotypic differences in reproductive success stem from inequalities in fertility and survival ability. In the case of fertility, the inequalities can arise through variability in the onset and duration of reproductive periods or through differences in mating success and in the number of functional gametes produced. Unequal survival ability, on the other hand, may occur through a host of different environmental effects, including differences in the ability of certain genotypes to compete for essential resources, to escape predation, or to withstand the rigors of their physical environment.

In addition to the selective process that occurs in nature, another form of selection, called *artificial selection,* has been practiced throughout a greater part of our history by humans on our crop plants and domesticated animals. Through careful breeding programs, in which those individuals with desirable qualities are saved for reproductive purposes, we have been able to modify wild species of plants and animals and develop numerous breeds and varieties to suit our needs and wishes. Since many domesticated stocks are unable to compete effectively with their wild relatives, they are permitted to flourish only through human intervention with the natural selective process. The topic of selective breeding is discussed at length in Chapter 16. We will limit our discussion here mainly to the forms of selection that occur in nature.

Reproductive Fitness

The reproductive success, or **fitness,** of a given genotype is expressed in the form of a net reproductive rate. The reproductive rate of each genotype is calculated by taking the average number of offspring produced per individual of the genotype and multiplying this number by the probability that each individual survives from birth through its reproductive period. The basis for these calculations is most clearly seen in terms of populations that reproduce in discrete, nonoverlapping generations, such as annual plants. Suppose, for example, that each parent reproduces once during its lifetime and dies before its offspring reach the reproductive age. Let

l_{ij} = the probability, for a given genotype ij, that an individual present at the start of the generation survives to the reproductive age;

m_{ij} = the average number of offspring produced per surviving individual of genotype ij at the reproductive age.

The net reproductive rate of the genotype in question is then $m_{ij}l_{ij}$. Note that by multiplying m_{ij} by l_{ij}, we obtain the projected average number of offspring produced per zygote of genotype ij. Suppose, for example, that four offspring are produced, on the average, by each adult of genotype *AA*. If an individual of this genotype has only a 50 percent chance of surviving from the zygote stage to the adult, each *AA* zygote is then expected eventually to contribute an average of $4(0.5) = 2$ offspring to the next generation.

We can apply the concept of a reproductive rate to population genetics if we consider that by forming offspring, an individual is contributing genes to the next generation. The more offspring the individual produces, the greater its contribution to the gene pool of future generations. Thus, if we multiply the reproductive rate of a genotype, which measures the average contribution of an individual of that genotype, by the frequency of the genotype in the population, we obtain a measure of the proportionate contribution of the individuals with that genotype as a group to the genetic constitution of the next generation. To illustrate, let us assume a random mating population in which the genotypes (AA, Aa, aa) have frequencies (p^2, $2pq$, q^2) at the zygote stage, and in which fertility rates for these genotypes (m_{AA}, m_{Aa}, m_{aa}) and their probabilities of survival (l_{AA}, l_{Aa}, l_{aa}) have values of (3, 4, 10) and (0.9, 0.7, 0.1) respectively. The net reproductive rates ($m_{AA}l_{AA}$, $m_{Aa}l_{Aa}$, $m_{aa}l_{aa}$) are then (2.7, 2.8, 1.0). If we multiply each of these values by the relative frequency in which the genotype occurs in the population, the proportionate genetic contributions of the three genotypes become $(2.7)p^2 : (2.8)2pq : (1.0)q^2$. Observe that the aa individuals in this example have the lowest reproductive rate, even though they are quite prolific as adults. This example illustrates the importance of taking both the survival ability and the fertility into account before making any judgment regarding the genetic contribution of each genotype.

The product $m_{ij}l_{ij}$ provides a measure of absolute fitness expressed in the form of a reproductive rate. Usually, though, we are not immediately concerned with the absolute values of these rates but only with how the reproductive ability of one genotype compares in a relative way to that of another. We therefore use a measure of *relative fitness* (designated w_{ij}), which we express as the reproductive rate of a specified genotype divided by the reproductive rate of the genotype with the greatest absolute fitness. Let us consider our numerical example once again. In this example, $m_{Aa}l_{Aa} = 2.8$ is the largest rate. The relative fitness values for the three genotypes then become

$$w_{AA} = \frac{2.7}{2.8} = 0.96, \; w_{Aa} = \frac{2.8}{2.8} = 1.0, \text{ and } w_{aa} = \frac{1.0}{2.8} = 0.36.$$

Thus, if mating is random so that the ratio of genotypes among the zygotes is $p^2 : 2pq : q^2$, the proportionate contributions at reproduction would be expressed as $(0.96)p^2 : (1.0)2pq : (0.36)q^2$.

Relative fitness, as we have defined it, can only have values ranging from 0 to 1. A fitness of 0 implies that the genotype in question fails to reproduce, as would be the case if it were lethal. When this happens, we say that there is *complete selection* operating against that particular genotype. A relative fitness between 0 and 1 signifies a less extreme form of selection, known as *partial selection.* In this case, the genotype is able to reproduce, but at a rate that is less than optimal. Most forms of selection that operate in nature would fall into this category. At the other extreme is a relative fitness of 1. This is the maximum fitness value that any genotype can have and is assigned to the genotype with the optimal reproductive capacity.

EXAMPLE 14.1. Studies have shown that the survival of *Drosophila* larvae in food that contains 0.1 percent octanoate depends, among other things, on the genotype of the fly at the glucose 6-phosphate dehydrogenase (G6PD) locus. Two alleles at this locus that affect survival are *F*, which codes for an allozyme that forms a fast-moving band during gel electrophoresis, and *S*, which codes for a slow-moving allozyme. In one experiment, larvae from *F/S* × *F/S* crosses were grown in the selective medium. The survivors were collected and analyzed, to determine the presence of the *F* and *S* alleles. Among 2400 survivors examined, 520 were *F/F* in genotype, 1250 were *F/S*, and 630 were *S/S*. Calculate from these results the relative fitness values for each of these genotypes in the selective medium.

Solution. Since the study was restricted to a single environment and to the offspring of a single genotype, the survival component of fitness in only this one selective medium can be evaluated. We can ignore other possible effects of G6PD on the total fitness or assume that these factors are constant and independent of genotype under the conditions employed in the experiment. We can determine the relative fitness values by first noting that the ratio of genotypes among the survivors (520 : 1250 : 630) corresponds to the ratio ($w_{FF}p^2N : w_{FS}2pqN : w_{SS}q^2N$), where $N = 2400$ (the total number) and p^2, $2pq$, and q^2 are the expected genotype frequencies, which are ($1/4$, $1/2$, $1/4$) for zygotes produced by a monohybrid cross. If we divide each observed number in the ratio by its expected value before selection (i.e., p^2N, $2pqN$, or q^2N), we should then obtain the ratio of relative fitness values ($520/600 : 1250/1200 : 630/600$) = (0.867 : 1.042 : 1.05), which reduces to approximately (0.83 : 1 : 1). The relative fitness values for the three genotypes are thus approximately $w_{SS} = w_{FS} = 1$ and $w_{FF} = 0.83$.

Selection operates on a genotype through its phenotype. The effectiveness of selection will therefore depend on the degree of dominance exhibited by a gene. In the sections that follow, we will consider three cases: (1) selection against recessive genotypes, (2) selection against dominant genotypes, and (3) selection against homozygous genotypes (selection favoring heterozygotes). The first two types of selection lead, in theory, to the eventual elimination of the disfavored allele. The third type, by contrast, gives rise to a state of equilibrium in which both alleles remain in the population. For now, we will assume that fitness values are constant and are independent of the size and genotypic composition of the population. Later in this chapter, we will relax this assumption and consider what happens when the fitness values themselves are functions of the genotype frequencies.

Selective Elimination of an Allele

Complete Selection. Many alleles result in premature lethality (death before reproductive age) or infertility when present in the homozygous state but have little or no effect on the fitness of heterozygotes. For these alleles, complete selection operates against the recessive homozygotes. Though obviously an extreme form of selection, complete selection serves as a useful model to demonstrate the consequences of the continued elimination of a disfavored type and can provide further insights into the selective process.

The basic selection model for a recessive lethal allele is developed in Table 14.1. This model assumes three genotypes (*AA*, *Aa*, *aa*) with relative fitness values of (1, 1, 0). Moreover, selection is assumed to operate after zygote formation but prior to the age of reproduction, so that no *aa* individuals survive to reach sexual maturity. If we start with a zygote population at generation 0 in which the ratio of (*AA* : *Aa* : *aa*) is ($p_0^2 : 2p_0q_0 : q_0^2$), and in which the ratio of (*A* : *a*) is ($p_0 : q_0$), the genotypic ratio of the adult population must then change to ($p_0^2 : 2p_0q_0 : 0$), owing to the loss of the recessive homozygotes. Although the recessive homozygotes are all eliminated in the first generation, the recessive allele is not, since it is still present in the heterozygous carriers among the breeding adults.

The consequences of a single generation of complete selection can be calculated in terms of the reduction in the frequency of the recessive allele. Since the recessive homozygotes fail to survive to reproductive maturity, all matings are restricted to crosses involving dominant genotypes (*A-* × *A-*). The frequency of the recessive allele among the zygotes in generation 1 (q_1) is thus equal to the probability that an offspring will receive a copy of the *a* allele from a dominant parent. In Chapter 13 we calculated this probability as

$$q_1 = \frac{\text{Frequency of } a \text{ among } A\text{- parents}}{\text{Total frequency of } A \text{ and } a \text{ among } A\text{- parents}} = \frac{p_0q_0}{p_0^2 + 2p_0q_0} = \frac{q_0}{1 + q_0}.$$

Table 14.1. Reduction in the frequency of a recessive lethal gene during one generation.

	ALGEBRAIC MODEL			ARITHMETIC EXAMPLE		
Generation 0						
Allele frequency:	$f(a) = q_0$			$f(a) = 1/4$		
Genotype frequencies:	*AA*	*Aa*	*aa*	*AA*	*Aa*	*aa*
Zygotes:	p_0^2	$2p_0q_0$	q_0^2	9/16	6/16	1/16
Fitness:	1	1	0	1	1	0
Adults:	$\frac{p_0^2}{\text{total}}$	$\frac{2p_0q_0}{\text{total}}$	0	9/15	6/15	0
	Adult total $= p_0^2 + 2p_0q_0$			Adult total = 15/16		
Generation 1						
Allele frequency:	$f(a) = q_1 = \frac{p_0q_0}{p_0^2 + 2p_0q_0} = \frac{q_0}{1 + q_0}$			$f(a) = \frac{1/4}{1 + 1/4} = 1/5$ or $f(a) = 0 + 1/2(6/15) = 1/5$		

Assumptions:
1. Recessive homozygotes are formed normally among zygotes but fail to reproduce.
2. The surviving dominant *AA* and *Aa* genotypes mate at random in the adult stage so that zygotes are produced according to random mating frequencies.

Furthermore, since the heterozygotes are indistinguishable in phenotype from the dominant homozygotes, mating should be random among the surviving adults. The frequency of *aa* zygotes in generation 1 can therefore be computed from the square of the allele frequency as

$$q_1^2 = \left(\frac{q_0}{1 + q_0}\right)^2.$$

The pattern is expected to be the same in the next generation. Hence, the frequency of the lethal allele among the zygotes after two generations of selection becomes

$$q_2 = \frac{q_1}{1 + q_1} = \frac{q_0/(1 + q_0)}{1 + [q_0/(1 + q_0)]} = \frac{q_0}{1 + 2q_0}$$

and so on, so that after t generations,

$$q_t = \frac{q_0}{1 + tq_0}. \tag{14.1}$$

As a numerical example, suppose we start with a population in which the genotypes (*AA*, *Aa*, *aa*) have frequencies of (¼, ½, ¼), so that $q_0 = ½$. The frequency of the recessive allele after it has been completely selected against for one generation ($t = 1$) will be $q_1 = ½/(1 + ½) = ⅓$. The incidence of recessive lethal homozygotes in generation 1 becomes $(⅓)^2 = ⅑$. After two generations ($t = 2$), the disfavored allele and genotype frequencies are further reduced to $q_2 = ½/(1 + 2 \cdot ½) = ¼$ and $(¼)^2 = 1/16$, respectively, and so on.

The number of generations that are required to reduce the allele frequency from q_0 to some value q_t can be derived from Eq. (14.1) by multiplying both sides by $1 + tq_0$ and rewriting it as $q_t + tq_0q_t = q_0$. Solving for t, we get

$$t = \frac{(q_0 - q_t)}{q_0q_t} = \frac{1}{q_t} - \frac{1}{q_0}. \tag{14.2}$$

From Eq. (14.2) we see that the time necessary to reduce the frequency of a lethal gene from, say, 0.1 to 0.05 would be $t = 1/0.05 - 1/0.1 = 20 - 10 = 10$ generations, and the time needed to reduce the allele frequency from 0.01 to 0.005 would be $t = 1/0.005 - 1/0.01 = 200 - 100 = 100$ generations. As this example shows, a reduction in the frequency of a lethal gene is followed by a corresponding decrease in the efficiency of selection against it. When q is small, most recessive genes are in heterozygous combinations and are not exposed to the effects of selection. In this condition, the value of q declines at a very low rate. The decline in the efficiency of selection limits the effectiveness of any program designed to eliminate harmful recessive genes from breeding stocks of crop plants and domesticated animals. Even when complete selection is practiced and the recessive homozygotes are not allowed to mate, *aa* individuals will continue to be produced for a very long time, as an occasional result of *Aa* × *Aa* crosses. Similar programs to get rid of recessive disorders in humans by sterilization or other eugenic measures are limited by the same genetic factors and are likely to be rather ineffective.

EXAMPLE 14.2. As we learned in Chapter 11, Tay-Sachs disease is a fatal recessive disorder, in which affected individuals are unable to produce the enzyme hexosaminidase A. The absence of this enzyme results in the accumulation of a specific fatty substance (ganglioside GM_2) within the nerve cells and the subsequent deterioration of the central nervous system. The allele for this disorder is most common in certain Jewish populations, in which it has been estimated to occur at a frequency as high as $1/80$. If we were to ignore any factors that might act to increase the incidence of this disease, how many generations would be required for selection against homozygous recessives to reduce the frequency of the Tay-Sachs allele in these populations to $1/800$?

Solution. We are asked to compute a value for t (time in generations) for the change from $q_0 = 1/80$ to $q_t = 1/800$. With complete selection against this recessive trait, we can calculate the value as $t = \dfrac{1}{1/800} - \dfrac{1}{1/80} = 800 - 80 = 720$ generations.

In contrast to recessive lethal genes, dominant lethals are removed from a population at a rapid rate. In the case of a dominant lethal gene, the genotypes (*AA*, *Aa*, *aa*) will have relative fitness values of (0, 0, 1), and only the recessive homozygotes will be able to reproduce. In theory, dominant lethal genes will be eliminated in a single generation.

Partial Selection. While the elimination of lethal alleles constitutes the most severe form of selection, the selective processes that occur in nature are usually less than complete, often with genotypes that differ only slightly from one another in terms of relative fitness. In cases of partial selection, the relative intensity of selection will then depend on the exact distribution of fitness values among the genotypes.

The effect of one generation of partial selection is usually expressed in terms of the change in allele frequency. This change can be shown either as $\Delta q = q' - q$, where q and q' designate the frequencies of the a allele in successive generations, or as $\Delta p = p' - p$, which gives the change in the frequency of the A allele. Since $p + q = 1$ and $p' + q' = 1$, we see that $(p' - p) + (q' - q) = 1 - 1 = 0$. Thus, $\Delta p = -\Delta q$, so that changes in the two allele frequencies will differ from one another only in direction.

The basic relationship between the change in allele frequency and the distribution of relative fitness values can be derived with the aid of the general selection model given in Table 14.2 and the following identity:

$$\Delta q = q' - q = q' - q'q - q + q'q = q'(1 - q) - q(1 - q') = q'p - qp'$$
$$= pq\left(\frac{q'}{q} - \frac{p'}{p}\right).$$

The values q'/q and p'/p are factor increases in q and p that measure the growth rates of the two alleles in the gene pool of the population. Both q'/q and p'/p are dependent upon the fitness values of the contributing genotypes. The exact dependence is derived in Table 14.2, from which we obtain

$$\frac{q'}{q} = \frac{qw_{aa} + pw_{Aa}}{\overline{w}} \quad \text{and} \quad \frac{p'}{p} = \frac{pw_{AA} + qw_{Aa}}{\overline{w}}.$$

Table 14.2. The general selection model.

	AA	Aa	aa	TOTAL
Frequency before selection:	p^2	$2pq$	q^2	1
Fitness:	w_{AA}	w_{Aa}	w_{aa}	
Proportionate contribution:	p^2w_{AA}	$2pqw_{Aa}$	q^2w_{aa}	$\overline{w}$
Frequency after selection:	$(p^2)\frac{w_{AA}}{\overline{w}}$	$(2pq)\frac{w_{Aa}}{\overline{w}}$	$(q^2)\frac{w_{aa}}{\overline{w}}$	1

Allele frequencies in the next generation:

$$p' = f(AA) + \tfrac{1}{2}f(Aa) = (p^2)\frac{w_{AA}}{\overline{w}} + (pq)\frac{w_{Aa}}{\overline{w}} = p\left[\frac{pw_{AA} + qw_{Aa}}{\overline{w}}\right]$$

$$q' = f(aa) + \tfrac{1}{2}f(Aa) = (q^2)\frac{w_{aa}}{\overline{w}} + (pq)\frac{w_{Aa}}{\overline{w}} = q\left[\frac{qw_{aa} + pw_{Aa}}{\overline{w}}\right]$$

The factor $\overline{w}$ constitutes the *average fitness* of the population, expressed as

$$\overline{w} = p^2w_{AA} + 2pqw_{Aa} + q^2w_{aa}.$$

The average contribution of the different genotypes of the selected group to the gametes is given by $\overline{w}$, which attains its maximum value of 1 within a nonselective environment, in which $w_{AA} = w_{Aa} = w_{aa} = 1$.

By substituting the expressions given above for q'/q and p'/p into the identity for Δq and rearranging the terms, we get the following general relationship:

$$\Delta q = pq\left[\frac{(w_{aa} - w_{Aa})q - (w_{AA} - w_{Aa})p}{\overline{w}}\right]. \quad \textbf{(14.3)}$$

The factors $(w_{aa} - w_{Aa})q$ and $(w_{AA} - w_{Aa})p$ in this equation designate the relative contributions of the two alleles to fitness, above the fitness value expressed by the heterozygote. For example, the difference $w_{aa} - w_{Aa}$ measures the effect that allele *a* would have on fitness if it were substituted for the *A* allele in the heterozygotes; the difference $w_{AA} - w_{Aa}$ gives the effect of the reverse substitution in the heterozygote ($Aa \rightarrow AA$).

Equation (14.3) consists of three fundamental parts.

1. The change in allele frequency is directly proportional to the product pq. Partial selection is therefore most effective at intermediate allele frequencies, that is, when the frequencies of the dominant and recessive alleles are similar. If one allele is much more frequent than the other, selection will be much less efficient, since Δq approaches 0 when p (or q) is close to 0 or 1.
2. The change in allele frequency is also directly proportional to the difference of the relative contributions of the two alleles to fitness: $(w_{aa} - w_{Aa})q - (w_{AA} - w_{Aa})p$. This difference in the contributions of the two alleles determines the direction of selective change. If the difference is positive, the *a* allele will increase in frequency; if it is negative, the *a* allele will decrease in frequency.

3. The change in allele frequency is inversely proportional to the average fitness of the population, $\overline{w}$. The change in allele frequency is therefore greatest in highly selective environments, in which $\overline{w}$ is small in comparison to its maximum value of 1.

EXAMPLE 14.3. One of the most spectacular evolutionary changes to occur in recent history has been the selective increase in the frequency of the dark (or melanic) form of the peppered moth, *Biston betularia*. The change occurred during the past century in urban areas of England, where the terrain had become darkened by the soot of industrial pollution. The dark form of the moth, which is determined largely by a single dominant gene, has an advantage over the recessive light-colored variety by virtue of its better camouflaged concealment from predators when it is resting on a sooty background. Let us assume for the purposes of calculation that the light-colored form of the moth makes up 49 percent of a random mating population when it emerges as an adult and has only half the reproductive potential of the melanic form. **(a)** Calculate $\overline{w}$, the average fitness of the moth population. **(b)** Determine the frequencies of both forms of the moth in the next generation.

Solution. **(a)** Since the frequency of the recessive form is $q^2 = 0.49$, the frequencies of the recessive and dominant alleles are $q = \sqrt{0.49} = 0.7$ and $p = 0.3$. Also, because the recessive form has only half the reproductive potential of the melanic form, the relative fitness values are $w_{AA} = w_{Aa} = 1$ and $w_{aa} = 0.5$. The average fitness becomes $\overline{w} = (1)(0.3)^2 + (1)(2)(0.3)(0.7) + (0.5)(0.7)^2 = 0.09 + 0.42 + (0.5)(0.49) = 0.755$. **(b)** The frequency of the recessive allele in the next generation is $q' = (w_{aa}q^2 + w_{Aa}pq)/\overline{w} = [(0.5)(0.49) + (1)(0.3)(0.7)]/0.755 = 0.6$. Thus, the frequency of the light-colored moth in the next generation is $q'^2 = 0.36$ and the frequency of the melanic form is $1 - 0.36 = 0.64$. Note that the light-colored form has decreased in frequency by $0.49 - 0.36 = 0.13$ in one generation.

A numerical example of the general selection model is given in Table 14.3, in which the solution is obtained by using Eq. (14.3). For the most part, we will not be concerned with exact solutions to Eq. (14.3); instead, we will obtain approximate solutions that describe the basic features of different types of selection without all the cumbersome algebraic detail. The major assumption that we shall make is that the average fitness of the population does not differ significantly from its maximum value of 1. Setting $\overline{w}$ equal to 1 is a suitable approximation for conditions of weak selection intensity, when both $w_{aa} - w_{Aa}$ and $w_{AA} - w_{Aa}$ have values of 0.01 or less (unlike the example given in Table 14.3).

One important case of partial selection that is covered by our general selection model is selection against recessive homozygotes. In this special case, only *aa* individuals are at a selective disadvantage so that $w_{AA} = w_{Aa} = 1$ and $w_{AA} - w_{Aa} = 0$. Also, since w_{aa} is less than w_{Aa}, we can further reduce algebraic clutter by means of the expression $w_{aa} - w_{Aa} = -s$, where s is a measure of selection intensity and has a value between 0 and 1. By making these substitutions in Eq. (14.3) and setting $\overline{w}$ equal to 1, we get the following approximate solution for

Table 14.3. A specific example of the selection model, with allele frequencies starting at $p = 1/4$ and $q = 3/4$.

	AA	Aa	aa	TOTAL
Frequency before selection:	$1/16$	$6/16$	$9/16$	1
Fitness:	1	1	$1/3$	
Proportionate contribution:	$1/16(1)$	$6/16(1)$	$9/16(1/3)$	$10/16$
Frequency after selection:	$1/10 = 0.1$	$6/10 = 0.6$	$9/10(1/3) = 0.3$	1

Allele frequency in the next generation:

$q' = 0.3 + 1/2(0.6) = 0.6$

Change in allele frequency:

$\Delta q = q' - q = 0.60 - 0.75 = -0.15$

or, directly obtained by using Eq. 14.3,

$$\Delta q = (1/4)(3/4)\frac{(1/3 - 1)(3/4) - (1 - 1)(1/4)}{10/16} = (3)\frac{-(2/3)(3/4)}{10} = -0.15.$$

selection against recessive homozygotes:

$$\Delta q = -s(1 - q)q^2 \tag{14.4}$$

in which the minus sign indicates that q is decreasing in value. Note that the effectiveness of selection in this case depends not only on the selection intensity, as measured by s, but also on q^2. Thus, when q is small, q^2 will be very small (e.g., when $q = 0.01$, $q^2 = 0.01^2 = 0.0001$), so that selection against recessive homozygotes becomes a very inefficient process. We can derive the same conclusion from Fig. 14.1, which shows the consequences of selection against a recessive

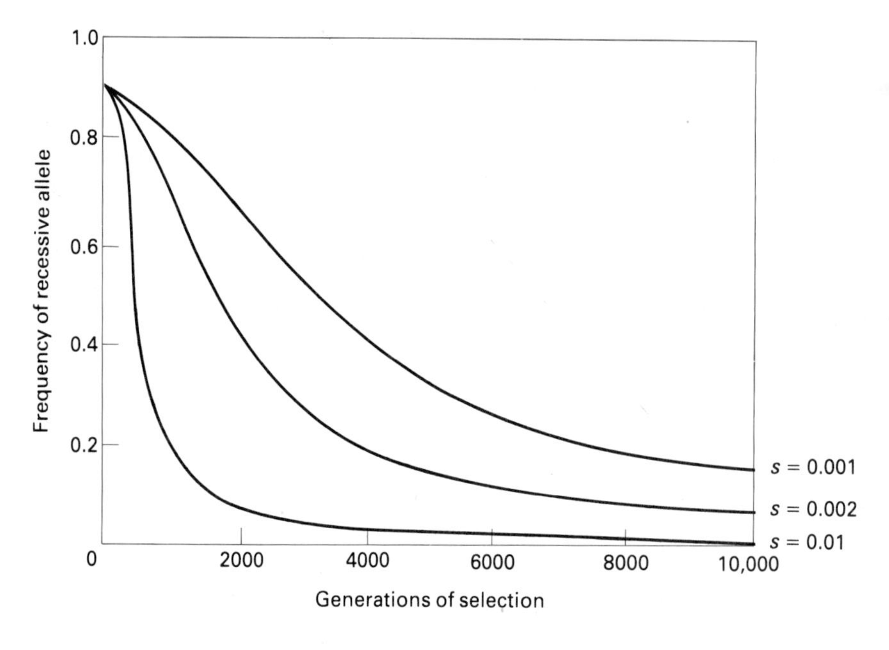

Figure 14.1. Different intensities of selection against a recessive detrimental allele that occurs initially (at generation 0) at a frequency of 0.9. At first, the reduction in allele frequency is comparatively rapid, because the allele is common, but it occurs at a very slow pace when the allele becomes rare. The elimination of the allele also proceeds at a faster rate with an increase in the strength of selection, measured by s.

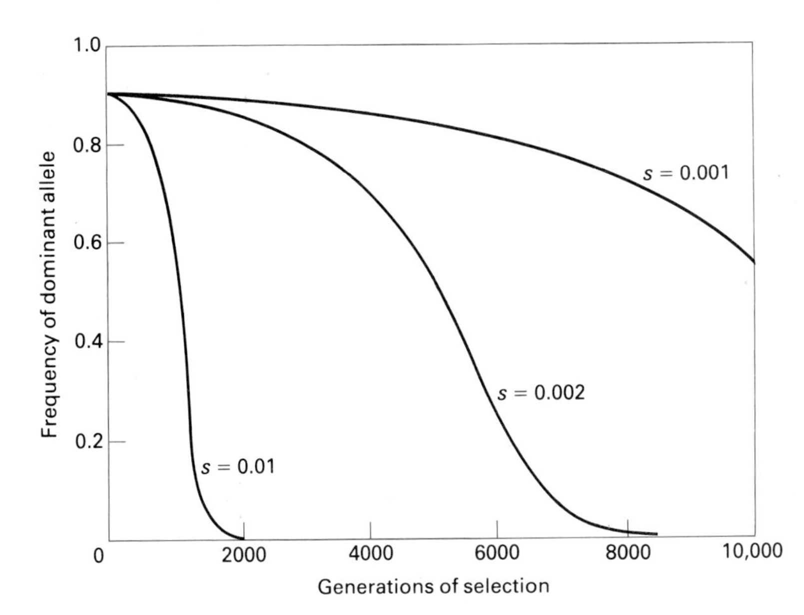

Figure 14.2. Different intensities of selection against a dominant detrimental allele that occurs initially (at generation 0) at a frequency of 0.9. The reduction in allele frequency is slow at first, because the allele is common, but it accelerates to a more rapid pace as the dominant allele declines in frequency. The elimination of the allele also proceeds at a faster rate as the strength of selection increases.

allele at various selection intensities. Observe that selection against recessives is characterized by rapid selection in the beginning, when q is large, and by a slow rate at the end, when q is small.

Another important case of partial selection that is included in our general model is selection against the dominant allele. If the A allele is completely dominant, w_{AA} and w_{Aa} are equal and $w_{AA} - w_{Aa}$ is again equal to 0. In this case, however, w_{aa} is greater than w_{Aa} so that $w_{aa} - w_{Aa} = +s$. If we make these substitutions in Eq. (14.3), assuming again that $\overline{w} = 1$ and replacing Δq with $-\Delta p$, we get the following approximate solution for *selection against dominants:*

$$\Delta p = -sp(1 - p)^2. \tag{14.5}$$

The effectiveness of selection in this case depends on the square of the frequency of the favored allele, $(1 - p)^2$. When p is large, with a value close to 1, the factor $(1 - p)^2$ will approach 0, so that selection against dominants becomes very inefficient. When p is small, $(1 - p)^2$ approaches 1, and the A allele is then reduced by the fraction $-\Delta p/p = s$ in each generation. These characteristics are shown graphically in Fig. 14.2. Observe that selection against dominants gets off to a slow start when p is large, but the rate of selection accelerates as the A allele declines in frequency.

Selection in Action

Several examples of partial selection have been documented in nature. Many are the result of human-induced changes in the environment. For the most part, they involve genetic mechanisms that are more complex than those that we assumed in the preceding models. Among the examples that are of special concern to human health and to the management of crops and domesticated animals are the rise in the resistance of bacteria to antibiotics and of insects to various insecticides.

Table 14.4. Some antibiotics and their mechanisms of action.

ANTIBIOTIC	MICROBIAL SOURCE	MODE OF ACTION
Ampicillin	*Penicillium* sp.	Inhibits cell wall synthesis
Bacitracin	*Bacillus subtilis*	Inhibits cell wall synthesis
Chloramphenicol	*Streptomyces venezuelae*	Blocks protein synthesis
Erythromycin	*Streptomyces erythreus*	Blocks protein synthesis
Novobiocin	*Streptomyces niveus* or *speroides*	Inhibits DNA synthesis
Penicillin G	*Penicillium chrysogenum*	Inhibits cell wall synthesis
Polymyxin B	*Bacillus polymyxa*	Destroys cell membrane
Streptomycin	*Streptomyces griseus*	Results in abnormal protein synthesis
Tetracycline	*Streptomyces aureofaciens*	Blocks protein synthesis

Resistance to Chemicals. Antibiotics are chemical substances that are produced by microorganisms and that inhibit the growth of other microorganisms. Some of the better-known antimicrobial agents and their mechanisms of action are listed in Table 14.4. With the advent of the widespread use of antibiotics, bacterial drug resistance has become a problem of major practical concern. Antibiotic resistance can arise in either of two ways. First, resistant bacteria can arise spontaneously, as a result of gene mutation in a normally sensitive strain. The mutant bacteria become resistant by acquiring the ability to destroy the antibiotic chemically or to avoid the killing effect of the drug in some other way. For example, resistance to penicillin is due to the formation of an enzyme called penicillinase, which modifies the structure of penicillin and thereby inactivates the toxic effect of the antibiotic on the bacteria. Drug resistance arising out of a mutation is rare and tends to have little impact on the genetic structure of bacterial populations under the conditions that they normally encounter. But in an environment that contains the drug, the resistant mutants are favored and will increase in frequency, eventually supplanting the sensitive strain. Fortunately, most mutations of this type confer resistance to only a single kind of antibiotic, leaving the bacteria susceptible to the actions of other antibiotics, so that they can be controlled. Nevertheless, the widespread and, in some cases, almost indiscriminate use of antibiotics in medicine and as supplements in livestock feed has served as a selective environment for the increase of certain resistant strains.

In recent years, a multiple drug resistance has become of major clinical concern, following the discovery that bacteria can become simultaneously resistant to more than a single antibiotic when the process is mediated by an R plasmid (see Chapter 4). R plasmids are extrachromosomal agents that carry genes for resistance to a variety of antibiotics, including sulfonamides, streptomycin, chloramphenicol, and tetracyclines, and can transfer these genes to other bacteria by means of conjugation. Plasmid-mediated transfer is thus the second mechanism by which drug resistance can arise in bacteria. This mechanism is particularly troublesome, since the resistance spreads rapidly in the presence of antibiotics and is very difficult to control. For example, in one hospital in Japan, multiple drug resistance increased in *Shigella* (the bacterium that produces dysentery) from 0.2 percent in 1954 to 52 percent in 1964.

Pesticide resistance is also becoming a serious problem, in that it has been recorded in over 200 species of insect pests during the past 30 years. Several adaptive mechanisms are involved in providing the different insect pests with greater immunity to insecticides. These mechanisms include greater resistance to the penetration of the poison through the integument, the acquisition of enzymes to detoxify the chemical, and the production of enzymes with increased resistance to the inhibitory action of the insecticide. Behavioral changes have also evolved, producing animals that are more likely to avoid contact with the poison. While none of these mechanisms will normally provide complete immunity to the insecticide, their combined effects can give rise to animals that show a high level of resistance (Fig. 14.3). For this reason, progressively greater doses of the chemicals are needed to overcome the increase in insecticide resistance, thus increasing the cost of pest control as well as posing a serious threat to human health and to the survival of beneficial wildlife.

Figure 14.3. Selection for resistance to insecticides. Resistance occurs through a combination of mechanisms, ranging from the development of avoidance behavior to the acquisition of changes at the molecular level. None of these mechanisms will normally provide complete immunity when acting alone. However, their combined actions can greatly reduce the potentially lethal effect of the insecticide (as indicated here by a reduction in the width of the arrows). *Source:* From Solbrig/Solbrig, *Introduction to Population Biology and Evolution*, © 1979. Addison-Wesley, Reading, MA. Reprinted with permission.

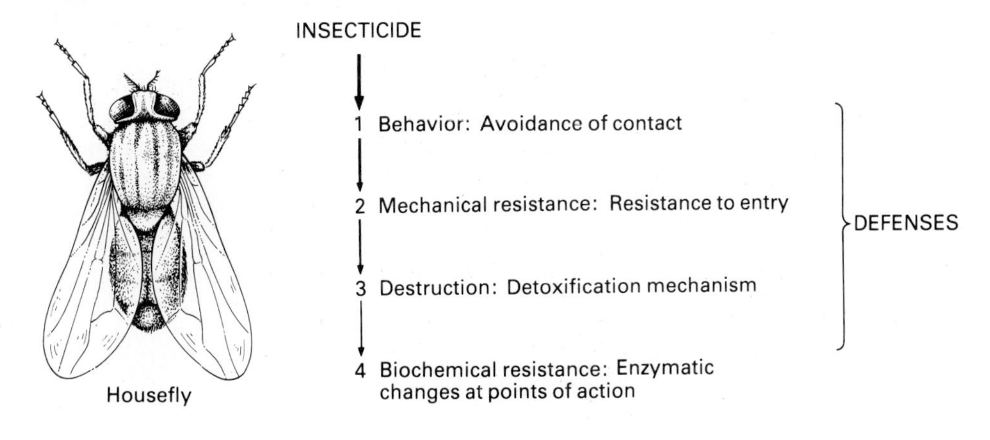

Resistance to Disease. In addition to providing greater resistance to chemicals, natural selection also acts to decrease susceptibility to disease. One interesting example concerns the host-parasite relationship that presently exists in the European wild rabbit, *Oryctolagus cuniculus*, in Australia. The rabbit was introduced to Australia from Europe in 1788 and subsequently spread throughout the greater part of the continent, becoming a serious economic pest. Beginning in 1950, the viral disease myxomatosis was used in an attempt to reduce the rabbit population. Rabbits were infected with the disease and released into densely populated areas. With the mosquito serving as a carrier for the disease, the virus was rapidly disseminated into most rabbit-infested areas. At first, in its original virulent form, the virus was very effective, proving fatal to some 98 percent of the rabbits. But the mortality rate subsequently declined to about 25 percent, owing to selection for mutant forms of the rabbit that were more resistant to viral infection.

The reduction in rabbit mortality was accompanied by selection for attenuated (less virulent) forms of the virus. Decreased virulence is selectively advantageous to the virus, since it increases the average lifespan of its host and thereby improves the chance that the virus will be transferred by the mosquito to uninfected rabbits. Thus, not only the rabbits but also the viruses evolved to the mutual benefit of both coexisting populations. This process of evolutionary change that occurs jointly in two or more interacting populations is called **coevolution;** it causes the host population to become more resistant to parasites and the parasites to become less apt to destroy their host.

Evidence seems to indicate that human populations have also participated in coevolution. When various pathogens were introduced to the American continent by immigrants from Europe, certain diseases such as tuberculosis were initially devastating to the native inhabitants. Yet, as the epidemics swept through the native populations, fewer fatalities occurred with each succeeding wave, so that these diseases are now less of a threat to the descendants than they were to their forebears who were originally exposed. In the case of tuberculosis, the decline in mortality that has been recorded is not attributable to any large extent to improvements in therapy or in sanitary conditions but apparently results from an increase in resistance by the host population and the attenuation of the virulence of the pathogen.

Balancing Selection

We tend to think of selection in terms of the development of new adaptive types and the directional change in allele frequency that this requires. But selection can also give rise to a state of equilibrium. In an environment that has remained comparatively stable for some time, the most fit combinations of genes are often the most frequently occurring types. The majority of selective changes in this type of environment are directed toward eliminating less fit genotypes on either end of the phenotypic spectrum that arise by segregation and recombination. This process of eliminating the adaptively inferior extremes provides a mechanism for preserving those genotypes that are best suited for the conditions that are prevalent at a particular time. Selection then acts to maintain the status quo. In this section, we will describe a few selective mechanisms that are believed to be important in the creation and preservation of this balanced state.

Heterozygote Advantage. As we pointed out in an earlier section, when selection favors the heterozygotes so that both homozygous types are at a selective disadvantage, an equilibrium is established in which both alleles are retained in the population. This process can be shown by a simple model in which complete selection is assumed to act against both *AA* and *aa* individuals. If we start with genotypes (*AA*, *Aa*, *aa*) at frequencies of (¼, ½, ¼) among the zygotes, the elimination of the homozygous genotypes results in a population of breeding adults with genotypic frequencies of (0, 1, 0). Since only heterozygotes remain in the adult population in each generation, only *Aa* × *Aa* matings can occur, so that the genotypic frequencies among the zygotes in the next generation will once again be (¼, ½, ¼). A balanced lethal condition is thus established in which the allele frequencies are constant at $p = 1/2$ and $q = 1/2$, even though half of the population is lost through selection in each generation.

We can use our general selection model to extend the case of heterozygote advantage to include partial selection against both *AA* and *aa* homozygotes. Let us assume that mating is random, so that the genotypes among the zygotes conform to Hardy-Weinberg frequencies of p^2, $2pq$, and q^2, and let us also assume that selection occurs prior to the reproductive age. Since heterozygotes have the selective advantage in this case, w_{Aa} will be greater than both w_{AA} and w_{aa}. We can therefore make the following designations: $w_{aa} - w_{Aa} = -s_a$ and $w_{AA} - w_{Aa} = -s_A$. If we substitute these values into Eq. (14.3) and assume that $\overline{w} = 1$, we get the approximate solution for *selection favoring heterozygotes:*

$$\Delta q = pq(s_A p - s_a q). \qquad \textbf{(14.6)}$$

Note that when $s_A p = s_a q$ in this relationship, $\Delta q = 0$. An equilibrium is then expected to develop in which the allele frequencies remain constant from one generation to the next. The equilibrium frequencies of the two alleles can be derived from the equality $s_A p = s_a q$ by replacing q with $1 - p$ and rearranging the terms: $s_A p = s_a(1 - p) = s_a - s_a p$. Solving for p (and q), we get *at equilibrium*

$$p = s_a/(s_A + s_a) \quad \text{and} \quad q = s_A/(s_A + s_a).$$

The allele frequencies are thus determined by the differences in relative fitness, designated by s_A and s_a.

For a numerical example, let us suppose that the fitness distribution (w_{AA}, w_{Aa}, w_{aa}) of a population is (0.9, 1, 0.8). The relevant differences in fitness are then $s_A = 1 - 0.9 = 0.1$ and $s_a = 1 - 0.8 = 0.2$. Thus, the allele frequencies at equilibrium become $p = 0.2/(0.1 + 0.2) = 2/3$ and $q = 1/3$. This example is considered further in Table 14.5, which shows the genotypic composition of the equilibrium population both prior to and following selection. Although the frequencies of *AA* and *aa* are p^2 and q^2 among the zygotes, their proportions decline among the adults as a result of selection. This seesaw process is expected to continue as long as the equilibrium conditions hold.

Table 14.5. Genotype frequencies before and after selection at the equilibrium established with $s_A = 0.1$ and $s_a = 0.2$.

Allele frequencies (before selection):

$$f(A) = p = 0.2/(0.1 + 0.2) = 2/3$$
$$f(a) = q = 0.1/(0.1 + 0.2) = 1/3$$

Gentotype frequencies:	AA	Aa	aa	Total
Before selection:	$(2/3)^2 = 4/9 = 0.444$	$2(2/3)(1/3) = 4/9 = 0.444$	$(1/3)^2 = 1/9 = 0.112$	1
Fitness:	$1 - 0.1 = 0.9$	1	$1 - 0.2 = 0.8$	
Contribution:	$4/9(0.9) = 0.400$	$4/9 = 0.444$	$1/9(0.8) = 0.089$	0.933
After selection:	$\frac{0.400}{0.933} = 0.429$	$\frac{0.444}{0.933} = 0.476$	$\frac{0.089}{0.933} = 0.095$	1

Allele frequencies (after selection):

$$f(A) = p = 0.429 + 1/2(0.476) = 2/3$$
$$f(a) = q = 0.095 + 1/2(0.476) = 1/3$$

EXAMPLE 14.4. About 1 child in 2500 Caucasians is born with the recessive disorder cystic fibrosis. The disease rarely affects other races. There is some evidence to suggest that the abnormally high incidence of cystic fibrosis in white populations is a result of a reproductive advantage by the heterozygotes. By what value must the reproductive potential of the heterozygotes exceed that of the dominant homozygotes in order to maintain the incidence of cystic fibrosis at its present level through heterozygote advantage? (*Note:* Assume that the fitness of affected individuals is zero.)

Solution. The frequency of the allele for cystic fibrosis must be $q = \sqrt{1/2500} = 1/50$. This frequency must also equal $s_A/(s_A + s_a)$, if the proposed mechanism of heterozygote advantage is correct. In this case $s_a = w_{Aa} - w_{aa} = 1 - 0 = 1$ and $s_A = w_{Aa} - w_{AA}$, which represents the fractional excess in fitness of the heterozygotes over the dominant homozygotes. Thus, $1/50 = s_A/(s_A + 1)$, which gives us $50s_A = s_A + 1$ or $49s_A = 1$. The reproductive potential of the heterozygotes must therefore exceed that of the dominant homozygotes by $1/49 = 0.02$, or 2 percent.

Heterozygote advantage represents one important mechanism by which selection can maintain two or more alleles in a population at reasonably high frequencies. The result is a stable variation pattern known as **balanced polymorphism,** in which the less-favored phenotypes are retained in the population along with the more adapted types. A well-known example of a balanced polymorphism in humans is sickle-cell anemia, which is prevalent in parts of Africa and Southern Asia. Recall from earlier chapters that homozygotes for the hemoglobin S allele (Hb^SHb^S) have the severe erythrocyte destruction that is associated with this anemia and tend to die in early childhood. They are clearly less fit than the heterozygotes (Hb^AHb^S), who have a milder form of sickling without anemia. But in regions of the world where falciparum malaria is prevalent, homozygotes for the normal hemoglobin allele (Hb^AHb^A) are also less fit than the heterozygotes, since they are less resistant to the malarial parasite and often die from the disease. Thus, the heterozygotes have a selective advantage and are more likely to reproduce than either homozygous type.

Many forms of evidence have been used to link the sickle-cell polymorphism to a differential resistance to malaria. One has been the close correlation that exists between the world-wide distribution of the sickle-cell allele and that of falciparum malaria (see Fig. 14.4). Hospital records have also revealed that the death rate from this form of malaria is much higher among homozygotes with normal hemoglobin than it is among heterozygotes. Of course, in nonmalarial regions of the world, heterozygotes do not have a selective advantage, and selection then occurs against the sickle-cell allele. In the black population of the United States, therefore, the frequency of the allele has fallen to a relatively low level, with heterozygous carriers currently making up about 9 percent of the American black population.

Frequency-Dependent Selection. Another type of selection that can result in a state of balanced polymorphism is *frequency-dependent selection.* This form of selection occurs when the fitness values are inversely related to the genotype frequencies, and it leads to a situation in which each genotype (or phenotype) gains a selective advantage when it is rare and is selected against when it is common. In this type of selection, no one genotype can completely eliminate any other genotype, and all of the affected genotypes will remain in the population at numbers below the saturation level of their environment.

There are several mechanisms that can give rise to a minority advantage, as this type of balance is called. One is competition for essential resources such as food or suitable living space. Since the resources are required for continued survival and reproduction, and since they are present in limited amounts, the fitness values of those genotypes that are directly involved in the competitive interaction will be adversely affected. If the genotypes differ sufficiently in their resource requirements (or preferences), competition will then be greater among members of the same genotype than among those of different genotypes. The fitness of each genotype will therefore decrease as its members increase in abundance so that the effects of competition become more severe.

A second mechanism that provides for a minority advantage is selective predation, in which the most frequently occurring phenotypes in a prey population are preferentially attacked by predators. This mechanism is thought to operate primarily in populations that serve as major food sources for vertebrate predators. Predators of this kind often develop a search image and learn preferentially to attack prey of the phenotype in greatest abundance. By selectively feeding on the more abundant types of prey, predators waste little time in search and pursuit of less abundant phenotypes. Selective predation also gives the less frequently occurring forms of prey a chance to reproduce and recover their losses until they too increase to levels high enough to be favored by the predator. Thus, each distinct type of prey within a population will be at an advantage when it is rare and at a disadvantage when it is common.

A third mechanism that has been thoroughly studied in *Drosophila* is the rare-mate advantage. In this form of sexual selection, males of the less common strains

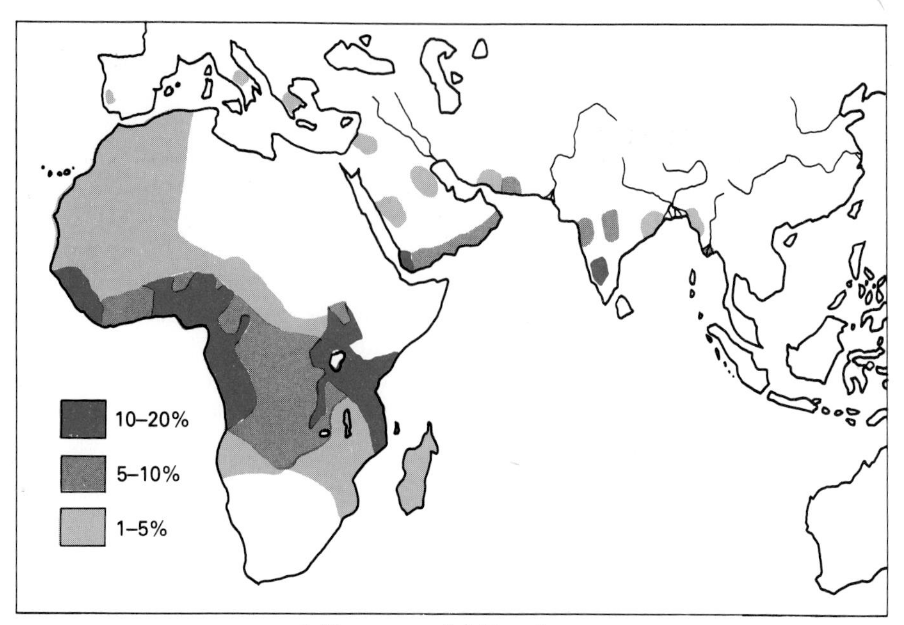

(a) Frequency of sickle cell gene

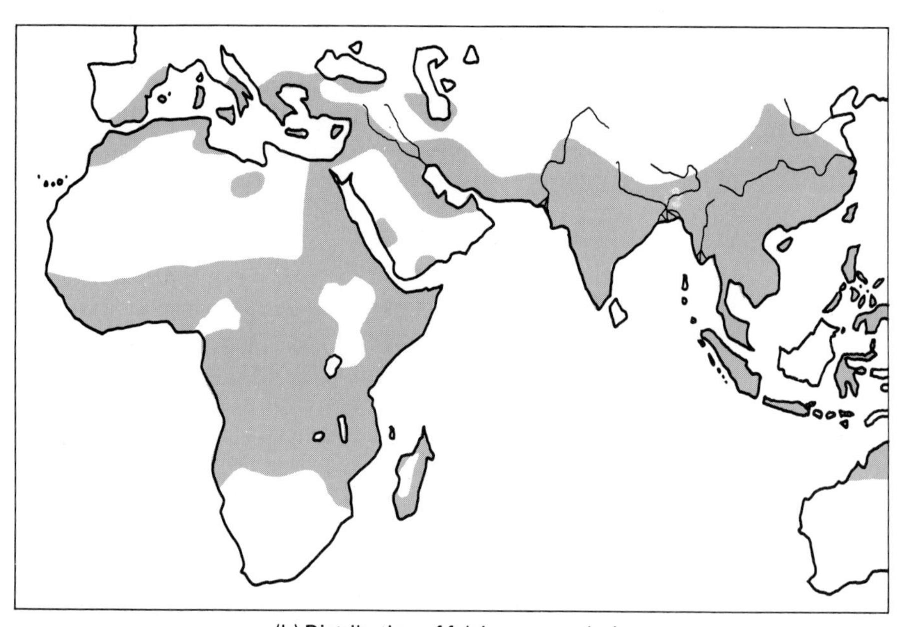

(b) Distribution of falciparum malaria

Figure 14.4. A comparison of the distribution patterns of (a) the sickle cell allele and (b) falciparum malaria. A good correlation exists in the joint occurrence of these diseases over much of their areas of distribution. *Source:* From L. S. Dillon, *Evolution: Concepts and Consequences*, 2nd ed., 1978. The C. V. Mosby Co., St. Louis.

mate disproportionately more often than males of the more abundant strains. The female flies apparently recognize different male genotypes by way of olfactory cues and give preference to the rare males.

TO SUM UP

1. Selection occurs when some genotypes produce more (or less) surviving offspring, on the average, than others. These differences in reproductive ability among the genotypes result from inequalities in fertility and survival ability.
2. The fitness of a given genotype is a measure of its reproductive success under a certain set of environmental conditions. The fitness can be expressed either in the form of an absolute reproductive rate or relative to the reproductive rate of the genotype with the largest reproductive ability.
3. Selection against the recessive homozygote is characterized by a rapid reduction in the frequency of the recessive phenotype when the frequency of the recessive allele is large and by a very slow decrease when q is small and most recessive alleles are "hidden" in heterozygotes. In contrast, selection against dominants gets off to a slow start when p is large but accelerates as the dominant allele declines in frequency.
4. Selection that favors the heterozygote will lead to an equilibrium condition known as a balanced polymorphism, in which both the selectively advantageous and the detrimental phenotypes are maintained in a population at relatively high frequencies. A balanced polymorphism can also arise from some form of frequency-dependent selection, in which the genotypes are at an advantage when they are rare and selected against when they are common.

Nonselective Changes in Gene Frequency

Three processes in addition to selection are known to alter the genetic structure of populations. They are mutation, migration, and random genetic drift. Like selection, all three can produce changes in gene frequency and thus all three constitute evolutionary forces. Unlike selection, however, none of the three has a direct and causal bearing on survival and fertility; they are therefore only indirectly related to adaptation. In this section, we shall consider each of these processes separately and shall describe their potential roles in promoting evolutionary change.

Mutation

As we learned in Chapter 11, a mutation is a heritable change in the genetic material that is detected by the sudden appearance of a new and different phenotype within a population or family group. In contrast with selection, mutations are random; no particular mutation can be directed by the environment in which it occurs. In other words, particular mutations occur strictly by chance, not as an adaptive response to environmental conditions. For example, streptomycin resistance does not arise in bacterial cells in response to their exposure to the antibiotic; instead, it occurs spontaneously prior to the addition of the drug to the culture medium. When added to the medium, the antibiotic simply serves as a selective agent by permitting only resistant cells to grow.

A clear demonstration of the randomness of mutation is provided by the *replica plating technique,* which was first used for this purpose by Joshua and Esther

Lederberg. The procedure is outlined in Fig. 14.5. It involves inoculating an agar growth medium in several petri dishes with bacteria taken from colonies grown in a separate petri dish, known as the master plate. The bacteria are transferred by first pressing the agar surface of the master plate to a sterile velvet-covered block. The fibers of the velvet act as small inoculating needles, which hold the bacterial samples in the identical spatial arrangement in which they grew. Several sterile plates are then pressed to the velvet surface. Each plate inoculated in this way becomes a replica of the master plate, with the same identifiable pattern of colony positions on its agar surface. When the replica plates are exposed to a selective agent such as streptomycin, the same colonies are resistant on all of the plates. This result is consistent with a prior spontaneous origin of streptomycin resistance on the master plate, since it is highly unlikely that the streptomycin would induce an identical pattern of resistant growth in each of several indepen-

Figure 14.5. Use of the replica plating technique to show the randomness of mutation. All of the colonies on the master plate that gave rise to colonies on the replica plates are found to contain streptomycin-resistant cells, whereas all of the colonies on the master plate that fail to yield colonies on the replica plates are found to contain only streptomycin-sensitive cells (Step 5).

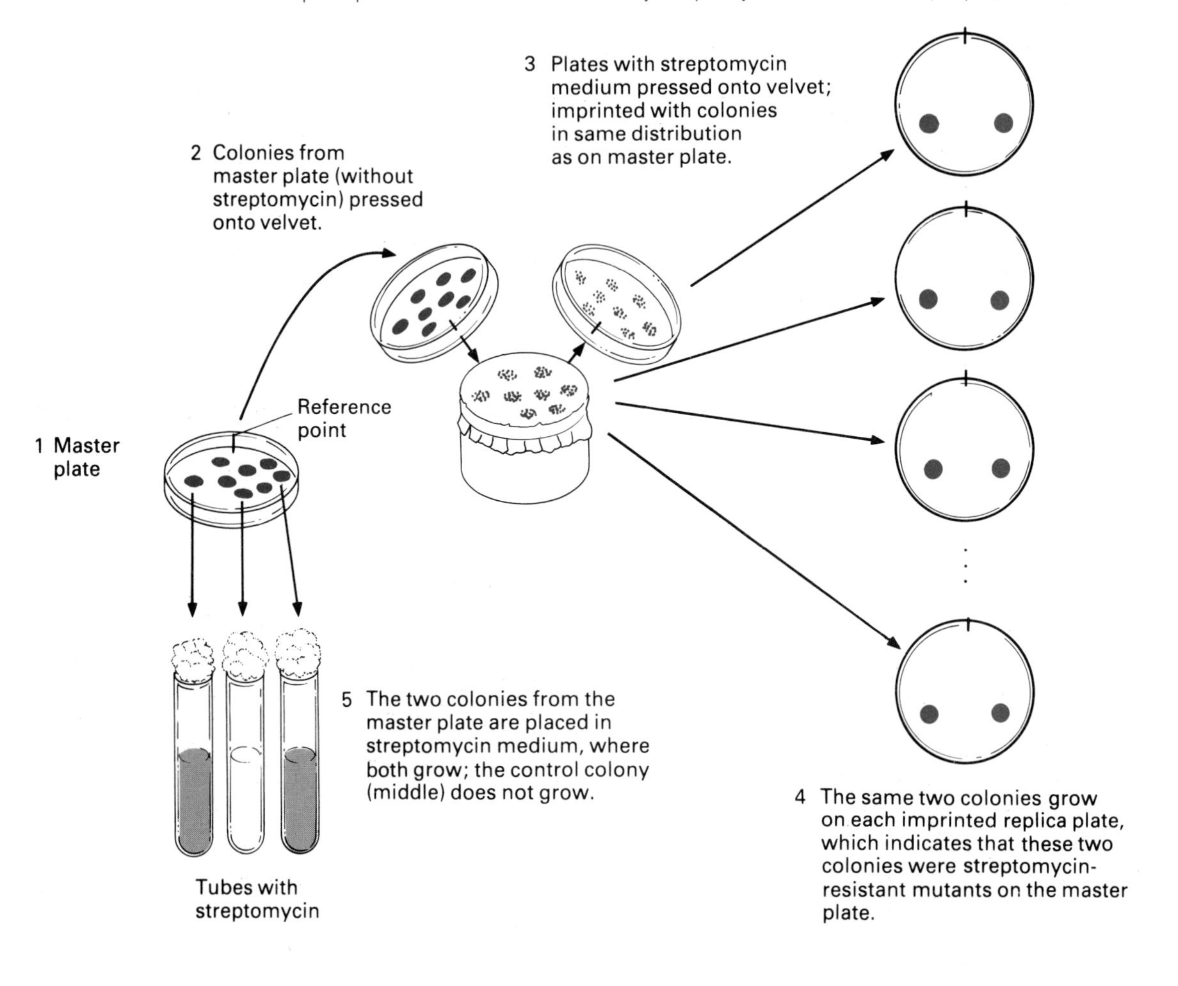
3 Plates with streptomycin medium pressed onto velvet; imprinted with colonies in same distribution as on master plate.
2 Colonies from master plate (without streptomycin) pressed onto velvet.
Reference point
1 Master plate
5 The two colonies from the master plate are placed in streptomycin medium, where both grow; the control colony (middle) does not grow.
Tubes with streptomycin
4 The same two colonies grow on each imprinted replica plate, which indicates that these two colonies were streptomycin-resistant mutants on the master plate.

dent samples. Moreover, when the investigators test the colonies on the master plate that gave rise to the resistant colonies on the streptomycin-treated replica plates, these colonies are always found to contain some streptomycin-resistant cells. In contrast, the other colonies on the master plate rarely contain such cells. Resistant colonies therefore arise from preexisting mutant cells; they do not develop as an adaptive change in response to their exposure to the chemical agent. Mutations are therefore not the result of directed change.

Because mutations are random, undirected events, we can infer that their occurrence in organisms that are already attuned to the environment in which they live is more likely to lower the fitness of the organisms than to improve it. The workings of an expensive clock can serve as an analogy. Most random changes in such an instrument are apt to make it run worse rather than better. A few beneficial mutations do occur, however, along with some that are by chance beneficial at a later time, and these mutations provide the genetic basis for adaptation.

Gene mutations play two different roles in evolution. First and most important, they serve as the initial source of all alleles, providing the raw materials on which natural selection can act. Those mutations that improve fitness are selected for and those that do not improve fitness tend to be eliminated from the population. Second, as we have already pointed out, mutations can alter gene frequencies and can therefore constitute an evolutionary force. It is this second role that concerns us here.

Effects of Recurrent Mutation. The mutational origin of each allele is a recurrent process that occurs at a certain relative rate. This process can be illustrated as

$$A \xrightarrow{\quad u \quad} a,$$

where u is the mutation rate from A to a, which is defined as the probability that a mutation to a occurs at the A locus per gamete per generation. If the starting population has p A alleles among the total, each of which has a probability u of undergoing mutation, then the frequency of a alleles appearing in the next generation as a result of mutation will be $(u)(p)$. In this case, we can treat p as the probability of selecting an A allele from the total and u as the probability that the A allele that is selected undergoes mutation to the alternative allelic state. Since the proportion of A alleles in the population will decline as A mutates to a, the factor $(u)(p)$ also designates the decrease in p from one generation to the next. The change in the allele frequency owing to mutation is therefore

$$\Delta p = -up.$$

Mutational changes in the allele frequency thus proceed at a rate that depends on the existing frequency of the allele being mutated. If we start with a population of all AA individuals at time zero and follow the population over many generations, we would observe that the A allele declines most rapidly at the beginning and that it declines at a progressively slower rate as the value of p gets smaller.

Some mutationally induced changes in gene frequency are shown in Fig. 14.6 for different values of u. Observe that although mutation does act as an evolu-

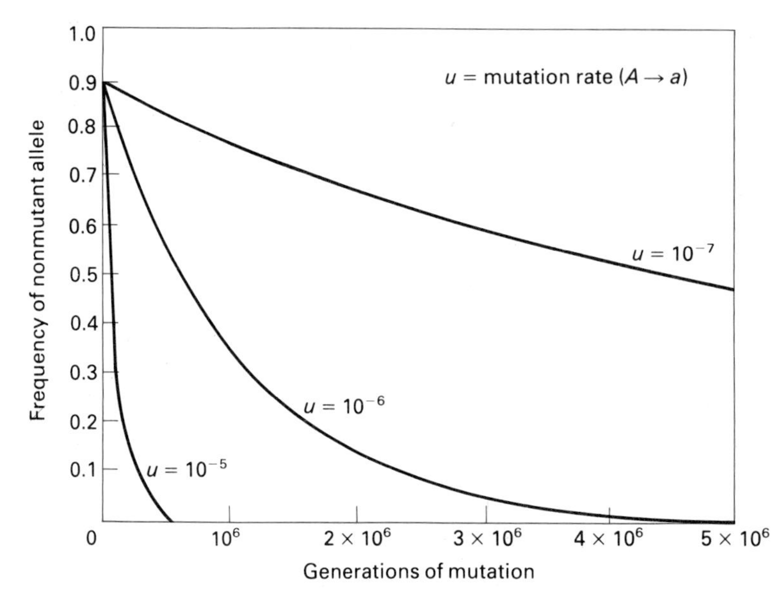

Figure 14.6. Reduction in the frequency of an allele through recurrent mutation. This reduction occurs at a faster rate when the probability of mutation is initially large. But compare this relatively slow rate of evolutionary change that occurs by means of mutation with the much faster rate of change that occurs as shown in Figures 14.1 and 14.2.

tionary force, the change in the gene frequency in each generation is likely to be very small, regardless of the value of p. In fact, mutation rates rarely exceed 10^{-4} (1 in 10,000) and usually range in order of magnitude between 10^{-5} and 10^{-6} per generation. Mathematically, the value of a mutation rate is roughly the reciprocal of the time needed to reduce the frequency of an allele to one-third of its starting value. Most mutations therefore require at least 10,000 generations (usually more) just to decrease the frequency of an allele by one-third.

Another factor that contributes to the ineffectiveness of mutation is the reversibility of the process. In other words, not only does A have a probability per unit time of mutating to its allele a, but once a is formed, it also has a chance of undergoing mutation back to the original state. We can show the reversibility of the mutation process by writing

$$A \underset{v}{\overset{u}{\rightleftarrows}} a,$$

where u is the rate of forward mutation (A to a), and v is the rate of back mutation (a to A). The values of u and v usually differ; u is most often numerically greater than v by a factor of at least 10. Forward mutation has a larger probability than back mutation, because, as pointed out in Chapter 11, a greater variety of coding changes are possible that can modify or destroy gene function than those that can specifically effect reversion to the original state by a particular mutational alteration.

Because of the reversible nature of mutation, an equilibrium between the frequencies of any two alleles will eventually be established, assuming that the mutation rates remain constant. At equilibrium, $(u)(p)$, which is the frequency of A alleles mutating to form a alleles per generation, must equal the frequency of a alleles mutating back to form A alleles, $(v)(q)$. Equating these changes and rearranging the terms, we get

$$up = vq = v(1 - p) = v - vp,$$

or

$$(u + v)p = v.$$

The frequencies of the two alleles *at mutational equilibrium* then become

$$p = v/(u + v) \qquad \text{and} \qquad q = 1 - p = u/(u + v).$$

For example, if the *A* allele mutates nine times more frequently than does the *a* allele (i.e., $u = 9v$), the allele frequencies at equilibrium are expected to be

$$p = v/(9v + v) = v/10v = 0.1 \qquad \text{and} \qquad q = 0.9.$$

Opposing Mutation and Selection. Although it is conceivable that some genes might eventually reach an equilibrium between forward and reverse mutation, most probably will not. As we pointed out earlier, most mutations that occur within a population are detrimental in character and will be selected against. Consequently, mutation and selection will often behave as opposing processes, with mutation acting to increase the number of detrimental alleles and selection serving to counteract this effect and to restore conditions to what they were prior to the introduction of the less fit type. If mutation and selection are acting in opposite directions, an equilibrium will develop when a balance exists between the mutation rate and the rate of selective elimination.

We can develop the mathematical basis for an equilibrium between opposing mutation and selection pressures by considering the example of partial selection against the recessive. From Eq. (14.4) we see that an approximate proportion $s(1 - q)q^2$ of *a* alleles will be lost in each generation as a result of selection against the recessive homozygotes. But if the *A* allele mutates to its recessive form at a rate of u per generation, a fraction $(u)(p) = u(1 - q)$ of *a* alleles will be restored to the population through mutation. A balance exists between these opposing processes when $u(1 - q) = s(1 - q)q^2$. (Back mutation is negligible at the low values of q expected in the vicinity of the equilibrium state.) The approximate condition *for equilibrium* is therefore

$$u = sq^2, \quad \text{or}$$
$$q^2 = u/s.$$

Thus, the frequency of the recessive genotype at equilibrium will depend directly on the mutation rate and will depend inversely on the strength of selection against it. For example, with complete selection against recessives, so that $s = 1$, and a mutation rate of $u = 10^{-4}$, the incidence of the lethal trait is expected to be $q^2 = u = 10^{-4}$, or 1 in 10,000 zygotes formed. The frequency of the lethal allele is thus $q = \sqrt{u} = 10^{-2}$. Note that the small values of mutation rates will tend to keep the frequency of the mutant gene quite low at equilibrium, even when the opposing selection pressures are small. This procedure provides a simple way of estimating mutation rates and has been used extensively in the past for this purpose.

The opposing effects of selection and mutation are illustrated graphically in Fig. 14.7. Observe that a relaxation of selection against a detrimental trait results

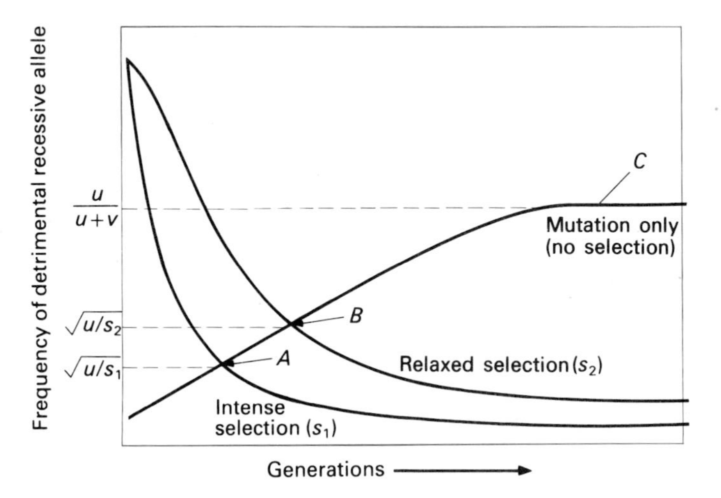

Figure 14.7. Effects of relaxing the selection intensity against a detrimental recessive allele. The graph shows idealized curves for the reduction in allele frequency that results from intense and relaxed selection and for the increase in frequency that occurs with mutation alone. An equilibrium between opposing mutation and intense selection will be established at point *A*. The relaxation of selection will result in an increase in the equilibrium frequency of the allele to point *B*. The elimination of selection entirely will eventually result in the mutational equilibrium at point *C*.

in the establishment of a new equilibrium with a higher frequency of the disfavored allele. Although the selection pressure has been reduced in this example, the mutation rate has not, which results in an increase in the frequency of the mutant allele. Such is the case in human populations when treatment of a recessive disorder, such as PKU, makes reproduction possible for many affected individuals. The recessive homozygotes can then marry and have children, thereby adding recessive mutant alleles to the genetic load of future generations.

EXAMPLE 14.5. Example 14.2 gives the frequency of the Tay-Sachs gene in certain Jewish populations as 1/80. If this detrimental gene is maintained through a balance between mutation and complete selection against the recessive trait, what is the mutation rate from the dominant normal allele to the recessive Tay-Sachs allele?

Solution. With complete selection against recessives, $s = 1$, so that the mutation rate (u) from the dominant to the recessive allele equals q^2. Thus, for Tay-Sachs disease $u = (1/80)^2 = 1/6400$, or 1.56×10^{-4} per generation. Because this value is unusually large for a mutation rate, geneticists have proposed alternative mechanisms, such as heterozygote advantage, to account for the high incidence of the allele in these Jewish populations.

Migration

Up until now, we have made the assumption of an isolated population that does not exchange genes with neighboring groups. Most breeding populations are not completely isolated from others of the same species and some exchange of genes normally takes place. If populations differ in genetic composition, the flow of genes from one population to another by way of migration can then serve as an evolutionary force by changing gene frequencies.

A simple model that describes the effects of migration is shown in Fig. 14.8. This model assumes one-way migration from a large source population, whose alleles (A, a) have frequencies of (p_m, q_m), to a smaller recipient population, with corresponding allele frequencies of (p, q). Since migration is in one direction only, p_m and q_m are treated as constants. A comparable situation existed among American blacks during their period of slavery. While a certain amount of racial mixing did occur, the flow of genes was in essence unidirectional, since subsequent descendants of mixed ancestry were almost always considered black.

Let us suppose that as a result of migration, a fraction m of all gametes in the recipient population are contributed by migrant individuals. What will be the frequency of the A allele in the next generation of the recipient population? Since a proportion m of all alleles are migrant, $1 - m$ must be nonmigrant. Therefore, p', which is the frequency of the A allele after one generation, is the average contribution of the two groups to the allele frequency. If we multiply p_m and p by m and $(1 - m)$, respectively, the frequency of the A allele in the next generation becomes

$$p' = mp_m + (1 - m)p. \tag{14.7}$$

In this equation, m and $1 - m$ are weighting factors that designate the relative contributions of migrants and nonmigrants to the genic composition of the progeny.

Figure 14.8. Model depicting unidirectional recurrent migration from a large source population to a smaller recipient group. Recurrent migration of this type gradually alters the genetic composition of the recipient population in the direction of the source population.

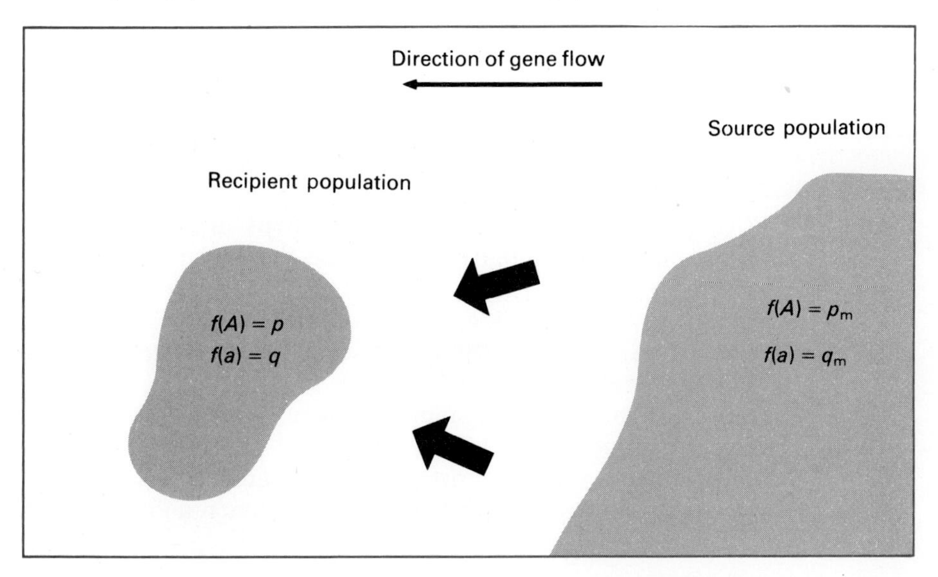

For a numerical example, let us assume that p_m and p are 1 and 0.1, respectively, and that 20 percent of the gametes contributed to the next generation are from migrants. The value of p in the next generation will then be

$$p' = (0.2)(1) + (0.8)(0.1) = 0.28.$$

Expressed in another way, there has been an increase in the frequency of the A allele of $\Delta p = 0.28 - 0.10 = 0.18$ during the course of one generation. Obviously, if this process were to continue, the recipient population would eventually approach the migrant population in genetic composition.

EXAMPLE 14.6. Because of a certain amount of racial admixture, United States blacks have European as well as African ancestry. Blood tests reveal that the frequency of the Fy^a allele at the Duffy blood group locus is equal to 0.081 in blacks in New York City. The allele has a frequency of 0.43 in Europeans but is essentially absent from African populations. Estimate the average fraction of genes of European descent among New York City blacks.

Solution. We can treat this situation as a case of one-way migration, in which the European genes have been incorporated into a base population of basically African ancestry. Let m in Eq. (14.7) represent the proportion of European (migrant) genes in the black population. Solving for m, we get

$$m = \frac{p' - p}{p_m - p}.$$

In our example, p', p, and p_m represent the respective frequencies of the Fy^a allele among New York City blacks, Africans, and Europeans. Making the appropriate substitutions, we obtain

$$m = \frac{0.081 - 0}{0.43 - 0} = 0.188.$$

On an average basis, therefore, about 19 percent of the ancestry of the black population sampled in this study is European.

Random Genetic Drift

Genetic drift refers to random fluctuations in gene frequency that arise as a consequence of sampling error in the reproductive process. It occurs to some extent in all finite populations, but it is especially important in small breeding groups (consisting of less than 100 individuals, as a general rule of thumb). In populations of this size, chance variation may constitute the most important factor involved in genetic change.

An example of fluctuations in the allele frequency is shown in Fig. 14.9 for three small populations. The populations in this example have the same initial allele frequency and consist of the same number of individuals throughout the period of observation. While all have the same number of A alleles to begin with,

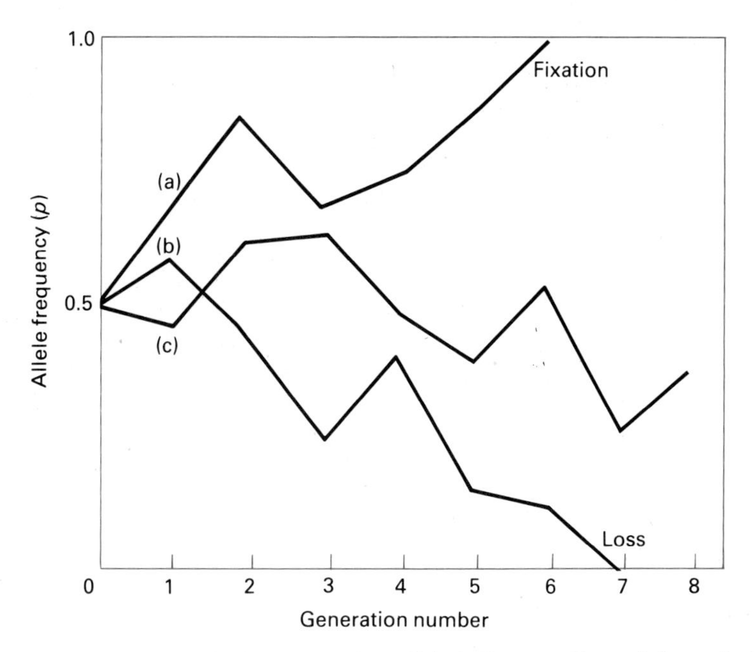

Figure 14.9. Random drift in the frequency of an allele in three small populations all started with $p = ½$. The ultimate outcome is homozygosity resulting from the random fixation (a) or loss (b) of one or the other allele. Population (c) has retained both alleles for the time being.

the populations start to diverge in genetic composition early in the process. Population (a) eventually reaches *fixation* of the *A* allele ($p = 1$) and consists of all *AA* individuals, while population (b) experiences *loss* of this allele ($p = 0$) and becomes strictly *aa* in genotype. In contrast, population (c) retains both alleles for the time being. When a population undergoes fixation or loss of an allele, its members are all homozygous and will no longer vary. The population then becomes trapped, so to speak, in a homozygous mode until the allele is again introduced through mutation or migration. A useful analogy is the path of a ball on a pinball machine (Fig. 14.10). Once the ball enters the trough on either side, the situation is comparable to the fixation or loss of an allele, since the path of the ball can no longer change. These simple examples serve to point out two important consequences of genetic drift: (1) an *increase in variation among populations* as the populations diverge in allele frequency and (2) a *decrease in variation within populations* as the members become homozygous in genotype.

Theory of Genetic Drift. We can gain a better understanding of what sampling error means in reference to Mendelian inheritance by considering the progeny produced by a large number of isolated groups with the same initial composition. Let us begin by assuming that each population in this series consists of only two individuals, both of whom are *Aa* in genotype, and that each population remains at this size through the next generation. Since the population size is assumed to be constant, *Aa* × *Aa* matings in all of the groups are restricted to two offspring

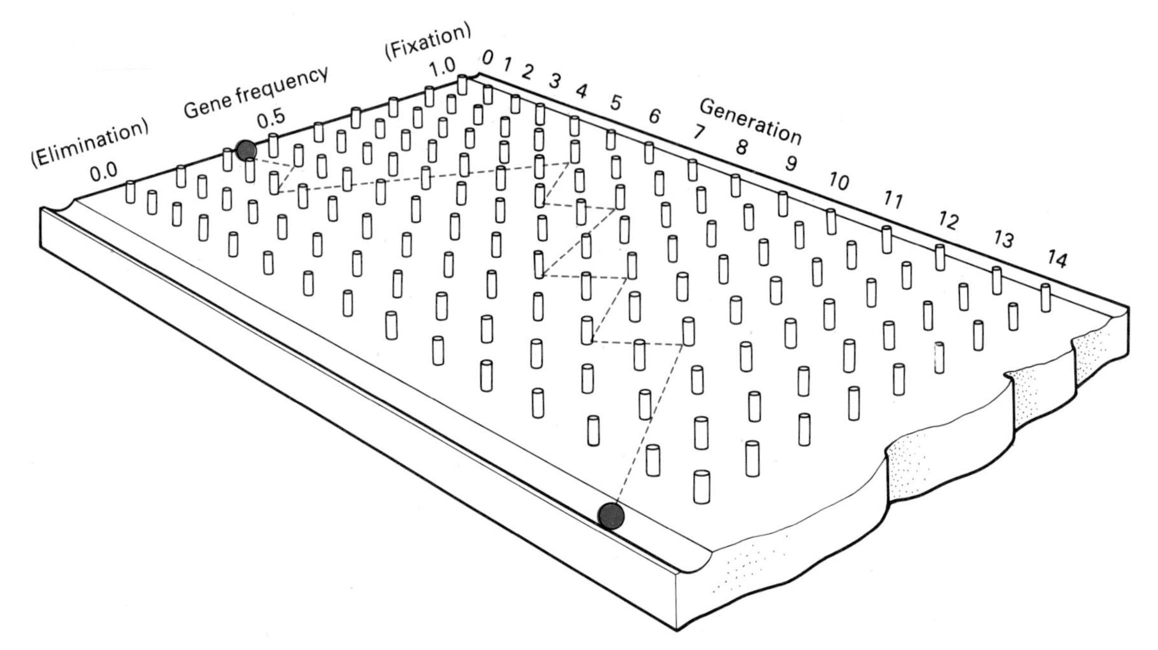

Figure 14.10. Pinball analogy of genetic drift. In this model, chance fluctuations in gene frequency are likened to the random path of a ball on a pinball machine. Once the ball enters the trough on either side, the situation is then comparable to the fixation or loss of an allele, since the path of the ball can no longer change. *Source:* L. S. Dillon, *Evolution: Concepts and Consequences*, 2nd ed., 1978. The C. V. Mosby Co., St. Louis.

each. The six sibling pairs that are possible among the progeny of the populations correspond to the six random paired combinations of *AA*, *Aa*, and *aa* genotypes. These combinations are shown in Table 14.6, along with the probability that each occurs from an *Aa* × *Aa* cross. Observe that when the results are arranged according to the frequency of the *A* allele in the progeny, the number of cross results reduces to five. These results, which represent the five possible values of p (0, $\frac{1}{4}$, $\frac{1}{2}$, $\frac{3}{4}$, and 1), have probabilities that can be computed from the terms of the binomial expansion $(\frac{1}{2} + \frac{1}{2})^4$. Hence, the probability that the *A* allele is lost ($p = 0$) in the next generation is the first term of this expansion, or $(\frac{1}{2})^4 = \frac{1}{16}$, while the probability that p is decreased to $\frac{1}{4}$ is the second term, or $4(\frac{1}{2})^4 = \frac{4}{16}$, and so on.

The size of the progeny population is the critical factor in determining the extent of chance fluctuation in our model. For example, if three offspring are produced instead of two, seven frequency classes are possible, with probabilities distributed according to the terms of the binomial expansion $(\frac{1}{2} + \frac{1}{2})^6$, and the probability of the fixation (or loss) of an allele is only $(\frac{1}{2})^6 = \frac{1}{64}$. In general, if N is the number of progeny, then the probability of fixation (or loss) of an allele can be calculated as $(\frac{1}{2})^{2N}$, which is the end term in the expansion $(\frac{1}{2} + \frac{1}{2})^{2N}$. Since $(\frac{1}{2})^{2N}$ declines exponentially with an increase in N, we would not expect in large populations to find substantial departures in gene frequency, such as those required to reach homozygosity.

EXAMPLE 14.7. Suppose that a population is started with two *Aa* heterozygotes, which produce four offspring in the next generation. What is the probability that the frequency of the *A* allele is increased to $p = 3/4$ among the progeny simply as a result of random drift?

Solution. To produce four offspring, the parents must contribute a total of eight alleles to the next generation. Of these eight alleles, six must be of type *A* in order for the frequency of the *A* allele to increase to 3/4. The probability of such a change is therefore

$$P(6A, 2a) = \frac{8!}{6!2!}(1/2)^6(1/2)^2 = 7/64.$$

Bottlenecks and Founder Effects. In addition to the pattern of drift that occurs whenever matings are restricted over time to very small groups, two other situations can lead to random changes in the allele frequency. One of these situations, known as the **bottleneck effect,** occurs in populations (even moderately large ones) that undergo periodic fluctuations in size. When population sizes are at their lowest values (caused, for example, by a natural catastrophe or some infectious disease), the populations are so depleted in numbers that sampling error becomes a distinct possibility. Random changes can then occur, altering the genetic composition of the population. These periodic reductions in numbers thus serve as bottlenecks that can lead to a much higher degree of sampling variation than would ever be expected in stable populations of comparable average size.

Another situation in which sampling error is believed to be important is a phenomenon known as the **founder effect.** This situation occurs whenever a population is being established by a small group of colonizers. The founders represent only a small fraction of the genetic variability in the species and may differ substantially in gene frequencies from the parent population from which they were derived. For example, an allele may be absent in the founding group

Table 14.6. Chance distribution of two offspring from heterozygous parents ($Aa \times Aa$).

DISTRIBUTION OF OFFSPRING			NUMBER OF ALLELES IN SAMPLE	FREQUENCIES OF ALLELES IN SAMPLE	PROBABILITY OF OFFSPRING
AA	*Aa*	*aa*			
2	0	0	4 *A*, 0 *a*	$p = 1, q = 0$	$\frac{4!}{4!0!}(1/2)^4(1/2)^0 = 1/16$
1	1	0	3 *A*, 1 *a*	$p = 3/4, q = 1/4$	$\frac{4!}{3!1!}(1/2)^3(1/2)^1 = 4/16$
0	2	0	2 *A*, 2 *a*	$p = 1/2, q = 1/2$	$\frac{4!}{2!2!}(1/2)^2(1/2)^2 = 6/16$
	or				
1	0	1			
0	1	1	1 *A*, 3 *a*	$p = 1/4, q = 3/4$	$\frac{4!}{1!3!}(1/2)^1(1/2)^3 = 4/16$
0	0	2	0 *A*, 4 *a*	$p = 0, q = 1$	$\frac{4!}{0!4!}(1/2)^0(1/2)^4 = 1/16$

Table 14.7. Blood group percentages in Pennsylvania Dunkers, compared with two control populations.

POPULATION	A	B	AB	O	M	MN	N	Rh_1	Rh_2	rh
	BLOOD GROUPS									
United States	40	11	4	45	29	50	21	54	15	15
Western Germany	45	10	5	41	30	50	20	—	—	—
Dunkers	59	3	2	36	45	42	14	58	16	11

Source: From Glass, Sacks, Jahn, and Hess. 1952. *Am. Nat.* 86:145–159. Copyright 1952 by The University of Chicago Press.

or may occur at such a low frequency that it is easily lost before the new population grows appreciably in size.

Founder effects in humans have been studied in various small religious isolates in which the members have tended to marry within the group and have remained more or less separated genetically from the surrounding population. A religious sect known as the Dunkers has provided some useful genetic information in this regard. The Dunkers were established from 27 families that came to the United States from western Germany over 200 years ago. They have since lived in small farming communities in eastern Pennsylvania. One small community with about 300 members was studied by a team of researchers headed by Bentley Glass. The researchers analyzed the relative blood group frequencies for the ABO, MN, and Rh systems, as well as some other easily identifiable traits, including the presence or absence of mid-digital hair (hair on the middle segments of the fingers) and the state of attached or free-hanging earlobes. They then compared the results with the corresponding data obtained from the surrounding American population and the general population in western Germany from which the Dunkers had emigrated. The frequencies that were obtained for the different blood groups are shown in Table 14.7. It is apparent from the ABO and MN blood group data that the Dunkers show significant differences in blood group frequencies from both of the control populations. At the present state of our knowledge, such differences can best be attributed to sampling variation that existed in the initial founding group and, to a certain extent, to the random genetic drift that probably occurred after the Dunkers had migrated to this country.

TO SUM UP

1. Gene mutations play two roles in evolution: (1) they provide the raw material for selection to act on and (2) they alter gene frequencies. Since mutation rates are extremely small, usually on the order of 10^{-5} to 10^{-6} per generation, mutation is generally regarded as a very weak evolutionary force.
2. The forward mutation of a given allele is usually opposed by back mutation, and it will approach a state of equilibrium if no other factors are operating to change the allele frequency. Most mutant alleles arising in a population that is already attuned to its environment are detrimental in character and are selected against. Thus, in addition to back mutation, gene mutation also tends to be opposed by selection.
3. Migration can lead to a flow of genes from one population to another and can thus serve as an evolutionary force by changing gene frequencies. If migration were to

proceed in the absence of any other evolutionary force, it would eventually eliminate any genetic differences existing between the affected populations.

4. Genetic drift refers to the random fluctuations in gene frequency that occur mainly in small populations. The likelihood of random drift is inversely related to the size of the population and can, by chance, lead to the fixation or loss of an allele in small breeding groups.

Differentiation of Populations

Much of the basic theory of population genetics is based on large, isolated populations in which all of the individuals in the population have an equal chance of mating with any member of the opposite sex, regardless of geographic location. While such idealized conditions aid in the development of comparatively simple models, they nevertheless give rise to inconsistencies. For instance, it is difficult to conceive of a very large population, especially one having a wide geographic distribution, that is also random-mating throughout. Organisms are not infinitely mobile, nor can they transmit their gametes over very large areas, so that widely separated individuals in the population can be isolated by distance and never get a chance to mate. Furthermore, individuals within the population often aggregate into groups, if for no other reason than to share a favorable habitat area. For example, many fish aggregate into schools, some birds form flocks, trees are often clumped into groves, and certain grasses are often clumped into meadows. While members of the same local group (or subpopulation) can mate at random, matings between members of different groups will occur less frequently and will depend on the migration rate. This breeding structure, which is that of a large population subdivided into many smaller local breeding populations, permits all of the evolutionary forces (including migration) to act concurrently throughout the entire assemblage. Subdivision is thus an important contributing factor to genetic variation, since it permits genetic differences to develop not only among individuals of the same local population but also among members of different subpopulations of the larger, more inclusive group.

Joint Action of Evolutionary Forces

When we consider all of the evolutionary forces at one time, it is often useful to combine and classify migration, mutation, and selection together as *systematic processes.* When acting singly or in some combination, the systematic processes tend to produce an equilibrium allele frequency in a population. For example, we saw this process take place in the opposing actions of mutation and selection. In contrast, genetic drift is a *dispersive process,* which acts through random fluctuations to scatter the allele frequencies away from their equilibrium values toward the limits of 0 or 1.

The joint actions of the systematic and dispersive pressures are pictured in Fig. 14.11(a). Note that the two major processes have opposing tendencies. Ultimately, a balance is reached when the dispersion of allele frequencies owing to genetic drift is held in check by the systematic forces. The equilibrium condition that then exists can be described by a probability curve, which gives the probability that a

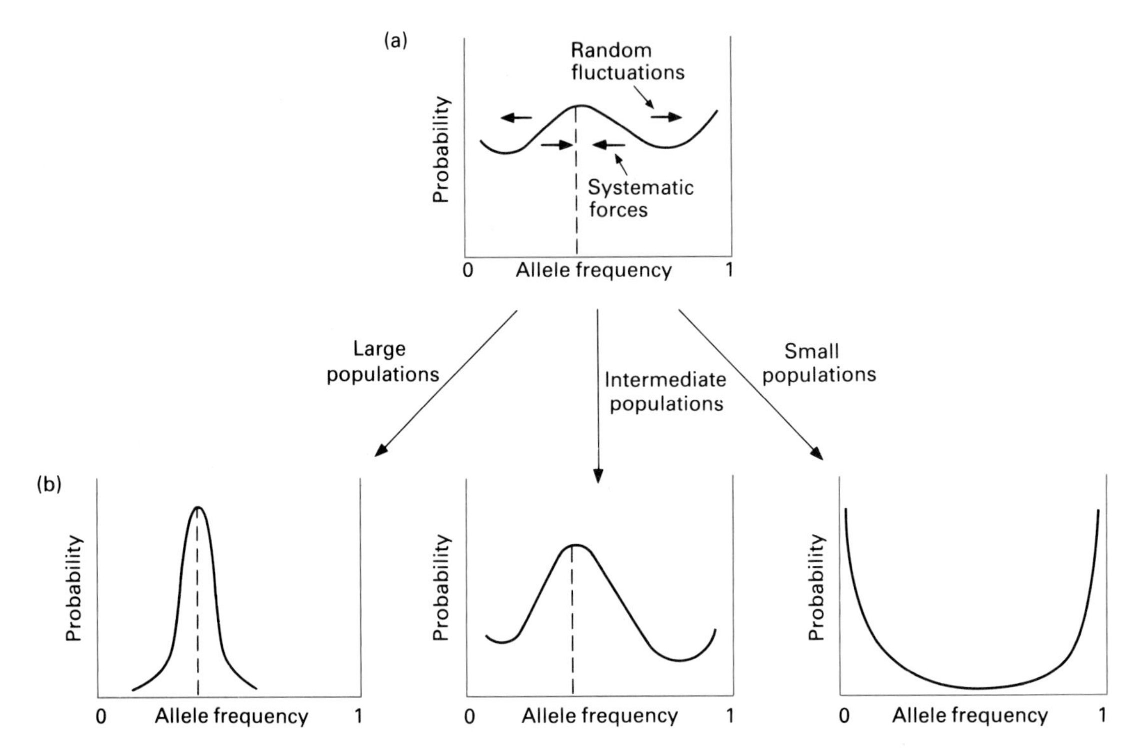

Figure 14.11. Opposing actions of the systematic evolutionary forces (selection, migration, mutation) and the dispersal force of random drift on allele frequencies. Each resulting graph depicts the probability distribution of allele frequencies in a population of a certain size. The dotted line designates the theoretical equilibrium frequency of the allele.

local population has some particular allele frequency. We can think of such a curve as designating the distribution of allele frequencies that we might observe at a single point in time in a large assemblage of local populations of the same size and under the same conditions of mutation, migration, and selection. We can also picture this curve as showing the variation in p (or q) that might occur under equilibrium conditions in the same population over a large number of generations.

Three basic types of probability curves are shown in Fig. 14.11(b) for large, intermediate, and small populations. Observe that when the population size is large, all breeding groups will tend to have allele frequencies at or near the equilibrium value. In large populations, the systematic forces predominate. Since there is little tendency for an allele to change in frequency from one generation to the next, continued evolutionary progress will depend on the rare occurrence of new beneficial mutations and on changes in the environmental conditions. When the local populations are small, however, the force of genetic drift predominates, so that most of these populations will undergo fixation or loss of the allele in question. Therefore, the curve for a small population is U-shaped, with the most probable allele frequency values being 0 and 1. These homozygous populations would be so depleted of genetic variation that they would be unable to respond to any change in selective pressures. The end result is extinction.

When populations are intermediate in size, there is appreciable scatter of allele frequencies, although a maximum probability occurs in the vicinity of the equilibrium value. In populations of intermediate size, then, the allele frequencies will tend to vary but will still remain under the influence of the systematic pressures. This situation provides for considerable genetic divergence without depleting the local populations of the allelic variation needed to form adaptive combinations of genes.

EXAMPLE 14.8. One important genetic effect that is associated with the formation and subsequent divergence of isolated subpopulations is an overall increase in the average frequency of homozygous genotypes in these subgroups above that expected for a single freely intermating population. The increase in homozygosity that accompanies the formation of isolates—a phenomenon called *Wahlund's principle*—can be reversed in certain cases by the fusion of the formerly isolated groups. The problem presented here illustrates the genetic effects that are associated with the formation and fusion of isolates. Suppose that two equal-sized isolates are formed from a large random mating population in which the allele frequencies are $p = q = 1/2$. Assume that the frequencies of the genotypes (AA, Aa, aa) are (0.64, 0.32, 0.04) in isolate 1 and (0.04, 0.32, 0.64) in isolate 2. Compare the average genotype frequencies within the two isolates with the frequencies that the genotypes would have if the two isolates were consolidated again into a single random mating population.

Solution. The average genotype frequencies for the two isolates are $f(AA) = (0.64 + 0.04)/2 = 0.34$, $f(Aa) = (0.32 + 0.32)/2 = 0.32$, and $f(aa) = (0.04 + 0.64)/2 = 0.34$. The average allele frequencies are then $f(A) = f(a) = 0.34 + 1/2(0.32) = 0.5$, the value expected for isolates that are random samples of the parent population. Therefore, if the isolates were to be consolidated into one random mating group, the genotype frequencies would become $f(AA) = (1/2)^2 = 0.25$, $f(Aa) = 2(1/2)(1/2) = 0.50$, and $f(aa) = (1/2)^2 = 0.25$. Note that the overall frequency of homozygotes is greater among the isolated groups than it is in a freely intermating population of corresponding genetic makeup. As we can see from this example, the breakdown of isolating barriers could thus lead to a reduction in the overall frequency of harmful recessive traits.

Geographic Differentiation

Spatial distribution adds another dimension to evolutionary change. Since local populations are geographically separated, they are apt to experience different environmental conditions. Systematic pressures will therefore differ as selection promotes adaptation to local conditions. Consequently, each breeding group within the overall population may have a different equilibrium allele frequency. These differences are comparatively slight between populations that inhabit similar portions of the geographic range of the species. Individuals within similar populations will tend to have a large number of characteristics in common. When these genetically similar populations become distinct from and readily recognized in comparison to other such groups in other portions of the species geographic range, they are classified as a **race.** Different races are thus defined as genetically distinct populations (or groups of populations) of the same species. If there is some

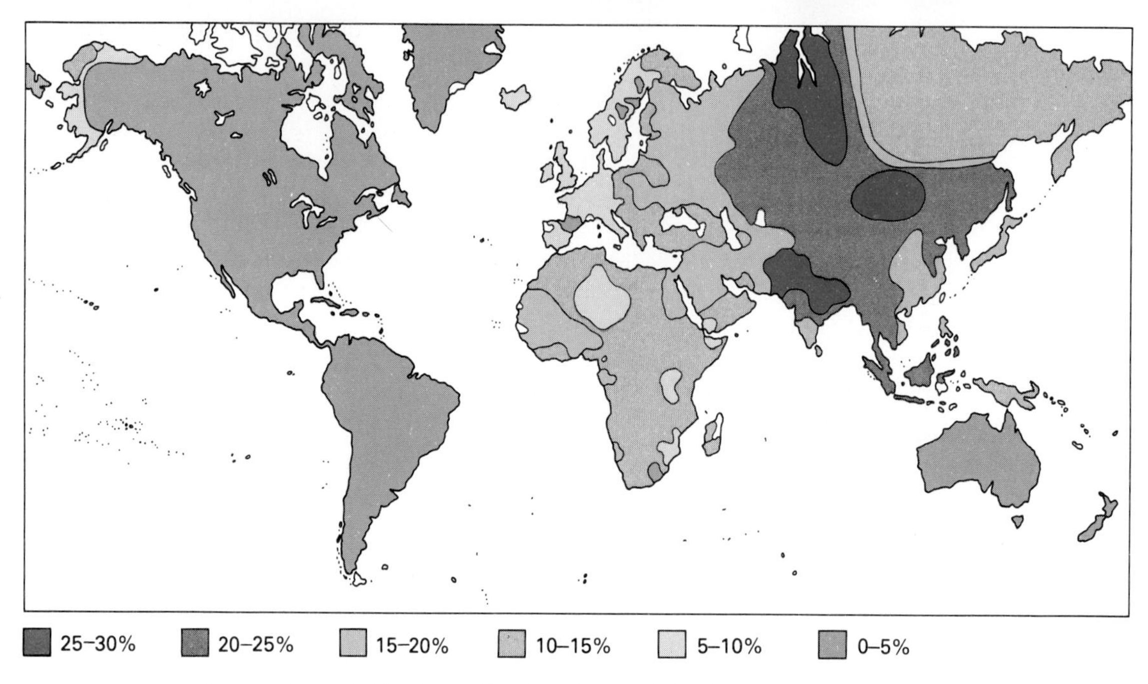

Figure 14.12. Distribution of the B allele (I^B) of the ABO blood group in the populations of the world. Note the east-west frequency gradient that occurs in Europe and western Asia. This gradual change (or cline) in gene frequency has been correlated with migrations (including invasions) from Asia westward.

restriction to gene exchange, the races will tend to retain their distinctive characteristics even with continued migration between them. Such is the case with the human races, in which unrestricted gene exchange has been prevented by social, religious, and cultural differences as well as by geographic separation.

Most anthropologists subdivide the human species into at least six geographic races: (1) Australoid, (2) Bushmen, (3) Caucasoid, (4) Mongoloid, (5) Negroid, and (6) Polynesian. In order to simplify the classification and characterization of races, human geneticists have used traits with a simple inheritance pattern for the purposes of comparison. One such trait is the presence of the A and B antigens in the ABO blood group system. The worldwide distribution of the I^B allele for the B antigen is shown in Fig. 14.12. This allele is either absent or present in very low frequencies (5 to 10 percent) in the native populations of Australia and North and South America. In Africa, it is present in rather high frequencies (15 to 20 percent) in the equatorial regions and decreases to 10 to 15 percent (and even lower in certain isolated localities) in both the northerly and southerly directions.

Of particular interest are the changes in the frequency of the I^B allele that occur along an east-west strip in Europe and Asia. A *cline* (or gradual geographic change) occurs in the allele frequency in this region; the cline is apparently a result of invasions of the European region by the Huns and other Asiatic tribes between the fifth and sixteenth centuries. The highest I^B frequencies occur in central Asia, with a peak frequency of about 30 percent. The frequency gradually declines to a low in western Europe of about 5 percent, and even less among the

Basques living in the Pyrenees between France and Spain. The presence of a cline, rather than sharp discontinuities in allele frequency, indicates a partial breakdown of the cultural barriers to gene exchange between natives and migrants. Such a breakdown often occurs during times of war, and to an extent, it depends on the length of contact between the native populations and the invading armies.

Researchers have studied other genes that show even greater differences than those of the ABO system in allele frequencies among the human races. In most of these cases, little is known about the evolutionary factors that are important in maintaining the polymorphisms in the different populations or about why the differences developed in the first place. Natural selection is undoubtedly responsible for at least part of this variation, but the effects of genetic drift may also have played a role.

Evolutionary Divergence

The evolutionary processes that promote differentiation of populations of the same species are also responsible in large part for the diversification of the different species. To a population geneticist, the species constitutes the most inclusive population. Each species is typically subdivided into numerous local populations; in sexually reproducing organisms these local populations are linked together by their potential for gene exchange. Whereas populations of the same species are capable of interbreeding, they do not ordinarily exchange genes with populations of other species. Different species are therefore said to be **reproductively isolated.**

Several mechanisms are known to prevent interbreeding between species. Usually, more than one of these mechanisms operates in any single case. As shown in Table 14.8, these isolating mechanisms act at two levels: *prezygotic* and *postzygotic.* The prezygotic isolating mechanisms include various anatomical, physiological, ecological, and behavioral barriers to mating or cross-fertilization between

Table 14.8. Summary of important reproductive isolating mechanisms.

A. Prezygotic mechanisms: Fertilization and zygote formation are prevented.
  1. Ecological: The populations live in the same region but occupy different habitats or become sexually mature at different times.
  2. Behavioral (in animals): Incompatible mating behavior prevents mating.
  3. Anatomical: Incompatible structure of reproductive organs (genitalia in animals, flowers in plants) prevents mating.
  4. Physiological: Gametes fail to survive in alien reproductive tracts.

B. Postzygotic mechanisms: Zygotes are formed but fail to survive or fail to produce viable and fertile offspring.
  1. Hybrid inviability.
  2. Hybrid sterility (partial or complete): Hybrids fail to reproduce because gonads fail to develop properly or chromosomes (or chromosome segments) segregate abnormally during meiosis.
  3. Inviability and/or sterility of F_2: F_1 hybrids are normal, but F_2 includes many weak or sterile individuals.

species; these barriers prevent the formation of hybrid zygotes. By contrast, the postzygotic isolating mechanisms act after fertilization has taken place and result in the formation of inviable, aberrant, or sterile hybrids. The postzygotic mechanisms are accidental byproducts of the genetic differentiation that occurs when populations adapt to differing conditions. As the populations diverge in genetic character, their genes are less likely to act harmoniously in a hybrid. Chromosome mutations can also arise that will lead to problems during meiosis in hybrid individuals (see Chapter 12). Although the production of unfit hybrids serves the same general purpose of preventing gene exchange as does the inability to form hybrids, it is nevertheless more wasteful of reproductive effort. Therefore, in regions where species have contact, we would expect natural selection to act by promoting the development of prezygotic barriers to gene exchange in populations that are already isolated from one another by postzygotic mechanisms.

Because of these intrinsic barriers to gene exchange, species will behave as separate evolutionary units, taking different evolutionary paths. Two major types of changes in species are generally recognized. They are **phyletic evolution** and **speciation.** Phyletic evolution is evolution within a lineage, or single line of descent. The phyletic component of species modification represents the gradual change in the overall genetic makeup of an already established species, and it is indicated by each separate line of the tree diagram in Fig. 14.13. As we have already seen, such changes come about by the joint action of the systematic and dispersive evolutionary forces. Speciation, on the other hand, is the splitting of a lineage into two or more separate lines of descent. Speciation thus results in a multiplication of the number of existing species. This branching of lineages occurs when a population of an established species becomes reproductively isolated from

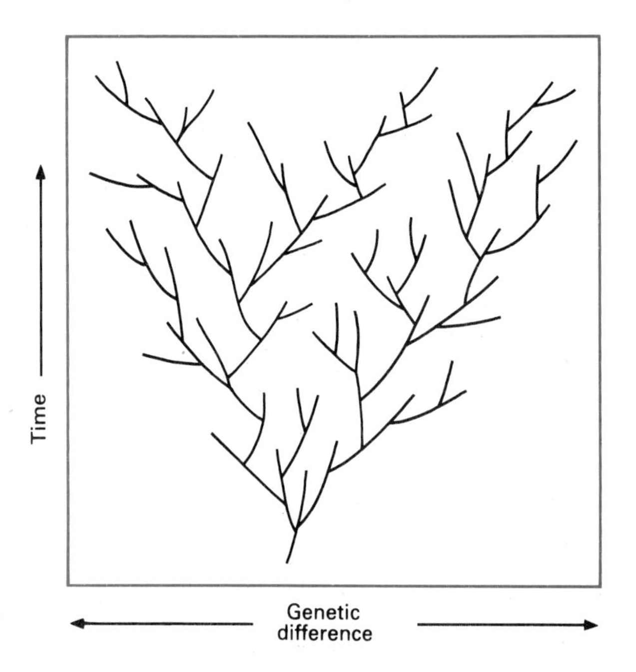

Figure 14.13. Hypothetical phylogeny showing the origin and diversification of species. Each branch point represents a speciation event, whereas the changes that occur within each line of descent constitute the phyletic component of evolution.

the other populations of that species. The prevailing view is that speciation can take place during periods when gene flow between the segments of a population is cut off or at least substantially reduced. The extrinsic barriers to gene flow could include geographic barriers (such as deserts, rivers, mountain ranges, and so on) or a simple switching of a parasite from one type of host to another. Whatever the mechanism, once the populations are isolated physically, they are then free to diverge until normal interbreeding is impossible. The two groups at this point can be regarded as separate species.

Molecular Phylogenies: DNA Hybridization. In recent years, geneticists have been able to study the phylogenetic relationships among different species by measuring their differences (and similarities) at the molecular level. One method that is particularly well suited for measuring the overall differences in DNA is **DNA hybridization.** In this technique, the DNA that is extracted from the different species is denatured, mixed, and allowed to renature. If the DNA strands from two different species (species A and B) were identical in base sequence, then the slow cooling of a mixture of their denatured DNA strands should produce three kinds of duplex molecules: renatured A helices, renatured B helices, and hybrid helices, in a ratio of 1:1:2, respectively. This process can be depicted as follows:

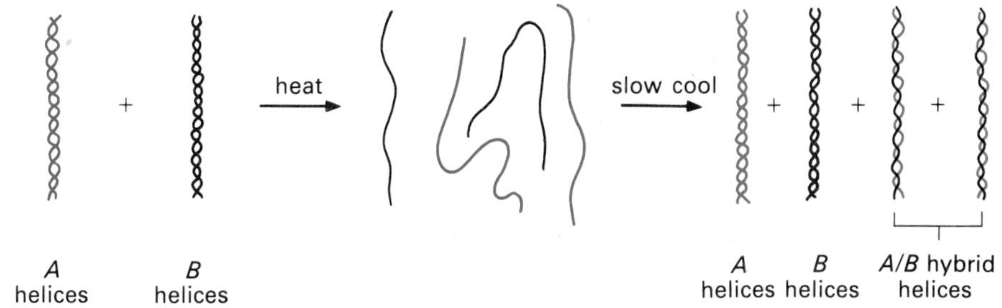

heat
slow cool
A helices
B helices
A helices
B helices
A/B hybrid helices

Of course, our theoretical example here is realistically an impossibility, since the DNA strands of even closely related species could not be perfectly complementary. The hybrid duplexes characteristically contain mismatched regions, where the nucleotide sequences of the two species are dissimilar enough to preclude hydrogen bonding. One measure of the proportion of noncomplementary bases in a hybrid DNA molecule is its melting temperature. Imperfectly paired DNA duplexes denature at a lower temperature than do the native molecules. Each 1°C difference in melting temperature between the native DNA and a hybrid duplex molecule corresponds to approximately one percent mismatched nucleotides. For example, native human DNA melts at a temperature that is 2.4°C higher than hybrid duplexes made up of human and chimpanzee DNA strands. We can therefore estimate that human and chimpanzee DNA strands differ by 2.4 percent of their nucleotide sequences. From these differences in melting temperature, we can directly estimate the genetic differentiation that has occurred within and among various phylogenetic groups.

There is a second hybridization technique that we can use to measure the overall degree of homology between the DNA sequences of two species. In this technique, the DNA of one of the species is immobilized in a matrix such as agar or on a filter. To illustrate this procedure, suppose that a large amount of denatured filter-bound DNA of species A is placed in a solution that contains a small amount of denatured DNA fragments of species A that are radioactively labeled. Also placed in the solution are increasing amounts of denatured DNA fragments of species B. The unlabeled DNA strands from species B in this mixture will compete with the labeled DNA strands from species A for complementary sequences on the filter-bound DNA from species A. The competition will be most severe when the concentration of DNA from species B is high. At very high concentrations of species B DNA, the only labeled species A DNA that becomes bound to the filter are those sequences that form a complex with sequences of the filter-bound DNA that are not complementary to any region in the competing DNA from species B (Fig. 14.14). The minimum level of labeled DNA that is able to hybridize thus indicates the proportion of DNA sequences that are not homologous in the two species. Figure 14.14 shows that the DNA of species B in this example differs in nucleotide sequence from that of species A by only about 5 percent, whereas the DNA of species A and species C share 50 percent of their sequences.

Figure 14.14. Homology between the DNA sequences of species A and those of species B and C. Homology is judged by the degree to which the denatured DNA from species B or C competes with the labeled DNA from species A. The vertical axis represents the relative amount of labeled denatured DNA from species A that is able to hybridize in the face of competition from increasing amounts of DNA from species B or C (horizontal axis). Only about 5% of the DNA sequences from species B are not homologous to those of species A, whereas about 50% of the DNA sequences from species C differ from those of species A.

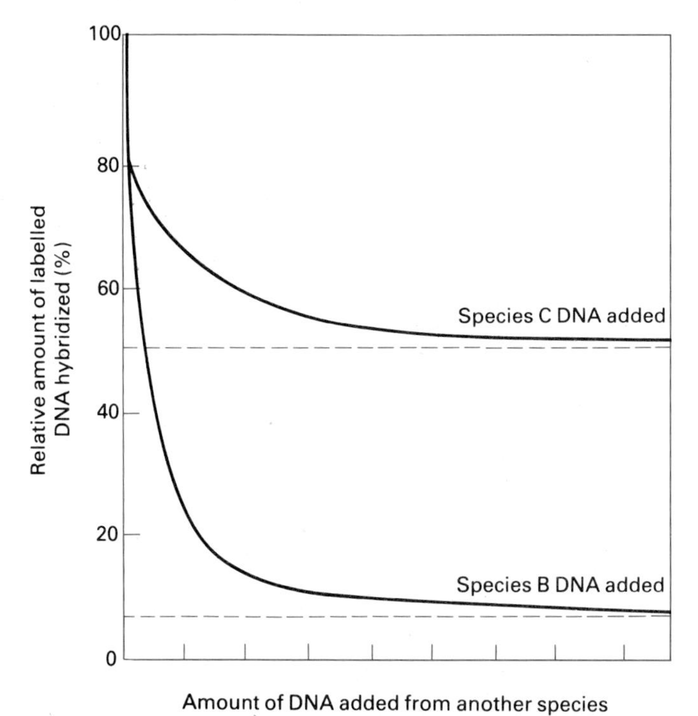

Table 14.9. Degree of similarity between human DNA and DNA from ten other species, expressed as the degree to which binding between labeled and unlabeled human DNA strands is inhibited.

ORGANISM	DEGREE OF TAXONOMIC DIFFERENTIATION	PERCENT INHIBITION OF HUMAN–HUMAN DNA BINDING
Human	—	100
Chimpanzee	family	100
Gibbon	family	94
Rhesus monkey	superfamily	88
Capuchin monkey	superfamily	83
Tarsier	suborder	65
Lemur	suborder	47
Tree shrew	suborder	28
Mouse	order	21
Hedgehog	order	19
Chicken	class	10

Source: From B. H. Hoyer and R. B. Roberts, 1967. In *Molecular Genetics,* Academic Press, New York.

The results of hybridization experiments with human DNA are given in Table 14.9. These results are consistent with the known phylogenetic relationships among the taxonomic groups. Note that according to the DNA hybridization criterion, chimpanzee DNA is identical to human DNA. This conclusion is erroneous and is merely an indication of the present limitations of the technique.

Molecular Phylogenies: Amino Acid Sequences. Various amino acid sequencing techniques have enabled geneticists to study evolutionary divergence at the molecular level by measuring the differences that exist among species in the primary structures of their proteins. This method is exemplified by studies on the differences in the amino acid sequences of cytochrome *c*.

Cytochrome *c* is a protein that plays a vital role in cellular respiration in the aerobic prokaryotes and eukaryotes. It is a relatively simple protein consisting of a single polypeptide chain that averages slightly longer than 100 amino acid sites. The amino acid sequences of cytochrome *c* from a number of different organisms are shown in Fig. 14.15. A comparison of these sequences shows that a high degree of similarity exists even among groups as diverse as vertebrates, molds, and higher plants (such as wheat). The similarities in sequence are most evident for the least divergent organisms shown and appear to decline with distance on an evolutionary scale. Thus, disparities in sequence tend to be greater for more distantly related species than for organisms that are grouped close together in most classification schemes. For instance, no amino acid differences exist between the cytochrome *c* of humans and chimpanzees, while only one amino acid difference (at position 66) exists between the cytochrome *c* of humans and rhesus monkeys. Greater differences appear in comparisons with other organisms; for example, the cytochrome *c* of humans and horses differ at 11 amino acid sites.

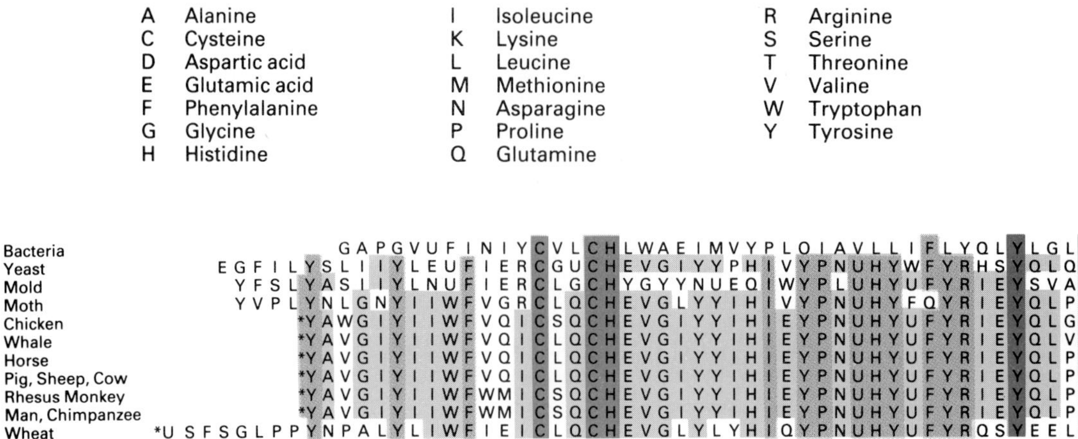

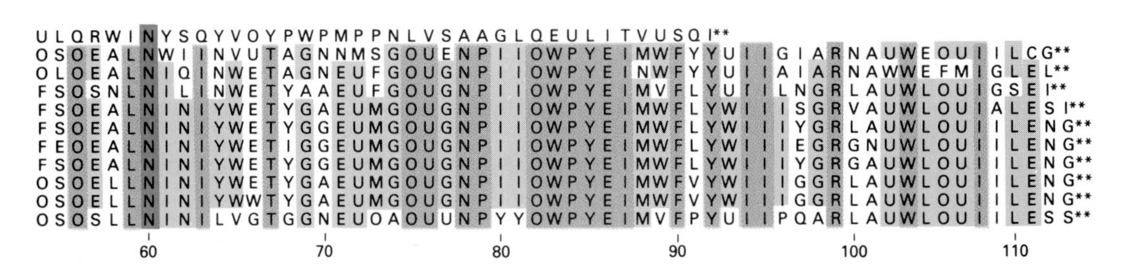

Figure 14.15. Amino acid sequences of cytochrome *c* isolated from different organisms. Invariable sites are indicated by the darkest shading. Increased variability at amino acid sites is indicated by a reduction in the intensity of shading. Note that many of the amino acids found in the vertebrates at certain locations are also shared by the yeast, mold, or wheat. *Source:* From L. S. Dillon, *Evolution: Concepts and Consequences*, 2nd ed., 1978. The C. V. Mosby Co., St. Louis.

Geneticists have used the sequence data for different proteins in conjunction with the genetic code (Chapter 9) to construct phylogenetic trees of the organisms studied. A phylogenetic tree that is based on sequence differences in cytochrome *c* is shown in Fig. 14.16. Trees such as this one use branching lines to suggest the evolutionary relationships, with the length of each branch drawn in direct proportion to the minimum number of base changes in the DNA needed to account for the particular amino acid differences in the protein. For example, a change from methionine (AUG) to isoleucine (AUU, AUC, or AUA) at the same amino acid site could occur with a minimum of one base change, whereas a change from methionine (AUG) to glutamine (CAA or CAG) would require at least two base substitutions. These molecular phylogenetic trees have the same topological features (order of branches, but not necessarily their length or angle) as phylogenetic trees that are derived by traditional methods. The phylogenetic relationships determined from the sequence data thus serve to confirm and extend the evolutionary record that is based on taxonomic and fossil studies.

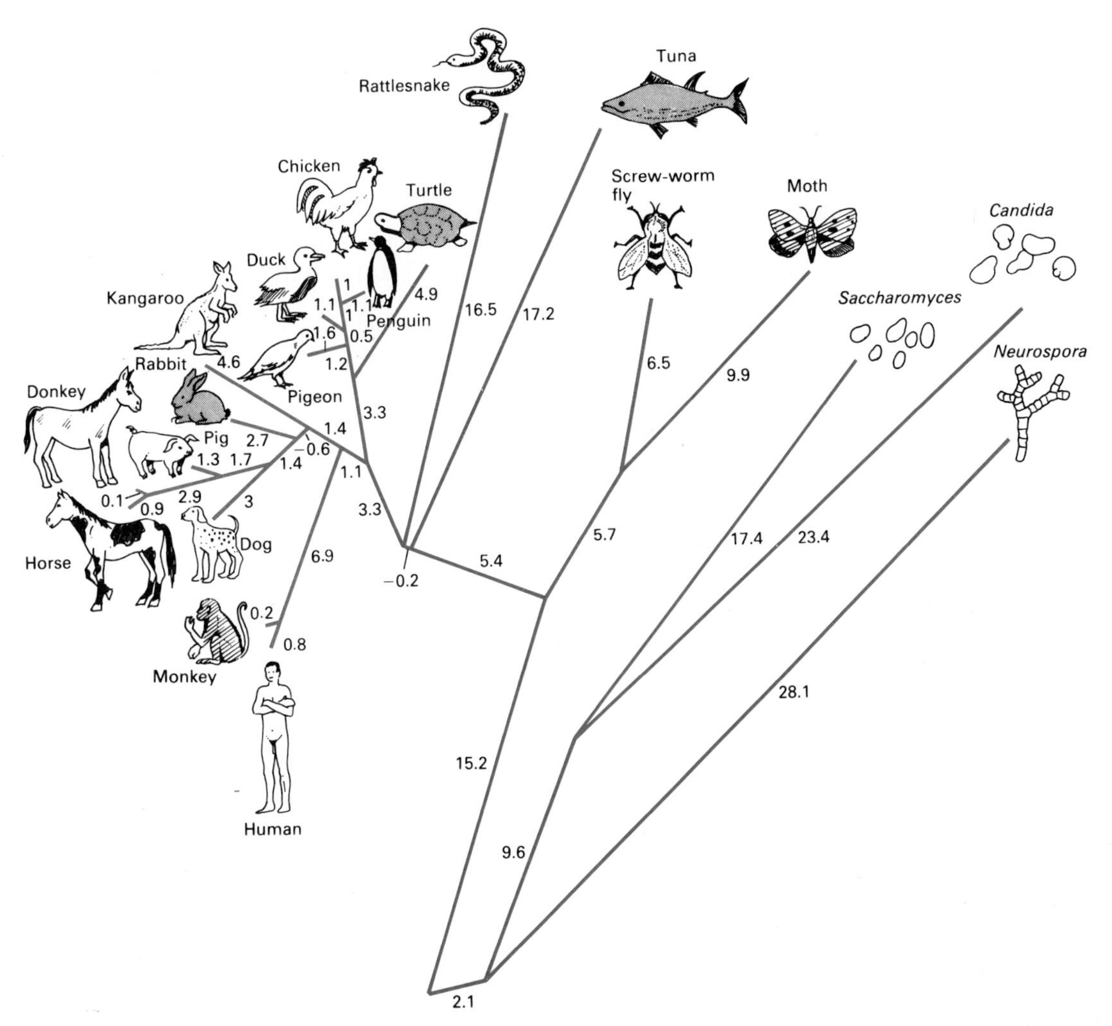

Figure 14.16. A phylogenetic tree based on differences in the amino acid sequence of cytochrome *c* from 20 organisms. Trees such as this one have the same topological features as those derived by traditional methods and serve to confirm and extend the evolutionary record. *Source:* From W. M. Fitch and E. Margoliash, *Science* 155:279–284, 1967. Copyright 1967 by the American Association for the Advancement of Science.

TO SUM UP

1. Migration, mutation, and selection are systematic processes that when acting alone or jointly, tend to produce an equilibrium gene frequency. Their effects are opposed by the dispersive process of genetic drift, which tends to scatter gene frequencies away from the potential equilibrium point toward values of 0 or 1. The combined effects of systematic and dispersive processes permit the development of genetic differences between local breeding populations.

2. When groups of geographically related populations develop distinct and readily recognized differences from other groups of the same species, they are classified as a race. Members of different races are capable of interbreeding but retain their distinctive characteristics because of geographic or, in the case of humans, cultural barriers to gene exchange.
3. A species consists of one or more populations that are actually or potentially interbreeding and are reproductively isolated from other such groups. New species can arise when populations of an existing species that are physically separated accumulate sufficient genetic differences to prevent interbreeding if they should come into contact with one another.
4. Amino acid sequencing techniques have enabled geneticists to detect evolutionary divergence in the protein structures of different species. Phylogenetic trees have been constructed based on the number of amino acid differences.

Questions and Problems

1. Distinguish among a race, a species, and a local population. Rank them in order of increasing complexity.
2. Some species, called sibling species, are so similar in appearance that they are difficult to distinguish by most morphological criteria. Why would a biologist classify such obviously similar organisms as different species and yet group others as diverse as the Chihuahua and the Saint Bernard breeds of dogs into one species?
3. Distinguish between the dispersive and systematic evolutionary forces. How does a combination of systematic and dispersive pressures provide for the development and maintenance of variation both within and between local breeding populations?
4. What is meant by the founder effect? Suppose that a group is founded by four individuals derived at random from a population that has genotypes in a ratio of 1 *AA* : 2 *Aa* : 1 *aa*. What is the probability that the starting group differs in allele frequency from the population from which it was derived?
5. Distinguish between the equilibrium associated with a balanced polymorphism (such as the one that involves the sickle-cell gene) and the random mating equilibrium described by the Hardy-Weinberg law. In what ways are they similar?
6. Assume that the forward mutation rate ($A \xrightarrow{u} a$) is 5×10^{-5} and the back mutation rate ($a \xrightarrow{v} A$) is 10^{-5}. What are the expected frequencies of the two alleles once an equilibrium is established between forward and back mutation?
7. In a population of 100,000,000 people, 10,000 have a serious genetic disease caused by a recessive allele. If these individuals are kept from reproducing, how many generations would it take for the frequency of this disorder to be reduced to one in a million persons? (Ignore the effects of back mutation.)
8. Calculate Δq in Eq. (14.4) and Δp in Eq. (14.5) for allele frequencies of 0, 0.2, 0.3, 0.4, 0.6, 0.7, 0.8, and 1, and with $s = 0.01$. Plot your results in graphical form as Δq plotted against q in the first case and as Δp against p in the second. Describe any differences in the selective efficiencies of the two models and indicate the range of allele frequencies at which each type of selection is most effective.
9. Kettlewell studied the differential effects of predation by birds on the light-colored and dark (or melanic) forms of the peppered moth, *Biston*

betularia, in a woodland near a heavily industrialized area of England. It was proposed that the melanic form would have the selective advantage against the soot-darkened tree trunks of this region, being less conspicuous to bird predators than the light-colored moths. To test this hypothesis, Kettlewell released a mixture of dark- and light-colored moths in the area and later recaptured the survivors. The following data were obtained:

	Released	Recaptured
Dark-colored moths:	154	82
Light-colored moths:	73	16

Calculate from these results the relative fitness values of the light- and dark-colored moths in terms of their differential survival from predation by birds in this industrialized area.

10. Suppose that a farmer decides to cull recessive homozygotes from his flock of sheep because of the poor quality of their wool, leaving only dominant homozygotes and heterozygotes free to mate. The farmer starts with a large flock of lambs consisting of $\frac{9}{16}$ *AA*, $\frac{6}{16}$ *Aa*, and $\frac{1}{16}$ *aa*. **(a)** What genotype frequencies are expected to occur among the adult sheep that are left for breeding purposes? **(b)** What genotype frequencies are expected to occur among the lambs in the next generation? **(c)** How many generations of culling are needed to reduce the frequency of the recessive allele in this flock to one-third of its initial value?

11. The farmer in Problem 10 discovers that about half of the heterozygotes for the recessive allele produce a slightly poorer quality of wool than do the other sheep with the dominant phenotype. **(a)** If the farmer can detect the poorer wool quality in time to cull these heterozygotes from the breeding flock along with the recessive homozygotes, what genotypic frequencies will appear among the breeding adults selected from a population of $\frac{9}{16}$ *AA*, $\frac{6}{16}$ *Aa*, and $\frac{1}{16}$ *aa*? What genotype frequencies will appear among the lambs of the next generation? **(b)** Show that the method of selection described in this problem will result in a reduction in the frequency of the recessive allele by one-half in each generation.

12. Ichthyosis congenita is a recessive lethal condition in humans that is characterized by abnormal leathery skin with deep, bleeding fissures. This abnormality arises with a mutation rate of approximately 10^{-5} per gamete per generation in human populations. What is the most likely frequency of the heterozygous carriers of the recessive lethal gene for this disorder?

13. Suppose that a program to control malaria is initiated in certain parts of Africa so that heterozygotes for the sickle-cell gene no longer have a selective advantage. Assume, in addition, that prior to the institution of malaria control, heterozygotes composed 20 percent of the adults in the affected regions. If the reproductive capacity of recessive homozygotes (individuals with sickle-cell anemia) is essentially zero in these regions, how many generations of malaria control are required to reduce the frequency of the heterozygotes to one-fifth of the initial value?

14. In natural populations, random changes in gene frequencies occur not only as a result of chance variation in the segregation ratios of alleles but also as a consequence of the random nature of survival. For example, suppose that the zygotes produced by a population consist of an equal mixture of *AA* and *Aa* genotypes and that individuals of both genotypes have a 10 percent chance of surviving to the reproductive age. Among 10 surviving adults, 6 are *AA* in genotype and 4 are *Aa*. **(a)** Calculate the change in allele frequency that has occurred between the zygote and adult stage. **(b)** Calculate the probability that this change has of occurring. **(c)** Does your answer in part (a) represent a selective or nonselective change in allele frequency? Explain.

15. In the text, models for the selective elimination of an allele were restricted to cases of complete dominance. Suppose that selection is acting instead on an allele that is incompletely dominant, so that the heterozygotes have a fitness value that is exactly intermediate between both homozygous genotypes (e.g., $w_{AA} - w_{Aa} = w_{Aa} - w_{aa} = s$). Using Eq. (14.3), show that when $\overline{w}$ is approximately equal to 1, selection against an allele that is incompletely dominant leads to the comparatively simple result $\Delta q = -sq(1 - q)$.

16. Dobzhansky studied changes in the frequencies of two gene arrangements, Standard (ST) and Chiricahua (CH), on chromosome III of *Drosophila pseudoobscura*. He observed the following numbers of each genotype present among egg samples and freshly emerging adults:

	CH/CH	CH/ST	ST/ST	TOTAL
Egg sample:	42	88	20	150
Young adults:	16	83	31	130

Assuming that the egg sample represents the frequencies before selection, determine the relative fitness values for young adults.

17. **(a)** Show that the average fitness of a population in which there is partial selection against the recessive homozygotes is equal to $1 - sq^2$, where $s = w_{Aa} - w_{aa}$. **(b)** Show that when selection against recessive homozygotes is complete so that $s = 1$, the average fitness of the population becomes $p(1 + q)$.

18. Would you expect a recessive lethal gene that is carried on the X chromosome to be maintained at a higher or lower frequency by the opposing mutation and selection pressures than a recessive lethal gene that is carried on an autosome? Explain.

19. Studies of the allele for sickle-cell hemoglobin (Hb^S) revealed that in a certain province in Africa, the genotypes ($Hb^A Hb^A$, $Hb^A Hb^S$, $Hb^S Hb^S$) had fitness values of (0.84, 1.0, 0). **(a)** What are the equilibrium frequencies of the Hb^A and Hb^S alleles? **(b)** Assuming that selection operates in this case by differential survival between birth and adulthood, what are the expected frequencies of the three genotypes among infants at birth? What about adults?

20. Studies on a local frog population reveal that 9 percent of the frogs exhibit the recessive spotted condition. In each generation, 5 percent of this population is derived by migration from a large neighboring population that is homozygous for the nonspotted phenotype. **(a)** If frogs mate at random for spotting, what is the expected frequency of spotted frogs in the local population after one generation of migration subsequent to the study? **(b)** What proportion of spotted frogs is expected to appear eventually in the local population if migration continues for a very long time with no other forces acting to change the gene frequency?

21. Suppose that a population is founded by a group that consists of four individuals: two males of genotype *AA* and one female each of genotypes *AA* and *Aa*. The females produce two offspring apiece that survive to adulthood. **(a)** What is the probability that the progeny will have the same allele frequencies as their parents? **(b)** If at every generation each mated couple in this population continues to produce only two offspring, what is the probability that the allele frequencies will remain the same at each of two consecutive generations? At each of *t* generations?

22. Neurofibromatosis, a syndrome dealt with in the film and play entitled "Elephant Man," is a dominant genetic disorder that is characterized by tumorlike formations on the skin and in the nervous tissue. The autosomal gene for this disorder has one of the highest known mutation rates in humans; this rate is approximately equal to 10^{-4} per gamete per generation. **(a)** Show that among 10^6 births, 200 new cases of neurofibromatosis are expected to occur as a result of mutation. (Assume that the trait, and hence the gene for this abnormality, is rare in the population, as in fact is the case.) **(b)** It has been estimated that individuals with neurofibromatosis have only half the reproductive potential of homozygotes for the normal allele. What is the expected frequency of the disorder, assuming a balance between the opposing effects of selection and mutation? (*Hint*: Since the frequency of the allele is rare, you can assume that essentially all individuals born with this disorder are heterozygous.)

23. Height in a particular plant is influenced strongly by the presence of the two alleles *A* and *a*. When the plants are fully grown, the average heights of the three genotypes are *AA* = 28 cm; *Aa* = 48 cm; and *aa* = 36 cm. A population of this plant species is subdivided into two isolates of equal size: I and II. In isolate I, the frequency of allele *A* is 0.1; in isolate II, it is 0.9. **(a)** Assuming random pollination within each isolate, calculate the average plant height in each isolate. **(b)** Calculate the average height for the subdivided population as a whole (i.e., the average of both isolates in the mixture). **(c)** Suppose that there is a breakdown of the barriers separating isolates I and II, so that they are combined into one large, random mating population. Calculate the average plant height for the combined population once random mating has occurred. **(d)** Explain the genetic basis for the difference in average plant height in the subdivided and freely intermating populations.

24. **(a)** Using Eq. (14.3), show that when selection is against the heterozygote (so that $w_{AA} - w_{Aa} = w_{aa} - w_{Aa} = s$), an equilibrium is theoretically possible at $q = \frac{1}{2}$. **(b)** Show that the equilibrium for $q = \frac{1}{2}$ in this case is unstable, with values of q increasing in magnitude above $\frac{1}{2}$ and decreasing below it. (*Hint*: Substitute values of q both greater than and less than $\frac{1}{2}$ into Eq. (14.3) and determine whether Δq is positive or negative.) **(c)** One example of selection against heterozygotes in the human population occurs in the maternal–fetal Rh incompatibility described in Chapter 6. Recall that Rh-positive offspring of Rh-negative mothers can develop a severe hemolyitc anemia known as erythroblastosis fetalis, which in the past has frequently been fatal. Each of these Rh-positive offspring must have received a recessive allele from his or her Rh-negative mother, so that the effect is restricted to heterozygotes. Given that selection against heterozygotes can lead to a stable state only when the value of q reaches either 0 or 1, suggest reasons why both of the Rh alleles are still present in Caucasian populations at relatively high frequencies. **(d)** In 1942, Haldane suggested that Rh-negative mothers might compensate for the loss of Rh-positive children through hemolytic disease by producing more offspring than they normally would. Describe how such compensation could serve to balance the losses from incompatibility and also to retain both alleles in the population.

Chapter 15

Quantitative Inheritance

In the previous chapters, we were mainly concerned with the inheritance of sharply defined, contrasting characteristics, such as round and wrinkled pea seeds, A, B, AB, and O blood types, and a host of other similarly inherited traits. As a general rule, these characters can be readily identified and easily categorized into discrete, nonoverlapping phenotypic classes. Such sharply discontinuous characteristics are known as *qualitative traits* and are usually determined by alleles at one or just a few loci. By contrast, the traits that are concerned with height, weight, and fertility, for example, vary in degree (rather than in kind) along some quantitative scale of measurement. These continuously varying characteristics are known as **quantitative** (or **polygenic**) **traits** and their pattern of inheritance is referred to as *quantitative* (or *polygenic*) *inheritance.*

The analysis of quantitative traits is extremely important in the study of human genetics and in agriculture. Most traits of economic importance, such as milk production in cattle, monthly weight gain in swine, and grain yield in cereal crops, are quantitative in nature. These traits result from the action and interaction of many genes and are highly susceptible to environmental modification. The effects of the individual genes are therefore obscured and cannot be measured directly. For this reason, in the study of quantitative characters, we must use techniques that depend on a statistical analysis of the combined actions of several genes rather than techniques that analyze genes individually on the basis of their phenotypic ratios. In this chapter, we will discuss these techniques and describe their theoretical and practical value in determining the nature of quantitative inheritance.

Continuous Variation and Quantitative Traits

The quantitative differences within a population are expressed in terms of the observed variability in a set of measurements (e.g., variation in centimeters of height, variation in grams of weight, and so on). Unlike qualitative differences in kind (such as red vs. yellow or tall vs. dwarf), which can often be classified on the basis of mere visual examination, quantitative differences require plotting techniques and statistical procedures for their analysis. We will therefore begin with a consideration of some of the methods and models used to describe quantitative variation.

Nature and Sources of Variation

When measurements are taken of quantitative traits, the data are usually analyzed by grouping the measured phenotypic values into designated classes and computing the frequency of individuals within each class. The values are then plotted in the form of a *frequency distribution*. This type of graph shows the frequency in each class as a function of the corresponding phenotypic value. One example of a frequency distribution is the histogram (bar graph) in Fig. 15.1. Observe that when a large number of phenotypic classes are defined, the frequency distribution approaches a continuous bell-shaped curve. The curve implies that regardless of how

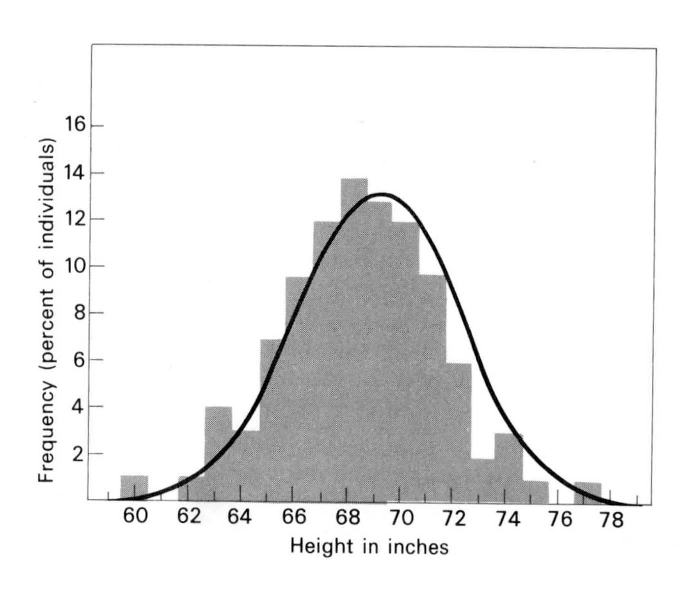

Figure 15.1. Frequency distribution of mature body height of human males. The histogram (bar graph) gives the heights of 100 men by grouping the measurements into 1-inch class intervals. Superimposed on the histogram is a continuous bell-shaped curve that corresponds to the limiting theoretical distribution of body height in a very large population. The continuous curve is formed in theory by increasing the number of classes to the point at which the class intervals are infinitesimally small.

narrow an interval is chosen on the scale of measurement, we will always expect to find some individuals in each phenotypic class if the population is large enough.

Two factors are responsible for the phenotypic differences that occur in a frequency distribution: differences in genotype and differences in environment. We can summarize this relationship quantitatively if we express the measured or observed value of a quantitative character as the sum

$$P = G + E,$$

where P is the phenotypic value, G is the genotypic value, and E is the environmental deviation. In this model, G is regarded as the theoretical value of P when all individuals of a given genotype are exposed to the same environment (i.e., when $E = O$). Furthermore, the environment is assumed to produce departures in phenotype from the value of G about as often in one direction as in the other, so that the average value of P for a given genotype, when measured under all environmental conditions to which the population is normally exposed, is also equal to G. This relationship is shown by Fig. 15.2, in which the differing degrees of environmental variation are assumed to be acting on the phenotypes expressed by alleles at a single locus. Observe that when the environmental departures are large enough, variability at even a single locus can obscure the boundaries between the phenotypic classes and can produce a single continuous distribution.

The environmental deviation (E) is actually a catchall term that includes any departure from the theoretical value of P that is not directly attributable to genetic effects. Among the important causes of environmental deviations are nutritional and climatic factors, such as differences in the diets of animals and differences in the light, temperature, moisture, and soil conditions for plants. The effects of these environmental deviations can be felt at all stages in life, but they are particularly critical during periods of active growth and development.

Most quantitative traits are determined by genes at more than a single locus. Several different genotypes are therefore likely to occur in a population, each with its own potential phenotype. Thus, in addition to the effects of the environment, phenotypic variability can also arise from genetic causes. The overall effects of genotypic differences on a quantitative trait are shown in Fig. 15.3, in which the environmental variation is assumed to be negligible. As the number of pairs of alleles that affect the trait increases, smaller differences occur between the phenotypic classes, ultimately producing a distribution that is indistinguishable by most experimental procedures from a continuous bell-shaped curve.

Statistical Analysis of Quantitative Traits

Mean and Variance. As we have just learned, when a large population is measured for a quantitative trait, the array of phenotypic values gives rise to a frequency distribution that approximates a normal curve. This curve is a theoretical distribution that is continuous and is symmetrical around a centrally occurring value. Two descriptive measures are used to characterize the general location of a normal curve on the horizontal axis and the breadth of its distribution; these measures are the *mean* and the *variance*. Their relationship to the normal curve is

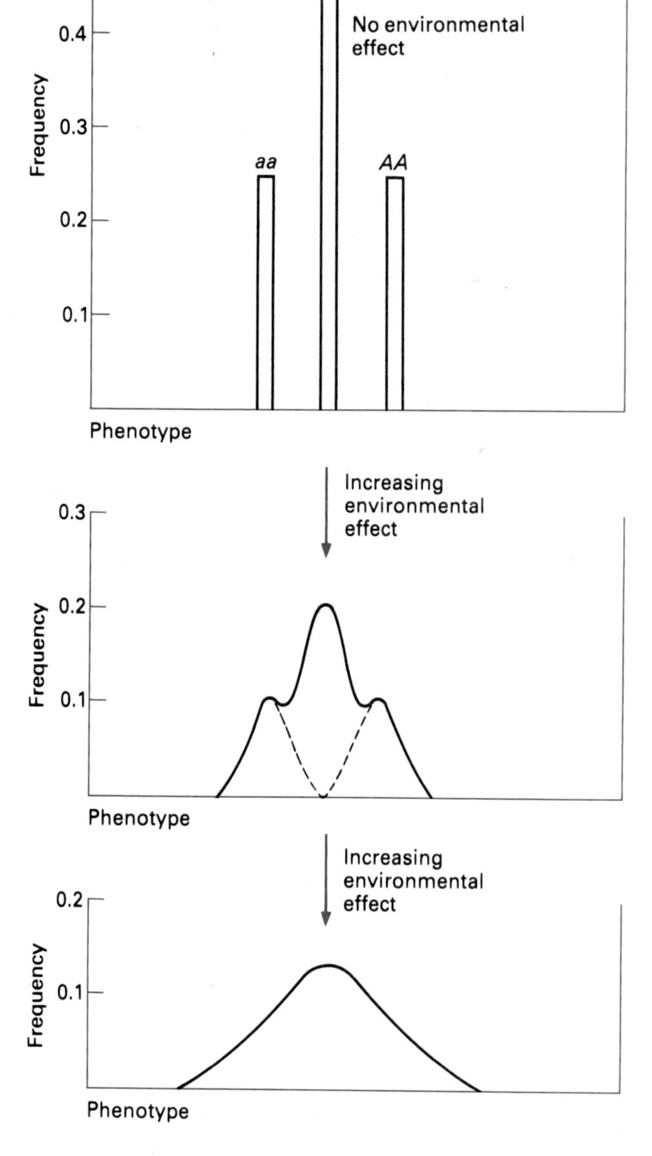

Figure 15.2. Effects of increasing environmental variation on the phenotypes expressed by alleles at a single locus. The genotypes *AA*, *Aa*, and *aa* are assumed to exist in the population at a ratio of 1:2:1. Note that when the environmental effect is large enough, class boundaries become obscured and the phenotypic distribution approaches a continuous bell-shaped curve.

shown in Fig. 15.4. Ordinarily, the mean is simply the average value and is thus one of several statistical measures that can be used to describe the location of the centrally occurring values in a distribution. In the case of the normal curve, however, the mean *is* the central value (or median, as it is called). It is also the most frequently occurring value (or mode). Although the mean equals the median and the mode in a symmetrical normal curve, this property does not necessarily hold true for actual distributions observed in nature, many of which are asymmetrical, or skewed, in appearance.

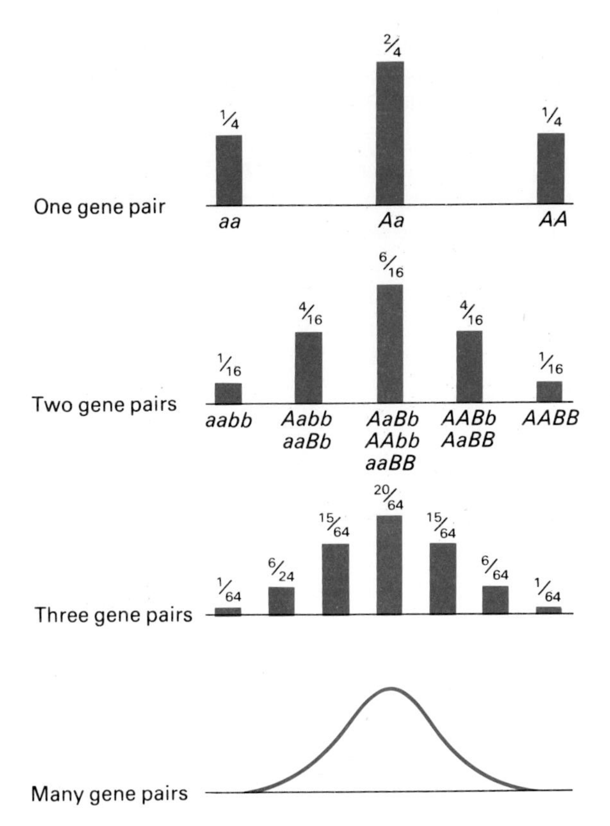

Figure 15.3. Effects of increasing the number of gene pairs that contribute to the expression of a quantitative trait. Each frequency distribution represents the results of a hybrid cross, in which genes segregate independently and the parents are heterozygous for the alleles at every locus. Note that an increase in the number of genes is equivalent to an increase in the number of classes and to a reduction in the size of each class interval. In the limiting case, class boundaries become obscured and the distribution approaches a continuous bell-shaped curve.

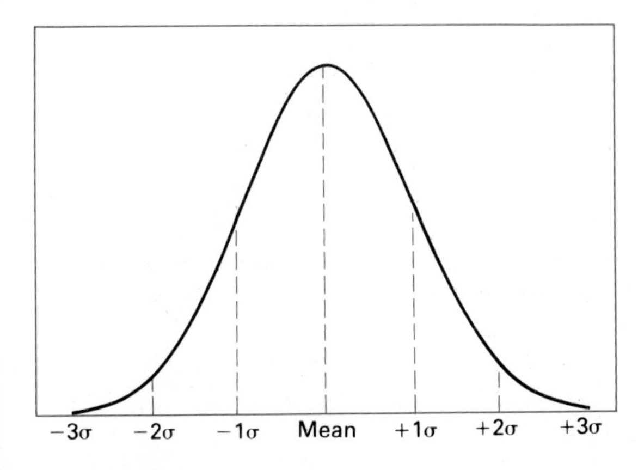

Figure 15.4. A normal frequency distribution, showing the mean and standard deviation (σ, which is equal to the square root of the variance). This is a theoretical distribution in which 68.3%, 95.4%, and 99.7% of all values are included within the intervals, which are $\pm 1\sigma$, $\pm 2\sigma$, and $\pm 3\sigma$, respectively, from the mean.

The variance, which is a measure of the degree of dispersion, or breadth, of the curve, is also illustrated in Fig. 15.4. The square root of the variance, commonly known as the *standard deviation,* measures the horizontal distance from the mean to each of the inflection points on the curve. Mathematical calculations have established that in any normal distribution, approximately 68.3 percent of all values fall within one standard deviation unit above and below the mean, about 95.4 percent fall within ± two standard deviation units of the mean, and about 99.7 percent fall within ± three standard deviation units of the mean. Hence, the larger the value for the standard deviation (and thus the larger the variance), the greater will be the total amount of variability within the population. This relationship is shown by the overlapping curves in Fig. 15.5. These curves are assumed to have the same mean but different variances.

When the mean and the variance are used to describe a normal curve, they have several useful properties that provide the theoretical basis for calculating their values from actual distributions of quantitative traits. First, the mean phenotypic value, designated $\bar{P}$, provides a measure of the average (or central tendency) of a population and can be calculated as follows:

$$\bar{P} = P_1f_1 + P_2f_2 + P_3f_3 + \ldots$$

or, in more compact form, as

$$\bar{P} = \sum P_if_i \qquad \textbf{(15.1)}$$

where Σ designates the sum, P_i is the phenotypic value of some specified class, and f_i is the frequency of that class. For example, if there are three phenotypic classes with P values of 2, 3, and 4, making up ¼, ½, and ¼ of the population, respectively, the mean phenotypic value is then $\bar{P} = (2)(1/4) + (3)(1/2) + (4)(1/4) = 3$.

If all of the members of a population are treated separately and are not grouped into classes, then f_i is the frequency of each individual, so that $f_i = 1/N$, where N is the number of individuals in the population (or in the sample selected from the population). The formula for the mean then reduces to the more familiar form $\bar{P} = \Sigma P_i/N$.

The phenotypic variance, symbolized V_P, provides a measure of dispersion and can be computed as the average of the squared deviations from the mean:

$$V_P = (P_1 - \bar{P})^2f_1 + (P_2 - \bar{P})^2f_2 + (P_3 - \bar{P})^2f_3 + \ldots$$

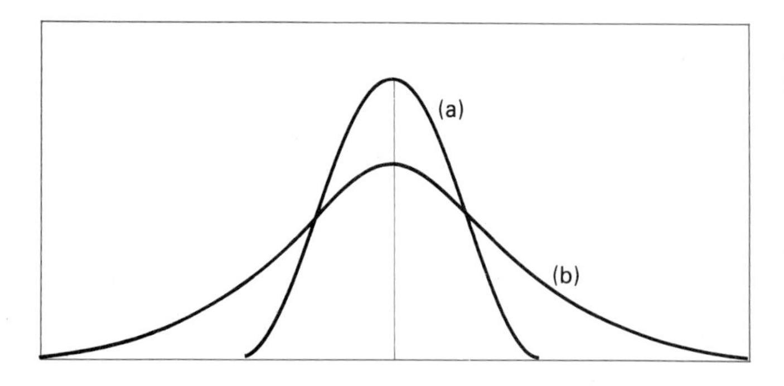

Figure 15.5. Frequency distributions with the same mean but with different variances. Curve (b) has a greater variance than curve (a) and includes a greater amount of variation.

which can be summarized as

$$V_P = \sum (P_i - \bar{P})^2 f_i. \qquad \textbf{(15.2)}$$

The values we used earlier to compute the sample mean can serve as a numerical example. Since $\bar{P} = 3$, then $V_P = (2 - 3)^2(1/4) + (3 - 3)^2(1/2) + (4 - 3)^2(1/4) = 1/4 + 1/4 = 1/2$.

We should note that when the value of $\bar{P}$ used in Eq. (15.2) is based on a selected sample of individuals rather than on the entire population, the formula yields a biased estimate of V_P, which tends to be smaller than the true value, on the average. The formula can be corrected for a sample by multiplying by the factor $N/(N - 1)$, where N is the sample size. When each member of the sample is treated separately, so that $f_i = 1/N$, an unbiased estimate of the sample can then be computed as

$$V_P = \frac{N}{N - 1} \sum \frac{(P_i - \bar{P})^2}{N} = \sum \frac{(P_i - \bar{P})^2}{N - 1}.$$

EXAMPLE 15.1. A representative sample of 10 body weights (in grams) of the Flemish breed of rabbits is shown below. Determine the range of weights that should in theory include 95.4 percent of the rabbits in the population.

3305	3667	3677	3696	3311
3442	3829	3561	3774	3738

Solution. The mean weight for this sample is $\bar{P} = \Sigma P_i/N = 36{,}000/10 = 3600$ g. The range of weights that would include 95.4 percent of the measurements is $\bar{P} \pm 2\sqrt{V_P}$, where $\sqrt{V_P}$ is the square root of the variance, known as the standard deviation. The standard deviation can be calculated as follows:

P_i	$P_i - \bar{P}$	$(P_i - \bar{P})^2$
3305	− 295	87025
3442	− 158	24964
3667	+ 67	4489
3829	+ 229	52441
3677	+ 77	5929
3561	− 39	1521
3696	+ 96	9216
3774	+ 174	30276
3311	− 289	83521
3738	+ 138	19044

$\Sigma P_i = 36{,}000 \qquad \Sigma(P_i - \bar{P}) = 0 \qquad \Sigma(P_i - \bar{P})^2 = 318{,}426$

$V_P = \Sigma(P_i - \bar{P})^2/(N - 1) = 318{,}426/9 = 35{,}380.7; \quad \sqrt{V_P} = \pm 188.1.$

The range is 3600 ± 376.2 g. The lower limit is 3223.8 g and the upper limit is 3976.2 g.

Genotypic Variation and Heritability. The underlying contributions of genes and the environment to the overall variation of a quantitative trait can be studied statistically by an *analysis of variance.* This procedure was originally applied to genetic study in the 1920s by the geneticist R. A. Fisher. The method entails partitioning the phenotypic variance (V_P) into the sum of the component variances, so that each component measures the amount of variation in a particular causal factor or combination of causal factors. At the most superficial level of analysis, V_P can be partitioned into three component parts as follows:

$$V_P = V_G + V_E + V_{GE}$$

where V_G is the *genetic variance,* which arises from the presence of different genotypes in the population, V_E is the *environmental variance,* which is due to environmental differences, and V_{GE} is the *genotype-environment interaction variance.* If all genotypes are randomly found in all environments and show the same general pattern of response to each environmental deviation (i.e., each environment affects all genotypes similarly), then the genetic and environmental sources of variation are independent and $V_{GE} = 0$. In this case, the phenotypic variance is simply $V_P = V_G + V_E$.

We are often more concerned with the relative contributions of each component of the variance rather than with their absolute values. Much of the relevant information regarding the sources of phenotypic variation can therefore be incorporated into one statistic, which is given by the ratio

$$H_B = V_G/V_P \tag{15.3}$$

where H_B is the **heritability**. The subscript B indicates that it is the heritability in the broad sense and is used to distinguish H_B from a similar but more narrowly defined statistic, designated H_N, which we will discuss in the next chapter.

The heritability H_B expresses the proportion of phenotypic variation that is a result of genotypic differences in the population. The value of H_B can vary from 0 to 1. If the environmental conditions are constant and uniform throughout the population (although this is realistically an impossibility), then $H_B = V_G/V_G = 1$, which indicates that all of the variation is genetic in origin. On the other hand, when the population is genetically uniform with regard to those genes that affect the trait in question, then $H_B = 0/V_E = 0$, which indicates that all of the variation is environmentally induced. Typically, the value of H_B lies somewhere between these extremes. Let us take human height as an example. Variation in height for a given age and sex will tend to be influenced by both genetic factors and environmental factors (such as diet). Since neither V_G nor V_E is then equal to 0, the analysis of a population becomes a problem of separating the genetic from the environmental contributions to the variance.

EXAMPLE 15.2. Data based on observations of 612 families by Miall and Oldham indicated that only 14.2 percent of the total variance in blood pressure (systolic pressure) could be ascribed to environmental variation. What is the heritability for this trait? How much greater is the effect of genetic differences than environmental differences on the total variability?

Solution. If we ignore the possible effects of genotype-environmental interactions, then $V_P = V_G + V_E$. Since $V_E/V_P = 0.142$, the heritability (H_B) must be $V_G/V_P = 1 - V_E/V_P = 1 - 0.142 = 0.858$. Note that by dividing V_G/V_P by V_E/V_P we get V_G/V_E, which is the ratio of the relative importance of genetic factors over environmental factors in determining phenotypic differences. In this case, $V_G/V_E = 0.858/0.142 = 6$. Thus, genetic variation is six times greater than environmental variation in this instance.

It is important to keep in mind that H_B does not measure the degree to which a trait is determined by genes; rather, it measures the relative contributions of genotypic and environmental differences to the total phenotypic variation. The heritability fraction therefore reflects differences within the population and does not apply to single individuals. The fact that H_B is expressed as a ratio can also lead to confusion about its meaning. For example, a high value of H_B does not necessarily imply that V_G is large. Nor does a low value of H_B indicate a high value for V_E. A trait may be influenced to only a small degree by environmental factors and yet be characterized by a low heritability if little genetic variation exists in the population. We should also emphasize that H_B is a variable that not only differs from trait to trait but also varies with the gene frequency and with the environment in which the population lives. Estimates of heritability are therefore specific for a certain population and only apply to that particular population under a particular set of environmental conditions.

There are several techniques that have been used to measure heritability for different traits. The technique known as the *twin method* yields estimates of H_B based on comparisons of variability among identical and fraternal twins. This approach, which has been used quite extensively in human genetics, is considered further in the following sections.

Twin Studies. Twins can be either monozygotic (MZ) or dizygotic (DZ). *Monozygotic* (or identical) *twins* are derived from a single zygote that separates into two embryos during an early developmental stage. Except for possible somatic mutations, MZ twins are genetically identical. *Dizygotic* (or fraternal) *twins* arise when two eggs are released at the same time and are fertilized by different sperm. Since DZ twins develop from two independent zygotes, they are no more similar genetically than are ordinary siblings.

We can calculate the proportions of MZ and DZ twins from the frequency of births involving twins of opposite sexes. The members of an MZ pair must be identical with respect to sex; by contrast, only half of the DZ pairs will be of the same sex, since the likelihood is equal that DZ twins will be of the same or of opposite sexes. The proportion of DZ twins must then be twice the proportion of twins of unlike sexes:

$$\text{Proportion of DZ pairs} = \frac{2(\text{number of pairs of unlike sexes})}{\text{all twin pairs}}$$

and the proportion of MZ twin pairs will be the difference:

$$\text{Proportion of MZ pairs} = 1 - \text{proportion of DZ pairs}$$

These formulas provide a simple method of estimating the frequencies of MZ and DZ twins in different human populations. When such calculations are made, they indicate that about 30 percent of all twin births are monozygotic. In terms of total pregnancies, this means about 4 MZ pairs occur in every 1000 births. This figure is about the same in all human populations throughout the world, which indicates that the occurrence of MZ twins is largely a random (or accidental) event. On the other hand, the frequency of DZ twins varies with ethnic origin (it is higher among blacks than whites, for example) and changes with maternal age (mothers in their mid-thirties have the greatest chance of producing DZ twins). These relationships and the tendency for DZ twins to run in families suggests that there is a genetic component involved in the occurrence of DZ twins.

When twins are used in genetic research, members of MZ pairs and members of same-sexed DZ pairs are usually compared, and the differences that exist between the members of each pair with respect to the trait in question are recorded. We can then calculate phenotypic variances based on these differences. Suppose that we wish to estimate the heritability of adult height among persons of the same sex. Let V_{MZ} = the variance in height between members of MZ twins and V_{DZ} = the variance in height between members of DZ twins. If no differences in height were observed between MZ twins, then V_{MZ} would equal zero. Even though MZ twins are genetically identical, however, they do show some phenotypic differences of environmental origin. For example, differences in height may arise from differences in nutrition (prenatal or otherwise) or from differences in exposure to certain childhood diseases. Whatever the cause, V_{MZ} should be proportional to the environmental variance, V_E. In contrast, V_{DZ} will include both the environmental and genetic components of the variation. But in DZ twins, the genetic component is only half of what the value of V_G would be for unrelated persons, since DZ twins are siblings and will have half of their genes in common, on the average. Thus, if we assume that V_{GE} is negligible, we can approximate $\frac{1}{2}V_G$ by the difference $V_{DZ} - V_{MZ}$. An estimate of H_B can then be obtained from the ratio

$$2(V_{DZ} - V_{MZ})/V_P$$

where V_P is the phenotypic variance of unrelated persons.

Heritability estimates based on twin studies are given for a number of human traits in Table 15.1. We should point out that heritability estimates calculated in

Table 15.1. Heritability estimates based on twin studies for a number of human traits.

TRAIT	HERITABILITY
Height	0.81
Weight	0.78
Cephalic index (head breadth/head length)	0.75
IQ	
Binet test	0.68
Otis test	0.80

this manner are subject to various sources of error. One problem with obtaining a reliable estimate of H_B by this method is that the environmental variance is assumed to be the same for both MZ and DZ twins. This need not be the case. The fact that MZ twins are so similar in appearance often leads to their being treated more alike. Exposure to more similar experiences can lead in turn to a reduction in V_{MZ} below the environmental variance for DZ twins.

TO SUM UP

1. Quantitative traits, including height, weight, and fertility, show a continuous variation pattern over their entire range of values. Since it is usually impossible to separate individuals into discrete phenotypic classes, quantitative traits are characterized by a statistical analysis of the individual measurements rather than by the ratios among the phenotypic classes.
2. The phenotypic variability of quantitative traits arises as a result of variations in genotype and in the environmental conditions. These effects are summarized mathematically by expressing the phenotypic value (P) of an individual as the sum of the potential genotypic value (G) and the environmental deviation (E), so that $P = G + E$.
3. The mean phenotypic value ($\bar{P}$) is a measure of the average (or central tendency) of a population, while the phenotypic variance (V_P) is a measure of the dispersion of values about the mean. The underlying contributions of genes and the environment to the overall variation of a quantitative trait can be measured by partitioning the phenotypic variance into three component parts: the genetic variance (V_G), the environmental variance (V_E), and the genotype-environment interaction variance (V_{GE}), so that $V_P = V_G + V_E + V_{GE}$.
4. The heritability of a trait, in the broad sense, is the proportion of the total phenotypic variation that is due to genetic differences, so that $H_B = V_G/V_P$. Heritability estimates vary with the allele frequency and with the environment; they therefore apply only to the specific populations in which they were measured.
5. Monozygotic twins originate from a single zygote and are genetically identical, whereas dizygotic twins are derived from independent zygotes and are genetically equivalent to ordinary siblings. Comparisons of the members of monozygotic and dizygotic twin pairs provide data for the calculation of heritability estimates and supply information concerning the relative importance of genetic and environmental influences on quantitative traits.

Inheritance of Quantitative Traits

Studies conducted during the period 1910 to 1913 by H. Nilsson-Ehle in Sweden and by E.M. East in the United States did much to establish a Mendelian basis for the inheritance of quantitative traits. One of these studies by East was based on crosses between strains of tobacco (*Nicotiana longiflora*) that differed in the length of the flowers. The results of this study are summarized in Fig. 15.6. East crossed two parental varieties with average flower lengths of $\bar{P}(\text{of } P_1) = 40.4$ mm and $\bar{P}(\text{of } P_2) = 93.1$ mm. Some variability existed in each strain. But if we assume that the parental strains are homozygous for their respective alleles that affect flower length, we can ascribe this observed variation to the environment. The parental cross yielded an intermediate F_1 with an average flower length of

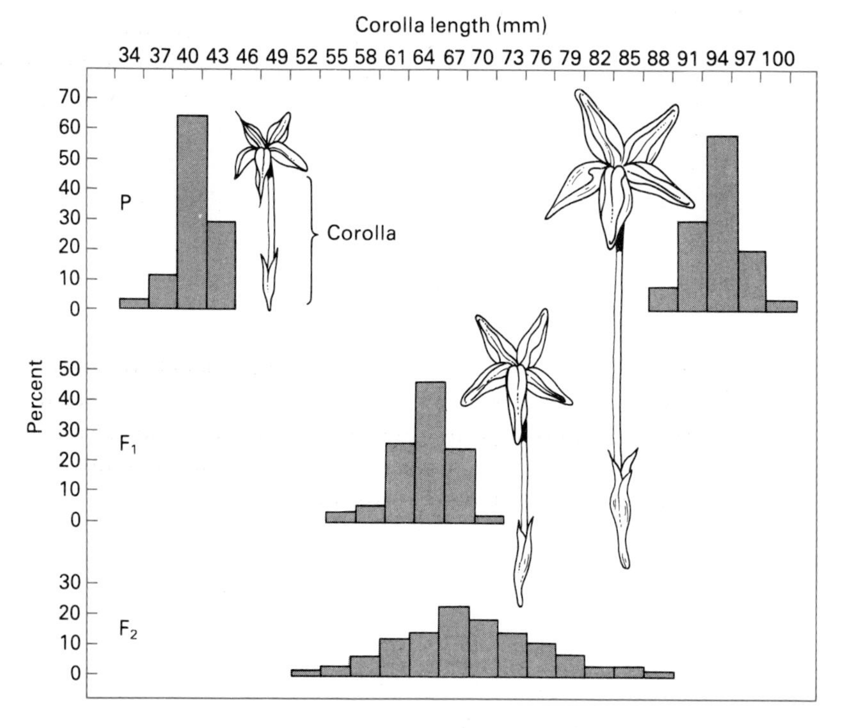

Figure 15.6. The results of crosses between strains of tobacco (*Nicotiana longiflora*) that differ in flower (corolla) length. Observe that the average lengths in the F_1 and F_2 are intermediate to those of the parents and that the F_2 distribution incorporates the greatest variability. *Source:* Reprinted by permission from F. J. Ayala and J. A. Kiger, Jr., *Modern Genetics*, Copyright © 1980, by The Benjamin/Cummings Publishing Company.

$\bar{P}(\text{of } F_1) = 64$ mm. Once again, some variability was observed among individuals, and the variance of the F_1 was similar to that of each of the parental strains. The F_1 were inbred, producing F_2 progeny with an average flower length of $\bar{P}(\text{of } F_2) = 68$ mm, which is similar to the mean of the F_1, except that the F_2 had a large amount of variation about the mean. The overall pattern of these results is typical of the pattern observed for other quantitative traits. We can summarize this pattern by the following two general statements:

1. The average phenotypic values of the F_1 and F_2 are similar in magnitude and intermediate between those of the parental strains.
2. The variance of the F_1 is similar to those of the parental strains, which indicates that the differences were derived from the environmental influence, whereas the variance of the F_2 is considerably greater, so that this variance apparently incorporates variability that is both genetic and environmental in origin.

Results such as those in the experiment on flower length led Nilsson-Ehle and East independently to propose the multiple-gene hypothesis. This hypothesis states that quantitative traits are a result of the cumulative effects of many genes, each of which contributes such a small amount to the overall phenotype that their individual actions are obscured by the genotype as a whole and by the effects of the environment. When many genes act together in this way to produce a quantitative trait, they are called *polygenes,* as they were named by Mather in 1941. While the biochemical actions of polygenes are still unknown, their statistical effects on quantitative inheritance have been studied intensively. We will discuss these effects in the following sections.

Polygenes and Additive Effects

The Additive Polygene Model. We can use a simple model to explain the cumulative effects of polygenes. In this model, two alleles are assumed to be present at each locus: a contributing (or active) allele, designated by a capital letter, and a noncontributing (or null) allele, symbolized by a lowercase letter. Let us also assume that all gene pairs assort independently and that all active alleles contribute an equal amount to the overall trait in an additive fashion.

We can clarify the additive action of polygenes by applying the model to East's observations on flower length in tobacco. Let us begin by supposing that only one gene pair, *A* and *a*, are segregating in the cross for flower length and that each *A* allele contributes 24 mm to a base value of 42 mm. The base value constitutes the contribution to this trait of all nonsegregating alleles that are always present in the homozygous state. The two parental strains would then be *aa* and *AA* in genotype, with average phenotypic values of 42 mm and (2)(24) + 42 = 90 mm, respectively. In theory, these average phenotypic values would be the only lengths expected for the two strains under uniform environmental conditions; they are calculated by multiplying the number of active alleles in a genotype by the contribution of each active allele (24 mm) and adding this result to the base value (42 mm). Thus, the *Aa* flowers of the F_1 plants are expected to average (1)(24) + 42 = 66 mm and the alleles for flower length in the F_2 plants should segregate into three classes:

ACTIVE ALLELES	GENOTYPE	FREQUENCY	VALUE (MM)
2	*AA*	¼	90
1	*Aa*	½	66
0	*aa*	¼	42

The proportions in the F_2 correspond to the terms of the binomial expansion $(p + q)^2 = p^2 + 2pq + q^2$, where p and q are the frequencies of the active allele and the null allele, respectively. Since both p and q equal ½ in this series of crosses, we get a ratio of 1:2:1 in the F_2.

We can make the model more realistic by assuming more gene pairs. For example, with two gene pairs (*A,a* and *B,b*), each capital-letter gene would be expected to contribute 12 mm above the base value of 42 mm. The parental strains would then be *aabb* and *AABB,* with average flower lengths of 42 mm and (4)(12) + 42 = 90 mm. The *AaBb* flowers of the F_1 would have the same mean value of (2)(12) + 42 = 66 mm. In this case, however, the genes for flower length in the

F_2 plants should segregate into five classes:

ACTIVE GENES	GENOTYPES	FREQUENCY	VALUE (MM)
4	$\frac{1}{16}$ *AABB*	$\frac{1}{16}$	90
3	$\frac{2}{16}$ *AABb* + $\frac{2}{16}$ *AaBB*	$\frac{4}{16}$	78
2	$\frac{4}{16}$ *AaBb* + $\frac{1}{16}$ *AAbb* + $\frac{1}{16}$ *aaBB*	$\frac{6}{16}$	66
1	$\frac{2}{16}$ *Aabb* + $\frac{2}{16}$ *aaBb*	$\frac{4}{16}$	54
0	$\frac{1}{16}$ *aabb*	$\frac{1}{16}$	42

The class proportions now follow the binomial expansion $(p + q)^4 = p^4 + 4p^3q + 6p^2q^2 + 4pq^3 + q^4$, where p and q are again equal to $\frac{1}{2}$.

We can generalize from the preceding pattern to predict that if n is the number of independently segregating gene pairs and R is the difference between the means of the parental strains (equal to $90 - 42 = 48$ mm in the example just given), the contribution of each active gene in the model will be $R/2n$. In the absence of environmental variation, the F_2 should segregate $2n + 1$ phenotypic classes, which correspond in frequency to the terms of the binomial expansion $(p + q)^{2n}$. For example, if we assume three gene pairs in our example, then each active gene should contribute $48/6 = 8$ mm to the total value, and the F_2 should segregate $2n + 1 = 7$ classes in a ratio of 1:6:15:20:15:6:1.

Since both p and q are $\frac{1}{2}$, it is possible to predict the expected proportion of each F_2 class by employing the binomial formula (Chapter 2) as follows:

$$P(x \text{ active genes}) = \frac{(2n)!}{x!(2n-x)!}(\tfrac{1}{2})^x(\tfrac{1}{2})^{2n-x}. \qquad \textbf{(15.4)}$$

Thus, the expected proportion of the F_2 that would have four active genes out of a total of $2n = 6$ is

$$P(4 \text{ active genes}) = \frac{6!}{4!2!}(\tfrac{1}{2})^4(\tfrac{1}{2})^2 = 15/64.$$

For three gene pairs in our example, this proportion is expected to have a phenotypic value of $(4)(8) + 42 = 74$ mm, ignoring environmental variation.

EXAMPLE 15.3. Suppose that the number of leaves on a particular plant is determined by four pairs of unlinked polygenes. Plants that are heterozygous at all four loci are crossed and produce many offspring. How many phenotypic classes are expected among the progeny? What fraction of the offspring is expected to have exactly the average number of leaves?

Solution. With $n = 4$ gene pairs, there will be $2n + 1 = 9$ phenotypic classes, assuming that there are no environmental effects. The mean number of leaves will correspond to the phenotypic class that has exactly four active genes. The expected frequency of this class is

$$P(4 \text{ active genes}) = \frac{8!}{4!4!}(\tfrac{1}{2})^4(\tfrac{1}{2})^4 = 70/256.$$

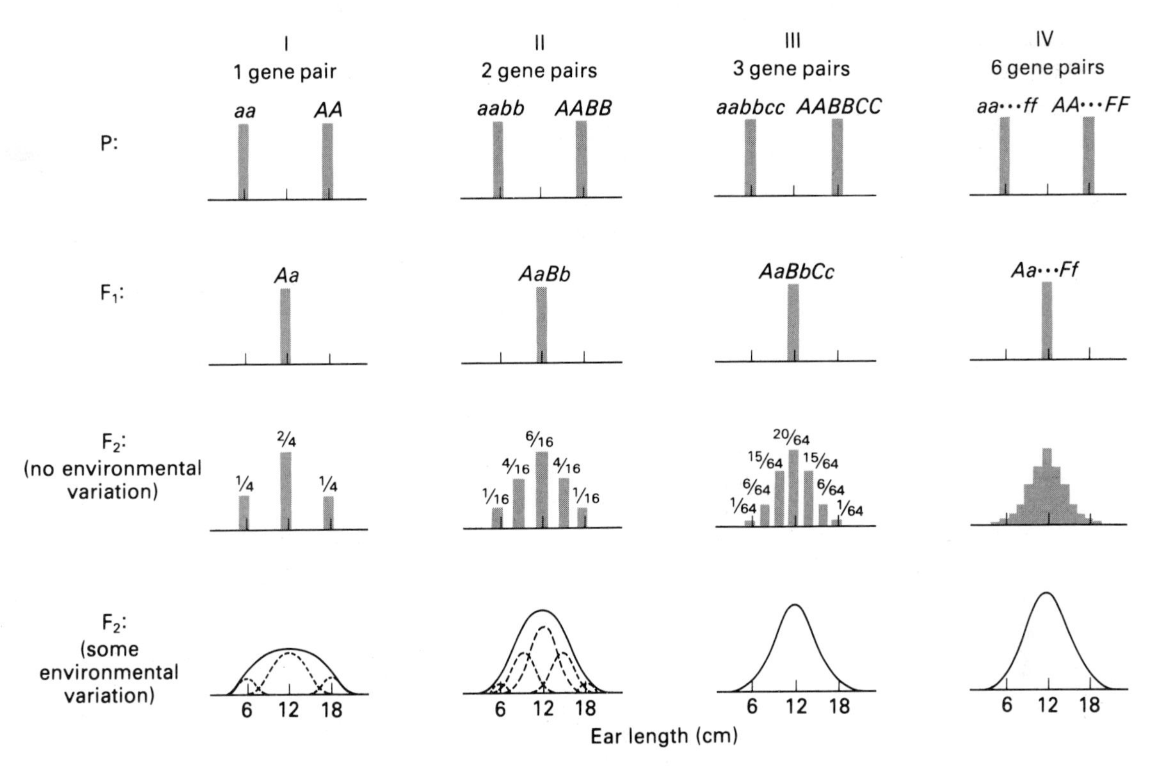

Figure 15.7. Diagrammatical summary of the additive polygene model for different numbers of gene pairs. In each case, homozygous parental lines are crossed to form an intermediate F_1, which in turn produces the F_2. The highly variable F_2 distributions are shown as they would appear both in the absence and in the presence of environmental variation. *Source:* Reprinted by permission from F. J. Ayala and J. A. Kiger, Jr., *Modern Genetics*, Copyright © 1980, by The Benjamin/Cummings Publishing Company.

The basic pattern just described is shown in Fig. 15.7 for increasing numbers of gene pairs. In addition to the expected distribution of the parental, F_1, and F_2 generations under uniform environmental conditions, the F_2 distribution is also modified to show how it would appear when the environment contributes significantly to the phenotypic variation. It is apparent that with a sufficient amount of environmental variation, the F_2 are not separable into discrete phenotypic classes, so that the class frequencies cannot be compared directly with the terms of the binomial distribution.

Although highly oversimplified, this model does provide insight into the behavior of polygenes and seems to fit the cross results of some traits rather well. For example, plant height and egg production in chickens are traits that yield results that are consistent with the model, as does the degree of skin pigmentation in humans (Fig. 15.8). In 1964, the English anthropologists Harrison and Owens used a reflectance spectrophotometer to estimate the amount of melanin pigment by measuring the percentage of light reflected by the skin at different wavelengths. Since reflectance and skin pigmentation are inversely related, darker skin will reflect less light. These investigators examined blacks, Caucasians, the F_1 from

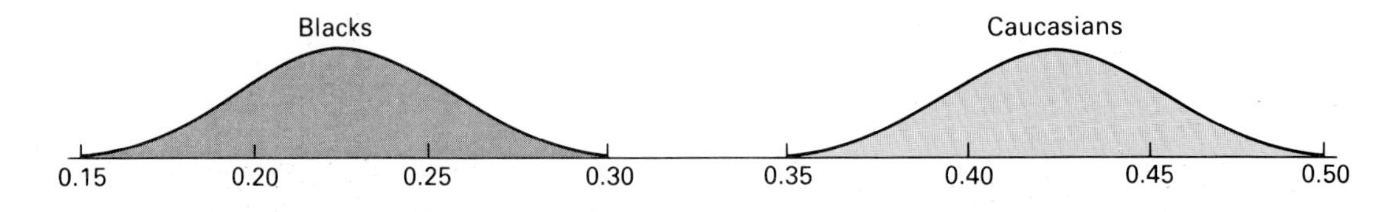

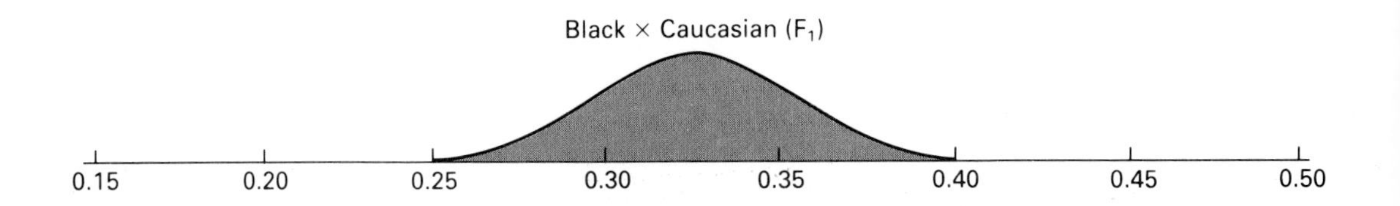

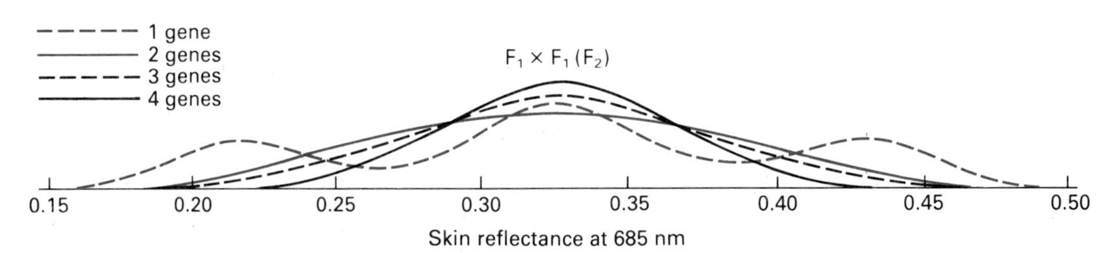

Figure 15.8. Distribution of human skin color as measured by the skin reflectance of red light at a wavelength of 685 nm. The F_2 distributions are theoretical curves based on the additive polygene model, assuming an environmental effect and different numbers of gene pairs involved in variations in skin color. *Source:* From *Genetics, Evolution, and Man* by W. F. Bodmer and L. L. Cavalli-Sforza. W. H. Freeman and Company. Copyright © 1976.

interracial matings, and the F_2 offspring from matings between the F_1, as well as children from $F_1 \times$ white and $F_1 \times$ black backcrosses. Their results clearly indicate a polygenic basis of inheritance, in that pigmentation ranged from black to white through various intermediate shades. A statistical analysis of their results suggests the likelihood that either three or four gene pairs are responsible for human skin color. Moreover, an estimate of 0.65 was obtained for heritability in the broad sense, which indicates that 65 percent of the total variance in skin color is genetically determined.

Estimating Gene Effects and Number of Loci. When a trait conforms to our simple model, it is a relatively easy task to measure the effects of individual genes and to determine the number of segregating gene pairs from the F_2 population. The number of phenotypic classes $(2n + 1)$ and the frequency of any extreme phenotype, $(1/2)^{2n}$, provide estimates of n. For example, the presence of nine phenotypic classes in the F_2 and a frequency of 1 out of 256 individuals falling in each of the parental (or extreme) classes would indicate the presence of $n = 4$ gene pairs, since $2(4) + 1 = 9$ and $(1/2)^{2(4)} = 1/256$. We can therefore estimate the contribution of each active gene from the formula $R/2n$, where R is the difference between the parental means.

This approach is limited in practice by the size of the F_2 population that is needed in order to observe at least one individual of each of the extreme phenotypes. It is therefore of little use when large numbers of gene pairs (usually five or more) are involved. The approach also requires that we be able to distinguish between phenotypic classes or, at the very least, to detect the homozygous individuals that show the extreme expression of the character. Obviously, other ways of estimating the number of genes must be used when the environment acts to obscure discrete phenotypic differences.

One alternative approach to estimating the number of genes is based on an analysis of variance, as shown in Table 15.2. Observe that by letting $R/2n$ represent the contribution of each active allele, the genetic variance of the F_2 becomes $R^2/8n$. Since the phenotypic variance of the F_2 (designated V_{F_2}) is equal to the sum of the genetic variance and the environmental variance, it can be written as

$$V_{F_2} = \frac{R^2}{8n} + V_{F_1}.$$

In this expression, the phenotypic variance of the F_1 (V_{F_1}) is taken as an estimate of the environmental variance, since the F_1 individuals are genetically uniform. Solving for n, we get $V_{F_2} - V_{F_1} = R^2/8n$, which gives

$$n = \frac{R^2}{8(V_{F_2} - V_{F_1})}. \qquad \textbf{(15.5)}$$

Table 15.2. Relating the number of genes with additive effects to the genetic variance of the F_2.

1. Start by considering only one gene pair:
   Contribution of the A locus:

GENOTYPES AT THE A LOCUS	FREQUENCY AMONG F_2	CONTRIBUTION TO GENOTYPIC VALUE
AA	¼	$2(R/2n)$*
Aa	½	$R/2n$
aa	¼	0

Mean contribution of A locus $= \frac{1}{4}(2R/2n) + \frac{1}{2}(R/2n) + \frac{1}{4}(0) = R/2n$

$$\text{Contribution to } V_G = \tfrac{1}{4}\left(\frac{2R}{2n} - \frac{R}{2n}\right)^2 + \tfrac{1}{2}\left(\frac{R}{2n} - \frac{R}{2n}\right)^2 + \tfrac{1}{4}\left(0 - \frac{R}{2n}\right)^2$$

$$= \tfrac{1}{4}\left(\frac{R}{2n}\right)^2 + \tfrac{1}{4}\left(\frac{R}{2n}\right)^2$$

$$= \frac{R^2}{8n^2}$$

2. Now extend the model to all n gene pairs:
   Computation of $\overline{P}$ and V_G for F_2 (take n times the contribution of a single locus);

$\overline{P} = n(R/2n) = R/2$
$V_G = n(R^2/8n^2) = R^2/8n$

* *Note*: $R/2n$ = the contribution of each active allele, where R = the difference between the phenotypic means of the parental lines and n = the total number of segregating pairs of polygenes.

For a numerical example, let us use East's data on flower length in tobacco (see Fig. 15.6). In this example, the difference in parental means is $R = 93.1 - 40.4 = 52.7$ mm. The phenotypic variance of the F_1 is 8.62 mm², while that of the F_2 is 42.37 mm². Substituting these values into Eq. (15.5), we get

$$n = \frac{(52.7)^2}{8(42.37 - 8.62)} = \frac{2777.29}{8(33.75)} = 10.3$$

or about 10 pairs of polygenes. The contribution of each active gene is $R/2n = 52.7/(2)(10) = 2.6$ mm.

EXAMPLE 15.4. The average body weight of the Flemish breed of rabbits is 3600 g. The average weight of the Himalayan breed is 1875 g. When crosses were made between these two breeds, Castle found the mean weights of the F_1 and F_2 to be intermediate between these extremes. The values of the variances were calculated as $V_{F_1} = 26{,}244$ g² and $V_{F_2} = 52{,}900$ g². Estimate the number of pairs of genes that contribute to variability in body weight in these crosses, and determine the average contribution of each active gene.

Solution. The number of gene pairs can be estimated using Eq. (15.5) as follows:

$$n = \frac{(3600 - 1875)^2}{8(52900 - 26244)} = 13.95, \text{ or 14 gene pairs.}$$

The average contribution of each active gene would be

$$\frac{R}{2n} = \frac{3600 - 1875}{2(14)} = 61.6 \text{ g.}$$

We should emphasize that in using Eq. (15.5), we have tacitly assumed a number of simplifying conditions that, if not correct, can lead to errors in estimation. One simplifying condition is that both parental strains are homozygous, with all of the contributing genes present in one strain and all of the noncontributing genes present in the other. When this is not the case, the extreme phenotypes in the segregating F_2 population may actually exceed the parental averages. For example, suppose that the parental cross is *AABBCCdd* × *aabbccDD*, rather than *AABBCCDD* × *aabbccdd*. Although the F_1 and F_2 generations will have the same distribution pattern in both cases, the extreme phenotypes in the F_2 will exceed those of the parental lines. The value of R in Eq. (15.5), which is based on the difference between the parental means, will then lead to an underestimate of the number of segregating loci. This situation, in which the extremes of the F_2 exceed those of the parents, is known as **transgressive variation.** It is a familiar occurrence in human pedigrees, for example, in which one or more children may be taller or shorter than either of their parents or any of their grandparents.

We also assumed in our model that all of the polygenes in the series assort independently of each other. This is highly unlikely when large numbers of genes are involved. Obviously, if there are more gene pairs acting on a quantitative trait than there are chromosome pairs in the organism, some of the genes must be linked. If two such genes are closely linked, then crossing over in the F_1 may not be sufficient to separate them in the F_2. As a result, they may be measured incorrectly in the model as one locus. Thus, linkage will also tend to produce underestimates of the number of segregating gene pairs.

A third oversimplification is that all of the contributing genes must have equal and additive effects on the trait in question. Polygenes, like other genes, do not always act in an additive fashion, nor are their actions always independent of the presence of genes at other loci. Inconsistencies do develop, with genes interacting to produce deviations from additivity. Because such departures are relatively common, we will consider them in detail in the following section. Suffice it to say for now that the unequal and nonadditive actions of certain polygenes tend to inflate the variance of the F_2 and thus contribute even further to the possibility of underestimating the number of loci.

Dominance and Nonallelic Gene Interaction

Interactions involving polygenes are of two basic types: (1) *dominance interactions,* or interactions between alleles, and (2) *interlocus interactions,* or interactions between nonallelic genes (e.g., epistasis). The two types of interactions can be expressed as components of the genotypic value of an individual, as follows:

$$G = A + D + I,$$

where A is the additive value of a genotype and D and I are the dominance and interlocus deviations, respectively. In this expression, D and I represent departures from the theoretical value of a genotype that is assumed in the additive model. Thus, when D and I are zero, dominance and interlocus interactions do not occur and the genotypic values are strictly additive.

We can clarify the concept that gene interactions produce a departure from the additive gene effect when we consider the case of two alleles at a single locus, one of which is completely dominant to the other. In this situation, heterozygotes and dominant homozygotes both have the same average phenotype. An example of this type of interaction is shown in Fig. 15.9, in which the phenotypic value is plotted against the number of active alleles. According to the additive model, the phenotypic value of a genotype should increase linearly with the number of contributing alleles. In other words, the difference between *AA* and *aa* should be twice as great as the difference between *Aa* and *aa*. Therefore, the three points on the straight line drawn on the graph correspond to the theoretical additive values of the different genotypes. The differences between these points and the real or observed values constitute the dominance deviations.

The effect of complete dominance on a quantitative character is shown in Fig. 15.10, in terms of the phenotypic distribution patterns that are expected for different cross results. Observe that with dominance, the phenotypic distribution of the F_1 will always be shifted in the direction of the dominant parent. The effect of dominance on the appearance of the F_2 distribution is not quite as obvious; this effect depends on the number of segregating loci. With one gene pair and complete dominance, a $\frac{3}{4}(A\text{-}) + \frac{1}{4}(aa)$ distribution is expected in the F_2, assuming no environmental variation. When two gene pairs are segregating independently in an $F_1 \times F_1$ cross, three phenotypes are expected, in the ratio $(\frac{3}{4} + \frac{1}{4})^2 = \frac{9}{16}(A\text{-}B\text{-}) + \frac{6}{16}(A\text{-}bb + aaB\text{-}) + \frac{1}{16}(aabb)$. In general, therefore, n inde-

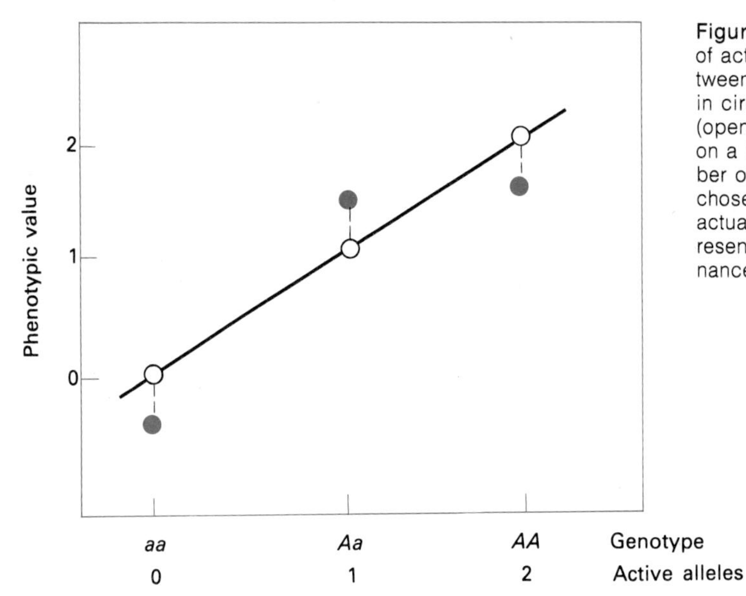

Figure 15.9. A graph of phenotype vs. number of active alleles, showing the relationship between the observed phenotypic values (filled-in circles) and the theoretical additive values (open circles). The additive values are placed on a line that increases linearly with the number of contributing alleles; these values were chosen so as to minimize departures from the actual phenotypic values. The departures, represented by the broken lines, are the dominance deviations.

Figure 15.10. The effect of dominance on a quantitative trait for different numbers of gene pairs. The phenotypic distributions of the F_1 and F_2 are shifted in the direction of the dominant parental line.

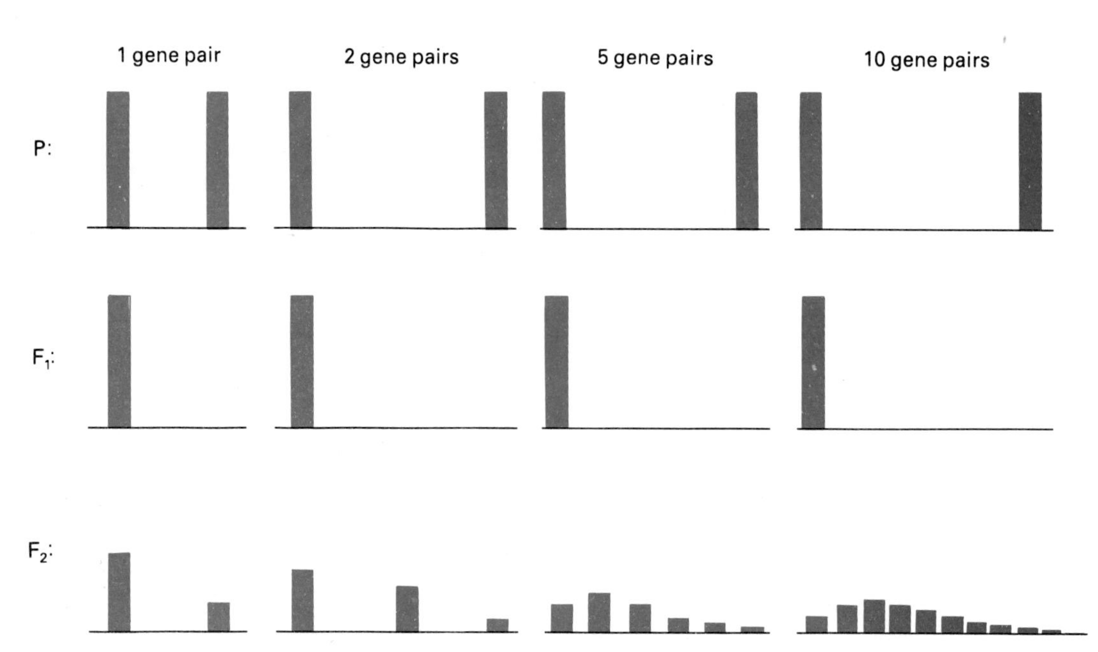

pendently segregating gene pairs with dominance will produce an F_2 distribution of $n + 1$ classes, which correspond in frequency to the terms of the binomial expansion $(3/4 + 1/4)^n$. Note that when n is small, we can easily recognize the presence of dominance by the asymmetrical appearance of the F_2 distribution. As n increases, however, the degree of asymmetry declines, ultimately producing an F_2 distribution that is difficult to distinguish from the distribution for a purely additive gene effect.

EXAMPLE 15.5. Suppose that height in a particular plant is determined by three independently segregating gene pairs (*A,a*; *B,b*; and *C,c*), with each active gene showing complete dominance and adding 2 centimeters in both heterozygous and homozygous combinations to a base height of 10 centimeters. The cross *AABBCC* × *aabbcc* is performed, and the F_1 are intercrossed to produce the F_2. Give the heights that are expected for the parents in the cross and for the F_1, and describe the distribution of heights in the F_2.

Solution. The phenotypes of the different generations can be summarized as follows:

P	*AABBCC*, 16 cm	×		*aabbcc*, 10 cm
F_1	*AaBbCc*, 16 cm			
F_2 **Heights:**	16 cm	14 cm	12 cm	10 cm
Contributing loci:	3	2	1	0
Frequencies:	27/64	27/64	9/64	1/64

The frequencies of the F_2 classes were computed using the formula

$$P(x \text{ contributing loci}) = \frac{n!}{x!(n-x)!}(3/4)^x(1/4)^{n-x}$$

In our illustrations of dominance so far, we have assumed that nonallelic genes not only segregate and assort independently but also express themselves in an independent and additive fashion. Our assumptions correspond to the gene contributions in the following system:

	¾ *A*– (α)	¼ *aa* (0)
¾ *B*– (β)	9/16 *A*–*B*– (α + β)	3/16 *aaB*– (β)
¼ *bb* (0)	3/16 *A*–*bb* (α)	1/16 *aabb* (0)

where α and β are the contributions of *A*- and *B*-, respectively. Although dominance interactions are assumed to occur at each of the loci, interlocus interactions are absent, since the different gene pairs contribute to the overall trait in an additive fashion. In the absence of interlocus interactions, $G = A + D$ for each genotype, because $I = 0$. But if, for example, the *b* allele in the *bb* homozygote were to mask the expression of the *A* gene, so that the *A-bb* genotype would have the same phenotypic value as *aabb,* then interlocus interaction is present in the form of epistasis, in addition to the dominance effects. If this were the case, then both dominance and nonallelic gene interactions would contribute to the departures from additivity.

We can incorporate the variability that is associated with each component of the genotypic value in our model $G = A + D + I$ by expressing the genetic variance as follows:

$$V_G = V_A + V_D + V_I$$

where V_A, V_D, and V_I are the additive genetic variance, the dominance variance, and the interlocus interaction variance, respectively. Of these components, V_A is usually the largest in value, V_D is next, and V_I is generally the smallest—often small enough to be ignored without producing a serious error in the calculations.

Since $V_P = V_G + V_E + V_{GE}$, it follows that

$$V_P = V_A + V_D + V_I + V_E + V_{GE}.$$

This is a general expression that incorporates many of the sources of variation in a quantitative trait. Complex statistical procedures have been developed to evaluate each component of the phenotypic variance. These procedures are not considered here, since they are beyond the scope of the text. Suffice it to say that in most studies, V_{GE} is assumed to be zero and, in many cases, the interlocus interaction effects are assumed to be negligible as well and are simply ignored. The evaluation of V_A, V_D, and V_E is therefore of major concern.

Multiplicative Gene Effects and Threshold Characters

Genes with Multiplicative Effects. The basic theory of polygenes was developed initially with the concept of additive gene action. In many quantitative traits, however, the contributing genes appear to act in a multiplicative rather than an additive fashion, increasing the phenotypic value by a constant percent rather than by the addition of a constant absolute amount. Suppose, for example, that the flower length in a particular plant is increased by a factor of 10 percent for each contributing gene. The incremental increase per added gene will then be multiplicative in overall effect, and the size of the increase will depend on the existing genotype. For instance, the flower length of a plant that already has the genotype for a potential length of 10 mm will only increase to $10 + (0.1)(10) = 11$ mm upon the addition of another active gene, since the phenotype would change by 10 percent of 10 mm—a 1-mm difference. But the flower length of a plant that has the genotype for a potential length of 100 mm would increase to $100 + (0.1)(100) = 110$ mm, for a change of 10 mm. In contrast, an active gene for an additive effect might add 1 mm to the flower length regardless of the existing genotype, so that it will not matter whether the plant already has the potential for a length of 10 mm or 100 mm.

A multiplying gene effect is often associated with traits that are the net result of a biological growth process (e.g., fruit weight in tomatoes). Since growth is a multiplicative process, quantitative characters of this type can be thought of as resulting from the combination of a large number of factors. This multiplying effect gives rise to phenotypic values that vary systematically along a geometric (or logarithmic) scale rather than an arithmetic scale. For example, suppose that fruit weight in a particular plant is doubled by the addition of each active gene. The net effect is to produce a geometric series of relative weights (i.e., 1, 2, 4, 8, 16, etc.). If these data are plotted on an arithmetic scale (1, 2, 3, 4, 5, etc.), the resulting distributions will be asymmetrical (skewed) in appearance, with a peak that is displaced toward the lower end of the curve and an upper tail that approaches zero very slowly (see Fig. 15.11a). Arithmetic plotting procedures also give the impression that the strains with the largest average phenotypes are also the most variable. But this is merely a scale effect that disappears when the data are plotted along a geometric or logarithmic horizontal axis (Fig. 15.11b). The scale effect is overcome by changing the increasing distances between the values in a geometric series (such as $2^0 \rightarrow 2^1 \rightarrow 2^2 \rightarrow 2^3 \rightarrow 2^4 \rightarrow \ldots$) to distances of equal length in the corresponding arithmetic series($0 \rightarrow 1 \rightarrow 2 \rightarrow 3 \rightarrow 4 \rightarrow \ldots$) by means of a logarithmic transformation. (Recall that the logarithm of 2^x is directly proportional to the exponent x.) If the gene effects are truly multiplicative, we would then expect the transformed data to fit a normal probability curve.

Figure 15.11. Distribution of a character that is determined by multiplicative gene effects. The distribution is asymmetrical in appearance when the phenotypic values are plotted along an arithmetic scale (graph a) but is symmetrical in appearance when the values are plotted along a logarithmic scale (graph b).

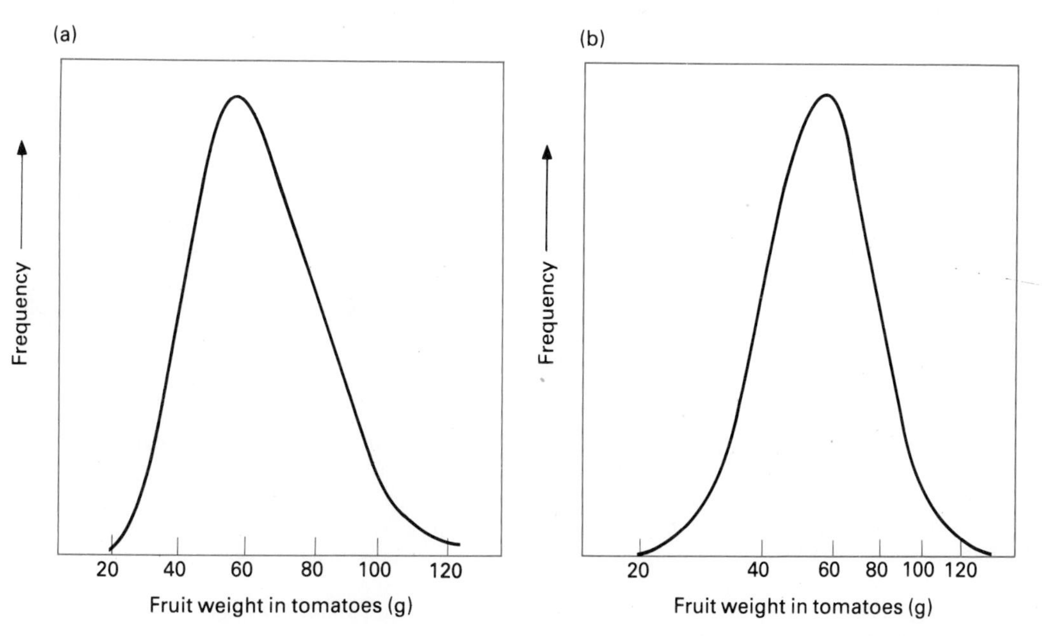

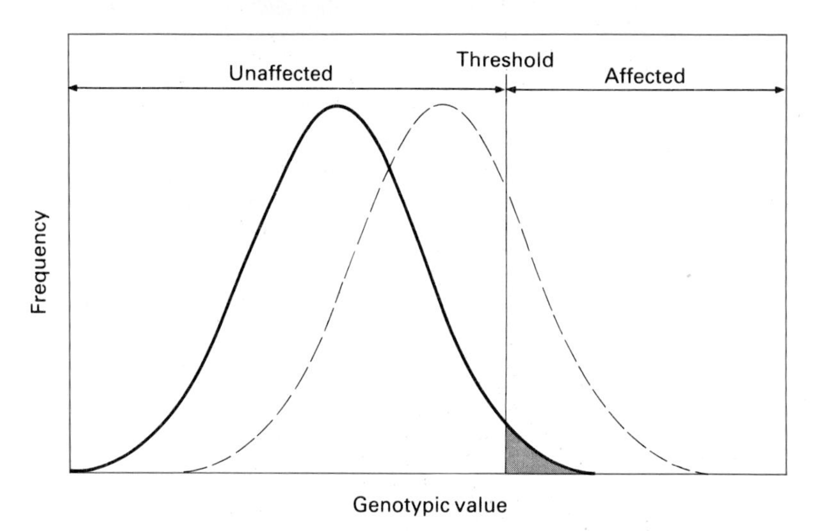

Figure 15.12. Relationship between polygenes and the expression of a discontinuous (threshold) trait. Polygenes are assumed to produce an underlying continuous distribution in the presence of environmental variation, but only two alternative traits are actually expressed. Individuals who exceed a particular threshold value will show one trait; those below this value will express the alternative. Solid line—distribution for the general population; broken line—distribution for close relatives of affected individuals, who are more likely than the general population to exceed the threshold.

Threshold Traits. Many discontinuous traits that can be characterized by the simple alternatives of being either present or absent are nevertheless thought to have an underlying polygenic basis. The discontinuities in this case are believed to arise as a result of **threshold effects.** Figure 15.12 illustrates a threshold effect in terms of a potentially continuous background of genotypic values. Although the underlying genetic effects are continuously distributed, the phenotypic distribution is discontinuous, since all of the individuals whose values fall below a certain threshold level will have one phenotype, while those whose values exceed the threshold level will have another phenotype.

The concept of a threshold effect was first proposed by the geneticist Sewall Wright in order to account for the inheritance of additional toes in guinea pigs. Although most guinea pigs have three toes on each foot, some animals are polydactylous, with four toes. Wright discovered that even though the polydactylous condition in guinea pigs has a genetic basis, its mode of inheritance does not conform to any simple Mendelian pattern. This discrepancy has since been shown to occur for certain other phenotypes, including such diverse traits as cleft lip, diabetes, and schizophrenia in humans. Table 15.3 lists a number of human disorders that are presently thought to be caused by threshold effects and gives a brief description of each.

Table 15.3. Some possible threshold traits in humans.

DISORDER	BRIEF DESCRIPTION
Cleft lip	Congenital malformation involving fissure of the lip.
Cleft lip and cleft palate	Cleft lip associated with fissure of the roof of the mouth (palate).
Club foot	Malformation involving turned-in, sometimes abnormally shaped foot.
Congenital hip	Congenital dislocation of the hip.
Diabetes (one type)	A deficiency of the hormone insulin.
Pyloric stenosis	A narrowing of the opening between the stomach and the small intestine.
Schizophrenia	Mental disorder that is sometimes accompanied by delusions and hallucinations.
Spina bifida	Open spine, caused by failure of embryonic neural tube to close.

We can measure the degree to which a genetic component is involved in a suspected threshold trait by testing for the presence or absence of the trait in both members of monozygotic and dizygotic twins. If the trait is either present or absent in both members of a twin pair, the twins are said to be **concordant** with respect to that character. The twins are **discordant** if one has the trait and the other does not. The *concordance frequency* is then calculated as the proportion of concordant twin pairs among all twins that include at least one member who has the trait. If genetic factors are the sole causes of variation, then all MZ twin pairs should be concordant, whereas DZ twin pairs are no more likely to be concordant than are siblings in general. Environmental differences tend to reduce the concordance frequencies, but they do so for both MZ and DZ twin pairs. A significantly larger concordance frequency in MZ than in DZ twins is therefore suggestive of a genetic predisposition for the trait in question.

For all of the disorders listed in Table 15.3, the concordance frequencies are greater in MZ than in DZ twins. This finding supports the view that a genetic component is involved in each of these characteristics. For conditions such as diabetes and cleft lip, for example, concordance is much larger for identical twins (over 40 percent, as compared to less than 10 percent among DZ twin pairs). These data alone do not prove a polygenic basis with a threshold effect, however, for they can also be interpreted in other ways, such as one or more dominant genes that show incomplete penetrance. The different interpretations have led to disagreements about the causes of these disorders.

EXAMPLE 15.6. The diagnosis of whether twins of the same sex are monozygotic or dizygotic is not always a simple matter. Although we expect identical twins to resemble each other very closely, fraternal twins may also look very much alike. One approach that is helpful in the diagnosis of zygosity is the similarity method, in which we calculate the probability that the twins are monozygotic, based on the known frequency of MZ twins in the population and on the likelihood of their both having an observed combination of certain traits. For simply inherited, fully penetrant traits, the probability of concordance is 1 for MZ twins. In this case, discordance establishes dizygosity. The overall probability that DZ twins will be alike in all of the traits studied is typically much less than 1; the precise value depends on the number of traits being considered and on the genotypes of the parents. The calculations used in the similarity method of diagnosing zygosity are illustrated by the problem presented here. Suppose that parents of blood types AB Rh^+ × O Rh^- give birth to same-sexed twins of blood type A Rh^-. If 30 percent of all twin births are monozygotic, what is the probability that these are MZ twins?

Solution. Knowing the blood type of the twins enables us to deduce that the Rh^+ parent is heterozygous for the Rh^- recessive allele. Two possibilities exist: either the twins are monozygotic, with a probability of 30 percent, or dizygotic, having the same sex and blood type. The probability of the latter occurrence would be 0.7 × ½(same sex) × ½(same ABO blood type) × ½(same Rh type) = 0.0875. Thus, the probability that these twins are monozygotic becomes 0.3/(total) = 0.3/(0.3 + 0.0875) = 0.774. (*Note*: In this calculation, 0.3 is divided by the sum of the probabilities of both mutually exclusive events, since the sum does not equal 1.)

TO SUM UP

1. The inheritance patterns that are associated with quantitative traits can be studied conveniently by crossing homozygous parental strains that differ significantly in their average phenotypic values. The phenotypic differences that occur within each homozygous strain are due solely to environmental effects.
2. When matings between homozygous parental strains are made and the F_1 are intercrossed to produce the F_2, the average phenotypes of the F_1 and F_2 populations tend to be similar to one another in value and intermediate between the values of the parents. The variance of the F_1 is strictly environmental in origin and is about the same as the variances of the parental varieties, whereas the variance of the F_2 is significantly greater in magnitude, as it includes both genotypic and environmental sources of variation.
3. The simplest model that can be used to account for quantitative inheritance assumes that a quantitative trait is determined by the additive contributions of a large number of independently segregating gene pairs, known as polygenes. Each gene pair is composed of a contributing (or active) allele and a noncontributing (or null) allele. The active genes are assumed to have actions that are indistinguishable from one another in a statistical sense. Each active gene contributes such a small amount to the overall trait that its effect is obscured by the genotype as a whole and by the environmental variation.
4. Several factors are known to be responsible for departures from the additive polygene model. These factors include close linkage of polygenes, dominance, and nonallelic gene interactions (epistasis). We can statistically measure the effects of dominance and epistasis by describing their actions in terms of deviations between the actual genotypic values and the values that we would expect on the basis of additivity.
5. In many quantitative traits, especially those that result from the biological process of growth, polygenes appear to act in a multiplicative rather than an additive fashion, increasing the phenotypic value by the product of a constant factor instead of by the addition of a constant amount. Other quantitative traits show a threshold effect. These traits have a polygenic basis but show a discontinuous form of variation, in which individuals whose values fall below a certain threshold level have one phenotype and those whose values exceed the threshold level have another.

Genetics and Behavior Patterns

An organism's behavior, like all of its other characteristics, is determined by its genotype interacting with the environment. For example, honey bees can communicate information about sources of food by dances they perform. Although the character of the dance changes with environmental variables such as the distance and direction of the food source, the general behavior patterns are sufficiently stereotyped so that we can use them, just as we would any anatomical trait, for identifying species of bees. These behavior patterns can even serve to distinguish among the different races within a species of bees.

In the higher animals, particularly in humans, behavioral traits are among the most difficult to study. Because of the problems of providing adequate environmental control and conducting objective measurements, it is difficult to access the relative importance of heredity and environment in behavioral traits. Precise genetic analyses are also complicated by the fact that the genes involved in producing such a trait can act on a variety of different body structures. Each organism is affected by stimuli that act on receptor organs both inside and outside of the body. The information received from these sensory structures is then processed by the nervous and endocrine systems. These systems, in turn, activate a variety of effector organs (muscles, glands, and so on), which enable the organism to meet the challenges of its environment. In this process, the genes may act to (1) alter the information input by affecting the sensory systems of the organisms, (2) modify the way in which the sensory information is processed by acting on the nervous and endocrine systems, or (3) change the response by acting on the effector systems.

One of the greatest conceptual problems we encounter when we are dealing with the development of behavioral traits is understanding the roles played by inheritance and learning. We can depict their actions in much the same way as we did the joint effects of genes and the environment on the development of a nonbehavioral characteristic. If we let P_t represent the phenotype at one stage of development and P_{t+1} represent the phenotype in the following stage, the actions of the genotype (G_t) and the environment (E_t) on growth and differentiation during the intervening period can be written as follows:

$$P_t + G_t + E_t \longrightarrow P_{t+1}.$$

All of these factors must be included in the basic equation, since every trait must develop in an environment, just as it must develop in the presence of genes. Thus, no phenotype—behavioral or otherwise—can be attributed solely to genetic factors or solely to environmental factors; rather, the phenotype is a result of the accumulated interactions of these two forces during development.

It is technically wrong to designate a form of behavior as being either strictly innate (genetically determined) or learned (environmental in origin). In theory, however, it is possible for the *differences* in behavior among individuals or the *differences* in behavior in the same individual at different stages of development to be due entirely to either genetic or environmental causes. Usually both types of differences are present. For example, the fact that one individual can speak a foreign language while another of the same age and educational background cannot may simply be attributed to learning. But the fact that an individual can speak the language at one age but not at a much earlier stage of development may involve a complex mixture of environmental and genetic determinants. In this case, the difference would depend on both learning and maturation during the developmental process.

Genetic Analysis of Behavioral Traits

There are basically three approaches used by researchers to evaluate the relative importance of heredity in determining behavioral differences. One aproach involves the use of *crossing experiments,* in which crosses are made between highly inbred (and presumably homozygous) lines or strains that differ significantly in some form of behavior. As we described previously, when these crosses are carried through to the F_2, they can sometimes provide information concerning the number of genes involved in the phenotypic differences and their degree of dominance.

A second approach uses *selection experiments,* in which a trait is modified in opposite directions by selective breeding. As illustrated in Fig. 15.13, this approach involves selection against the intermediate types and for the extreme homozygous genotypes. If the behavioral differences are at least partly hereditary, both "high" and "low" lines should be established among the progeny, and these lines should continue to diverge in average character during the subsequent generations of selection.

The third approach is that of *twin studies,* in which comparisons are made between monozygotic and dizygotic twins. As we indicated previously, twin analysis provides data that can be used in the calculation of heritability estimates. Since controlled genetic crosses and selection experiments are clearly inappropriate for human studies, this approach is the method of choice in the genetic analysis of human behavior.

The most convincing studies on the role of heredity in determining behavior have involved differences that are attributable to only a few gene pairs. One clear-cut example concerned the studies of W.C. Rothenbuhler on nest-cleaning behavior in honeybees. The Brown strain of honeybees (hygienic strain) is resistant to an infection known as American foulbrood, which is caused by the bacterium *Bacillus larvae.* Resistance is due to the cleaning behavior of worker bees, who uncap the cells in the hive that contain dead pupae and remove the dead before the disease can spread to other compartments. The Van Scoy strain (nonhygienic strain) is not resistant; the bees in this strain do not open the compartments, so that the dead pupae remain in their cells indefinitely. When the two strains were crossed, the F_1 bees produced were nonhygienic. When the F_1 drones were backcrossed to queen bees of the recessive Brown line, four kinds of offspring were formed in about equal proportions: (1) hygienic bees, (2) uncappers only, who opened cells but did not remove dead pupae, (3) removers only, who removed dead pupae after the cells were uncapped by humans, and (4) nonhygienic bees, who neither uncapped cells nor removed dead pupae. These behavioral differences are consistent with the hypothesis of two unlinked gene pairs: one involving the recessive

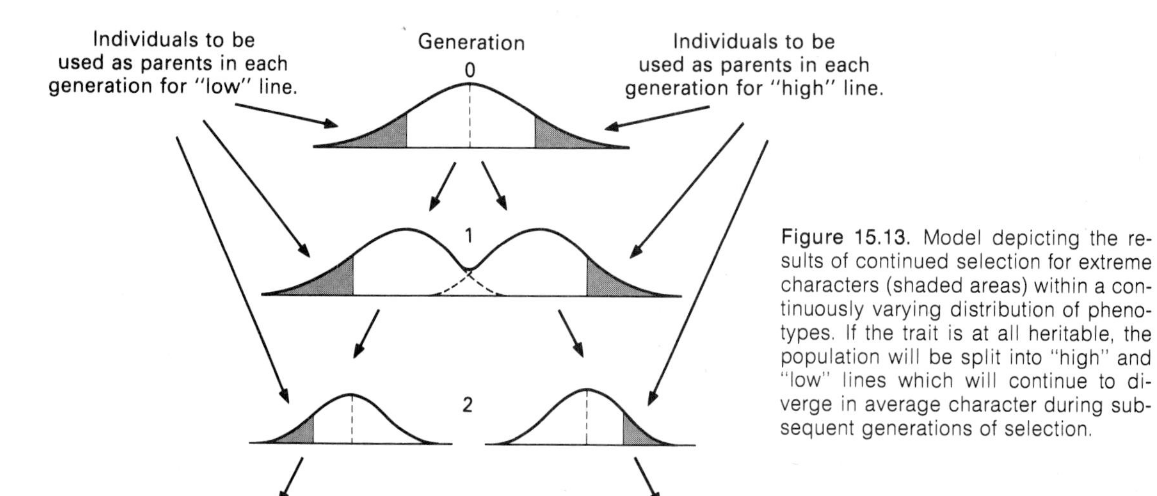

Figure 15.13. Model depicting the results of continued selection for extreme characters (shaded areas) within a continuously varying distribution of phenotypes. If the trait is at all heritable, the population will be split into "high" and "low" lines which will continue to diverge in average character during subsequent generations of selection.

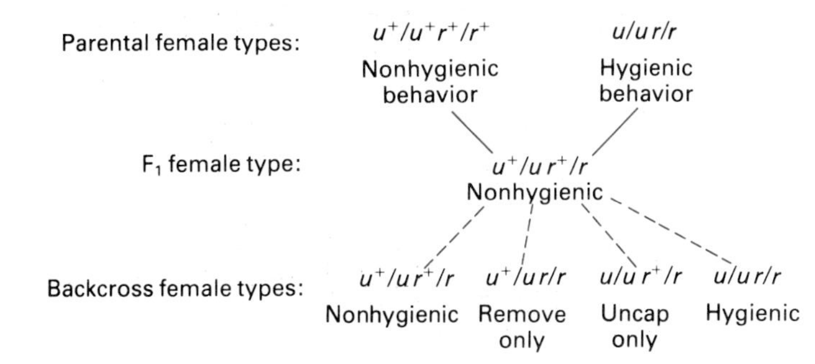

Figure 15.14. Genetics of hygienic behavior in honeybees, as illustrated by the female genotypes of the F_1 and backcross progeny. The behavioral differences are consistent with the hypothesis of two unlinked gene pairs: one involving the recessive allele *u* for uncapping of cells and another involving the recessive allele *r* for removal of dead larvae.

allele *u* for uncapping and another involving the recessive allele *r* for removal. The backcross, which is diagrammed in Fig. 15.14, is equivalent genetically to a dihybrid testcross, which should give rise to the observed results.

While studies on traits that have a simple inheritance pattern provide the most convincing examples of the role of genes in determining behavior, most behavioral characteristics involve many genes, each of which has only a small individual effect. These traits must therefore be studied by the methods used in quantitative genetics. In this section, we will provide two examples that serve to illustrate this approach to behavioral genetic analysis.

Geotaxis in *Drosophila*. Fruit flies vary in their response to gravity. Some show positive geotaxis (with a tendency to move downward when placed in a vertical tube), while others show negative geotaxis (moving upward against gravity). The geotactic response of flies can be studied using a vertical maze that contains a series of vertical alleys linked together by T openings, similar to the maze shown in Fig. 15.15. Each fly enters the apparatus through the initial T opening, attracted

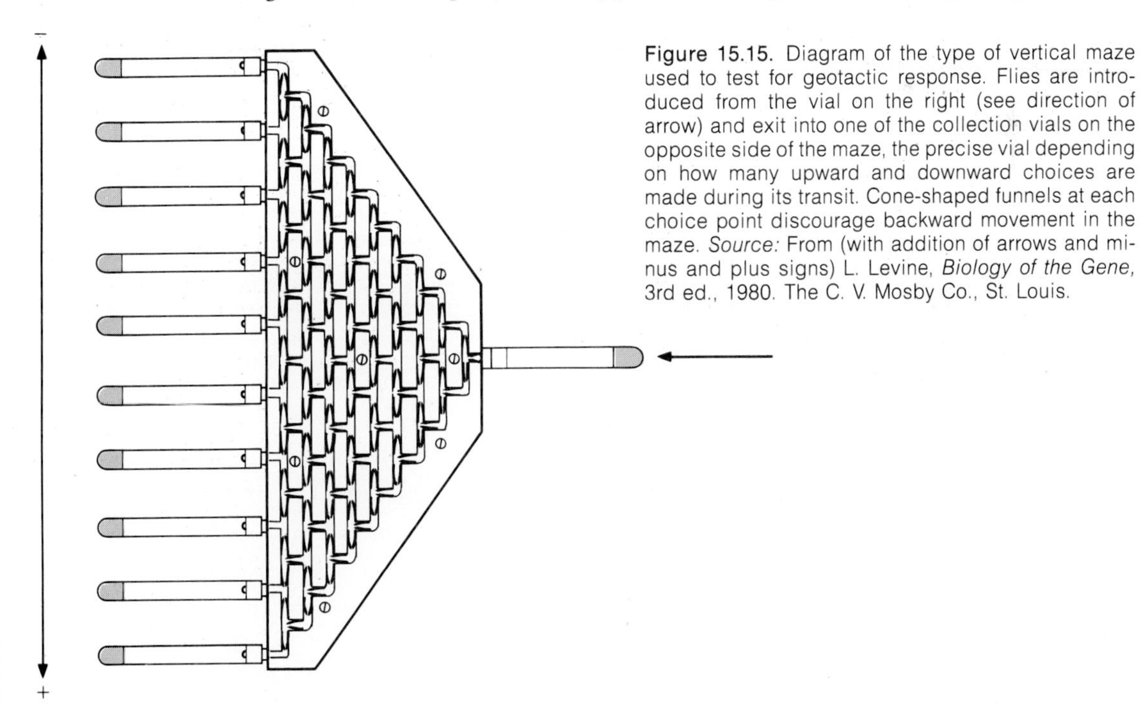

Figure 15.15. Diagram of the type of vertical maze used to test for geotactic response. Flies are introduced from the vial on the right (see direction of arrow) and exit into one of the collection vials on the opposite side of the maze, the precise vial depending on how many upward and downward choices are made during its transit. Cone-shaped funnels at each choice point discourage backward movement in the maze. *Source:* From (with addition of arrows and minus and plus signs) L. Levine, *Biology of the Gene*, 3rd ed., 1980. The C. V. Mosby Co., St. Louis.

by light, and eventually reaches one of the collecting vials on the opposite side of the maze, the precise vial depending on how many upward and downward choices are made during its transit. A positively geotactic fly will eventually enter one of the lower vials, whereas a negatively geotactic fly will tend to arrive at one of the upper vials.

In one study published by Erlenmeyer-Kimling and colleagues, flies were selected for both positive and negative geotaxis. The selection was accomplished by mating the flies in the lower vials to one another for a positive geotactic response and by mating those in the upper vials together for the reverse response. Selection then proceeded over many generations by using the most positively geotactic flies in the low line and the most negatively geotactic flies in the high line as parents at each reproductive interval.

The average geotactic responses of both selected populations during 65 generations of repeated selection are shown in Fig. 15.16. Note that both the upper and lower lines continue to diverge for several generations, which indicates that a strong genetic component is involved in geotaxis. Starting at generation 25, Hirsch and Erlenmeyer-Kimling crossed samples of flies from the original unselected stock and from each of the selected lines to a special tester strain that carried dominant gene markers on the X chromosome and on chromosomes II and III, as well as inversions that served to maintain the integrity of these chromosomes by suppressing crossing over (see Chapter 12). Female flies in the F_1 progeny that were heterozygous for the dominant markers Bar eyes (*B*) Curly wings (*Cy*), and Stubble bristles (*Sb*) were then backcrossed to males from the samples to be tested.

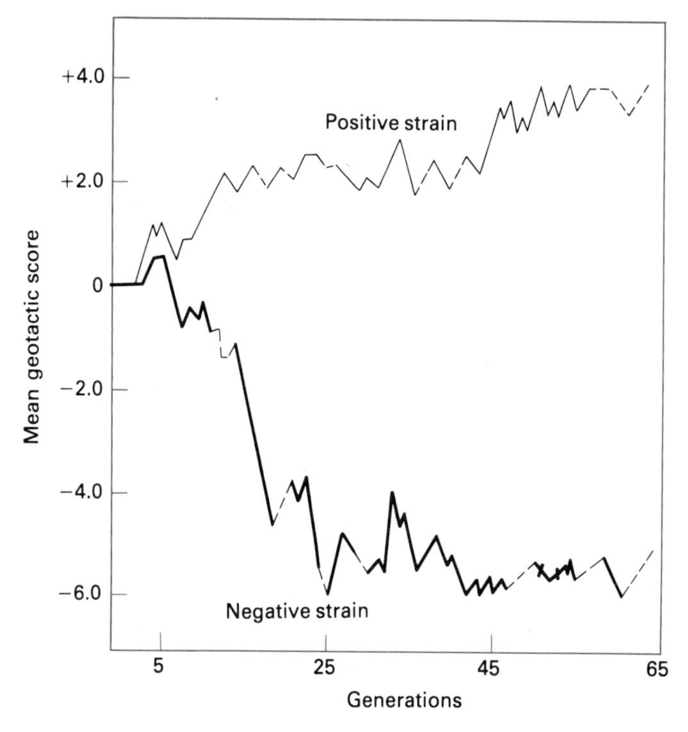

Figure 15.16. Response to selection for negative and positive geotaxis in *Drosophila* during a period of 65 generations. In the selection for negative geotaxis, only males and females from the upper tubes were used as parents for each and every generation, while in the selection for positive geotaxis, only males and females from the lower tubes were chosen as parents. The mean geotactic score, which ranges from −6 to +4, represents an arbitrary measurement of geotactic response. *Source:* From L. Erlenmeyer-Kimling, J. H. Hirsch, and J. M. Weiss, *J. Comparative and Physiological Psychology* 55:722, 1962. Copyright 1962 by the American Psychological Association. Reprinted by permission of the publisher and author.

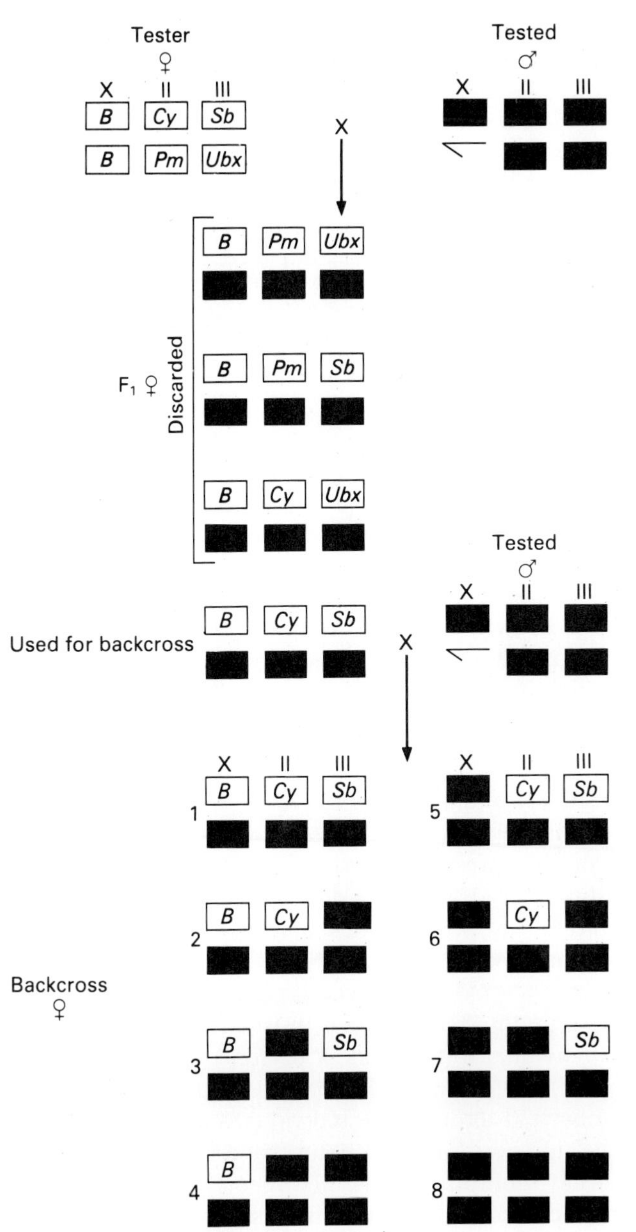

Figure 15.17. Mating design used for chromosomal analysis of geotaxis in *Drosophila*. Flies are initially mated to a tester strain which carries dominant marker genes on chromosomes X, II, and III, as well as inversions to suppress crossing over. The backcross progeny are scored as to geotactic response and are used to evaluate the contribution of each of the three chromosomes to positive or negative geotaxis. *Source:* From L. Erlenmeyer-Kimling, J. H. Hirsch and J. M. Weiss, *J. Comparative and Physiological Psychology* 55:722, 1962. Copyright 1962 by the American Psychological Association. Reprinted by permission of the publisher and author.

This cross yielded eight (2^3) chromosome types among the backcross progeny. The backcross results are shown in Fig. 15.17, along with the rest of the crossing sequence.

The geotactic responses of each type of heterozygote in the backcross progeny were then compared with the responses of the corresponding homozygous types. These comparisons enabled the researchers to determine the cumulative effects of the genes located on each of the three major chromosomes on the geotaxis traits.

Table 15.4. Net contribution to positive (+) and negative (−) geotactic score by genes on the three major chromosomes of *Drosophila melanogaster.*

GROUP STUDIED	CHROMOSOMES X	II	III
Positive strain	+1.39	+1.81	+0.12
Unselected population	+1.03	+1.74	−0.29
Negative strain	+0.47	+0.33	−1.08

Source: From L. Erlenmeyer-Kimling, J.H. Hirsch, and J.M. Weiss, *J. Comparative and Physiological Psychology* 55: 732-739, 1962.

The results of these comparisons are listed in Table 15.4. It was found that the X chromosome and chromosome II contribute to positive geotaxis and that chromosome III contributes to negative geotaxis in the unselected stock. Selection for a negative geotactic response results in marked reductions in the contributions of the X chromosome and chromosome II to positive geotaxis and an enhancement of the contribution of chromosome III to negative geotaxis. In contrast, selection for a positive geotactic response produces little change in the effects of the X chromosome and chromosome II, while the contribution of chromosome III changes from negative to positive. Although the researchers were unable to determine the number of genes involved and their individual effects, the fact that genes present on all three of the major chromosomes contributed to geotaxis in *Drosophila* clearly demonstrated a polygenic basis for this behavioral trait.

The Inheritance of IQ. The IQ (or intelligence quotient) is a quantitative measure of intelligent behavior that is closely correlated with achievement in school and academic success in general. It represents the ratio of an individual's mental age to chronological age, multiplied by 100, and is determined by a standardized testing procedure. Studies indicate that the IQ is actually a composite score that measures the average of a number of primary abilities, such as verbal ability, reasoning ability, and the abilities to memorize and visualize objects in space. Since primary abilities are not necessarily interrelated, their distribution of relative importance may vary from one individual to the next. A person's IQ is thus clearly a multidimensional character, which includes several somewhat independent abilities; ideally, it should be expressed in terms of the individual's profile of primary abilities rather than by a single number. A case in point is the lowered IQ that is sometimes associated with Turner syndrome. While first reports suggested a possible slight mental retardation, later studies on affected individuals revealed normal to superior verbal IQ. The reduction in overall score was largely caused by the difficulty that some of these individuals have with a form of space perception that requires right–left orientation.

When the total IQ scores are plotted in the form of a frequency distribution, they tend to follow the bell-shaped pattern that is expected for a normal curve (Figure 15.18). The distribution has a mean score of 100 (for whites) and a standard deviation of approximately 15. About 95 percent of the population are therefore included within the range of IQ scores of $100 \pm 2(15)$, or 70 to 130. As

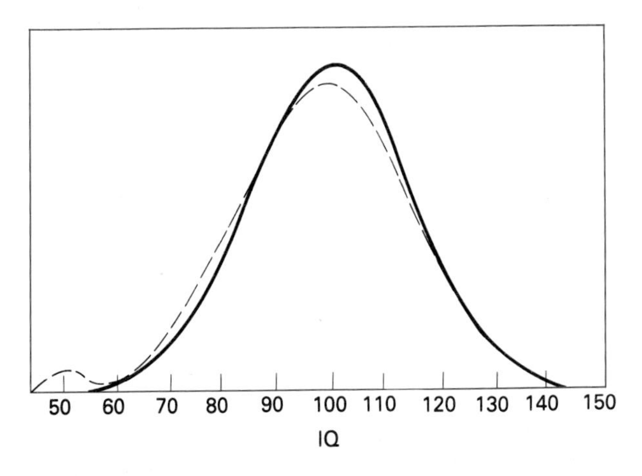

Figure 15.18. Distribution of IQ scores among whites (broken line). The distribution of observed scores is compared with a theoretical normal curve (solid line). Note that the observed distribution has a bulge on the low side.

we have seen, frequency distributions of this general shape are often observed for quantitative traits and while such curves cannot be taken by themselves as evidence for a particular genetic mechanism, they are suggestive of a polygenic mode of inheritance. There is some departure from the normal curve, however, with the extreme low-score end of the IQ distribution being larger than expected. This departure has been interpreted to mean that many individuals who are classified as being mentally retarded have values that lie outside of the normal range of IQ values that result from multiple gene inheritance. These extreme low-score values are possibly a result of single gene effects (e.g., phenylketonuria), chromosome aberrations (e.g., Down syndrome), and developmental accidents rather than the normal random segregation of polygenes.

Evidence for a genetic basis of variation in IQ scores has come from heritability (H_B) estimates, which range in value from about 0.6 to 0.8, and from comparisons of IQ measurements in various groups of individuals reared together (unrelated individuals, parent and child, siblings, MZ twins, and DZ twins) and reared separately (unrelated persons, siblings, and MZ twins). Several of these IQ comparisons are summarized in Fig. 15.19 by means of correlation coefficients. Each correlation coefficient measures the degree of correspondence in IQ scores between the persons concerned. The correlation coefficient is equal to 1 if the trait in question (IQ in this case) is the same for the members of all pairs of individuals or if the phenotypic value of one member of a pair always varies in direct proportion to the phenotypic value of the other. The coefficient is zero if the phenotypic values vary independently in all pairs of individuals. The expected values are the phenotypic values that are predicted for strictly additive gene effects. The genetic correlation between each pair of relatives is then equal to the proportion of segregating genes that the members of the pair have in common. For instance, first-degree relatives (siblings or a parent and child) are, on the average, identical for one-half of their genes. Thus, if the environmental influences are nonexistent, siblings are expected to show a correlation coefficient of 0.5 regardless of whether they are reared together or apart. Monozygotic twins are expected to show a correlation coefficient of 1, with members of each twin pair having identical phenotypic values. No correlation is expected between unrelated persons. While

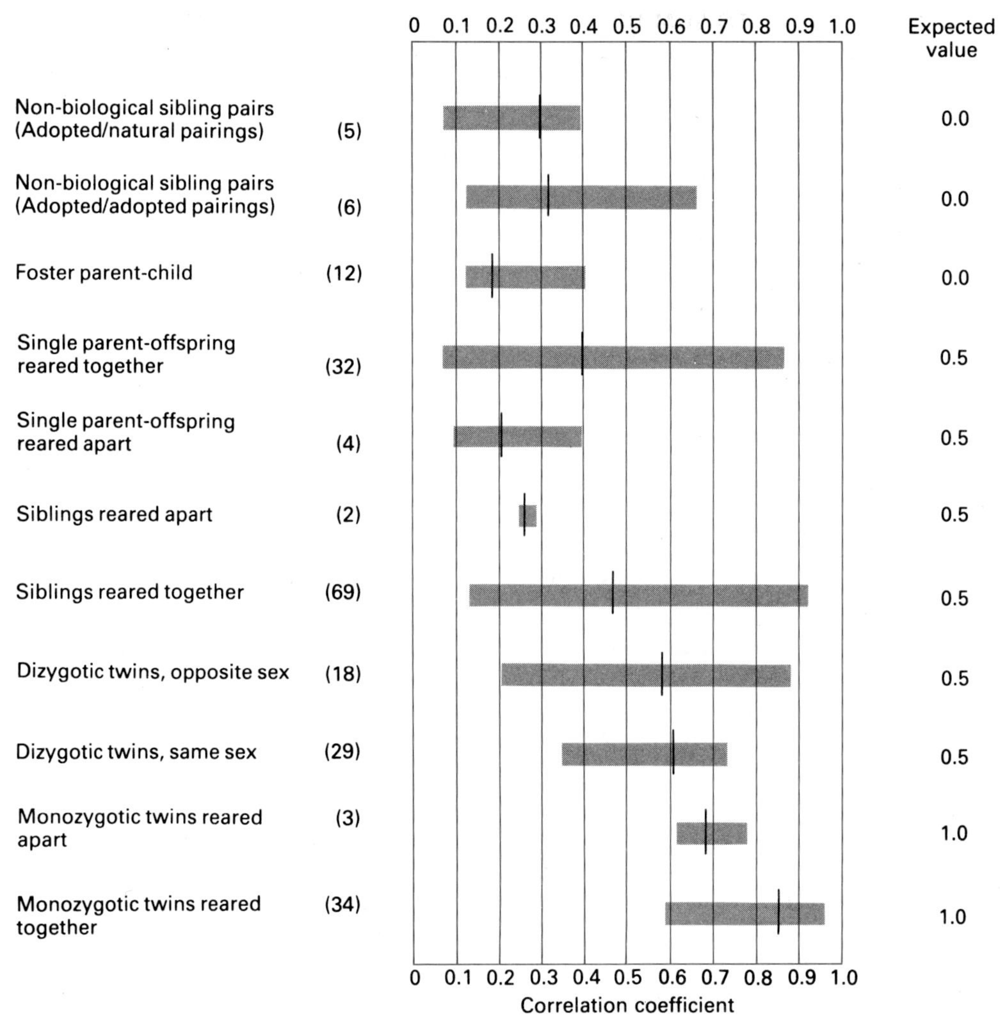

Figure 15.19. Summary of correlation coefficients on a variety of IQ tests between individuals with different degrees of genetic and environmental relationship. The bar represents the range of correlation in each case, and the vertical line designates the median correlation coefficient. The number of studies included in each sample is given in parentheses. *Source:* After T. J. Bouchard, Jr., and M. McGue, *Science* 212:1055–1059, 1981. Copyright 1981 by the American Association for the Advancement of Science.

the correspondence with the expected values is good in the case of siblings, significant departures occur among MZ twins and unrelated persons. The very high correlation in test scores between MZ twins supports the view that there is a high degree of genetic determination for IQ. But note that rearing apart reduces the resemblance of MZ twins on this trait. This finding and the fact that raising children together in the same orphanage or foster home significantly increases the correlation between unrelated persons provide evidence for a substantial influence of home environment on IQ score.

TO SUM UP

1. The behavior of an organism is determined by its genotype interacting with the environment. The genes for behavioral traits are among the most difficult to study since they can act in a complex manner on a number of different body structures. These structures are involved in the detection of environmental stimuli and the subsequent processing of sensory information.
2. Three important sources of information are used for the genetic analysis of behavioral traits: controlled genetic crosses, selection experiments, and twin studies. These sources provide useful data concerning the role of genes in behavior and the relative importance of heredity in determining behavioral differences.
3. Traits whose inheritance pattern is simple (e.g., nest-cleaning behavior in honeybees) provide the most clearcut evidence of the role of genes in behavior. But most behavioral traits involve many gene pairs and must therefore be studied by the methods of quantitative genetics. For example, the geotactic responses of fruit flies are influenced by genes on three major chromosomes, some of which contribute to negative geotaxis while others contribute to positive geotaxis.
4. The IQ (or intelligence quotient) is a quantitative measure of human intelligence. The correlations in IQ between pairs of individuals who are related to varying degrees support the view that there is a high degree of genetic determination for this trait. But because of the complexity of the trait and the problems of providing adequate environmental control, no definite statements can be made at present concerning the relative importance of heredity and environment in contributing to differences in IQ.

Questions and Problems

1. Define the following terms:
   heritability
   phenotypic mean
   polygenes
   threshold characters
   transgressive variation
   variance, standard deviation
2. Distinguish between the members of the following paired terms:
   (a) Additive vs. multiplicative gene effects
   (b) Concordant vs. discordant twins
   (c) Continuous vs. discrete variation
   (d) Dominance vs. interlocus interactions
   (e) Genetic vs. environmental variance
   (f) Monozygotic vs. dizygotic twins
   (g) Quantitative vs. qualitative traits
3. In what respects are the genes that are responsible for quantitative characters different from the traditional Mendelian genes that specify qualitative traits? Is the difference fundamental or only one of degree? Explain.
4. Indicate whether the following factors would serve to increase or decrease the heritability of a quantitative trait: **(a)** An increase in the homozygosity of the relevant genes. **(b)** A reduction in the environmental variability.
5. Twin concordance has been determined for a number of different traits that are suspected of having a genetic basis. Three examples are listed below:

TRAIT	CONCORDANCE FREQUENCY MZ TWINS	DZ TWINS
Arterial hypertension	25.0	6.6
Death from acute infection	7.9	8.8
Diabetes mellitus	47.0	9.7

   (a) Which of these traits gives evidence for the greatest degree of genetic determination? Which gives evidence for the least degree of genetic determination? Explain. **(b)** Why are the concordance frequencies for MZ twins significantly less than 100 percent in every case?
6. What proportion of the F_2 should be homozygous for all of the contributing genes if the homozygous parental strains differ in genotype at **(a)** 2, **(b)** 3, **(c)** 4, **(d)** n gene loci?

7. What proportion of the F_2 should resemble a parental strain that is homozygous for completely dominant alleles at each of **(a)** 2, **(b)** 3, **(c)** 4, **(d)** n gene loci?
8. Guernsey cows produce milk with about 30 percent higher butterfat content than Holsteins. From crosses between Guernseys and Holsteins, none of the 2000 offspring in the F_2 had a butterfat percentage as high as the Guernseys. What conclusions can be drawn concerning the minimum number of genes that are involved in determining the difference in butterfat percentage in these breeds of cattle?
9. The number of leaves per stem in a particular plant is known to be a quantitative trait. Suppose that a cross between plants that each have 14 leaves per stem yields offspring in the following ratio: 1 with 10 leaves : 8 with 11 leaves : 28 with 12 leaves : 56 with 13 leaves : 70 with 14 leaves : 56 with 15 leaves : 28 with 16 leaves : 8 with 17 leaves : 1 with 18 leaves. Using the first letter of the alphabet to designate the first gene pair and as many more letters as necessary, give the genotypes of the parents of this offspring distribution.
10. Among 20,000 pregnancies in Sweden, there were approximately 240 twin births, of which there were 75 male pairs, 75 female pairs, and 90 pairs of mixed sex. **(a)** Calculate the percentages of MZ twins and DZ twins among all twin births in Sweden. **(b)** Determine the numbers of MZ twins and DZ twins born in Sweden per 1000 pregnancies.
11. In a recent study on twins of whom at least one member had the mental disorder schizophrenia, there were 11 concordant and 11 discordant MZ twin pairs and 3 concordant and 30 discordant DZ twin pairs. Calculate the concordance frequencies for schizophrenia among MZ and DZ twin pairs in this study.
12. Emerson and East found the F_1 from a cross between Black Mexican corn and a variety of popcorn to have the following ear lengths:

Ear length (cm):	9	10	11	12
Number of ears:	1	12	12	14
Ear length (cm):	13	14	15	
Number of ears:	17	9	4	

(Total = 69 ears)

(a) Calculate the mean and variance of ear length in this sample. What is the probable cause of variation? **(b)** If ear length is assumed to follow a normal distribution, within what range of values can we expect 95.4 percent of the measurements to fall?

13. In Problem 16 of Chapter 12, we learned that three pairs of independently assorting genes, (R_1 and r_1, R_2 and r_2, and R_3 and r_3) are responsible for kernel color in wheat. Kernels are white in the $r_1r_1r_2r_2r_3r_3$ strain and red in all other strains. It was not mentioned in Chapter 12 that active R genes contribute to red coloration in an additive fashion, so that kernel color in wheat forms an almost continuous gradation from very light red in strains with only one active gene to very dark red in the completely homozygous $R_1R_1R_2R_2R_3R_3$ strain. Suppose that we make the cross $R_1R_1r_2r_2R_3R_3 \times r_1r_1R_2R_2r_3r_3$ and carry it into the F_2 by intercrossing the F_1. Ignore the effects of environmental variation in answering the following questions regarding the results of this cross: **(a)** How would the kernel color of the F_1 compare to that of each of the parents? **(b)** Compare the mean of the F_1 with the mean of the F_2. **(c)** What proportion of the F_2 will have kernels that are white? **(d)** What proportion of the F_2 will have kernels of the same color as the $R_1R_1r_2r_2R_3R_3$ parental strain? **(e)** What proportion of the F_2 should breed true for the kernel color of the $R_1R_1r_2r_2R_3R_3$ strain? **(f)** What proportion of the F_2 will have kernels of the same color as the F_1? **(g)** What proportion of the F_2 should breed true for the kernel color of the F_1? **(h)** If it were possible to distinguish between the color produced by different numbers of contributing genes, how many classes of kernel color would be expected in the F_2?
14. Although the inheritance of eye color in humans is complex and incompletely understood, at least seven different eye colors can be identified: light blue, blue, blue-green, hazel, light brown, brown, and dark brown. **(a)** Assuming that eye color is a quantitative trait, propose genotypes for these classes of eye color. **(b)** Use your genetic model in part (a) to predict the distribution of eye color that you would observe among the offspring of hazel-eyed couples, where both the husband and wife are heterozygous for all of the contributing genes.
15. In certain breeds of cattle, solid color is due to

the dominant gene *S*, while white spotting is caused by its recessive allele *s*. The degree of spotting appears to be a quantitative trait, with *ss* animals ranging from those of almost-solid color, with just a few white patches, to those that are almost white, with just a trace of color on a white background. Suppose that crosses are made between *Ss* cattle that are also heterozygous for polygenes that modify the degree of spotting. What fraction of the offspring is expected to be almost white if there are **(a)** 2, **(b)** 3, **(c)** 4, **(d)** *n* pairs of polygenes that affect spotting in these crosses?

16. In wheat, the number of days that elapse between planting and the time that heads of grain appear (known as the heading date) is a quantitative trait. Crosses between two true-breeding varieties of wheat, with heading dates of 56 and 72 days, respectively, produced a quite uniform F_1, with an average heading date of 64 days. An $F_1 \times F_1$ cross resulted in a highly variable F_2, with heading dates that varied symmetrically between the extremes of the early- and late-heading true-breeding parental strains. Among 3000 F_2 plants examined, 12 had a heading date of 56 days. Estimate the number of gene loci that are involved in determining the heading date, and calculate the average contribution of each active allele at these loci.

17. Two other true-breeding strains of wheat were crossed, with heading dates of 60 and 68 days, respectively. The average heading date of the F_1 produced was again 64 days, as was the average of the F_2. In this case, however, the F_2 distribution extended beyond the extremes in heading date that were exhibited by either of the true-breeding parental strains. Out of 1000 F_2 plants, three had a heading date of 56 days, five had a heading date of 72 days, and the others had heading dates between these extremes. What genotypes are possible for the true-breeding parental plants in this series of crosses? (Start with the letter a for the first gene locus, and use as many more letters as necessary.)

18. The average weight of a mature male of the Bantam breed of chickens is 1.4 lb, while that of the Plymouth Rock male is 6.6 lb. Crosses between the Bantam and Plymouth Rock breeds produce an F_1 with a mean weight of 3.4 lb and a variance of 0.3 lb^2. The F_2 has a mean weight of 3.6 lb and a variance of 1.2 lb^2. **(a)** Ignoring dominance and interlocus interactions, estimate the number of gene loci involved in determining the difference in weight between these breeds of chickens. **(b)** Calculate the heritability in the broad sense (H_B) for weight in these chickens.

19. In a cross between two diverse types of corn, Emerson and East obtained the distribution of earlength for the parental strains, F_1, and F_2 as shown at the bottom of this page. **(a)** Calculate the mean and variance of each distribution. **(b)** Using Eq. (14.5), estimate the number of loci that determine ear length in these crosses. **(c)** How does the value calculated in part (b) compare with the number you would get if you based your estimate on the fraction of the F_2 with the longest (or shortest) length? Explain the reason for any difference.

20. Suppose that variation in height in a particular plant is determined by three gene pairs (*A* and *a*, *B* and *b*, and *C* and *c*) that exhibit complete dominance. Assume that each active gene, in either the homozygous or heterozygous state, contributes 4 inches of height to the base height of 10 inches. If we ignore variation owing to the environment, an *AABBCC* plant would then be 22 inches high and an *aabbcc* plant would be 10 inches. The cross *AABBCC* × *aabbcc* is made and is carried into the F_2 by intercrossing the F_1. **(a)** What is the height of the F_1? **(b)** Determine the height distribution in the F_2. **(c)** Calculate the mean and variance of height in the F_2. How do they compare in value to those of the F_1?

21. Reconsider Problem 20, but now assume that interlocus interactions of the following type occur. Suppose that the *cc* gene pair exerts a mask-

EAR LENGTH (CM)

	5	6	7	8	9	10	11	12	13	14	15	16	17	18	19	20	21
Parental strains P60:	4	21	24	8													
Parental strains P54:						3	11	12	15	26	15	10	7	2			
F_1:						1	12	12	14	17	9	4					
F_2:			1	10	19	26	47	73	68	68	39	25	15	9	1		

ing effect over the B gene, so that whenever BB or Bb occurs together with cc, the plant will exhibit the same height as a plant of the $bbcc$ genotype. A plant of genotype $AABBcc$ would then be 14 inches high, as would a plant of genotype $AAbbcc$. Any C- plant will respond as it did in Problem 20. Again, the cross $AABBCC \times aabbcc$ is made and is carried into the F_2. **(a)** Determine the height distribution in the F_2. **(b)** Calculate the mean and variance of height in the F_2. How do these values compare with the corresponding values computed in Problem 20?

22. Scott and Fuller have studied the mode of inheritance of various behavioral differences among breeds of dogs. In one experiment, they attempted to measure the inheritance pattern for barking tendency. They used the African Basenji breed, with a high threshold of stimulation, and the American Cocker Spaniel breed, with a low threshold. The tendencies to bark were assessed by measuring the percentage of dogs who barked in 10-minute test periods while pairs of litter mates were allowed to compete for a bone. The results obtained for the Basenji and Cocker Spaniel breeds and for their F_1 and F_2 offspring are presented below.

	% OF DOGS BARKING IN A 10-MINUTE PERIOD
Basenji	19.6
Cocker Spaniel	68.2
F_1	60.1
F_2	55.5

Compare the F_1 and F_2 results with what is expected if the lower threshold (greater tendency to bark) of the Cocker Spaniel is inherited as the expression of a dominant gene at a single locus.

23. The average fruit weight of the Red Currant tomato is 1 g, while that of the Putnam's Forked variety is 58 g. When crossed, the two varieties produce an F_1 that has an average fruit weight of only 7.6 g, far less than the arithmetic average of the two parental strains. One plausible explanation for the low average of the F_1 is to assume that, say, 10 gene pairs are determining fruit weight, with each active gene exerting a geometric effect when present by multiplying the value of the residual genotype by a constant amount of 1.5 g. Thus, genotypes $aabbccddee$, $AaBbCcDdEe$, and $AABBCCDDEE$ would have phenotypic values of 1, $(1.5)^5 = 7.6$, and $(1.5)^{10} = 58$ g, respectively. Employing the above model, predict the distribution of fruit weight among the F_2 of a cross between the Red Currant and Putnam's Forked strains.

24. Twin boys were born who had the same blood type with regard to the ABO, Rh, and MN blood groups. The twins were A Rh^+ MN, while the father and mother of the twins were O Rh^- M and AB Rh^+ MN, respectively. Blood tests also disclosed that the maternal grandmother of the twins was A Rh^- M. Given that MZ twins make up 30 percent of all twin births, calculate the probability that these twins are monozygotic.

25. The heritability (H_B) of a quantitative trait is measured in two populations, A and B, in each of two different environments, x and y. Assume that population A has twice the genotypic variation for the trait in question as does population B in both environments. Also assume that environment x, as measured by the total phenotypic response of the two populations, is three times as variable as environment y. **(a)** Arrange the four population–environment combinations (A in x, A in y, B in x, and B in y) in order of increasing heritability. (Ignore genotype–environment interaction.) **(b)** Repeat part (a) of this problem, but now assume that genotype–environment interaction occurs such that the variability of environment x, as measured by the phenotypic response of population B, is the same as the variability of environment y. Population A still responds as if the variability of environment x were three times that of environment y. Has the relative order of heritability estimates changed? If so, in what way?

26. Consider a single locus with two alleles, A and a, that occur at frequencies of p and q, respectively, in a random mating population. Assume that the genotypes at this locus contribute to a quantitative trait in the following manner:

Genotype:	AA	Aa	aa
Frequency:	p^2	$2pq$	q^2
Contribution:	2	1	0

Using the definitions of the mean and variance given in the text, show that the mean and variance of the phenotypic contribution of this locus are $2p$ and $2pq$, respectively. (Note that both the mean and the variance of a trait in a population depend on the allele frequency.)

Chapter 16

Mating Systems and Selective Breeding

The only mating pattern that we have dealt with so far has been random mating. But individuals do not always interbreed at random; they may select their mates on the basis of phenotype. Even when mating preferences are not evident, the dispersal ability of organisms is often quite limited, so that individuals tend to be restricted to matings with others close at hand. This restriction can result in inbreeding, in which the mates have a higher degree of genetic relatedness than they would if they were selected strictly at random from anywhere within the population. Such nonrandom mating patterns by themselves do not alter gene frequencies and should not be regarded as evolutionary forces. Only the genotype frequencies are theoretically changed. Nonetheless, if a difference in fertility—either natural or artificially imposed—is associated with one or more of the genotypes, nonrandom mating can greatly accelerate the process of selection. Plant and animal breeders have successfully used this fact to develop new and more productive strains of plants and animals and to adapt existing types to a variety of conditions.

The sections that follow cover a broad range of topics dealing with nonrandom mating and artificial selection. We will start with a discussion of inbreeding and its genetic consequences, along with a brief consideration of matings that are based on phenotypic similarity. We will then examine outcrossing and hybrid vigor, and finish with a description of the types and effects of selective mating.

Mating Systems

The mating system of a population is the pattern by which mating occurs. The basic types of mating patterns are outlined as follows:

A. **Random Mating:** No correlation between the genotypes (or phenotypes) of mates.

B. **Nonrandom Mating:** Positive or negative correlation between the genotypes (or phenotypes) of mates.

   1. Positive correlation: Mates tend to be more closely related **(inbreeding)** or more similar in phenotype **(assortative mating)** than randomly chosen individuals.
   2. Negative correlation: Mates tend to be less closely related **(outbreeding)** or less similar in phenotype **(disassortative mating)** than randomly chosen individuals.

Each form of nonrandom mating can vary in the degree to which it departs from randomness, depending on the population and, in certain cases, on the trait being studied. Nonrandom mating is said to be complete when all mates are either identical in phenotype (or genotype) or the reverse.

As a general rule, the two alternative forms of nonrandom mating tend to have opposite genetic effects. Inbreeding and assortative mating, which are characterized by a positive correlation between mates, act to increase the frequency of homozygous genotypes present in a population above the frequency that is expected for random mating and to reduce the frequency of heterozygotes. In contrast, outbreeding and disassortative mating serve to maintain heterozygosity within the breeding population.

Inbreeding and Degree of Relationship

Inbreeding is said to occur when offspring are produced by genetically related parents. Matings between close relatives are called **consanguineous matings.** Inbreeding can result either from the systematic choice of relatives as mates or through a reduction in population size to a level at which randomly selected mates have a high probability of having common ancestry; the consequences are similar in either case.

The Inbreeding Coefficient. Several measures of inbreeding intensity have been proposed. One that is commonly used by geneticists is the **inbreeding coefficient** (designated as F), which was first defined by the population geneticist Sewall Wright. The inbreeding coefficient is the probability that two genes at a given locus are identical by descent. Stated in another way, it is the probability that the two alleles are replicates of the same allelic gene in a common ancestor.

We can further clarify the definition of F by examining the pedigree of a mating between half-siblings, shown in Fig. 16.1. Assume that none of the grandparents of offspring I in the pedigree are related and that the common ancestor, which is the grandfather in this case, is not inbred. Offspring I can have two possible genotypes in which the alleles are replicates of the same gene in the grandfather. They are A_3A_3 and A_4A_4, where the subscripts are used to differentiate between alleles. Genotypes in which the alleles are identical by descent are called **autozygous** genotypes. All of the other genotypes that are possible in this case (A_1A_3, A_2A_4, A_3A_5, etc.) are said to be **allozygous** genotypes. Unlike autozygotes, who must be homozygous, allozygotes can be either homozygous or heterozygous. A homozygous allozygote would possess alleles that are identical in function but are independent in origin. Thus, F is the probability that the individual is autozygous at the locus in question.

The computation of F is quite simple. In the case of our pedigree, the inbreeding coefficient of individual I is

$$F_{\mathrm{I}} = P(A_3A_3 \text{ or } A_4A_4) = P(A_3A_3) + P(A_4A_4).$$

The probability of getting two A_3 alleles from the grandfather, designated $P(A_3A_3)$, is equal to the probability that both parents are heterozygous for the A_3 allele (computed numerically as $1/2 \times 1/2 = 1/4$), times the probability that they both transmit the A_3 allele to their offspring (also $1/2 \times 1/2 = 1/4$). The total probability becomes $P(A_3A_3) = (1/4)(1/4) = 1/16$. A similar procedure would yield the same numerical result for the probability of getting two A_4 alleles from the common ancestor, or $P(A_4A_4)$. Adding the two probabilities, we get

$$F_{\mathrm{I}} = 1/16 + 1/16 = 1/8,$$

which is the inbreeding coefficient for the offspring of a half-sib mating.

Figure 16.1. Pedigree of a mating between half-siblings. The offspring of this mating can have two possible genotypes in which the alleles are identical by descent from the common ancestor (the grandfather in this case). They are A_3A_3 and A_4A_4, where the subscripts are used to differentiate between alleles. Inside the symbols for the individuals, A = common ancestor and I = inbred individual.

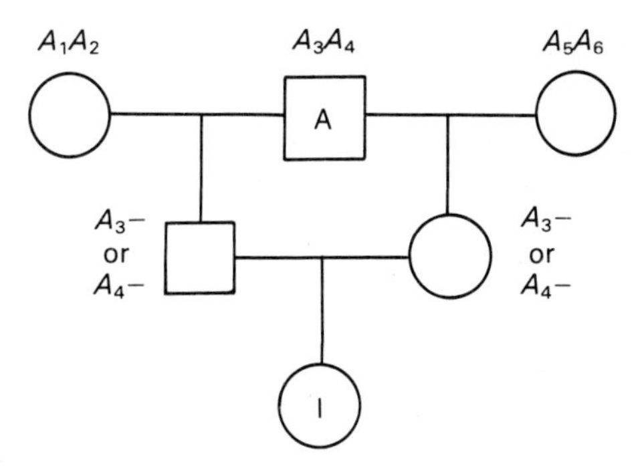

The inbreeding coefficient not only serves as an indicator of inbreeding intensity; it can also be used to measure the degree of genetic relationship between mates. In general, the closer the relationship between the mated individuals, the greater the likelihood of their carrying genes at equivalent loci that are replicates of the same DNA in a common ancestor. The inbreeding coefficient therefore measures relatedness by specifying the probability that genes selected at random from equivalent loci in different individuals are identical by descent. For example, the probability of identical genes for the half-siblings in Fig. 16.1 is $\frac{1}{8}$, which is equal to the chance that genes selected at random at the *A* locus from each of two half-siblings are both copies of the same allele in the relative whom the siblings have in common.

Table 16.1 lists the inbreeding coefficients for the offspring of couples who have varied degrees of relatedness. It is assumed in every case that the common ancestors are not inbred. Note that the value of F ranges from $\frac{1}{4}$ for the offspring of full siblings to 0 for the offspring of parents who are unrelated.

Measuring Inbreeding from Pedigrees. There are certain rules that help to simplify the computation of inbreeding coefficients from pedigree data. The most basic rule is that *if the parents of an individual have only one ancestor in common, the inbreeding coefficient of the individual can be expressed as*

$$F = (\tfrac{1}{2})^n \quad \textbf{(16.1)}$$

where n is the number of ancestors on the path from the individual through one of the parents to the common ancestor and back through the other parent. To use this equation, we must assume that the common ancestor is not inbred. For a specific example, let us consider the modified pedigree in Fig. 16.2, which is a *path diagram* of the pedigree in Fig. 16.1 for a half-sibling mating. In the path diagram, only the individuals who are directly involved in the transmission of genes from the common ancestor to offspring I are included. Note that only three individuals need to be included in the path. Consequently, $n = 3$ and as we determined previously, $F_{\mathrm{I}} = (\frac{1}{2})^3 = \frac{1}{8}$.

For a simple proof of this formula, let us consider the gametes in Fig. 16.2, which are labeled *a, b, c,* and *d.* Since individual I is formed from the fusion of gametes *c* and *d,* then F_{I}, which is the inbreeding coefficient of individual I, is the

Table 16.1. Inbreeding coefficients for offspring of couples of varied degrees of relatedness.

DEGREE OF RELATEDNESS	INBREEDING COEFFICIENT OF OFFSPRING
First-degree relatives (parent-child, full sibling)	$\frac{1}{4}$
Second-degree relatives (aunt-nephew, uncle-niece, half-sibling)	$\frac{1}{8}$
Third-degree relatives (first cousins)	$\frac{1}{16}$
Unrelated individuals	0

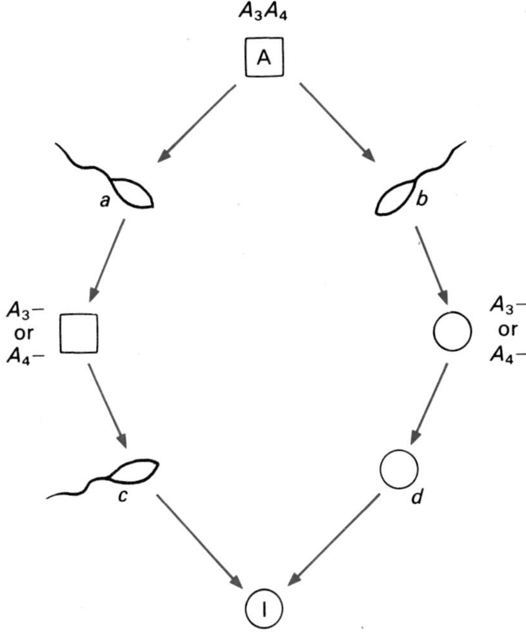

Figure 16.2. Path diagram of a half-sib mating. Only the one common ancestor and those individuals in the direct line of descent are included in the modified pedigree. Gametes are included for the sake of clarity and are labeled *a*, *b*, *c*, and *d*. The common ancestor is not inbred.

probability that the allele carried by gamete c is identical by descent to the allele carried by gamete d. If we use an equals sign to indicate that the included alleles are identical by descent, then

$$F_I = P(c = d) = P(c = a) \cdot P(a = b) \cdot P(d = b).$$

There are two possible ways that gametes a and b both carry identical alleles: either $a = A_3$ and $b = A_3$ or $a = A_4$ and $b = A_4$. Thus, $P(a = b) = P(a = A_3, b = A_3) + P(a = A_4, b = A_4) = 1/4 + 1/4 = 1/2$. By referring to Fig. 16.2, we can also see that $P(c = a) = 1/2$ and $P(d = b) = 1/2$. We can therefore write $F_I = (1/2)^3 = 1/8$, as in the formula.

Equation 16.1 can also be modified for use in more complicated situations. Two examples follow.

1. *Common ancestor is inbred.* If the common ancestor is also inbred, with an inbreeding coefficient of F_A, the inbreeding coefficient of the offspring becomes

$$F_I = (1/2)^n(1 + F_A). \qquad \textbf{(16.2)}$$

In this case, the two alleles in the common ancestor (e.g., A_3 and A_4) have an F_A chance of being identical by descent from an earlier generation, thereby increasing the probability that they will be identical in the offspring. For a numerical example, let us suppose that the common ancestor in Figs. 16.1 and 16.2 is also the result of a mating between half-siblings. The inbreeding coefficient of the common ancestor is then $F_A = 1/8$ and that of offspring I is

$$F_I = (1/2)^3(1 + 1/8) = 9/64.$$

2. *More than one common ancestor.* When the parents of an individual have more than one ancestor in common, the inbreeding coefficient for that individual is the sum of the F values derived from each of the common (but unrelated) ancestors separately. In symbolic form,

$$F = \sum(1/2)^n(1 + F_A) \qquad \textbf{(16.3)}$$

where Σ designates taking the sum. For example, let us consider the pedigree for the mating between full siblings given in Fig. 16.3. Figure 16.3(a) gives the conventional pedigree notation and Fig. 16.3(b) shows the corresponding path diagram, depicting the two mutually exclusive paths taken by genes derived from each of the grandparents of the individual. Now, if we assume that neither common ancestor is inbred, so that $F_A = 0$ for both, offspring I will have a $(1/2)^3$ chance of receiving alleles that are identical by descent from the grandmother plus a $(1/2)^3$ chance of receiving such alleles from the grandfather. The total inbreeding coefficient of individual I is then

$$F_I = (1/2)^3 + (1/2)^3 = 1/8 + 1/8 = 1/4.$$

A pedigree need not be restricted to two mutually exclusive paths but can be broken up into as many circular paths as there are ancestors in common. This concept is illustrated by the pedigree involving half first cousins and second cousins shown in Fig. 16.4. In this pedigree, there are three common ancestors and, therefore, three mutually exclusive paths leading to offspring I: one with $n = 5$ and two others with $n = 7$ each. The total inbreeding coefficient of individual I is then

$$F_I = (1/2)^5 + (1/2)^7 + (1/2)^7 = 3/64.$$

Figure 16.3. Pedigree of a mating between full-siblings. (a) Conventional pedigree, showing two common ancestors (grandparents in this case). (b) Path diagram of full-sib mating, showing the two mutually exclusive paths taken by genes derived from each of the grandparents of individual I.

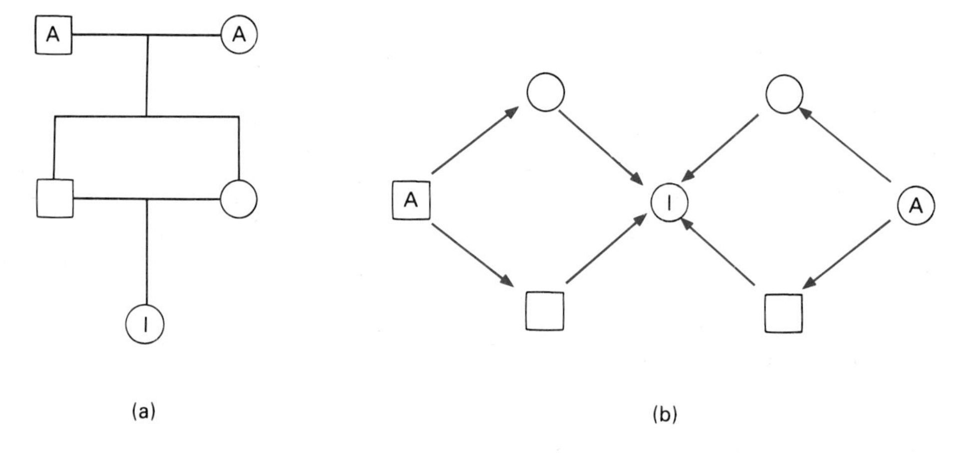

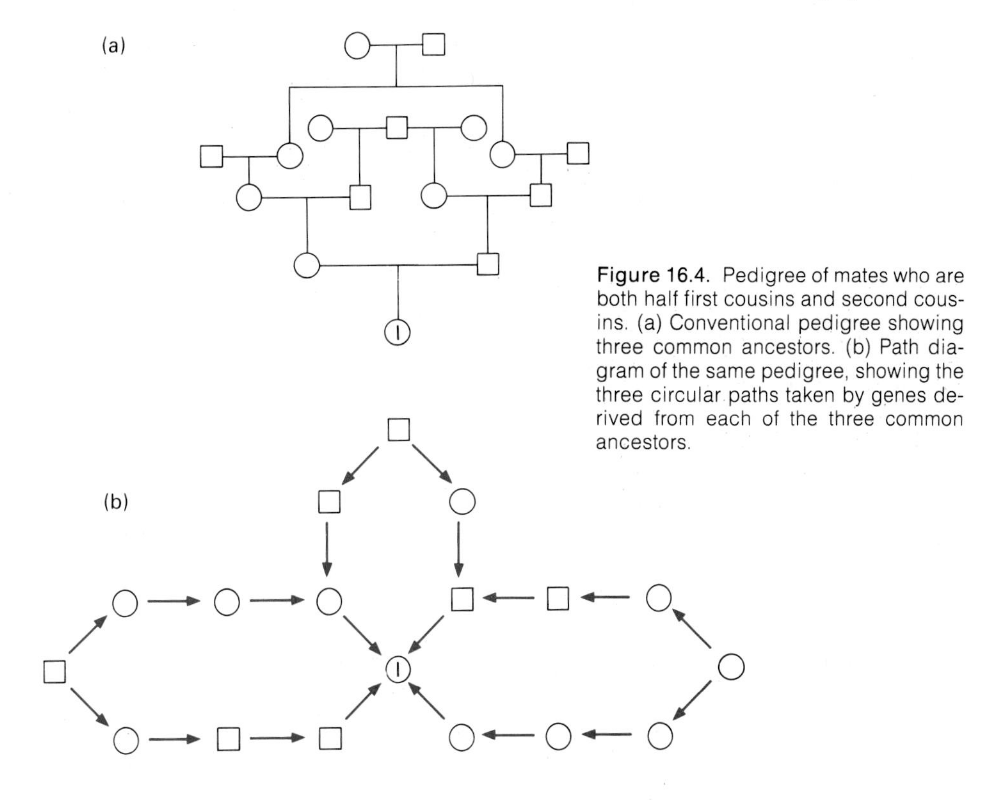

Figure 16.4. Pedigree of mates who are both half first cousins and second cousins. (a) Conventional pedigree showing three common ancestors. (b) Path diagram of the same pedigree, showing the three circular paths taken by genes derived from each of the three common ancestors.

Effects of Inbreeding on Populations

Genotype Frequencies with Inbreeding. Consanguineous matings will alter the genetic structure of a population from the structure that is predicted for random mating; the extent of alteration depends on the intensity with which the inbreeding occurs. To illustrate, let us assume a population in which the individuals have an average inbreeding coefficient F. When applied to an entire breeding group, F is the probability that an individual selected at random from the population is autozygous, with alleles at a particular locus that are identical by descent. Thus, for the two alleles A and a, a population with inbreeding will consist of a proportion F of the AA and aa autozygotes in which the alleles are identical by descent, and a proportion $1 - F$ of the AA, Aa, and aa allozygotes in which the alleles are independent in origin and are not replicas of a single ancestral gene.

The subdivision of a population with inbreeding into its autozygous and allozygous components is illustrated in Fig. 16.5. Since autozygous individuals must be homozygous, the proportions of the AA and aa genotypes among the autozygotes are simply equal to the frequencies of their respective alleles, p and q. In determining the proportions of the autozygous genotypes, therefore, we need to compute the frequency of only one allele in each case; once the first allele is specified, we immediately know what the second allele in a given pair will be.

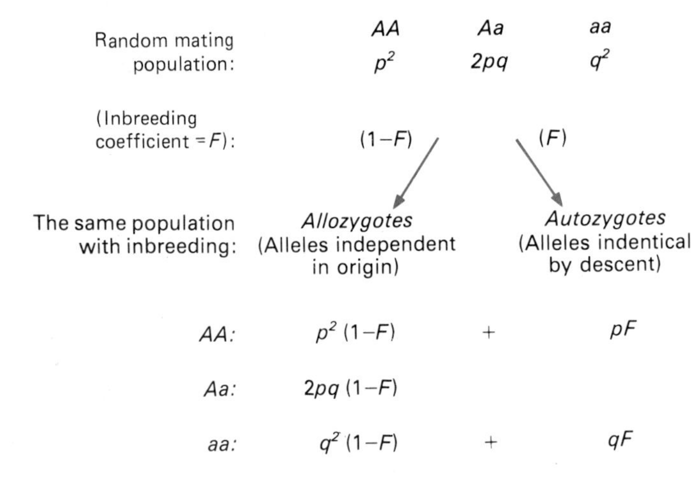

Figure 16.5. Subdivision of a random mating population into autozygous and allozygous components upon inbreeding. A fraction F of the population has alleles that are identical by descent; a fraction $1 - F$ inherits alleles of independent origin.

Allozygotes, on the other hand, are independent in origin, in that they are formed from a random combination of alleles. The frequencies of the *AA*, *Aa*, and *aa* genotypes among the allozygotes are therefore equal to their Hardy-Weinberg proportions p^2, $2pq$, and q^2. The total frequencies of the three genotypes *AA*, *Aa*, and *aa* within the inbred population are then $p^2(1 - F) + pF$, $2pq(1 - F)$, and $q^2(1 - F) + qF$, respectively.

As we can see from this example, the genotype frequencies under the conditions of inbreeding will depend directly on the intensity of inbreeding, as measured by F. In the absence of inbreeding, $F = 0$ and the genotypes will conform to their Hardy-Weinberg frequencies. With inbreeding, F will be greater than 0. Thus, *inbreeding will produce a reduction in the frequency of heterozygotes* from $2pq$ to $2pq(1 - F)$. This reduction is offset by an increase in the proportions of the homozygous genotypes.

We should emphasize that in this example, F represents the average inbreeding coefficient, with a value that depends not only on the kinds of consanguineous matings in the population but also on the frequencies at which these matings occur. For instance, we can see this dual dependence in a population in which 90 percent of all individuals are of random parentage ($F = 0$) and 10 percent are from matings between first cousins ($F = 1/16$). The average inbreeding coefficient for such a population would be $(0)(0.9) + (1/16)(0.1) = 1/160$.

We should also stress that although the genotype frequencies have been altered, *the allele frequencies are not changed as a result of inbreeding*. For example, the frequency of the *A* allele in a population with inbreeding will equal $f(AA) + \frac{1}{2}f(Aa) = p^2(1 - F) + pF + pq(1 - F) = p(p + q)(1 - F) + pF = p - pF + pF = p$, regardless of the value of F.

EXAMPLE 16.1. Assuming that inbreeding is the only factor contributing to changes in the genetic structure of a population, what would be the value of the inbreeding coefficient that is attained once the genotype frequencies become $f(AA) = 0.752$, $f(Aa) = 0.096$, and $f(aa) = 0.152$?

Solution. First note that the frequency of heterozygotes under the conditions of inbreeding can be expressed as $f(Aa) = 2pq(1 - F)$. Solving for F, we get

$$1 - F = \frac{f(Aa)}{2pq} \quad \text{or} \quad F = 1 - \frac{f(Aa)}{2pq}$$

where $f(Aa)$ in this relationship equals 0.096 and the allele frequencies can be calculated as

$$p = 0.752 + \tfrac{1}{2}(0.096) = 0.8 \quad \text{and} \quad q = 1 - 0.8 = 0.2.$$

Substituting these values, we get

$$F = 1 - \frac{0.096}{2(0.8)(0.2)} = 1 - 0.3 = 0.7.$$

The inbreeding coefficient is thus 0.7.

Inbreeding and Homozygosis. If inbreeding continues to occur over an extended period of time, the magnitude of the inbreeding coefficient will gradually increase, since it becomes progressively more likely that two alleles are identical by descent from some previous generation. This increase in F will be accompanied by a rise in the frequency of homozygotes and by a corresponding reduction in the frequency of heterozygotes. The increase in homozygosity that accompanies inbreeding is most easily shown in terms of a population undergoing *self-fertilization* (or *selfing*). This extreme form of inbreeding is relatively common among plants. Table 16.2 lists some examples of economically important plants that are largely self-pollinated, along with some examples of cross-pollinated crops.

The results of several generations of selfing are shown in Fig. 16.6. Note that when a homozygous parent (AA or aa) is selfed, it produces only homozygous offspring that are identical in genotype to the parent. In contrast, heterozygotes form a progeny of ¼ AA, ½ Aa, and ¼ aa. The proportion of heterozygotes is therefore reduced in each generation by a half. If this mode of reproduction were

Table 16.2. Examples of economically important plants that are largely self-pollinated and those that are largely cross-pollinated.

Largely self-pollinated plants:			
	apricots	peaches	soybeans
	barley	peanuts	tobacco
	beans	peppers	tomatoes
	citrus fruits	rice	wheat
	cotton	sorghum	
Largely cross-pollinated plants:			
	apples	corn	raspberries
	almonds	cucumbers	rye
	cabbage	grapes	strawberries
	carrots	orchids	sugar beets
	celery	plums	

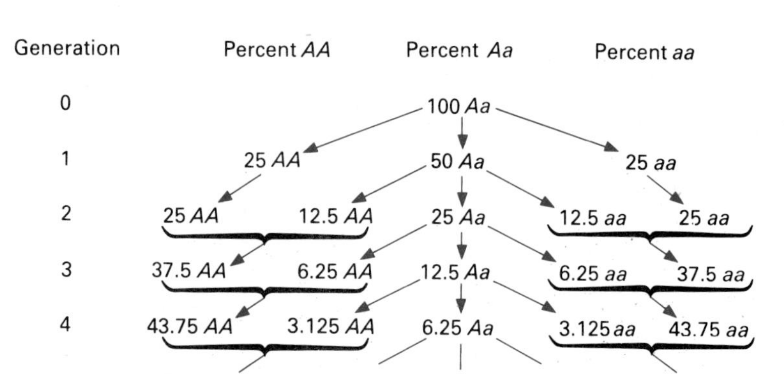

Figure 16.6. Reduction in the frequency of heterozygotes during four generations of self-fertilization. The population is started with all heterozygotes. Note that homozygotes produce only homozygous offspring of the same genotype, whereas heterozygotes produce all three genotypes in a 1:2:1 ratio. The result is a reduction in the frequency of heterozygotes by a factor of ½ in each generation.

to continue, the frequency of heterozygotes after t generations of selfing, $f(Aa)_t$, would become

$$f(Aa)_t = (1/2)^t f(Aa)_0 \tag{16.4}$$

where $f(Aa)_0$ is the frequency of heterozygotes when $t = 0$. Thus, when $t = 1$, $f(Aa)_1 = 1/2 f(Aa)_0$; when $t = 2$, $f(Aa)_2 = 1/4 f(Aa)_0$; and so on. Ultimately, the frequency of heterozygotes will approach zero. Once the heterozygotes are eliminated from the population, all of the individuals must be homozygous for one allele or the other.

EXAMPLE 16.2. How many generations of selfing are required to reduce the frequency of heterozygotes in a plant population to just under 0.001 of the initial value?

Solution. Applying Eq. (16.4), we see that $(1/2)^t$ has a value just under 0.001 when $t = 10$. After ten generations of selfing, the frequency of heterozygotes is $f(Aa)_{10} = (1/1024)f(Aa)_0$.

We can easily transform Eq. (16.4) into a relationship that shows the time-dependent increase in F that occurs during selfing. First, if we let $f(Aa)_0 = 2pq$, then $f(Aa)_t = (2pq)(1/2)^t$ in Eq. (16.4). Since $f(Aa)_t$ also equals $2pq(1 - F)$, we see that

$$1 - F = (1/2)^t \quad \text{or} \quad F = 1 - (1/2)^t$$

for a self-fertilizing system. In the first generation of selfing, $F = 1 - 1/2 = 1/2$. In the second generation, F is increased to $1 - (1/2)^2 = 1 - 1/4 = 3/4$. This increase continues until eventually F approaches 1. At this time, inbreeding is said to be complete, and only the AA and aa autozygotes remain in the population, at frequencies of p and q, respectively.

An increase in homozygosity is not restricted to self-fertilization but can occur with other, less extreme forms of inbreeding as well. Figure 16.7 compares the approach toward homozygosity that is produced by different kinds of consanguineous matings with the homozygosity that results from continued selfing. Full-sibling (brother-sister) mating, which is the most intense form of inbreeding in higher animals, produces homozygosity at a significantly slower rate than self-fertilization. It takes approximately nine generations of repeated sibling mating to attain an inbreeding coefficient of 90 percent, whereas only about three generations of selfing are needed to achieve the same result. As we would expect, the less intense forms of inbreeding produce an even slower approach to homozygosity.

Fixation of Genetic Characters. The preceding model focused attention on a single pair of alleles. But when consanguineous matings occur, the mates possess genetic similarities that extend over all gene pairs. Inbreeding must therefore affect all segregating loci. The ultimate outcome of close inbreeding is then complete homozygosity, which results in the production of different *pure inbred lines*. Each inbred line will be uniform in phenotype (except for environmentally induced variation) and will differ from other inbred lines on the basis of the alleles that

Figure 16.7. Increase in the value of the inbreeding coefficient under different systems of mating that involve recurrent inbreeding. Continued selfing subdivides a population into lines, each of which is perpetuated by a single parent. In sibling mating, two parents (a brother and sister) are used to perpetuate each line in every generation. While two pairs of parents are used in first cousin matings, four pairs are used in second cousin matings.

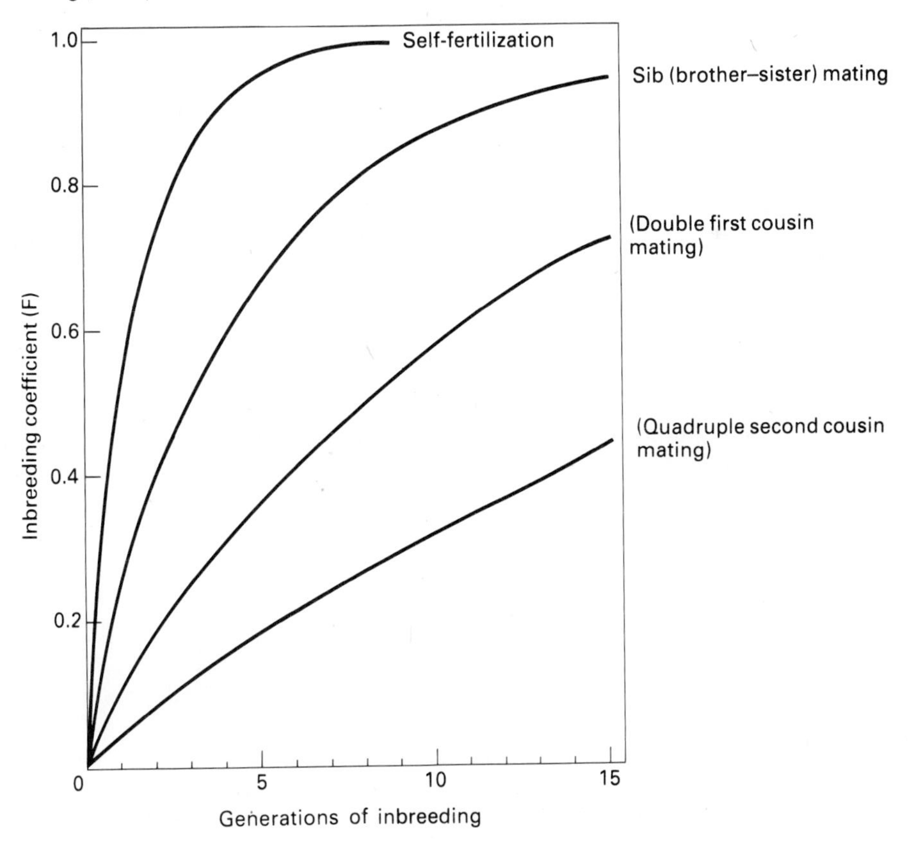

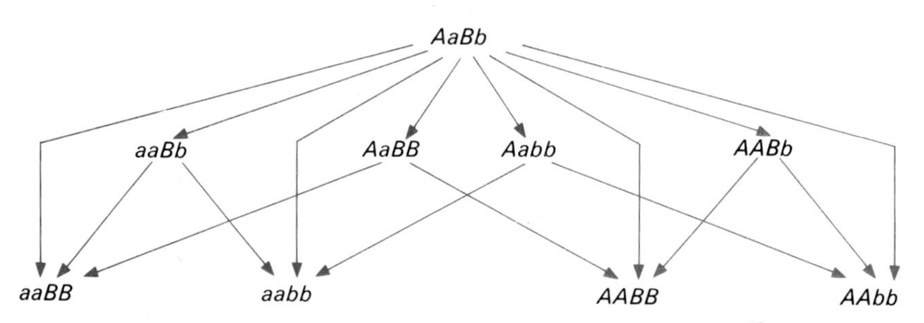

Figure 16.8. Formation of pure lines upon continued selfing of a tetrahybrid plant. Four true-breeding lines will be produced in this case once all of the pertinent loci are homozygous.

are fixed at its various loci. For example, if we consider only four gene loci, one line may have the genotype *AAbbccdd*, while another might be *aaBBccDD*.

The subdivision of a hybrid plant population into different pure-breeding lines through continued selfing is illustrated in Fig. 16.8. The genotypes of the lines represent the different possible combinations of homozygous gene pairs. For this reason, the number of pure-breeding lines will depend simply on the number of gene pairs concerned. One pair of genes, *A* and *a*, will produce two pure inbred lines: *AA* and *aa*. With two pairs of genes, *A,a* and *B,b*, four lines are possible: *AABB*, *AAbb*, *aaBB*, and *aabb*. If we extend this pattern to n gene pairs, we can use the general formula 2^n to predict the number of pure inbred lines. Thus, with ten gene pairs, more than a thousand (2^{10}) pure-breeding lines are theoretically possible, and with 20 gene pairs, more than a million pure-breeding lines can be formed.

If the subdivision of a hybrid population proceeds in the absence of an evolutionary force, all of the inbred lines are equally probable. For this reason, the allele frequencies do not change in the collective inbred population. Only the genotype frequencies are altered.

Increase in Deleterious Recessive Traits. Estimates have been made for the human species that predict that each of us carries, on the average, two recessive genes that would be lethal if they were in the homozygous state. Fortunately, such deleterious genes are rare and, so long as mates are selected at random from a large population, the vast majority of these genes remain hidden in heterozygous combinations. A problem develops when there are matings between close relatives, since consanguineous matings lead to an increase in the homozygosity of the offspring. We can picture this effect of close inbreeding in more explicit terms by dividing the expected frequency of recessive homozygotes under conditions of close inbreeding, $q^2(1 - F) + qF$, by the expected frequency for this class under conditions of random mating. We get

$$\frac{q^2(1 - F) + qF}{q^2} = 1 - F + (1/q)F = 1 + \frac{F(1 - q)}{q}.$$

This formula represents the factor by which consanguineous matings increase the frequency of recessive homozygotes above the frequency expected for a random mating population. Note that the ratio increases in direct proportion to the factor

$(1 - q)/q$. Thus, inbreeding will have a greater effect if the trait in question is rare (i.e., when q is much less than 1) than if the trait is common.

For a numerical example, let us suppose that a deleterious recessive gene occurs with a frequency of $q = 10^{-2}$ in a population. The frequency of recessive homozygotes under conditions of random mating would then be $q^2 = 10^{-4}$. Now assume that inbreeding occurs, with an inbreeding coefficient of $1/16 = 0.0625$ (the value that is expected for matings between first cousins). The factor increase in recessive individuals under inbreeding then becomes

$$1 + \frac{F(1 - q)}{q} = 1 + \frac{(0.0625)(0.99)}{(0.01)} = 7.2$$

or about seven times greater than in a random mating population. Similar calculations reveal that when a recessive gene occurs at a frequency as low as 10^{-3}, this same intensity of inbreeding would increase the frequency of recessive homozygotes to $1 + (0.0625)(0.999)/(0.001)$, or over 60 times greater than from random mating.

We should bear in mind that inbreeding in itself is not necessarily harmful. If, by chance, the initial population were substantially free of detrimental recessive genes, close inbreeding could occur without substantial risk of expressing a recessive disorder. This was apparently the case among the pharaohs of ancient Egypt, among whom brother-sister matings were the rule (Fig. 16.9). The opposite effect is also possible. If rare, beneficial recessive genes occur in a population, close inbreeding would act to improve their chances of being expressed in homozygous combinations.

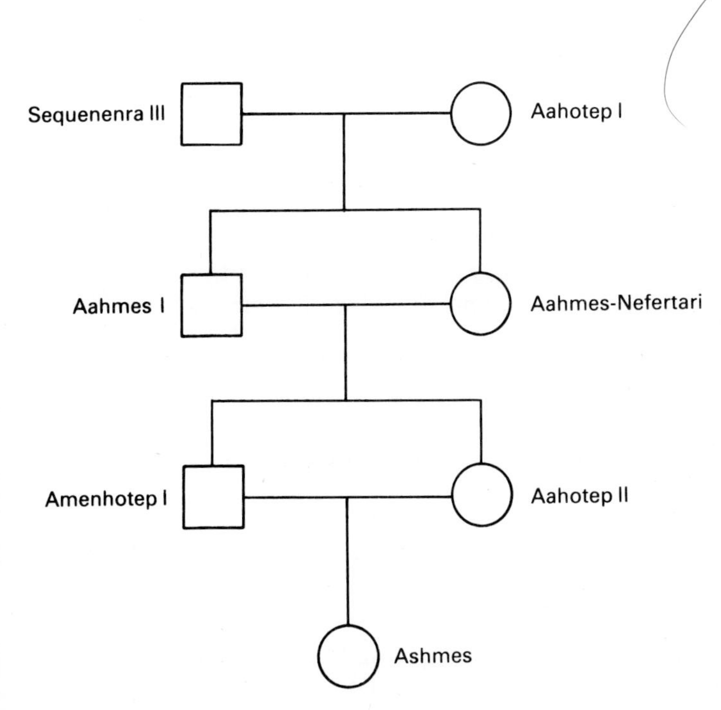

Figure 16.9. Partial pedigree of the Eighteenth Dynasty of Egypt, starting at 1580 B.C., illustrating the practice of brother-sister matings.

Inbreeding in Isolates. In bisexual organisms, each individual has two parents, four grandparents, eight great-grandparents, and so on, for a total of 2^t ancestors t generations back. It should be apparent from the exponential nature of this progression that t does not have to be very large before the number of separate ancestors that are theoretically possible exceeds the size of any real population. Therefore, all potential mates in a finite population must share at least one common ancestor at some point in the more or less distant past, and that the smaller the size of the population, the closer will be their probable relationship. Thus, in small populations, close inbreeding is largely unavoidable and may be quite common, even when mates are selected at random with regard to genotype.

The effects of population size on the severity of inbreeding are particularly evident in small religious isolates. One group that has been intensively studied is the Amish sect, which originated as an offshoot of the Mennonite Church. The Amish live in relatively isolated communities in the eastern United States and maintain a rural lifestyle that rejects most modern conveniences. Because of the closed nature of the group, marriages have been restricted largely to members of the Amish sect. This restriction has led to a high degree of inbreeding, even though the marriages do not always involve close relatives. A medical survey revealed that certain recessive disorders occur within the group at levels that are considerably higher than in the noninbred neighboring populations. One example is the Ellis-van Creveld syndrome, a rare form of dwarfism that is characterized by extra fingers and disproportionately shortened limbs. The incidence of this disorder is extraordinarily high among the Amish of Lancaster County, Pennsylvania, where it exists at a frequency of approximately 5×10^{-3}. Elsewhere, the frequency is extremely low.

Assortative Mating

One common departure from random mating is assortative mating, in which an increase in phenotypic resemblance enhances the choice of mate. For example, humans frequently exhibit assortative mating for height, intelligence, and many traits associated with race. We therefore tend to see a positive correlation between husband and wife regarding height, IQ, and certain racial characteristics.

We can best see the genetic effects of mating according to phenotypic resemblance by considering an example of complete assortative mating for a simply inherited trait, in which the matings are restricted to individuals of the same phenotype. If dominance is complete, the mating combinations are then either dominant $\times$ dominant (A- $\times$ A-) or recessive $\times$ recessive ($aa \times aa$). In this case, assortative mating has effectively subdivided the population into two components, A- and aa, based on the phenotypic expression of a single pair of alleles.

In order to compare the effects of this mating system with the effects of random mating, let us assume that the initial population has alleles (A,a) at frequencies (p, q) and genotypes (AA, Aa, aa) at frequencies (p^2, $2pq$, q^2). The allele frequencies in the dominant (A-) mating component are then $f(A) = p$ and $f(a) = pq/(p^2 + 2pq) = q/(1 + q)$, and those in the recessive (aa) mating component are $f(A) = 0$ and $f(a) = q - q/(1 + q) = q^2/(1 + q)$. During complete assortative mating, heterozygotes can be formed in this population only by A- $\times$ A- matings.

Table 16.3. Approach to homozygosity under complete phenotypic assortative mating with (AA, Aa, aa) at initial frequencies of (1/4, 1/2, 1/4).

GENERATION	AA	Aa*	aa
0	1/4	1/2	1/4
1	2/6	1/3	2/6
2	3/8	1/4	3/8
3	4/10	1/5	4/10
4	5/12	1/6	5/12
.			
.			
.			
limit	1/2	0	1/2

* In this example, $f(Aa)_t = f(Aa)_0/[1 + tf(Aa)_0]$.

Therefore, if mates are selected at random with regard to the AA and Aa genotypes within the dominant component, the frequency of heterozygotes in the next generation becomes $(2)(p)[q/(1 + q)] = 2pq/(1 + q)$. The frequency of heterozygotes has thus been reduced in complete assortative mating by a factor of $1 + q$ during the first generation. For example, if $p = q = 1/2$, the heterozygotes would decline in frequency during the first generation from $2(1/2)(1/2) = 1/2$ to $2(1/2)(1/2)/(1 + 1/2) = 1/3$. The homozygotes would increase correspondingly from 1/2 to 2/3. If this process continued over several generations, it would lead to the changes shown in Table 16.3. As in the case of close inbreeding, the ultimate outcome is the complete elimination of heterozygotes from the population.

Despite the similar genetic consequences of continued inbreeding and complete assortative mating, there are some important differences between these mating systems. First, the rate at which the heterozygotes decline in frequency tends to be slower for matings that are based solely on phenotypic similarity. Given the same starting population as in Table 16.3, for example, heterozygotes would be reduced to a frequency of $(1/2)^9 = 1/512$ after 8 generations of selfing (see Eq. (16.4)), but just to 1/10 after complete assortative mating for the same number of generations. Second, unlike inbreeding, in which homozygosity is not confined to a single locus but extends to all segregating gene pairs, phenotypic assortative mating affects only those genes that are responsible for the trait in question, so long as the matings are restricted to unrelated individuals. Thus, even though the A,a locus might segregate out only homozygous types as a result of assortative mating, other gene pairs (B,b; C,c; etc.) could, in theory, be maintained in random mating equilibrium.

Disassortative Mating and Outbreeding

Disassortative mating and outbreeding represent other ways in which the pattern of mate selection can depart from randomness. In both of these mating systems, a negative correlation occurs between mates; but in disassortative mating, the negative correlation results from the tendency of mates to be dissimilar in phenotype, while in outbreeding, they tend to be dissimilar in ancestry. In general, both systems act to maintain the heterozygosity in a population. This effect is illustrated by an extreme form of disassortative mating, in which only dominant × recessive crosses occur. In this system, two kinds of matings are possible with regard to genotype: $AA \times aa$ and $Aa \times aa$. The first type of mating can produce only *Aa* offspring, while the second type gives rise to *Aa* and *aa* progeny in equal numbers. In the next generation, and in all subsequent generations, therefore, the matings are restricted to $Aa \times aa$, which produces an equilibrium condition of ½ *Aa* : ½ *aa*.

One example of this situation is the maintenance of dimorphism for style shape in many flowering plants (Fig. 16.10). The flower morphology is typically of two types: (1) long pistils and short stamens, called the "pin" type, and (2) short pistils and long stamens, called the "thrum" type. This difference in flower structure, known as *heterostyly* (which means "different styles"), helps to prevent selfing. The trait is controlled by a gene complex, or supergene, that consists of at least seven very closely linked loci involved in the determination of flower morphology and pollen tube growth. In some species, the pin types behave as though they were recessive (*ss*), while the thrum types seem to be dominant (*S*-). Pollinating insects tend to deliver pollen from *ss* plants to *S*- plants and vice versa, thus perpetuating through disassortative mating a dimorphism between the pin and thrum types at a 1 : 1 ratio.

Another mechanism has evolved among higher plants that acts to prevent self-fertilization and to encourage outcrossing. This mechanism is an incompatibility system involving *self-sterility* alleles. The self-sterility alleles (designated S_1, S_2, S_3, etc.) inhibit the growth of pollen tubes in styles with which they have an allele in common (the styles probably cause this inhibition by producing an

Figure 16.10. Two types of flower morphology in the primrose (*Primula*). (a) The pin type, showing a highly placed stigma and low-placed anthers. (b) The thrum type, with the reverse morphology. This difference in style morphology, known as heterostyly, helps to prevent selfing.

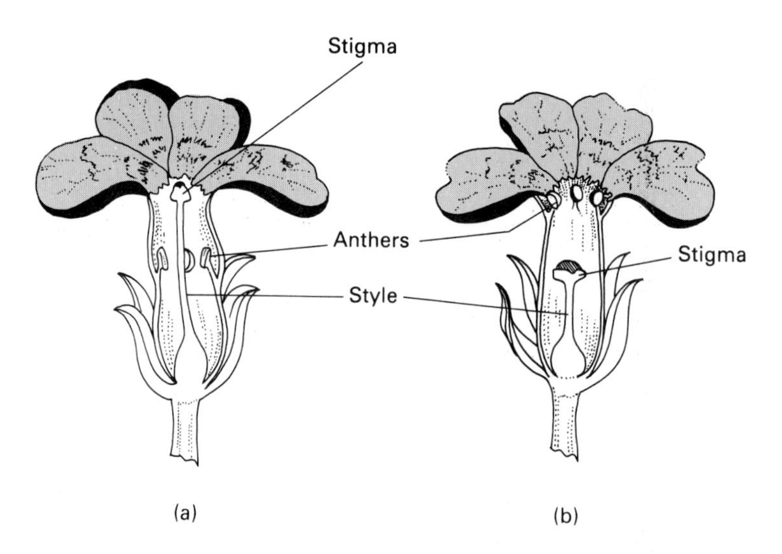

antibody-like substance). For example, pollen that contains the S_1 allele cannot fertilize plants with genotypes S_1S_1, S_1S_2, S_1S_3, and so on, but it can fertilize plants without the S_1 allele (e.g., S_2S_2, S_2S_3, etc.). Plants that have this mechanism are obviously self-incompatible with regard to pollination, and plants that are homozygous for the same sterility alleles are cross-incompatible. Plants that have only one allele in common can cross-pollinate, but with reduced fertility. Thus, S_3-containing pollen from an S_1S_3 plant can fertilize S_1S_2 plants, although the pollen from the same plant that contains the S_1 allele cannot. Because this genetic system is extremely common among cultivated plants, it has been necessary to include pollinator plants in various fruit orchards, for example, to ensure an adequate supply of compatible pollen.

TO SUM UP

1. Nonrandom mating occurs whenever there is a correlation between the ancestry or phenotypes of mates. The correlation is positive in the case of inbreeding and assortative mating; it is negative for outbreeding and disassortative mating.
2. Inbreeding occurs through the mating of relatives (called consanguineous matings), which results either from a systematic choice of relatives as mates or from a reduction in population size to a level at which randomly selected mates have a high probability of having common ancestry. The basic effect of inbreeding is to increase the probability that an inbred individual is homozygous for alleles that are identical by descent (i.e., replicates of the same gene in a common ancestor).
3. The inbreeding coefficient, F, measures the probability that two alleles are identical by descent. Its value varies with the degree of relatedness between the parents and with the number of generations of inbreeding. The value of F is 0 for a noninbred individual and is equal to 1 when inbreeding is complete.
4. Consanguineous matings lead to an increase in the homozygosity of the offspring. For this reason, there is an increased risk that a rare recessive trait will appear among inbred persons as compared with the members of a random mating population.
5. The smaller the population, the greater is the chance that two randomly selected individuals have an ancestor in common. Inbreeding can therefore occur in small populations even if mating is random, and it is one of the factors responsible for the abnormally high frequencies of rare recessive disorders in religious isolates.
6. Assortative mating according to phenotypic similarity produces homozygosity among the offspring, but it tends to be less efficient in this respect than close inbreeding. In contrast, disassortative mating tends to maintain the heterozygosity in a population.

Heterosis

A common observation is that the hybrids that are produced upon crossing different inbred lines tend to show an increase in commercially desirable characteristics relating to size, growth rate, and fertility. This increase in performance by hybrids is called **hybrid vigor** or **heterosis.** While heterosis is well known among cultivated plants (e.g., corn, tomatoes, onions, sorghum), in which the increases in yield are of particular economic benefit, it has also been reported for a variety of other, less familiar plant and animal species, so that it appears to be a widespread phenomenon. In this section, we will discuss the genetic basis for heterosis and will consider some of the procedures that contribute to its commercial use.

Heterosis and Inbreeding Depression

We can think of heterosis as the complement of the phenomenon known as **inbreeding depression.** Inbreeding depression is the decline in average performance (or vigor) that is associated with inbreeding. This decline results from the gradual loss of heterozygosity during inbreeding, as shown for corn in Fig. 16.11.

For a numerical example of the relationship between inbreeding depression and heterosis, let us consider a certain variety of self-fertilizing plants in which the recessive homozygotes grow to only half the size of the *AA* and *Aa* genotypes. The relative performances of each genotype in terms of vegetative growth can then be assigned the following values:

Genotypes:	*AA*	*Aa*	*aa*
Relative performance:	1.0	1.0	0.5

Suppose that the parents consist of only heterozygotes (*Aa*), with an average performance of 1. One generation of selfing would then produce an F_1 composed of ¼ *AA*, ½ *Aa*, and ¼ *aa*, with a performance value as follows:

$$\text{Average performance of } F_1 = (1)\tfrac{1}{4} + (1)\tfrac{1}{2} + (0.5)\tfrac{1}{4} = 0.875.$$

Figure 16.11. Heterosis and inbreeding depression in corn. The highly vigorous F_1 is produced by crossing the two inbred lines P_1 and P_2. Selfing over several generations (F_2 to F_8) leads to a reduction in vigor back to an approximate average of the two parental lines. *Source:* Reprinted by permission from F. J. Ayala and J. A. Kiger, Jr., *Modern Genetics*. Copyright © 1980, by The Benjamin/Cummings Publishing Company.

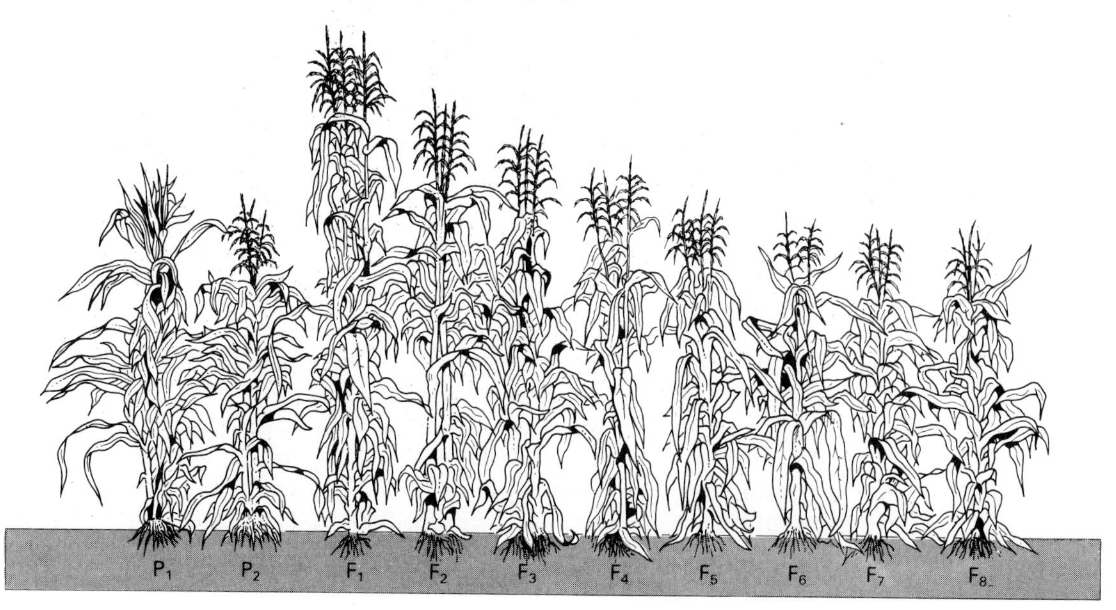

The relative vigor, as determined from the yield of an average plant, has thus declined by $1 - 0.875 = 0.125$ in just one generation. If selfing were to continue in the absence of any selection pressures, the average performance of the group would gradually decline over many generations of inbreeding to a minimum value of $(1)\frac{1}{2} + (0.5)\frac{1}{2} = 0.75$. This value would be reached once the plant population consists solely of the two inbred lines, *AA* and *aa*. But note that the performance that is lost during several generations of close inbreeding can be regained through heterosis in a single generation of outbreeding by simply crossing the inbred lines, $AA \times aa$, since the population would then consist of only heterozygotes.

Measuring Heterosis. It is possible to quantify the concept of heterosis by defining it as the excess in vigor of an F_1 hybrid over the midpoint of the inbred parental line. In the preceding numerical example, for instance, the relative performance of the F_1 heterozygotes is 1 whereas the midpoint performance of the two inbred *AA* and *aa* parents is $\frac{1}{2}(1 + 0.5) = 0.75$. Heterosis in this example is thus $1 - 0.75 = 0.25$.

When performance is determined by a single locus (usually this is not the case), one of the alleles must possess some degree of dominance if heterosis is to occur. This dependence on dominance is shown by the following scale of phenotypic values:

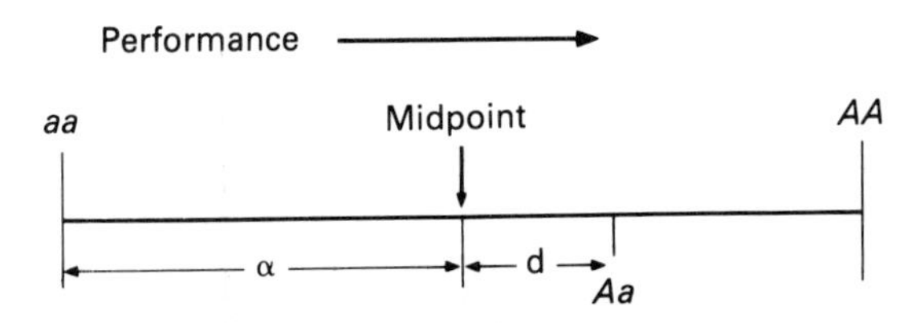

On this scale of relative performance, α is the midpoint value of the two homozygotes, and d is the excess in performance of the heterozygote over this midpoint value. From our earlier definition of heterosis, we see that heterosis $= d$ for a trait determined by a single locus. The relationship between d and α determines the degree of dominance. When $d = \alpha$, dominance is complete, with the heterozygote showing the same phenotypic value as the *AA* homozygote. When $d > \alpha$, overdominance occurs, with the heterozygote outperforming either homozygote. When $d = 0$ for a single locus, the allele contributions are additive. Some dominance effect is therefore essential for heterosis.

EXAMPLE 16.3. The average plant heights of two inbred lines of corn are 63.5 inches and 57.3 inches. The amount of heterosis exhibited by hybrids of these lines (the F_1) is 32.8 inches. Predict the average plant heights of the F_1 and F_2 generations.

Solution. The average plant height of the F_1 can be calculated as follows:

$$\begin{aligned}\text{Average value of } F_1 &= \text{Heterosis} + \text{Midpoint value of parents}\\ &= 32.8 + \tfrac{1}{2}(63.5 + 57.3)\\ &= 93.2 \text{ inches}\end{aligned}$$

Selfing the F_1 to form the F_2 will result in an average decline of heterozygous loci by one-half (see Eq. 16.4). Because of inbreeding depression, the average height of the F_2 should decrease below that of the F_1 by half the initial gain through heterosis:

$$\begin{aligned}\text{Average value of } F_2 &= \text{Average value of } F_1 - \tfrac{1}{2}(\text{heterosis}) \\ &= 93.2 - \tfrac{1}{2}(32.8) \\ &= 76.8 \text{ inches}\end{aligned}$$

Since characteristics such as yield are usually polygenic in nature, our model must be extended to include the sum (Σ) of d values over all segregating loci:

$$\text{Heterosis} = \sum d.$$

Implicit in this expression is the assumption that the parental varieties are homozygous at all of the loci that contribute to the trait. We should also point out that even though heterosis is dependent on the sum of the d values, the absence of heterosis should not imply an absence of complete or partial dominance. If some of the loci are dominant in one direction (such as increased size at maturity), while others are dominant in the opposite direction, it is possible that their contributions might cancel each other out. Heterosis would then equal zero despite the dominance effects of the individual loci.

Genetic Basis for Heterosis

One of the first theories developed to explain heterosis proposed that the increases in vigor in heterozygotes are caused by the combined dominance effects of polygenes. According to the so-called dominance theory of heterosis, various contributing and noncontributing alleles are fixed in different homozygous combinations within each inbred line. Because of dominance, these gene combinations tend to be less than optimal in total performance. But crossing the pure-breeding lines would redistribute the contributing alleles into the more optimal array that is present within the fully heterozygous hybrids. The net result is an increase in vigor.

For an illustrative example of the dominance model of heterosis, let us suppose there are four contributing genes (*A, B, C,* and *D*), which because of complete dominance individually add one unit to the total yield in both heterozygous and homozygous combinations (e.g., *Aa* and *AA*). Let us further assume that the corresponding small-letter alleles (*a, b, c,* and *d*) are noncontributing. One possible cross between two pure-breeding lines is the following:

Inbred lines: *AAbbCCdd* × *aaBBccDD*
F_1 hybrids: *AaBbCcDd*

Even though each pure-breeding line has the same number of contributing genes as the F_1 hybrids, heterosis would still occur, with the following numerical results:

Inbred lines: 2 × 2
F_1 hybrids: 4

As we can see from this example, heterozygosity is only incidental to the expression of heterosis in the dominance theory, since a homozygous combination of contributing genes (e.g., *AABBCCDD*) is assumed to add as much (or possibly more) to the overall performance as the heterozygous combination.

Overdominance and Complementary Gene Interactions. Alternative models have been proposed for heterosis that account for the increase in heterozygote performance in terms of overdominance at individual loci or complementary gene effects at separate loci or both mechanisms. For instance, the overdominance theory holds that the heterozygous state is essential for the full expression of heterosis. According to this view, the value of d is greater than α at one or more loci, so that the performance of the heterozygote is superior to both the dominant and recessive homozygotes. For example, heterozygotes may produce a more rapid and luxuriant growth under a wider range of conditions, or they may be more resistant to disease.

Many hypotheses have been proposed to account for the expression of overdominance by allelic genes. A few of these hypotheses have been supported by experimental evidence. Recent explanations have attempted to provide a molecular interpretation of overdominance. One possible mechanism is the formation of hybrid proteins, such as allozymes, by the combination of the polypeptide products of different alleles. For example, let us assume two alleles, A_1 and A_2, which produce polypeptides 1 and 2, respectively. Suppose that any two of the polypeptide chains can combine to form a functional dimeric enzyme. Three allozymes are then possible in the A_1A_2 heterozygote: 1-1, 2-2, and the hybrid enzyme 1-2. Only the 1-1 form will be present in the A_1A_1 homozygote and only the 2-2 form will be present in the A_2A_2 homozygote. This system could give rise to overdominance through several different mechanisms. It may be possible that the heterodimer 1-2 is a more efficient catalyst than either the 1-1 or 2-2 homodimer, or that it can form a metabolic product in amounts that are more optimal for growth. If the allozymes have different optimal environmental conditions (e.g., different optimal temperatures), the greater variety of enzyme forms in the heterozygote may serve to protect or "buffer" this genotype against environmental disturbances. The hybrid would thus tend to react less violently to changes in growth conditions, giving rise (as is often observed) to a smaller environmental variance than either of its inbred parents.

Cooperative interactions of this type are not restricted to alleles but may occur between nonallelic genes as well. Many oligomeric proteins (such as hemoglobin) are known to be constructed from a combination of the polypeptide products of different genes. When allelic variation is present, a complementary gene interaction of this type could generate a large potential number of enzyme forms in the heterozygotes, some of which may be more active in catalysis than others. In the case of two gene pairs, for example, the enzymes produced by *AaBb* dihybrids might be superior in performance to those formed by any other genotype at these loci. When two (or more) loci are involved in the production of the same end product, epistatic effects are also possible. When epistasis is present, heterosis could involve the cooperative (or synergistic) actions of alleles at different loci, so

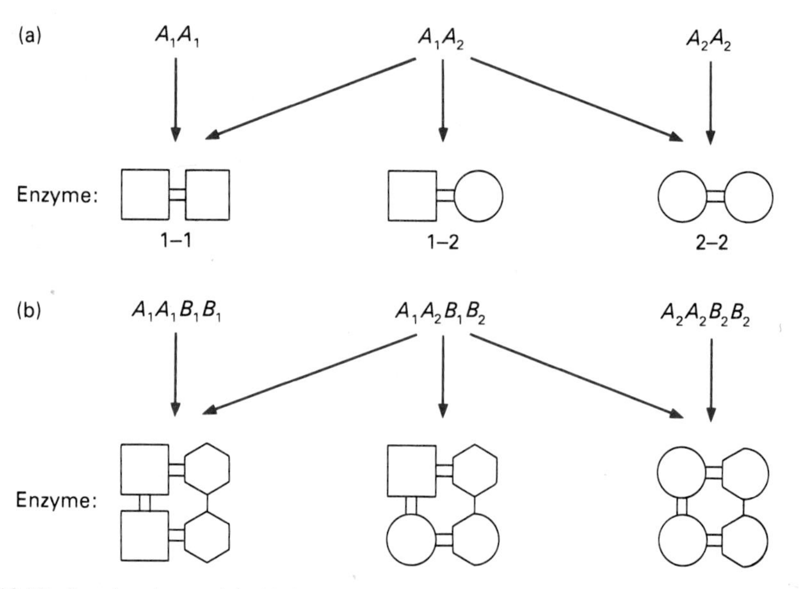

Figure 16.12. A molecular model of hybrid vigor through overdominance. (a) A single-locus model, in which alleles are assumed to interact to produce a dimeric enzyme. Heterosis may then be possible through the production of a more efficient hybrid catalyst or through the flexibility provided by the presence of a greater variety of enzyme forms. (b) A multiple-locus model, in which two gene pairs interact to produce an oligomeric enzyme (only 3 of the 9 possible genotypes are shown). Nonallelic gene interaction can provide a greater potential for variation than overdominance alone. These interactions can contribute to heterosis by causing the dihybrids to be more active in overall catalysis than the other genotypes.

that the total effect is greater than the sum of the separate contributions taken independently. Such a model for the molecular basis of overdominance is illustrated in Fig. 16.12, in which it is compared with a model that involves allelic interactions at a single locus.

Practical Uses of Inbreeding and Hybridization

Inbreeding and its opposite, outbreeding, have been used extensively by plant and animal breeders for the development of new and more productive varieties. The use of inbreeding in these breeding programs has a two-fold purpose. Its major value is in the fixation of desirable characteristics through homozygosity. For example, certain breeds of cattle have been selected and inbred for type over a period of several generations. In any such program, there is always the danger of inbreeding deterioration, caused by the exposure of harmful recessive genes. Thus, in the case of the Hereford breed of beef cattle, inbreeding has resulted in a rise in the incidence of births of dwarf calves. This trait is caused by a recessive gene that was originally present in certain ancestors of the breed (see Chapter 1). Intense inbreeding in other breeds of animals has also revealed recessive disorders, such as reduced vitality and lethal or crippling effects. A case in point is the high incidence of hip-dysplasia in the German shepherd dog. The recessive gene for this defect was introduced into the breed during its development.

Another purpose of inbreeding, which is closely related to the first purpose, is to produce genetic uniformity. Nowhere is this characteristic more desirable than in the experimental animals used for biological research. For example, highly inbred strains of laboratory mice have been developed by close inbreeding over many generations. Because of homozygosity, the members of each strain are, for all practical purposes, genetically identical (isogenic). Since the individuals of each strain are genetically alike, they tend to respond uniformly to different experimental treatments. This lack of genetic variation provides greater precision in measuring responses to treatments and permits the researchers to use fewer animals for a given test. One particular use of isogenic strains of animals has been in the study of tissue transplantation. Skin grafts, as well as other types of grafts, can be successfully transplanted in isogenic animals, since the highly inbred hosts do not reject tissue from donors of the same isogenic strain.

Outbreeding and Crossbreeding. Like inbreeding, the crossing of two different inbred lines (or outbreeding) produces genetic uniformity. But the uniformity created by hybridization is only temporary, since the F_1 hybrids cannot breed true. Therefore, the primary value of outbreeding lies in the short-term advantages of heterosis.

Numerous cases are known in which the advantages of heterosis have been exploited in animal breeding. In poultry, for instance, considerable increases in meat and egg yields have been obtained through outbreeding, and millions of hybrid chicks are now produced each year. Substantial increases in weight are also obtained through outbreeding in pigs, as are improvements in meat and wool production in sheep and in milk and butterfat production in cattle. But the slowness of reproduction of the larger domestic animals (such as cattle) makes the formation of the pure lines that are needed for outbreeding less practical in these species.

Inbreeding and subsequent outbreeding do not introduce any new genetic information into a population. They simply sort out and redistribute the genes that were originally present. To introduce a new characteristic into a strain or breed requires a mutation, which is extremely rare, or matings with individuals from an entirely different race or species. Such matings, known as crossbreeding, constitute an extreme form of outbreeding. This process has been used quite successfully in plants for the introduction of hardiness, disease resistance, and other commercially desirable characteristics into cultivated varieties (see Chapter 17).

Crossbreeding is also possible in animals, but to a more limited extent. While the hybrids produced from interspecific crosses tend to be more vigorous than their parents, they are usually sterile and cannot reproduce. A good example is the mule, which is formed from a cross between a mare (female horse) and a jack (male donkey). As a beast of burden, the mule is superior in many ways to either of its parents. It is larger and stronger than the donkey and more hardy and disease resistant than the horse. Yet because of hybrid sterility, these animals must be formed anew in each generation from crosses between horses and donkeys.

Occasionally, it is possible to form new intrafertile breeds despite the reduced fertility of hybrids. Such is the case when the bison, or American buffalo, is crossed with beef cattle. Hybrids of such crosses are usually sterile, but a few offspring occur that are fertile among themselves and are even fertile with cattle when large quantities of semen from the buffalo-cattle hybrids are used to inseminate the cattle artificially. Subsequent selection of the offspring has resulted in a breed known as the beefalo, which produces more usable meat and is ready for market sooner than domestic beef.

The hybrids of cattle and species such as the yak also show hybrid vigor, as do the hybrids of crosses between the standard breeds of cattle and the zebu, a separate race of cattle that is native to India. One breed of beef cattle (known as the Santa Gertrudis) has been developed in the United States from crosses between Shorthorn cattle and the American Brahman, which consists of about seven-eighths zebu and one-eighth domestic breed. The Santa Gertrudis breed resulted from several years of backcrossing, selection, and inbreeding; it combines the heat-tolerance and resistance to subtropical animal pests exhibited by the Brahman with the superior meat production of the Shorthorn (Fig. 16.13).

Figure 16.13. The Santa Gertrudis breed and the breeds used in its development. (a) A Brahman bull. (b) A Shorthorn cow. (c) A Santa Gertrudis bull.

(a)

(b)

(c)

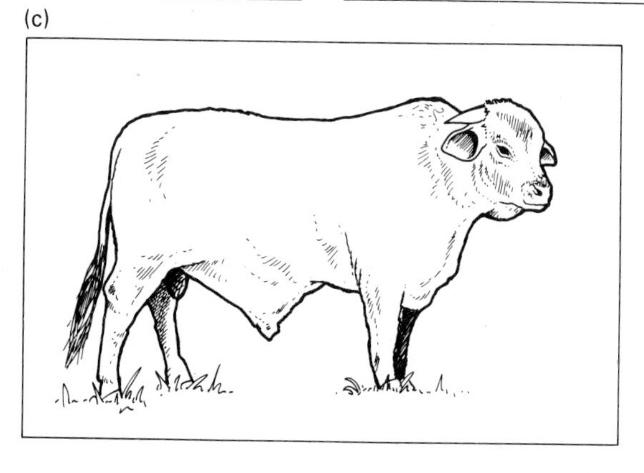

Production of Hybrid Seed Corn. The practical advantages of heterosis have been exploited most successfully with self-fertilizing plants. In these plants, the inbred lines with the desired characteristics are obtained at a comparatively rapid rate and the space and labor that are required for producing and testing the lines are not prohibitive. One of the greatest successes has been the development of hybrid corn. Hybrid vigor is so marked in this species that together with improved farming methods, hybrid corn has contributed to an approximate tripling of average per-acre yields in the United States during the past 30 years. Most corn now planted in the United States is grown from hybrid seed.

The usual method for producing hybrid seed for use in farming is to begin by establishing inbred lines through controlled pollination. The lines with the most desirable characteristics are kept and are crossed in paired combinations to produce the F_1 hybrids. In order for the plant breeders to cross the inbred lines with each other, the seed-bearing parental plants must either have their tassels removed or have male sterile cytoplasm (see Chapter 8). [There was a strong move to cytoplasmic male sterility in the production of hybrid corn until 1970. During that year a new race of Southern corn leaf blight (*Helminthosporium maydis*) appeared that preferentially attacked plants with "T" ("Texas") cytoplasm, the type being used. In the following year seed producers reverted to detasselling, although other blight-resistant sterile cytoplasms are now being used to some extent. Reliable cytoplasmic male sterility mechanisms are currently being researched, since male sterility has important advantages in ensuring purity in addition to cost savings for both the farmer and the seed producer.] Modern detasselling methods are not always manual. Special machines can be employed that pull most (70 to 95 percent) of the tassels, and a crew then follows the machine and detassells the rest. These detasselled plants are subsequently fertilized by the wind-borne pollen from the other parental line.

Generally, the yields of hybrid seed in pounds or bushels from the inbred lines are low, since the inbred plants are small and have little vigor. Hybrid seed was therefore produced commercially (prior to 1960) from crosses between the more uniform and vigorous F_1 hybrids of these inbred lines, which serves to reduce the cost of commercial seed. A procedure that once was widely used in the production of hybrid seed is diagrammed in Fig. 16.14. In this example, crosses are made between inbred lines A and B and between inbred lines C and D to produce two single-cross F_1 hybrids, designated AB and CD. The hybrids are then crossed to produce double-cross AB × CD hybrid seed, which is sold to farmers for planting.

We should stress that the double-cross procedure that is used for the production of hybrid corn does not improve heterosis above the level that is manifested in the single-cross hybrids. This procedure merely lowers the cost of hybrid seed by having the seed produced on the large and uniform ears of the single-cross plants. In the mid-1960s, several seed companies, led by DEKALB, began moving into single-cross methods to produce seed corn. The plant breeders recognized that even though the yield in pounds or bushels from a female inbred plant is much lower than that from a female single-cross plant, the yield in terms of the *number* of seeds produced is about the same, since seeds produced on inbred plants are smaller. Despite their smaller size, these seeds germinate as well as the seeds produced by hybrid plants and they actually grow a more productive plant (in terms of yield per acre of seed produced) than does the double-cross hybrid seed, especially at the higher yield levels (e.g., above 100 bu/A) possible with modern farming methods. Single-cross hybrids respond to fertilizer, high plant population density, and other improved management factors better than double-cross hybrids.

Figure 16.14. The double-cross method used for producing hybrid corn. Crosses between inbred lines A and B and between inbred lines C and D yield the single-cross hybrids AB and CD. These hybrids are then crossed to produce the double-cross hybrid seed (AB × CD), which is sold to the farmer for planting. *Source:* Reprinted by permission from F. J. Ayala and J. A. Kiger, Jr., *Modern Genetics*. Copyright © 1980, by The Benjamin/Cummings Publishing Company.

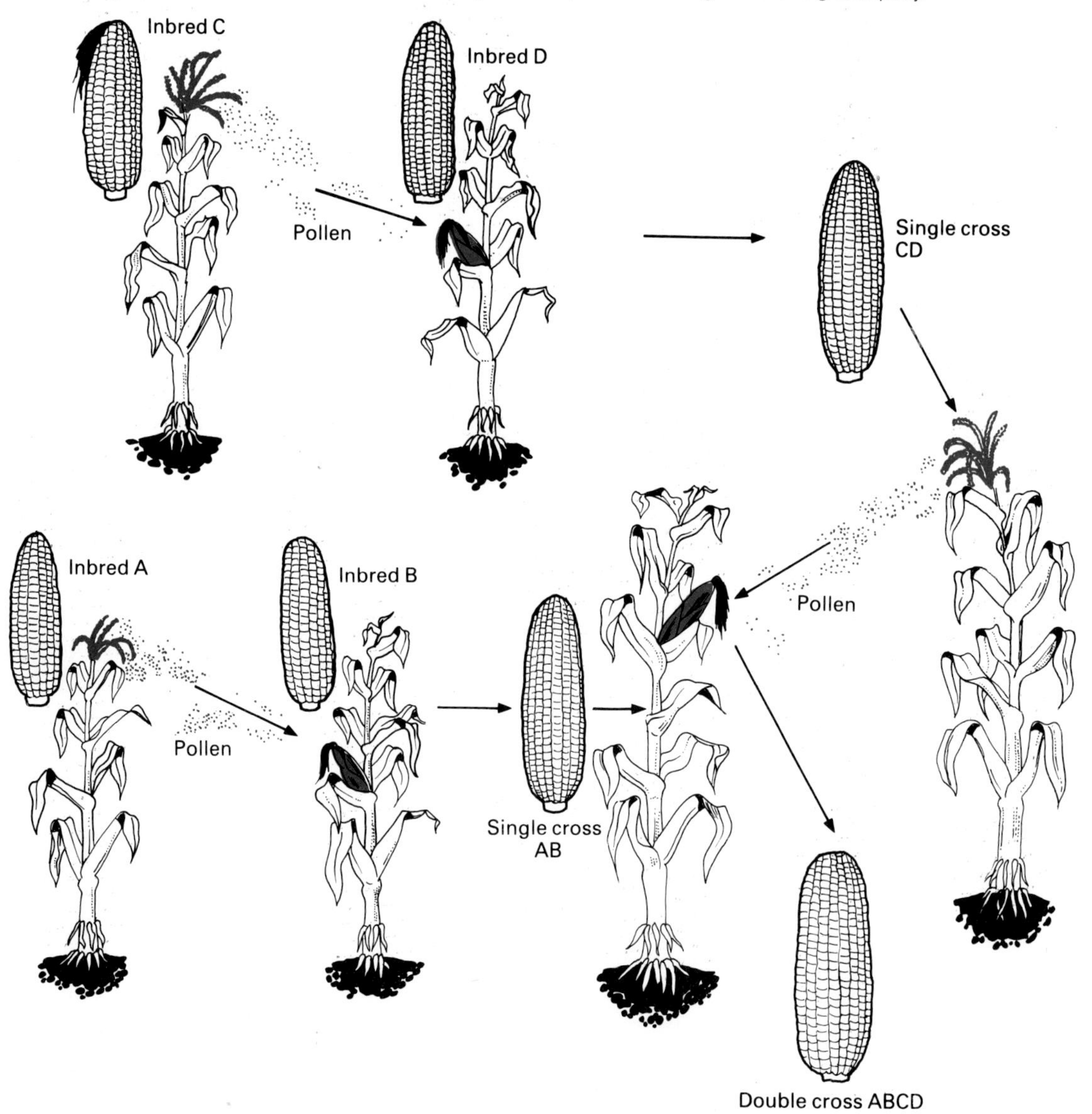

Therefore, even though single-cross seed prices are somewhat higher than those for double-cross seed, the agronomic management level of corn farming has improved to the point that single-cross hybrids now have a much higher return on investment to the farmer than double-cross seed. As a result, double-cross hybrid seed is now rare in the United States. Only in countries such as Argentina, where rainfall amounts and distribution are poor, are double crosses still the rule.

TO SUM UP

1. The increase in the performance of hybrids that is produced by crossing different inbred lines is known as heterosis or hybrid vigor. The complement of heterosis is inbreeding depression, which is the reduction in average performance that is associated with inbreeding.
2. We can quantify the concept of hybrid vigor by expressing heterosis as the excess in vigor of an F_1 hybrid over the midpoint of the inbred parental lines.
3. The dominance theory of heterosis proposes that the increase in vigor of the heterozygotes is caused by the combined dominance effects of polygenes. According to this view, the heterozygous condition is not a requirement for increased vigor, since a homozygous combination of dominant genes at every locus is assumed to add as much (if not more) to the overall performance as one that is completely heterozygous.
4. The overdominance theory holds that the heterozygote state is essential for the full expression of heterosis. This theory assumes that the performance of the heterozygote is intrinsically superior to that of either homozygous genotype.
5. Both inbreeding and outbreeding have practical applications. Inbreeding is generally used for the fixation of desirable characteristics through homozygosity and for the production of genetic uniformity in experimental animals and plants. Outcrossing, as exemplified by the production of hybrid seed corn, is used primarily to exploit the practical advantages of heterosis.
6. Crossbreeding is the mating of individuals from entirely different races or species. This method is commonly employed to introduce certain desirable characteristics (such as disease resistance) into a breed of domestic animal or a variety of cultivated plant.

Genetic Advance Through Selection

There are two general ways in which a plant or animal breeder can alter the genotype frequencies within a population. One is to control the way in which individuals are mated. As we have seen, the use of this method controls the mating system of the population and determines whether the individuals undergo inbreeding or outbreeding. A second and more basic approach is to choose which individuals are to be used as parents (assuming that all are not free to mate). This method, which alters the genotype frequencies by changing the gene frequencies, constitutes the process of artificial selection. It is this second approach to plant and animal breeding that we shall examine in the present section.

Response to Individual Selection

The most direct approach to deciding which members of a population are to be used as parents is to choose or reject individuals solely on the basis of their own merits, without reference to pedigree or progeny. This selection process can happen in one of two ways. The breeder may simply weed out the undesirable types in each generation to prevent their reproduction. Or the breeder may proceed in a more positive manner by choosing individuals with certain outstanding qualities and then confining reproduction to this select group. The end result is much the same with either method. In both cases, there is an improvement in the average phenotype among the parents of the next generation, and if the trait is at all heritable, there is also a change in the gene frequencies in the desired direction. This method of selection, which is based strictly on the performance of each individual, is known as *individual selection.* It is also called *mass selection,* especially when the individual matings are not controlled within the selected group. Mass selection is a common plant-breeding practice; seeds are often selected from only desirable plants and are planted en masse for reproduction.

Selection Differential and Regression. The method of individual (or mass) selection is summarized diagrammatically in Fig. 16.15. Since most characters of economic importance are quantitative traits, the distribution of phenotypic values in the population is pictured in terms of a continuously varying bell-shaped curve. Included within this curve is the corresponding distribution for the selected parents, with an average value that is greater than the mean of the original population. The difference between the mean of the selected parents and the mean of the base population from which the parents were selected is called the **selection differential.** This measure of selection intensity is commonly employed when selection acts on a continuously varying trait.

The response to selection, as measured by the improvement in the mean phenotypic value of the base population, is illustrated in Fig. 16.16. Observe that different degrees of response are possible, depending on the importance of the environmental variation. If the character under study is determined entirely by

Figure 16.15. Effects of the process of individual (or mass) selection on a continuously varying trait. The merit of individual phenotypes is assumed to vary linearly from least desirable on the left to most desirable on the right. The selection differential is the difference between the population mean and the mean of those selected and is the measure of selection intensity. (a) In truncate selection, all of the individuals above a certain value are chosen as parents regardless of the merit of other characteristics. (b) In selection on the basis of total merit, weight is given to the desirability of other characteristics in addition to the one of major concern.

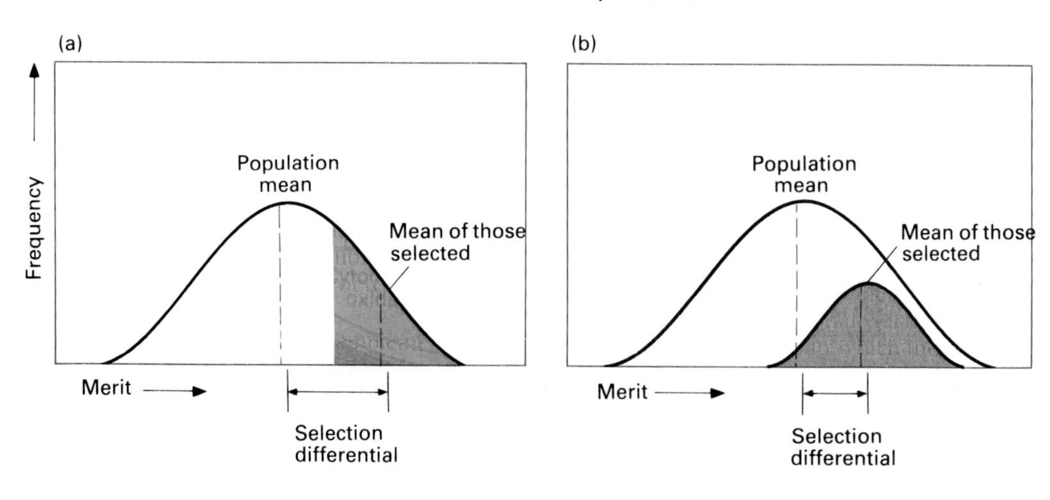

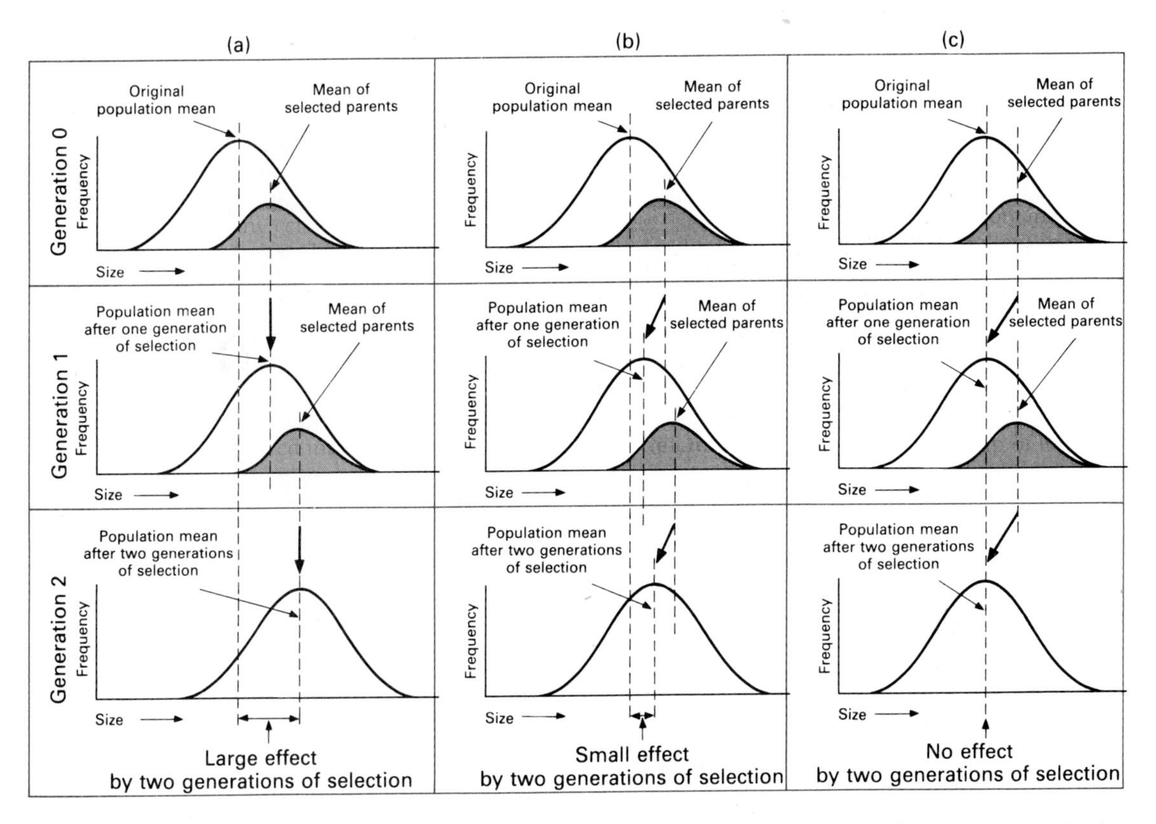

Figure 16.16. Effects of two generations of selection on a quantitative trait. The response to selection is measured by the improvement of the mean phenotype of the base population. (a) The phenotypic variability is caused entirely by additive gene effects. The mean of the progeny is equal to the mean of the selected parents. (b) The phenotypic variability is caused by variation in both heredity and environment. The progeny regress toward the original population mean. (c) The phenotypic variability is caused entirely by environmental variation. The progeny mean is now the same as the mean of the population from which their parents came.

additive gene effects (Fig. 16.16a), the progeny mean equals the mean of the selected parents. The selection response is then equal to the selection differential.

When the selected trait is an expression of additive gene effects, it is easy to demonstrate the equality of the progeny and parental means by considering the distributions of progeny that are produced by various matings among the selected parents. For example, consider a trait that is determined by three gene pairs. Let us suppose that each active gene contributes one unit to the overall trait and that each of the parents that are selected has five contributing genes in total. A representative mating might be as follows:

GENOTYPIC CONTRIBUTION	PHENOTYPIC VALUE
AaBBCC × *AaBBCC*	5 × 5
↓	↓
¼ *AABBCC* + ½ *AaBBCC* + ¼ *aaBBCC*	¼(6) + ½(5) + ¼(4) = 5

The average phenotypic value of the progeny is thus the same as that of their parents. (Convince yourself that this result is also true for other possible parental mating combinations.)

Departures from the one-to-one relationship between the means of the progeny and parental distributions will occur when the gene effects are subject to environmental variation. The appearance, or phenotype, of an individual then becomes an unreliable indicator of the genotype. For example, a dairy farmer who is selecting for increased milk yields may by chance choose a cow with only a mediocre genotype but with an outstanding production record that is attributable mainly to a favorable environment. Conversely, the dairy farmer may fail to select a cow with a superior genotype that has a production record that is less than favorable because of poor environmental circumstances. The net result of such errors is the tendency of the progeny to average less than the mean of their selected parents (Fig. 16.16b). We can most clearly picture the basis for this reduction in selection efficiency by considering a situation in which all of the variation in individual performance is of environmental origin (Fig. 16.16c). In this situation, selection for increased performance will be based strictly on environmental advantage, since the genotypes are all the same. As long as the overall environment is not improved, the progeny produced under these conditions will average no better than the population from which their parents came.

The tendency on the part of the offspring of selected parents to slip back to the average of the population from which the parents were chosen is known as *regression to the mean* (or simply *regression*). Regression obviously acts to impede any changes made in average performance; it therefore constitutes a major factor that limits the rate of genetic advance through selection.

Environmentally induced variation is only one of several factors that act to promote regression to the mean. Another factor is dominance. With dominant genes, the heterozygotes will equal or approach the dominant homozygotes in performance level. In this case, errors can result from mistaking a heterozygote with a superior phenotype, but with low transmitting ability, for the desired true-breeding homozygous strain. Here, too, the phenotype is an unreliable indicator of genotype. For a numerical example, let us assume that a certain trait is determined by three gene pairs, all of which show complete dominance, and that each active gene in either a homozygous or a heterozygous combination (e.g., *AA* or *Aa*) contributes one unit to the overall phenotype. Suppose that in selecting parents with the optimum phenotype, we choose individuals that have only five contributing genes, as follows:

GENOTYPIC CONTRIBUTION	PHENOTYPIC VALUE
AaBBCC × *AaBBCC*	3 × 3
↓	↓
¼ *AABBCC* + ½ *AaBBCC* + ¼ *aaBBCC*	¼(3) + ½(3) + ¼(2) = 2.75

Regression, as evidenced by the reduction in the average phenotype, is in this case a consequence of allelic segregation at heterozygous loci. Note that the contribution of dominance to regression would be lacking in the cross *AABBCC* × *AABBCC*, as might be the case if we were able to select the parental genotypes without ambiguity.

Nonallelic gene interaction is another factor that serves to lower the predictability of the selection response when the parents are chosen solely on the basis of phenotype. These interactions are therefore a further cause of regression. We can illustrate this relationship by the following example involving epistasis. Let us consider a trait that is determined by the additive effects of two gene pairs, with each active allele at the *B,b* and *C,c* loci contributing one unit to the overall phenotype. Let us also assume a type of interlocus interaction in which the *B* and *C* genes are expressed when the genotype includes *AA* or *Aa* but are not expressed in the presence of the *aa* gene pair. A representative mating between parents of optimum phenotype might then be as follows:

GENOTYPIC CONTRIBUTION	PHENOTYPIC VALUES
AaBBCC × *AaBBCC*	4 × 4
↓	↓
¼ *AABBCC* + ½ *AaBBCC* + ¼ *aaBBCC*	¼(4) + ½(4) + ¼(0) = 3

Again, we see a reduction in the average phenotype. We can envision numerous other examples of epistasis, in which the interlocus interactions have more subtle effects on the phenotype, with the gene loci interacting in pairs, in threes, or in higher numbers.

Selection Response and Heritability. As we have seen, a population's response to selection is enhanced by additive gene effects. The predictability of the response is maximum when all of the variation is additive, so that the mean of the progeny equals the mean of the parents. Predictability declines with dominance, interlocus interactions, and environmental deviations, as the phenotype becomes a progressively less reliable indicator of an individual's breeding value. We can quantify these observations by relating the selection response (designated by *R*) to the selection differential (symbolized as *S*) by means of the following equation:

$$R = (H_N)S. \tag{16.5}$$

In this expression, H_N is the heritability in the narrow sense: a factor that measures the fraction of the total variation that is a result of additive gene effects. The value of H_N serves a predictive role in breeding programs by expressing the reliability of the phenotype as a guide to an individual's genotype. It is defined as the ratio of additive genetic variance to phenotypic variance:

$$H_N = \frac{V_A}{V_P}. \tag{16.6}$$

Thus, when all of the phenotypic variation is a result of additive gene effects, $H_N = V_A/V_A = 1$. Then $R = S$ and the gain in selection is equal to the selection differential. On the other hand, where there is no additive genetic variance, $H_N = 0$ and this method of selection is totally ineffective.

By considering each of the variables in Eq. (16.5) in greater detail, we can use this relationship to illustrate the major factors that are important in a breeding program. First, the selection differential (*S*) depends on the amount of phenotypic variability in the base population and on the proportion of the population that is selected. This dependence is shown in Fig. 16.17 in terms of three distributions of phenotypic values. For example, if only the top 30 percent in each of two base populations is kept for breeding (Figs. 16.17a and b), the selection differential

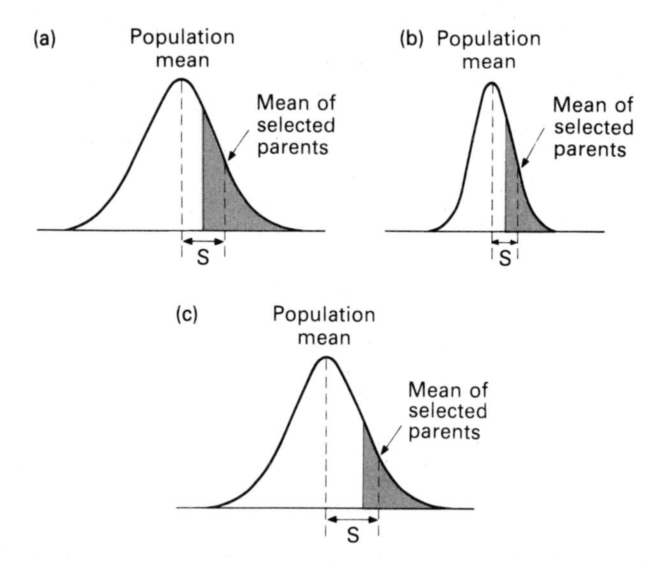

Figure 16.17. Dependence of the selection differential (S) on the amount of variability of the character and on the fraction of the population that is selected. (a) 30% selected; standard deviation 2 units. (b) 30% selected; standard deviation 1 unit. (c) 20% selected; standard deviation 2 units.

(and thus the selection response) will be greater for the more variable population. The selection intensity could be strengthened even further if only the top 20 percent of the more variable group is chosen (Fig. 16.17c). Second, the heritability (H_N) designates that only the additive component of the total variability is important for genetic improvement. A breeder would thus maximize the chances of success by selecting the most desirable individuals from a highly variable base population in which there is a substantial heritability for the trait in question.

The heritability values for various characteristics in domestic plants and animals are listed in Table 16.4. We must emphasize that the H_N values are not fixed

Table 16.4. Heritability in the narrow sense for various traits among domestic plants and animals.

CHARACTERISTIC	HERITABILITY (H_N)
Slaughter weight in cattle	0.85
Plant height in corn	0.70
Root length in radishes	0.65
Egg weight in poultry	0.60
Thickness of back fat in pigs	0.55
Fleece weight in sheep	0.40
Milk production in cattle	0.30
Yield in corn	0.25
Egg production in poultry	0.20
Ear length in corn	0.17
Conception rate in cattle	0.05

properties of the listed characteristics but tend to vary with the population and the environment. For instance, a low heritability value may simply mean that there is a large environmental component of variation in the population in which it was measured. The effectiveness of selection on such a character might then be improved by lowering V_E by employing more carefully controlled methods of husbandry.

EXAMPLE 16.4. The heritability (H_N) of milk production in cattle is 0.3. The average annual milk production in a large dairy herd is 14,000 lb. The average annual milk production of individuals selected to be parents for the next generation is 18,000 lb. Determine the average annual milk production expected among the progeny.

Solution. Recall that $R = (H_N)S$. In this case, $H_N = 0.3$, $S = 18{,}000 - 14{,}000 = 4000$ lb/yr, and R = average value of progeny minus 14,000. Therefore, the average value of the progeny = $14{,}000 + (0.3)(4000) = 15{,}200$ lb/yr.

Selection for More Than One Character. Genetic progress is most rapid when a plant or animal breeder can restrict selection to a single trait. This is seldom the case, however. Breeders usually cannot afford to select rigidly for a single character and ignore all other characters; rather, they must take other factors into consideration. For example, in selecting for breeding animals, a dairy farmer is not only concerned with total milk production but also with butterfat percentage and a number of other characteristics that must conform to breed or type, such as color, spotting pattern, and the conformation of the back and udder. While the appearance of the individual is not necessarily related to milk production, selecting for characteristics that are considered superior by show-ring standards is of obvious economic benefit, since it further improves the cash value of the animal.

As a general rule, genetic progress varies inversely with the number of criteria used in judging the merit of an individual. We can most readily see this relationship when we select for characters that are independently inherited. Suppose that there is a 1 in 10 chance of selecting an individual with a superior genotype for each of a number of desired characteristics. There is then a 1 in 100 chance of simultaneously selecting the proper combination of genes for two of these traits, and a 1 in 1000 chance for three. Obviously, no breeder who selects on the basis of two or more components independently can expect to achieve success without compromising the initial objectives to some extent. For example, a poultry breeder may find it necessary to choose hens with a lower-than-desired body weight at maturity in order to ensure a better-than-average egg production. By selecting in this way, the breeder reduces the effectiveness of selection for body weight alone.

A breeder reaches an impasse in attempting to select simultaneously for negatively correlated characters. Negatively correlated characters are different traits within the same organism (such as milk yield and butterfat percentage in cattle and egg size and egg production in chickens) for which an improvement in one is normally associated with a decline in the other. Certain traits may also show positive correlation. For example, traits such as milk yield and total butterfat production in cattle tend to increase simultaneously. While the degree of correlation is less than perfect in this case, choosing parents on the basis of this combination of traits would not seriously impair genetic progress, since selection for one is, in effect, selection for the other.

Other Methods of Selection

We have seen that genetic gains under selection depend largely on the accuracy with which a breeder can judge the genetic merit of individuals. Since the appearance of an individual is only at best an uncertain guide to its breeding value, individual selection tends to be accompanied by errors of a genetic nature and is of questionable usefulness when the heritability of a trait is low. Several methods have been devised that help to correct errors in selection based on phenotype alone. In this section, we will briefly describe a few of these procedures.

Family Selection. One approach that is taken in animal breeding is to supplement the phenotype as the sole guide to an animal's breeding worth with information from the relatives and descendants of the individual. The advantage gained from considering an individual's pedigree and progeny is greatest when the heritability of a trait is low, especially when environmental deviations constitute a large part of the phenotypic variance. By averaging the phenotypic values of the family members, environmental "noise" tends to cancel out, and a better estimate of the genotypic value can be made.

Information from relatives decreases rapidly in importance as their degree of relationship to the individual declines. Full sibs have both parents in common and share half of their genes, on the average; half sibs have only one parent in common and share only one-quarter of their genes; and so on. The most useful information regarding the genotype of an individual can be obtained from (1) the phenotypes of its siblings and (2) the phenotypes of its progeny. Information from full or half sibs is commonly used when we cannot measure the breeding worth of an individual directly. For example, the genetic potential of a dairy bull for milk production cannot be determined directly, but it can be estimated from the production records of the female relatives.

Progeny testing, the second source of information, can also be used to achieve the same end. Recall from Chapter 1 that a progeny test is a method of estimating the genotype of an individual from the phenotypes of its offspring. When this method is used, the breeding worth of a male for such decidedly female characters as egg and milk production can be expressed mathematically by using some quantitative measure of merit. One measure, known as the midpoint index, is a comparatively simple measure of breeding worth that apportions equal weight to the transmitting abilities of the male and female parents. By assuming that the production averages of the mothers and daughters are accurate measures of their genetic values, then the average of the daughters = ½(average of dams) + ½(transmitting potential average of sire), or

$$\text{Transmitting potential of sire} = 2(\text{average of daughters}) - (\text{average of dams}).$$

One drawback of the progeny test is that the information concerning a parent's breeding value must await the measurement of its offspring. This delay slows the selection process, since the selection of the parents cannot be carried out until after the offspring have reached sexual maturity.

Pedigree Selection and Line Breeding. Some of the guesswork involved in selection can be eliminated by the use of inbreeding. One method, known as pedigree selection, has been used quite successfully among self-fertilizing plants. In this method, plants are allowed to self-pollinate to form inbred lines. Different lines are then isolated, and the promising ones are tested for a variety of desired characteristics with the goal of developing superior, true-breeding strains. Pedigree selection differs from mass selection in that the homozygous varieties are produced by this method through inbreeding rather than exclusively through selection. Pedigree selection is generally more expensive than mass selection, since it requires the testing of several inbred lines and can also lead to inbreeding depression. When loss of vigor is a problem, mass selection without inbreeding is preferable, since the homozygosity that is needed for the formation of uniform lines can then be restricted to only those characteristics that are subjected to the selection process.

Another form of selective inbreeding that is commonly used by animal breeders is line breeding. In line breeding, animals are selected in part on the basis of pedigree, in such a way that the parents chosen are always related by descent from one or more outstanding ancestors. The degree of relationship between mates need not be high and is often kept at a minimum so as to avoid the undesirable consequences of close inbreeding. This form of breeding conserves valuable characteristics by concentrating them within a line of descent and thus maximizes the contribution of an individual (or individuals) whose characteristics are admired.

TO SUM UP

1. In individual (or mass) selection, reproduction is confined to a select few who are chosen solely on the basis of individual merit, without reference to pedigree or progeny. The intensity of the selection process is measured in terms of the selection differential, which is the difference between the mean of the selected parents and the mean of the base population from which the parents were chosen.
2. When a selected character is determined entirely by additive gene effects, the mean of the progeny equals the mean of the selected parents. The selection response, as measured by the improvement in the mean phenotypic value of the base population, is then equal to the selection differential.
3. Variation in the form of dominance, interlocus interactions, and environmental deviations produces a phenomenon known as regression to the mean, in which the offspring of selected parents tend to slip back to the average of the population from which the parents were chosen. The probability that the progeny will not regress is measured by the heritability in the narrow sense, which is expressed as the ratio of additive genetic variance to phenotypic variance: $H_N = V_A/V_P$.
4. Genetic gains under individual (or mass) selection depend on the accuracy with which a breeder can judge the genetic merit of an individual from its phenotype. Information concerning the relatives and descendants of the individual helps to reduce the number of errors made in selecting desirable breeding stock. Of particular value in this regard are the phenotypes of the siblings and progeny of the individual concerned.

Questions and Problems

1. Define the following terms:
   Consanguineous matings
   Crossbreeding
   Inbreeding coefficient
   Line breeding
   Regression to the mean
   Selection differential
2. Distinguish between the members of the following paired terms:
   (a) Assortative vs. disassortative mating
   (b) Autozygous vs. allozygous genotypes
   (c) Inbreeding vs. outbreeding
   (d) Inbreeding depression vs. heterosis
   (e) Individual vs. pedigree selection
3. How is it possible for the members of a small population to select mates at random and still undergo inbreeding at the same time?
4. Calculate the inbreeding coefficient for individual I in each of the following pedigrees:

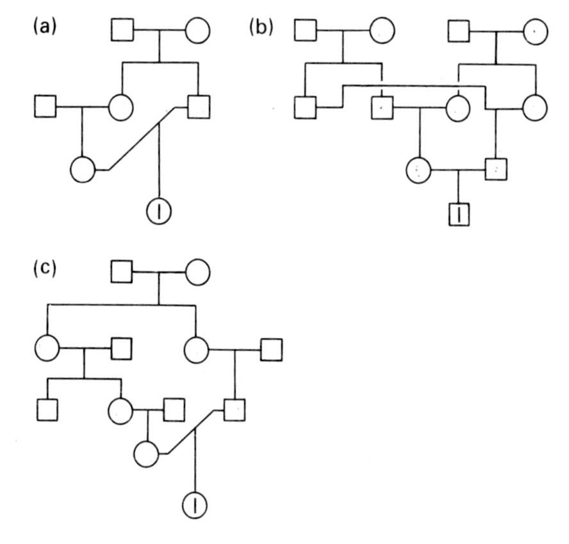

5. A population has allele frequencies of $p = 0.4$ and $q = 0.6$. What are the expected frequencies of the genotypes (*AA, Aa, aa*) in the population when the inbreeding coefficient equals 0.5?
6. A strictly self-fertilizing plant population is started with ¼ *AA*, ½ *Aa*, and ¼ *aa* individuals. What will be the genotype frequencies after **(a)** 1, **(b)** 2, **(c)** 3, and **(d)** *t* generations of selfing?
7. An autosomal recessive disorder occurs at a frequency of 1/10,000 in a large random mating population. The incidence of this disorder is 50 times higher in a neighboring religious isolate. If the allele frequencies are the same in both populations, what is the inbreeding coefficient for the isolate?
8. When hybrid seed for field corn is produced through a double cross, why are the progeny of the double cross generally more variable than the progeny of either single cross?
9. Suppose that a breeder has two strains of barley that produce about the same amount of grain per acre under similar field conditions. The breeder has tried various selection procedures in the past and has been unsuccessful in improving grain production by either strain. Crossbreeding alone has also proved unsuccessful, since no improvement in yield was observed with the hybrid that was formed by crossing the two strains. When the hybrid was subjected to several generations of mass selection, however, the breeder was able to increase the yield considerably. Provide an explanation for these results.
10. In the following pedigree, individual I is an offspring of a marriage between first cousins.

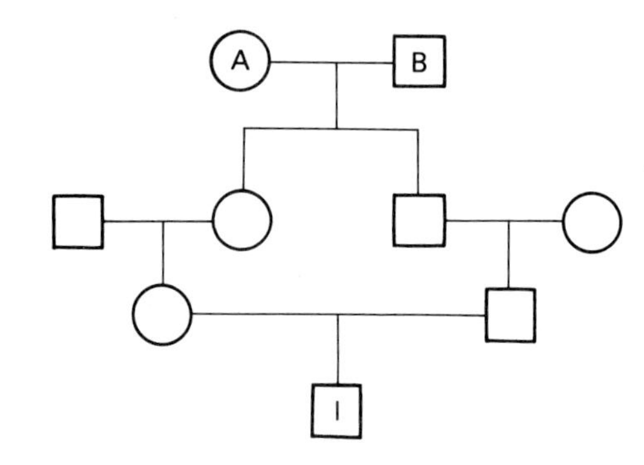

Determine the probability that individual I is affected by a rare autosomal recessive disorder **(a)** if his great-grandmother (designated A in the pedigree) is affected; **(b)** if his maternal grandmother is affected; **(c)** if there are no affected relatives and the frequency of the recessive allele in the population is 0.01.
11. Suppose that individual A in the pedigree in Problem 10 is heterozygous for defective tooth enamel (a rare X-linked dominant disorder) and that individual B is colorblind (a rare X-linked recessive disorder). Determine the

probability that individual I **(a)** is colorblind; **(b)** has defective teeth; **(c)** has normal teeth and normal color vision.

12. How many generations of self-fertilization are required to reduce the frequency of the heterozygotes in a plant population to 0.4 percent of its initial value?

13. Show algebraically that the frequencies of genotypes in a population with inbreeding, which are given in Fig. 16.5, can be expressed in the following modified forms: $f(AA) = p^2 + pqF$, $f(Aa) = 2pq - 2pqF$, and $f(aa) = q^2 + pqF$.

14. A sample that is selected at random from a population consists of 52 *AA*, 28 *Aa*, and 20 *aa* individuals. If inbreeding is the only process that contributes to changes in genotype frequency in this population, what is the value of the inbreeding coefficient?

15. Because of random genetic drift, the frequency of a deleterious recessive gene in a small religious isolate is 0.05, which is 50 times higher than in the surrounding population. **(a)** What is the frequency of the recessive disorder among children of unrelated parents in the isolate? In the surrounding population? **(b)** What is the frequency of the recessive disorder among children of marriages between first cousins in the isolate? In the surrounding population? **(c)** By what factor do marriages between first cousins increase the frequency of the disorder above that of random mating in the isolate? In the surrounding population? Explain the difference in values.

16. Starting with a cross between two inbred strains of tobacco, H. H. Smith obtained an F_1 with an average height of 43.2 inches and an F_2 with an average height of 40.6 inches. Although the F_1 was shorter than the tallest parental strain by an average of 4.6 inches, there was heterosis, evidenced by the fact that both the F_1 and F_2 generations exceeded the midpoint height of the parental plants. From these results, predict the average plant heights of both inbred parental strains. Compare your estimates with the actual measured values of 47.8 and 28.7 inches.

17. The average monthly egg production of a poultry flock is 15 eggs per hen. Individual hens that average 18 eggs per month are selected from the flock for breeding purposes. If the heritability of egg production is 0.2, calculate **(a)** the selection differential, **(b)** the expected gain in egg production (the selection response), and **(c)** the predicted average monthly egg production in the next generation.

18. The annual milk production records of three daughters of each of two bulls are compared in the following table with those of their dams:

Bull A:	Daughter	Dam
	15,155 lb	14,154 lb
	17,350 lb	16,768 lb
	12,160 lb	13,756 lb

Bull B:	Daughter	Dam
	13,112 lb	11,716 lb
	12,283 lb	9,947 lb
	18,350 lb	13,154 lb

Calculate the midparent index for each bull. Which is the better bull?

19. Suppose that a plant that is initially heterozygous for three independently assorting gene loci (*A,a; B,b;* and *C,c*) undergoes repeated self-fertilization. **(a)** What is the probability that the *A,a* locus is still heterozygous after two generations of selfing? **(b)** What is the probability that any two of the loci are still heterozygous after two generations of selfing? **(c)** What is the probability that none of the loci will be heterozygous after two generations of selfing?

20. Certain researchers have objected to the dominance theory of heterosis on the basis that actual experiments have failed to find pure-breeding (homozygous) lines that perform as well as the F_1 hybrids. Proponents of the dominance theory counter by pointing out that quantitative traits that show heterosis are usually determined by genes at many loci. The probability of obtaining highly vigorous true-breeding lines is thus extremely small. For example, suppose that heterosis is determined by the presence of completely dominant alleles at each of 10 independently assorting gene loci. Show that when the F_1 heterozygotes are intercrossed, the probability of obtaining a pure-breeding line for the phenotype expressed by the hybrid is, in this case, less than one in a million.

21. According to the dominance theory, another factor that reduces the chances of obtaining a pure-breeding line that performs as well as the hybrid types is the presence of linked genes. Consider the following hybrid arrangement of genes as an example:

$$\frac{A\ b\ C\ d\ E\ f}{a\ B\ c\ D\ e\ F}$$

Suppose that heterosis is determined by the completely dominant (capital-letter) allele at each of these loci. If each neighboring pair of loci is separated by 10 map units, what is the probability that selfing the hybrid will produce a pure-breeding line with the same phenotype as that expressed by the hybrid arrangement of genes?

22. Assume that height in a particular plant is determined by five independently assorting gene pairs (*A,a; B,b; C,c; D,d;* and *E,e*), with each capital-letter gene contributing 2 inches when in the heterozygous or homozygous condition, to a base height of 10 inches. For example, if we ignore the environmental effects, the heights of the *aabbccddee* and *A-B-C-D-E-* plants are thus 10 inches and 20 inches, respectively. Assume that you have a plant of genotype *AaBbCCDDEE* at your disposal. **(a)** If this plant is selfed, what average height is expected among the plants of the F_1 generation? **(b)** Calculate the inbreeding depression that would be observed in this case. **(c)** Suppose that you select all 20-inch plants from the F_1 population produced in this experiment to serve as parents for the next generation (the F_2). If you self these selected F_1 plants (en masse), what is the average height expected in the F_2? **(d)** Calculate the selection response, the selection differential, and the heritability in the narrow sense (H_N) for height in this selection experiment.

23. Suppose that a fraction c of a population practices first-cousin matings, while the rest $(1 - c)$ mates at random. Show that the fraction of all recessive homozygotes in the population that are produced by first-cousin matings can be expressed as

$$\frac{c(16q^2 + pq)}{16q^2 + cpq} = \frac{c(15q + 1)}{16q + cp}.$$

24. The following is a pedigree of Roan Gauntlet, a famous Shorthorn bull.

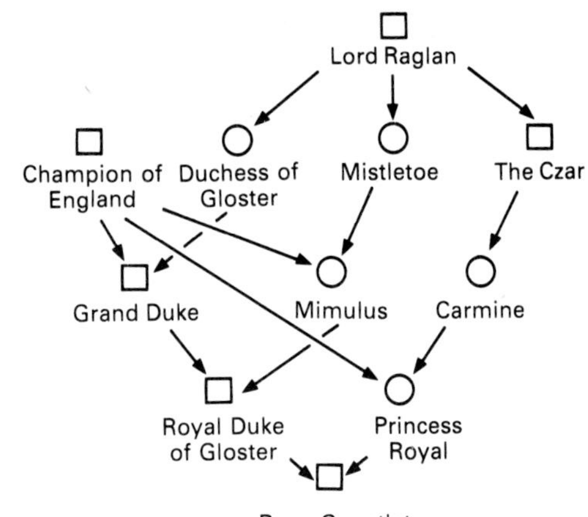

Calculate the inbreeding coefficient of Roan Gauntlet. (*Note:* the pedigree consists of four different circular paths, two containing Lord Raglan and two containing Champion of England.)

25. One extreme form of line breeding that helps to maximize the genetic contribution of an outstanding individual involves performing repeated backcrosses between the individual (usually a male) and its offspring from successive generations. This breeding pattern is illustrated by the following pedigree, in which a male (A) is mated successively to his daughter (C) and then to his granddaughter (D):

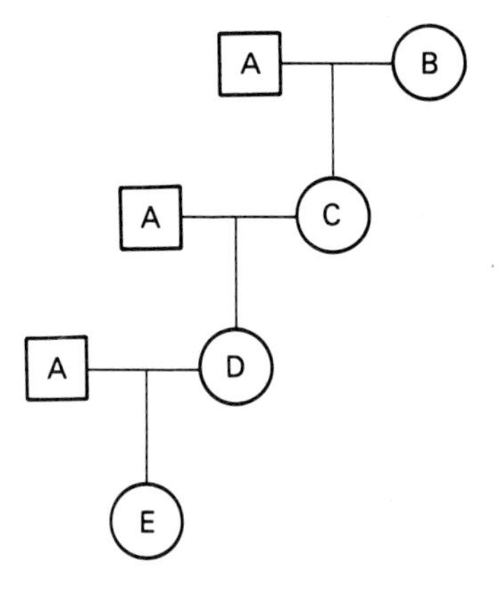

(a) Assuming that A and B in the pedigree are unrelated and not inbred, calculate the inbreeding coefficient of the daughter (C). Of the granddaughter (D). Of the great-granddaughter (E). **(b)** What percentage of the genes of individual A does the daughter contain? The granddaughter? The great-granddaughter?

Suggested Readings / Part VI

Chapter 13

Clowes, R. C. The molecule of infectious drug resistance. *Scientific American* 228:18–27, 1973.

Crow, J. F. and M. Kimura. *An Introduction to Population Genetics Theory,* Harper & Row, New York, 1970.

Elseth, G. D. and K. D. Baumgardner. *Population Biology,* D. Van Nostrand, New York, 1981.

Hardy, G. H. Mendelian proportions in a mixed population. *Science* 28:49–50, 1908.

Hartl, D. L. *Principles of Population Genetics,* Sinauer, Sunderland, Mass., 1980.

Pirchner, F. *Population Genetics in Animal Breeding,* Freeman, San Francisco, 1969.

Wallace, B. *Basic Population Genetics,* Columbia University Press, New York, 1981.

Chapter 14

Allison, A. C. Protection afforded by sickle-cell trait against subtertian malarial infection. *British Med. J.* 1:290–294, 1954.

Ayala, F. J. (ed.) *Molecular Evolution,* Sinauer, Sunderland, Mass., 1976.

Bodmer, W. F. and L. L. Cavalli-Sforza. *Genetics, Evolution and Man,* Freeman, San Francisco, 1976.

Crow, J. F. and M. Kimura. *An Introduction to Population Genetics Theory,* Harper & Row, New York, 1970.

Dobzhansky, Th., F. J. Ayala, G. L. Stebbins and J. W. Valentine. *Evolution,* Freeman, San Francisco, 1977.

Eigen, M., W. Gardiner, P. Schuster and R. Winker-Oswatitsch. The origin of genetic information. *Scientific American* 244:88–118, 1981.

Elseth, G. D. and K. D. Baumgardner. *Population Biology,* D. Van Nostrand, New York, 1981.

Mettler, L. E. and T. G. Gregg. *Population Genetics and Evolution,* Prentice-Hall, Englewood Cliffs, N.J., 1969.

Roughgarden, J. *Theory of Population Genetics and Evolutionary Ecology: An Introduction,* Macmillan, New York, 1979.

Wallace, B. *Basic Population Genetics,* Columbia University Press, New York, 1981.

White, M. J. D. *Modes of Speciation,* Freeman, San Francisco, 1978.

Zuckerkandl, E. The evolution of hemoglobin. *Scientific American* 212:110–118, 1965.

Chapter 15

Bodmer, W. F. and L. L. Cavalli-Sforza. *Genetics, Evolution and Man,* Freeman, San Francisco, 1976.

Brues, A. M. A genetic analysis of human eye color. *Am. J. Physical Anthropology* (New Series) 4:1–36, 1946.

Cavalli-Sforza, L. L. and W. F. Bodmer. *The Genetics of Human Populations,* Freeman, San Francisco, 1971.

Ehrman, L. and P. A. Parsons. *Behavior Genetics and Evolution,* McGraw-Hill, New York, 1981.

Elseth, G. D. and K. D. Baumgardner. *Population Biology,* D. Van Nostrand, New York, 1981.

Falconer, D. S. *Introduction to Quantitative Genetics,* Longman, Harlow, Essex, UK, 1981.

Mather, K. and J. L. Jinks. *Introduction to Biometrical Genetics,* Cornell University Press, Ithaca, N.Y., 1977.

Chapter 16

Beadle, G. W. The ancestry of corn. *Scientific American* 242:112–119, 1980.

Bodmer, W. F. and L. L. Cavalli-Sforza. *Genetics, Evolution and Man,* Freeman, San Francisco, 1976.

Cavalli-Sforza, L. L. and W. F. Bodmer. *The Genetics of Human Populations,* Freeman, San Francisco, 1971.

Crow, J. F. and M. Kimura. *An Introduction to Population Genetics Theory,* Harper & Row, New York, 1970.

Falconer, D. S. *Introduction to Quantitative Genetics,* Longman, Harlow, Essex, UK, 1981.

Sprague, G. F. The genetics of corn breeding. *Stadler Sympos.* 4:69–81, 1972.

Tatum, L. A. The southern corn leaf blight epidemic. *Science* 171:1113–1116, 1971.

Part VII

Genetic Engineering

Chapter 17

Approaches to Genetic Engineering

The genetic alteration of plants and animals has been practiced by humans for centuries. It has only been in recent years, however, that we have developed techniques that enable us to modify the genotypes of organisms directly by incorporating genes from entirely different species. We shall therefore limit the meaning of the term "genetic engineering" here to include only those genetic alterations that are accomplished by unusual cytogenetic procedures or by direct manipulation of genes or gametes.

Genetic engineering promises to have a major impact on the future. The new genetic technologies open up new approaches to the synthesis of drugs and other chemicals and to pollution control. The techniques also provide new tools that can be used to supplement the traditional breeding practices for the improvement of crop plants and domesticated animals. Many of the new technologies carry with them possible risks to human health and to the quality of the environment. Some potential risks, such as those associated with manipulating the genes of known toxins or of pathogenic organisms, are certain and quantifiable. Others, such as the inadvertent creation of some new "super-pathogen" are purely hypothetical and without experimental foundation. Certain of the techniques pose special problems of moral concern, relating in particular to how and if these methods should be applied to humans.

In this chapter, we will describe many of the new techniques that promise to be useful in genetic engineering. We will consider not only their commercial applications but also the ways in which they are used in basic genetic research. We begin with a discussion of recombinant DNA techniques.

Recombinant DNA Technology

Prior to the 1970s, geneticists studying the genetic organization of eukaryotic chromosomes were hindered by the large amounts of noncoding DNA sequences present within the genomes of higher organisms. Over the past decade, however, the development of recombinant DNA technology has revolutionized the study of eukaryotic genetics. By providing the tools needed to isolate unlimited quantities of purified single genes from eukaryotic organisms, recombinant DNA technology has permitted scientists to examine eukaryotic genes at a level of detail comparable to that available for bacterial and viral genes. Important findings have already been made concerning the molecular structure of eukaryotic genes, the mapping of gene locations in chromosomes, and the discovery and characterization of sequences that are important for gene regulation.

The application of recombinant DNA techniques is also having a tremendous impact on the emphasis placed by the pharmaceutical and chemical industries on biotechnology. Microorganisms are being used to manufacture substances that have previously been available only from natural sources. In this section, we will discuss the procedures used to construct recombinant DNA molecules and the applications of these procedures to both genetic research and industry.

Restriction Endonucleases

Underlying all of the recombinant DNA technology is the experimental discovery of a group of bacterial enzymes referred to as **restriction endonucleases.** The role of these enzymes in nature is to defend a bacterium from the invading DNA of a phage or of another bacterial strain. As early as 1953, researchers knew that if the DNA from one strain of *E. coli* is introduced into a different strain, the foreign DNA is almost always broken down by certain nucleases within the recipient cell. The only way that the foreign DNA can escape fragmentation is if it is methylated in a way that is specific to each strain. It was not until the late 1960s, however, that the restriction endonucleases that catalyze this degradation were identified.

Certain restriction enzymes cleave the DNA at specific recognition sites. These endonucleases are named by the first letter of the bacterial genus and the first two letters of the species from which the enzymes are obtained; in many cases, a letter follows indicating the antigenic type or strain of the bacterium, and a Roman numeral is included if the cell contains more than one such enzyme. For example, *Hin*dII and *Hin*dIII are restriction endonucleases isolated from *Haemophilus influenzae,* serotype d. Each of the two enzymes cuts both strands of the DNA double helix, but within a different sequence of six base pairs, as follows:

*Hind*II: 5′ G-T-Py↓Pu-A-C 3′ / 3′ C-A-Pu↑Py-T-G 5′

*Hind*III: 5′ A↓A-G-C-T-T 3′ / 3′ T-T-C-G-A↑A 5′

The arrows indicate the points of cleavage of the enzymes, and Pu and Py designate unspecified purine and pyrimidine bases.

Restriction nucleases that cleave the DNA only at a specific sequence are very valuable tools in molecular genetics, because they generate a defined set of DNA fragments, all of which end in the same nucleotide sequence. Furthermore, a wide variety of these enzymes have been isolated from over 230 bacterial strains, and over 70 different recognition sites for cleavage have been identified. Table 17.1

Table 17.1. Some restriction enzymes and their recognition sequences (arrows indicate cleavage sites).

MICROORGANISM	ENZYME ABBREVIATION	SEQUENCE
Arthrobacter luteus	*Alu*I	5′ A G↓C T 3′ 3′ T C↑G A 5′
Bacillus amyloliquefaciens	*Bam*HI	5′ G↓G A T C C 3′ 3′ C C T A G↑G 5′
Escherichia coli	*Eco*RI	5′ G↓A A T T C 3′ 3′ C T T A A↑G 5′
Haemophilus aegyptius	*Hae*II	5′ Pu G C G C↓Py 3′ 3′ Py↑C G C G Pu 5′
Haemophilus aegyptius	*Hae*III	5′ G G↓C C 3′ 3′ C C↑G G 5′
Haemophilus haemolyticus	*Hha*I	5′ G C G↓C 3′ 3′ C↑G C G 5′
Haemophilus influenzae	*Hind*II	5′ G T Py↓Pu A C 3′ 3′ C A Pu↑Py T G 5′
Haemophilus influenzae	*Hind*III	5′ A↓A G C T T 3′ 3′ T T C G A↑A 5′
Haemophilus parainfluenzae	*Hpa*I	5′ G T T↓A A C 3′ 3′ C A A↑T T G 5′
Haemophilus parainfluenzae	*Hpa*II	5′ C↓C G G 3′ 3′ G G C↑C 5′
Serratia marcescens	*Sma*I	5′ G G G↓C C C 3′ 3′ C C C↑G G G 5′
Streptomyces albus	*Sal*I	5′ G↓T C G A C 3′ 3′ C A G C T↑G 5′
Xanthomonas oryzae	*Xor*II	5′ C G A T C↓G 3′ 3′ G↑C T A G C 5′

lists a few of these enzymes, along with the restriction sequence (recognition site) of each enzyme. The restriction sequences are typically four or six base pairs in length and are *palindromes,* a term that in general refers to a group of letters or words that reads the same forward and backward. In this case, the sequence of bases on the two strands are the same when each is read in the $5' \rightarrow 3'$ direction.

Each restriction enzyme will cleave DNA from any source (viral, bacterial, plant, or animal), as long as that DNA contains one or more copies of its specific recognition sequence. The more such sequences are contained in the DNA, the shorter will be the lengths of the fragments that are produced by enzymatic digestion. Furthermore, most types of DNA can be cleaved by a number of different restriction enzymes, since a strand of DNA typically has a variety of restriction sequences. For example, suppose that the sequence of base pairs along a molecule of DNA is determined strictly at random. Any one of four base pairs (AT, TA, GC, or CG) then has an equal chance of occupying a particular residue site. Each restriction sequence of four base pairs is thus expected to occur at a frequency of $\frac{1}{4} \times \frac{1}{4} \times \frac{1}{4} \times \frac{1}{4} = (\frac{1}{4})^4 = \frac{1}{256}$; in other words, each sequence is expected to be present, on the average, once in every 256 base pairs. Even a small viral chromosome comprising only 5000 nucleotides would thus contain approximately 20 copies of each possible sequence that is four base pairs in length.

Some of the restriction endonucleases, such as *Hin*dII, cleave the DNA into blunt-ended fragments. Other enzymes, such as *Hin*dIII or *Eco*RI (the first restriction endonuclease to be isolated from *E. coli*), make staggered cuts, yielding fragments that have single-stranded complementary (cohesive) ends. These ends are of crucial importance in the generation of hybrid or recombinant DNA molecules, as we shall see shortly.

Construction of Recombinant DNA Molecules. In 1972, investigators at Stanford University found that any two DNA fragments produced by the *Eco*RI restriction enzyme, regardless of their origin, will form hydrogen bonds with one another at their complementary ends. The fragments can then be joined permanently by the action of a DNA ligase. The result is an artificially created recombinant DNA molecule. To create a recombinant molecule that is also capable of replicating itself, the DNA of a plasmid is used in its construction. The plasmid, which carries one or more restriction sequences, is cut open with a restriction enzyme, such as *Eco*RI (Fig. 17.1). The plasmid pSC101 (see Fig. 11.25) was initially used because it contains only a single *Eco*RI recognition site. The now-linear DNA of the plasmid is mixed with DNA fragments from another source (referred to here as the donor DNA), which also has cohesive ends generated by *Eco*RI. In the original experiments conducted in 1973 by H. Boyer and S. Cohen, the donor DNA was that of another plasmid. In general, however, the donor DNA can be obtained from any microbial, animal, or plant source. The plasmid chromosome forms hydrogen bonds with the donor fragment, then reforms a circle, and is finally sealed with ligase. The outcome is a hybrid plasmid chromosome that constitutes the recombinant DNA molecule.

The recombinant plasmid can be introduced into cells of the bacterium *E. coli* after the bacteria are treated with calcium to increase their permeability to DNA.

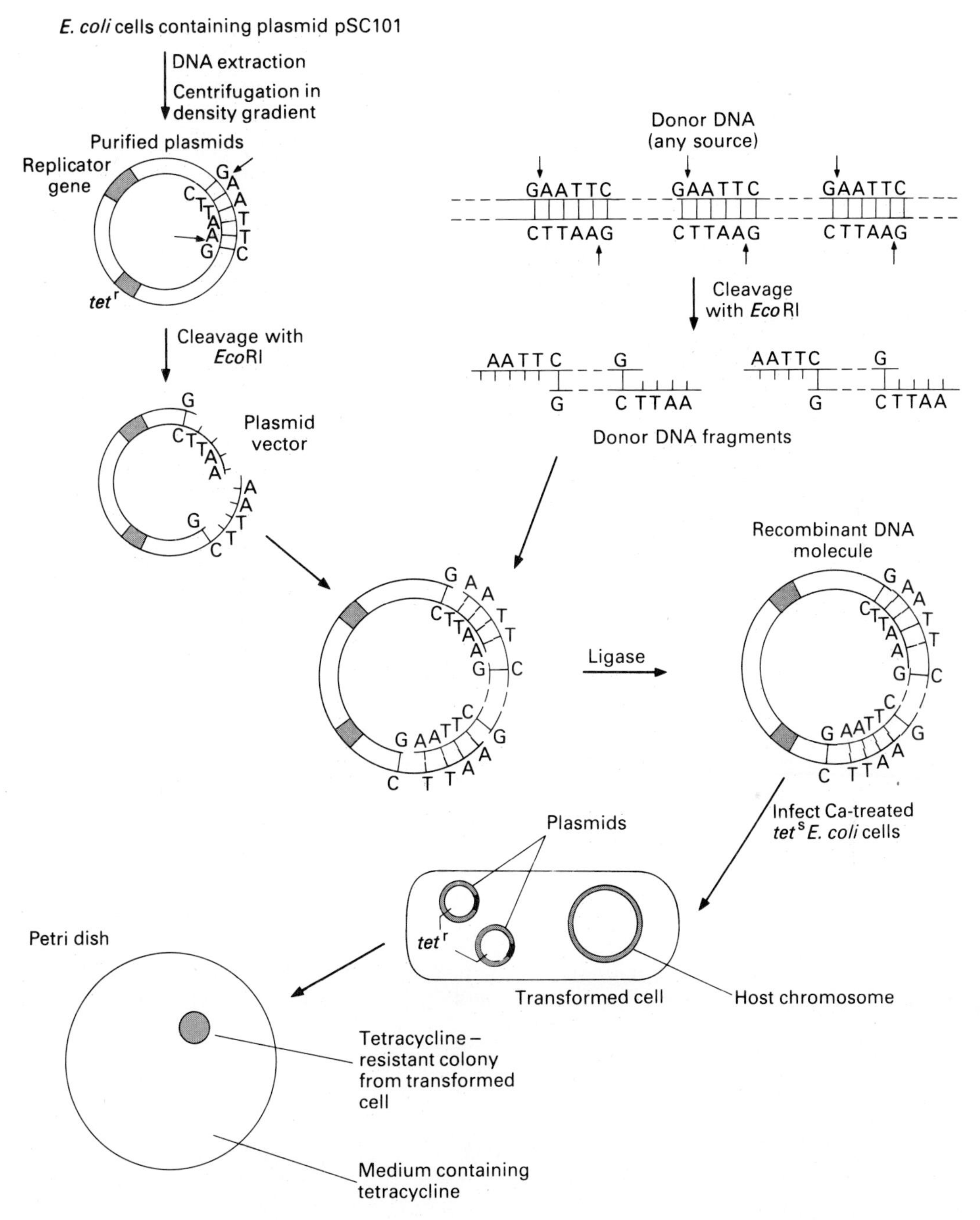

Figure 17.1. Construction of a recombinant DNA molecule, and its use in transforming bacterial cells with donor DNA from any prokaryotic or eukaryotic source. Transformed cells are selected by their resistance to an antibiotic (in this case, tetracycline), a property that is conferred by the recombinant plasmid.

The hybrid molecule multiplies, as would the natural plasmid, within the cytoplasm of the host cell, conveying upon the cell a totally new genetic property. The plasmid is thus said to serve as the vector for the transfer and replication of the donor genetic material. Many different bacterial plasmids as well as viruses (e.g., phage lambda) are currently being used as vectors in recombinant DNA work with bacterial cells as the recipients. A few of these cloning vectors are listed in Table 17.2. The vectors carry a variety of restriction sequences.

Until recently, few plasmids had been identified and characterized in plant and animal cells. As plasmids or viral vectors are discovered in these organisms, greater opportunities for engineering the cells of higher organisms will become available. Genetic engineering is presently being done on eukaryotic cells using a

Table 17.2. Some recombinant DNA cloning vectors.

TYPE	VECTOR	RESTRICTION SEQUENCES	FEATURES
Plasmid (*E. coli*)	pBR322	*Bam*HI, *Eco*RI, *Hae*III, *Hin*dIII, *Pst*I, *Sal*I, *Xor*II	Carries genes for tetracycline and ampicillin resistance.
Plasmid (yeast-*E. coli* hybrid)	pYe(CEN3)41	*Bam*HI, *Bgl*II, *Eco*RI, *Hin*dIII, *Pst*I, *Sal*I	Multiplies in *E. coli* or yeast cells.
Cosmid (artificially constructed *E. coli* plasmid carrying λ *cos* site)	pJC720	*Hin*dIII	Can be packaged into λ phage particles for efficient introduction into bacteria; replicates as a plasmid; useful for cloning large DNA inserts.
Virus	Charon phage	*Eco*RI, *Hin*dIII, *Sst*I	Constructed using restriction nucleases and ligase, having foreign DNA as central portion, with λ DNA at each end; carries β-galactosidase gene; packaged into λ phage particles; useful for cloning large DNA inserts.
Virus	M13	*Eco*RI	Single-stranded DNA virus; useful in studies employing single-stranded DNA insert.
Virus	SVGT5	*Bam*HI, *Hin*dIII	SV40 virus fragment; infects animal cells in culture.
Transposable genetic element	P element	*Ava*II, *Nin*dIII, *Sal*I	*Drosophila* transposable element, into which DNA segments can be inserted; exhibits highly efficient transfer of the inserted segment into the *Drosophila* germ line.
Plasmid	Ti	under study	Maize plasmid
Transposable genetic element	Ds element	under study	Maize transposable element

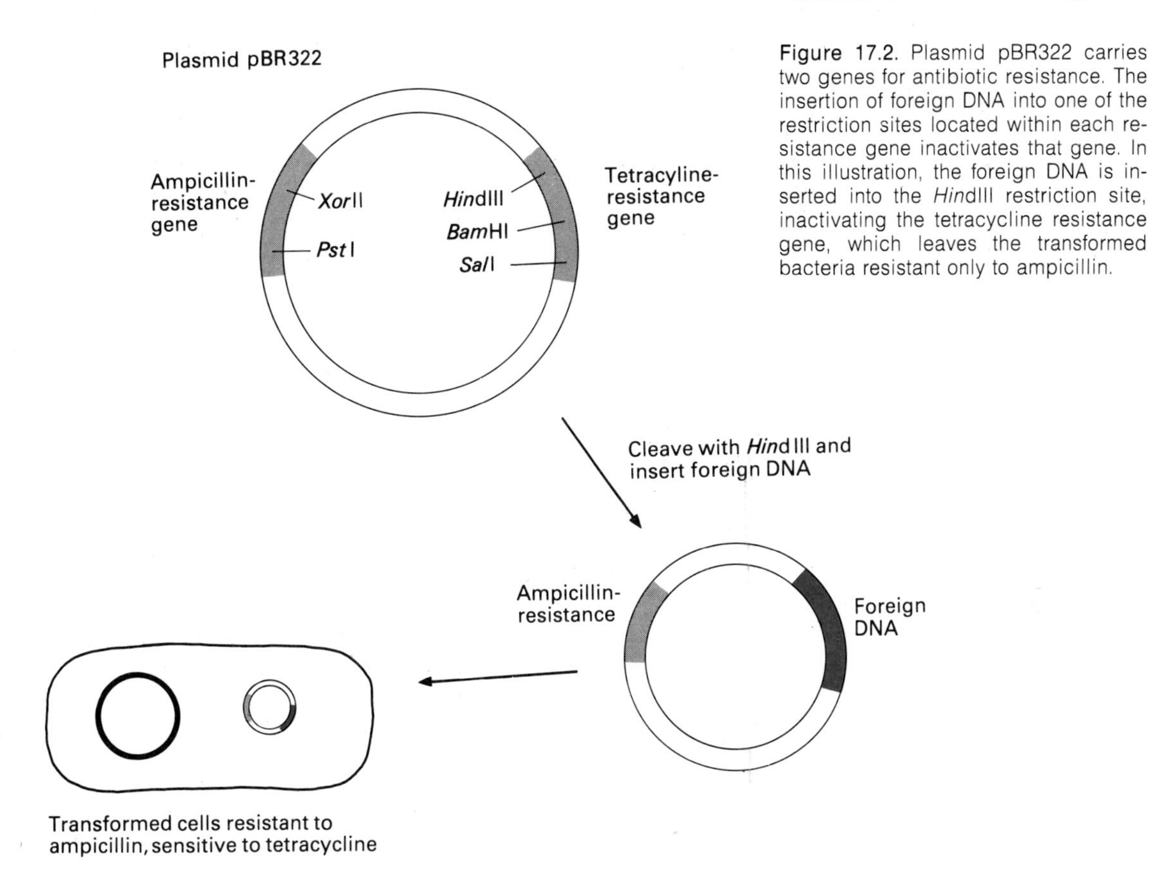

Figure 17.2. Plasmid pBR322 carries two genes for antibiotic resistance. The insertion of foreign DNA into one of the restriction sites located within each resistance gene inactivates that gene. In this illustration, the foreign DNA is inserted into the *Hind*III restriction site, inactivating the tetracycline resistance gene, which leaves the transformed bacteria resistant only to ampicillin.

variety of methods for gene transfer, including the use of plasmids, viruses, and transposable genetic elements as vectors and DNA-mediated gene transfer (see Chapter 7).

The primary advantage of using plasmids as recombinant DNA vectors stems from their characteristic property of carrying one or more genes for antibiotic resistance. Because of this feature, the selection of bacteria that have taken up a hybrid plasmid becomes a simple matter. Following incubation with recombinant plasmids, the recipient bacteria (which previously carried no plasmids of their own) are placed on a medium containing one or more antibiotics. Only those cells that have acquired a plasmid are able to survive. In order to be certain that the plasmid is a hybrid molecule carrying the donor DNA, the investigators use a particular type of plasmid, such as plasmid pBR322, that carries two different genes for antibiotic resistance (Fig. 17.2). Each resistance gene contains a different set of restriction enzyme sequences. Using a particular restriction endonuclease (*Hin*dIII is shown in Fig. 17.2), the investigators splice the donor DNA into one of the resistance genes (the tetracycline resistance gene in this case). This event inactivates the gene, so that any bacteria that receive the hybrid plasmid will become resistant only to the other antibiotic. This highly selective system ensures that the plasmid that is received by a bacterial cell is a recombinant DNA molecule of defined construction.

Applications in Genetic Research

With the advent of recombinant DNA technology, research in molecular biology is proceeding at such a rapid pace that genetic knowledge is said to be doubling every two years. Most advances have come about because of three main research applications of recombinant molecules: DNA cloning, restriction enzyme mapping, and DNA sequencing.

DNA Cloning. The cloning of a DNA fragment refers to the amplification of a defined DNA restriction fragment by means of its replication within the cells of a recipient bacterial population, so that large amounts of that particular DNA fragment are obtained in pure form. The first step in this procedure is to identify those few cells in the recipient bacterial population that have taken up a hybrid plasmid carrying the gene we wish to clone. Digestion by a restriction enzyme releases many different DNA fragments; finding the fragment that carries the gene we wish to study is something like finding the proverbial needle in a haystack. For example, suppose we wish to clone the human β hemoglobin gene. Treatment of the human DNA complement with a restriction enzyme releases thousands of fragments, only one of which carries the β hemoglobin gene. After these fragments are inserted into plasmids, the plasmids are allowed to infect *E. coli* cells, at concentrations of plasmids and bacteria such that each cell gets no more than one plasmid. We would obtain from this procedure a recipient cell population whose members carry a variety of DNA fragments. We must next probe this population for the cells that carry the desired gene.

To identify the cells that carry the gene in question, we first dilute the recipient cell population and spread samples over the surface of nutrient agar plates, so that colonies form after incubation overnight. Each colony represents a pure culture derived from a single recipient cell and can be considered to be a single-cell isolate. The replica-plating technique is then used to transfer the colonies onto nitrocellulose filter paper, where the cells are lysed and the released DNA is denatured (Fig. 17.3). The single-stranded DNA binds to the paper at the point at which each colony was transferred. A **nucleic acid probe** is then used to screen the DNA isolates, in order to identify those colonies that contain the desired gene. The probe (designated **cDNA**) consists of a small amount of radioactive denatured DNA that is complementary in sequence to the strands that make up the gene in question. (We will discuss the method by which this probe is prepared in the next section.) When the cDNA probe is added to the DNA isolates, it hybridizes only with those DNA samples that contain the gene in question. A solution that washes away all single-stranded DNA is next poured through the filter, leaving only double-stranded DNA bound to the nitrocellulose paper. Upon autoradiography, the radioactive label on the cDNA identifies the DNA isolates (and their corresponding single-cell isolates on the master plate) that contain the desired gene.

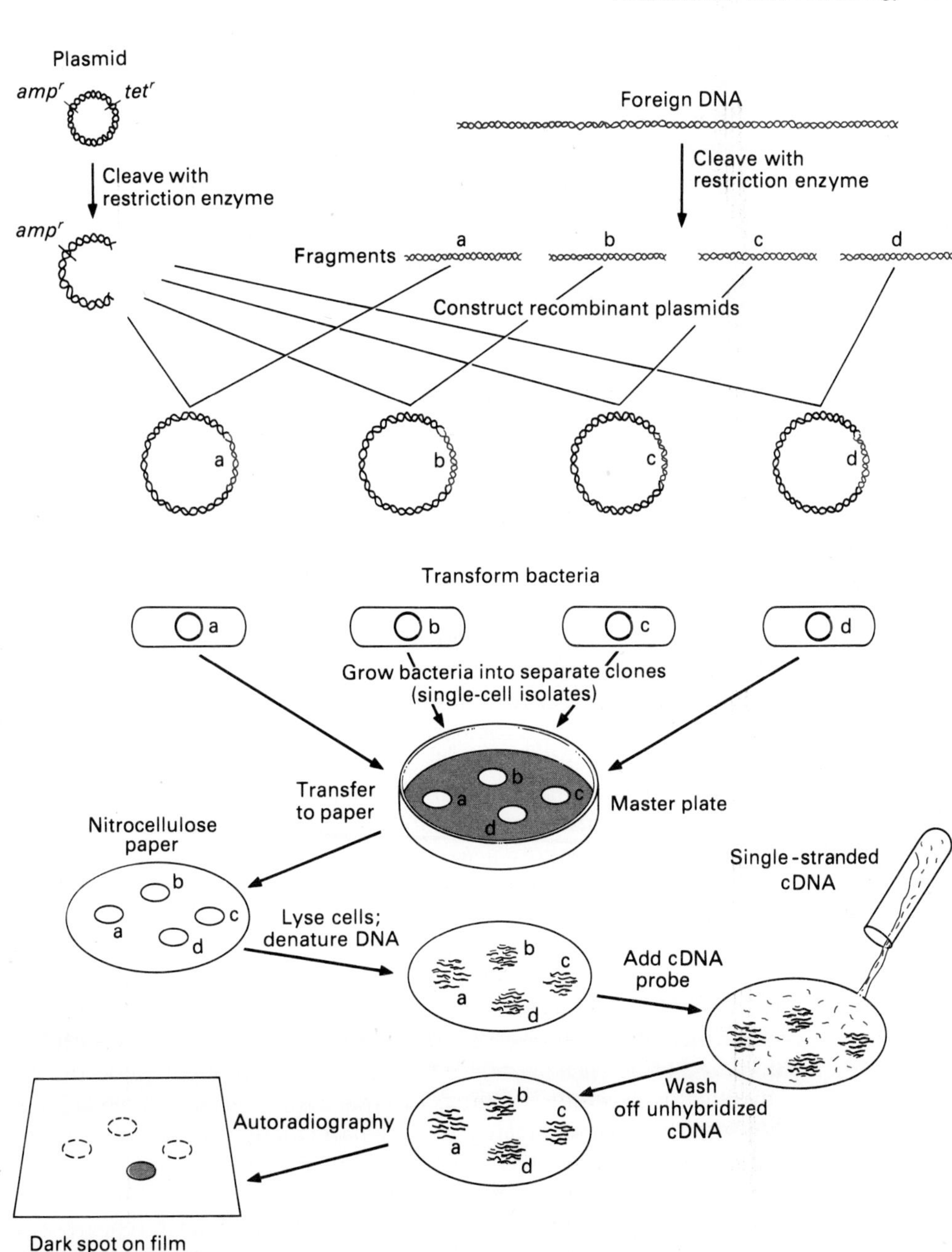

Figure 17.3. Cloning of foreign DNA fragments in bacterial cells and the procedure used to identify the clones that carry a specific fragment. The desired fragment is identified using a DNA probe, which is a small amount of DNA that is complementary to the desired DNA fragment.

Because of the high sensitivity of this procedure, a single technician can screen several hundred thousand bacterial isolates in a single operation.

Once the bacterial colonies that carry the desired gene have been identified, the gene can be cloned. These bacteria are allowed to grow in mass culture, producing up to 10^9 cells per ml, with a corresponding number of copies of the cloned gene. (Some mutant strains of *E. coli* that are used as recipients replicate the new plasmid 20 to 40 times within each cell, thereby producing very high yields of the cloned gene.) The DNA is extracted from the cells of the culture and is digested by restriction enzymes to yield the desired gene, which can then be separated from the rest of the DNA by any of several experimental procedures. Since Kornberg's in vitro DNA replication system is unable to replicate DNA faithfully in large quantities, the development of recombinant DNA technology has meant that it is possible, for the first time, to obtain unlimited amounts of purified DNA. The cloning of a gene by means of its replication within a growing bacterial population thus allows geneticists quickly and inexpensively to obtain very large amounts of that gene in pure form. This capability opens the door to a variety of studies of gene structure and chromosome organization. One of the major applications is nucleic acid sequencing, which we will discuss shortly. Other uses of the technology in research have included the study of how control molecules bind to particular regions of DNA, the analysis of gene structure, the examination of the eukaryotic genome for chromosomal regions that are important in control or in the structural organization of the chromosome, and the study of selected regions of chromosomes from tumor viruses in order to understand better how these viruses transform cells to a cancerous state. Discoveries that have already been made include the split nature of eukaryotic genes, the existence of clustered gene families for many of the more abundant proteins (e.g., hemoglobin, histones, and antibodies), the presence of spacer regions between the genes within a cluster, and the finding that movable genetic elements are a generalized phenomenon.

In the approach that we have considered thus far, recombinant DNA experiments begin with donor DNA that represents a random sample of all the fragments produced by restriction enzyme digestion. Since we do not know which gene is being incorporated into any particular plasmid, these studies are called "shotgun" experiments. One possible danger in this approach is that if the DNA of eukaryotic cells harbors tumor proviruses (as is currently thought to be the case by many investigators), then the geneticists might be engineering a potentially cancer-causing bacterium. Researchers were particularly concerned about this possibility because of the widespread distribution of *E. coli* in nature and because the laboratory strains are descendants of the *E. coli* that occurs naturally in the human intestine (even though laboratory strains of *E. coli* have repeatedly been shown no longer to be capable of survival in the human gut). The National Institutes of Health ultimately issued guidelines for conducting recombinant DNA experiments after public attention was drawn to the matter by molecular biologists. Work is now proceeding at a rapid pace under various levels of physical and biological containment. The higher levels of containment include special research facilities and the use of *E. coli* strains so disabled by mutations that it would be impossible for them to live outside of the highly artificial laboratory conditions.

Construction of cDNA Probes. Even though a population of recipient bacteria may contain an entire eukaryotic genome, geneticists can identify only those genes for which they have specific DNA probes. Much effort has therefore gone into the production of a variety of different cDNA sequences. The procedure basically involves two steps: (1) synthesis of the probe itself and (2) identification of the specific gene carried by the probe. We will first discuss how a pure DNA probe is constructed.

The mRNA of a eukaryotic cell can be easily differentiated from other kinds of cellular RNA, since mRNA molecules contain a unique poly-A tail at their 3′ end. The poly-A tail will bind to a filter on which poly-T segments have been fixed, yielding mRNA molecules with a double-stranded tail of AT base pairs. The enzyme reverse transcriptase, which synthesizes a strand of DNA using RNA as a template, is able to use the poly-T section as a primer to polymerize a strand of DNA that is complementary to the message (Fig. 17.4). The enzyme DNA polymerase is then used to copy this DNA into a complementary strand, displacing the

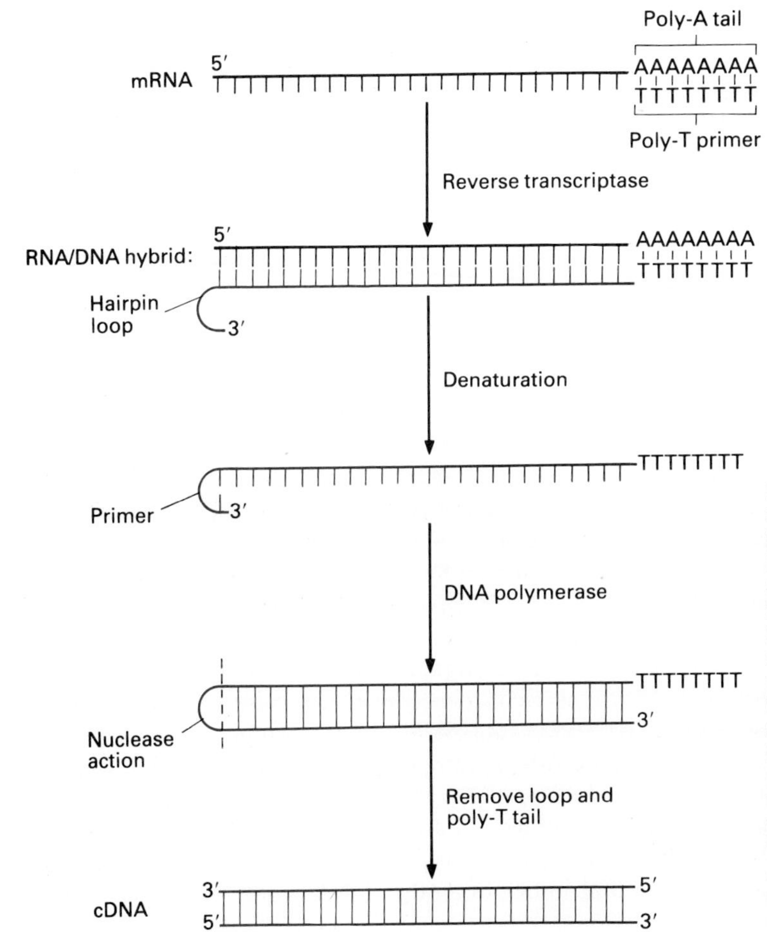

Figure 17.4. Construction of a double-stranded DNA probe. A mRNA molecule is bound to a filter by means of hybridization of its poly-A tail to the poly-T segment. The enzyme reverse transcriptase is then employed to synthesize a DNA strand that is complementary to the mRNA, using the poly-T segment as a primer. Reverse transcriptase leaves a hairpin loop on the 3′ end of the DNA strand. Following denaturation of the RNA/DNA duplex, this loop provides a primer for DNA polymerase, which synthesizes a strand of DNA that is complementary to the first strand. Nucleases then remove the single-stranded loop and the poly-T tail, leaving a DNA molecule that is complementary to the original mRNA.

mRNA. The result is a population of double-stranded DNA molecules (cDNA) that are complementary to the mRNA molecules in the cellular extract.

At this point in the procedure, we will have as many different types of cDNA molecules as there are types of mRNA in the eukaryotic cell. This is a major problem, because it is not easy to differentiate between the various mRNA or cDNA molecules. Most recombinant DNA work has therefore been restricted to certain genes in highly specialized cells that synthesize predominantly one kind of protein (and therefore produce mostly one kind of mRNA). This restriction makes it much easier to locate the cDNA probe of interest within the mixture of probes produced as a result of the procedures outlined in the preceding paragraph. Cells that have been used as sources of nucleic acid for probes have included reticulocytes (in which 90 percent of the protein made is hemoglobin), plasma cells (which synthesize only antibody protein), and hormone-stimulated chicken oviduct cells (over 50 percent of whose protein is ovalbumin, the protein in egg whites). For our discussion here, let us assume that we are using mRNA that was extracted from chicken oviduct cells. Using reverse transcriptase, we obtain a mixture of cDNA probes, of which about 50 percent are copies of the ovalbumin gene. It now remains specifically to identify the ovalbumin cDNA molecules.

To identify a DNA probe, we must first insert it into a plasmid and clone the hybrid plasmid in *E. coli*. This procedure provides us with sufficient amounts of pure DNA with which to work. Recombination with a plasmid requires that cohesive ends be attached to the cDNA molecule. We can construct cohesive ends on DNA with the enzyme called terminal transferase, which catalytically adds nucleotides to the 3′ end of each polynucleotide chain (Fig. 17.5). For example, by adding poly-A to the 3′ ends of the cDNA and poly-T to the 3′ ends of the plasmid, the two DNA segments can join together to form a hybrid plasmid, which can then be cloned.

We must now identify which of the *E. coli* clones carries the ovalbumin cDNA. For this purpose, we extract the DNA from each clone and denature the DNA on a nitrocellulose filter. We then pass the mRNA extract from oviduct cells through the filter (Fig. 17.6). Only those mRNA molecules that are complementary to a given cDNA are retained on the filter, hybridized to the cDNA probe. Each isolate of bound mRNA is then itself washed off and concentrated, to yield pure samples of individual types of mRNA molecules that correspond to the various cDNA probes present on the filters. To identify each mRNA sample, we add them to separate test tubes containing a cell-free translation system (the tubes contain all the components needed for protein synthesis, including ribosomes, tRNA molecules, amino acids, protein initiation, elongation, and termination factors, GTP, and so on). Through established biochemical procedures, we are subsequently able to identify the protein made in each test tube. In our example case, we would find that about 50 percent of the mRNA samples produce ovalbumin in the test tube protein synthesizing system. The cDNA molecules that correspond to these mRNA molecules are thereby identified, as are the bacterial clones that contain the ovalbumin recombinant DNA. By growing these clones in mass culture, we can then collect a large potential source of a pure cDNA probe.

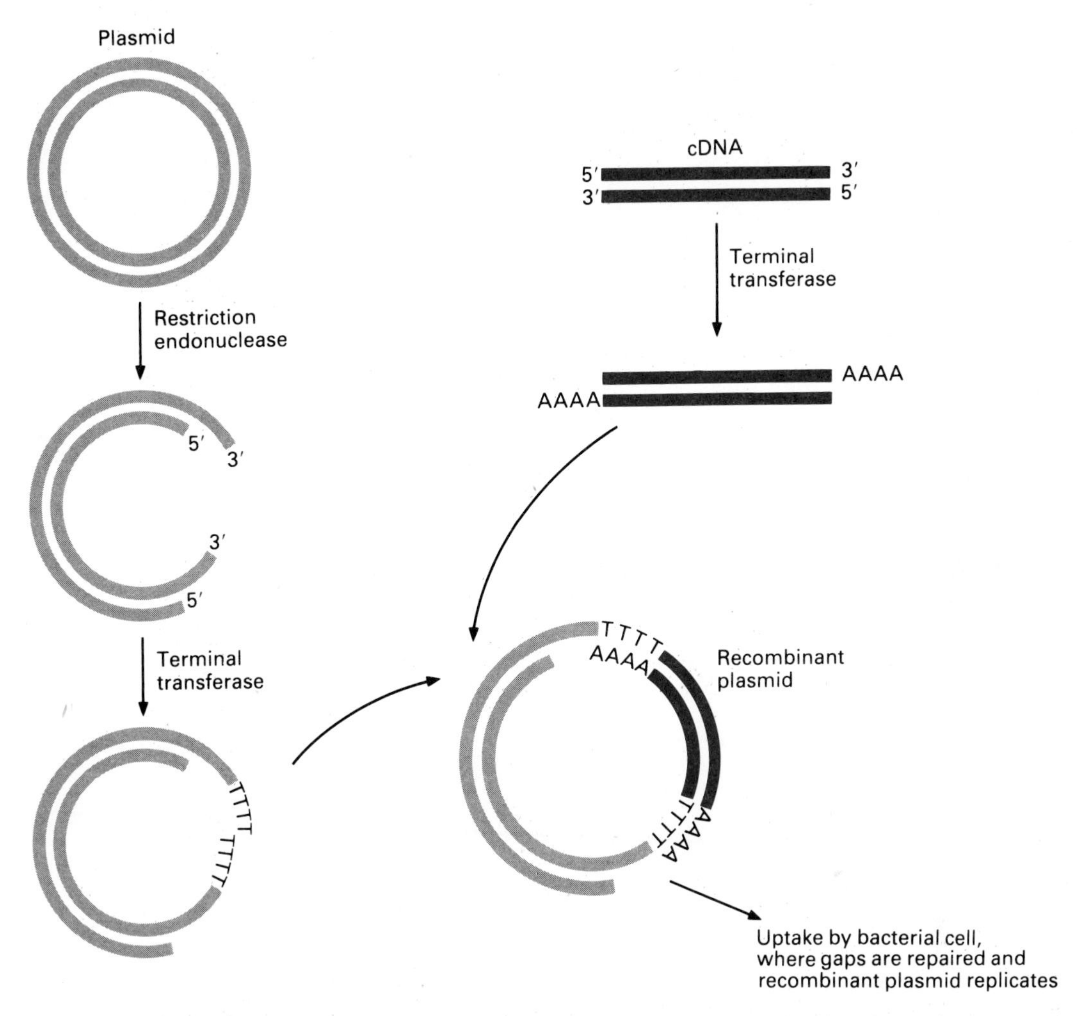

Figure 17.5. Cloning of a cDNA molecule. The enzyme terminal transferase is used to construct complementary single-stranded 3′ ends on a cleaved plasmid and on the cDNA. The hybrid plasmid that subsequently forms is inserted into *E. coli*, where bacterial enzymes repair the gaps. The cDNA-recombinant plasmid then replicates within the replicating bacterial cells, which results in large quantities of the cDNA.

Once the geneticists have a known probe, they can quickly screen large numbers of bacterial colonies for the desired gene. If the geneticists had available an entire "library" of probes, they could screen an entire eukaryotic genome by identifying the restriction fragments that carry each gene of the organism's genetic complement. Investigators are at least a decade away from being able to screen the entire restriction complement of a higher organism, however, because of the

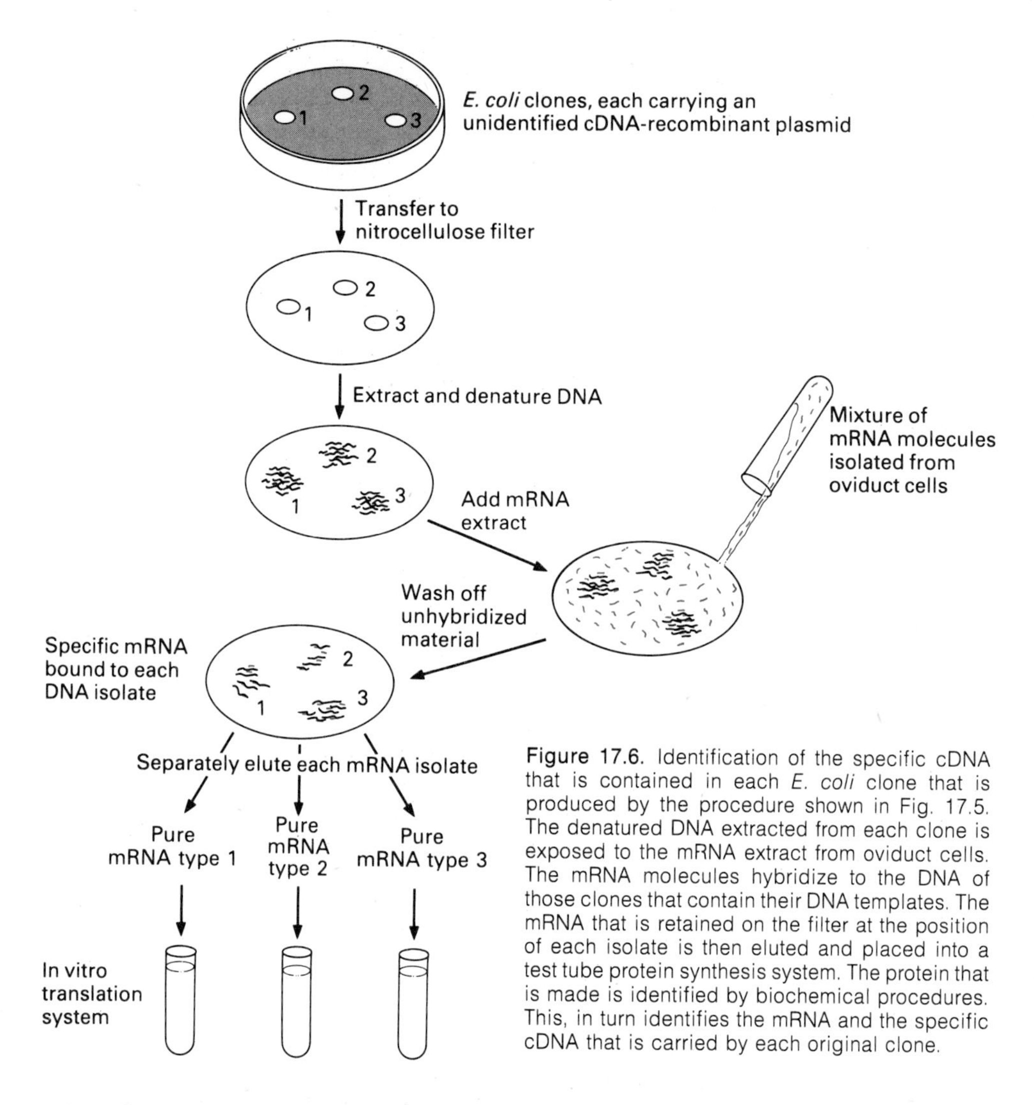

Figure 17.6. Identification of the specific cDNA that is contained in each *E. coli* clone that is produced by the procedure shown in Fig. 17.5. The denatured DNA extracted from each clone is exposed to the mRNA extract from oviduct cells. The mRNA molecules hybridize to the DNA of those clones that contain their DNA templates. The mRNA that is retained on the filter at the position of each isolate is then eluted and placed into a test tube protein synthesis system. The protein that is made is identified by biochemical procedures. This, in turn identifies the mRNA and the specific cDNA that is carried by each original clone.

complex procedure that is used in identifying cDNA probes. Better techniques are needed for the separation of mRNA, since most cells express many different genes, each of which synthesizes a different kind of protein in relatively small amounts. Recent advances in the technology of amino acid sequencing of small proteins are providing an alternative method by which specific cDNA probes can be identified. Investigators determine the partial amino acid sequence of a small amount of the desired gene product and then, working with the genetic code dictionary, they artificially synthesize a DNA probe corresponding to the amino acid sequence. These synthetic probes can then be used to screen a mixture of cDNA molecules.

Restriction Mapping. Another use of recombinant DNA technology in genetic research is to map individual DNA molecules on the basis of the specific locations of their restriction sequences. In order to obtain a restriction map, researchers use various restriction endonucleases, both separately and in combination, to cleave the DNA into different-sized fragments. The fragments are separated by gel electrophoresis and are analyzed to determine the number and kinds of restriction sites included in each. By comparing the sequences that are included within the different overlapping fragments, the geneticists can then deduce the proper arrangement of restriction sites within the DNA molecule.

For example, suppose that complete digestion of a sample of DNA by two restriction endonucleases (labeled x and y) yields four fragments: A, B, C, and D. These fragments are identified on the basis of their migration patterns in gel electrophoresis. Digestion of the DNA sample by restriction endonuclease x alone yields two fragments: one containing A and B and the other containing C and D. Digestion by restriction enzyme y alone also produces two fragments: one containing A and C and the other containing B and D. By piecing the overlapping fragments together, we see that the map is circular:

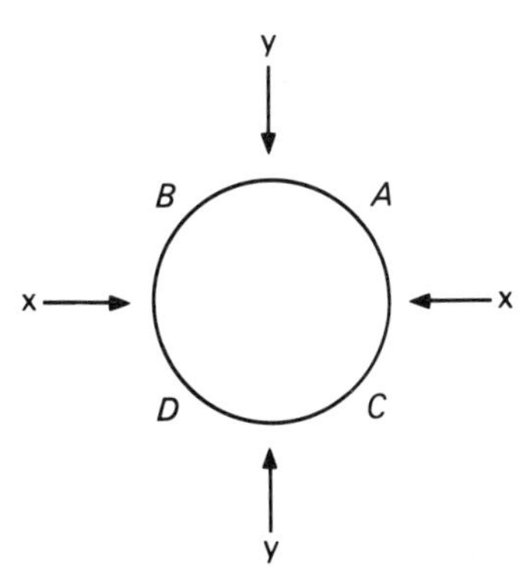

with restriction endonuclease x cleaving at sites in the intervals A-C and B-D, while restriction endonuclease y cleaves at sites within the intervals A-B and C-D.

Different approaches can be taken in the development of a restriction map. For example, in the construction of the first restriction map in 1971, researchers used the enzyme *Hin*dII to cut the DNA of monkey virus SV40 into 11 fragments. The order in which these fragments occur in the DNA was deduced by studying the pattern of fragmentation over time. The first cut breaks the circular DNA molecule into a linear one, which is then cut into progressively smaller fragments that are overlapped by the fragments of previous cuts. Repetition of the fragmentation process with other enzymes yields a more detailed map, which gives the locations of several different restriction sequences (Fig. 17.7).

EXAMPLE 17.1. Suppose that complete digestion of a DNA molecule by the restriction enzyme *Hin*dII produces eight segments: A, B, C, D, E, F, G, and H. These segments can be separated by gel electrophoresis. Partial digestion by the enzyme yields DNA pieces of larger size; when isolated and further analyzed, these pieces are shown to include fragments of the following composition: (A, B, D, F, G), (A, D, E, F), (B, G, H), (C, E, H), and (B, F), where the parentheses indicate that the order of the enclosed segments

(A, B, C, etc.) is unknown. Construct a restriction map of the DNA molecule from these data, giving the correct order of the segments A through H.

Solution. Note that the fragments overlap with respect to their included segments. It is thus possible to arrange them in the following manner:

(1) A-D-F-B-G
(2) E-A-D-F
(3) B-G-H
(4) H-C-E
(5) F-B

The fact that E maps at both ends indicates that the map is circular:

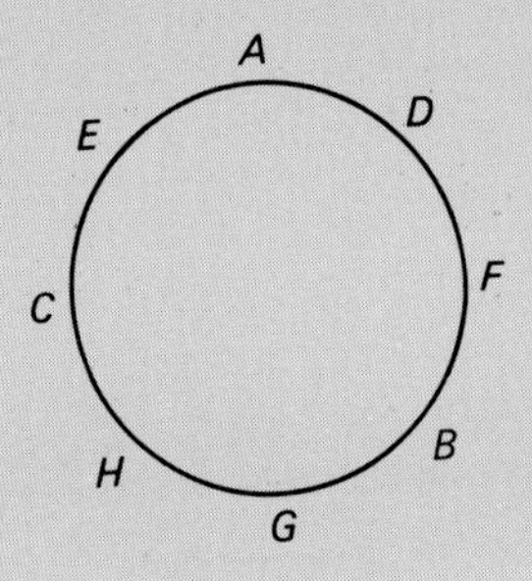

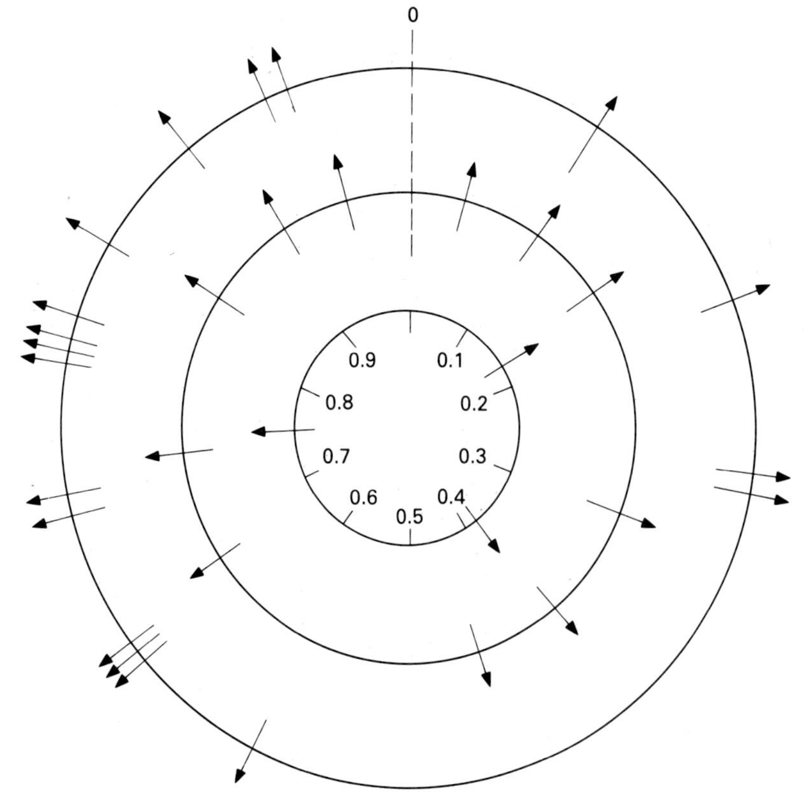

Figure 17.7. Restriction map of the SV40 genome, showing the positions of cleavage by *Hae*III (outer circle), *Hin*dII and *Hin*dIII (middle circle), and *Hpa*I (inner circle. The point marked 0 is the position of the single cut made by *Eco*RI. Maps of the cleavage sites on this genome by other restriction enzymes are also available. Map units are given as: distance from *Eco*RI site/length of SV40 genome. *Source:* From Yang *et al.*, *European J. Biochem.* 61:119, 1976, and from K. J. Danna *et al.*, *J. Mol. Biol.* 78:374, 1973.

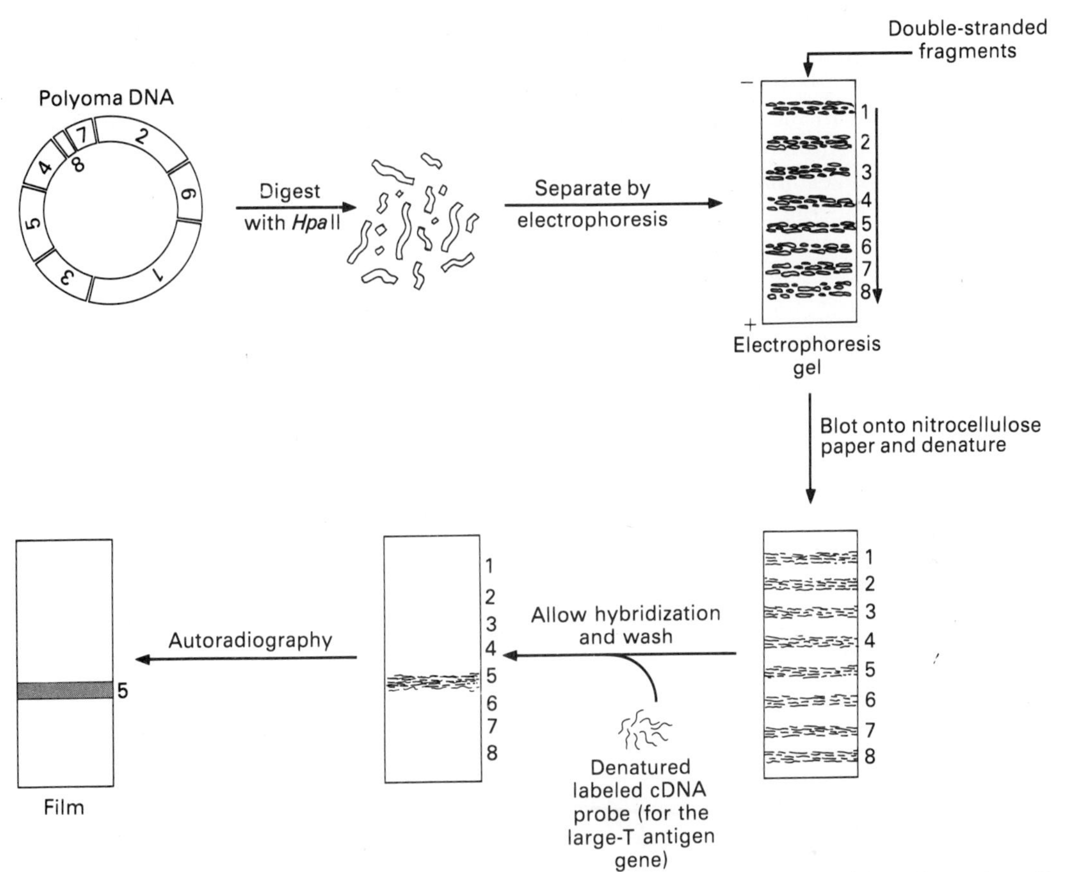

Figure 17.8. The use of the Southern blot technique to identify the location of the polyoma large-T antigen gene. The polyoma DNA is first fragmented into eight sections by the restriction enzyme *Hpa*II. The fragments are separated on the basis of size by electrophoresis, yielding pure fragments. Each fragment class is transferred to nitrocellulose paper, where the DNA is denatured. A small amount of pure cDNA for the large-T antigen gene is applied, after which the DNA is washed to remove the single-stranded material. Autoradiography reveals hybridization of the probe for the large-T antigen gene to band 5, thus identifying fragment 5 as the location of the large-T antigen gene. The investigators then refer to the restriction map of polyoma DNA to identify the chromosomal position of this gene.

The construction of a restriction map enables the geneticists to locate regions of genetic importance on a chromosome. Once the geneticists have identified a DNA fragment that carries a particular gene, they can quickly locate the chromosomal position of that gene using a restriction map. To locate a specific gene on a particular restriction fragment, the *Southern blot* technique is used (Fig. 17.8). In this procedure, the fragmented DNA is subjected to gel electrophoresis, which separates the fragments on the basis of size. Each electrophoretic band therefore corresponds to a pure isolate of one particular restriction fragment. The DNA in each band is applied (blotted) onto nitrocellulose paper and is denatured. A labeled cDNA probe of the desired gene is then applied to the paper; through hybridization and autoradiography, this probe identifies the location on the gel of the class of restriction fragments that carries the gene in question. The location of this fragment on the restriction map thus identifies the chromosomal position of the gene relative to the restriction sequences. Restriction mapping has been used in combination with more traditional genetic analysis (see Chapter 8) to construct maps of the DNA of both the human and yeast mitochondrial genomes (Fig. 17.9). These maps reveal that mitochondria from the two organisms have essentially the same genes, but the organization of these genes on the mitochondrial chromosome is quite

Figure 17.9. Maps of the DNA molecules of human (inner circles) and yeast (outer circles) mitochondria. Dark areas represent genes coding for known proteins and for rRNA, and unassigned reading frames (U.R.F.), which are presumably genes coding for proteins that have not yet been identified. Lighter areas code for tRNA; each is labeled with its specific amino acid. Dashed lines indicate stretches of yeast mitochondrial DNA that have not yet been mapped. Note that several genes are split, as are the majority of eukaryotic nuclear genes. *Source:* From L. A. Grivell, *Scientific American* 248:78–89, 1983. Copyright © 1983 by Scientific American, Inc. All rights reserved.

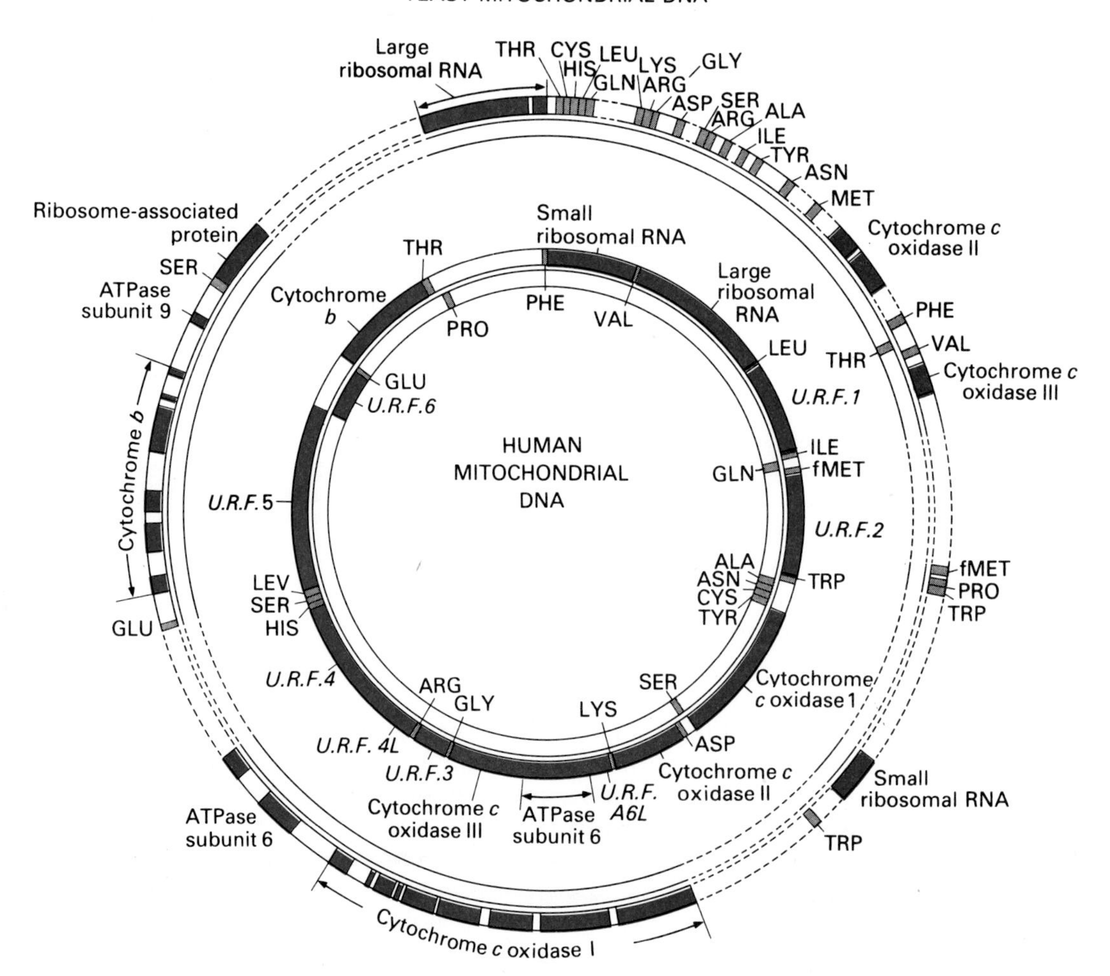

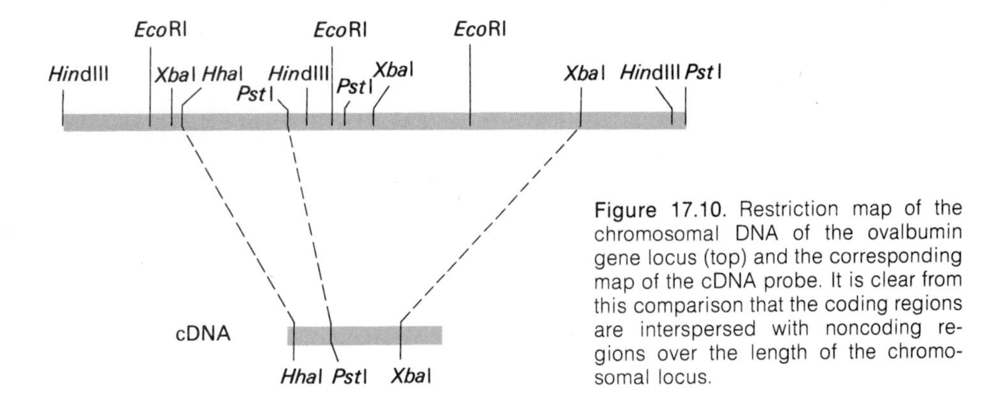

Figure 17.10. Restriction map of the chromosomal DNA of the ovalbumin gene locus (top) and the corresponding map of the cDNA probe. It is clear from this comparison that the coding regions are interspersed with noncoding regions over the length of the chromosomal locus.

different. Restriction mapping has also been used in mapping those regions of the DNA of tumor viruses that correspond to particular viral functions. Because tumor viruses are difficult to work with in traditional genetic analysis, restriction mapping has been extremely useful in providing geneticists with information about the locations of genes on tumor virus chromosomes.

One of the first major discoveries made by restriction mapping and the Southern blot technique was the split nature of eukaryotic genes. When investigators studying the structure of the ovalbumin gene applied the Southern blot procedure, they found that the cDNA probe for this gene hybridized with several different restriction fragments, rather than just one. Because cDNA is derived from mRNA, it contains only coding sequences (exons). Therefore, this finding suggested that the chromosomal coding sequences for ovalbumin might be spread over an extended length of DNA, encompassing several restriction sequences. When the researchers prepared a restriction map of the chromosomal locus for the ovalbumin gene and compared it with the restriction map of the cDNA (Fig. 17.10), it was clear that most of the nucleotide sequences that are present in the chromosomal DNA are not contained in the cDNA. This in itself was not a surprising finding, since it had been known for some time that mRNA is derived from a much longer RNA transcript. It had been thought, however, that processing of the hnRNA into mRNA simply involved the removal of long noncoding sequences at the ends of the molecule. Figure 17.10 demonstrates that the situation is more complex than had previously been thought, in that coding regions are interspersed with noncoding regions over the length of the chromosomal locus.

EXAMPLE 17.2. The technique of amniocentesis makes it possible for us to diagnose many genetic disorders prenatally, by examining the protein deficiencies in fetal cells cultured from the amniotic fluid (refer to Table 11.2). But several important disorders cannot be detected by this technique. One such condition is phenylketonuria, which cannot be detected in the discarded fetal cells because the enzyme whose absence causes the disease is produced only in liver cells. The gene that is responsible for this enzyme is inactive in other types of cells. Another example concerns sickle-cell anemia and other hemoglobin disorders. Until recently, these diseases had not been detectable through amniocentesis, because the amniotic fluid does not contain fetal blood cells. But a development related to recombinant DNA technology is now making it possible to detect hemoglobin disorders prenatally. Investigators have found that the DNA of the chromosomal region that codes for the β-polypeptide chain of normal hemoglobin contains an *Hpa*I restriction sequence. They have also discovered that 80 percent of the individuals with hemoglobin S lack this cleavage site. Based on this information, outline a procedure that will allow the prenatal detection of sickle-cell anemia.

Solution. The technique would use Southern blot analysis to detect the position of the fragment from digestion by restriction endonuclease *Hpa*I that carries the β-globin gene. You would begin by noting the normal position of the nitrocellulose band that carries this gene. You would then compare this position with the position of the β-gene on the gel in bands that are produced by the digestion of DNA from fetal amniotic cells. In sickle-cell anemics, the mutant β-gene is included in a fragment that is larger than normal; this larger fragment gives a band that does not migrate as far across the gel as does the band for the normal β-gene.

DNA Sequencing. Until recently, nucleic acid sequencing had been limited to short polynucleotide chains that were less than 100 nucleotides in length. In 1964, for example, Holley's laboratory reported the primary structure of the tRNA for alanine in yeast, comprising 77 nucleotides in length. Several other tRNAs of comparable size were subsequently sequenced, but little progress was made in sequencing the longer DNA molecules until the mid-1970s. The major difficulty with DNA sequencing is that only four different subunits (A, T, G, and C) make up DNA molecules, which may range in size from 5000 base pairs in the smaller phages to several billion base pairs in mammals, and even longer in many kinds of plants. The lack of diversity of the subunits makes direct sequencing of such long molecules impossible. One of the recent developments that has contributed to an advance in sequencing technology was the discovery of restriction endonucleases that cleave the DNA into short, defined fragments, each of which can be cloned to obtain large amounts of pure material. The other development has been the establishment of methods that allow the sequencing of DNA lengths up to 500 nucleotides in a time interval of just one or two days. The individual restriction fragments are short enough to be sequenced by these methods. Once their nucleotide sequences are known, the individual fragments can be pieced together to give the entire sequence of the total length of DNA.

The method of DNA sequencing that we shall consider here was developed by A. Maxam and W. Gilbert in the late 1970s. To illustrate the procedure, suppose that we cleave the DNA of a virus by using the *Hha*I restriction endonuclease (see Table 17.1 for its recognition sequence), yielding fragments that average 256 base pairs in length. The fragments are then denatured. Assume that these fragments begin with the sequence 5′ CGCTCCACGTA . . . 3′. One end of the sequence, in this case the 5′ end, is next labeled with ^{32}P-adenine. The DNA strands are incubated with labeled adenosine triphosphate and a specific enzyme that adds the labeled adenine (*A) to the 5′ end, to produce the sequence 5′ *ACGCTCCACGTA . . . 3′. The labeled strands are then subjected to chemical reagents that break the DNA at defined points. For example, the combination of hydrazine at high salt concentration followed by piperidine cleaves the DNA strands to the 5′ side of the C residues. This action yields fragments of varying lengths, each having a labeled 5′ end and having lost a C at the 3′ end: *ACGCTCCA, *ACGCTC, *ACGCT, *ACG, and *A. Other sets of cleavage products are similarly generated by the use of reagents that break the DNA strands at the A residues (dimethylsulfate at neutral pH and high temperature), at the G residues (dimethylsulfate at low pH and low temperature), and at both the C and T residues (hydrazine at low salt concentration, followed by piperidine). No reagent has been found that breaks the DNA at only the T residues.

The next step in the procedure is to separate the fragments on the basis of size by means of gel electrophoresis. After the electrophoretic separation of the fragments, their positions are detected as dark bands on an autoradiographic film (Fig. 17.11). Note that the positions of the fragments on the gel directly indicate the nucleotide sequence. The smallest (fastest moving) fragment is *A; it is produced by cleavage at C (columns 1 and 4). Therefore, C is next to *A and is the first nucleotide on the original DNA strand. The next smallest fragment is produced by cleavage at G (column 3), which means that G is the second nucleotide along the strand. The third fastest fragment is produced by cleavage at C, so C is the third nucleotide. The fourth fastest fragment is the result of cleavage at C and T, but not at C alone, which gives T as the fourth nucleotide, and so on. Thus, we can read the sequence from the bottom to the top of the gel by simply noting where cleavage has occurred in order to produce the appropriate band on the film.

Once each restriction fragment has been sequenced in this manner, we are then faced with organizing the fragments into the overall chromosomal DNA sequence. One way that we can accomplish this task is by using another restriction enzyme to generate an additional set of fragments, which are also sequenced by the Maxam-Gilbert method. For example, the restriction enzyme *Hae*III cuts the DNA at a sequence of four base pairs. The *Hae*III sequence, like the sequence for *Hha*I, is expected to occur every 256 nucleotide pairs along the DNA (but not at the same points, since the recognition sequence is different). The *Hha*I and *Hae*III sites should thus alternate regularly, so that most *Hha*I fragments contain *Hae*III sites, and vice versa (Fig. 17.12). The overlap in the nucleotide sequences of the two sets of restriction fragments would then allow us to place them together into a linear sequence representing the entire length of the DNA.

With the methods that are presently available, it is feasible to perform a sequence analysis of the segments that are produced by the digestion by restriction enzymes of a region 5000 to 50,000 base pairs long on a viral or eukaryotic chromosome. But the analysis of longer chromosomal regions has not been accomplished. Even the relatively simple *E. coli* chromosome, with its four million base pairs, has about 15,600 copies of each restriction sequence of four base pairs listed in Table 17.1. Cleavage produces far too many fragments to make sequence analysis practicable in terms of time and cost. Current efforts are therefore being directed at sequencing individual genes and particular regions of chromosomes, rather than complete genomes.

One of the first applications of DNA sequencing was in determining the nucleotide arrangements within the control regions of bacterial and phage operons. Knowledge of the sequences of operators and promoters has contributed greatly to our understanding of transcription and of the way in which effector molecules interact with DNA. The additional possibility exists that once a gene or control region is isolated and sequenced, nucleotides at specific positions can be altered by chemical treatment (so-called site-specific or directed mutation) and the effect of each mutation on gene expression can be noted. Geneticists are beginning to dissect a gene physically into regions whose importance differs with regard to the normal activity of the gene. Sequencing data will also be invaluable in elucidating

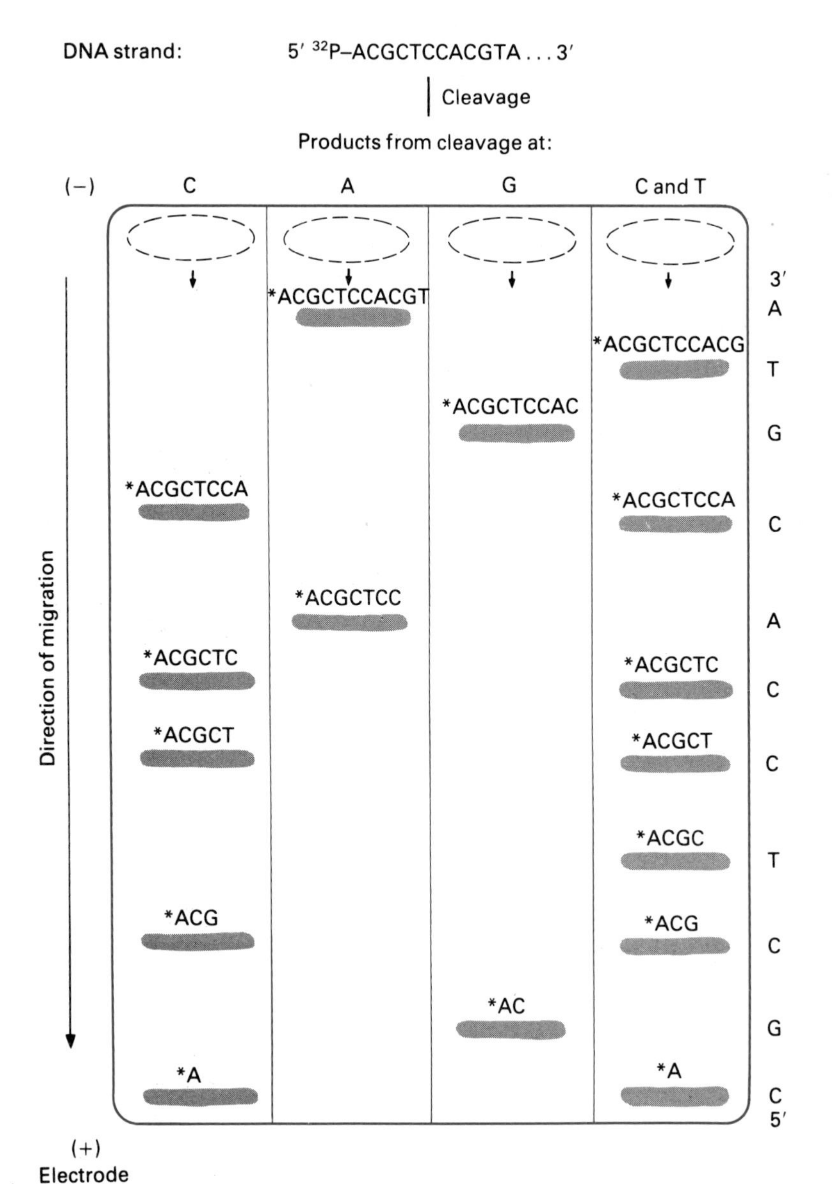

Figure 17.11. The Maxam-Gilbert method for DNA sequencing, using gel electrophoresis to separate DNA fragments of varying lengths that have been produced by selective cleavage. The distance traveled by a fragment depends inversely on its length. The DNA sequence can be read directly from the gel, bottom to top.

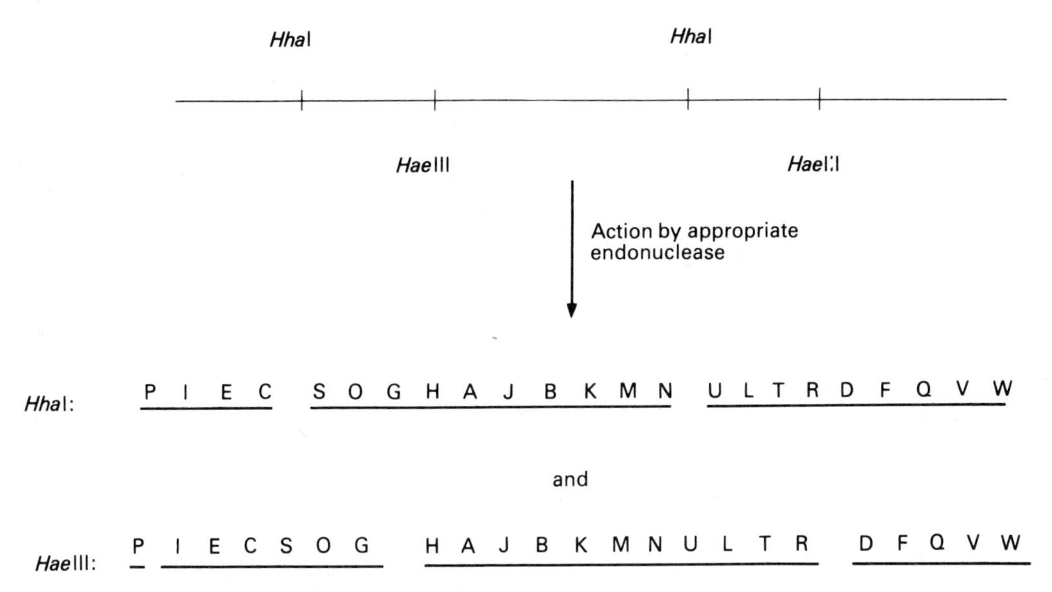

Figure 17.12. Spacing of two different restriction sites along the DNA breaks the molecule into overlapping segments (letter sequences). Once the nucleotide sequence of each segment is determined by the Maxam-Gilbert procedure, the overlapping regions are matched. In this manner, the nucleotide sequence of the entire length of DNA is determined: P I E C S O G H A J B K M N U L T R D F Q V W.

the organizaton of DNA in eukaryotic chromosomes and in deducing the evolutionary relationships among genes and among organisms. DNA sequencing is also the best method for determining the amino acid sequence of many proteins. In most cases, it is more efficient to sequence a gene and determine the amino acid sequence of its protein from knowledge of the genetic code than it is to sequence the protein directly by chemical methods.

Expression of Eukaryotic Genes in Bacteria. So far, we have considered only the cloning of DNA and its applications in genetic research; we have said nothing about the expression of that DNA. Bacteria are not able to express eukaryotic genes with introns, so that techniques have had to be devised to obtain eukaryotic genes that are composed only of coding sequences. This task can be accomplished in several ways. One possibility is to splice cDNA directly into a plasmid; because the cDNA is a complementary copy of the processed mRNA, it lacks the intervening sequences. Another possibility, which is feasible in the case of very short genes, is the chemical synthesis of the gene. For example, the researchers can use a *gene machine,* which is programmed with knowledge of the amino acid sequence of the encoded polypeptide (Fig. 17.13). The genes for human insulin and the growth hormone somatostatin have been synthesized artificially. Generally, artificial synthesis has been very slow and expensive, so that few genes have been made in this manner. Progress in gene synthesis technology is proceeding at a rapid rate however.

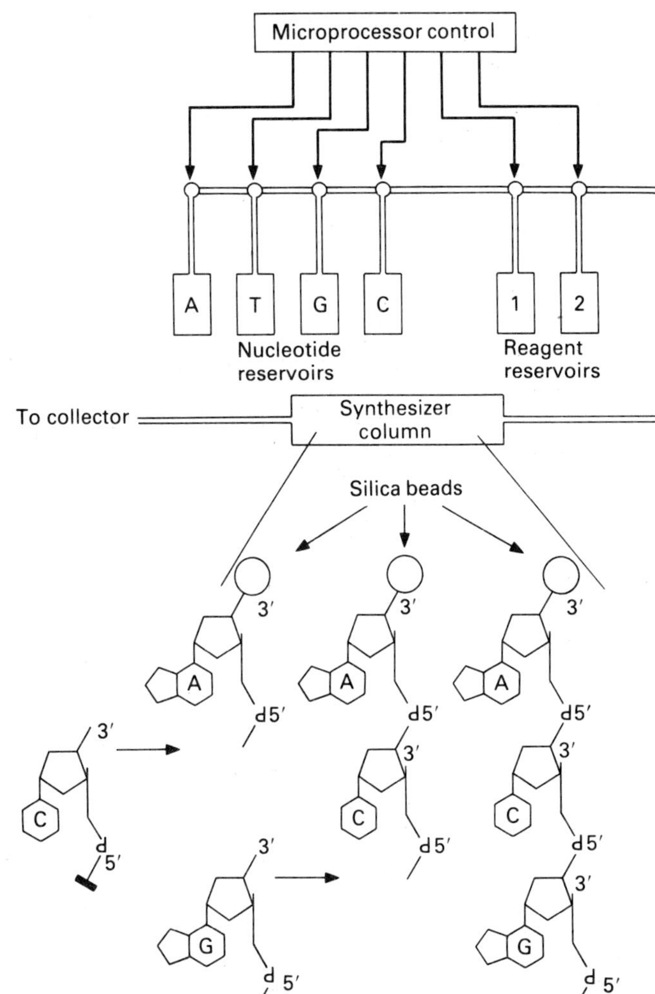

Figure 17.13. The workings of a "gene machine." The desired sequence of nucleotides is entered on a keyboard. The microprocessor controls valves that allow the nucleotides and reagents needed at each step to be pumped through the column. Silica beads within the column serve as supports for the synthesis of short single-stranded DNA segments. The 3′ end of the first nucleotide of the segment is fixed to a bead. The 5′ end of that nucleotide is free to bond to the 3′ end of the second nucleotide. Each nucleotide that enters the column has its 5′ end blocked with a chemical reagent. The reagent is removed following the formation of a phosphodiester bond, leaving the 5′ end free to accept the next nucleotide. In this manner, nucleotides are added at the rate of about one every 30 minutes.

Neither cDNA nor artificial genes contain the control regions that are normally adjacent to the coding sequences on the chromosomes. Without the influence of a promoter region that will combine with bacterial RNA polymerase, an inserted gene cannot be transcribed. To overcome this problem, geneticists splice the cDNA or artificial gene into a specially constructed plasmid that carries bacterial control signals. For instance, Fig. 17.14 shows the insertion of the somatostatin gene into a plasmid that contains the control regions and β-galactosidase gene of the *lac* operon. Somatostatin is then synthesized by the recipient cells as a short polypeptide attached to the end of β-galactosidase, from which it is cleaved and isolated in pure form.

Another problem with bacterial synthesis of the products of eukaryotic genes is that bacteria often degrade foreign protein. For this reason, geneticists frequently modify the host cell by mutation so that it has lost its ability to recognize a protein as foreign. Another approach that has been used successfully with the somatostatin gene is to add a methionine codon to the beginning of its coding sequence. This experimental "trick" serves the same purpose as modifying the host bacterium,

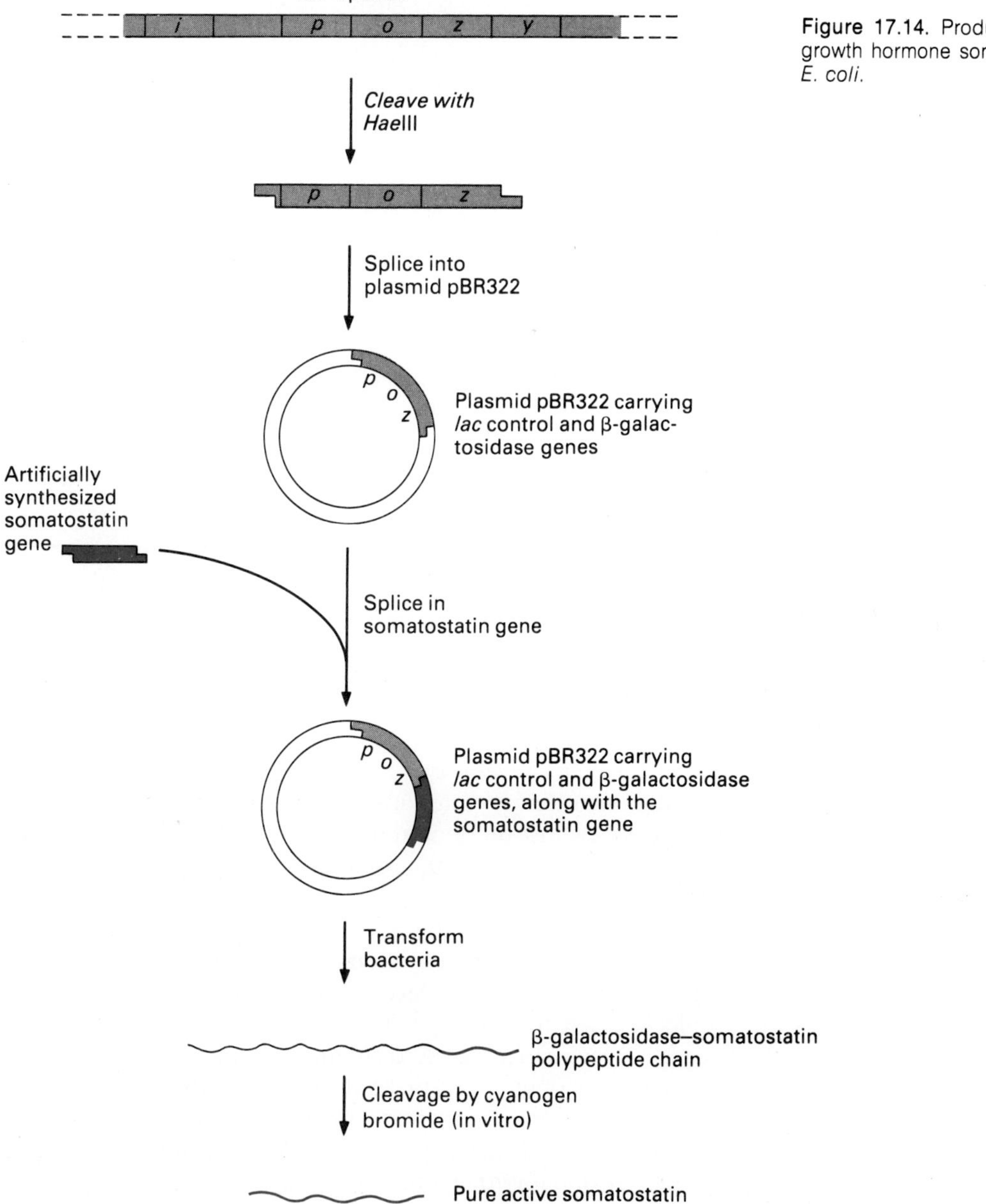

Figure 17.14. Production of the growth hormone somatostatin by *E. coli*.

since the presence of an initial methionine prevents the recipient cell from degrading the polypeptide product of the inserted eukaryotic gene.

Applications in Industry

Biotechnology involves the use of living organisms or their components (e.g., enzymes) in industrial processes. The art of fermentation is the oldest form of biotechnology. Nearly 9000 years ago, ancient societies made use of the microbial fermentation process in the conversion of sugar to alcohol in order to make beer. A variety of microbial products have since served humankind, providing food, nutritional supplements such as vitamins, beverages, and products for medicine (e.g., antibiotics) and for industry.

Recombinant DNA technology is making a substantial impact on biotechnology, because it allows us to manipulate the genetic material of microorganisms directly, in order to produce the desired characteristics. Using this technology, geneticists are able not only to improve the efficiency with which microbes carry out natural fermentation but also to program microbes to produce substances that they never synthesized in nature. The first commercial applications of these advances have been in the pharmaceutical and chemical industries. Potential applications include food processing, mineral leaching and recovery, oil recovery, and pollution control. We will discuss some of these uses of recombinant DNA methods in the following sections.

The Pharmaceutical Industry. The pharmaceutical industry has been the first to make widespread practical use of recombinant DNA techniques (Fig. 17.15). Bacterial cells are now making a variety of pharmacologically active proteins, including human *insulin,* to replace the insulin that is isolated from the pancreas glands of cattle and swine (which causes allergic reactions in some patients and does not prevent the deterioration of the kidneys and retinas that accompany diabetes). Human insulin made by bacterial cells was released for sale in the United States, United Kingdom, the Netherlands, and West Germany in 1982. Other products now under clinical or animal testing are human *growth hormone,* to replace its natural source—human cadavers; several types of *interferon,* an antiviral protein that is available naturally from human foreskins in very small amounts and at a staggering cost; and *vaccines* against the foot-and-mouth and hepatitis viruses. In all of these cases, the protein produced by the bacteria is of comparable structure, purity, and activity to natural sources. Future prospects for synthesis by bacteria include silk, blood coagulation factors (for therapy in diseases such as hemophilia), new varieties of antibiotics (to replace the traditionally used ones, to which many bacteria have become resistant), numerous vaccines and hormones, and a variety of enzymes that could be used to replace defective catalysts that cause certain human genetic diseases. Recombinant DNA techniques, coupled with the new technology of *protein engineering,* will even make it possible in the near future to create proteins with novel properties not found in nature. Investigators would modify the gene that codes for a particular protein to alter its coding sequence in a predictable fashion, designed to improve some functional property of the protein.

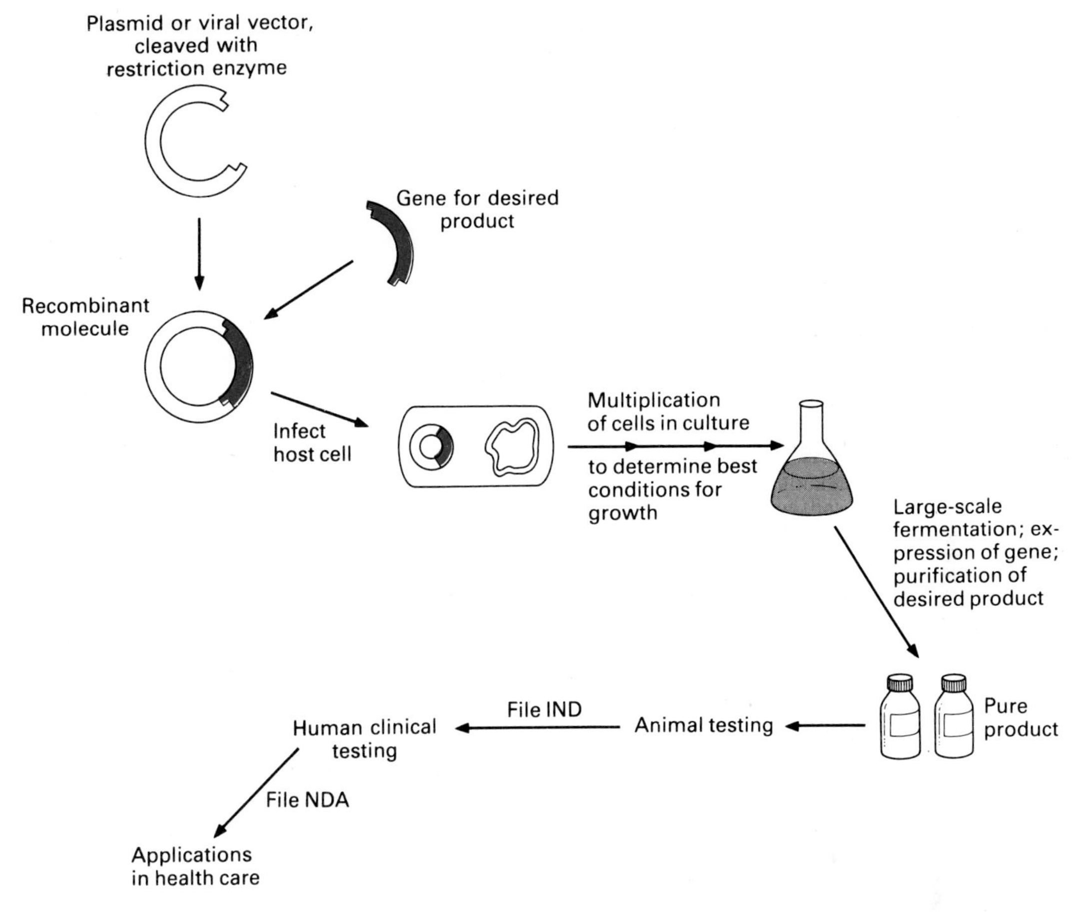

Figure 17.15. The process used in developing a pharmaceutical product from a genetically engineered microorganism. A recombinant DNA molecule is constructed carrying the gene that codes for the desired product. (This gene can be obtained either from a biological source or by organic synthesis.) The gene is cloned in a microbial host cell. Large-scale fermentation then yields large amounts of the desired product, which is purified and packaged for testing. Animal testing is followed by submission of an investigational new drug application (IND), which, if approved, allows clinical trials with humans. Following clinical testing, a new drug application (NDA) is filed with the FDA. If approved by the FDA, the product may then be marketed in the United States.

The modification of a gene can be accomplished by using an automated gene synthesizer to make a mutant form of a gene by introducing a specific alteration (e.g., a particular base pair substitution) at a specific site along the gene.

By making a pharmaceutical product widely available at a reasonable cost, recombinant DNA technology has two types of impact. First, substances that have known medical promise will be available for testing (e.g., interferon can be tested for therapeutic use against cancer and viruses). Second, pharmacologically active substances that have no present use can be explored for potential new therapeutic applications. It is very likely that recombinant DNA methods represent the next great advance in clinical medicine.

The Chemical Industry. Recombinant DNA technology is expected to make a major contribution in the future to the production of organic chemicals, such as alcohols and other organic solvents, acids, plastics, synthetic fibers and rubbers, agricultural chemicals, and cosmetics. At present, approximately 80 percent of the raw material needed to produce these chemicals comes from petroleum and natural gas. Because the world's supply of petroleum is threatened by dwindling resources, politics, and increased costs, and because chemical synthesis requires high energy input and generates unwanted by-products and other pollutants, there is increasing interest in using biomass (natural renewable resources) as raw material in the biological synthesis of organic materials. In biological synthesis, enzymes replace chemical catalysts, which require high temperatures. Enzymes have essentially 100 percent conversion efficiency, producing none of the undesirable by-products of chemical synthesis. Biosynthesis thus drastically reduces chemical pollution and, with it, the costs of pollution control and waste disposal. In fact, wastes that are created biologically are often themselves valuable as sources of nutrients.

In principle, all organic compounds could be produced by biological systems. In practice, however, biological synthesis is limited by two main factors: (1) the availability of the organism or enzymes for the desired synthesis and (2) the availability and cost of the raw material. Scientists are therefore attempting to engineer microorganisms genetically so that they are able to utilize readily available substances as raw materials. With respect to the available technology, food-related biomass sources (such as the starch in corn and potatoes, and various sugars) are the best candidates at present to serve as the raw material of biological conversions; yet, the use of sugars and starches in fermentation processes is not currently as profitable as is their direct use as food. Cellulose biomass sources (wood, agricultural wastes, municipal wastes) are more promising candidates to serve as raw materials in the future, but technological barriers concerned with the collection, storage, pretreatment (to generate fermentable substances), and waste disposal must first be overcome.

As an example of the biological synthesis of an organic compound, let us consider the industrial production of ethanol, one of the most important industrial compounds. Ethanol has traditionally been made from petroleum or natural gas by a reaction sequence that requires a large input of water and very high temperatures. In an alternative process, yeast can be employed to catalyze the conversion of crude sugar (from sugarcane or beet molasses) to ethanol biologically, at much lower temperatures (Fig. 17.16). This method already is cost-competitive with industrial ethanol produced from fossil fuels. But the sugar supplies fluctuate in price and are themselves needed as food. Therefore, researchers are presently working to develop a strain of microorganism that can break down cornstarch into its component sugar units, which can then be fermented into ethanol or other organic materials. Even if this possibility is realized, the problem of availability of biomass raw material remains. For instance, replacing all of the gasoline used in

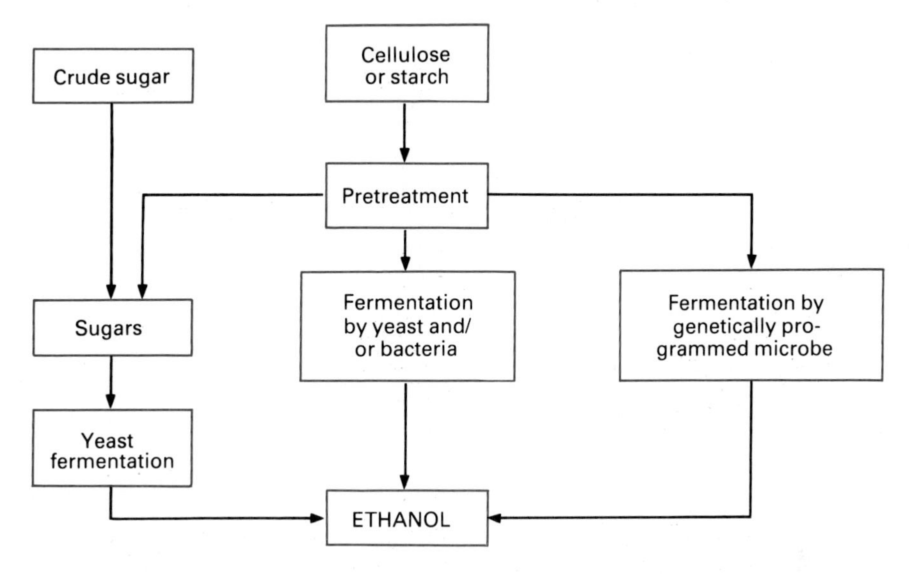

Figure 17.16. Biological synthesis of ethanol. Currently, crude sugar (sugarcane or molasses) is converted into sugars that are used as substrates by fermenting yeast to produce ethanol. Investigators are presently working to isolate microorganisms that can degrade cellulose or starch into their component sugars and can then ferment these sugars into ethanol. Work is also underway to program known yeast and bacterial species genetically to produce ethanol from cellulose in a single operation.

this country with gasohol (9 parts gasoline : 1 part ethanol) would require over 100 times more ethanol for use in fuel than our current level of usage. Excess corn and other grains simply cannot provide such large amounts of biomass. For this reason, scientists are presently working to develop a microorganism that can feed directly on cellulose.

Pollution Control. Another important application of recombinant DNA technology is in the detoxification and degradation of sewage and industrial wastes by microorganisms. The use of organic wastes as substrates for biological conversion into industrial compounds is an exciting possibility that would help overcome shortages of raw materials, while at the same time it would dispose of noxious substances.

Microorganisms are now being made available to degrade specific compounds, such as polychlorinated biphenyls (PCBs). One bacterium, *Pseudomonas putida,* has been engineered with a plasmid that carries genes coding for four different enzymes that degrade the octane, hexane, decane, xylene, toluene, camphor, and naphthalene components of oil. The bacterium works well in the laboratory under controlled environmental conditions, but it has yet to be proved effective in cleaning up oil spills. This particular bacterial strain made legal history in 1980, when A. M. Chakrabarty, its programmer, was issued the first patent ever granted for a genetically manipulated microorganism.

TO SUM UP

1. Restriction endonucleases are bacterial enzymes that catalyze the breakdown of DNA. Many restriction endonucleases cleave DNA at specific recognition sites four or six base pairs in length that are palindromes, in that their base sequence reads the same forward and backward. Some restriction endonucleases make staggered cuts at their recognition sites, to yield DNA fragments that have single-stranded complementary (cohesive) ends.
2. Recombinant DNA molecules can be artificially created by inserting foreign genes into the DNA of a vector such as a plasmid or virus. The splicing together of two different DNA molecules is accomplished by joining the DNA fragments at their cohesive ends, which are formed by treatment with the same restriction endonuclease. The DNA fragments first join through hydrogen bonding and are subsequently sealed with DNA ligase. The hybrid plasmids or viruses that are produced in this manner are then replicated by their incorporation, through infection, into a suitable bacterial host.
3. When a hybrid plasmid (recombinant DNA) replicates within a bacterium, the piece of foreign DNA is said to be cloned. Cloning permits geneticists to obtain substantial quantities of the foreign genes, the RNA transcripts, or their polypeptide products for practical purposes. The cloned foreign genes are identified by specific hybridization tests; these tests use a nucleic acid probe made of cDNA, which consists of radioactive DNA that is complementary in base sequence to the gene or genes in question.
4. The practical uses of cloned recombinant DNA in research include restriction mapping (which involves ordering the genes in a DNA molecule relative to the arrangement of cleavage sites of various restriction endonucleases) and DNA sequencing (which determines the primary structure of the DNA). So far, the applications of recombinant DNA technology in industry have involved introducing genes into bacteria for the production of pharmacologically active proteins such as peptide hormones, enzymes, and vaccines. Numerous other potential applications can be envisioned and are presently being explored.

Advances in Plant and Animal Improvement

The successful genetic manipulation of microorganisms by molecular geneticists led researchers in the agricultural sciences to consider applying similar techniques of genetic engineering to higher plants and animals. But the vastly greater complexity of eukaryotes has impeded major progress in this area. Although it still holds great promise for the future, the impact of the new methods of genetic engineering on agriculture has, until now, been minimal.

In this section, we will discuss some promising genetic technologies that are available for plant and animal improvement. These techniques, as they are presently envisaged, will not replace the established breeding practices; rather, they will serve as *additional* tools for the development of new and more productive varieties.

Incorporating Wild Gene Resources into Crop Plants

In order for the technologies that have been developed for both classical plant breeding and the new genetic engineering techniques to be applied successfully, there must be a continued source of genetic diversity. Ordinarily, the availability of diverse plant types is limited by the amount of variation in the gene pool of the population. The potential for genetic variation in most plant populations is large. A rapid erosion of this variability usually takes place during selective breeding, though, leaving the selected stocks more genetically uniform than the original population and thus more vulnerable to new diseases and adverse changes in climate. As new strains are produced from existing varieties, less variability remains for further improvement. Eventually, a limit is approached, at which the existing gene pool of the plant population has been exploited to its fullest potential. Additional improvements in disease resistance or some other characteristic must then be induced or must come from other sources.

One important source of genetic variabiliy for crop plants is from wild related species. There have been numerous cases in which the wild, weedy relatives of cultivated plants (such as melons, strawberries, tobacco, and wheat, to name a few) have provided genetic traits of economic value. By all indications, this source of variation will also be important in the future, enabling plant geneticists to make improvements with regard to factors such as increased pest and drought resistance, tolerance to poor soils and high salt content (as in heavily irrigated lands), increased nutritional (e.g., protein) content, requirement for less fertilizer, and better growth under minimum tillage systems.

In addition to desirable genes (such as those for disease resistance), the chromosomes of a wild species also have many genes that produce an undesirable effect when introduced into the genome of a cultivated plant. To avoid the often undesirable consequences of mixing whole genomes, geneticists rely on techniques that permit a transfer of only a single chromosome or chromosome segment. We will describe one set of procedures that has been used successfully in wheat.

Chromosome and Gene Transfer in Wheat. E. R. Sears has been instrumental in the development of techniques for the transfer of chromosomes and chromosome segments into wheat from its wild diploid relatives. Recall that the common wheat is a hexaploid, with a diploid number of 14 chromosomes in each of the three genomes A, B, and D. This plant thus has a total of $3 \times 14 = 42$ chromosomes, which can be represented by the genomic formula AABBDD.

Two types of plant lines that carry a pair of chromosomes from a different (alien) species can be established in wheat. One type is an *alien-addition line* which has a chromosome pair from an alien species in addition to all of its own chromosomes. The second type is an *alien-substitution line,* which carries a pair of alien chromosomes in place of a corresponding (homeologous) pair of wheat chromosomes. The alien-addition line is formed through a series of crosses, starting with the creation of a fertile amphidiploid between common wheat and the alien species (Fig. 17.17). Backcrossing the amphidiploid to common wheat will produce progeny with only a haploid set of the alien genome. Members of this haploid set are randomly dispersed into gametes, so that repeated backcrosses with common wheat can eventually produce various alien-addition lines that are monosomic for one or more different alien chromosomes. Selfing the members of each monosomic line can then lead to the formation of some disomic offspring in which the alien chromosome is paired. All of the chromosomes in the disomic alien-addition line thus have a regular pairing partner.

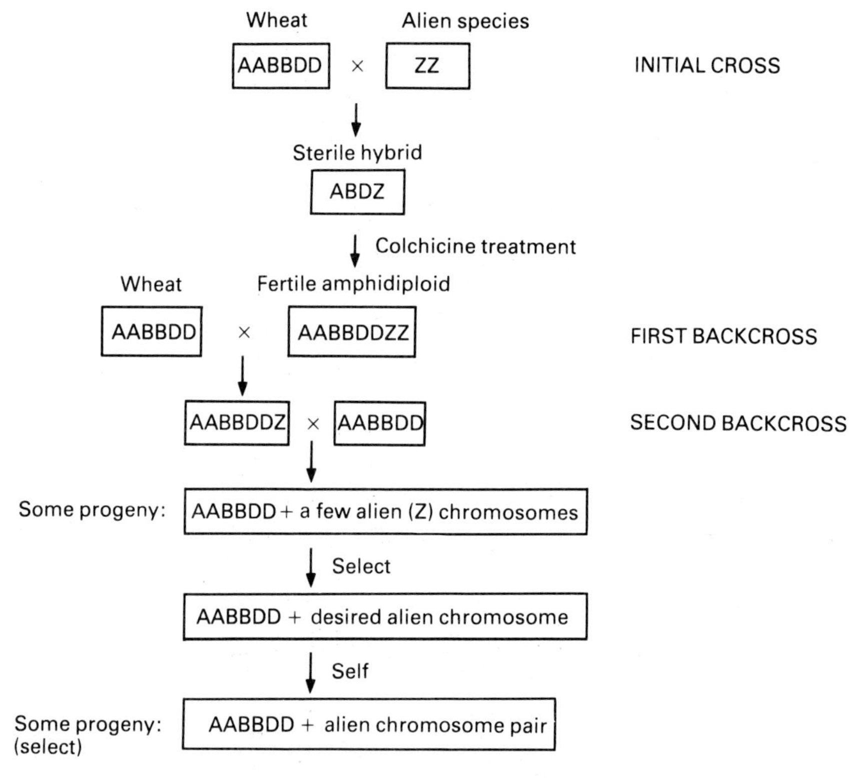

Figure 17.17. Production of a wheat line that carries an alien chromosome pair as an addition to the normal chromosome complement. A fertile amphidiploid that carries the diploid complement of alien chromosomes is first produced. Repeated backcrossing to common wheat yields some progeny that are monosomic for just a few alien chromosomes. Selfing of these plants and selection techniques can then be used to isolate some offspring that carry a particular alien chromosome pair.

One procedure for forming an alien-substitution line starts with a cross between a member of an alien-addition line and a monosomic wheat that is missing a chromosome homeologous to the alien pair (Fig. 17.18). Offspring are selected that are monosomic for both the wheat and the alien chromosome. Selfing these double-monosomic offspring can result in several different kinds of progeny with regard to karyotype. A few of these progeny plants will be disomic for the alien chromosome and nullisomic for the corresponding homeologous wheat chromosome. These progeny plants constitute an alien-substitution line.

The addition or substitution of an entire chromosome or chromosome pair from an alien species seldom results in a plant with characteristics that are suitable for immediate practical use. Additional chromosome engineering is usually necessary in order to reduce further the number of detrimental alien genes, while still retaining the genes that produce the desired effect. One approach is to use ionizing

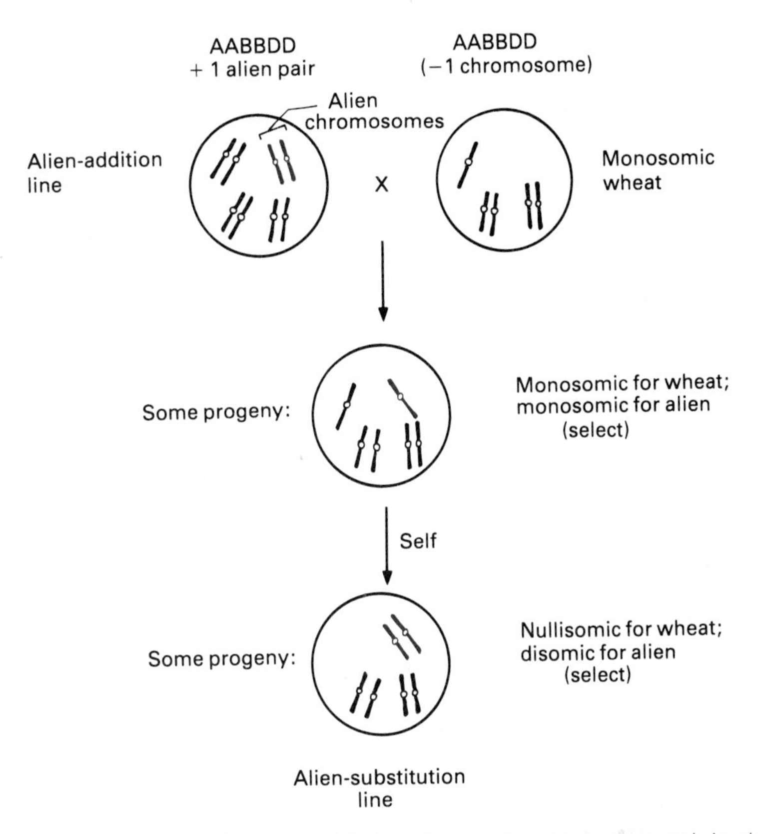

Figure 17.18. Production of a line of wheat that carries an alien chromosome pair in place of the corresponding pair of wheat chromosomes. A cross between an alien-addition line and a monosomic wheat that is missing a chromosome homeologous to the alien pair yields some offspring that are monosomic for both the wheat and the alien chromosome. Selfing of such a plant produces some progeny that contain a pair of alien chromosomes while lacking completely the homeologous wheat chromosomes.

radiation to induce translocations in the pollen of an alien-addition line (Fig. 17.19). This technique was applied by Sears in transferring the dominant *R* gene for rust resistance from the wild grass *Triticum umbellulatum* into common wheat. In this case, cytological studies revealed that the translocation homozygote that was produced contained little more genetic material from its wild grass relative than the gene for rust resistance.

New Genetic Technologies for Plant Breeding

The major goals of crop breeding are to improve the quantity and quality (e.g., nutritional value) of crop yields and to reduce the production costs. In the past, these goals were achieved by a repertoire of genetic techniques that included various selection and breeding methods and the artificial induction of polyploids

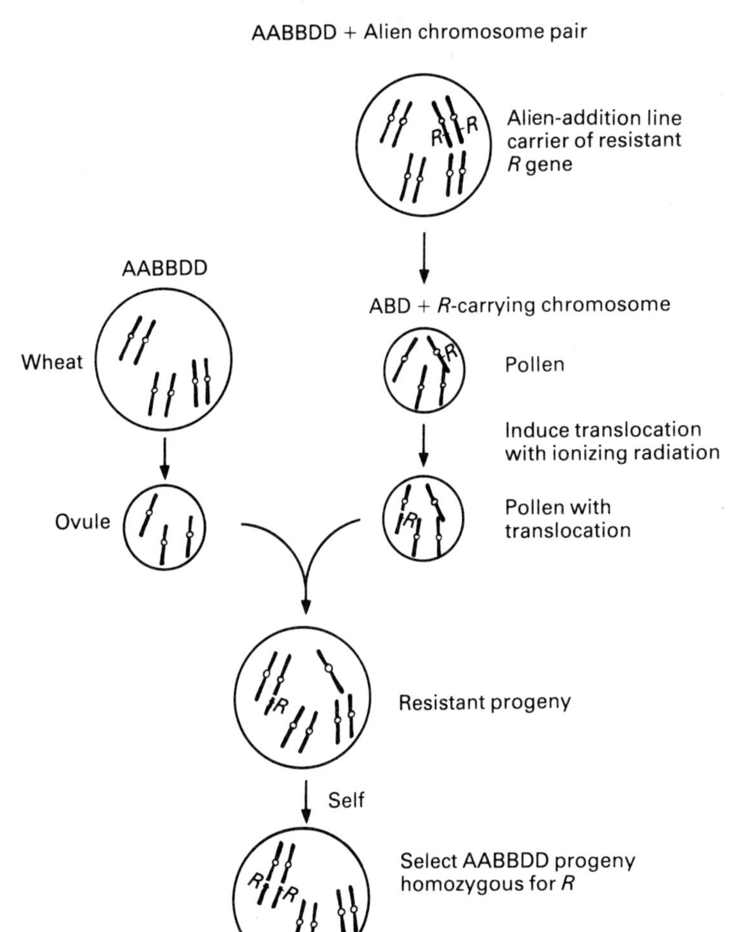

Figure 17.19. Production of a wheat line that carries only a particular gene derived from an alien line. Pollen from an alien-addition line is irradiated to introduce a translocation of a chromosome segment that includes gene *R* to one of the wheat chromosomes. Selection of resistant offspring and their selfing yields some progeny that are homozygous for gene *R*.

with colchicine. Plant geneticists now have available new technologies that enable them to culture and to manipulate the cells of higher plants genetically in a manner similar to the methods used with microorganisms. The genetically altered cells can then be grown into mature plants with the desired characteristics.

Cloning, Protoplast Fusion, and Plant Regeneration. The development of a different plant variety through the use of the new genetic engineering technologies is basically a four-step process (Fig. 17.20). The first step is *the growth of isolated plant cells in tissue culture.* Single isolated cells can be induced to divide in broth or on agar medium to produce clones that consist of millions of genetically identical descendants. The cultured cells can be kept alive by transferring them periodically to new culture medium or, in some cases, by storing them in a frozen state in liquid nitrogen.

The second step is *the genetic manipulation of the cultured cells.* Various genetic procedures are (or soon will be) available for the introduction of genes into cultured cells. One such procedure involves the transfer of foreign DNA that is attached to the DNA of a plasmid or virus (this method will be described in the next section). Another approach, which was already considered briefly in Chapter 12, is the fusion of genetically different cells that have been stripped of their cell walls. When two of these protoplasts fuse, they form a single hybrid cell that contains the genetic information of the two different cell types. In this way, the genes of even distantly related species can be combined without the restrictions of natural breeding barriers, and there is a potential for the creation of novel plants.

The third step is *the screening of the genetically modified cells for useful characteristics*. For example, if the desired trait is resistance to the toxin that is produced by a certain plant pathogen, then screening might simply involve growing the cells in the presence of the toxin and selecting the resistant clones for further testing. The growth of haploid cells by means of anther culture promises to be useful in the induction and selection of genetic changes (see Chapter 12). Through the use of anther culture, recessive alterations in the genome that are produced by mutation or some other method are expressed immediately without the masking effect of dominance. Colchicine can then be applied to promote the doubling of the chromosome number, in order to produce diploid plants that are homozygous for the altered gene.

Figure 17.20. The development of new plant varieties regenerated from genetically engineered cells growing in culture.

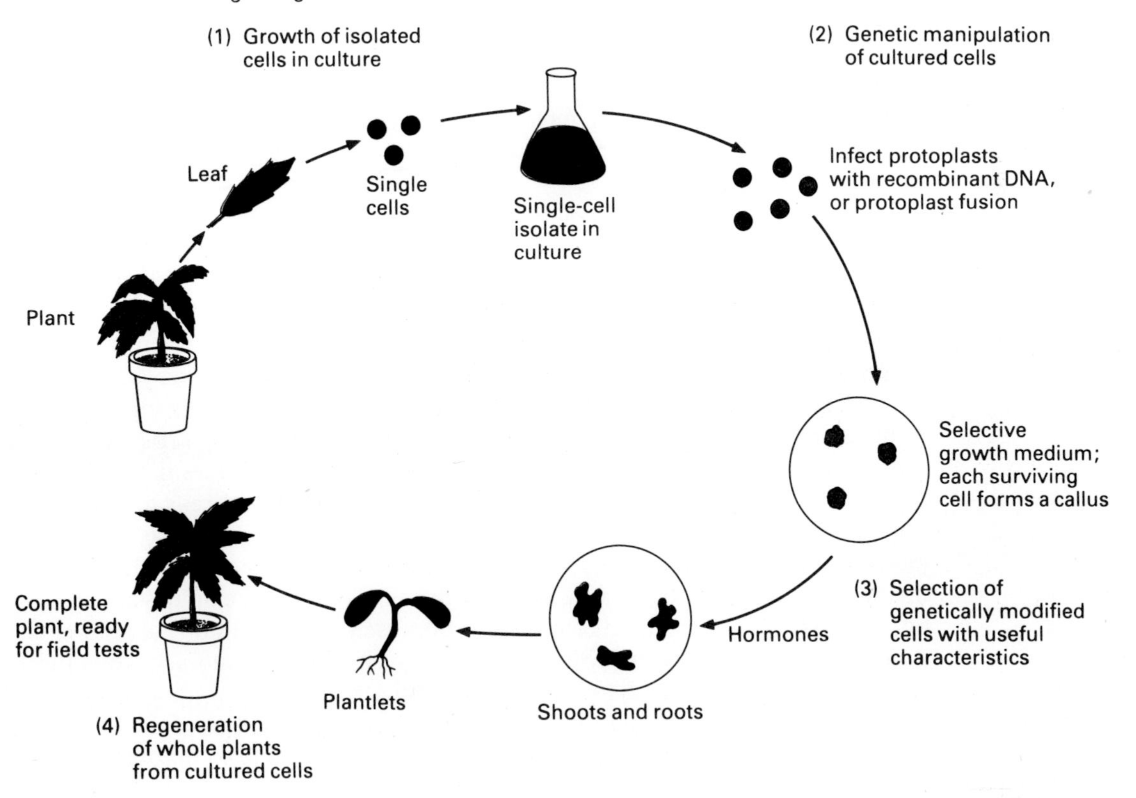

The fourth step is *the regeneration of whole plants from cultured cells.* Protoplasts can be induced to regenerate cell walls and proliferate into a cell mass, or callus. By adjusting the plant hormone levels in the growth medium, the researchers can cause the callus tissue to develop into small plantlets with immature roots, stems, and leaves. Eventually, the plantlets are transferred to soil, where they can grow to form complete normal plants.

The procedure just described has several applications in addition to its potential use in genetic engineering. For example, this method is presently being applied in the selection and mass propagation of plants for commercial use, such as trees for reforestation. It also has commercial importance in the development of virus-free varieties of plants.

Plants as Recipients of Recombinant DNA. Investigators are now working to adapt recombinant DNA procedures to plant cells as recipients. One major barrier to accomplishing the engineering of plant cells with a novel gene is our limited understanding of plant molecular genetics. Genes for most important plant characteristics have not yet been identified. In addition, our understanding of the molecular bases of gene expression in plants, as in all eukaryotes, is very limited, since regulatory processes in eukaryotes differ from those of prokaryotes. In the late 1970s, it became possible to insert replicating yeast genes into yeast protoplasts. Each yeast donor gene is first purified by splicing it into a yeast-*E. coli* hybrid plasmid and cloning the plasmid in bacteria. The *E. coli* section of the hybrid plasmid allows it to replicate in bacterial cells; the yeast portion allows it to replicate in yeast cells. Yeast is likely to become increasingly important in the applications of recombinant DNA technology to genetic research because it is a eukaryote, its genes are fairly well known, it contains plasmids that can be used as vectors, it is not a pathogen, and it can be easily contained since it has no form of aerial dispersion.

The potential applications of recombinant DNA technology in higher plants are enormous. The major obstacle to using these procedures has been a paucity of vectors that will allow the donor material to be replicated in recipient plant cells. The few known DNA plant viruses are unable to multiply if foreign DNA is inserted into them. Plant geneticists have so far focused their attention on a bacterial plasmid called Ti, which is found in *Agrobacterium tumefaciens. A. tumefaciens* causes crown gall tumors in wounded dicotyledonous plants, by transferring a segment of the Ti plasmid (called T DNA) into plant cells. The T DNA inserts itself into a randomly selected position on a plant chromosome. Certain genes that are carried on the T DNA prevent cell differentiation, causing the infected cells to grow in an uncontrolled manner into a tumor. Other genes on the T DNA cause the infected plant cells to produce the enzyme opine synthetase, which catalyzes the synthesis of a special class of nitrogen-rich amino acids called opines. The opines are required by *A. tumefaciens* as a source of nitrogen. Because the Ti plasmid is able to confer a new genetic property on the infected plant cells, investigators have come to regard it as a promising vector for recombinant DNA work.

In 1982, it was reported that the tumor-producing activities of the Ti plasmid had been separated from its capability for transferring genes, so that the recipient cells do not become cancerous. The genes that normally block the differentiation of infected plant cells are inactivated by mutation; the plasmid itself still remains active, though, as shown by the presence of opines in infected cells. Investigators are now concentrating their attention on splicing donor DNA into these altered plasmids and using them to introduce foreign genes into plant cells (Fig. 17.21). Early in 1983, researchers reported that the Ti plasmid had been used to introduce modified bacterial genes coding for antibiotic resistance into plant cells. The introduced genes were expressed, conferring resistance to certain antibiotics on the recipient cells. (The donor bacterial genes were modified by fusing their protein-coding sequences with the control signals of a Ti gene; the modified genes were then spliced into the Ti plasmid vector.)

One further problem that is encountered in this research is the restricted specificity of *A. tumefaciens,* which infects only dicotyledonous plants (dicots). Many important crop plants, such as the cereal grains, are monocots. Scientists have recently isolated and cloned a transposable *Ds* element from maize that may permit the development of a gene transfer system for this crop plant. One possible gene transfer system utilizes cells that lack the enzyme alcohol dehydrogenase (ADH) as recipients; these cells are unable to grow anaerobically. Researchers are attempting to splice the *adh* gene for this enzyme together with some other desired donor gene into the *Ds* element, creating a recombinant DNA molecule that includes the *Ds* element as the vector. Protoplasts that are deficient in ADH would then be incubated with the recombinant DNA and the ability of the recipient cells to grow anaerobically would be used to select those cells that have acquired the desired genes. As soon as techniques are developed that will enable investigators to grow whole corn plants from protoplasts, as is possible with some other plants, this system may provide the means to engineer corn and other monocots with recombinant DNA.

Perhaps the most widely publicized application of recombinant DNA technology in plants is the engineering of corn and other crop plants, so as to enable them to fix their own nitrogen from the atmosphere, thus eliminating the need for large amounts of nitrogen fertilizer. Nitrogen-fixing plants acquire this ability from bacteria that infect the roots of some legumes and fix nitrogen from the atmosphere. The nitrogen-fixing bacteria have established a complex symbiotic relationship with their host: In return for usable nitrogen, the plant supplies the bacteria with a source of carbon. In order for a stable symbiotic relationship to develop, researchers must therefore engineer both the bacterium (so that it can infect corn, for example) and the corn cells. As the first step in this process, researchers have recently assembled a bacterial plasmid that carries all 17 of the known genes involved in nitrogen fixation in the bacterium *Klebsiella pneumoniae.* By infecting *E. coli,* this recombinant plasmid confers upon its bacterial host the ability to fix nitrogen. But infection of yeast cells with the engineered plasmid does not result in the acquisition of the ability to fix nitrogen. Apparently, the yeast cells are unable to express the incorporated prokaryotic genes correctly. So little is known of the molecular genetics of plants that it is difficult at present to determine why the nitrogen-fixing bacterial genes fail to be expressed in yeast cells.

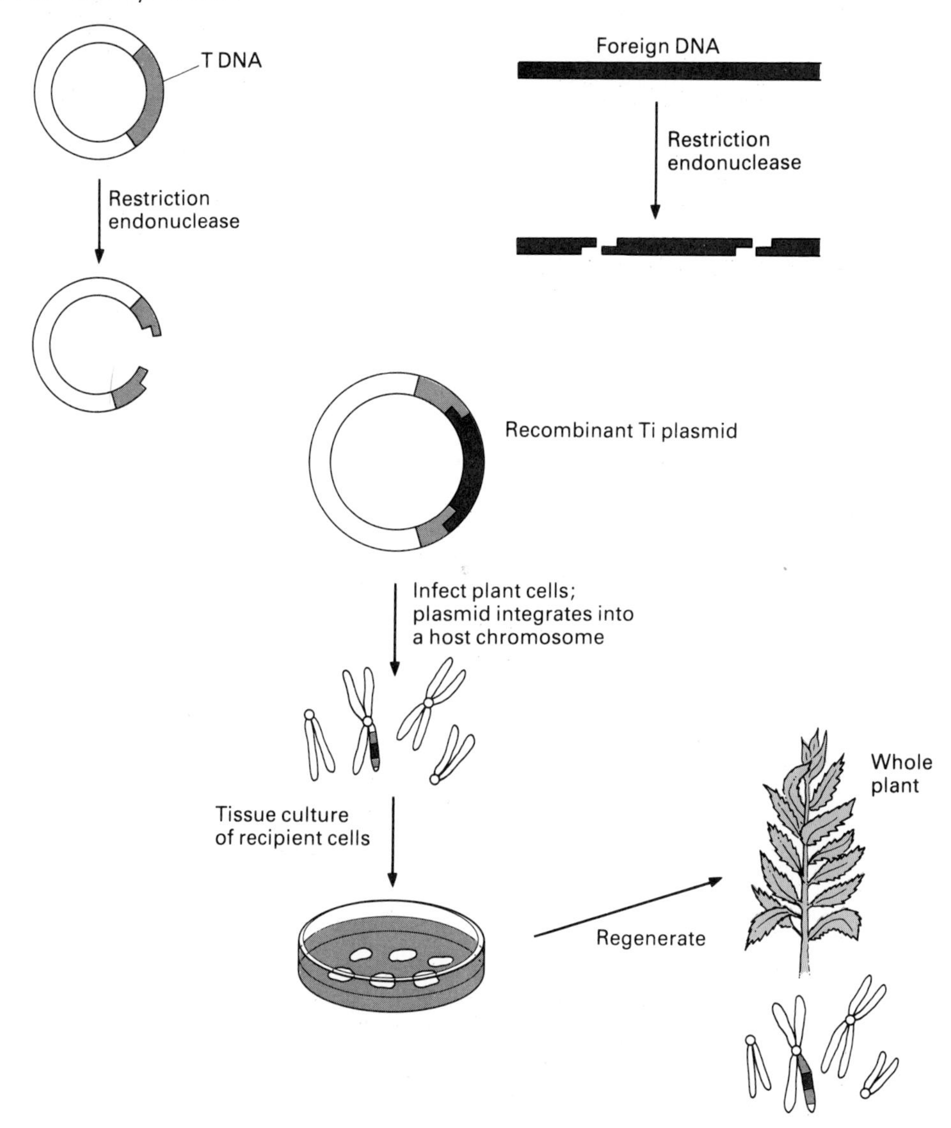

Figure 17.21. Possible means of introducing foreign DNA into plant cells. A recombinant plasmid is constructed from a modified Ti plasmid (that does not transform recipient cells to the tumorous state) and a foreign DNA. Upon infection of a plant cell, the recombinant plasmid integrates into a host chromosome. (Integration can be measured by the production of the enzyme opine synthetase by the host cells.) After the cells are grown in culture, plants are regenerated from the cultured cells. If the cells of these plants retain the Ti plasmid and its associated foreign DNA it becomes possible to obtain a new plant strain through genetic engineering.

Although fundamental problems remain to be solved, it appears that recombinant DNA technology will have an important impact on agriculture. The new technology is not a panacea, and it is doubtful that it will ever replace the traditional breeding methods. Yet it does hold promise of providing additional variability from which to select. One concern in this regard, however, is the ongoing loss of irreplaceable genetic variation from plants in the wild, which is caused particularly by the clearing of land in the tropical rain forests for farming. Ironically, now that recombinant DNA methods may soon make it possible to screen this wealth of variation, the variation is being lost at a rapid rate.

New Technologies for Animal Improvement

Several techniques have been developed over the past 30 years that enable breeders selectively to control and manipulate the reproduction of their animals to a degree that was impossible with traditional breeding methods. One of the first and most important techniques to be successfully applied to animal breeding was artificial insemination. This advance allowed the widespread use of sperm from sires of proven genetic quality. The general application of this procedure became possible with the development of a successful means to store sperm in the frozen state on a long-term basis. For example, the sperm of cattle can be kept in the frozen state (−196°C) for an indefinite period without a major loss of viability.

Additional advances have been made in recent years that permit breeders to increase greatly the reproductive efficiency of genetically superior animals. One is a method known as *superovulation,* in which hormones are used to induce the release of a greater number of eggs than is normally released by a female at ovulation. This technique has definite commercial potential for cattle breeders. A cow typically releases only one egg at ovulation, so as to produce less than 10 calves during her normal lifespan. With superovulation, however, the number of eggs that are released at a single ovulation can be increased to as many as 8 to 10. When superovulation is coupled with *embryo transfer,* which is another technique that has proved feasible in cattle, each embryo that is formed upon fertilization can be recovered from the animal that produced the eggs and can be implanted in the uterus of another animal. This procedure enables a breeder to obtain large numbers of offspring from a blue-ribbon female, by using hormonally prepared, but genetically less desirable animals to support the pregnancy.

Other reproductive technologies that are likely to become important in the future are sex selection, in vitro fertilization, and cloning. In *sex selection,* the sex of the offspring is controlled. This reproductive capability would be particularly beneficial in the dairy industry, since the profit in this industry is derived mainly from milk, which is a product of only the female sex. No reliable method currently exists for controlling the sex at fertilization, although research has been done on selecting sperm cells that carry either the X or the Y chromosome. The methods that researchers have attempted to use in distinguishing between the X- and Y-bearing sperm have met with varying degrees of success; these methods include separating the sperm cells according to their swimming abilities, their sedimentation rates in a centrifuge or under gravity, and their rates of migration in an electric field.

In vitro fertilization is the fusion of egg and sperm outside the reproductive tract. This form of fertilization has been accomplished in different animals and in humans. But no reliable method has been developed for its widespread application to farm animals. This technique would be particularly useful in overcoming the infertility of a highly prized animal that is unable to sustain a normal pregnancy. For instance, eggs that are obtained from a superovulating female could be selectively fertilized in vitro with X-bearing sperm and then transferred to the reproductive tracts of other females for the completion of normal development.

As we learned in an earlier chapter, *cloning* is the production of genetically identical individuals. Clones are produced naturally by vegetative reproduction, which is a common method of reproduction in plants and microorganisms. Members of monozygotic twins in higher animals are also identical in genotype, but they are not clones of their parents, since they are produced by sexual reproduction. Cloning would be particularly advantageous in animal breeding, since most desirable traits that are of economic importance depend on many gene loci (polygenes), which tend to recombine during the formation of the progeny. Through the use of cloning, the breeders would retain the genotype of a highly prized animal *intact,* without the risk of producing an offspring with a less desirable combination of genes. Two procedures have been successful in the experimental production of identical offspring. One involves the mechanical separation of cells in early embryos. In effect, this method serves to produce monozygotic twins, triplets, and so on, by artificial means. For example, researchers have successfully divided two-cell sheep embryos in half to produce identical twins; they have also been able to separate a four-cell sheep embryo into four parts, to produce identical quadruplets.

Another technique that can give rise to individuals with identical nuclear genes is *nuclear transplantation.* This procedure involves the removal or destruction of the nucleus of an egg and the subsequent insertion of a different nucleus by means of a micropipet. The offspring that develop will then have the *nuclear* genotype of the donor individual. This method has been applied successfully using the comparatively large eggs of frogs and toads (Fig. 17.22). It has also recently been accomplished in mice, even with their much smaller eggs (less than one-tenth the size of frog eggs). We should point out, however, that the chance of successful nuclear transfer tends to decline with the developmental age of the cell from which the donor nucleus is derived. For example, Briggs and King discovered in the 1950s that nuclei that are derived from cells at the blastula stage of development in the frog are fully capable of supporting normal development when transferred to enucleated eggs, but they lose this ability by the later gastrula stage. Such results would indicate that many cells may become irreversibly differentiated as development proceeds, so that their nuclei are no longer able to direct the formation of a complete individual from the zygote. Not all nuclei of differentiated cells lose this capability, though. For instance, some nuclei that had been taken from certain specialized cells of a fully developed toad (*Xenopus*) were able to sustain normal development when transplanted into eggs.

Genetic Engineering of Animal Cells. Procedures have been developed for inserting a particular DNA fragment that carries a single gene into cultured animal cells. In one such technique, the gene is first linked to the DNA of an animal virus; the recombinant virus is then used to infect the recipient cells. For example, this method has been employed successfully to insert the β hemoglobin gene of rabbits into monkey cells. To accomplish the transfer, the β hemoglobin gene was

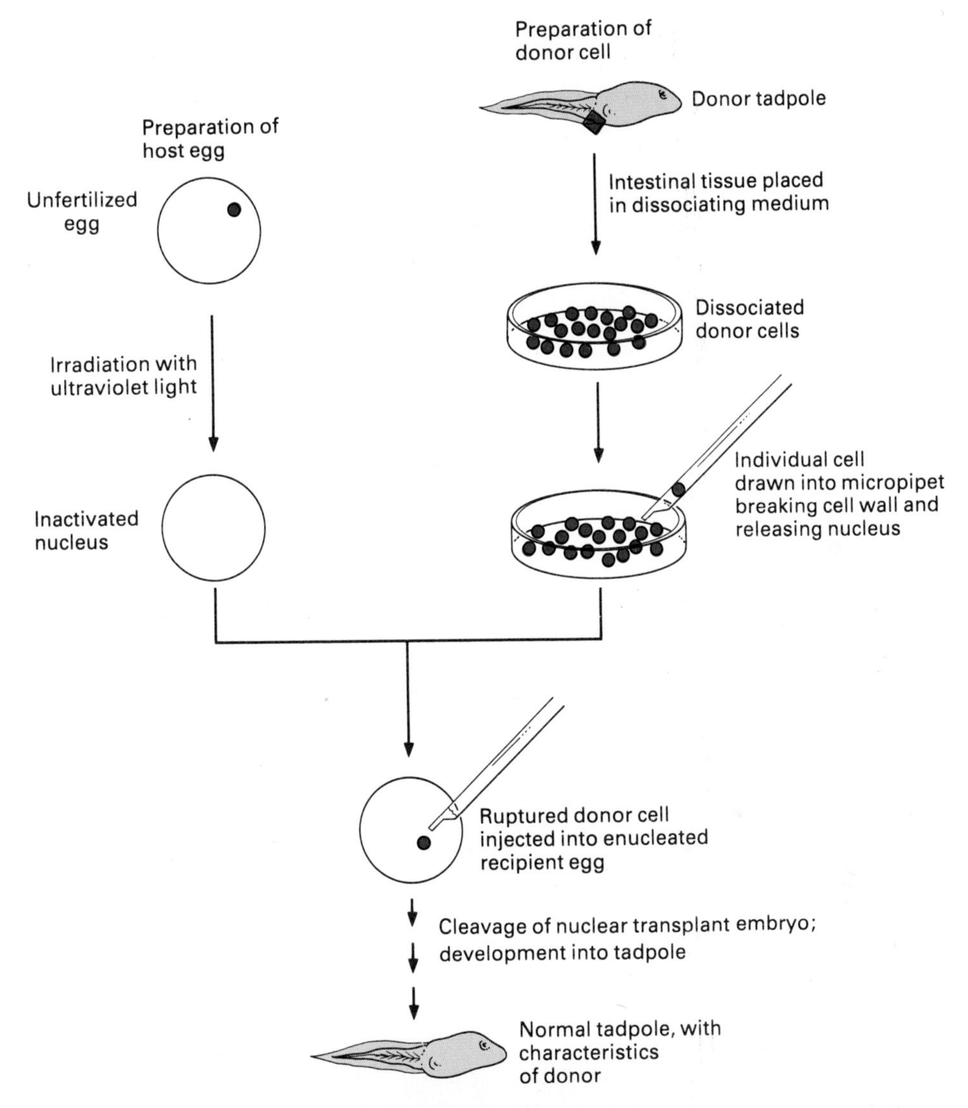

Figure 17.22. Nuclear transplantation in *Xenopus laevis*. A nucleus from an intestinal epithelial cell is sucked into a micropipet and is then injected into a recipient egg, the nucleus of which has been inactivated with ultraviolet light. A normal tadpole, which expresses the phenotypic characteristics of the donor individual, develops from this egg.

first spliced into the DNA of the monkey tumor virus SV40, using recombinant DNA techniques (Fig. 17.23). A few of the cells that had been infected with the recombinant virus expressed the foreign gene by making both the mRNA of the gene and its encoded protein.

A major drawback of this recombinant DNA system is that only a few cells incorporate the desired gene on a stable basis, even though very large populations of target cells are employed in the process. To improve the efficiency of gene transfer, geneticists have been working to develop a procedure that would enable

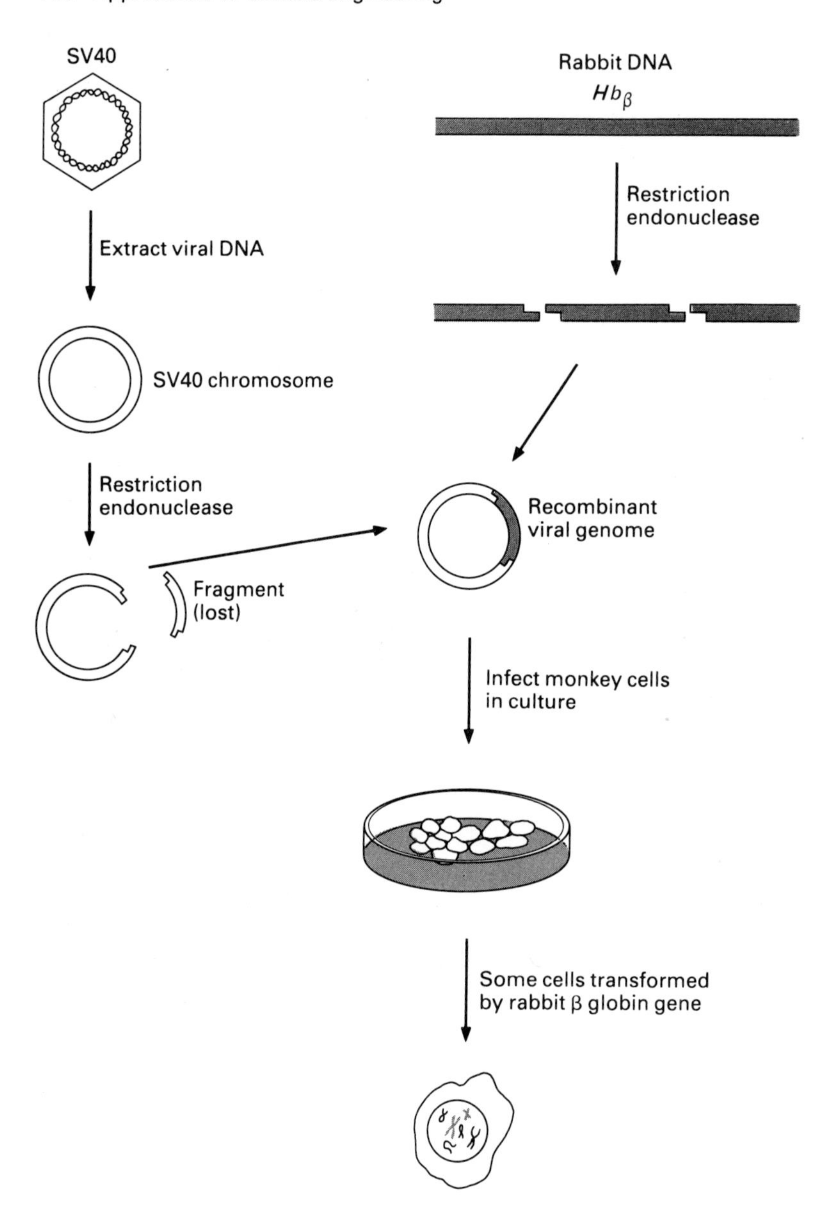

Figure 17.23. Technique used to transfer the rabbit β hemoglobin gene into monkey cells. The β hemoglobin gene is spliced into a vector, which in this case is the DNA of the monkey virus SV40. Some monkey cells that have been infected with the recombinant viral chromosome express the rabbit β globin gene.

the insertion of a specific gene directly into the nucleus of a single recipient cell. The cell could then be propagated into a clone of genetically altered cells. For example, in one reported experiment, two recombinant plasmids—one carrying a thymidine kinase (*TK*) gene from a herpes simplex virus and the other carrying a human β hemoglobin gene—were both inserted into the nuclei of TK^- mouse cells by *microinjection*, as shown in Fig. 17.24. The colonies formed from the microinjected cells expressed the herpes *TK* gene, which indicates that the TK^- defect in the mouse cells had been corrected by the functional viral gene. In this experiment, the genetically engineered mouse cells did not synthesize any β hemoglobin chains, but they did produce a few mRNA molecules that coded for the human β hemoglobin protein.

The transformation of mouse cells by the *TK* gene from the herpes simplex virus is an example of successful genetic engineering at the level of a single mammalian cell. The next step in such procedures is to extend the transformation of cells that are growing in laboratory culture to the genetic alteration of cells in an experimental animal, with the object of correcting a genetic defect that is expressed by a multicellular organism. Gene therapy of this type may be possible in the near future, as shown by the report in late 1981 that fertilized mouse eggs that had been microinjected with a rabbit hemoglobin gene were successfully implanted in female mice. These females subsequently delivered several offspring whose bone marrow cells were found to be synthesizing rabbit hemoglobin. Furthermore, two of these offspring, when later intermated, passed on the rabbit hemoglobin gene to some of their progeny.

Benefits of Recombinant DNA Technology in Animals. There is reason to believe that the first major benefit of recombinant DNA technology to animal husbandry will probably come from its use in the commercial production of animal feed supplements and various pesticides, herbicides, and antibiotics. The amino acids and vitamins that are used to supplement feed are quite expensive and contribute in large part to the total cost of the feed to the farmer. For example, feed costs for vitamins alone ran between $180 and $200 million in the world market in 1981. Some of these vitamins, such as vitamin B_{12}, are extremely costly to produce by the conventional methods. If microorganisms can be engineered to synthesize these compounds more efficiently, the costs for animal feed and feed products could be substantially reduced.

Figure 17.24. Microinjection of a gene from a herpes simplex virus and a human β hemoglobin gene into the nucleus of a mouse cell. A coverslip carrying mouse cells that are mutant for the *TK* gene that is needed for DNA synthesis (genotype TK^-) is inverted over a slide to form a chamber. The chamber is filled with culture medium and is sealed with silicon oil. A micropipet filled with a solution containing copies of the normal (TK^+) gene from herpes virus and the human β hemoglobin gene is inserted into a mouse cell nucleus, and one or more copies of both donor genes are injected.

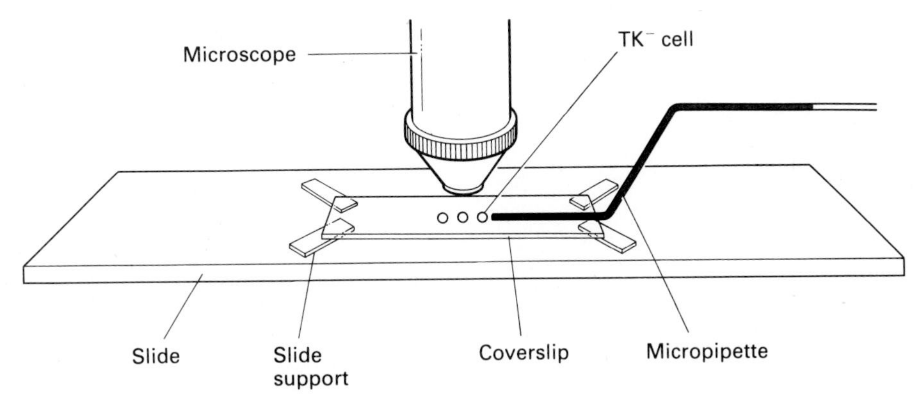

Other, less immediate benefits would include the use of recombinant DNA technology to improve the phenotypes of farm animals. But we must learn more about the genetics of these animals if such techniques are to become a reality. We must be able to identify the major genes that affect economically important traits and must then determine their chromosomal locations. While progress is being made in this area, several problems remain. One major obstacle to applying recombinant DNA technology to animal breeding is that most economically useful characteristics are quantitative traits, which are determined by many genes with comparatively small, cumulative effects. These gene loci are very difficult to identify as separate entities and to work with on an individual basis. In contrast, rare mutant genes that have a major impact on the phenotype of an individual, such as those associated with certain metabolic diseases in humans, lend themselves more readily to the manipulative procedures of recombinant DNA research. For example, the normal allele for the gene that causes Lesch-Nyhan syndrome in humans (the *HPRT* gene) has already been cloned and has been used successfully in the laboratory to correct the defect in human cells with a virus that inserts the normal allele into them. Normal alleles for other genes involved in genetic diseases are also available, thus providing the basic material for gene replacement therapy in the not-too-far-distant future.

Suggested Readings / Part VII

Chapter 17

Brill, W. J. Agricultural microbiology. *Scientific American* 245:198–215, 1981.

Chambon, P. Split genes. *Scientific American* 244:60–71, 1981.

Chilton, M. A vector for introducing new genes into plants. *Scientific American* 248:51-59, 1983.

Cohen, S. The manipulation of genes. *Scientific American* 233:24–33, 1975.

Diacumakos, E. G. Genetic engineering of mammalian cells. *Scientific American* 245:106–121, 1981.

Feldman, M. and E. R. Sears. The wild gene resources of wheat. *Scientific American* 244:102–113, 1981.

Gilbert, W. and L. Villa-Komaroff. Useful proteins from recombinant bacteria. *Scientific American* 242:74–97, 1980.

Impacts of Applied Genetics, Office of Technology Assessment, Congress of the United States, U.S. Government Printing Office, 1980.

Messel, H. (ed.). *The Biological Manipulation of Life,* Pergamon Press, Sydney, Australia, 1981.

Motulsky, A. G. Impact of genetic manipulation on society and medicine. *Science* 219:135–140, 1983.

Novick, R. P. Plasmids. *Scientific American* 243:103–127, 1980.

Pestka, F. The purification and manufacture of human interferon. *Scientific American* 249:36–43, 1983.

Shepard, J. F. Regeneration of potato plants from leaf cell protoplasts. *Scientific American* 246:154–166, 1982.

Watson, J. D. and J. Tooze. *The DNA Story, A Documentary History of Gene Cloning,* Freeman, San Francisco, 1981.

Watson, J. D., J. Tooze, and D. T. Kurtz. *Recombinant DNA: A Short Course,* Scientific American Books, Freeman, New York, 1983.

Glossary

Acentric chromosome A chromosome without a centromere.

Acrocentric chromosome A chromosome that has its centromere very near one end.

Active site The region of a protein that has a specific shape that is responsible for the function of the protein; the region of an enzyme to which the substrate binds.

Allele frequency The proportion of all alleles of a particular gene in a population that are of a specific type.

Alleles (allelic genes) Two or more alternative forms of a gene, any one of which can exist at a particular chromosomal locus.

Allopolyploidy Euploidy that involves multiple sets of nonhomologous chromosomes.

Allosteric site A region of a protein (other than the active site) that binds a specific effector molecule and thereby influences the functioning of the active site.

Allozygote (allozygous genotype) A homozygous or heterozygous genotype in which the alleles are not identical by descent.

Allozymes Alternative forms of an enzyme that are coded by the different alleles at a particular gene locus.

Ames test A test to detect mutagenicity of chemical substances, based on the ability of the chemicals to induce mutation in the bacterium *Salmonella*.

Amino acid A monomer unit of a polypeptide chain.

Aminoacyl-tRNA synthetase One of 20 or more enzymes that can catalyze the covalent attachment of a particular amino acid to its corresponding tRNA.

Amniocentesis Technique of prenatal diagnosis of fetal abnormalities, based on karyotype and biochemical analyses of fetal cells obtained from the amniotic fluid.

Amphidiploid An allotetraploid having a diploid chromosome complement from each of two separate ancestral species.

Anaphase Stage of mitosis and meiosis in which sister chromatids (mitosis, meiosis II) or homologs (meiosis I) are pulled toward the opposite poles of the spindle.

Aneuploidy Chromosome mutation involving the gain or loss of less than an entire set of chromosomes.

Annealing A process whereby two single strands of DNA come together to form a double helix.

Anticodon A triplet nucleotide sequence on a tRNA molecule that recognizes a particular mRNA codon at the ribosome; it functions to ensure that the amino acid carried by the aminoacyl-tRNA is inserted at its proper position in a growing polypeptide chain.

Antiparallel The opposite orientations of the two strands of a DNA double helix, so that the 5′ end of one strand aligns with the 3′ end of the other strand.

Aporepressor An inactive repressor molecule that must be activated by combining with the corepressor before it can bind to the operator and turn off the transcription of an operon.

Ascospore A haploid product of a cross between fungi of opposite mating type, found in certain fungal species in which spores are produced in a sac called an ascus.

Ascus A sac that encloses the products (ascospores) of a cross between fungi of opposite mating types.

Assortative mating Mating in which mates tend

to be more similar in phenotype than randomly chosen individuals.

Asymmetric replication Duplication of DNA via the unwinding of the parental helix and the synthesis of daughter strands, proceeding in one direction from a fixed starting point.

Autopolyploidy Euploidy that involves multiple sets of homologous chromosomes.

Autoradiography A process whereby radioactive materials are incorporated into a cellular structure, which is then placed next to a photographic film; the decay of the radioisotope forms a pattern on the film that corresponds to the distribution of the radioactive material over the structure.

Autosome Any chromosome that is not a sex chromosome.

Autozygote (autozygous genotype) A homozygous genotype in which the alleles are identical by descent.

Auxotroph A mutant microorganism that will grow only if the medium is supplemented with a specific organic substance that is not required by the wild-type (prototrophic) organism.

Backcross A mating of an F_1 individual to an individual having one of the parental genotypes.

Back mutation The change of a mutant allele back to wild-type through reversal of the original mutational change.

Bacteriophage (phage) A virus that infects bacteria.

Balanced polymorphism A stable genetic polymorphism that is maintained by natural selection.

Barr body A condensed, genetically inactive mammalian X chromosome that stains darkly in interphase.

Base analog A compound that is similar in molecular structure to one of the bases in DNA; because of this similarity, the analog may be mistakenly incorporated into DNA and therein acts as a mutagen.

Binomial event An event consisting of a total number of objects, or series of trials, partitioned into just two different classes (successes vs. failures).

Bivalent A structure formed from the synapsis of homologous duplicated chromosomes in prophase I of meiosis.

Bottleneck effect Random genetic drift as a consequence of periodic reductions in population size.

Callus An undifferentiated clump of plant cells derived through cell culture of a protoplast.

Carcinogen Any substance that increases the likelihood of cancer.

Catabolite repression Inhibition of transcription of an inducible operon by glucose or a breakdown product of glucose.

cDNA *See* Complementary DNA.

Cell hybridization *See* Somatic cell hybridization.

Centric fusion Fusion of the long arms of two acrocentric chromosomes to make a single chromosome.

Centromere The constricted region of a chromosome to which spindle fibers attach during cell division.

Chiasma (plural, **chiasmata**) A microscopically visible point of crossing over observed in late meiotic prophase I.

Chromatid One of two identical longitudinal halves of a duplicated chromosome; the two sister chromatids are held together at the region of the centromere.

Chromatin Fibrous material containing DNA (and RNA and protein) that is found in interphase nuclei, when discrete chromosomes are not visible with the light microscope; the structural material of eukaryotic chromosomes.

Chromosomal mosaic An individual composed of two or more chromosomally different types of cells.

Chromosome A structure that carries genes and other DNA, usually associated with RNA and specific proteins; the eukaryotic chromosome is visible as an entity only under the light microscope during mitosis or meiosis, when the chromatin exists in a highly condensed form.

Chromosome arm The segment between the centromere and one of the ends of a chromosome; most chromosomes have two arms, whose lengths are determined by the position of the centromere.

Chromosome bands Regions of chromosomes that retain a stain upon certain chemical treatments.

Chromosome puff A swelled region originating from a site of active transcription on a polytene chromosome.

Chromosome set The group of chromosomes that carries the basic complement of genetic information for a particular species.

Cis arrangement Arrangement of linked nonallelic genes in a heterozygote where both mutant genes are located on one chromosome and their wild-type alleles are located on the homologous chromosome.

Cis dominant The dominance of a gene over genes adjacent to it on the same chromosome, but not to genes located on the homologous chromosome.

Cistron Functional unit of a gene defined on the basis of a complementation test as that region of a gene within which two mutations cannot complement; that functional unit that codes for one polypeptide chain.

Clone A group of genetically identical cells or individuals derived by way of asexual reproduction from a single ancestor.

Cloning *See* DNA cloning.

Codominance Situation where contrasting alleles of a heterozygote are both expressed in full.

Codon Sequence of three bases in DNA or mRNA that serves as a code word for a particular amino acid.

Coefficient of coincidence A measure of the degree of interference in multiple crossing over, expressed as the ratio of the observed frequency of double recombinants to the frequency of double recombinants expected in the absence of interference.

Coevolution The process of evolutionary change that occurs jointly in two or more interacting populations, such as predator and prey populations or host and parasite populations.

Cohesive ends Complementary single-stranded regions at the ends of a DNA molecule or DNA fragment.

Colchicine A drug that interferes with the formation of the spindle.

Colinearity Linear correspondence between the amino acid sequence of a polypeptide and the coding sequence of its gene.

Colony A visible clone of bacterial cells.

Competence A physiological property of transformable cells that allows them to take up naked DNA.

Complementary DNA (cDNA) Synthetic DNA synthesized from a specific RNA template through the action of the enzyme reverse transcriptase.

Complementation The production of a wild-type phenotype when two different mutations are present together with their wild-type alleles in the same cell.

Complementation test (1) A test to determine whether two different mutations affecting the same trait are in the same or in different genes—a test for allelism; (2) a test to determine whether two mutations within a gene locus are in the same cistron.

Complete dominance The expression of one allele or characteristic in a heterozygote to the exclusion of a contrasting (recessive) allele or characteristic.

Concatamer A long DNA molecule that is an intermediate in the replication of some viral genomes; contains repeated copies of the basic viral genome.

Concordant twins Both members either have or do not have a particular trait.

Condensation A term used to describe the coiling and supercoiling of interphase chromatin into the highly coiled fibers of chromosomes.

Conditional mutant A mutant gene that has the wild-type phenotype under certain (permissive) environmental conditions and a mutant phenotype under other (restrictive) environmental conditions.

Conditional probability The situation where the probability with which an event occurs is dependent on whether another particular event has occurred previously.

Conjugation A mode of transfer of genetic material from a donor bacterial cell to a recipient cell in which there is direct physical contact between the two cells.

Consanguineous mating A mating between relatives.

Contact inhibition The density dependent inhibition of the growth of normal (non-cancerous) cells, mediated by cells touching one another; loss of contact inhibition is an obvious distinguishing feature of cancerous cells.

Corepressor A molecule (often the end product of a metabolic pathway) that combines allosterically with an aporepressor to activate it.

Cot curve A graph describing the renaturation of single-stranded DNA fragments with time; used to determine the degree of repetitiveness in the DNA base sequence.

Coupling gene arrangement *See* Cis arrangement.

Cross-fertilization Union of gametes that are derived from different individuals.

Crossbreeding A mating between individuals of different races or species.

Crossing over The exchange of parts between homologous segments of nonsister chromatids by breakage and reunion.

Cytokinesis Division of the parental cytoplasm into daughter cells, through the formation of a cleavage furrow (in animals) or cell plate (in plants) to separate the daughter cells.

Cytoplasmic inheritance Inheritance via genes found in cytoplasmic factors or cytoplasmic organelles, such as mitochondria or chloroplasts.

Cytoplasmic sterility Male or female sterility in plants caused by a factor transmitted through the cytoplasm.

Degeneracy A property of the genetic code whereby two or more codons specify the same amino acid.

Deletion (1) Chromosome structural mutation involving the loss of a chromosome segment from a chromosome set; (2) a gene (point) mutation involving the loss of a single base pair.

Deletion mapping The localization of the points of mutation along a chromosome by a series of matings of point mutants with known deletion mutants.

Denaturation Loss of the secondary structure of a DNA or protein molecule without breakage of the major bonds of its constituent chains; denaturation of DNA results in separation of the two strands.

Deoxyribose The five-carbon sugar found in each nucleotide in DNA.

Dicentric chromosome A chromosome having two centromeres.

Dihybrid cross A mating that considers differences at two gene loci; in a classical dihybrid cross, the two individuals mated are heterozygous for each of the gene pairs.

Dioecious plant A plant species in which the male and female organs are found on separate individuals.

Diploid Possessing two complete sets of chromosomes.

Disassortative mating A mating in which mates tend to be less similar in phenotype than randomly chosen individuals.

Discordant twins One member has, and the other lacks, a particular trait.

Disulfide bond S–S linkage formed between SH (sulfhydryl) side groups of two cysteine residues located in the same or in different polypeptide chains.

DNA cloning The production of many copies of a pure DNA fragment through its incorporation into a plasmid or viral vector, and replication of the recombinant vector molecule in the cells of recipient bacteria.

DNA hybridization Annealing of complementary DNA strands derived from different sources; used to determine the degree of similarity of the base sequences of the two DNAs.

DNA ligase An enzyme whose sole activity is to catalyze the formation of a broken 5′–3′ phosphodiester bond in DNA.

DNA polymerase An enzyme that can catalyze the synthesis of a DNA strand using the complementary strand as a template.

Dominance *See* Complete dominance.

Dominance relationship The relationship between alleles in a heterozygous genotype that determines the phenotypic expression of the heterozygote. *See* Complete dominance, Incomplete dominance, Codominance.

Dosage compensation The mechanism by which males and females produce equal quantities of a product coded for by an X-linked gene; in mammals, this is accomplished by inactivation of one of the X chromosomes in the cells of the female.

Drift *See* Genetic drift.

Duplication A chromosome structural mutation involving an extra copy of a particular chromosome segment in a chromosome set.

Eclipse period Early phase of intracellular lytic growth, prior to progeny phage assembly; defined by the absence of infective phage particles.

Electrophoresis A technique for separating the components of a mixture of DNA or protein molecules in an electric field.

Embryoid A multicellular plant structure derived from hormonally stimulated growth of a haploid product of meiosis.

Endogenote That section of DNA of a recipient cell that is homologous to a donor DNA segment (exogenote). *See also* Merozygote.

Endonuclease A type of enzyme that catalyzes the breakage of a 5′–3′ phosphodiester bond in the interior of a polynucleotide strand, producing a nick.

Episome A genetic element that can replicate autonomously in the cytoplasm of the cell that harbors the element or can be inserted into the chromosome of the host cell.

Epistasis Interaction between nonallelic genes in which an allele of one gene masks the expression of the alleles of the other gene.

Euchromatin Chromatin that stains lightly and is not as highly condensed as heterochromatin; thought to contain transcriptionally active genes, as opposed to heterochromatin, in which the genes are inactive.

Eugenics Selective human breeding based on ideas of desirable and undesirable genotypes; conscious attempt to alter the human gene pool by selective breeding.

Eukaryotic cell Any cell other than a bacterium; a cell having a distinct nucleus separated from the cytoplasm by a nuclear envelope and having various cellular organelles.

Euploidy A chromosome mutation that involves the gain or loss of one or more entire chromosome sets.

Exogenote A DNA segment from a donor cell that has been incorporated into a recipient cell having a DNA segment (endogenote) homologous to the donor fragment. *See also* Merozygote.

Exon A coding segment within a split gene locus; a portion of a split gene that is transcribed into mRNA.

Exonuclease A type of enzyme that catalyzes the sequential removal of nucleotides from an exposed end of a polynucleotide chain through the hydrolysis of phosphodiester bonds.

Expressivity The degree of expression of a completely penetrant genotype as a particular phenotype.

Feedback inhibition The allosteric inactivation of an enzyme in a metabolic pathway by the end product of the pathway.

Fine-structure mapping The mapping of different mutation sites within a single gene.

First-division segregation pattern A linear arrangement of ascospores in an ordered tetrad that reflects the segregation of alleles at the first meiotic division, showing that no recombination has occurred between that allele pair and the centromere.

Fitness The reproductive success of a genotype, measured as some form of a reproductive rate.

Fixation Attainment of an allele frequency value of one, resulting in a population that consists solely of one homozygous genotype.

Forward mutation A mutation that converts a wild-type gene into a mutant allele.

Founder effect A form of genetic drift occurring when a new population is established by a small group of colonizers whose alleles may differ substantially in frequency from the parent population from which the colonizers were derived.

Frameshift mutation The addition or deletion of a nucleotide pair causing a shift in the frame of reading of subsequent DNA code words.

Gamete A mature reproductive cell produced by meiosis, with only half the number of chromosomes present in the body cells of the same organism; it is an egg in females and a sperm (or pollen) in males.

Gametophyte The haploid gamete-producing stage in the life cycle of plants.

Gene The fundamental unit of inheritance that transmits a specification from one generation to the next and is present in the cell as a segment of DNA that codes for one particular product.

Gene cluster A closely linked group of functionally related genes.

Gene conversion A meiotic process by which one allele directs the change of another allele to its own form.

Gene family A group of genes that have similar nucleotide sequences and are thought to have evolved from a common ancestral gene.

Gene frequency *See* Allele frequency.

Gene locus The specific location on a chromosome of a particular gene.

Genetic code The set of correspondences between nucleotide pair triplets in mRNA and amino acids incorporated into a polypeptide chain.

Genetic distance A measure of the distance separating two linked loci based on the percent crossing over that occurs between them.

Genetic drift A change in gene frequency that occurs in small populations as a result of random sampling error during reproduction.

Genetic map A diagram showing the relative positions of genes along a chromosome, and some measure of distance between them.

Genetic transformation The modification of a bacterial or plant or animal cell by the uptake of DNA from a cell of a different genotype.

Genome The entire complement of genes carried by a cell.

Genotype The gene constitution of an individual.

Genotype frequency The proportion of individuals of a particular genotype in a population.

Germinal mutation A mutation that occurs in a cell that is destined to become a gamete.

Gynandromorph A sexual mosaic.

Haploid A cell having one set of chromosomes, or an organism composed of such cells.

Haploidization The formation of the haploid

condition through progressive loss of chromosomes as a result of mitotic nondisjunction.

Hemizygote A diploid genotype or individual that has only a single copy of a particular gene—for example, an X-linked gene in the heterogametic sex.

Heritability in the broad sense The fraction of the total phenotypic variance for a particular trait that is due to genetic differences among individuals.

Heritability in the narrow sense The proportion of the total phenotypic variance for a trait that can be attributed to the additive genetic variance.

Hermaphrodite A single animal having both male and female reproductive parts.

Heterochromatin Chromatin that is very tightly coiled and stains darkly; it is thought to be genetically inactive.

Heteroduplex A DNA duplex molecule that contains one or more incorrectly matched base pairs.

Heterogametic sex The sex that has heteromorphic sex chromosomes and therefore produces two different kinds of gametes with respect to the sex chromosomes.

Heterogeneous nuclear RNA The primary trancription product of eukaryotic split genes; includes mRNA precursors and other types of RNA.

Heterokaryon A single cell containing two genetically different nuclei, formed as a result of cell fusion without accompanying nuclear fusion.

Heterokaryon test A test for cytoplasmically inherited genes, based on new associations of phenotypes in cells derived from specially marked heterokaryons.

Heterokaryosis The process of formation of a heterokaryon.

Heterosis Increased vigor of a hybrid offspring derived from inbred (homozygous) parental lines.

Heterozygote An individual or genotype that carries a pair of contrasting allelic genes.

Highly repetitive DNA Eukaryotic DNA that consists of short base sequences that repeat themselves more than 10^6 times in the genome.

Histones Positively charged (basic) proteins that complex with DNA in eukaryotic cells to form chromatin fibers.

Holandric traits Traits determined by Y-linked genes.

Homeologous chromosomes Chromosomes having partial homology, indicative of some original ancestral homology.

Homogametic sex The sex that has homologous sex chromosomes and therefore produces only one type of gamete with respect to the sex chromosomes.

Homologous chromosomes Members of a chromosome pair (one maternally derived and the other paternally derived) that synapse during meiosis, match in size and structure, and carry alleles of the same genes.

Homozygote An individual or genotype that carries a pair of like alleles.

hnRNA *See* Heterogeneous nuclear RNA.

Hybrid (1) A heterozygote, (2) a progeny individual from a cross between two species or varieties, (3) a duplex nucleic acid molecule made up of strands derived from different sources.

Hybrid cell The result of somatic cell fusion followed by nuclear fusion.

Hybridization (1) Formation of a hybrid by performing a cross between two species or varieties, (2) formation of a duplex nucleic acid molecule from strands derived from different sources (for example, *see* DNA hybridization and Molecular hybridization).

Hybrid vigor *See* Heterosis.

Hydrogen bond A weak chemical bond formed between a hydrogen atom and an electronegative atom such as oxygen or nitrogen.

Hydrophobic bond Attraction between water-insoluble regions of two molecules or of two parts of a single molecule.

Inbreeding Mating in which mates tend to be more closely related than randomly chosen individuals.

Inbreeding coefficient The probability that two alleles at a given locus are identical by descent, being replicates of the same allele in a common ancestor.

Inbreeding depression The decline in vigor that accompanies inbreeding if the heterozygote shows superior performance to at least one of the homozygous genotypes.

Incomplete dominance The situation where contrasting alleles in a heterozygote show a phenotype that is intermediate between the two corresponding homozygous genotypes.

Independent assortment The segregation of nonallelic genes during meiosis so as to give all possible gametic gene combinations in equal frequency; typical of genes located on different chromosome pairs, in which case the segregation of one pair of chromosomes has no influence on the segregation of any other chromosome pair.

Independent events In a group of events, the

occurrence of one event does not influence the chance of occurrence of any other event.

Inducer An environmental substance (effector molecule) that indirectly triggers the transcription of an operon by allosteric inactivation of the repressor molecule.

Induction of an operon The process by which the transcription of an operon is triggered by an inducer substance.

Induction of prophage The process by which a prophage leaves the bacterial chromosome, thereby initiating the lytic cycle of phage growth.

Insertion sequence (IS element) A transposable genetic element found in bacteria.

In situ "In place"; in the natural or original position.

Interference The lack of independence of multiple crossover events from one another. *See* Negative interference and Positive interference.

Interphase All periods of the cell life cycle other than nuclear division; a "resting" or nondividing stage where chromosome duplication occurs.

Intron A noncoding region within a split gene locus; it is transcribed into hnRNA but is not found in mature mRNA.

Inversion A chromosome structural mutation involving the reversal of the gene order within a segment of a chromosome.

Inverted repeat (IR) sequences Base pair sequences that read the same if one is read in the opposite order from the other.

In vitro "In glass"; an experimental situation outside the organism.

In vivo "In life"; within a living organism or a cell.

Karyotype The chromosomal complement of a eukaryotic organism or cell, usually viewed during mitotic metaphase.

Latent period The period of intracellular lytic growth of viruses.

Lethal gene A gene whose expression results in the death of the individual.

Ligase *See* DNA ligase.

Linkage The tendency of nonallelic genes located on the same chromosome to assort into gametes in the same combinations as they were inherited, rather than independently.

Linkage equilibrium Condition in which the frequency of a gamete carrying a particular combination of nonallelic genes is equal to the product of the corresponding gene frequencies.

Linkage group A group of genes that show linkage with one another and correspond to a particular chromosome.

Linkage map *See* Genetic map.

Linked genes Nonallelic genes that show linkage, measured as a departure from independent assortment.

Locus *See* Gene locus.

Lyonization The process in mammals by which an X chromosome undergoes genetic inactivation through condensation into a Barr (sex chromatin) body.

Lysogenic bacterium A bacterial cell that contains a prophage, so that the cell is capable of spontaneous lysis due to the uncoupling of the prophage from the bacterial chromosome.

Lysogenic cycle The method of phage reproduction through existence as a prophage, with replication of the prophage occurring in synchrony with replication of the host chromosome.

Lytic cycle The method of viral reproduction consisting of the production of progeny viruses and lysis of the infected cell.

Mainband DNA The largest band of DNA obtained by density gradient equilibrium centrifugation of the DNA complement of an organism.

Map unit A unit of distance in a genetic map; the classical map unit is equivalent to one percent recombination; in bacteria, one map unit is equivalent to one minute of conjugation time.

Maternal effect The determination of the phenotype of the offspring by cytoplasmic elements coded by the nuclear genotype of the mother.

Maternal inheritance The extranuclear inheritance of a trait exclusively through the female parent. *See* Cytoplasmic inheritance.

Meiocyte A primordial germ cell that undergoes meiosis to yield gametes.

Meiosis Period of the life cycle of a meiocyte where two successive nuclear divisions occur resulting in the formation of gametes (or spores), each containing one-half of the number of chromosome sets of the original cell.

Meiotic nondisjunction *See* Nondisjunction.

Melting temperature The temperature at which a DNA molecule denatures and becomes single stranded; it is measured at the midpoint of the transition between native double-stranded DNA and denatured single-stranded DNA.

Merozygote A partially diploid bacterial cell, containing a segment of donor chromosome (exogenote) in addition to the homologous segment of recipient chromosome (endogenote).

Messenger RNA (mRNA) The end product of the transcription of a structural gene; contains the genetic information that specifies cellular synthesis of a particular polypeptide chain.

Metabolic block A nonfunctioning reaction in a metabolic pathway, which is the result of a defect in the enzyme that normally catalyzes the reaction.

Metacentric chromosome A chromosome that has its centromere located in a median position.

Metaphase Stage of mitosis and meiosis when chromosomes (mitosis and meiosis II) or bivalents (meiosis I) align on the equatorial plane of the spindle.

Missense mutation A change in DNA that results in an amino acid substitution in the encoded polypeptide.

Mitosis The period of the cell life cycle of a eukaryotic cell when nuclear division occurs resulting in daughter cells that each contain the same complement of chromosomes as the original cell.

Mitotic nondisjunction *See* Nondisjunction.

Moderately repetitive DNA Eukaryotic DNA that consists of base sequences that repeat themselves from 10 to 10^5 times in the genome.

Molecular hybridization Hybridization of DNA strands derived from different sources (see DNA hybridization), or of DNA with RNA.

Monoecious plant A single plant that contains both male and female reproductive parts.

Monohybrid cross A mating that involves differences at a single gene locus; in a classical monohybrid cross, both of the individuals mated are heterozygous for the allele pair.

Monomorphic trait A trait for which all individuals in a population are of the same phenotype.

Monoploid A cell having a single set of chromosomes, or an organism composed of such cells.

Monosomy An aneuploid condition involving the loss of a single chromosome; in a diploid organism, this gives a chromosome number of $2n - 1$.

Mosaic An individual composed of two or more genetically different types of cells.

mRNA *See* Messenger RNA.

Multiple allelism The existence of three or more allelic states of a gene.

Mutable gene A gene having a relatively high rate of mutation.

Mutagen Any agent that is capable of increasing the rate of mutation.

Mutant A cell or any organism that carries a mutant gene.

Mutation (1) A process that produces a gene or a chromosome complement that is different from wild-type, (2) the gene or chromosome complement that results from such a process.

Mutation rate The probability of mutation at a given gene locus per gamete or cell per unit of time (e.g., per generation).

Mutator gene A gene that increases the rate of mutation of one or more other genes in the genetic complement.

Mutually exclusive events A series of alternative events only one of which can occur at any one time.

Negative interference A phenomenon by which the occurrence of a crossover event enhances the chance that additional crossing over occurs nearby, making double crossovers more common than expected without interference.

Neutral mutation A missense mutation that does not affect protein function.

Nick A broken 5′–3′ phosphodiester bond within a strand of DNA.

Nicking enzyme *See* Endonuclease.

Nonalleles (nonallelic genes) Genes that occupy different chromosomal positions.

Nondisjunction The failure of homologous chromosomes (meiosis I) or sister chromatids (meiosis II, mitosis) to separate at anaphase.

Nonhistone proteins All proteins other than histones that are found associated with chromatin.

Nonparental ditype tetrad An ascus that contains only ascospores having recombinant gene arrangements.

Nonrandom mating *See* Assortative mating, Inbreeding, Outbreeding, and Disassortative mating.

Nonsense mutation A change in DNA that results in a codon that does not specify an amino acid, leading to premature termination of polypeptide synthesis.

Nuclear transplantation Replacement of the nucleus of an egg with the nucleus of another cell, usually a partially differentiated somatic cell.

Nucleic acid probe Purified radioactive complementary DNA (cDNA) carrying a known gene; used to identify the particular restriction fragment that carries the same gene.

Nucleoid A DNA mass within a prokaryotic cell, chloroplast, or mitochondrion.

Nucleolar organizer region Chromosomal region or regions in a eukaryotic cell that contain the genes that code for ribosomal RNA.

Nucleolus An organelle in a eukaryotic nucleus that contains ribosomal RNA and multiple copies of the genes coding for rRNA.

Nucleoside A nitrogenous base bound to a sugar molecule, but lacking a phosphate group.

Nucleosome The structural subunit of eukaryotic chromatin, composed of a core of eight histone molecules about which is wrapped two full turns (166 base pairs) of double helical DNA.

Nucleotide The monomer unit of nucleic acid, consisting of a nitrogenous base, a sugar, and a phosphate group.

Null allele A gene whose effect is either an absence of cellular gene product, or an absence of normal function at the phenotypic level.

Nullisomy An aneuploid condition involving the complete loss of one type of chromosome, thus producing a chromosome number such as $n - 1$ or $2n - 2$.

Okazaki fragment Short segment of DNA formed by the discontinuous nature of replication of one of the parental DNA strands.

Oncogene A potential cancer producing gene, of host or viral origin.

Oncogenic virus A cancer-causing virus.

Oogenesis The process of development and maturation of the female gamete or egg in animals.

Operator site A DNA region at one end of an operon that serves as the binding site for active repressor protein.

Operon Two or more adjacent structural genes that are transcribed into a single molecule of mRNA, plus the adjacent transcriptional control sites (operator and promoter).

Ordered tetrad An ascus in which the meiotic products are arranged in a linear order reflecting the linear order of meiotic segregation events.

Outbreeding Mating in which mates tend to be less closely related genetically than randomly chosen individuals.

Overlapping genes Genes whose positions along the DNA overlap.

Paracentric inversion An inversion that does not involve the centromere.

Parental type gamete A gamete that contains the parental arrangement of nonallelic genes.

Parental ditype tetrad An ascus that contains only ascospores with parental gene arrangements.

Particulate theory A theory that proposes that discrete hereditary units ("particles") are transmitted unchanged from one generation to the next, not being altered by mixing with units from another individual in forming offspring.

Pedigree A diagrammatic representation of two or more generations of related individuals, showing the pattern of transmission of particular genetic traits.

Penetrance The percentage of individuals of a certain genotype that show the expected phenotype.

Peptide bond The chemical linkage formed between the carboxyl group (—COOH) of one amino acid and the amino group (—NH_2) of the adjacent amino acid in a polypeptide chain.

Pericentric inversion An inversion that includes the centromere within the inverted segment.

Phenocopy An environmentally induced phenotype that resembles one known to be genetically determined.

Phenotype The characteristic expressed by a genotype; the appearance of an individual.

Phosphodiester bond The chemical linkage formed between a phosphate group and the sugars of two adjacent nucleosides in a strand of DNA or RNA.

Phyletic evolution The gradual evolutionary change within a single line of descent, or lineage.

Plaque A clear area on a bacterial lawn, left by lysis of the bacteria through successive infections by a single original phage particle and its descendants.

Plasmid A small genetic element that replicates autonomously in the cytoplasm of a prokaryotic or eukaryotic cell.

Pleiotropy Phenomenon whereby a single mutant gene is observed to influence several different traits.

Point mutation A gene mutation that is caused by a change (substitution or addition or deletion) of a single base pair.

Polar body A tiny product of meiosis I or meiosis II in females that will not become a functional egg.

Polygenic trait *See* Quantitative trait.

Polymorphic trait Trait for which at least two phenotypic forms exist in a population.

Polypeptide Polymer composed of a linear sequence of amino acids linked to one another through peptide bonds.

Polyploidy A euploid condition produced by the gain of one or more entire sets of chromosomes.

Polysome Complex of mRNA and ribosomes involved in the synthesis of multiple copies of a polypeptide chain.

Polytene chromosomes Large (multistranded) chromosomes found in the secretory tissues of diptera; formed by successive chromosome replication without accompanying chromatid separation or cell division.

Position effect Variation in the expression of a gene that depends on its location in the genome.

Positive interference The phenomenon where the occurrence of a crossover event decreases the chance that additional crossing over occurs nearby, giving fewer double crossovers than expected on the basis of independence of crossover events.

Primary constriction The centromere region of a chromosome.

Primary structure The particular sequence of monomer units (amino acids or nucleotides) making up a macromolecule (protein or nucleic acid).

Primase An enzyme that catalyzes the DNA-dependent synthesis of a short RNA primer. *See also* Primer.

Primer A segment of a DNA or RNA chain that is elongated with deoxyribonucleotides by DNA polymerase during DNA replication.

Probability The theoretical value of the relative frequency of an event.

Prokaryotic cell A bacterial cell; a cell having no nuclear envelope and hence no separate nucleus, and lacking various cellular organelles.

Promoter A regulatory site at one end of a gene or an operon that serves as the binding site for RNA polymerase in the initiation of transcription.

Prophage An intracellular phage chromosome that is integrated into the chromosome of the host bacterial cell.

Prophase The initial stage of mitosis and meiosis in which the duplicated chromosomes condense, the nuclear envelope and nucleolus disappear, and the spindle apparatus forms.

Protein A macromolecule composed of one or more polypeptide chains folded together in a specific fashion.

Protoplast A cell whose wall has been removed.

Prototroph A wild-type microorganism that will grow in minimal medium, being able to synthesize for itself all organic compounds required for growth and thus requiring no growth supplements in the medium.

Provirus A virus chromosome that is integrated into a chromosome of its host cell.

Pseudodominance The phenomenon whereby a single copy of a recessive gene is expressed due to the deletion of its dominant allele from a heterozygote.

Pseudogene A DNA sequence that is homologous with a known gene, but is never transcribed.

Pure-breeding *See* True-breeding.

Purine A nitrogenous base that consists of a double heterocyclic ring structure; in DNA and RNA, adenine and guanine are purine bases.

Pyrimidine A nitrogenous base that consists of a single heterocyclic ring structure; in DNA and RNA, cytosine, thymine, and uracil are pyrimidine bases.

Pyrimidine dimer Two adjacent pyrimidine bases on the same DNA strand that are covalently bonded together.

Quantitative trait A trait that is determined by the cumulative effects of many genes (called polygenes), whose action typically is very sensitive to environmental factors; usually characterized by a continuous distribution of phenotypes in a population, rather than discrete phenotypic classes.

Quaternary structure The multimeric constitution of a protein.

Race A genetically distinct population characterized by gene and genotype frequencies that differ from those of other races of the same species.

Random genetic drift *See* Genetic drift.

Random mating Selection of mates without regard to genotype or phenotype.

Random sampling error The chance deviation of observed results from those expected.

Recessive The allele whose expression is masked by a contrasting dominant allele in a heterozygote; the phenotype that is not expressed by the heterozygote.

Reciprocal crosses A pair of crosses of the type female genotype A × male genotype B and female genotype B × male genotype A; for example, *AA* male × *aa* female and *aa* male × *AA* female are reciprocal crosses.

Recombinant DNA A novel DNA molecule constructed from segments of DNA derived from two or more different sources.

Recombinant-type gamete A gamete with a gene combination that does not reflect the parental arrangement of nonallelic genes.

Recombination Any process that generates new nonallelic gene combinations.

Relative frequency The number of ways that a

particular event can occur divided by the total number of possible outcomes.

Renaturation *See* Annealing.

Repetitive DNA *See* Highly repetitive DNA, and Moderately repetitive DNA.

Replicon Unit of replication of a eukaryotic chromosome, consisting of a length of DNA under the influence of a single replication-initiation site.

Repressor protein A regulatory molecule that binds to an operator gene and prevents transcription of an operon.

Reproductive isolation The genetic divergence of two populations to the extent that their members are unable to successfully interbreed.

Repulsion gene arrangement *See* Trans arrangement.

Restriction endonuclease One of many bacterial endonucleases that cleave DNA strands within certain recognition sequences.

Restriction fragments The segments of double-stranded DNA produced by the action of a restriction enzyme.

Restriction map The diagrammatic representation of a chromosome showing the locations of the recognition sequences of one or more restriction enzymes.

Retrovirus An RNA tumor virus that employs the enzyme reverse transcriptase to synthesize a DNA copy of the viral genome within the host cell.

Reverse transcriptase An enzyme that is able to catalyze the synthesis of DNA using an RNA strand as the template.

Reversion The production of a wild-type gene from a mutant allele by means of back mutation.

Ribose The five-carbon sugar found in each nucleotide in RNA.

Ribosomal RNA (rRNA) The product of transcription of ribosomal RNA genes; complexes with ribosomal proteins to make ribosomes.

Ribosome Cellular organelle that is the site at which translation of mRNA into protein occurs; composed of rRNA plus proteins.

RNA polymerase An enzyme that catalyzes the synthesis of a strand of RNA using a DNA strand as the template.

Robertsonian translocation *See* Centric fusion.

rRNA *See* Ribosomal RNA.

Satellite Chromosomal material that is attached to the tips of some chromosomes by a narrow constriction.

Satellite DNA DNA that bands at a different position than the bulk of cellular DNA when subjected to equilibrium density gradient centrifugation, based on its different base composition.

Secondary constriction A narrowed region on a chromosome other than the primary constriction (centromere).

Secondary structure The three-dimensional conformation formed by the regular folding of a polynucleotide or polypeptide chain; often in the form of a helix.

Second-division segregation pattern A linear arrangement of ascospores in an ordered tetrad that reflects the segregation of alleles at the second meiotic division, showing that crossing over has occurred between that allele pair and the centromere.

Selection differential The numerical difference between the mean of the parents selected to produce a progeny population, and the mean of the base population from which the parents were selected.

Self-fertilization (selfing) The union of gametes derived from a single individual.

Semisterility A characteristic of individuals that are heterozygous for certain kinds of chromosomal mutations, expressed as a reduced number of viable gametes and hence reduced fertility of such individuals.

Sense strand The DNA strand that serves as the template for transcription of a given gene.

Sex chromatin body *See* Barr body.

Sex chromosome A chromosome whose presence or absence is correlated with the sex of an individual.

Sex-influenced trait Trait determined by alleles (usually autosomal) whose dominance relationship is influenced by the sex hormones.

Sex-limited trait Trait expressed exclusively in one sex, usually determined by an autosomal gene.

Sex-linkage Inheritance pattern caused by genes located exclusively on the X chromosome or exclusively on the Y chromosome; usually taken to be synonymous with X-linkage.

Sex-linked gene Gene located on either the X chromosome or the Y chromosome, but not on both; usually taken to mean an X-linked gene.

Sex pilus An appendage of a donor bacterial cell that becomes modified into a conjugation tube through which donor DNA passes into a recipient cell.

Siblings (sibs) Brothers and/or sisters; offspring of the same parents.

Silent mutation A DNA base pair substitution that does not alter the amino acid assignment of the affected code word.

Sister chromatids The two chromatids that are derived from the duplication of one chromosome during interphase.

Solenoid structure The condensed chromosome fiber formed by the coiling of the nucleosome fiber.

Somatic cell Any cell that is not a constituent of the germinal (reproductive) tissue of an organism.

Somatic mutation A mutation that takes place in a somatic cell.

Somatic cell hybridization The fusion of somatic cells from different species to form a hybrid cell whose nucleus contains the chromosomes of each species.

Somatic doubling The doubling of the chromosome number in a cell as a result of chromosome duplication without the subsequent segregation of sister chromatids into separate daughter cells.

Speciation The splitting of a single lineage into two or more genetically distinct lines of descent as a consequence of the development of reproductive isolation.

Spermatogenesis The process of development and maturation of the male gamete or sperm in animals.

Spindle The organized system of microtubules that attach to the centromere regions of duplicated chromosomes and lead to opposite poles of the cell; responsible for chromosome movement during mitosis and meiosis.

Splicing The excision of introns from the primary RNA transcripts of eukaryotic genes, and the joining together of the exon regions into mature mRNA.

Split gene A gene locus that is subdivided into coding and noncoding nucleotide sequences.

Spores (1) The haploid reproductive cells formed by sexual reproduction in plants and in fungi; in plants, mitotic division of a spore forms the gametophyte; (2) haploid somatic cells that are produced by asexual reproduction in fungi; they can act as gametes or as the initial cells for a new haploid generation.

Sporophyte The diploid sexually reproducing stage in the life cycle of plants that produces the spores by meiosis.

Structural gene A non-regulatory gene that codes for the amino acid sequence of a polypeptide.

Submetacentric chromosome A chromosome that has its centromere located in a submedian position (neither at the center nor near one end).

Supercoiling Higher level coiling of the double helical DNA molecule.

Suppression The phenomenon whereby a second mutation restores the wild-type phenotype that was lost by a previous mutation at a different site; not to be confused with true back mutation.

Symmetrical replication The duplication of DNA via unwinding and synthesis of daughter strands in a bidirectional fashion, proceeding in both directions from a fixed starting point.

Synapsis The pairing of homologous chromosomes during meiotic prophase I.

Synaptinemal complex The nucleoprotein complex that is formed between homologous chromosomes undergoing synapsis.

Syntenic genes Nonallelic genes that are located on the same chromosome pair; syntenic genes may or may not show genetic linkage, depending on the distance separating them.

Tautomerism The spontaneous and reversible change in the distribution of protons in a nitrogenous base, resulting in altered hydrogen-bonding properties of that base.

Telocentric chromosome A chromosome that has its centromere at one end.

Telophase The stage of mitosis and meiosis in which the nuclear envelope reassembles, spindle fibers disappear, and chromosomes uncoil.

Temperate phage A bacterial virus that is capable of forming a prophage state.

Template strand A nucleotide sequence of DNA that acts as a "mold" or "pattern" to determine the nucleotide sequence of a daughter DNA strand or of a strand of RNA.

Teratogen Any environmental agent that interferes with embryonic or fetal development.

Terminalization The shift of chiasmata from their original positions to points closer to the end of the synapsed chromosomes.

Terminal redundancy A description of certain chromosomes that end in the same sequence of genes or nucleotides with which they began.

Tertiary structure The three-dimensional conformation formed by the folding of the secondary structure of a nucleic acid or protein.

Testcross A cross of a dominant individual (usually a heterozygote, when the genotype is known) to a recessive individual; this cross can be used to determine the genotype of the dominant individual.

Tetrad The four haploid spores (or spore pairs)

produced by meiosis in certain fungi and contained within a single ascus.

Tetraploid A cell containing four sets of chromosomes, or an organism composed of such cells.

Tetratype tetrad An ascus containing four types of spores, two of the parental gene arrangements and two of the recombinant gene arrangements.

Threshold trait Situation where a trait is characterized by discrete phenotypic classes, but with an underlying polygenic basis in which phenotypic differences are the result of particular combinations of polygenes that either exceed or fall below certain threshold values.

Thymine dimer A pyrimidine dimer involving two thymine bases.

Trans arrangement Arrangement of linked nonallelic genes in a heterozygote where one wild-type gene and a mutant nonallele are located on each member of the chromosome pair.

Transcription The synthesis of RNA using a DNA template.

Transduction The transfer of DNA from one cell to another by way of a viral or plasmid vector, with no physical contact between donor and recipient cells.

Transfer RNA (tRNA) The product of transcription of transfer RNA genes; each tRNA molecule binds to and carries a specific amino acid to the ribosomes, and its anticodon recognizes the specific mRNA codon for that amino acid.

Transformation (1) Treatment of cells with purified DNA for the purpose of altering their genetic constitution—*See* Genetic transformation; (2) the change of cells into a cancerous (tumerous) state.

Transgressive variation The occurrence of phenotypic classes in the F_2 generation that are more extreme than the phenotypes of the P or F_1 generations.

tRNA *See* Transfer RNA.

Transition A base pair substitution mutation in DNA that involves the replacement of a purine with another purine and of a pyrimidine with another pyrimidine; for example, AT $\rightarrow$ GC.

Translation The production at a ribosome of a polypeptide whose amino acid sequence is determined by the sequence of codons along a mRNA molecule; the transfer of genetic information from mRNA to protein.

Translocation (1) A chromosome structural mutation involving the relocation of a chromosome segment to a nonhomologous chromosome; (2) the movement of peptidyl-tRNA from the A site to the P site of a ribosome during protein synthesis, accompanied by a shift of the mRNA by one codon.

Transposable genetic element (transposon) Any DNA segment that is capable of moving from one chromosomal location to another.

Transversion A base pair substitution mutation in DNA that involves the replacement of a purine with a pyrimidine and of a pyrimidine with a purine; for example, AT $\rightarrow$ CG.

Triploid A cell containing three sets of chromosomes, or an organism composed of such cells.

Trisomy An aneuploid condition that involves the gain of a single chromosome; in a diploid organism, this gives a chromosome number of $2n + 1$.

True-breeding line or strain A group of genetically identical individuals that, when intercrossed, always produce offspring that are identical to their parents.

Unequal crossing over The occurrence of a crossover between homologous chromosomes that are not perfectly aligned.

Unineme theory The theory that states that a single double-stranded molecule of DNA is included within each eukaryotic chromosome.

Unique-sequence DNA DNA that contains a nucleotide sequence that lacks repetition.

Unordered tetrad An ascus in which the meiotic products are randomly placed.

Variable expressivity See Expressivity.

Variegation The occurrence within a tissue of sectors with different phenotypes.

Virion The mature, nonreplicating form of a virus.

Virulent phage A bacterial virus that is limited to the lytic cycle of growth, being incapable of forming a prophage.

Wild-type The allele or genotype or phenotype that is found in nature or in the standard laboratory strain for a given organism.

X-linkage The presence of a gene on the X chromosome, but not on the Y chromosome.

Y-linkage The presence of a gene on the Y chromosome, but not on the X chromosome.

Zygote A fertilized egg.

Zygotic combinations The various gene combinations produced by fertilization; entries within a Punnett square.

Answers to Problems

Chapter 1

2. (a) All *Aa*. (b) 1 *AA* : 1 *Aa*. (c) 1 *AA* : 2 *Aa* : 1 *aa*. (d) All *AaBb*. (e) 1 *AABB* : 2 *AaBB* : 1 *aaBB* : 1 *AABb* : 2 *AaBb* : 1 *aaBb*.
3. White. Two dominant parents, if they are both heterozygous, can produce a recessive offspring, but two recessive parents cannot produce a dominant offspring.
4. (a) Wire-haired texture. (b) 3 wire-haired : 1 smooth-haired.
5. Black coat color is dominant to white. Using gene symbols *B-* = black and *bb* = white, the parental genotypes are, in order, *Bb* × *Bb* → 3 *B-* : 1 *bb*, *BB* × *Bb* (or *BB* × *BB*) → all *B-*, *BB* × *bb* → all *Bb*, *Bb* × *bb* → 1 *Bb* : 1 *bb*, and *bb* × *bb* → all *bb*.
6. Cross 1: *AA* × *aa*; cross 2: *Aa* × *Aa*; cross 3: *aa* × *aa*; cross 4: *Aa* × *aa*; cross 5: *AA* × *AA* (or *AA* × *Aa*).
7. (a) *Rr*. (b) 1 black and white : 1 red and white.
8. (a) 1 palomino : 1 chestnut. (b) 1 palomino : 1 cremello. (c) All cremello.
9. (a) 3 round : 1 wrinkled. (b) 1 large, numerous : 2 medium, several : 1 small, few.
10. Three-fourths.
11. The Mexican Hairless condition is the heterozygous expression of an allele that is lethal when homozygous. Thus, Mexican Hairless parents: *Aa* × *Aa* → Zygotes: 1 *AA* : 2 *Aa* : 1 *aa* (dies) → Surviving pups: 1 *AA* (haired) : 2 *Aa* (hairless).
12. (a) Incomplete dominance. (b) Half will be normal. None will be severely affected.
13. (a) *AA* = silver, *Aa* = platinum, and *aa* = white (a lethal trait). (b) Breed silver and platinum foxes: *AA* × *Aa* → 1 *AA* : 1 *Aa*.
14. (a) 3 green : 1 albino among seedlings; all green plants at maturity. (b) Two-thirds.
15. (a) All F_1 progeny have purple flowers and long stems. (b) $\frac{9}{16}$ purple, long : $\frac{3}{16}$ purple, short : $\frac{3}{16}$ white, long : $\frac{1}{16}$ white, short.
16. *Aabb* × *aaBb*.
17. (a) 1:1:1:1. (b) 1:1:1:1. (c) 1:1:2. (d) 2:1:1. (e) 1:3. (f) 3:1. (g) 3:1. (h) 1:2:1.
18. Cross 1: *CcLl* × *CcLl*. Cross 2: *Ccll* × *ccLl*. Cross 3: *CcLl* × *Ccll*. Cross 4: *CCll* × *ccLl*. Cross 5: *CCLl* × *ccLl*.
19. Cross moderately resistant and slightly resistant plants and select for highly resistant *AaBb* progeny. Make *AaBb* × *AaBb* cross and select for *AABB* plants that breed true for high resistance.
20. (a) *AaBb* female × *Aabb* male. (b) *aaBb* male × *AAbb* female.
21. (a) Blue Andalusian is the heterozygous expression of a pair of alleles showing incomplete dominance. (b) No, because Blue Andalusians are heterozygous.
22. 2 black, creeper : 4 Blue Andalusian, creeper : 2 splashed white, creeper : 1 black, normal : 2 Blue Andalusian, normal : 1 splashed white, normal.
23. (a) Nine phenotypes, constituting the nine combinations of black, Blue Andalusian, and splashed white with extreme frizzled, mildly frizzled, and normal feathers. (b) 1:2:1:2:4:2:1:2:1.

24. (a) Two gene pairs, because the F_2 ratio is a modification of the 9:3:3:1 dihybrid phenotypic ratio. (b) White-fruited parent = *AAbb*, yellow-fruited parent = *aaBB*, F_1 = *AaBb*, F_2 = *A-B-* and *A-bb* = white fruited plants, *aaB-* = yellow-fruited plants, and *aabb* = green-fruited plants.
25. (a) $\frac{9}{16}$ normal hearing. (b) One example would be *AAbb* × *aaBB* → all *AaBb*.
26. (a) Recessive. (b) Modifier gene interaction; complementary gene interaction. (c) White Wyandotte = *aaCCii*, White Leghorn = *AACCII*, White Silkie = *AAccii*. Cross 1—P: *aaii* × *AAII* → F_1: *AaIi* (white) → F_2: $\frac{13}{16}$ (*A-I-* + *aaI-* + *aaii*) white : $\frac{3}{16}$ (*A-ii*) colored. Cross 2—P: *aaCC* × *AAcc* → F_1: *AaCc* (colored) → F_2: $\frac{9}{16}$ (*A-C-*) colored : $\frac{7}{16}$ (*A-cc* + *aaC-* + *aacc*) white.
27. (a) Two gene pairs. (b) Parents: red (*AABB*) × white (*aabb*) → F_1: red (*AaBb*) → F_2: $\frac{15}{16}$ (about 94%) red (*A-B-* + *A-bb* + *aaB-*) : $\frac{1}{16}$ (about 6%) white (*aabb*).
28. (a) $\frac{12}{16}$ gray : $\frac{3}{16}$ chestnut : $\frac{1}{16}$ sorrel. (b) *EeGg* mare × *eegg* stallion.
29. (a) $\frac{3}{4}$ colored : $\frac{1}{4}$ colorless. (b) $\frac{9}{16}$ colored : $\frac{7}{16}$ colorless. (c) $\frac{1}{8}$ colored : $\frac{7}{8}$ colorless. (d) $\frac{27}{64}$ colored : $\frac{37}{64}$ colorless.

Chapter 2

2. (a) 4 + 4 = 8. (b) 4 × 4 = 16. (c) 4 × 3 × 2 × 1 = 24. (d) 3 (2 or 4 or 6). (e) 1 (4_1 and 4_2) + 1 (5_1 and 3_2) + 1 (3_1 and 5_2) + 1 (6_1 and 2_2) + 1 (2_1 and 6_2) = 5.
3. (a) 8/52 = 2/13. (b) (1/13)(1/13) = 1/169. (c) (4/52)(3/51)(2/50)(1/49) = 1/270,725. (d) 1/2. (e) 5/36.
4. (a) 7!/5!2! = 21. (b) $\frac{7!}{4!3!}(\frac{1}{2})^4(\frac{1}{2})^3 = 35/128$.
5. (a) $\frac{1}{4}$. (b) ($\frac{1}{4}$)($\frac{1}{4}$) = 1/16. (c) (1/4)(3/4) = 3/16. (d) 3/8.
6. (a) 3/4. (b) 3/4 × 1/2 = 3/8. (c) 1/2 × 1/2 = 1/4.
7. (a) 2^3 = 8. (b) 2^4 = 16. (c) 2^5 = 32. (d) = 2^n.
8. (a) 3^3 = 27. (b) 3^4 = 81. (c) 3^5 = 243. (d) 3^n.
9. (a) 8. (b) 16. (c) 32. (d) 2^n.
10. (a) 1/16. (b) 1/16. (c) 3/32. (d) 1/32.
11. (a) $\frac{6!}{4!2!}(\frac{1}{2})^6 = 15/64$. (b) 1/64. (c) 15/64 + 6/64 + 1/64 = 11/32. (d) 1/32.
12. (a) 1/4. (b) 27/256. (c) 27/64. (d) 81/4096. (e) 294/4096.
13. (a) 1/3. (b) 8/81.
14. (a) 24. (b) 9/128.
15. $\frac{5!}{2!3!}(\frac{3}{16})^2(\frac{13}{16})^3 = 0.18857$.
16. 1/8.
17. (a) 1. (b) 2. (c) 3. (d) 5. (e) 8.
18. (a) $\chi^2 = 12$. (b) 9:7 ratio with complementary gene interactions.
19. $\chi^2 = 0.199$.
20. $\chi^2 = 0.451$, $\chi^2 = 0.618$.

Chapter 3

2. (a) False. Transformation involves naked DNA. (b) True. (c) False. It indicates that the protein part of the virus does not enter the cell. (d) False. Cytosine is a pyrimidine. (e) False. Either to the 5′ or 3′ carbon. No ester linkage can be formed with the 2′ carbon since it lacks an oxygen. (f) False. DNA molecules can vary in length and as to whether they are single- or double-stranded. (g) False. An (A + T):(G + C) ratio can vary depending on the organism. (h) True. (i) False. It would contain a higher proportion of GC base pairs. (j) True. (k) True. (l) True. (m) True. (n) False. A primer is a segment of a polynucleotide chain onto which nucleotides are added by DNA polymerase. (o) False. It would block the synthesis of RNA primers. (p) False. It has been implicated in the degradation of RNA primers and subsequent repair synthesis.
3. (a) Adenine, guanine, and cytosine. (b) Thymine, uracil. (c) Complementary. (d) A = T and G = C. (e) Hydrogen, phosphodiester. (f) Hydrogen, complementary strands. (g) Information storage, replication, and mutation. (h) Ligases. (i) 1000, 200. (j) Cytosine. (k) Guanine.
5. (a) $(10^3)(2^{10}) = 10^6$. (b) $(10^3)(2^{10})(2^{10}) = 10^9$. (*Note*: 2^{10} is approximately equal to 10^3.)
6. (1) Double-stranded DNA. (2) Double-stranded DNA. (3) Double-stranded RNA. (4) Single-stranded RNA. (5) Single-stranded DNA.
7. %GC = (86.5 -- 69.3)/0.41 = 42%.

8. Buoyant density (1) = 1.660 + 0.00098(45) = 1.704. Buoyant density (2) = 1.714.
10. (a) 1, 0, 0. (b) 1/4 intermediate and 3/4 light. (c) If conservative, 1/8 heavy and 7/8 light. If dispersive, all within a broad intermediate band.
11. 200,000.
12. 3.8×10^5, 2.5×10^8.
13. Thymine = 30%, guanine = 20%, cytosine = 20%.
14. A = 15%, T = 15%, G = 35%, C = 35%.
15. 2.7×10^3 times greater in volume.
16. 2×10^{-16} cm^3.
22. 0.2 percent.
24. (a) (A + G):(T + C) = 2/3. (b) A:T = 1/2, G:C = 2, (A + T):(G + C) = 1/4, (A + G):(T + C) = 3/2. (c) A:T = 1, G:C = 1, (A + T):(G + C) = 1/4, (A + G):(T + C) = 1.
25. Because the strands of a DNA molecule are complementary and have opposite polarity, G-G = C-C, A-G = C-T, and C-T = A-G. The other pairs of dinucleotides need not be equal.

Chapter 4

4. Acrocentric: (c) and (f); metacentric: (a); submetacentric: (b) and (d); telocentric: (e). Only chromosome (c) has satellites.
5. (a) Telophase. (b) Prophase. (c) Anaphase.
6. (a) 20, 20, and 40. (b) 20, 20, and 40. (c) 40, 40, and 0. (d) 20, 20, and 0.
7. (a) S. (b) Prophase. (c) Anaphase.
8. (a) 93 days.
9. (a) 11.2 picograms, 11.2 picograms, 5.6 picograms. (b) 0.12 picogram, 0.12 picogram, 0.12 picogram. (c) 0.24 picogram, 0.12 picogram, 0.12 picogram.
10. 4×10^6 base pairs.
11. 40 minutes.
12. 40,000 base pairs per minute.
14. $32 \times 12 = 384$.
15. (a) 142 times longer. (b) 7.2 times larger in volume.
16. (a) 46%, 33%, 17%, 4%. (b) 0.15. (c) 24, 18, 12, and 6 minutes, respectively.
17. $2^{50} = 10^{15}$.
18. $(10^2)(2^{30}) = 10^{11}$.
19. (a) 35/128. (b) 1/16.

Chapter 5

2. (a) Both. (b) Meiosis. (c) Both. (d) Meiosis. (e) Mitosis.
4. Metaphase II.
5. (a) 40. (b) 20.
7. (a) 20, 20, 40. (b) 20, 20, 40. (c) 10, 10, 20. (d) 20, 20, 0. (e) 10, 10, 0.
8. (a) 4. (b) 8. (c) 2^n.
9. (a) G–I and G–J. (b) D–E–I and D–E–J. (c) A, B, C, D, E, F, G, H; I, J, K. (d) A, B, C, F, H, K.
10. 21.
11. 1 flinty : 1 floury; all floury.
13. 79.5% FDS and 20.5% SDS.
14. (a) (1) PD, (2) NPD, (3) T, (4) T, (5) PD, (6) NPD, (7) T. (b) PD (13 + 30) and NPD (12 + 33) tetrad types are about equal in frequency. (c) For *A,a*, %SDS = 61%. For *D,d*, %SDS = 73%.
15. (a) $2^3 = 8$. (b) $(1/2)^3 = 1/8$. (c) 1/2.
18. (a) $(0.5)^3(0.7)^3 = 0.043$. (b) $(0.043)^2 = 0.0018$.
19. (a) 8. (b) *AbCD* and *ABcd* = 0.765, *Abcd* and *ABCD* = 0.135, *AbCd* and *ABcD* = 0.085, and *AbcD* and *ABCd* = 0.015.
20. (a) Female. (b) Male. (c) Metamale.
22. (a) Female (pistillate). (b) Monoecious plants that are male sterile.
28. (a) 5 male : 3 female. (b) *Tra/tra* female × *tra/tra* male.

Chapter 6

2. (a) Dominant. (b) 3/4.
3. Two complementary gene pairs.
4. *P*(3 *A*- : 1 *aa*) = 0.42.
5. Autosomal dominant: (c), (d), (e), (f); autosomal recessive: (b), (c), (d), (e); X-linked dominant: (a), (b), (c), (e), (f); X-linked recessive: (a), (b), (d), (e), (f).
6. (a) 1/4, regardless of sex. (b) 1/4 (in half the male children only).
7. (1) Autosomal dominant, (2) X-linked dominant (cannot exclude autosomal dominant), (3) autosomal dominant, (4) autosomal recessive.
8. (b) Tortoiseshell female × yellow male. (c) Male.
9. $X^{w^e}X^w \times X^{w^+}Y$.
10. The short-bristled phenotype is the heterozygous expression of an X-linked gene that is

lethal when present in males or in the homozygous condition in females.

11. (a) ½ barred males : ½ nonbarred females. (b) $Z^BZ^b \times Z^BW$. (c) 1/3 barred males : 1/3 barred females : 1/3 nonbarred females.
12. (a) 1/2. (b) 5/16.
13. (a) 1/2. (b) 1/2. (c) 1/8.
14. 1/4.
15. (a) Male. (b) 3 horned : 1 hornless among males, and 3 hornless : 1 horned among females.
17. (a) *hh* male × *Hh* female. (b) ½ hen-feathered, nonbarred females : 3/8 hen-feathered, barred males : 1/8 cock-feathered, barred males.
18. (a) 1/9. (b) 1/4.
19. (a) 55, if all alleles are codominant or incompletely dominant. (b) 10, if the alleles show a hierarchy of complete dominance.
20. (a) 6, 15. (b) 90. (c) 4. (d) 6.
22. (a) Wild-type > steelblu > silverblu. (b) 3. (c) 2, 1. (d) wild-type × steelblu.
23. Student 2 = O, student 3 = AB, student 4 = A, student 5 = B.
24. (a) $I^Ai \times I^Bi$. (b) $I^AI^B \times I^Bi$. (c) $I^AI^B \times I^AI^B$. (d) $I^Ai \times I^Ai$.
25. AB; AB; B, O; A, O; A, O; B, O.
26. (a) AB. (b) A. (c) B.
27. No; AB parent cannot have an O child.
28. I. *iiRr* and I^AiRr. II. *iirr*, I^AiRr, and I^BiRr. III. I^AI^Brr and I^AiR-.
29. Pedigree (1): (a) and (d) consistent; Pedigree (2): (d) and (e) consistent.
30. Pedigree (1): (a) and (e) consistent; Pedigree (2): (b), (d), and (e) consistent; Pedigrees (1) and (2): only (e) consistent.
31. Only (c) is consistent.

Chapter 7

2. Seven pairs.
4. *a* 8 *b* 6 *c* 2 *d* 4 *e*
5. No, about 25% of the offspring are colored, as expected for independent assortment.
6. The genes for fruit shape and pubescence are linked, located 8 map units apart on the chromosome. The F_1 heterozygotes produced in the separate crosses differ in their arrangement of genes: in cross 1, the F_1 gene arrangement is cis; in cross 2, the F_1 gene arrangement is trans.
7. *a b c d* and *e*.
8. (a) Cis. (b) 35 map units. (c) *BC*/*BC* × *bc*/*bc*.
9. (a) 0.09. (b) 0.025. (c) 0.9.
10. (a) I. *EeR*- and *eerr*. II. *eeRr*, *EeRr*, and *eerr*. III. *eerr*, *Eerr*, *EeRr*, *eerr*, *eeRr*, *eerr*, *EeRr*, and *EeRr*. (b) Cis. (c) III-2 and III-5. (d) 25%.
11. (a) 0.25. (b) 0.00625. (c) 0.2625.
12. (a) 0.48 : 0.48 : 0.02 : 0.02. (b) 0.04%. (c) 24.96%. (d) 46.16%.
13. (a) 0.5 wild-type : 0.25 red, black : 0.25 purple, gray.
14. In *AaBb*, genes are linked in trans and separated by 20 map units. In *BbCc*, genes assort independently. In *CcDd*, genes are linked in cis and separated by 40 map units.
15. (a) *r* and *s* are linked, *p* assorts independently of the others. (b) Trans. (c) 5 map units.
16. (b) *rd* 6.5 *cn* 9.5 *vg*. (d) 0.65.
17. (a) *lg* 19.0 *gl* 41.3 *v*. (b) $lg\ gl^+\ v/lg^+\ gl\ v^+$.
18. (Only chromosomes received from the trihybrid are shown.) (a) *ABC* + *abc* = 0.846, *Abc* + *aBC* = 0.094, *ABc* + *abC* = 0.054, *AbC* + *aBc* = 0.006. (b) *ABC* + *abc* = 0.843, *Abc* + *aBC* = 0.097, *ABc* + *abC* = 0.057, *AbC* + *aBc* = 0.003.
19. (a) *p* 5 *r* 8 *q*. (b) Parental-types = 0.875, single recombinant-types (*p*-*r*) = 0.045, single recombinant-types (*q*-*r*) = 0.075, double recombinant-types = 0.005. (c) 1.48.
20. (a) 7. (b) 3 singles, 3 doubles, and 1 triple. (c) 10^6.
22. (a) 29.5 map units. (b) 8.5 map units. (c) 19.3 map units.
23. The genes are located 18.5 map units apart on the same chromosome. The gene-centromere distances are *c*-centromere = 6.5 map units and *d*-centromere = 11.2 map units.
24. The *c* and *leu* markers are separated by 7.9 map units.
25. The genes are located 20 map units apart on the same chromosome.
27. *a* 3.6 *b* 6.8 *c*.
28. *e* 9.7 *d* 17.4 *f*.
29. *PGK* 88 *HPRT* 75 *G6PD*.
30. (a) 1 − E_3, 11 − E_1, 17 − E_4, X − E_2. (b) E_4. (c) E_1 = LDH, E_2 = PGK, and E_3 = 6PGD.

Chapter 8

2. (a) 200. (b) 200.
3. (b) 12 map units.

5. The map is circular:

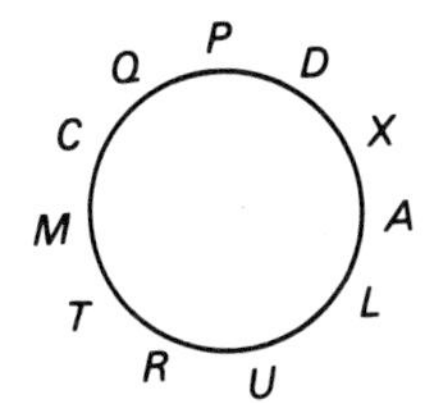

6. The map is circular:

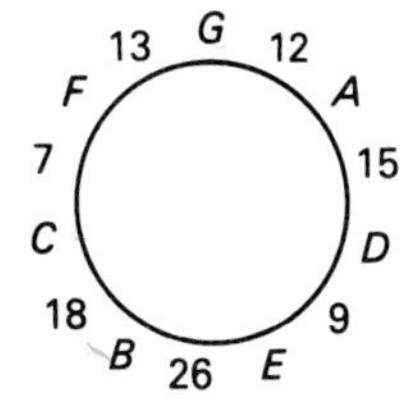

7. (a) Cytoplasmic inheritance. (b) The *arg* gene is on a nuclear chromosome.
8. (a) Normal *Vv* offspring. (b) Normal L_1Vv offspring. (c) Normal *Vv* offspring. (d) Virescent L_2Vv offspring. (e) 3/4 normal (*V*-) : 1/4 virescent (*vv*) offspring. (f) Virescent L_2 offspring.
10. (a) All *Rr* male fertile. (b) 3 male fertile : 1 male sterile offspring.
12. (a) 6. (b) 10. (c) $N(N - 1)/2$.
14. (a) Cotransduction frequency for $a^+b^+ = 54\%$. Cotransduction frequency for $a^+c^+ = 86\%$. (b) *a c b*.
16. (a) *a*, *b*, and *d*. (b) *a 40 d 50 b*.
17. (a) All s^+s in genotype but sinistral in phenotype, like the maternal parent. (b) 1 s^+s^+ : 2 s^+s : 1 *ss*, but all dextral in phenotype. (c) $s^+s^+ \rightarrow$ all s^+s^+ (dextral), $s^+s \rightarrow 1\ s^+s^+$: 2 s^+s : 1 *ss* (all dextral), $ss \rightarrow$ all *ss* (sinistral).
19. 2.5 nucleotides.
20. *c a b*.
21.

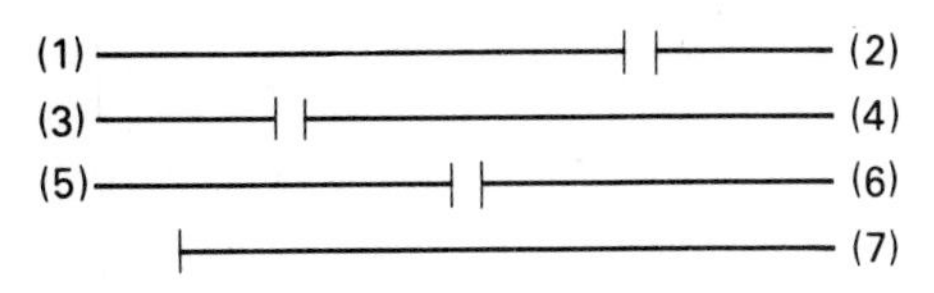

22. *b e a c d*.

Chapter 9

2. (a) False. Amino acids occur in proteins. (b) True. (c) False. Some proteins serve as antibodies, others in O_2 transport, etc. (d) True. (e) True. (f) False. Amino acid changes that affect the three-dimensional structure of proteins can also destroy enzymatic activity. (g) False. The strands, while complementary, possess a different sequence of codons. (h) False. The code is (almost) universal. (i) False. Three amino acids have six codons each. (j) True. (k) True.
3. (a) Oligomeric (or multimeric). (b) Secondary. (c) Peptide. (d) Allosteric, active. (e) Inborn errors of metabolism. (f) Degeneracy. (g) Exons, introns. (h) Translation. (i) Nucleolar organizer. (j) Transfer. (k) Aminoacyl-tRNA. (l) Amino acid, mRNA codon. (m) Inosine.
5. (a) (5′) G-U-C-U-U-A-C-G-C-U-A-G (3′). (b) CAG, AAU, GCG. (c) Val-leu-arg.
6. 3.75%.
8. (a) 9. (b) 17.
9. (a) 15. (b) 432.
10. (a) 400. (b) 8×10^3. (c) 1.6×10^5. (d) $(20)^n$.
11. $\xrightarrow{(1)}$ α-ketoisocaproic acid $\xrightarrow{(2)}$ leucine.
12. B $\xrightarrow{(3)}$ D $\xrightarrow{(1)}$ A $\xrightarrow{(4)}$ E $\xrightarrow{(2)}$ C.
13. A(1, 5), B(3, 4), C(2).
14.

T A C	G A G	A A T	A T G	A C T
A T G	C T C	T T A	T A C	T G A
A U G	C U C	U U A	U A C	U G A
U A C	G A G	A A U	A U G	— — —
met	leu	leu	tyr	none

15. 3.
16. 7.
20. UUU = 21.6%; UUC, UCU, and CUU = 14.4% each; UCC, CUC, and CCU = 9.6% each; CCC = 6.4%.
21. Phe, 1; leu, 2/3; ser, 2/3; pro, 4/9.
22. (a) $(\text{ile-tyr})_n$. (b) $(\text{asp})_n$ and $(\text{met})_n$. (c) $(\text{asn})_n$ and $(\text{ile})_n$.

Chapter 10

4. $R = 1$, so that $\frac{1}{2} = C/C_0 = 1/(1 + KC_0t_{1/2})$. Rearranging, we get $1 + KC_0t_{1/2} = 2$, or $KC_0t_{1/2} = 1$. Thus, $K = 1/C_0t_{1/2}$.
5. (b) +, +, +, +. (c) −, −, −, −. (d) +, +, −, −. (e) −, +, −, +. (f) +, +, −, +. (g) −, +, +, +. (h) −, −, −, −. (i) −, −, −, −.
6. *a* = structural gene, *b* = operator, *c* = regulatory gene.

7. (a) 0.2, 0.3, 0.5. (b) 10^{-3}, 10^{-1}, 10^{3}. (c) 2×10^{-4}, 3×10^{-2}, 5×10^{2}. (d) 2.5×10^{6}, 1.67×10^{4}, 1. (e) 80.
8. $R^{o}o^{+}S^{+}$ = constitutive (a) for I, repressed (c) for II. $R^{+}o^{-}S^{+}$ = constitutive (a) for I, repressed (c) for II. $R^{i}o^{+}S^{+}/R^{+}o^{+}S^{+}$ = repressed (c) for I, inducible (b) for II. $R^{o}o^{+}S^{+}/R^{+}o^{+}S^{+}$ = inducible (b) for I, inducible (b) for II. $R^{+}o^{-}S^{+}/R^{+}o^{+}S^{-}$ = constitutive (a) for I, repressed (c) for II.
10. (a) 35/128. (b) 1/128.

Chapter 11

3. Only arginine in the case of a transition. Either arginine, glycine, serine, leucine, or cysteine in the case of a transversion.
4. Either CTT → TTT or CTC → TTC.
5. (a), (c), (d), (f).
6. UUA, UUG, UCA, UCG, UAU, UAC, UGU, UGC, UGG, CAA, AAA, GAA, CAG, AAG, GAG, CGA, AGA, GGA.
7. (a) Silent. (b) Missense. (c) Nonsense. (d) Missense. (e) Silent. (f) Missense. (g) Nonsense.
8. 4.3×10^{-6}.
9. The products of somatic mutation in plants can be propagated vegetatively (i.e., by asexual means).
12. Wild-type: A-C-C-T-T-C-C-G-A-G-C-T-T-G-C-C-A-C. Mutant 1 (transition): G-C-C-T-T-C-C-G-A-G-C-T-T-G-C-C-A-C. Mutant 2 (transversion): A-C-C-T-A-C-C-G-A-G-C-T-T-G-C-C-A-C. Mutant 3 (transition): A-C-C-T-T-C-C-A-A-G-C-T-T-G-C-C-A-C. Mutant 4 (transition): A-C-C-T-T-C-C-G-A-A-C-T. Mutant 5 (addition): A-C-C-T-T-C-A-C-G-A-G-C-T-T-G-C-C-A. Mutant 6 (deletion of mutant 5): A-C-C-T-T-C-A-C-G-A-G-C-T-G-C-C-A-C. (c) 54.
14. (a) 3%. (b) 12%. (c) 15%.
15. Strain 1: AT → GC transition. Strain 2: Transversion. Strain 3: Addition-deletion mutation. Strain 4: GC → AT transition.
17. The parents are homozygous for different recessive mutant genes.
18. HbS and HbC are allelic mutations in the β gene. Ho-2 is not allelic to HbS; rather, it is caused by a mutation in the α gene.
19. (a) Enzyme 2, enzyme 1. (b) *SSbb* (brown) × *ssBB* (scarlet) → *SsBb* (red). (c) 9/16 *S-B-* (red) : 3/16 *S-bb* (brown) : 3/16 *ssB-* (scarlet) : 1/16 *ssbb* (colorless).
20. (a) White. (b) Blue. (c) Purple. (d) 9 purple : 3 blue : 4 white.
23. 9 pigmented (*A-B-*) : 7 colorless (*A-bb*, *aaB-*, and *aabb*).
24. 13 colorless (*C-D-*, *C-dd*, and *ccdd*) : 3 pigmented (*ccD-*).
25. (a) 9 dark (*E-F-*) : 6 intermediate (*E-ff* and *eeF-*) : 1 white (*eeff*). (b) 15 colored (*E-F-*, *E-ff*, and *eeF-*) : 1 colorless (*eeff*).
26. *A* = receptor element; *B* = target gene; *C* = regulatory gene, causing mutational instability.

Chapter 12

2. 23.
3. 10.
4. (a) 1/2. (b) Anaphase II of oogenesis.
5. 45; the mother is probably a translocation carrier.
6. The cell with 69 chromosomes is triploid ($3n$). The cell with 52 chromosomes is aneuploid ($2n + 6$).
7. 31.
8. (a) 2 colorless : 1 red. (b) 1 colorless : 1 red.
9. The loss of a Y chromosome during the initial cleavage stages of an XY zygote, followed by the separation of the cells to form an XY and XO twin pair.
11. (a) With $x = 7$, $2x = 14$, $3x = 21$, $4x = 28$, and $7x = 49$. (b) By means of polyploidy.
12. 16.
13. 14.
14. In the mother during anaphase II of oogenesis.
15. 5 dominant : 1 recessive.
16. 63 red : 1 white.
17. 1/1296.
19. Band 7 in region 2 of the long arm of chromosome 4. It is lightly stained.
21. See below.
23. (a) Centric fusion. (b) Pericentric inversion.
24. (a) 1 normal : 1 translocation heterozygote. (b) Cross to normal stock. All offspring will be

21. *ABCDEF* $\xrightarrow{\text{inversion } BCD}$ *ADCBEF* $\xrightarrow{\text{inversion } BE}$ *ADCEBF* $\xrightarrow{\text{inversion } EBF}$ *ADCFBE*.
ADCBEF $\xrightarrow{\text{inversion } ADCB}$ *BCDAEF*

translocation heterozygotes.

26. (a) About 1/32. (b) About 1/1920. (c) 1/175.
28. *c-e-a-d-b-f*.
29. 24 map units.
30. (a) 6 *AA* : 16 *Aa* : 6 *aa*, or 3:8:3. (b) 187 dominant : 9 recessive.

Chapter 13

3. $f(\mathrm{M}) = 0.291$, $f(\mathrm{MN}) = 0.496$, $f(\mathrm{N}) = 0.213$; $f(L^M) = 0.539$, $f(L^N) = 0.461$.
4. (a) 0.4 and 0.6. (b) $f(A) = 0.4$, $f(a) = 0.6$; $f(AA) = 0.16$, $f(Aa) = 0.48$, $f(aa) = 0.36$.
5. 18%.
6. (a) 16%. (b) 32%.
7. (a) $p = 0.7$ and $q = 0.3$. (c) 91% were purple and 9% were yellow.
8. (c) Black is the recessive trait. Therefore, $q^2 = 0.64$, so that $q = 0.8$.
9. $q^2 = 1/9$, so that $q = 1/3$ and $2pq = 4/9$.
10. $q = 1/3$.
12. (a) Population 1: $p = 0.74$, $q = 0.26$; population 2: $p = 0.4$, $q = 0.6$; population 3: $p = 0.1$, $q = 0.9$; population 4: $p = 0.5$, $q = 0.5$. (b) Population 1: no—0.55, 0.38, 0.07; population 2: no—0.16, 0.48, 0.36; population 3: yes; population 4: no—0.25, 0.50, 0.25.
13. Herd 2 conforms to a random mating population.
15. (a) $q = 1/50$, so that $2pq = 98/2500$, or about 4%. (b) 0.16%.
16. (a) 0.16%. (b) About 8%. (c) 92%.
17. (a) Disorder A is recessive; disorder B is dominant. (b) Disorder A: $p = 0.99$, $q = 0.01$; disorder B: $p = 0.02$, $q = 0.98$.
18. 3; 6; $N(N + 1)/2$.
19. Yes, there is a close agreement between observed numbers (483 and 278) and expected numbers (495 and 266). $\chi^2 = 0.83$.
20. (a) $f(\mathrm{M}) = 0.16$. Assuming random mating, $f(\mathrm{MN}) = 0.48$ and $f(\mathrm{N}) = 0.36$. (b) $f(\mathrm{M}) = 0.2$, $f(\mathrm{MN}) = 0.4$, $f(\mathrm{N}) = 0.4$.
21. (a) $p = 0.3$, $q = 0.3$, $r = 0.4$. (b) 198 A, 198 B, 108 AB, 96 O.
22. 0.16.
23. (a) About 10%. (b) 19%.
24. (a) $f(AB) = 0.18$. $f(Ab) = 0.42$, $f(aB) = 0.12$, $f(ab) = 0.28$. (b) 0.2352.
26. (a) $f(AB) = f(ab) = 5/12$, $f(Ab) = f(aB) = 1/12$. (b) $p_1 = q_1 = ½$, $p_2 = q_2 = ½$. (c) *AABB* : *AaBB* : *aaBB* : *AABb* : *AaBb* : *aaBb* : *AAbb* : *Aabb* : *aabb* = 25:10:1:10:52:10:1:10:25. (d) No, yes. (e) *AABB* : *AaBB* : *aaBB* : *AABb* : *AaBb* : *aaBb* : *AAbb* : *Aabb* : *aabb* = 1:2:1:2:4:2:1:2:1.
27. $f(AB) = f(ab) = 0.458$, $f(Ab) = f(aB) = 0.042$; $p_1 = q_1 = 0.5$, $p_2 = q_2 = 0.5$.
28. (a) Females: $f(AA) = 0.12$, $f(Aa) = 0.56$, $f(aa) = 0.32$; Males: $f(A) = 0.2$, $f(a) = 0.8$. (b) Females: $f(AA) = 1/9$, $f(Aa) = 4/9$, $f(aa)$ 4/9; Males: $f(A) = 1/3$, $f(a) = 2/3$.

Chapter 14

4. 0.727.
6. $p = 1/6$, $q = 5/6$.
7. 900 generations.
9. w(for dark-colored moths) = 1, w(for light-colored moths) = 0.41.
10. (a) $f(AA) = 9/15$, $f(Aa) = 6/15$. (b) $f(AA) = 16/25$, $f(Aa) = 8/25$, $f(aa) = 1/25$. (c) 8 generations.
11. (a) $f(AA) = 3/4$, $f(Aa) = 1/4$; $f(AA) = 49/64$, $f(Aa) = 14/64$, $f(aa) = 1/64$.
12. 0.0063.
13. 40 generations.
14. (a) 0.05. (b) 105/512. (c) Nonselective.
16. w(for *CH/CH*) = 0.245, w(for *CH/ST*) = 0.607, w(for *ST/ST*) = 1.
19. (a) $f(Hb^A) = 0.862$, $q = f(Hb^S) = 0.138$. (b) Infants: $f(Hb^AHb^A) = 0.743$, $f(Hb^AHb^S) = 0.238$, $f(Hb^SHb^S) = 0.019$; Adults: $f(Hb^AHb^A) = 0.724$, $f(Hb^AHb^S) = 0.276$, $f(Hb^SHb^S) = 0$.
20. (a) 8.1%. (b) 0%.
21. (a) 1/2. (b) 1/4, $(1/2)^t$.
22. (b) 4×10^{-4}.
23. (a) Isolate I: 38.08 cm; isolate II: 31.68 cm. (b) 34.88. (c) 40 cm.

Chapter 15

6. (a) 1/16. (b) 1/64. (c) 1/256. (d) $(½)^{2n}$ or $(¼)^n$.
7. (a) 9/16. (b) 27/64. (c) 81/256. (d) $(3/4)^n$.
8. Probably greater than five gene pairs.
9. *AaBbCcDd*.
10. (a) MZ twins = 25%, DZ twins = 75%. (b) MZ twins = 3/1000 births, DZ twins = 9/1000 births.
11. 0.5 for MZ twins, 0.09 for DZ twins.
12. (a) Mean length = 12.1 cm; variance = 2.3 cm². (b) 9.1–15.1 cm.
13. (a) Intermediate in color. (b) The mean would be the same in the F_1 and F_2. (c) 1/64. (d)

15/64. (e) 3/64. (f) 20/64. (g) 0. (h) 7, including white.

14. (a)

Eye color	Contributing gene
light blue	0
blue	1
blue green	2
hazel	3
light brown	4
brown	5
dark brown	6

(b) 1 light blue : 6 blue : 15 blue-green : 20 hazel : 15 light brown : 6 brown : 1 dark brown.

15. (a) 1/64. (b) 1/256. (c) 1/1024. (d) $(\frac{1}{4})^{n+1}$.
16. Four gene loci with each active allele contributing two days to heading date.
17. *AAbbccdd* (or any genotype homozygous for 1 active gene pair), 60 days; *aaBBCCDD* (or any genotype complementary to the other parent and homozygous for 3 active gene pairs), 68 days.
18. (a) About 4 gene loci. (b) 0.75.
19. (a) Mean (P60) = 6.63 cm, mean (P54) = 16.80 cm, mean (F_1) = 12.12 cm, mean (F_2) = 12.89 cm, variance (F_1) = 2.28 cm^2, variance (F_2) = 5.06. (b) 5 gene loci.
20. (a) 22 inches. (b) 27 (22 inches) : 27 (18 inches) : 9 (14 inches) : 1 (10 inches). (c) Mean height of F_2 = 19 inches, variance of F_2 = 9 $inches^2$.
21. (a) 27 (22 inches) : 18 (18 inches) : 15 (14 inches) : 4 (10 inches). (b) Mean of F_2 = 18.25 inches, variance of F_2 = 14.44 $inches^2$.
22. F_1: 68.2, F_2: 51.15.
23. 1 (1 g) : 10 (1.5 g) : 45 (2.25 g) : 120 (3.38 g) : 210 (5.06 g) : 252 (7.59 g) : 210 (11.39 g) : 120 (17.08 g) : 45 (25.63 g) : 10 (38.44 g) : 1 (57.66 g).
24. 0.873.
25. (a) B in *x*, A in *x*, B in *y*, A in *y*.

Chapter 16

4. (a) 1/8. (b) 1/8. (c) 1/32.
5. $f(AA) = 0.28, f(Aa) = 0.24, f(aa) = 0.48$.
6. (a) 3/8, 2/8, 3/8. (b) 7/16, 2/16, 7/16. (c) 15/32, 2/32, 15/32. (d) $\frac{1}{2}[1 - (\frac{1}{2})^{t+1}]$, $(\frac{1}{2})^{t+1}$, $\frac{1}{2}[1 - (\frac{1}{2})^{t+1}]$.
7. $F = 0.49$.
10. (a) 1/16. (b) 1/12. (c) 7.2×10^{-4}.
11. (a) 1/4. (b) 1/8. (c) 5/8 (or slightly less if crossing over has a significant effect).
12. 8.
14. 0.376.
15. (a) 0.0025, 10^{-6}. (b) 0.0055, 6.3×10^{-5}. (c) 2.2, 63.
16. Tallest parent = 47.8 inches, shortest parent = 28.2 inches.
17. (a) 3 eggs per hen. (b) 0.6 eggs per hen. (c) 15.6 eggs per hen.
18. Bull A: 14,884 lb, bull B: 17,558 lb.
19. (a) 1/4. (b) 9/64. (c) 27/64.
20. Frequency of pure-breeding heterotic line = $(\frac{1}{4})^{10} = (\frac{1}{2})^{10}(\frac{1}{2})^{10} = (1/1024)(1/1024) < 10^{-6}$.
21. 2.5×10^{-11}.
22. (a) 19 inches. (b) 1 inch. (c) 19.33 inches. (d) Selection response = 0.33 inch, selection differential = 1 inch, heritability = 0.33.
24. 9/64.
25. (a) 0, 1/4, 3/8. (b) 1/2, 3/4, 7/8.

Index

107 heteroduplex

105 tautomeric shift

? 90 density gradient equilibrium centri.

118 concatamers

129 genome

138 uninemic

134 nucleosomes

142 solenoid

146 nucleoids

154 meiocytes

158 bivalent

158 terminalization